Kishore Vaigyanik Protsahan Yojana

KVPY

11 Years' SOLVED PAPERS

2019-2009

Stream SB/SX

Included 5 Practice Sets

Kishore Vaigyanik Protsahan Yojana

KVPY

11 Years' SOLVED PAPERS

2019-2009

Stream SB/SX

Authors
Lakshman Prasad *(Mathematics)*
Deepak Paliwal, Mansi Garg *(Physics)*
Neha Minglani Sachdeva *(Chemistry)*
Sanubia Saleem *(Biology)*

ARIHANT PRAKASHAN (Series), MEERUT

ARIHANT PRAKASHAN (Series), MEERUT

卐 **Administrative & Production Offices**

Regd. Office

'Ramchhaya' 4577/15, Agarwal Road, Darya Ganj, New Delhi -110002
Tele: 011- 47630600, 43518550; Fax: 011- 23280316

卐 **Head Office**

Kalindi, TP Nagar, Meerut (UP) - 250002
Tel: 0121-7156203, 7156204

卐 **Sales & Support Offices**

Agra, Ahmedabad, Bengaluru, Bareilly, Chennai, Delhi, Guwahati, Hyderabad, Jaipur, Jhansi, Kolkata, Lucknow, Nagpur & Pune.

卐 **ISBN** : 978-93-24195-38-8

PO No : TXT-XX-XXXXXXX-X-XX

Published by Arihant Publications (India) Ltd.

For further information about the books published by Arihant
log on to www.arihantbooks.com or email to info@arihantbooks.com

Follow us on   

Kishore Vaigyanik Protsahan Yojana

ABOUT THE EXAM

KVPY i.e. Kishore Vaigyanik Protsahan Yojana is a National Level Fellowship (scholarship) Program in Basic Science (Physics, Chemistry, Mathematics & Biology) upto Pre-Phd Level, run by Department of Science & Technology, Government of India and Conducted by IISC (Indian Institute of Science) Bangalore, Karnataka Annually.

It Was Started in 1999 to Encourage Basic Sciences Students to take up Research Career in Sciences. The Objective of the Exam is to Encourage Talented Students for Research Career in Sciences.

ELIGIBILITY CRITERIA

KVPY scholarships are given only to Indian Nationals to study in India. There are three streams in KVPY; SA, SB & SX. Eligibility criteria for different streams is discussed below;

- **For SA** Class 11 Students who passed class 10 with minimum 75% (65% for SC/ST/PWD) marks in Mathematics & Science.

  The fellowship of students selected in SA will be activated only if they pursue undergraduate courses in Basic Sciences (B.Sc./B.S./B.Stat./B.Math/Integrated M.Sc. or M.S.) and have secured a minimum of 60% (50% for SC/ST/PWD) marks in science subjects in class 12th.

- **For SX** Class 12 Students aspiring to pursue undergraduate program (B.Sc. etc) with basic sciences (Physics, Chemistry, Mathematics & Biology) who passed class 10 with minimum 75% (65% for SC/ST/PWD) marks in Mathematics & Science.

  The fellowship of students selected in SX will be activated only if they pursue undergraduate courses in Basic Sciences (B.Sc./B.S./B.Stat./B.Math/Integrated M.Sc. or M.S.) and have secured a minimum of 60% (50% for SC/ST/PWD) marks in science subjects in class 12th.

- **For SB** B.Sc. Ist year Students who passed class 12 with 60% marks in Maths & Sciences (PCMB) & class 10 with minimum 75% marks in Mathematics & Science.

  In order to activate fellowship, in the first year of undergraduate course they should secure minimum 60% (50% for SC/ST/PWD) marks.

Those students who are intending or pursuing undergraduate program under distance education scheme or correspondence course of any university are not eligible.

SYLLABUS OF KVPY

There is no prescribed syllabus for KVPY aptitude test, it aims to assess the understanding & analytical ability of the students than his/her factual knowledge. However questions are framed from syllabus upto class 10/12/Ist Year of Undergraduate Courses in basic sciences, as applicable. There are two Questions Papers in KVPY; one for stream SA & Other for SB/SX (Question Paper is same for SB & SX).

QUESTION PAPERS PATTERN

There are two Questions Papers in KVPY; one for stream SA & Other for SX/SB (Question Paper is same for SB & SX).

- **Question Paper for SA Stream** caries 80 Questions for 100 marks. There are Two Parts in the Question Paper; Part I has 15 Questions of 1 mark each for Mathematics, Physics, Chemistry & Biology while Part II has 5 Questions of 2 marks each for Mathematics, Physics, Chemistry & Biology.
- **Question Paper for SB/SX Stream** caries 120 Questions for 160 marks. There are Two Parts in the Question Paper; Part I has 20 Questions of 1 mark each for Mathematics, Physics, Chemistry & Biology while Part II has 10 Questions of 2 marks each for Mathematics, Physics, Chemistry & Biology.

MODE OF EXAM

KVPY is conducted in Online Mode in English & Hindi Medium.

TIME OF EXAM

- Normally the notification or advertisement for KVPY appear in National Newspapers on May 11 (Technology Day) and Second Sunday of July every year.
- Generally the exam is conducted in the month of November.

SELECTION PROCESS

After scrutiny of application forms on the basis of eligibility criteria for various streams all eligible students are called for Aptitude Test conducted in English & Hindi Medium at different centers across the country. On the basis of performance in aptitude test shortlisted students are called for an interview, which is the final stage of selection procedure.

FELLOWSHIPS

The selected students are eligible to receive KVPY fellowship after class 12th/Ist Year of Undergraduate course only if they pursue Undergraduate Courses in Basic Science, upto Pre-PhD or 5 Years whichever is earlier.

Details of fellowships are listed below;

Basic Science	Monthly Fellowship	Annual Contingency Grant
SA/SX/SB during Ist to IIIrd year of B.Sc./B.S./BB.Stat./B.Math/ Integrated M.Sc or M.S.	Rs. 5000	Rs. 20000
SA/SX/SB during M.Sc. / IVth to Vth years of Integrated M.Sc /M.S./ M.Math/ M.Stat.	Rs. 7000	Rs. 28000

CONTINUATION / RENEWAL OF FELLOWSHIP

- The fellow should continue to study basic science and should also maintain a minimum level of academic performance as Ist division or 60% (50% for SC/ST/PWD) marks in aggregate. Also the fellow has to pass all the subjects prescribed for that particular year.
- In each year marks are to be certified by the Dean or Head of the Institution.
- The fellowship will be discontinued if above marks are not obtained. However if fellow passed all the subjects & obtain marks more than 60% (50% for SC/ST/PWD) in subsequent year, the fellowship can be renewed only for that year onwards.
- If KVPY fellow opts out of the basic science at any stage then monthly fellowship and contingency grant will be forfeited from him.

KVPY Timeline 2020

IMPORTANT DATES

Opening of Application Portal	:	2nd Week of July 2020
Last Date of Submission of Online Application	:	1st Week of September 2020
KYPY Aptitude Test	:	1st Week of November 2020

APPLICATION FEE

For General Category	:	Rs. 1000/-
For SC/ST/ Persons with Disabilities	:	Rs. 500/-

For more details visit:www.kvpy.iisc.ernet.in

CONTENTS

QUESTION PAPER 2019

Stream : SB/SX

MM : 160

Instructions

1. There are 120 questions in this paper.
2. The question paper contains two parts; Part I (1 Mark Questions) and Part II (2 Marks Questions).
3. There are four sections in each part; Mathematics, Physics, Chemistry and Biology.
4. Out of the four options given with each question, only one is correct.

PART-I (1 Mark Questions)

MATHEMATICS

1. The number of four-letter words that can be formed with letters a, b, c such that all three letters occur is

(a) 30 (b) 36 (c) 81 (d) 256

2. Let $A = \left\{\theta \in R : \left(\frac{1}{3}\sin\theta + \frac{2}{3}\cos\theta\right)^2 = \frac{1}{3}\sin^2\theta + \frac{2}{3}\cos^2\theta\right\}$.

Then

(a) $A \cap [0, \pi]$ is an empty set

(b) $A \cap [0, \pi]$ has exactly one point

(c) $A \cap [0, \pi]$ has exactly two points

(d) $A \cap [0, \pi]$ has more than two points.

3. The area of the region bounded by the lines $x = 1, x = 2$, and the curves $x(y - e^x) = \sin x$ and $2xy = 2\sin x + x^3$ is

(a) $e^2 - e - \frac{1}{6}$ (b) $e^2 - e - \frac{7}{6}$

(c) $e^2 - e + \frac{1}{6}$ (d) $e^2 - e + \frac{7}{6}$

4. Let AB be a line segment with mid-point C and D be the mid-point of AC. Let C_1 be the circle with diameter AB and C_2 be the circle with diameter AC. Let E be a point on C_1 such that EC is perpendicular to AB. Let F be a point on C_2 such that DF is perpendicular to AB and E and F lie on opposite sides of AB. Then, the value of $\sin \angle FEC$ is

(a) $\frac{1}{\sqrt{10}}$ (b) $\frac{2}{\sqrt{10}}$

(c) $\frac{1}{\sqrt{13}}$ (d) $\frac{2}{\sqrt{13}}$

5. The number of integers x satisfying

$$-3x^4 + \det\begin{bmatrix} 1 & x & x^2 \\ 1 & x^2 & x^4 \\ 1 & x^3 & x^6 \end{bmatrix} = 0 \text{ is equal to}$$

(a) 1 (b) 2

(c) 5 (d) 8

6. Let P be a non-zero polynomial such that $P(1+x)=P(1-x)$ for all real x and $P(1)=0$. Let m be the largest integer such that $(x-1)^m$ divides $P(x)$ for all such $P(x)$. Then, m equals

(a) 1 (b) 2
(c) 3 (d) 4

7. Let

$$f(x)=\begin{cases} x\sin\left(\frac{1}{x}\right) \text{ when } x\neq 0 \\ 1 \text{ when } x=0 \end{cases}$$

and $A=\{x\in R: f(x)=1\}$. Then, A has

(a) exactly one element
(b) exactly two elements
(c) exactly three elements
(d) infinitely many elements

8. Let S be a subset of the plane defined by $S=\{(x,y):|x|+2|y|=1\}$. Then, the radius of the smallest circle with centre at the origin and having non-empty intersection with S is

(a) $\frac{1}{5}$ (b) $\frac{1}{\sqrt{5}}$
(c) $\frac{1}{2}$ (d) $\frac{2}{\sqrt{5}}$

9. The number of solutions of the equation $\sin(9x)+\sin(3x)=0$ in the closed interval $[0,2\pi]$ is

(a) 7 (b) 13
(c)19 (d) 25

10. Among all the parallelograms whose diagonals are 10 and 4, the one having maximum area has its perimeter lying in the interval

(a) (19, 20] (b) (20, 21]
(c) (21, 22] (d) (22, 23]

11. The number of ordered pairs (a,b) of positive integers such that $\frac{2a-1}{b}$ and $\frac{2b-1}{a}$ are both integers is

(a) 1 (b) 2
(c) 3 (d) more than 3

12. Let $z=x+iy$ and $w=u+iv$ be complex numbers on the unit circle such that $z^2+w^2=1$. Then the number of ordered pairs (z,w) is

(a) 0 (b) 4
(c) 8 (d) infinite

13. Let E denote the set of letters of the English alphabet, $V=\{a,e,i,o,u\}$ and C be the complement of V in E. Then, the number of four-letter words (where repetitions of letters are allowed) having at least one letter from V and at least one letter from C is

(a) 261870 (b) 314160
(c) 425880 (d) 851760

14. Let $\sigma_1,\sigma_2,\sigma_3$ be planes passing through the origin. Assume that σ_1 is perpendicular to the vector $(1,1,1)$, σ_2 is perpendicular to a vector (a,b,c) and σ_3 is perpendicular to the vector (a^2,b^2,c^2). What are all the positive values of a,b and c so that $\sigma_1\cap\sigma_2\cap\sigma_3$ is a single point?

(a) Any positive value of $a,b,$ and c other than 1
(b) Any positive values of a,b and c where either $a\neq b, b\neq c$ or $a\neq c$
(c) Any three distinct positive values of a,b and c
(d) There exist no such positive real numbers $a,b,$ and c.

15. Ravi and Rashmi are each holding 2 red cards and 2 black cards (all four red and all four black cards are identical). Ravi picks a card at random from Rashmi and then Rashmi picks a card to random from Ravi. This process is repeated a second time. Let p be the probability that both have all 4 cards of the same colour. Then, p satisfies

(a) $p\leq 5\%$ (b) $5\%<p\leq 10\%$
(c) $10\%<p\leq 15\%$ (d) $15\%<p$

16. Let A_1, A_2 and A_3 be the regions on R^2 defined by

$$A_1=\{(x,y): x\geq 0, y\geq 0, 2x+2y-x^2-y^2>1>x+y\}$$
$$A_2=\{(x,y): x\geq 0, y\geq 0, x+y>1>x^2+y^2\},$$
$$A_3=\{(x,y): x\geq 0, y\geq 0, x+y>1>x^3+y^3\}$$

Denote by $|A_1|, |A_2|$ and $|A_3|$ the areas of the regions A_1, A_2 and A_3 respectively. Then,

(a) $|A_1|>|A_2|>|A_3|$ (b) $|A_1|>|A_3|>|A_2|$
(c) $|A_1|=|A_2|<|A_3|$ (d) $|A_1|=|A_3|>|A_2|$

17. Let $f:R\to R$ be a continuous function such that $f(x^2)=f(x^3)$ for all $x\in R$. Consider the following statements.

I. f is an odd function.
II. f is an even function.
III. f is differentiable everywhere.

Then,

(a) I is true and III is false (b) II is true and III is false
(c) Both I and III are true (d) Both II and III are true

18. Suppose a continuous function $f:(0,\infty)\to R$ satisfies

$$f(x)=2\int_0^x tf(t)dt+1, \forall\ x\geq 0.$$

Then, $f(1)$ equals

(a) e (b) e^2 (c) e^4 (d) e^6

19. Let $a>0, a\neq 1$. Then, the set S of all positive real numbers b satisfying $(1+a^2)(1+b^2)=4ab$ is

(a) an empty set
(b) a singleton set
(c) a finite set containing more than one element
(d) $(0,\infty)$

20. Let $f : R \to R$ be a function defined by

$$f(x) = \begin{cases} \dfrac{\sin(x^2)}{x} \text{ if } x \neq 0 \\ 0 \text{ if } x = 0 \end{cases}$$

Then, at $x = 0$, f is

(a) not continuous
(b) continuous but not differentiable
(c) differentiable and the derivative is not continuous
(d) differentiable and the derivative is continuous

PHYSICS

21. In a muonic atom, a muon of mass of 200 times of that of electron and same charge is bound to the proton. The wavelengths of its Balmer series are in the range of

(a) X-rays (b) infrared rays
(c) γ-rays (d) microwave

22. We consider the Thomson model of the hydrogen atom in which the proton charge is distributed uniformly over a spherical volume of radius 0.25 Å. Applying the Bohr condition in this model, the ground state energy (in eV) of the electron will be close to

(a) $-13.6/4$ (b) -13.6
(c) $-13.6/2$ (d) -2×13.6

23. A spherical rigid ball is released from rest and starts rolling down an inclined plane from height $h = 7$ m, as shown in the figure. It hits a block at rest on the horizontal plane (assume elastic collision). If the mass of both the ball and the block is m and the ball is rolling without sliding, then the speed of the block after collision is close to

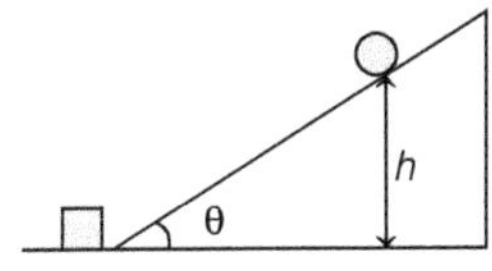

(a) 6 m/s (b) 8 m/s
(c) 10 m/s (d) 12 m/s

24. A girl drops an apple from the window of a train which is moving on a straight track with speed increasing with a constant rate. The trajectory of the falling apple as seen by the girl is

(a) parabolic and in the direction of the moving train
(b) parabolic and opposite to the direction of the moving train
(c) an inclined straight line pointing in the direction of the moving train
(d) an inclined straight line pointing opposite to the direction of the moving train

25. A train is moving slowly at 2m/s next to a railway platform. A man, 1.5 m tall, alights from the train such that his feet are fixed on the ground. Taking him to be a rigid body, the instantaneous angular velocity (in rad/s) is

(a) 1.5 (b) 2.0
(c) 2.5 (d) 3.0

26. A point mass M moving with a certain velocity collides with a stationary point mass $M/2$. The collision is elastic and in one-dimension. Let the ratio of the final velocities of M and $M/2$ be x. The value of x is

(a) 2 (b) 3
(c) 1/2 (d) 1/4

27. A particle of mass $2/3$ kg with velocity $v = -15$ m/s at $t = -2$ s is acted upon by a force $F = k - \beta t^2$. Here, $k = 8\text{N}$ and $\beta = 2\text{N/s}^2$. The motion is one-dimensional. Then, the speed at which the particle acceleration is zero again, is

(a) 1 m/s (b) 16 m/s
(c) 17 m/s (d) 32 m/s

28. A certain stellar body has radius $50 R_s$ and temperature $2T_s$ and is at a distance of 2×10^{10} AU from the earth. Here, AU refers to the earth-sun distance and R_s and T_s refer to the sun's radius and temperature, respectively. Take, both star and sun to be ideal black bodies. The ratio of the power received on earth from the stellar body as compared to that received from the sun is close to

(a) 4×10^{-20} (b) 2×10^{-6}
(c) 10^{-8} (d) 10^{-16}

29. As shown in the schematic below, a rod of uniform cross-sectional area A and length l is carrying a constant current i through it and voltage across the rod is measured using an ideal voltmeter. The rod is stretched by the application of a force F.

Which of the following graphs would show the variation in the voltage across the rod as function of the strain ε when the strain is small. Neglect Joule heating.

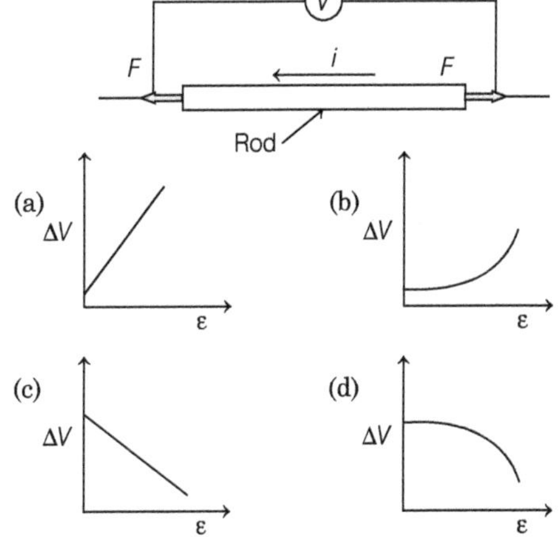

30. Two identical coherent sound sources R and S with frequency f are 5 m apart. An observer standing equidistant from the source and at a perpendicular distance of 12 m from the line RS hears maximum sound intensity.

When he moves parallel to RS, the sound intensity varies and is a minimum when he comes directly in front of one of the two sources. Then, a possible value of f is close to (the speed of sound is 330 m/s)

(a) 495 Hz (b) 275 Hz
(c) 660 Hz (d) 330 Hz

31. A photon falls through a height of 1 km through the earth's gravitational field. To calculate the change in its frequency, take its mass to be $h\nu / c^2$. The fractional change in frequency ν is close to

(a) 10^{-20} (b) 10^{-17}
(c) 10^{-13} (d) 10^{-10}

32. 0.02 moles of an ideal diatomic gas with initial temperature 20 °C is compressed from 1500 cm^3 to 500 cm^3. The thermodynamic process is such that $pV^2 = \beta$, where β is a constant. Then, the value of β is close to (the gas constant, $R = 8.31$ J/K/mol).

(a) 7.5×10^{-2} Pa - m^6 (b) 1.5×10^{2} Pa - m^6
(c) 3×10^{-2} Pa - m^6 (d) 2.0×10^{1} Pa - m^6

33. A heater supplying constant power P watts is switched ON at time $t = 0$ min to raise the temperature of a liquid kept in a calorimeter of negligible heat capacity. A student records the temperature of the liquid $T(t)$ at equal time intervals. A graph is plotted with $T(t)$ on the Y-axis *versus* t on the X-axis. Assume that there is no heat loss to the surroundings during heating. Then,

(a) the graph is a straight line parallel to the time axis
(b) the heat capacity of the liquid is inversely proportional to the slope of the graph
(c) if some heat were lost at a constant rate to the surroundings during heating, the graph would be a straight line but with a larger slope
(d) the internal energy of the liquid increases quadratically with time

34. Unpolarised red light is incident on the surface of a lake at incident angle θ_R. An observer seeing the light reflected from the water surface through a polariser notices that on rotating the polariser, the intensity of light drops to zero at a certain orientation. The red light is replaced by unpolarised blue light. The observer sees the same effect with reflected blue light at incident angle θ_B.

Then,

(a) $\theta_B < \theta_R < 45°$ (b) $\theta_B = \theta_R$
(c) $\theta_B > \theta_R > 45°$ (d) $\theta_R > \theta_B > 45°$

35. A neutral spherical copper particle has a radius of 10 nm (1 nm $= 10^{-9}$ m). It gets charged by applying the voltage slowly adding one electron at a time.

Then, the graph of the total charge on the particle *versus* the applied voltage would look like

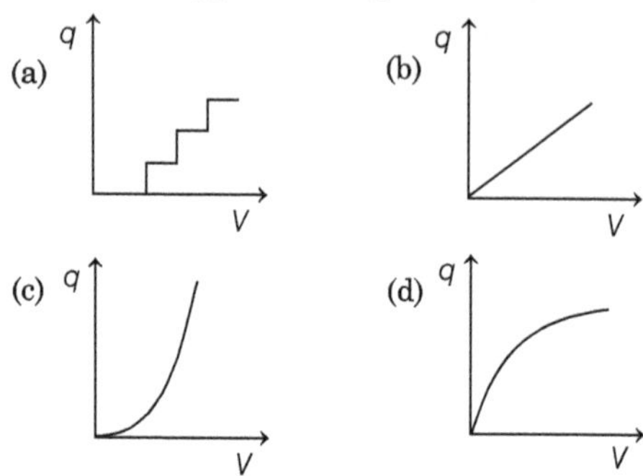

36. A charge $+q$ is distributed over a thin ring of radius r with line charge density $\lambda = q\sin^2\theta/(\pi r)$. Note that the ring is in the XY- plane and θ is the angle made by $\mathbf{r}$ with the X-axis. The work done by the electric force in displacing a point charge $+Q$ from the centre of the ring to infinity is

(a) equal to $qQ / 2\pi\varepsilon_0 r$
(b) equal to $qQ / 4\pi\varepsilon_0 r$
(c) equal to zero only, if the path is a straight line perpendicular to the plane of the ring
(d) equal to $qQ / 8\pi\varepsilon_0 r$

37. Originally the radioactive beta decay was thought as a decay of a nucleus with the emission of electrons only (Case I) . However, in addition to the electron, another (nearly) massless and electrically neutral particle is also emitted (Case II). Based on the figure below, which of the following is correct?

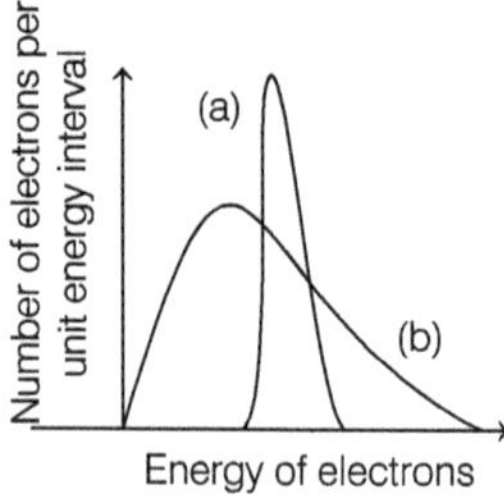

(a) (a) in both cases I and II
(b) (a) in case I and (b) in case II
(c) (a) in case II and (b) in case I
(d) (b) in both cases I and II

38. One gram mole of an ideal gas A with the ratio of constant pressure and constant volume specific heats $\gamma_A = 5/3$ is mixed with n gram moles of another ideal gas B with $\gamma_B = 7/5$. If the γ for the mixture is 19/13, then what will be the value of n?

(a) 0.75 (b) 2 (c) 1 (d) 3

39. How will the voltage (V) between the two plates of a parallel plate capacitor depend on the distance (d) between the plates, if the charge on the capacitor remains the same?

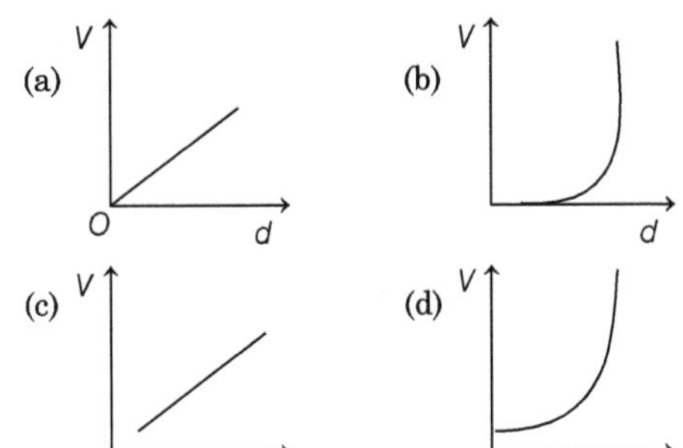

40. Three large identical plates are kept parallel to each other. The outer two plates are maintained at temperatures T and $2T$, respectively. The temperature of the middle plate in steady state will be close to

(a) 1.1 T (b) 1.3 T (c) 1.7 T (d) 1.9 T

CHEMISTRY

41. The major products of the following reaction.

$PhCOCBr_3 \xrightarrow{(aq)\ NaOH}$

are

(a) $Br_3C—OH$ and $PhCHO$

(b) $PhCOONa$ and $CHBr_3$

(c) $PhCOCHBr_2$ and NaBr

(d) PhH and CBr_3CO_2Na

42. Among the following,

I: 1-bromocyclohexene (Br) II: tert-butyl bromide (Br) III: MeO–C_6H_4–CH_2Br IV: 1-bromobicyclo[2.2.1]heptane (Br)

the compounds which can undergo an S_N1 reaction in an aqueous solution, are

(a) I and IV only (b) II and IV only
(c) II and III only (d) II, III and IV only

43. The major product of the following reaction.

$EtO–CO–(CH_2)_6–CN \xrightarrow[\text{Toluene, –78°C then } H_3O^+]{\text{Excess DIBAL-H}}$

is

(a) $H–CO–(CH_2)_6–CHO$

(b) $H–CO–(CH_2)_6–CN$

(c) $EtO–CO–(CH_2)_6–CHO$

(d) $HO–CO–(CH_2)_6–CHO$

44. Permanent hardness of water can be removed by

(a) heating
(b) treating with sodium acetate (CH_3COONa)
(c) treating with calcium hydrogen carbonate $Ca(HCO_3)_2$
(d) treatment with sodium hexametaphosphate ($Na_6P_6O_{18}$)

45. Alkali metals (M) dissolve in liquid NH_3 to give

(a) MNH_2
(b) MH
(c) $[M(NH_3)_x]^+ + [e(NH_3)_y]^-$
(d) M_3N

46. The absolute configurations of the following compounds

H, CH_2SH, H_3C, CH_2OH ; HO, H, H_3C, CH_2SH

respectively, are

(a) R and R (b) S and S
(c) R and S (d) S and R

47. The diamagnetic species among the following is

(a) O_2^+ (b) O_2^-
(c) O_2 (d) O_2^{2-}

48. Among the following transformations, the hybridisation of the central atom remains unchanged in

(a) $CO_2 \longrightarrow HCOOH$ (b) $BF_3 \longrightarrow BF_4^-$
(c) $NH_3 \longrightarrow NH_4^+$ (d) $PCl_3 \longrightarrow PCl_5$

49. For an octahedral complex MX_4Y_2 (M = a transition metal, X and Y are monodentate achiral ligands), the correct statement among the following is

(a) MX_4Y_2 has 2 geometrical isomers, one of which is chiral
(b) MX_4Y_2 has 2 geometrical isomers both of which are achiral
(c) MX_4Y_2 has 4 geometrical isomers, all of which are achiral
(d) MX_4Y_2 has 4 geometrical isomers, two of which are chiral

50. The values of the Henry's law constant of Ar, CO_2, CH_4 and O_2 in water at 25 °C are 40.30, 1.67, 0.41 and 34.86 kbar, respectively. The order of their solubility in water at the same temperature and pressure is

(a) $Ar > O_2 > CO_2 > CH_4$
(b) $CH_4 > CO_2 > Ar > O_2$
(c) $CH_4 > CO_2 > O_2 > Ar$
(d) $Ar > CH_4 > O_2 > CO_2$

51. Thermal decomposition of N_2O_5 occurs as per the equation below:

$$2N_2O_5 \longrightarrow 4NO_2 + O_2$$

The correct statement is

(a) O_2 production rate is four times the NO_2 production rate
(b) O_2 production rate is the same as the rate of disappearance of N_2O_5
(c) rate of disappearance of N_2O_5 is one-fourth of NO_2 production rate
(d) rate of disappearance of N_2O_5 is twice the O_2 production rate

52. For a first order chemical reaction,

(a) the product formation rate is independent of reactant concentration.
(b) the time taken for the completion of half of the reaction ($t_{1/2}$) is 69.3% of the rate constant (k)
(c) the dimension of Arrhenius pre-exponential factor is reciprocal of time.
(d) the concentration *vs* time plot for the reactant should be linear with a negative slope

53. The boiling point of 0.001 M aqueous solutions of $NaCl, Na_2SO_4, K_3PO_4$ and CH_3COOH should follows the order.

(a) $CH_3COOH < NaCl < Na_2SO_4 < K_3PO_4$
(b) $NaCl < Na_2SO_4 < K_3PO_4 < CH_3COOH$
(c) $CH_3COOH < K_3PO_4 < Na_2SO_4 < NaCl$
(d) $CH_3COOH < K_3PO_4 < NaCl < Na_2SO_4$

54. An allotrope of carbon which exhibits only two types of C — C bond distance of 143.5 pm and 138.3 pm, is

(a) charcoal (b) graphite
(c) diamond (d) fullerene

55. Nylon-2-nylon-6 is a co-polymer of 6-aminohexanoic acid and

(a) glycine (b) valine
(c) alanine (d) leucine

56. A solid is hard and brittle. It is an insulator in solid state, but conducts electricity in molten state. The solid is a

(a) molecular solid
(b) ionic solid
(c) metallic solid
(d) covalent solid

57. The curve that best describes the adsorption of a gas (x)g on 1.0 g of a solid substrate as a function of pressure (p) at a fixed temperature

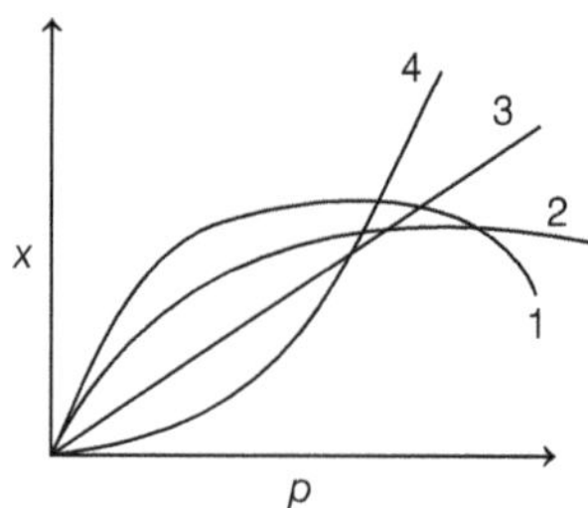

is

(a) 1 (b) 2 (c) 3 (d) 4

58. The octahedral complex $[Co(NH_3)_5SO_4]Cl$ exists in two isomeric forms X and Y. Isomer X reacts with $AgNO_3$ to give a white precipitate, but does not react with $BaCl_2$. Isomer Y gives white precipitate with $BaCl_2$ but does not react with $AgNO_3$.
Isomers X and Y are

(a) ionisation isomers (b) linkage isomers
(c) coordination isomers (d) solvate isomers

59. The correct order of basicity of the following amines

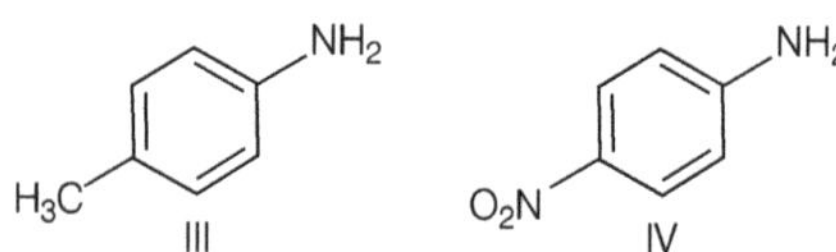

is

(a) I > II> III> IV (b) I > III> II> IV
(c) III > II> I> IV (d) IV > III> II> I

60. Electrolysis of concentrated aqueous solution of NaCl results in

(a) increase in pH of the solution
(b) decrease in pH of the solution
(c) O_2 liberation at the cathode
(d) H_2 liberation at the anode

BIOLOGY

61. Ethanol is used to treat methanol toxicity because ethanol

(a) is a competitive inhibitor of alcohol dehydrogenase
(b) is a non-competitive inhibitor of alcohol dehydrogenase
(c) activates enzymes involved in methanol metabolism
(d) inhibits methanol uptake by cells

62. Given below is a diagram of the stomatal apparatus. Match the labels with the corresponding names of the compounds.

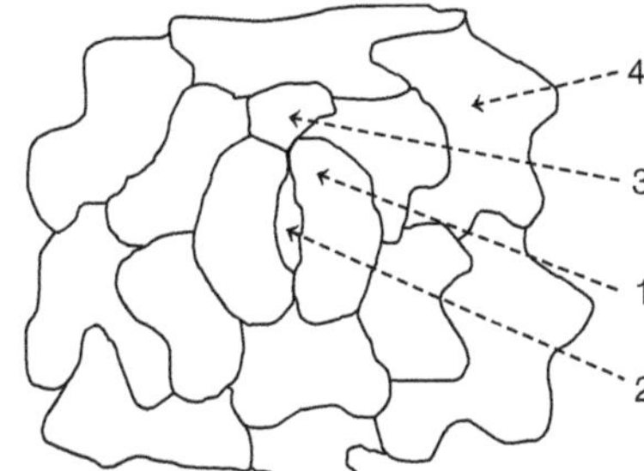

Choose the correct combination.

(a) 1. Stomatal pore;	2. Guard cell;	3. Epidermal cell;	4. Subsidiary cell
(b) 1. Guard cell;	2. Stomatal pore;	3. Subsidiary cell;	4. Epidermal cell
(c) 1. Subsidiary cell;	2. Guard cell;	3. Stomatal pore;	4. Epidermal cell
(d) 1.Guard cell;	2. Stomatal pore;	3. Epidermal cell;	4. Subsidiary cell

63. Which one of the following pairs was excluded from Whittaker's five kingdom classification?
(a) Viruses and lichens
(b) Algae and *Euglena*
(c) Lichens and algae
(d) *Euglena* and viruses

64. When a plant species is grown in shade tends to produce thinner leaves with more surface area, and when grown under abundant sunlight starts producing thicker leaves with reduced surface area. This phenomenon is an example of
(a) character displacement
(b) phenotypic plasticity
(c) natural selection
(d) genotypic variation

65. Sacred groves found in several regions in India are an example of
(a) *in situ* conservation (b) *ex situ* conservation
(c) reintroduction (d) restoration

66. Which one of the following immune processes is most effectively controlled by anti-histamines?
(a) Cell-mediated auto-immunity
(b) IgE-mediated exaggerated immune response
(c) IgG-mediated humoral immune response
(d) IgM-mediated humoral immune response

67. Which one of the following is explained by the endosymbiotic theory?
(a) The interaction between bacteria and viruses
(b) The symbiosis between plants and animals
(c) The origin of mitochondria and chloroplast
(d) The evolution of multicellular organisms from unicellular ones

68. According to the logistic population growth model, the growth rate is independent of
(a) per capita birth rate (b) per capita death rate
(c) resource availability (d) environmental fluctuations

69. A violent volcanic eruption wiped out most of the life forms in an island. Over different forms of simple organisms colonised this region, followed by the emergence of other organisms such as shrubs, woody plants, invertebrates and mammals. This ecological process is referred to as
(a) generation (b) replacement
(c) succession (d) turnover

70. Which one of the following microbial product is called "clot buster"?
(a) Cyclosporin-A (b) Paracetamol
(c) Statins (d) Streptokinase

71. Which one of the following elements is not directly involved in transcription?
(a) Promoter (b) Terminator
(c) Enhancer (d) *Ori C*

72. Which one of the following phyla is a pseudocoelomate?
(a) Cnidaria (b) Nematoda
(c) Mollusca (d) Chordates

73. Which one of the following glands does not secrete saliva?
(a) Submaxillary gland (b) Lacrimal gland
(c) Parotid gland (d) Sublingual gland

74. Which one of the following options correctly represents the tissue arrangement in roots?
(a) Cortex, pericycle, Casparian strip, vascular bundle
(b) Pericycle, cortex, Casparian strip, vascular bundle
(c) Cortex, Casparian strip, pericycle, vascular bundle
(d) Casparian strip, pericycle, cortex, vascular bundle

75. During fermentation of glucose to ethanol, glucose is
(a) first reduced and then oxidised
(b) only oxidised
(c) neither oxidised nor reduced
(d) only reduced

76. Which of the following is/are the product(s) of cyclic photophosphorylation?
(a) Both NADPH and H^+ (b) NADPH
(c) ATP (d) Both ATP and NADPH

77. Which one of the following amino acids is least likely to be in the core of a protein?
(a) Phenylalanine (b) Valine
(c) Isoleucine (d) Arginine

78. Which one of the following statements is a general feature of global species diversity?
(a) It increases from high to low altitudes
(b) It increases from low to high altitudes
(c) It changes over time but not spatially
(d) It changes randomly across space and time

79. Which one of the following conditions is not responsible for the presence of deoxygenated blood in the arteries of a newborn?

(a) Pneumonia
(b) Atrial septal defect
(c) Shunt between pulmonary artery and aorta
(d) Phenylketonuria

80. *Rhizobium* forms symbiotic association with roots in legumes and fixes atmospheric nitrogen. Which one of the following statement is correct about this process?

(a) Activity of nitrogenase is sensitive to oxygen
(b) Activity of nitrogenase is insensitive to oxygen
(c) Anaerobic conditions allow ATP independent conversion of nitrogen to ammonia
(d) Under aerobic conditions, atmospheric nitrogen can be converted to nitrates by *Rhizobium*

PART–II (2 Marks Questions)

MATHEMATICS

81. The points C and D on a semicircle with AB as diameter are such that $AC=1, CD=2$ and $DB=3$. Then, the length of AB lies in the interval.

(a) $[4, 4.1)$ (b) $[4.1, 4.2)$
(c) $[4.2, 4.3)$ (d) $[4.3, \infty)$

82. Let ABC be a triangle and let D be the mid-point of BC. Suppose $\cot(\angle CAD):\cot(\angle BAD)=2:1$. If G is the centroid of ΔABC, then the measure of $\angle BGA$ is

(a) 90° (b) 105°
(c) 120° (d) 135°

83. Let $f(x)=x^6-2x^5+x^3+x^2-x-1$ and $g(x)=x^4-x^3-x^2-1$ be two polynomials. Let a, b, c and d be the roots of $g(x)=0$. Then, the value of $f(a)+f(b)+f(c)+f(d)$ is

(a) –5 (b) 0
(c) 4 (d) 5

84. Let $\mathbf{a}=\hat{\mathbf{i}}+\hat{\mathbf{j}}+\hat{\mathbf{k}}, \mathbf{b}=2\hat{\mathbf{i}}+2\hat{\mathbf{j}}+\hat{\mathbf{k}}$ and $\mathbf{c}=5\hat{\mathbf{i}}+\hat{\mathbf{j}}-\hat{\mathbf{k}}$ be three vectors. The area of the region formed by the set of points whose position vectors $\vec{r}$ satisfy the equations $\mathbf{r}\cdot\mathbf{a}=5$ and $|\mathbf{r}-\mathbf{b}|+|\mathbf{r}-\mathbf{c}|=4$ is closest to the integer.

(a) 4 (b) 9
(c) 14 (d) 19

85. The number of solutions to

$$\sin(\pi\sin^2\theta)+\sin(\pi\cos^2\theta)=2\cos\left(\frac{\pi}{2}\cos\theta\right)$$

satisfying $0\le\theta\le 2\pi$ is

(a) 1 (b) 2
(c) 4 (d) 7

86. Let

$$J=\int_0^1\frac{x}{1+x^8}dx.$$

Consider the following assertions:

I. $J>\frac{1}{4}$ II. $J<\frac{\pi}{8}$

Then,

(a) only I is true (b) only II is true
(c) both I and II are true (d) neither I nor II is true

87. Let $f:(-1,1)\to R$ be a differentiable function satisfying

$(f'(x))^4=16(f(x))^2$ for all $x\in(-1,1)$

$f(0)=0$

The number of such functions is

(a) 2 (b) 3
(c) 4 (d) more than 4

88. For $x\in R$, let $f(x)=|\sin x|$ and $g(x)=\int_0^x f(t)dt$. Let $p(x)=g(x)-\frac{2}{\pi}x$. Then

(a) $p(x+\pi)=p(x)$ for all x
(b) $p(x+\pi)\ne p(x)$ for at least one but finitely many x
(c) $p(x+\pi)\ne p(x)$ for infinitely many x
(d) p is a one-one function

89. Let A be the set of vectors $\mathbf{a}=(a_1, a_2, a_3)$ satisfying

$$\left(\sum_{i=1}^{3}\frac{a_i}{2^i}\right)^2=\sum_{i=1}^{3}\frac{a_i^2}{2^i}$$

Then,

(a) A is empty
(b) A contains exactly one element
(c) A has 6 elements
(d) A has infinitely many elements

90. Let $f:[0,1]\to[0,1]$ be a continuous function such that $x^2+(f(x))^2\le 1$ for all $x\in[0,1]$ and $\int_0^1 f(x)dx=\frac{\pi}{4}$

Then, $\int_{\frac{1}{2}}^{\frac{1}{\sqrt{2}}}\frac{f(x)}{1-x^2}dx$ equals

(a) $\frac{\pi}{12}$ (b) $\frac{\pi}{15}$ (c) $\frac{\sqrt{2}-1}{2}\pi$ (d) $\frac{\pi}{10}$

PHYSICS

91. A metal rod of cross-sectional area 10^{-4}m^2 is hanging in a chamber kept at 20 °C with a weight attached to its free end. The coefficient of thermal expansion of the rod is $2.5 \times 10^{-6}\ \text{K}^{-1}$ and its Young's modulus is $4 \times 10^{12}\text{N/m}^2$. When the temperature of the chamber is lowered to T, then a weight of 5000 N needs to be attached to the rod, so that its length is unchanged. Then, T is

(a) 15 °C (b) 12°C
(c) 5°C (d) 0°C

92. A short solenoid (length l and radius r with n turns per unit length) lies well inside and on the axis of a very long, coaxial solenoid (length L, radius R and N turns per unit length, with $R > r$). Current I follows in the short solenoid. Choose the correct statement.

(a) There is uniform magnetic field $\mu_0 nI$ in the long solenoid.
(b) Mutual inductance of the solenoids is $\pi\mu_0 r^2 nNl$.
(c) Flux through outer solenoid due to current I in the inner solenoid is proportional to the ratio R/r.
(d) Mutual inductance of the solenoids is $\pi\mu_0 rRnNlL/(rR)^{1/2}$.

93. Consider the wall of a dam to be straight with height H and length L. It holds a lake of water of height $h(h < H)$ on one side. Let the density of water be ρ_w. Denote the torque about the axis along the bottom length of the wall by τ_1. Denote also a similar torque due to the water up to height $h/2$ and wall length $L/2$ by τ_2. Then, τ_1/τ_2 (ignore atmospheric pressure) is

(a) 2 (b) 4
(c) 8 (d) 16

94. Two containers C_1 and C_2 of volumes V and $4V$ respectively, hold the same ideal gas and are connected by a thin horizontal tube of negligible volume with a valve which is initially closed. The initial pressures of the gas in C_1 and C_2 are p and $5p$, respectively. Heat baths are employed to maintain the temperatures in the containers at 300 K and 400 K, respectively.

The valve is now opened. Select the correct statement.

(a) The gas will flow from the hot container to the cold one and the process is reversible.
(b) The gas will flow from one container to the other till the number of moles in two containers are equal.
(c) A long time after the valve is opened, the pressure in both the containers will be $3p$.
(d) A long time after the valve is opened, number of moles of gas in the hot container will be thrice that of the cold one.

95. Four electrons, each of mass m_e are in a one dimensional box of size L. Assume that, the electrons are non-interacting, obey the Pauli exclusion principle and are described by standing de Broglie waves confined within the box. Define $\alpha = h^2/8m_e L^2$ and U_0 to be the ground state energy. Then,

(a) the energy of the highest occupied state is 16α
(b) $U_0 = 30\alpha$
(c) the total energy of the first excited state is $U_0 + 9\alpha$
(d) the total energy of the second excited state is $U_0 + 8\alpha$

96. A rope of length L and uniform linear density is hanging from the ceiling. A transverse wave pulse, generated close to the free end of the rope, travels upwards through the rope. Select the correct option.

(a) The speed of the pulse decreases as it moves up.
(b) The time taken by the pulse to travel the length of the rope is proportional to $\sqrt{L}$.
(c) The tension will be constant along the length of the rope.
(d) The speed of the pulse will be constant along the length of the rope.

97. A circuit consists of a coil with inductance L and an uncharged capacitor of capacitance C. The coil is in a constant uniform magnetic field such that the flux through the coil is ϕ. At time $t = 0$ min, the magnetic field is abruptly switched OFF. Let $\omega_0 = 1/\sqrt{LC}$ and ignore the resistance of the circuit.

Then,

(a) current in the circuit is $I(t) = (\phi/L)\cos\omega_0 t$
(b) magnitude of the charge on the capacitor is $|Q(t)| = 2C\omega_0|\sin\omega_0 t|$
(c) initial current in the circuit is infinite
(d) initial charge on the capacitor is $C\omega_0\phi$

98. Consider the configuration of a stationary water tank of cross-section area A_0 and a small bucket as shown in figure below;

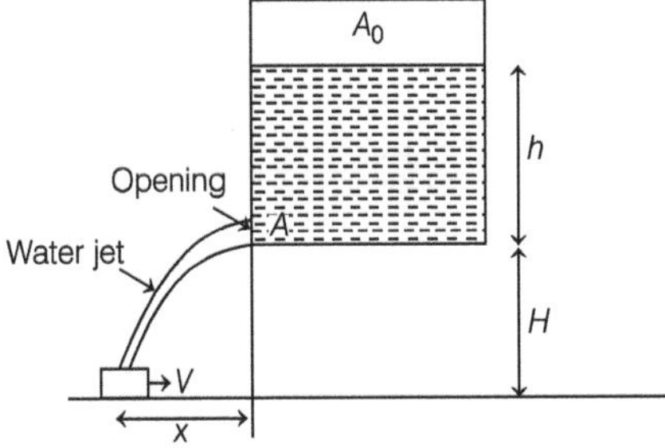

What should be the speed v of the bucket, so that the water leaking out of a hole of cross-section area A (as shown) from the water tank does not fall outside the bucket? (Take, $h = 5\text{m}$, $H = 5\text{m}$, $g = 10\text{m/s}^2$, $A = 5\text{cm}^2$ and $A_0 = 500\text{cm}^2$).

(a) 1 m/s (b) 0.5 m/s (c) 0.1 m/s (d) 0.05 m/s

99. The circuit below is used to heat water kept in a bucket.

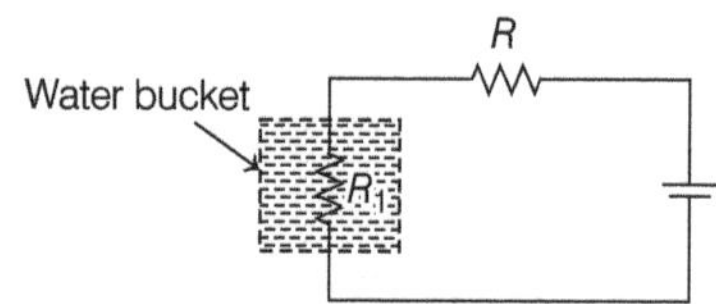

Assuming heat loss only by Newton's law of cooling, the variation in the temperature of the water in the bucket as a function of time is depicted by

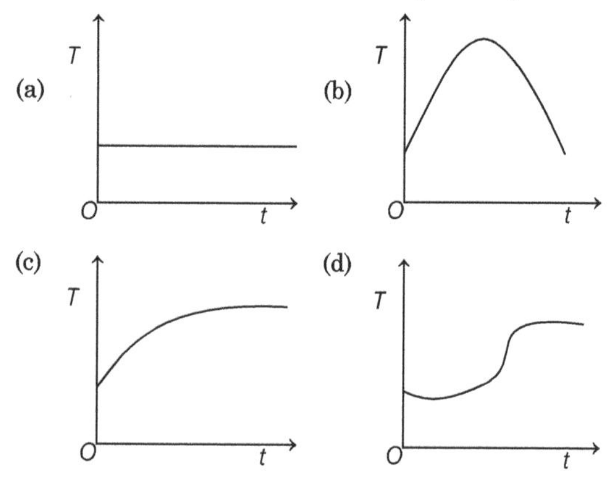

100. A bubble of radius R in water of density ρ is expanding uniformly at speed v. Given that water is incompressible, the kinetic energy of water being pushed is

(a) zero (b) $2\pi\rho R^3 v^2$
(c) $2\pi\rho R^3 v^2/3$ (d) $4\pi\rho R^3 v^2/3$

CHEMISTRY

101. The product of which of the following reactions forms a reddish brown precipitate when subjected to Fehling's test?

(a) Benzene $\xrightarrow[\text{anh. AlCl}_3\text{, CuCl}]{\text{CO, HCl}}$

(b) $CH_3CH_2COCl + (CH_3CH_2)_2Cd \longrightarrow$

(c) p-Toluic acid (COOH) $\xrightarrow[\text{2. H}_2\text{, Pd-BaSO}_4]{\text{1. PCl}_5}$

(d) 1-Methylcyclohexene $\xrightarrow[\text{2. Zn/H}_2\text{O}]{\text{1. O}_3}$

102. The major products X, Y and Z in the following sequence of transformations

Aniline (NH_2) $\xrightarrow{(CH_3CO)_2O} X \xrightarrow[\text{conc. H}_2\text{SO}_4\text{, 15°C}]{\text{conc.HNO}_3} Y \xrightarrow{aq.\text{NaOH}} Z$

are

(a) X= 4-aminoacetophenone (NH_2), Y= 4-amino-3-nitroacetophenone (NH_2, NO_2), Z= HO–CO– aryl with NH_2, NO_2

(b) X= acetanilide (H–N–COCH_3), Y= 4-nitroacetanilide (O_2N), Z= 4-nitroaniline (NH_2, O_2N)

(c) X= 2-aminoacetophenone (NH_2), Y= O_2N-substituted 2-aminoacetophenone, Z= O_2N-substituted 2-aminobenzoic acid (NH_2, OH)

(d) X= acetanilide, Y= 2-nitroacetanilide (NO_2), Z= 2-nitroaniline (NO_2, NH_2)

103. In the following reaction, P gives two products Q and R each in 40% yield.

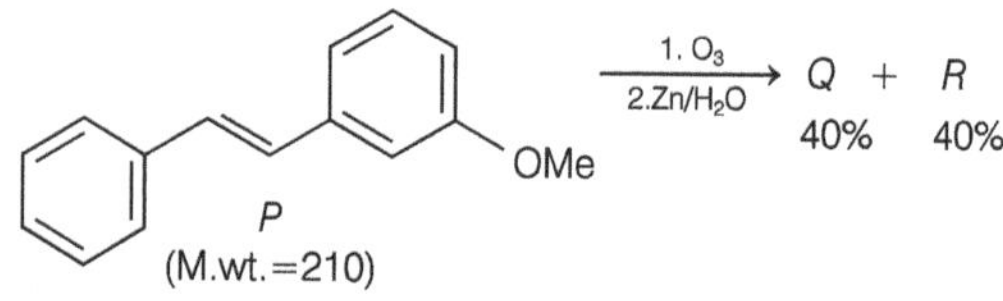

If the reaction is carried out with 420 mg of P, the reaction yields 108.8mg of Q. The amount of R produced in the reaction is closest to

(a) 97.6 mg (b) 108.8 mg
(c) 84.8 mg (d) 121.6 mg

104. Solubility products of CuI and Ag_2CrO_4 have almost the same value ($\sim 4\times10^{-12}$). The ratio of solubilities of the two salts (CuI: Ag_2CrO_4) is closest to

(a) 0.01 (b) 0.02 (c) 0.03 (d) 0.10

105. Given that, the molar combustion enthalpy of benzene, cyclohexane and hydrogen are x, y, and z, respectively, the molar enthalpy of hydrogenation of benzene to cyclohexane is

(a) $x - y + z$ (b) $x - y + 3z$
(c) $y - x + z$ (d) $y - x + 3z$

106. Among the following, the pair of paramagnetic complexes is

(a) $K_3[Fe(CN)_6]$ and $K_3[CoF_6]$
(b) $K_3[Fe(CN)_6]$ and $[Co(NH_3)_6]Cl_3$
(c) $K_4[Fe(CN)_6]$ and $K_3[CoF_6]$
(d) $K_4[Fe(CN)_6]$ and $[Co(NH_3)_6]Cl_3$

107. The major products X and Y in the following sequence of transformations

Benzene $\xrightarrow[\text{2. Molten NaOH,}\Delta \quad \text{3. } H_3O^+]{\text{1. Oleum}}$ X $\xrightarrow[\text{2. } CO_2 \quad \text{3. } H_3O^+]{\text{1. NaOH}}$ Y

are

(a) X= (benzene ring with HO and SO_3H), Y= (benzene ring with HO_2C and SO_3H)

(b) X= (benzene ring with OH), Y= (benzene ring with CO_2H)

(c) X= (benzene ring with SO_3H and OH), Y= (benzene ring with CO_2H and OH)

(d) X= (benzene ring with OH), Y= (benzene ring with CO_2H and OH)

108. 3.0g of oxalic acid $[(CO_2H)_2 \cdot 2H_2O]$ is dissolved in a solvent to prepare a 250mL solution. The density of the solution is 1.9 g/mL. The molality and normality of the solution, respectively, are closest to

(a) 0.10 mol kg^{-1} and 0.38 N
(b) 0.10 mol kg^{-1} and 0.19 N
(c) 0.05 mol kg^{-1} and 0.19 N
(d) 0.05 mol kg^{-1} and 0.09 N

109. In a titration experiment, 10 mL of an $FeCl_2$ solution consumed 25 mL of a standard $K_2Cr_2O_7$ solution to reach the equivalent point. The standard $K_2Cr_2O_7$ solution is prepared by dissolving 1.225g of $K_2Cr_2O_7$ in 250 mL water. The concentration of the $FeCl_2$ solution is closest to

[Given : molecular weight of $K_2Cr_2O_7$ = 294g mol^{-1}]

(a) 0.25 N (b) 0.50 N
(c) 0.10 N (d) 0.04 N

110. Atoms of an element Z form hexagonal closed packed (hcp) lattice and atoms of element X occupy all the tetrahedral voids. The formula of the compound is

(a) XZ (b) XZ_2
(c) X_2Z (d) X_4Z_3

BIOLOGY

111. In a population N_{AA} and N_{aa} are the numbers of homozygous individuals of allele 'A' and 'a', respectively and N_{Aa} is the number of heterozygous individuals. Which one of the following options is the allele frequency of 'A' and 'a' in a population with $N_{AA} = 90$, $N_{Aa} = 40$ and $N_{aa} = 70$?

(a) $A = 0.55$ and $a = 0.45$
(b) $A = 0.40$ and $a = 0.60$
(c) $A = 0.35$ and $a = 0.65$
(d) $A = 0.25$ and $a = 0.75$

112. A newly discovered organism possesses a genetic material with a new base composition consisting of the sugar and phosphate backbone as found in existing natural DNA. The five novel bases in this genetic material-namely, P, Q, R, S, T are heterocyclic structures with 1, 1, 2, 2 and 3 rings, respectively. Assuming the new DNA forms a double helix of uniform width, which one of the following would be the most appropriate base pairing?

(a) P with Q; R with T; S with T
(b) P with T; R with S; Q with T
(c) P with S; Q with R; S with T
(d) P with Q; R with S; S with T

113. Amino acid analysis of two globular protein samples yielded identical composition per mole. Which one of the following characteristics is necessarily identical for the two proteins?

(a) Disulphide bonds
(b) Primary structure
(c) Molecular mass
(d) Three-dimensional structure

114. Which of the following conversions in glycolysis is an example of substrate level phosphorylation?

(a) Glyceraldehyde -3-phosphate to 1, 3-bisphosphoglycerate
(b) 1, 3-bisphosphoglycerate to 3-phosphoglycerate
(c) Fructose 6-phosphate to fructose-1, 6-bisphosphate
(d) Glucose-6-phosphate to fructose-6-phosphate

115. A plant heterozygous for height and flower colour (TtRr) are selfed and 1600 of the resulting seeds are planted. If the distance between the loci controlling height and flower colour is 1 centiMorgan then how many offsprings are expected to be short with white flower (ttrr)?

(a) 1 (b) 10 (c) 100 (d) 400

116. Which one of the following will be the ratio of heavy, intermediate and light bands in Meselson and Stahl's experiment after two generations if DNA replication were conservative?

(a) 0 : 2 : 2 (b) 1 : 0 : 3
(c) 2 : 2 : 0 (d) 2 : 0 : 2

117. Given the graphs below, the interaction between species 1 and 2 can be classified as

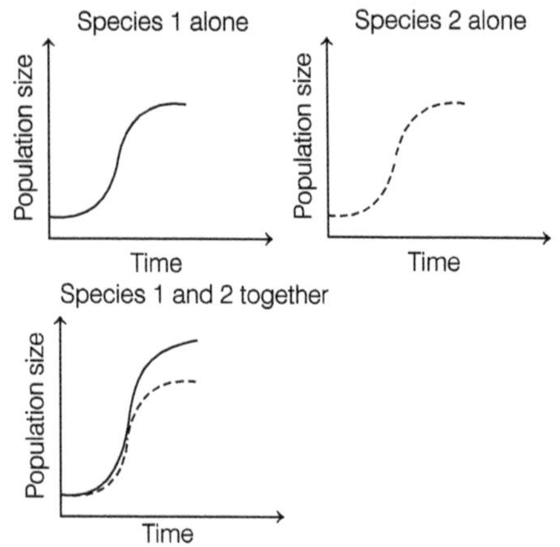

(a) amensalism
(b) commensalism
(c) mutualism
(d) competition

118. The additional nuclear ploidy levels found in a diploid angiosperm species in full bloom compared to its vegetative stage are

(a) 1N and 2N (b) 2N and 3N
(c) 3N and 4N (d) 1N and 3N

119. The bill sizes in a bird species of seed crackers from West Africa shows a bimodal distribution. Their most abundant food sources are two types of marsh plants that produce hard and soft seeds, consumed preferentially by the large and small billed birds, respectively. This bimodal distribution of bill sizes is a likely consequence of

(a) directional selection (b) stabilising selection
(c) disruptive selection (d) sexual selection

120. The containers X and Y have 1 litre of pure water and 1 litre of 0.1 M sugar solution, respectively. Which one of the following statements would be correct regarding their water potential (Ψ) and osmotic potential (Ψs)?

(a) Both Ψ and Ψs are zero in X
(b) Both Ψ and Ψs are zero in Y
(c) Ψ in X is zero and Ψs in Y is negative
(d) Ψ in X is negative and Ψs in Y is zero

Answers

PART-I

1	(b)	2	(b)	3	(b)	4	(a)	5	(b)	6	(b)	7	(a)	8	(b)	9	(b)	10	(c)
11	(c)	12	(c)	13	(a)	14	(c)	15	(a)	16	(c)	17	(d)	18	(a)	19	(a)	20	(d)
21	(a)	22	(b)	23	(c)	24	(d)	25	(b)	26	(d)	27	(c)	28	(d)	29	(a)	30	(a)
31	(c)	32	(a)	33	(b)	34	(c)	35	(a)	36	(b)	37	(c)	38	(b)	39	(c)	40	(c)
41	(b)	42	(c)	43	(a)	44	(d)	45	(c)	46	(d)	47	(d)	48	(c)	49	(b)	50	(c)
51	(d)	52	(c)	53	(a)	54	(d)	55	(a)	56	(b)	57	(b)	58	(a)	59	(b)	60	(a)
61	(a)	62	(b)	63	(a)	64	(b)	65	(a)	66	(b)	67	(c)	68	(d)	69	(c)	70	(d)
71	(d)	72	(b)	73	(b)	74	(c)	75	(c)	76	(c)	77	(d)	78	(a)	79	(d)	80	(a)

PART-II

81	(b)	82	(a)	83	(b)	84	(a)	85	(d)	86	(a)	87	(d)	88	(a)	89	(b)	90	(a)
91	(a)	92	(b)	93	(d)	94	(d)	95	(d)	96	(b)	97	(a)	98	(c)	99	(c)	100	(b)
101	(d)	102	(b)	103	(c)	104	(b)	105	(b)	106	(a)	107	(d)	108	(c)	109	(a)	110	(c)
111	(a)	112	(b)	113	(c)	114	(b)	115	(c)	116	(d)	117	(b)	118	(d)	119	(c)	120	(c)

Solutions

1. *(b)* Out of given three letters a, b, c, we will repeat exactly one of the letter and we can do this in $^3C_1$ ways.

Now, we have four letters and out of these four letters two are identical, so number of ways to arrange these four letters is $\frac{4!}{2!}$.

So, the number of four letter words that can be formed with letters a, b, c such that all three letters occur is

$$^3C_1 \times \frac{4!}{2!} = 36$$

2. *(b)* Given trigonometric relation is

$$\left(\frac{1}{3}\sin\theta + \frac{2}{3}\cos\theta\right)^2 = \frac{1}{3}\sin^2\theta + \frac{2}{3}\cos^2\theta$$

$$\Rightarrow \quad \frac{1}{9}\sin^2\theta + \frac{4}{9}\cos^2\theta + \frac{4}{9}\sin\theta\cos\theta = \frac{1}{3}\sin^2\theta + \frac{2}{3}\cos^2\theta$$

$$\Rightarrow \frac{2}{9}\sin^2\theta + \frac{2}{9}\cos^2\theta - \frac{4}{9}\sin\theta\cos\theta = 0$$

$$\Rightarrow \quad \sin^2\theta + \cos^2\theta - 2\sin\theta\cos\theta = 0$$

$$\Rightarrow \quad \sin 2\theta = 1$$

$$\Rightarrow \quad 2\theta = 2n\pi + \frac{\pi}{2}, n \in I$$

$$\Rightarrow \quad \theta = n\pi + \frac{\pi}{4}, n \in I$$

$$\therefore \quad A = \left\{\theta \in R : \theta = n\pi + \frac{\pi}{4}, n \in I\right\}$$

$$\therefore A \cap [0, \pi] = \left\{\frac{\pi}{4}\right\}$$

$\therefore$ $A \cap [0, \pi]$ has exactly one point.

3. *(b)* The area of region bounded by the curves

$$x(y - e^x) = \sin x \text{ or } y = \frac{\sin x}{x} + e^x \text{ and}$$

$$2xy = 2\sin x + x^3$$

or

$$y = \frac{\sin x}{x} + \frac{x^2}{2}$$

between lines $x = 1$ and $x = 2$ is

$$A = \left|\int_1^2 \left\{\left(\frac{\sin x}{x} + e^x\right) - \left(\frac{\sin x}{x} + \frac{x^2}{2}\right)\right\} dx\right|$$

$$= \left|\int_1^2 \left(e^x - \frac{x^2}{2}\right) dx\right| = \left|\left[e^x - \frac{x^3}{6}\right]_1^2\right| = e^2 - e - \frac{7}{6}$$

4. *(a)* According to given informations in the question, if radius of circle C_1 is $2a$, centre c is at origin (0, 0) and AB is along X-axis, then $B(2a, 0)$, $A(-2a, 0)$, $E(0, 2a)$ and $F(-a, -a)$.

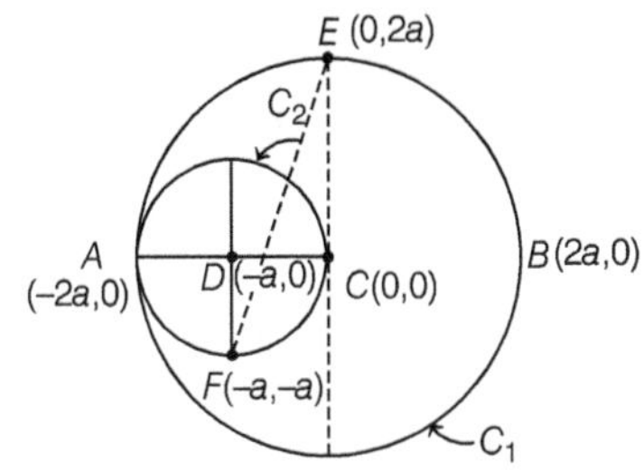

Let $\angle FEC = \theta$

Then, slope of line $EF = m = \tan\left(\frac{\pi}{2} - \theta\right)$

$$\Rightarrow \cot\theta = \frac{2a - (-a)}{0 - (-a)} = 3 \qquad \left[\because m = \frac{y_2 - y_1}{x_2 - x_1}\right]$$

$$\therefore \sin(\angle FEC) = \sin\theta = \frac{1}{\sqrt{1 + \cot^2\theta}} = \frac{1}{\sqrt{1 + 9}} = \frac{1}{\sqrt{10}}$$

5. *(b)* Given, $-3x^4 + \det\begin{bmatrix} 1 & x & x^2 \\ 1 & x^2 & x^4 \\ 1 & x^3 & x^6 \end{bmatrix} = 0$

$$\Rightarrow \quad x^8 + x^5 + x^5 - x^4 - x^7 - x^7 = 3x^4$$

$$\Rightarrow \quad x^8 - 2x^7 + 2x^5 - 4x^4 = 0$$

$$\Rightarrow \quad x^4[x^4 - 2x^3 + 2x - 4] = 0$$

$$\Rightarrow \quad x^4[x^3(x - 2) + 2(x - 2)] = 0$$

$$\Rightarrow \quad x^4(x^3 + 2)(x - 2) = 0$$

$\because x$ is an integer, so $x = 0, 2$.

6. *(b)* It is given that P is a non-zero polynomial such that, $P(1 + x) = P(1 - x)$

On differentiating both sides w.r.t. x, we get

$$P'(1 + x) = -P'(1 - x)$$

Now on putting $x = 0$, we get

$$P'(1) = -P'(1)$$

$$\Rightarrow \quad P'(1) = 0$$

i.e the polynomial P touches the X-axis at $x = 1$ and $P(1) = 0$ (given)

$$\therefore \quad P(x) = (x - 1)^2 Q(x)$$

$\therefore$ Largest integral value of 'm' such that $(x - 1)^m$ divides $P(x)$ is 2.

7. *(a)* Given function

$$f(x) = \begin{cases} x\sin\left(\frac{1}{x}\right), & \text{when } x \neq 0 \\ 1, & \text{when } x = 0 \end{cases}$$

Now, for $x = 0$, $f(x) = 1$

and for $x \neq 0$, $f(x) = 1 \Rightarrow \sin\frac{1}{x} = \frac{1}{x}$ has no solution.

$\therefore$ The set $A = \{x \in R : f(x) = 1\}$ has exactly one element.

8. *(b)* The radius of the smallest circle with centre at the origin and having non-empty intersection with $S = [(x, y) : |x| + 2|y| = 1]$ is radius of inscribed circle of rhombus having sides represented by equation $|x| + 2|y| = 1$.

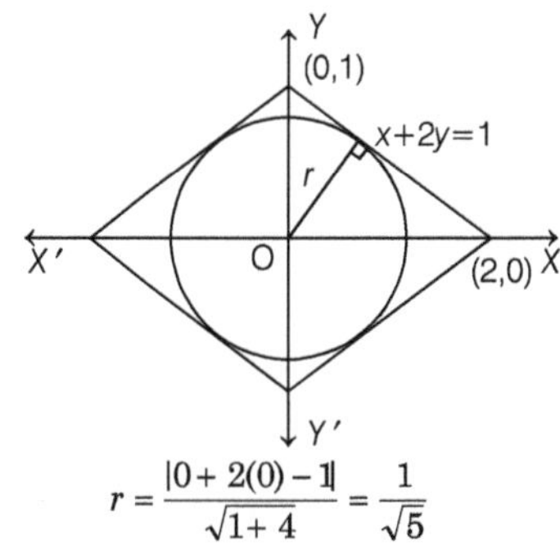

$$r = \frac{|0 + 2(0) - 1|}{\sqrt{1 + 4}} = \frac{1}{\sqrt{5}}$$

9. *(b)* Given trigonometric equation is

$$\sin(9x) + \sin(3x) = 0$$

$$\Rightarrow \quad 2\sin 6x \cos 3x = 0$$

$$\therefore \quad \text{either } \sin 6x = 0$$

$$\text{or} \quad \cos 3x = 0$$

for $x \in [0, 2\pi]$

$\sin 6x = 0$

$$\Rightarrow \quad x = 0, \frac{\pi}{6}, \frac{\pi}{3}, \frac{\pi}{2}, \frac{2\pi}{3}, \frac{5\pi}{6},$$

$$\pi, \frac{7\pi}{6}, \frac{4\pi}{3}, \frac{3\pi}{2}, \frac{5\pi}{3}, 2\pi$$

and $3x = 0$

$$\Rightarrow \quad x = \frac{\pi}{6}, \frac{\pi}{2}, \frac{5\pi}{6}, \frac{7\pi}{6}, \frac{3\pi}{2}, \frac{11\pi}{6}$$

So, number of solution of given equation is 13.

10. *(c)* Area of parallelogram whose diagonals are 10 and 4 and if angle between adjacent side is θ is

$$A = \frac{10 \times 4}{\sin\theta}$$

The area will be maximum if $\theta = \frac{\pi}{2}$, so the parallelogram must be rhombus

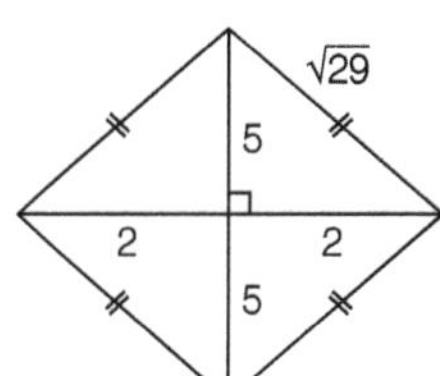

Perimeter of rhombus

$$= 4\sqrt{29} = \sqrt{464} \text{ and } \sqrt{464} \in (21, 22]$$

11. *(c)* For positive integers 'a' and 'b' the numbers $\frac{2a-1}{b}$ and $\frac{2b-1}{a}$ are integers if 'a' and 'b' and odd integers because '$2a-1$' and '$2b-1$' are odd integers.

Now, let $\frac{2a-1}{b}=\alpha$ and $\frac{2b-1}{a}=\beta$,

where α, β are integers.

then $2a-1=\alpha b$ and $2b-1=\beta a$

so $4a-2=\alpha(\beta a+1)$

$$\Rightarrow \quad a=\frac{\alpha+2}{4-\alpha\beta}$$

$\because a$ is an integer, then $0<\alpha\beta<4$

$\therefore$ Possible value of $\alpha=1, 2$ or 3 and $\beta=1, 2$ or 3

such that $0<\alpha\beta<4$

Now, when $(\alpha, \beta)=(1, 1)$, then$(a, b)=(1, 1)$

When $(\alpha, \beta)=(1, 2)$, then (a, b) have no value

When $(\alpha, \beta)=(1, 3)$, then $(a, b)=(3, 5)$

and similarly when $(\alpha, \beta)=(3, 1)$, then $(a, b)=(5, 3)$

So, number of ordered pairs (a, b) is 3.

12. *(c)* It is given that complex numbers $z=x+iy$ and $w=u+iv$ are on the unit circle such that $z^2+w^2=1$ ...(i)

So, $\overline{(z^2)}+\overline{(w^2)}=1$

$\Rightarrow (\bar{z})^2+(\bar{w})^2=1$

$\Rightarrow \left(\frac{\bar{z}z}{z}\right)^2+\left(\frac{\bar{w}w}{w}\right)^2=1$

$\Rightarrow \frac{1}{z^2}+\frac{1}{w^2}=1 \Rightarrow z^2+w^2=z^2w^2$

$\Rightarrow z^2w^2=1$ ...(ii)

$\because$ The number of solution of equations $x^2+y^2=1$ and $x^2y^2=1$ is eight

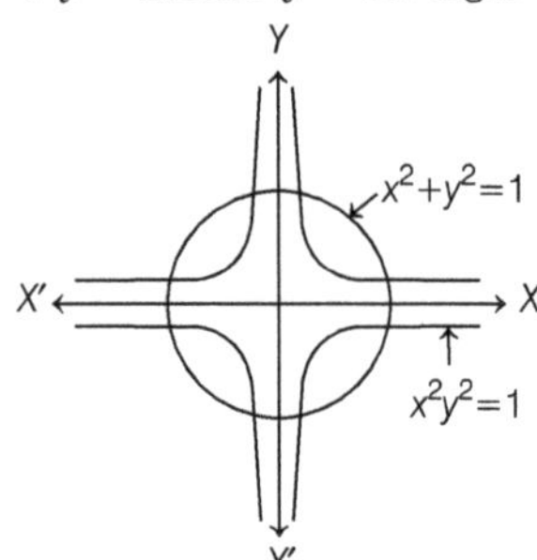

$\therefore$ Number of ordered pair (z, w) is also eight.

13. *(a)* The number of all four-letter words $=26^4$

The number of all four-letter words from set $V=\{a, e, i, o, u\}$ i.e vowels $=5^4$

Similarly, the number of all four-letter words from set of consonants $=21^5$

$\therefore$ Number of words which contains atleast one vowel and atleast one consonant

$=26^4-21^4-5^4$

$=(26^2+21^2)(26^2-21^2)-5^4$

$=(676+441)(5)(47)-5^4$

$=5[(1117\times 47)-125]$

$=5\times(52374)=261870$

14. *(c)* According to given information equation of planes

$\sigma_1; x+y+z=d_1,$

$\sigma_2: ax+by+cz=d_2$

and $\sigma_3: a^2x+b^2y+c^2z=d_3$

Now, for unique solution

$\Delta \neq 0.$

and $\Delta=\begin{vmatrix}1 & 1 & 1\\ a & b & c\\ a^2 & b^2 & c^2\end{vmatrix}$

On applying $c_2 \to c_2-c_1$ and $c_3 \to c_3-c_1$

$$\Delta=\begin{vmatrix}1 & 0 & 0\\ a & b-a & c-a\\ a^2 & b^2-a^2 & c^2-a^2\end{vmatrix}=(b-a)(c-a)\begin{vmatrix}1 & 0 & 0\\ a & 1 & 1\\ a^2 & b+a & c+a\end{vmatrix}$$

On applying $c_3 \to c_3-c_2$

$$(b-a)(c-a)\begin{vmatrix}1 & 0 & 0\\ a & 1 & 0\\ a^2 & b+a & c-b\end{vmatrix}$$

$=(a-b)(b-c)(c-a)$

So, if a, b and c has any three distinct values then $\Delta \neq 0$.

15. *(a)* It is given that, Ravi and Rashmi are each holding 2 red cards and 2 black cards (all four red and all four black cards are identical).

Let Ravi picks a Red card from Rashmi and Rashmi picks black card from Ravi and in second time also Ravi picks a red card from Rashmi and Rashmi picks black card from Ravi, then finally Ravi have four red cards and Rashmi have four black cards, then probability of such event is

$$\frac{{}^2C_1}{{}^4C_1}\times\frac{{}^2C_1}{{}^5C_1}\times\frac{{}^1C_1}{{}^4C_1}\times\frac{{}^1C_1}{{}^5C_1}=\frac{4}{400}=\frac{1}{100}$$

Similarly, the probability if at the end Ravi have four black and Rashmi have four red cards is $\frac{1}{100}$.

So, required probability

$=\frac{1}{50}=2\%=p\leq 5\%$

16. *(c)* Area of region defined by

$$A_1=\left\{(x, y): x\geq 0, y\geq 0, 2x+2y-x^2-y^2>1>x+y\right\}$$

is $|A_1|$.

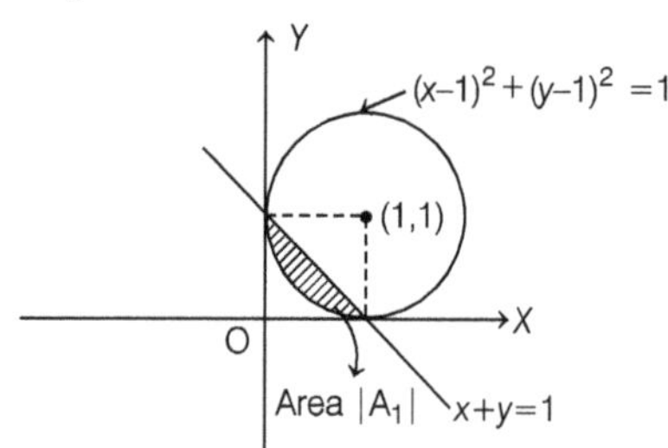

Similarly, $|A_2|$ is the area of region defined by

$A_2=\{(x, y): x\geq 0, y\geq 0, x+y>1>x^2+y^2\}$

Here, $|A_1|=|A_2|$

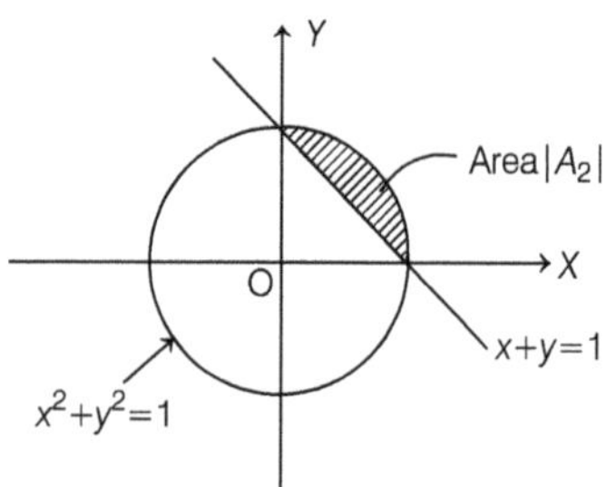

and area $|A_3|$ is the area of region defined by

$A_3=\{(x, y): x\geq 0, y\geq 0, x+y>1>x^3+y^3\}$

then $|A_3|>|A_2|=|A_1|$

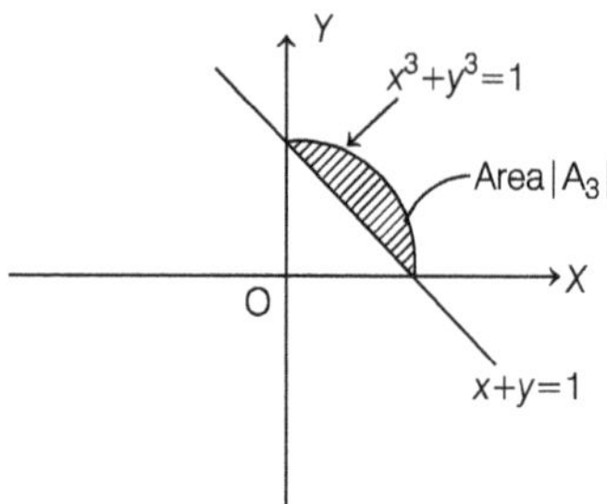

17. *(d)* Given function $f: R \longrightarrow R$ be a continuous function such that

$f(x^2)=f(x^3)\,\forall x\in R$

then $f(x)=f(x^{2/3})$ [on replacing x by $x^{1/3}$]

Similarly,

$f(x)=f(x^{2/3})=f(x^{4/9})=f(x^{8/27})=$

$\ldots=f(x^{(2/3)^n})$

$=f(x^0)$ [as x tends to infinity] $=f(1)$

$\therefore f(x)=f(1)=$ constant

The function $f(x)=$ constant is even and differentiable everywhere.

18. *(a)* It is given that a continuous function

$f : [0, \infty) \longrightarrow R$ satisfies

$$f(x) = 2\int_0^x tf(t)dt + 1, \forall x \geq 0$$

$$\therefore \quad f'(x) = 2xf(x)$$

$$\Rightarrow \quad \frac{f'(x)}{f(x)} = 2x$$

On integrating both sides, we get

$\ln|f(x)| = x^2 + c$

$\Rightarrow f(x) = Ke^{x^2}$, where $K = e^c$

$\because \; f(0) = 2\int_0^0 t\,f(t)dt + 1 = 1$

$\therefore \quad K = 1$

$\therefore \quad f(1) = e$

19. *(a)* Given relation

$$(1 + a^2)\;(1 + b^2) = 4ab$$

$$\Rightarrow \quad a^2 + b^2 - 2ab = 2ab - 1 - a^2b^2$$

$$\Rightarrow \quad (a - b)^2 = -(1 - ab)^2$$

$\because a > 0, a \neq 1$ and b is a positive real number

$\therefore (a - b)^2 \neq 0 \neq -(1 - ab)^2$, because $(a - b)^2$ and $(1 - ab)^2$ are non-negative real numbers.

$\therefore$ Set S is an empty set.

20. *(d)* Given function

$$f(x) = \begin{bmatrix} \dfrac{\sin(x^2)}{x} & , x \neq 0 \\ 0 & , \text{if } x = 0 \end{bmatrix}$$

then $\lim\limits_{x \to 0} f(x) = \lim\limits_{x \to 0} \dfrac{\sin x^2}{x} = \lim\limits_{x \to 0}$

$$x\frac{\sin x^2}{x^2} = 0 = f(0)$$

Hence, $f(x)$ is continuous at $x = 0$

Now, for differentiability

RHD (at $x = 0$)

$$= \lim_{h \to 0} \frac{f(0 + h) - f(0)}{h} = \lim_{h \to 0} \frac{\sin h^2}{h^2} = 1$$

and LHD (at $x = 0$)

$$= \lim_{h \to 0} \frac{f(0 - h) - f(0)}{-h} = \lim_{h \to 0} \frac{\sin h^2}{h^2} = 1$$

So, $f(x)$ is differentiable at $x = 0$

$$\therefore \quad f'(x) = \begin{bmatrix} 2\cos(x^2) - \dfrac{\sin x^2}{x^2} & , \text{if } x \neq 0 \\ 1 & , \text{if } x = 0 \end{bmatrix}$$

$$\because \lim_{x \to 0} f'(x) = \lim_{x \to 0} \left[2\cos(x^2) - \frac{\sin x^2}{x^2}\right]$$

$$= 2 - 1 = 1$$

$\therefore \lim\limits_{x \to 0} f'(x) = f'(0)$

So, $f(x)$ is differentiable and the derivative is continuous

21. *(a)* A muon is an unstable elementary particle of mass nearly 200 m_e and charge $\pm$ e.

Here, a negative muon is given bound to a proton.

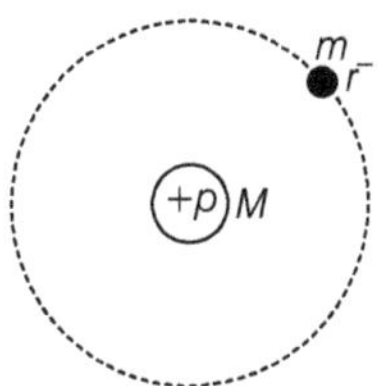

So, $m = 200\ m_e$ and $M = 1836\ m_e$

(as mass of a proton is 1836 times of mass of electron)

So, reduced mass of system,

$$m' = \frac{mM}{m + M} = \frac{200\ m_e \times 1836 m_e}{200\ m_e + 1836\ m_e} \approx 180\ m_e$$

As mass of muon is comparable to mass of proton, we have to take account of motion of nucleus. That's way we are calculating reduced mass.

Now, as energy of an electron in nth orbit is of H-atom,

$$E_n = \frac{me^4}{8\varepsilon_0^2 h^2 n^2}$$

So, energy of a muon of muonic atom is

$$E_n' = \frac{m'e^4}{8\varepsilon_0^2 h^2 n^2} \Rightarrow E_n' = 180\ E_n$$

Now, considering a Balmer transition, $n = 3$ to $n = 2$.

$$\therefore \quad \Delta E' = \frac{hc}{\lambda}$$

$$\text{or} \quad \Delta E' = E'_{n=3} - E'_{n=2} = \frac{hc}{\lambda}$$

$$\Rightarrow \quad 180\ (E_{n=3} - E_{n=2}) = \frac{hc}{\lambda}$$

$$\Rightarrow \quad 180\left(\frac{-13.6}{3^2} - \frac{-13.6}{2^2}\right) = \frac{hc}{\lambda}$$

$$\Rightarrow \quad 180 \times 1.89 \text{eV} = \frac{1240(\text{eV-nm})}{\lambda(\text{nm})}$$

$$\Rightarrow \quad \lambda(\text{nm}) = 3.6\,\text{nm}$$

This emitted radiation is X-rays.

Note X-rays have a wavelength range of 0.01 nm to 10 nm.

22. *(b)* As per the condition given in question, we have following picture of atom,

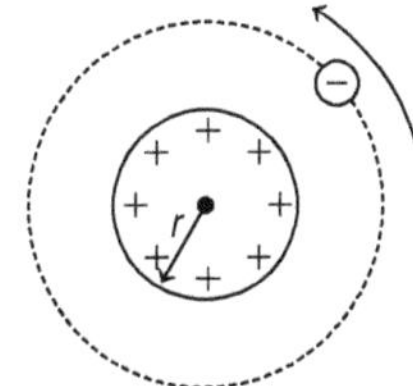

As a spherical charge distribution acts like a point charge concentrated at centre, above model is equivalent to Bohr's model.

Hence, we expert same energy value of electron as that of Bohr's atom.

$\therefore$ Ground state energy of electron $= -13.6$ eV.

Note Question is incorrect as Thomson's model does not have any ground state configuration. Thomson's model of H-atom is as shown below;

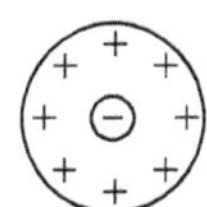

Now, if we calculate energy is above case, we get

$$U = V_{\text{centre}} = \frac{3}{2} \times \frac{1}{4\pi\varepsilon_0} \times \frac{e}{\varepsilon} \times -e$$

$$= -86.4\,\text{eV}$$

Not matching with any option.

23. *(c)* As collision is elastic and masses of colliding objects are equal, so there is exchange of total energy of moving mass to stationary mass.

Hence, velocity of block = velocity of sphere just before collision.

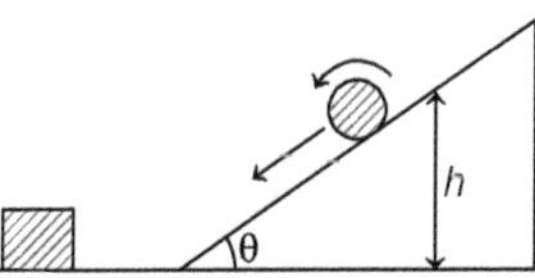

For sphere, if v = velocity of translation of centre of mass at the bottom of plane and ω = angular speed, then by energy conservation, we have

$$mgh = \frac{1}{2}mv^2 + \frac{1}{2}I\omega^2$$

Here, $\omega = \dfrac{v}{R}$ and $\quad I = \dfrac{2}{5}mR^2$

Substituting and solving for velocity, we have

$$v = \sqrt{\frac{10}{7}gh} = \sqrt{\frac{10 \times 10 \times 7}{7}} = 10\ \text{m/s}$$

$\therefore$ Velocity of block after collision $= 10$ m/s.

24. *(d)* As the apple is released, it has two accelerations,

a_1 = pseudo acceleration due to acceleration to train

and a_2 = acceleration due to gravity.

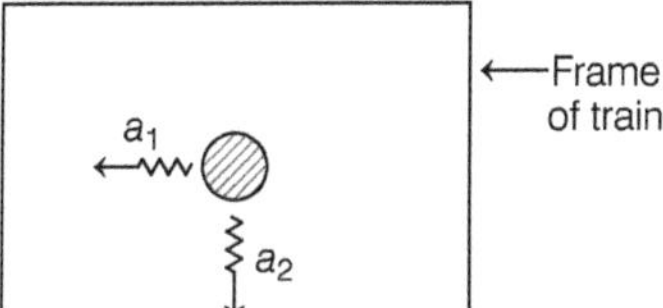

Resultant of these accelerations are as shown below.

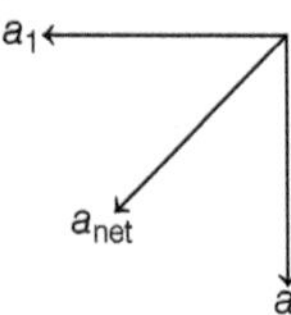

Hence, trajectory of falling apple is an inclined straight line opposite to the direction of train.

25. *(b)* As man alight, its angular momentum about a point on platform is conserved.

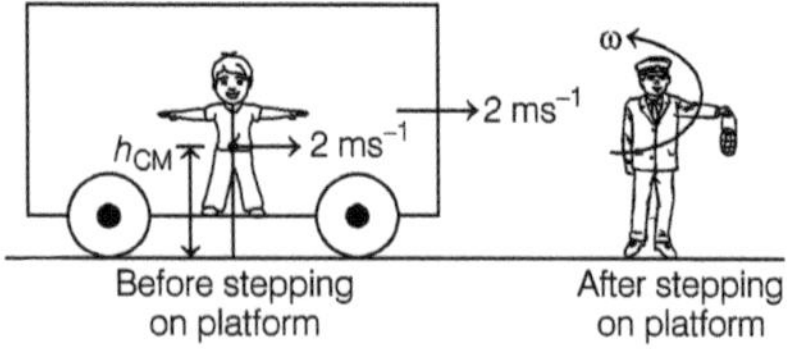

By angular momentum conservation,

$$L_{\text{initial}} = L_{\text{final}} \Rightarrow mvr = I\omega$$

$$\Rightarrow \quad mvr = \frac{ml^2}{3} \cdot \omega \Rightarrow \frac{3vr}{l^2} = \omega$$

or $$\omega = \frac{3 \times 2 \times \left(\frac{1.5}{2}\right)}{(1.5)^2} = 2 \text{rad/s}$$

So, instantaneous angular speed of man = 2 rad/s.

26. *(d)* As collision is elastic both linear momentum and kinetic energy are conserved.

We have, following given condition,

M →u_1 (M/2) u_2=0

Before collision

M →v_1 (M/2) →v_2

After collision

Momentum conservation gives,

$$Mu_1 = Mv_1 + \frac{M}{2}v_2 \Rightarrow 2u_1 = 2v_1 + v_2 \quad \text{... (i)}$$

Energy conservation gives,

$$\frac{1}{2}Mu_1^2 = \frac{1}{2}Mv_1^2 + \frac{1}{2}\frac{M}{2}v_2^2$$

$$\Rightarrow \quad 2u_1^2 = 2v_1^2 + v_2^2 \quad \text{... (ii)}$$

Now, substituting for u_1 from Eq. (i) in Eq. (ii), we get

$$2\left(\frac{2v_1 + v_2}{2}\right)^2 = 2v_1^2 + v_2^2$$

$$\Rightarrow \quad 4v_1^2 + v_2^2 + 4v_1v_2 = 4v_1^2 + 2v_2^2$$

$$\Rightarrow \quad v_2^2 = 4v_1v_2 \text{ or } \frac{v_1}{v_2} = \frac{1}{4}$$

$\therefore$ Ratio of final velocities of M and $\frac{M}{2}$ is $\frac{1}{4}$.

27. *(c)* Force on the object is $F = k - \beta t^2$

v=–15 ms^{-1} F m=2/3 kg t=–2s

Acceleration of the particle,

$$a = \frac{F}{m} = \frac{k - \beta t^2}{m}$$

Acceleration is zero when $k = \beta t^2$ or

$$t^2 = \frac{k}{\beta} \text{ or } t^2 = \frac{8}{2} \Rightarrow t = 2\text{s}$$

Now, $$a = \frac{dv}{dt} = \frac{k - \beta t^2}{m} \Rightarrow dv = \frac{k - \beta t^2}{m} \cdot dt$$

Integrating between given limits, we have

$$\Rightarrow \int_{v=-15}^{v(t=2s)} dv = \int_{t=-2s}^{t=2s} \frac{k - \beta t^2}{m}\, dt$$

$$= \frac{3}{2}\int_{t=-2}^{t=2} (8 - 2t^2)dt$$

$$v(t = 2s) - (-15) = \frac{3}{2}\left[8t - \frac{2t^3}{3}\right]_{-2}^{2}$$

$$\Rightarrow v_{\text{at } t=2s} = -15 + \frac{3}{2} \times \left[8(2-(-2)) - \frac{2}{3}(8-(-2)^3)\right]$$

$$\Rightarrow v_{\text{at } t=2s} = -15 + \frac{3}{2}\left(32 - \frac{32}{3}\right)$$

$$= -15 + \frac{3}{2}\left(\frac{64}{3}\right)$$

$$= -15 + 32 = 17 \text{ ms}^{-1}$$

28. *(d)* Intensity of radiation of sun on earth or solar constant,

$$S_1 = \frac{P}{4\pi R_0^2} = \frac{4\pi R_s^2 \cdot \sigma \cdot T_s^4}{4\pi R_0^2}$$

$$= \sigma \frac{R_s^2}{R_0^2} \cdot T_s^4 \quad \text{...(i)}$$

where, R_s = radius of sun,
R_o = radius of orbit,
σ = Stefan-Boltzmann constant and
T_s = temperature of sun.

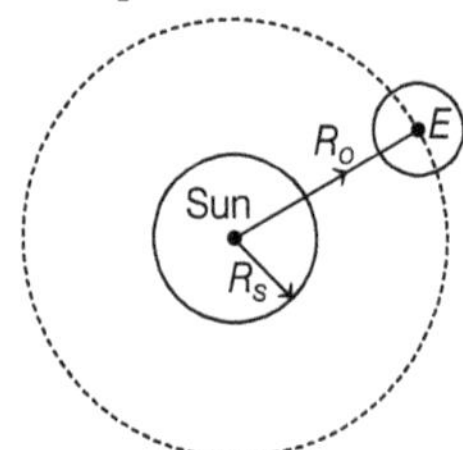

Now, intensity of radiation received from steller body on earth's surface,

$$S_2 = \frac{\sigma(50R_s)^2}{(2 \times 10^{10} R_0)^2} \cdot (2T_s)^4$$

$$\Rightarrow \quad S_2 = \frac{2500 \times 16}{4 \times 10^{20}} \times \sigma . \frac{R_s^2}{R_0^2} . T_s^4$$

$$\Rightarrow \quad S_2 = 10^{-16} S_1 \quad \text{(from Eq. (i))}$$

$$\Rightarrow \quad \frac{S_2}{S_1} = 10^{-16}$$

29. *(a)* When rod is stretched, its length increases. Potential drop across the rod also increases due to increase in resistance of rod.

Resistance of rod,

$$R = \frac{\rho l}{A} = \frac{\rho l^2}{X}$$

($\because$ volume of rod, $X = Al$)

Change in resistance of rod,

$$\Delta R = \frac{\rho \,.\, 2l\Delta l}{X}$$

Change in potential drop across rod,

$$\Delta V = i\Delta R = \frac{i \cdot \rho \cdot 2l\Delta l}{X} = \frac{2i\rho l^2}{X} . \frac{\Delta l}{l} = \frac{2i\rho l^2}{X} . \varepsilon$$

or $$\Delta V \propto \varepsilon$$

As change in potential drop is directly proportional to strain, voltage as a function of strain is as shown below;

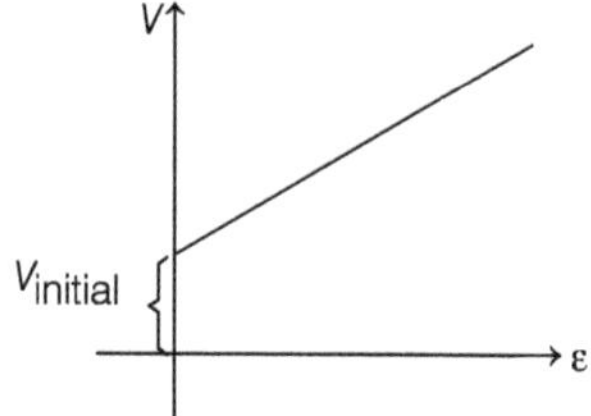

30. *(a)*

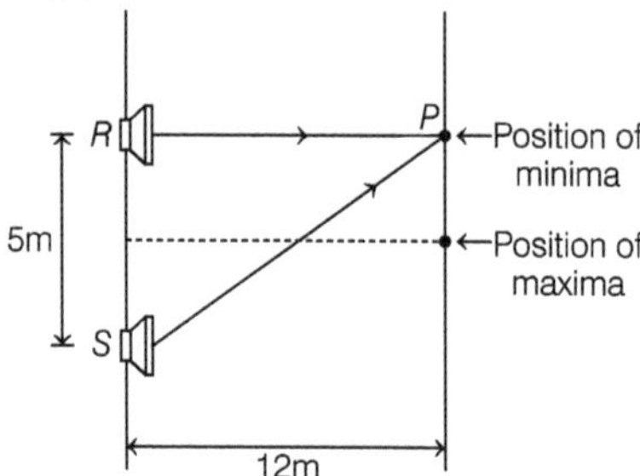

For a minima at P, path difference of sounds reaching P must be an odd multiple of half wavelength.

So, $$SP - RP = (2n + 1)\frac{\lambda}{2} \quad \text{... (i)}$$

where, $n = 0, 1, 2, 3...$

From above figure,

$$SP = \sqrt{(RP)^2 + (RS)^2}$$

$$SP = \sqrt{12^2 + 5^2} = 13$$

So, path difference,

$$SP - RP = 13 - 12 = 1 \text{m}$$

Hence, from Eq. (i),

$$1 = \frac{(2n + 1)v}{2f}$$

or frequency of sound, $f = \left(\frac{2n + 1}{2}\right)v$

Possible values of frequency of sound are for $n = 0$,

$$f_1 = \frac{v}{2} = \frac{330}{2} = 165\text{ Hz}$$

For $n = 1$,

$$f_2 = \frac{3v}{2} = 495\text{ Hz}, \ldots\ldots, \text{ etc.}$$

Hence, option 'a' matches with f_2.

31. *(c)* A photon of mass m and frequency ν falls through a height H through the earth's gravitational field as shown in figure below.

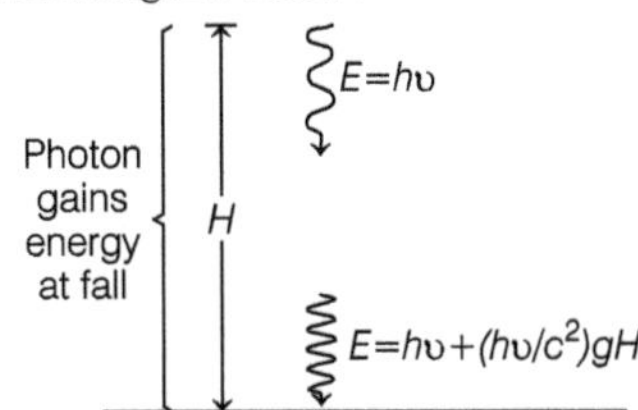

When the photon falls through the earth's gravitational field, then it is gains extra energy. Therefore,

Final photon energy

= Initial photon energy + Increase in energy

$$\Rightarrow \quad h\nu' = h\nu + mgH$$

$$= h\nu + \frac{h\nu}{c^2} \cdot gH = h\nu\left(1 + \frac{gH}{c^2}\right)$$

So, $\frac{f' - f}{f}$ = fractional change in frequency

$$= \frac{gH}{c^2} = \frac{10 \times 1000}{(3 \times 10^8)^2} = \frac{10}{9} \times 10^{-13}.$$

32. *(a)* Process equation is

$$pV^2 = \beta \qquad \ldots \text{(i)}$$

As gas is ideal, it obeys gas equation,

$$pV = nRT \qquad \ldots \text{(ii)}$$

From Eqs. (i) and (ii), gives

$$(nRT) \cdot V = \beta$$

Here, $n = 0.02$ moles,

$R = 8.31\,\text{JK}^{-1}\text{mol}^{-1}$,

$T = 20°\text{C} + 273 = 293\text{ K}$

and $V = 1500\text{ cm}^3 = 1.5 \times 10^{-3}\text{ m}^3$

$$\therefore \quad \beta = 0.02 \times 8.31 \times 293 \times 1.5 \times 10^{-3}$$

$$= 7.3 \times 10^{-2}\text{ Pa-m}^6$$

$$\approx 7.5 \times 10^{-2}\text{ Pa-m}^6$$

33. *(b)* Assuming no heat loss,

Heat gained by liquid in calorimeter = Heat supplied by heater

$$\Rightarrow mS(T_f - T_i) = Pt$$

$$\Rightarrow \quad T_f = \frac{P}{mS}t + T_i$$

Comparing above equation with

$$y = mx + c,$$

Slope of line $= \frac{P}{mS}$

or slope $\propto \frac{1}{S}$

So, graph of T *versus* t is as shown below.

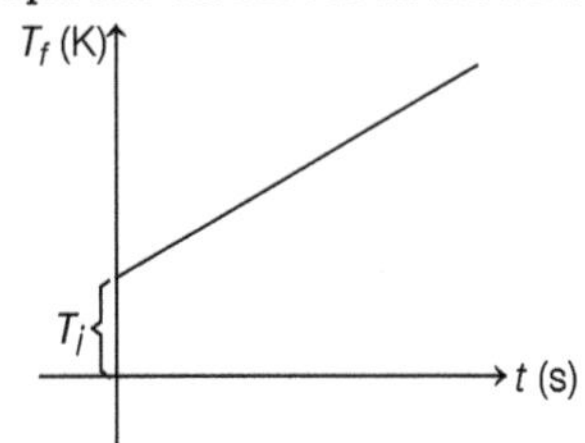

34. *(c)* As intensity of reflected light changes on rotating the polaroid this means reflected beam is polarised.

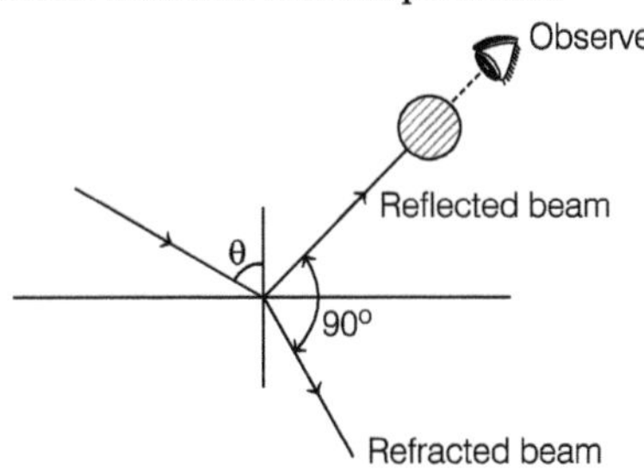

By Brewster's law,

$$n = \tan\theta$$

$$n_{\text{Blue}} > n_{\text{Red}}$$

$$\Rightarrow \quad \tan\theta_B > \tan\theta_R$$

$$\Rightarrow \quad \theta_B > \theta_R$$

Also, from geometry of figure, $\theta > 45°$.

So, we have $\theta_B > \theta_R > 45°$.

35. *(a)* From $V = \frac{kq}{r}$,

$$q = \frac{V \cdot r}{k} = C \cdot V$$

As, $q = ne$, hence graph of q *versus* V is a step function.

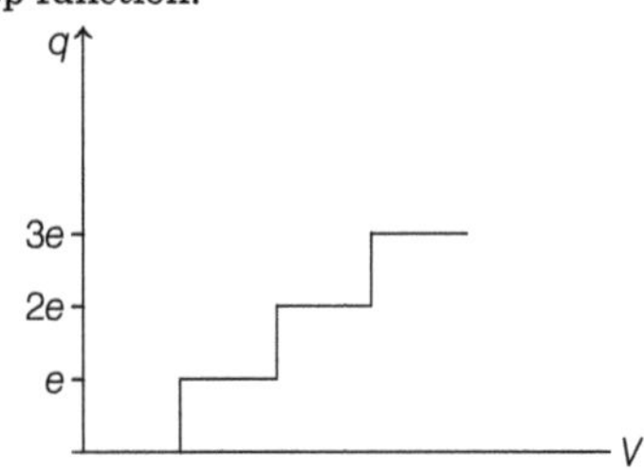

36. *(b)*

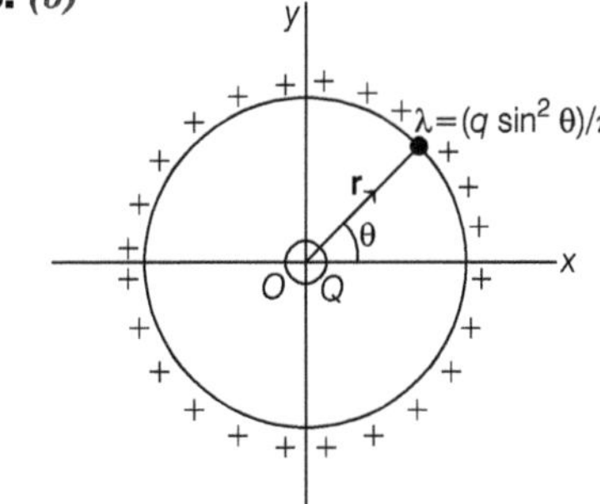

As charge distribution is over a circle, potential due to charge over ring is

$$V = \frac{k}{r} \cdot q_{\text{total}}$$

where, $q_{\text{total}} = \int_0^{2\pi} \frac{q\sin^2\theta}{\pi r}(r d\theta)$

$$= \frac{q}{\pi}\int_0^{2\pi} \frac{1 - \cos 2\theta}{2} . d\theta$$

$$= \frac{q}{\pi}\left[\left(\frac{\theta}{2} - \frac{\sin 2\theta}{4}\right)\right]_0^{2\pi} = q$$

So, potential at centre of ring = V

$$= \frac{kq_{\text{total}}}{r}$$

$$\Rightarrow \quad V = \frac{kq}{r}$$

Also, work done in taking a charge Q from centre of ring to infinity = potential energy of system

$$= V \cdot Q = \frac{kqQ}{r} = \frac{qQ}{4\pi\varepsilon_0 r}$$

37. *(c)* β^- decay occurs as

$${}^A_Z X \longrightarrow {}^A_{Z+1} Y + {}^0_{-1}\beta + \bar{\nu}$$

Most of the decay energy is carried by both emitted electron β^- and antineutrino $(\bar{\nu})$. So, β^-- particles emerges with varrying energies and energy spectrum of emitted β^- is as shown;

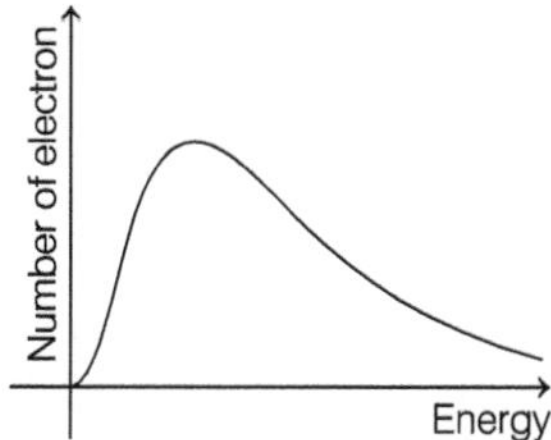

If there are no antineutrinos, energy is carried mainly by β^-- particles, so energy spectrum will be

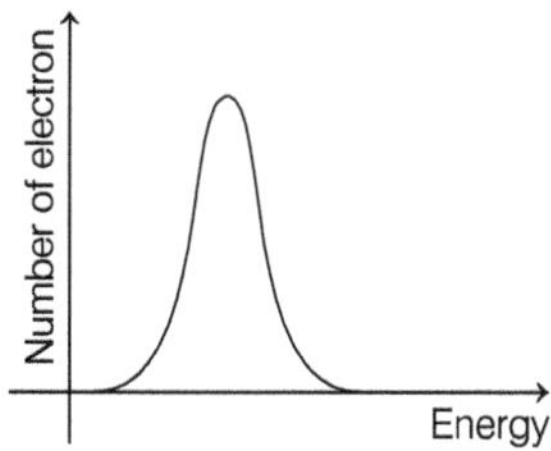

38. *(b)* By using formula,

$$\gamma_{\text{mixture}} = \frac{\left(\frac{\mu_1 Cp_1 + \mu_2 Cp_2}{\mu_1 + \mu_2}\right)}{\left(\frac{\mu_1 Cv_1 + \mu_2 Cv_2}{\mu_1 + \mu_2}\right)}$$

$$\gamma_{\text{mixture}} = \frac{\mu_1 Cp_1 + \mu_2 Cp_2}{\mu_1 Cv_1 + \mu_2 Cv_2}$$

$$= \frac{\mu_1\left(\frac{\gamma_1}{\gamma_1 - 1}\right)R + \mu_2\left(\frac{\gamma_2}{\gamma_2 - 1}\right)R}{\mu_1\left(\frac{R}{\gamma_1 - 1}\right) + \mu_2\left(\frac{R}{\gamma_2 - 1}\right)}$$

or $\gamma_{\text{mixture}} = \dfrac{\mu_1\gamma_1(\gamma_2 - 1) + \mu_2\gamma_2(\gamma_1 - 1)}{\mu_1(\gamma_2 - 1) + \mu_2(\gamma_1 - 1)}$

It is given, $\gamma_{\text{mixture}} = \dfrac{19}{13}$

$$\gamma_1 = \frac{5}{3}, \gamma_2 = \frac{7}{5}$$

$$n_1 = 1, n_2 = n$$

Substituting values, we have

$$\frac{19}{13} = \frac{\frac{5}{3}\left(\frac{7}{5} - 1\right) + n\frac{7}{5}\left(\frac{5}{3} - 1\right)}{\left(\frac{7}{5} - 1\right) + n\left(\frac{5}{3} - 1\right)}$$

$$\Rightarrow \quad \frac{19}{13} = \frac{\frac{5}{3}\times\frac{2}{5} + n\frac{7}{5}\times\frac{2}{3}}{\frac{2}{5} + n\frac{2}{3}}$$

$$\Rightarrow 19\left(\frac{2}{5} + n\frac{2}{3}\right) = 13\left(\frac{2}{3} + n\frac{14}{15}\right)$$

$$\Rightarrow \quad n = 2$$

$\therefore$ Number of moles of gas $B = 2$.

39. *(c)* From $C = \dfrac{q}{V}$

We have, $V = \dfrac{q}{C} = \dfrac{qd}{\varepsilon_0 A}$ or $V \propto d$

Hence, graph is a straight line.

When both plates are touched $(d = 0)$, system still have some capacity $(C \neq 0)$.

So, from $C = \dfrac{q}{V}$ we can see that at very small seperations $V \neq \infty$ or graph does not have any values for very small d.

$\therefore V$ *versus* d graph is as shown below :

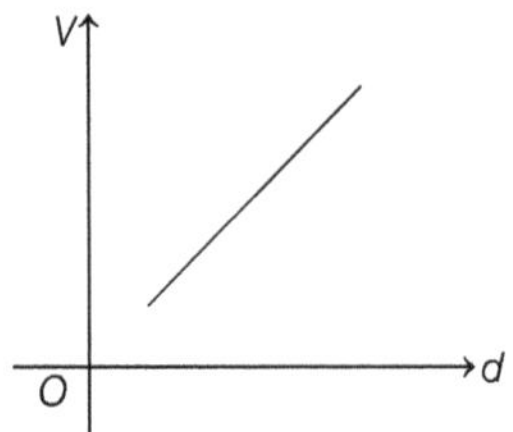

40. *(c)* Plates does not touch each other, so exchange of heat takes place in form of radiations.

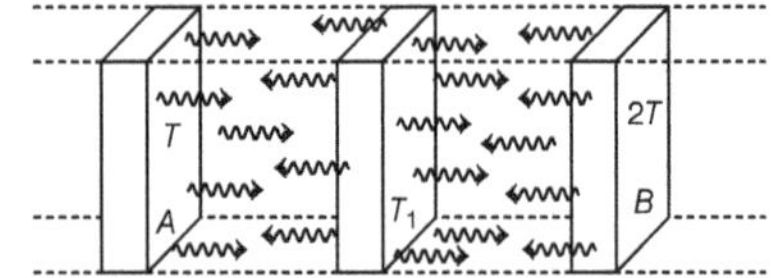

In steady state, heat gained by middle plate = heat lost by it.

$\Rightarrow$ Heat radiations from plate A + Heat radiations from plate B = Heat radiations emitted from both surfaces

$\Rightarrow A\varepsilon\sigma T^4 + A\varepsilon\sigma(2T)^4 = 2A\varepsilon\sigma(T_1)^4$

$$\Rightarrow \quad T^4 + 2^4T^4 = 2T_1^4 \Rightarrow \frac{17}{2}T^4 = T_1^4$$

$$\Rightarrow \quad T_1^4 = \left(\frac{17}{2}\right)^{\frac{1}{4}} \cdot T \text{ or } T_1 = 1.7\,T$$

41. *(b)* Ph–CO–ONa and $CHBr_3$

Hint Reaction given is the last stage of haloform reaction.

OH^- nucleophile attacks carbonyl carbon, resulting in the leaving of CBr_3^-.

Ph–CO–CBr_3 $\xrightarrow{:\ddot{O}H^-}$ Ph–C(O^-)(OH)–CBr_3 $\longrightarrow$ Ph–CO–OH + CBr_3^-

Ph–CO–O^- + $CHBr_3$ $\xleftarrow[\text{Proton exchange}]{\text{Intermolecular}}$

42. *(c)* S_N1 reaction involves carbocation, and therefore only those alkyl halides which can give stable carbocations undergo S_N1 reaction.

I. (1-bromocyclohexene) $\longrightarrow$ (vinylic cyclohexenyl cation)

Unstable Vinylic carbocations are unstable.

II. (tert-butyl bromide) $\longrightarrow$ (tert-butyl cation)

Stable *tert*-butyl carbocation has great stability due to hyper conjugation.

III. MeO–C_6H_4–CH_2Br $\longrightarrow$ MeO–C_6H_4–$\overset{+}{C}H_2$

Stable Resonance and $+R$ effect.

IV. (1-bromonorbornane) $\longrightarrow$ (bridgehead cation)

Unstable Bridge-head carbocations are very unstable.

43. *(a)* H–CO–$(CH_2)_6$–CO–H

DIBAL-H reduces ester and nitrile functional groups to aldehyde. It does not reduce aldehyde. DIBAL-H or diisobutyl aluminium hydride is

$[(CH_3)_2CHCH_2]_2$Al—H

44. *(d)* Sodium hexametaphosphate $(Na_6P_4O_{18})$ (or Calgon) coverts Ca^{2+} and Mg^{2+} ions in water (which cause hardness) into a soluble complex.

$$Na_6P_6O_{18} \longrightarrow 2Na^+ + Na_4P_6O_{18}^{2-}$$

$$Ca^{2+} + Na_4P_6O_{18}^{2-} \longrightarrow 2Na^+ + CaNa_2P_6O_{18}^{2-}$$

45. *(c)* When alkali metals are dissolved in liquid ammonia, they exist in solution as ammoniated cations after releasing valence electrons, and the released electrons also stay in solution in an ammoniated state. Ammoniation is solvation by ammonia.

46. *(d)* (H, CH_2SH, CH_3, CH_2OH) ; (OH, H, H_3C, CH_2SH)

Using CIP (Cahn-Ingold-Prelog) ranking method, $CH_2SH > CH_2OH > CH_3 > H$ is the ranking order.

$\therefore$ ④ ① ③ ② $\equiv$ ① ④ ② ③ *S*- configuration

① OH, ④ H, ③ H_3C, ② CH_2SH ; ① ④ ③ ② = *R*-configuration

47. *(d)* To be diamagnetic, no unpaired electron must be present.

Hint Write molecular orbital electronic configuration of each species.

For all options given,

$\sigma_{1s}^2\,\sigma_{1s}^{*2}\,\sigma_{2s}^2\,\sigma_{2s}^{*2}\,\sigma_{2p_z}^2\,\pi_{2p_x}^2 = \pi_{2p_y}^2$ is common.

After that, $O_2^+ : \pi_{2p_x}^{*1}$,

$O_2^- : \pi_{2p_x}^{*2} = \pi_{2p_y}^{*1}$,

and $O_2 : \pi_{2p_x}^{*1} = \pi_{2p_y}^{*1}$,

all contain unpaired electron. They are paramagnetic, but $O_2^{2-} : \pi_{2p_x}^{*2} = \pi_{2p_y}^{*2}$ has no unpaired electrons.

$\therefore$ It is diamagnetic.

48. *(c)*

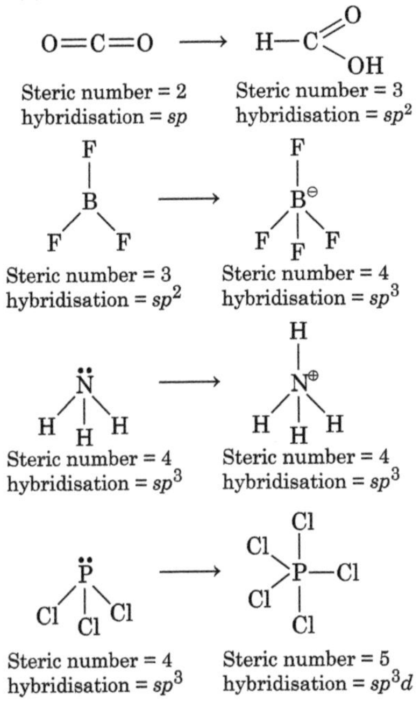

$O=C=O \longrightarrow H-C(=O)-OH$

Steric number = 2, hybridisation = sp → Steric number = 3, hybridisation = sp^2

$BF_3 \longrightarrow BF_4^{\ominus}$

Steric number = 3, hybridisation = sp^2 → Steric number = 4, hybridisation = sp^3

$NH_3 \longrightarrow NH_4^{\oplus}$

Steric number = 4, hybridisation = sp^3 → Steric number = 4, hybridisation = sp^3

$PCl_3 \longrightarrow PCl_5$

Steric number = 4, hybridisation = sp^3 → Steric number = 5, hybridisation = sp^3d

Thus, the hybridisation of the central atom remains unchanged in

$$NH_3 \to NH_4^+.$$

49. *(b)* Two geometrical isomers (*cis* and *trans* with respect to ligand *Y*) are possible. The *trans* isomer has both centre of symmetry and several planes of symmetry.

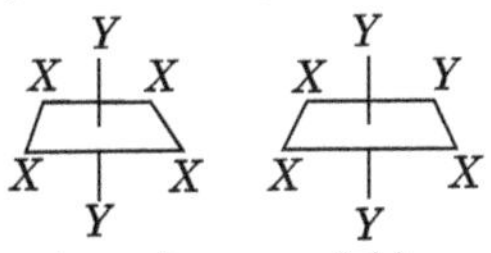

(*trans*) (*cis*)

and, therefore it is achiral.

The *cis* isomer has two planes of symmetry and, therefore it is also achiral.

50. *(c)* By Henry's law,

$p_{gas} \propto X_{gas}$ (in liquid solution)

and $p_{gas} = K_H \times \chi_{gas}$

K_H = Henry's law constant.

∴ At a given partial pressure of gas,

$$K_H \propto \frac{1}{\chi_{gas}}$$

∴ The greater the value of K_H, the less its solubility.

Here, K_H of $Ar > O_2 > CO_2 > CH_4$.

∴ Solubility of $CH_4 > CO_2 > O_2 > Ar$.

51. *(d)* Rate of disappearance of N_2O_5 and rate of formation of O_2 must be in 2 : 1 ratio as per the stoichiometry of the reaction,

$$2N_2O_5 \longrightarrow 4NO_2 + O_2$$

52. *(c)* Arrhenius equation for rate constant, *k* is $k = Ae^{-E_a/RT}$, where *A* is the pre-exponential factor. Since, $e^{-E_a/RT}$ is dimensionless, dimension of $A = k$.

For first order reaction, dimension of $k = \text{time}^{-1}$

∴ Also same for *A*.

As for the other options,

For a first order reaction, the product formation rate is directly proportional to reactant concentration, half-life $= \frac{0.693}{k}$ or 69.3% of $\left(\frac{1}{k}\right)$, and concentration of reactant changes with time by the equation $[A]_t = [A]_0 e^{-kt}$, which shows exponential decreases, not linear.

53. *(a)* Elevation in boiling point, $\Delta T_b = i \times K_b \times m$. ($K_b$ is constant for a solvent)

If molality (*m*) is same, $\Delta T_b \propto$ van't Hoff factor (*i*) CH_3COOH only partially dissociates,

∴ $i < 2$

$NaCl$, Na_2SO_4 and K_3PO_4 almost completely dissociate to produce 2, 3 and 4 ions per formula respectively,

∴ $i \approx 2$, 3, and 4 respectively

∴ The order of *i* = the order of boiling points

$= CH_3COOH < NaCl < Na_2SO_4 < K_3PO_4$

54. *(d)* In fullerene, each carbon is bonded to three other carbon atoms. Two types of those bonds are of greater single bond character and greater length (143.5 pm) and one is of greater double bond character and shorter length (138.3 pm).

55. *(a)*

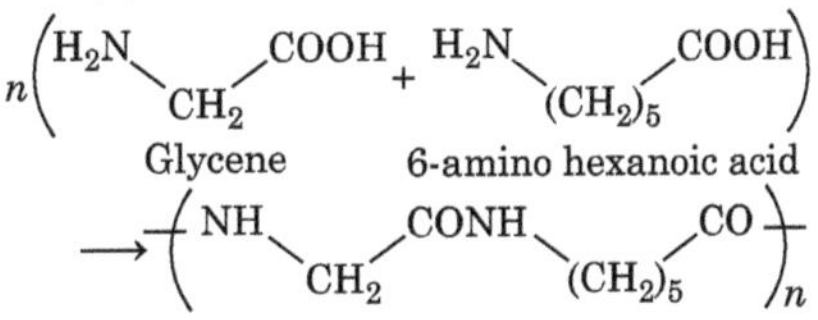

$$n\left(H_2N-CH_2-COOH + H_2N-(CH_2)_5-COOH\right) \longrightarrow \left(NH-CH_2-CONH-(CH_2)_5-CO\right)_n$$

Glycene; 6-amino hexanoic acid

Nylon-2-nylon-6

56. *(b)* Characteristics given are that of ionic solids.

Hard, brittle, insulator as solid and conductor in molten state.

57. *(b)* Variation of mass of gas adsorbed (*x*) per mass of adsorbent (*m*) with increase in pressure is given by

$\frac{x}{m} \propto p^{1/n}$ (Freundlich isotherm)

which is graphically represented by graph 2.

58. *(a)* To give precipitate with $AgNO_3$, the isomer must be $[Co(NH_3)_5SO_4]Cl$ and to give precipitate with $BaCl_2$, it must be $[Co(NH_3)_5 \cdot Cl]SO_4$.

They are ionisation isomers.

59. *(b)* I is most basic, because the lone pair of nitrogen is not delocalised by resonance in that, but in all others, there is resonance with benzene ring.

III is more basic than II due to + *R* (or + *M*) effect of $-CH_3$, and IV is less basic than II because of −*R* (or− *M*) effect of $-NO_2$, i.e. correct order is I > III > II > IV.

60. *(a)* The cathode half-reaction in the process is

$$2H_2O + 2e^- \longrightarrow H_2 + 2OH^-$$

which increases $[OH^-]$ and, therefore pH and the anode half-reaction.

$2Cl^- \longrightarrow Cl_2 + 2e^-$ does not produce or consume H^+ or OH^-.

Overall, the reaction can be written as

$2NaCl + 2H_2O \longrightarrow 2NaOH + H_2 + Cl_2$

61. *(a)* Ethanol is used to treat methanol toxicity because it is a competitive inhibitor of alcohol dehydrogenase competitive inhibitor competes with substrates for the active site of enzyme.

Ethanol has high binding affinity then method for alcohol dehydrogenase. This enzyme converts methane to formaldehyde into formate and formate accumulation causes blindness and affect central nervous system. Ethanol ingestion slows down the metabolism of methanol, so that kidneys, got sufficient time to filter out methanol.

62. *(b)*

1. **Guard cell** Stomatal guard cells control the pore to balance CO_2 entry into the leaf for photosynthesis with water loss through transpiration. It also regulates membrane transport, signalling and homeostasis.

2. **Stomatal pore** It allows CO_2 to diffuse into leaf for photosynthesis.

3. **Subsidiary cell** These are accessory cells to guard cells. They play an important role in ion-channel mediated opening and closing of guard cells.

4. **Epidermal cell** Provides protection to the plants from the external environment.

63. *(a)* Viruses and lichens were excluded from **Whittaker's five kingdom classification** because Whittaker's classification was based on certain character like cell structure, mode and source of nutrition, body organisation and reproduction. Viruses and lichens does not fulfil the classification criteria as viruses are acellular organism and lichens are symbiotic association of algac and fungi.

64. *(b)* Plant species when grown in shade produces thin leaves with more surface area to capture more of the limited light and when grown in light produce thick leaves to maximise photosynthesis in light. This phenomenon is an example of phenotypic plasticity. It is the ability of organism to alter their phenotype in response to environmental factor without changing its genome.

65. *(a)* Sacred groves found in Westren Ghats, central India and North East India are example of *in situ* conservation of living organisms mainly plants and animals.

Sacred groves are diversity rich regions protected by local people for religious and cultural beliefs. Endangered, threatend and endemic species are conserved in these regions.

66. *(b)* IgE mediated immunity is most effectively controlled by anti-histamines in immune response. Anti-histamines are used to prevent the symptoms of hay fever and allergy. IgE class antibodies along with antigen and mast cell triggres the allergic reaction by releasing histamines, which in turn cause inflammatory response. Anti-histamines prevents the effects of histamines by affecting IgE mediate response.

67. *(c)* The origin of mitochondria and chloroplast are explained by endosymbiotic theory. This theory first postulated by **Lynn Margulis** in 1967. Mitochondria and chloroplast are eukaryotic cell organelles with bacterial characteristics. According to this theory eukaryotic mitochondria evolved from a small, prokaryotic autotrophic bacterium that was engulfed by large primitive heterotrophic eukaryotic cell. Endosymbiosis of chloroplast occurs when a cell engulfed a photosynthetic cyanobacterium.

68. *(d)* Growth rate is independent of environmental fluctuation as population exhibits logistic growth when resources are limited. Population expansion decreases as resources become scarce. After reaching to the carrying capacity there is little or no change in population size over time.

Birth rate It is the ability of individuals of a population to produce new individuals. Natality is the scientific term for birth rate.

Death rate Scientific term for death rate is mortality. Mortality refers to the death of individuals in population.

Resource availability A resource is an substance in environment required by an organism for growth and reproduction. It is one of the main factor which determines the ecological dynamics of species or population.

69. *(c)* Natural events like volcanic eruptions, wildfires and flood, etc., wipes out most of the life form and leaves the barren land.

Succession starts when organism colonise on barren land. Algae, fungi and simple plants such as lichens and mosses grows firstly on barren land upon successful colonisation of plants, animals and small invertebrates starts to colonise themselves and become stable.

70. *(d)* Streptokinase in known as **'clot bluster'**. It is produced by beta - hemolytic streptocci. Streptokinase cleaves the Arg/val bond in plasminogen to form proteolytic enzyme plasmin. Plasmin breaks the main component of blood clot 'fibrin'.

Cyclosporin-A is an immunosuppressant used to prevent organ rejection after kidney, heart and liver transplant.

Paracetamol is used to treat pain and fever.

Statins lowers the cholesterol level in blood.

71. *(d)* *Ori* C is not involved in transcription. Replication starts from sequences found on chromosome called origin of replication the point at which DNA opens, helicase unwinding the DNA double helix resulting in the formation of replication fork.

Promoter is a region of DNA from where transcription of gene is initiated.

Terminator is a sequence of DNA that causes RNA polymerase to terminate transcription.

Enhancers are DNA sequences that cause an increase in the level of expression of gene.

72. *(b)* Nematoda is a pseudocoelomate. Pseudocoelomate animals possess pseudocoeloms. It is a fluid filled body cavity between the endoderm and mesoderm.

73. *(b)* Lacrimal gland does not secrete saliva. It secretes aqueous layer of tear film. Submaxillary, partoid and sublingual glands are associated with saliva secretion.

74. *(c)* Cortex, Casparian strip, pericycle, vascular bundle represents the tissue arrangement in roots.

Cortex It conducts water and minerals across the root.

Casparian strip It prevents water from entering the pericycle. Casparian strip are made up of suberin.

Pericycle It gives rise to lateral roots.
Vascular bundles It consists of xylem and phloem. Xylem carries water and dissolved minerals and phloem carries food.

75. *(c)* In alcoholic fermentation, conversion of glucose into ethanol occurs. The NADH generated by the oxidation of glyceraldehyde 3-phosphate is consumed in the reduction of acetaldehyde to ethanol. Thus, there is no net oxidation - reduction in the conversion of glucose into ethanol.

76. *(c)* Product of cyclic photophosphorylation is ATP. In cyclic photophosphorylation only one photosystem (PS-I) is involved when light is absorbed by PS-I excited electrons enters into electron transport chain to produce ATP.

Both ATP and NADPH are produced in non-cyclic photophosphorylation, which includes both photosystems (PS-I and PS-II).

77. *(d)* Arginine is a hydrophilic amino acid found at the surface of protein.

78. *(a)* Global species diversity increases from high to low altitudes. High altitudes are less diverse then lower altitudes because atmosphere becomes less dense and due to the lack of oxygen few species are able to sustain their life at higher altitudes.

79. *(d)* Phenylketonuria is not responsible for the presence of deoxygenated blood in the arteries of a newborn. It is a genetic disorder in which level of phenylalanine is increased in blood. This occurs because of defect in PAH gene.

Phenylalanine hydroxylase, which converts phenylalanine to tyrosine. In the absence of PAH gene enzyme PAH is not synthesised and breakdown of phenylalanine is completely inhibited.

Symptoms of PKU includes seizures tremors, stunted growth, hyperactivity, etc.

Pneumonia is an infection of lungs can be caused by bacteria, virus and fungi. Air sacs present in the lungs get inflamed and filled with fluid resulting in cough, fever and breath shortness.

80. *(a) Rhizobium* is a Gram negative bacteria fixes atmosphoric nitrogen by infecting roots of leguminous plants and leads to the formation of root nodules. Nitrogen fixation is catalysed by enzyme nitrogenase.

Nitrogenase is very sensitive to oxygen. If oxygen is present, nitrogenase is not able to perform its function.

Leg haemoglobin scavanges O_2 if present for proper functioning of enzyme. It is a ATP dependent process and required anaerobic condition.

81. *(b)* Let the diameter $AB = x$

$\because \quad \angle ACB = \frac{\pi}{2}$

and $\quad \angle ADB = \frac{\pi}{2}$

$\therefore \quad BC = \sqrt{x^2 - 1}$

and $\quad AD = \sqrt{x^2 - 9}$

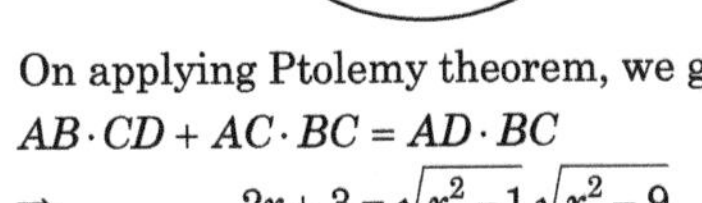

On applying Ptolemy theorem, we get

$AB \cdot CD + AC \cdot BC = AD \cdot BC$

$\Rightarrow \quad 2x + 3 = \sqrt{x^2-1}\sqrt{x^2-9}$

On squaring both sides, we get

$4x^2 + 9 + 12x = x^4 - 10x^2 + 9$

$\Rightarrow x^4 - 14x^2 - 12x = 0$

$\Rightarrow \quad x^3 - 14x - 12 = 0$

Let $\quad f(x) = x^3 - 14x - 12$

$\because \quad f(4) = -4$

$f(4.1) = -0.479$

$f(4.2) = 3.288$

$\because \quad f(4.1) \cdot f(4.2) < 0$

$\Rightarrow \quad x \in [4.1, 4.2)$

82. *(a)* Let ΔABC and D is the mid- point of side BC,

then

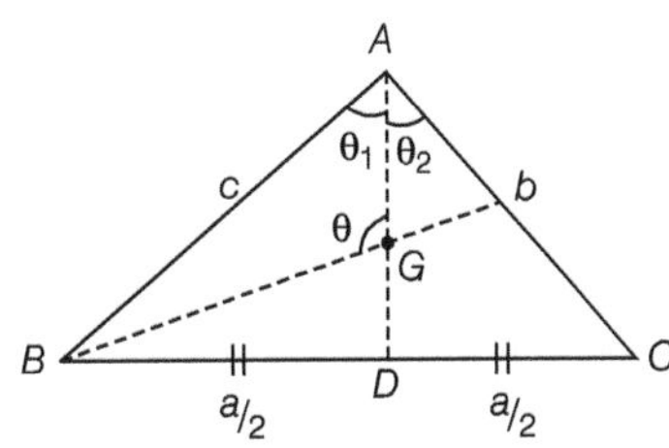

According to Apollonius theorem,

$$AD^2 = \frac{1}{2}\left\{b^2 + c^2 - \frac{a^2}{4}\right\}$$

$$\Rightarrow \quad AD = \frac{1}{2}\sqrt{2b^2 + 2c^2 - a^2}$$

$$\because \quad \frac{AG}{GD} = \frac{2}{1}$$

$$\Rightarrow \quad AG = \frac{2}{3}AD = \frac{1}{3}\sqrt{2b^2 + 2c^2 - a^2}$$

Similarly $BG = \frac{1}{3}\sqrt{2c^2 + 2a^2 - b^2}$

So, $\cos\theta = \dfrac{BG^2 + AG^2 - AB^2}{2(BG)(AG)}$,

where $\theta = \angle AGB$

$$= \frac{\frac{1}{9}(2c^2 + 2a^2 - b^2) + \frac{1}{9}(2b^2 + 2c^2 - a^2) - c^2}{2(BG)(AG)}$$

$$= \frac{a^2 + b^2 - 5c^2}{18(BG)(AG)} \quad \ldots(i)$$

Now, as $\quad \cot\theta_2 = 2\cot\theta_1$

$$\Rightarrow \quad \frac{\cos\theta_2}{\sin\theta_2} = 2\frac{\cos\theta_1}{\sin\theta_1}$$

$$\Rightarrow \frac{AD^2 + b^2 - \frac{a^2}{4}}{2AD \cdot b\sin\theta_2} = 2\frac{AD^2 + c^2 - \frac{a^2}{4}}{2AD \cdot c\sin\theta_1}$$

$$\Rightarrow \frac{AD^2 + b^2 - \frac{a^2}{4}}{4(\text{area}\,\Delta ADC)} = 2\frac{AD^2 + c^2 - \frac{a^2}{4}}{4(\text{area}\,\Delta ADB)}$$

$$\Rightarrow \quad AD^2 + b^2 - \frac{a^2}{4} = 2AD^2 + 2c^2 - \frac{a^2}{2}$$

$[\because \Delta ADC = \Delta ADB]$

$$\Rightarrow \quad AD^2 = b^2 + \frac{a^2}{4} - 2c^2$$

$$\Rightarrow \frac{1}{4}(2b^2 + 2c^2 - a^2) = b^2 + \frac{a^2}{4} - 2c^2$$

$$\Rightarrow \quad 2b^2 + 2a^2 = 10c^2$$

$$\Rightarrow \quad a^2 + b^2 = 5c^2 \quad \ldots(ii)$$

From Eqs. (i) and (ii), we get

$$\cos\theta = 0$$

$$\Rightarrow \quad \theta = \frac{\pi}{2}$$

83. *(b)* Given,

$f(x) = x^6 - 2x^5 + x^3 + x^2 - x - 1$

$= x^2(x^4 - x^3 - x^2 - 1)$

$\quad -x(x^4 - x^3 - x^2 - 1) + 2x^2 - 2x - 1$

$\Rightarrow f(x) = (x^2 - x)g(x) + 2x^2 - 2x - 1$

$\therefore \quad f(a) = 2a^2 - 2a - 1$

$\therefore \quad \Sigma f(a) = f(a) + f(b) + (c) + f(d)$

$= 2\Sigma a^2 - 2\Sigma a - \Sigma 1$

$= 2((\Sigma a)^2 - 2\Sigma ab) - 2\Sigma a - 4$

$\because a, b, c, d$ are roots of equation

$g(x) = x^4 - x^3 - x^2 - 1 = 0$

then $\Sigma a = 1, \Sigma ab = -1$

$\therefore \quad \Sigma f(a) = 2[(1)^2 - 2(-1)] - 2(1) - 4$

$= 6 - 2 - 4 = 0$

84. *(a)* Since, $|\mathbf{r} - \mathbf{b}| + |\mathbf{r} - \mathbf{c}| = 4$ is equation of ellipsoid in plane $\mathbf{r} \cdot \mathbf{a} = 5$ or $x + y + z = 5$.

$\therefore$ Distance between foci $= |\mathbf{b} - \mathbf{c}|$

$= |3\hat{\mathbf{i}} - \hat{\mathbf{j}} - 2\hat{\mathbf{k}}| = \sqrt{14} = 2ae$

and $2a = 4$

$$\therefore \quad e = \frac{\sqrt{14}}{4} \Rightarrow e^2 = 1 - \frac{b^2}{a^2}$$

$$\Rightarrow \quad \frac{14}{16} = 1 - \frac{b^2}{4} \Rightarrow \frac{b^2}{4} = \frac{2}{16}$$

$$b^2 = \frac{1}{2}$$

$\therefore$ Area of ellipsoid

$$= \pi ab = \pi(2)\left(\frac{1}{\sqrt{2}}\right) = \sqrt{2}\,\pi$$

And from the given options the closest integer to $\sqrt{2}\pi$ is 4.

85. *(d)* The given equation

$$\sin(\pi\sin^2\theta) + \sin(\pi\cos^2\theta) = 2\cos\left(\frac{\pi}{2}\cos\theta\right)$$

$$\Rightarrow 2\sin\left(\frac{\pi(\sin^2\theta + \cos^2\theta)}{2}\right)$$

$$\cos\left(\frac{\pi(\sin^2\theta - \cos^2\theta)}{2}\right) = 2\cos\left(\frac{\pi}{2}\cos\theta\right)$$

$$\Rightarrow \cos\left(\frac{\pi}{2}\cos 2\theta\right) = \cos\left(\frac{\pi}{2}\cos\theta\right)$$

$$\Rightarrow \frac{\pi}{2}\cos 2\theta = 2n\pi \pm \frac{\pi}{2}\cos\theta,\ n \in \text{Integers}$$

$$\Rightarrow \quad \cos 2\theta = 4n \pm \cos\theta,\ n \in I.$$

Case I

If $\cos 2\theta = 4n + \cos\theta \Rightarrow \cos 2\theta - \cos\theta = 4n$

The above equation will hold only if $n = 0$, so $\quad \cos 2\theta = \cos\theta$

$$\Rightarrow 2\cos^2\theta - \cos\theta - 1 = 0 \Rightarrow \cos\theta = 1, -\frac{1}{2}$$

$$\Rightarrow \quad \theta = 2k\pi,\ 2k\pi \pm \frac{2\pi}{3},\ k \in I$$

$\because \quad \theta \in [0, 2\pi]$

So, $\theta = 0, 2\pi, \frac{2\pi}{3}, \frac{4\pi}{3}$

Case II

If $\cos 2\theta = 4n - \cos\theta$

$\Rightarrow \quad \cos 2\theta + \cos\theta = 4n$

The above equation will hold only if $n = 0$,

so $\quad \cos 2\theta + \cos\theta = 0$

$\Rightarrow 2\cos^2\theta + \cos\theta - 1 = 0$

$\Rightarrow \quad \cos\theta = -1, \frac{1}{2}$

$\Rightarrow \quad \theta = (2m+1)\pi,\ 2m\pi \pm \frac{\pi}{3},\ m \in I.$

$\because \quad \theta \in [0, 2\pi]$

So, $\theta = \pi, \frac{\pi}{3}, \frac{5\pi}{3}$

$\therefore$ There are total 7 solutions.

86. *(a)* As we know, $1 + x^8 < 2$□, $\forall x \in (0, 1)$

$\Rightarrow \quad \frac{1}{2} < \frac{1}{1+x^8}, \forall x \in (0, 1)$

$\Rightarrow \quad \frac{x}{2} < \frac{x}{1+x^8}, \forall x \in (0, 1)$

$\Rightarrow \quad \int_0^1 \frac{x}{2}\, dx < \int_0^1 \frac{x}{1+x^8} dx$

$\Rightarrow \quad \int_0^1 \frac{x}{1+x^8} dx > \frac{1}{4}$

$\Rightarrow \quad J > \frac{1}{4}$

Similarly, as we know, $1 + x^8 < 1 + x^4$, $\forall x \in (0, 1)$

$\Rightarrow \int_0^1 \frac{x}{1+x^4} dx < \int_0^1 \frac{x}{1+x^8} dx$

$\Rightarrow J > \frac{1}{2}[\tan^{-1}(x^2)]_0^1$

$\Rightarrow J > \frac{\pi}{8}$

$\therefore$ Statement $J = \int_0^1 \frac{x}{1+x^8} dx > \frac{1}{4}$ is true,

while the statement $J = \int_0^1 \frac{x}{1+x^8} dx < \frac{\pi}{8}$ is not true.

87. *(d)* It is given that,

$(f'(x))^4 = 16(f(x))^2, \forall x \in (-1, 1)$

$\Rightarrow \quad (f'(x))^2 = \pm 4f(x)$

Case I

If $f(x) \geq 0 \forall x \in (-1, 1)$

Then, $f'(x) = \pm 2\sqrt{f(x)}$

$\Rightarrow \quad \frac{f'(x)}{\sqrt{f(x)}} = \pm 2$

$\Rightarrow \quad 2\sqrt{f(x)} = \pm 2x, \qquad [\because f(0) = 0]$

$\Rightarrow \quad f(x) = x^2 \forall x \in (-1, 1)$

Case II

If $f(x) < 0 \forall x \in (-1, 1)$

Then, $f'(x) = \pm 2\sqrt{-f(x)}$

$\therefore \quad f(x) = -x^2 \ \forall x \in (-1, 1)$

$\therefore \quad f_1(x) = \begin{cases} x^2, & 0 \le x < 1 \\ x^2, & -1 < x < 0 \end{cases}$

$f_2(x) = \begin{cases} x^2, & 0 \le x < 1 \\ -x^2, & -1 < x < 0 \end{cases}$

$f_3(x) = \begin{cases} -x^2, & 0 \le x < 1 \\ x^2, & -1 < x < 0 \end{cases}$

$f_4(x) = \begin{cases} -x^2, & 0 \le x < 1 \\ -x^2, & -1 < x < 0 \end{cases}$

$f_5(x) = 0$

$f_6(x) = \begin{cases} 0, & 0 \le x < 1 \\ x^2, & -1 < x < 0 \end{cases}$, ...and so on.

Hence, there are more than 4 functions possible.

88. *(a)* Given, for $x \in R$

$f(x) = |\sin x|$

$g(x) = \int_0^x f(t)dt$, and

$p(x) = g(x) - \frac{2}{\pi}x$

$\because p(x+\pi) = g(x+\pi) - \frac{2}{\pi}(x+\pi)$

$= \int_0^{x+\pi} |\sin(t)|\, dt - \frac{2}{\pi}x - 2$

$= \int_0^{\pi} |\sin t|\, dt + \int_{\pi}^{\pi+x} |\sin t|\, dt - \frac{2}{\pi}x - 2$

$= 2\int_0^{\pi/2} \sin t\, dt + \int_0^x |\sin t|\, dt - \frac{2}{\pi}x - 2$

[$\because |\sin t|$ is periodic function having period π]

$= 2[-\cos t]_0^{\pi/2} + g(x) - \frac{2}{\pi}x - 2$

$= 2(0 - (-1)) + g(x) - \frac{2}{\pi}x - 2$

$= 2 + g(x) - \frac{2}{\pi}x - 2 = g(x) - \frac{2}{\pi}x = p(x)$

$\therefore p(x+\pi) = p(x)$ for all x.

89. *(b)* Given, $\left(\sum_{i=1}^{3} \frac{a_i}{2^i}\right)^2 = \sum_{i=1}^{3} \frac{a_i^2}{2^i}$

$\Rightarrow \left(\frac{a_1}{2} + \frac{a_2}{4} + \frac{a_3}{8}\right)^2 = \frac{a_1^2}{2} + \frac{a_2^2}{4} + \frac{a_3^2}{8}$

$\Rightarrow \frac{a_1^2}{4} + \frac{a_2^2}{16} + \frac{a_3^2}{64} + \frac{a_1 a_2}{4} + \frac{a_2 a_3}{16} + \frac{a_3 a_1}{8} = \frac{a_1^2}{2} + \frac{a_2^2}{4} + \frac{a_3^2}{8}$

$\Rightarrow \frac{a_1^2}{4} + \frac{3a_2^2}{16} + \frac{7a_3^2}{64} = \frac{a_1 a_2}{4} + \frac{a_2 a_3}{16} + \frac{a_3 a_1}{8}$

$\Rightarrow 16a_1^2 + 12a_2^2 + 7a_3^2$

$= 16a_1 a_2 + 4a_2 a_3 + 8a_3 a_1$

$\Rightarrow (8a_1^2 + 8a_2^2 - 16a_1 a_2) + (4a_2^2 + a_3^2 - 4a_2 a_3) + (2a_3^2 + 8a_1^2 - 8a_3 a_1) + 4a_3^2 = 0$

$\Rightarrow 8(a_1 - a_2)^2 + (2a_2 - a_3)^2 + 2(a_3 - 2a_1)^2 + 4a_3^2 = 0$

The above relation will be true if

$a_1 = a_2,\ 2a_2 = a_3,\ a_3 = 2a$ and $a_3 = 0$

So, $a_1 = a_2 = a_3 = 0$

Then, set A contains exactly one element.

90. *(a)* For a continuous function $f : [0, 1] \longrightarrow [0, 1]$, it is given that

$x^2 + (f(x))^2 \le 1$

$\Rightarrow (f(x))^2 \le 1 - x^2$

$\Rightarrow \quad f(x) \le \sqrt{1-x^2} \qquad \{\because f(x) \in [0, 1]\}$

$\therefore \int_0^1 f(x)dx \le \int_0^1 \sqrt{1-x^2}\, dx$

$\Rightarrow \int_0^1 f(x)dx \le \left[\frac{x}{2}\sqrt{1-x^2} + \frac{1}{2}\sin^{-1} x\right]_0^1$

$\Rightarrow \int_0^1 f(x)dx \le \frac{\pi}{4}$

So, $f(x) = \sqrt{1-x^2}$, because it is given that $\int_0^1 f(x) = \frac{\pi}{4}$.

Now,

$\int_{1/2}^{1/\sqrt{2}} \frac{f(x)}{1-x^2} dx = \int_{1/2}^{1/\sqrt{2}} \frac{dx}{\sqrt{1-x^2}} = [\sin^{-1} x]_{1/2}^{1/\sqrt{2}}$

$= \frac{\pi}{4} - \frac{\pi}{6} = \frac{\pi}{12}$

91. *(a)* Length of rod given remains unchanged. This means contraction due to cooling is equals to elongation due to hanging of weight.

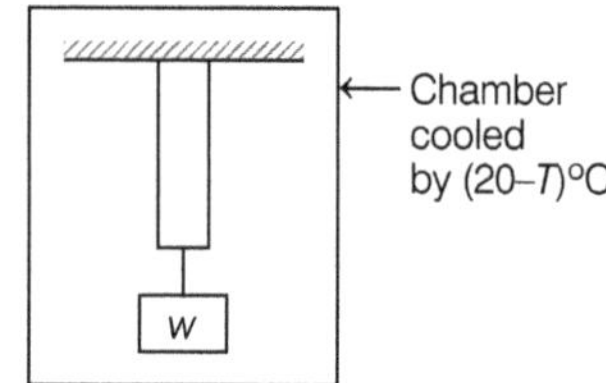

As, thermal strain = strain caused by weight

$\Rightarrow \quad \alpha\Delta\theta = \frac{\Delta l}{l}$

$\Rightarrow \quad \Delta\theta = \frac{\Delta l / l}{\alpha} = \frac{F}{YA\alpha} \quad \left(\because Y = \frac{F/A}{\Delta l / l}\right)$

$\Rightarrow \Delta\theta = \frac{5000}{4 \times 10^{12} \times 10^{-4} \times 2.5 \times 10^{-6}} = 5°C$

Note $\Delta\theta$ in K is same as $\Delta\theta$ in °C.

$\therefore 20 - T = 5°C$ or $T = 15°C$

92. *(b)* We use reciprocatory theorem to calculate mutual inductance of the system of solenoids.

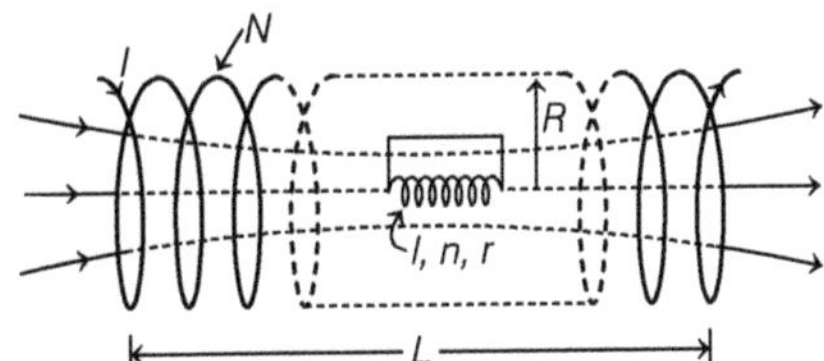

Let outer solenoid carries a current I_1. Flux linked with small solenoid (2) is proportional to the current in large solenoid (1).

$\Rightarrow \phi_{21} \propto I_1 \Rightarrow (n\ l)(\mu_0 N I_1)(\pi r^2) \propto I_1$

Proportionality constant is the coefficient of mutual inductance M of system.

Note that $M_{12} = M_{21}$ in all cases.

So, $nl\mu_0 NI_1 \pi r^2 = MI_1$ or $M = \pi\mu_0 r^2 nNl$

93. *(d)*

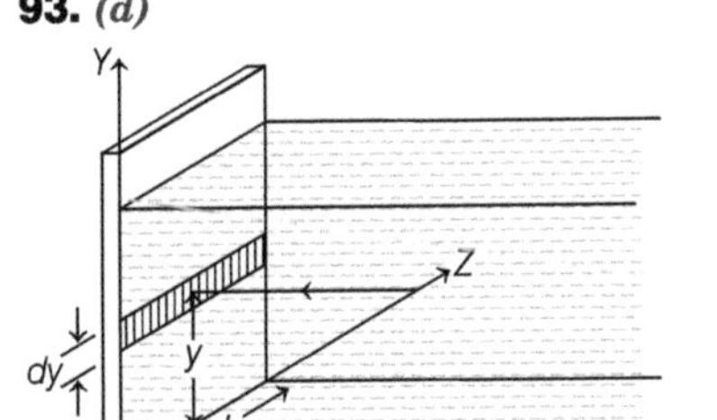

Consider a strip of width dy at a depth $(h - y)$ below the surface.

Torque on bottom length (OZ in diagram) due to force of water on strip of width dy is

$d\tau$ = Force × Perpendicular distance

= Pressure × Area × Distance

$d\tau = \rho g(h - y) \cdot (L \cdot dy) \cdot y$

So, torque on bottom length due to complete volume of water,

$$\tau_1 = \int_0^h L\rho g(h - y) y \cdot dy$$

$$= L\rho g\left[\int_0^h (hy - y^2) dy\right]$$

$$= L\rho g\left[\frac{hy^2}{2} - \frac{y^3}{3}\right]_0^h$$

$$= L\rho g\left[\frac{h^3}{2} - \frac{h^3}{3}\right]$$

$$\text{or } \tau_1 = \frac{L\rho g h^3}{6} \qquad \ldots \text{(i)}$$

and torque due to water upto height $\frac{h}{2}$ and wall length $\frac{1}{2}$,

$$\tau_2 = \int_0^{\frac{h}{2}} \frac{L}{2}\rho g\left(\frac{h}{2} - y\right) y\, dy$$

$$= \frac{L}{2}\rho g\left[\left(\frac{hy^2}{4} - \frac{hy^3}{3}\right)\right]_0^{\frac{h}{2}}$$

$$= \frac{L}{2}\rho g\left[\frac{h^3}{16} - \frac{h^3}{24}\right] = \frac{L\rho g h^3}{16 \times 6}$$

So, $\frac{\tau_1}{\tau_2} = 16$

94. *(d)* Let n_1 and n_2 are number of moles of gas present in container C_1 and C_2 respectively, before the value is opened.

Then, using $pV = nRT$.

$$n_1 = \frac{pV}{R(300)}$$

and $$n_2 = \frac{5p(4V)}{R(400)}$$

5p, 4V
p, V
C_1 →
← C_2
300 K → 400 K

When value is opened gas flows from C_2 to C_1 till pressure in C_1 and C_2 is equal.

Let after equalisation of pressure in both C_1 and C_2, its value is p_0,

Then, using $pV = nRT$.

$p_0 V = n_1{}'R(300)$(Container C_1)

and $p_0 4V = n_2{}'R(400)$ (Container C_2)

So, $$n_1{}' = \frac{p_0 V}{R(300)}$$

and $$n_2{}' = \frac{p_0(4V)}{R(400)}$$

As no gas is leaked from containers,

$n_1 + n_2 = n_1{}' : n_2{}'$

$$\Rightarrow \frac{pV}{R(300)} + \frac{20pV}{R(400)} = \frac{p_0 V}{R(300)} + \frac{4p_0 V}{R(400)}$$

So, $$p\left(\frac{1}{300} + \frac{20}{400}\right) = p_0\left(\frac{1}{300} + \frac{4}{400}\right)$$

$\Rightarrow \quad 4p = p_0$

Now, $$\frac{n_2{}'}{n_1{}'} = \frac{\frac{4p_0 V}{R(400)}}{\frac{p_0 V}{R(300)}} = 3$$

Finally number of moles of C_2 is thrice of C_1.

95. *(d)* For an electron described by standing waves confined within a box of length L, the wave can occupy only states for which

$$L = \frac{n\lambda}{2}, \text{ where } n = 1, 2, 3 \ldots$$

Now, by de Broglie wavelength of this electron,

$$\lambda = \frac{h}{p} = \frac{h}{\sqrt{2mE}}$$

From above two equations, we have

$$E_n = \left(\frac{h^2}{9\,mL^2}\right) n^2$$

(where, $n = 1, 2, 3 \ldots$)

Now, ground state energy is obtained by substituting $n = 1$.

$$\therefore \quad U_0 = E_1 = \frac{h^2}{8mL^2}$$

Energy of first excited state,

$$E_2 = \frac{h^2}{8mL^2} \times 2^2$$

$$= 4U_0 = U_0 + 3\alpha$$

Also, energy for second excited state,

$$E_3 = \frac{h^2}{8mL^2} \times 3^2$$

$$= 9U_0 = U_0 + 8\alpha$$

96. *(b)* Tension at a section x distance from free end is

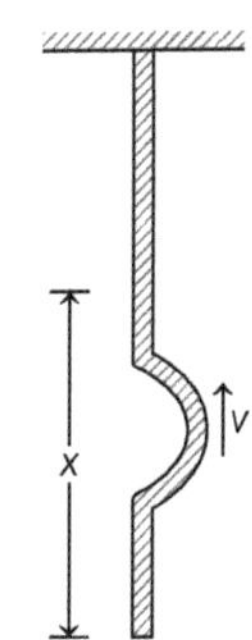

$$T = mg = \mu \cdot x \cdot g$$

where, μ = mass per unit length of rope.

Wave speed over the rope is given by

$$v = \sqrt{\frac{T}{\mu}} = \sqrt{\frac{\mu \cdot x \cdot g}{\mu}} = \sqrt{g \cdot x}$$

$\Rightarrow \quad v^2 = gx$

Comparing above equation with $v^2 - u^2 = 2as$, we get

$\Rightarrow \quad v^2 = 2ax = gx$

$\Rightarrow$ Acceleration, $a = \frac{g}{2}$

Now, if t = time for wave pulse to cover a distance x,

then from

$$s = ut + \frac{1}{2}at^2,$$

we have

$$x = \frac{1}{2} \times \frac{g}{2} \times t^2$$

or $$t = 2\sqrt{\frac{x}{g}}$$

$\Rightarrow$ $$t \propto \sqrt{x}$$

or time to travel complete length of rope $t \propto \sqrt{L}$.

97. *(a)* At $t = 0$ capacitor is uncharged and flux of inductor is ϕ.

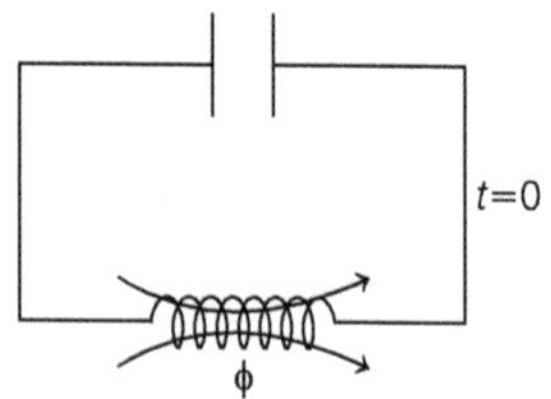

Now, using $\phi = LI$, at $t = 0$ current in circuit is

$$I_0 = \frac{\phi}{L}$$

Instantaneous current in circuit is $\frac{dq}{dt}$.

where, q is solution of

$$L\frac{d^2q}{dt^2} + \frac{q}{C} = 0$$

or $$\frac{d^2q}{dt^2} + \frac{q}{LC} = 0$$

At $t = 0$, current in circuit is non-zero, so instantaneous current is given by

$$= I_0 \cos\omega t = \frac{\phi}{L} \cdot \cos\omega t$$

98. *(c)*

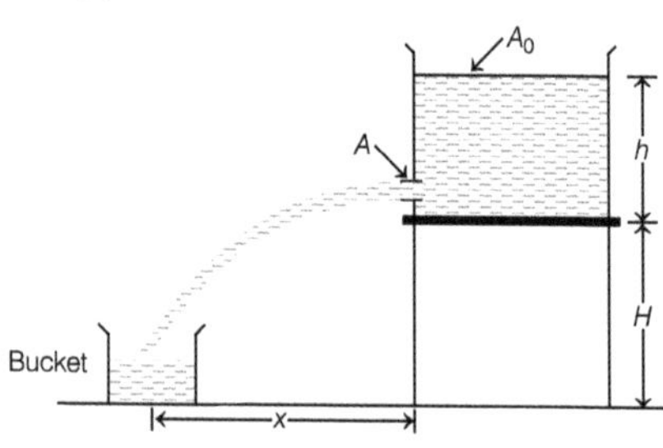

Velocity of outflow (efflux) from hole $= \sqrt{2gh}$.

Range of water jet,

$$x = \sqrt{2gh} \times \text{Time of fall}$$

$$= \sqrt{2gh} \times \sqrt{\frac{2H}{g}} = 2\sqrt{hH}$$

Velocity v of bucket is

$$v = \frac{dx}{dt} = \frac{d}{dt}2\sqrt{hH}$$

$$= 2\sqrt{H} \cdot \frac{d\sqrt{h}}{dt}$$

$\Rightarrow$ $$v = 2\sqrt{H} \cdot \frac{1}{2\sqrt{h}} \cdot \frac{dh}{dt}$$

$$= \sqrt{\frac{H}{h}} \cdot \frac{dh}{dt} \qquad \text{...(i)}$$

Now, using equation of continuity at area A_0 and area A, we have

$$Av_1 = A_0V$$

where, $v_1 = \sqrt{2gh}$

and $V = \frac{dh}{dt}$.

$\therefore$ $$A\sqrt{2gh} = A_0\left(\frac{dh}{dt}\right)$$

Substituting for $\frac{dh}{dt}$ from Eq (i), we get

$$v = \sqrt{\frac{H}{h}} \cdot \frac{A}{A_0} \cdot \sqrt{2gh}$$

$$= \sqrt{\frac{5}{5}} \cdot \left(\frac{5}{500}\right) \cdot \sqrt{2 \times 10 \times 5}$$

$$= \frac{1}{10}\ \text{ms}^{-1}$$

$$= 0.1\ \text{ms}^{-1}$$

99. *(c)* In given condition,

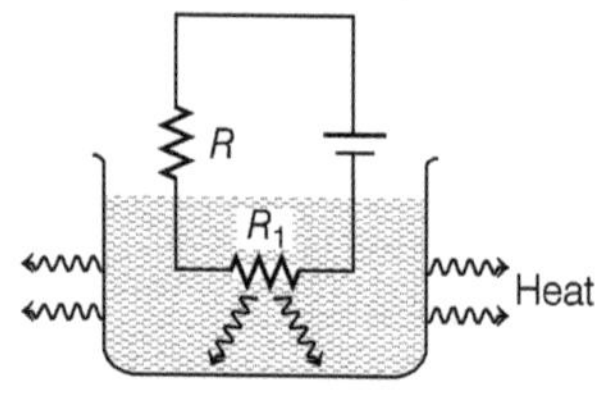

Heat gained by water heater – Heat lost governed by Newton's law of cooling = Net heat gained by water

$$\Rightarrow i^2R_1 - K(T - T_0) = mS\left(\frac{\Delta T}{\Delta t}\right)$$

Here, T = temperature and t = time.

$\Rightarrow$ $$\frac{dT}{dt} = \frac{i^2R_1}{mS} - \frac{K}{mS}(T - T_0)$$

$$= \frac{i^2R_1}{mS} + \frac{KT_0}{mS} - \frac{K}{mS}T$$

Above equation can be written as,

$$\frac{dT}{dt} = A + BT$$

or $$\frac{dT}{A + BT} = dt$$

Integrating, we get

$\Rightarrow$ $$\frac{1}{B}.\log_e(A + BT) = t$$

$\Rightarrow$ $$A + BT = e^{tB}$$

$\Rightarrow$ $$T = \frac{1}{B}(e^{tB} - A)$$

Hence, temperature T depends on time t exponentially.

$\therefore$ Best suitable graph is

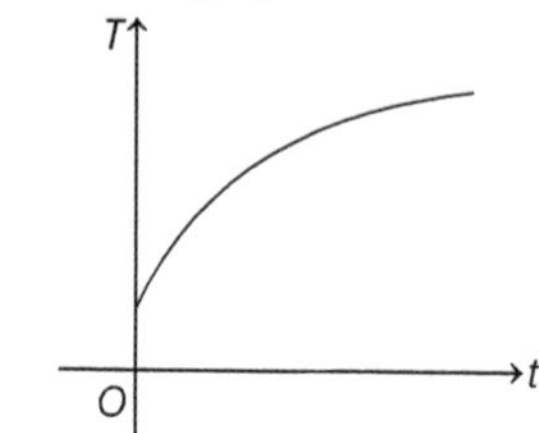

100. *(b)*

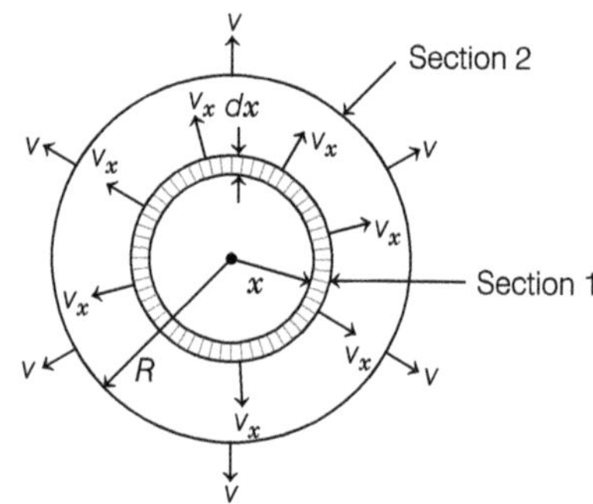

Let bubble of radius x expands and pushes water of density ρ by a distance dx. Then, kinetic energy of water of strip dx is

$$dK = \frac{1}{2}dm \cdot v_x^2 \Rightarrow dK = \frac{1}{2}4\pi x^2 \cdot dx \cdot \rho \cdot v_x^2$$

or $$dK = 2\pi R^2 \cdot \rho \cdot v_x^2 \cdot dx \qquad \text{... (i)}$$

Considering laminar flow, for section 1 and 2 shown in figure, by equation of continuity

$$A_1v_1 = A_2v_2 \Rightarrow 4\pi x^2 v_x = 4\pi R^2 v$$

$\Rightarrow$ $$v_x = \frac{R^2}{x^2}.v \qquad \text{... (ii)}$$

From Eqs. (i) and (ii), we have

$$dK = 2\pi x^2\rho\left(\frac{R^2}{x^2} \cdot v \cdot\right)^2 dx$$

$$\Rightarrow dK = 2\pi R^4\rho v^2.\left(\frac{dx}{x^2}\right)$$

So, kinetic energy of complete volume of water,

$$K = \int_R^\infty dK = 2\pi R^4\rho v^2\int_R^\infty \frac{dx}{x^2}$$

$\Rightarrow$ $$K = 2\pi R^4 \cdot \rho \cdot v\left[-\frac{1}{x}\right]_R^\infty$$

$$= 2\pi\rho R^3 v^2$$

101. *(d)*

Cyclohexene derivative $\xrightarrow[\text{2. Zn/H}_2\text{O}]{\text{1. O}_3}$ product with C=O and CHO

The product has an aldehyde group (which is not aromatic aldehyde)

$\therefore$ Gives Fehling's test.

Products of options (a), (b) and (c) are

CHO (Aromatic aldehyde)

(Ketone)

CHO (Aromatic aldehyde)

and none of them give Fehling's test.

102. *(b)*

(X)

(amine + acid annydride $\longrightarrow$ amide)

conc.HNO_3 / conc.H_2SO_4 15°C

NO_2 (Y)

$(-NH-CO-NH_3$ is a *para* directing group)

aq. NaOH

O_2N (Z)

(amide is hydrolysed to amine).

103. *(c)*

1. O_3 2. Zn/H_2O (Ozonolysis)

P, OMe, CHO + OHC, OMe, Q and R

Number of moles of $P = \dfrac{420 \times 10^{-3}}{210}$

$= 2 \times 10^{-3}$

Number of moles of Q and R = 40% of

$P = \dfrac{40}{100} \times 2 \times 10^{-3}$

$= 8 \times 10^{-4}$

mass of CHO (M = 106)

$= 8 \times 10^{-4} \times 10^{6} = 84.8 \times 10^{-3}$

$= 84.8$ mg

mass of CHO, OMe (M = 136)

$= 8 \times 10^{-4} \times 136$

$= 108.8 \times 10^{-3}$

$= 108.8$ mg

which is Q.

$\therefore$ Mass of R = 84.8 mg

104. *(b)* $CuI(s) \rightleftharpoons Cu^{+}(aq) + I^{-}(aq)$

$K_{sp} = [Cu^{+}][I^{-}]$

If its solubility = x, at eqbm., $[Cu^{+}] = x$ and $[I^{-}] = x$, $K_{sp} = x^2$.

Since, $K_{sp} = 4 \times 10^{-12}$, $x = 2 \times 10^{-6}$ M

$Ag_2CrO_4(s) \rightleftharpoons 2Ag^{+}(aq) + CrO^{2-}(aq)$

$$K_{sp} = [Ag^{+}]^2[CrO_4^{2-}]$$

If its solubility = y, at eqbm., $[Ag^{+}] = 2y$ and $[CrO_4^{2-}] = y$

$\therefore \quad K_{sp} = (2y)^2 y = 4y^3$

Since, $K_{sp} = 4 \times 10^{-12}$,

$4y^3 = 4 \times 10^{-12}$,

$y = 10^{-4}$ M

$\therefore \quad x : y = 2 \times 10^{-6} ; 10^{-4}$

$= 2 \times 10^{-2} = 0.02$

105. *(b)* Hydrogenation of benzene is given by

$$C_6H_6 + 3H_2 \longrightarrow C_6H_{12} \quad \ldots (a)$$

Enthalpies are given for,

$$C_6H_6 + \frac{15}{2}O_2 \longrightarrow 6CO_2 + 3H_2O ; \Delta H = x \quad \ldots (b)$$

$$C_6H_{12} + 9O_2 \longrightarrow 6CO_2 + 6H_2O ; \Delta H = y \quad \ldots (c)$$

$$H_2 + \frac{1}{2}O_2 \longrightarrow H_2 ; \Delta H = z \quad \ldots (d)$$

Reaction (a) = (b) − (c) + 3(d),

$\therefore \Delta H$ of a $= x - y + 3z$

106. *(a)* $K_3[Fe(CN)_6]$

Central metal ion is Fe^{3+}, which has [Ar] $3d^5$ configuration as free ion.

In strong ligand field of CN^-, its splits to become $t_{2g}^{5} e_g^{0}$.

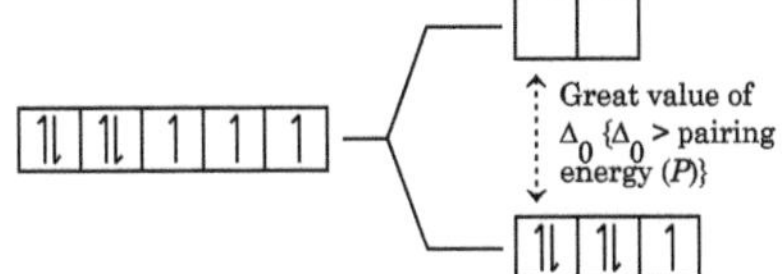

One unpaired electron will be there, and $\therefore$ It is paramagnetic.

$K_3[CoF_6]$: Central metal ion is Co^{3+}.

Free metal ion configuration [Ar] $3d^6$.

In weak ligand field of F^-, its splits and becomes $t_{2g}^4 e_g^2$.

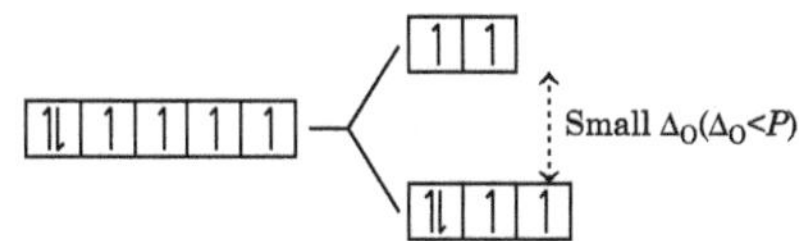

Four unpaired electrons, therefore paramagnetic.

In $K_4[Fe(CN)_6]$, central ion is Fe^{2+} and in $[Co(NH_3)_6]Cl_3$, central ion is Co^{3+}.

In both of them, free metal ion is $[Ar]3d^6$.

Ligands NH_3 and CN^- are of strong field, $\Delta_0 > P$

So, d^6 splits to $t_{2g}^6 e_g^0$

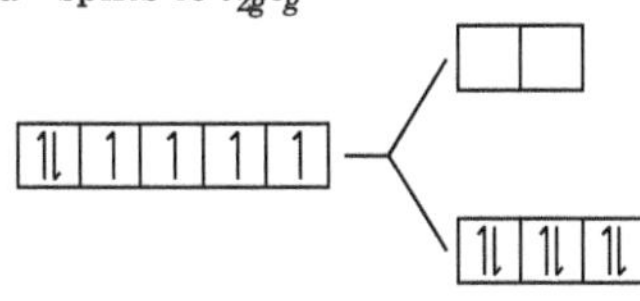

No unpaired electrons, therefore diamagnetic.

107. *(d)*

SO_3H; 1. Oleum; 2. Molten NaOH, D; ONa; 3. H_3O^+; OH (X)

OH (X) 1. NaOH ONa

ONa, COOH; 2. CO_2, 3. H_3O^+ (Y)

X = (phenol, OH),

Y = (CO_2H, OH)

108. *(c)* Molality

$$= \frac{\text{Mass of solute} \times 1000}{\text{Molar mass of solute} \times \text{Mass of solvent}}$$

Mass of solvent

= mass of solution − Mass of solute

↓ ↓

Volume × density 3 g

$= 250\,mL \times 1.9\,g/mL = 475\,g$

$\therefore$ Mass of solvent = 472 g

Molar mass of $(CO_2H)_2 . 2H_2O = 126\,g\,mol^{-1}$

$$\therefore \text{Molality} = \frac{3 \times 1000}{472 \times 126}$$

$$= 0.05\ \text{mol kg}^{-1}$$

Normality

$$= \frac{\text{Number of equivalents of solute}}{\text{Volume of solution } (l)}$$

$$= \frac{\text{Mass of solute } (g) \times 1000}{\text{Equivalent mass of solute} \times \text{Volume of solution (mL)}}$$

Equivalent mass of oxalic acid

$$= \frac{\text{Molar mass}}{2}$$

$$= 63\text{g / equi.}$$

$$\therefore \text{Normality} = \frac{3 \times 1000}{63 \times 250}$$

$$= 0.19\ \text{(equivalents)/l}$$

or 0.19 N

109. *(a)* In titration at equivalence point, number of equivalents of both reactants must be same.

$\therefore N_1V_1 = N_2V_2$

N (normality) of $FeCl_2$

$$= \frac{(N \times V) \text{ of } K_2Cr_2O_7}{V \text{ of } FeCl_2}$$

N (normality) of $K_2Cr_2O_7$

$$= \frac{\text{Mass} \times 1000}{\text{Equivalent mass } \times \text{Volume of solution prepared}}$$

$$\text{Equivalent mass} = \frac{\text{Molar mass}}{n \text{ factor}}$$

$$= \frac{294}{6}$$

$$= 49\ \text{g/equi.}$$

($K_2Cr_2O_7$ as oxidising agent is reduced to $2\,Cr^{3+}$).

Oxidation state of 2 Cr atoms change from + 6 to + 3.

Total change is $3 \times 2 = 6$, ($\therefore n$-factor = 6)

$$\therefore N \text{ (of } K_2Cr_2O_7) = \frac{1225 \times 1000}{49 \times 250} = 0.1\text{N}$$

$$\therefore N \text{ (of } FeCl_2) = \frac{0.1 \times 25}{10}$$

$$= 0.25\ \text{N}$$

110. *(c)* Number of lattice points in a hcp lattice : number of tetrahedral voids is in 1 : 2 ratio.

$\therefore$ Number of Z atoms : X atoms = 1:2

$\therefore$ Formula of compound = X_2Z

111. *(a)* Allele A frequency

$$= \frac{2N_{AA} + A_a}{2N}$$

Total number of individuals in population $= N$

$N = 90 + 40 + 70 = 200$

$N = 200$

$$A = \frac{2 \times 90 + 40}{2 \times 200}$$

$$= \frac{180 + 40}{400}$$

$$= \frac{220}{400} = 0.55$$

$$\text{Allele '}a\text{' frequency} = \frac{2\,N_{aa} + A_a}{2N}$$

$$a = \frac{2 \times 70 + 40}{2 \times 200}$$

$$= \frac{140 + 40}{400}$$

$$= \frac{180}{400} = 0.45$$

So, $A = 0.55$

$a = 0.45$

112. *(b)* P = 1 ring

Q = 1 ring

R = 2 ring

S = 2 ring

T = 3 ring

P binds with T from four structured ring.

Q binds with T from four structured ring.

R binds with S from four structured ring.

113. *(c)* Amino acid analysis of two globular protein yield identical composition per mole it means molecular weight of two protein is same because molecular weight of a protein can be predicted on the basis of amino acid composition.

114. *(b)* The conversion of 1, 3-bisphosphoglycerate to 3-phosphoglycerate is an example of substrate level phosphorylation. ATP is produced in this reaction.

$$\text{1, 3-bisphosphoglycerate} + \text{ADP} \underset{\text{Kinase}}{\overset{\text{Phosphoglycerate}}{\rightleftharpoons}} \text{3-phosphoglycerate} + \text{ATP}$$

In this process, high energy phosphate group is transferred from 1, 3-bisphosphoglycerate which results in the formation of ATP. This ATP formation is referred as substrate level phosphorylation because the phosphate donor 1, 3-bisphosphoglycerate is a substrate.

115. *(c)* Plant is heterozygous for the following traits

T – Tall plant

t – Short plant

R – Red colour flower

r – White colour flower

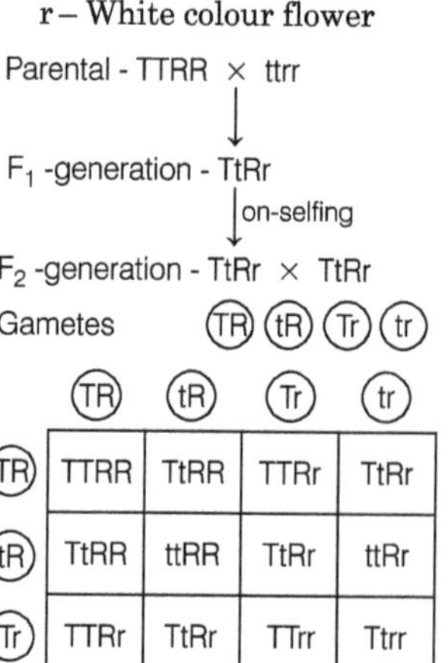

	TR	tR	Tr	tr
TR	TTRR	TtRR	TTRr	TtRr
tR	TtRR	ttRR	TtRr	ttRr
Tr	TTRr	TtRr	TTrr	Ttrr
tr	TtRr	ttRr	Ttrr	ttrr

Ratio is - 9 : 3 : 3 : 1

out of 1600 seeds ratio is

900 : 300 : 300 : 100

100 offsprings are completely homozygous for trait short and white flower.

116. *(d)* The Meselson and Stahl's experiment showed that DNA in replicated by semi-conservative mechanism. In conservative replication, on daughter duplex consist of two newly synthesised strand and parental start is conserved. Parental DNA grown is ^{15}N medium (heavy isotope of nitrogen) when centrifuged in first and second generation produces only light and heavy band because parental DNA contains ^{15}N in DNA and newly synthesised contain ^{14}N in bases. No intermediate density band is formed. Intermediate band is formed in semi-conservative and dispersive replication, where one DNA strand contains ^{15}N and other strand contains ^{14}N.

Separation of DNA by centrifugation

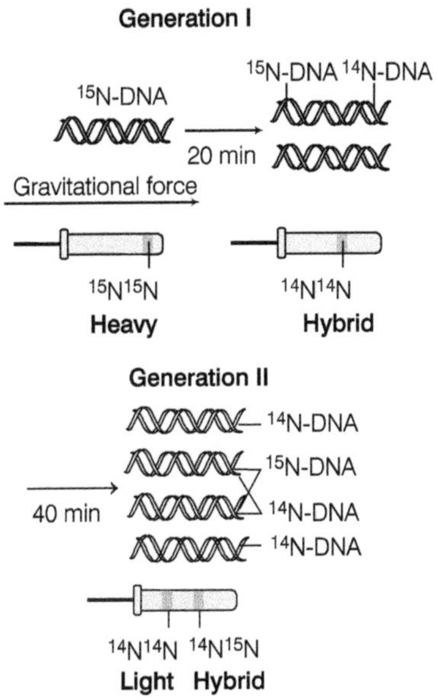

Generation I

^{15}N-DNA

20 min

^{15}N-DNA ^{14}N-DNA

Gravitational force

^{15}N^{15}N
Heavy

^{14}N^{14}N
Hybrid

Generation II

40 min

^{14}N-DNA
^{15}N-DNA
^{14}N-DNA
^{14}N-DNA

^{14}N^{14}N ^{14}N^{15}N
Light Hybrid

So, ratio is 2 : 0 : 2

117. *(b)* The graph shows commensalism type of interaction between species 1 and 2. Species grows with time when grows alone. When species 1 and 2 grow together, species 1 grows more whereas no change is shown by species 2. In commensalism, one organism gets benefits and other derives niether benefit nor harm. In amensalism, one species harms another species without getting any kind of benefits. In mutualism, both species benefits each other. In competition both organisms are harmed.

118. *(d)* In diploid species ploidy level is $2N$, while additional ploidy level are $1N$ and $3N$.

$1N \rightarrow$ found in gametes and $3N \rightarrow$ found in endosperm (triploid).

119. *(c)* Smaller billed birds feed more efficiently on soft seeds whereas large billed birds feed on hard seed. Birds having intermediate bill size cannot feed on hard and soft seeds efficiently. Bimodial distribution of bill sizes is a result of disruptive selection which favours larger smaller bill size individuals over intermediate sized bills.

120. *(c)* Water potential and osmotic potential of pure water is zero because water is open to atmosphere, Hydrostatic pressure of water is same as atmospheric pressure and there are no solutes in water. Container X does not contain any solute. So, ψ for container X is zero. Container Y contains 0.1 M sugar solution, so ψ and ψ_s would be negative because solute lowers the water potential of a solution.

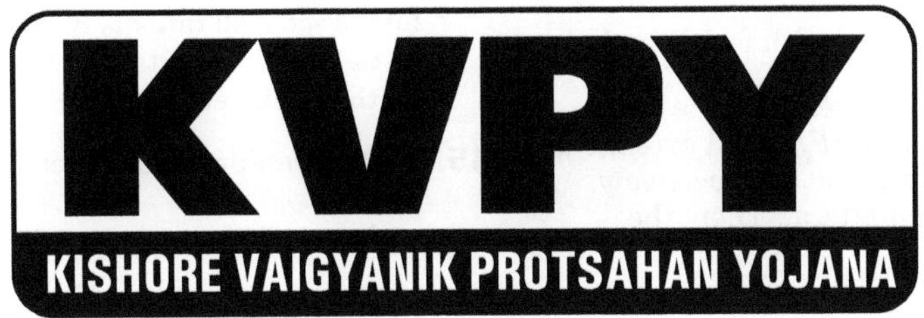

QUESTION PAPER 2018

Stream : SB/SX

MM : 160

Instructions

1. There are 120 questions in this paper.
2. The question paper contains two parts; Part I (1 Mark Questions) and Part II (2 Marks Questions).
3. There are four sections in each part; Mathematics, Physics, Chemistry and Biology.
4. Out of the four options given with each question, only one is correct.

PART-I (1 Mark Questions)

MATHEMATICS

1. Suppose $A = \begin{bmatrix} a & b \\ c & d \end{bmatrix}$ is a real matrix with non-zero entries, $ad - bc = 0$ and $A^2 = A$. Then, $a + d$ equals

(a) 1 (b) 2 (c) 3 (d) 4

2. On any given arc of positive length on the unit circle $|z| = 1$ in the complex plane.

(a) there need not be any root of unity
(b) there lies exactly one root of unity
(c) there are more than one but finitely many roots of unity
(d) there are infinitely many roots of unity

3. For $0 < \theta < \frac{\pi}{2}$, four tangents are drawn at the four points $(\pm 3\cos\theta, \pm 2\sin\theta)$ to the ellipse $\frac{x^2}{9} + \frac{y^2}{4} = 1$. If $A(\theta)$ denotes the area of the quadrilateral formed by these four tangents, the minimum value of $A(\theta)$ is

(a) 21 (b) 24
(c) 27 (d) 30

4. Let $S = \{x \in R : \cos(x) + \cos(\sqrt{2}x) < 2\}$, then

(a) $S = \phi$
(b) S is a non-empty finite set
(c) S is an infinite proper subset of $R - \{0\}$
(d) $S = R - \{0\}$

5. On a rectangular hyperbola $x^2 - y^2 = a^2$, $a > 0$, three points A, B, C are taken as follows : $A = (-a, 0)$; B and C are placed symmetrically with respect to the X-axis on the branch of the hyperbola not containing A. Suppose that the ΔABC is equilateral. If the side length of the ΔABC is ka, then k lies in the interval

(a) (0, 2] (b) (2, 4]
(c) (4, 6] (d) (6, 8]

6. The number of real solutions x of the equation $\cos^2(x\sin(2x)) + \frac{1}{1 + x^2} = \cos^2 x + \sec^2 x$ is

(a) 0 (b) 1
(c) 2 (d) infinite

7. Let $\frac{x^2}{a^2}+\frac{y^2}{b^2}=1$, $a>b$ be an ellipse with foci F_1 and F_2. Let AO be its semi-minor axis, where O is the centre of the ellipse. The lines AF_1 and AF_2, when extended, cut the ellipse again at points B and C respectively. Suppose that the ΔABC is equilateral. Then, the eccentricity of the ellipse is

(a) $\frac{1}{\sqrt{2}}$ (b) $\frac{1}{\sqrt{3}}$ (c) $\frac{1}{3}$ (d) $\frac{1}{2}$

8. Let $a=\cos 1^\circ$ and $b=\sin 1^\circ$. We say that a real number is algebraic if it is a root of a polynomial with integer coefficients. Then,

(a) a is algebraic but b is not algebraic
(b) b is algebraic but a is not algebraic
(c) both a and b are algebraic
(d) neither a nor b is algerbraic

9. A rectangle with its sides parallel to the X-axis and Y-axis is inscribed in the region bounded by the curves $y=x^2-4$ and $2y=4-x^2$. The maximum possible area of such a rectangle is closest to the integer.

(a) 10 (b) 9 (c) 8 (d) 7

10. Let $f(x)=x|\sin x|, x\in R$. Then,

(a) f is differentiable for all x, except at $x=n\pi$, $n=1,2,3,\ldots$
(b) f is differentiable for all x, except at $x=n\pi$, $n=\pm1,\pm2,\pm3,\ldots$
(c) f is differentiable for all x, except at $x=n\pi$, $n=0,1,2,3$
(d) f is differentiable for all x, except at $x=n\pi$, $n=0,\pm1,\pm2,\pm3,\ldots$

11. $f:[-1,1]\to R$ be a function defined by

$$f(x)=\begin{cases} x^2\left|\cos\left(\frac{\pi}{x}\right)\right| & \text{for } x\neq 0 \\ 0 & \text{for } x=0.\end{cases}$$

The set of points where f is not differentiable is

(a) $\{x\in[-1,1]: x\neq 0\}$
(b) $\left\{x\in[-1,1]: x=0 \text{ or } x=\frac{2}{2n+1}, n\in Z\right\}$
(c) $\left\{x\in[-1,1]: x=\frac{2}{2n+1}, n\in Z\right\}$
(d) $[-1,1]$

12. The value of the integral $\int_0^\pi (1-|\sin 8x|)dx$ is

(a) 0 (b) $\pi-1$ (c) $\pi-2$ (d) $\pi-3$

13. Let $\ln x$ denote the logarithm of x with respect to the base e. Let $S\subset R$ be the set of all points where the function $\ln(x^2-1)$ is well-defined. Then, the number of functions $f:S\to R$ that are differentiable, satisfy $f'(x)=\ln(x^2-1)$ for all $x\in S$ and $f(2)=0$, is

(a) 0 (b) 1 (c) 2 (d) infinite

14. Let S be the set of real numbers p such that there is no non-zero continuous function $f:R\to R$ satisfying $\int_0^x f(t)dt = p\,f(x)$ for all $x\in R$. Then, S is

(a) the empty set
(b) the set of all rational numbers
(c) the set of all irrational numbers
(d) the whole set R

15. The probability of men getting a certain disease is $\frac{1}{2}$ and that of women getting the same disease is $\frac{1}{5}$. The blood test that identifies the disease gives the correct result with probability $\frac{4}{5}$. Suppose a person is chosen at random from a group of 30 males and 20 females, and the blood test of that person is found to be positive. What is the probability that the chosen person is a man?

(a) $\frac{75}{107}$ (b) $\frac{3}{5}$ (c) $\frac{15}{19}$ (d) $\frac{3}{10}$

16. Number of functions $f:[0,1]\to[0,1]$ satisfying $|f(x)-f(y)|=|x-y|$ for all x, y in $[0,1]$ is

(a) exactly 1 (b) exactly 2
(c) more than 2, but finite (d) infinite

17. Suppose A is 3×3 matrix consisting of integer entries that are chosen at random from the set $\{-1000,-999,\ldots 999,1000\}$. Let P be the probability that either $A^2=-I$ or A is diagonal, where I is the 3×3 identity matrix. Then,

(a) $P<\frac{1}{10^{18}}$ (b) $P=\frac{1}{10^{18}}$
(c) $\frac{5^2}{10^{18}}\le P\le\frac{5^3}{10^{18}}$ (d) $P\le\frac{5^4}{10^{18}}$

18. Let x_k be real numbers such that $x_k\ge k^4+k^2+1$ for $1\le k\le 2018$. Denote $N=\sum_{k=1}^{2018}k$. Consider the following inequalities.

I. $\left(\sum_{k=1}^{2018}kx_k\right)^2\le N\left(\sum_{k=1}^{2018}kx_k^2\right)$ II. $\left(\sum_{k=1}^{2018}kx_k\right)^2\le N\left(\sum_{k=1}^{2018}k^2x_k^2\right)$

Then,

(a) both I and II are true (b) I is true and II is false
(c) I is false and II is true (d) both I and II are false

19. Let $x^2=4ky$, $k>0$ be a parabola with vertex A. Let BC be its latusrectum. An ellipse with centre on BC touches the parabola at A, and cuts BC at points D and E such that $BD=DE=EC$ (B,D,E,C in that order). The eccentricity of the ellipse is

(a) $\frac{1}{\sqrt{2}}$ (b) $\frac{1}{\sqrt{3}}$ (c) $\frac{\sqrt{5}}{3}$ (d) $\frac{\sqrt{3}}{2}$

20. Let $f:[0,1]\to[-1,1]$ and $g:[-1,1]\to[0,2]$ be two functions such that g is injective and $gof:[0,1]\to[0,2]$ is surjective. Then,

(a) f must be injective but need not be surjective
(b) f must be surjective but need not be injective
(c) f must be bijective
(d) f must be a constant function

PHYSICS

21. A table has a heavy circular top of radius 1 m and mass 20 kg placed on four light (considered massless) legs placed symmetrically on its circumference. The maximum mass that can be kept anywhere on the table without toppling it is close to

(a) 20 kg (b) 34 kg (c) 47 kg (d) 59 kg

22. Air (density ρ) is being blown on a soap film (surface tension T) by a pipe of radius R with its opening right next to the film. The film is deformed and a bubble detaches from the film when the shape of the deformed surface is a hemisphere. Given that the dynamic pressure on the film due to the air blown at speed v is $\frac{1}{2}\rho v^2$, the speed at which the bubble formed is

(a) $\sqrt{\frac{T}{\rho R}}$ (b) $\sqrt{\frac{2T}{\rho R}}$ (c) $\sqrt{\frac{4T}{\rho R}}$ (d) $\sqrt{\frac{8T}{\rho R}}$

23. For an ideal gas, the internal energy is given by $U = 5pV/2 + C$, where C is a constant. The equation of the adiabats in the pV-plane will be

(a) p^5V^7 = constant (b) p^7V^5 = constant

(c) p^3V^5 = constant (d) p^5V^2 = constant

24. An ideal gas undergoes change in its state from the initial state I to the final state F *via* two possible paths as shown below. Then,

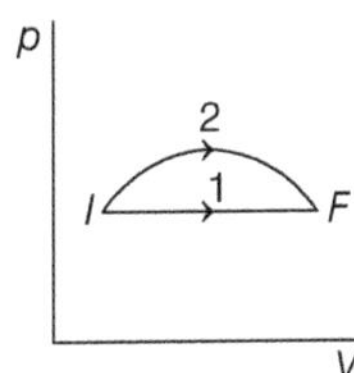

(a) there is no change in internal enorgy along path 1

(b) heat is not absorbed by the gas in both paths

(c) the temperature of the gas first increases and then decreases for path 2

(d) work done by the gas is larger in path 1

25. A thermally insulated rigid container of 1 L volume contains a diatomic ideal gas at room temperature. A small paddle installed inside the container is rotated from the outside, such that the pressure rises by 10^5 Pa. The change in internal energy is close to

(a) zero (b) 67 J

(c) 150 J (d) 250 J

26. In Young's double slit experiment, the amplitudes of the two waves incident on the two slits are A and $2A$. If I_0 is the maximum intensity, then the intensity at a spot on the screen, where the phase difference between the two interfering waves is ϕ.

(a) $I_0 \cos^2(\phi/2)$ (b) $\frac{I_0}{3}\sin^2(\phi/2)$

(c) $\frac{I_0}{9}(5 + 4\cos\phi)$ (d) $\frac{I_0}{9}(5 + 8\cos\phi)$

27. Figures below show water flowing through a horizontal pipe from left to right.
Note that the pipe in the middle is narrower. Choose the most appropriate depiction of water levels in the vertical pipes.

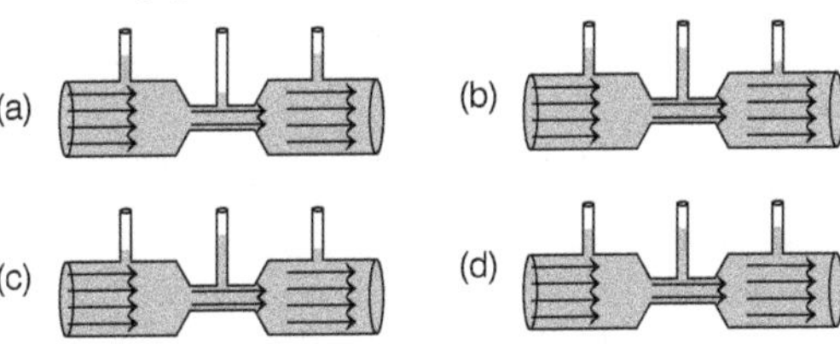

28. A plank is moving in a horizontal direction with a constant acceleration $a\hat{\mathbf{i}}$. A uniform rough cubical block of side l rests on the plank and is at rest relative to the plank.

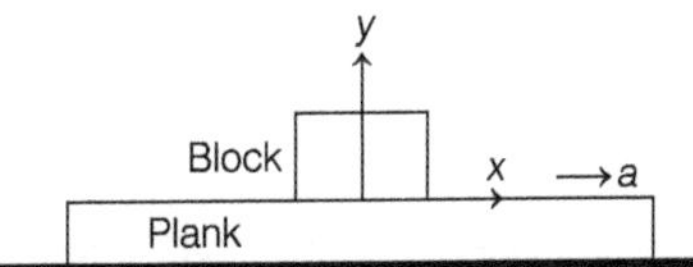

Let the centre of mass of the block be at $(0, l/2)$ at a given instant. If $a = g/10$, then the normal reaction exerted by the plank on the block at that instant acts at

(a) (0, 0) (b) $(-l/20, 0)$

(c) $(-l/10, 0)$ (d) $(l/10, 0)$

29. Using the Heisenberg uncertainty principle, arrange the following particles in the order of increasing lowest energy possible.

(I) An electron in H_2 molecule

(II) A hydrogen atom in a H_2 molecule

(III) A proton in the carbon nucleus

(IV) A H_2 molecule within a nanotube

(a) I < III < II < IV

(b) IV < II < I < III

(c) II < IV < III < I

(d) IV < I < II < III

30. The current in flowing along the path $ABCD$ of a cube (shown in the left figure) produces a magnetic field at the centre of cube of magnitude B. Dashed line depicts the non-conducting part of the cube.

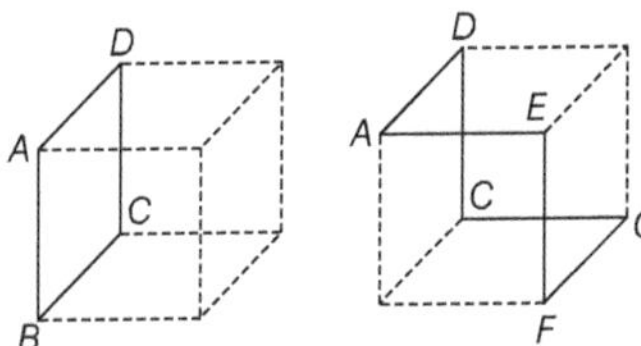

Consider a cubical shape shown to the right which is identical in size and shape to the left. If the same current now flows in along the path $DAEFGCD$, then the magnitude of magnetic field at the centre will be

(a) zero (b) $\sqrt{2}B$ (c) $\sqrt{3}B$ (d) B

31. A thin metallic disc is rotating with constant angular velocity about a vertical axis that is perpendicular to its plane and passes through its centre. The rotation causes the free electrons in the disc to redistribute. Assume that, there is no external electric or magnetic field. Then,

(a) a point on the rim of the disc is at a higher potential than its centre

(b) a point on the rim of the disc is at a lower potential than its centre

(c) a point on the rim of the disc is at the same potential as its centre

(d) the potential in the material has an extremum between centre and the rim

32. One mole of a monoatomic gas and one mole of a diatomic gas are initially in the same state. Both gases are expanded isothermally and then adiabatically, such that they acquire the same final state. Choose the correct statement.

(a) Work done by diatomic gas is more than that by monoatomic gas

(b) Work done by monoatomic gas is more than that by diatomic gas

(c) Work done by both the gases are equal

(d) Change in internal energies of both the gases are equal

33. An ideal gas is made to undergo the cyclic process shown in the figure below.

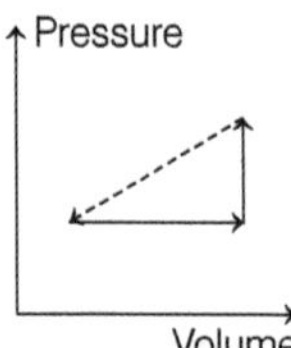

Let ΔW depict the work done, ΔU be the change in internal energy of the gas and Q be the heat added to the gas. Sign of each of these three quantities for the whole cycle will be (0 refers to no change)

(a) – , 0, – (b) + , 0, + (c) 0, 0, 0 (d) +,+,+

34. Two balls of mass M and $2M$ are thrown horizontally with the same initial velocity v_0 from top of a tall tower and experience a drag force of $-kv$ $(k > 0)$, where v is the instantaneous velocity.

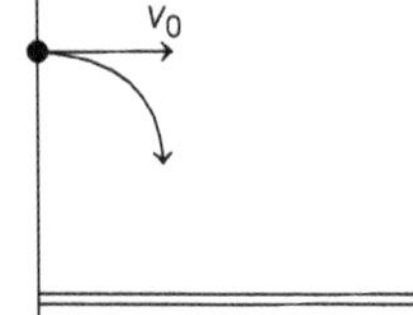

Then,

(a) the heavier ball will hit the ground further away than the lighter ball

(b) the heavier ball will hit the ground closer than the lighter ball

(c) both balls will hit the ground at the same point

(d) both balls will hit the ground at the same time

35. Consider a glass cube slab of dielectric bound by the planes $x = 0, x = a; y = 0, y = b; z = 0, z = c$; with $b > a > c$. The slab is placed in air and has a refractive index of n. The minimum value for n, such that all rays entering the dielectric at $y = 0$ reach $y = b$ is

(a) 1 (b) $\sqrt{2}$
(c) $\sqrt{3}$ (d) 2

36. The graph shows the log of activity $\log R$ of a radioactive material as a function of time t in minutes.

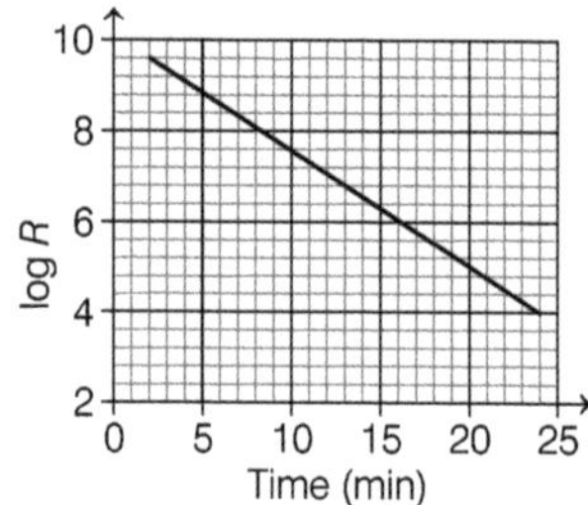

The half-life (in minute) for the decay is closest to

(a) 2.1 (b) 3.0
(c) 3.9 (d) 4.4

37. The magnetic field is uniform for $y > 0$ and points into the plane. The magnetic field is uniform and points out of the plane for $y < 0$. A proton denoted by filled circle leaves $y = 0$ in the $-y$-direction with some speed as shown below.

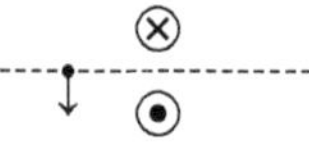

Which of the following best denotes the trajectory of the proton?

(a) 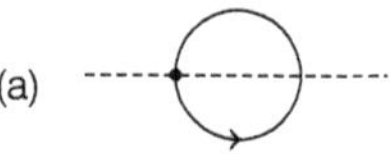 (b) 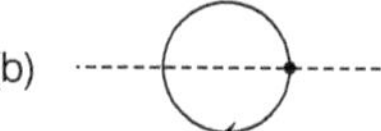

(c) 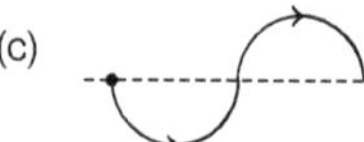 (d) 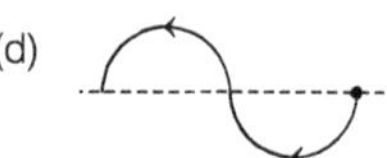

38. The Hitomi satellite recently observed the Lyman alpha emission line ($n = 2$ to $n = 1$) of hydrogen-like iron ion (atomic number of iron is 26) from the Perseus galaxy cluster. The wavelength of the line is closest to

(a) 2 Å (b) 1 Å
(c) 50 Å (d) 10 Å

39. Assume that, the drag force on a football depends only on the density of the air, velocity of the ball and the cross-sectional area of the ball. Balls of different sizes but the same density are dropped in an air column. The terminal velocity reached by balls of masses 250 g and 125 g are in the ratio

(a) $2^{1/6}$ (b) $2^{1/3}$
(c) $2^{1/2}$ (d) $2^{2/3}$

40. An electrostatic field line leaves at an angle α from point charge q_1 and connects with point charge $-q_2$ at an angle β (q_1 and q_2 are positive) see figure below. If $q_2 = \frac{3}{2} q_1$ and $\alpha = 30°$, then

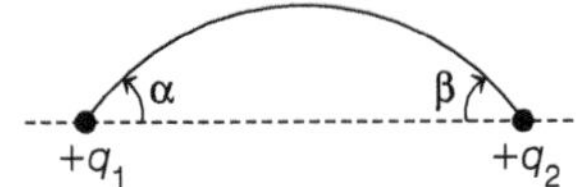

(a) $0° < \beta < 30°$ (b) $\beta = 30°$
(c) $30° < \beta \leq 60°$ (d) $60° < \beta \leq 90°$

CHEMISTRY

41. The amount (in mol) of bromoform ($CHBr_3$) produced when 1.0 mole of acetone reacts completely with 1.0 mole of bromine in the presence of aqueous NaOH is
(a) $\frac{1}{3}$ (b) $\frac{2}{3}$ (c) 1 (d) 2

42. The following compound

can readily be prepared by Williamson ether synthesis by reaction between

(a) [OH-bicyclic] and [benzyl–Cl] (b) [OH-bicyclic] and [benzyl–I]
(c) [Cl-bicyclic] and [benzyl–OH] (d) [I-bicyclic] and [benzyl–OH]

43. X and Y

X: H, H, Cl, Cl, H_3C, CH_3
Y: Cl, H_3C, H, H, CH_3, Cl

are
(a) enantiomers (b) diastereomers
(c) constitutional isomers (d) conformers

44. The higher stabilities of *tert*-butyl cation over isopropyl cation and *trans*-2- butene over propene, respectively, are due to orbital interactions involving

(a) $\sigma \to \pi$ and $\sigma \to \pi^*$
(b) $\sigma \to$ vacant p and $\sigma \to \pi$
(c) $\sigma \to \sigma^*$ and $\sigma \to \pi$
(d) $\sigma \to$ vacant p and $\sigma \to \pi^*$

45. Benzaldehyde can be converted to benzyl alcohol in concentrated aqueous NaOH solution using
(a) acetone (b) acetaldehyde
(c) formic acid (d) formaldehyde

46. The major product of the following reaction,

$PhCO(CH_2)_4CO_2H \xrightarrow{NaBH_4}$

is

(a) $PhCH(OH)(CH_2)_4CO_2H$
(b) $PhCH(OH)(CH_2)_4CH_2OH$
(c) $PhCO(CH_2)_4CH_2OH$
(d) $Ph(CH_2)_5CO_2H$

47. Among the following species, the H—X—H angle (X = B, N or P) follows the order.
(a) $PH_3 < NH_3 < NH_4^+ < BF_3$
(b) $NH_3 < PH_3 < NH_4^+ < BF_3$
(c) $BF_3 < PH_3 < NH_4^+ < NH_3$
(d) $BF_3 < NH_4^+ < NH_3 < PH_3$

48. The ionic radii of Na^+, F^-, O^{2-}, N^{3-} follow the order
(a) $O^{2-} > F^- > Na^+ > N^{3-}$ (b) $N^{3-} > Na^+ > F^- > O^{2-}$
(c) $N^{3-} > O^{2-} > F^- > Na^+$ (d) $Na^+ > F^- > O^{2-} > N^{3-}$

49. The oxoacid of phosphorus having the strongest reducing property is
(a) H_3PO_3 (b) H_3PO_2
(c) H_3PO_4 (d) $H_4P_2O_7$

50. Among C, S and P, the element (s) that produce (s) SO_2 on reaction with hot conc. H_2SO_4 is / are
(a) Only S (b) Only C and S
(c) Only S and P (d) C, S and P

51. The complex that can exhibit linkage isomerism is
(a) $[Co(NH_3)_5(H_2O)]Cl_3$ (b) $[Co(NH_3)_5(NO_2)]Cl_2$
(c) $[Co(NH_3)_5(NO_3)](NO_3)_2$ (d) $[Co(NH_3)_5Cl]SO_4$

52. The tendency of X in BX_3 (X = F, Cl, OMe, NMe) to form a π-bond with boron follows the order
(a) $BCl_3 < BF_3 < B(OMe)_3 < B(NMe_2)_3$
(b) $BF_3 < BCl_3 < B(OMe)_3 < B(NMe_2)_3$
(c) $BCl_3 < B(NMe_2)_3 < B(OMe)_3 < BF_3$
(d) $BCl_3 < BF_3 < B(NMe_2)_3 < B(OMe)_3$

53. Consider the following statements about Langmuir isotherm.

I. The free gas and adsorbed gas are in dynamic equilibrium.
II. All adsorption sites are equivalent.
III. The initially adsorbed layer can act as a substrate for further adsorption.
IV. The ability of a molecule to get adsorbed at a given site is independent of the occupation of neighboring sites.

The correct statements are
(a) I, II, III and IV (b) I, II and IV
(c) I, III, and IV (d) I, II, and III

54. Among the following, the plot that correctly represents the conductometric titration of 0.05 M H_2SO_4 with 0.1 M NH_4OH is

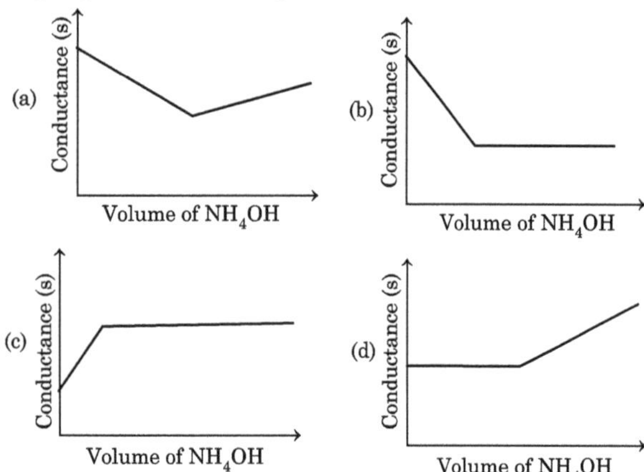

55. The correct representation of wavelength intensity relationship of an ideal black body radiation at two different temperatures T_1 and T_2 is

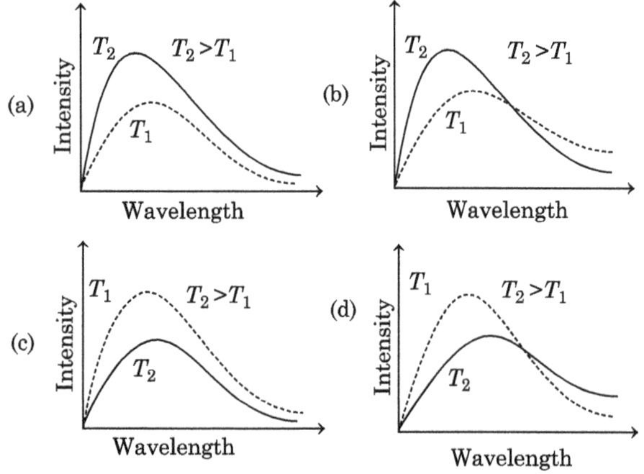

56. The pressure (p) and volume (V) isotherm of a van der Waals' gas, at the temperature at which it undergoes gas to liquid transition, is correctly represented by

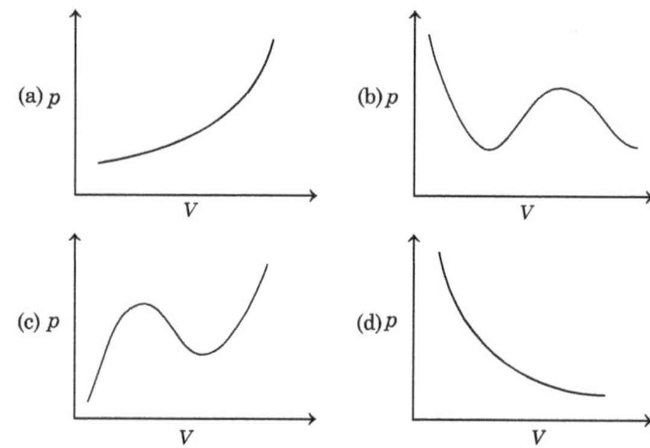

57. A buffer solution can be prepared by mixing equal volumes of
(a) 0.2 M NH_4OH and 0.1 M HCl
(b) 0.2 M NH_4OH and 0.2 M HCl
(c) 0.2 M NaOH and 0.1 M CH_3COOH
(d) 0.1 M NH_4OH and 0.2 M HCl

58. The plot of total vapour pressure as a function of mole fraction of the components of an ideal solution formed by mixing liquids X and Y is

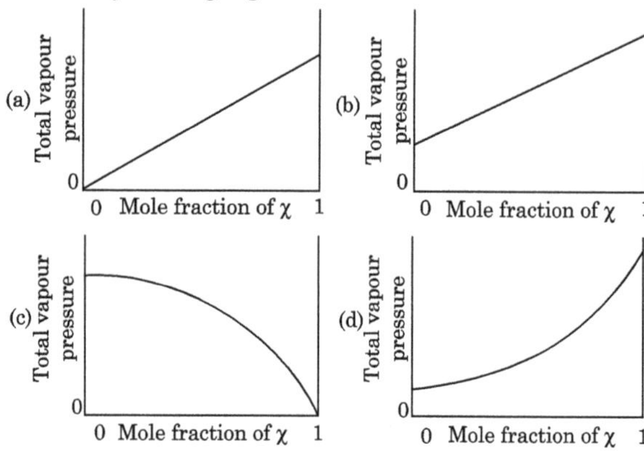

59. On complete hydrogenation, natural rubber produces
(a) polyethylene
(b) ethylene propylene copolymer
(c) polyvinyl chloride (d) polypropylene

60. The average energy of each hydrogen bond in A—T pair is x kcal mol^{-1} and that in G—C pair is y kcal mol^{-1}. Assuming that no other interaction exists between the nucleotides, the approximate energy required in kcal mol^{-1} to split the following double stranded DNA into two single strands is

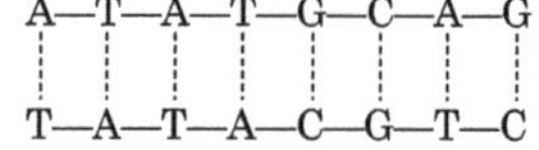

[Each dashed line may represent more than one hydrogen bond between the base pairs]

(a) $10x + 9y$ (b) $5x + 3y$ (c) $15x + 6y$ (d) $5x + 4.5y$

BIOLOGY

61. What is the maximum number of oxygen atoms that a molecule of haemoglobin can bind?

(a) 2 (b) 4 (c) 8 (d) 16

62. *Bt* toxin produced by *Bacillus thuringiensis* does not kill the producer because the toxin is

(a) in an inactive protoxin form
(b) rapidly secreted outside
(c) inactivated by an antitoxin
(d) in unfolded form

63. An angiosperm was identified with its endosperm of $6n$. Assuming that this is a self-pollinating species, which one of the following is the correct ploidy of the parent?

(a) $3n$ (b) $4n$ (c) $6n$ (d) $8n$

64. Which one of the following statements is true about viruses?

(a) All viruses possess a protein coat around its genetic material at all stages of their life cycle
(b) All viruses contain RNA as genetic material
(c) All viruses contain DNA as genetic material
(d) All viruses replicate only within the host cell

65. Mitochondrial cristae are infoldings of the

(a) outer membrane and they increase the surface area
(b) outer membrane and they decrease the surface area
(c) inner membrane and they increase the surface area
(d) inner membrane and they decrease the surface area

66. In biological nitrogen-fixation, the enzyme nitrogenase converts

(a) nitrate to nitrite
(b) atmospheric nitrogen to nitrite
(c) nitrite to ammonia
(d) atmospheric nitrogen to ammonia

67. The graph below represents the absorption spectrum of a major pigment contributing to photosynthesis.

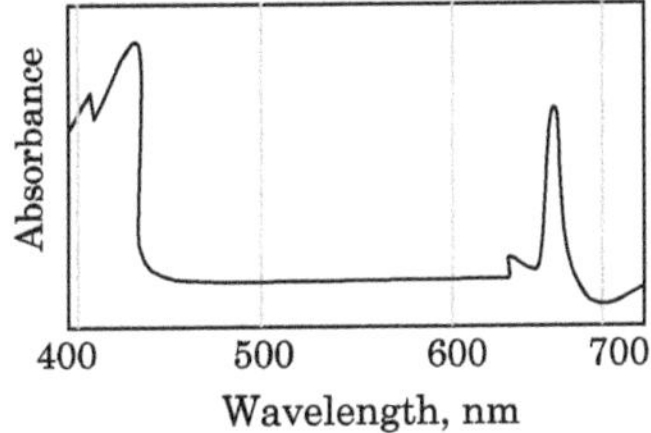

Which one of the following best represents the photosynthetic efficiency of the pigment?

(a)
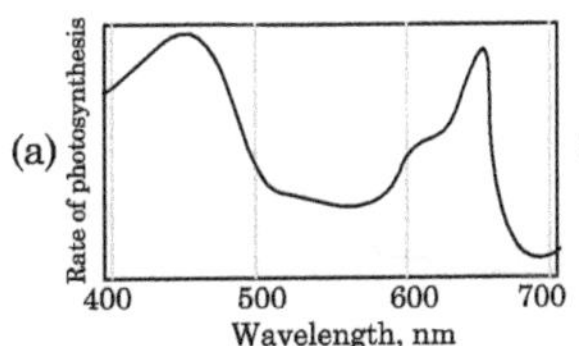

(b)
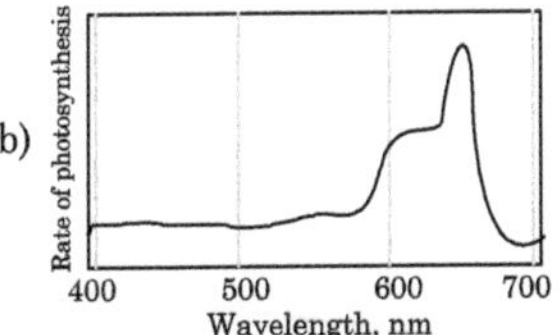

(c)
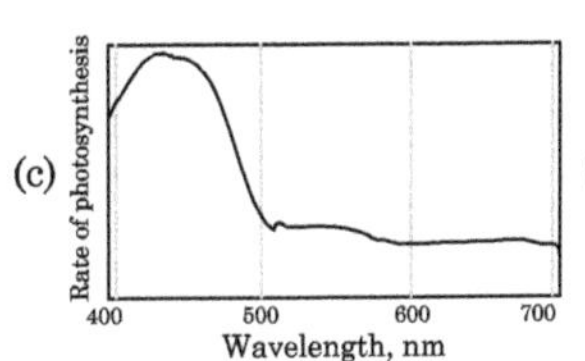

(d)
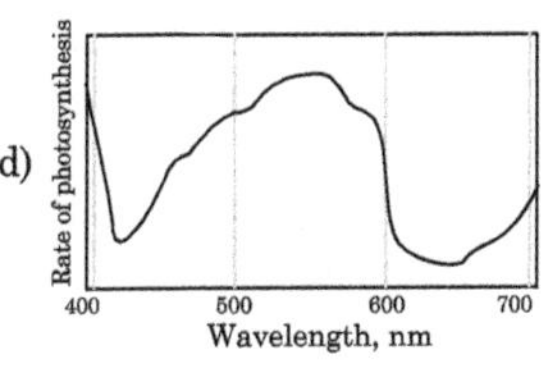

68. Which one of the following properties of normal cell is lost during its transition to cancerous cell?

(a) Glutamine utilisation (b) Contact inhibition
(c) Glucose utilisation (d) Membrane fluidity

69. Which one of the following gases is produced during fermentation by yeast ?

(a) CO_2 (b) O_2 (c) H_2 (d) N_2

70. Serine proteases are called so because they

(a) require free serine for their activity
(b) cleave after serine residues in the substrate
(c) are inhibited by the presence of free serine
(d) have a serine residue at their active site

71. The maximum number of genotypes of the pollens produced by a tall pea plant with round, yellow, seeds of the genotype TtRrYY, if the three loci are unlinked, would be

(a) 1 (b) 2 (c) 4 (d) 8

72. Which one of the following statements is true with respect to human ovary?

(a) Oestrogen is secreted by Graafian follicles and progesterone by corpus luteum
(b) Oestrogen is secreted by corpus luteum and progesterone by Graafian follicles
(c) Both oestrogen and progesterone are secreted by Graafian follicles
(d) Both oestrogen and progesterone are secreted by corpus luteum

73. Which one of the following statements is incorrect with respect to human antibodies?

(a) They can neutralise microbes
(b) They are synthesised by T-cells
(c) They are made up of four polypeptide chains
(d) Milk contains antibodies

74. Concentration (%) of NaCl isotonic to human blood is

(a) 0.085 - 0.09% (b) 1.7 - 1.8%
(c) 3.4 - 3.6% (d) 0.85 - 0.9%

75. Which one of the following statements is true about the Golgi apparatus?

(a) It is found only in animals
(b) It is found only in prokaryotes
(c) It modifies and targets proteins to the plasma membrane
(d) It is a site for ATP production

76. Creutzfeldt Jakob Disease (CJD) is a transmissible disease caused by a

(a) virus (b) bacterium
(c) fungus (d) misfolded protein

77. A researcher found petrified dinosaur faeces. Which one of the following is unlikely to be found in this fossil?
(a) Decayed conifer wood
(b) Bamboo
(c) Cycad
(d) Giant fern

78. Which one of the pairs of amino acids contains two chiral centres?
(a) Isoleucine and threonine
(b) Leucine and valine
(c) Valine and isoleucine
(d) Threonine and leucine

79. In photosynthetic carbon-fixation, which one of the following reacts with CO_2?
(a) Phosphoglycolate
(b) 3-phosphoglycerate
(c) Ribulose-1, 5-bisphosphate
(d) Ribose-5-phosphate

80. Match the diseases in Column I with the routes of infection in Column II. Choose the correct combination.

	Column I	Column II
A.	Tuberculosis	1. Contaminated food and water
B.	Dysentery	2. Inhalation of aerosol
C.	Filariasis	3. Contact *via* skin
D.	Syphilis	4. Sexual intercourse
		5. Mosquito bite

	A	B	C	D
(a)	2	1	5	4
(b)	2	1	3	5
(c)	1	3	5	4
(d)	2	3	4	5

PART-II (2 Marks Questions)

MATHEMATICS

81. Let R be a rectangle, C be a circle, and T be a triangle in the plane. The maximum possible number of points common to the perimeters of R, C and T is
(a) 3 (b) 4 (c) 5 (d) 6

82. The number of different possible values for the sum $x+y+z$, where x, y, z are real number such that $x^4 + 4y^4 + 16z^4 + 64 = 32xyz$ is
(a) 1 (b) 2 (c) 4 (d) 8

83. Let Γ be a circle with diameter AB and centre O. Let l be the tangent to Γ at B. For each point M on Γ different from A, consider the tangent t at M and let interest l at P. Draw a line parallel to AB through P intersecting OM at Q. The locus of Q as M varies over Γ is
(a) an arc of a circle
(b) a parabola
(c) an arc of an ellipse
(d) a branch of a hyperbola

84. The number of solutions x of the equation $\sin(x+x^2) - \sin(x^2) = \sin x$ in the interval [2, 3] is
(a) 0 (b) 1 (c) 2 (d) 3

85. The number of polynomials $p: R \to R$ satisfying $p(0) = 0$, $p(x) > x^2$ for all $x \neq 0$ and $p''(0) = \frac{1}{2}$ is
(a) 0 (b) 1
(c) more than 1, but finite (d) infinite

86. Suppose the limit
$$L = \lim_{n\to\infty} \sqrt{n} \int_0^1 \frac{1}{(1+x^2)^n} dx$$
exists and is larger than $\frac{1}{2}$. Then,
(a) $\frac{1}{2} < L < 2$ (b) $2 < L < 3$
(c) $3 < L < 3$ (d) $L \geq 4$

87. Consider the set A_n of points (x, y) such that $0 \leq x \leq n, 0 y \leq n$ where n, x, y are integers. Let S_n be the set of all lines passing through at least two distinct points from A_n. Suppose we choose a line l at random from S_n. Let P_n be the probability that l is tangent to the circle $x^2 + y^2 = n^2\left(1 + \left(1 - \frac{1}{\sqrt{n}}\right)^2\right)$.

Then, the limit $\lim_{n \to \infty} P_n$ is
(a) 0 (b) 1 (c) $\frac{1}{\pi}$ (d) $\frac{1}{\sqrt{2}}$

88. Let $f: [0, 1] \to R$ be an injective continuous function that satisifes the condition $-1 < f(0) < f(1) < 1$. Then, the number of functions $g: [-1, 1] \to [0, 1]$ such that $(gof)(x) = x$ for all $x \in [0, 1]$ is
(a) 0
(b) 1
(c) more than 1, but finite
(d) infinite

89. The maximum possible area bounded by the parabola $y = x^2 + x + 10$ and a chord of the parabola of length 1 is
(a) $\frac{1}{12}$ (b) $\frac{1}{6}$ (c) $\frac{1}{3}$ (d) $\frac{1}{2}$

90. Suppose z is any root of $11z^8 + 21iz^7 + 10iz - 22 = 0$ where $i = \sqrt{-1}$. Then, $S = |z|^2 + |z| + 1$ satisfies
(a) $S \leq 3$ (b) $3 < S < 7$
(c) $7 \leq S < 13$ (d) $S \geq 13$

PHYSICS

91. In steady state heat conduction, the equations that determine the heat current $\mathbf{j}(\mathbf{r})$ [heat flowing per unit time per unit area] and temperature $T(\mathbf{r})$ in space are exactly the same as those governing the electric field $\mathbf{E}(\mathbf{r})$ and electrostatic potential $V(\mathbf{r})$ with the equivalence given in the table below.

Heat flow	Electrostatics
$T(\mathbf{r})$	$V(\mathbf{r})$
$\mathbf{j}(\mathbf{r})$	$\mathbf{E}(\mathbf{r})$

We exploit this equivalence to predict the rate $\dot{Q}$ of total heat flowing by conduction from the surfaces of spheres of varying radii, all maintained at the same temperature. If $\dot{Q} \propto R^n$, where R is the radius, then the value of n is

(a) 2 (b) 1
(c) -1 (d) -2

92. An arrangement of spring, strings, pulley and masses is shown in the figure below.

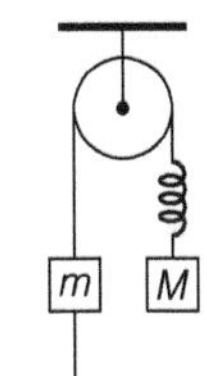

The pulley and the strings are massless and $M > m$. The spring is light with spring constant k. If the string connecting m to the ground is detached, then immediately after detachment

(a) the magnitude of the acceleration of m is zero and that of M is g
(b) the magnitude of the acceleration of m is $(M - m)g/m$ and that of M is zero
(c) the accelerations of both masses are same
(d) the elongation in the spring is $(M - m)g/k$

93. The potential due to an electrostatic charge distribution is $V(r) = \dfrac{qe^{-\alpha r}}{4\pi\varepsilon_0 r}$, where α is positive.

The net charge within a sphere centred at the origin and of radius $1/\alpha$ is

(a) $2q/e$ (b) $(1 - 1/e)q$
(c) q/e (d) $(1 + 1/e)q$

94. A wheel of radius R is trapped in a mud pit and spinning. As the wheel is spinning, it splashes mud blobs with initial speed u from various points on its circumference. The maximum height from the centre of the wheel, to which a mud blob can reach is

(a) $u^2/2g$ (b) $\dfrac{u^2}{2g} + \dfrac{gR^2}{2u^2}$
(c) 0 (d) $R + \dfrac{u^2}{2g}$

95. Two rods of copper and iron with the same cross-sectional area are joined at S and a steady current I flows through the rods as shown in the figure.

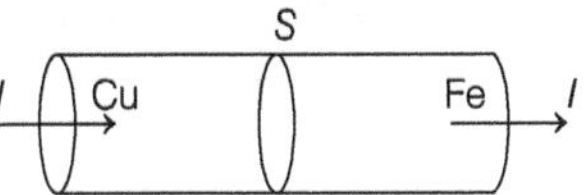

Choose the most appropriate representation of charges accumulated near the junction S.

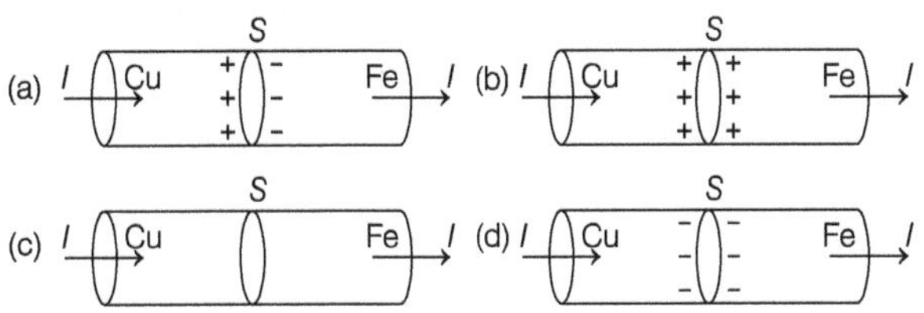

96. Graphs below show the entropy *versus* energy U of two systems 1 and 2 at constant volume. The initial energies of the systems are indicated by $U_{1,i}$ and $U_{2,i}$, respectively. Graphs are drawn to the same scale. The systems are then brought into thermal contact with each other. Assume that, at all times the combined energy of the two systems remains constant. Choose the most appropriate option indicating the energies of the two systems and the total entropy after they achieve the equilibrium.

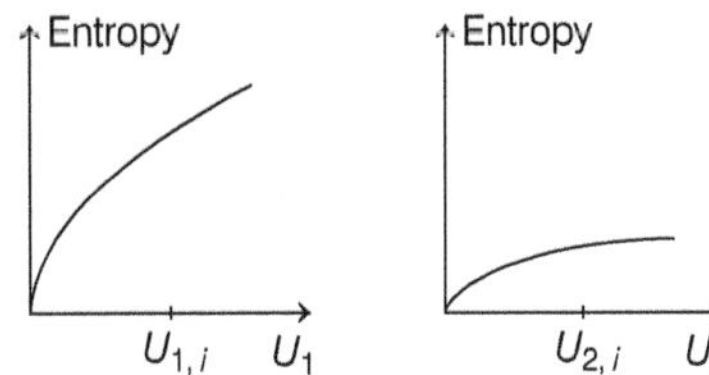

(a) U_1 increases and U_2 decreases and the total entropy remains the same
(b) U_1 decreases and U_2 increases and the total entropy remains the same
(c) U_1 increases and U_2 decreases and the total entropy increases
(d) U_1 decreases and U_2 increases and the total entropy increases

97. The image of an object O due to reflection from the surface of a lake is elongated due to the ripples on the water surface caused by a light breeze. This is because the ripples act as tilted mirrors as shown below. Consider the case, where O and the observer E are at the same height above the surface of the lake. If the maximum angle that the ripples make with the horizontal is α, then the angular extent δ of the image will be

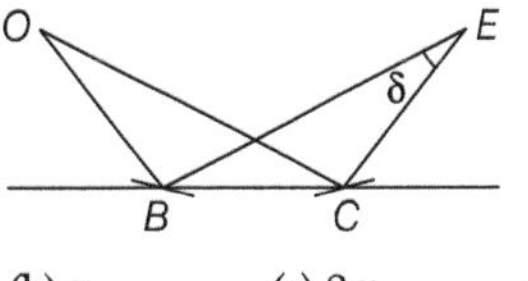

(a) $\dfrac{\alpha}{2}$ (b) α (c) 2α (d) 4α

98. A spiral galaxy can be approximated as an infinitesimally thin disc of a uniform surface mass density (mass per unit area) located at $z = 0$. Two stars A and B start from rest from heights $2z_0$ and z_0 (z_0 << radial extent of the disc), respectively and fall towards the disc, cross over to the other side and execute periodic oscillations. The ratio of time periods of A and B is

(a) $2^{-1/2}$ (b) 2 (c) 1 (d) $2^{1/2}$

99. Two mutually perpendicular infinitely long straight conductors carrying uniformly distributed charges of linear densities λ_1 and λ_2 are positioned at a distance r from each other.

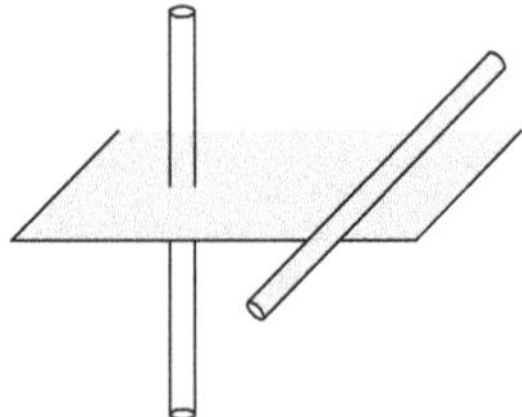

Force between the conductors depends on r as

(a) I/r (b) I/r^2 (c) r (d) r^0

100. The graph below shows the variation of a force F with time t on a body which is moving in a straight line. Dependence of force on time is $F \propto t^n$. Initially body is at rest.

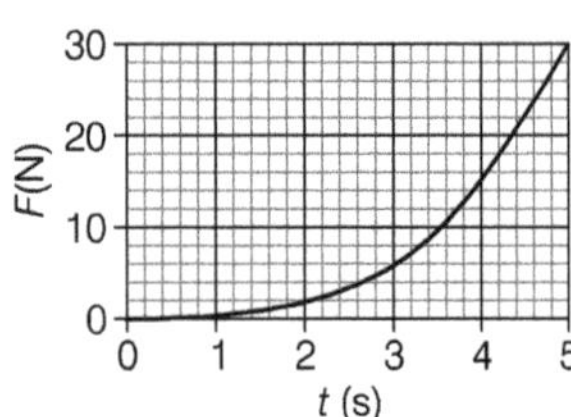

If the speed of the object is 2 m/s at 3 s, then the speed at 4 s will be approximately (in m/s)

(a) 2.5 (b) 6.5
(c) 7.8 (d) 3.1

CHEMISTRY

101. For the electrochemical cell shown below

$$\text{Pt}|\text{H}_2\ (p = 1\text{ atm})|\text{H}^+\ (aq., x\text{ M})|$$
$$|\text{Cu}^{2+}\ (aq., 1.0\text{M})|\text{Cu}(s)$$

The potential is 0.49 V at 298 K. The pH of the solution is closest to

[Given, standard reduction potential, $E°$ for Cu^{2+}/Cu is 0.34 V.

Gas constant, R is 8.31 J K^{-1} mol^{-1}

Faraday constant, F is 9.65×10^4 J V^{-1} mol^{-1}]

(a) 1.2 (b) 8.3
(c) 2.5 (d) 3.2

102. Consider the following reversible first order reaction of X at an initial concentration $[X]_0$. The values of the rate constants are $K_f = 2s^{-1}$ and $K_b = 1s^{-1}$

$$X \underset{K_b}{\overset{K_f}{\rightleftharpoons}} Y$$

A plot of concentration of X and Y as function of time is

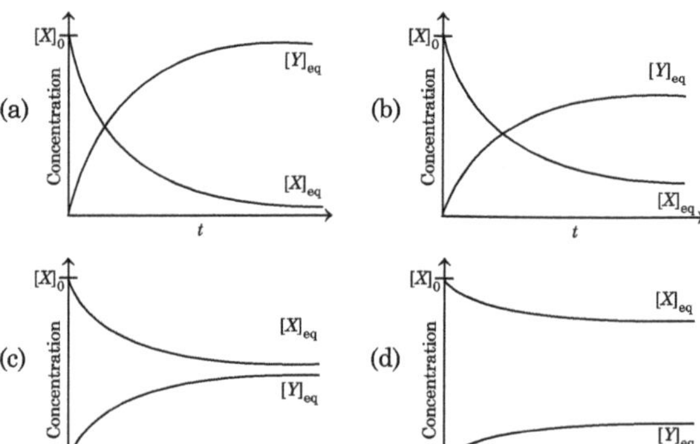

103. Nitroglycerine (MW = 227.1) detonates according to the following equations,

$$2C_3H_5(NO_3)_3(l) \longrightarrow 3N_2(g) + \frac{1}{2}O_2(g) + 6CO_2(g) + 5H_2O(g)$$

The standard molar enthalpies of formation, ΔH_f° for all the compounds are given below.

$\Delta H_f^\circ[C_3H_5(NO_3)_3] = -364$ kJ/mol
$\Delta H_f^\circ[CO_2(g)] = -393.5$ kJ/mol
$\Delta H_f^\circ[H_2O(g)] = -241.8$ kJ/mol
$\Delta H_f^\circ[N_2(g)] = 0$ kJ/mol
$\Delta H_f^\circ[O_2(g)] = 0$ kJ/mol

The enthalpy change when 10 g of nitroglycerine is detonated is

(a) −100.5 kJ/mol (b) −62.5 kJ/mol
(c) −80.3 kJ/mol (d) −74.9 kJ/mol

104. The heating of $(NH_4)_2Cr_2O_7$ produces another chromium compound along with N_2 gas. The change of the oxidation state of Cr in the reaction is

(a) +6 to +2 (b) +7 to +4
(c) +8 to +4 (d) +6 to +3

105. The complex having the highest spin only magnetic moment is

(a) $[Fe(CN)_6]^{3-}$ (b) $[Fe(H_2O)_6]^{2+}$
(c) $[MnF_6]^{4-}$ (d) $[NiCl_4]^{2-}$

106. Among Ce($4f^1\,5d^1\,6s^2$), Nd ($4f^1 6s^2$), Eu ($4f^7 6s^2$) and Dy($4f^{10} 6s^2$), the elements having highest and lowest 3rd ionisation energies, respectively are

(a) Nd and Ce (b) Eu and Ce
(c) Eu and Dy (d) Dy and Nd

107. The major product of the following reaction sequence,

Ph, Me; (i) B_2H_6 (ii) H_2O_2/NaOH (iii) conc. H_2SO_4

is

(a) Ph, Me (b) Ph (c) Ph, Me (d) Ph, OH

108. Among the following reactions, a mixture of diastereomers is produced from

(a) Me, H, HBr (b) Me, H, H_2/Pt

(c) Me, H, HBr ROOR, $h\nu$ (d) Me, H, B_2H_6 H_2O_2/NaOH

109. Reaction of phenol with NaOH followed by heating with CO_2 under high pressure, and subsequent acidification gives compound X as the major product, which can be purified by steam distillation. When reacted with acetic anhydride in the presence of a trace amount of conc. H_2SO_4, compound X produces Y as the major product. Compound Y is

(a) O, O, CO_2H (b) OH, O, O, O

(c) O, O, CO_2H (d) OH, O, O, O

110. A tetrapeptide is made of naturally occurring alanine, serine, glycine and valine. If the C-terminal amino acid is alanine and the N-terminal amino acid is chiral, the number of possible sequences of the tetrapeptide is

(a) 12 (b) 8 (c) 6 (d) 4

BIOLOGY

111. What is the probability that a human individual would receive the entire haploid set of chromosomes from his/her grandfather?

(a) $\frac{1}{2}$ (b) $\left(\frac{1}{2}\right)^{23}$ (c) $\left(\frac{1}{2}\right)^{2}$ (d) $\left(\frac{1}{2}\right)^{46}$

112. Which one among the following primer pairs would amplify the fragment of DNA given below?

5′-CTAGTCGTCGAT-$(N)_{300}$-GACTGAGCTGAGCTG-3′

3′-GATCAGCAGCTA-$(N)_{300}$-CTGACTCGACTCGAC-5′

(a) 5′-CTAGTCGTCGAT-3′ and 5′-GACTGAGCTGAGCTG-3′

(b) 5′ - CTGACTCGACTCGAC - 3′ and 5′ -CTAGTCGTCGAT - 3′

(c) 5′ - CTAGTCGTCGAT - 3′ and 5′ - CAGCTCAGCTCAGTC - 3′

(d) 5′ - CTAGTCGTCGAT - 3′ and 5′ - GTCGAGTCGAGTCAG - 3′

113. The following graphs with the solid and dotted lines correspond to the reactions without and with enzyme, respectively. Which of the following graph(s) correctly represents the concept of activation energy?

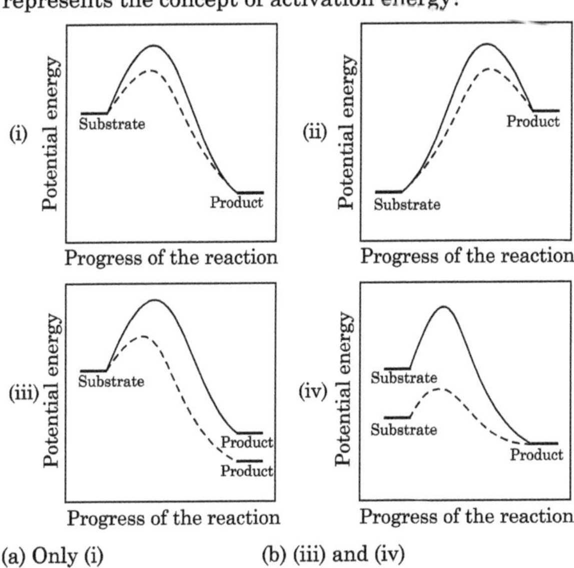

(a) Only (i) (b) (iii) and (iv)

(c) Only (ii) (d) (i) and (ii)

114. A novel species with double-stranded genetic material consists of 5 bases namely P, Q, R, S, T, with percentage given below.

	P	Q	R	S	T
Percentage	22	28	22	12	16

Based on the above information, which one of the following inferences is not supported by the observations?

(a) S base pairs with T and Q base pairs with R
(b) S base pairs with Q and T base pairs with Q
(c) P base pairs with R and S base pairs with Q
(d) P base pairs with R and T base pairs with Q

115. How many different blood groups are possible in a diploid species with ABCO blood grouping system involving I^A, I^B, I^C and I^O alleles (I^O is recessive and others are codominant)?
(a) 4 (b) 6 (c) 7 (d) 8

116. Within the exponential phase of growth, if the initial surface area and the growth rate of a leaf are 10 mm^2 and 0.015 mm^2/hour respectively, the area of the leaf after 4 days would range from
(a) 10-12 mm^2 (b) 20-24 mm^2
(c) 30-36 mm^2 (d) 40-48 mm^2

117. If the acidic, basic and hydrophobic residues of proteins are considered to be red, green and blue in colour, respectively, then a globular protein in aqueous solution would have
(a) red and blue on the surface and green at the core
(b) red and green on the surface and blue at the core
(c) blue on the surface and red and green at the core
(d) blue and green on the surface and red at the core

118. A lysosome vesicle of 1 μm diameter has an internal pH of 5.0. The total number of H^+ ions inside this vesicle would range from
(a) 10^3-10^4 (b) 10^4-10^5
(c) 10^5-10^{10} (d) 10^{10}-6.023×10^{23}

119. Match the vitamins listed in Column I with their respective coenzyme forms in Column II. Choose the correct combination.

	Column I		Column II
A.	Vitamin-B_1	1.	Thiamine pyrophosphate
B.	Vitamin-B_2	2.	Flavine adenine dinucleotide
C.	Vitamin-B_6	3.	Methylcobalamin
D.	Vitamin-B_{12}	4.	Coenzyme-A
		5.	Pyridoxal phosphate

	A	B	C	D		A	B	C	D
(a)	5	3	1	4	(b)	3	4	2	1
(c)	1	2	5	3	(d)	1	4	2	3

120. Two independent experiments related to photosynthesis were conducted, one with ^{18}O-labelled water (experiment P) and the other ^{14}C-labelled CO_2 (experiment Q). Which one of the following options lists the first labelled products in experiments P and Q, respectively?
(a) P : O_2 Q : 3- phosphoglycerate
(b) P : 3-phosphoglycerate Q : NADPH
(c) P : O_2 Q : NTP
(d) P : 3-phosphoglycerate Q : 3-phosphoglycerate

Answers

PART-I

1 (a)	2 (d)	3 (b)	4 (d)	5 (b)	6 (b)	7 (d)	8 (c)	9 (b)	10 (b)
11 (c)	12 (c)	13 (d)	14 (d)	15 (a)	16 (b)	17 (a)	18 (a)	19 (c)	20 (b)
21 (c)	22 (d)	23 (a)	24 (a, c)	25 (d)	26 (c)	27 (a)	28 (b)	29 (b)	30 (c)
31 (b)	32 (a)	33 (a)	34 (a)	35 (b)	36 (b)	37 (d)	38 (a)	39 (a)	40 (a)
41 (a)	42 (b)	43 (d)	44 (d)	45 (d)	46 (a)	47 (*)	48 (c)	49 (b)	50 (d)
51 (b)	52 (a)	53 (b)	54 (b)	55 (a)	56 (b)	57 (a)	58 (b)	59 (b)	60 (a)
61 (c)	62 (a)	63 (b)	64 (d)	65 (c)	66 (d)	67 (a)	68 (b)	69 (a)	70 (d)
71 (c)	72 (a,d)	73 (b)	74 (d)	75 (c)	76 (d)	77 (b)	78 (a)	79 (c)	80 (a)

PART-II

81 (d)	82 (c)	83 (b)	84 (c)	85 (a)	86 (a)	87 (a)	88 (d)	89 (b)	90 (b)
91 (a)	92 (b)	93 (a)	94 (b)	95 (b)	96 (c)	97 (c)	98 (b)	99 (d)	100 (b)
101 (c)	102 (b)	103 (b)	104 (d)	105 (c)	106 (b)	107 (c)	108 (a)	109 (a)	110 (d)
111 (b)	112 (c)	113 (d)	114 (a)	115 (c)	116 (d)	117 (b)	118 (a)	119 (c)	120 (a)

** No options are correct.*

Solutions

1. *(a)* We have,

$$A=\begin{bmatrix}a & b\\ c & d\end{bmatrix}$$

$$A^2=\begin{bmatrix}a & b\\ c & d\end{bmatrix}\begin{bmatrix}a & b\\ c & d\end{bmatrix}$$

$$A^2=\begin{bmatrix}a^2+bc & ab+bd\\ ac+cd & bc+d^2\end{bmatrix}$$

Given, $A^2=A$ and $ad-bc=0$

$$\therefore \begin{bmatrix}a^2+bc & ab+bd\\ ac+cd & bc+d^2\end{bmatrix}=\begin{bmatrix}a & b\\ c & d\end{bmatrix}$$

$$ab+bd=b$$

$$b(a+d)=b$$

$$\Rightarrow \quad a+d=1$$

2. *(d)* We have,

$$|z|=1$$

$$\Rightarrow \quad z^n=1$$

$$\Rightarrow \quad z^n=e^{i2k\pi};\ k\in I$$

$$\Rightarrow \quad k\in(0,n-1)$$

$$\Rightarrow \quad z=e^{\frac{i2k\pi^2}{n}}$$

$$z=1, e^{\frac{i2\pi}{n}}, e^{\frac{i4\pi}{n}},\ldots, e^{\frac{i2(n-1)\pi}{n}}$$

All roots of unity lie on arc of circle.

$\therefore$ There are infinitely many roots of unity.

3. *(b)* We have,

Equation of ellipse

$$\frac{x^2}{9}+\frac{y^2}{4}=1$$

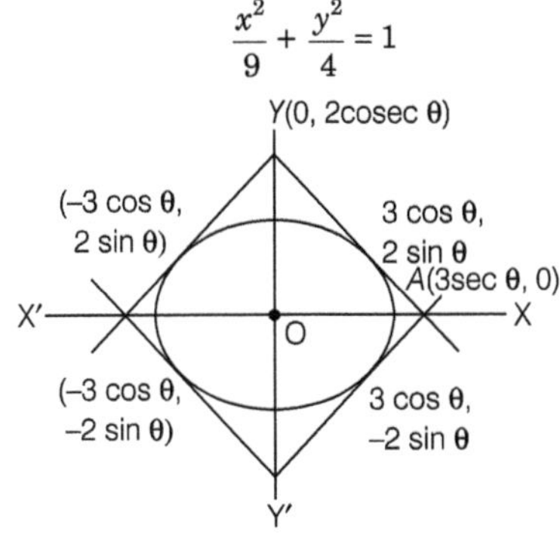

Equation of tangent at $(3\cos\theta, 2\sin\theta)$ is

$$\frac{x}{3}\cos\theta+\frac{y}{2}\sin\theta=1$$

Intercept of tangent is $(3\sec\theta, 0)$ and $(0, 2\operatorname{cosec}\theta)$.

Area of quadrilateral

$$=4\times\frac{1}{2}3\sec\theta\times 2\operatorname{cosec}\theta$$

$$=12\sec\theta\operatorname{cosec}\theta=\frac{24}{\sin 2\theta}$$

Minimum area = 24

4. *(d)* We have,

$S=\{x\in R: \cos x+\cos\sqrt{2}x<2\}$

Maximum value of $\cos x$ and $\cos\sqrt{2}x$ is 1 at $x=0$.

$\therefore \quad \cos x+\cos\sqrt{2}x=2$ at $x=0$

Hence, $\quad S=R-\{0\}$.

5. *(b)* We have rectangular hyperbola

$$x^2-y^2=a^2$$

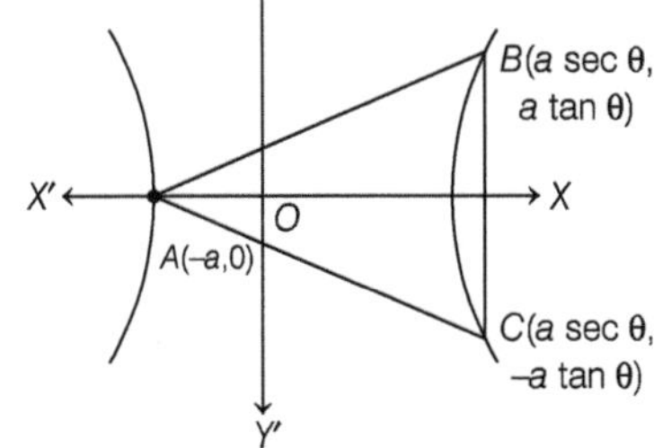

Given ABC is an equilateral triangle.

$$\therefore \quad AB=BC=AC$$

$$AB^2=BC^2$$

$$a^2(\sec\theta+1)^2+a^2\tan^2\theta=4a^2\tan^2\theta$$

$$\Rightarrow \quad (\sec\theta+1)^2=3\tan^2\theta$$

$$\Rightarrow \quad (\sec\theta+1)^2=3(\sec^2\theta-1)$$

$$\Rightarrow \quad (\sec\theta+1)^2=3(\sec\theta+1)(\sec\theta-1)$$

$$\Rightarrow \quad \sec\theta+1=3\sec\theta-3$$

$$\Rightarrow \quad \sec\theta=2$$

$$\Rightarrow \quad \theta=60^\circ$$

$\because$ Side $BC=2a\tan\theta$

$$=2a\tan 60^\circ=2a\sqrt{3}$$

But side of triangle is ka.

$$\therefore \quad ka=2a\sqrt{3}$$

$$k=2\sqrt{3}$$

Hence, $k\in(2,4]$.

6. *(b)* We have,

$$\cos^2(x\sin(2x))+\frac{1}{1+x^2}=\cos^2x+\sec^2x$$

$$\text{LHS}=\cos^2(x\sin 2x)+\frac{1}{1+x^2}$$

$$\cos^2(x\sin 2x)+\frac{1}{1+x^2}\le 2$$

$$\text{RHS}=\cos^2x+\sec^2x\ge 2$$

$$\text{LHS}=\text{RHS}$$

$\therefore \quad \cos^2x+\sec^2x=2$ at $x=0$

Hence, only one solution.

7. *(d)* Given, $\dfrac{x^2}{a^2}+\dfrac{y^2}{b^2}=1$

ABC is an equilateral traingles.

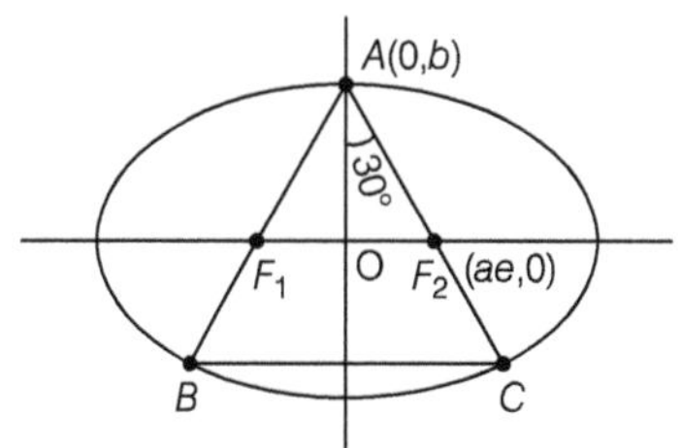

In ΔAOF_2,

$$\tan 30^\circ=\frac{OF_2}{OA}$$

$$\frac{1}{\sqrt{3}}=\frac{ae}{b}$$

$$\Rightarrow \quad b=\sqrt{3}ae$$

$$\Rightarrow \quad b^2=3a^2e^2$$

$$\Rightarrow \quad \frac{b^2}{a^2}=3e^2$$

$$\Rightarrow \quad 1-e^2=3e^2 \qquad \left[\because e=\sqrt{1-\frac{b^2}{a^2}}\right]$$

$$\Rightarrow \quad 4e^2=1$$

$$\Rightarrow \quad e^2=\frac{1}{4} \Rightarrow e=\frac{1}{2}$$

8. *(c)* We have,

$$a=\cos 1^\circ \text{ and } b=\sin 1^\circ$$

$$(\cos x+i\sin x)^n=\cos nx+i\sin nx$$

$$(\cos x-i\sin x)^n=\cos nx-i\sin nx$$

$$\Rightarrow \quad 2\cos nx=(\cos x+i\sin x)^n+(\cos x-i\sin x)^n$$

Put $x=1^\circ$ and $n=60$

$$=2\cos 60^\circ=2(\cos^{60}1^\circ+{}^{60}C_2\cos^{58}1^\circ\, i^2\sin^2 1^\circ\ldots)$$

Change all $\sin 1^\circ$ to $\cos 1^\circ$ using the identity $\cos^2 1^\circ=1-\sin^2 1^\circ$ equation with root $\cos 1^\circ$ so it is algebraic.

Similarly, for $b=\sin 1^\circ$

9. *(b)* Equation of parabola

$$y=x^2-4 \text{ and } 2y=4-x^2$$

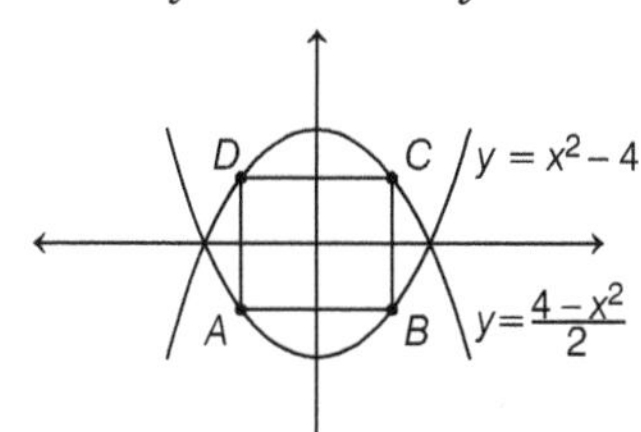

Let point $B = (h, \frac{4-h^2}{2})$

$$A = (-h, \frac{4-h^2}{2})$$
$$C = (h_1\ h^2 - 4)$$
$$D = (-h, h^2 - 4)$$

Area of rectangle

$$ABCD = AB \times BC = 2h \times \left(\frac{4-h^2}{2} - h^2 + 4\right)$$

$$\Rightarrow \quad A = 12h - 3h^3 \Rightarrow \frac{dA}{dh} = 12 - 9h^2$$

For maxima or minima

$$\frac{dA}{dh} = 0 \quad \Rightarrow 12 - 9h^2 = 0$$

$$h = \pm\frac{2}{\sqrt{3}} \text{ maximum at } h = \frac{2}{\sqrt{3}}$$

$$\therefore \quad A = 12\left(\frac{2}{\sqrt{3}}\right) - 3\left(\frac{2}{\sqrt{3}}\right)^3 = \frac{24}{\sqrt{3}} - \frac{8}{\sqrt{3}}$$

$$= \frac{16}{\sqrt{3}} = \frac{16\sqrt{3}}{3}$$

$$A - \frac{16 \times 1.73}{3} = 9.22$$

$$[A] = [9.22] = 9$$

10. *(b)* We have, $f(x) = x|\sin x|, x \in R$

$$f(x) = \begin{cases} x\sin x, & x \in (2n\pi, (2n+1)\pi) \\ -x\sin x, & x \in ((2n+1)\pi, 2n\pi) \end{cases}$$

$$f'(n\pi) = \lim_{x \to n\pi} \frac{f(x) - f(n\pi)}{x - n\pi}$$

$$= \lim_{x \to n\pi} \frac{x|\sin x|}{x - n\pi}$$

Clearly, $f(x)$ is differentiable for all x except $x = n\pi, n = \pm 1, \pm 2, \pm 3, \ldots$

11. *(c)* We have,

$$f(x) = \begin{cases} x^2\left|\cos\left(\frac{\pi}{x}\right)\right| & \text{for } x \neq 0 \\ 0 & \text{for } x = 0 \end{cases}$$

$f(x)$ is not differentiable at

$$\left|\cos\frac{\pi}{x}\right| = 0$$

$$\cos\frac{\pi}{x} = 0 \Rightarrow \frac{\pi}{x} = (2n+1)\frac{\pi}{2}$$

$$\Rightarrow \quad x = \frac{2}{2n+1}$$

$\because f(x)$ is not differentiable at

$$x \in [-1, 1] : x = \frac{2}{2n+1}, n \in Z$$

12. *(c)* Let $I = \int_0^{\pi} (1 - |\sin 8x|)\,dx$

$$I = \int_0^{\pi} dx - \int_0^{\pi} |\sin 8x|\,dx$$

$$I = \pi - \int_0^{\frac{\pi}{8} \times 8} |\sin 8x|\,dx$$

$$I = \pi - 8\int_0^{\frac{\pi}{8}} \sin 8x\,dx \ [\because \sin 8x\ B \text{ periodic with } \tfrac{\pi}{8}]$$

$$I = \pi - 8\left[\frac{-\cos 8x}{8}\right]_0^{\pi/8}$$

$$I = \pi + (\cos\pi - \cos 0) = \pi - 2$$

13. *(d)* We have,

$$f'(x) = \ln(x^2 + 1)$$
$$f(x) = \int \ln(x^2 - 1)dx$$
$$f(x) = x\ln(x^2 - 1) - \int \frac{2x^2}{x^2 - 1}\,dx$$
$$f(x) = x\ln(x^2 - 1) - 2\int\left(\frac{x^2 - 1}{x^2 - 1} + \frac{1}{x^2 - 1}\right)dx$$
$$f(x) = x\ln(x^2 - 1) - 2x - \ln\left(\frac{x-1}{x+1}\right) + C$$
$$f(2) = 2\ln(3) - 4 - \ln\left(\frac{1}{3}\right) + C = 0$$

$[\because f(2) = 0]$

$$\Rightarrow C = 4 - 3\ln 3$$
$$\therefore f(x) = x\ln(x^2 - 1) - 2n\ln\frac{(x-1)}{(x+1)} + 4 - 3\ln 3$$

defined for S infinite C values possible in set S such that $f'(x) = \ln(x^2 - 1)$.

14. *(d)* Given,

$$\int_0^x f(t)dt = p\,f(x)$$

$$\Rightarrow \quad f(x) = p\,f'(x)$$
$$\Rightarrow \quad \frac{f'(x)}{f(x)} = \frac{1}{p}$$
$$\Rightarrow \quad \log f(x) = \frac{x}{p} + C$$
$$\Rightarrow \quad f(x) = Ae^{x/p}$$

Putting $x = 0$

$$f(0) = Ae^0 \Rightarrow A = 0 \quad [\because f(0) = 0]$$
$$\because \quad f(x) = 0$$
$$\therefore \quad pf(0) = 0$$

Case I $\quad f(0) = 0, p \neq 0$

There is no non-zero continuous to $f(x)$.

Case II $\quad p = 0$

$$\therefore \quad \int_0^x f(t)\,dt = 0, \forall x \in R$$

If is possible when $f(x) = 0$.

Hence, $\forall\, p \in R$. There is no non-zero continous function. Hence, $S \in R$.

15. *(a)* We have 30 males and 20 females.

$$P(\text{males}) = \frac{30}{50} = \frac{3}{5}$$

$$P(\text{females}) = \frac{20}{50} = \frac{2}{5}$$

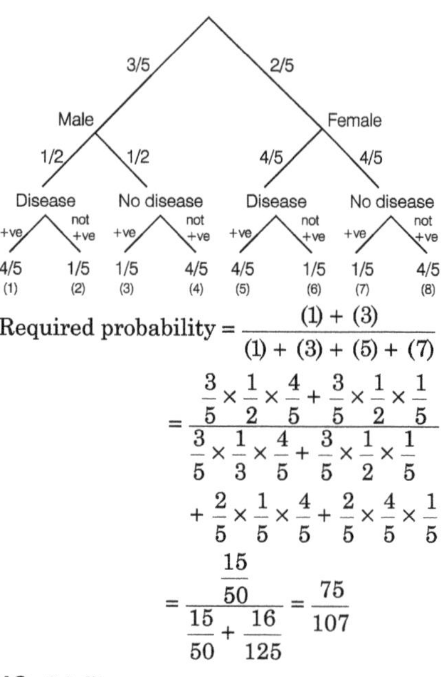

$$\text{Required probability} = \frac{(1) + (3)}{(1) + (3) + (5) + (7)}$$

$$= \frac{\frac{3}{5} \times \frac{1}{2} \times \frac{4}{5} + \frac{3}{5} \times \frac{1}{2} \times \frac{1}{5}}{\frac{3}{5} \times \frac{1}{3} \times \frac{4}{5} + \frac{3}{5} \times \frac{1}{2} \times \frac{1}{5} + \frac{2}{5} \times \frac{1}{5} \times \frac{4}{5} + \frac{2}{5} \times \frac{4}{5} \times \frac{1}{5}}$$

$$= \frac{\frac{15}{50}}{\frac{15}{50} + \frac{16}{125}} = \frac{75}{107}$$

16. *(b)* Given,

$$|f(x) - f(y)| = |x - y| \text{ for all in } [0, 1]$$

$$\Rightarrow \quad \frac{|f(x) - f(y)|}{|x - y|} = 1$$

$$\Rightarrow \lim_{x \to y} \frac{|f(x) - f(y)|}{x - y} = 1 \Rightarrow |f'(x)| = 1$$

$$\Rightarrow f'(x) = \pm 1 \Rightarrow f(x) = \pm x$$
$$x \in [0, 1]$$

$\therefore f(x)$ has exactly 2 function.

17. *(a)* We have, A is 3×3 matrix consisting of integer entries from $\{-1000, -999, \ldots 999, 1000\}$

Total entries = 2001

Total number of outcomes = $(2001)^9$

Given, $A^2 = -I, A = -A^{-1}$ (which is not possible)

A is diagonal matrix.

$$A = \begin{bmatrix} a & 0 & 0 \\ 0 & a & 0 \\ 0 & 0 & a \end{bmatrix}$$

Total number of favourable outcomes $= (2001)^3$

Required probability

$$= \frac{(2001)^3}{(2001)^9}(2001)^{-6} = (2000 + 1)^{-6}$$

$$p = (2000)^{-6}\left(1 + \frac{1}{2000}\right)^{-6}$$

$$P = \frac{1}{2^6 \times 10^{18}}\left(1 + \frac{1}{2000}\right)^{-6}$$

$$P < \frac{1}{2^6 \times 10^{18}} \quad \left[\because 1 + \frac{1}{2000} \text{ is decreasing}\right]$$

$$P < \frac{1}{10^{18}}$$

18. *(a)* We have,

$x_k \geq k^4 + k^2 + 1, \forall\, k \in [1, 2018]$

$$N \sum_{k=1}^{2018} k$$

$$\text{I. } I\left(\sum_{k=1}^{2018} kx_k\right)^2 \leq N\left(\sum_{k=1}^{2018} kx_k^2\right)$$

We know

$$\left(\frac{x_1 + x_2 + x_3 + \ldots + x_n}{n}\right) \leq \frac{x_1^2 + x_2^2 + x_3^2 + \ldots + x_n^2}{n}$$

$$\therefore \left[\frac{x_1 + (x_2 + x_2) + (x_3 + x_3 + x_3) + \ldots (x_{2018} + x_{2018\ldots\ldots}\ 2018 \text{ times})}{1 + 2 + 3 + 4 + \ldots + 2018}\right]^2$$

$$\leq \frac{x_1^2 + (x_2^2 + x_2^2) + (x_3^2 + x_3^2 + x_3^2)\ldots}{1 + 2 + 3 + 4 + \ldots + 2018}$$

$$\Rightarrow \left(\frac{x_1 + 2x_2 + 3x_3 + \ldots + 2018x_{2018}}{1 + 2 + 3 + 4 + \ldots + 2018}\right)$$

$$\leq \frac{x_1^2 + 2x_2^2 + \ldots + 2018x_{2018}^2}{1 + 2 + 3 + \ldots + 2018}$$

$$\Rightarrow \frac{\left(\sum_{k=1}^{2018} k x_k\right)^2}{\left(\sum_{k=1}^{2018} k\right)^2} \leq \frac{\sum_{k=1}^{2018} k\, x_k^2}{\sum_{k=1}^{2018} k} = \left(\sum_{k=1}^{2018} kx_k\right)^2 \leq N \sum_{k=1}^{2018} kx_k^2$$

Hence, I is true.

$$\text{II. } \left(\sum_{k=1}^{2018} kx_k\right)^2 \leq N\left(\sum_{k=1}^{2018} k^2 x_k^2\right)$$

Similarly for I

$$\left(\frac{x_1 + 2x_2 + 3x_3 + \ldots + 2018x_{2018}}{2018}\right)^2$$

$$\leq \left(\frac{x_1^2 + (2x_1) + \ldots (2018x_{2018})}{2018}\right)$$

$$\Rightarrow \frac{\left(\sum_{k=1}^{2018} kx_k\right)^2}{(2018)^2} \leq \frac{\sum_{k=1}^{2018} k^2 x_{k^2}}{2018}$$

$$= \left(\sum_{k=1}^{2018} kx_k\right)^2 \leq 2018 \sum_{k=1}^{2018} k^2 x_{k^2}$$

$$N = \sum_{k=1}^{2018} k = \frac{2018 \times 2019}{2}$$

$\because \left(\sum_{k=1}^{2018} kx_k\right)^2$ is always less than or equal to $2018 \sum_{k=1}^{2018} k^2 x_{k^2}$.

$\therefore$ It will always be less than $N\left(\sum_{k=1}^{2018} k^2 x_{k^2}\right)$

Hence, I and II both are true.

19. *(c)* Given, $x^2 = 4ky$

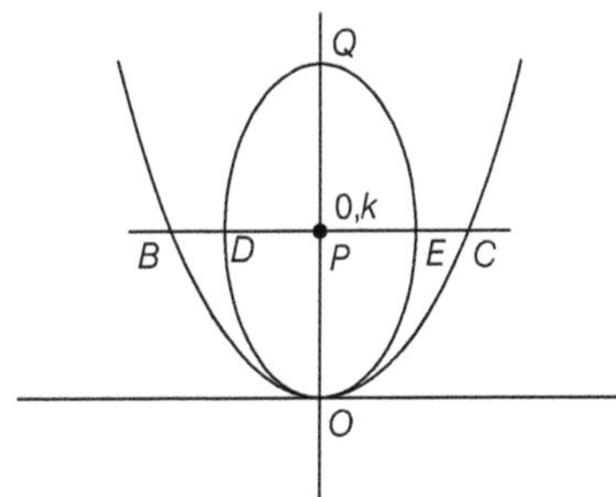

BC is latusrectum.

$$\therefore \quad BC = 4k$$

$$BD = DE = EC$$

$$\therefore \quad DE = \frac{BC}{3} = \frac{4k}{3}$$

P is centre of ellipse.

$$\therefore \quad PE = \frac{2k}{3}$$

$$OP = k$$

$\because$ Eccentricity of ellipse

$$\sqrt{1 - \frac{PE^2}{OP^2}} = \sqrt{1 - \frac{4k^2}{9k^2}}$$

$$\Rightarrow \quad e = \sqrt{\frac{9-4}{9}} = \frac{\sqrt{5}}{3}$$

20. *(b)* Let $h(x) = g(f(x))$ where, $g(x)$ is injective and $h(x)$ is surjective.

$h(x)$ is surjective.

$\therefore$ Codomain of $h(x)$ = Range of $h(x)$.

Range of $h(x) = [0, 2]$ which is also codomain of g. So, must be surjective.

Now, domain of $g = [-1, 1]$ which must be range of f. But codomain of $f = [-1, 1]$

So, it must be surjective.

21. *(c)*

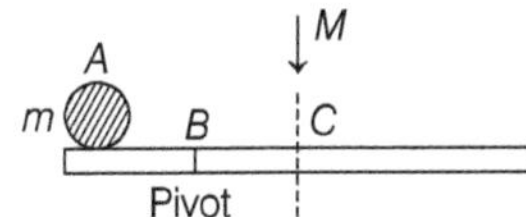

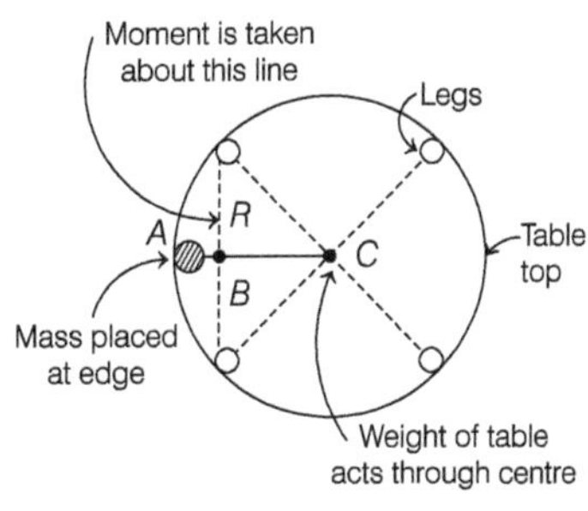

From the above figure, we can see that due to mass placed at edge of table, it tends to rotate about a line joining two legs. Hence, we take pivot in balancing of weight of mass m and weight of mass M of table at point B.

In critical condition,

Moment of weight of mass m about B = Moment of weight of table about B

$$\Rightarrow \quad mg \cdot (AB) = Mg \cdot (BC)$$

$$\Rightarrow \quad m = \frac{M \cdot (BC)}{AB} \quad \ldots \text{(i)}$$

From geometry of table,

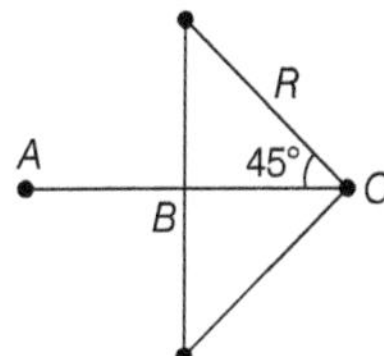

$$BC = R\cos 45° = \frac{R}{\sqrt{2}}$$

So, $$AB = R - \frac{R}{\sqrt{2}}$$

$$= R\left(1 - \frac{1}{\sqrt{2}}\right)$$

Hence, from Eq. (i), we have

$$m = \frac{M \times R/\sqrt{2}}{R\left(1 - \frac{1}{\sqrt{2}}\right)}$$

Here, $M = 20\,\text{kg}$

$$\therefore \quad m = \frac{20}{\sqrt{2} - 1} \approx \frac{20}{0.4} \approx 50\,\text{kg}$$

Nearest value is $m = 47\,\text{kg}$.

22. *(d)* Bubble detached from the look at ends of circumference, when force of dynamic pressure exceeds force of surface tension.

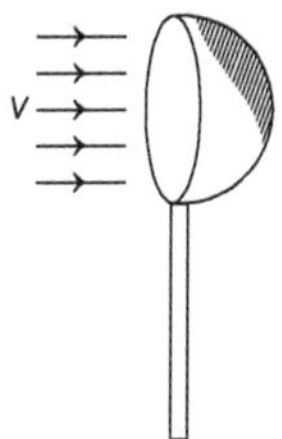

$\Rightarrow p_{\text{dynamic}} \times$ Area $\geq$ Force of surface tension

$$\Rightarrow \frac{1}{2}\rho v^2 \times \pi R^2 \geq 2\,(2\pi RT)$$

Note that factor 2 on right hand side appears as there are two sides of film surface.

$$\Rightarrow \quad v \geq \sqrt{\frac{8T}{\rho R}}$$

$$\Rightarrow \quad v_{\min} = \sqrt{\frac{8T}{\rho R}}$$

23. *(a)* For an ideal gas,

$$C_V = \left.\frac{\partial U}{\partial T}\right|_{V = \text{constant}}$$

or $$C_V = \frac{dU}{dT}$$

Also, for 1 mole of gas,

$$C_V = \frac{f}{2}\cdot R$$

where, f = degree of freedom.

Hence, we have

$$\frac{f}{2}R = \frac{dU}{dT}$$

Here, $U = \frac{5}{2}pV + C = \frac{5}{2}RT + C$

[$\because$ one mole of gas is considered]

So, $$\frac{f}{2}R = \frac{d}{dT}\left(\frac{5}{2}RT + C\right)$$

$$\Rightarrow \frac{f}{2}R = \frac{5}{2}R \Rightarrow f = 5$$

Now, using $\gamma = 1 + \frac{2}{f}$

where, γ = ratio of specific heats = adiabatic index.

We have, $\gamma = 1 + \frac{2}{5} \Rightarrow \gamma = \frac{7}{5}$

So, equation of adiabats can be written as

$pV^\gamma = \text{constant} \Rightarrow pV^{7/5} = \text{constant}$

$\Rightarrow p^5V^7 = \text{constant}$

24. *(a,c)* Given paths over *p-V* graph are

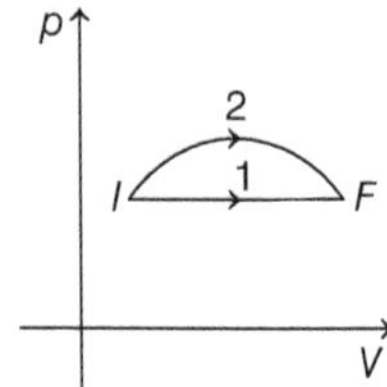

- As in both processes, initial and final points are same, change in internal energy (which is not a path function) is same. So, option (a) is correct.
- Both processes are expansion process. So, heat is absorbed in both 1 and 2. Hence, option (b) is incorrect.
- Now, consider isotherms over *p-V* graph given, we clearly see that for path 2 temperature increases and then decreases. So, option (c) is correct.

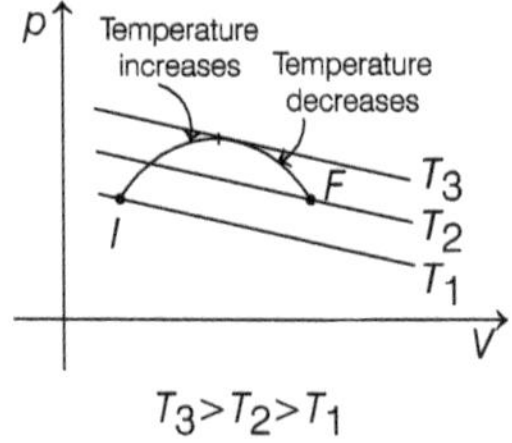

- Area enclosed by path 2 in *p-V* graph is larger.

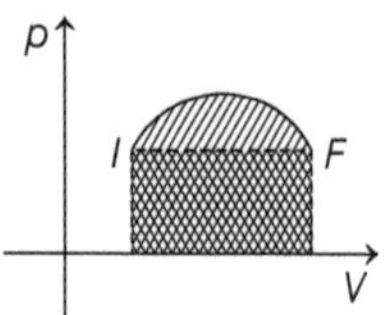

So, work done is more in path 2. Hence, option (d) is incorrect.

25. *(d)* Change in internal energy of n moles of an ideal gas, when temperature changes by ΔT is

$$\Delta U = n\cdot\frac{f}{2}\cdot R\cdot\Delta T$$

$$= \frac{f}{2}(nR\Delta T) = \frac{f}{2}(\Delta p\cdot V)$$

[$\therefore$ for ideal gas $pV = nRT$]

Here, $f = 5$ (gas is diatomic)

$\Delta p = 10^5$ Pa, $V = 1\text{L} = 10^{-3}\ \text{m}^3$

So, $\Delta U = \frac{5}{2}\times 10^5 \times 10^{-3} = 250\,\text{J}$

26. *(c)* Resultant intensity when two waves with phase difference ϕ interfere is

$$I = I_1 + I_2 + 2\sqrt{I_1}\sqrt{I_2}\cos\phi$$

As, intensity $I \propto A^2$ or $I = kA^2$, where k = a constant and A is amplitude.

So, $I = A_1^2 + A_2^2 + 2A_1A_2\cos\phi$

Here, $A_1 = A$ and $A_2 = 2A$

$\Rightarrow I = A^2 + (2A)^2 + 2(A)(2A)\cos\phi$

$= A^2(5 + 4\cos\phi)$ ... (i)

Intensity is maximum (I_0), when $\cos\phi = 1$

$\Rightarrow I_0 = A^2(5 + 4\times 1) = 9A^2$

$\Rightarrow A^2 = I_0/9$ ...(ii)

So, resultant intensity,

$I = A^2(5 + 4\cos\phi)$

[From Eqs. (i) and (ii)]

$$\Rightarrow I = \frac{I_0}{9}(5 + 4\cos\phi)$$

27. *(a)* By Bernoulli's equation for streamline flow,

$$p + \frac{1}{2}\rho v^2 + \rho gh = \text{a constant}$$

In given situation h = constant

$$\therefore p + \frac{1}{2}\rho v^2 = \text{constant}$$

When velocity of flow in narrow section increases, pressure decreases.

So, current level in vertical tubes is as shown below.

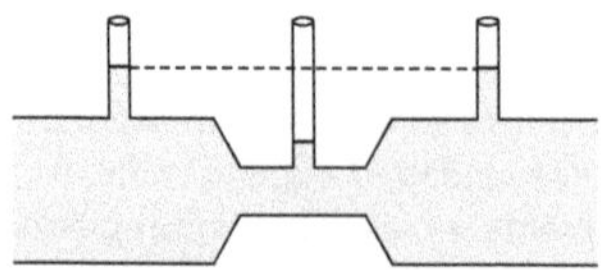

28. *(b)* Given situation is

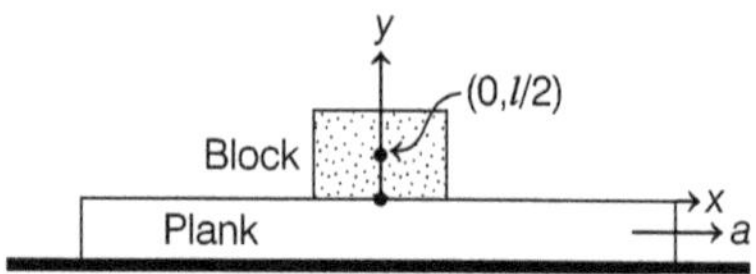

Forces acting on the block are

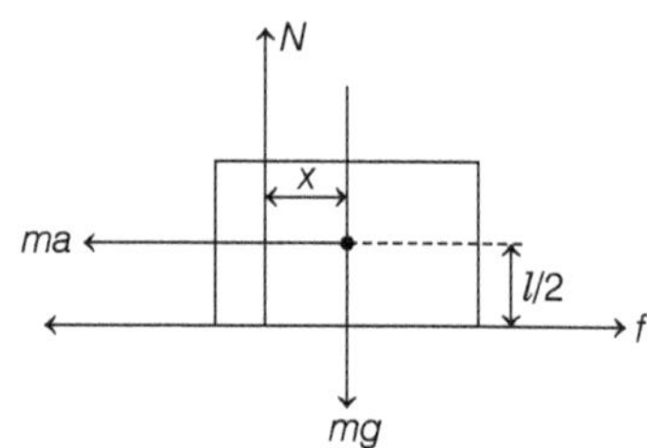

In above diagram, ma = pseudo force due to acceleration of plank,

f = force of static friction, mg = weight of block and N = normal reaction.

Here, we must note that due to motion of plank, normal reaction force does not passes through centre of mass.

If line of action of normal force at a distance x from centre, then for equilibrium of block, net torque about centre of mass must be zero.

$$\Rightarrow \tau_{\substack{\text{Normal}\\\text{reaction}}} = \tau_{\text{Friction}}$$

mg and ma does not produces any torque about centre of mass as their line of action passes through centre of mass.

$\Rightarrow N\cdot x = f\cdot l/2$

$\Rightarrow mgx = ma\frac{l}{2}$

[$\because N = mg$ and $f = ma$]

$\Rightarrow x = \left(\frac{a}{g}\right)\frac{l}{2}$

or $x = \frac{1}{10}\cdot\frac{l}{2}$ $\left[\because a = \frac{g}{10}\right]$

or $x = \frac{l}{20}$

So, coordinates of point from which reaction passes are $\left(-\frac{l}{20}, 0\right)$.

29. *(b)* A H_2 molecule in a nanotube is bound only by adhesive forces. So, its energy is least.

A hydrogen atom in a H_2 molecule is bound by interatomic forces. So, its energy is less than the energy of an electron in H_2 molecule.

Proton in a carbon nucleus is bound because of strong nuclear forces. So, its energy is largest.

$\therefore$ Correct order is IV < II < I < III.

30. *(c)* Given magnetic field due to a square loop *DABCA* at centre of cube is **B**.

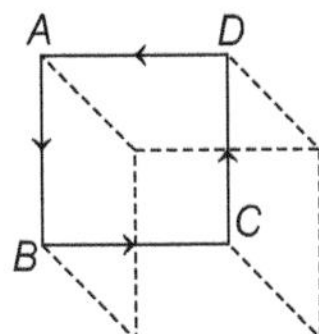

Now, loop *DAEFGCD* can be viewed as super position of three square loops as shown below.

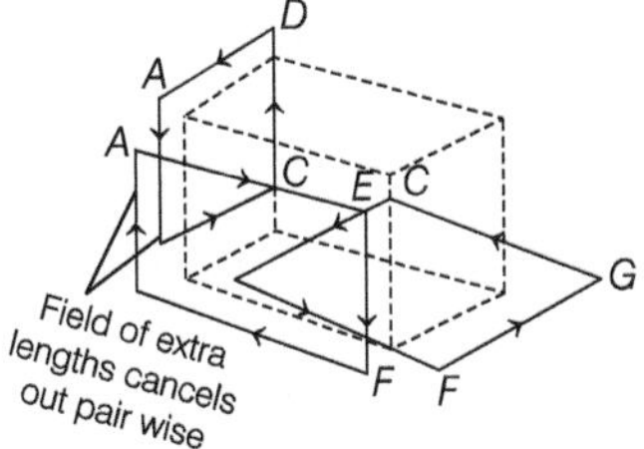

So, net field at centre = resultant of fields of these three loops

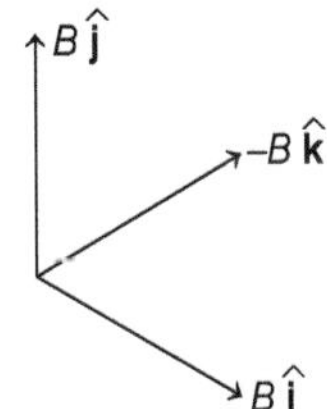

$\therefore \qquad \mathbf{B}_{\text{net}} = B\hat{\mathbf{i}} + B\hat{\mathbf{j}} - B\hat{\mathbf{k}}$

Magnitude of resultant field is

$$|\mathbf{B}_{\text{net}}| = \sqrt{B^2 + B^2 + B^2} = \sqrt{3} \cdot B$$

31. *(b)* Centrifugal action causes electrons to accumulate around rim of disc.

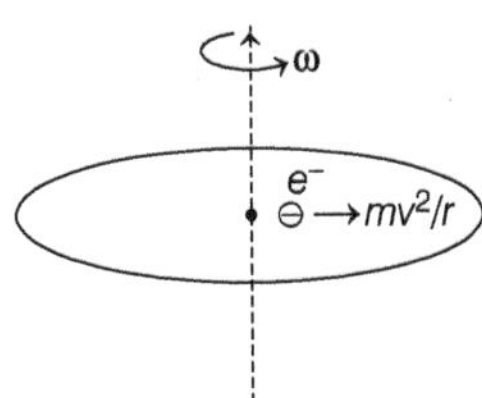

So, rim of disc is at a negative potential with respect to centre of disc. So, points of rim are at lower potential than its centre.

32. *(a)* Expansion curves are as shown below.

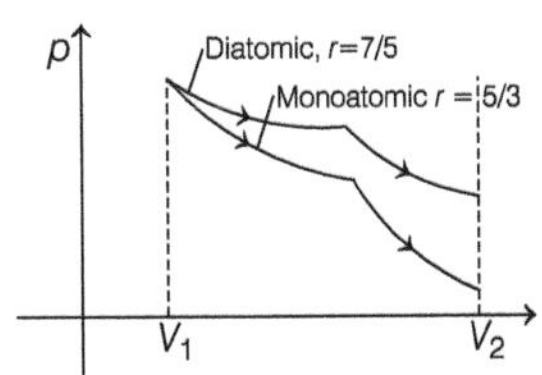

As, slope of an adiabatic curve $= -\gamma\left(\frac{p}{V}\right)$

Also, $\gamma_{\text{diatomic}}\left(=\frac{7}{5}\right) < \gamma_{\text{monoatomic}}\left(=\frac{5}{3}\right)$

So, adiabatic expansion will be more steep for monoatomic gas, whereas isotherms for both gases are nearly same.

As, work done = area under *p*-*V* graph

$\therefore$ Work done is more for diatomic gas.

33. *(a)* Given cyclic process is

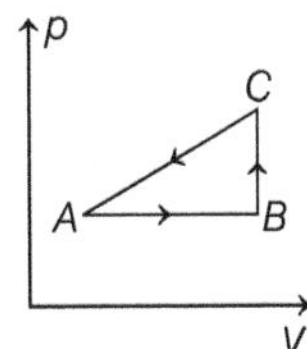

Area under compression process **CA** is more than area under expansion process **AB**. So, net work done is negative.

i.e. $\qquad \Delta W < 0$

Also, in a cyclic process, change in internal energy is zero.

i.e. $\qquad \Delta U = 0$

Now, by using first law of thermodynamics, we have

$$\Delta Q = 0 + \Delta W$$

we see that, $\qquad \Delta Q < 0$

34. *(a)* Only force resisting motion of particle in horizontal direction is drag force given by

$$F = -kv$$

But $\qquad F = m \cdot a_x$

where, a_x = acceleration in *x*-direction caused by drag force.

$\therefore \qquad ma_x = -kv \Rightarrow a_x = -\frac{kv}{m}$

$\therefore$ Acceleration of particle is *x*-direction is inversely proportional to mass.

So, decreases in velocity in *x*-direction for lighter particle will be more.

Hence, heavier mass will hit the ground further away from the lighter ball.

35. *(b)* Planes bounding the dielectric are as shown below.

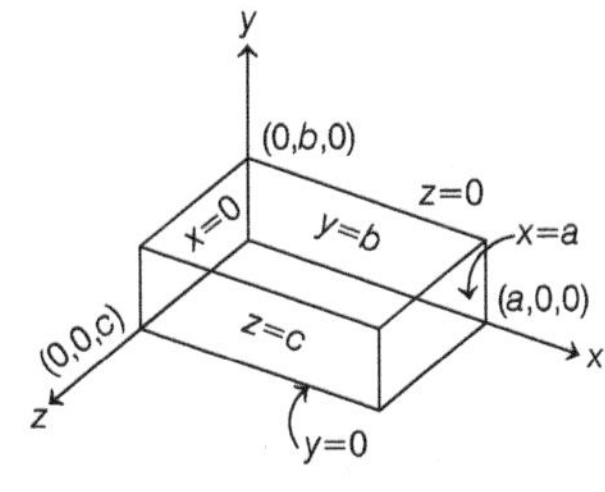

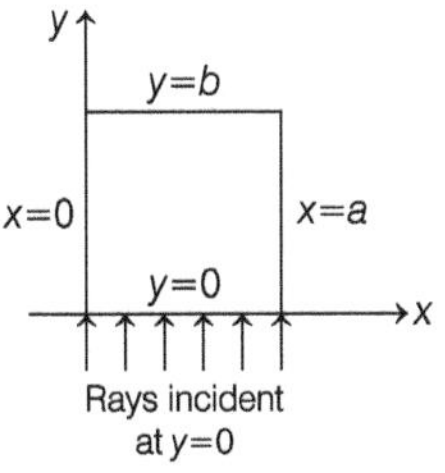

Due to dielectric, a light ray incident at $y = 0$, at some angle θ bends towards normal and it strikes plane $x = a$.

If angle of incidence at $x = a$, is more than angle of critical incidence, TIR occurs and light ray, then emerges out of plane $y = b$.

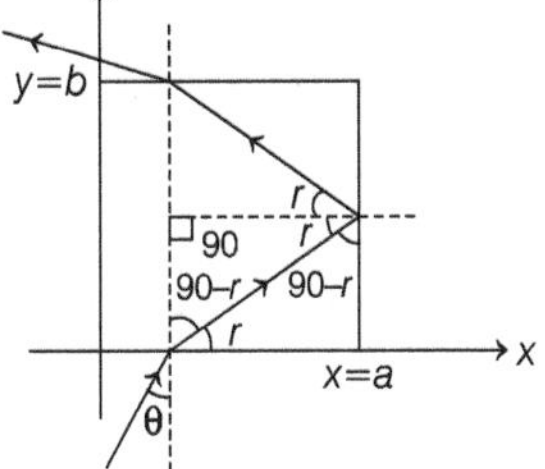

For TIR at $x = a$,

$$\sin(90° - r) = \frac{1}{n} \text{ or } \cos r = \frac{1}{n} \qquad \ldots \text{(i)}$$

Also, $\qquad n = \frac{\sin\theta}{\sin r}$

Now θ at $y = 0$, varies from 0° to a maximum of 90°.

So, for limiting incidence,

$$n = \frac{\sin 90°}{\sin r} \Rightarrow n = \frac{1}{\sin r} \qquad \ldots\text{(ii)}$$

From Eqs. (i) and (ii), we get

$$\sin r = \cos r \Rightarrow r = 45°$$

Substituting $r = 45°$ in Eq. (i) or Eq. (ii),

we get $n_{\min} = \frac{1}{\cos 45°} = \sqrt{2}$

36. *(b)* Activity of a radioactive sample is given by

$$R = -\frac{dN}{dt} = -\frac{d}{dt}N_0e^{-\lambda t} = \lambda N_0 \cdot e^{-\lambda t}$$

So, $\log R = \log(\lambda N_0) + \log(e^{-\lambda t})$

$\Rightarrow \quad \log R = -\lambda t + \log(\lambda N_0)$

This equation is form of $y = mx + C$

So, absolute value of slope of $\log R$ *versus* *t* graph gives decay constant λ.

Now, from graph,

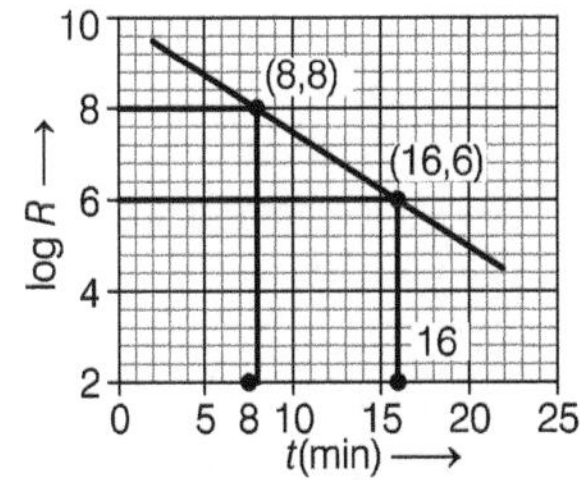

We get, slope $= \left|\dfrac{8-6}{8-16}\right| = \dfrac{1}{4} = \lambda$

So, half-life time period of sample is

$$T_{1/2} = \frac{\log 2}{\lambda} = \frac{0.693}{1/4} \approx 3.0 \text{ min}$$

37. *(d)* Following Fleming's left hand rule, proton initially rotates clockwise and it enters region $y > 0$ with its velocity in positive y-direction.

Here in region $y > 0$, magnetic field is into plane of paper. So, proton again follows left hand rule and now tend to rotate anti-clockwise as shown below.

38. *(a)* Using Balmer's formula, wavelength λ of emission from ion is

$$\frac{1}{\lambda} = RZ^2\left(\frac{1}{n_f^2} - \frac{1}{n_i^2}\right)$$

Here,

Rydberg constant, $R = 1.09 \times 10^7 \text{ m}^{-1}$

Atomic number of iron, $Z = 26$

$$n_i = 2, n_f = 1$$

$$\therefore \quad \frac{1}{\lambda} = 1.09 \times 10^7 \times (26)^2 \times \left(\frac{1}{1^2} - \frac{1}{2^2}\right)$$

$\Rightarrow \quad \lambda = 1.8 \times 10^{-10}$ Å

or $\quad \lambda \approx 2$ Å

39. *(a)* Given, drag force F_d depends on air density σ, velocity of ball v and area of cross-section of ball A.

$$\therefore \quad F_d \propto \sigma^a v^b A^c$$

Substituting dimensions of different physical quantities, we have

$$[MLT^{-2}] = [ML^{-3}]^a [LT^{-1}]^b [L^2]^c$$

Equating dimensions of fundamental quantities, we have

$$a = 1 \quad \ldots(i)$$

$$-3a + b + 2c = 1 \quad \ldots(ii)$$

$$-b = -2 \Rightarrow b = 2 \quad \ldots(iii)$$

Putting the values of a and b in Eq (ii), we get

$\Rightarrow \quad -3 \times 1 + 2 + 2c = 1 \Rightarrow c = 1$

$\because \; a = 1, b = 2, c = 1$

So, drag force on football is

$$F_d = k \cdot \sigma \cdot v^2 \cdot A$$

When football reaches terminal speed, its weight is balanced by drag force.

$$\Rightarrow \quad mg = F_d$$

$$\Rightarrow \quad mg = k\sigma v_T^2 A$$

where, v_T = terminal speed of football.

$$\Rightarrow \quad mg = k\sigma v_T^2 \pi R^2$$

$$\Rightarrow \quad mg = k\sigma v_T^2 \pi \left(\frac{m}{\frac{4}{3}\pi\rho}\right)^{\frac{2}{3}}$$

[from $m = \frac{4}{3}\pi R^3 \rho$, where ρ = density of football]

$$\Rightarrow \quad v_T \propto m^{1/6}$$

$$\Rightarrow \quad \frac{v_1}{v_2} = \left(\frac{m_1}{m_2}\right)^{\frac{1}{6}} = \left(\frac{250}{125}\right)^{\frac{1}{6}} = 2^{1/6}$$

40. *(a)* Consider another symmetric field line below line joining centres of charges q_1 and q_2.

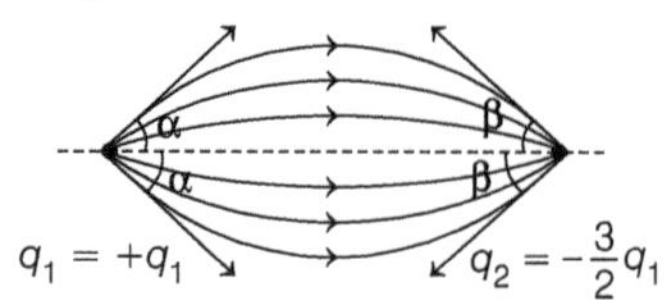

In given situation, flux (field lines) leaving charge q_1 at a solid angle 2α = flux terminating over charge q_2 at a solid angle 2β.

Clearly, in given situation

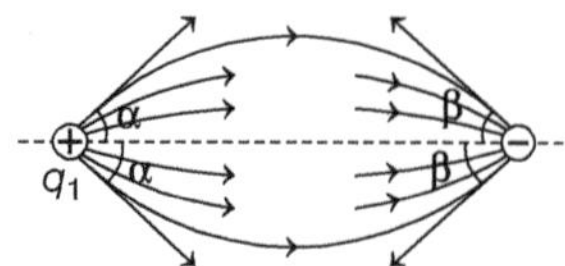

We can say that, flux leaving charge q_1 through a cone of semi-vertical angle α = flux terminating on charge q_2 through a cone of semi-vertical angle β.

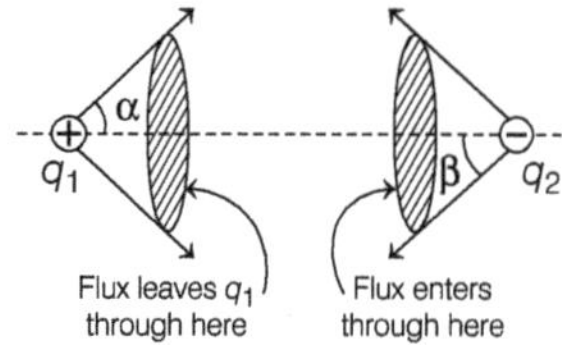

To calculate flux, we first find flux through an elemental ring of base of cone and then we integrate to get total flux.

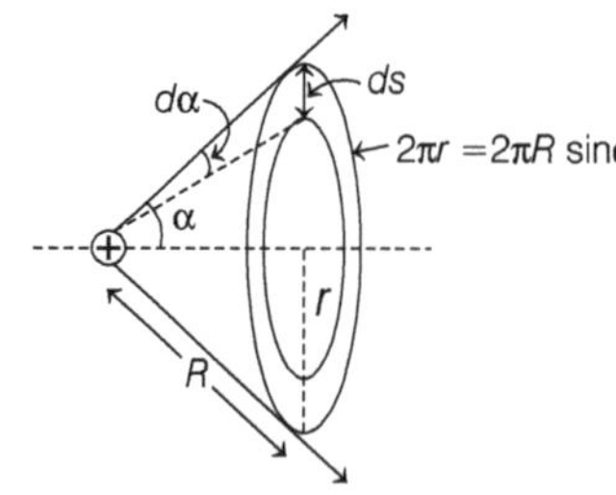

Area of elemental ring,

$$dA = 2\pi r ds = 2\pi R \sin\alpha \cdot R d\alpha$$

Flux through elemental ring (**E**||**dA**)

$$d\phi = \frac{kQ}{R^2} \cdot 2\pi R^2 \sin\alpha \, d\alpha$$

Total flux through base of cone

$$\phi = \int_0^\alpha \frac{kQ}{R^2} \cdot 2\pi R^2 \sin\alpha \, d\alpha$$

$$= kQ \, 2\pi \int_0^\alpha \sin\alpha \, d\alpha$$

$$\phi = \frac{Q}{2\varepsilon_0} \cdot (1 - \cos\alpha)$$

So, equating flux of both cones, we get

$$\frac{q_1}{2\varepsilon_0}(1 - \cos\alpha) = \frac{q_2}{2\varepsilon_0}(1 - \cos\beta)$$

$$\Rightarrow \quad q_1(1 - \cos\alpha) = \frac{3}{2}q_1(1 - \cos\beta)$$

Substituting $\alpha = 30°$ in above equation, we get

$$\Rightarrow \quad \frac{2}{3}(1 - \cos 30°) = 1 - \cos\beta$$

$$\Rightarrow \quad \frac{2}{3}\left(1 - \frac{\sqrt{3}}{2}\right) = 1 - \cos\beta$$

$$\Rightarrow \quad \cos\beta = 1 - \frac{2}{3}\left(1 - \frac{\sqrt{3}}{2}\right)$$

$$\Rightarrow \quad \cos\beta \approx 0.9$$

So, angle β is not more than 30°.

41. *(a)*

$$\underset{\text{Acetone}}{CH_3-\overset{\overset{O}{||}}{C}-CH_3} + 3Br_2 + 4NaOH \longrightarrow CH_3\overset{\overset{O}{||}}{C}O\overset{-}{}\overset{+}{Na} + 3NaBr + 3H_2O + CHBr_3$$

3 moles of bromine produces 1 mole of $CHBr_3$.

$\therefore$ 1 mole of bromine produces $\frac{1}{3}$ mole of $CHBr_3$.

42. *(b)* Williamson ether synthesis is the reaction in which an alkyl halide is allowed to react with alkoxide to give ether. The reaction involves S_N2 attack of an alkoxide ion on primary alkyl halide. In option (a) and (b) there are alkoxide and alkyl halide, but option (b) is preferred because I^- is better leaving group than Cl^-.

43. *(d)*

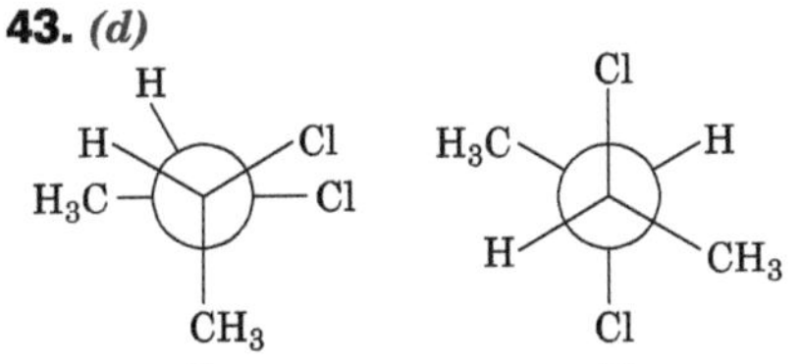

X: Newman projection with H, H_3C, Cl, Cl, CH_3; Y: Newman projection with Cl, H_3C, H, H, CH_3, Cl

X *Y*

Both *X* and *Y* are conformers of each other, where *X* is eclipsed and *Y* is staggered form.

Conformers are the spatial arrangements of atoms, which can be converted into one another by rotation around a C—C single bond. The conformation in which the substitutents are as closed together as possible is called eclipsed conformation, whereas in staggered form the substituents are far apart as possible.

44. *(d)* *Tert*-butyl cation is more stable than isopropyl cation, because of better hyper conjugation between σ-vacant and *p*-orbital.

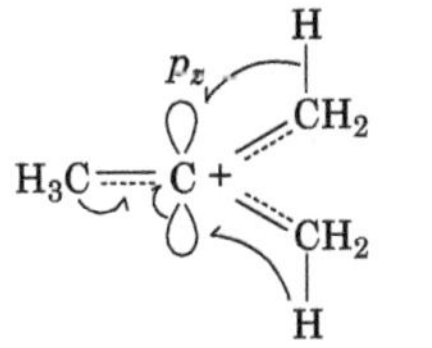

(Hyper-conjugation in *tert*-butyl carbocation)

whereas, *trans*-2-butene is more stable than propene because of hyper-conjugation σ-π* -orbitals.

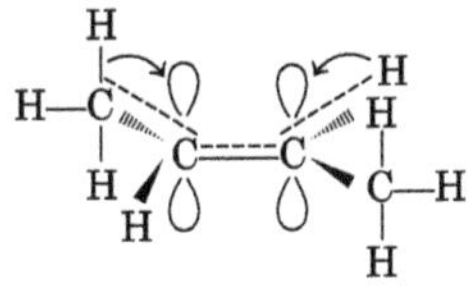

(Hyper-conjugation in *trans*-2-butene)

45. *(d)* Benzaldehyde can be converted to benzyl alcohol in concentrated aq. NaOH solution using formaldehyde.

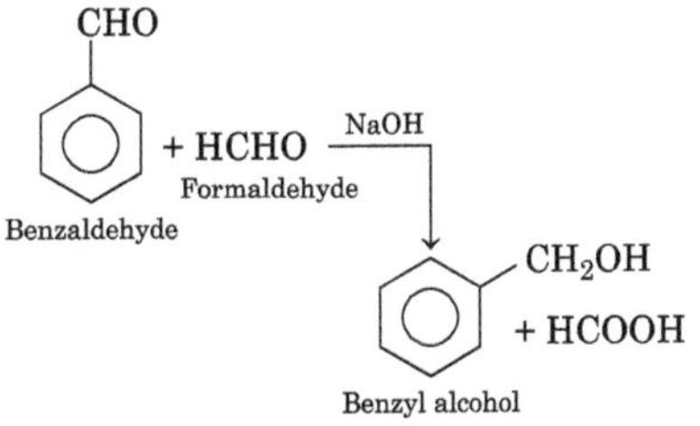

Benzaldehyde (CHO) + HCHO (Formaldehyde) $\xrightarrow{NaOH}$ Benzyl alcohol (CH_2OH) + HCOOH

This reaction is known as Cannizzaro reaction, where one molecule of benzaldehyde is reduced to alcohol while formaldehyde is oxidised to carboxylic acid salt. Aldehydes, which do not have an α-hydrogen atom undergo this type of reaction.

46. *(a)*

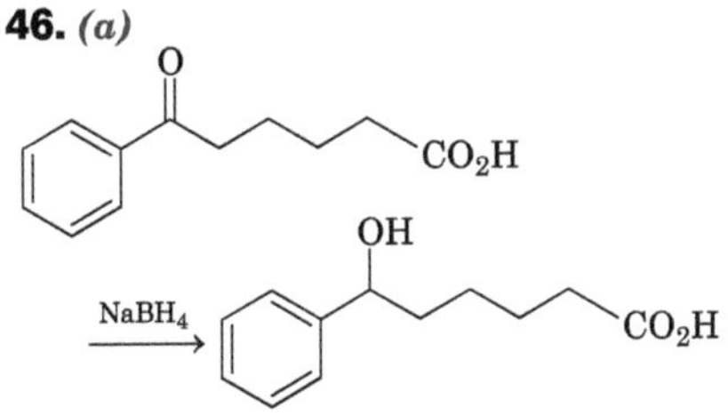

$C_6H_5CO(CH_2)_4CO_2H \xrightarrow{NaBH_4} C_6H_5CH(OH)(CH_2)_4CO_2H$

$NaBH_4$ is a selective reducing agent which reduces carboxyl group of aldehyde and ketone to alcohol, but does not reduces COOH group.

47. *(*)* As the electronegativity of the central atom increases the bond angle also increases. So, NH_3 will have high bond angle than PH_3. The bond angle of NH_4^+ is highest because of its symmetrical geometry, also bond pair and lone pair on N of NH_3 suffer repulsion, which decreases its bond angle in comparison to NH_4^+. Thus, the correct order of bond angle is

$$PH_3 < NH_3 < NH_4^+$$

But in the given question none of the option is correct because there is no H—*X*—H bond angle in BF_3.

48. *(c)* The given species are isoelectronic (same number of electrons) to each other. In case of isoelectronic species, cation with greater positive charge will have a smaller radius because of the greater attraction of the electrons to nucleus. Anion with greater negative charge will have the largest radius. This is because the net repulsion of electrons will out weight the nuclear charge and the ion will expand in size.

Thus, the order of ionic radii will be,

$$N^{3-} > O^{2-} > F^- > Na^+$$

49. *(b)* The extent of reducing property is determined by the oxidation state of central atom. More is the oxidation state lesser will be the reducing property. The oxidation states of P in given oxoacids are as follows :

(i) H_3PO_3

$$3(+1) + x + 3(-2) = 0$$
$$3 + x - 6 = 0$$
$$x = +3$$

(ii) H_3PO_2

$$3(+1) + x + 2(-2) = 0$$
$$3 + x - 4 = 0$$
$$x = +1$$

(iii) H_3PO_4

$$3(+1) + x + 4(-2) = 0$$
$$3 + x - 8 = 0$$
$$x = +5$$

(iv) $H_4P_2O_7$

$$4(+1) + 2x + 7(-2) = 0$$
$$4 + 2x - 14 = 0$$
$$x = +5$$

As the oxidation state of P is least in H_3PO_2. Thus, it has a strongest reducing property.

50. *(d)* The reactions between given elements, S, C and P with conc. H_2SO_4 are as follows:

$C + H_2SO_4 \longrightarrow CO_2 + SO_2 + H_2O$
$S + H_2SO_4 \longrightarrow SO_2 + H_2O$
$P_4 + H_2SO_4 \longrightarrow H_3PO_4 + SO_2 + H_2O$

Thus, all the given elements produces SO_2 on reaction with hot concentration.

51. *(b)* Linkage isomerism arises in a coordination compound containing ambidentate ligand. Among the given ligands only, NO_2 is an ambidentate ligand as it can be bonded from both N or O-atom to metal atom. Thus, $[Co(NH_3)_5(NO_2)]Cl_2$ show linkage isomerism.

52. *(a)* Boron atom has an incomplete octet. It can accept a lone pair from *X* given in the question and can also donate electron to it, thus it can form a π-bond also known as back bonding.

The extent of back bonding depends upon the electronegativity of *X* and on the size of the valence shell. Lesser is the electronegativity of *X* and similar is the size of valence shell of *X* and B, more will be the extend of back bonding.

Cl and F are more electronegative than OMe and NMe group, so there extent of back bonding will be least. Among BCl_3 and BF_3, Cl will have least tendency to form π - bond with boron, because of large size of its 3*p*-orbital.

Among $B(OMe)_3$ and $B(NMe_2)_3$, (NMe) will have the maximum tendency to form π-bond because N in NMe is less electronegative than O in OMe group. Thus, the tendency to form π-bond with boron follows the order.

$BCl_3 < BF_3 < B(OMe)_3 < B(NMe_2)_3$.

53. *(b)* Langmuir isotherm gives the relationship between the release of adsorbed gas and pressure. The Langmuir isotherm assumes that the gas attaches to the surface of adsorbent. The free gas and adsorbed gas are in dynamic equilibrium. The surface of the adsorbent is uniform that is, all the adsorption sites are equivalent. The ability of a molecule to get adsorbed at a given site is independent of the occupation of neighbouring sites. Thus, statements (i), (ii) and (iv) are correct.

54. *(b)* The principle of conductometric titration is based on the fact that during the titration one of the ions is replaced by other and invariably these two ions differ in the ionic conductivity.

In the conductometric titration of 0.05 M H_2SO_4 with 0.1 M NH_4OH, initially the fast moving H^+ ion get neutralised as H_2O and is replaced by slow moving NH_4^+ ion upto neutralisation point. After neutralisation point, weak electrolyte NH_4OH is added gradually, which do not affect the conductance. Thus, plot (b) is the correct one.

55. *(a)* The ideal body which emits and adsorbs radiations of all frequencies is called a black body and the radiation emitted by such a body is called black body radiation. The exact frequency distribution of the emitted radiation from black body depends only on its temperature. At a given temperature intensity of radiation emitted increases with decrease of wavelength, reaches a maximum value at a given wavelength and then starts decreasing with further decrease of wavelength. Thus, correct representation is shown in option (a).

56. *(b)* van der Waals' gases are thoses' gases which shows positive or negative deviation from ideal gas equation. At constant temperature, the pressure of the van der Waals' gas, which undergo gas to liquid transition is inversely proportional to its volume, i.e. as the volume increases pressure decreases. Thus, pressure volume isotherm given in option (b) is correct.

57. *(a)* The solutions which resist change in pH on dilution or with the addition of small amounts of acid or alkali are called buffer solution. It is a mixture of weak base with its salt or weak acid with its salt.

Thus, buffer solution can be prepared by mixing equal volumes of 0.2 M NH_4OH (weak base) and 0.1 M HCl (strong acid). It results in the formation of NH_4OH (weak base) + NH_4Cl (salt of weak base NH_4OH), which is a basic buffer mixture.

58. *(b)* According to Raoult's law,

$$p = p^0\chi$$

According to Dalton's law of partial pressure,

$$p_T = p_x + p_y$$

$$p_T = p^0{}_x\chi_x + p^0{}_y\chi_y$$

$$= p^0{}_x\chi_x + p^0{}_y(1-\chi_x)$$

$$p_T = p^0{}_Y + (p^0{}_x - p^0{}_y)\,\chi_x$$

Thus, total vapour pressure of solution varies linearly with mole fraction of component *X*.

Case I, at $\chi_X = 0$

$$p_T = p^0{}_Y$$

Case II, at $X_X = 1$

$$p_T = p^0{}_Y + p^0{}_X - p^0{}_Y$$

$$p_T = p^0_X$$

Thus, correct graph is given in option (b).

59. *(b)* Natural rubber is polyisoprene.

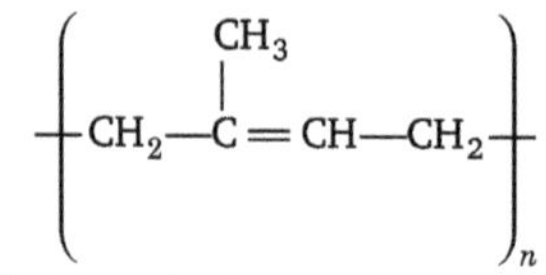

$$\left[\!-CH_2-\overset{\overset{\displaystyle CH_3}{|}}{C}=CH-CH_2-\!\right]_n$$

Which on hydrogenation gives polyisobutyl group.

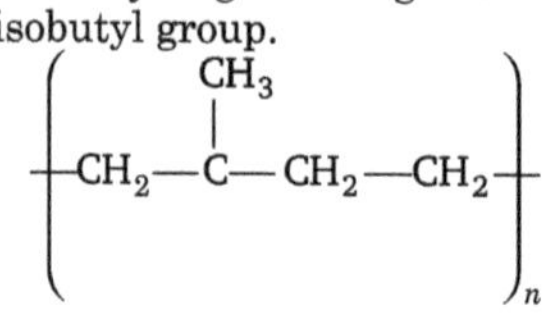

$$\left[\!-CH_2-\overset{\overset{\displaystyle CH_3}{|}}{C}-CH_2-CH_2-\!\right]_n$$

Which can be considered as ethylene propylene copolymer $\left[\!-CH_2-\overset{\overset{\displaystyle CH_3}{|}}{C}-\!\right]_n$

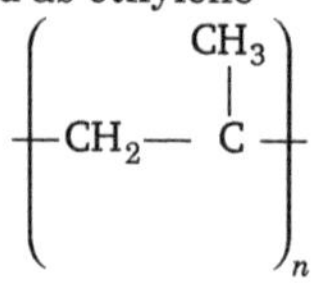

$$-\!\!(CH_2-CH_2\!)\!\!-_n$$

$$\left[\!-CH_2-\overset{\overset{\displaystyle CH_3}{|}}{C}=CH-CH_2-\!\right]_n \xrightarrow{H_2} \left[\!-CH_2-\overset{\overset{\displaystyle CH_3}{|}}{CH}-CH_2-CH_2-\!\right]_n$$

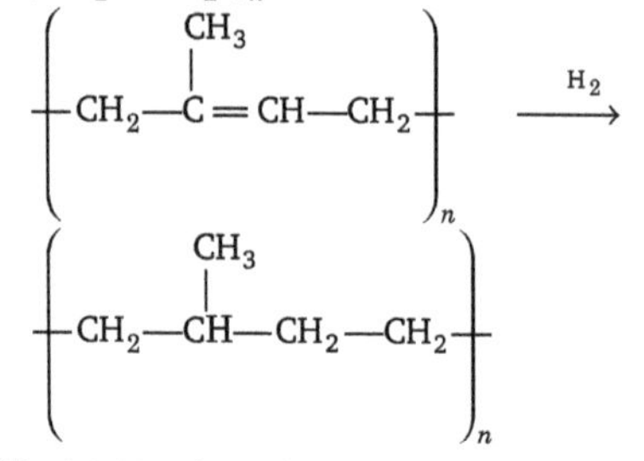

60. *(a)* Number of H-bonds in each A—T pair = 2.

Number of H-bonds in G—C pair = 3.

The given double stranded DNA has 5 A—T and 3 G —C pairs.

∴ Total number of H-bonds in 5 A—T pair

$$= 5 \times 2 = 10$$

Total number of H-bonds in 3G — C pair

$$= 3 \times 3 = 9$$

⇒ Average energy of 10 H-bonds in A—T pair = $10x$.

Average energy of 9 H-bonds in G—C pair = $9y$.

Thus, the approximate energy required to split the double stranded DNA in two single strands = $10x + 9y$.

61. *(c)* The mammalian haemoglobin molecule can bind (carry) up to four oxygen (O_2) molecules which means eight oxygen atoms. Haemoglobin has an oxygen binding capacity of 1.34 mL O_2 per gram, which increases the total blood oxygen capacity seventy fold compared to dissolved oxygen in blood.

62. *(a)* *Bt* toxin is produced by *Bacillus thuringiensis* (soil bacterium). This toxin does not kill the bacteria because when it is present in the bacteria, it is in an inactive and crystalline form. It becomes active and toxic when it is consumed by insects such as lepidopterans (armyworm), coleopterans (beetles) and dipterans (flies/mosquitoes).

63. *(b)* The ploidy of endosperm in angiosperm is $3n$ (central cell fused with one male gamete, i.e. $2n + n$) and the ploidy of an angiosperm is $2n$. Thus, if endosperm $= 6n$, the plant will be $4n$.

64. *(d)* Option (a) is correct. Other statements can be corrected as viruses are acellular entities and contain either DNA or RNA. They replicate using the replication proteins of their host cell. The protein shell of virus is called capsid, it encloses the genetic material of the virus. It protects the nucleic acid from digestion by enzymes. The polymerase genes of virus are expressed early in the cycle of virus but capsid and tail proteins are expressed later.

65. *(c)* Mitochondria are membrane bound organelles and like the nucleus, have a double membrane. The outer membrane is fairly smooth but the inner membrane is highly convoluted, forming folds called cristae. The cristae greatly increase the inner membrane's surface area. The increased surface area helps in faster production of ATP.

66. *(d)* Biological nitrogen-fixation occurs when atmospheric nitrogen is converted to ammonia by an enzyme called nitrogenase. This enzyme is used by some organisms to fix atmospheric nitrogen gas (N_2).

67. *(a)* The given graph represents absorption spectrum of chlorophyll-*a*. The photosynthetic efficiency graph can also be referred to as action spectrum. The set of wavelengths absorbed by a pigment is its absorption spectrum. Chlorophyll-*a* gives absorption peaks at 430 nm and 662 nm.

The absorption spectrum of chlorophyll-*a* includes wavelengths of blue and orange-red light as is indicated by their peaks, around 450-475 and 650-675 nm. Thus, option (a) is correct.

68. *(b)* Contact inhibition is a process of arresting cell growth when cells come in contact with each other. As a result, normal cells stop proliferating when they form a monolayer in a culture dish. Contact inhibition is powerful anticancer mechanism that gets lost in cancer cells, which helps in rapid spread of concreous cells.

69. *(a)* Yeast contains enzymes that catalyse the breakdown of glucose during fermentation to produce ethanol and carbon dioxide.

$$\underset{\text{Glucose}}{C_6H_{12}O_6} \xrightarrow[\text{Fermentation}]{\text{Yeast}} \underset{\text{Ethanol}}{2C_2H_5OH} + \underset{\text{Carbon dioxide}}{2CO_2}$$

70. *(d)* Proteases are enzymes that cleave peptide bonds in proteins, in which serine serves as the nucleophilic amino acid at the enzymes active site. They are found ubiquitously in both eukaryotes and prokaryotes.

71. *(c)* Pollen grains are male gametes. The number of different gametes produced by a plant Tt Rr YY is 2^n where n = number of heterozygous alleles. Thus, $2^2 = 4$.

72. *(a, d)* Both statements *a* and *d* are correct. Because Follicle Stimulating Hormone (FSH) stimulates the ovarian production of oestrogen by the granulosa cells of the ovarian (Graafian) follicles and corpus lutea. Progesterone is secreted by corpus luteum which regulates the condition of the inner lining of the uterus.

73. *(b)* An antibody is a protein that is synthesised by human in response to the presence of a foreign substance in our body called as antigen. Antibodies are created by white blood cells or more specifically by B-cells.

74. *(d)* A 0.85 – 0.9% NaCl solution is said to be isotonic when blood cells present in such a medium, the intracellular and extracellular fluids are in osmotic equilibrium across the cell membrane and there is no net influx or efflux of water.

75. *(c)* Golgi bodies are membrane bound organelle of eukaryotic cells that is made up of a series of flattened stacked pouches called cisternae. These are responsible for transporting, modifying and packaging proteins and lipids into vesicles for delivery to targeted destination like plasma membrane.

76. *(d)* Creutzfeldt Jakob Disease (CJD) is caused by an abnormal infectious misfolded protein in the brain called as prion. Normally these misfolded prion proteins are recycled by the body, but they can build up in the brain if they are not recycled.

77. *(b)* Dinosaurs lived in the Mesozoic era (66-252 million years ago) which is divided into Triassic, Jurassic and Cretaceous periods. Bamboo species that we know today is about 30-40 million years ago. Thus, bamboo cannot be present in the fossil of dinosaur faeces as they were not present at that time.

78. *(a)* Isoleucine, an essential amino acid is one of the three amino acids having branched hydrocarbon side chains. It is usually interchangeable with leucine and occasionally with valine in proteins. These two amino acids isoleucine and threonine have in common the fact that they have two chiral centres.

79. *(c)* Ribulose-1,5-bisphosphate reacts with carbon dioxide to produce 3-keto 2-carboxyarabinitol, 1, 5-bisphosphate as the first step in the biochemical process of carbon-fixation. It needs an enzyme to do this, i.e. RuBisCO.

80. *(a)* Tuberculosis is acquired after inhalation of aerosol droplets containing the bacilli from the cough of infected individuals.

Dysentery is an intestinal inflammation caused by contaminated food and water consumptions.

Filariasis is a parasitic disease caused by an infection with roundworms. They spread by blood-feeding diptera as black flies and mosquitoes.

Syphilis is a bacterial infection usually spreads by sexual contact that starts as a painless sore.

81. *(d)* Maximum possible number of points common to the perimeter of rectangle, circle and triangle is 6.

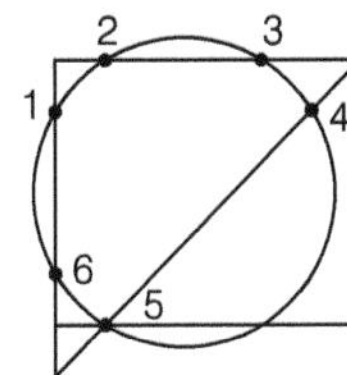

82. *(c)* We have,

$$x^4 + 4y^4 + 16z^4 + 64 = 32xyz$$

We know $\quad$ AM ≥ GM

$$\Rightarrow \frac{x^4 + 4y^4 + 16z^4 + 64}{4} \geq (x^4 \times 4y^4 \times 16z^4 \times 64)^{1/4}$$

$$\Rightarrow \quad x^4 + 4y^4 + 16z^4 + 64 \geq 32xyz$$

But given $x^4 + 4y^4 + 16z^4 + 64 = 32xyz$

$$\therefore \quad x^4 = 4y^4 = 16z^4 = 64$$

$$\Rightarrow \quad x = \pm 2\sqrt{2},\ y = \pm 2,\ z = \pm\sqrt{2}$$

For x, y, z

For $x^4 + 4y^4 + 16z^4 + 64 = 32xyz$

Either each of x, y, z is positive → 1 case or two of x, y, z are negative → 3 cases

∴ 4 cases of different (x, y, z) triplets

4 possible $x + y + z$ values $(x \neq y \neq z)$.

83. *(b)* Let the equation of circle

$$x^2 + y^2 = r^2$$

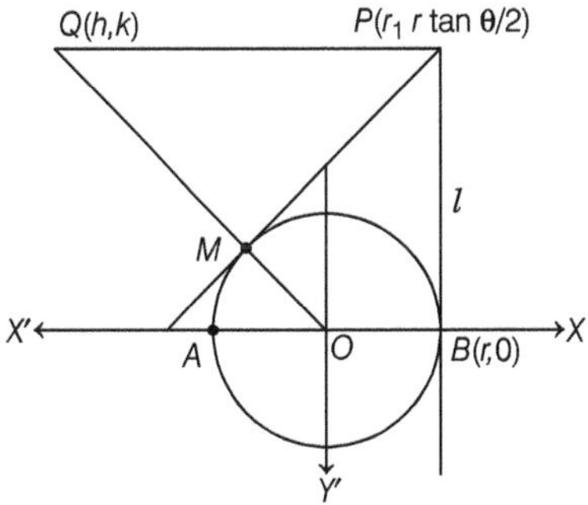

Let $M(r\cos\theta, r\sin\theta)$

Equation of tangents at $M(r\cos\theta, r\sin\theta)$ is $rx\cos\theta + ry\sin\theta = r^2$

$$\Rightarrow \quad x\cos\theta + y\sin\theta = r$$

Tangent is passes through P.

$$\therefore \quad x\cos\theta + y\sin\theta = r$$

$$\Rightarrow \quad y = \frac{r(1-\cos\theta)}{\sin\theta} = r\tan\frac{\theta}{2}$$

PQ is parallel to AB.

$$\therefore \quad k = r\tan\frac{\theta}{2}$$

OQ is normal of circle.

$$\therefore \text{Slope of normal} = \frac{k}{h} = \tan\theta$$

$$\Rightarrow \quad \frac{k}{h} = \frac{2\tan\frac{\theta}{2}}{1-\tan^2\frac{\theta}{2}}$$

$$\Rightarrow \quad \frac{k}{h} = \frac{2\frac{k}{r}}{1-\frac{k^2}{r^2}} \quad \left[\because \tan\frac{\theta}{2} = \frac{k}{r}\right]$$

$$\Rightarrow \quad 2hr = r^2 - k^2 \Rightarrow k^2 = r^2 2hr$$

$$\therefore \text{ Locus of } Q \text{ is } y^2 = -2r\left(x - \frac{r}{2}\right)$$

which is parabola.

84. *(c)* Given,

$$\sin(x + x^2) - \sin(x^2) = \sin x$$

$$\Rightarrow \quad \sin(x + x^2) = \sin(x^2) + \sin x$$

$$\Rightarrow 2\sin\left(\frac{x+x^2}{2}\right)\cos\left(\frac{x+x^2}{2}\right)$$

$$= 2\sin\left(\frac{x^2+x}{2}\right)\cos\left(\frac{x^2-x}{2}\right)$$

$$\sin\left(\frac{x+x^2}{2}\right) = 0$$

or $\cos\left(\frac{x^2+x}{2}\right) - \cos\left(\frac{x^2-x}{2}\right) = 0$

$$\frac{x^2+x}{2} = x\pi \text{ or } 2\sin\frac{x^2}{2}\sin\frac{x}{2} = 0$$

$$\Rightarrow \frac{x^2+x}{2} = 0,\ \pi,\ 2\pi \text{ or } \frac{x^2}{2} = 0,\ \pi,\ \frac{n}{2} = 0,\ \pi,\ 2\pi$$

$$\Rightarrow x^2 + x = 2\pi \text{ or } x^2 = 2\pi$$

$$\Rightarrow x^2 + x - 2\pi = 0 \text{ or } x = \sqrt{2\pi}$$

$$\Rightarrow \quad x = \frac{-1 \pm \sqrt{1+8\pi}}{2} \text{ or } 2 < \sqrt{2\pi} < 3$$

$$\Rightarrow \quad \sqrt{1+8\pi} \approx \sqrt{25.14}$$

$$\therefore \quad x = \frac{5.2-1}{2} = \frac{4.2}{2} = 2.1$$

$\because$ Total numbers of solution lies between (2, 3) = 2

85. *(a)* We have,

$$p(x) > x^2, p(0) = 0,\ p''(0) = \frac{1}{2}$$

Let $g(x) = p(x) - x^2$

$g(x) > 0,\ \forall\ x \neq 0$

and $g(0) = p(0) - 0 = 0$

$\Rightarrow x = 0$ should be minima.

$\because g''(x)$ should be ≥ 0 at $x = 0$

Now, $g'(x) = p'(x) - 2x$

$g''(x)^2 = p''(x) - 2$

$$g''(0) = p''(0) - 2 = \frac{1}{2} - 2 = -\frac{3}{2}$$

But $g''(0) \geq 0$

$\because$ No polynomial exists.

86. *(a)* We have,

$$L = \lim_{n\to\infty} \sqrt{n} \int_0^1 \frac{1}{(1+x^2)^n} dx$$

$$(1+x^2)^n > 1 + nx^2$$

$$\Rightarrow \quad \frac{1}{(1+x^2)^n} < \frac{1}{1+nx^2}$$

$$\Rightarrow \quad \int_0^1 \frac{dx}{(1+x^2)^n} < \int_0^1 \frac{dx}{1+nx^2}$$

$$\Rightarrow \quad \int_0^1 \frac{dx}{(1+x^2)^n} < \frac{1}{\sqrt{n}} [\tan^{-1}\sqrt{n}\ x]_0^1$$

$$\Rightarrow \quad \int_0^1 \frac{dx}{(1+x^2)^n} < \frac{1}{\sqrt{n}} \tan^{-1}\sqrt{n}$$

$$\Rightarrow \quad L = \lim_{n\to\infty} \sqrt{n} \int_0^1 \frac{1}{(1+x^2)^n} dx$$

$$< \lim_{n\to\infty} \frac{\sqrt{n}}{\sqrt{n}} (\tan^{-1}\sqrt{n})$$

$$\Rightarrow \quad L < \tan^{-1}\infty = \frac{\pi}{2}$$

$$\therefore \quad \frac{1}{2} < L < 2 \qquad \left[\because \frac{\pi}{2} < 2\right]$$

87. *(a)* Equation of line passing through two points (x_1, y_1) and (x_2, y_2) is

$$y - y_1 = \frac{y_2 - y_1}{x_2 - x_1}(x - x_1)$$

$$\Rightarrow (x_2 - x_1)y + (y_1 - y_2)x + y_1(x_1 - x_2) + x_1(y_2 - y_1) = 0$$

$\Rightarrow ax + by + c = 0$, where $a, b, c \in I$

$a = y_1 - y_2,\ b = x_2 - x_1$

$c \in y(x_1 - x_2) + x_1(y_2 - y_1)$

Perpendicular distance from origin to line $ax + by + c = 0$ is $\frac{c}{\sqrt{a^2+b^2}}$ and $\frac{c^2}{a^2+b^2}$ is rational.

Case I n is not pefect square and square of radius $= n^2\left(1\left(1-\frac{1}{\sqrt{n}}\right)^2\right)$ = rational

$$\Rightarrow \quad r^2 \neq \frac{c^2}{a^2+b^2}$$

$\Rightarrow ax + by + c = 0$ never be tangent to give n circle

$$\Rightarrow \quad \lim_{n\to\infty} P_n = 0$$

Case II If n is perfect square.

In this case number of tangents passing through two points from given set are few, but total number of lines are in much quantity when n approaches to infinite.

$$\Rightarrow \quad \lim_{n\to\infty} P_n = 0$$

88. *(d)* We have,

$$-1 < f(0) < f(1) < 1$$

$$g = [-1, 1] \to [0, 1]$$

$$gof(x) = x,\ \forall\ x \in [0, 1]$$

Only condition that $g(x)$ should satisfy for $gof(x) = x,\ \forall\ x \in [0, 1]$ is that $g(x)$ should attain all values in [0, 1] when range of $f(x)$ a subset of (–1, 1) is used as image for $g(x)$. Thus, there can be infinite such functions $g(x)$ with domain [–1, 1] and range [0, 1].

89. *(b)* We have,

$$y = x^2 + x + 10$$

$$\Rightarrow \quad x^2 + x + \frac{1}{4} = y - 10 + \frac{1}{4}$$

$$\Rightarrow \quad \left(x + \frac{1}{2}\right)^2 = \left(y - \frac{39}{4}\right)$$

Latusrectum of parabola is 1.

$\therefore$ Length of chord is also 1.

Area is maximum when chord is latusrectum.

Area of shaded region =

Area of rectangle $OABC - 2\int_{-1/2}^{0}$ parabola

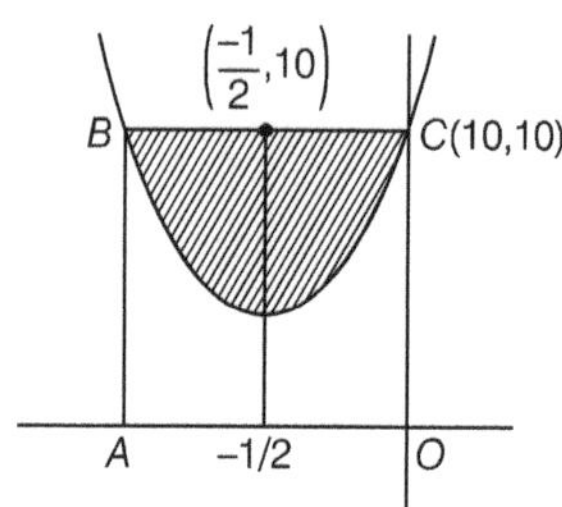

$$= 10 - 2\int_{-\frac{1}{2}}^{0} (x^2 + x + 10)dx$$

$$= 10 - 2\left[\frac{x^3}{3} + \frac{x^2}{2} + 10\right]_{-\frac{1}{2}}^{0}$$

$$= 10 - 2\left[\frac{-1}{24} + \frac{1}{8} - \frac{10}{2}\right] = \frac{1}{6}$$

90. *(b)* We have,

$$11z^8 + 20iz^7 + 10iz - 22 = 0$$

$$\Rightarrow \quad z^8 + \frac{20}{11}iz^7 + \frac{10i}{11}z - 2 = 0$$

Root of equation are $z_1, z_2, z_3, z_4, z_5, \ldots, z_8$

$$\sum_{i=1}^{8} zi = \frac{20}{11}i$$

$$z_1 z_2 z_3 z_4 z_5 z_6 z_7 z_8 = 2$$

$$\Rightarrow \quad |z_1 z_2 z_3 z_4 z_5 z_6 z_7 z_8| = 2$$

$$\Rightarrow |z_1||z_2||z_3||z_4||z_4||z_5||z_6||z_7||z_8| = 2$$

$$\because \quad 1 < |z_i| < 2$$

$$\therefore \quad S = |z|^2 + |z| + 1$$

$$\Rightarrow \quad 1 + 1 + 1 < |z|^2 + |z| + 1 < 2^2 + 2 + 1$$

$$\Rightarrow \quad 3 < |z|^2 + |z| + 1 < 7$$

$$\therefore \quad 3 < S < 7$$

91. *(a)* Temperature difference causes heat to flow.

So, potential difference is equivalent to temperature difference.

As, $\frac{dV}{dR} = -E = \frac{-kQ}{R^2}$

We can write for heat flux,

$$\frac{dT}{dR} = \frac{-kQ}{R^2}$$

Here, Q = heat transferred.

$$\Rightarrow \quad Q \propto R^2$$

when $\frac{dT}{dR}$ = constant.

$$\therefore \quad n = 2$$

92. *(b)*

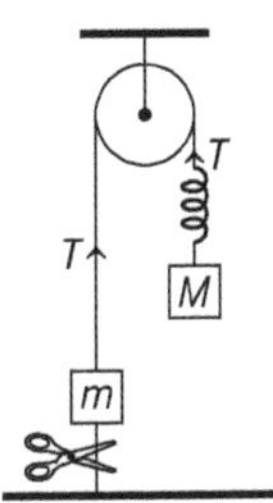

At $t = 0$, $T = mg$

When string is cut, mass M remains static and spring snaps. For mass m at the instant string is cut, forces on m are

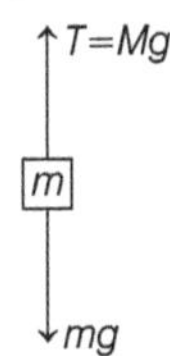

Acceleration of m is upwards as spring snaps,

$$a_m = \frac{F_{\text{net}}}{m} = \frac{Mg - mg}{m} = \left(\frac{M - m}{m}\right) g$$

93. *(a)* Electric field due to given charge distribution is

$$E = -\frac{dV}{dr} = -\frac{d}{dr}\frac{qe^{-\alpha r}}{4\pi\varepsilon_0 r}$$

$$= \frac{-q}{4\pi\varepsilon_0}\frac{d}{dr}\left(\frac{e^{-\alpha r}}{r}\right)$$

$$= \frac{-q}{4\pi\varepsilon_0}\left(\frac{-\alpha r e^{-\alpha r} - e^{-\alpha r}}{r^2}\right)$$

$$= \frac{q}{4\pi\varepsilon_0}\cdot e^{-\alpha r}\cdot\left(\frac{\alpha r + 1}{r^2}\right)$$

Electric field at $r = \frac{1}{\alpha}$ is

$$E\left(r = \frac{1}{\alpha}\right) = \frac{q}{4\pi\varepsilon_0}\cdot e^{-\alpha \times \frac{1}{\alpha}}\left(\frac{\alpha \times \frac{1}{\alpha} + 1}{1/\alpha^2}\right)$$

$$= \frac{(q/e)}{4\pi\varepsilon_0}\cdot 2\alpha^2$$

Flux through a sphere of radius $\frac{1}{\alpha}$ is

$$\phi = \int \mathbf{E}\cdot d\mathbf{A}$$

For spherical distribution,

$\mathbf{E}\cdot d\mathbf{A} = EdA\cos 0^\circ = EdA$

and E = uniform.

So, we have

$$\therefore \quad \phi = E\int dA$$

$$= \frac{q/e}{4\pi\varepsilon_0}\cdot 2\alpha^2\cdot 4\pi\left(\frac{1}{\alpha}\right)^2$$

or $\quad \phi = \dfrac{2q}{\varepsilon_0 e}$

From Gauss' law, we have $\phi = \dfrac{q_{\text{enclosed}}}{\varepsilon_0}$

So, charge enclosed in given sphere is $\dfrac{2q}{e}$.

94. *(b)* Let a mud blob is detached from the circumference of wheel with initial speed u at angle θ as shown below.

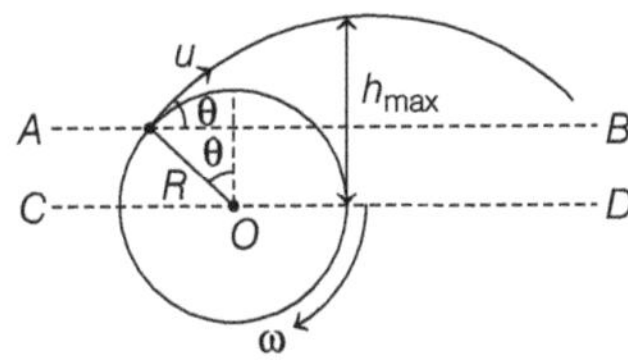

Height upto which mud blob can be thrown is

h = Maximum height of projectile + Height at which mud blob is thrown

$$\therefore \quad h = \frac{u^2\sin^2\theta}{2g} + R\cos\theta \quad \text{...(i)}$$

h is maximum when $\dfrac{dh}{d\theta} = 0$

$$\Rightarrow \frac{d}{d\theta}\left(\frac{u^2\sin^2\theta}{2g} + R\cos\theta\right)$$

$$\Rightarrow \frac{u^2}{2g}.\,2\sin\theta\cos\theta - R\sin\theta = 0$$

$$\Rightarrow \quad \frac{u^2}{g}.\cos\theta = R \text{ or } \cos\theta = \frac{Rg}{u^2}$$

$$\Rightarrow \quad \sin^2\theta = 1 - \frac{R^2g^2}{u^2}$$

Substituting these values in Eq (i), we get

$$\lambda_{\max} = \frac{u^2}{2g}\left(1 - \frac{R^2g^2}{u^4}\right) + R\left(\frac{Rg}{u^2}\right)$$

$$= \frac{u^2}{2g} - \frac{R^2g}{2u^2} + \frac{R^2g}{u^2} = \frac{u^2}{2g} + \frac{R^2g}{2u^2}$$

95. *(b)* When a current flows through a junction or a contact between two metals of different conductivities a charge is accumulated in the vicinity of contact. The charge accumulation is directly proportional to the current. Charge accumulation occurs in a region whose thickness is nearly equals to the mean free path in the metal.

In case current flows from a metal of large conductivity to a metal of small conductivity, charge accumulation is of positive type and *vise-versa*. So, correct option is (b).

96. *(c)* By second law of thermodynamics, entropy of a system either increases or remains constant (only in reversible quasistatic processes).

Also, from graphs given

$$S_2 > S_1 \Rightarrow \frac{Q}{T_2} > \frac{Q}{T_1}$$

where, S is entropy, Q is heat and T is the temperature.

or $\quad T_1 < T_2$

Hence, temperature of system 1 is less than system 2. So, heat flows from system 2 to system 1.

Hence, U_1 increases and U_2 decreases.

97. *(c)*

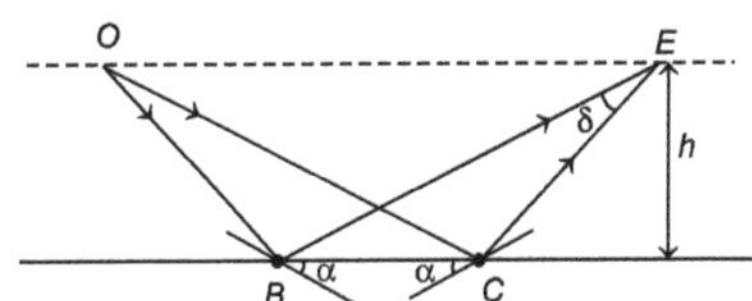

We can consider ripples as plane mirrors reflecting light to form image with angular extent δ.

It is given O and E are at same level, so $OEBC$ is a trapezium with $OE \parallel BC$.

Now, different angles of trapezium $OEBC$ are as shown below.

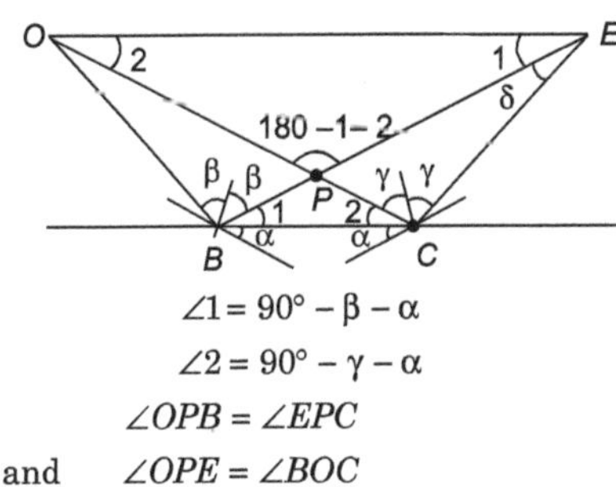

$$\angle 1 = 90^\circ - \beta - \alpha$$

$$\angle 2 = 90^\circ - \gamma - \alpha$$

$$\angle OPB = \angle EPC$$

and $\quad \angle OPE = \angle BOC$

$\Rightarrow \; 90^\circ - \beta - \alpha = 90^\circ - \gamma - \alpha$

$\Rightarrow \quad \beta = \gamma \quad$ ...(i)

As, $\quad \angle OPC = 180^\circ$

$\Rightarrow \quad 180^\circ - \beta + \gamma + 2\alpha + 180 - 2\gamma - \delta$

$\Rightarrow \quad 2\alpha + \beta - \gamma - \delta = 0$

$\Rightarrow \quad \delta = 2\alpha \quad [\because \beta = r \text{ from Eq. (i)}]$

98. *(b)* Spiral galaxies consists of flat rotating disc containing stars, gas and dust.

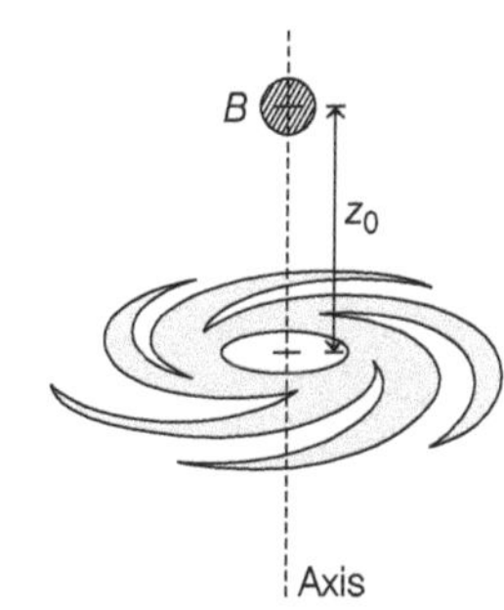

Now, we thought of spiral galaxy in form of a disc.

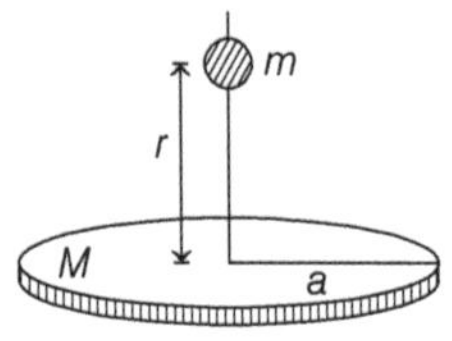

Gravitational intensity at distance r over axis is

$$I_g = \frac{2GM}{a^2}\left(1 - \frac{r}{(r^2 + a^2)^{1/2}}\right)$$

$$= \frac{2GM}{a^2}\left(1 - \frac{1}{\left(1 + \frac{a^2}{r^2}\right)^{1/2}}\right)$$

$$= \frac{2GM}{a^2}\left(1 - \left(1 + \frac{a^2}{r^2}\right)^{-\frac{1}{2}}\right)$$

Using binomial approximation, we have

$$I_g \approx \frac{2GM}{a^2}\left(1 - 1 + \frac{1}{2}\frac{a^2}{r^2}\right) \approx \frac{2GM}{r^2}$$

So, forces on planet A and B are

$$F_A = \frac{2GMm_A}{4z_0^2}, F_B = \frac{2GMm_B}{z_0^2}$$

Accelerations of A and B respectively, are

$$a_A = \frac{GM}{2z_0^2} \text{ and } a_B = \frac{2GM}{z_0^2}$$

If T_A and T_B are time periods of oscillations of A and B, then

$$\frac{T_A}{T_B} = \sqrt{\frac{a_B}{a_A}}$$

[we use $T \propto \frac{1}{\sqrt{a}}$ as in case of a linear oscillator]

$$= \sqrt{\frac{4}{1}} \Rightarrow \frac{T_A}{T_B} = 2$$

99. *(d)* Consider the arrangement given,

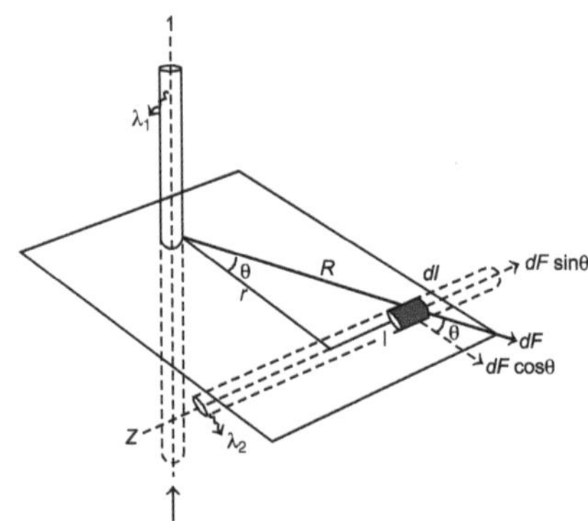

Force on a differentiable length element (dl) of conductor 2 due to field of conductor 1 is

$$dF = E_1 dq$$

where, E_1 = field of wire 1 at location of dl.

This force can be resolved into components $dF\cos\theta$ and $dF\sin\theta$

net force is sum of all $dF\cos\theta$ components, whereas $\Sigma dF\sin\theta = 0$.

So, net force on conductor 2 is

$$F = \int dF = \int E_1 dq$$

Here,

$$E_1 = \frac{2k\lambda_1}{R} = 2k\lambda_1$$

$$dq = \lambda_2 dl$$

$$\frac{l}{r} = \tan\theta \text{ and } \frac{r}{R} = \cos\theta$$

$$\Rightarrow \quad dl = r\sec^2\theta\, d\theta$$

Also, θ varies from $\frac{-\pi}{2}$ to $\frac{\pi}{2}$ for conductor 2.

So, $F = \int_{-\pi/2}^{\pi/2} \frac{2k\lambda_1}{r\sec\theta}\cdot\lambda_2\cdot r\sec^2\theta\, d\theta$

$$= 2k\lambda_1\lambda_2 \int_{-\pi/2}^{\pi/2} \sec\theta\, d\theta$$

As $\int_{-\pi/2}^{\pi/2} \sec\theta\, d\theta$ = a constant

We can say that,

$F_{net} = C\cdot\lambda_1\lambda_2$, where C = some constant.

$\therefore \; F_{net} \propto r^0$

100. *(b)* Given force varies as

$$F \propto t^n \Rightarrow F = kt^n \qquad \ldots(i)$$

From graph, we observe

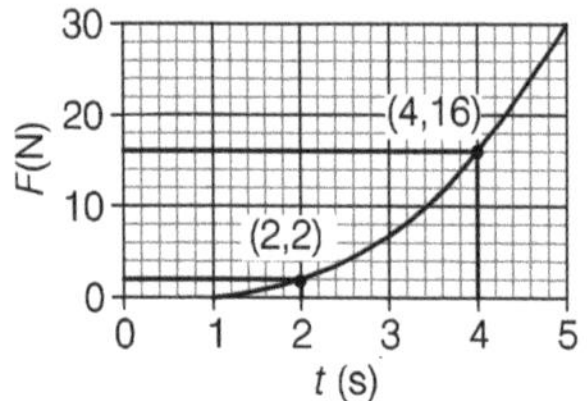

at $t = 2$ s, $F = 2$ N

and at $t = 4$ s, $F = 16$ N

So, putting the values in Eq (i), we have

$$16 = k(4)^n \text{ and } 2 = k(2)^n$$

$$\Rightarrow \quad 8 = \frac{2^{2n}}{2^n} \Rightarrow n = 3$$

Also, $2 = k(2)^n \Rightarrow 2 = k(2)^3 \Rightarrow k = \frac{1}{4}$

So, force is given by $F = \frac{t^3}{4}$

Now, $F = \frac{dp}{dt}$, (where, p = momentum)

$$\Rightarrow \quad dp = Fdt \Rightarrow \int_{v_i}^{v_f} dp = \int_{t_i}^{t_f} F\,dt$$

$$\Rightarrow \quad m(v_f - v_i) = \int_{t_i}^{t_f} \frac{t^3}{4}\cdot dt$$

$$\Rightarrow \quad m(v_f - v_i) = \frac{t_f^4 - t_i^4}{16} \qquad \ldots(ii)$$

Now, when $t_i = 0$ and $t_f = 3$ s,

We have, $v_i = 0$ and $v_f = 2\text{ ms}^{-1}$

So, from Eq. (ii), we have

$$2m = \frac{81}{16} \qquad \ldots(iii)$$

and when we take $t_i = 0$, $t_f = 4$ s, $v_i = 0$ and $v_f = v$ (let), then again from Eq (ii), we gives

$$mv = \frac{256}{16} \qquad \ldots(iv)$$

Dividing Eq. (iv) by Eq. (iii), we get

$$\frac{v}{2} = \frac{256}{81}$$

$$\Rightarrow \quad v = \frac{256 \times 2}{81} \approx 6.5\text{ ms}^{-1}$$

101. *(c)* According to Nernst equation,

$$E_{cell} = E^\circ_{cell} - \frac{0.0591}{n}\log\frac{[P]}{[R]}$$

For the given electrochemical cell

Pt | H_2 (p = 1 atm) |

H^+ (*aq.* xM) || Cu^{2+} (*aq.* 1.0 M)/Cu(s)

$$E_{cell} = E^\circ_{cell} - \frac{0.0591}{n}\log\frac{[H^+]^2}{[Cu^{2+}]}$$

$$= 0.49 = 0.37 - \frac{0.0591}{2}\log\frac{x^2}{1}$$

$$0.49 = 0.37 - \frac{0.0591}{2} \times 2\log x$$

$0.15 = -0.0591 \times \log x$

$2.53 = -\log x$

Also, $-\log[H^+] = $ pH

Here, $[H^+] = x$

$\therefore$ pH $= -\log x$

$\Rightarrow$ pH = 2.53

102. *(b)* For first order reaction,

$$X \rightleftharpoons Y$$

Given, $K_f = 2\text{ s}^{-1}$

$K_b = 1\text{ s}^{-1}$

$$\therefore \quad K_{eq} = \frac{K_f}{K_b} = \frac{[Y]}{[X]}$$

$$K_{eq} = \frac{2}{1} = \frac{[Y]}{[X]}$$

$$\Rightarrow \quad 2[X] = [Y]$$

or

$$2[X]_{eq} = [Y]_{eq}$$

$$\therefore \quad [Y] > [X]$$

Thus, as the rate of reaction increases, there will be increased in the concentration Y_{eq}, but it will be less than initial concentration of $X (X_0)$, whereas the concentration of X_{eq} decreases. Thus, plot (b) is correct.

103. *(b)* $2C_3H_5(NO_3)_3\,(l) \longrightarrow$

$$\downarrow$$
$$\Delta H_f^\circ = -364$$

$$3N_2(g) + \frac{1}{2}O_2(g) + 6CO_2(g) + 5H_2O(g)$$

$$\downarrow \quad \downarrow \quad \downarrow \quad \downarrow$$

$$\Delta H_f^\circ = 0 \;\; \Delta H_f^\circ = 0 \;\; \Delta H_f^\circ = -393.5 \;\; \Delta H_f^\circ = -241.8$$

$$\Delta H^\circ_{\text{reaction}} = \Sigma H^\circ_{f(P)} - \Sigma H^\circ_{f(R)}$$

$$\Delta H^\circ_{\text{reaction}} = 3 \times 0 + \frac{1}{2} \times 0 + 6 \times -393.5 + 5 \times -241.8 - (2 \times -364)$$

$$= -2361 - 1209 + 728$$

$$= -2842 \text{ kJ/mol}$$

For 2 moles of nitroglycerine enthalpy – 2842 kJ/mol.

$\therefore$ For 1 mole of nitroglycerine enthalpy

$$= \frac{-2842}{2} = -1421 \text{ kJ/mol.}$$

227.1g of $C_3H_5(NO_3)_3$ has enthalpy – 1421 kJ/mol.

$\therefore$ 10 g of $C_3H_5(NO_3)_3$ has enthalpy

$$= -\frac{1421}{227.1} \times 10$$

$$= -62.5 \text{ kJ/mol}$$

104. *(d)* The heating of $(NH_4)_2Cr_2O_7$ produces Cr_2O_3 compound along with N_2.

$$(NH_4)_2Cr_2O_7 \xrightarrow{\Delta} N_2 + Cr_2O_3 + 4H_2O$$

Let the oxidation state of Cr be x.

In $(NH_4)_2Cr_2O_7$,

$$= 2(+1) + 2(x) + 7(-2) = 0$$
$$2 + 2x - 14 = 0$$
$$2x = 12 \Rightarrow x = +6$$

In Cr_2O_3,

$$2(x) + 3(-2) = 0$$
$$2x + (-6) = 0$$
$$2x = +6 \Rightarrow x = +3$$

$\therefore$ The change of the oxidation state of Cr in the reaction is from + 6 to + 3.

105. *(c)* The spin only magnetic moment for a complex can be calculated using formula

$$\mu = \sqrt{n(n+2)}$$

(a) $[Fe(CN)_6]^{3-}$

The oxidation number Fe in $[Fe(CN)_6]^{3-}$ is +3. The electronic configuration of Fe^{3+} is – (Ar) $3d^5\,4s^0$

3d					4s	4p		
1	1	1	1	1				

As CN^- is a strong ligand pairing will occur.

$[Fe(CN)_6]^{3-}$

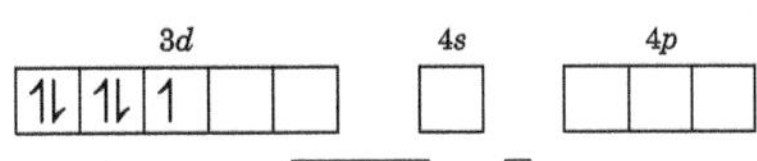

$\therefore n = 1 \Rightarrow \mu = \sqrt{1(1+2)} = \sqrt{3}$

(b) $[Fe(H_2O)_6]^{2+}$

The oxidation number of Fe in $[Fe(H_2O)_6]^{2+}$ is +2. The electronic configuration of Fe^{2+} is – [Ar] $3d^6\,4s^0$

As H_2O is weak ligand, so pairing will not occur.

$[Fe(H_2O)_6]^{2-}$

3d					4s	4p		
1L	1	1	1	1				

$\therefore n = 4$

$$\mu = \sqrt{4(4+2)} = \sqrt{24}$$

(c) $[MnF_6]^{4-}$

The oxidation number of Mn is +2. It's configuration is – [Ar] $3d^5\,4s^0$

As F^- is a weak field ligand, so pairing will not occur.

$[MnF_6]^{4-}$

3d					4s	4p		
1	1	1	1	1				

$\therefore n = 5$

$$\mu = \sqrt{5(5+2)} = \sqrt{35}$$

(d) $[NiCl_4]^{2-}$

In $NiCl_4^{2-}$, the oxidation number of Ni will be +2. It's configuration is – [Ar] $3d^8\,4s^0$

As Cl^- is a weak field ligand, so pairing will not occur.

$[NiCl_4]^{2-}$

3d					4s	4p		
1L	1L	1L	1	1				

$n = 2$

$\mu = \sqrt{2(2+2)} = \sqrt{8}$

Thus, $[MnF_6]^{4-}$ will have the highest spin only magnetic moment.

106. *(b)* For 3rd ionisation energy, the electron must be removed from + 2 oxidation state of given metals.

The electronic configuration of Ce is

$$4f^1\,5d^1\,6s^2$$

The electronic configuration of Ce^{2+} is

$$4f^1\,5d^1$$

The electronic configuration of Nd is

$$4f^4\,6s^2$$

The electronic configuration of Nd^{2+} is

$$4f^4$$

The electronic configuration of Eu is

$$4f^7\,6s^2$$

The electronic configuration of Eu^{2+} is

$$4f^7$$

The electronic configuration of Dy is

$$4f^{10}\,6s^2$$

The electronic configuration of Dy^{2+} is $4f^{10}$.

Eu will have the maximum 3rd ionisation enthalphy, because the electron has to be removed from stable half-filled $(4f^7)$ Eu^{2+} configuration, whereas Ce will have the lowest 3rd ionisation enthalpy as the electron has to be removed from 5*d*-orbital.

107. *(c)*

Step (i) and (ii) is the hyboration reaction, where diborane adds to an alkane in presence of H_2O_2. This is according to anti-Markownikoff's rule.

Step (iii) is the elimination reaction, where H_2O is eliminated and leads to formation of 2 alkanes. The major product is according to Satzeff rule.

108. *(a)* Diastereomers are optical isomers, which are not mirror images of one each other.

The product obtained from the given reactions are as follows.

(a)

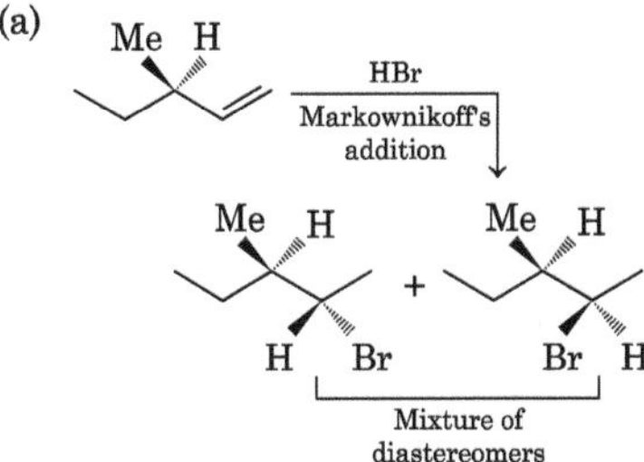

(b)

Me H $\xrightarrow{H_2/Pt}$ Me H

(c)

Me H $\xrightarrow[\text{ROOR, } h\nu]{\text{HBr}}$ Me H, Br, H

Anti-Markownikoff's addition

(d)

Me H $\xrightarrow[H_2O_2/NaOH]{B_2H_6}$ Me H, OH

Thus, reaction given in option (a) is correct.

109. *(a)* Reaction of phenol with NaOH followed by heating with CO_2 under high pressure and subsequent acidification gives salicyclic acid (*X*) as the major product, which can be purified by steam distillation. When reacted with acetic anhydride in presence of a trace amount of conc. H_2SO_4, compound *X* produces methyl salicyclate (*Y*) as the major product.

OH $\xrightarrow[\text{(ii) } H_3O^+]{\text{(i) } CO_2, \text{ NaOH}}$ Salicylic acid (*X*) $\xrightarrow[CH_3-\overset{O}{\overset{\|}{C}}-\overset{O}{\overset{\|}{C}}-CH_3]{\text{conc. } H_2SO_4}$ Methyl salicylate (*Y*)

110. *(d)* A tetra-peptide is made of naturally occuring alanine serine glycine and valine.

It is given that C-terminal amino acid is alanine and N-terminal amino acid is chiral, thus glycine can't be N-terminal amino acid as it is an achiral amino acid. Therefore possible sequence of tetra-peptide are as follows.

(i) Alanine, glycine, serine, valine

(ii) Alanine, glycine, valine, serine

(iii) Alanine, serine, glycine, valine

(iv) Alanine, valine, glycine, serine

Thus it can have 4 possible sequences of the tetrapeptide.

111. *(b)* Your father has 23 chromosomes from his father and 23 from his mother. You get one of each of the 23 pairs from your father. Since, there is a 50% chance of getting your paternal grandmother's chromosome or your paternal grandfather's chromosome for each of the 23 in the cell, the probability of getting every one of the 23 from one grandparent, is $(0.5)^{23}$ or $\left(\frac{1}{2}\right)^{23}$ or 1 in 8 million.

112. *(c)* A primer is a short nucleic acid sequence that provides a starting point for DNA synthesis.

These have opposite polarity to template DNA.

Thus, the correct option is (c) because

Template DNA -

5′ - CTAGTCGTCGAT - $(N)_{300}$- GACTGAGCTGAGCTG - 3′

3′ - GATCAGCAGCTA - $(N)_{300}$- CTGACTCGACTCGAC - 5′

Primer on Template

5′ - CTAGTCGTCGAT - $(N)_{300}$- GACTGAGCTGAGCTG - 3′

3′ - GATCAGCAGCTA - $(N)_{300}$- CTGACTCGACTCGAC - 5′

5′ - CTAGTCGTCGAT - $(N)_{300}$- GACTGAGCTGAGCTG - 3′

3′ - GATCAGCAGCTA - $(N)_{300}$- CTGACTCGACTCGAC - 5′

113. *(d)* An enzyme is a biochemical catalyst that lowers the activation energy of the reaction without affecting the substrate or product. Also, they are not consumed in the reaction. Therefore, only graph (i) and (ii) depict corrrectly the reaction with an enzyme and without an enzyme.

114. *(a)* In a double-stranded genetic material, if two bases are joined by bonds their percentage should be same.

Therefore, according to the information given in question, percentages of S and T are almost same and Q and R are same. Thus, S pairs with T and Q pairs with R.

115. *(c)* For a diploid species with ABCO, blood grouping can be represented as

Genotype	I^CI^O	I^AI^B	I^AI^O	I^BI^O	I^OI^O	I^AI^A	I^BI^B
Phenotype or Blood group	Blood group C	Blood group AB	Blood group A	Blood group B	Blood group O	Blood group A	Blood group B
Genotype	I^AI^C	I^BI^C	I^CI^C				
Phenotype or Blood group	Blood group AC	Blood group BC	Blood group C				

Thus, there are 10 genotypes but only 7 different types of blood group.

116. *(d)* Equation of exponential growth in plants is $W_t = W_0 \times e^{rt}$

$W_0 = 10\text{mm}^2$

$r = 0.015\text{mm}^2 / \text{hour}$

$t = 4\text{days} \longrightarrow 4 \times 24 = 96 \text{ hours}$

$W_t = 10 \times e^{0.015 \times 96} = 42.2$

117. *(b)* Globular proteins are folded such that their tertiary structure consists of the polar or hydrophilic amino acids (both acidic and basic) arranged on the outside surface and non-polar or hydrophobic amino acids on the inside or core of the three dimensional shape. Thus, the globular protein mentioned in the question would have red and green colour on the surface and blue colour at the core in aqueous solution.

118. *(a)* Lysosome radius is

$$0.5\mu\text{m} = 0.5 \times 10^{-6} \text{m}$$

Spherical lysosome volume is $\frac{4}{3}\pi r^3$

$\frac{4}{3}\pi\frac{1}{8} \times 10^{-18} \text{m}^3$ or $\frac{4}{3}\pi\frac{1}{8} \times 10^{-15}$ L

$[H^+] = 10^{-pH}$ mol / L

Number of moles of H^+ ions

$$= \frac{4}{3}\pi\frac{1}{8} \times 10^{-15} \times 10^{-5} \text{ mol}$$

Number of H^+ ions $= n \times NA$

n = number of moles

NA = Avogadro number

$[H^+]$ number $= \frac{4}{3}\pi\frac{1}{8} \times 10^{-20} \times 6.023 \times 10^{23}$

$[H^+]$ number = in between 10^3–10^4

119. *(c)* Vitamin-B complex are necessary for vital functions and essential for the formation of coenzyme involved in enzymatic reaction. **Example** Vitamin-B_1 also called thiamine required for TPP (Thiamine Pyrophosphate). Vitamin-B_2 also called riboflavin, required for FMN/FAD formation, vitamin-B_3 also called niacin or nicotinic acid required for the formation of NAD, NADP.

120. *(a)* By the use of O^{18} radioisotope, it was confirmed that O_2 exists from H_2O not from CO_2 during light reaction. By the use of C^{14} radioisotope.

QUESTION PAPER 2017

Stream : SB/SX (Nov 19)

MM : 160

Instructions

1. There are 120 questions in this paper.
2. Tho question paper contains two parts; Part I (1 Mark Questions) and Part II (2 Marks Questions).
3. There are four sections in each part; Mathematics, Physics, Chemistry and Biology.
4. Out of the four options given with each question, only one is correct.

PART-I (1 Mark Questions)

MATHEMATICS

1. Consider a rigid square $ABCD$ as in the figure with A and B on the X and Y-axes, respectively.

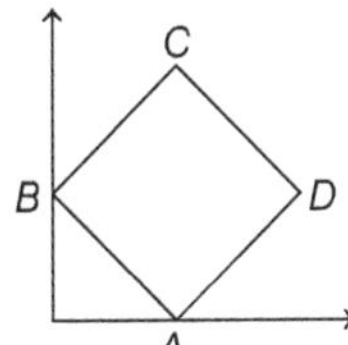

When A and B slide along their respective axes, the locus of C forms a part of

(a) a circle
(b) a parabola
(c) a hyperbola
(d) an ellipse which is not a circle

2. Among the inequalities below, which ones are true for all natural numbers n greater than 1000?

I. $n! \le n^n$ II. $(n!)^2 \le n^n$

III. $10^n \le n!$ IV. $n^n \le (2n)!$

(a) I and IV (b) I, III and IV
(c) II and IV only (d) I, II, III and IV

3. Let $S = \left\{\frac{a^2+b^2+c^2}{ab+bc+ca} : a, b, c \in R, ab+bc+ca \neq 0\right\}$, where R is the set of real numbers. Then, S equals

(a) $(-\infty, -1] \cup [1, \infty)$ (b) $(-\infty, 0) \cup (0, \infty)$
(c) $(-\infty, -1] \cup [2, \infty)$ (d) $(-\infty, -2] \cup [1, \infty)$

4. Let S be the infinite sum given by $S = \sum_{n=0}^{\infty} \frac{a_n}{10^{2n}}$ where $(a_n)_{n \ge 0}$ is a sequence defined by $a_0 = a_1 = 1$ and $a_j = 20a_{j-1} - 108a_{j-2}$ for $j \ge 2$. If S is expressed in the form $\frac{a}{b}$, where a, b are coprime positive integers, then a equals

(a) 2017 (b) 2020 (c) 2023 (d) 2025

5. Define a function $f(x) = \frac{16x^2 - 96x + 153}{x-3}$ for all real $x \neq 3$. The least positive value of $f(x)$ is

(a) 16 (b) 18 (c) 22 (d) 24

6. Let $n > 2$ be an integer and define a polynomial

$p(x) = x^n + a_{n-1}x^{n-1} + \dots + a_1 x + a_0$

where $a_0, a_1, \dots, a_{n-1}$ are integers. Suppose we know that $np(x) = (1 + x)p'(x)$. If $b = p(1)$, then

(a) b is divisible by 10 (b) b is divisible by 3
(c) b is a power of 2 (d) b is a power of 5

7. The number of 5-tuples (a, b, c, d, e) of positive integers such that

I. a, b, c, d, e are the measures of angles of a convex pentagon in degrees

II. $a \le b \le c \le d \le e$

III. a, b, c, d, e are in arithmetic progression is

(a) 35 (b) 36 (c) 37 (d) 126

8. Thirty two persons $X_1, X_2, \dots, X_{32}$ are randomly seated around a circular table at equal intervals. Two persons X_i and X_j are said to be within earshot of each other if there are at most three persons between them on the minor arc joining X_i and X_j. The probability that X_1 and X_2 are within earshot of each other is, $\left(\text{Here, } {}^nC_r = \frac{n!}{(n-r)!r!}\right)$

(a) $\frac{\binom{32}{2}30!}{8(32!)}$ (b) $\frac{\binom{32}{2}30!}{4(32!)}$ (c) $\frac{8}{31}$ (d) $\frac{4}{31}$

9. Let n be the smallest positive integer such that $1 + \frac{1}{2} + \frac{1}{3} + \dots + \frac{1}{n} \ge 4$. Which one of the following statements is true?

(a) $20 < n \le 60$ (b) $60 < n \le 80$
(c) $80 < n \le 100$ (d) $100 < n \le 120$

10. A pair of 12-sided fair dice with faces numbered 1, 2, 3, ..., 12 is rolled. The probability that the sum of the numbers appearing has remainder 2 when divided by 9 is

(a) $\frac{7}{72}$ (b) $\frac{5}{48}$ (c) $\frac{11}{144}$ (d) $\frac{1}{9}$

11. Let $x_1, x_2, \dots, x_6$ be the roots of the polynomial equation $x^6 + 2x^5 + 4x^4 + 8x^3 + 16x^2 + 32x + 64 = 0$.

Then,

(a) $|x_i| = 2$ for exactly one value of i
(b) $|x_i| = 2$ for exactly two values of i
(c) $|x_i| = 2$ for all values of i
(d) $|x_i| = 2$ for no value of i

12. In the complex plane, let $z_1 = \sqrt{3} + i$ and $z_2 = \sqrt{3} - i$ be two adjacent vertices of an n-sided regular polygon centered at the origin. Then, n equals

(a) 4 (b) 6 (c) 8 (d) 12

13. Let $A^{-1} = \begin{bmatrix} 1 & 2017 & 2 \\ 1 & 2017 & 4 \\ 1 & 2018 & 8 \end{bmatrix}$. Then, $|2A| - |2A^{-1}|$ is equal to

(a) 3 (b) −3 (c) 12 (d) −12

14. An ellipse with its minor and major axis parallel to the coordinate axes passes through (0, 0), (1, 0) and (0, 2). One of its foci lies on the Y-axis. The eccentricity of the ellipse is

(a) $\sqrt{3} - 1$ (b) $\sqrt{5} - 2$ (c) $\sqrt{2} - 1$ (d) $\frac{\sqrt{3} - 1}{2}$

15. Let $I_n = \int_0^1 e^{-y} y^n \, dy$, where n is a non-negative integer. Then, $\sum_{n=1}^{\infty} \frac{I_n}{n!}$ is

(a) 1 (b) $1 - \frac{1}{e}$ (c) $\frac{1}{e}$ (d) $1 + \frac{1}{e}$

16. The number of solutions of the equation $\sin\theta + \cos\theta = \sin 2\theta$ in the interval $[-\pi, \pi]$ is

(a) 1 (b) 2 (c) 3 (d) 4

17. Let $z_1, z_2, \dots, z_7$ be the vertices of a regular heptagon that is inscribed in the unit circle with centre at the origin in the complex plane. Let $w = \sum_{1 \le i < j \le 7} z_i z_j$, then $|w|$ is equal to

(a) 0 (b) 1 (c) 2 (d) 3

18. The sound of a cannon firing is heard one second later at a position B that at position A. If the speed of sound is uniform, then

(a) the positions A and B are foci of a hyperbola, with cannon's position on one branch of the hyperbola
(b) the position A and B are foci of an ellipse with cannon's position on the ellipse
(c) one of the positions A, B is focus of a parabola with cannon's position on the parabola
(d) it is not possible to describe the positions of A, B and the cannon with the given information

19. A spherical ball is kept at the corner of a rectangular room such that the ball touches two (perpendicular) walls and lies on the floor. If a point on the sphere is at distance of 9, 16, 25 from the two walls and the floor, then a possible radius of the sphere is

(a) 13 (b) 15 (c) 26 (d) 36

20. Let m, n be two distinct integers chosen randomly from the set {0, 1, 2, ..., 99}. Then, the probability that $4^m + 4^n + 3$ is divisible by 5 lies in the interval

(a) (0, 0.25] (b) (0.25, 0.5] (c) (0.5, 0.75] (d) (0.75, 1)

PHYSICS

21. The distance s travelled by a particle in time t is $s = ut - \frac{1}{2}gt^2$. The initial velocity of the particle was measured to be $u = 1.11 \pm 0.01$ m/s and the time interval of the experiment was $t = 1.01 \pm 0.1$ s. The acceleration was taken to be $g = 9.8 \pm 0.1$ m/s^2. With these measurements, the student estimates the total distance travelled. How should the student report the result?

(a) 1121 ± 0.1 m (b) 1.1 ± 0.1 m
(c) 1.12 ± 0.07 m (d) 1.1 ± 0.07 m

22. A massive black hole of mass m and radius R is spinning with angular velocity ω. The power P radiated by it as gravitational waves is given by $P = Gc^{-5}m^x R^y \omega^z$, where c and G are speed of light in free space and the universal gravitational constant, respectively. Then,

(a) $x = -1, y = 2, z = 4$ (b) $x = 1, y = 1, z = 4$
(c) $x = -1, y = 4, z = 4$ (d) $x = 2, y = 4, z = 6$

23. Consider the following statements for air molecules in an air tight container :

(I) The average speed of molecules is larger than root mean square speed.
(II) Mean free path of molecules is larger than the mean distance between molecules.
(III) Mean free path of molecules increases with temperature.
(IV) The rms speed of nitrogen is smaller than oxygen molecule.

Which of the above statements are correct?

(a) Only statement II is correct
(b) Statements II and III are correct
(c) Statements II and IV are correct
(d) Statements I, II and IV are correct

24. Three circularly shaped linear polarisers are placed coaxially. The transmission axis of the first polariser is at 30°, the second one is at 60° and the third at 90° to the vertical all in the clockwise sense. Each polariser additionally absorbs 10% of the light. If a vertically polarised beam of light of intensity $I = 100\ \text{W/m}^2$ is incident on this assembly of polarisers, then the final intensity of the transmitted light will be close to

(a) $10\ \text{W/m}^2$ (b) $20\ \text{W/m}^2$
(c) $30\ \text{W/m}^2$ (d) $50\ \text{W/m}^2$

25. One end of a rod of length L is fixed to a point on the circumference of a wheel of radius R. The other end is sliding freely along a straight channel passing through the centre O of the wheel as shown in the figure below. The wheel is rotating with a constant angular velocity ω about O. Taking $T = \frac{2\pi}{\omega}$, the motion of the rod is

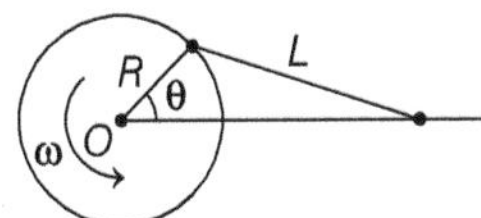

(a) simple harmonic with a period of T
(b) simple harmonic with a period of $T/2$
(c) not simple harmonic but periodic with a period of T
(d) not simple harmonic but periodic with a period of $T/2$

26. A rope of mass 5 kg is hanging between two supports as shown alongside. The tension at the lowest point of the rope is close to (take, $g = 10\ \text{m/s}^2$)

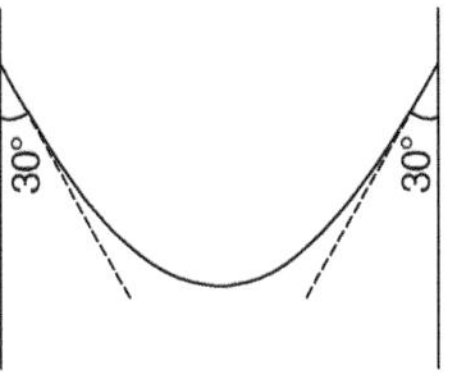

(a) 22 N (b) 44 N (c) 28 N (d) 14 N

27. A uniform rope of total length l is at rest on a table with fraction f of its length hanging (see figure). If the coefficient of friction between the table and the chain is μ, then

(a) $f = \mu$ (b) $f = 1/(1+\mu)$
(c) $f = 1/(1+1/\mu)$ (d) $f = 1/(\mu + 1/\mu)$

28. A light beam travelling along the X-axis with planar wavefront is incident on a medium of thickness t. In the region, where light is falling the refractive index can be taken to be varying such that $(dn/dy) > 0$. The light beam on the other side of the medium will emerge

(a) parallel to the X-axis
(b) bending downward
(c) bending upward
(d) split into two or more beams

29. Let the electrostatic field E at distance r from a point charge q not be an inverse square but instead an inverse cubic, e.g. $\mathbf{E} = k\frac{q}{r^3}\hat{\mathbf{r}}$, here k is a constant.

Consider the following two statements:

(I) Flux through a spherical surface enclosing the charge is $\phi = q_{\text{enclosed}}/\varepsilon_0$.
(II) A charge placed inside uniformly charged shell will experience a force.

Which of the above statements are valid?

(a) Only statement I is valid
(b) Only statement II is valid
(c) Both statements I and II are invalid
(d) Both statements I and II are valid

30. A star of mass M and radius R is made up of gases. The average gravitational pressure compressing the star due to gravitational pull of the gases making up the star depends on R as

(a) $1/R^4$ (b) $1/R$ (c) $1/R^2$ (d) $1/R^6$

31. The black shapes in the figure below are closed surfaces. The electric field lines are in red. For which case, the net flux through the surfaces is non-zero?

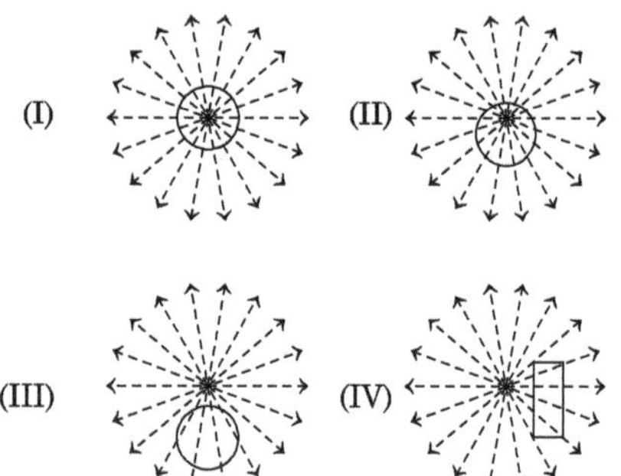

(a) In all cases net flux is non-zero
(b) For III and IV cases
(c) For I and II cases
(d) For II, III and IV cases

32. A particle of charge q and mass m enters a region of a transverse electric field of $E_0\hat{\mathbf{j}}$ with initial velocity $v_0\hat{\mathbf{i}}$. The time taken for the change in the de-Broglie wavelength of the charge from the initial value of λ_0 to $\lambda_0/3$ is proportional to

(a) $\frac{q}{m}$ (b) $\frac{m}{q}$ (c) $\sqrt{\frac{q}{m}}$ (d) $\sqrt{\frac{m}{q}}$

33. Consider the following nuclear reactions:

I. ${}^{14}_{7}N + {}^{4}_{2}He \longrightarrow {}^{17}_{8}O + X$

II. ${}^{9}_{4}Be + {}^{4}_{2}H \longrightarrow {}^{12}_{6}He + Y$

Then,

(a) X and Y are both protons
(b) X and Y are both neutrons
(c) X is a proton and Y is a neutron
(d) X is neutron and Y is a proton

34. Consider a plane parallel beam of light incident on a plano-cylindrical lens as shown below. Which of the following will you observe on a screen placed at the focal plane of the lens?

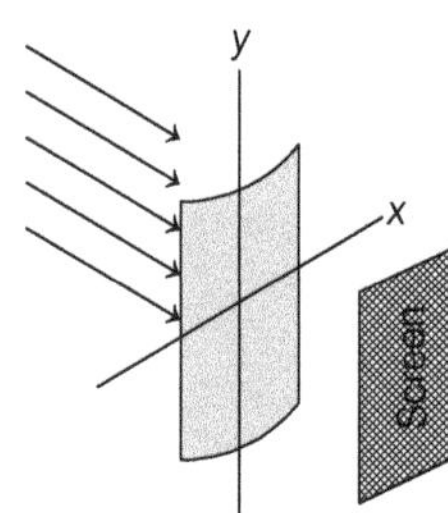

(a) The screen will be uniformly illuminated
(b) There will be a single bright spot on the screen
(c) There will be a single bright line on the screen parallel to the X-axis
(d) There will be a single bright line on the screen parallel to the Y-axis

35. The n-side of the depletion layer of a p-n junction

(a) always has same width as of the p-side
(b) has no bound charges
(c) is negatively charged
(d) is positively charged

36. A small ring is rolling without slipping on the circumference of a large bowl as shown below in the figure. The ring is moving down at P_1, comes down to the lower most point P_2 and is climbing up at P_3. Let $\mathbf{v}_{CM}$ denote the velocity of the centre of mass of the ring. Choose the correct statement regarding the frictional force on the ring.

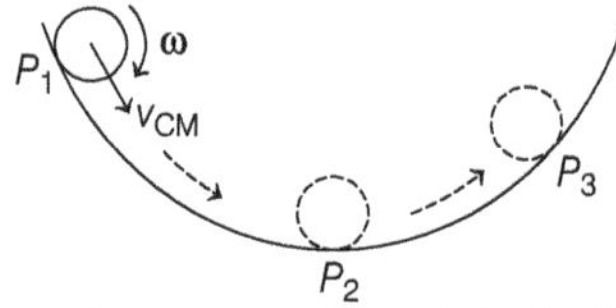

(a) It is opposite to $\mathbf{v}_{CM}$ at the points P_1, P_2 and P_3
(b) It is opposite to $\mathbf{v}_{CM}$ at P_1 and in the same direction as $\mathbf{v}_{CM}$ at P_3
(c) It is in the same direction as $\mathbf{v}_{CM}$ at P_1 and opposite to $\mathbf{v}_{CM}$ at P_3
(d) It is zero at the points P_1, P_2 and P_3

37. A bomb explodes at time $t = 0$ in a uniform, isotropic medium of density ρ and releases energy E, generating a spherical blast wave. The radius R of this blast wave varies with time t as

(a) t (b) $t^{2/5}$ (c) $t^{1/4}$ (d) $t^{3/2}$

38. A closed pipe of length 300 cm contains some sand. A speaker is connected at one of its ends. The frequency of the speaker at which the sand will arrange itself in 20 equidistant piles is close to (velocity of sound is 300 m/s)

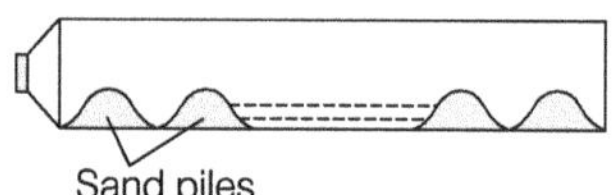

(a) 10 kHz (b) 5 kHz (c) 1 kHz (d) 100 kHz

39. A planet of radius R_p is revolving around a star of radius R^*, which is at temperature T^*. The distance between the star and the planet is d. If the planet's temperature is $f\,T^*$, then f is proportional to

(a) $\sqrt{R^*/d}$ (b) R^*/d (c) $R^* R_p/d^2$ (d) $(R^*/d)^4$

40. Some of the wavelengths observed in the emission spectrum of neutral hydrogen gas are 912, 1026, 1216, 3646, 6563 Å. If broad band light is passing through neutral hydrogen gas at room temperature, then the wavelength that will not be absorbed strongly is

(a) 1026 Å (b) 1216 Å
(c) 912 Å (d) 3646 Å

CHEMISTRY

41. The major product formed in the following reaction is

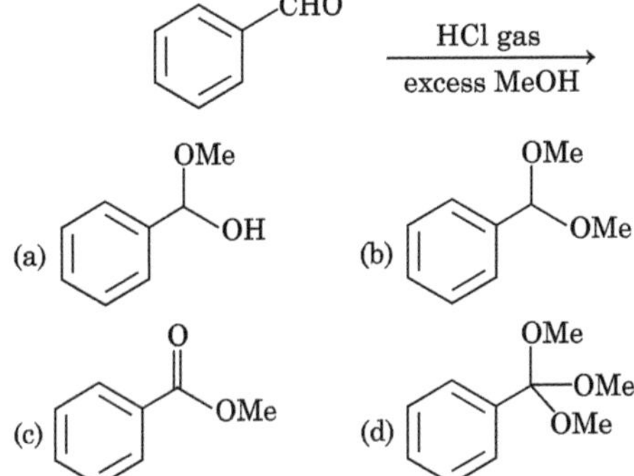

42. Which among the following is a non-benzenoid aromatic compound?
(a) *o*-xylene (b) Phenanthrene
(c) Indole (d) Thiophene

43. Natural rubber is a polymer of
(a) neoprene (b) chloroprene
(c) isoprene (d) styrene

44. The following tripeptide

can be represented as
(a) tyr — val — thr (b) phe — ala — ser
(c) phe — leu — cys (d) lys — ala — ser

45. The sugar units present in natural DNA and RNA, respectively are
(a) D-2-deoxyribose and L-ribose
(b) L-2-deoxyribose and D-ribose
(c) D-2-deoxyribose and D-ribose
(d) L-2-deoxyribose and L-ribose

46. The major product formed in the following reaction is, $CH_3Br + CH_3CH_2ONa \longrightarrow$
(a) $CH_3CH_2CH_2OH$ (b) CH_3OCH_3
(c) $CH_3CH_2OCH_3$ (d) $CH_3CH_2OCH_2Br$

47. The most abundant metal ion present in the human body is
(a) Zn^{2+} (b) Ca^{2+} (c) Na^+ (d) Fe^{2+}

48. Phosphorus reacts with chlorine gas to give a colourless liquid, which fumes in moist air to produce HCl and
(a) $POCl_3$ (b) H_3PO_3 (c) PH_3 (d) H_3PO_4

49. The oxidising ability of the given anions follows the order
(a) $TiO_4^{4-} < VO_4^{3-} < CrO_4^{2-} < MnO_4^-$
(b) $VO_4^{3-} < CrO_4^{2-} < MnO_4^- < TiO_4^{4-}$
(c) $CrO_4^{2-} < MnO_4^- < VO_4^{3-} < TiO_4^{4-}$
(d) $VO_4^{3-} < TiO_4^{4-} < CrO_4^{2-} < MnO_4^-$

50. The complete hydrolysis of XeF_6 results in the formation of
(a) XeO_2F_2 (b) $XeOF_4$ (c) XeO_3 (d) XeO_2

51. The reactivity of the following compounds toward water is in the order
(a) $Cl_2O_7 < P_2O_5 < B_2O_3$
(b) $B_2O_3 < P_2O_5 < Cl_2O_7$
(c) $P_2O_5 < B_2O_3 < Cl_2O_7$
(d) $B_2O_3 < Cl_2O_7 < P_2O_5$

52. Among the following complexes, the one that can exist as facial (*fac*) and meridional (*mer*) isomers is
(a) $[Co(NO_2)_3(NH_3)_3]$ (b) $K_3[Fe(CN)_6]$
(c) $[Co(H_2O)_2(NH_3)_4]Cl_3$ (d) $[CoCl(NH_3)_5]Cl_2$

53. An excess of $Ag_2CrO_4(s)$ is added to a 5×10^{-3} M K_2CrO_4 solution. The concentration of Ag^+ in the solution is closest to
[Solubility product for $Ag_2CrO_4 = 1.1 \times 10^{-12}$]
(a) 2.2×10^{-10} M (b) 1.5×10^{-5} M
(c) 1.0×10^{-6} M (d) 5.0×10^{-3} M

54. The packing efficiency in a body centered cubic (bcc) structure is closest to
(a) 74% (b) 63% (c) 68% (d) 52%

55. The consecutive reaction $X \longrightarrow Y \longrightarrow Z$ takes place in a closed container. Initially, the container has A_0 moles of X (and no Y and Z). The plot of total moles of the constituents in the container as a function of time will be

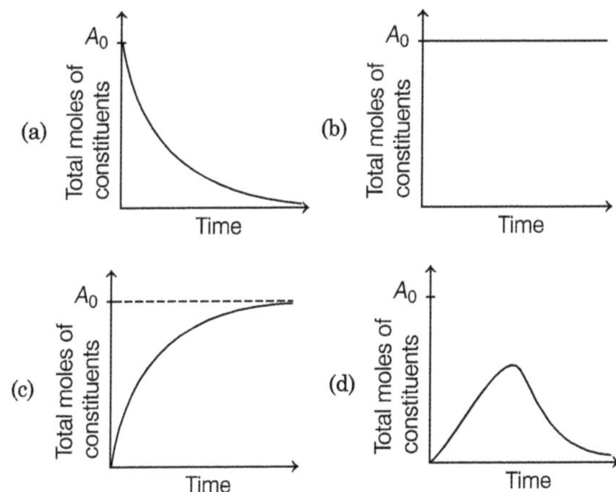

56. The particles emitted during the sequential radioactive decay of $^{238}U_{92}$ to $^{206}Pb_{82}$ are
(a) 5 α and 6 β (b) 6 α and 8 β
(c) 8 α and 4 β (d) 8 α and 6 β

57. The allowed set of quantum numbers for an electron in a hydrogen atom is
(a) $n = 4, l = 2, m_l = 0, m_s = 0$
(b) $n = 3, l = 1, m_l = -3, m_s = -1/2$
(c) $n = 3, l = 3, m_l = -1, m_s = 1/2$
(d) $n = 2, l = 1, m_l = -1, m_s = 1/2$

58. The plot that best represents the relationship between the extent of adsorption (x/m) and pressure (p) is

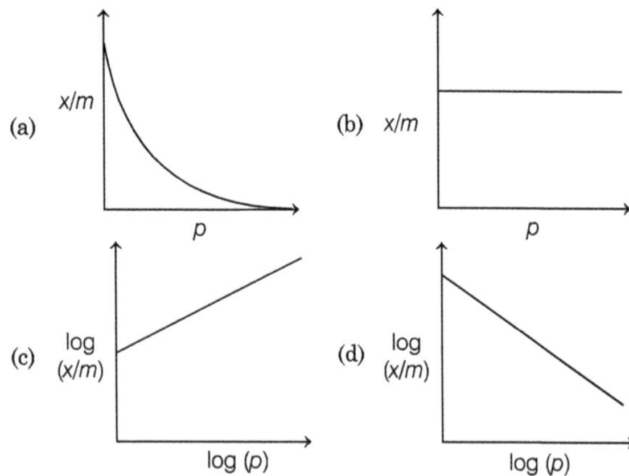

59. The pH of 0.1 M acetic acid solution is closest to [Dissociation constant of the acid, $K_a = 1.8 \times 10^{-5}$]
(a) 2.87 (b) 1.00 (c) 2.07 (d) 4.76

60. The limiting molar conductivities of the given electrolytes at 298 K follow the order
$\lambda^{\circ}_{(K^+)} = 73.5$, $\lambda^{\circ}_{(Cl^-)} = 76.3$,
$\lambda^{\circ}_{(Ca^{2+})} = 119.0$,
$\lambda^{\circ}_{(SO_4^{2-})} = 160.0 \, S \, cm^2 \, mol^{-1}$
(a) $KCl < CaCl_2 < K_2SO_4$ (b) $KCl < K_2SO_4 < CaCl_2$
(c) $K_2SO_4 < CaCl_2 < KCl$ (d) $CaCl_2 < K_2SO_4 < KCl$

BIOLOGY

61. Resting membrane potential of a neuron is approximately
(a) −70 mV (b) + 70 mV
(c) −0.7 V (d) + 0.7 V

62. Amphimixis is
(a) a fusion of pronuclei of male gametes
(b) a fusion of pronuclei from male and female gametes
(c) a fusion of pronuclei of female gametes
(d) the development of a somatic cell into an embryo

63. Activation of sympathetic nervous system
(a) decreases blood pressure
(b) causes pupil contraction
(c) increases heart rate
(d) causes bronchoconstriction

64. At physiological temperature, sterols in biological membranes
(a) increase their fluidity
(b) decrease their fluidity
(c) increase their permeability to water
(d) decrease their permeability to water

65. Which one of the following is a heteropolysaccharide?
(a) Glycogen (b) Starch
(c) Cellulose (d) Hyaluronic acid

66. Bacterial plasmids are genetic entities that
(a) are non-transferable to the same bacterial species
(b) are capable of independent replication
(c) have RNA as genetic material
(d) always require integration in the genome for their replication

67. Skin-prick test on the forearm is conducted to identify the responsible allergen. This is because
(a) of the presence of mast cells under the skin
(b) lymphocytes migrate rapidly from the blood to the skin
(c) hair follicles can enhance the reaction
(d) neutrophils migrate rapidly from the blood to the skin

68. Which one of the following processes in *E. coli* does not directly involve RNA?
(a) DNA replication (b) Transcription
(c) Translation (d) DNA repair

69. Which one of the following statements is incorrect for translation in cytoplasm?
(a) One codon codes for only one amino acid
(b) One amino acid may be coded by many codons
(c) More than one amino acids are coded by one specific codon
(d) There are some codons that do not code for any amino acid

70. Two homozygous parents harbouring two different alleles of a gene, exhibiting incomplete dominance for flower colour were used for a genetic experiment. Which one of the following statements is incorrect?
(a) The F_2-generation will consist of plants of three different flower colours
(b) The genotypic and phenotypic ratios obtained in the F_2-generation will be different
(c) The F_1-generation will be of a different flower colour compared to both the parents
(d) The genotypic ratio obtained in the F_2-generation will be the same irrespective of whether it is complete dominance or incomplete dominance

71. Which one of the following is an essential condition for a population to be at Hardy-Weinberg equilibrium?
(a) Random mating
(b) Immigration
(c) Emigration
(d) Geographical isolation

72. Inbreeding in a population leads to
(a) decrease in recessive disorders
(b) heterosis
(c) increase in homozygosity
(d) increase in heterozygosity

73. Which one of the following molecules serves as a substrate for direct synthesis of ATP?
(a) 1, 3-bisphosphoglycerate (b) Glucose-6-phosphate
(c) Pyruvate (d) Fructose-1,6-bisphosphate

74. If a pure chlorophyll solution is illuminated with ultraviolet light, the solution appears
(a) green (b) violet (c) red (d) black

75. Botanical names of plants are given in Column I and the family/order name in Column II. Choose the appropriate combination from the options below.

	Column I		Column II
A.	*Tamarindus indica*	1.	Arecaceae
B.	*Cocos nucifera*	2.	Liliaceae
C.	*Colchicum autumnale*	3.	Solanaceae
D.	*Withania somnifera*	4.	Papilionaceae

	A	B	C	D		A	B	C	D
(a)	4	1	2	3	(b)	4	2	3	1
(c)	1	2	4	3	(d)	4	1	3	2

76. Nitrogen-fixation is inhibited by oxygen. However, in aerobic nitrogen-fixing bacteria, nitrogen is fixed in the presence of oxygen. Nitrogenase in such organisms is protected by which one of the following mechanisms?

(a) Channelising oxygen to form ozone
(b) Removal of oxygen by metabolic activity
(c) Utilising oxygen for membrane remodelling
(d) Utilising oxygen for synthesis of pentapeptide chain in peptidoglycan

77. Frederick Griffith performed an experiment where mice were killed when injected with a mixture of Heat-Killed S-type *Streptococcus* (HKS) and Live R-type *Streptococcus* (LRS) but not with HKS or LRS separately. Mice were killed because

(a) lipids from HKS made LRS virulent
(b) RNA from HKS transformed LRS and made it virulent
(c) proteins from HKS made LRS virulent
(d) DNA from HKS transformed LRS and made it virulent

78. In diabetic patients, the pH of blood plasma can decrease leading to acidosis. This is because tissues catabolise

(a) amino acids leading to loss of buffering capacity of the blood
(b) stored glycogen leading to the accumulation of pyruvic acid
(c) stored fatty acids leading to the accumulation of beta hydroxybutyric acid and acetoacetic acid
(d) nucleic acid pool leading to decrease in blood pH

79. If the number of alveoli in an individual is doubled without changing the total alveolar volume, the gas exchange capacity of the lungs will

(a) increase for both O_2 and CO_2
(b) decrease for both O_2 and CO_2
(c) remain unaltered for both O_2 and CO_2
(d) increase for O_2 and decrease for CO_2

80. In an experiment, bacteria were infected with ^{32}P labelled virus in a ratio of 5 : 1. The culture was rigorously shaken followed by centrifugation. Radioactivity was

(a) lost due to metabolic activity
(b) detected in supernatant as inorganic phosphate
(c) detected in the supernatant in association with viral capsid
(d) detected in bacterial cell pellet

PART-II (2 Marks Questions)

MATHEMATICS

81. Let AB be the latusrectum of the parabola $y^2 = 4ax$ in the XY-plane. Let T be the region bounded by the finite arc AB of the parabola and the line segment AB. A rectangle $PQRS$ of maximum possible area is inscribed in T with P, Q on line AB, and R, S on arc AB. Then, area $(PQRS)$/area (T) equals

(a) $\frac{1}{2}$ (b) $\frac{1}{3}$ (c) $\frac{1}{\sqrt{2}}$ (d) $\frac{1}{\sqrt{3}}$

82. Let S be the set of all permutations $a_1, a_2, \ldots, a_6$ of 1, 2, ..., 6 such that $a_1, a_2, \ldots, a_k$ is not a permutation of 1, 2, ..., k for any k, $1 < k < 5$. Then, the number of elements in A is

(a) 192 (b) 408 (c) 312 (d) 528

83. The area bounded by the curves $y = \frac{1}{4}|4 - x^2|$ and $y = 7 - |x|$ is

(a) 18 (b) 32 (c) 36 (d) 64

84. An ellipse $\frac{x^2}{a^2} + \frac{y^2}{b^2} = 1$, $a > b$ and the parabola $x^2 = 4(y + b)$ are such that the two foci of the ellipse and the end points of the latusrectum of parabola are the vertices of a square. The eccentricity of the ellipse is

(a) $\frac{1}{\sqrt{13}}$ (b) $\frac{2}{\sqrt{13}}$ (c) $\frac{1}{\sqrt{11}}$ (d) $\frac{2}{\sqrt{11}}$

85. A sector is removed from a metallic disc and the remaining region is bent into the shape of a circular conical funnel with volume $2\sqrt{3}\pi$. The least possible diameter of the disc is

(a) 4 (b) 6 (c) 8 (d) 12

86. Let $g(x) = \int_0^{|x|^{3/4}} t^{2/3} \sin\frac{1}{t}\, dt$, for all real x. Then, $\lim_{x \to 0} \frac{g(x)}{x}$ is equal to

(a) ∞ (b) $-\infty$ (c) 0 (d) $\frac{3}{4}$

87. Let $a_n = \int_{-\pi}^{\pi} |x-1| \cos nx \, dx$ for all natural numbers n. Then, the sequence $(a_n)_{n \geq 0}$ satisfies

(a) $\lim_{n \to \infty} a_n = \infty$

(b) $\lim_{n \to \infty} a_n = -\infty$

(c) $\lim_{n \to \infty} a_n$ exists and is positive

(d) $\lim_{n \to \infty} a_n = 0$

88. Let $f(x)$ be a polynomial with integer coefficients satisfying $f(1) = 5$ and $f(2) = 7$. The smallest possible positive value of $f(12)$ is

(a) 5 (b) 7 (c) 27 (d) 15

89. Suppose four balls labelled 1, 2, 3, 4 are randomly placed in boxes B_1, B_2, B_3, B_4. The probability that exactly one box is empty is

(a) $\frac{8}{256}$ (b) $\frac{9}{16}$ (c) $\frac{27}{256}$ (d) $\frac{9}{64}$

90. Let $f(x) = \log(1 + x^2)$ and A be a constant such that $\frac{|f(x) - f(y)|}{|x - y|} \leq A$ for all x, y real and $x \neq y$. Then, the least possible value of A is

(a) equal to 1

(b) greater than 1 but less than 2

(c) greater than 0 but less than 1

(d) greater than 2

PHYSICS

91. One mole of an ideal monoatomic gas undergoes the following four reversible processes:

Step 1 It is first compressed adiabatically from volume 8.0 m^3 to 1.0 m^3.

Step 2 Then expanded isothermally at temperature T_1 to volume 10.0 m^3.

Step 3 Then expanded adiabatically to volume 80.0 m^3.

Step 4 Then compressed isothermally at temperature T_2 to volume 8.0 m^3.

Then, T_1/T_2 is

(a) 2 (b) 4 (c) 6 (d) 8

92. A solid cube of wood of side $2a$ and mass M is resting on a horizontal surface as shown below.

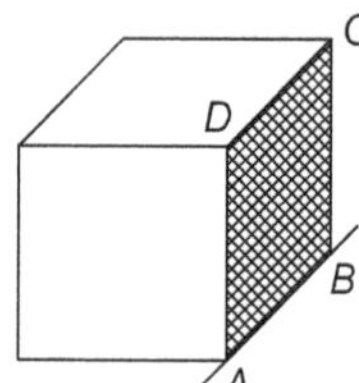

The cube is free to rotate about the fixed axis AB. A bullet of mass $m(<< M)$ and speed v is shot horizontally at the face opposite to $ABCD$ at a height h above the surface to impart the cube an angular speed ω_c, so that the cube just topples over. Then, ω_c is (**Note** the moment of inertia of the cube about an axis perpendicular to the face and passing through the centre of mass is $2Ma^3/3$)

(a) $\sqrt{3gM/2ma}$ (b) $\sqrt{3g/4h}$

(c) $\sqrt{3g(\sqrt{2}-1)/2a}$ (d) $\sqrt{3g(\sqrt{2}-1)/4a}$

93. A uniform thin wooden plank AB of length L and mass M is kept on a table with its B end slightly outside the edge of the table. When an impulse J is given to the end B, the plank moves up with centre of mass rising a distance h from the surface of the table. Then,

(a) $h > 9J^2/8M^2g$

(b) $h = J^2/2M^2g$

(c) $J^2/2M^2g < h < 9J^2/8M^2g$

(d) $h < J^2/2M^2g$

94. A square shaped wire loop of mass m, resistance R and side a moving speed v_0, parallel to the X-axis, enters a region of uniform magnetic field B, which is perpendicular to the plane of the loop. The speed of the loop changes with distance $x\,(x < a)$ in the field, as

(a) $v_0 - \frac{B^2a^2}{Rm}x$ (b) $v_0 - \frac{B^2a^2}{2Rm}x$

(c) $v_0 - \frac{B^2a}{Rm}x^2$ (d) v_0

95. The emission series of hydrogen atom is given by

$$\frac{1}{\lambda} = R\left(\frac{1}{n_1^2} - \frac{1}{n_2^2}\right)$$

where, R is the Rydberg constant. For a transition from n_2 to n_1, the relative change $\Delta\lambda/\lambda$ in the emission wavelength, if hydrogen is replaced by deuterium (assume that, the mass of proton and neutron are the same and approximately 2000 times larger than that of electrons) is

(a) 0.025% (b) 0.005%

(c) 0.0025% (d) 0.05%

96. When light shines on a p-n junction diode, the current I *versus* voltage V is observed as in figure below.

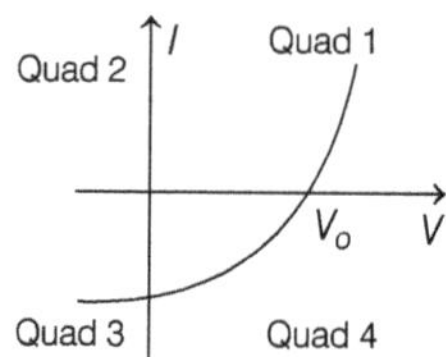

In which quadrant(s) does the diode generate power, so that it can be used as a solar cell?

(a) Quad 1 only (b) Quad 1 and 3 only

(c) Quad 4 only (d) Quad 1 and 4 only

97. Four identical beakers contain same amount of water as shown below.

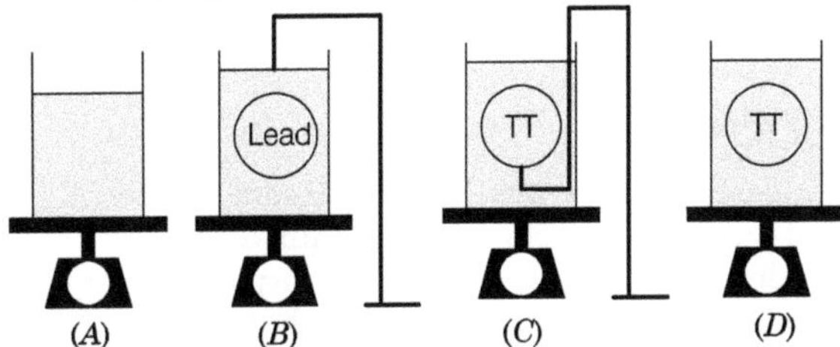

Beaker A contains only water. A lead ball is held submerged in the beaker B by string from above. A same sized plastic ball, say a table tennis (TT) ball, is held submerged in beaker C by a string attached to a stand from outside. Beaker D contains same sized TT ball which is held submerged from a string attached to the bottom of the beaker. These beakers (without stand) are placed on weighing pans and register readings w_A, w_B, w_C and w_D for A, B, C and D, respectively. Effects of the mass and volume of the stand and string are to be neglected.

(a) $w_A = w_B = w_C = w_D$
(b) $w_B = w_C > w_D > w_A$
(c) $w_B = w_C > w_A > w_D$
(d) $w_B > w_C > w_D > w_A$

98. Back surface of a glass (refractive index n and thickness t) is polished to work as a mirror as shown below. A laser beam falls on it and is partially reflected and refracted at the air-glass interface and fully reflected at the mirror surface, respectively. A pattern of discrete spots of light is observed on the screen.

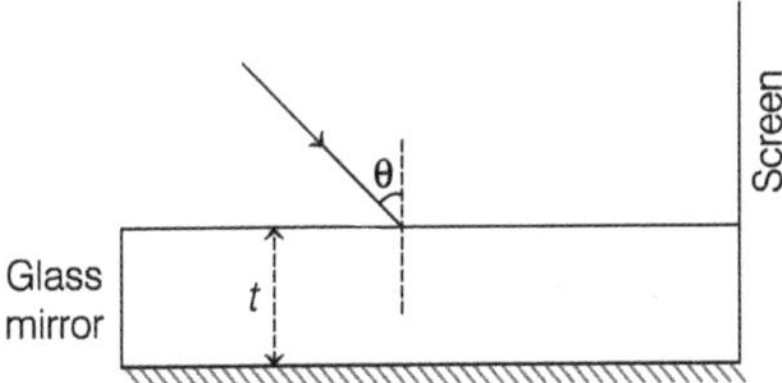

The spacing between the spots on the screen will be

(a) $\dfrac{2t\cos\theta}{\sqrt{n^2-\sin^2\theta}}$
(b) $\dfrac{2t\sin\theta}{\sqrt{n^2-\sin^2\theta}}$
(c) $\dfrac{2t\tan\theta}{\sqrt{n^2-\sin^2\theta}}$
(d) $\dfrac{2t\sin\theta}{\sqrt{1-\dfrac{\sin^2\theta}{n^2}}}$

99. Consider the following statements regarding the photoelectric effect experiment:

(I) Photoelectrons are emitted as soon as the metal is exposed to light.

(II) There is a minimum frequency below which no photo-current is observed.

(III) The stopping potential is proportional to the frequency of light.

(IV) The photo-current varies linearly with the intensity of the light.

Which of the above statements indicate that light consists of quanta (photons) with energy proportional to frequency?

(a) Statements I and III are correct
(b) Statements II and III are correct
(c) Statements II, III and IV are correct
(d) Statements I, II and III are correct

100. Consider the L-C-R circuit given below. The circuit is driven by a 50 Hz, AC source with peak voltage 220 V. If $R = 400\ \Omega$, $C = 200\ \mu F$ and $L = 6$ H, the maximum current in the circuit is closest to

(a) 0.12 A (b) 0.55 A (c) 1.2A (d) 5.5 A

CHEMISTRY

101. In the reaction,

O, Cl —(i) X, (ii) Y→ CO_2H

X and Y are

(a) $X = H_2$, $Pd/BaSO_4$; $Y =$ NaOAc, Ac_2O
(b) $X = LiAlH_4$; $Y =$ NaOAc, Ac_2O
(c) $X = H_2$, Pd/C; $Y =$ NaOH, Ac_2O
(d) $X = LiAlH_4$; $Y =$ NaOH, Ac_2O

102. In the following reaction,

CN —(i) $SnCl_2/HCl$, (ii) H_3O^+→ X —Dil. NaOH, H_3C–C_6H_4–CO–CH_3→ Y

X and Y are

(a) $X =$ Cl $Y =$ O, H_3C
(b) $X =$ Cl $Y =$ O, CH_3
(c) $X =$ CHO $Y =$ O, H_3C
(d) $X =$ CHO $Y =$ O, CH_3

103. Acetophenone ($PhCOCH_3$) reacts with perbenzoic acid to produce a compound X. Reaction of X with excess CH_3MgBr followed by treatment with aqueous acid predominantly produces

(a) $Ph-C(=O)-OH$ (b) $Ph-C(=O)-OCH_3$

(c) $(H_3C)_2C(CH_3)-OH$ (d) $(H_3C)(Ph)C(CH_3)-OH$

104. The fusion of chromite ore ($FeCr_2O_4$) with Na_2CO_3 in air gives a yellow solution upon addition of water. Subsequent treatment with H_2SO_4 produces an orange solution. The yellow and orange colours, respectively, are due to the formation of

(a) Na_2CrO_4 and $Na_2Cr_2O_7$
(b) $Cr(OH)_3$ and $Na_2Cr_2O_7$
(c) $Cr_2(CO_3)_3$ and $Fe_2(SO_4)_3$
(d) $Cr(OH)_3$ and Na_2CrO_4

105. Hybridisation and geometry of $[Ni(CN)_4]^{2-}$ are

(a) sp^2d and tetrahedral
(b) sd^3 and square planar
(c) sp^3 and tetrahedral
(d) dsp^2 and square planar

106. The total number of geometrical isomers possible for an octahedral complex of the type $[MA_2B_2C_2]$ is
(M = transition metal; A, B and C are monodentate ligands)

(a) 3 (b) 4 (c) 5 (d) 6

107. The maximum work (in $kJ\ mol^{-1}$) that can be derived from complete combustion of 1 mole of CO at 298 K and 1 atm is
[Standard enthalpy of combustion of $CO = -283.0 kJ\ mol^{-1}$; standard molar entropies at $298\ K : S_{O_2} = 205.1\ J\ mol^{-1}$, $S_{CO} = 197.7\ J\ mol^{-1}$, $S_{CO_2} = 213.7\ J\ mol^{-1}$]

(a) 257 (b) 227 (c) 57 (d) 127

108. 18 g of glucose ($C_6H_{12}O_6$) dissolved in 1 kg of water is heated to boiling. The boiling point (in K) measured at 1 atm pressure is closest to [Ebulioscopic constant, K_b for water is $0.52\ K\ kg\ mol^{-1}$. Consider absolute zero to be $-273.15\ C°$]

(a) 373.15 (b) 373.10
(c) 373.20 (d) 373.25

109. Polonium (atomic mass = 209) crystallises in a simple cubic structure with a density of $9.32\ g\ cm^{-3}$ Its lattice parameter (in pm) is closest to

(a) 421 (b) 334 (c) 481 (d) 193

110. The following reaction takes place at 298 K in an electrochemical cell involving two metals A and B,
$A^{2+}(aq) + B(s) \rightarrow B^{2+}(aq) + A(s)$
with $[A^{2+}] = 4 \times 10^{-3}$ M and $[B^{2+}] = 2 \times 10^{-3}$ M in the respective half-cells, the cell EMF is 1.091 V.
The equilibrium constant of the reaction is closest to

(a) 4×10^{36} (b) 2×10^{37}
(c) 2×10^{34} (d) 4×10^{37}

BIOLOGY

111. Suppose the three non-linked autosomal genes A, B and C control coat colour in an animal and the dominants alleles A, B and C are responsible for dark colour and the recessive alleles a, b and c are responsible for light colour. If a cross between a male of $AABBCC$ genotype and a female of $aabbcc$ genotype produces 640 offsprings in the F_2-generation, how many of them are likely to be of the parental genotype?

(a) 10 (b) 20 (c) 160 (d) 640

112. In a population of families having three children each, the percentage of population of families having both boys and girls is

(a) 10 (b) 25 (c) 50 (d) 75

113. As indicated in the gel image, lanes X and Y represent samples obtained from a circular plasmid DNA after complete digestion using restriction enzyme X or Y with different recognition sites, respectively. How many sites for X and Y are present in the plasmid (sizes of the bands in kilo base pairs (kb) is shown)?

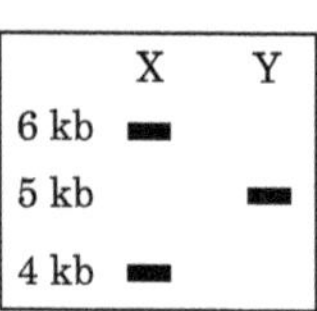

(a) 1 for X, 1 for Y (b) 2 for X, 1 for Y
(c) 1 for X, 2 for Y (d) 2 for X, 2 for Y

114. Matthew Meselson and Franklin Stahl grew *E.coli* (doubling time is 20 min) in medium containing $^{15}NH_4Cl$ for many generations. Then the *E.coli* was transferred to medium containing $^{14}NH_4Cl$. After 40 minutes, the cells were harvested and DNA was extracted and subjected to cesium chloride density gradient centrifugation. The proportion of light and hybrid DNA densities will be

(a) 50% light and 50% hybrid DNA
(b) 100% light DNA
(c) 100% hybrid DNA
(d) 25% light and 75% hybrid DNA

115. In a population interaction between the species X and the species Y, which one of the following statements is correct?

(a) When X benefits and Y is disadvantaged, it is competition
(b) When both X and Y benefit, it is mutualism
(c) When both X and Y are disadvantaged, it is predation
(d) When both X and Y are disadvantaged, it is parasitism

116. The protein P, the oligosaccharide O, and the oligonucleotide N are composed of 100 amino acid residues, 100 hexose residues and 100 nucleotides, respectively. Which one of the following orders of molecular weights is correct?

(a) P > O > N
(b) P > N > O
(c) N > O > P
(d) O > P > N

117. An octapeptide (NH_2-Asn-Glu-Tyr-Lys-Trp-Met-Glu-Gly) is subjected to complete protease and chemical digestion. Based on the results obtained, choose the incorrect option from below.

(a) Trypsin generates mixtures of dimer and trimer
(b) Trypsin generates tetramers only
(c) Cyanogen bromide generates a hexamer and a dimer
(d) Chymotrypsin generates mixture of dimer and trimers

118. Match the enzymes in Column I with their respective biochemical reactions in column II. Choose the correct combination from below

Column I	Column II
A. Transaminases	1. Removal of phosphoryl group from a specific amino acid
B. Protein kinases	2. Removal of α-amino group from a specific amino acid
C. Protein phosphatases	3. Addition of phosphoryl group to a specific amino acid
D. Dehydrogenases	4. Interconversion of optical isomers
	5. Oxidation and reduction of substrates

	A	B	C	D		A	B	C	D
(a)	4	2	3	5	(b)	2	1	2	4
(c)	2	3	1	5	(d)	5	2	3	1

119. Which one of the following graphs best describes the Blood Pressure (BP) change when blood moves from aorta to capillaries?

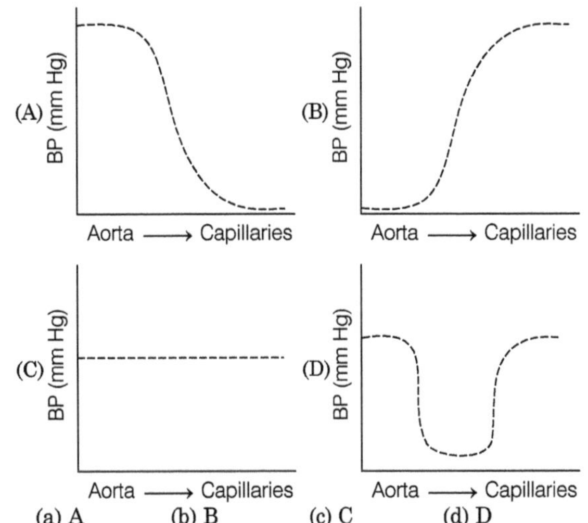

(a) A (b) B (c) C (d) D

120. The following two pedigrees describe the autosomal genetic disorders P and Q in Family 1 and Family 2, respectively.

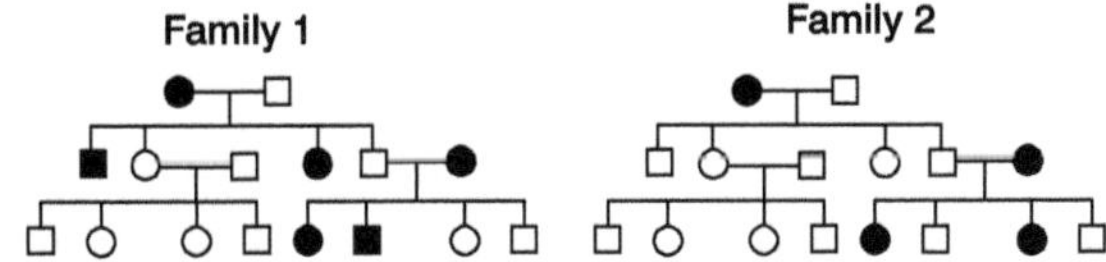

Choose the correct statement from the following options.

(a) Both P and Q are dominant traits
(b) P is a dominant trait and Q is a recessive trait
(c) Both P and Q are recessive traits
(d) P is a recessive trait and Q is a dominant trait

Answers

PART-I

1 (d)	2 (b)	3 (d)	4 (d)	5 (d)	6 (c)	7 (b)	8 (c)	9 (a)	10 (d)
11 (c)	12 (b)	13 (c)	14 (c)	15 (c)	16 (b)	17 (a)	18 (a)	19 (a)	20 (a)
21 (b)	22 (d)	23 (a)	24 (c)	25 (c)	26 (d)	27 (c)	28 (c)	29 (b)	30 (a)
31 (c)	32 (b)	33 (c)	34 (d)	35 (d)	36 (b)	37 (b)	38 (c)	39 (a)	40 (d)
41 (b)	42 (d)	43 (c)	44 (b)	45 (c)	46 (c)	47 (b)	48 (b)	49 (a)	50 (c)
51 (b)	52 (a)	53 (b)	54 (c)	55 (b)	56 (d)	57 (d)	58 (c)	59 (a)	60 (a)
61 (a)	62 (b)	63 (c)	64 (a)	65 (d)	66 (b)	67 (a)	68 (d)	69 (c)	70 (b)
71 (a)	72 (c)	73 (a)	74 (c)	75 (a)	76 (b)	77 (d)	78 (c)	79 (a)	80 (d)

PART-II

81 (d)	82 (d)	83 (b)	84 (b)	85 (b)	86 (c)	87 (d)	88 (c)	89 (b)	90 (a)
91 (b)	92 (d)	93 (c)	94 (a)	95 (d)	96 (c)	97 (b)	98 (a)	99 (d)	100 (a)
101 (a)	102 (d)	103 (c)	104 (a)	105 (d)	106 (c)	107 (a)	108 (c)	109 (b)	110 (b)
111 (b)	112 (d)	113 (d)	114 (a)	115 (b)	116 (c)	117 (a)	118 (c)	119 (a)	120 (b)

Solutions

1. *(d)* Given $ABCD$ is a square A and B. A and B slide along theirs respective axes.

Let $AB = BC = CD = AD = a$

In ΔBMC,

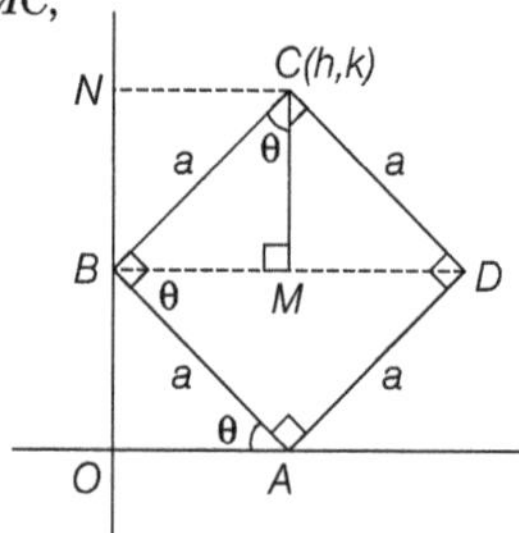

$$\sin\theta = \frac{BM}{BC}$$

$$\sin\theta = \frac{h}{a}$$

$$h = a\sin\theta \quad \text{...(i)}$$

$$k = OB + NB$$

$$k = a\sin\theta + a\cos\theta$$

$$[\because OB = a\sin\theta,\ NB = a\cos\theta]$$

$$\Rightarrow \quad k = h + a\cos\theta$$

$$\Rightarrow \quad k - h = a\cos\theta \quad \text{...(ii)}$$

On squaring and adding Eqs. (i) and (ii), we get

$$h^2 + (k - h)^2 = a^2$$

$\because$ Locus of CB

$$x^2 + (y - x)^2 = a^2$$

$$\Rightarrow \quad 2x^2 + y^2 - 2xy = a^2$$

which is ellipse not a circle.

2. *(b)* We have,

I. $n! \le n^n$

$$\frac{n^n}{n!} = \frac{n}{n} \times \frac{n}{(n-1)} \times \frac{n}{(n-2)} \times \ldots \times \frac{n}{1}$$

Clearly $\frac{n^n}{n!} \ge 1$

$\therefore n^n \ge n!$ True

II. $(n!)^2 \le n^n$ is False

III. $10^n \le n!$

$$\frac{n!}{10^n} = \frac{n}{10} \times \frac{n-1}{10} \times \frac{n-2}{10} \times \ldots \times \frac{1}{10}$$

Given, $n > 1000$. Clearly, $\frac{n!}{10^n} \ge 1$

Hence, it is also true.

IV. $n^n \le (2n)!$

$$\frac{2n!}{n^n} = \frac{1 \cdot 2 \cdot 3 \cdot 4 \ldots n(n+1)(n+2)\ldots(n+n)}{n \times n \times n \ldots n \text{ times}}$$

$$= \frac{n!(n+1)(n+2)(n+3)\ldots(n+1)}{n \times n \times n \times \ldots n \text{ times}}$$

$$= n!\left(1+\frac{1}{n}\right)\left(1+\frac{2}{n}\right)\left(1+\frac{3}{n}\right)\ldots\left(1+\frac{n}{n}\right)$$

Clearly $n \ge 1$.

$\therefore$ It is also true.

3. *(d)* We have,

$$S = \frac{a^2 + b^2 + c^2}{ab + bc + ca}$$

We know that,

$$(a-b)^2 + (b-c)^2 + (c-a)^2 \ge 0$$

$$\Rightarrow \quad a^2 + b^2 + c^2 - (ab + bc + ca) \ge 0$$

$$\Rightarrow a^2 + b^2 + c^2 \ge ab + bc + ca$$

$$\Rightarrow \quad \frac{a^2 + b^2 + c^2}{ab + bc + ca} \ge 1 \quad [\because ab + bc + ca > 0]$$

$$\therefore \quad S \ge 1$$

We know that,

$$(a + b + c)^2 \ge 0$$

$$\therefore a^2 + b^2 + c^2 + 2(ab + bc + ca) \ge 0$$

$$ab + bc + ca < 0$$

$$\frac{a^2 + b^2 + c^2 + 2(ab + bc + ca)}{ab + bc + ca} < 0$$

$$\frac{a^2 + b^2 + c^2}{ab + bc + ca} + 2 \le 0$$

$$\frac{a^2 + b^2 + c^2}{ab + bc + ca} \le -2$$

$$\therefore \quad S \in (-\infty, -2] \cup [1, \infty)$$

4. *(d)* We have, $S = \sum_{n=0}^{\infty} \frac{a_n}{10^{2n}}$

$a_0 = a_1 = 1$ and $a_j = 20a_{j-1} - 108a_{j-2}, j \ge 2$

$a_n = 20a_{n-1} - 108a_{n-2}$

$$\frac{a_n}{10^{2n}} = \frac{20a_{n-1}}{10^{2n}} - \frac{108a_{n-2}}{10^{2n}}$$

$$= \frac{20}{100}\frac{a_{n-1}}{10^{2(n-1)}} - \frac{108}{10^4}\frac{a_{n-2}}{10^{2(n-2)}}$$

$$\sum_{n=2}^{\infty} \frac{a_n}{10^{2n}} = \frac{1}{5}\frac{\Sigma a_{n-1}}{10^{2(n-1)}} - \frac{27}{2500}\frac{\Sigma a_{n-2}}{10^{2(n-2)}}$$

$$S - 1 - \frac{1}{10} = \frac{1}{5}(S - 1) - \frac{27}{2500}S$$

$$S\left(1 - \frac{1}{5} + \frac{27}{2500}\right) = 1 + \frac{1}{100} - \frac{1}{5}$$

$$S\left(\frac{2500 - 500 + 27}{2500}\right) = \frac{100 + 1 - 20}{100}$$

$$S = \frac{81 \times 25}{2027}$$

$$S = \frac{2025}{2027}$$

$$\therefore \quad a = 2025$$

5. *(d)* Given,

$$f(x) = \frac{16x^2 - 96x + 153}{x - 3}.$$

Let $\quad f(x) = y$

$$\therefore \quad y = \frac{16x^2 - 96x + 153}{x - 3}$$

$$\Rightarrow xy - 3y = 16x^2 - 96x + 153$$

$$\Rightarrow 16x^2 - x(96 + y) + 153 + 3y = 0$$

$$x \in R$$

$$\therefore \quad D \ge 0$$

$$\because (96 + y)^2 - 4 \times 16(153 + 3y) \ge 0$$

$$(96)^2 + 192y + y^2 - 64(153) - 192y \ge 0$$

$$y^2 \ge 64(153) - 96^2$$

$$y^2 \ge 9792 - 9216$$

$$y^2 \ge 576$$

$$\therefore \quad y \in (-\infty, -24] \cup [24, \infty)$$

$\therefore$ Least positive value of $f(x) = 24$

6. *(c)* We have,

$$p(x) = x^n + a_{n-1}x^{n-1} + a_{n-2}x^{n-2} + \ldots + a_1x + a_0$$

$$np(x) = (1 + x)p'(x)$$

and $\quad p(1) = b$

$$p'(x) = nx^{n-1} + (n-1)a_{n-1}x^{n-2} + (n-2)a_{n-2}x^{n-3} + \ldots + a_1$$

$$xp'(x) = nx^n + (n-1)a_{n-1}x^{n-1} + (n-2)a_{n-2}x^{n-2} + \ldots + a_1x$$

$$np(x) = nx^n + na_{n-1}x^{n-1} + a_{n-2}x^{n-2} + \ldots + na_1x + na_0$$

$$np(x) - xp'(x) = a_{n-1}x^{n-1} + 2a_{n-2}x^{n-2} + \ldots + (n-1)a_1x + na_0$$

$$np(x) - xp'(x) = p'(x)$$

$$\therefore \quad n = a_{n-1},\ (n-1)a_{n-1} = 2a_{n-2} \ldots na_0 = a_1$$

$$\therefore \quad a_{n-2} = \frac{(n-1)n}{2} = {}^nC_2$$

$$a_{n-3} = \frac{n(n-1)(n-2)}{3} = {}^nC_3$$

$$\because p(x) = x^n + {}^nC_1x^{n-1} + {}^nC_2x^{n-2} + \ldots + {}^nC_x$$

$$p(1) = 1 + {}^nC_1 + {}^nC_2 + {}^nC_3 + \ldots + {}^nC_x$$

$$p(1) = 2^n$$

$$\therefore \quad b = 2^n$$

7. *(b)* We have,

I. $a, b, ,c, d, e$ are angle of convex pentagon in degree.

II. $a \le b \le c \le d \le e$

III. a, b, c, d, e are in AP

$$a + b + c + d + e = 540°$$

Let $a = a$, common difference $= D$

$\therefore \quad \frac{5}{2}(2a + 4D) = 540°$

$a + 2D = 108$ and $a + 4D < 180°$

[$\because$ interior angle of polygon is less than 180°]

$\therefore \quad 108° - 2D + 4D < 180°$

$2D < 180° - 108°$

$0 < D < 36$

$\therefore$ Total 36 types are possible.

8. *(c)* We have,

Total number of person = 32

they seated in a around a circular table

***Case* I** No person between X_1 and X_2

Probability $= \frac{30! \times 2!}{31!} = \frac{2}{31}$

***Case* II** One person between X_1 and X_2

Probability $= \frac{30C_1 \times 29! \times 2!}{31!} = \frac{2}{31}$

***Case* III** Two persons between X_1 and X_2

Probability $\frac{30C_2 \times 28! \times 2! \times 2!}{31!} = \frac{2}{31}$

***Case* IV** Three persons between X_1 and X_2

Probability $= \frac{30C_3 \times 27! \times 2! \times 3!}{31!} = \frac{2}{31}$

Total probability $= \frac{2}{31} + \frac{2}{31} + \frac{2}{31} + \frac{2}{31} = \frac{8}{31}$

9. *(a)* We know,

$$\frac{\log_e(1+x)}{x} < 1$$

$$\therefore \quad \frac{\log_e\left(1+\frac{1}{x}\right)}{1/x} < 1$$

$$\log_e\left(\frac{x+1}{x}\right) < \frac{1}{x}$$

$$\log_e(x+1) - \log_e x < \frac{1}{x}$$

$$\log_e 2 - \log_e 1 < 1$$

$$\log_e 3 - \log_e 2 < \frac{1}{2}$$

$$\log_e(n+1) - \log_e n < \frac{1}{n}$$

$$\log_e(n+1) - \log_e 1 < 1 + \frac{1}{2} + \frac{1}{3} + \ldots + \frac{1}{n}$$

$$\log_e(n+1) \le 4 \quad \left[\because 1 + \frac{1}{2} + \frac{1}{3} + \ldots + \frac{1}{n} \ge 4\right]$$

$$n + 1 < e^4$$

$$n < e^4 - 1$$

$$n < (2.7)^4 - 1 \quad [\because e = 2.718]$$

$$n < 54.59 - 1$$

$$n < 53.59$$

$\therefore$ Hence, option (a) is correct.

10. *(d)* We have a pair of 12-sided fair dices with number (1, 2, 3, ..., 12)

Total sample space $= (12)^2 = 144$.

Favourable outcome = Sum of number appearing remainder 2 when divided by 9 is

***Case* I** $x_1 + x_2 = 2\{(1, 1)\}$

***Case* II** $x_1 + x_2$
$= 11\{(1, 10), (2, 9), (3, 8), (4, 7)$
$(5, 6) (6, 5) (7, 4) (8, 3) (9, 2) (10, 1)\}$

***Case* III** $x_1 + x_2 = 20\{(8, 12), (9, 11),$
$(10, 10), (11, 9), (12, 8)\}$

Total favourable outcomes
$= 1 + 10 + 5 = 16$

$\therefore$ Required probability $= \frac{16}{144} = \frac{1}{9}$

11. *(c)* We have,

$$x^6 + 2x^5 + 4x^4 + 8x^3 + 16x^2 + 32x + 64 = 0$$

$$\Rightarrow x^6\left(1 + \frac{2}{x} + \left(\frac{2}{x}\right)^2 + \left(\frac{2}{x}\right)^3 + \left(\frac{2}{x}\right)^4 + \left(\frac{2}{x}\right)^5 + \left(\frac{2}{x}\right)^6\right) = 0$$

$$\Rightarrow \frac{1 - \left(\frac{2}{x}\right)^7}{1 - \frac{2}{x}} = 0 \qquad [x^6 \ne 0]$$

$$\Rightarrow \left(\frac{2}{x}\right)^7 = 1 \Rightarrow x^7 = 2^7$$

$$\therefore \quad x = 2$$

$\therefore |x_i| = 2$ for all value of i

12. *(b)* Given, $z_1 = \sqrt{3} + i$ and $z_2 = \sqrt{3} - i$ are two adjacent vertices of n-sided regular polygon centred of origin.

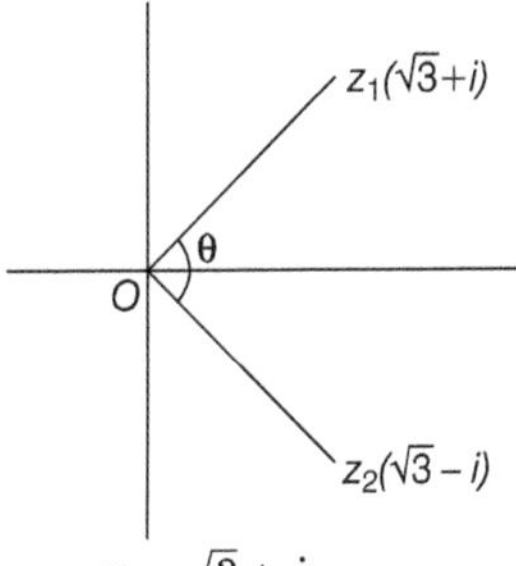

$$z_1 = \sqrt{3} + i$$

$$\arg(z_1) = \tan^{-1}\left(\frac{1}{\sqrt{3}}\right) = \frac{\pi}{6}$$

$$z_2 = \sqrt{3} - i$$

$$\arg(z_2) = \tan^{-1}\left(-\frac{1}{\sqrt{3}}\right) = -\frac{\pi}{6}$$

$$\angle z_2Oz_1 = \frac{\pi}{3}$$

$$\therefore \quad \frac{n\pi}{3} = 2\pi$$

$$\Rightarrow \quad n = 6$$

13. *(c)* Given,

$$A^{-1} = \begin{bmatrix} 1 & 2017 & 2 \\ 1 & 2017 & 4 \\ 1 & 2018 & 8 \end{bmatrix}$$

$$|A^{-1}| = \begin{vmatrix} 1 & 2017 & 2 \\ 1 & 2017 & 4 \\ 1 & 2018 & 8 \end{vmatrix}$$

Apply $R_1 \to R_1 - R_2$

$$|A^{-1}| = \begin{vmatrix} 0 & 0 & -2 \\ 1 & 2017 & 4 \\ 1 & 2018 & 8 \end{vmatrix}$$

Expand along R_1

$$|A^{-1}| = -2(2018 - 2017) = -2$$

$$AA^{-1} = I$$

$$|A||A^{-1}| = 1$$

$$|A| = \frac{1}{|A^{-1}|^2} = -\frac{1}{2}$$

Now, $|2A| - |2A^{-1}| = 8|A| - 8|A^{-1}|$

$$= 8\left[\frac{-1}{2} + 2\right] = 12$$

14. *(c)* An ellipse passes through (0, 0), (1, 0) and (2, 0). Its minor and major axis are parallel to coordinates axes. One of its foci lies on Y-axis.

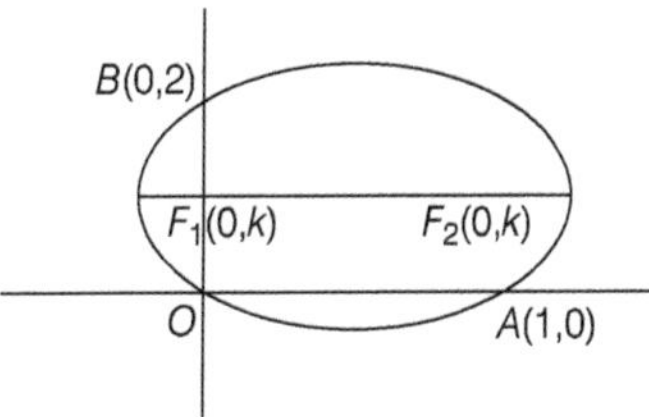

Let one foci is $(0, k)$.

$\therefore OB$ is the latusrectum of ellipse $F_1(0, k)$ is mid-point of OB

$F_1(0, 1) \qquad \left[\because k = \frac{2+0}{2} = 1\right]$

Now $F_2(h, k) = F_2(h, 1)$

Now by definition of ellipse

$$BF_1 + BF_2 = 2a = OF_1 + OF_2$$

$$\therefore \quad BF_1 + BF_2 = OF_1 + OF_2$$

$$\sqrt{1+1} + \sqrt{(h-1)^2 + 1} = \sqrt{0+1} + \sqrt{h^2 + 1}$$

$$\sqrt{2} + \sqrt{(h-1)^2 + 1} = 1 + \sqrt{h^2 + 1}$$

$$\Rightarrow \quad h = 1$$

$\therefore F_1(0, 1)$ and $F_2(1, 1)$

$$F_1F_2 = 2ae = 1$$

$$OF_1 + OF_2 = 2a = 1 + \sqrt{2}$$

$$\therefore \quad e = \frac{1}{\sqrt{2}+1} \Rightarrow e = \frac{\sqrt{2}-1}{2-1}$$

$$\Rightarrow \quad e = \sqrt{2} - 1$$

15. *(c)* We have, $I_n = \int_0^1 e^{-y} y^n dy$

$$I_n = [-y^n e^{-y}]_0^1 + \int_0^1 n y^{n-1} e^{-y} dy$$

$$I_n = -\frac{1}{e} + nI_{n-1}$$

$$\Rightarrow \frac{I_n}{n!} = \frac{-1}{n!e} + \frac{I_{n-1}}{(n-1)!}$$

$$\Rightarrow \frac{I_n}{n!} - \frac{I_{n-1}}{(n-1)!} = \frac{-1}{n!e}$$

$$n = 1 \quad \frac{I_1}{1!} - \frac{I_0}{0!} = -\frac{1}{e}$$

$$n = 2 \quad \frac{I_2}{2!} - \frac{I_1}{1!} = -\frac{1}{2!e}$$

$$n = 3 \quad \frac{I_3}{3!} - \frac{I_1}{2!} = -\frac{1}{3!e}$$

$$n = n \quad \frac{I_n}{n!} - \frac{I_n}{n-1} = \frac{1}{n!e}$$

Adding these form, we get

$$\frac{I_n}{n!} - I_0 = -\frac{1}{e}\left(1 + \frac{1}{1!} + \frac{1}{2!} + \ldots + \frac{1}{n!}\right)$$

$$\frac{I_n}{n!} = I_0 - \frac{1}{e}\left(1 + \frac{1}{1!} + \frac{1}{2!} + \ldots + \frac{1}{n!}\right)$$

$$\sum_{n=1}^{\infty} \frac{I_n}{n!} = \frac{1}{e} + 1 - \frac{1}{e}(e) \quad \left[\because e = 1 + \frac{1}{1!} + \frac{1}{2!} + \ldots\right]$$

$$= \frac{1}{e} + 1 - 1 = \frac{1}{e}$$

16. *(b)* We have, $\sin\theta + \cos\theta = \sin 2\theta$

On squaring both sides, we get

$$\sin^2\theta + \cos^2\theta + \sin 2\theta = \sin^2 2\theta$$

$$\Rightarrow \sin^2 2\theta - \sin 2\theta - 1 = 0$$

$$\Rightarrow \sin 2\theta = \frac{1 \pm \sqrt{5}}{2}$$

$$\Rightarrow \sin 2\theta = \frac{1 - \sqrt{5}}{2} \Rightarrow \sin 2\theta \neq \frac{\sqrt{5} + 1}{2}$$

$\therefore$ Two solutions exist interval $[-\pi, \pi]$.

17. *(a)* We have, $z_1, z_2, z_3, \ldots, z_7$ be the vertices of regular heptagon inscribed in the unit circle.

$\therefore z_1, z_2, z_3, \ldots, z_7$ are seventh roots of units

$$z_1 + z_2 + z_3 + \ldots + z_7 = 0$$

We know that,

$$(x_1 + x_2 + x_3 + \ldots + x_n)^2 = \sum_{i=1}^{n} x_i^2 + 2 \sum_{i \le i < j \le n} x_i x_j$$

$$\therefore (z_1 + z_2 + z_3 + \ldots + z_7)^2 = \sum_{i=1}^{7} z_i^2 + 2 \sum_{1 \le i < j \le 7} z_i z_j$$

$$0 = 0 + 2|w| \quad \left[\because \sum_{i=1}^{7} z_i^2 = 0\right]$$

$$\Rightarrow \quad |w| = 0 \quad \left[\because |w| = \sum_{1 \le i < j \le 7} z_i z_j\right]$$

18. *(a)* Let the position of cannon = P

speed of sound = S

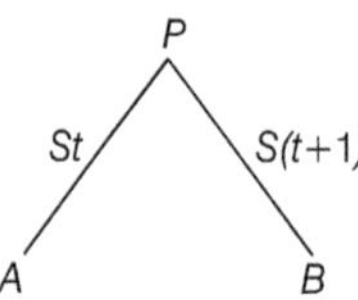

Time of heard of sound of cannon firing of at $A = t$

$\therefore$ at $B = t + 1$

Hence, PA = Speed × Time

$$PA = St$$

and $$PB = S(t + 1) = St + S$$

$$PB - PA = St + S - St$$

$$PB - PA = S$$

$$PB - PA = \text{constant}$$

$\therefore A$ and B are foci of hyperbola.

19. *(a)* We have, point on sphere from coordinates axes is (9, 16, 25)

Sphere touches the coordinates axis

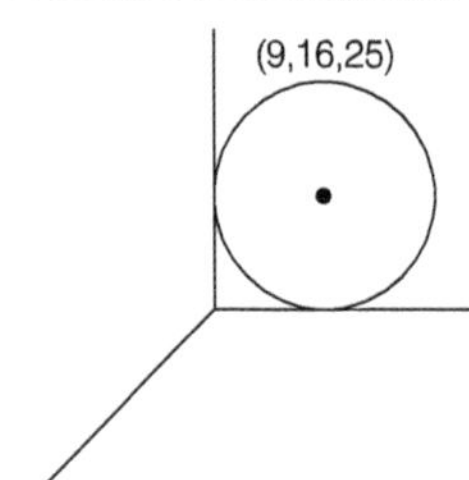

$\therefore$ Centre (r, r, r) and radius is r

$$\therefore (r - 9)^2 + (r - 16)^2 + (r - 25)^2 = r^2$$

$$3r^2 - 2r(9 + 16 + 25) + 9^2 + 16^2 + 25^2 = r^2$$

$$2r^2 - 100r + 962 = 0$$

$$r^2 - 50r + 481 = 0$$

$$(r - 37)(r - 13) = 0$$

$$r = 13, 37$$

$\therefore$ Possible radius of sphere = 13

20. *(a)* We have,

m, n be two distinct integer chosen randomly from the set {0, 1, 2, 3, ..., 99}

$\therefore$ Total number of outcomes = $100\,C_2$

$$4^m + 4^n + 3$$

$$(5 - 1)^m + (5 - 1)^n + 3$$

$$5K + (-1)^m + 5\lambda + (-1)^n + 3$$

$$5(K + \lambda) + (-1)^m + (-1)^n + 3$$

$4^m + 4^n + 3$ is divisible by 5 if m and n both are even.

$\therefore$ Favourable outcomes = $50\,C_2$

$$\text{Required probability} = \frac{50\,C_2}{100\,C_2}$$

$$= \frac{50 \times 49}{100 \times 99} = \frac{49}{198} = 0.247$$

$$0.247 \in (0, 0.25]$$

21. *(b)* In multiplication and division number of significant figure in result = smallest number of significant figures (as that of numbers involved in product or division).

In addition and subtraction number of significant figures = smallest number of significant decimal places (as that of numbers used in sum and difference)

So, from given data

$$s = \underbrace{ut}_{\text{Three significant figures}} + \underbrace{\frac{1}{2}at^2}_{\text{One digit after decimal}}$$

One digit after decimal

So, result contains only one digit after decimal.

22. *(d)* Given, $P = Gc^{-5} m^x R^y \omega^z$ ... (i)

Here, dimensions of various physical quantities are

Angular speed, $\omega = [T^{-1}]$

Power, $P = [ML^2T^{-3}]$

Mass, $m = [M]$

Radius, $R = [L]$

Speed, $c = [LT^{-1}]$

Universal gravitational constant, $G = [M^{-1}L^3T^{-2}]$

Substituting dimensions in Eq. (i), we have

$$[ML^2T^{-3}] = [M^{-1}L^3T^{-2}][L^{-5}T^5][M]^x [L]^y [T^{-z}]$$

Equating dimensions of same quantity, we get $1 = -1 + x \Rightarrow x = 2$

$$2 = 3 - 5 + y \Rightarrow y = 4$$

$$-3 = -2 + 5 - z \Rightarrow z = 6$$

23. *(a)* Average speed is related to the rms speed as

$$\bar{c} = \sqrt{\frac{8}{3\pi}} \cdot c_{\text{rms}}$$

$$\Rightarrow \quad \bar{c} < c_{\text{rms}}$$

So, statement I is incorrect.

Mean free path of molecules is larger than mean distance between molecules.

So, statement II is correct.

Mean free path decreases with temperature as collision frequency increases with temperature. So, statement III is incorrect.

As, $c_{\text{rms}} \propto \frac{1}{\sqrt{M}} \Rightarrow$ rms speed of nitrogen molecule is higher than that of oxygen molecule at same temperature. So, statement IV is incorrect.

24. *(c)*

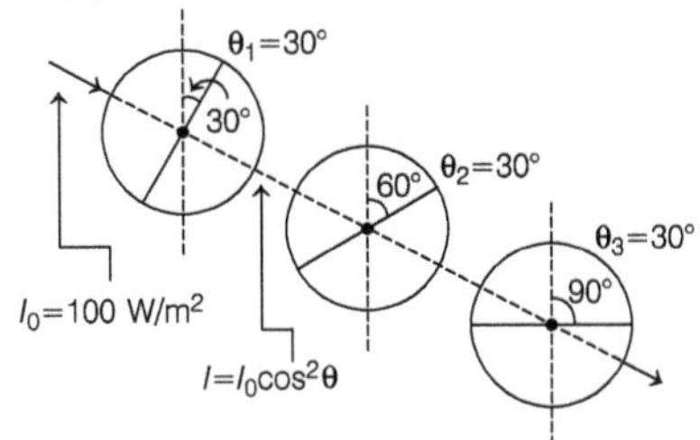

When an already polarised beam passes through a polaroid, intensity I of light obtained is given by Malus' law

$$I = I_0 \cos^2\theta$$

where, I_0 = intensity of incident polarised light and θ = angle between plane of polarisation (of light) and transmission axis of polaroid.

As incident light is vertically polarised, angle θ for polaroids are

For I polaroid angle is $\theta_1 = 30°$

For II polaroid angle is $\theta_2 = 30°$

For III polaroid angle is $\theta_3 = 30°$

So, intensity obtained after first polaroid is $I_1 = kI_0 \cos^2\theta_1$.

where, k = absorption coefficient.

$$= 0.9 \times 100 \times \cos^2 30°$$

$$= 0.9 \times 100 \times \frac{3}{4} = 67.5 \frac{\text{W}}{\text{m}^2}$$

Intensity after second polaroid is

$$I_2 = kI_1 \cos^2\theta_2 = 0.9 \times 67.5 \times \frac{3}{4}$$

$$= 45.56 \frac{\text{W}}{\text{m}^2}$$

Intensity after third polaroid is

$$I_3 = kI_2 \cos^2\theta_3$$

$$I_3 = 0.9 \times 45.56 \times \cos^2 30°$$

$$\approx 30\ \text{W/m}^2$$

25. *(c)* Rod is connected to wheel, so it oscillates following rotation of pivot P. Hence, its motion is forced oscillatory motion with time period T.

26. *(d)* At point of contact P tension T can be resolved into vertical and horizontal components.

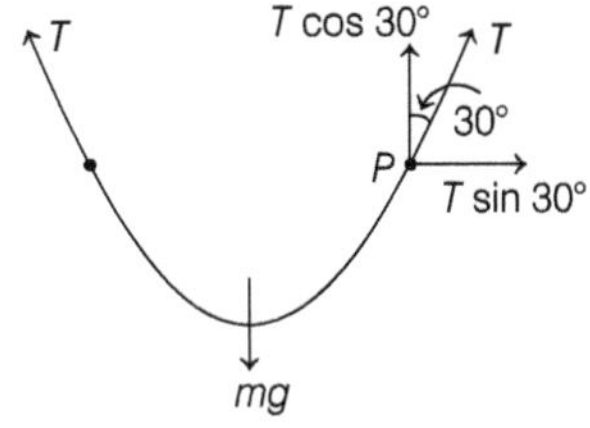

For equilibrium,

$$2T \cos 30° = mg$$

$$\therefore \quad T = \frac{mg}{2\cos 30°}$$

If tension at lowest point is T_1, then for horizontal equilibrium,

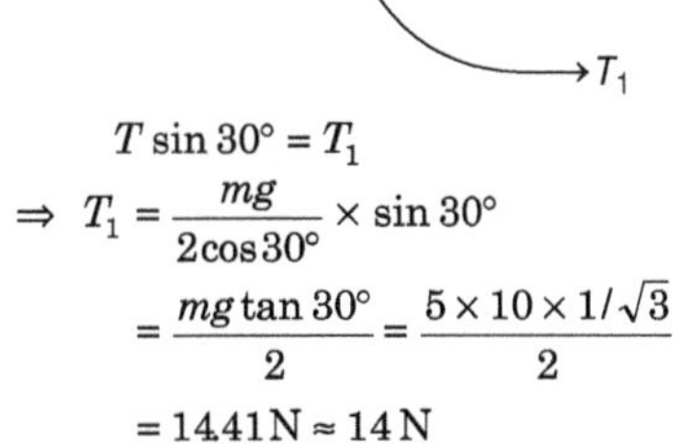

$$T \sin 30° = T_1$$

$$\Rightarrow \quad T_1 = \frac{mg}{2\cos 30°} \times \sin 30°$$

$$= \frac{mg \tan 30°}{2} = \frac{5 \times 10 \times 1/\sqrt{3}}{2}$$

$$= 14.41\ \text{N} \approx 14\ \text{N}$$

27. *(c)*

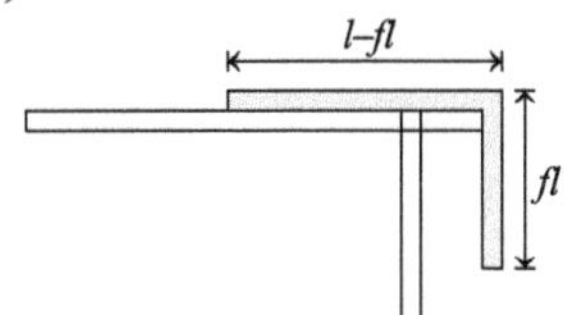

Let mass per unit length of chain is x.

Then, weight pulling the chain on table is

$$F = flxg$$

Weight of chain on table

$$= \text{Normal reaction of table}$$

$$= l(1-f)xg$$

Friction acting on chain,

$$f = \mu N = \mu l\,(1-f)\,xg$$

In equilibrium,

$$f = F \Rightarrow \mu l(1-f)xg = flxg$$

$$\Rightarrow \quad \mu(1-f) = f \Rightarrow \mu = f(1+\mu)$$

$$\Rightarrow \quad f = \frac{\mu}{1+\mu} \Rightarrow f = \frac{1}{\left(1+\frac{1}{\mu}\right)}$$

28. *(c)* Given, $\frac{dn}{dy} > 0$

This means refractive index increases in y-direction.

Let AB is incident wavefront.

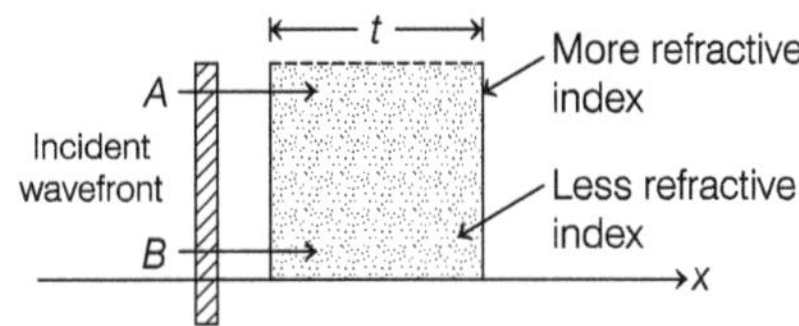

As, B is in rare medium and A is in denser medium, B travels faster than A.

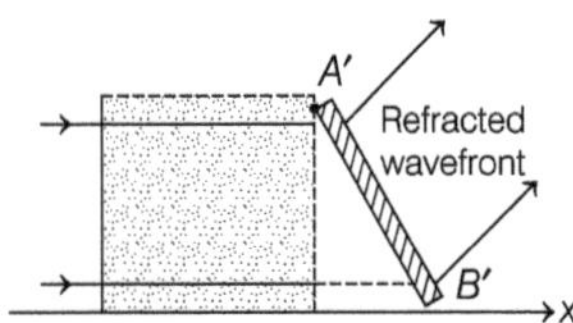

So, new wavefront $A'B'$ is bend upwards.

29. *(b)* If $E = \frac{kq}{r^3} \cdot \hat{\mathbf{r}}$, then flux through a shell enclosing a charge q is

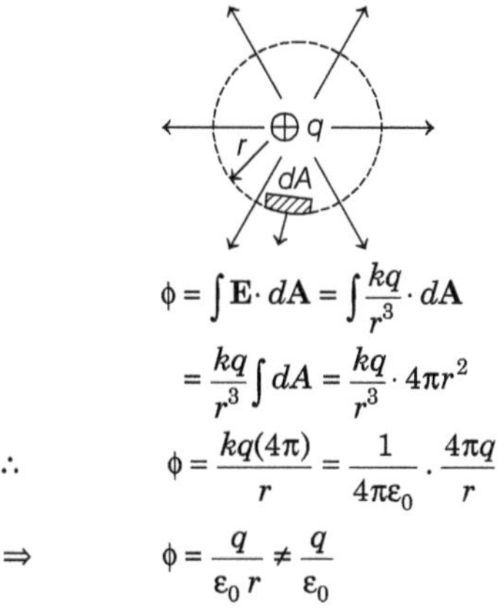

$$\phi = \int \mathbf{E} \cdot d\mathbf{A} = \int \frac{kq}{r^3} \cdot d\mathbf{A}$$

$$= \frac{kq}{r^3} \int dA = \frac{kq}{r^3} \cdot 4\pi r^2$$

$$\therefore \quad \phi = \frac{kq(4\pi)}{r} = \frac{1}{4\pi\varepsilon_0} \cdot \frac{4\pi q}{r}$$

$$\Rightarrow \quad \phi = \frac{q}{\varepsilon_0 r} \neq \frac{q}{\varepsilon_0}$$

Also, when $E = \frac{kq}{r^3}$, Gauss' law is not valid and electric field at an interior point of a charged shell is non-zero.

$\therefore$ Charge placed inside shell experience a force.

30. *(a)* For a spherical shell of radius r and thickness dr, weight of layer is balanced by pressure on layer.

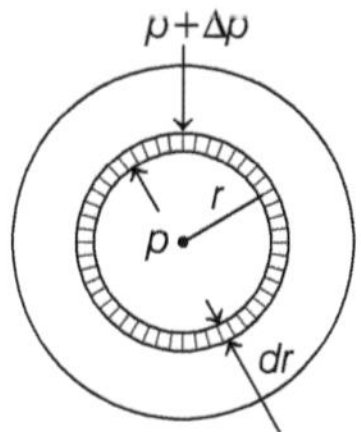

$$\therefore \quad \text{Weight} = \{p - (p + \Delta p)\}\, 4\pi r^2$$

$$\Rightarrow \quad mg = -\,\Delta p \cdot 4\pi r^2$$

$$\Rightarrow 4\pi r^2 \cdot dr \cdot \rho \cdot g = -\,\Delta p \cdot 4\pi r^2 \Rightarrow -\Delta p = \rho g dr$$

$$\Rightarrow \quad -\,\Delta p = \rho^2 \cdot \frac{4}{3}\pi R\, dr$$

So, $$p_{\text{av}} = \frac{1}{R}\int_0^R \Delta p = \frac{4}{3}\pi\rho^2 R \int_0^R dr$$

$$= \frac{4}{3}\frac{\pi MR \cdot R}{(V)^2}$$

$$= \frac{4}{3}\pi \cdot \frac{MR^2}{\left(\frac{4}{3}\pi R^3\right)^2}$$

$$\Rightarrow \quad p_{\text{av}} = \frac{3}{4\pi} \cdot \frac{M}{R^4}$$

$$\Rightarrow \quad p_{\text{av}} \propto \frac{1}{R^4}$$

31. *(c)* Flux through a closed surface is non-zero only, when charge is enclosed by the surface. So, net flux through III and IV is zero.

32. *(b)* Initial value of de-Broglie wavelength is λ_0.

$\therefore \quad \lambda_0 = \dfrac{h}{mv_0}$

where, v_0 = initial speed and m = mass of particle.

So, when de-Broglie wavelength is $\dfrac{\lambda_0}{3}$, velocity of particle will be $3v_0$.

Now, acceleration of the particle is

$$a = \frac{qE_0}{m} \cdot \hat{\mathbf{j}}$$

So, velocity of particle after t seconds is

$$v = u + at = v_0\hat{\mathbf{i}} + \left(\frac{qE_0 t}{m}\right)\hat{\mathbf{j}}$$

When velocity is $3v_0$, then

$$3v_0 = \sqrt{v_0^2 + \left(\frac{qE_0 t}{m}\right)^2}$$

$$\Rightarrow \quad 9v_0^2 = v_0^2 + \left(\frac{qE_0 t}{m}\right)^2$$

$$\Rightarrow \quad t^2 = \frac{8v_0^2 m^2}{q^2 E_0^2} \Rightarrow t = \frac{2\sqrt{2}\, v_0}{E_0} \cdot \left(\frac{m}{q}\right)$$

$$\therefore \quad t \propto \frac{m}{q}$$

33. *(c)* Balancing given nuclear reactions, we get

$$^{14}_{7}\text{N} + {}^{4}_{2}\text{He} \longrightarrow {}^{17}_{8}\text{O} + {}^{1}_{1}X$$

and $\quad {}^{9}_{4}\text{Be} + {}^{4}_{2}\text{H} \longrightarrow {}^{12}_{6}\text{He} + {}^{1}_{0}Y$

So, X is a proton and Y is a neutron.

34. *(d)* Consider a section (a slice) of cylindrical plano-convex lens,

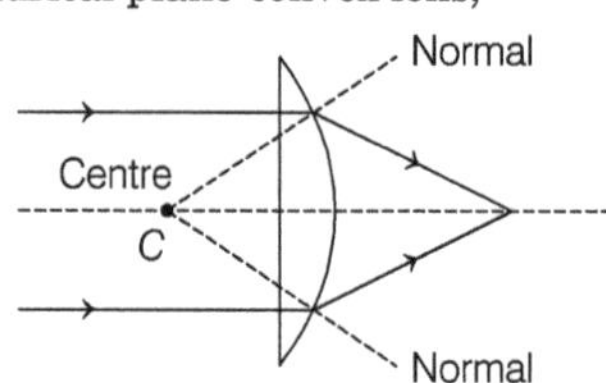

Following refraction laws, parallel beam converges to focus.

So, for complete cylindrical lens, parallel beam converges at focal plane. Hence, on screen, we get a single bright line on screen.

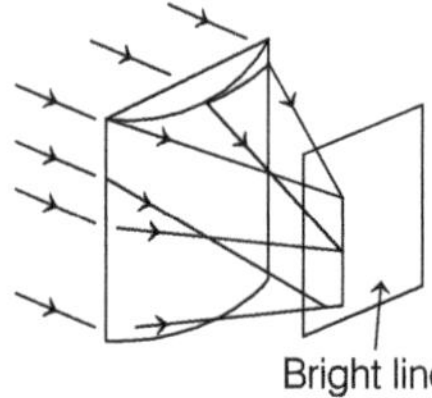

35. *(d)* Depletion layer is formed due to diffusion of electrons from n-side to p-side. There they combine with holes of p-side.

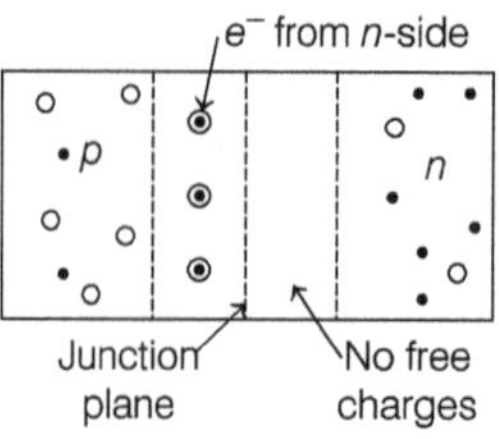

In a p-n junction, there are no free charges in depletion region. It contains only bound charges. Width of depletion region is inversely proportional to dopant concentration. So, it may have different width in p and n-sides. Due to diffusion of electrons, p-side is negatively charged and n-side is positively charged.

36. *(b)*

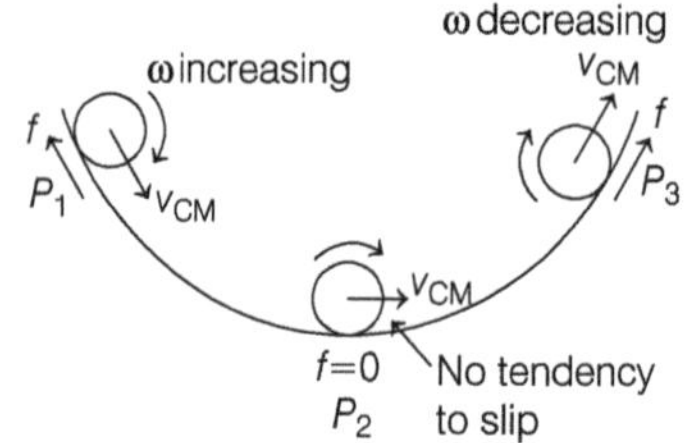

At P_1, ring has a tendency to slip downwards. So, friction acts upward producing rolling torque.

At P_2, friction is of static nature.

At P_3, ball has a tendency to slide down.

$\therefore$ Friction acts parallel to v_{CM}.

37. *(b)* Velocity of propagation of blast wave is

$$v = \sqrt{\frac{\gamma p}{\rho}} \quad \text{... (i)}$$

$\gamma = \dfrac{C_p}{C_V}$, where p = pressure

and ρ = density.

As, $\quad p = \dfrac{nRT}{V} = \dfrac{nRT}{\frac{4}{3}\pi r^3}$

and $\quad v = \dfrac{dr}{dt}$

So, by Eq. (i), we have

$$\frac{dr}{dt} = \sqrt{\frac{\gamma nRT}{\frac{4}{3}\pi r^3 \cdot \rho}}$$

$$\Rightarrow \quad \frac{dr}{dt} = k \cdot r^{-3/2}$$

$$\Rightarrow \quad kdt = r^{3/2} dr$$

Integrating above equation, we get

$$\Rightarrow \quad k\int dt = \int r^{3/2} dr$$

$$\Rightarrow \quad kt = \frac{2}{5} r^{5/2}$$

So, $\quad r \propto t^{\frac{2}{5}}$

38. *(c)* Given is a Kundt's tube experiment. It is used today for demonstrating standing waves. A standing wave is generated in tube causes small amount of chalk powder to pile up at nodes.

Distance between the piles is one half wavelength ($\lambda / 2$) of longitudinal waves.

Now, given in 300 cm, there are 20 piles of powder.

$$\therefore \quad 20 \times \frac{\lambda}{2} = 300 \Rightarrow \lambda = \frac{300}{10 \times 100}\text{ m}$$

So, frequency of sound is

$$f = \frac{v}{\lambda} = \frac{300}{\left(\frac{300}{1000}\right)} = 1\text{kHz}$$

39. *(a)* In equilibrium,

Energy radiated by planet

= Energy received by planet

By Kirchhoff's rule, energy radiated by planet is proportional to fourth power of its temperature.

$$\Rightarrow \quad E_1 = kT^4 = k(fT')^4$$

where, temperature of planet is fT' (given).

Energy received by planet

= Solar constant of planet

$$= E_2 = \frac{R^2 \sigma (T')^4}{d^2}$$

where, R = radius of star,

T' = temperature of star,

σ = Stefan's constant

and $\quad d$ = distance of planet and star.

In equilibrium,

$$E_1 = E_2 \Rightarrow kf^4\, (T')^4 = \sigma\left(\frac{R^2}{d^2}\right) \cdot T'^4$$

$$\Rightarrow \quad f \propto \sqrt{\frac{R}{d}}$$

40. *(d)* Energy of a photon with wavelength λ is

$$E = \frac{12400\ (\text{eV} \cdot \text{Å})}{\lambda(\text{Å})}$$

Energies of wavelengths (of photons) observed in emission spectrum are

$$E_1 = \frac{12400}{912} = 13.5\text{ eV}$$

$$E_2 = \frac{12400}{1026} = 12.08\text{ eV}$$

$$E_3 = \frac{12400}{1216} = 10.2\text{ eV}$$

$$E_4 = \frac{12400}{3646} = 3.4\text{ eV}$$

When these radiations pass through neutral hydrogen gas at room temperature, as all of atoms are probably in their ground states, absorbed energy must be more than or equal to 10.2 eV.

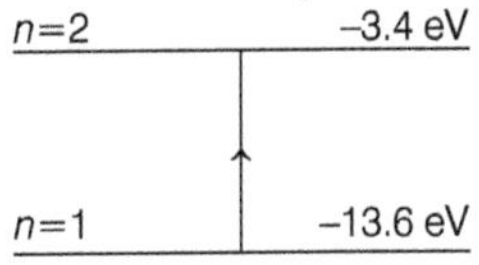

n=2 −3.4 eV

n=1 −13.6 eV

So, $\lambda = 3646\,\text{Å}$ wavelength is not strongly absorbed by the given sample.

41. *(b)*

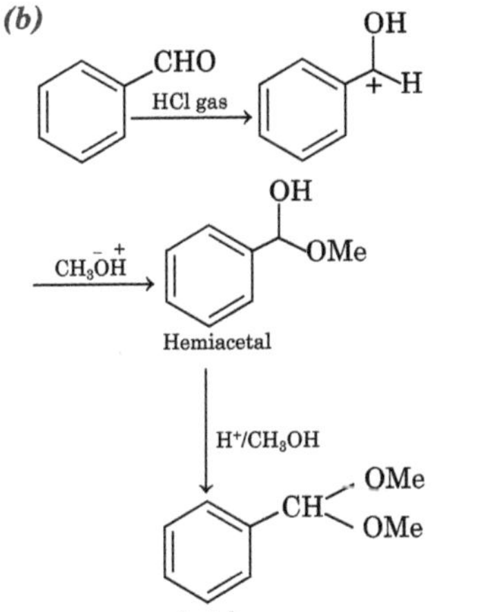

CHO —HCl gas→ OH, +H → $CH_3\bar{O}\overset{+}{H}$ → OH, OMe (Hemiacetal) —H^+/CH_3OH→ CH(OMe)(OMe) (Acetal)

This reaction is known as nucleophilic addition reaction, where CH_3O^- from CH_3OH group act as a $Nu^{\ominus}$ and attack at electro-deficient side of carbonyl compound. Aldehydes react with one equivalent of monohydric alcohol in the presence of dry HCl gas to yield alkoxy alcohol intermediate known as hemiacetal which further react with one more molecule of alcohol to give a gem-dialkoxy compound known as acetal.

42. *(d)* Non-benzenoid aromatic compounds are those compounds which do not have benzene ring. They exhibit aromaticity due to alternate π-bonds in the molecule.

The structure of compounds given in options are as follows :

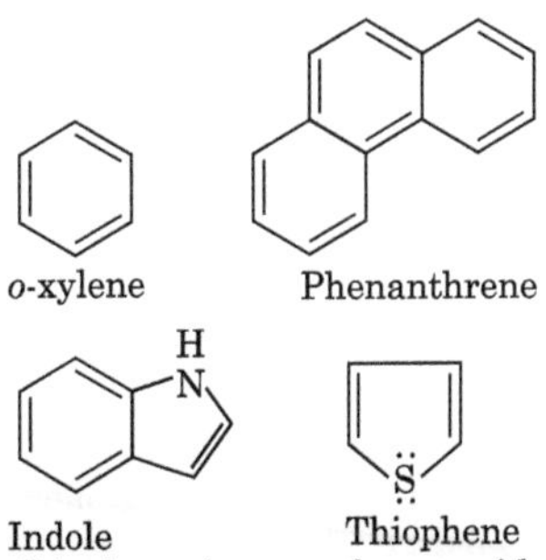

o-xylene Phenanthrene

Indole Thiophene

Thus, thiophene is a non-benzenoid compound.

43. *(c)* Natural rubber is a linear polymer of isoprene (2-methyl-1, 3-butadiene) and is also called as *cis*-1, 4-polyisoprene.

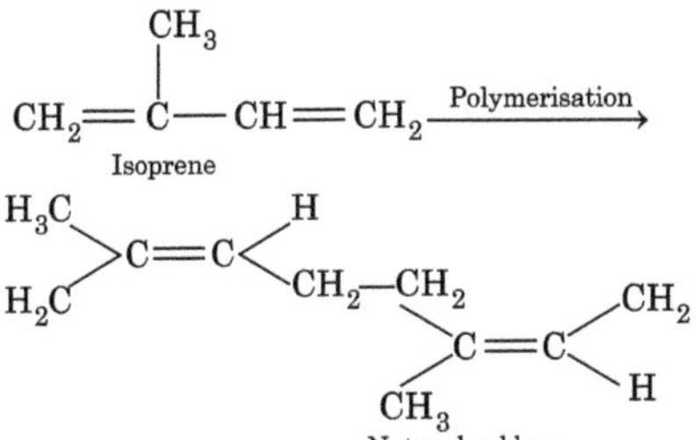

$$CH_2{=}\underset{\text{Isoprene}}{\overset{CH_3}{C}}{-}CH{=}CH_2 \xrightarrow{\text{Polymerisation}}$$

Natural rubber

44. *(b)* The following tripeptide is made from phenyl alanine (on left), alanine (middle) and serine (right).

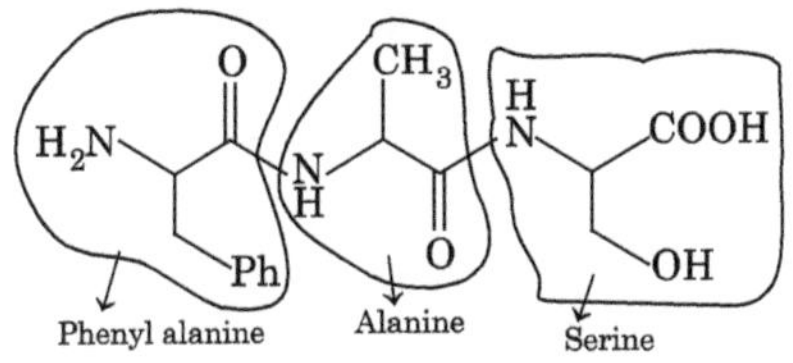

Phenyl alanine Alanine Serine

Thus, it can be represented as, phe–ala–ser.

45. *(c)* Deoxyribo nucleic acid or DNA contains 2-deoxy-D-ribose sugar unit, whereas ribo nucleic acid or RNA contains D-ribose sugar unit.

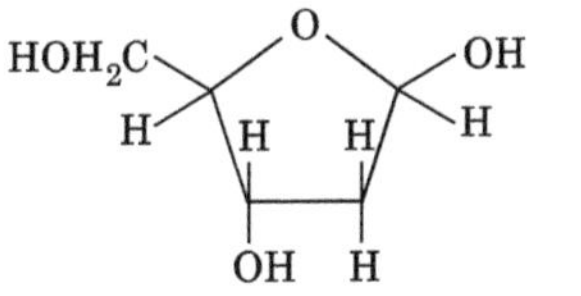

D-2 deoxyribose in DNA

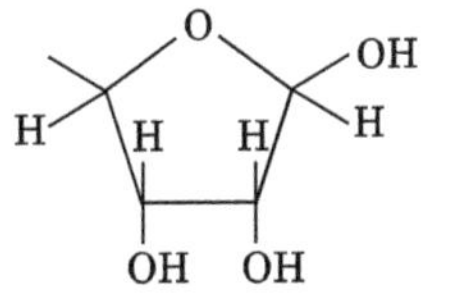

D-ribose in RNA

46. *(c)* $CH_3Br + CH_3CH_2ONa \longrightarrow CH_3CH_2OCH_3$

This reaction is also known as Williamson synthesis reaction. In this method, an alkyl halide is allowed to react with sodium alkoxide to give ether. The reaction involves S_N2 attack of an alkoxide ion on primary alkyl halide.

47. *(b)* The most abundant metal ion present in human body is Ca^{2+}. This essential mineral is required in large quantities and makes up between 1-2% of human body mass with more than 99% found in bones and teeth.

48. *(b)* Phosphorus reacts with chlorine gas to give phosphorus trichloride, PCl_3 which is a colourless liquid. Fumes of PCl_3 in moist air produces HCl and phosphorous acid, H_3PO_3.

$$P_4 + Cl_2 \longrightarrow \underset{\text{Colourless liquid}}{PCl_3} \xrightarrow[\text{Moisture}]{H_2O} H_3PO_3 + HCl$$

49. *(a)* Oxidising ability of a substance means the ability to oxidise other and reduce itself.

As the oxidation number on central atom increases the oxidising ability of given anions also increases.

The order of oxidation number of central atom in given anions are as follows

$$Ti^{4+} + < V^{5+} < Cr^{6+} < Mn^{7+}$$

Thus, the order of oxidising ability will also be same, i.e.

$$TiO_4^{4-} < VO_4^{3-} < CrO_4^{2-} < MnO_4^-$$

50. *(c)* The complete hydrolysis of XeF_6 results in the formation of xenon trioxide XeO_3. This XeO_3 is highly explosive and act as a powerful oxidising agent in aqueous solution.

$$XeF_6 + 3H_2O \longrightarrow XeO_3 + 6HF$$

51. *(b)* All the given oxides are acidic in nature. Thus, more is acidity of oxide faster will be the reaction or more will be the reactivity.

As the electronegativity of the central atom increases, the acidic strength of these oxides increases. The order of electronegativity of central atom is

$$B < P < Cl$$

So, the acidic strength of the oxides follows order

$$B_2O_3 < P_2O_5 < Cl_2O_7.$$

Therefore, the reactivity of given oxide towards water follows the above given order.

52. *(a)* Facial (*fac*) and meridional (*mer*) isomers are type of geometrical isomers which occurs in octahedral coordination entities of the type Ma_3b_3. The type of complexes given in the options are as follows :

Complexes	Type of complexes
$[Co(NO_2)_3(NH_3)_3]$	$[Ma_3b_3]$
$K_3[Fe(CN)_6]$	$[Ma_6]$
$[Co(H_2O)_2(NH_3)_4]Cl$	$[Ma_2b_4]$
$[CoCl(NH_3)_5]Cl_2$	$[Mab_5]$

Thus, $[Co(NO_2)_3(NH_3)_3]$ shows *fac* and *mer* isomers.

fac *mer*

53. *(b)* Given,
solubility product of $Ag_2CrO_4 = 1.1 \times 10^{-12}$
Concentration of CrO_4^{2-} ions, $[CrO_4]^{2-}$
$= 5 \times 10^{-3}$ M
For reaction,

$$Ag_2CrO_4 \rightleftharpoons 2Ag^+ + CrO_4^{2-}$$

$$K_{sp} = [Ag^{2+}]^2 [CrO_4^{2-}]$$

$$1.1 \times 10^{-12} = [Ag^{2+}]^2 [5 \times 10^{-13}]$$

$$\therefore \quad [Ag^+] = 1.48 \times 10^{-5} \text{ M} \approx 1.5 \times 10^{-5} \text{ M}$$

54. *(c)* For bcc lattice,

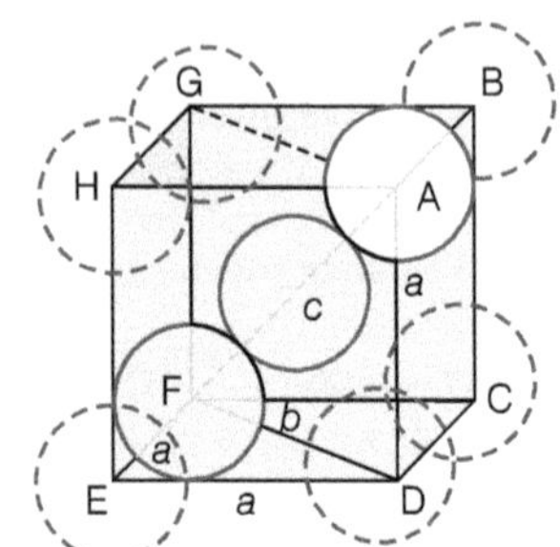

In ΔEFD

$$b^2 = a^2 + a^2 = 2a^2$$
$$b = \sqrt{2}a$$

Now, in ΔAFD

$$c^2 = a^2 + b^2$$
$$= a^2 + 2a^2 = 3a^2$$
$$c = \sqrt{3}a$$

Also, $\sqrt{3}a = 4r$

$$\frac{\sqrt{3}}{4}a = r$$

In bcc, $Z = 2$

$$\therefore \text{Volume of cube} = a^3 = \left(\frac{4}{\sqrt{3}}r\right)^3$$

Packing effeciency

$$= \frac{\text{Volume occupied by two spheres in a unit cell}}{\text{Total volume of unit cell}} \times 100$$

$$= \frac{2 \times \frac{4}{3}\pi r^3 \times 100}{[4/\sqrt{3}r]^3} = 68\%$$

55. *(b)*

	$X \longrightarrow$	$Y \longrightarrow$	Z
Initial, $t = 0$	A_0	0	0
conc., $t = t$	$A_0 - x$	y	z

where, y and z are number of moles of Y and Z, respectively formed at time t and $x = y + z$.

$\therefore$ Total number of moles at

$$t = t = A_0 - x + y + z$$

This total number of moles at time $= t$ will be more than that, at $t = 0$, i.e.

$$(A_0 - x + y + z) > A_0$$

As, it is a closed container, so total number of moles will remain constant throughout the time. Thus, plot (b) is correct option.

56. *(d)* Let number of α-particles emitted during radioactive decay $= n_1$
Let number of β-particles emitted during radioactive decay $= n_2$

$$^{238}U_{92} \longrightarrow {}^{206}Pb_{82} + n_1\alpha^{+1} + n_2\beta^{-1}$$

The α-particle represents a helium atom. In α-decay the atomic mass is reduced to four units.

$$\therefore \quad 238 - 206 = 4n_1$$

$$\Rightarrow \quad \frac{238 - 206}{4} = n_1$$

$$n_1 = 8$$

In β decay the atomic number is reduced to two units.

$$\therefore \quad 92 - 82 = 2n_2$$

$$\Rightarrow \quad \frac{12}{2} = n_2$$

$$n_2 = 6$$

57. *(d)* The allowed set of quantum numbers for an electron in an element must follow the given conditions
(i) the value of l ranges from 0 to $n - 1$
(ii) the value of m_l ranges from $-l$ to $+l$
(iii) value of m_s of an electron $= +\frac{1}{2}$ or $-\frac{1}{2}$

Thus, the allowed set of quantum numbers for an electron in a hydrogen atom is $n = 2, l = 1, m_l = -1, m_s = 1/2$.

58. *(c)* The relation between the extent of adsorption (x/m) and pressure (p) is given by Freundlich adsorption is other m.
According to this,

$$\frac{x}{m} \propto p^{1/n} \; (n > 1)$$

where, x is the mass of gas adsorbed on mass m of the adsorbent at pressure p, k and n are constant.

$$\therefore \quad \frac{x}{m} = kp^{1/n}$$

Taking log both sides

$$\log\frac{x}{m} = \frac{1}{n}\log p + \log k$$

Comparing the above equation with the equation of straight line, $y = mx + c$.
$\therefore$ The graph would be linear, where

$$\text{slope} = \frac{1}{n}, \; c = \log k$$

Thus, correct plot is (c).

59. *(a)* Given,
dissociation constant, $K_a = 1.8 \times 10^{-5}$
Concentration of $[H^+] = 0.1$ M
For dissociation of acetic acid

$$CH_3COOH \rightleftharpoons CH_3COO^- + H^+$$

$$H^+ = \sqrt{k_a c}$$
$$= \sqrt{1.8 \times 10^{-5} \times 0.1}$$
$$= \sqrt{1.8} \times 10^{-3}$$
$$= 1.34 \times 10^{-3}$$
$$pH = -\log[H^+]$$
$$= -\log(1.34 \times 10^{-3})$$
$$= 3 - \log 1.34$$
$$= 2.87$$

60. *(a)* The limiting molar conductivities of the given electrolytes can be calculated as follows :

$$\lambda^o_{KCl} = \lambda^o_{K^+} + \lambda^o_{Cl^-}$$
$$= 73.5 + 76.3 = 149.8 \text{ Scm}^2\text{ mol}^{-1}$$
$$\lambda^o_{CaCl_2} = \lambda^o_{Ca^{2+}} + 2\lambda^o_{Cl^-}$$
$$= 119 + 2 \times 76.3 = 271.6 \text{ Scm}^2\text{mol}^{-1}$$
$$\lambda^o_{K_2SO_4} = 2\lambda^o_{K^+} + \lambda^o_{SO_4^{2-}}$$
$$= 2 \times 73.5 + 160.0$$
$$= 307 \text{ Scm}^2\text{ mol}^{-1}$$

Thus, the limiting molar conductivities follow the order,

$$\lambda^o_{KCl} < \lambda^o_{CaCl_2} < \lambda^o_{K_2SO_4}.$$

61. *(a)* Resting membrane potential is potential difference across the plasma membrane when neuron is at rest. It is -70 mV.

62. *(b)* Amphimixis is the process of merging of the nuclei of the sperm (male gamete) and egg cells (female gamete) during sexual reproduction.

63. *(c)* The Sympathetic Nervous System (SNS) releases the hormones (catecholamines-epinephrine and norepinephrine) to accelerate the heart rate. The Sympathetic Nervous System (SNS) is a part of the Autonomic Nervous System (ANS), alongwith the Parasympathetic Nervous System (PNS). The SNS activates what is often termed as the body's fight or flight response (during emergency situations).

64. *(a)* Cholesterol (sterol) acts as a bidirectional regulator of membrane fluidity because at high temperature (above 40°C) it stabilises the membrane and raises its melting point, whereas at low temperatures (below 40°C) it intercalates between phospholipids and prevents them from clustering together and stiffening. Thus, option (a) is correct, as physiological temperature is the average temperature of human body is 37°C.

65. *(d)* Hyaluronic acid, formed by thousands of alternative units of N-acetyl glucosamine and glucuronic acid is an example of heteropolysaccharide. A heteropolysaccharide contain two or more different monosaccharide units.

66. *(b)* Every plasmid has its own 'origin of replication'-a stretch of DNA that ensures it gets replicated (copied) by the host bacterium.

For this reason, plasmids can copy themselves independently of the bacterial chromosome, so there can be many copies of a plasmid with in one bacterial cell.

67. *(a)* Prick testing involves introducing a needle into the upper layers of the skin through a drop of allergen extract and gently lifting the epidermis up. The release of preformed histamine from mast cells (which are formed due to allergy) causes increased vascular permeability *via* smooth muscle contraction and development of a wheal, inflammatory mediators initiate a neural reflex causing vasodilation, leading to erythema. Prick testing methods are the initial technique for detecting the IgE.

68. *(d)* DNA repair is defined as the cellular responses that are associated with the restoration of the normal base-pair sequence and structure of damaged DNA. Thus, it does not involve RNA directly other processes involves RNA as follows

(i) Transcription has its first step in which a particular segment of DNA is copied into *m*RNA.

(ii) DNA replication used RNA primers.

(iii) Translation involves decoding of *m*RNA to produce specific amino acid chain.

69. *(c)* Statement (c) is incorrect because one codon can code for a single amino acid only. But some amino acids are encoded by more than one codon.

70. *(b)* Lets take RR and rr as two homozygous parents. Thus, both genotypic and phenotypic ratios are same.

71. *(a)* The conditions to maintain the Hardy-Weinberg equilibrium are no mutation, no gene flow, large population size, random mating and no natural selection.

The Hardy-Weinberg equilibrium states that allele and genotype frequencies in a population will remains constant from generation to generations in the absence of other evolutionary influences.

72. *(c)* Inbreeding results in homozygosity which can increase the chances of offspring being affected by delecterious or recessive traits. Inbreeding is the production of offspring from the mating or breeding of individuals or organisms that are closely related genetically.

73. *(a)* 1, 3-bisphosphoglyceric acid is a 3-carbon organic molecule present in most, if not all, living organisms. It primarily exists as a metabolic intermediate in both glycolysis during respiration and the Calvin cycle during photosynthesis. It serves as a substrate for direct synthesis of ATP.

1, 3 bi PGA ⟶ 3 PGA

ADP → ATP

74. *(c)* If a pure chlorophyll solution is illuminated with ultraviolet light, the solution appears red. This happens due to the phenomenon called as fluorescence. In this, absorption of a photon causes a transition of the chlorophyll molecule from its ground state to its excited state. Chlorophyll absorbs red and blue light.

75. *(a)* *Tamarindus indica* (tamarind) is a leguminous tree in the family–Fabaceae or Papilionaceae indigenous to tropical Africa. The coconut tree (*Cocos nucifera*) is a member of the palm tree family–Arecaceae. *Colchicum autumnale* is a toxic autumn blooming flowering plant that belongs to family– Liliaceae. *Withania somnifera* (ashwagandha) is known as poison gooseberry belonging to Solanaceae family.

76. *(b)* Nitrogenases are enzymes used by some organisms to fix atmospheric nitrogen gas (N_2). It is sensitive to oxygen thus is protected by removal of excess oxygen by organisms in their various metabolic activities.

77. *(d)* Frederick Griffith concluded that the R-strain bacteria must have taken up what he called a 'transforming principle' (i.e. DNA) from the heat-killed S-bacteria (HKS), which allowed them to transform into smooth-coated bacteria and become virulent.

78. *(c)* Diabetic acidosis occurs when cells in the body do not receive the sugar they need for energy. This happen due to plenty of glucose in the bloodstream, but not enough insulin to help convert glucose for use in the cells. The body recognises this and start breaking down muscle and fat for energy. This breakdown produces amino acids (β-hydroxybutyric acid and acetoacetic acid), which cause a decrease in pH of blood plasma.

79. *(a)* Alveoli are air sacs for gaseous exchange in the body. If the number of alveoli in an individual is doubled without changing the total alveolar volume, the gas exchange capacity of the lungs will increase for both O_2 and CO_2 due to increase in the surface area of gaseous exchange.

80. *(d)* Centrifugation separates cell components on the basis of size and density. The larger and denser components experience the greatest centrifugal force and move most rapidly. After centrifugation the nuclei and cytoskeleton are present in the pellet. Thus in the experiment, since the bacteria were infected with ^{32}P-labelled virus it means ^{32}P gets incorporated in the DNA and the DNA is detected in the bacterial cell pellet.

81. *(d)* Area of region *OBAO*

$$= \text{area } (T) = 2\int_0^a \sqrt{4ax}\, dx$$

$$= 4\sqrt{a}\left[\frac{2}{3}x^{3/2}\right]_0^a = 4\sqrt{a}\left[\frac{2}{3}a^{3/2} - 0\right] = \frac{8}{3}a^2$$

Area of rectangle = (*PQRS*)

$$= RS \times RQ = 2y(a - x)$$

$$A = 2y\left(a - \frac{y^2}{4a}\right) \qquad \left[\because x = \frac{y^2}{4a}\right]$$

$$\frac{dA}{dy} = 2a - \frac{3y^2}{2a}$$

For maxima or minima $\frac{dA}{dy} = 0$

$$2a - \frac{3y^2}{2a} = 0 \Rightarrow y = \frac{2a}{\sqrt{3}}$$

$\therefore$ Area of rectangle

$$PQRS = 2 \times \frac{2a}{\sqrt{3}}\left(a - \frac{a}{3}\right) = \frac{8a^2}{3\sqrt{3}}$$

$$\frac{\text{Area of rectangle } PQRS}{\text{Area of rectangle } (T)} = \frac{8a^2/3\sqrt{3}}{8a^2/3} = \frac{1}{\sqrt{3}}$$

82. *(d)* Let S be the set of all permutation in $a_1, a_2, a_3, a_4, a_5, a_6$ of 1, 2, 3, 4, 5, 6

$\therefore \quad S = 6! = 720$

Let S' be the set of all permutation containing (2, 3, 4).

Total number $= (2! + 3! + 4!)3!$

$= (2 + 6 + 24) \times 6$

$= 32 \times 6 = 192$

Number of element in A = Number of element in S – Number of element in S'

$= 720 - 192 = 528$

83. *(b)* We have,

$$y = \frac{1}{4}|4 - x^2| \text{ and } y = 7 - |x|$$

Graph of given curves

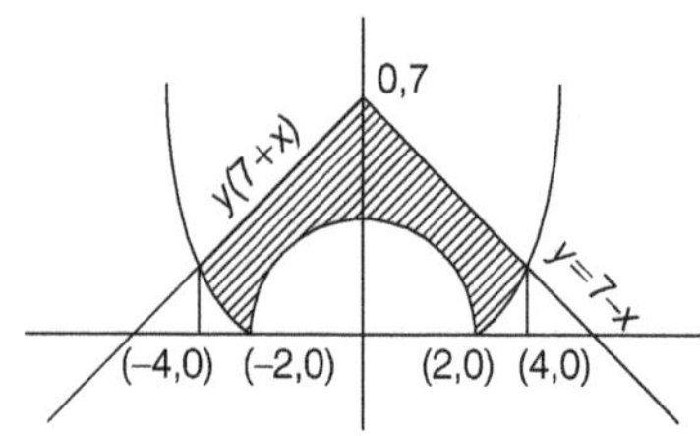

Area of shaded region

$$= 2\left[\int_0^4 (7 - x)dx - \int_0^4 \frac{1}{4}|4 - x^2|dx\right]$$

$$= 2\left[\int_0^4 (7 - x)dx - \frac{1}{4}\left[\int_0^2 (4 - x^2)dx + \int_2^4 (x^2 - 4)dx\right]\right]$$

$$= 2\left[\left[7x - \frac{x^2}{2}\right]_0^4 - \frac{1}{4}\left[4x - \frac{x^3}{3}\right]_0^2 - \frac{1}{4}\left[\frac{x^3}{3} - 4x\right]_2^4\right]$$

$$= 2\left[(28 - 8) - \frac{1}{4}\left(8 - \frac{8}{3}\right) - \frac{1}{4}\left(\left(\frac{16}{3} - 16\right) - \left(\frac{8}{3} - 8\right)\right)\right]$$

$$= 2\left[\left(20 - 2 + \frac{2}{3} - \frac{4}{3} - 4 + \frac{2}{3} + 2\right)\right]$$

$= 2(16) = 32$

84. *(b)* Equation of ellipse

$$\frac{x^2}{a^2} + \frac{y^2}{b^2} = 1, a > b$$

Equation of parabola $x^2 = 4(y + b)$

Foci of ellipse $(\pm ae, 0)$.

End of latusrectum of parabola

$= (\pm 2, 1 - b)$

D(–2,(1–b)) C(2,(1–b))

A(–ae,0) B(ae,0)

$ABCD$ is a square

$AB = CD \Rightarrow 2ae = 4 \Rightarrow ae = 2$

$BC = CD \Rightarrow BC^2 = CD^2$

$(2 - ae)^2 + (1 - b)^2 = 4^2$

$\Rightarrow \quad 0 + (1 - b)^2 = 4^2$

$\Rightarrow \quad 1 - b = \pm 4$

$b = 5, b = -3 \Rightarrow a = \sqrt{29}$ or $\sqrt{13}$

$\therefore \quad e = \sqrt{1 - \frac{b^2}{a^2}}$

$\Rightarrow \quad e = \sqrt{1 - \frac{9}{13}} = \frac{2}{\sqrt{13}}$ or 2

85. *(b)* We have,

A sector is removed from a metallic disc and remaining region is bent into the shape of circular conical funnel with volume $2\sqrt{3}\pi$

Let radius of disc $= r$

$\therefore$ r is the slant height of cone volume of cone $= \frac{1}{3}\pi x^2 h$

$$\Rightarrow \quad 2\sqrt{3}\pi = \frac{1}{3}\pi x^2 h$$

r r

$$x^2 h = 6\sqrt{3} \Rightarrow x^2 = \frac{6\sqrt{3}}{h}$$

$$r^2 = x^2 + h^2$$

$$r^2 = \frac{6\sqrt{3}}{h} + h^2$$

r h r x

Let $\quad r^2 = l$

$\therefore \quad l = \frac{6\sqrt{3}}{h} + h^2$

$$\frac{dl}{dh} = \frac{-6\sqrt{3}}{h^2} + 2h$$

For maxima or minima $\frac{dl}{dh} = 0$

$\because \quad h^3 = 3\sqrt{3}$

$\Rightarrow \quad h = \sqrt{3}$

$\Rightarrow \quad x^2 = \frac{6\sqrt{3}}{3} = 6$

$\therefore \quad r^2 = 6 + 3 = 9$

$\Rightarrow \quad r = 3$

$\therefore$ Diameter of disc

$= 2r = 2 \times 3 = 6$

86. *(c)* We have,

$$g(x) = \int_0^{|x|^{3/4}} t^{2/3} \sin\frac{1}{t}\,dt$$

$$g'(x) = |x|^{\frac{1}{2}} \sin\frac{1}{|x|^{3/4}}$$

$$\lim_{x \to 0}\frac{g(x)}{x} = \lim_{x \to 0}\frac{g'(x)}{1}$$

$$= \lim_{x \to 0}|x|^{1/2}\sin\frac{1}{|x|^{3/4}}$$

$= 0$

87. *(d)* We have,

$$a_n = \int_{-\pi}^{\pi}|x - 1|\cos nx\,dx$$

$$a_n = \int_{-\pi}^{1} -(x - 1)\cos nx\,dx - \int_1^{\pi}(x - 1)\cos nx\,dx$$

$$a_n = -\left[\frac{(x - 1)\sin nx}{n}\right]_{-\pi}^{1} - \left[\frac{\cos n\pi}{n^2}\right]_{-\pi}^{1} + \left[\frac{(x - 1)\sin nx}{n}\right]_1^{\pi} - \left[\frac{\cos n\pi}{n^2}\right]_1^{\pi}$$

$$a_n = \frac{2\pi\sin n\pi}{n} + \frac{2}{n^2}\cos n\pi - \frac{2}{n^2}\cos x\pi$$

$$a_n = \frac{2\pi\sin n\pi}{n}$$

$$\lim_{x \to \infty} a_n = \lim_{n \to \infty}\frac{2\pi\sin n\pi}{n} = 0$$

88. *(c)* Let $f(x) = ax + b$

$f(1) = a + b = 5 \quad \ldots$(i)

$f(2) = 2a + b = 7 \quad \ldots$(ii)

On solving Eqs. (i) and (ii), we get

$a = 2, b = 3$

$\therefore \quad f(x) = 2x + 3$

$f(12) = 2(12) + 3 = 27$

89. *(b)* Four balls 1, 2, 3, 4 are randomly placed in boxes B_1, B_2, B_3, B_4.

Probability of exactly one box is empty is

$$\frac{{}^4C_1 \times \frac{4!}{1! \times 2!} \times \frac{1}{2!} \times 3!}{4^4} = \frac{4 \times 6 \times 6}{4 \times 4 \times 4 \times 4} = \frac{9}{16}$$

90. *(a)* We have, $f(x) = \log(1+x^2)$

$$f'(x) = \frac{2x}{1+x^2}$$

Range of $f'(x) = [-1, 1]$

$$\frac{|f(x) - f(y)|}{|x - y|} \leq A$$

Maximum value of $\frac{|f(x) - f(y)|}{|x - y|}$ always less than A.

So, least possible value of A is 1.

91. *(b)*

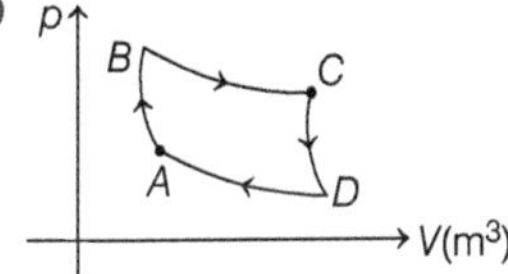

Given,

Step 1 A to B adiabatic compression, $V_A = 8\text{ m}^3, V_B = 1\text{ m}^3$

Step 2 B to C isothermal expansion,

$$T_C = T_B = T_1, V_C = 10\text{m}^3$$

Step 3 C to D adiabatic expansion,

$$V_D = 80\text{ m}^3.$$

Step 4 D to A isothermal compression,

$$T_D = T_A = T_2, V_A = 8\text{ m}^3$$

Now, for processes A to B,

$$T_A V_A^{\gamma-1} = T_B V_B^{\gamma-1}$$

Substituting values in above equation, we get

$$T_2 (8)^{\frac{5}{3}-1} = T_1 (1)^{\frac{5}{3}-1}$$

$\therefore$ For ideal monoatomic gas, $\gamma = \frac{5}{3}$

So, $$\frac{T_1}{T_2} = 8^{\frac{2}{3}} = 4$$

92. *(d)* Bullet provides cube an angular impulse which causes it to topple.

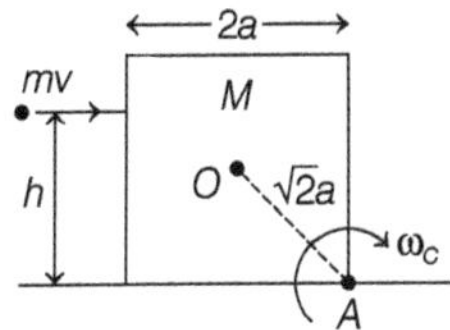

If I_A = moment of inertia of cube about point A and ω_C = initial angular speed about A.

Then, initial rotational kinetic energy of cube is

$$K_i = \frac{1}{2} I_A \omega_c^2$$

Moment of inertia of cube about point A, using parallel axes theorem is

$$I_A = I_0 + M(OA)^2$$

$$[\because I_0 = \frac{2}{3} Ma^2 \text{ and } OA = \sqrt{2}a]$$

$$= \frac{2Ma^2}{3} + M(\sqrt{2}a)^2 = \frac{8}{3} Ma^2$$

Note In question paper I_0 is misprinted as $\frac{2}{3} Ma^3$.

$\therefore$ Initial rotational kinetic energy,

$$K_i = \frac{1}{2}\left(\frac{8}{3} Ma^2\right)\omega_c^2 = \frac{4}{3} Ma^2 \omega_c^2$$

This rotation energy must be greater than or equal to potential energy of cube when its centre is at highest level above ground during toppling.

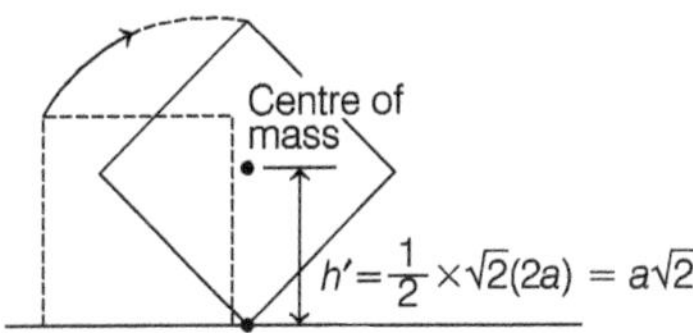

As, maximum height of centre of mass is $a\sqrt{2}$, potential energy of cube in this position is

$$U_f = Mg(h') = Mga\sqrt{2}$$

Now, by conservation of energy, we have

$$U_i + K_i = U_f + K_f$$

For critical (minimum) angular speed ω_c, $K_f = 0$.

So,

$$Mg\left(\frac{2a}{2}\right) + \frac{4}{3} Ma^2 \omega_c^2 = Mga\sqrt{2} + 0$$

$$\Rightarrow \quad \omega_c = \sqrt{\frac{3g(\sqrt{2}-1)}{4a}}$$

93. *(c)* When an impulse J is given to wooden plank AB, then there are two possibilities

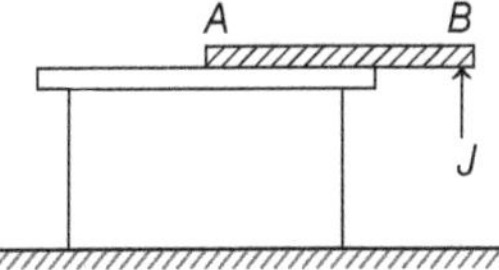

(i) Plank started to rotate about A as soon as impulse is imparted.

(ii) Plank moves vertically without rotation.

Now, we consider both cases one by one.

Case I

In case rotation occurs about A, angular momentum about A is conserved.

$$\Rightarrow \quad I_A \omega = JL$$

where, I_A = moment of inertia of plank about $A = \frac{ML^2}{3}$

and ω = angular speed of plank about A.

$$\frac{ML^2}{3}\omega = LJ$$

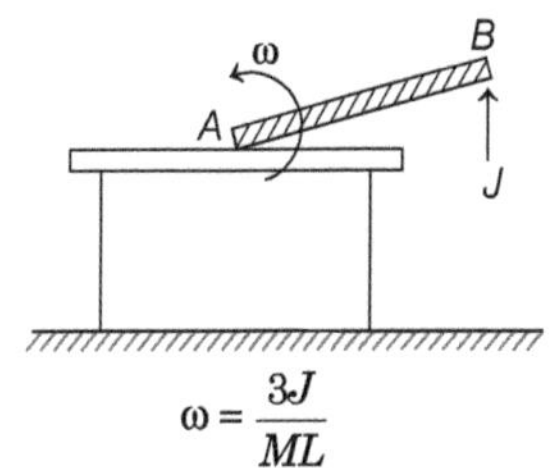

$$\Rightarrow \quad \omega = \frac{3J}{ML}$$

So, linear velocity of centre of mass,

$$v = \left(\frac{L}{2}\right) \times \omega \Rightarrow v = \frac{3J}{2M}$$

If h = maximum height attained by centre of mass, then

By energy conservation, we have

$$\frac{1}{2} mv^2 = mgh \Rightarrow h = \frac{v^2}{2g} = \frac{9J^2}{8M^2 g}$$

***Case* II** If there is no rotation of plank, then by conservation of linear momentum, we have

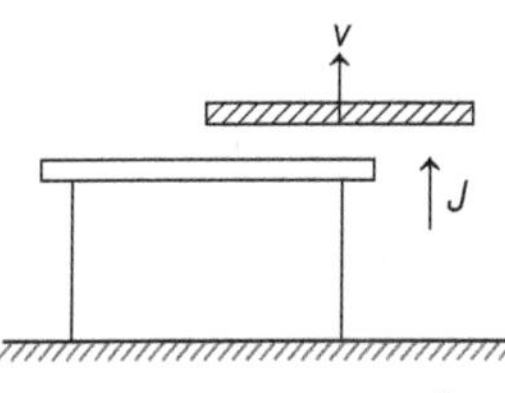

$$M \cdot v = J \Rightarrow v = \frac{J}{M}$$

and from conservation of energy, we have

$$\frac{1}{2} mv^2 = mgh \Rightarrow h = \frac{J^2}{2M^2 g}$$

Clearly, maximum height attained by centre of mass of plank is

$$\frac{J^2}{2M^2 g} < h < \frac{9J^2}{8M^2 g}$$

94. *(a)* As wire loop enters the region of magnetic field, an emf is induced in the wire loop. The current of induced emf causes an opposing force on the wire loop.

$$\times \quad \times \quad \times \quad \times B$$

$F \leftarrow$ [loop] $\rightarrow v$, $\rightarrow X$

This force is given by

$$F = -BIa = -B\left(\frac{E}{R}\right) \cdot a$$

$$= -B\left(\frac{Bav}{R}\right) a = \frac{-B^2 a^2 v}{R}$$

(Negative sign shows retarding force)

So, deacceleration A of loop is

$$A = \frac{F}{m} = \frac{-B^2 a^2 v}{mR}$$

$$\Rightarrow \quad \frac{dv}{dt} = \frac{-B^2 a^2}{mR} v \Rightarrow \frac{dv}{dx} \cdot \frac{dx}{dt} = \frac{-B^2 a^2 v}{mR}$$

$$\Rightarrow \quad v\frac{dv}{dx} = \frac{-B^2a^2}{mR}\cdot v$$

$$\Rightarrow \quad dv = \frac{-B^2a^2}{mR}\cdot dx$$

Integrating between limits, we have

$$\int_{v_0}^{v} dv = \int_0^x -\frac{B^2a^2}{mR}\cdot dx$$

$$v - v_0 = \frac{-B^2a^2}{mR}\cdot x$$

So, velocity of loop at distance x is

$$\Rightarrow \quad v = v_0 - \frac{B^2a^2}{mR}\cdot x$$

95. *(d)* In Bohr's theory, it is assumed that nucleus of H-atom is so heavy that it remained fixed at the circular orbit, while electron revolves around it. This assumption is true only when mass of nucleus is infinite compared to mass of electron.

Infact the nuclear mass is finite and in H-atom, nucleus is only about 2000 times (~ 1836 times) heavier than orbiting electron.

So, the assumption that nucleus is fixed is not justified.

The nucleus and electron must be revolving around their common centre of mass and orbit must be elliptical.

Let m_e is mass of electron and m_n is mass of nucleus, then distances of electron and nucleus from centre of mass must be

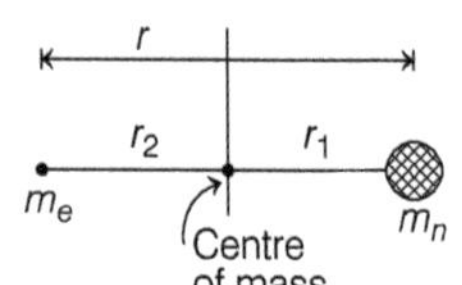

$$r_1 = \left(\frac{m_e}{m_n + m_e}\right) r \text{ and } r_2 = \left(\frac{m_n}{m_n + m_e}\right) r$$

So, orbital angular momentum of atom

$$= L_{\text{electron}} + L_{\text{proton}}$$

$$= m_e r_2^2\omega + m_n r_1^2\omega$$

$$= \left(\frac{m_e m_n}{m_e + m_n}\right) r^2\omega = \mu r^2\omega$$

where, $\frac{m_e m_n}{m_e + m_n} = \mu$ = reduced mass.

Now, by Bohr's postulate, we have

$$\mu\omega r^2 = \frac{nh}{2\pi} \Rightarrow \mu v r = \frac{nh}{2\pi}$$

$\therefore$ Energy of electron in nth orbit of a one electron atom is

$$E_n = -\left(\frac{\mu Z^2 e^4}{8n^2h^2\varepsilon_0^2}\right)$$

When electron jumps from n_2 to n_1 state, radiated wavelength will be

$$\frac{1}{\lambda} = \frac{\mu e^4 Z^2}{8h^3 c\varepsilon_0}\left(\frac{1}{n_1^2} - \frac{1}{n_2^2}\right)$$

Keeping other parameters same, for a given transition n_2 to n_1, we can say

$$\frac{1}{\lambda} \propto \mu Z^2 \Rightarrow \frac{1}{\lambda} = C\mu Z^2 \quad [\because C = \text{constant}]$$

or $$\frac{1}{\lambda} = C\left(\frac{m_e m_n}{m_e + m_n}\right)\cdot Z^2$$

Now, for H-atom ($Z = 1, A = 1$),

$$\frac{1}{\lambda_1} = C\cdot\left(\frac{m_e \times 2000\, m_e}{2001\, m_e}\right)\times 1$$

$$\Rightarrow \quad \frac{1}{\lambda_1} = C\cdot\frac{2000}{2001}\cdot m_e$$

$$\Rightarrow \quad \lambda_1 = \frac{2001}{2000C\cdot m_e} \qquad \ldots\text{(i)}$$

For deuterium atom ($Z = 1, A = 2$),

$$\frac{1}{\lambda_2} = C\cdot\left(\frac{m_e \times 4000\, m_e}{4001\, m_e}\right)\times 1$$

$$\Rightarrow \frac{1}{\lambda_2} = C\cdot\frac{4000\, m_e}{4001} \Rightarrow \lambda_2 = \frac{4001}{4000\,C\cdot m_e} \qquad \ldots\text{(ii)}$$

Subtraction Eq. (ii) by Eq. (i), we have

$$\Rightarrow |\lambda_2 - \lambda_1| = \left|\frac{4001}{4000\,C\cdot m_e} - \frac{2001}{2000C\cdot m_e}\right|$$

$$= \frac{2}{4000\,C\cdot m_e}$$

$$\Rightarrow \left|\frac{\lambda_2 - \lambda_1}{\lambda_1}\right| = \frac{\left(\dfrac{2}{4000\,C\cdot m_e}\right)}{\left(\dfrac{2001}{2000\,C\cdot m_e}\right)} = \frac{1}{2001}$$

So, % change in wavelength is

$$\frac{\Delta\lambda}{\lambda_1}\times 100 = \frac{1}{2001}\times 100 \approx 0.05\%$$

96. *(c)* In a solar cell, when light falls at the junction ($E_g < hf$). Electron hole pairs generated are moved to opposite side by junction field. Electrons towards n-side and holes towards p-side.

If no load is connected they accumulate on n and p-sides, giving rise to a photo-voltage.

When external circuit is complete, a photo-current I_L flows.

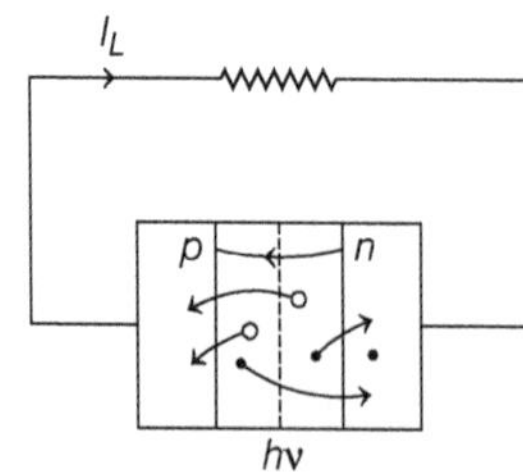

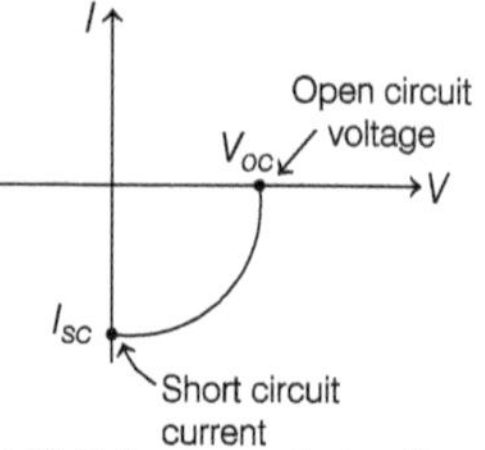

Note that V-I characteristics lies in fourth quadrant of coordinate axes.

97. *(b)* ***Case* A** Here only force acting on weighing pan is weight of water.

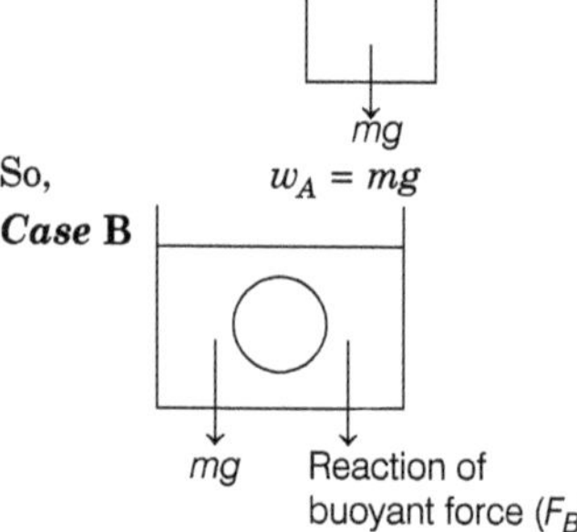

So, $\quad w_A = mg$

***Case* B**

In this case, downward forces are weight and reaction of buoyant force.

$\therefore \quad w_B = mg + F_B$

***Case* C**

mg Reaction of buoyant force (F_B)

In this case, downward acting forces are weight and reaction of buoyant force which is same is as that of case B as balls are of same size.

$\therefore \quad w_C = mg + F_B$

***Case* D**

T F_B mg m′g

In this case, forces acting on bottom of beaker are

$m'g$: weight of ball

mg : weight of water

F_B : reaction of buoyant force

T : tension in string

Also, $\quad T = F_B$

So, net downward force on bottom of beaker is $mg + m'g$.

$\therefore \quad w_D = mg + m'g$

So, $\quad w_B = w_C > w_D > w_A$

98. *(a)* Following ray diagram shows path of reflected and refracted rays.

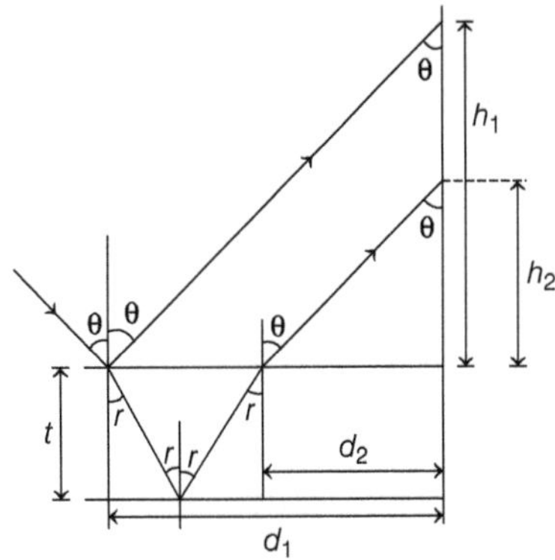

From above diagram, we have

$$\tan\theta = \frac{d_1}{h_1} \text{ and } \tan\theta = \frac{d_2}{h_2}$$

So, $$h_1 - h_2 = \frac{d_1}{\tan\theta} - \frac{d_2}{\tan\theta}$$

$$\Rightarrow \quad h_1 - h_2 = \frac{d_1 - d_2}{\tan\theta}$$

Now, as $$\frac{\sin\theta}{\sin r} = n$$

$$\Rightarrow \quad \sin r = \frac{\sin\theta}{n}$$

So, $$\tan r = \frac{d_1 - d_2}{2t}$$

$$\Rightarrow \quad d_1 - d_2 = 2t \cdot \tan r$$

$$\Rightarrow \quad d_1 - d_2 = \frac{2t \cdot \sin r}{\sqrt{1 - \sin^2 r}} = \frac{\frac{2t}{n}\sin\theta}{\sqrt{1 - \left(\frac{\sin^2\theta}{n^2}\right)}} = \frac{2t\sin\theta}{\sqrt{n^2 - \sin^2\theta}}$$

Hence, $$h_1 - h_2 = \frac{d_1 - d_2}{\tan\theta}.$$

or $$h_1 - h_2 = \left(\frac{2t\sin\theta}{\sqrt{n^2 - \sin^2\theta}} \times \frac{1}{\tan\theta}\right)$$

$$h_1 - h_2 = \frac{2t\cos\theta}{\sqrt{n^2 - \sin^2\theta}}$$

99. *(d)* Photoemission is an instantaneous process. So, statement I is correct. Threshold frequency exists for every metal. So, statement II is correct.

Energy of incident photon ∝ Frequency ∝ Stopping potential. So, statement III is correct.

Photo-current ∝ Intensity of light but only above threshold frequency.

So, statement IV is not fully correct.

100. *(a)* Current in a series *L-C-R* circuit is given by

$$I = \frac{V}{Z} = \frac{V}{\sqrt{(X_L - X_C)^2 + R^2}}$$

$$= \frac{200}{\sqrt{\left(600\pi - \frac{50}{\pi}\right)^2 + (400)^2}} = 0.12\text{ A}$$

101. *(a)*

Benzoyl chloride (PhCOCl) $\xrightarrow[\text{Rosenmund reaction}]{\text{(i) } H_2, Pd/BaSO_4 \ (X)}$ Benzaldehyde (PhCHO) $\xrightarrow[\text{Perkin reaction}]{\text{(ii) NaOAc, } Ac_2O}$ Cinnamic acid (Ph—CH=CH—COOH)

In step (i), benzoyl chloride undergoes reduction in the presence of $H_2, Pd/BaSO_4$, to give benzaldehyde. This reaction is known as rosenmund reduction. The reagent is selective reducing agent, which reduces acetyl chloride to aldehyde group.

In step (ii), benzaldehyde undergoes aldol condensation with acid anhydride in the presence of an alkali salt of the acid (NaOAc) to give cinnamic acid (α, β-unsaturated aromatic acid). This reaction is known as Perkin's reaction.

102. *(d)*

PhCN $\xrightarrow[\text{Stephen reduction}]{\text{(i) } SnCl_2/HCl}$ Ph—CH=NH (Imine compound) $\xrightarrow{\text{(ii) } H_3O^+}$ PhCHO (*X*)

(*X*) + (iii) H_3C—C_6H_4—CO—CH_3 $\xrightarrow{\text{NaOH (Aldol condensation)}}$ H_3C—C_6H_4—CO—CH=CH—Ph (*Y*)

In step (i), the nitrile group is reduced to imine group, which then undergo hydrolysis to give aldehyde group. This reaction is known as stephen reduction.

In step (iii), the formed benzaldehyde then undergoes aldol condensation to give α, β-unsaturated aromatic compound.

103. *(c)* Acetophenone reacts with perbenzoic acid to produce compound phenylacetate, Ph—O—C(=O)—CH_3.

This reaction is known as Baeyer Villiger oxidation, where a ketone group gets converted an ester in presence of a peroxide group.

Reaction of *X* with excess of CH_3MgBr followed by treatment with aqueous acid predominantly produces *tert*-butyl alcohol.

Ph—C(=O)—CH_3 $\xrightarrow[\text{Baeyer Villiger oxidation}]{PhCO_3H \text{ (Peroxy benzoic acid)}}$ Ph—O—C(=O)—CH_3 (*X*)

Ph—O—C(=O)—CH_3 $\xrightarrow{CH_3MgBr}$ CH_3—C(=O)—CH_3 $\xrightarrow{CH_3MgBr}$ CH_3—C(CH_3)($\overset{+}{O}MgBr$)—CH_3 $\xrightarrow{H_3O^+}$ CH_3—C(CH_3)(OH)—CH_3

104. *(a)* The fusion of chromite ore ($FeCr_2O_4$) with Na_2CO_3 in air to produce Na_2CrO_4, a yellow coloured solution in aqueous solution. Subsequent treatment of, it with H_2SO_4 produces $Na_2Cr_2O_7$, which is orange coloured solution.

$$8Na_2CO_3 + 4FeCr_2O_4 + 7O_2 \longrightarrow \underset{\text{Yellow colour}}{Na_2CrO_4} + 2Fe_2O_3 + 8O_2$$

$$2Na_2CrO_4 \xrightarrow{2H^+} \underset{\text{Orange}}{Na_2Cr_2O_7} + 2Na^+ + H_2O$$

105. *(d)* The oxidation state of Ni in $[Ni(CN)_4]^{2-}$ is +2. The electronic configuration of Ni^{2+} is [Ar] $3d^8\ 4s^0$.

Ni^{2+} Ground state: 3d [⇅ | ⇅ | ⇅ | ↑ | ↑] 4s [ ] 4p [| |]

AsCN⁻ is a strong field ligand, so it will pair up electrons.

$[Ni(CN)_4]^{2-}$: 3d [⇅ | ⇅ | ⇅ | ⇅ |] 4s [ ] 4p [| |] — dsp^2

The hybridisation of Ni in $[Ni(CN)_4]^{2-}$ is dsp^2 and thus, it will a square planar geometry.

$$\left[\begin{matrix} NC & & CN \\ & Ni & \\ NC & & CN \end{matrix}\right]^{2-}$$

106. *(c)* Geometrical isomers arises in heteroleptic complexes due to different possible geometric arrangements of ligand.

There are 5 geometrical isomers possible for an octahedral complex of the type $[MA_2B_2C_2]$, which are as follows:

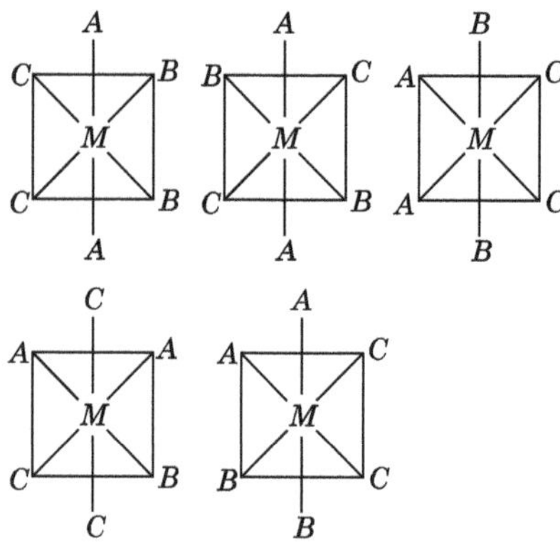

107. *(a)* For the reaction,

$$CO(g) + \frac{1}{2}O_2(g) \longrightarrow CO_2(g),$$

$$\Delta H_C = -283.5$$

$$\Delta S = \Sigma S_P - \Sigma S_R$$
$$= 213.7 - \left(197.7 + \frac{1}{2} \times 205.1\right)$$
$$= -86.5 \text{ J mol}^{-1}$$
$$\Delta G = \Delta H - T\Delta S$$
$$= -283 - 298 \times -86.5 \times 10^{-3}$$
$$= -257.0 \text{ kJ mol}^{-1}$$
$$w_{max} = -\Delta G = -(-2570.0) = 257 \text{ kJ mol}^{-1}$$

108. *(c)* Given, weight of glucose ($C_6H_{12}O_6$), $w_2 = 18$ g

Weight of water, $w_1 = 1$kg

K_b of water $= 0.52$ K kg mol^{-1}

Molar mass of glucose

$$= 12 \times 6 + 12 + 16 \times 6$$
$$= 72 + 12 + 96 = 180$$

Number of moles of glucose $= 18/180 = \frac{1}{10}$

As we know, $\Delta T_b = K_b \times m$

$$\therefore \quad \Delta T_b = 0.52 \times \frac{\frac{1}{10}}{1}$$
$$\Delta T_b = 0.052$$

Also, $\Delta T_b = T_b - T_b^0$

$$0.052 = T_b - (373.15)$$
$$T_b = 0.052 + 373.15 = 373.2 \text{ K}$$

109. *(b)* Given,

density of polonium $= 9.32$ g cm^{-3}

As we know, $\rho = \frac{M \times Z}{V \times N_A}$

where, ρ = Density

M = Mass of the element

Z = Number of atom in a unit lattice

V = Volume

N_A = Avogadro's number

For scc, $Z = 1$

$$\therefore \quad 9.32 = \frac{209 \times 1}{a^3 \times 6.022 \times 10^{23}}$$
$$a^3 = 37.2 \times 10^{-24}$$
$$a = 3.33 \times 10^{-8} \text{ cm} \approx 334 \text{ pm}$$

110. *(b)* For the reaction,

$$A^{2+}(aq) + B(s) \longrightarrow B^{2+}(aq) + A(s)$$

According to Nernst equation,

$$E_{cell} = E^\circ_{cell} - \frac{0.0591}{n}\log\frac{[P]}{[R]}$$
$$1.091 = E^\circ_{cell} - \frac{0.0591}{2}\log\frac{[B^{2+}]}{[A^{2+}]}$$
$$1.091 = E^\circ_{cell} - \frac{0.0591}{2}\log\frac{2 \times 10^{-3}}{4 \times 10^{-3}}$$
$$E^\circ_{cell} = 1.099$$
$$\Delta G = -2.303\, RT \log K_{eq} \quad \ldots(i)$$

Also, $\Delta G = -nFE^\circ$ ...(ii)

Equating Eqs. (i) and (ii)

$$-nFE^\circ = -2.303\, RT\log K_{eq}$$
$$= -2 \times 96500 \times 1.099$$
$$= -2.303 \times 8.314 \times 298 \log K_{eq}$$
$$\log K_{eq} = \frac{2 \times 96500 \times 1.099}{2.303 \times 8.314 \times 298}$$
$$\log K_{eq} = 37.17$$
$$\therefore \quad K_{eq} = 2 \times 10^{37}$$

111. *(b)* In a Mendelian trihybrid cross between *AABBCC* and *aabbcc* parents the F_2 progenies produced are 640.

$\therefore$ The number of parental combinations

$$= \frac{2}{64} \times 640 = 20$$

112. *(d)* Probability of having 2 boys and 1 girl $= \frac{3}{8}$

Probability of 1 boy and 2 girls $= \frac{3}{8}$

Total probability $= \frac{3}{8} + \frac{3}{8} = \frac{6}{8} = 75\%$

113. *(d)* There will be 2 restriction sites for X and 2 restriction sites for Y. This is because there occurs two bands for X enzymes of 6Kb and 4Kb. It means the plasmid is of 10Kb. And also there occurs one thick for Y enzymes of 5Kb. The 5Kb band can be thick due to the fact that the Y enzymes cut the plasmid at 5Kb and 5Kb sites.

114. *(a)* Since, the DNA replication is semiconservative (i.e., one strand of the parent double helix is conserved in each new DNA molecule), the proportion of light and hybrid bands in the experiment would be 50% light and 50% hybrid DNA.

115. *(b)* Mutualism describes the ecological interaction between two species where both are benefitted.

When X benefits and Y is disadvantaged it is predation or parasitism.

When both X and Y are disadvantaged it is competition.

116. *(c)* Proteins are made up of amino acids, oligosaccharide of hexose and oligonucleotide of nucleotides. We known that the molecular weight of nucleotide is 500 Da approx, hexose is 180 Da approx and amino acid is 110 Da approx.

Thus, correct order of molecular weights is N > O > P

Necleotide > Oligosaccharide > Protein

117. *(a)* The octapeptide NH_2-Asn-Glu-Tyr-Lys-Trp-Met-Glu-Gly on complete digestion with protease will give following results. On digestion with trypsin

NH_2-Asn-Glu-Try-Lys (Pentamer)

and Trp-Met-Glu-Gly (Tetramer)

On digestion with chymotrypsin

NH_2-Asn-Glu-Tyr (Tetramer) Lys-Trp (Dimer) Met-Glu-Gly (Trimer)

On digestion with cyanogen bromide

NH_2-Asn-Glu-Tyr-Lys-Trp-Met (Hexamer) Glu-Gly (Dimer)

118. *(c)* **Transaminase** are enzymes that catalyses a transamination reaction between an amino acid and an α-keto acid. **Protein kinase** modifies other molecules, mostly proteins by chemically adding phosphate groups to them. **Protein phosphatases** is an enzyme that removes a phosphate group from a protein. A **dehydrogenase** is an enzyme belonging to the group of oxidoreductases that oxidises a substrate by reducing an electron acceptor, usually NAD^+ / $NADP^+$ or flavin coenzyme such as FAD or FMN.

119. *(a)* Blood pressure is highest within the large arteries (such as the aorta) because they are connected directly to the ventricle of the heart. Compared to aorta blood pressure is low in capillaries. Thus, graph (A) is correct.

120. *(b)* Autosomal recessive traits are typically not seen in every generation of an affected family. An autosomal dominant is seen in every generation and affects males and females in the same way. Thus, in the family 1 pedigree, the trait is autosomal dominant (because present in each generation) and in the family 2 pedigree the trait is autosomal recessive (because it is not present in each generation).

QUESTION PAPER 2017

Stream : SB/SX (Nov 5)

MM : 160

Instructions

1. There are 120 questions in this paper.
2. The question paper contains two parts; Part I (1 Mark Questions) and Part II (2 Marks Questions).
3. There are four sections in each part; Mathematics, Physics, Chemistry and Biology.
4. Out of the four options given with each question, only one is correct.

PART-I (1 Mark Questions)

MATHEMATICS

1. Let BC be a fixed line segment in the plane. The locus of a point A such that the ΔABC is isosceles, is (with finitely many possible exceptional points)

(a) a line
(b) a circle
(c) the union of a circle and a line
(d) the union of two circles and a line

2. The number of solution pairs (x, y) of the simultaneous equations $\log_{1/3}(x+y)+\log_3(x-y)=2$ $2^{y^2}=512^{x+1}$ is

(a) 0 (b) 1
(c) 2 (d) 3

3. The value of the limit $\lim_{x\to-\infty}(\sqrt{4x^2-x}+2x)$ is

(a) $-\infty$ (b) $-\frac{1}{4}$
(c) 0 (d) $\frac{1}{4}$

4. Let R be a relation on the set of all natural numbers given by $a\,R\,b \Leftrightarrow a$ divides b^2.

Which of the following properties does R satisfy?

I. Reflexivity II. Symmetry
III. Transitivity

(a) I only (b) III only
(c) I and III only (d) I and II only

5. The fractional part of a real number x is $x-[x]$, where $[x]$ is the greatest integer less than or equal to x. Let F_1 and F_2 be the fractional parts of $(44-\sqrt{2017})^{2017}$ and $(44+\sqrt{2017})^{2017}$, respectively. Then, F_1+F_2 lies between the numbers

(a) 0 and 0.45 (b) 0.45 and 0.9
(c) 0.9 and 1.35 (d) 1.35 and 1.8

6. The number of real solutions of the equation $2\sin 3x+\sin 7x-3=0$, which lie in the interval $[-2\pi, 2\pi]$ is

(a) 1 (b) 2 (c) 3 (d) 4

7. Suppose p, q, r are real numbers such that
$q = p(4-p),\ r = q(4-q),\ p = r(4-r)$.
The maximum possible value of $p+q+r$ is
(a) 0 (b) 3 (c) 9 (d) 27

8. The parabola $y^2 = 4x+1$ divides the disc $x^2+y^2 \le 1$ into two regions with areas A_1 and A_2. Then, $|A_1 - A_2|$ equals
(a) $\frac{1}{3}$ (b) $\frac{2}{3}$ (c) $\frac{\pi}{4}$ (d) $\frac{\pi}{3}$

9. A shooter can hit a given target with probability $\frac{1}{4}$.
She keeps firing a bullet at the target until she hits it successfully three times and then she stops firing. The probability that she fires exactly six bullets lies in the interval
(a) (0.5272, 0.5274) (b) (0.2636, 0.2638)
(c) (0.1317, 0.1319) (d) (0.0658, 0.0660)

10. Consider the following events :
E_1 : Six fair dice are rolled and at least one die shows six.
E_2 : Twelve fair dice are rolled and at least two dice show six.
Let p_1 be the probability of E_1 and p_2 be the probability of E_2. Which of the following is true?
(a) $p_1 > p_2$ (b) $p_1 = p_2 = 0.6651$
(c) $p_1 < p_2$ (d) $p_1 = p_2 = 0.3349$

11. For how many different values of a does the following system have at least two distinct solutions?

$$ax + y = 0$$
$$x + (a+10)y = 0$$

(a) 0 (b) 1
(c) 2 (d) Infinitely many

12. Let R be the set of real numbers and $f : R \to R$ be defined by $f(x) = \frac{\{x\}}{1+[x]^2}$, where $[x]$ is the greatest integer less than or equal to x, and $\{x\} = x - [x]$. Which of the following statements are true?

I. The range of f is a closed interval.
II. f is continuous on R.
III. f is one-one on R.

(a) I only (b) II only
(c) III only (d) None of I, II and III

13. Let $x_n = (2^n + 3^n)^{1/2n}$ for all natural numbers n. Then,
(a) $\lim_{n\to\infty} x_n = \infty$ (b) $\lim_{n\to\infty} x_n = \sqrt{3}$
(c) $\lim_{n\to\infty} x_n = \sqrt{3} + \sqrt{2}$ (d) $\lim_{n\to\infty} x_n = \sqrt{5}$

14. One of the solutions of the equation $8\sin^3\theta - 7\sin\theta + \sqrt{3}\cos\theta = 0$ lies in the interval
(a) (0°, 10°] (b) (10°, 20°] (c) (20°, 30°] (d) (30°, 40°]

15. Let a, b, c, d, e be real numbers such that $a+b<c+d$, $b+c<d+e$, $c+d<e+a$, $d+e<a+b$. Then,
(a) the largest is a and the smallest is b.
(b) the largest is a and the smallest is c.
(c) the largest is c and the smallest is e.
(d) the largest is c and the smallest is b.

16. If a fair coin is tossed 5 times, the probability that heads does not occur two or more times in a row is
(a) $\frac{12}{2^5}$ (b) $\frac{13}{2^5}$ (c) $\frac{14}{2^5}$ (d) $\frac{15}{2^5}$

17. Consider the following parametric equation of a curve :

$$x(\theta) = |\cos 4\theta| \cos\theta$$
$$y(\theta) = |\cos 4\theta| \sin\theta$$

for $0 \le \theta \le 2\pi$

Which one of the following graphs represents the curve?

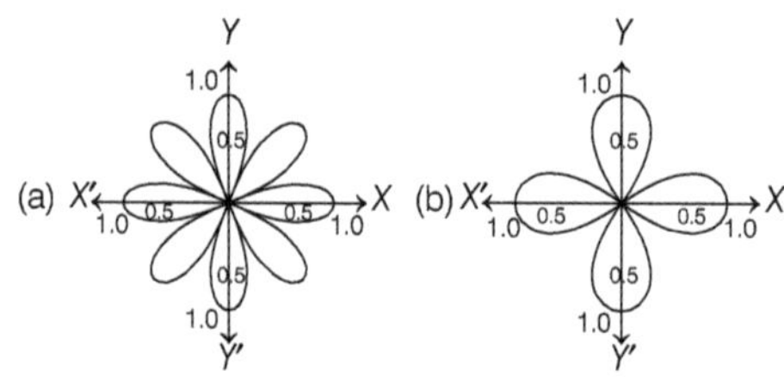

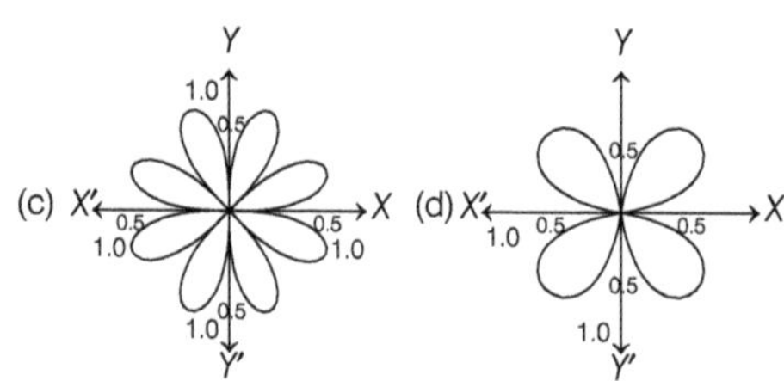

18. Let $A = (a_1, a_2)$ and $B = (b_1, b_2)$ be two points in the plane with integer coordinates. Which one of the following is not a possible value of the distance between A and B?
(a) $\sqrt{65}$ (b) $\sqrt{74}$ (c) $\sqrt{83}$ (d) $\sqrt{97}$

19. Let $f(x) = \max\left\{3, x^2, \frac{1}{x^2}\right\}$ for $\frac{1}{2} \le x \le 2$. Then, the value of the integral $\int_{1/2}^{2} f(x)dx$ is
(a) $\frac{11}{3}$ (b) $\frac{13}{3}$ (c) $\frac{14}{3}$ (d) $\frac{16}{3}$

20. Let $a_1 = i + \frac{1}{i}$ for $i = 1, 2, \ldots, 20$. Put

$p = \frac{1}{20}(a_1 + a_2 + \ldots + a_{20})$ and

$q = \frac{1}{20}\left(\frac{1}{a_1} + \frac{1}{a_2} + \ldots + \frac{1}{a_{20}}\right)$. Then,

(a) $q \in \left(0, \frac{22-p}{21}\right)$ (b) $q \in \left(\frac{22-p}{21}, \frac{2(22-p)}{21}\right)$
(c) $q \in \left(\frac{2(22-p)}{21}, \frac{22-p}{7}\right)$ (d) $q \in \left(\frac{22-p}{7}, \frac{4(22-p)}{21}\right)$

PHYSICS

21. The magnitude of acceleration of the electron in the *n*th orbit of hydrogen atom is a_H and that of singly ionised helium atom is a_{He}. The ratio of $a_H : a_{He}$ is

(a) 1 : 8 (b) 1 : 4
(c) 1 : 2 (d) dependent on n

22. A carrot looks orange in colour because of the β carotene molecule in it. This means that the β carotene molecule absorbs light of wavelengths

(a) longer than 550 nm (b) shorter than 550 nm
(c) longer than 700 nm (d) shorter than 700 nm

23. If some charge is given to a solid metallic sphere, the field inside remains zero and by Gauss's law all the charge resides on the surface. Now, suppose that Coulomb's force between two charges varies as $1/r^3$. Then, for a charged solid metallic sphere

(a) field inside will be zero and charge density inside will be zero
(b) field inside will not be zero and charge density inside will not be zero
(c) field inside will not be zero and charge density inside will be zero
(d) field inside will be zero and charge density inside will not be zero

24. Using dimensional analysis, the resistivity in terms of fundamental constants $h, m_e, c, e, \varepsilon_0$ can be expressed as

(a) $\frac{h}{\varepsilon_0 m_e c e^2}$ (b) $\frac{\varepsilon_0 m_e c e^2}{h}$
(c) $\frac{h^2}{m_e c e^2}$ (d) $\frac{m_e \varepsilon_0}{c e^2}$

25. Consider a bowl filled with water on which some black pepper powder have been sprinkled uniformly. Now, a drop of liquid soap is added at the centre of the surface of water. The picture of the surface immediately after this will look like

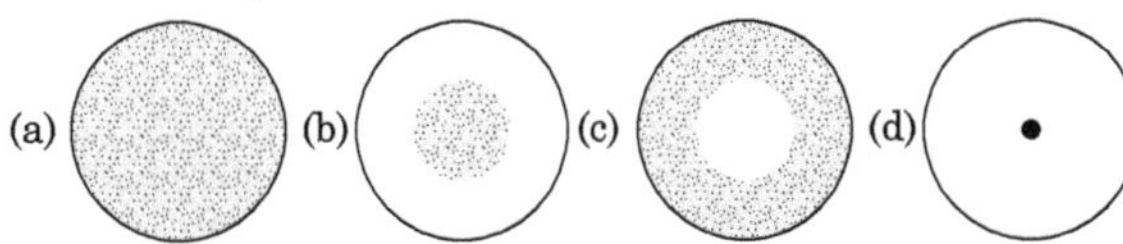

26. It was found that the refractive index of material of a certain prism varied as $1.5 + 0.004/\lambda^2$, where λ is the wavelength of light used to measure the refractive index. The same material was then used to construct a thin prism of apex angle 10°. Angles of minimum deviation δ_m of the prism were recorded for the sources with wavelengths λ_1 and λ_2, respectively. Then,

(a) $\delta_m(\lambda_1) < \delta_m(\lambda_2)$ if $\lambda_1 < \lambda_2$
(b) $\delta_m(\lambda_1) > \delta_m(\lambda_2)$ if $\lambda_1 > \lambda_2$
(c) $\delta_m(\lambda_1) > \delta_m(\lambda_2)$ if $\lambda_1 < \lambda_2$
(d) δ_m is the same in both cases

27. Two circularly shaped linear polarisers are placed coaxially. The transmission axis of the first polariser is at 30° from the vertical while the second one is at 60°, both in the clockwise sense. If an unpolarised beam of light of intensity $I = 20\ \text{W/m}^2$ is incident on this pair of polarisers, then the intensities I_1 and I_2 transmitted by the first and second polarisers respectively, will be close to

(a) $I_1 = 10.0\ \text{W/m}^2$ and $I_2 = 7.5\ \text{W/m}^2$
(b) $I_1 = 20\ \text{W/m}^2$ and $I_2 = 15\ \text{W/m}^2$
(c) $I_1 = 10.0\ \text{W/m}^2$ and $I_2 = 8.6\ \text{W/m}^2$
(d) $I_1 = 15.0\ \text{W/m}^2$ and $I_2 = 0.0\ \text{W/m}^2$

28. An electron in an electron microscope with initial velocity $v_0\hat{i}$ enters a region of a stray transverse electric field $E_0\hat{j}$. The time taken for the change in its de-Broglie wavelength from the initial value of λ to λ/3 is proportional to

(a) E_0 (b) $\frac{1}{E_0}$ (c) $\frac{1}{\sqrt{E_0}}$ (d) $\sqrt{E_0}$

29. A bird sitting on a single high tension wire does not get electrocuted because

(a) the circuit is not complete
(b) the bird feet has an insulating covering
(c) capacitance of the bird is too small and the line frequency is too small
(d) resistance of the bird is too high

30. A positive charge q is placed at the centre of a neutral hollow cylindrical conducting shell with its cross-section as shown in the figure below.

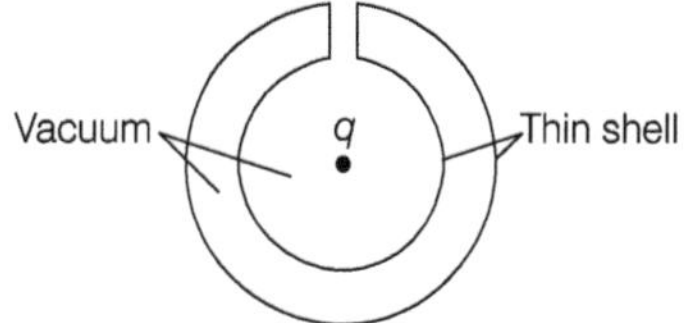

Which one of the following figures correctly indicates the induced charge distribution on the conductor? (Ignore edge effects)

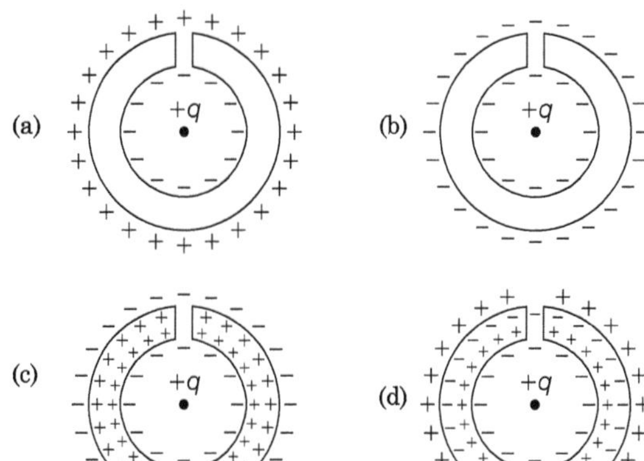

31. A transverse wave of frequency 500 Hz and speed 100 m/s is travelling in the positive x-direction on a long string. At time $t = 0$ s, the displacements at $x = 0.0$ m and at $x = 0.25$ m are 0.0 m and 0.02 m, respectively. The displacement at $x = 0.2$ m at $t = 5 \times 10^{-4}$ s is

(a) – 0.04 m (b) – 0.02 m (c) 0.04 m (d) 0.02 m

32. A thin piece of thermal conductor of constant thermal conductivity insulated on the lateral sides connects two reservoirs which are maintained at temperatures T_1 and T_2 as shown in the figure alongside. Assuming that the system is in steady state, which of the following plots best represents the dependence of the rate of change of entropy on the ratio of T_1 / T_2?

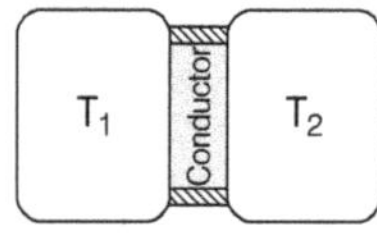

(a)
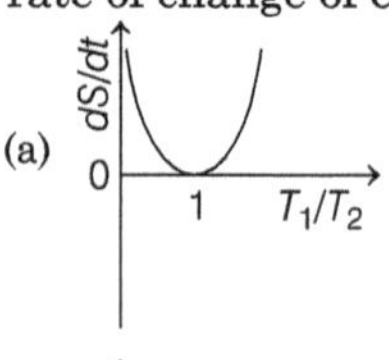

(b)
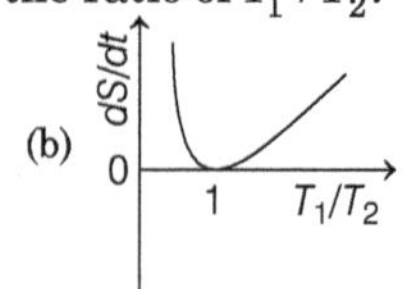

(c)
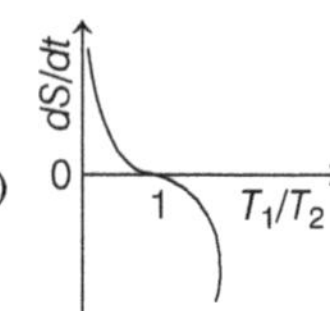

(d)
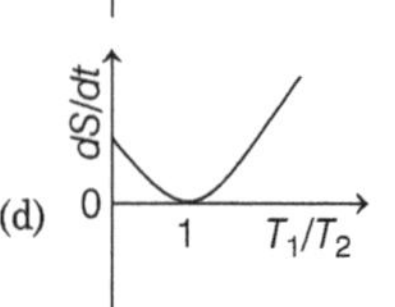

33. Which of the following plots represents schematically the dependence of the time period of a pendulum, if measured and plotted as a function of the amplitude of its oscillations? (**Note** amplitude need not be small)

(a)
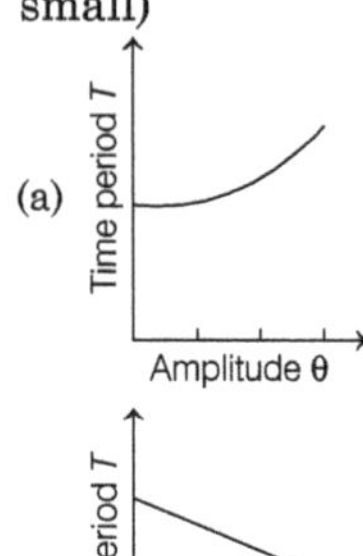

(b)
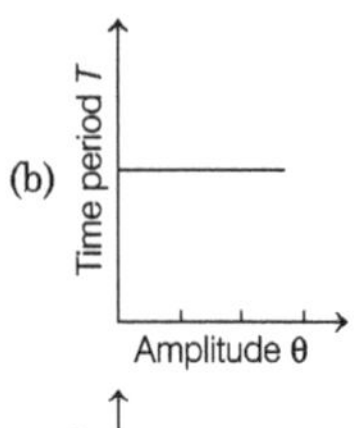

(c)
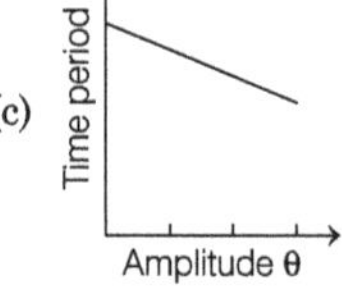

(d)
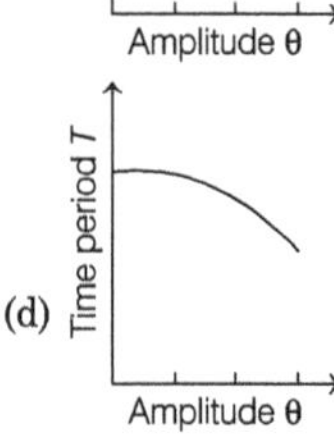

34. On a pulley of mass M hangs a rope with two masses m_1 and m_2 $(m_1 > m_2)$ tied at the ends as shown in the figure below.

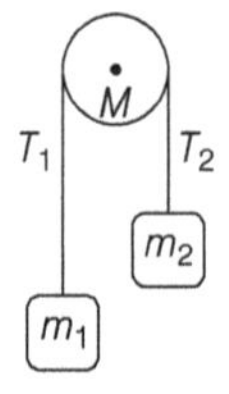

The pulley rotates without any friction, whereas the friction between the rope and the pulley is large enough to prevent any slipping. Which of the following plots best represents the difference between the tensions in the rope on the two sides of the pulley as a function of the mass of the pulley?

(a)
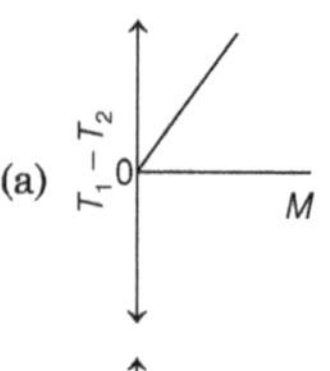

(b)
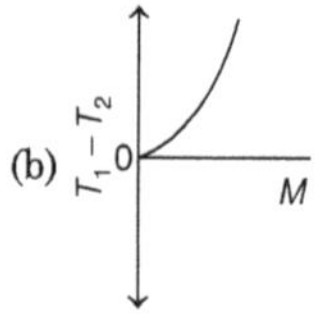

(c)
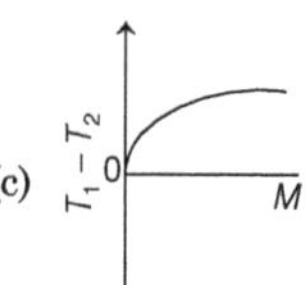

(d)
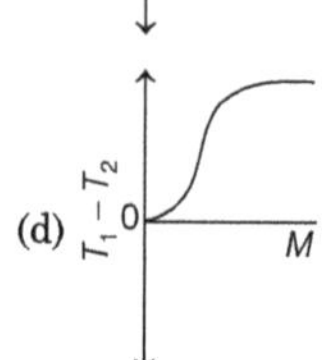

35. Two satellites S_1 and S_2 are revolving around a planet in the opposite sense in coplanar circular concentric orbits. At time $t = 0$, the satellites are farthest apart. The periods of revolution of S_1 and S_2 are 3 h and 24 h, respectively. The radius of the orbit of S_1 is 3×10^4 km. Then, the orbital speed of S_2 as observed from

(a) the planet is $4\pi \times 10^4$ km h^{-1}, when S_2 is closest from S_1
(b) the planet is $2\pi \times 10^4$ km h^{-1}, when S_2 is farthest from S_1
(c) S_1 is $\pi \times 10^4$ km h^{-1}, when S_2 is closest from S_1
(d) S_1 is $3\pi \times 10^4$ km h^{-1}, when S_2 is closest to S_1

36. A rectangular region of dimensions $\omega \times l$ $(\omega << l)$ has a constant magnetic field into the plane of the paper as shown in the figure below. On one side, the region is bounded by a screen. On the other side, positive ions of mass m and charge q are accelerated from rest and towards the screen by a parallel plate capacitor at constant potential difference $V < 0$ and come out through a small hole in the upper plate. Which one of the following statements is correct regarding the charge on the ions that hit the screen?

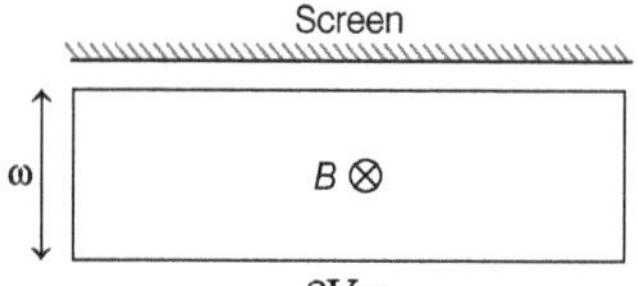

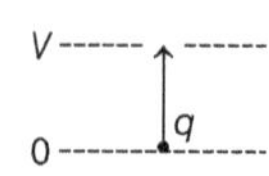

(a) Ions with $q > \dfrac{2Vm}{B^2\omega^2}$ will hit the screen

(b) Ions with $q < \dfrac{2Vm}{B^2\omega^2}$ will hit the screen

(c) All ions will hit the screen

(d) Only ions with $q = \dfrac{2Vm}{B^2\omega^2}$ will hit the screen

37. Force **F** applied on a body is written as $\mathbf{F} = (\hat{\mathbf{n}} \cdot \mathbf{F})\hat{\mathbf{n}} + \mathbf{G}$, where $\hat{\mathbf{n}}$ is a unit vector. The vector **G** is equal to

(a) $\hat{\mathbf{n}} \times \mathbf{F}$ (b) $\hat{\mathbf{n}} \times (\hat{\mathbf{n}} \times \mathbf{F})$
(c) $(\hat{\mathbf{n}} \times \mathbf{F}) \times \mathbf{F} / |\mathbf{F}|$ (d) $(\hat{\mathbf{n}} \times \mathbf{F}) \times \hat{\mathbf{n}}$

38. A particle of mass m moves around the origin in a potential $\frac{1}{2}m\omega^2r^2$, where r is the distance from the origin. Applying the Bohr's model in this case, the radius of the particle in its nth orbit in terms of $a = \sqrt{h/(2\pi m\omega)}$ is

(a) $a\sqrt{n}$ (b) an
(c) an^2 (d) $an\sqrt{n}$

39. Two bottles A and B have radii R_A and R_B and heights h_A and h_B respectively, with $R_B = 2R_A$ and $h_B = 2h_A$. These are filled with hot water at 60°C. Consider that heat loss for the bottles takes place only from side surfaces. If the time, the water takes to cool down to 50°C is t_A and t_B for bottles A and B, respectively. Then, t_A and t_B are best related as

(a) $t_A = t_B$ (b) $t_B = 2t_A$
(c) $t_B = 4t_A$ (d) $t_B = t_A/2$

40. The number of gas molecules striking per second per square metre of the top surface of a table placed in a room at 20°C and 1 atmospheric pressure is of the order of ($k_B = 1.4 \times 10^{-23}$ JK^{-1} and the average mass of an air molecule is 5×10^{-27} kg)

(a) 10^{27} (b) 10^{23}
(c) 10^{25} (d) 10^{29}

CHEMISTRY

41. The major product formed in the following reaction is

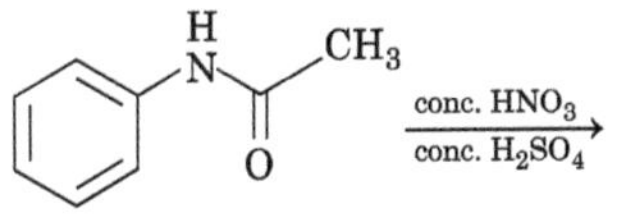

C6H5–NH–CO–CH3 $\xrightarrow[\text{conc. } H_2SO_4]{\text{conc. } HNO_3}$

(a) 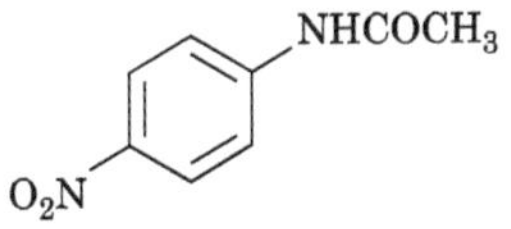 p-O$_2$N–C$_6$H$_4$–NHCOCH$_3$ (b) m-NO$_2$–C$_6$H$_4$–NHCOCH$_3$
(c) 2,4-(NO$_2$)$_2$C$_6$H$_3$–NHCOCH$_3$ (d) 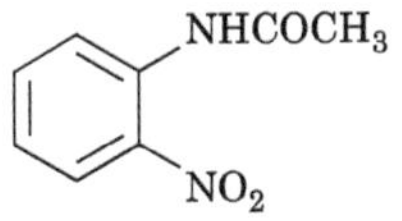 o-NO$_2$–C$_6$H$_4$–NHCOCH$_3$

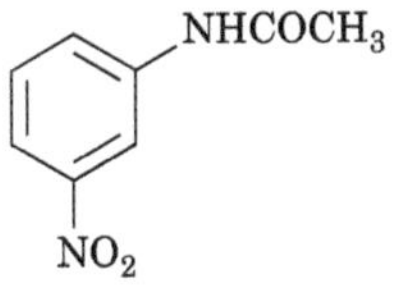

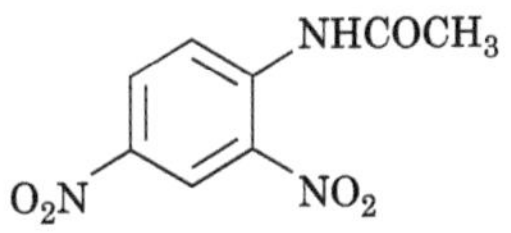

42. Among the α-amino acids - threonine, tyrosine, methionine, arginine and tryptophan, those which contain an aromatic group in their side chain are

(a) threonine and arginine
(b) tyrosine and tryptophan
(c) methionine and tyrosine
(d) arginine and tryptophan

43. The number of stereoisomers possible for the following compound is

$CH_3—CH=CH—CH(OH)—CH_3$

(a) 1 (b) 2 (c) 3 (d) 4

44. In electrophilic aromatic substitution reactions of chlorobenzene, the *ortho/para*-directing ability of chlorine is due to its

(a) positive inductive effect (+*I*)
(b) negative inductive effect (–*I*)
(c) positive resonance effect (+*R*)
(d) negative resonance effect (–*R*)

45. Among the following,

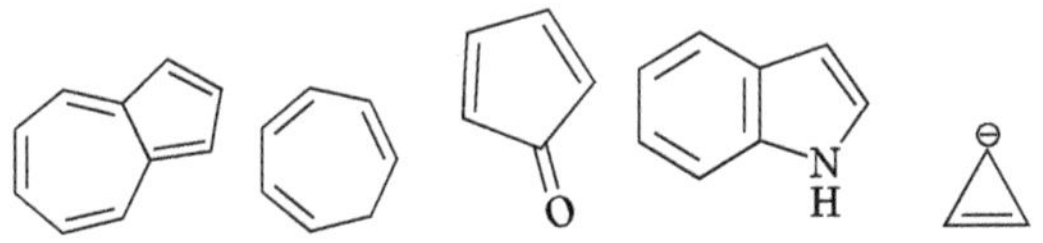

I II III IV V

the antiaromatic compounds are

(a) I and IV (b) III and V
(c) II and V (d) I and III

46. Upon reaction with CH_3MgBr followed by protonation, the compound that produces ethanol is

(a) CH_3CHO (b) HCOOH
(c) HCHO (d) $(CHO)_2$

47. Which of the following is not an oxidation-reduction reaction?

(a) $H_2 + Br_2 \longrightarrow 2HBr$
(b) $NaCl + AgNO_3 \longrightarrow NaNO_3 + AgCl$
(c) $2Na_2S_2O_3 + I_2 \longrightarrow Na_2S_4O_6 + 2NaI$
(d) $Cl_2 + H_2O \longrightarrow HCl + HOCl$

48. The thermal stability of alkaline earth metal carbonates $MgCO_3$, $CaCO_3$, $SrCO_3$ and $BaCO_3$, follows the order

(a) $BaCO_3 > SrCO_3 > CaCO_3 > MgCO_3$
(b) $CaCO_3 > SrCO_3 > BaCO_3 > MgCO_3$
(c) $MgCO_3 > CaCO_3 > SrCO_3 > BaCO_3$
(d) $SrCO_3 > CaCO_3 > MgCO_3 > BaCO_3$

49. When a mixture of diborane and ammonia is heated, the final product is

(a) BH_3 (b) NH_4BH_4 (c) NH_2NH_2 (d) $B_3N_3H_6$

50. Among the following metals, the strongest reducing agent is

(a) Ni (b) Cu (c) Zn (d) Fe

51. The molecule, which is not hydrolysed by water at 25°C is

(a) $AlCl_3$ (b) $SiCl_4$ (c) BF_3 (d) SF_6

52. Among the following compounds, the one which does not produce nitrogen gas upon heating is

(a) $(NH_4)_2Cr_2O_7$ (b) NaN_3
(c) NH_4NO_2 (d) $(NH_4)_2(C_2O_4)$

53. Chlorine has two naturally occurring isotopes, ^{35}Cl and ^{37}Cl. If the atomic mass of Cl is 35.45, the ratio of natural abundance of ^{35}Cl and ^{37}Cl is closest to

(a) 3.5 : 1 (b) 3 : 1 (c) 2.5 : 1 (d) 4 : 1

54. The reaction $C_2H_6(g) \rightleftharpoons C_2H_4(g) + H_2(g)$ is at equilibrium in a closed vessel at 1000 K. The enthalpy change (ΔH) for the reaction is 137.0 kJ mol^{-1}. Which one of the following actions would shift the equilibrium to the right?
(a) Decreasing the volume of the closed reaction vessel
(b) Decreasing the temperature at which the reaction is performed
(c) Adding an inert gas to the closed reaction vessel
(d) Increasing the volume of the closed reaction vessel

55. The enthalpy (H) of an elementary exothermic reaction, $A \rightleftharpoons B$ is schematically plotted against the reaction coordinate. The plots in the presence and absence of a catalyst are shown in dashed and solid lines, respectively. Identify the correct plot for the reaction.

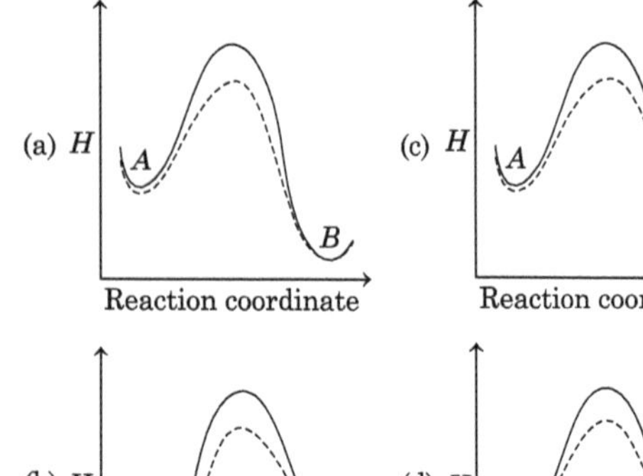

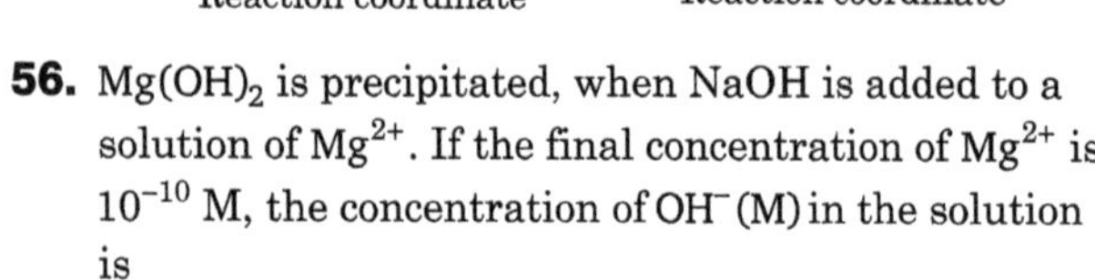

56. $Mg(OH)_2$ is precipitated, when NaOH is added to a solution of Mg^{2+}. If the final concentration of Mg^{2+} is 10^{-10} M, the concentration of OH^- (M) in the solution is
[Solubility product for $Mg(OH)_2 = 5.6 \times 10^{-12}$]
(a) 0.056 (b) 0.12 (c) 0.24 (d) 0.025

57. A constant current (0.5 amp) is passed for 1 hour through (i) aqueous $AgNO_3$, (ii) aqueous $CuSO_4$ and (iii) molten AlF_3, separately. The ratio of the mass of the metals deposited on the cathode is
[M_{Ag}, M_{Cu}, M_{Al} are molar masses of the respective metals.]
(a) $M_{Ag} : 2M_{Cu} : 3M_{Al}$ (b) $M_{Ag} : M_{Cu} : M_{Al}$
(c) $6M_{Ag} : 3M_{Cu} : 2M_{Al}$ (d) $3M_{Ag} : 2M_{Cu} : M_{Al}$

58. A reaction has an activation energy of 209 kJ mol^{-1}. The rate increases 10-fold when the temperature is increased from 27°C to X°C. The temperature X is closest to
[Gas constant, $R = 8.134$ J mol^{-1} K^{-1}]
(a) 35 (b) 40 (c) 30 (d) 45

59. A mineral consists of a cubic close-packed structure formed by O^{2-} ions, where half the octahedral voids are occupied by Al^{3+} and one-eighth of the tetrahedral voids are occupied by Mn^{2+}. The chemical formula of the mineral is
(a) $Mn_3Al_2O_6$ (b) $MnAl_2O_4$
(c) $MnAl_4O_7$ (d) $Mn_2Al_2O_5$

60. For a $4p$-orbital, the number of radial and angular nodes, respectively are
(a) 3, 2 (b) 1, 2 (c) 2, 4 (d) 2, 1

BIOLOGY

61. Interferons combat viral infection by
(a) inhibiting viral packaging directly
(b) increasing the binding of antibodies to viruses
(c) binding to the virus and agglutinating them
(d) restricting viral spread to the neighbouring cells

62. Leydig cells synthesise
(a) insulin
(b) growth hormone
(c) testosterone
(d) oestrogen

63. Glucagon increases the blood glucose concentration by
(a) promoting glycogenolysis
(b) increasing the concentration of fructose 2,6-bisphosphate
(c) increasing the concentration of pyruvate kinase
(d) inhibiting gluconeogenesis

64. Which one of the following is not essential for Polymerase Chain Reaction (PCR)?
(a) Restriction enzyme
(b) Denaturation of DNA
(c) Primers
(d) DNA polymerase

65. CO_2 acts as a greenhouse gas because
(a) it is transparent to heat but traps sunlight
(b) it is transparent to sunlight but traps heat
(c) it is transparent to both sunlight and heat
(d) it traps both sunlight and heat

66. A graph of species richness *vs* area on log-log axes is
(a) linear (b) sigmoidal
(c) oscillatory (d) parabolic

67. Concentration of Na^+ ions outside a nerve cell is ~100 times more than inside. The concentration of K^+ ions is more inside the cells. The levels of Na^+ ions and K^+ ions are maintained by
(a) free diffusion of Na^+ ions and pumping of K^+ ions across the membrane
(b) Na^+ and K^+ pumps in the membrane
(c) free diffusion of K^+ ions and pumping of Na^+ ions across the membrane
(d) water channels formed by lipids in the membrane

68. In a chemical reaction, enzymes catalyse the reaction by
(a) lowering the activation energy
(b) increasing the activation energy
(c) decreasing the free energy change between reactants and products
(d) increasing the free energy change between reactants and products

69. The rigidity of cellulose is due to
(a) coiled structure of glucose polymer
(b) $\beta(1 \rightarrow 4)$ glycosidic linkage
(c) hydrogen bonding with adjacent glucose polymer
(d) cross-linking between glucose and peptides

70. Antigen-antibody reactions
(a) always result in precipitation of the complex
(b) depend only on covalent interactions
(c) are irreversible
(d) depend on ionic and hydrophobic interactions

71. Which one of the following combinations of molecular masses of polypeptides are obtained from purified human IgM when analysed on Sodium Dodecyl Sulphate Polyacrylamide Gel Electrophoresis (SDS-PAGE) under reducing conditions?
(a) 55 kDa, 15 kDa
(b) 70 kDa, 25 kDa, 15 kDa
(c) 55 kDa, 25 kDa
(d) 155 kDa

72. For a particular gene that determines the coat colour in a diploid organism, there are three different alleles that are codominant. How many different skin colours are possible in such an organism?
(a) 9 (b) 6
(c) 4 (d) 3

73. Two genetic loci controlling two different traits are linked. During the inheritance of these traits, the Mendelian laws that would be affected is/ are
(a) law of dominance, law of segregation and law of independent assortment
(b) law of segregation and law of independent assortment
(c) only law of independent assortment
(d) only law of segregation

74. Which one of the following statements is incorrect?
(a) Alleles are different forms of the same gene
(b) Alleles are present at the same locus
(c) Alleles code for different isoforms of a protein
(d) Alleles are non-heritable

75. Which one of the following statements is incorrect about restriction endonucleases?
(a) They serve as primitive form of immune system in bacteria
(b) They digest the DNA non-randomly
(c) They digest the DNA at specific location
(d) They digest the DNA from free ends

76. The number of net ATP molecules produced from 1 glucose molecule during glycolysis is
(a) 1 (b) 2 (c) 3 (d) 4

77. Which one of the following coenzymes is required for the conversion of L-alanine to a racemic mixture of D and L-alanine?
(a) Pyridoxal-6-phosphate
(b) Thiamine pyrophosphate
(c) Coenzyme-A
(d) Flavin adenine dinucleotide

78. The cyclic electron flow during photosynthesis generates
(a) NADPH alone (b) ATP and NADPH
(c) ATP alone (d) ATP, NADPH and O_2

79. Match the type of cells given in Column I with organisms given in Column II. Choose the appropriate combination from the options below.

	Column I		Column II
A.	Flame cells	1.	Sponges
B.	Collar cells	2.	*Hydra*
C.	Stinging cells	3.	*Planaria*

(a) A–3, B–1, C–2 (b) A–3, B–2, C–1
(c) A–1, B–2, C–3 (d) A–2, B–3, C–1

80. Compared to the atmospheric air, the alveolar air has
(a) more pO_2 and less pCO_2
(b) less pO_2 and more pCO_2
(c) more pO_2 and more pCO_2
(d) less pO_2 and less pCO_2

PART-II (2 Marks Questions)

MATHEMATICS

81. Let x, y, z be positive integers such that HCF $(x, y, z) = 1$ and $x^2 + y^2 = 2z^2$. Which of the following statements are true?

I. 4 divides x or 4 divides y.
II. 3 divides $x + y$ or 3 divides $x - y$.
III. 5 divides $z(x^2 - y^2)$.

(a) I and II only (b) II and III only
(c) II only (d) III only

82. How many different (mutually non-congruent) trapeziums can be constructed using four distinct side lengths from the set {1, 2, 3, 4, 5, 6}?
(a) 5 (b) 11 (c) 15 (d) 30

83. A solid hemisphere is mounted on a solid cylinder, both having equal radii. If the whole solid is to have a fixed surface area and the maximum possible volume, then the ratio of the height of the cylinder to the common radius is
(a) $1:1$ (b) $1:2$ (c) $2:1$ (d) $\sqrt{2}:1$

84. Let ABC be an acute scalene triangle, and O and H be its circumcentre and orthocentre respectively. Further, let N be the mid-point of OH. The value of the vector sum $\overrightarrow{NA}+\overrightarrow{NB}+\overrightarrow{NC}$ is

(a) $\vec{0}$ (zero vector) (b) $\overrightarrow{HO}$
(c) $\frac{1}{2}\overrightarrow{HO}$ (d) $\frac{1}{2}\overrightarrow{OH}$

85. The quotient when $1+x^2+x^4+x^6+\ldots+x^{34}$ is divided by $1+x+x^2+x^3+\ldots+x^{17}$ is

(a) $x^{17}-x^{15}+x^{13}-x^{11}+\ldots+x$
(b) $x^{17}+x^{15}+x^{13}+x^{11}+\ldots+x$
(c) $x^{17}+x^{16}+x^{15}+x^{14}+\ldots+1$
(d) $x^{17}-x^{16}+x^{15}-x^{14}+\ldots-1$

86. Let R be the region of the disc $x^2+y^2\le 1$ in the first quadrant. Then, the area of the largest possible circle contained in R is

(a) $\pi(3-2\sqrt{2})$ (b) $\pi(4-3\sqrt{2})$
(c) $\frac{\pi}{6}$ (d) $\pi(2\sqrt{2}-2)$

87. Let R be the set of real numbers and $f: R\to R$ be given by $f(x)=\sqrt{|x|}-\log(1+|x|)$. We now make the following assertions:

I. There exists a real number A such that $f(x)\le A$ for all x.

II. There exists a real number B such that $f(x)\ge B$ for all x.

(a) I is true and II is false
(b) I is false and II is true
(c) I and II both are true
(d) I and II both are false

88. Define $g(x)=\int_{-3}^{3} f(x-y)f(y)dy$, for all real x, where

$$f(t)=\begin{cases}1, & 0\le t\le 1\\ 0, & \text{elsewhere}\end{cases}$$

Then,

(a) $g(x)$ is not continuous everywhere
(b) $g(x)$ is continuous everywhere but differentiable nowhere
(c) $g(x)$ is continuous everywhere and differentiable everywhere except at $x=0, 1$
(d) $g(x)$ is continuous everywhere and differentiable everywhere except at $x=0, 1, 2$

89. The integer part of the number $\sum_{k=0}^{44}\frac{1}{\cos k^\circ \cos(k+1)^\circ}$ is

(a) 50 (b) 52
(c) 57 (d) 59

90. The number of continuous functions $f:[0,1]\to R$ that satisfy $\int_0^1 xf(x)dx=\frac{1}{3}+\frac{1}{4}\int_0^1 (f(x))^2dx$ is

(a) 0 (b) 1
(c) 2 (d) infinity

PHYSICS

91. One end of a rod of length $L=1$ m is fixed to a point on the circumference of a wheel of radius $R=1/\sqrt{3}$ m. The other end is sliding freely along a straight channel passing through the centre O of the wheel as shown in the figure below.

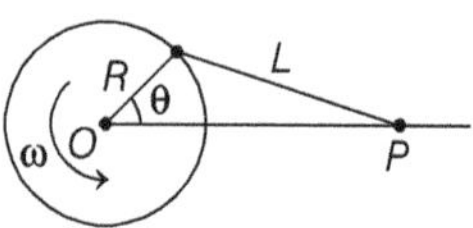

The wheel is rotating with a constant angular velocity ω about O. The speed of the sliding end P, when $\theta=60^\circ$ is

(a) $\frac{2\omega}{3}$ (b) $\frac{\omega}{3}$ (c) $\frac{2\omega}{\sqrt{3}}$ (d) $\frac{\omega}{\sqrt{3}}$

92. One mole of an ideal monoatomic gas undergoes the following four reversible processes:

Step 1 It is first compressed adiabatically from volume V_1 to 1 m^3.

Step 2 Then expanded isothermally to volume 10 m^3.

Step 3 Then expanded adiabatically to volume V_3.

Step 4 Then compressed isothermally to volume V_1.

If the efficiency of the above cycle is 3 / 4, then V_1 is

(a) 2 m^3 (b) 4 m^3 (c) 6 m^3 (d) 8 m^3

93. A neutron star with magnetic moment of magnitude m is spinning with angular velocity ω about its magnetic axis. The electromagnetic power P radiated by it is given by $\mu_0^x m^y \omega^z c^u$, where μ_0 and c are the permeability and speed of light in free space, respectively. Then,

(a) $x=1, y=2, z=4$ and $u=-3$
(b) $x=1, y=2, z=4$ and $u=3$
(c) $x=-1, y=2, z=4$ and $u=-3$
(d) $x=-1, y=2, z=4$ and $u=3$

94. A solid cube of wood of side $2a$ and mass M is resting on a horizontal surface as shown in the figure below.

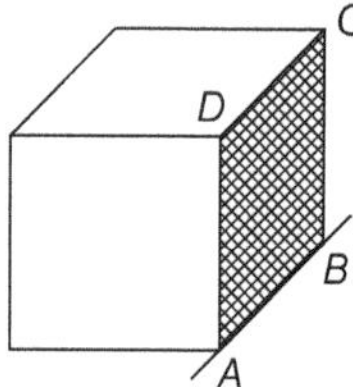

The cube is free to rotate about a fixed axis AB. A bullet of mass m ($<< M$) and speed v is shot horizontally at the face opposite to $ABCD$ at a height of $4a/3$ from the surface to impart the cube an angular speed ω. It strikes the face embeds in the cube. Then, ω is close to (**Note** : The moment of inertia of the cube about an axis perpendicular to the face and passing through the centre of mass is $2Ma^2/3$)

(a) Mv/ma (b) $Mv/2ma$ (c) mv/Ma (d) $mv/2Ma$

95. A gas obeying the equation of state $pV = RT$ undergoes a hypothetical reversible process described by the equation, $pV^{5/3}\exp\left(-\frac{pV}{E_0}\right) = C_1$, where C_1 and E_0 are dimensioned constants. Then, for this process, the thermal compressibility at high temperature

(a) approaches a constant value

(b) is proportional to T

(c) is proportional to $T^{1/2}$

(d) is proportional to T^2

96. To calculate the size of a hydrogen ion using the Bohr's model, we assume that its two electrons move in an orbit such that they are always on diametrically opposite sides of the nucleus. With each electron having the angular momentum $\hbar = h/2\pi$ and taking electron interaction into account the radius of the orbit in terms of the Bohr's radius of hydrogen atom $a_B = \frac{4\pi\varepsilon_0 h^2}{me^2}$ is

(a) a_B (b) $\frac{4}{3}a_B$ (c) $\frac{2}{3}a_B$ (d) $\frac{3}{2}a_B$

97. A square-shaped conducting wire loop of dimension a moving parallel to the X-axis approaches a square region of size b ($a < b$), where a uniform magnetic field B exists pointing into the plane of the paper (see figure). As the loop passes through this region, the plot correctly depicting its speed v as a function of x is

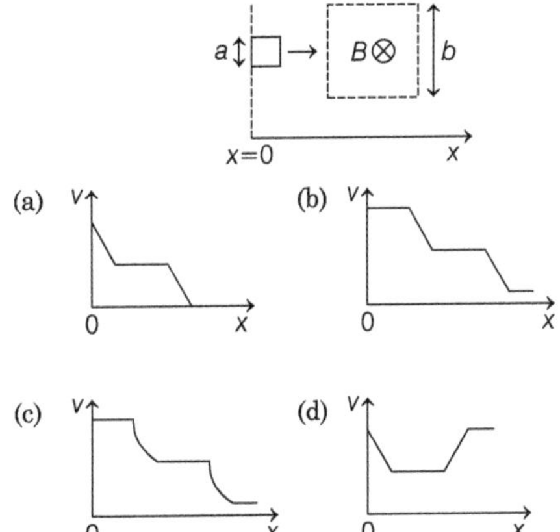

98. The figure of a centimetre scale below shows a particular position of the Vernier calipers. In this position, the value of x shown in the figure is (figure is not to scale)

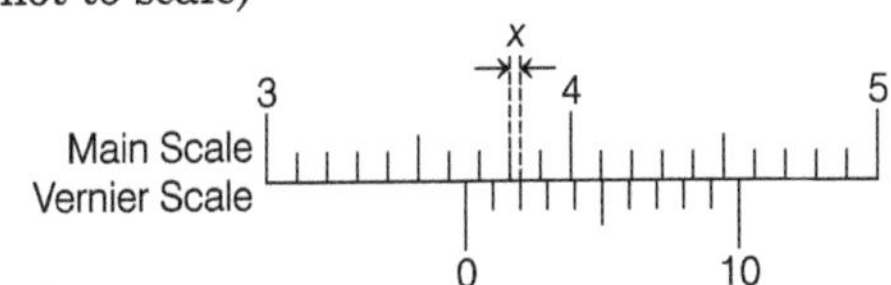

(a) 0.02 cm (b) 3.65 cm (c) 4.15 cm (d) 0.03 cm

99. A parallel beam of light is incident on a tank filled with water up to a height of 61.5 mm as shown in the figure below.

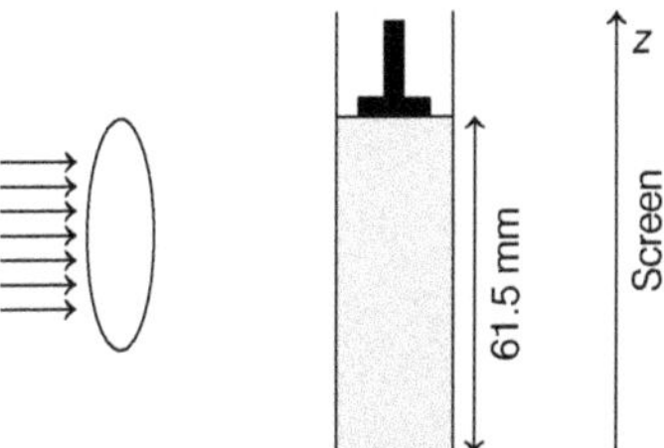

Ultrasonic waves of frequency 0.5 MHz are sent along the length of the water column using a transducer placed at the top and they form longitudinal standing waves in the water. Which of the schematic plots below best describes the intensity distribution of the light as seen on the screen? (Take, the speed of sound in water to be 1500 m/s)

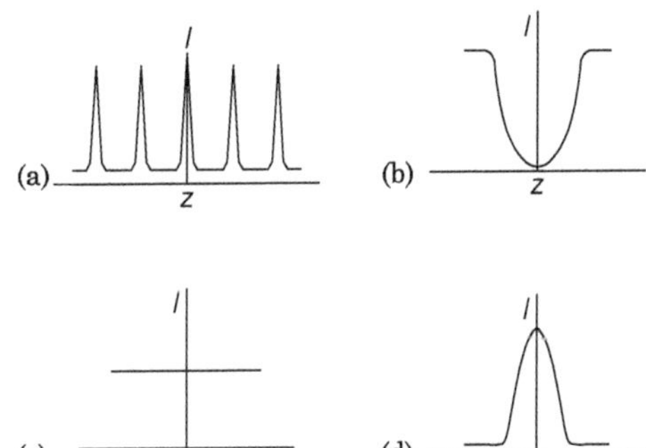

100. A star of mass M (equal to the solar mass) with a planet (much smaller than the star) revolves around the star in a circular orbit. The velocity of the star with respect to the centre of mass of the star-planet system is shown below.

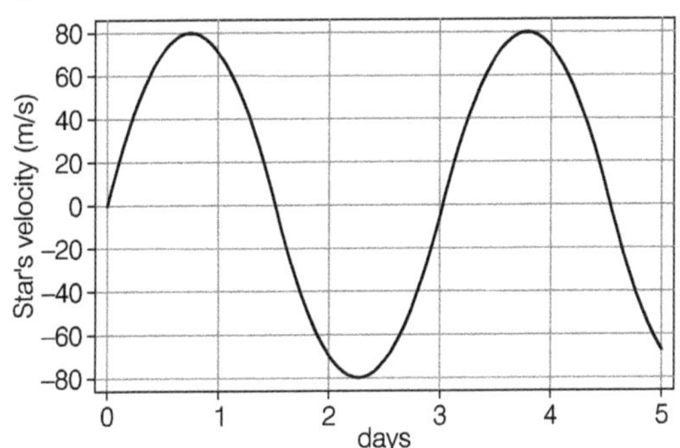

The radius of the planet's orbit is closest to (1 AU = Earth-Sun distance)

(a) 0.004 AU

(b) 0.008 AU

(c) 0.04 AU

(d) 0.12 AU

101. In the following reaction sequence,

$\xrightarrow[\text{(ii) } H_2O_2/NaOH]{\text{(i) } B_2H_6} X \xrightarrow{CrO_3/H_2SO_4} Y$

X and Y are

(a) $X =$ OH $Y =$ CHO

(b) $X =$ OH $Y =$ COOH

(c) $X =$ OH $Y =$

(d) $X =$ OH $Y =$

102. In the following reactions,

$H_3C\text{–}CO\text{–}CH_3 \xrightarrow[\text{(ii) } D_3O^+]{\text{(i) } NaBH_4} X$; $H_3C\text{–}CO\text{–}CH_3 \xrightarrow[\text{(ii) } H_3O^+]{\text{(i) } NaBD_4} Y$

X and Y are

(a) $X =$ H, OD, H_3C, CH_3 $Y =$ D, OH, H_3C, CH_3

(b) $X =$ D, OH, H_3C, CH_3 $Y =$ H, OD, H_3C, CH_3

(c) $X = Y =$ D, OH, H_3C, CH_3

(d) $X = Y =$ H, OH, H_3C, CH_3

103. Which of the following alkenes can generate optically active compounds upon hydrogenation?

I II

III IV

(a) I, III and IV (b) II and III
(c) I and III (d) II and IV

104. When heated in air, brown copper powder turns black. This black powder would turn brown again when heated with

(a) CO (b) O_2 (c) H_2 (d) NH_3

105. The geometry and magnetic property of $[NiCl_4]^{2-}$, respectively are

(a) tetrahedral, paramagnetic
(b) tetrahedral, diamagnetic
(c) square planar, paramagnetic
(d) square planar, diamagnetic

106. Among (i) $[Cr(en)_3]^{3+}$, (ii) *trans* – $[Cr(en)_2Cl_2]^+$, (iii) *cis* – $[Cr(en)_2Cl_2]^+$, (iv) $[CO(NH_3)_4Cl_2]^+$, the optically active complexes are

(a) (i) and (ii) (b) (i) and (iii)
(c) (ii) and (iii) (d) (ii) and (iv)

107. ^{227}Ac has a half-life of 22 years with respect to radioactive decay. The decay follows two parallel paths: $^{227}Ac \longrightarrow {}^{227}Th$ and $^{227}Ac \longrightarrow {}^{223}Fr$. If the percentage of the two daughter nuclides are 2.0 and 98.0, respectively, the decay constant (in year^{-1}) for $^{227}Ac \longrightarrow {}^{227}Th$ path is closest to

(a) 6.3×10^{-2} (b) 6.3×10^{-3}
(c) 6.3×10^{-1} (d) 6.3×10^{-4}

108. A system consisting of 1 mole of an ideal gas undergoes a reversible process, $A \rightarrow B \rightarrow C \rightarrow A$ (schematically indicated in the figure below). If the temperature at the starting point A is 300 K and the work done in the process $B \rightarrow C$ is 1 L atm, the heat exchanged in the entire process in L atm is

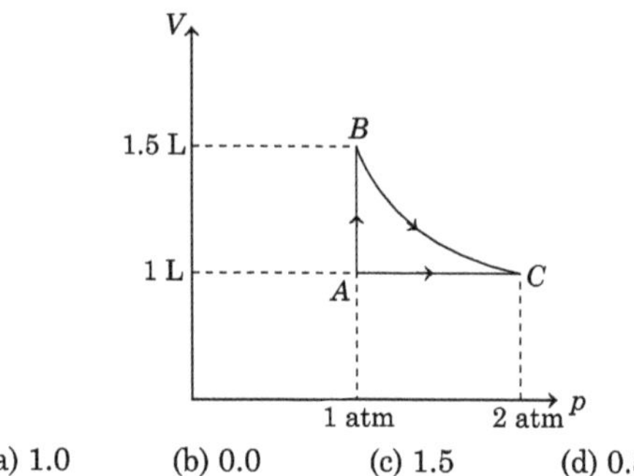

(a) 1.0 (b) 0.0 (c) 1.5 (d) 0.5

109. A mixture of toluene and benzene boils at 100°C. Assuming ideal behaviour, the mole fraction of toluene in the mixture is closest to

[Vapour pressures of pure toluene and pure benzene at 100°C are 0.742 and 1.800 bar, respectively. 1 atm = 1.013 bar]

(a) 0.824 (b) 0.744 (c) 0.544 (d) 0.624

110. A two-dimensional solid pattern formed by two different atoms X and Y is shown below.

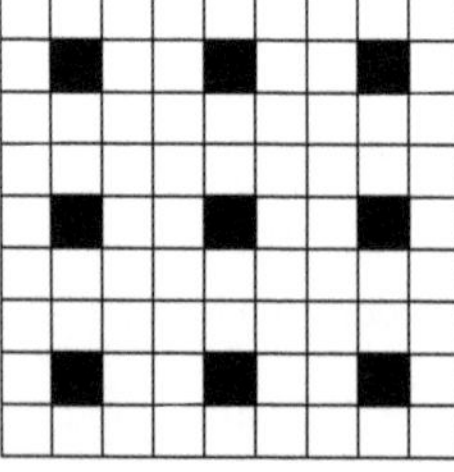

The black and white squares represent atoms X and Y, respectively. The simplest formula for the compound based on the unit cell from the pattern is

(a) XY_8 (b) X_4Y_9
(c) XY_2 (d) XY_4

BIOLOGY

111. The genetic distance between genes A and B is 10 cm. An organism with Ab combination of the alleles is crossed with the organism with aB combination of alleles. What will be the percentage of the gametes with AB allele combination produced by an F_1 individual?

(a) 1 (b) 5
(c) 10 (d) 50

112. Proteins P, Q, and R are associated with intact organellar membrane in a cell. If the intact organelle is treated with a high ionic strength buffer, only protein R remained associated with the membrane fraction. Based on this, one could conclude that

(a) P and Q are peripheral membrane proteins
(b) R is a peripheral membrane protein
(c) P and Q are integral membrane bound proteins
(d) P is peripheral and Q is integral membrane protein

113. In photosynthesis, oxygen is produced by

(a) photosystem-I from carbon dioxide
(b) photosystem-II from carbon dioxide
(c) photosystem-I from water
(d) photosystem-II from water

114. How many different proteins consisting of 100 amino acids can be formed from 20 different amino acids?

(a) 20^{100} (b) 100^{20}
(c) 2^{20} (d) 20×100

115. Molecular weight of *E. coli* DNA is 3.1×10^9 g/mol. Average molecular weight of nucleotide pair is 660 g/mol and each nucleotide pair contributes to 0.34 nm to the length of DNA. The length of *E. coli* DNA molecule will be approximately

(a) 0.8 nm (b) 1.6 nm
(c) 1.6 μm (d) 1.6 mm

116. Which one of the following options is true with respect to emigration?

(a) It is the difference between the births and deaths in a population
(b) It is the difference between individuals who have come to a habitat and who have left the habitat
(c) It involves individuals of different species coming to a habitat from elsewhere during the period under consideration
(d) It involves individuals of a population leaving a habitat during the time period under consideration

117. Choose the correct combination of statements given below related to cysteine residue in proteins.

I. Cysteine can be linked to tyrosine by S—O bond.
II. Cysteine can be linked to another cysteine by S—S bond.
III. Cysteine can complex with Zn^{2+}.
IV. Cysteine can be linked to methionine by S-S bond.

(a) I and II (b) II and III
(c) III and IV (d) I and IV

118. The minimum number of plants to be screened to obtain a plant of the genotype AabbCcDd from a cross between plants of genotypes AaBbCcDd and AABbCCDd is

(a) 8 (b) 16 (c) 32 (d) 64

119. When a pure bred, red flower producing plant of genotype RR is crossed with a pure bred, white flower producing plant of genotype rr, all the F_1 plants produced pink flowers. If all the plants in each generation from F_1 to F_6 are selfed, what will be the percentage of plants with red and white flowers in the final population consisting of a large number of individuals? (Consider that flower colour has no effect on reproduction and survival.)

(a) 3-4 (b) 12-13
(c) 49-51 (d) 97-100

120. The schematic below describes the status of *lac* operon in the absence of lactose. Which one of the following happens when lactose is present in the cell?

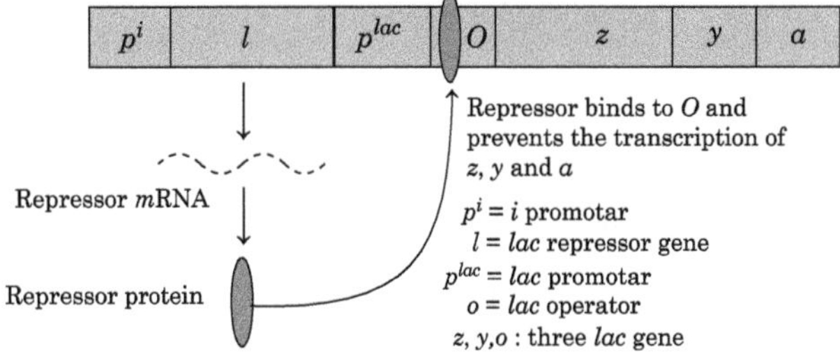

(a) Lactose binds to p^i and stops the transcription of i
(b) Lactose is converted to allolactose, which binds to p^{lac} and results in the displacement of the repressor from O
(c) Lactose is converted to allolactose, which binds to the repressor protein and prevents its interaction with O.
(d) Lactose has no effect on the status of the *lac* operon.

Answers

PART-I

1 (d)	2 (b)	3 (d)	4 (a)	5 (c)	6 (b)	7 (c)	8 (b)	9 (d)	10 (a)
11 (c)	12 (d)	13 (b)	14 (b)	15 (a)	16 (b)	17 (a)	18 (c)	19 (c)	20 (a)
21 (a)	22 (b)	23 (d)	24 (c)	25 (c)	26 (c)	27 (a)	28 (b)	29 (*)	30 (a)
31 (b)	32 (b)	33 (a)	34 (c)	35 (d)	36 (b)	37 (d)	38 (a)	39 (b)	40 (a)
41 (a)	42 (b)	43 (d)	44 (c)	45 (b)	46 (c)	47 (b)	48 (a)	49 (d)	50 (c)
51 (d)	52 (d)	53 (b)	54 (d)	55 (a)	56 (c)	57 (c)	58 (a)	59 (b)	60 (d)
61 (d)	62 (c)	63 (a)	64 (a)	65 (b)	66 (a)	67 (b)	68 (a)	69 (c)	70 (d)
71 (b)	72 (b)	73 (c)	74 (d)	75 (d)	76 (b)	77 (a)	78 (c)	79 (a)	80 (b)

PART-II

81 (b)	82 (b)	83 (a)	84 (c)	85 (d)	86 (a)	87 (b)	88 (d)	89 (c)	90 (b)
91 (a)	92 (d)	93 (a)	94 (d)	95 (a)	96 (*)	97 (b)	98 (*)	99 (a)	100 (c)
101 (b)	102 (a)	103 (c)	104 (c)	105 (a)	106 (c)	107 (d)	108 (d)	109 (b)	110 (a)
111 (b)	112 (a)	113 (d)	114 (a)	115 (d)	116 (d)	117 (b)	118 (c)	119 (d)	120 (c)

** No options are correct.*

Solutions

1. *(d)* Given in ΔABC, BC is fixed, A is variable and ABC is an isosceles triangle.

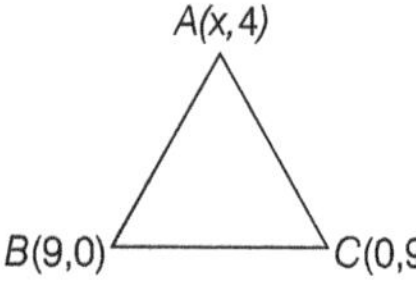

***Case* I** In ΔABC,

If $\quad AB = AC$

Then, locus of A is perpendicular bisector of BC.

$\therefore$ Locus of A is a straight line.

***Case* II** When $AB = BC$

BC fixed $B(a, 0)$, $C(0, a)$, locus of A is $(x-a)^2 + y^2 = 2a^2$, which represents the equation of circle.

Similarly, when $AC = BC$ also locus of A is circle.

$\therefore$ Locus of A is the union of two circles and the lines.

2. *(b)* We have,

$$\log_{1/3}(x+y) + \log_3(x-y) = 2$$

$$2y^2 = 512^{x+1}$$

$\Rightarrow \log_{3^{-1}}(x+y) + \log_3(x-y) = 2$

$\Rightarrow -\log_3(x+y) + \log_3(x-y) = 2$

$\Rightarrow \quad \dfrac{x-y}{x+y} = 3^2 = 9$

$\Rightarrow \quad x - y = 9x + 9y$

$\Rightarrow \quad -8x = 10y \Rightarrow -4x = 5y \quad \ldots$(i)

and $\quad 2^{y^2} = 2^{9(x+1)}$

$\Rightarrow \quad y^2 = 9(x+1) \quad \ldots$(ii)

From Eqs. (i) and (ii), we get

$$\left(\frac{-4x}{5}\right)^2 = 9(x+1)$$

$\Rightarrow \quad 16x^2 = 225x + 225$

$\Rightarrow \quad 16x^2 - 225x - 225 = 0$

$\Rightarrow \quad (16x + 15)(x - 15) = 0$

$\Rightarrow \quad x = 15, -\dfrac{15}{16}$

When $x = 15$, $y = -12$

$$x = -\frac{15}{16}, y = \frac{3}{4} \text{ (not possible)}$$

$\therefore$ Only one solution $(15, -12)$.

3. *(d)* We have, $\lim\limits_{x \to -\infty} (\sqrt{4x^2 - x} + 2x)$

$$= \lim_{x \to -\infty} \frac{4x^2 - x - 4x^2}{\sqrt{4x^2 - x} - 2x}$$

$$= \lim_{x \to -\infty} \frac{-x}{-x\left(\sqrt{4 - \dfrac{x}{x^2}} + 2\right)}$$

$$= \lim_{x \to -\infty} \frac{1}{\sqrt{4 - \dfrac{1}{x}} + 2} = \frac{1}{2+2} = \frac{1}{4}$$

4. *(a)* We have, $aRb : a$ divides b^2

For reflexive : $(a, a) \in R$

$\therefore \quad aRa : a$ divides a^2.

Hence, R is reflexive.

For symmetric : $(a, b) \in R \Rightarrow (b, a) \in R$

a divides b^2 and b not divides a^2.

Hence, it is not symmetric.

For transitive : $(a, b) \in R$, $(b, c) \in R$

$\Rightarrow (a, c) \in R$

$\Rightarrow (8, 4) : 8$ divides 4^2

$\Rightarrow (4, 2) : 4$ divides 2^2

But $(8/2) : 8$ not divides 2^2

$\therefore$ It is not transitive.

5. *(c)* We have,

F_1 and F_2 be the fractional part of $(44 - \sqrt{2017})^{2017}$ and $(44 + \sqrt{2017})^{2017}$

$\Rightarrow \quad (44 + \sqrt{2017})^{2017} = I + F_2$

$\Rightarrow \quad (44 - \sqrt{2017})^{2017} = F_1$

$[\because 0 < 44 - \sqrt{2017} < 1]$

$\therefore I + F_1 + F_2$

$= 2({}^{2017}C_0(44)^{2017} + {}^{2017}C_2(44)^{2015}(2017) + \ldots]$

$\Rightarrow I + F_1 + F_2 = 2\,$(integer)

$\therefore \quad I + F_1 + F_2$ is an even integer.

$\therefore \quad F_1 + F_2$ is also integer.

But, $0 < F_1 + F_2 < 2$

But, $F_1 + F_2$ is an integer.

$\therefore \quad F_1 + F_2 = 1$

Hence, $F_1 + F_2$ is lie between 0.9 and 1.35

6. *(b)* Given, $2\sin 3x + \sin 7x - 3 = 0$

$\Rightarrow \quad 2\sin 3x + \sin 7x = 3$

It is possible only $\sin 3x = 1$ and $\sin 7x = 1$

$\because \quad \sin 3x = 1$

$\therefore \quad 3x = n\pi + (-1)^n \frac{\pi}{2}$

$x = \frac{n\pi}{3} + (-1)^n \frac{\pi}{6}$

$\because \quad \sin 7x = 1$

$\therefore \quad 7x = m\pi + (-1)^m \frac{\pi}{2}$

$x = \frac{m\pi}{7} + (-1)^m \frac{\pi}{14}$

Common solution are $-\frac{\pi}{2}, \frac{3\pi}{2}$ in $[-2\pi, 2\pi]$.

Hence, only two solution exists.

7. *(c)* We have,

$q = p(4 - p)$ ...(i)

$r = q(4 - q)$ ...(ii)

$p = r(4 - r)$ ...(iii)

On adding Eqs. (i), (ii) and (iii), we get

$p + q + r = 4(p + q + r) - (p^2 + q^2 + r^2)$

$\Rightarrow \quad 3(p + q + r) = p^2 + q^2 + r^2$

$\Rightarrow \quad p + q + r = \frac{p^2 + q^2 + r^2}{3}$

Maximum value of $p + q + r$ at

$p = q = r = 3$

$\therefore$ Maximum value of $p + q + r$

$= \frac{3^2 + 3^2 + 3^2}{3} = 9$

8. *(b)* We have, parabola $y^2 = 4x + 1$

and circle $x^2 + y^2 \le 1$

Area of shaded region,

$$A_1 = 2\left(\int_{-1/4}^{0} \sqrt{4x + 1}\,dx + \int_{0}^{1} \sqrt{1 - x^2}\,dx\right)$$

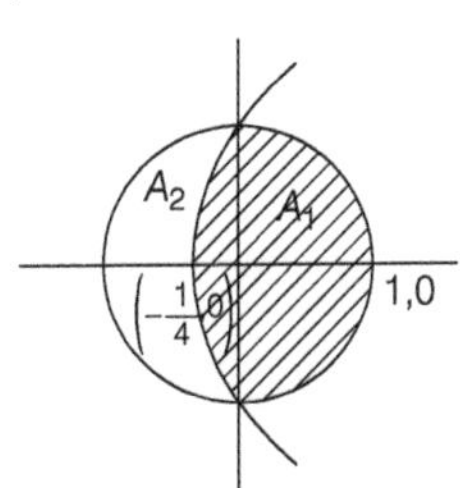

$\Rightarrow \quad A_1 = 2\left[\frac{2}{3}(4x + 1)^{3/2}\right]_{-1/4}^{0} + 2 \times \frac{\pi}{2}$

$\Rightarrow \quad A_1 = \frac{1}{3} + \frac{\pi}{2}$

$A_2 = \text{Area of circle} - A_1$

$= \pi - \left(\frac{1}{3} + \frac{\pi}{2}\right) = \frac{\pi}{2} - \frac{1}{3}$

$\therefore \quad |A_1 - A_2| = \left|\frac{1}{3} + \frac{\pi}{2} - \frac{\pi}{2} + \frac{1}{3}\right| = \frac{2}{3}$

9. *(d)* $n = 5, p = \frac{1}{4}, q = \frac{3}{4}$

$\therefore$ Required probability $= {}^5C_2 (p)^2 q^3 \times p$

$= {}^5C_2 \left(\frac{1}{4}\right)^2 \left(\frac{3}{4}\right)^3 \times \frac{1}{4} = \frac{270}{4096} = 0.6591$

10. *(a)* E_1 = Six fair dice are rolled and at least one die shows six

$P(E_1) = p_1 = 1 - (\text{no die show six})$

$= 1 - \left(\frac{5}{6}\right)^6 = 0.6651$

E_2 = Twelve fair dice rolled and at least two dice show six.

Here, $n = 12, p = \frac{1}{6}, q = \frac{5}{6}$

$P(E_2) = P(r \ge 2) = 1 - [P(r = 0) + P(r = 1)]$

$\Rightarrow P(E_2) = p_2 = 1 - \left[\left(\frac{5}{6}\right)^{12} + {}^{12}C_1 \left(\frac{5}{6}\right)^{11}\left(\frac{1}{6}\right)\right]$

$= 0.61866$

$\therefore \quad p_1 > p_2$

11. *(c)* We have, $ax + y = 0$ ...(i)

$x + (a + 10)y = 0$ ...(ii)

From Eqs. (i) and (ii), we get

$\frac{a}{1} = \frac{1}{a + 10} \Rightarrow a^2 + 10a - 1 = 0$

$a = \frac{-10 \pm \sqrt{104}}{2}$

$\therefore$ Two values of a for systems has at least two distinct solution.

12. *(d)* We have,

$f(x) = \frac{\{x\}}{1 + [x]^2}$

$\Rightarrow \quad f(x) = \frac{x - [x]}{1 + [x]^2}$

Range of $f(x) = [0, 1)$.

$\therefore$ Range of f is semi-closed interval f is discontinuous on integer.

Clearly, f is not one-one function.

$\therefore$ Option (d) is correct.

13. *(b)* We have,

$x_n = (2^n + 3^n)^{\frac{1}{2n}}$

$\Rightarrow \quad \lim_{n \to \infty} x_n = \lim_{n \to \infty} (2^n + 3^n)^{\frac{1}{2n}}$

$= \lim_{n \to \infty} 3^{n/2n} \left[\left(\frac{2}{3}\right)^n + 1\right]^{\frac{1}{2n}}$

$= 3^{1/2} \times 1 = \sqrt{3}$

14. *(b)* Given,

$8\sin^3\theta - 7\sin\theta + \sqrt{3}\cos\theta = 0$

$\Rightarrow \quad 2(4\sin^3\theta) - 7\sin\theta + \sqrt{3}\cos\theta = 0$

$\Rightarrow \quad 2(3\sin\theta - \sin 3\theta) - 7\sin\theta + \sqrt{3}\cos\theta = 0$

$\Rightarrow \quad 6\sin\theta - 2\sin 3\theta - 7\sin\theta + \sqrt{3}\cos\theta = 0$

$\Rightarrow \quad \sqrt{3}\cos\theta - \sin\theta - 2\sin 3\theta = 0$

$\Rightarrow \quad \frac{\sqrt{3}}{2}\cos\theta - \frac{1}{2}\sin\theta = \sin 3\theta$

$\Rightarrow \quad \sin\left(\frac{\pi}{3} - \theta\right) = \sin 3\theta$

$\Rightarrow \quad \frac{\pi}{3} - \theta = 3\theta \Rightarrow 4\theta = 60° \Rightarrow \theta = 15°$

$\therefore \quad \theta \in (10°, 20°]$

15. *(a)* Given,

$a + b < c + d$ ...(i)

$b + c < d + e$ ...(ii)

$c + d < e + a$ ...(iii)

$d + e < a + b$ ...(iv)

From Eqs. (i) and (iii), we get

$a + b + c + d < c + d + e + a$

$\Rightarrow \quad b < e$ ...(v)

From Eqs. (ii) and (iv), we get

$b + c + d + e < d + c < a + b$

$\Rightarrow \quad c < a$ ...(vi)

From Eqs. (i) and (iv), we get

$a + b + d + e < c + d + a + b$

$e < c$ ...(vii)

From Eqs. (v), (vi), (vii), we get

$a > c > e > b$

$\therefore$ Largest value is a and smallest value is b.

16. *(b)* A fair coin is tossed 5 times.

Total number of sample space $= 2^5$

Probability of head does not occur two or more time in rows.

Possible cases

***Case* I** All tails $= \left(\frac{1}{2}\right)^5 = \frac{1}{2^5}$

***Case* II** 4 tails and one head

${}^5C_4 \left(\frac{1}{2}\right)^5 = \frac{5}{2^5}$

***Case* III** 3 tails and 2 head $\times T \times T \times T \times$

${}^4C_2 \times \left(\frac{1}{2}\right)^5 = \frac{6}{2^5}$

***Case* IV** 2 tails and 3 heads $\times T \times T \times$

${}^3C_3 \times \left(\frac{1}{2}\right)^5 = \frac{1}{2^5}$

$\therefore$ Required probability

$= \frac{1}{2^5} + \frac{5}{2^5} + \frac{6}{2^5} + \frac{1}{2^5} = \frac{13}{2^5}$

17. *(a)* We have,

$x(\theta) = |\cos 4\theta|\cos\theta$ and $y(\theta) = |\cos 4\theta|\sin\theta$

θ	0	45°	90°	135°	180°	225°
$x(\theta)$	1	$\frac{1}{\sqrt{2}}$	0	$-\frac{1}{\sqrt{2}}$	−1	$-\frac{1}{\sqrt{2}}$
$y(\theta)$	0	$\frac{1}{\sqrt{2}}$	1	$\frac{1}{\sqrt{2}}$	0	$-\frac{1}{\sqrt{2}}$

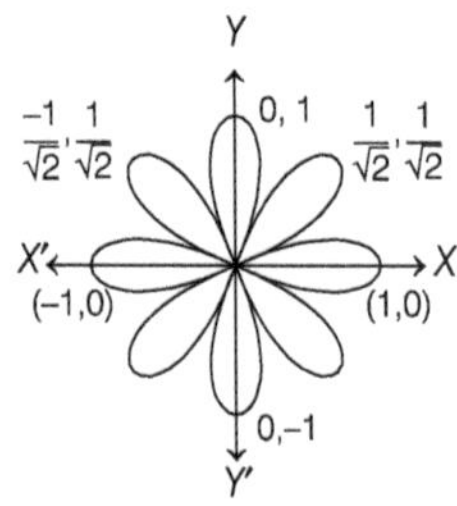

18. *(c)* Given,

$A = (a_1, a_2)$, $B = (b_1, b_2)$, where a_1, a_2, b_1, b_2 are integers.

Distance between AB

$$= \sqrt{(b_1 - a_1)^2 + (b_2 - a_2)^2}$$

$\therefore AB = \sqrt{\text{sum of squares of two number}}$

$\sqrt{65} = \sqrt{64 + 1}$, it is possible.

$\sqrt{74} = \sqrt{49 + 25}$, it is also possible.

$\sqrt{97} = \sqrt{81 + 16}$, it is also possible.

83 is not a sum of squares of two number.

Hence, option (c) is correct.

19. *(c)* We have,

$$f(x) = \max\left\{3, x^2, \frac{1}{x^2}\right\} \text{ for } x \in \left[\frac{1}{2}, 2\right]$$

y=x²
y=3
y=3
y=$\frac{1}{x^2}$
$\left(\frac{1}{\sqrt{3}},0\right)$ $(\sqrt{3},0)$

$$\int_{1/2}^{2} f(x)dx = \int_{1/2}^{1/\sqrt{3}} \frac{1}{x^2}dx + \int_{1/\sqrt{3}}^{\sqrt{3}} 3dx + \int_{\sqrt{3}}^{2} x^2 dx$$

$$\left[-\frac{1}{x}\right]_{1/2}^{1/\sqrt{3}} + [3x]_{1/\sqrt{3}}^{\sqrt{3}} + \left[\frac{x^3}{3}\right]_{\sqrt{3}}^{2}$$

$$= -\sqrt{3} + 2 + 3\sqrt{3} - 3\sqrt{3} + \frac{8}{3} - \frac{3\sqrt{3}}{3} = \frac{14}{3}$$

20. *(a)* Given $a_i = i + \frac{1}{i}$ for $i = 1, 2, 3, \ldots, 20$

$$p = \frac{1}{20}(a_1 + a_2 + a_3 + \ldots + a_{20})$$

$$q = \frac{1}{20}\left(\frac{1}{a_1} + \frac{1}{a_2} + \frac{1}{a_3} \ldots + \frac{1}{a_{20}}\right)$$

Clearly, $q > 0$.

Let $\quad q < \frac{22 - p}{21}$

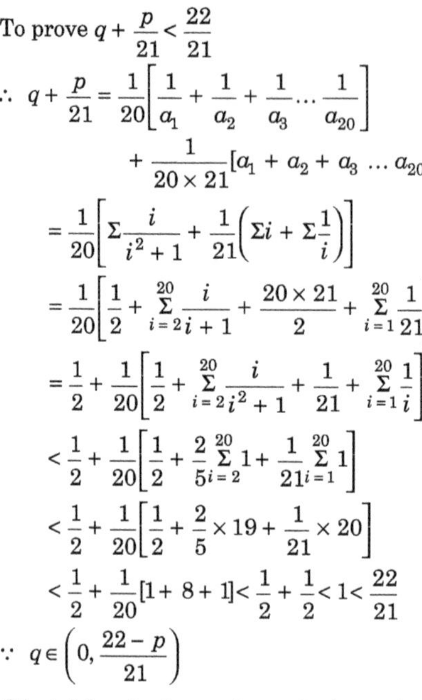

To prove $q + \frac{p}{21} < \frac{22}{21}$

$$\therefore q + \frac{p}{21} = \frac{1}{20}\left[\frac{1}{a_1} + \frac{1}{a_2} + \frac{1}{a_3} \ldots \frac{1}{a_{20}}\right] + \frac{1}{20 \times 21}[a_1 + a_2 + a_3 \ldots a_{20}]$$

$$= \frac{1}{20}\left[\Sigma\frac{i}{i^2+1} + \frac{1}{21}\left(\Sigma i + \Sigma\frac{1}{i}\right)\right]$$

$$= \frac{1}{20}\left[\frac{1}{2} + \sum_{i=2}^{20}\frac{i}{i+1} + \frac{20 \times 21}{2} + \sum_{i=1}^{20}\frac{1}{21}\right]$$

$$= \frac{1}{2} + \frac{1}{20}\left[\frac{1}{2} + \sum_{i=2}^{20}\frac{i}{i^2+1} + \frac{1}{21} + \sum_{i=1}^{20}\frac{1}{i}\right]$$

$$< \frac{1}{2} + \frac{1}{20}\left[\frac{1}{2} + \frac{2}{5}\sum_{i=2}^{20} 1 + \frac{1}{21}\sum_{i=1}^{20} 1\right]$$

$$< \frac{1}{2} + \frac{1}{20}\left[\frac{1}{2} + \frac{2}{5} \times 19 + \frac{1}{21} \times 20\right]$$

$$< \frac{1}{2} + \frac{1}{20}[1 + 8 + 1] < \frac{1}{2} + \frac{1}{2} < 1 < \frac{22}{21}$$

$$\because q \in \left(0, \frac{22 - p}{21}\right)$$

21. *(a)* In a hydrogen ion, velocity and radius of nth orbit are

$$v_n = C_1 \cdot \left(\frac{Z}{n}\right) \text{ and } r_n = C_2 \cdot \left(\frac{n^2}{Z}\right)$$

where, C_1 and C_2 are constants. So, acceleration of electron in nth orbit is

$$a_n = \frac{v_n^2}{r_n} = \frac{C_1^2}{C_2} \cdot \frac{\left(\frac{Z^2}{n^2}\right)}{\left(\frac{n^2}{Z}\right)} \Rightarrow a_n \propto \frac{Z^3}{n^4}$$

So, ratio of magnitude of acceleration of electron in nth orbit of hydrogen atom and that of singly ionised helium atom is given by

$$\frac{a_{\text{H}}}{a_{\text{He}}} = \frac{\left(\frac{Z_{\text{H}}^3}{n^4}\right)}{\left(\frac{Z_{\text{He}}^3}{n^4}\right)} = \frac{Z_{\text{H}}^3}{Z_{\text{He}}^3}$$

where, Z = atomic number.

$\Rightarrow \quad Z_{\text{H}} = 1$ and $Z_{\text{He}} = 2$

$\Rightarrow \quad \frac{a_{\text{H}}}{a_{\text{He}}} = \left(\frac{1}{2}\right)^3 = \frac{1}{8}$

22. *(b)* Carrot appears orange in white light because it reflects orange portion and absorbs remaining part of spectrum. Visible part of spectrum is

This part is absorbed: Violet ~ 450 nm, Blue, Green, Yellow ~ 550 nm

This part makes orange colour: Yellow ~ 550 nm, Orange, Red ~ 750 nm

Clearly, an orange object must absorbs wavelengths shorter than 550 nm.

23. *(d)* As Coulomb's force,

$$F \propto \frac{1}{r^3}$$

$\Rightarrow$ Electric field, $E \propto \frac{1}{r^3} \Rightarrow E = \frac{kq}{r^3}$

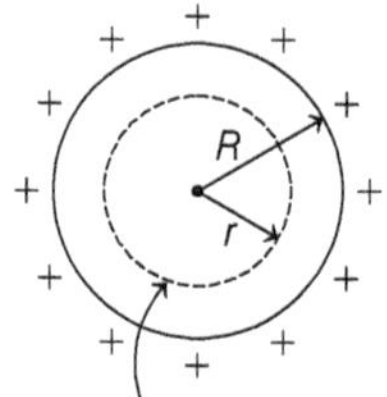

Gaussian surface

Now, for Gaussian surface $(r < R)$.

Electric flux linked with surface is

$$\phi = \int \frac{kq}{r^3} \cdot 2\pi r\, dr$$

$$= 2\pi kq \int \frac{dr}{r^2} = 2\pi kq\left(-\frac{1}{r}\right)$$

As flux is non-zero, charge density is also non-zero. Also, by symmetry of charge distribution electric field is zero.

24. *(c)* Let resistivity depends on given fundamental constants.

$$\rho = h^a m_e^b c^c e^d \varepsilon_0^f$$

where, k = a numeric constant.

Now, substituting dimensions of different physical constants, we have

$$[\text{ML}^3\text{T}^{-3}\text{A}^{-2}] = k[[\text{ML}^2\text{T}^{-1}]^a\, [\text{M}]^b\, [\text{LT}^{-1}]^c\, [\text{AT}]^d\, [\text{M}^{-1}\text{L}^{-3}\text{T}^4\text{A}^2]^f]$$

Equating dimensions, we have

$$1 = a + b - f$$
$$3 = 2a + c - 3f$$
$$-3 = -a - c + d + 4f$$
$$-2 = d + 2f$$

Solving these, we get

$$a = 2$$
$$b = -1$$
$$c = -1$$
$$d = -2$$
$$f = 0$$

So, resistivity can be expressed as

$$\rho = k\left(\frac{h^2}{m_e c e^2}\right)$$

25. *(c)* Soap solution lowers the surface tension of water.

So, water surface is punctured and is pulled back near circumference of vessel due to surface tension.

26. *(c)* Given, $\mu = 1.5 + \frac{0.004}{\lambda^2}$

So, deviation is given by

$$\delta = \frac{-d\mu}{d\lambda} = \frac{-d}{d\lambda}\left(1.5 + \frac{0.004}{\lambda^2}\right)$$

$$\Rightarrow \quad \delta = \frac{0.008}{\lambda^3} \Rightarrow \delta \propto \frac{1}{\lambda^3}$$

If $\lambda_1 < \lambda_2$, then $\delta(\lambda_1) > \delta(\lambda_2)$.

$$\Rightarrow \quad \delta_m(\lambda_1) > \delta_m(\lambda_2)$$

Alternative method

Deviation produced by a thin prism is

$$\delta = (\mu - 1)A$$
$$= \left(1.5 + \frac{0.004}{\lambda^2} - 1\right)A$$
$$\Rightarrow \quad \delta = \left(0.5 + \frac{0.004}{\lambda^2}\right)A$$

Clearly, δ is less for larger λ.

$\therefore$ When $\lambda_1 < \lambda_2$,

$$\delta(\lambda_1) > \delta(\lambda_2)$$
$$\Rightarrow \quad \delta_m(\lambda_1) > \delta_m(\lambda_2)$$

27. *(a)*

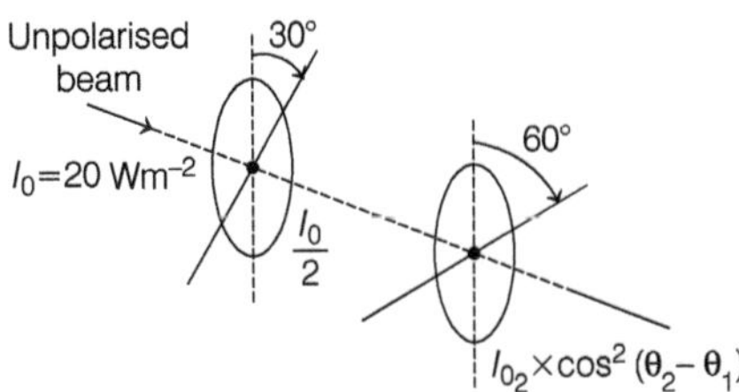

As beam incident over first polaroid is unpolarised, its intensity is reduced to half (Malus' law is not applicable).

So, intensity I_1 after first polaroid is

$$I_1 = \frac{I_0}{2} = \frac{20}{2} = 10 \text{ Wm}^{-2}$$

As light from first polaroid is polarised, Malus' law is now applicable.

So, intensity I_2 obtained after second polaroid is

$$I_2 = I_1 \cdot \cos^2\theta$$

where, θ = angle between transmission axis of first and second polaroid.

So, $\quad I_2 = 10 \times \cos^2 30°$

$$= 10 \times \left(\frac{\sqrt{3}}{2}\right)^2 = 7.5 \text{ Wm}^{-2}$$

28. *(b)* de-Broglie wavelength of electron is

$$\lambda = \frac{h}{p} = \frac{h}{mv}$$

Initially, $\quad \lambda = \frac{h}{mv_0}$

Finally, $\quad \frac{\lambda}{3} = \frac{h}{mv'}$

$$\Rightarrow \quad \frac{\lambda}{\lambda/3} = \frac{h/mv_0}{h/mv'}$$

So, final velocity is $v' = 3v_0$.

Acceleration of electron due to electric field is in y-direction, its magnitude is

$$a = \frac{eE_0}{m}\hat{\mathbf{j}}$$

Final velocity v' of electron after time t is

$$\mathbf{v}' = \mathbf{u} + \mathbf{a}t$$
$$\Rightarrow \quad v' = v_0\hat{\mathbf{i}} + \frac{eE_0t}{m}\hat{\mathbf{j}}$$

Magnitude of final velocity is

$$|\mathbf{v}'| = \sqrt{v_0^2 + \left(\frac{eE_0t}{m}\right)^2}$$

As $\quad |\mathbf{v}'| = 3v_0$

We have,

$$9v_0^2 = v_0^2 + \left(\frac{eE_0}{m}.t\right)^2$$
$$\Rightarrow \quad t = \sqrt{\frac{8mv_0^2}{e^2E_0^2}}$$
$$\Rightarrow \quad t \propto \frac{1}{E_0}$$

29. *(*)* When bird sits over a high tension wire, its both feet are

(As $\Delta V \approx 0$, both feets are placed nearby over wire)

nearly at same potential. So, no current flows through bird's body.

30. *(a)* As no field exists inside the shell body or in the cavity, correct charge distribution is as shown below in the figure,

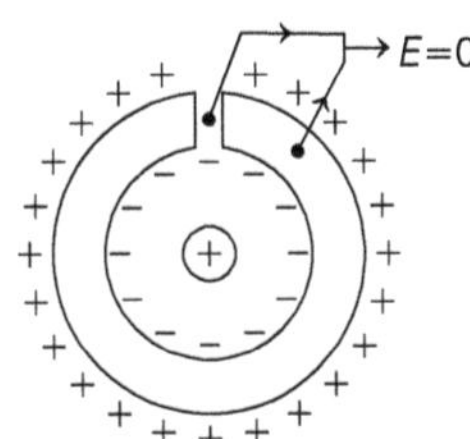

31. *(b)* Equation for a travelling wave is

$$y = a\sin(\omega t - kx + \phi_0)$$

As at $t = 0$ at $x = 0$ and $y = 0$,

$$\Rightarrow \quad \phi_0 = 0$$

So, equation of wave over string is

$$y = a\sin(\omega t - kx)$$

Here, $\quad \omega = 2\pi f = 2\pi \times 500$

$$= 1000\pi\left(\frac{\text{rad}}{\text{s}}\right)$$

Angular wave number,

$$k = \frac{\omega}{v} = \frac{1000\pi}{100} = 10\pi(\text{m}^{-1})$$

So, wave equation is

$$y = a\sin(1000\pi \cdot t - 10\pi \cdot x)$$

Given at $t = 0$ at $x = 0.25$ m and $y = 0.02$ m.

So, $\quad 0.02 = a\sin(-10\pi \times 0.25)$

$$= -a\sin\left(\frac{5}{2}\pi\right)$$

$[\because \sin(-\theta) = -\sin\theta]$

$$\Rightarrow \quad 0.02 = -a\sin\left(2\pi + \frac{\pi}{2}\right)$$
$$\Rightarrow \quad 0.02 = -a\sin\left(\frac{\pi}{2}\right) \text{ or } a = -0.02$$

Here note that a is amplitude and its positive and negative values are same. When we are getting a negative value this means particle is displaced below mean position.

So, we have

$$y = -0.02\sin(1000\pi t - 10\pi x)$$

Now, at $t = 5 \times 10^{-4}$ s and $x = 0.2$ m, value of displacement of particle is

$$y = -0.02\sin(1000\pi \times 5 \times 10^{-4} - 10\pi \times 0.2)$$
$$= -0.02\sin\left(\frac{\pi}{2} - 2\pi\right)$$
$$= 0.02\sin\left(2\pi - \frac{\pi}{2}\right)$$

$\left[\because \sin\left(\frac{\pi}{2} - 2\pi\right) = -\sin\left(2\pi - \frac{\pi}{2}\right).\right.$ Also, $\left.\sin(2\pi - \theta) = -\sin\theta\right]$

$$= -0.02 \text{ m}$$

32. *(b)* Entropy change for a system or body is

$$\Delta S = \frac{Q}{T}$$

Now, for given system to be in steady state, heat lost by resorvoir at temperature T_1 = heat gained by resorvoir at temperature T_2 (= Q say).

So, change in entropy for heat conduction process is

$$\Delta S = \frac{-Q}{T_1} + \frac{(+Q)}{T_2}$$
$$\Rightarrow \quad \Delta S = Q\left(\frac{1}{T_2} - \frac{1}{T_1}\right)$$

So, time rate of change of entropy is

$$\frac{dS}{dt} = \frac{d}{dt}\left\{Q\left(\frac{1}{T_2} - \frac{1}{T_1}\right)\right\}$$
$$= \left(\frac{1}{T_2} - \frac{1}{T_1}\right)\cdot\left(\frac{dQ}{dt}\right)$$
$$\frac{dS}{dt} = \left(\frac{T_1}{T_2} - 1\right)\cdot T_1\left(\frac{dQ}{dt}\right)$$

As, $\frac{dQ}{dt} = -kA\left(\frac{dT}{dx}\right)$

So, $\frac{dS}{dt} = -kA\frac{dT}{dx}\cdot T_1\left(\frac{T_1}{T_2} - 1\right)$

$$\Rightarrow \frac{dS}{dt} = kAT_1 \cdot \left(1 - \frac{T_1}{T_2}\right)\cdot \frac{dT}{dx}$$

$$= \frac{kAT_1}{x}\left(1 - \frac{T_1}{T_2}\right)(T_1 - T_2)$$

$$= \frac{kAT_1^2}{x}\left(1 - \frac{T_1}{T_2}\right)\left(1 - \frac{T_2}{T_1}\right)$$

Clearly, at $\frac{T_1}{T_2} = 1, \frac{dS}{dt} = 0.$

Also, graph is asymmetrical. So, correct option is (b).

33. *(a)* Time period of a simple pendulum depends on its amplitude as,

$$T = 2\pi\sqrt{\frac{l}{g}}\cdot\left(1 + \frac{\theta_0^2}{16}\right)$$

where, θ_0 = angular amplitude from mean position. Clearly, T increases with θ and its relation is parabolic. Its variation is as shown below in the figure.

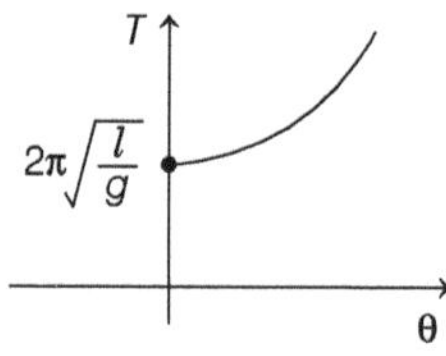

34. *(c)* As pulley is massive, tensions on both sides of pulley are not same.

From free body diagram, we have

T_1 m_1 a m_1g T_2 m_2 a m_2g

$$m_1g - T_1 = m_1a \quad \text{... (i)}$$

$$T_2 - m_2g = m_2a \quad \text{... (ii)}$$

and $(T_1 - T_2)R = I\alpha = \frac{Ia}{R}$ ...(iii)

From Eq. (iii), we have

$$T_1 - T_2 = \left(\frac{MR^2}{2}\cdot\frac{a}{R}\right)\times\frac{1}{R}$$

$$\Rightarrow T_1 - T_2 = M\cdot\frac{a}{2} \quad \text{...(iv)}$$

So, $T_1 - T_2$ increases as mass M of pulley increases.

Now, substituting for T_1 and T_2 from Eqs. (i) and (ii) in Eq. (iii), we get

$$m_1(g - a) - m_2(g + a) = \frac{Ma}{2}$$

$$\Rightarrow (m_1 - m_2)g = a\left\{\frac{M}{2} + (m_1 + m_2)\right\}$$

$$\Rightarrow a = \frac{(m_1 - m_2)g}{\frac{M}{2} + (m_1 + m_2)}$$

Substituting for a in Eq. (iv), we have

$$T_1 - T_2 = \frac{M(m_1 - m_2)g}{M + 2(m_1 + m_2)}$$

$$\Rightarrow T_1 - T_2 = \frac{(m_1 - m_2)g}{1 + \frac{2}{M}(m_1 + m_2)}$$

As $M \to \infty$, $T_1 - T_2$ = constant.

Also, when M increases from 0, $T_1 - T_2$ increases.

So, correct graph is option (c).

35. *(d)* From Kepler's law, we have

$$T^2 \propto R^3$$

So, for satellites S_1 and S_2 is

$$\Rightarrow \left(\frac{T_1}{T_2}\right)^2 = \left(\frac{R_1}{R_2}\right)^3$$

Radius of orbit of satellite S_2 around planet is

$$R_2 = \left(\frac{T_2^2 \times R_1^3}{T_1^2}\right)^{\frac{1}{3}}$$

$$= \left(\frac{24 \times 24 \times (3 \times 10^4)^3}{3 \times 3}\right)^{\frac{1}{3}}$$

$$= 4 \times 3 \times 10^4 \text{ km}$$

$$= 12 \times 10^4 \text{ km}$$

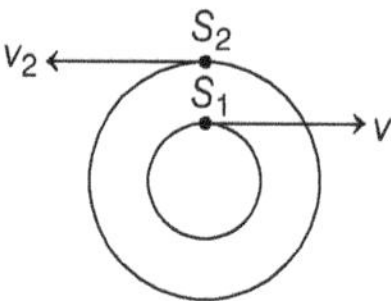

When satellites are closest to each other, orbital speed of S_2 as observed from S_1 is

$$v_{\text{relative}} = v_1 + v_2$$

$$= R_1\omega_1 + R_2\omega_2$$

$$= R_1 \times \frac{2\pi}{T_1} + R_2 \times \frac{2\pi}{T_2}$$

$$= 3 \times 10^4 \times \frac{2\pi}{3} + 12 \times 10^4 \times \frac{2\pi}{24}$$

$$= 2\pi \times 10^4 + \pi \times 10^4$$

$$= 3\pi \times 10^4 \text{ kmh}^{-1}$$

36. *(b)* Trajectory of charged particle in region of perpendicular magnetic field to its velocity is a circle.

So, path of charged particle will be as shown below in the figure,

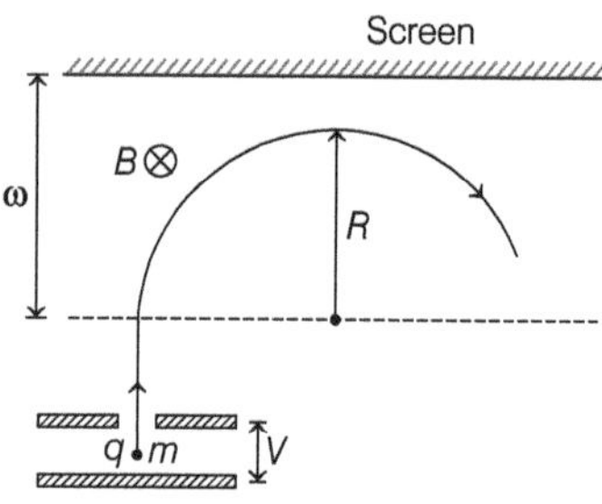

Clearly, charged particle will hit the screen, if $R > \omega$.

$$\Rightarrow \frac{mv}{Bq} > \omega \quad \left[\text{from, } Bqv = \frac{mv^2}{R}\right]$$

$$\Rightarrow \frac{\sqrt{2mK}}{Bq} > \omega$$

$$[\because \text{momentum, } p = mv = \sqrt{2mK}]$$

$$\Rightarrow \frac{\sqrt{2mqV}}{Bq} > \omega$$

or $\frac{2mV}{B^2q} > \omega^2$ or $q < \frac{2mV}{B^2\omega^2}$

37. *(d)* $(\hat{\mathbf{n}} \times \mathbf{F}) \times \hat{\mathbf{n}}$

$$= -\hat{\mathbf{n}} \times (\hat{\mathbf{n}} \times \mathbf{F})$$

$$[\because \mathbf{A} \times \mathbf{B} = -\mathbf{B} \times \mathbf{A}]$$

$$= -\{\hat{\mathbf{n}}(\hat{\mathbf{n}} \cdot \mathbf{F}) - \mathbf{F}(\hat{\mathbf{n}} \cdot \hat{\mathbf{n}})\}$$

∴ Vector triple product is defined as,

$$\mathbf{A} \times (\mathbf{B} \times \mathbf{C}) = \mathbf{B}(\mathbf{A} \cdot \mathbf{C}) - \mathbf{C}(\mathbf{A} \cdot \mathbf{B})$$

$$= \mathbf{F} - \hat{\mathbf{n}}(\hat{\mathbf{n}} \cdot \mathbf{F}) \quad [\because \hat{\mathbf{n}} \cdot \hat{\mathbf{n}} = 1]$$

So, $(\hat{\mathbf{n}} \times \mathbf{F}) \times \hat{\mathbf{n}} = \mathbf{F} - \hat{\mathbf{n}}(\hat{\mathbf{n}} \cdot \mathbf{F})$

$\Rightarrow \mathbf{F} = \hat{\mathbf{n}}(\hat{\mathbf{n}} \cdot \mathbf{F}) + (\hat{\mathbf{n}} \times \mathbf{F}) \times \hat{\mathbf{n}}$

So, $\mathbf{G} = (\hat{\mathbf{n}} \times \mathbf{F}) \times \hat{\mathbf{n}}$

38. *(a)* Energy of particle is

$$\frac{1}{2}m\omega^2r^2 = \frac{1}{2}mv^2$$

where, v = velocity of particle around the path.

$$\Rightarrow v = r\omega$$

Now, angular momentum of particle will be

$$L = mvr = mr^2\omega$$

By Bohr's model, we have

$$L = \frac{nh}{2\pi}$$

$$\Rightarrow mr^2\omega = \frac{nh}{2\pi} \Rightarrow r^2 = \frac{nh}{2\pi m\omega}$$

or $r = \sqrt{\frac{h}{2\pi m\omega}} \times \sqrt{n}$

$$\Rightarrow r = a\sqrt{n}$$

$$\left[\because \text{given, } \sqrt{\frac{h}{2\pi m\omega}} = a\right]$$

39. *(b)* Rate of heat loss for two bodies of same specific heat's and same density for which temperature difference (of bodies and surroundings) is same is given by

$$\frac{\Delta T}{\Delta t} = k \cdot \frac{A}{V}$$

where, ΔT = temperature change $(T_f - T_i)$, Δt = time interval, k = a constant depending on shape of bodies, A = surface area and V = volume.

As in given case, bottles are identical and they cool down by same temperatures,

$$\Delta t \propto \frac{V}{A}$$

So, $$\Delta t \propto \frac{A \cdot h}{A} \text{ or } \Delta t \propto h$$

$$\therefore \quad \frac{t_A}{t_B} = \frac{h_A}{h_B} \Rightarrow \frac{t_A}{t_B} = \frac{h_A}{2h_A} \quad [\because h_B = 2h_A]$$

$$\Rightarrow \quad t_B = 2t_A$$

40. *(a)* Using, $v_{rms} = \sqrt{\frac{3kT}{m}}$

and $$p = N \times 2mv_{rms}$$

We have,

$$N = \frac{p}{2m \cdot v_{rms}} = \frac{p \times \sqrt{m}}{2m\sqrt{3kT}} = \frac{(p/2)}{\sqrt{3mkT}}$$

$$= \frac{1.01 \times 10^5}{2 \times \sqrt{(3 \times 5 \times 10^{-27} \times 1.4 \times 10^{-23} \times 293)}}$$

$$= 6.4 \times 10^{27}$$

41. *(a)*

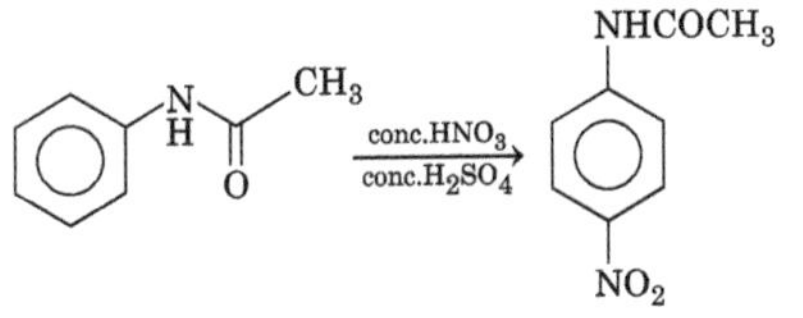

Acetanilide (Ph–NH–CO–CH_3) $\xrightarrow[\text{conc.} H_2SO_4]{\text{conc.} HNO_3}$ p-nitroacetanilide ($NHCOCH_3$ / NO_2)

In the above given reaction acetylated aniline is nitrated to get mononitro substituted product. In case of electrophilic substitution of aniline the reactivity is very high. Substitution tends to occur at all *ortho* and *para*-position. In order to get monosubstituted product the NH_2 group of aniline is protected by acetylation with acetic anhydride, then carrying out the nitration.

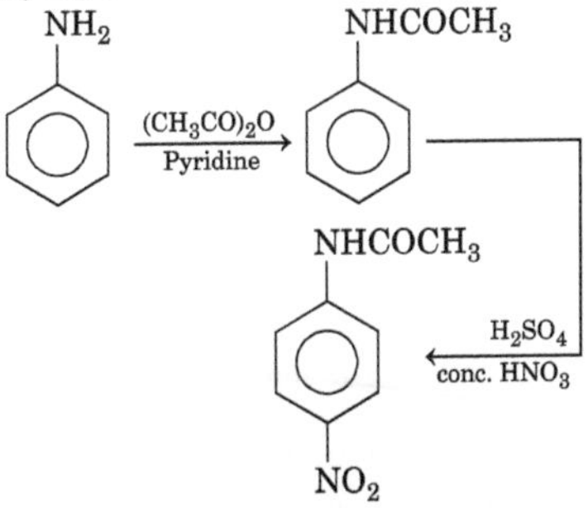

Aniline (NH_2) $\xrightarrow[\text{Pyridine}]{(CH_3CO)_2O}$ Acetanilide ($NHCOCH_3$) $\xrightarrow[\text{conc. } HNO_3]{H_2SO_4}$ p-nitroacetanilide ($NHCOCH_3$, NO_2)

42. *(b)* The structure of given α-amino acids are as follows :

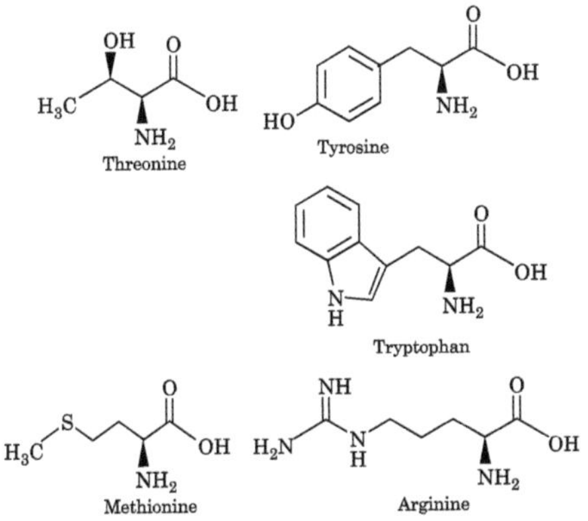

Threonine, Tyrosine, Tryptophan, Methionine, Arginine

Thus, tyrosine and tryptophan contain an aromatic group in their side chain.

43. *(d)* Number of stereoisomers = Number of optical isomers + number of geometrical isomers.

In the given compound,

$$CH_3 - CH = CH - \overset{*}{C}H(OH) - CH_3$$

There are 2 optical isomers as there is only 1 chiral (*) centre. Also, the compound is unsaturated (contains double bond), so it will have two geometrical isomers also, *cis* and *trans* form.

∴Number of stereoisomers, 2 + 2 = 4.

44. *(c)* In electrophilic aromatic substitution reaction of chlorobenzene, the *ortho/para* directing ability of chlorine is due to its + *R*-effect. The resonating structures of chlorobenzene are as follows :

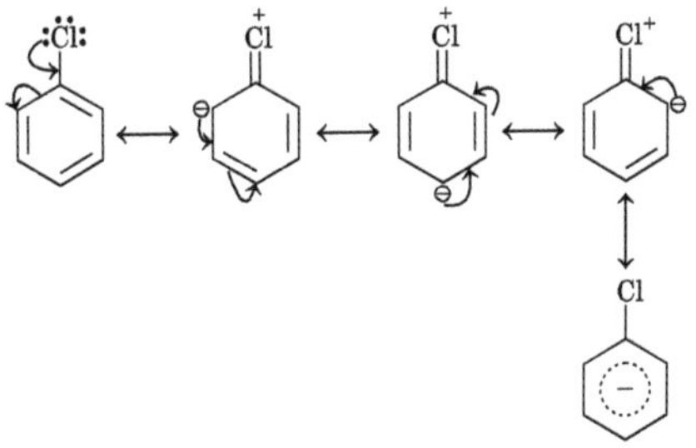

Cl ⟷ $\overset{+}{Cl}$ ⟷ $\overset{+}{Cl}$ ⟷ Cl^+ ↕ Cl (resonance hybrid)

Due to this effect, the electron density increases more at *ortho* and *para*-positions than at *meta*- position.

45. *(b)* One of the criteria to know anti-aromaticity is the molecule is that it should have $4n\pi$-electrons and all the carbon atoms should have sp^2-hybridisation.

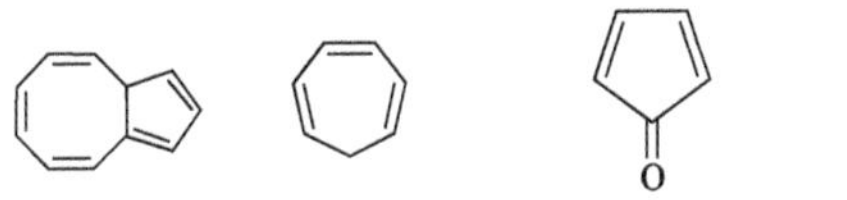

(I) 10π-electrons. It has $(4n+2)$ π-electrons ∴ It is aromatic.

(II) 6π-electrons. It has sp^3 hybridised atom with $(4n+2)$ π-electrons ∴ It is non-aromatic.

(III) 4π-electrons. All C-atoms has sp^2 hybridisation with electrons ∴ It is C-atom anti-aromatic.

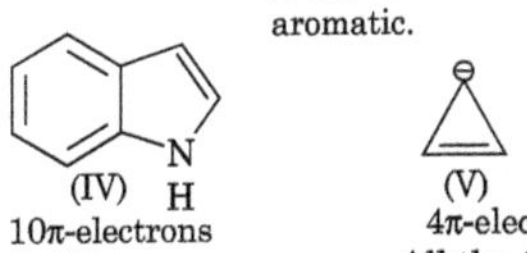

(IV) 10π-electrons. It follows $(4n+2)$ π-electrons rule, so it is aromatic.

(V) 4π-electrons. All the C-atoms are sp^2-hybridised ∴ It is anti-aromatic.

Thus, compound III and IV are anti-aromatic.

46. *(c)* As ethanol is primary alcohol, so it can be produced only by formaldehyde with Grignard's reagent. The reaction is as follows :

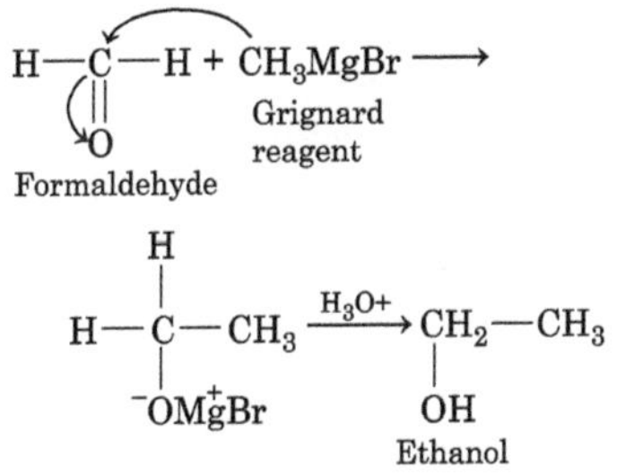

H–C(=O)–H (Formaldehyde) + CH_3MgBr (Grignard reagent) ⟶ H–C(H)(CH_3)–$\bar{O}\overset{+}{M}gBr$ $\xrightarrow{H_3O^+}$ $CH_2(OH)-CH_3$ (Ethanol)

47. *(b)* Oxidation-reduction reactions are those reactions in which there is a change in oxidation state of any element.

(a) $\overset{0}{H_2} + \overset{-2}{Br_2} \longrightarrow 2\overset{+1}{H}\overset{-1}{Br}$

(b) $\overset{+1}{Na}\overset{-1}{Cl} + \overset{+1}{Ag}\overset{-1}{NO_3} \longrightarrow \overset{+1}{Na}\overset{-1}{NO_3} + \overset{+1}{Ag}\overset{-1}{Cl}$

(c) $2\overset{+2}{Na_2}\overset{-6}{S_2O_3} + I_2 \longrightarrow \overset{+2}{Na_2}\overset{+12}{S_4O_6} + 2\overset{+1}{Na}\overset{-1}{I}$

(d) $\overset{-2}{Cl_2} + \overset{+2}{H_2}\overset{-2}{O} \longrightarrow \overset{+1}{H}\overset{-1}{Cl} + \overset{+1}{H}\overset{-2}{O}\overset{-1}{Cl}$

As there is no change in oxidation state in reaction given in (b). Thus, it is not an oxidation-reduction reaction.

48. *(a)* Thermal stability of alkaline earth metal carbonates increases down the group. This is because, down the group the size of metal increases, thus the carbonates require more heating to decompose. Thus, the order of thermal stability is

$BaCO_3 > SrCO_3 > CaCO_3 > MgCO_3$

49. *(d)* When a mixture of diborane and ammonia is heated, the final product is borazine, which is also known as inorganic benzene.

Most of its chemical properties is similar to that of benzene.

$B_2H_6 + NH_3 \longrightarrow$ (Diborane) Borazine $(B_3N_3H_6)$

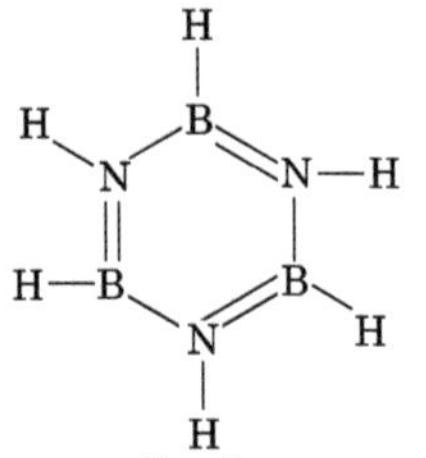

50. *(c)* The order of reducing agent can be determined with the help of electrochemical series. It is a series of electrodes or half-cells arranged in order of their decreasing standard reduction potential.

As the reduction potential decreases the strength of reducing agent also decreases.

According to this series, the order of increasing reducing agent,

$$\text{Fe} < \text{Cu} < \text{Ni} < \text{Zn}.$$

Thus, among the given metals, Zn is the strongest reducing agent.

51. *(d)* The moelcule, which is not hydrolysed by water at 25°C is SF_6. This is because it is kinetically stabilised, attack at S is impossible due to steric Hindrance and also because of the rigidity of F– SF angles.

The hydrolysed product of other molecules are as follows :

$AlCl_3 + H_2O \longrightarrow Al(OH)_3 + HCl$

$SiCl_4 + 2H_2O \longrightarrow 4HCl + SiO_2$

$BF_3 + H_2O \longrightarrow H_3BO_3 + HBF_4$

52. *(d)* The thermal products of given compounds are as follows :

(a) $(NH_4)_2Cr_2O_7 \xrightarrow{\Delta} Cr_2O_3 + N_2\uparrow + H_2O$

(b) $2NaN_3 \xrightarrow{\Delta} 2Na + 3N_2\uparrow$

(c) $NH_4NO_2 \xrightarrow{\Delta} N_2\uparrow + 2H_2O$

(d) $(NH_4)_2(C_2O_4) \xrightarrow{\Delta} 2NH_3 + \begin{matrix}COOH \\ | \\ COOH\end{matrix}$

Thus, among the given compounds, the one which does not produce nitrogen gas upon heating is $(NH_4)_2(C_2O_4)$.

53. *(b)* Let the natural abundance of ^{35}Cl be x.

Let the natural abundance of ^{37}Cl be y.

$$M_{av} = \frac{M_x x + M_y y}{x + y}$$

$$35.45 = \frac{35x + 37y}{x + y}$$

$$\therefore \quad \frac{x}{y} = \frac{3}{1}$$

54. *(d)* $C_2H_6(g) \rightleftharpoons C_2H_4(g) + H_2(g)$,

$$\Delta H = 137.0 \text{ kJ mol}^{-1}$$

Increasing the volume of the closed reaction vessel would shift the equilibrium to the right. As the volume increases pressure decreases, so number of gaseous molecules also decreases. Thus, equilibrium will shift to that direction where there is increase in the number of gaseous molecules, i.e. in forward direction. This is inaccordance to Le-Chatelier principle, which states that a system in chemical equilibrium changes towards a new equilibrium when subjected to a change in pressure, temperature, volume and concentration. The equilibrium shift partly counter acts the change that was applied to the system initially.

55. *(a)* For exothermic reaction,

$$A \rightleftharpoons B$$

$$\Delta H < 0.$$

$$\Delta H = E_{\text{Product}} - E_{\text{Reactant}}$$

$$\therefore \quad E_B - E_A < 0$$

$$\Rightarrow \quad E_B < E_A$$

This relation can be seen in plot A and C, but correct option is (a) as the catalyst does not change the initial and final state of reactants and products of a reaction.

56 *(c)* Given, K_{sp} $Mg(OH)_2 = 5.6 \times 10^{-12}$

Concentration of $Mg^{2+} = 10^{-10}$ M

For the reaction,

$$Mg(OH)_2 \rightleftharpoons Mg^{2+} + 2OH^-$$

$$K_{sp} = [Mg^{2+}][OH^-]^2$$

$$5.6 \times 10^{-12} = [10^{-10} \text{M}]\,[OH^-]^2$$

$$\frac{5.6 \times 10^{-12}}{10^{-10}} = [OH^-]^2$$

$$= \sqrt{5.6 \times 10^{-2}} = [OH^-]$$

$$\therefore \quad [OH^-] = 0.24 \text{ M}$$

Note Due to common ion effect, the concentration of OH^- gets suppressed.

57. *(c)* According to Faraday's second law of electrolysis, when the same quantity of electricity is passed through several electrolytes, the mass of the substances deposited are directly proportional to their respective chemical equivalent. i.e.

Mass of substance metal deposited $\propto$ equivalent weight.

Where, equivalent weight

$$= \frac{\text{Atomic mass}}{\text{Number of electrons transferred}}$$

Let the number of the equivalent of metal deposited be M.

(i) **For $AgNO_3$**

Number of moles equivalent of Ag^+ deposited $= \dfrac{M_{Ag}}{1}$

(ii) **For $CuSO_4$**

Number of moles equivalent of Cu^{2+} deposited $= \dfrac{M_{Cu}}{2}$

(iii) **For AlF_3**

Number of moles equivalent of Al^{3+} deposited $= \dfrac{M_{Al}}{3}$

$\therefore$ Ratio of metals deposited on the cathode, i.e.

Moles of $Ag^+ : Cu^{2+} : Al^{3+}$

$$= \frac{M_{Ag}}{1} : \frac{M_{Cu}}{2} : \frac{M_{Al}}{3}$$

Or $6M_{Ag} : 3M_{Cu} : 2M_{Al}$

58. *(a)* Given,

activation energy = 209 kJ mol^{-1}

$$K_2 = 10K_1$$

According to Arrhenius equation,

$$K = Ae^{-E_aRT}$$

Or $$\log\frac{K_2}{K_1} = \frac{E_a}{R} 2.303 \log\left[\frac{1}{T_1} - \frac{1}{T_2}\right]$$

$$\log 10 \frac{K_2}{K_1} = \frac{209 \times 10^3}{2.303 \times 8.314}\left[\frac{1}{300} - \frac{1}{X}\right]$$

$$\log 10 = \frac{209 \times 10^3}{2.303 \times 8.314}\left[\frac{1}{300} - \frac{1}{X}\right]$$

$$= \frac{2.303 \times 8.314}{209 \times 10^3} = \frac{1}{300} - \frac{1}{X}$$

$$9.16 \times 10^{-5} = 3.33 \times 10^{-5} - \frac{1}{T}$$

$$T = 308.4 \text{ K}$$

$$T = 308.4 - 273 = 35.4 \text{ K} \approx 35 \text{ K}$$

59. *(b)* Number O^{2-} ions

$$= 8 \times \frac{1}{8} + 6 \times \frac{1}{2} = 4$$

Number of O^{2-} ions = Number of octahedral voids = 4

Also, number of tetrahedral voids
= 2 × number of octahedral voids
= 8

Number of $Al^{3+} = \dfrac{1}{2} \times$ octahedral voids

$$= \frac{1}{2} \times 4 = 2$$

Number of Mn^{2+} ions

$$= \frac{1}{8} \times \text{tetrahedral voids} = \frac{1}{8} \times 8 = 1$$

Thus, the chemical formula of the mineral is $MnAl_2O_4$.

60. *(d)* For any orbital,

number of radical nodes = $n - l - 1$

number of angular nodes = l

For 4p-orbital, $n = 4, l = 1$

Number, of radical nodes = $4 - 1 - 1 = 2$

Number of angular nodes = 1

61. *(d)* Interferon is secreted by cells in response to stimulation by a virus or other foreign substance, but it does not directly inhibit the virus' multiplication. Rather, it stimulates the infected cells and those nearby to produce proteins that prevent the virus from replicating within them.

62. *(c)* Testosterone is produced by the gonads (by the Leydig cells in testes in men and by the ovaries in women) and small quantities are also produced by the adrenal glands in both sexes. It is an androgen, meaning that it stimulates the development of male characteristics.

63. *(a)* Glucagon elevates the concentration of glucose in the blood by promoting gluconeogenesis and glycogenolysis. When glucagon binds to the glucagon receptors, the liver cells convert the glycogen into individual glucose molecules and release them into the bloodstream, in a process known as glycogenolysis.

64. *(a)* For PCR, there are five chemical components needed, including a DNA template, DNA polymerase enzyme, primers, nucleotides and reaction buffer. The Polymerase Chain Reaction (PCR) is used to make millions of copies of a target piece of DNA. Thus, option (a) is correct as restriction enzyme is used for recombinant DNA technology.

65. *(b)* A greenhouse gas absorbs and emits solar radiant energy within the thermal infrared range. The primary greenhouse gases in earth's atmosphere are carbon dioxide, water vapour, methane, nitrous oxide and ozone. These are responsible for greenhouse effect.

66. *(a)* A graph of species richness *vs* area on log-log axes is linear.

Its equation is $\log S = \log C + Z \log A$

Where S = Number of species

A = Habitat area

Z = The slope of the species-area relationship in log-log space.

Species-area relationship on log-log axes

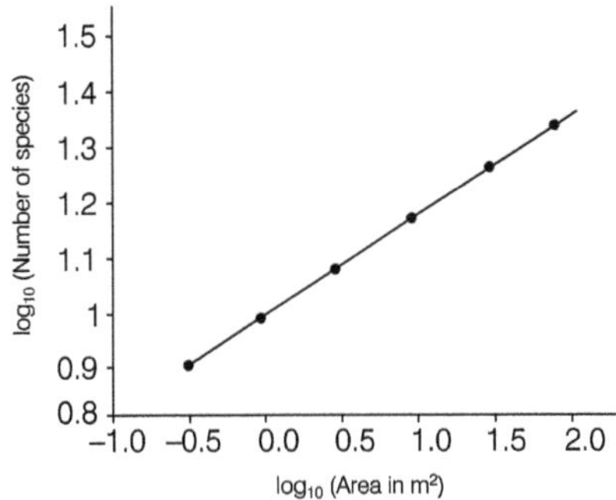

67. *(b)* The sodium-potassium ($Na^+ - K^+$) pump is found in many cell (plasma) membranes. Powered by ATP, the pump moves sodium and potassium ions in opposite directions, each against its concentration gradient. In a single cycle of the pump, three sodium ions are extruded from and two potassium ions are imported into the cell.

68. *(a)* The lower the activation energy for a reaction, the faster the rate. Thus, enzymes speed up reactions by lowering activation energy.

The activation energy is the energy which must be provided to a chemical reaction to result in order to start it.

69. *(c)* Cellulose is a straight chain polymer, without any coiling or branching. There occurs multiple hydroxyl groups on the glucose from one chain which forms hydrogen bonds with oxygen atoms on the same or on a neighbour chain, holding the chains firmly together side by side forming microfibrils with high tensile strength. Thus, option (c) is correct.

70. *(d)* Antigen-antibody reaction is the basis of humoral immunity or antibody -mediated immune response. Antigen-antibody interactions depend on four types of non-covalent interactions, i.e. hydrogen bonds, ionic bonds, hydrophobic interactions and van der Wall's interactions.

71. *(b)* IgM is the major antibody produced in the primary immune response and it exists as part of the B-cell antigen receptor on the surface of B-cells and as a secreted glycoprotein. Sodium Dodecyl Sulphate-Polyacrylamide Gel Electrophoresis (SDS-PAGE) is used to obtain the combinations of molecular masses of polypeptides from purified human IgM. The combinations obtained from SDS - PAGE are 70 kDa, 25 kDa and 15 kDa. So, the answer is (b).

72. *(b)* Skin colour is a good example of polygenic (multiple genes) inheritance. Skin colour is determined by the amount of melanin in the skin. This trait is controlled by at least three genes. The genes that determine skin colour have two alleles each and are found on different chromosomes. For example, three genes that are known to influence skin colour, each gene has one allele for dark skin and one for light skin colour. The allele for dark skin colour (D) is dominant to the allele for light skin colour (d). Skin colour is determined by the number of dark alleles a person has. Individuals who inherit no dark alleles will have very light skin colour, while those that inherit only dark alleles will have very dark skin colour. Individuals who inherit different combinations of light and dark alleles will have phenotypes of varying skin shades. Hence, with 3 genes for skin colour and each gene having 2 alleles, the number of skin colours possible $3 \times 2 = 6$. So, the answer is (b).

73. *(c)* Linked genes, violate the law of independent assortment. The segregation of alleles into gametes can be influenced by linkage, in which genes that are located physically close to each other on the same chromosome are more likely to be inherited as pair.

74. *(d)* An individual gene may code for a specific physical trait, that gene can exist in different forms or alleles. One allele for every gene in an organism is inherited from each of that organism's parents. Alleles produce phenotypes (or physical versions of a trait) that are either dominant or recessive.

75. *(d)* Statement (d) is incorrect, because restriction endonuclease enzymes cut within the DNA strands at specific sites called recognition sites. DNA from the ends are cut by exonuclease enzymes.

76. *(b)* The net ATP yield during glycolysis of one glucose molecule is 2 ATP. Glycolysis is the metabolic pathway that converts glucose into pyruvate.

77. *(a)* Pyridoxal-6-phosphate is the active form of vitamin-B_6. It acts as coenzyme in many enzymatic reactions. One of the reaction in which it functions as coenzymes is the conversion of L-alanine to a racemic mixture of D and L - alanine. This reaction is catalysed by enzyme alanine racemase. Thus, the correct answer is (a).

78. *(c)* During cyclic photophosphorylation, electrons flow a circular path in which only ATP (no NADPH) is produced. This type of photophosphorylation usually occurs in the thylakoid membrane.

79. *(a)* A flame cell is a specialised excretory cell found in the simplest freshwater invertebrates including flatworms, e.g. *Planaria*. Choanocytes (also known as collar cells) filter particles of food from the water and digest it. These are found in sponges.

All cnidarians (e.g. *Hydra*)have tentacles with stinging cells in their tips, which are used to capture and subdue prey.

80. *(b)* The partial pressure of oxygen in atmospheric air is higher than that of oxygen in alveolar air. In atmospheric air, pO_2 is about 159 mm Hg. In alveolar air, it is about 104 mm Hg.

The partial pressure of CO_2 in atmospheric air is lesser than that of CO_2 in alveolar air. In atmospheric air, pCO_2 is about 0.3 mm Hg. In alveolar air, it is about 40 mm Hg.

81. *(b)* We have

$$x^2 + y^2 = 2z^2 \text{ and HCF } (x, y, z) = 1$$

$x = 1, y = 7, z = 5$

Then, $\quad 1 + 49 = 50$

or $\quad x = 7, y = 1, z = 5$

$\quad 49 + 1 = 50$

$\therefore$ 4 not divides x or y.

$\therefore$ Ist statement is wrong.

$$x + y = 7 + 1 = 8,$$
$$x - y = 7 - 1 = 6$$

3 not divides $x + y$ but 3 divides $x - y$.

$\therefore$ IInd statement is TRUE

$$z(x^2 - y^2) = 5(49 - 1)$$

$\therefore$ 5 divides $z(x^2 - y^2)$.

IIIrd statement is also TRUE.

82. *(b)*

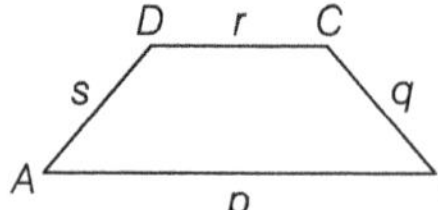

Length of sides of trapezium from {1, 2, 3, 4, 5, 6}.

From non-congruent trapezium

$$|r - p| < q + s < r + p$$

Possible combination are

$(r, p), (s, q) \equiv \{(5, 6), (1, 3)\}, \{(5, 6), (2, 4)\}$
$\{(5, 6), (1, 4)\}, \{(5, 6), (3, 4)\}, \{(6, 4), (1, 3)\},$
$\{(6, 4), (1, 5)\}, \{(6, 4), (2, 3)\}, \{(6, 4), (3, 5)\},$
$\{(4, 5), (1, 3)\} \{(4, 5), (1, 6)\} \{(4, 5), (2, 6)\}$

Total 11 combination is possible.

83. *(a)* Surface area of solid

i.e. $\quad S = 2\pi r^2 + 2\pi rh + \pi r^2$

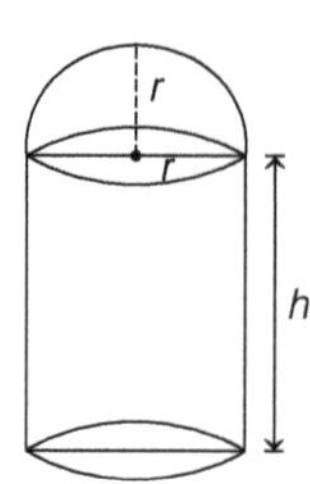

$$S = 3\pi r^2 + 2\pi rh$$
$$h = \frac{S - 3\pi r^2}{2\pi r}$$

Volume of solid

i.e. $\quad V = \pi r^2 h + \frac{2}{3}\pi r^3$

$\Rightarrow \quad V = \pi r^2\left(\frac{S - 3\pi r^2}{2\pi r}\right) + \frac{2}{3}\pi r^3$

$\Rightarrow \quad V = \frac{1}{2}(Sr - 3\pi r^3) + \frac{2}{3}\pi r^3$

$\Rightarrow \quad \frac{dV}{dr} = \frac{S}{2} - \frac{9\pi r^2}{2} + 2\pi r^2$

For maximum or minimum $\frac{dV}{dr} = 0$

$\therefore \quad S = 5\pi r^2$

$\Rightarrow \quad h = \frac{S - 3\pi r^2}{2\pi r} = \frac{5\pi r^2 - 3\pi r^2}{2\pi r} = r$

$\therefore \quad h = r \Rightarrow h : r = 1 : 1$

84. *(c)* Let position vector of ΔABC are $A(\vec{a}), B(\vec{b})$ and $C(\vec{c})$

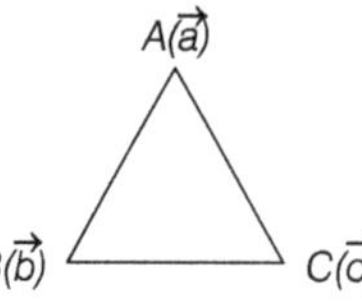

Let circumcentre of ΔABC, O (origin).

Centroid of ΔABC is $\frac{\vec{a} + \vec{b} + \vec{c}}{3}$

We know that centroid divide orthocentre and circumcentre in 2 : 1

i.e.

$$\underset{H}{} \overset{2}{\text{———}} \underset{G}{} \overset{1}{\text{———}} \underset{O}{}$$

$$HG : GO = 2 : 1$$

$$\overrightarrow{OG} = \left(\frac{\vec{a} + \vec{b} + \vec{c}}{3}\right)$$

$$\overrightarrow{OH} = \vec{a} + \vec{b} + \vec{c}$$

$\vec{N}$ is mid-point of $\overrightarrow{OH}$

$\therefore \quad \vec{N} = \frac{\vec{a} + \vec{b} + \vec{c}}{2}$

$\overrightarrow{NA} + \overrightarrow{NB} + \overrightarrow{NC}$

$$= \vec{a} - \left(\frac{\vec{a} + \vec{b} + \vec{c}}{2}\right) + \vec{b} - \left(\frac{\vec{a} + \vec{b} + \vec{c}}{2}\right) + \vec{c} - \left(\frac{\vec{a} + \vec{b} + \vec{c}}{2}\right)$$

$$= (\vec{a} + \vec{b} + \vec{c}) - \frac{3(\vec{a} + \vec{b} + \vec{c})}{2}$$

$$= -\frac{1}{2}(\vec{a} + \vec{b} + \vec{c}) = -\frac{1}{2}\overrightarrow{OH} = \frac{1}{2}\overrightarrow{HO}$$

85. *(d)* Let

$$p(x) = 1 + x^2 + x^4 + x^6 + \ldots + x^{34}$$

$\Rightarrow \quad p(x) = \frac{x^{36} - 1}{x^2 - 1}$

$\Rightarrow \quad q(x) = 1 + x + x^2 + x^3 + \ldots + x^{17}$

$\Rightarrow \quad q(x) = \frac{x^{18} - 1}{x - 1}$

$\therefore \quad \frac{p(x)}{q(x)} = \left(\frac{x^{36} - 1}{x^2 - 1}\right)\left(\frac{x - 1}{x^{18} - 1}\right)$

$\Rightarrow \quad \frac{p(x)}{q(x)} = \frac{(x^{18} + 1)(x^{18} - 1)(x - 1)}{(x + 1)(x - 1)(x^{18} - 1)}$

$\Rightarrow \quad \frac{p(x)}{q(x)} = \frac{x^{18} + 1}{x + 1}$

$\Rightarrow \frac{p(x)}{q(x)} = \frac{(x + 1)(x^{17} - x^{16} + x^{15} - x^{14} \ldots - 1)}{x + 1}$

$\therefore$ Quotient is $x^{17} - x^{16} + x^{15} - x^{14} \ldots -1$

86. *(a)* Given

$$x^2 + y^2 \le 1$$

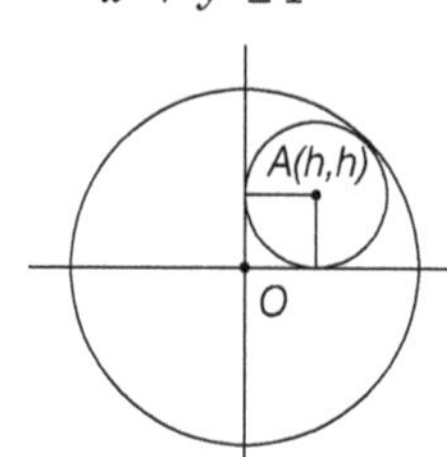

Largest and smallest circle touch internally.

$\therefore \quad |r_1 - r_2| = OA$

$\quad (h - 1)^2 = 2h^2$

[radius of smallest circle = h]

$\Rightarrow \quad h^2 - 2h + 1 = 2h^2$

$\Rightarrow \quad h^2 + 2h - 1 = 0$

$\Rightarrow \quad h = \frac{-2 \pm \sqrt{8}}{2}$

$\Rightarrow \quad h = -1 \pm \sqrt{2}$

$\Rightarrow \quad h = \sqrt{2} - 1$

$\because \quad r = \sqrt{2} - 1$

Area $= \pi(\sqrt{2} - 1)^2 = \pi(3 - 2\sqrt{2})$

87. *(b)* We have,

$$f(x) = \sqrt{|x|} - \log(1 + |x|)$$

Let $\quad g(x) = \sqrt{x} - \log(1 + x)$

$$g'(x) = \frac{1}{2\sqrt{x}} - \frac{1}{1+x}$$

$$\Rightarrow \quad g'(x) = \frac{1 + x - 2\sqrt{x}}{2(1+x)\sqrt{x}}$$

$$\Rightarrow \quad g'(x) = \frac{(\sqrt{x}-1)^2}{2(1+x)\sqrt{x}} > 0,\ x > 0$$

$\therefore$ Graph of $g(x)$ is

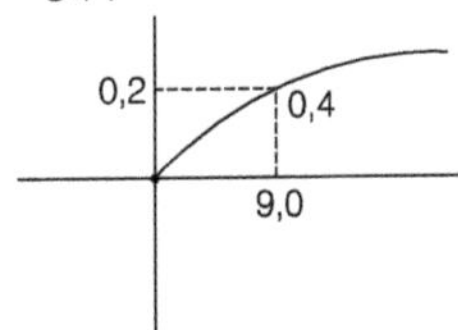

Graph of $f(x) = g(|x|)$

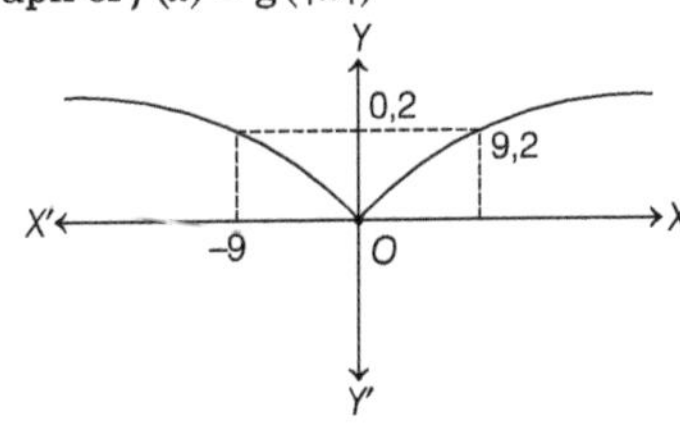

Clearly from graph:

I is false and II is true.

88. *(d)* We have,

$$g(x) = \int_{-3}^{3} f(x-y) f(y)\, dy$$

$$f(t) = \begin{cases} 1, & 0 \le t \le 1 \\ 0, & \text{elsewhere} \end{cases}$$

$$\Rightarrow \quad g(x) = \int_{-3}^{0} g(x-y) f(y)\, dy + \int_0^1 f(x-y) f(y)\, dy + \int_1^3 f(x-y) f(y)\, dy$$

$$\Rightarrow \quad g(x) = 0 + \int_0^1 f(x-y)\, dy + 0$$

$$\Rightarrow \quad g(x) = \int_0^1 f(x-y)\, dy$$

Put $x - y = t \Rightarrow -dy = dt$

$$\Rightarrow \quad g(x) = -\int_x^{x-1} f(t)\, dt \Rightarrow g(x) = \int_{x-1}^{x} f(t)\, dt$$

$$\Rightarrow \quad g(x) = \begin{cases} 0, & x \le 0 \\ x, & 0 < x < 1 \\ 2 - x, & 1 \le x \le 2 \\ 0, & x > 2 \end{cases}$$

Clearly, $g(x)$ is continues at for all x

$$g'(x) = \begin{cases} 0, & x \le 0 \\ 1, & 0 < x < 1 \\ -1, & 1 \le x \le 2 \\ 0, & x > 2 \end{cases}$$

$g(x)$ is not differentiable at 0, 1, 2.

89. *(c)* Let $S = \sum_{k=0}^{44} \frac{1}{\cos k^\circ \cos(k+1)^\circ}$

$$\Rightarrow \quad S = \frac{1}{\sin 1^\circ} \sum_{k=0}^{44} \frac{\sin(k+1-k)^\circ}{\cos k^\circ \cos(k+1)^\circ}$$

$$\Rightarrow \quad S = \frac{1}{\sin 1^\circ} \sum_{k=0}^{44} \left[\frac{\sin(k+1)^\circ \cos k^\circ}{\cos k^\circ \cos(k+1)^\circ} - \frac{\cos(k+1)^\circ \sin k^\circ}{\cos k^\circ \cos(k+1)^\circ} \right]$$

$$\Rightarrow \quad S = \frac{1}{\sin 1^\circ} \sum_{k=0}^{44} (\tan(k+1)^\circ - \tan k^\circ)$$

$$\Rightarrow \quad S = \frac{1}{\sin 1^\circ} [\tan 1^\circ - \tan 0^\circ + \tan 2^\circ - \tan 1^\circ + \ldots + \tan 45^\circ - \tan 44^\circ]$$

$$\Rightarrow \quad S = \frac{1}{\sin 1^\circ} \tan 45^\circ$$

$$\Rightarrow \quad S = \frac{1}{\sin 1^\circ} = \frac{1}{0.0174} = 57.29$$

$[\because \sin 1^\circ = 0.017]$

$$[S] = [57.29] = 57$$

90. *(b)* We have,

$$\int_0^1 x f(x)\, dx = \frac{1}{3} + \frac{1}{4} \int_0^1 (f(x))^2\, dx$$

$$\Rightarrow \quad -\frac{1}{3} = \frac{1}{4} \left[\int_0^1 [(f(x))^2 - 4x f(x)] dx \right]$$

$$\Rightarrow -\frac{1}{3} = \frac{1}{4} \left[\int_0^1 (f(x))^2 - 4x f(x) + (2x)^2 - (2x)^2 \right] dx$$

$$\Rightarrow -\frac{1}{3} = \frac{1}{4} \left[\int_0^1 [f(x) - 2x]^2 dx - \int_0^1 4x^2 dx \right]$$

$$\Rightarrow -\frac{4}{3} = \int_0^1 [f(x) - 2x]^2 dx - \left[\frac{4x^3}{3} \right]_0^1$$

$$\Rightarrow -\frac{4}{3} = \int_0^1 [f(x) - 2x]^2 dx - \frac{4}{3}$$

$$\Rightarrow \quad \int_0^1 [f(x) - 2x]^2 dx = 0$$

$\therefore$ Only one continuous function.

91. *(a)* Let rod OA makes angle θ at some instant t with OP as shown below in the figure.

Then, by applying cosine formula to ΔOAP, we get

$$\cos\theta = \frac{R^2 + x^2 - L^2}{2Rx}$$

$$= R^2 + x^2 - L^2 - 2Rx\cos\theta$$

Differentiating above equation with respect to time t, we get

$$\Rightarrow \quad \frac{dx^2}{dt} = 2R\left(x\frac{d}{dt}\cos\theta + \cos\theta \frac{dx}{dt} \right)$$

$$\Rightarrow 2x\frac{dx}{dt} = 2R\left(-x\sin\theta \frac{d\theta}{dt} + \cos\theta \frac{dx}{dt} \right)$$

$$\Rightarrow \quad xv = -xR\omega\sin\theta + Rv\cos\theta$$

$$\therefore \quad \frac{d\theta}{dt} = \omega \text{ and } \frac{dx}{dt} = v$$

$$\Rightarrow \quad v = \frac{Rx\omega \cdot \sin\theta}{x - R\cos\theta} \quad \ldots\text{(i)}$$

Now, when angle $\theta = 60°$, then

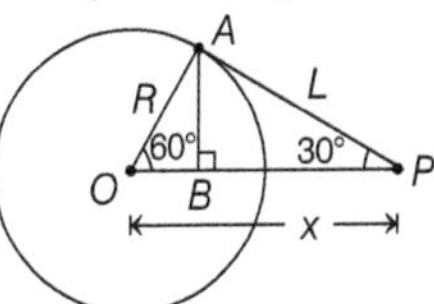

In ΔOAP, we have

$$\angle A = 90°,\ L = 1\,\text{m and } R = \frac{1}{\sqrt{3}}\,\text{m}$$

So, $x = \sqrt{L^2 + R^2} = \sqrt{1 + \frac{1}{3}} = \frac{2}{\sqrt{3}}$ m

Substituting values in Eq. (i), we get

$$v = \frac{\left(\frac{1}{\sqrt{3}} \times \frac{2}{\sqrt{3}} \times \sin 60° \right) \cdot \omega}{\left(\frac{2}{\sqrt{3}} - \frac{1}{\sqrt{3}} \cos 60° \right)}$$

$$\Rightarrow \quad v = \left(\frac{\frac{2}{3} \times \frac{\sqrt{3}}{2}}{\frac{2}{\sqrt{3}} - \frac{1}{2\sqrt{3}}} \right) \omega \Rightarrow v = \frac{2}{3}\omega$$

92. *(d)* Given p-V cycle is

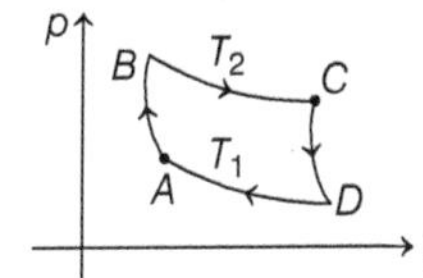

AB : Adiabatic compression,

$$V_A = V_1,\ V_B = 1\,\text{m}^3$$

BC : Isothermal expansion,

$$V_C = V = 10\,\text{m}^3$$

CD : Adiabatic expansion,

$$V_D = V_3$$

DA : Isothermal compression,

$$V_A = V_1$$

Cycle efficiency is given, $\eta = \frac{3}{4}$

For given Carnot's cycle,

$$\eta = 1 - \frac{T_1}{T_2} = 1 - \left(\frac{V_2}{V_1} \right)^{\gamma - 1}$$

[$\because$ Process AB is adiabatic, $\gamma = \frac{5}{3}$ for monoatomic gas]

$$\Rightarrow \quad \frac{3}{4} = 1 - \left(\frac{1}{V_1} \right)^{\frac{5}{3} - 1} \Rightarrow \frac{1}{V_1^{2/3}} = \frac{1}{4}$$

$$\Rightarrow \quad V_1 = 8\,\text{m}^3$$

93. *(a)* Given, power radiated P is

$$P = \mu_0^x m^y \omega^z c^u$$

Substituting dimensions of different physical quantities involved, we have

$$[ML^2T^{-3}] = [MLT^{-2}A^{-2}]^x [L^2A]^y [T^{-1}]^z [LT^{-1}]^u$$

Equating powers of fundamental quantities, we have

$$x = 1 \quad \text{...(i)}$$
$$x + 2y + u = 2 \quad \text{...(ii)}$$
$$-2x - z - u = -3 \quad \text{...(iii)}$$
$$-2x + y = 0 \quad \text{...(iv)}$$

From Eq. (i), putting the value of x in Eq. (iv), we get

$$\Rightarrow \quad -2\times 1 + y = 0 \Rightarrow y = 2 \quad \text{...(v)}$$

Now, from Eqs. (i) and (v), putting the values of x and y in Eq. (ii), we get

$$\Rightarrow \quad 1 + 2\times 2 + u = 2 \Rightarrow u = -3 \quad \text{...(vi)}$$

Now, again from Eqs. (i) and (vi), putting the values of x and u in Eq. (iii), we get

$$\Rightarrow -2\times 1 - z + 3 = -3 \Rightarrow z = 4$$

So, $x = 1$, $y = 2$, $z = 4$, $u = -3$.

94. *(d)*

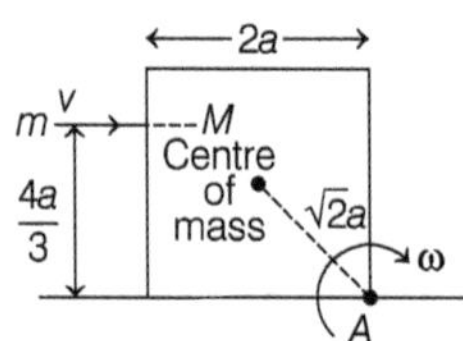

As there is no external torque on the system of bullet and block, angular momentum about A remains constant.

Now, moment of inertia of block about A using parallel axes theorem is

$$I_A = I_{CM} + Mh^2$$
$$= \frac{2}{3}Ma^2 + M(\sqrt{2}a)^2 = \frac{8}{3}Ma^2$$

Applying angular momentum conservation about A, we have

$$L_i = L_f \Rightarrow mvr = I_A\omega$$
$$\Rightarrow \quad \omega = \frac{mvr}{I_A} \quad \text{...(i)}$$

where, m = mass of bullet and ω = initial angular speed of block.

Substituting values in Eq. (i), we have

$$\omega = \frac{mv\left(\frac{4a}{3}\right)}{\frac{8}{3}Ma^2} \Rightarrow \omega = \frac{mv}{2Ma}$$

95. *(a)* Process equation is

$$pV^{\frac{5}{3}} \cdot \left(e^{\frac{-pV}{E_0}}\right) = C_1$$

$$\Rightarrow \quad pV^{\frac{5}{3}} = C_1 e^{\frac{pV}{E_0}}$$

Taking log to e base, we have

$$\log(p) + \frac{5}{3}\log(V) = \log(C_1) + \frac{pV}{E_0}$$

Differentiating with respect to volume, we have

$$\frac{1}{p}\cdot\frac{dp}{dV} + \frac{5l}{3V} = 0 + \frac{1}{E_0}\left(p + \frac{Vdp}{dV}\right)$$

$$\Rightarrow \quad \frac{dp}{p} + \frac{5}{3}\frac{dV}{V} = \frac{1}{E_0}(pdV + Vdp)$$

$$\Rightarrow \quad \frac{-dV}{dp} = \frac{\left(\frac{1}{p} - \frac{V}{E_0}\right)}{\left(\frac{p}{E_0} - \frac{5}{3V}\right)}$$

As, $\quad pV = RT$

$$\Rightarrow \quad -\frac{dV}{dp} = \frac{\left(\frac{V}{RT} - \frac{V}{E_0}\right)}{\left(\frac{p}{E_0} - \frac{5p}{3RT}\right)}$$

$$\Rightarrow \quad -\frac{1}{V}\frac{dV}{dp} = \frac{\left(\frac{1}{RT} - \frac{1}{E_0}\right)}{p\left(\frac{1}{E_0} - \frac{5}{3RT}\right)}$$

But this is compressibility of gas, so compressibility,

$$k = -\frac{1}{V}\frac{dV}{dp}$$
$$= \frac{\left(\frac{1}{RT} - \frac{1}{E_0}\right)}{p\left(\frac{1}{E_0} - \frac{5}{3RT}\right)}$$

As $T \to \infty$, $\frac{1}{RT} \to 0$

So, k approaches a constant value at higher temperatures.

96. *(*)* Given situation is

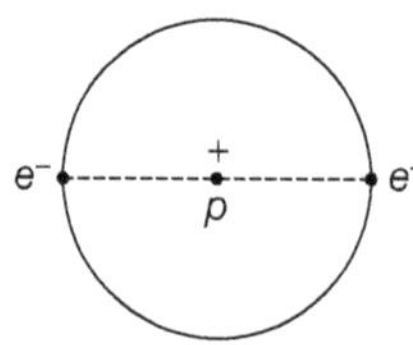

So, net centripetal force on a rotating electron is

$$F = \frac{mv^2}{r} = \frac{ke^2}{r^2} - \frac{ke^2}{(2r)^2}$$

$$\Rightarrow \quad mv^2r = \frac{3}{4}ke^2 \quad \text{...(i)}$$

From Bohr's postulate, we have

$$mvr = \frac{nh}{2\pi} \quad \text{...(ii)}$$

Squaring Eq. (ii) and dividing by Eq. (i), we have

$$\frac{m^2v^2r^2}{mv^2r} = \frac{\left(\frac{n^2h^2}{4\pi^2}\right)}{\left(\frac{3}{4}ke^2\right)}$$

$$\Rightarrow \quad r = \frac{n^2h^2}{3kme^2} = \frac{4\pi\varepsilon_0 n^2h^2}{3me^2}$$

For $n = 1$,

$$r = \frac{4\pi\varepsilon_0 h^2}{3me^2} = \frac{1}{3}\left(\frac{4\pi\varepsilon_0 h^2}{me^2}\right) = \frac{1}{3}a_B$$

(No option is matching)

97. *(b)* Opposing force on loop is

$$F = BIa$$

where, I = induced current in loop.

Also, $I = \frac{E}{R}$

where, E = induced emf.

$$\Rightarrow \quad I = \frac{-Bav}{R}$$

So, opposing force is

$$F = \frac{-B^2a^2}{R}\cdot v \Rightarrow mA = \frac{-B^2a^2}{R}\cdot v$$

where, A = acceleration of loop and m = mass of loop.

$$m\frac{dv}{dt} = \frac{-B^2a^2v}{R}$$

$$m.\frac{dv}{dx}\cdot\frac{dx}{dt} = \frac{-B^2a^2v}{R}$$

$$\Rightarrow \quad dv = \frac{-B^2a^2}{Rm}\cdot dx \quad \left[\therefore v = \frac{dx}{dt}\right]$$

Integrating above equation, we have

$$\Rightarrow \quad v = \frac{-B^2a^2x}{Rm} + C$$

where, C = constant of integration.

$\therefore v$ *versus* x is a straight line with negative slope. Also, emf is induced only when the loop is going in or emerging out of the region of magnetic field. So, correct graph is (b) .

98. *(*)* Given, position of Vernier and main scale is as shown below in the figure.

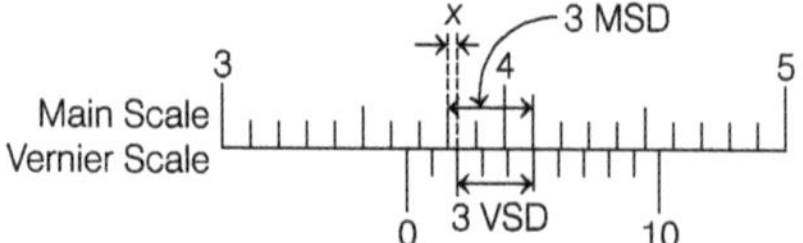

From above figure,

$$x = 3\text{MSD} - 3\text{VSD}$$
$$= 3\times 1\text{MSD} - 3\times 1\text{VSD}$$
$$= \left(3 - 3\times\frac{1}{10}\right)\text{mm}$$
$$= 3 - 0.3\text{ mm} = 2.7\text{ mm} = 0.27\text{ cm}$$

(No option is matching)

99. *(a)* Due to standing waves, density of water column is more at nodes and less at antinodes.
So, optical path length of light beams is different. This causes interference in light beams when they overlaps over screen. Hence, correct graph of intensity variation on screen is of (a).

100. *(c)* From given graph, time period of rotation of planet around star is obtained 3 days.

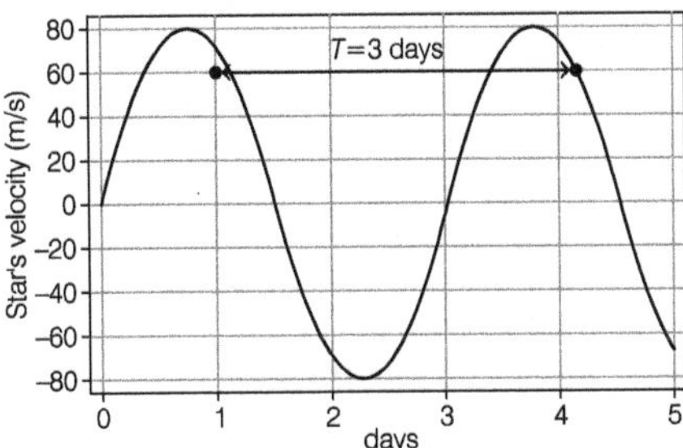

Now, we use

$$T^2 = \frac{4\pi^2}{GM} . r^3 \quad ...(i)$$

If we take time period in years and radius in AU, then $\frac{4\pi^2}{GM} = 1$

As for earth, $r = 1\text{AU}, T = 1\text{yr}$

So, $T^2 = \frac{4\pi^2}{GM} . r^3$

$\Rightarrow \frac{4\pi^2}{GM} = \frac{1^2}{1^3} = 1$

Now, in given question is

$$T = 3\text{ days} = \frac{3}{365}\text{yr}$$

So, from Eq. (i),

$$r = T^{2/3} \qquad \left[\because \frac{4\pi^2}{GM} = 1\right]$$

$$= \left(\frac{3}{365}\right)^{\frac{2}{3}} \approx \left(\frac{8}{1000}\right)^{\frac{2}{3}} = \left(\frac{64}{10^6}\right)^{1/3}$$

$$= \frac{4}{10^2} = 0.04\,\text{AU}$$

101. *(b)*

Methylenecyclohexane $\xrightarrow[\text{(ii) } H_2O_2/\text{ NaOH}]{\text{(i) } B_2H_6}$ cyclohexylmethanol (CH_2OH) (X) $\xrightarrow{CrO_3/H_2SO_4}$ cyclohexanecarboxylic acid ($COOH$) (Y)

Whenever B_2H_6 adds in the presence of peroxide to an alkene, an alcohol is formed. This addition reaction is in accordance to anti- Markownikoff's rule.
CrO_3 in the presence of H_2SO_4 is known as Jones reagent is selective oxidising agent. It converts primary alcohol to carboxylic and secondary alcohol to ketones.

102. *(a)*

$CH_3COCH_3 \xrightarrow{\text{(i) NaBH}_4} (CH_3)_2C(H)OH \xrightarrow{\text{(ii) } D_3O^+} (CH_3)_2C(H)OD$ (X)

$CH_3COCH_3 \xrightarrow{\text{(i) NaBD}_4} (CH_3)_2C(H)OD \xrightarrow{\text{(ii) } H_3O^+} (CH_3)_2C(D)OH$ (Y)

In case I, $NaBH_4$ is used as a reducing agent, where H^- gets transferred to ketonic group and alcohol is formed. In step (ii), H of the OH group gets substituted by D^+, when hydrolysis occurs in the presence of D_3O^+.
In case II, $NaBD_4$ is used as a reducing agent where D^- gets transferred to ketonic group and alcohol is formed. In step (ii), D of OD group gets substituted by H^+, when hydrolysis occurs in the presence of H_3O^+.

103. *(c)* The hydrogenation products of given alkenes are as follows

I $\xrightarrow{H_2/Ni}$ Optically active

II $\xrightarrow{H_2/Ni}$

III $\xrightarrow{H_2/Ni}$ Optically active

IV $\xrightarrow{H_2/Ni}$

Thus, alkene I and III generate optically active compounds upon hydrogenation.

104. *(c)* When heated in air, brown copper turns black. This black powder would turn brown again when heated with hydrogen (H_2). This is because here, H_2 will act as air reducing agent that reduces CuO into Cu, whereas air (O_2) will act as an oxidising agent.

$$\underset{\text{Brown}}{Cu} + O_2 \xrightarrow{\Delta} \underset{\text{Black}}{CuO}$$

$$\underset{\text{Black}}{CuO} + H_2 \xrightarrow{\Delta} Cu + H_2O$$

105. *(a)* The oxidation state of Ni in $[NiCl_4]^{2-}$ is +2. The electronic configuration of Ni^{2+} is $[Ar]3d^8 4s^0$.

Ni^{2+} ground state: 3d [⥮][⥮][⥮][↿][↿] 4s [ ] 4p [ ][ ][ ]

As Cl^- is a weak field ligand, pairing of electron will not occur.

$[NiCl_4]^{2-}$ = 3d [⥮][⥮][⥮][↿][↿] 4s [ ] 4p [ ][ ][ ] (sp^3-hybridisation)

As the hybridisation of Ni in $[NiCl_4]^{2-}$ is sp^3. Thus, the geometry will be tetrahedral. It is paramagnetic in nature as there are 2 unpaired electrons.

106. *(c)* Optically active complexes are those complexes in which the mirror images are non-superimposable on each other. Complexes of type $[M(AA)_3]$, *cis* $[M(AA)_2B_2]$ are optically active. The type of complexes given in the options are as follows :

Complexes	Types of complexes
$[Cr(en)_3]^{3+}$	$[M(AA)_3]$
trans-$[Cr(en)_2Cl_2]^+$	*trans*-$[M(AA)_2Cl_2]$
cis-$[Cr(en)_2Cl_2]^+$	*cis*-$[M(AA)_2Cl_2]$
$[Co(NH_3)_4Cl_2]^+$	$[MA_4B_2]$

Thus, among the given options $[Cr(en)_3]^{3+}$ and *cis* $[Cr(en)_2Cl_2]^+$ are optically active.

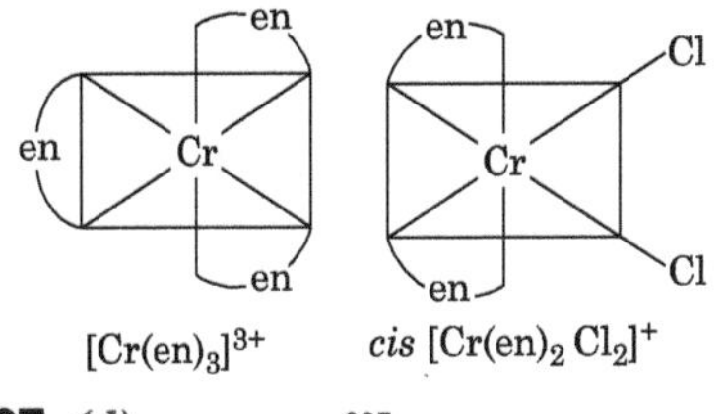

$[Cr(en)_3]^{3+}$ *cis* $[Cr(en)_2Cl_2]^+$

107. *(d)*

$$^{227}Ac \xrightarrow{k_1} {}^{227}Th, \quad ^{227}Ac \xrightarrow{k_2} {}^{223}Fr$$

As radioactive compounds are first order reactions

$\therefore \quad t_{1/2}$ of Ac $= \frac{0.693}{k}$ Or $k = \frac{0.693}{22}$

% of Th $\frac{k_1}{k_T} = \frac{2}{100}$

% of Fr $= \frac{k_2}{k_T} = \frac{98}{100}$

$\therefore \quad \frac{k_1}{k_2} = \frac{2}{98}$

Also, $k = k_1 + k_2$

$\frac{0.693}{22} = k_1 + \frac{98}{2}k_1$

$\therefore \quad k_1 = 6.3 \times 10^{-4}$

108. *(d)* For process, $A \longrightarrow B$

$$W = -p\Delta V = -1(1.5-1) = -0.5 \text{ L atm}$$
$$\Delta U = C_V \Delta T$$
$$\Delta U = 0 \text{ (isothermal process)}$$

$\therefore$ From first law of thermodynamics,

$$\Delta U = W_1 + q_1$$
$$0 = -0.5 + q$$
$$q_1 = 0.5 \text{ L atm}$$

For porcess, $B \longrightarrow C$

$$W = 1 \text{ L atm (given)}$$
$$\Delta U = 0 \text{ (isothermal process)}$$
$$\therefore \quad \Delta U = W_2 + q_2$$
$$0 = 1 + q_2$$
$$q_2 = -1 \text{ L atm}$$

For process, $A \longrightarrow C$

$$\Delta U = 0 \text{ (isothermal process)}$$
$$W_3 = -p\Delta V = 0 \text{ (volume is constant)}$$
$$\therefore \quad q_3 = 0$$

$\therefore$ Heat exchanged in entire reaction,

$$q = q_1 + q_2 + q_3$$
$$q = +0.5 - 1 + 0 = -0.5 \text{ L atm}$$

Here, –ve sign represents that heat is released.

109. *(b)* According to Raoult's law,

$$p = p_1^0 \chi_1$$
$$p_T = p_1^0 \chi_1 + p^0 \chi_2$$
$$= p_1^0 \chi_1 + p_2^0 (1 - \chi_1)$$
$$p_1^0 \chi_1 + p_2^0 - p_2^0 \chi_1$$
$$(p_1^0 - p_2^0)\chi_1 + p_2^0$$
$$p_T = p_{\text{toluene}} + p_{\text{benzene}}$$
$$1.013 = (0.742 - 1.8)\chi_{\text{toluene}} + 1.8$$
$$\Rightarrow \quad 1.013 = -1.058 \times \chi_{\text{toluene}} + 1.8$$
$$-0.787 = -1.058 \times \chi_{\text{toluene}}$$
$$x_{\text{toluene}} = 0.744$$

110. *(a)* Represents a unit cell

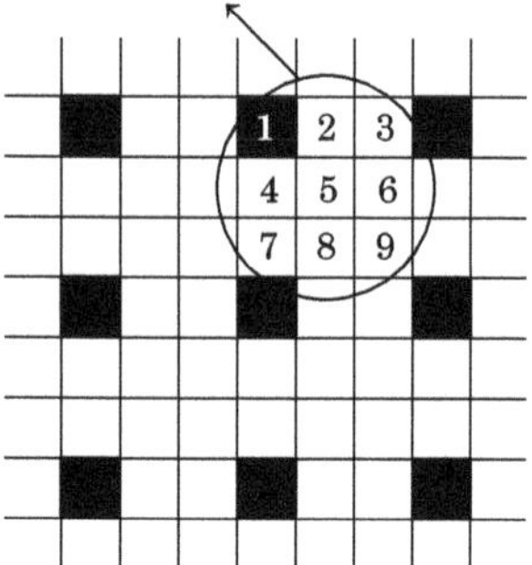

The whole solid is a matrix of 9×9, so unit cell will be matrix of 3×3.

Thus, from above figure, it is can be seen that a unit cell have 8 white square and 1 black square.

$\therefore$ Number of X-atoms in a unit cell $= 1$

Number of Y-atoms in a unit cell = 8

Thus, the formula for the compound based on the unit cell from pattern is XY_8.

111. *(b)* The genetic distance between genes A and B is 10 cm. It means, the recombination frequency is 10%. If the cross is made between the two parents with genotype Ab and aB, out of their progeny, 10% will be recombinant type (that is AB and ab) and 90% will be parental type (that is aB and Ab). AB allele represents half of the recombination type, that is 5%. Thus, the correct answer is option (b).

112. *(a)* A cell membrane is a double layer of lipids and proteins that surrounds a cell, and separates the cytoplasm from its surrounding environment. It consists of various proteins embedded in this lipid bilayer i.e. receptor proteins, integral proteins and peripheral proteins. Out of these peripheral membrane proteins are temporarily attached either to the lipid bilayer or to integral proteins by a combination hydrophobic, electrostatic and other non-covalent interactions. Thus on treating the cell membrane with high ionic strength buffer these can be easily removed. Thus we can conclude that in the experiment P and Q are peripheral membrane proteins.

113. *(d)* Photosystem-II is the first membrane protein complex in oxygenic photosynthetic organisms in nature. It producess atmospheric oxygen to catalyse the photooxidation of water by using light energy. Two molecules of water into one molecule of molecular oxygen occurs as

$$2H_2O \xrightarrow[\text{Chlorophyll}]{\text{Light}} 4H^+ + 4e^- + O_2 \uparrow$$

114. *(a)* Proteins are among the most important chemical to all life on the planet. Each protein is made up of many of the 20 different amino acids. So, the number of proteins which can be formed these different 20 amino acids if each protein contains 100 amino acids $= 20\text{th}^{100}$.

115. *(d)* 660 g/mol is the average molecular weight of nucleotide pair. Each nucleotide pair = 0.34 nm of the length of DNA.

The length of *E.coli* DNA molecule will be

$$= \frac{0.34 \times 3.1 \times 10^9}{660}$$
$$= 1.6 \times 10^6 \text{ nm}$$
$$= 1.6 \text{ mm}$$

116. *(d)* Emigration refers to the number of individuals of the population who left the habitat and gone elsewhere during a given time period.

Emigration thus leads to decrease in the number of individuals in a habitat.

117. *(b)* Cysteine is a sulphur containing amino acid. It is unstable in the air. In proteins it usually exists as cystine by forming a disulphide (S—S) bond between two cysteine residue.

Zinc ions have the ability to be chelated to cysteine residues within protein scaffolds. Thus, option (b) is correct.

118. *(c)* The minimum number of plants to be screened to obtain a plant of the genotype AabbCcDd is 32.

Because AaBbCcDd $\times$ AABbCCDd

AabbCcDd

$$\frac{1}{2} + \frac{1}{4} + \frac{1}{2} + \frac{1}{2} = \frac{1}{32}$$

119. *(d)* A cross between red flower producing plant and white flower producing plant is shown below

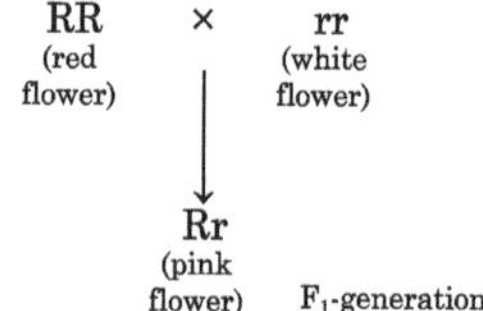

When F_1 to F_6 individuals during each generation are sefed, i.e.

Rr × Rr

RR	Rr Rr	RR
25%	50%	25%

This shows that 50% plants are homozygous with red and white flowers.

We know that homozygous plants will produce homozygous offsprings. Thus during each generation the heterozygous progeny will give further 50% heterozygous and 50% homozygous progeny. This will reduce the number of heterozygous successively.

$\therefore$ The final population will have 97-100% of homozygous progenies, i.e. red and white flowers plants.

120. *(c)* In the *lac* operon, when lactose is present, the *lac* genes are expressed, because lactose is converted to allolactose, which binds to the repressor protein and prevents its interaction with operator.

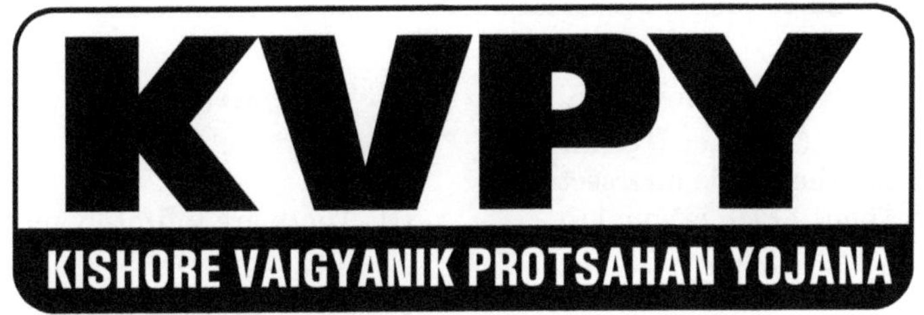

QUESTION PAPER 2016

Stream : SB/SX

MM : 160

Instructions

1. There are 120 questions in this paper.
2. The question paper contains two parts; Part I (1 Mark Questions) and Part II (2 Marks Questions).
3. There are four sections in each part; Mathematics, Physics, Chemistry and Biology.
4. Out of the four options given with each question, only one is correct.

PART-I (1 Mark Questions)

MATHEMATICS

1. The number of triples (x, y, z) of real numbers satisfying the equation $x^4 + y^4 + z^4 + 1 = 4xyz$ is

(a) 0 (b) 4
(c) 8 (d) more than 8

2. Let $P(x)$ be a polynomial with real coefficients such that $P(\sin^2 x) = P(\cos^2 x)$ for all $x \in [0, \pi/2)$. Consider the following statements:

I. $P(x)$ is an even function.

II. $P(x)$ can be expressed as a polynomial in $(2x-1)^2$.

III. $P(x)$ is a polynomial of even degree.

Then,

(a) all are false (b) only I and II are true
(c) only II and III are true (d) all are true

3. For any real number r, let $A_r = \{e^{i\pi rn} : n$ is a natural number$\}$ be a set of complex numbers. Then,

(a) $A_1, A\frac{1}{\pi}, A_{0.3}$ are all infinite sets

(b) A_1 is a finite set and $A\frac{1}{\pi}, A_{0.3}$ are infinite sets

(c) $A_1, A\frac{1}{\pi}, A_{0.3}$ are all finite sets

(d) $A_1, A_{0.3}$ are finite sets and $A\frac{1}{\pi}$ is an infinite set

4. The number of integers k for which the equation $x^3 - 27x + k = 0$ has at least two distinct integer roots is

(a) 1 (b) 2
(c) 3 (d) 4

5. Suppose the tangent to the parabola $y = x^2 + px + q$ at (0, 3) has slope -1. Then, $p + q$ equals

(a) 0 (b) 1
(c) 2 (d) 3

6. Let $O = (0,0)$: let A and B be points respectively on X-axis and Y-axis such that $\angle OBA = 60°$. Let D be a point in the first quadrant such that OAD is an equilateral triangle. Then, the slope of DB is

(a) $\sqrt{3}$ (b) $\sqrt{2}$
(c) $\frac{1}{\sqrt{2}}$ (d) $\frac{1}{\sqrt{3}}$

7. Suppose the parabola $(y-k)^2 = 4(x-h)$, with vertex A, passes through $O=(0,0)$ and $L=(0,2)$. Let D be an end point of the latusrectum. Let the Y-axis intersect the axis of the parabola at P. Then, $\angle PDA$ is equal to

(a) $\tan^{-1}\frac{1}{19}$ (b) $\tan^{-1}\frac{2}{19}$ (c) $\tan^{-1}\frac{4}{19}$ (d) $\tan^{-1}\frac{8}{19}$

8. In a circle with centre O, suppose A, P, B are three points on its circumference such that P is the mid-point of minor arc AB. Suppose when $\angle AOB = \theta$, $\frac{\text{area }(\Delta AOB)}{\text{area }(\Delta APB)} = \sqrt{5}+2$.

If $\angle AOB$ is doubled to 2θ, then the ratio $\frac{\text{area }(\Delta AOB)}{\text{area }(\Delta APB)}$ is

(a) $\frac{1}{\sqrt{5}}$ (b) $\sqrt{5}-2$ (c) $2\sqrt{3}+3$ (d) $\frac{\sqrt{5}-1}{2}$

9. Let $X = \{x \in R : \cos(\sin x) = \sin(\cos x)\}$. The number of elements in X is

(a) 0 (b) 2 (c) 4 (d) not finite

10. A sphere with centre O sits on the top of a pole as shown in the figure.

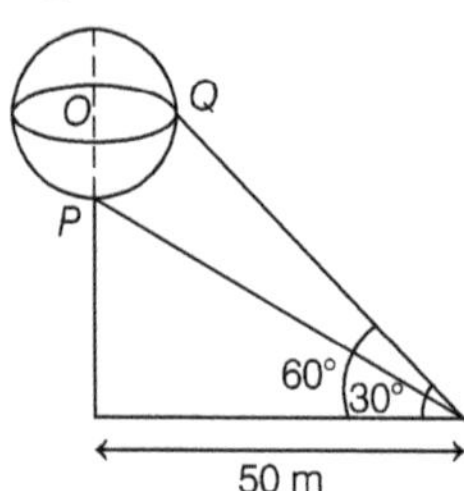

An observer on the ground is at a distance 50 m from the foot of the pole. She notes that the angles of elevation from the observer to points P and Q on the sphere are 30° and 60°, respectively. Then, the radius of the sphere in metres is

(a) $100\left(1-\frac{1}{\sqrt{3}}\right)$ (b) $\frac{50\sqrt{6}}{3}$

(c) $50\left(1-\frac{1}{\sqrt{3}}\right)$ (d) $\frac{100\sqrt{6}}{3}$

11. The graph of the function $f(x) = x + \frac{1}{8}\sin(2\pi x)$, $0 \le x \le 1$ is shown below. Define $f_1(x) = f(x)$, $f_{n+1}(x) = f(f_n(x))$, for $n \ge 1$.

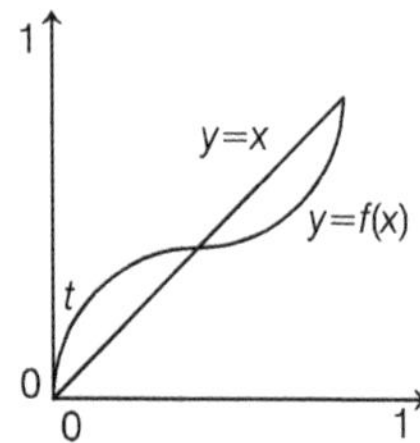

Which of the following statements are true?

I. There are infinitely many $x \in [0,1]$ for which $\lim_{n\to\infty} f_n(x) = 0$

II. There are infinitely many $x \in [0,1]$ for which $\lim_{n\to\infty} f_n(x) = \frac{1}{2}$

III. There are infinitely many $x \in [0,1]$ for which $\lim_{n\to\infty} f_n(x) = 1$

IV. There are infinitely many $x \in [0,1]$ for which $\lim_{n\to\infty} f_n(x)$ does not exist.

(a) I and III only (b) II only
(c) I, II, III only (d) I, II, III and IV

12. The limit $\lim\limits_{x\to\infty} x^2 \int_0^x e^{t^3-x^3}\,dt$ equals

(a) $\frac{1}{3}$ (b) 2 (c) ∞ (d) $\frac{2}{3}$

13. The polynomial equation $x^3 - 3ax^2 + (27a^2+9)x + 2016 = 0$ has

(a) exactly one real root for any real a
(b) three real roots for any real a
(c) three real roots for any $a \ge 0$, and exactly one real root for any $a < 0$
(d) three real roots for any $a \le 0$, and exactly one real root for any $a > 0$

14. The area of the region bounded by the curve $y = |x^3 - 4x^2 + 3x|$ and the X-axis, $0 \le x \le 3$, is

(a) $\frac{37}{6}$ (b) $\frac{9}{4}$ (c) $\frac{37}{12}$ (d) 0

15. The number of continuous function $f : [0,1] \to [0,1]$ such that $f(x) < x^2$ for all x and $\int_0^1 f(x)\,dx = \frac{1}{3}$ is

(a) 0 (b) 1 (c) 2 (d) infinite

16. On the real line R, we define two functions f and g as follows:

$f(x) = \min\{x-[x], 1-x+[x]\}$,
$g(x) = \max\{x-[x], 1-x+[x]\}$,

where $[x]$ denotes the largest integer not exceeding x. The positive integer n for which $\int_0^n (g(x)-f(x))\,dx = 100$ is

(a) 100 (b) 198 (c) 200 (d) 202

17. Let $\vec{v}$ be a vector in the plane such that $|\mathbf{v}-\mathbf{i}| = |\mathbf{v}-2\mathbf{i}| = |\mathbf{v}-\mathbf{j}|$. Then, $|\mathbf{v}|$ lies in the interval

(a) (0,1] (b) (1, 2] (c) (2, 3] (d) (3, 4]

18. A box contains b blue balls and r red balls. A ball is drawn randomly from the box and is returned to the box with another ball of the same colour. The probability that the second ball drawn from the box is blue, is

(a) $\frac{b}{r+b}$ (b) $\frac{b^2}{(r+b)^2}$ (c) $\frac{b+1}{r+b+1}$ (d) $\frac{b(b+1)}{(r+b)(r+b+1)}$

19. The number of non-congruent integer-sided triangles whose sides belong to the set {10, 11, 12, ..., 22} is

(a) 283 (b) 446
(c) 448 (d) 449

20. Suppose we have to cover the XY-plane with identical tiles such that no two titles overlap and no gap is left between the tiles. Suppose that we can choose tiles of the following shapes : equilateral triangle, square, regular pentagon, regular hexagon. Then, the tiling can be done with tiles of

(a) all four shapes
(b) exactly three of the four shapes
(c) exactly two of the four shapes
(d) exactly one of the four shapes

PHYSICS

21. Physical processes are sometimes described visually by lines. Only the following can cross.

(a) Streamlines in fluid flow
(b) Lines of forces in electrostatics
(c) Rays in geometrical optics
(d) Lines of force in magnetism

22. A uniform ring of radius R is moving on a horizontal surface with speed v, then climbs up a ramp of inclination 30° to a height h. There is no slipping in the entire motion. Then, h is

(a) $v^2/2g$ (b) v^2/g (c) $3v^2/2g$ (d) $2v^2/g$

23. A gas at initial temperature T undergoes sudden expansion from volume V to $2V$. Then,

(a) the process is adiabatic
(b) the process is isothermal
(c) the work done in this process is $nRT\ln_e(2)$, where n is the number of moles of the gas
(d) the entropy in the process does not change

24. Photons of wavelength λ are incident on a metal. The most energetic electrons ejected from the metal are bent into a circular are of radius R by a perpendicular magnetic field having a magnitude B. The work function of the metal is (where, symbols have their usual meanings)

(a) $\frac{hc}{\lambda} - m + \frac{q^2B^2R^2}{2m}$ (b) $\frac{hc}{\lambda} + 2m\left(\frac{qBR}{2m}\right)^2$

(c) $\frac{hc}{\lambda} - mc^2 - \frac{q^2B^2R^2}{2m}$ (d) $\frac{hc}{\lambda} - 2m\left(\frac{qBR}{2m}\right)^2$

25. A container is divided into two equal parts I and II by a partition with a small hole of diameter d. The two partitions are filled with same ideal gas, but held at temperatures $T_I = 150$ K and $T_{II} = 300$ K by connecting to heat reservoirs. Let λ_I and λ_{II} be the mean free paths of the gas particles in the two parts, such that $d >> \lambda_I$ and $d >> \lambda_{II}$. Then, the λ_I / λ_{II} is close to

(a) 0.25 (b) 0.5 (c) 0.7 (d) 1.0

26. A conducting bar of mass m and length l moves on two frictionless parallel rails in the presence of a constant uniform magnetic field of magnitude B directed into the page as shown in the figure.

The bar is given an initial velocity v_0 towards the right at $t = 0$.

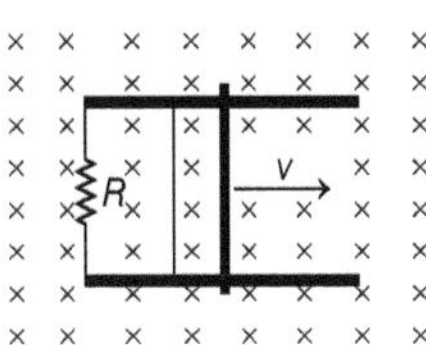

Then, the

(a) induced current in the circuit is in the clockwise direction
(b) velocity of the bar decreases linearly with time
(c) distance the bar travels before it comes to a complete stop is proportional to R
(d) power generated across the resistance is proportional to l

27. A particle with total mechanical energy, which is small and negative. Its under the influence of a one dimensional potential $U(x) = x^4/4 - x^2/2$ J, where x is (in metre). At time $t = 0$ s, it is at $x = -0.5$ m. Then, at a later time it can be found

(a) anywhere on the X-axis
(b) between $x = -1.0$ m to $x = 1.0$ m
(c) between $x = -1.0$ m to $x = 0.0$ m
(d) between $x = 0.0$ m to $x = 1.0$ m

28. A nurse measures the blood pressure of a seated patient to be 190 mm of Hg.

(a) The blood pressure at the patient's feet is less than 190 mm of Hg
(b) The actual pressure is about 0.25 times of atmospheric pressure
(c) The blood pressure at the patient's neck is more than 190 mm of Hg
(d) The actual pressure is about 1.25 times of atmospheric pressure

29. A particle at a distance of 1 m from the origin starts moving, such that $dr / d\theta = r$, where r and θ are polar co-ordinates. Then, the angle between resultant velocity and tangential velocity is

(a) 30°
(b) 45°
(c) 60°
(d) dependent on where the particle is

30. Electrons accelerated from rest by an electrostatic potential are collimated and sent through a Young's double slit experiment. The fringe width is ω. If the accelerating potential is doubled, then the width is now close to

(a) $0.5\,\omega$ (b) $0.7\,\omega$ (c) $1.0\,\omega$ (d) $2.0\,\omega$

31. A metallic sphere is kept in between two oppositely charged plates. The most appropriate representation of the field lines is

(a) 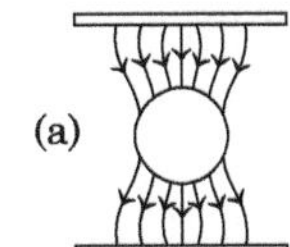 (b) 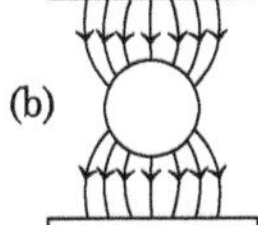 (c) 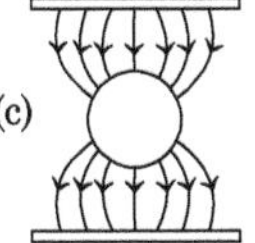 (d) 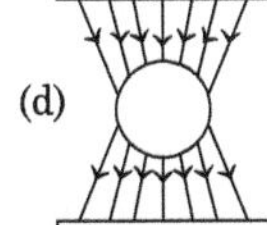

32. An electron with kinetic energy E collides with a hydrogen atom in the ground state. The collision will be elastic

(a) for all values of E
(b) for $E < 10.2\,\text{eV}$
(c) for $10.2\,\text{eV} < E < 13.6\,\text{eV}$ only
(d) for $0 < E < 3.4\,\text{eV}$ only

33. The continuous part of X-ray spectrum is a result of the

(a) photoelectric effect (b) raman effect
(c) compton effect (d) inverse photoelectric effect

34. Thermal expansion of a solid is due to the

(a) symmetric characteristic of the inter atomic potential energy curve of the solid
(b) asymmetric characteristic of the inter atomic potential energy curve of the solid
(c) double well nature of the inter atomic potential energy curve of the solid
(d) rotational motion of the atoms of the solid

35. An electron and a photon have same wavelength of 10^{-9} m. If E is the energy of the photon and p is the momentum of the electron, then the magnitude of E/p (in SI unit) is

(a) 1.00×10^{-9} (b) 1.50×10^{8}
(c) 3.00×10^{8} (d) 1.20×10^{7}

36. If one takes into account finite mass of the proton, then the correction to the binding energy of the hydrogen atom is approximately (take, mass of proton $= 1.60 \times 10^{-27}$ kg and mass of electron $= 9.10 \times 10^{-31}$ kg)

(a) 0.06% (b) 0.0006% (c) 0.02% (d) 0.00%

37. A monochromatic light source S of wavelength 440 nm is placed slightly above a plane mirror M as shown below. Image of S in M can be used as a virtual source to produce interference fringes on the screen. The distance of source S from O is 20.0 cm and the distance of screen from O is 100.0 cm (figure is not to scale). If the angle $\theta = 0.50 \times 10^{-3}$ radians, then the width of the interference fringes observed on the screen is

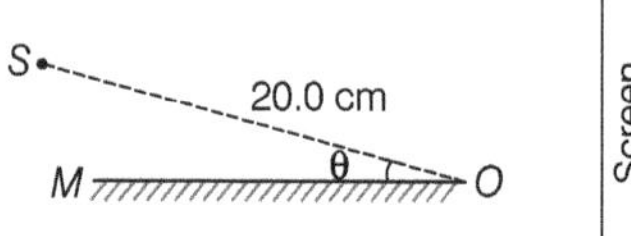

(a) 2.20 mm (b) 2.64 mm (c) 1.10 mm (d) 0.55 mm

38. A nuclear fuel rod generates energy at a rate of $5 \times 10^{8}\ \text{W/m}^3$. It is in the shape of a cylinder of radius 4.0 mm and length 0.20 m. A coolant of specific heat $4 \times 10^{3}\ \text{Jkg}^{-1}\text{K}^{-1}$ flows past it at a rate of $0.2\ \text{kgs}^{-1}$. The temperature rise in this coolant is approximately

(a) 2°C (b) 6°C (c) 12°C (d) 30°C

39. A hearing test is conducted on an aged person. It is found that her thresold of hearing is 20 dB at 1 kHz and it rises linearly with frequency to 60 dB at 9 kHz. The minimum intensity of sound that the person can hear at 5 kHz is

(a) 10 times than that at 1 kHz
(b) 100 times than that at 1 kHz
(c) 0.5 times than that at 9 kHz
(d) 0.05 times than that at 9 kHz

40. Two infinitely long parallel wires carry currents of magnitude I_1 and I_2 are at a distance 4 cm apart. The magnitude of the net magnetic field is found to reach a non-zero minimum value between the two wires and 1 cm away from the first wire. The ratio of the two currents and their mutual direction is

(a) $\frac{I_2}{I_1} = 9$, anti-parallel (b) $\frac{I_2}{I_1} = 9$, parallel
(c) $\frac{I_2}{I_1} = 3$, anti-parallel (d) $\frac{I_2}{I_1} = 3$, parallel

CHEMISTRY

41. The shape of SCl_4 is best described as a

(a) square (b) tetrahedron
(c) square pyramid (d) *see-saw*

42. Among the following atomic orbital overlaps, the non-bonding overlap is

(a) 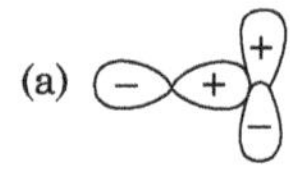 (b) 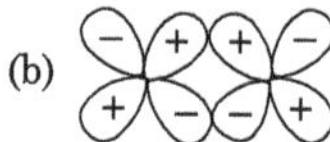

(c) 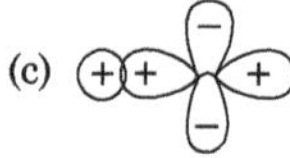 (d) 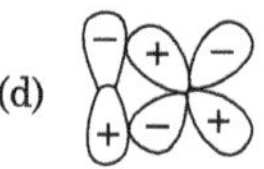

43. Among the following complexes, the one that can exhibit optical activity is

(a) $[CoCl_6]^{3-}$ (b) $[Co(en)Cl_4]^{-}$
(c) *cis* $-[Co(en)_2Cl_2]^{+}$ (d) *trans* $-[Co(en)_2Cl_2]^{+}$

44. The pK_a of oxoacids of chlorine in water follows the order

(a) $HClO < HClO_3 < HClO_2 < HClO_4$
(b) $HClO_4 < HClO_3 < HClO_2 < HClO$
(c) $HClO_4 < HClO_2 < HClO_3 < HClO$
(d) $HClO_2 < HClO < HClO_3 < HClO_4$

45. The packing efficiency of the face centered cubic (fcc), body centered cubic (bcc) and simple/primitive cubic (pc) lattices follows the order

(a) fcc > bcc > pc (b) bcc > fcc > pc
(c) pc > bcc > fcc (d) bcc > pc > fcc

46. The ratio of root mean square velocity of hydrogen at 50 K to that of nitrogen at 500 K is closest to

(a) 1.18 (b) 0.85 (c) 0.59 (d) 1.40

47. The molecule with the highest dipole moment among the following is

(a) NH_3 (b) NF_3 (c) CO (d) HF

48. The most stable Lewis acid-base adduct among the following is

(a) $H_2O \rightarrow BCl_3$ (b) $H_2S \rightarrow BCl_3$
(c) $H_3N \rightarrow BCl_3$ (d) $H_3P \rightarrow BCl_3$

49. The reaction of D-glucose with ammoniacal $AgNO_3$ produces

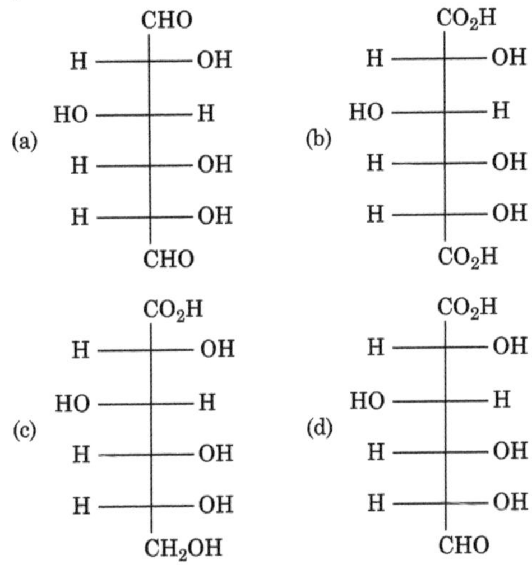

50. The reagent(s) used for the conversion of benzene diazonium hydrogensulphate to benzene is/are

(a) H_2O (b) $H_3PO_2 + H_2O$
(c) $H_2SO_4 + H_2O$ (d) CuCl/HCl

51. The major product obtained in the reaction of toluene with 1 - bromo-2- methyl propane in the presence of anhyd. $AlCl_3$ is

(a) H_3C (b) CH_3
(c) H_3C (d) H_3C

52. The major product in the following reaction is

OH, OH, O $\xrightarrow[(CH_3COOH)]{(CH_3CO)_2O}$

(a) OH, O, CH_3, C, O (b) O, CH_3, O, OH, O

(c) H_3C, OH, OH, O (d) H_3C, O, OH, OH, O

53. The compounds containing *sp*-hybridised carbon atom are

(I) H_3C—N(—CHO)—CH_3 (II) N (pyridine)

(III) H_3C—CN (IV) H_2C—C—$CHCH_3$

(a) I and II (b) III and IV
(c) II and III (d) I and IV

54. Upon heating with acidic $KMnO_4$, an organic compound produces hexan - 1,6-dioic acid as the major product. The starting compound is

(a) benzene
(b) cyclohexene
(c) 1-methylcyclohexene
(d) 2-methylcyclohexene

55. It takes 1 h for a first order reaction to go to 50% completion. The total time required for the same reaction to reach 87.5% completion will be

(a) 1.75 h (b) 6.00 h
(c) 3.50 h (d) 3.00 h

56. A unit cell of calcium fluoride has four calcium ions. The number of fluoride ions in the unit cell is

(a) 2 (b) 4
(c) 6 (d) 8

57. The equilibrium constant of a 2 electron redox reaction at 298 K is 3.8×10^{-3}. The cell potential E° (in V) and the free energy change ΔG° (in kJ mol^{-1}) for this equilibrium, respectively are

(a) −0.071, − 13.8 (b) −0.071, 13.8
(c) 0.71, − 13.8 (d) 0.071, − 13.8

58. The number of stereoisomers possible for the following compound is

CH_3—CH═CH—CH(Br)—CH_2—CH_3

(a) 2 (b) 3
(c) 4 (d) 8

59. In the radioactive disintegration series $^{232}_{90}Th \rightarrow ^{208}_{82}Pb$, involving α and β decay, the total number of α and β particles emitted are

(a) 6 α and 6 β
(b) 6 α and 4 β
(c) 6 α and 5 β
(d) 5 α and 6 β

60. In the following compressibility factor (*Z*) *versus* pressure graph at 300 K, the compressibility of CH_4 at pressure < 200 bar deviates from ideal behaviour because

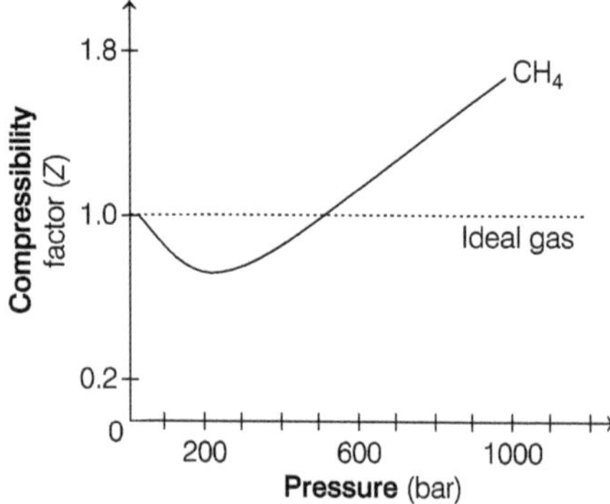

(a) the molar volume of CH_4 is less than its molar volume in the ideal state
(b) the molar volume of CH_4 is same as that in its ideal state
(c) inrermolecular interactions between CH_4 molecules decreases
(d) the molar volume of CH_4 is more than its molar volume in the ideal state

BIOLOGY

61. Which of the following molecules is a primary acceptor of CO_2 in photosynthesis?
(a) Pyruvate (b) 3-phosphoglycerate
(c) Phosphoenol pyruvate (d) Oxaloacetate

62. Which one of the following pairs of molecules, never forms a hydrogen bond between them?
(a) Water and water (b) Water and glucose
(c) Water and ethanol (d) Water and octane

63. Lactase hydrolyses lactose into
(a) glucose + glucose (b) glucose + galactose
(c) galactose + galactose (d) galactose + fructose

64. Which of the following statements is incorrect regarding biological membrane ?
(a) It is composed of lipids and proteins
(b) Peripheral proteins are loosely associated with the membrane
(c) Integral proteins span the lipid bilayer
(d) Lipids and membrane proteins do not provide structural and functional asymmetry

65. The percentage of sunlight captured by plants is
(a) 2-10% (b) 10-20%
(c) 60-80% (d) 100%

66. The hard outer layer of pollens, named exine, is made of
(a) cellulose (b) tapetum
(c) sporopollenin (d) pectin

67. Insectivorous plants such as Venus fly trap catch and digest insects in order to supplement the deficiency of
(a) sulphur (b) nitrogen
(c) potassium (d) phosphorus

68. Which of the following statements about nucleosome is true?
(a) It consists of only DNA
(b) It is nucleus-like structure found in prokaryotes
(c) It consists of DNA and proteins
(d) It consists of only histone proteins

69. Epithelial cells in animals are held by specialised junctions one of them being gap junction. Function of a 'gap junction' is to
(a) facilitate cell-cell communication by rapid transfer of small molecules
(b) cement the neighbouring cells
(c) stop substances from leaking
(d) provide gaps in the tissue to facilitate uptake of small molecules across tissues

70. Which of the following statements is true about glandular epithelium in salivary gland?
(a) It consists of isolated single cells
(b) It consists of multicellular cluster of cells
(c) Its secretions are endocrine
(d) It consists of squamous epithelial cells

71. Which one of the following ion pairs is involved in nerve impulse?
(a) $Na^+.K^+$ (b) $Na^+.Cl^-$ (c) $K^+.Cl^-$ (d) $K^+.Ca^{2+}$

72. Which of the following hormones that controls blood pressure is secreted by human heart?
(a) Erythropoietin (b) Atrial Natriuretic Factor
(c) ACTH (d) Glucocorticoid

73. Oxytocin and vasopressin are synthesised in
(a) hypothalamus (b) adrenal gland
(c) pituitary gland (d) ovary

74. If you exhale multiple times into a conical flask containing lime water through a single inlet fixed through a stop cork, lime water will
(a) become cooler (b) turn milky
(c) remain unchanged (d) turn yellow

75. The path of passage of stimulus when you accidentally touch a hot plate is
(a) receptor $\rightarrow$ brain $\rightarrow$ muscles
(b) muscles $\rightarrow$ spinal cord $\rightarrow$ receptor
(c) muscles $\rightarrow$ brain $\rightarrow$ receptor
(d) receptor $\rightarrow$ spinal cord $\rightarrow$ muscles

76. In the presence of glucose and lactose, *Escherichia coli* utilises glucose. However, lactose also enters the cells because
(a) low level of permease is always present in the cell
(b) it uses the same transporter as glucose
(c) it diffuses through the bacterial cell membrane
(d) it is transported through porins

77. Passive immunisation is achieved by administering
(a) heat-killed vaccines
(b) toxoids
(c) live attenuated vaccines
(d) antibodies

78. Which of the following anions neutralises the acidic pH of the chyme that enters into the duodenum from the stomach?
(a) $H_2PO_4^-$ (b) HSO_4^-
(c) HCO_3^- (d) CH_3COO^-

79. If $^{14}CO_2$ is added to a suspension of photosynthesising chloroplasts, which of the following will be the first compound to be radioactive?
(a) ATP (b) NADPH
(c) NADH (d) 3-phosphoglycerate

80. Which of the following species makes the largest true flower in the world?
(a) *Amorphophallus titanium* (b) *Rafflesia arnoldii*
(c) *Nelumbo nucifera* (d) *Helianthus annuus*

PART-II (2 Marks Questions)

MATHEMATICS

81. The remainder when the polynomial $1+x^2+x^4+x^6+\ldots+x^{22}$ is divided by $1+x+x^2+x^3+\ldots+x^{11}$ is
(a) 0 (b) 2
(c) $1+x^2+x^4+\ldots+x^{10}$ (d) $2(1+x^2+x^4+\ldots+x^{10})$

82. The range of the polynomial $P(x)=4x^3-3x$ as x varies over the interval $\left(-\frac{1}{2},\frac{1}{2}\right)$ is
(a) $[-1,1]$ (b) $(-1,1]$ (c) $(-1,1)$ (d) $\left(-\frac{1}{2},\frac{1}{2}\right)$

83. Ten ants are on the real line. At time $t=0$, the kth ant starts at the point k^2 and travelling at uniform speed, reaches the point $(11-k)^2$ at time $t=1$. The number of distinct times at which at least two ants are at the same location is
(a) 45 (b) 11 (c) 17 (d) 9

84. A wall is inclined to the floor at an angle of 135°. A ladder of length l is resting on the wall. As the ladder slides down, its mid-point traces an arc of an ellipse. Then, the area of the ellipse is

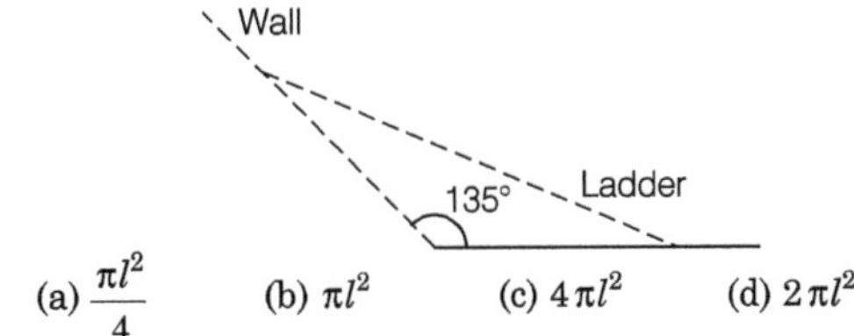

(a) $\frac{\pi l^2}{4}$ (b) πl^2 (c) $4\pi l^2$ (d) $2\pi l^2$

85. Let AB be a sector of a circle with centre O and radius d. $\angle AOB=\theta\left(<\frac{\pi}{2}\right)$ and D be a point on OA such that BD is perpendicular OA. Let E be the mid-point of BD and F be a point on the arc AB such that EF is parallel to OA. Then, the ratio of length of the arc AF to the length of the arc AB is
(a) $\frac{1}{2}$ (b) $\frac{\theta}{2}$ (c) $\frac{1}{2}\sin\theta$ (d) $\frac{\sin^{-1}\left(\frac{1}{2}\sin\theta\right)}{\theta}$

86. Let $f(x)$ be a non-negative differentiable function on $[0,\infty)$ such that $f(0)=0$ and $f'(x)\le 2f(x)$ for all $x>0$. Then, on $[0,\infty)$.
(a) $f(x)$ is always a constant function
(b) $f(x)$ is strictly increasing
(c) $f(x)$ is strictly decreasing
(d) $f'(x)$ changes sign

87. For each positive real number λ. Let A_λ be the set of all natural numbers n such that $|\sin(\sqrt{n+1})-\sin(\sqrt{n})|<\lambda$. Let A_λ^c be the complement of A_λ in the set of all natural numbers. Then,
(a) $A_{1/2}, A_{1/3}, A_{2/5}$ are all finite sets
(b) $A_{1/3}$ is a finite set but $A_{1/2}, A_{2/5}$ are infinite sets
(c) $A_{1/2}^c, A_{1/3}^c, A_{2/5}^c$ are all finites sets
(d) $A_{1/3}, A_{2/5}$ are finite sets and $A_{1/2}$ is an infinite set

88. Let f be a continuous function defined on [0, 1] such that $\int_0^1 f^2(x)\,dx=\left(\int_0^1 f(x)dx\right)^2$. Then, the range of f
(a) has exactly two points
(b) has more than two points
(c) is the interval [0, 1]
(d) is a singleton

89. Three schools send 2, 4 and 6 students, respectively to a summer camp. The 12 students must be accommodated in 6 rooms numbered 1, 2, 3, 4, 5, 6 in such a way that each room has exactly 2 students and both from the same school. The number of ways, the students can be accommodated in the rooms is
(a) 60 (b) 45
(c) 32400 (d) 2700

90. Let a be a fixed non-zero complex number with $|a|<1$ and $w=\left(\frac{z-a}{1-\bar{a}z}\right)$, where z is a complex number. Then,
(a) there exists a complex number z with $|z|<1$ such that $|w|>1$
(b) $|w|>1$ for all z such that $|z|<1$
(c) $|w|<1$ for all z such that $|z|<1$
(d) there exists z such that $|z|<1$ and $|w|=1$

PHYSICS

91. A light balloon filled with helium of density ρ_{He} is tied to a long light string of length l and the string is attached to the ground. If the balloon is displaced slightly in the horizontal direction from the equilibrium and released. Then,

(a) the balloon undergoes simple harmonic motion with period $2\pi\sqrt{\left(\frac{\rho_{air}}{\rho_{air}-\rho_{He}}\right)\frac{l}{g}}$

(b) the balloon undergoes simple harmonic motion with period $2\pi\sqrt{\left(\frac{\rho_{air}-\rho_{He}}{\rho_{air}}\right)\frac{l}{g}}$

(c) the balloon undergoes simple harmonic motion with period $2\pi\sqrt{\left(\frac{\rho_{He}}{\rho_{air}-\rho_{He}}\right)\frac{l}{g}}$

(d) the balloon undergoes conical oscillations with period $2\pi\sqrt{\left(\frac{\rho_{air}+\rho_{He}}{\rho_{air}-\rho_{He}}\right)\frac{l}{g}}$

92. Consider a cube of uniform charge density ρ. The ratio of electrostatic potential at the centre of the cube to that at one of the corners of the cube is

(a) 2 (b) $\sqrt{3}/2$ (c) $\sqrt{2}$ (d) 1

93. Two infinitely long wires each carrying current I along the same direction are made into the geometry as shown in the figure below.

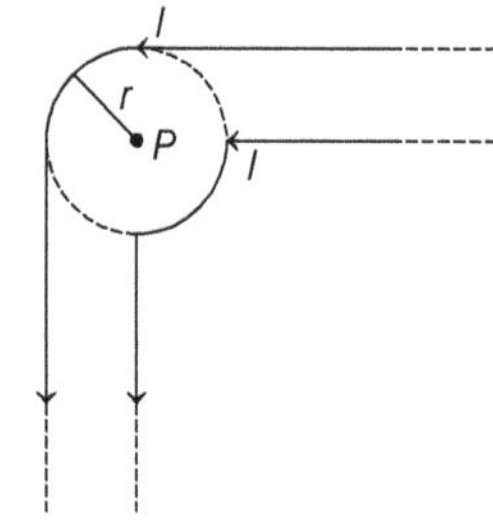

The magnetic field at the point P is

(a) $\frac{\mu_0 I}{\pi r}$ (b) $\frac{\mu_0 I}{r}\left(\frac{1}{\pi}+\frac{1}{4}\right)$ (c) zero (d) $\frac{\mu_0 I}{2\pi r}$

94. A photon of wavelength λ is absorbed by an electron confined to a box of length $\sqrt{(35h\lambda/8mc)}$. As a result, the electron makes a transition from state $k=1$ to the state n. Subsequently the electron transist from the state n to the state m by emitting a photon of wavelength $\lambda'=1.75\,\lambda$. Then,

(a) $n=4, m=2$ (b) $n=5, m=3$
(c) $n=6, m=4$ (d) $n=3, m=1$

95. Consider two masses with $m_1>m_2$ connected by a light inextensible string that passes over a pulley of radius R and moment of inertia I about its axis of rotation. The string does not slip on the pulley and the pulley turns without friction. The two masses are released from rest separated by a vertical distance $2h$. When the two masses pass each other, the speed of the masses is proportional to

(a) $\sqrt{\frac{m_1-m_2}{m_1+m_2+\frac{I}{R^2}}}$ (b) $\sqrt{\frac{(m_1+m_2)(m_1-m_2)}{m_1+m_2+\frac{I}{R^2}}}$

(c) $\sqrt{\frac{m_1+m_2+\frac{I}{R^2}}{m_1-m_2}}$ (d) $\sqrt{\frac{\frac{I}{R^2}}{m_1+m_2}}$

96. An ideal gas is taken reversibly around the cycle a-b-c-d-a as shown on the temperature T- entropy S diagram.

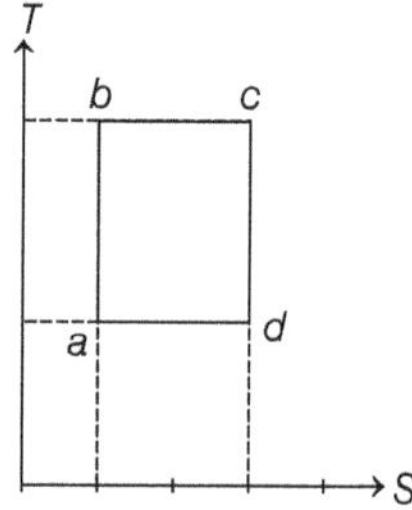

The most appropriate representation of above cycle on a internal energy U- volume V diagram is

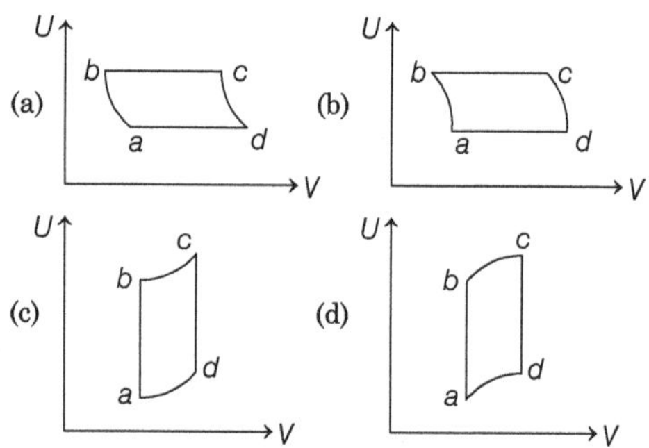

97. The heat capacity of one mole an ideal is found to be $C_V=3R(1+aRT)/2$, where a is a constant. The equation obeyed by this gas during a reversible adiabatic expansion is

(a) $TV^{3/2}e^{aRT}$ = constant (b) $TV^{3/2}e^{3aRT/2}$ = constant
(c) $TV^{3/2}$= constant (d) $TV^{3/2}e^{2aRT/3}$ = constant

98. If the input voltage V_i to the circuit below is given by $V_i(t)=A\cos(2\pi f t)$ and the output voltage is given by $V_o(t)=B\cos(2\pi f t+\phi)$.

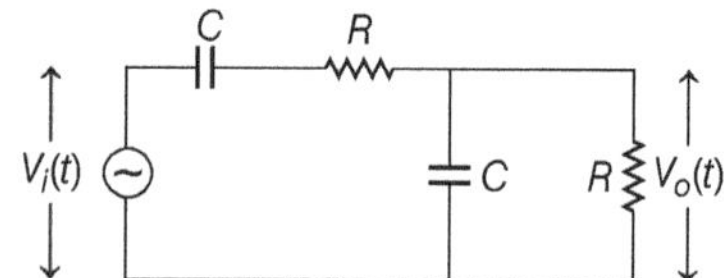

Which one of the following four graphs best depict the variation of ϕ *versus* f?

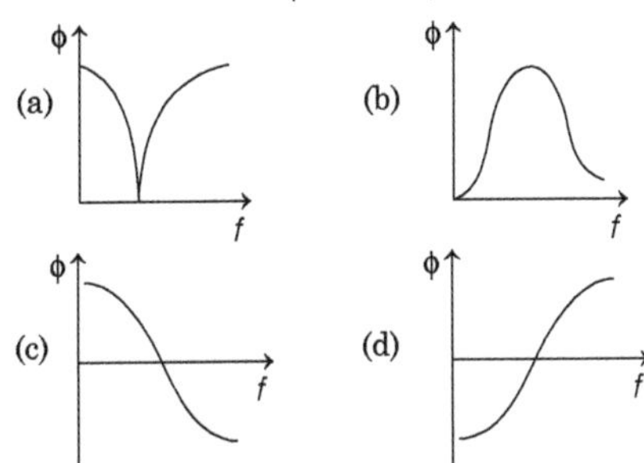

99. A glass prism has a right-triangular cross-section ABC, with $\angle A = 90°$. A ray of light parallel to the hypotenuse BC and incident on the side AB emerges grazing the side AC. Another ray, again parallel to the hypotenuse BC, incident on the side AC suffers total internal reflection at the side AB. Which one of the following must be true about the refractive index μ of the material of the prism?

(a) $\sqrt{\frac{3}{2}} < \mu < \sqrt{2}$ (b) $\mu > \sqrt{3}$

(c) $\mu < \sqrt{\frac{3}{2}}$ (d) $\sqrt{2} < \mu < \sqrt{3}$

100. A smaller cube with side b (depicted by dashed lines) is excised from a bigger uniform cube with side a as shown below, such that both cubes have a common vertex P. Let $X = a / b$. If the centre of mass of the remaining solid is at the vertex O of smaller cube, then X satisfies

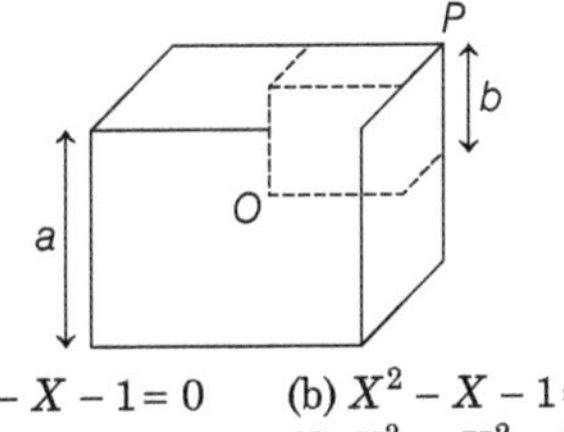

(a) $X^3 - X^2 - X - 1 = 0$ (b) $X^2 - X - 1 = 0$

(c) $X^3 + X^2 - X - 1 = 0$ (d) $X^3 - X^2 - X + 1 = 0$

CHEMISTRY

101. X, Y and Z in the following reaction sequence are

$\xrightarrow{O_2} X \xrightarrow[H_2O]{H^-} Y + Z$

(a) $X =$ OOH $Y =$ $Z =$ OH

(b) $X =$ OOH $Y =$ OH $Z =$

(c) $X =$ OH $Y =$ OOH $Z =$ O

(d) $X =$ OOH $Y =$ OH $Z =$ O

102. The reagents required for the following two-step transformation are

CN

(a) (i) HBr, benzoyl peroxide; (ii) CH_3CN

(b) (i) HBr; (ii) NaCN

(c) (i) Br_2; (ii) NaCN

(d) (i) NaBr; (ii) NaCN

103. In the reaction sequence,

CHO, CHO $\xrightarrow[2.\ H_3O^+]{1.\ \text{Conc. KOH}} X \xrightarrow[\Delta]{H^+} Y$

the major products X and Y respectively are

(a) CH_2OH, CO_2H and O, O

(b) CO_2H, CO_2H and O, O, O

(c) CH_2OH, CH_2OH and O

(d) CHO, CH_2OH and O, O

104. In the following reactions,

\+ O, O, O $\xrightarrow{\text{Anhyd. } AlCl_3} X \xrightarrow[\Delta]{\text{Phosphoric acid}} Y$

X and Y, respectively, are

(a) O, CO_2H and O, O

(b) and

(c) and

(d) and

105. Copper (atomic mass = 63.5) crystallises in a fcc lattice and has density 8.93 g cm^{-3}. The radius of copper atom is closest to

(a) 361.6 pm (b) 511.4 pm (c) 127.8 pm (d) 102.8 pm

106. Given, the standard potentials $E^\circ_{(Cu^{2+}/Cu)}$ and $E^\circ_{(Cu^{+}/Cu)}$ as 0.340 V and 0.522 V respectively, the value of $E^\circ_{(Cu^{2+}/Cu^{+})}$ is

(a) 0.364 V (b) 0.158 V (c) − 0.182 V (d) − 0.316 V

107. For electroplating 1.5 A current is passed for 250 s through 250 mL of 0.15 M solution of MSO_4. Only 85% of the current was utilised for electrolysis. The molarity of MSO_4 solution after electrolysis is closest to [Assume that the volume of the solution remained constant]

(a) 0.14 (b) 0.014 (c) 0.07 (d) 0.035

108. The hybridisation of the central atom and the shape of $[IO_2F_5]^{2-}$ ion, respectively are

(a) sp^3d^3

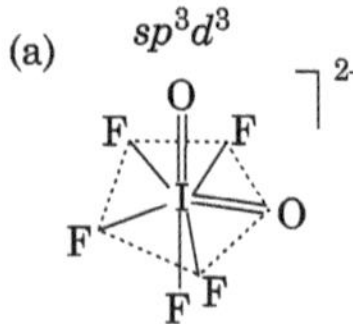

(b) sp^2d^4

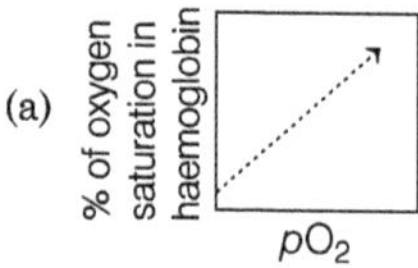

(c) sp^2d^4

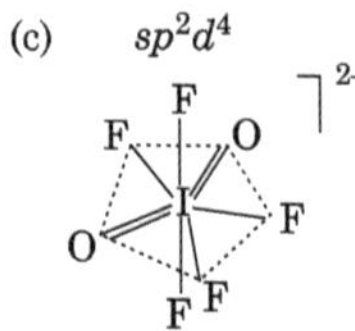

(d) sp^3d^3

109. 2.33 g of compound X (empirical formula $= CoH_{12}N_4Cl_3$) upon treatment with excess $AgNO_3$ solution produces 1.435 g of a white precipitate. The primary and secondary valences of cobalt in compound X, respectively are

[Given, atomic mass : Co = 59, Cl = 35.5, Ag = 108]

(a) 3, 6 (b) 3, 4 (c) 2, 4 (d) 4, 3

110. The specific conductance (κ) of 0.02 M aqueous acetic acid solution at 298 K is 1.65×10^{-4} S cm^{-1}. The degree of dissociation of acetic acid is

[Given equivalent conductance at infinite dilution of H^{-} = 349.1 S cm^2 mol^{-1} and CH_3COO^{-} = 40.9 S cm^2 mol^{-1}]

(a) 0.021 (b) 0.21

(c) 0.012 (d) 0.12

BIOLOGY

111. Match the following organelles in Group I with the structure in Group II. Choose the correct combination.

Group I	Group II
A. Mitochondrion	1. Cisternae
B. Golgi apparatus	2. Cristae
C. Chloroplast	3. Thylakoids
D. Centrosome	4. Radial spokes

Codes

(a) A–2, B–1, C–3, D–4 (b) A–3, B–1, C–2, D–4

(c) A–4, B–1, C–2, D–3 (d) A–4, B–2, C–1, D–3

112. A human population containing 200 individuals has two alleles at the 'T' locus, named T and t. T which produces tall individuals, is dominant over t, which produces short individuals. If the population has 90 TT, 40 Tt and 70 tt genotypes, what will be the frequencies of these two alleles in this population?

(a) T, 0.50 : t, 0.50 (b) T, 0.55 : t, 0.45

(c) T, 0.45 : t, 0.35 (d) T, 0.90 : t, 0.10

113. Which of the following graphs best describes the oxygen dissociation curve where pO_2 is the partial pressure of oxygen?

(a) (b) (c) (d)

114. Which of the following best describes the DNA content and the number of chromosomes at the end of S and M -phases of the cell cycle in mitosis, if the DNA content of the cell in the beginning of cell cycle (G_1-phase) is considered as C and the number of chromosomes 2N?

(a) 2C and 2N for S-phase : 2C and 2N for M-phase

(b) 2C and N for S-phase : 2C and N for M-phase

(c) 2C and 2N for S-phase : C and 2N for M-phase

(d) C and N for S-phase : C and 2N for M-phase

115. Study the following graph of metabolic rate of various terrestrial mammals as a function of their body mass and choose the correct option below.

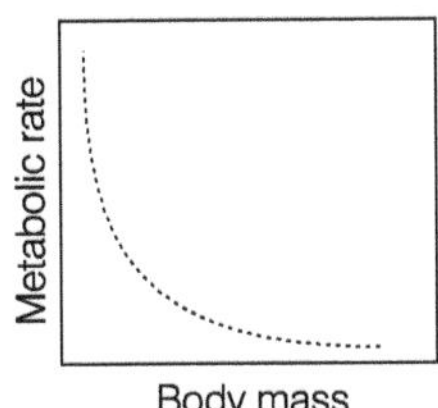

(a) Animals are distributed throughout the curve with the smaller animals towards the left and progressively bigger animals towards the right
(b) The smaller animals below a certain critical mass cluster at the left end of the curve and the larger animals above the critical mass cluster on the right end.
(c) Animals are distributed throughout the curve with the larger animals towards the left and progressively smaller animals towards the right
(d) The larger animals above a certain critical mass cluster at the left end of the curve and the smaller animals below the critical mass cluster on the right end

116. Match the human disorders shown in Group I with the biochemical processes in Group II. Choose the correct combination.

Group I	Group II
A. Phenylketonuria	1. Melanin synthesis
B. Albinism	2. Conversion of phenylalanine to tyrosine
C. Homocystinuria	3. Tyrosine degradation
D. Argininemia	4. Methionine metabolism
	5. Urea synthesis

Codes
(a) A–2, B–1, C–4, D–5 (b) A–1, B–4, C–2, D–5
(c) A–2, B–1, C–5, D–3 (d) A–5, B–3, C–1, D–2

117. An *m*RNA is transcribed from a DNA segment having the base sequence 3′-TACATGGGTCCG-5′. What will be the correct order of binding of the four amino acyl-*t*RNA complexes given below during translation of this *m*RNA?

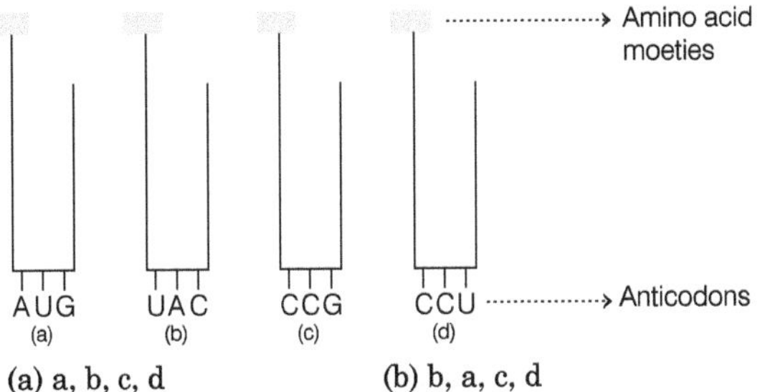

(a) a, b, c, d (b) b, a, c, d
(c) c, d, a, b (d) b, a, d, c

118. If the initial number of template DNA molecules in a PCR reaction is 1000. The number of product DNA molecules at the end of 20 cycles will be closest to
(a) 10^3
(b) 10^6
(c) 10^9
(d) 10^{12}

119. The allele for black hair (B) is dominant over brown hair (b) and the allele for brown eye (E) is dominant over blue eye (e). Out of the offsprings obtained upon mating a black-haired and brown-eyed individual (BbEe) with a brown-haired and brown-eyed individual (bbEE), the ratio of brown-haired and brown-eyed individuals to black-haired and brown-eyed individuals is
(a) 2 : 1
(b) 3 : 1
(c) 1 : 1
(d) 1 : 2

120. In an experiment represented in the schematic below, a plant species was grown in different day and night cycles and its photoperiodic flowering behaviour was noted. This species is a

Light	Dark	
16 hrs	6 hrs	No flower
16 hrs	7 hrs	No flower
16 hrs	8 hrs	No flower
16 hrs	9 hrs	Flower
16 hrs	10 hrs	Flower
16 hrs	11 hrs	No flower
8 hrs	10 hrs	Flower
10 hrs	10 hrs	Flower
12 hrs	10 hrs	Flower
8 hrs	8 hrs	No flower
10 hrs	8 hrs	No flower
12 hrs	8 hrs	No flower

(a) short day plant and actually measures day length to flower
(b) short day plant and actually measures night length to flower
(c) long day plant and actually measures night length to flower
(d) long day plant and actually measures day length to flower

Answers

PART-I

Q	Ans	Q	Ans	Q	Ans	Q	Ans	Q	Ans	Q	Ans	Q	Ans	Q	Ans	Q	Ans	Q	Ans
1	(b)	2	(c)	3	(d)	4	(b)	5	(c)	6	(d)	7	(b)	8	(a)	9	(a)	10	(c)
11	(b)	12	(a)	13	(a)	14	(c)	15	(a)	16	(c)	17	(c)	18	(a)	19	(c)	20	(b)
21	(c)	22	(b)	23	(a)	24	(d)	25	(c)	26	(c)	27	(c)	28	(d)	29	(b)	30	(b)
31	(b)	32	(b)	33	(d)	34	(b)	35	(c)	36	(a)	37	(b)	38	(b)	39	(b)	40	(a)
41	(d)	42	(a)	43	(c)	44	(b)	45	(a)	46	(a)	47	(d)	48	(c)	49	(c)	50	(b)
51	(c)	52	(b)	53	(b)	54	(b)	55	(d)	56	(d)	57	(b)	58	(c)	59	(b)	60	(a)
61	(c)	62	(d)	63	(b)	64	(d)	65	(a)	66	(c)	67	(b)	68	(c)	69	(a)	70	(b)
71	(a)	72	(b)	73	(a)	74	(b)	75	(d)	76	(a)	77	(d)	78	(c)	79	(d)	80	(b)

PART-II

Q	Ans	Q	Ans	Q	Ans	Q	Ans	Q	Ans	Q	Ans	Q	Ans	Q	Ans	Q	Ans	Q	Ans
81	(d)	82	(c)	83	(c)	84	(a)	85	(d)	86	(a)	87	(c)	88	(d)	89	(c)	90	(c)
91	(c)	92	(a)	93	(d)	94	(c)	95	(a)	96	(a)	97	(*)	98	(c)	99	(d)	100	(a)
101	(d)	102	(b)	103	(a)	104	(a)	105	(c)	106	(b)	107	(a)	108	(d)	109	(a)	110	(a)
111	(a)	112	(b)	113	(d)	114	(c)	115	(a)	116	(a)	117	(d)	118	(c)	119	(c)	120	(b)

** No options are correct.*

Solutions

1. *(b)* We have,

$$x^4 + y^4 + z^4 + 1 = 4xyz$$

$$\text{AM} \geq \text{GM}$$

$$\therefore \frac{x^4 + y^4 + z^4 + 1}{4} \geq (x^4 \cdot y^4 \cdot z^4 \cdot 1)^{1/4}$$

$$\Rightarrow \frac{4xyz}{4} \geq |xyz| \Rightarrow xyz > 0$$

It is possible $(1, 1, 1)$ $(-1, -1, 1)$ $(-1, 1, -1)$, $(1, -1, -1)$.

So number of triplet $(x, y, z) = 4$

2. *(c)* We have,

$$P(\sin^2 x) = P(\cos^2 x),\ x \in \left[0, \frac{\pi}{2}\right)$$

$$P(\sin^2 x) = P(1 - \sin^2 x)$$

$$P(x) = P(1 - x),\ x \in [0,1]$$

$$P'(x) = -P'(1 - x)$$

So, $P'(x)$ is symmetric about line $x = \frac{1}{2}$

So, $P'(x)$ has highest degree is odd.

$\Rightarrow P(x)$ has highest degree is even.

Hence, option (c) is correct.

3. *(d)* Given,

$$A_r = \{e^{j\pi rn} : n \text{ is a natural number}\}$$

$A_1 = e^{jn\pi}$ which is finite.

$A_{0.3} = e^{j0 \cdot 3n\pi}$ is also finite

$A_{\frac{1}{\pi}} = e^{ni}$ is infinite set

$\therefore$ Option (d) is correct.

4. *(b)* We have,

$$x^3 - 27x + k = 0$$

Let $f(x) = x^3 - 27x$

$$f'(x) = 3x^2 - 27 = 3(x^2 - 9)$$

Now, sum of roots of $x^3 - 27x + k = 0$ is zero.

$\therefore$ Third root is also integer.

Now, put $x = 6t$

$$216t^3 - 27(6t) + k = 0$$

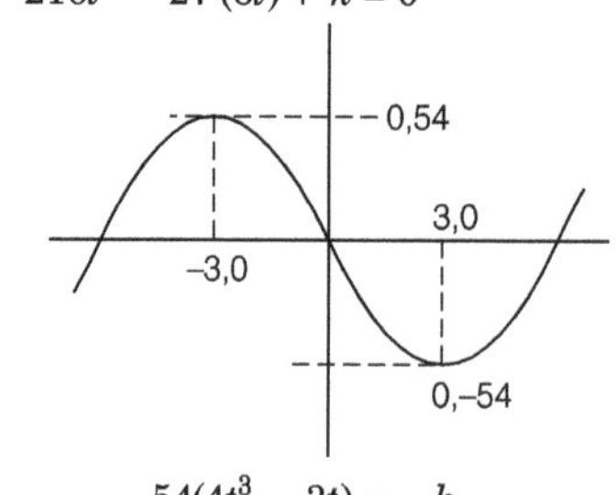

$$\Rightarrow 54(4t^3 - 3t) = -k$$

$$\Rightarrow 54\cos 3\theta = -k \quad [\because t = \cos\theta]$$

$$\Rightarrow \cos 3\theta = -\frac{k}{54}$$

Now, we get integral solution on $\theta = 0, \pi$

$\therefore$ Two values of k are possible.

5. *(c)* We have equation of parabola

$$y = x^2 + px + q$$

Its passes through $(0, 3)$.

$$\therefore \quad 3 = q \Rightarrow \frac{dy}{dx} = 2x + p$$

$$\Rightarrow \left(\frac{dy}{dx}\right)_{(0,3)} = p = -1$$

[$\because$ slope of tangent $= -1$]

$$\Rightarrow p + q = -1 + 3 = 2$$

6. *(d)* Given,

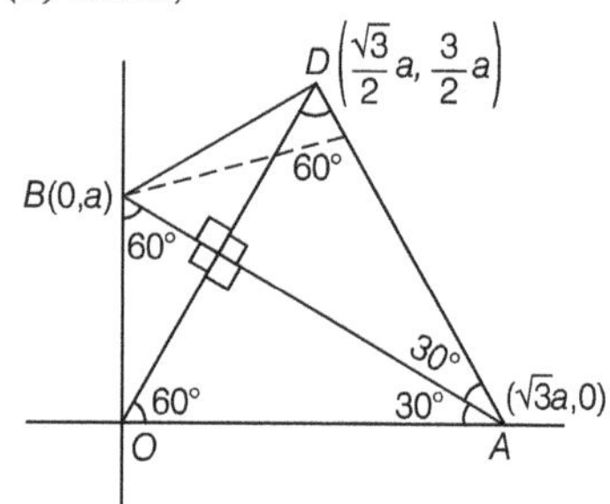

$$\angle OBA = 60°$$

$$\therefore \quad \angle OAB = 30°$$

OAD is an equilateral triangle.

$$\therefore \quad \angle AOD = \angle ODA = \angle OAD = 60°$$

$$OA = AD = OD$$

Let $B(0, a)$

$\therefore \quad A(\sqrt{3}a, 0)$

$D\left(\frac{\sqrt{3}}{2}a, \frac{3}{2}a\right)$

Slope of $BD = \frac{\frac{3}{2}a - a}{\frac{\sqrt{3}}{2}a} = \frac{1}{\sqrt{3}}$

7. *(b)* Given parabola $(y-k)^2 = 4(x-h)$

This is passes through (0, 0) and (0, 2).

$\therefore \quad k^2 = -4h \quad \ldots$(i)

and $\quad (2-k)^2 = -4h \quad \ldots$(ii)

From Eqs. (i) and (ii), we get

$k = 1, h = -\frac{1}{4}$

$\therefore$ Equation of parabola becomes

$(y-1)^2 = 4x + 1$

Focus of parabola $= \left(\frac{3}{4}, 1\right)$

End of latusrectum $D = \left(\frac{3}{4}, 3\right)$

Axis of parabola is $y - 1 = 0$

Point P intersection of axis of parabola and Y-axis.

$\therefore \quad P(0, 1)$

$D\left(\frac{3}{4}, 3\right)$

$A\left(-\frac{1}{4}, 1\right) \quad$ [$\because A$ is vertex of parabola]

Slope of $PD = \frac{3-1}{3/4 - 0} = 8/3 = m_1$

Slope of $DA = \frac{3-1}{\frac{3}{4} + \frac{1}{4}} = 2 = m_2$

$\therefore \quad \angle PDA = \theta = \tan^{-1}\left(\frac{m_1 - m_2}{1 + m_1 m_2}\right)$

$= \tan^{-1}\left(\frac{8/3 - 2}{1 + \frac{16}{3}}\right) = \tan^{-1}\left(\frac{2}{19}\right)$

8. *(a)* In circle with centre O, APB are three points on its circumference such that P is mid-point of minor arc AB

$\frac{\text{Area }(\Delta AOB)}{\text{Area }(\Delta APB)} = \frac{\sqrt{5}+2}{1}$

Area of $\Delta AOB = \frac{1}{2}\sin\theta$

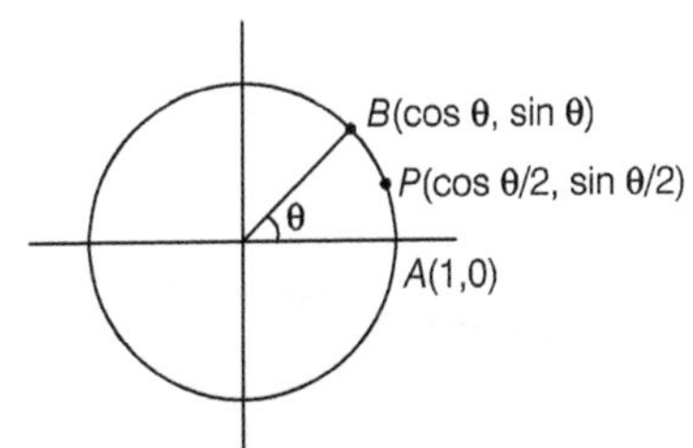

Area of $\Delta APB = \frac{1}{2}\begin{vmatrix} 1 & 0 & 1 \\ \cos\theta/2 & \sin\theta/2 & 1 \\ \cos\theta & \sin\theta & 1 \end{vmatrix}$

$= \frac{1}{2}(1 - \cos\theta/2)\, 2\sin\theta/2$

$\therefore \quad \frac{\text{Area of }\Delta AOB}{\text{Area of }\Delta APB} = \sqrt{5} + 2$

$\frac{\frac{1}{2}\sin\theta}{\frac{1}{2}2\sin\frac{\theta}{2}(1-\cos\frac{\theta}{2})} = \sqrt{5} + 2$

$= \frac{\cos\theta/2}{1 - \cos\theta/2} = \sqrt{5} + 2$

$\Rightarrow \cos\theta/2 = \frac{\sqrt{5}+1}{4} \Rightarrow \cos\theta = \frac{\sqrt{5}-1}{4}$

If $\theta \to 2\theta$

$\therefore \quad \frac{\text{Area }(\Delta AOB)}{\text{Area }(\Delta APB)} = \frac{\cos\theta}{1 - \cos\theta}$

$= \frac{\frac{\sqrt{5}-1}{4}}{1 - \left(\frac{\sqrt{5}-1}{4}\right)} = \frac{1}{\sqrt{5}}$

9. *(a)* We have,

$X = \{x \in R : \cos(\sin x) = \sin(\cos x)\}$

$\cos(\sin x) = \sin(\cos x)$

$\sin\left(\frac{\pi}{2} \pm \sin x\right) = \sin(\cos x)$

$\Rightarrow \quad \cos x = \frac{\pi}{2} \pm \sin x$

$\Rightarrow \cos x = n\pi + (-1)^n\left(\frac{\pi}{2} + \sin x\right), n \in I$

$\Rightarrow \cos x \pm \sin x = n\pi + (-1)^n \frac{\pi}{2}, n \in I$

As LHS $\in [-\sqrt{2}, \sqrt{2}]$, and it does not satisfy the RHS.

$\therefore$ No solution exists.

10. *(c)* Given, $OP = OQ = r = AR$

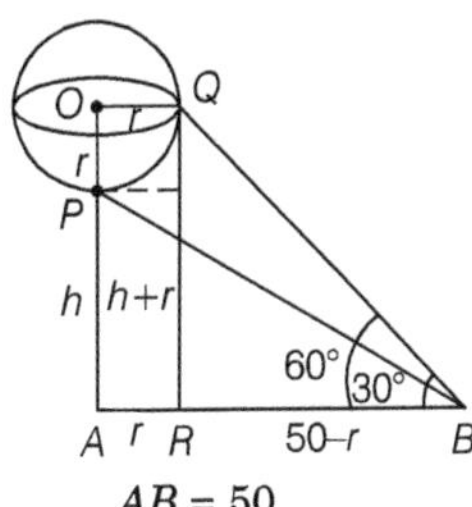

$AB = 50$

In ΔBRQ, $\tan 60° = \frac{RQ}{BR} = \frac{h+r}{50-r}$

$\Rightarrow \quad \sqrt{3}(50 - r) = h + r$

$\Rightarrow \quad h = 50\sqrt{3} - r(\sqrt{3}+1) \quad \ldots$(i)

In ΔAPB,

$\tan 30° = \frac{AP}{AB} = \frac{h}{50} \Rightarrow h = \frac{50}{\sqrt{3}} \quad \ldots$(ii)

From Eqs. (i) and (ii), we get

$\frac{50}{\sqrt{3}} = 50\sqrt{3} - r(\sqrt{3}+1)$

$\Rightarrow \quad r(\sqrt{3}+1) = \frac{100}{\sqrt{3}}$

$\Rightarrow \quad r = \frac{100}{\sqrt{3}(\sqrt{3}+1)} = \frac{100(3-\sqrt{3})}{9-3}$

$= 50\left(1 - \frac{1}{\sqrt{3}}\right)$

11. *(b)* We have,

$f(x) = x + \frac{1}{8}\sin(2\pi x), x \in [0, 1]$

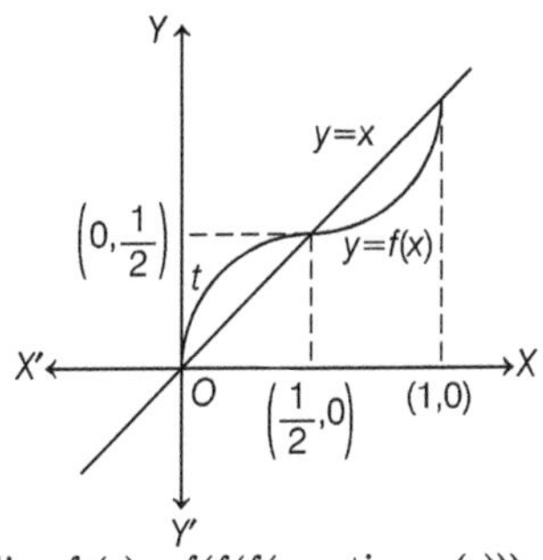

$\lim_{n\to\infty} f_n(x) = f(f(f(\ldots\infty \text{ times }(x)))$

Now, for $x_1 \in \left(0, \frac{1}{2}\right)$

$f(x_1) > x_1$ as $f(x)$ is concave-downward

Thus, $f_n \to \frac{1}{2} n \to \infty$

Similarly, for $x_1 \in \left(\frac{1}{2}, 1\right)$

$f(x_1) < x_1$ as $f(x)$ is concave upward.

Thus, $f_n \to \frac{1}{2}$ as $n \to \infty$

$\therefore$ Only II is true.

Hence, option (b) is correct.

12. *(a)* We have,

$\lim_{x\to\infty} x^2 \int_0^x e^{t^3 - x^3} dt = \lim_{x\to\infty} x^2 e^{-x^3} \int_0^x e^{t^3} dt$

$= \lim_{x\to\infty} \frac{x^2 \int_0^x e^{t^3} dt}{e^{x^3}}$

Apply L Hospital's rule

$= \lim_{x\to\infty} \frac{2x\int_0^x e^{t^3} dt + x^2 e^{x^3}}{3x^2 e^{x^3}}$

$= \lim_{x\to\infty} \frac{2\int_0^x e^{t^3} dt + x^2 e^{x^3}}{3x^2 e^{x^3}}$

Again Apply L Hospital's rule, we get

$= \lim_{x\to\infty} \frac{2e^{x^3} + e^{x^3} + 3x^3 e^{x^3}}{3e^{x^3} + 9x^3 e^{x^3}}$

$= \lim_{x\to\infty} \frac{e^{x^3}(3 + 3x^3)}{e^{x^3}(3 + 9x^3)} = \frac{1}{3}$

13. *(a)* We have,

$$x^3 - 3ax^2 + (27a^2 + 9)x + 2016 = 0$$

Let

$$f(x) = x^3 - 3ax^2 + (27a^2 + 9)x + 2016$$
$$f'(x) = 3x^2 - 6ax + 27a^2 + 9$$
$$f'(x) = 3(x^2 - 2ax + 9a^2 + 3)$$
$$f'(x) = 3((x-a)^2 + 8a^2 + 3)$$
$$f'(x) > 0, \forall x \in R$$

$\therefore f(x)$ is increasing function, $\forall a \in R$.

$\therefore f(x)$ has exactly one real root for any real a.

14. *(c)* We have, $y = |x^3 - 4x^2 + 3x|$

Area of region bounded by $y \le f(x)$, X-axis and $0 \le x \le 3$ is $\int_0^3 |x^3 - 4x^2 + 3x|\, dx$

$$\int_0^1 (x^3 - 4x^2 + 3x) - \int_1^3 (x^3 - 4x^2 + 3x)\, dx$$
$$= \left[\frac{x^4}{4} - \frac{2x^3}{3} + \frac{3x^2}{2}\right]_0^1 - \left[\frac{x^4}{4} - \frac{4x^3}{3} + \frac{3x^2}{2}\right]_1^3$$
$$= \left[\left(\frac{1}{4} - \frac{4}{3} + \frac{3}{2}\right) - (0)\right] - \left[\left(\frac{81}{4} - \frac{108}{3} + \frac{27}{2}\right) - \left(\frac{1}{4} - \frac{4}{3} + \frac{3}{2}\right)\right]$$
$$= \frac{37}{12}$$

15. *(a)* We have,

$$f : [0, 1] \to [0, 1]$$
$$f(x) < x^2$$

$f(x)$ is always positive $x \in [0, 1]$

$$\int_0^1 f(x) dx < \int_0^1 x^2$$
$$\int_0^1 f(x)\, dx < \left[\frac{x^3}{3}\right]_0^1$$
$$\int_0^1 f(x)\, dx < \frac{1}{3}$$

But $\int_0^1 f(x)\, dx = \frac{1}{3}$ which is not possible.

Hence, no solution exists.

16. *(c)* We have,

$$f(x) = \min\{x - [x], 1 - x + [x]\}$$
$$g(x) = \max\{x - [x], 1 - x + [x]\}$$

Graph of $f(x)$ is

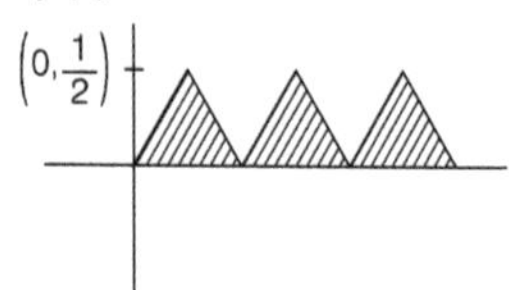

Graph of $g(x)$ is

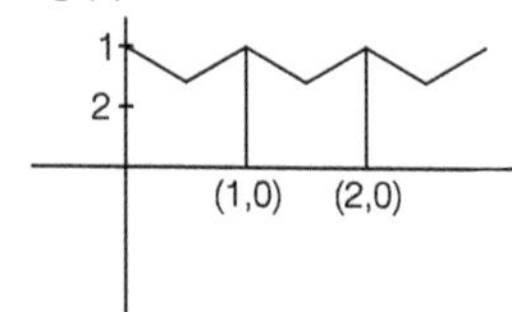

$$\int_0^n g(x)\, dx = \int_0^n g(x)\, dx = \frac{3n}{4}$$
$$\int_0^n f(x)\, dx = \frac{n}{4}$$
$$\because \int_0^n (g(x) - f(x))\, dx = \frac{3n}{4} - \frac{n}{4} = 100$$
$$\Rightarrow \frac{n}{2} = 100 \Rightarrow n = 200$$

17. *(c)* We have,

$$|\mathbf{v} - \mathbf{i}| = |\mathbf{v} - 2\mathbf{i}| = |\mathbf{v} - \mathbf{j}|$$

Clearly, $\mathbf{v}$ is circumcentre of ΔABC.

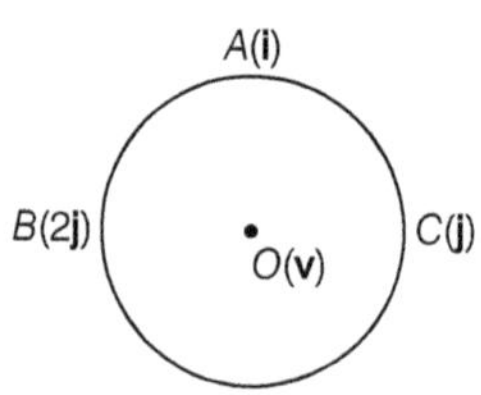

where $A(1, 0)$, $B(2, 0)$, $C(0, 1)$ and $O(x, y)$

$$\therefore \quad OA^2 = OB^2 = OC^2$$
$$(x-1)^2 + (y-0)^2 = (x-2)^2 + y^2 = x^2 + (y-1)^2$$
$$\Rightarrow x^2 - 2x + 1 + y^2 = x^2 - 4x + 4 + y^2 = x^2 + y^2 - 2y + 1$$
$$\Rightarrow 2x = 3 \Rightarrow x = \frac{3}{2}$$
$$\Rightarrow x^2 - 2x + 1 + y^2 = x^2 + y^2 - 2y + 1$$
$$\Rightarrow x = y = 3/2$$
$$\therefore (x, y) = \left(\frac{3}{2}, \frac{3}{2}\right)$$
$$\Rightarrow \vec{v} = \frac{3}{2}\hat{i} + \frac{3}{2}\hat{j}$$

$|\vec{v}| = \sqrt{\frac{9}{4} + \frac{9}{4}} = \frac{3\sqrt{2}}{2}$ lie in (2, 3].

18. *(a)* Consider the event

A = Blue ball is drawn from first draw

B = Red ball is drawn from first draw

C = Blue ball is drawn from second draw

$$P(A) = \frac{b}{b+r}, P\left(\frac{C}{A}\right) = \frac{b+1}{b+r+1}$$
$$P(B) = \frac{r}{b+r}, P\left(\frac{C}{B}\right) = \frac{b}{b+r+1}$$

Required probability

$$P(C) = P(A) \cdot P(C/A) + P(B) \cdot P(C/B)$$
$$= \frac{b}{b+r}\frac{b+1}{b+r+1} + \frac{r}{(b+r)}\frac{b}{b+r+1}$$
$$= \frac{b}{(b+r)(b+r+1)}(b+1+r) = \frac{b}{b+r}$$

19. *(c)* We have,

Set {10, 11, 12, ..., 22}

Number of scalene triangle

$$= {}^{13}C_3 - 3 = 283$$

Number of isosceles triangle

$$= ({}^{13}C_2 \times 2) - 4 = 152$$

Number of equilateral triangle

$$= {}^{13}C_1 = 13$$

So, total number of triangle

$$= 283 + 152 + 13 = 448$$

20. *(b)* We have,

four shape of tiles

(1) Equilateral triangle

(2) Square

(3) Regular pentagon

(4) Regular hexagon

Cover the XY-plane with four shapes such that no two tiles are overlaps.

$\therefore$ We use equilateral triangular, regular hexagon is also made of equilateral triangle but pentagon cannot cover the plane because of its shape.

$\therefore$ Exactly three of four shapes, i.e. equilateral triangle, square and regular hexagon.

21. *(c)* Rays can cross-each other while forming image etc.

- Light rays are path of light waves and two waves can cross without effecting each other's characteristics.
- In streamline flow different layers of flowing fluid does not mix with each other. So, streamlines never crosses in entire flow.
- Electric field lines and magnetic field lines never cross as there is only one direction of field at a point in space.

22. *(b)*

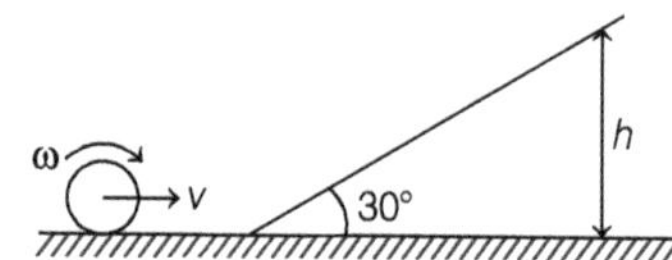

As ring is rolling without any slippage,

Total initial kinetic energy = Total final potential energy at height h

$\Rightarrow$ Kinetic energy of translation + Kinetic energy of rotation = Potential energy of ring at height h

$$\Rightarrow \frac{1}{2}mv^2 + \frac{1}{2}I\omega^2 = mgh$$

For ring, $I = mR^2$ and $v = R\omega$

$$\therefore \frac{1}{2}mv^2 + \frac{1}{2}mR^2 \times \frac{v^2}{R^2} = mgh$$
$$\Rightarrow mv^2 = mgh$$
$$\Rightarrow h = \frac{v^2}{g}$$

23. *(a)* In sudden expansion, time duration is small. Thus, heat does not leave or enter the gas mass.

Process of sudden expansion is adiabatic process, $\Delta Q = 0$.

24. *(d)*

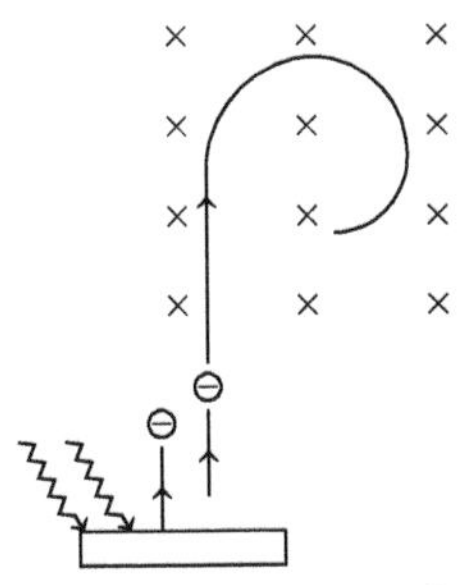

Let v = velocity of most energetically emitted electron.

Then, by Einstein's equation, we have

$$K_{max} = \frac{hc}{\lambda} - \phi_0$$

$\Rightarrow$ Work function ϕ_0 is

$$\phi_0 = \frac{hc}{\lambda} - K_{max}$$

$$= \frac{hc}{\lambda} - \frac{1}{2}mv^2 \quad \dots (i)$$

As, this electron describes a circular path of radius R in magnetic field B, then

$$R = \frac{mv}{qB}$$

$$\Rightarrow \quad v = \frac{qBR}{m}$$

Substituting v in Eq. (i), we have

$$\phi_0 = \frac{hc}{\lambda} - \frac{1}{2}m\left(\frac{q^2B^2R^2}{m^2}\right)$$

$$= \frac{hc}{\lambda} - 2m\left(\frac{q^2B^2R^2}{4m^2}\right)$$

$$\Rightarrow \quad \phi_0 = \frac{hc}{\lambda} - 2m\left(\frac{qBR}{2m}\right)^2$$

25. *(c)*

I II
Hole
λ_I T_I=150 K
λ_{II} T_{II}=300 K

In steady state, rate of diffusion of gases must be same from both sides.

$$\Rightarrow \quad r_1 = r_2$$

$$\Rightarrow \quad \frac{P_I}{\sqrt{T_I}} = \frac{P_{II}}{\sqrt{T_{II}}}$$

Now, mean free path of a gas molecule is

$$\lambda = \frac{k_B T}{\sqrt{2}\, \pi d^2 \cdot P}$$

or $$\lambda \propto \frac{T}{P}$$

So, $$\frac{\lambda_I}{\lambda_{II}} = \frac{T_I / P_I}{T_{II} / P_{II}}$$

$$= \frac{T_I}{T_{II}} \times \frac{P_{II}}{P_I} = \frac{T_I}{T_{II}} \times \frac{\sqrt{T_{II}}}{\sqrt{T_I}}$$

$$\Rightarrow \quad \frac{\lambda_I}{\lambda_{II}} = \frac{\sqrt{T_I}}{\sqrt{T_{II}}} = \sqrt{\frac{150}{300}}$$

$$\Rightarrow \quad \frac{\lambda_I}{\lambda_{II}} = 0.7$$

26. *(c)* In given arrangement, sliding bar is given an initial speed v towards left.

As, flux linked with loop is increasing induced current must be flowing counter clockwise to oppose the increasing flux.

During induction electrical energy generated is dissipated as heat. Power generated is given by

$$P = \frac{B^2l^2v^2}{R}$$

or $$P \propto l^2$$

Acceleration of sliding bar is

$$a = \frac{F}{m}$$

$$a = \frac{B^2l^2v}{Rm}$$

$$\Rightarrow \quad \frac{dv}{dt} = \frac{B^2l^2v}{Rm}$$

$$\Rightarrow \quad \frac{dv}{dx}\cdot\left(\frac{dx}{dt}\right) = \frac{B^2l^2v}{Rm}$$

$$\Rightarrow \quad dv = \frac{B^2l^2}{Rm}dx \quad \left[\because \frac{dx}{dt} = v\right]$$

$$\Rightarrow \quad v = \frac{B^2l^2}{Rm} . x$$

or $$x = \frac{Rmv}{B^2l^2} \Rightarrow x \propto R$$

27. *(c)* Given, potential energy of particle is

$$U = \frac{x^4}{4} - \frac{x^2}{2}$$

Differentiating with respect to x, we get

$$\frac{dU}{dx} = \frac{4x^3}{4} - \frac{2x}{2}$$

$$= x^3 - x$$

$$= x(x^2 - 1)$$

For maxima or minima,

$$\frac{dU}{dx} = 0$$

$$\Rightarrow \quad x(x^2 - 1) = 0$$

$$\Rightarrow \quad x = 0 \text{ or } x = \pm 1$$

Differentiating again, we have

$$\frac{d^2U}{dx^2} = 3x^2 - 1$$

So, $$\frac{d^2U}{dx^2} = \begin{cases} -1 \text{ at } x = 0 \\ +2 \text{ at } x = \pm 1 \end{cases}$$

So, by second derivative test, potential energy maxima occurs at $x = \pm 1$ and potential energy minima is at $x = 0$.

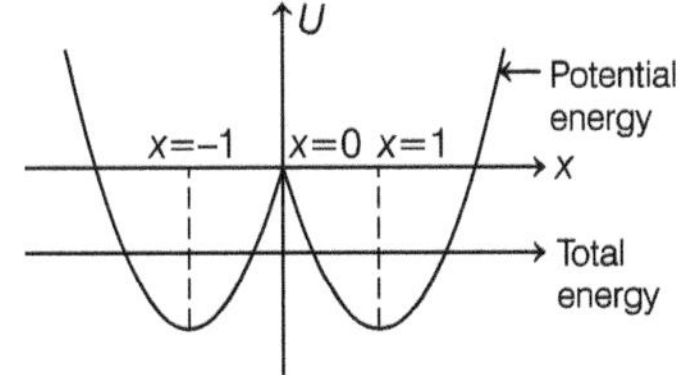

Now, as at $t = 0$, $x = -0.5$ m, so the particle is region $x = -1$ to $x = 0$. As total energy of particle is negative, so particle cannot crosses $x = 0$ mark.

28. *(d)* Blood pressure is gauge pressure.

Also,

Actual pressure = Atmospheric pressure + Gauge pressure

$\Rightarrow$ Actual pressure = 760 mm of Hg + 190 mm of Hg

= 950 mm of Hg

= 1.25×760 mm of Hg

$\therefore$ Actual pressure is 1.25 times of atmospheric pressure.

29. *(b)* Velocity v of a particle moving around a curved path can be resolved into radial $\left(v_r = \frac{dr}{dt}\right)$ and tangential (or transverse, $v_\theta = r\frac{d\theta}{dt}\right)$ velocity components.

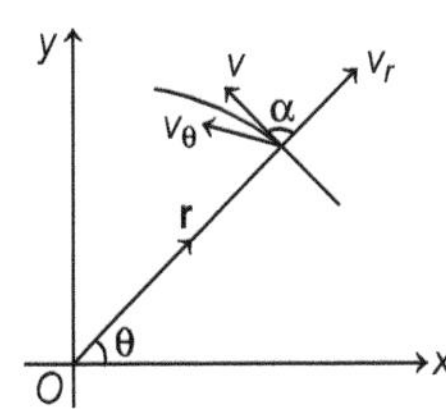

Angle α between velocity **v** and tangential velocity $\mathbf{v}_r$ is given by

$$\tan\alpha = \frac{v_\theta}{v_r}$$

$$\Rightarrow \quad \tan\alpha = \frac{\left(r\dfrac{d\theta}{dt}\right)}{\left(\dfrac{dr}{dt}\right)} = r\cdot\left(\frac{d\theta}{dr}\right)$$

$$= \frac{r}{(dr/d\theta)}$$

Here, given $\dfrac{dr}{d\theta} = r$

$$\therefore \quad \tan\alpha = \frac{r}{r} = 1$$

$$\Rightarrow \quad \alpha = \tan^{-1}(1) = 45°$$

30. *(b)* When electrons are accelerated through a potential of V volts, momentum gained by electrons is

$$\Delta p = mv = \sqrt{2mK}$$
$$= \sqrt{2meV}$$

where, K = kinetic energy, m = mass of electron, v = velocity of electron and e = charge of electron.

Wavelength λ associated with these moving electrons is

$$\lambda = \frac{h}{p} = \frac{h}{\sqrt{2meV}}$$

where, h is the Planck's constant.

Fringe width β shown by electrons in Young's double slit experiment, is

$$\beta = \frac{\lambda D}{d}$$

where, D is the distance of screen from fringes and d is the separation of slits.

$$\therefore \quad \beta = \frac{hD}{d\sqrt{2meV}}$$

$\Rightarrow \beta \propto \dfrac{1}{\sqrt{V}}$, keeping other parameters constant.

So, ratio of initial and final fringe widths is

$$\Rightarrow \quad \frac{\beta_i}{\beta_f} = \frac{\sqrt{V_f}}{\sqrt{V_i}} = \frac{\sqrt{2V_i}}{\sqrt{V_i}}$$

$$\Rightarrow \quad \beta_f = \frac{\beta_i}{\sqrt{2}} = 0.7\,\beta_i$$

$$\Rightarrow \quad \beta_f = 0.7\omega \qquad (\text{given, } \beta_i = \omega)$$

31. *(b)* Electric field lines starts from positive surface and terminates over negative surface. Electric field lines are perpendicular to the surface at point of contact. Electric field between two charged plates in uniform.

So, correct option is (b).

32. *(b)* First excitation energy of a hydrogen atom in ground state is 10.2 eV.

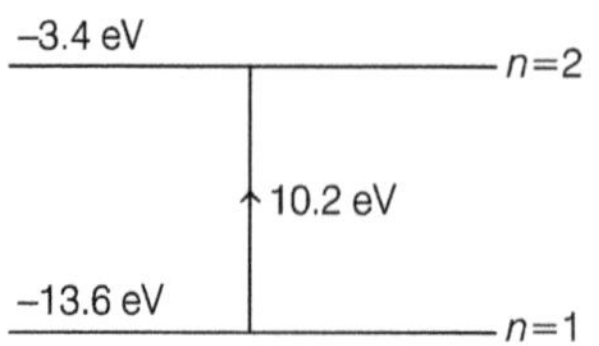

So, if energy of colliding electron is more than 10.2 eV, then hydrogen atom absorbs 10.2 eV from kinetic energy of electron and collision will be inelastic.

If energy of electron is less than 10.2 eV, then hydrogen atom does not absorbs energy and total energy before and after collision remains same and collision will be elastic.

33. *(d)* Continuous X-ray spectrum is produced when an missile electron is scattered from a target atom and energy is released in form of a photon.

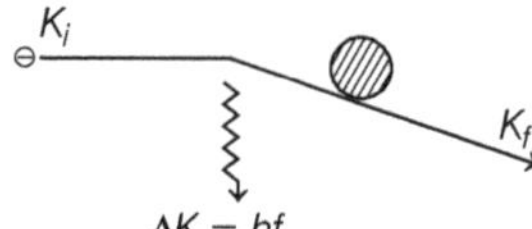

So, this is just opposite of photoelectric effect in which a photon releases an electron.

34. *(b)* An approximate potential energy curve for two particles is as shown below.

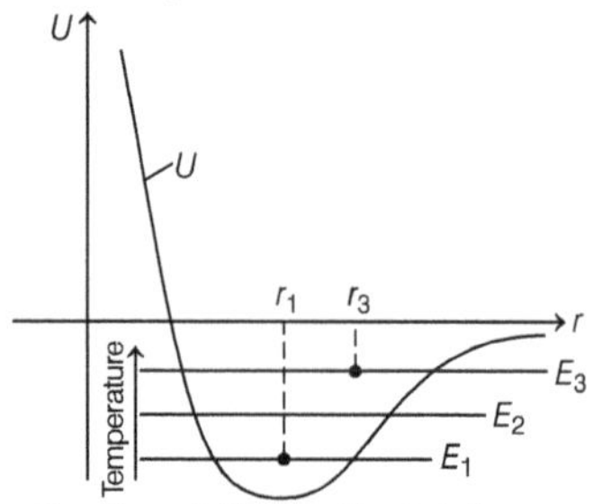

Here E_1, E_2 and E_3 are the total energy values of particles. As temperature increases, total energy E increases.

Hence, mean separation r also increases.

Note That mean separation r is mid point of length of E intercepted by potential energy curve.

35. *(c)* For a photon energy is

$$E = hf = \frac{hc}{\lambda}$$

For an momentum of electron is $p = \dfrac{h}{\lambda}$

So, $\dfrac{E}{p} = \dfrac{hc/\lambda}{h/\lambda} = c$

here, λ is same for both photon and electron.

$$\Rightarrow \quad \frac{E}{p} = 3\times10^8\ \text{ms}^{-1}$$

36. *(a)* If finite mass of proton is taken into account, then in expression of energy of electron, we must introduce reduced mass instead of mass of electron. So, expression for energy will be

$$E_n = -\left(\frac{m_p m_e}{m_e + m_p}\right)\cdot\frac{e^4}{8n^2\epsilon_0^2 h^2}$$

where, m_e = mass of electron, m_p = mass of proton.

We can rewrite above expression as

$$E_n = \frac{-m_p}{m_e + m_p}\cdot\left(\frac{m_e\cdot e^4}{8n^2\epsilon_0^2 h^2}\right)$$

$$= \left(\frac{m_p}{m_e + m_p}\right)(-13.6)$$

So, binding energy is nearly 0.06% less than 13.6 eV.

37. *(b)* Given arrangement is

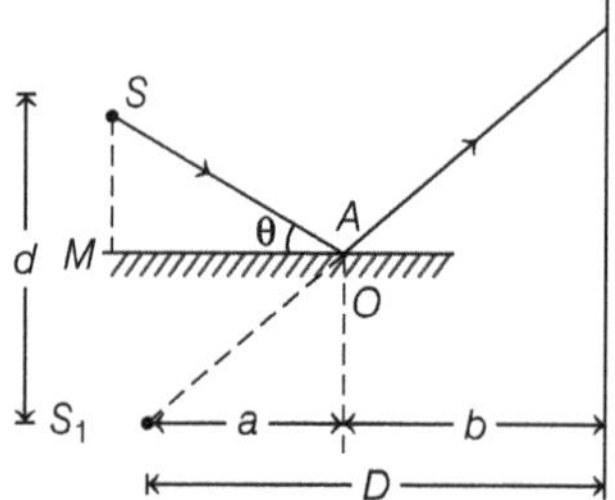

Angle $\theta = 0.5\times10^{-3}$ radians.

So, distance between source S and its image S_1 is

$$d = 2r\theta = 2\times SO\times\theta$$
$$= 2\times20\times0.5\times10^{-3}\ \text{cm}$$
$$= 20\times10^{-3}\ \text{cm} = 2\times10^{-4}\ \text{m}$$

Distance of sources S and S_1 from screen,

$$D = a + b = 20 + 100 = 120\,\text{cm}$$
$$= 120\times10^{-2}\ \text{m}$$

Wavelength of light used is

$$\lambda = 440\,\text{nm} = 440\times10^{-9}\ \text{m}$$

Both S and S_1 acts like slits (sources of coherent light) and produces fringes over screen.

Width of a fringe, $\beta = \dfrac{\lambda D}{d}$

$$\therefore \quad \beta = \frac{440\times10^{-9}\times120\times10^{-2}}{2\times10^{-4}}$$

$$= 2.64\,\text{mm}$$

38. *(b)* Energy generated per unit volume of fuel rod $= 5\times10^8\ \text{Wm}^{-2}$

Total energy generated per second by fuel rod $= 5\times10^8\times$ Volume of rod

$$= 5\times10^8\times\pi\times r^2\times h$$
$$= 5\times10^8\times\pi\times(4\times10^{-3})^2\times0.2$$
$$= 1600\,\pi\ \text{W}$$

Energy absorbed by coolant for per degree rise of temperature is

$= \text{Specific heat} \times \text{Mass flow rate}$

$= 4 \times 10^3 \times 0.2 \frac{W}{K}$

$= 8 \times 10^2 \frac{W}{K}$

If ΔT is rise of temperature of coolant, then equating heat absorbed by coolant and energy generated by fuel rod, we have

$8 \times 10^3 \times \Delta T = 1600\,\pi$

$\Rightarrow \quad \Delta T = \frac{1600 \times \pi}{8 \times 10^2} = 2\pi$

$= 6.28°C \approx 6°C$

39. *(b)* Sound level (in decibel) is defined as

$$\beta = 10 \log_{10}\left(\frac{I}{I_0}\right)$$

where, I = intensity of sound and I_0 = reference intensity ($\sim 10^{-12}\ W/m^2$).

Now, taking antilog, we have

$$\frac{I}{I_0} = 10^{\left(\frac{\beta}{10}\right)} \Rightarrow I = I_0 \times 10^{\left(\frac{\beta}{10}\right)} \quad \text{... (i)}$$

Here, $\beta = 20$ at 1 kHz and $\beta = 60$ at 9 kHz.

Taking a linear relation, $\beta = kf + c$

where, k and c are constants and f = frequency.

So, we have

$20 = k \times 1 + c \quad \text{...(ii)}$

and $\quad 60 = k \times 9 + c \quad \text{...(iii)}$

Subtracting, Eqs. (ii) and (iii), we get

$60 - 20 = 9k - k + c - c$

$40 = 8k + 0$

$8k = 40$

$k = 5$

Putting the value of k in Eq. (ii), we have

$20 = 5 + c \Rightarrow c = 15$

$\Rightarrow \quad c = 15$

$\therefore$ At $f = 5\,\text{kHz}$,

$\beta = kf + c = 5(5) + 15$

$= 40$

Now, width $\beta = 20$ and $\beta = 40$ from Eq. (i), we have

Intensity, $\quad I = I_0 \cdot 10^{\left(\frac{\beta}{10}\right)}$

$\Rightarrow \quad I_{1\,\text{kHz}} = I_0 (10)^{\frac{20}{10}} = I_0 \cdot 10^2$

$I_{5\,\text{kHz}} = I_0 (10)^{\frac{40}{10}} = I_0 \cdot 10^4$

$\Rightarrow \quad \frac{I_{5\,\text{kHz}}}{I_{1\,\text{kHz}}} = \frac{10^4}{10^2} = 100$

40. *(a)* Let magnetic field is minimum at some point P, distant x from first wire.

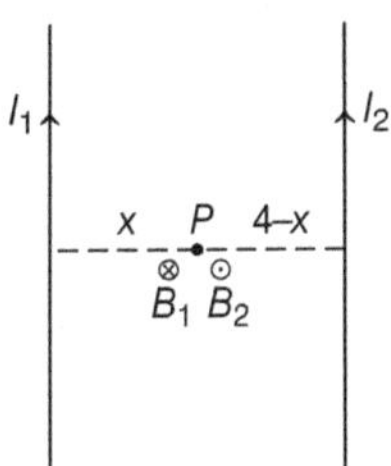

Net magnetic field at P is

$$B = \frac{\mu_0 I_1}{2\pi x \times 10^{-2}} - \frac{\mu_0 I_2}{2\pi (4 - x) \times 10^{-2}}$$

$$= \frac{\mu_0}{2\pi \times 10^{-2}}\left(\frac{I_1}{x} + \frac{I_2}{x - 4}\right)$$

For B to be minimum,

$$\frac{dB}{dx} = 0 \Rightarrow \frac{d}{dx}\left(\frac{I_1}{x} + \frac{I_2}{x - 4}\right) = 0$$

$$\Rightarrow -\frac{I_1}{x^2} - \frac{I_2}{(x-4)^2} = 0 \Rightarrow -\frac{I_1}{x^2} = \frac{I_2}{(x-4)^2}$$

$$\Rightarrow \quad \frac{I_1}{I_2} = -\left(\frac{x}{x - 4}\right)^2$$

with $x = 1\,\text{cm}$, we have

$$\frac{I_1}{I_2} = -\left(\frac{1}{1 - 4}\right)^2 = -\frac{1}{9} \text{ or } \frac{I_2}{I_1} = -9$$

Here, negative sign shows that they are anti-parallel.

41. *(d)* Sulphur has 6 valence electrons. Out of which it forms 4 bond pair with chlorine and one pair remain unbonded. Thus, it has 4 bond pairs and 1 lone pair. So, the shape of SCl_4 is best described as a *see-saw*.

Cl—S—Cl
Cl Cl
(See-saw)

42. *(a)* When orbitals of two atoms come close to form bond, their overlap may be positive or zero depending upon the sign and the direction of orientation of amplitude of orbital wave function in space.

In case of non-bonding overlap, there is no effective overlap where +ve and – ve overlap simultaneously. It can occur in case of p_x and p_z overlap as shown in option (a).

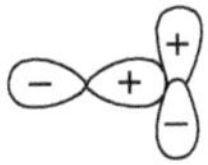

Non-bonding overlap between p_z and p_x orbital

The type of overlaps given in the other options are as follows:

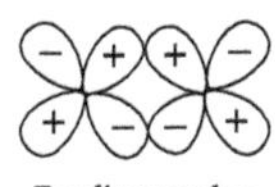

Bonding overlap between $d_{x^2-y^2}$ and $d_{x^2-y^2}$

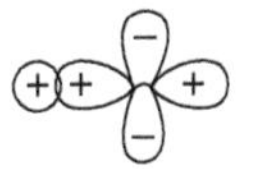

No type of overlap is possible as the combining orbitals do not have comparable energies

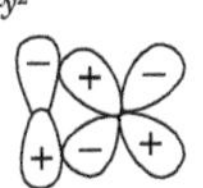

Antibonding overlap between p_x

43. *(c)* Optical activity is shown by those complexes, which have optical isomers. Optical isomers are mirror images that cannot be superimposed on each other. Complexes of type $[M(AA)_3]$ and $cis[M(AA)_2B]_2$ shows optical isomerism.

The type of complexes given in the options are as follows :

Complex	Type
$[CoCl_6]^{3-}$	$[MA_6]$
$[Co(en)Cl_4]^-$	$[M(AA)B_4]$
cis $[Co(en)_2Cl_2]^+$	*cis* $[M(AA)_2B_2]$
trans $[Co(en)_2\,Cl_2]^+$	*trans* $[M(AA)_2B_2]$

Thus, *cis* $[Co(en)_2Cl_2]^+$ shows optical activity.

44. *(b)* pK_a value of an acid is inversely proportional to acidic strength, i.e. stronger is the acid, lesser will be its pK_a value.

Also, the acidic strength increases with increase in the oxidation number of central atom.

The oxidation number of central atom in given acids are given below.

Acid	Oxidation Number of central atom
HClO	+1
$HClO_3$	+5
$HClO_2$	+3
$HClO_4$	+7

The acidic strength follows the order,

$HClO < HClO_2 < HClO_3 < HClO_4$

Thus, the pK_a of oxoacids of chlorine in water,

$HClO_4 < HClO_3 < HClO_2 < HClO$

45. *(a)* The packing efficiency of cubic lattice is given by formula

$$PE = \frac{\text{Volume occupied by spheres in lattice} \times 100}{\text{Total volume of the unit cell}}$$

Packing efficiency of fcc = 74%

Packing efficiency of bcc = 68%

Packing efficiency of pc = 52.4%

Thus, packing efficiency follow the order

fcc > bcc > pc

46. *(a)* Root mean square velocity of a gas is given by formula, $v_{rms} = \sqrt{\frac{3RT}{M}}$

where, R is gas constant

T is temperature

M is molar mass of gas

For hydrogen gas (H_2)

$M = 2g, T = 50\,K$

$\therefore (v_{rms})_{H_2} = \sqrt{\frac{3 \times R \times 50}{2}}$

For nitrogen gas (N_2)

$M = 28g, T = 500$

$(v_{rms})_{N_2} = \sqrt{\frac{3 \times R \times 500}{28}}$

$$\therefore \frac{(v_{rms})_{H_2}}{(v_{rms})_{N_2}} = \frac{\sqrt{\frac{3 \times R \times 50}{2}}}{\sqrt{\frac{3 \times R \times 500}{28}}} = 1.18$$

47. *(d)* The dipole moment of a molecule depends upon the electronegativity difference of the different atoms and on the geometry of a molecule.

Symmetrical molecules have zero dipole moment. As in all the given cases molecules are unsymmetrical, so they will have non-zero dipole moment. Therefore, here dipole moment depends upon the electronegativity difference of atoms.

Among the given molecules HF has the highest dipole moment, because there is large difference in electronegativity between H and F atom.

NH_3: $\mu = 1.47$ D NF_3: $\mu = 0.23$ D $C\equiv O$: $\mu = 0.122$ D $H—F$: $\mu = 1.78$ D

[The orbital dipole is in the direction opposite to the resultant dipole moment of three N—F]

48. *(c)* A complex formed by the dative bond between Lewis acid and Lewis base is called Lewis acid-base adduct. In all the given cases Lewis acid, i.e. BCl_3 is same. So, stability depends on Lewis base.

Among the given options $H_3N \rightarrow BCl_3$ form stable Lewis base adduct, because N of NH_3 contains a lone pair and is less electronegative, thus it can donate its lone pair into empty orbital of BCl_3 (Lewis acid) to form effective dative bond ($p\pi$-$p\pi$).

In option (a), O is highly electronegative, so it will not donate its lone pair to BCl_3.

In option (b) and (d), the formed dative bond $d\pi$-$p\pi$ is less effective than $p\pi$-$p\pi$ in case of $NH_3 \rightarrow BCl_3$.

49. *(c)* The reaction of D-glucose with ammoniacal $AgNO_3$, which is tollen's reagent produces D-gluconic acid. This reaction show that glucose contains free aldehydic group, which oxidises into carboxylic acid and thus regarded as reducing sugar.

D-glucose (CHO; H—OH; HO—H; H—OH; H—OH; CH_2OH) $\xrightarrow{AgNO_3}$ D-gluconic acid (COOH; H—OH; HO—H; H—OH; H—OH; CH_2OH)

50. *(b)* The reagent used for the conversion of benzene diazonium hydrogen sulphate to benzene is aqueous solution of phosphinic acid, which is a reducing agent.

Benzene diazonium sulphate $[C_6H_5\overset{+}{N}\equiv N]HSO_4^-$ $\xrightarrow{H_3PO_2 + H_2O}$ Benzene $+ N_2\uparrow + H_2O$

51. *(c)*

Toluene + $CH_3—CH(CH_3)—CH_2Br$ → p-$CH_3C_6H_4C(CH_3)_3$

This reaction is known as Friedel-Crafts alkylation, where alkyl group.

$(CH_3—\overset{+}{C}(CH_3)—CH_3)$

Which acts as an electrophile is added to toluene by an electrophilic aromatic substitution.

The mechanism of this reaction is as follows :

$CH_3—CH(CH_3)—CH_2\ddot{Br}: + AlCl_3 \rightarrow (CH_3)_2CHCH_2\overset{+}{Br}—\overset{-}{AlCl_3}$ $\xrightarrow{-AlCl_3Br^-}$ $(CH_3)_2CH\overset{+}{C}H_2$ (1° carbocation) $\xrightarrow{H^+ \text{ shift}}$ $(CH_3)_2\overset{+}{C}—CH_3$ (3° carbocation (more stable))

Toluene + $(CH_3)_3\overset{+}{C}$ (Electrophile) $\xrightarrow[ArS_E]{Slow}$ p-tert-butyltoluene

52. *(b)*

Salicylic acid $\xrightarrow[CH_3COOH]{(CH_3CO)_2O}$ Acetyl salicyclic acid

(Salicyclic acid) (Acetyl salicyclic acid)

The reactant given in above reaction is salicyclic acid, which undergoes acetylation to produce acetyl salicyclic acid, which is commonly known as aspirin. It is a drug used to treat pain, fever or inflammation.

53. *(b)* The hybridisation of carbon in given compounds are as follows :

(I) $\overset{sp^3}{CH_3}—N(\overset{sp^3}{CH_3})—\overset{sp^2}{C}(=O)—H$ (II) Pyridine (sp^2)

(III) $\overset{sp^3}{H_3C}—\overset{sp}{C}\equiv N$ (IV) $H_2\overset{sp^2}{C}=\overset{sp}{C}=\overset{sp^2}{C}H—\overset{sp^3}{CH_3}$

Thus, compounds (III) and (IV) contain sp-hybridised carbon atom.

54. *(b)* The starting compound is cyclohexene, which upon heating with acidic $KMnO_4$ produces hexan-1,6-dioic acid as a major product.

Cyclohexene $\xrightarrow[{[O]}]{\text{Acidic } KMnO_4}$ OHC—$(CH_2)_4$—CHO $\xrightarrow{[O]}$ HOOC—$(CH_2)_4$—COOH

Note Alkene on reaction with cold, aqueous solution of $KMnO_4$ produces *vicinal* glycols whereas, hot acidic solution of $KMnO_4$ oxidises alkenes to ketone which further gets oxidised to acids.

55. *(d)* For Ist order reaction,

$$t_{1/2} = \frac{0.693}{k}$$

$$1 = \frac{0.693}{k}$$

$$\therefore \quad k = \frac{0.693}{1}$$

Also,

$$t = \frac{2.303}{k} \log \frac{a}{a-x}$$

For 87.5% completion

$$t = \frac{2.303}{0.693} \log \frac{a}{a-0.875}$$

$$= \frac{2.303}{0.693} \log 8$$

$$= \frac{2.303}{0.693} \times 3 \times 0.3010 = 3 \text{ h}$$

56. *(d)* Calcium fluoride, CaF_2 is a type of AB_2 type crystal lattice. The calcium ions form the ccp arrangement and fluoride ions occupy tetrahedral sites. Thus, the coordination number of CaF_2 is 8 : 4. So, number of F^- ions will be 8 as for Ca^{2+} is surrounded by two F^- ions.

57. *(b)* Given,

$$K_{eq} = 3 \times 10^{-3}$$
$$n = 2$$
$$T = 298 \text{ K}$$

$$\Delta G = -RT \ln K_{eq}$$

or $\Delta G = -2.303 \times RT \log K_{eq}$

$$= -2.303 \times 8.314 \times 298 \log 3 \times 10^{-3}$$
$$= 13809.3876 \text{ J}$$
$$= 13.81 \text{ kJ}$$

Also, $\Delta G = -nFE°$

$$13809.387 = -2 \times 96500 \times E°$$
$$E° = -0.071 \text{ V}$$

58. *(c)* Total number of possible stereoisomers in an unsymmetrical compound = Number of optical isomers + Number of geometrical isomers.

In given compound, there is one chiral carbon.

$$CH_3—CH{=}CH—\overset{*}{C}H(Br)—CH_2—CH_3$$

$$n = 1$$

$\therefore$ Number of optical isomers = 2

Also, the compound contains a double bond, thus it will have two geometrical isomers also (*cis* and *trans* form).

Thus, total number of stereoisomers = 2 + 2 = 4

59. *(b)* An α-particle corresponds to helium atom. It *trans*-forms or decays into a different atomic nucleus with mass number that is reduced by four and an atomic number is reduced by two. An α-particle is identical to the nucleus of a helium-4-atom. In β-disintegration the atomic number is reduced by one unit.

For the reaction,

$$^{232}_{90}Th \rightarrow {}^{208}_{82}Pb + {}^{4}_{2}x\alpha + {}^{0}_{-1}y\beta$$

Let the number of α-particles be x and the number of β-particles by y.

$$\therefore \quad 232 = 208 + 4x$$
$$x = 6$$

and

$$90 = 82 + 2x - y$$
$$90 = 82 + 12 - y$$
$$y = 4$$

60. *(a)* Compressibility factor, $Z = \frac{pV}{nRT}$

For ideal gas $Z = 1$ (at all temperature and pressures)

For real gas, $Z > 1$ (At high pressures) or $Z < 1$ (At low pressures)

For real gas, $Z = \frac{pV_{real}}{nRT}$ ...(i)

If the gas shows ideal behaviour, then

$$V_{ideal} = \frac{nRT}{p}$$

or $\frac{p}{nRT} = \frac{1}{V_{ideal}}$ ...(ii)

Putting (ii) in (i)

$$Z = \frac{V_{real}}{V_{ideal}}$$

If we consider molar volume, then

$$Z = \frac{(V_m)_{real}}{(V_m)_{ideal}}$$

As in given case

$$Z = \frac{(V_m)_{real}}{(V_m)_{ideal}} < 1 \text{ at } p < 200 \text{ bar}$$

$\therefore (V_m)_{CH_4}$ real $< (V_m)_{CH_4}$ ideal

61. *(c)* The primary carbon dioxide acceptor in C_4-plant is phosphoenol pyruvate, it contains 3 carbon atoms, e.g. maize, sugarcane, etc. In C_3-plants, Ribulose 1, 5 Bisphosphate (RuBP) is the primary CO_2 acceptor, it is a 5 carbon compound, e.g. in beans, rice, etc. Thus, according to question, option (c) is correct.

62. *(d)* Octane is considered to be non-polar, it will not be soluble in water. Since, water is a polar solvent. This will happen because octane (hydrocarbons in general) contains neither ionic groups, non-polar functional groups that can interact with water molecules. Thus, option (d) is correct for the fact that water and octane does not form hydrogen bond between them.

63. *(b)* Lactase is an enzyme produced by many organisms. Lactase (also known as Lactase Phlorizin Hydrolase, or LPH) a part of the β-galactosidase family of enzymes, is a glycoside hydrolase involved in the hydrolysis of the disaccharide lactase into its constituents galactose and glucose monomers.

64. *(d)* Only statement (d) is incorrect and can be corrected as lipids and membrane proteins provide structural and functional asymmetry to biological membrane. This asymmetry allows the membrane to be rigid and permits a different intra-cellular environment from the existing extra-cellular environment.

65. *(a)* Most solar energy occurs at wavelengths unsuitable for photosynthesis. Between 90–98% of solar energy reaching on the earth is reflected from leaves and other surfaces and absorbed by other molecules, which convert it, to heat. Thus, only 2–10% of sunlight is available to be captured by plants.

66. *(c)* Sporopollenin is one of the most chemically inert biological polymers. It is a major component of the tough outer (exine) walls of plant spores and pollen grains. It is chemically very stable and is usually well-preserved in soils and sediments.

67. *(b)* Insectivorous plants trap, digest and absorb insects in order to supplement their nitrogen deficiency. Since, these plants live in marshy areas, they do not get nitrogen from the soil. Their nitrogen need is fulfilled by sucking the juice of insects.

68. *(c)* A structural unit of a eukaryotic chromosome, consisting of a length of DNA coiled around a core of histone proteins (of four types, i.e. H_2A, H_2B, H_3 and H_4) is called a nucleosome.

69. *(a)* Gap junction is an organised collection of protein channels in cell membranes that allows ions and small molecules to pass between adjacent cells. The protein channels that make up gap junctions consist of two connexons. One connexon resides in the membrane of one cell.

70. *(b)* Glandular epithelium are modified columnar epithelial tissues. They are large cells that are present in the tear glands, sweat glands, salivary glands, etc. Their main function is secretion. The glandular epithelium can have a cluster of either unicellular cells or multicellular cells.

71. *(a)* Ions that are important in the formation of a nerve impulse include sodium (Na^+) and potassium (K^+) ions. An action potential is a very rapid change in membrane potential that occurs when a nerve cell membrane is stimulated.

72. *(b)* Atrial Natriuretic Peptide (ANP) or Atrial Natriuretic Factor (ANF) is a natriuretic peptide hormone secreted from the cardiac atria. The main function of ANP is causing a reduction in expanded Extracellular Fluid (ECF) volume by increasing renal sodium excretion, i.e. it controls blood pressure.

73. *(a)* Oxytocin and Antidiuretic Hormone (ADH) or vasopressin are hormones synthesised by the magnocellular neurons located in the supraoptic and paraventricular nuclei of the hypothalamus delivered to the nerve terminals in posterior pituitary before secretion.

74. *(b)* The air, we breathe out or exhale contains CO_2

$$\underset{\text{Lime water}}{Ca(OH)_2} + \underset{\text{Exhaled air}}{CO_2} \longrightarrow \underset{\text{Milky white ppt}}{CaCO_3} + H_2O$$

On reaction with CO_2, the lime water will turn milky.

75. *(d)* The nerve pathway followed by a reflex action is called a reflex arc, e.g. a simple reflex arc happens if we accidentally touch something hot. Receptor in the skin detects a stimulus (the changes in temperature) and sends it to spinal cord. Effector produces a response (muscles contract to move hand away). Thus, the path is

receptor → spinal cord → muscles.

76. *(a)* Whenever glucose is present, *E. coli* metabolises it before using arabinose energy source such as lactose, arabinose, galactose and maltose. The enzymes to metabolise glucose are made constantly by *E.coli*. When both glucose and lactose are available, the genes for lactose metabolism are transcribed at low levels because of the presence of permease in the cell.

77. *(d)* Passive immunity is the transfer of active humoral immunity of readymade antibodies. Passive immunisation is used when there is a high risk of infection and insufficient time for the body to develop its own immune response or to reduce the symptoms of ongoing or immunosuppressive diseases.

78. *(c)* Sodium bicarbonate (HCO_3^-) present in pancreatic juice neutralises the acidity of the chyme to prevent damage to the walls of the duodenum and provides a neutral pH environment for the digestion of chyme.

79. *(d)* Photosynthesis occurs in two steps, i.e. light reaction (Calvin cycle) and dark reaction. During light reaction, CO_2 is accepted in the first step and reacts with RuBisCO to form 3-phosphoglycerate (3-PG). Thus, if $^{14}CO_2$ is added in the suspension, the first compound to be radioactive is 3-phosphoglycerate.

80. *(b)* *Rafflesia arnoldii*, commonly called the 'corpse lily', is a species of flowering plant in the parasitic genus *Rafflesia*. It is noted for producing the largest flower on the earth. It has a very strong and unpleasant odour of decaying flesh, earning it the nickname 'corpse flower'.

81. *(d)* Let

$$P(x) = 1 + x^2 + x^4 + x^6 + \dots + x^{22}$$
$$Q(x) = 1 + x + x^2 + x^3 + \dots + x^{11}$$
$$P(x) = \frac{x^{24}-1}{x^2-1}$$
$$\left[\because a + ar + ar^2 + \dots + ar^{n-1} = \frac{a(r^n-1)}{r-1}\right]$$

Similarly, $Q(x) = \dfrac{x^{12}-1}{x-1}$

$$\therefore \quad \frac{P(x)}{Q(x)} = \left(\frac{x^{24}-1}{x^2-1}\right)\left(\frac{x-1}{x^{12}-1}\right) = \frac{x^{12}+1}{x+1}$$

Now, $(1+x^{12})$ is divided by $(x+1)$, then by remainder theorem remainder is 2.

Now, $P(x)$ is divided by $Q(x)$, then remainder $= 2(1+x^2)(1+x^4+x^8)$

$$= 2(1 + x^2 + x^4 + \dots + x^{10})$$

82. *(c)* We have,

$$P(x) = 4x^3 - 3x \; x \in \left(-\frac{1}{2}, \frac{1}{2}\right)$$
$$P'(x) = 12x^2 - 3 = 3(4x^2 - 1)$$
$$P'(x) < 0 \; x \in \left(-\frac{1}{2}, \frac{1}{2}\right)$$

$\Rightarrow P(x)$ is decreasing $\left(-\frac{1}{2}, \frac{1}{2}\right)$

Range $\in P(-1), P(1)$

$$P(-1) = -4 + 3 = -1$$
$$P(1) = 4 - 3 = 1$$

$\because$ Range $(-1, 1)$.

83. *(c)* At time $t = 0$, kth ant starts at point k^2 and reaches at time $t = 0$ at the point $(11-k)^2$.

$\therefore$ Velocity of kth ant

$$= \frac{x_2 - x_1}{t_2 - t_1} = \frac{(11-k)^2 - k^2}{1-0}$$
$$u = 121 - 12k$$

Now, two ants are at the same location

$$\therefore \quad x_i = x_j$$
$$x_i = x + ut$$
$$k_i^2 - 22k_it + 121t = k_j - 22k_jt + 121t$$
$$\Rightarrow \quad t = \frac{k_j^2 - k_i^2}{22(k_j - k_i)} = \frac{k_j + k_i}{22} \quad [k_i \neq k_j]$$

Now, for $i = 1$,

values of t will be $\frac{3}{22}, \frac{4}{22}, \frac{5}{22}, \dots, \frac{11}{22}$ (9 values)

When $i = 2$,

value of t will be $\frac{4}{22}, \frac{5}{22}, \dots, \frac{11}{22}, \frac{12}{22}$

Only one distinct value.

Similarly, for $i = 3, 4, 5, 6, 7, 8, 9$, we get only 1 distinct value.

So, in all there are 17 distinct value of t.

84. *(a)*

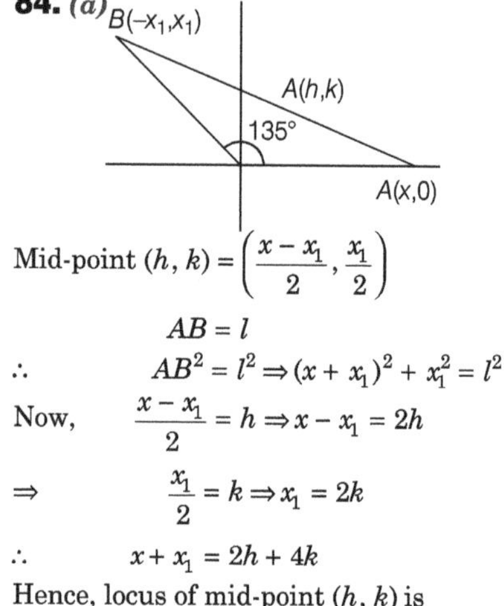

Mid-point $(h, k) = \left(\frac{x - x_1}{2}, \frac{x_1}{2}\right)$

$$AB = l$$
$$\therefore \quad AB^2 = l^2 \Rightarrow (x + x_1)^2 + x_1^2 = l^2$$
$$\text{Now,} \quad \frac{x - x_1}{2} = h \Rightarrow x - x_1 = 2h$$
$$\Rightarrow \quad \frac{x_1}{2} = k \Rightarrow x_1 = 2k$$
$$\therefore \quad x + x_1 = 2h + 4k$$

Hence, locus of mid-point (h, k) is

$$(2h + 4k)^2 + 4k^2 = l^2$$
$$\Rightarrow \quad x^2 + 5y^2 + 4xy = \frac{l^2}{4}$$
$$\therefore \quad \text{Area of ellipse} = \frac{\pi l^2}{4}$$

85. *(d)* Given,

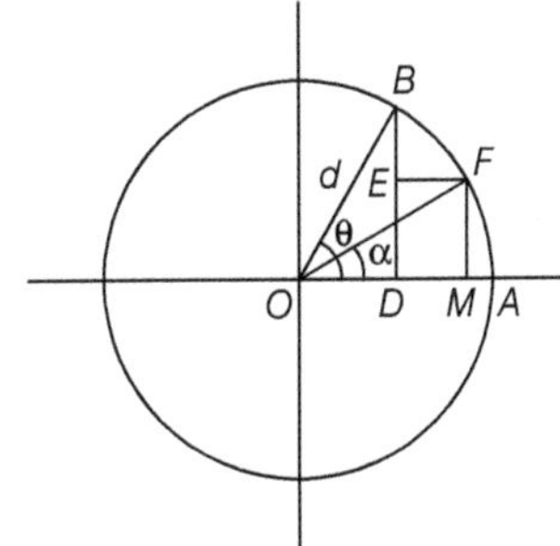

$\angle AOB = \theta$

$OA = OB = OF = d$

OA is perpendicular to BD.

E is mid-point of BD.

$\therefore \quad BE = ED$

EF is parallel to OA.

In ΔODB,

$$\sin\theta = \frac{BD}{d}$$

$$\Rightarrow \quad BD = d\sin\theta$$

$$\Rightarrow \quad DE = FM = \frac{1}{2}BD = \frac{d}{2}\sin\theta$$

In ΔOFM,

$$\sin\alpha = \frac{FM}{OF} = \frac{\frac{d}{2}\sin\theta}{d}$$

$$\sin\alpha = \frac{\sin\theta}{2}$$

$$\Rightarrow \quad \alpha = \sin^{-1}\left(\frac{\sin\theta}{2}\right)$$

$\therefore$ Ratio of length of arc AF to length

$$AB = \frac{\alpha}{\theta} = \frac{\sin^{-1}\left(\frac{1}{2}\sin\theta\right)}{\theta}$$

86. *(a)* We have,

$f(x)$ is non-negative differentiable function on $[0, \infty)$

$$f(0) = 0, f'(x) \le 2f(x)$$

$$\Rightarrow \quad f'(x) \le 2f(x)$$

$$\Rightarrow \quad \log f(x) \le 2x + c$$

$$\Rightarrow \quad f(x) \le Ae^{2x}$$

$$\Rightarrow \quad f(0) \le A$$

$$\Rightarrow \quad A = 0 \quad [f(0) = 0 \text{ and } f(x) \ge 0]$$

$$\therefore \quad f(x) = 0$$

Hence, $f(x)$ is always a constant function.

87. *(c)* We have,

$$|\sin\sqrt{n+1} - \sin\sqrt{n}| < \lambda, \ \lambda \in R$$

When $n \to \infty$

$$|\sin\sqrt{n+1} - \sin\sqrt{n}| \to 0$$

$\therefore$ There exist infinite natural number for which $|\sin\sqrt{n+1} - \sin\sqrt{n}| < \lambda, \forall \lambda \in 0$

Hence, $A_{1/2}, A_{1/3}, A_{2/5}$ are infinite sets.

$\therefore A^c_{1/2}, A^c_{1/3}, A^c_{2/5}$ are finite sets.

88. *(d)* We have,

$$\int_0^1 f^2(x)\,dx = \left(\int_0^1 f(x)\,dx\right)^2, x \in [0, 1]$$

We know Cauchy Schwartz inequality

$$\left[\int_a^b f(x)\cdot g(x)\,dx\right]^2 \le \int_a^b (f(x))^2 dx \int_a^b (g(x))^2 dx$$

Here, $g(x) = 1$ and equality holds only when $\frac{f(x)}{g(x)} = \lambda$

So, $f(x)$ is constant function.

$\therefore f(x)$ is a singleton.

89. *(c)* We have 2, 4, 6 students from three schools respectively.

6 rooms allotted for 1, 2, 3, 4, 5, 6 such that each room has exactly two students for same schools.

$\therefore$ Total number of ways student can be accommodated in the rooms

$$= \frac{4!}{2! \times 2! \times 2!} \times \frac{6!}{2! \times 2! \times 2! \times 3!} \times 6!$$

$$= 3 \times 15 \times 720$$

$$= 32400$$

90. *(c)* We have,

$$w = \left(\frac{z-a}{1-\bar{a}z}\right), |a| < 1$$

$$\Rightarrow \quad w(1-\bar{a}z) = z - a$$

$$\Rightarrow \quad (w - w\bar{a}z) = z - a$$

$$\Rightarrow \quad w + a = z(1 + w\bar{a})$$

$$\Rightarrow \quad z = \frac{w+a}{1+w\bar{a}}$$

When $|z| < 1$

$$\left|\frac{w+a}{1+w\bar{a}}\right| < 1$$

$$\Rightarrow \quad |w + a| < |1 + w\bar{a}|$$

$$\Rightarrow (w+a)(\bar{w}+\bar{a}) < (1+w\bar{a})(1+\bar{w}a)$$

$$\Rightarrow \quad w\bar{w} + w\bar{a} + a\bar{w} + a\bar{a} < 1 + \bar{w}a + w\bar{a} + w\bar{w}a\bar{a}$$

$$\Rightarrow \quad |w|^2 + |a|^2 < 1 + |w|^2|a|^2$$

$$\Rightarrow \quad |w|^2|a|^2 - |w|^2 - |a|^2 + 1 > 0$$

$$\Rightarrow \quad (|w|^2 - 1)(|a|^2 - 1) > 0$$

$$\Rightarrow \quad |w|^2 - 1 < 0 \quad [\because |a| < 1]$$

$$\therefore \quad |w| < 1, |z| < 1$$

91. *(c)* When balloon is slightly displaced horizontally component of buoyant force produces a torque about the end of string attached to ground. This torque produces side ways oscillations of balloon.

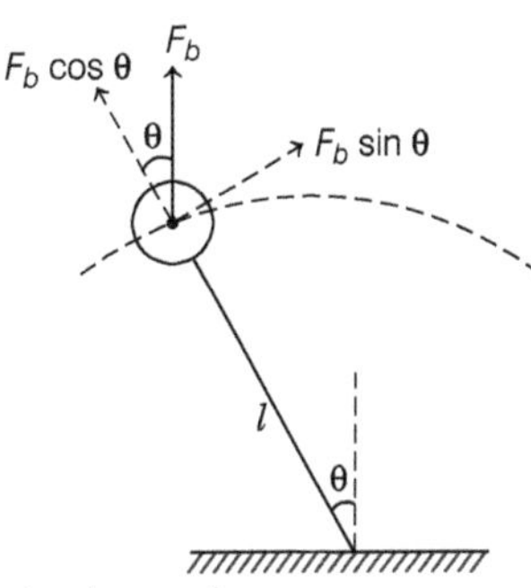

Restoring torque is

$$\tau_1 = F_b \sin\theta \times l = V(\rho_{air} - \rho_{He})\, gl\sin\theta$$
$$= V(\rho_{air} - \rho_{He})\, gl\theta$$

For small angular displacements, $\sin\theta \approx \theta$.

And the inertial torque an balloon is

$$\tau_2 = I\alpha = ml^2\alpha = V\cdot\rho_{He}\cdot l^2\cdot\alpha$$

Helium balloon can be viewed as a mass tied to end of a string.

Equating both torques, we have

$$V\cdot\rho_{He}\cdot l^2\cdot\alpha = -V(\rho_{air} - \rho_{He})l\theta g$$

(Negative sign appears as both torques are in opposite directions)

$$\alpha = -\left(\frac{\rho_{air} - \rho_{He}}{\rho_{He}}\right)\cdot\frac{g}{l}\cdot\theta$$

So, if ω is angular frequency, then

$$\omega^2 = \left(\frac{\rho_{air} - \rho_{He}}{\rho_{He}}\right)\cdot\frac{g}{l}$$

$\therefore$ Time period of oscillations of balloon is

$$T = \frac{2\pi}{\omega} = 2\pi\sqrt{\left(\frac{\rho_{He}}{\rho_{air} - \rho_{He}}\right)\cdot\frac{l}{g}}$$

92. *(a)*

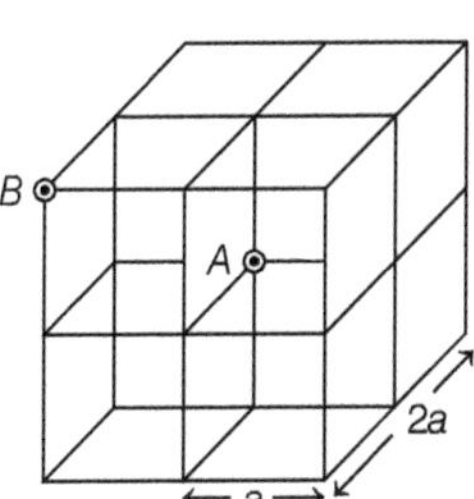

For any charge distribution,

$$\text{Potential} \propto \frac{\text{Charge}}{\text{Distance}}$$

So, potential due to a cube at its corner is

$$V = \frac{C\cdot Q}{a}$$

where, $a =$ side length, $Q =$ charge and $C =$ constant.

Now, consider point A is centre of cube.

Potential at point $A = 8 \times$ Potential due to a cube of side a and charge density ρ.

$\therefore \quad V_A = \dfrac{8C\cdot Q}{a} = \dfrac{8C\cdot\rho\cdot a^3}{a}$

where, ρ = charge density.

$\Rightarrow \quad V_A = 8C\rho\cdot a^2 \quad \text{... (i)}$

Now, potential at point B = potential due to a cube of side $2a$ and charge density ρ.

$\Rightarrow \quad V_B = \dfrac{C\cdot Q}{2a} = \dfrac{C\cdot\rho\cdot(2a)^3}{2a}$

$\Rightarrow \quad V_B = 4C\cdot\rho\cdot a^2 \quad \text{... (ii)}$

So, required ratio is $\dfrac{V_A}{V_B} = \dfrac{8C\rho a^2}{4C\rho a^2} = 2.$

93. *(d)*

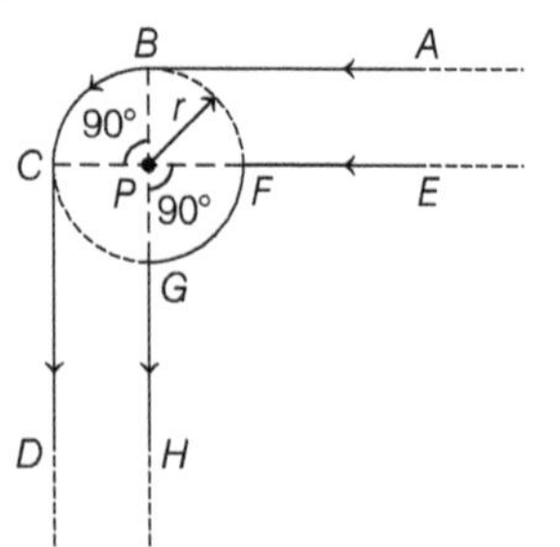

In given arrangement, net magnetic field at point P is vector sum of magnetic fields of wire AB, arc BC, wire CD, wire EF, arc FG and wire GH.

So, $B_{net} = B_{AB}\odot + B_{BC}\odot + B_{CD}\odot + B_{EF} + B_{FG}\otimes + B_{GH}$

As, P lies on axis of wires FE and GH.

So, $B_{FE} = B_{GH} = 0$

Also, magnetic fields of equal arcs BC and FG are in opposite directions.

So, $B_{BC}\odot + B_{FG}\otimes = 0$

So, $B_{net} = (B_{AB} + B_{CD}) = \dfrac{\mu_0 I}{4\pi r} + \dfrac{\mu_0 I}{4\pi r} = \dfrac{\mu_0 I}{2\pi r}$ T

94. *(c)* For a particle like electron, confined to a box of side length a is identical to a standing wave on a string of length a.

So, wavelength λ associated with electrons must have a value,

$$\frac{1}{2}\lambda = a, \frac{a}{2}, \frac{a}{3} \dots \frac{a}{n} \text{ or } \lambda = \frac{2a}{n}$$

Possible values of momentum of electron are $p = \dfrac{h}{\lambda} = \dfrac{hn}{2a}$

So, energy of electron corresponding to these values of momentum is

$$E = \frac{p^2}{2m} = \frac{n^2h^2}{2m\cdot 4a^2}$$

Here, given $\quad a^2 = \dfrac{35h\lambda}{8mc}$

So, $\quad E = \dfrac{n^2h^2}{8m\left(\dfrac{35\,h\lambda}{8mc}\right)} = \dfrac{n^2hc}{35\lambda} \quad \text{...(i)}$

Now, given when this electron absorbs a photon of wavelength $\lambda \left(\text{or energy} = \dfrac{hc}{\lambda}\right)$, it transits from state 1 to n.

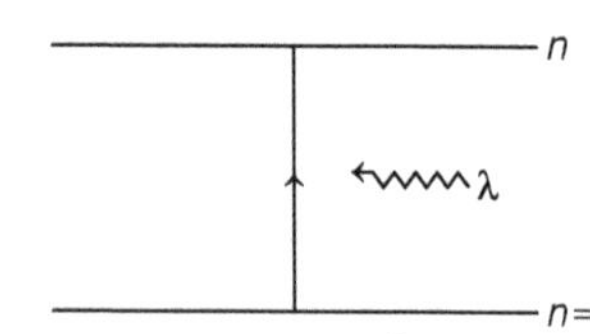

So, for electron $\quad \Delta E = \dfrac{hc}{\lambda}$

$\Rightarrow \quad E_n - E_1 = \dfrac{hc}{\lambda}$

$\Rightarrow \quad \dfrac{(n^2-1)hc}{35\lambda} = \dfrac{hc}{\lambda} \quad$ [from Eq. (i)]

$\Rightarrow \quad n^2 - 1 = 35 \Rightarrow n^2 = 36$

$\therefore \quad n = 6$

Now, given that electron makes transition $6 \to m$ by emitting photon of wavelength $\lambda' = 175\lambda$.

So, $\quad E_{n=6} - E_m = \dfrac{hc}{\lambda'}$

$\dfrac{36hc}{35\lambda} - \dfrac{m^2hc}{35\lambda} = \dfrac{hc}{175\lambda} \quad$ [Using (i)]

$\Rightarrow \quad \dfrac{36-m^2}{35} = \dfrac{1}{1.75}$

$\Rightarrow \quad 36 - m^2 = \dfrac{35}{175}$

$\Rightarrow \quad 36 - m^2 = 20 \Rightarrow m^2 = 16$

$\therefore \quad m = 4$

95. *(a)*

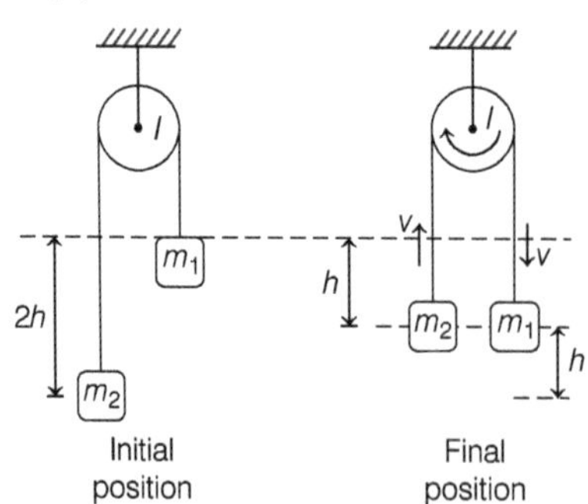

Loss of potential energy of m_1 appears as kinetic energies of m_1, m_2 and pulley and also as potential energies of m_1 and m_2. So, by energy conservation, we get

$$m_1gh = \frac{1}{2}m_1v^2 + \frac{1}{2}m_2v^2 + \frac{1}{2}I\omega^2 + m_2gh$$

Here, m_1 falls by distance h, m_2 rises by distance h, v = speed of m_1 = speed of m_2 as they passes each other and ω = angular speed of pulley = $\dfrac{v}{R}$.

So, $(m_1 - m_2)gh = \dfrac{v^2}{2}\left(m_1 + m_2 + \dfrac{I}{R^2}\right)$

$\therefore \quad v = \dfrac{2gh\,(m_1 - m_2)}{\left(m_1 + m_2 + \dfrac{I}{R^2}\right)}$

or $\quad v \propto \sqrt{\left(\dfrac{m_1 - m_2}{m_1 + m_2 + \dfrac{I}{R^2}}\right)}$

96. *(a)* In given cycle

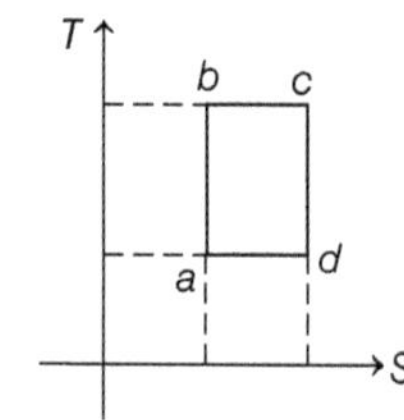

bc and ad are isothermal processes (T and U are constants). ab and cd are isentropic processes (S = constant).

So, plot of above cycle on U-V diagram is

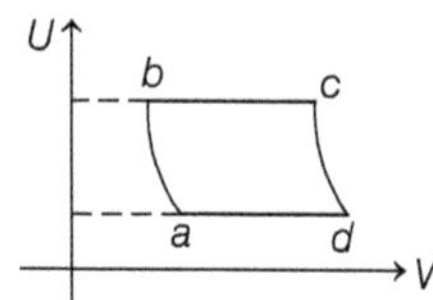

97. *(*)* In an adiabatic process, specific heat at constant volume is given by

$$C_V = \frac{f}{2}.R$$

Here, $\quad C_V = \dfrac{3}{2}R\,(1 + aRT)$

$\Rightarrow \quad \dfrac{fR}{2} = \dfrac{3R}{2}(1 + aRT)$

$\Rightarrow \quad f = 3\,(1 + aRT)$

Now, $\quad \gamma = 1 + \dfrac{2}{f}$

$\therefore \quad \gamma = 1 + \dfrac{2}{3 + 3\,aRT}$

So, process equation is $TV^{\gamma-1}$ = constant.

$\Rightarrow \quad TV^{\left(\frac{2}{3+3aRT}\right)}$ = constant

98. *(c)* Given, circuit is a captive resistive circuit. So, current leads the voltage drop across capacitor.

Also, phase difference ϕ of output voltage V_0 decreases with increasing frequency.

As, $X_c \propto \dfrac{1}{\omega}$ and $\tan\phi \propto X_c$

So, correct graph is

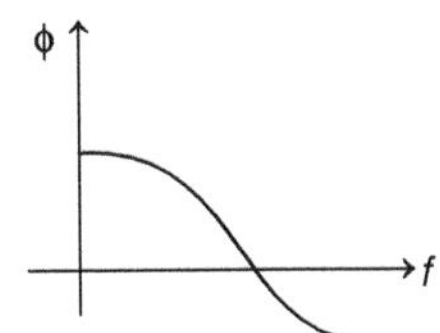

99. *(d)* When light is made incident over face AB, refraction occurs as shown below.

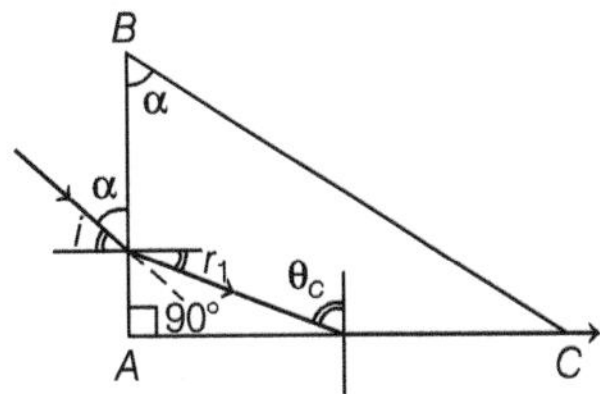

From geometry of figure,

$$r_1 = 90 - \theta_c$$
$$i = 90 - \alpha$$

Refractive index μ of prism is

$$\mu = \frac{1}{\sin\theta_c} = \frac{\sin i}{\sin r_1}$$

$$\Rightarrow \quad \mu = \frac{1}{\sin\theta_c} = \frac{\sin(90-\alpha)}{\sin(90-\theta_c)}$$

$$\Rightarrow \quad \mu = \frac{1}{\sin\theta_c} = \frac{\cos\alpha}{\cos\theta_c} \quad \ldots(\text{i})$$

Also, when incidence is made over face AC, refraction occurs as shown below.

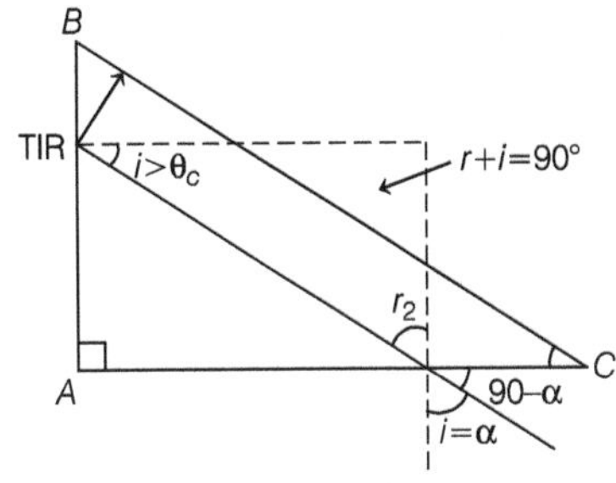

Again, from geometry of figure,

At surface AB, TIR occurs

$$\Rightarrow \quad i > \theta_c$$
$$\Rightarrow \quad 90 - r > \theta_c$$
$$\Rightarrow \quad 90 - \theta_c > r_2$$
$$\Rightarrow \quad \sin(90-\theta_c) > \sin r_2$$
$$\Rightarrow \quad \cos(\theta_c) > \frac{\sin\alpha}{\mu}$$

$$\left[\because \mu = \frac{\sin i}{\sin r_2} = \frac{\sin\alpha}{\sin r_2}\right]$$

$$\Rightarrow \quad \cos\theta_c > \frac{\sqrt{1-\cos^2\alpha}}{\mu}$$

$$\Rightarrow \quad \cos\theta_c > \frac{\sqrt{1-\mu^2\cos^2\theta_c}}{\mu}$$

$$\Rightarrow \quad 2\mu^2\cos^2\theta_c > 1 \Rightarrow 2\mu^2(1-\sin^2\theta_c) > 1$$

$$\Rightarrow \quad 2\mu^2\left(1-\frac{1}{\mu^2}\right) > 1$$

$$\Rightarrow \quad 2\mu^2 - 2 > 1$$

$$\Rightarrow \quad 2\mu^2 > 3$$

$$\Rightarrow \quad \mu > \sqrt{\left(\frac{3}{2}\right)}$$

This matches with option (d).

100. *(a)* We choose origin at point P.

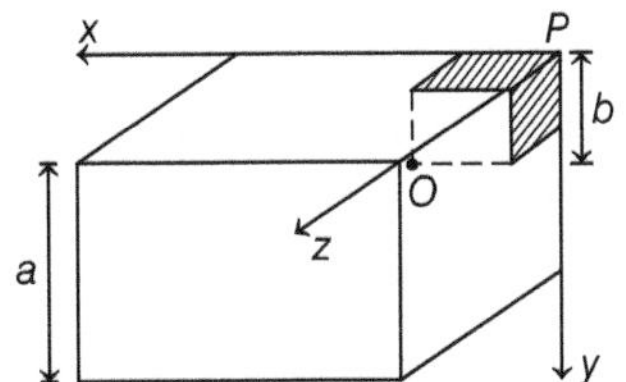

Now, given centre of mass of remaining solid is at point O. Coordinates of O as per axes choosen are $O \equiv (b, b, b)$

i.e. $x = b, y = b$ and $z = b$

Centre of mass of complete large cube lies at its centre.

$\therefore$ Coordinates of centre of mass of large cube are $x_1 = \frac{a}{2}, y_1 = \frac{a}{2}, z_1 = \frac{a}{2}$.

And centre of mass of removed cube of side b is $x_2 = \frac{b}{2}, y_2 = \frac{b}{2}, z_2 = \frac{b}{2}$.

Treating removed mass as negative mass, we have

$$x_{CM} \text{ (of remaining part)} = \frac{m_1x_1 - m_2x_2}{m_1 - m_2}$$

$$\Rightarrow \quad b = \frac{\rho a^3 \times \frac{a}{2} - \rho b^3 \times \frac{b}{2}}{\rho a^3 - \rho b^3}$$

where, ρ = mass density.

$$\Rightarrow \quad a^3 b - b^4 = \frac{a^4}{2} - \frac{b^4}{2}$$

$$\Rightarrow \quad \frac{a^3}{b^3} - 1 = \frac{a^4}{2b^4} - \frac{1}{2}$$

As $X = \frac{a}{b}$, we have

$$X^3 - 1 = \frac{X^4}{2} - \frac{1}{2}$$

$$\Rightarrow \quad 2X^3 - 1 = X^4$$

$$\Rightarrow \quad 2(X^3 - 1) = X^4 - 1$$

$$\Rightarrow 2(X-1)(X^2+1+X) = (X-1)(X+1)(X^2+1)$$

$$\Rightarrow \quad X^3 - X^2 - X - 1 = 0$$

101. *(d)*

Cumene $\xrightarrow{O_2}$ X (cumene hydroperoxide, $C_6H_5C(CH_3)_2$—O—OH) $\xrightarrow{H^+/H_2O}$ CH_3COCH_3 (Z, Acetone) + C_6H_5OH (Y, Phenol)

This method is used for the preparation of phenol from cumene. Cumene is isopropyl benzene, which is oxidised in the presence of air to cumene hydroperoxide. It is then converted to phenol and acetone by treating with dilute acid. Acetone, a by product of this reaction is also obtained in large quantities.

102. *(b)*

Styrene $\xrightarrow[H^+]{HBr}$ $C_6H_5\overset{+}{C}H—CH_3$ $\xrightarrow[\text{Markownikoff addition (Step 1)}]{Br^{\ominus}}$ $C_6H_5CH(Br)CH_3$ $\xrightarrow[\text{Step 2}]{NaCN, S_N2}$ $C_6H_5CH(CN)CH_3$

In step 1, the styrene reacts with HBr to give 2-bromo-2-phenyl ethane. This addition takes place by Markownikoff's rule. This formed halo derivative then undergo nucleophile substitution reaction with NaCN to give the desired product.

103. *(a)*

Phthalaldehyde (o-CHO, CHO) $\xrightarrow[\text{(2) } H_3O^+ \text{ Cannizzaro reaction}]{\text{(1) Conc. KOH}}$ o-(COO^-K^+)(CH_2OH)benzene $\xrightarrow{H^+}$ X (o-(COOH)(CH_2OH)benzene) $\xrightarrow[\text{Esterification}]{\Delta, -H_2O}$ Y (phthalide)

As the starting compound does not contain α-hydrogen thus in the presence of strong base, KOH it undergoes intramolecular Cannizzaro reaction.

In this reaction, one molecule of the aldehyde is reduced to alcohol while another molecule is oxidised to carboxylic acid salt. On heating, loss of water takes place and a lactone (cyclic ester) is formed by process of esterification.

104. *(a)*

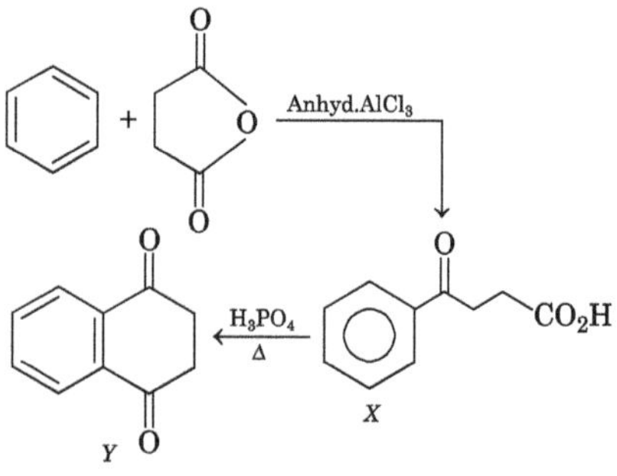

Anhyd.$AlCl_3$

H_3PO_4, Δ

CO_2H

Y X

The mechanism of the above reaction is as follows :

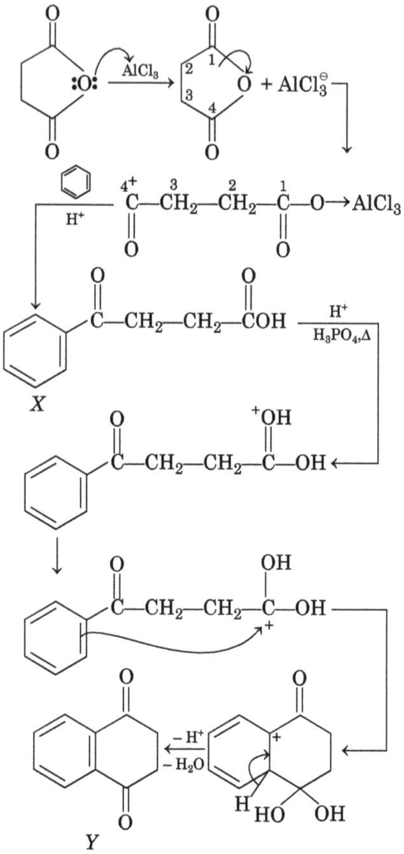

$AlCl_3$ $+ AlCl_3^{\ominus}$

$\overset{4+}{C}-\overset{3}{C}H_2-\overset{2}{C}H_2-\overset{1}{C}-O \rightarrow AlCl_3$

H^+

$C-CH_2-CH_2-COH$ $\xrightarrow[H_3PO_4,\Delta]{H^+}$

X

$C-CH_2-CH_2-\overset{+OH}{C}-OH$

$C-CH_2-CH_2-\overset{OH}{C}-OH$

$-H^+$, $-H_2O$

Y

105. *(c)* Given, density of copper lattice $= 8.93 \text{ g cm}^{-3}$

Number of atoms in fcc lattice, $Z = 4$

As we know,

$$\rho = \frac{M \times Z}{N_A \times a^3}$$

$$a^3 = \frac{\rho \times N_A}{M \times Z} = \frac{8.93 \times 6.023 \times 10^{23}}{4 \times 63.5}$$

$$a = (47.2 \times 10^{-24})^{1/3}$$

$$a = 3.61 \times 10^{-10} \text{ m}$$

$$= 361 \text{ pm}$$

Also, in fcc lattice

$$a = 2\sqrt{2}\, r$$

$$r = \frac{a}{2\sqrt{2}} = \frac{361}{2\sqrt{2}} = 127.8 \text{ pm}$$

106. *(b)* Given,

$$E^\circ_{Cu^{2+}/Cu} = 0.340 \text{ V}$$

$$E^\circ_{Cu^{+}/Cu} = 0.522 \text{ V}$$

$Cu^{2+} + 2e^- \longrightarrow Cu, E^\circ_1 = 0.340$ V ...(i)

$Cu^{+} + e^- \longrightarrow Cu, E^\circ_2 = 0.522$ V ...(ii)

$Cu^{2+} + e^- \longrightarrow Cu^+, E^\circ_3 = ?$ ...(iii)

In order to get Eq. (iii), we have two subtract Eq. (ii) from Eq. (i), but we cannot simplify subtract E°_2 from E°_1 as it is an intensive property, so we have to consider Gibb's free energy, which is an extensive property.

$$\therefore \Delta G^\circ_3 = \Delta G^\circ_1 - \Delta G^\circ_2$$

$$= -n_3FE^\circ_3 = -n_1FE^\circ_1 + n_2FE^\circ_2$$

$$-E^\circ_3 = \frac{-n_1E^\circ_1 + n_2E^\circ_2}{n_3}$$

$$-E^\circ_3 = \frac{-2 \times 0.34 + 0.522}{1}$$

$$E^\circ_3 = 0.158 \text{ V}$$

$\therefore$ Thus, $E^\circ_{Cu^{2+}/Cu^+} = 0.158$ V

107. *(a)* Given,

Current = 1.5 A

Time = 250 s

Volume = 250 mL

Molarity = 0.15 M

$$\text{Molarity} = \frac{\text{Number of moles of solute}}{\text{Volume of solution (L)}}$$

$\therefore$ Number of initial moles of metals

= Molarity × volume

$= 0.15 \times 0.250$

$= 0.0375$

According to Faraday's second law,

$$\frac{W}{M} = \frac{Q}{96500 \times n}$$

$$Q = It$$

$n = 2$ (as two electrons are transferred)

$$\therefore \frac{W}{M} = \frac{It}{96500 \times 2}$$

$$\frac{W}{M} = \frac{1.5 \times 0.85 \times 250}{96500 \times 2}$$

Number of moles deposited after electroplating $= \frac{W}{M} = 0.00165$

$\therefore$ Number of moles left after electroplating $= 0.0375 - 0.00165$

$= 0.03585$

Thus, molarity $= \frac{0.03585}{0.25} = 0.143$ M

108. *(d)* The hybridisation of central atom in a compound can be calculated as

$$H = \frac{1}{2}[V + MA \pm \text{anion/cation}]$$

where, H = Hybridisation

V = Valency of central atom

MA = Monovalent atom

In $[IO_2F_5]^{2-}$

$$H = \frac{1}{2}[7 + 5 + 2] = 7, sp^3d^3$$

$\therefore$ The hybridisation of I in $[IO_2F_5]^{2-}$ ion is sp^3d^3 and the shape corresponding to this hybridisation is pentagonal bipyramidal. Here, double bond would be on axial positions, at an angle 180° to each other, so as to minimize the repulsion caused by them.

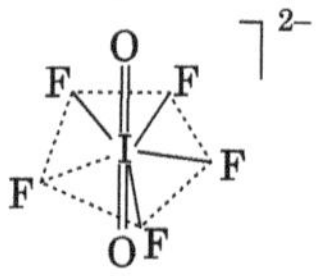

$[IO_2F_5]^{2-}$

sp^3d^3, pentagonal bipyramidal

109. *(a)* $\underset{\substack{X \\ 2.33 \text{ g}}}{CoH_{12}N_4Cl_3} + AgNO_3 \rightarrow \underset{\substack{\text{White ppt.} \\ 1.435 \text{ g}}}{AgCl}$

Empirical formula mass of $CoH_{12}N_4Cl_3$

$= 59 + 12 \times 1 + 14 \times 4 + 35.5 \times 3$

$= 233.5$ g

Number of moles of $CoH_{12}N_4Cl_3 = \frac{2.33}{233.5}$

$= 0.01$ mol

Number of moles of AgCl $= \frac{1.435}{143.32}$

$= 0.01$ mol

0.01 mole of compound X produce 0.01 mole of AgCl.

$\therefore$ Only one Cl^- is present outside the coordination complex. So, the formula of complex X will $[Co(NH_3)_4Cl_2]$ Cl

Primary valency of Co = It's oxidation number.

Let the oxidation number of Co in $[Co(NH_3)_4Cl_2]$ Cl be x

$$\therefore \quad x + 0 + 2(-1) = +1$$

$$x - 2 = +1$$

$$x = +3$$

Thus, the primary valency of Co will be +3.

Secondary valency of
Co = Coordination number of Co = 6

Thus, the secondary valency of cobalt in k is 6.

110. *(a)* Given, specific conductance of acetic acid = 1.65×10^{-4} S cm^{-2}

Molarity of acetic acid = 0.02 M

As we know, molar conductance can be calculated using formula

$$\lambda_m = \frac{1000 \times k}{M}$$

$$\lambda_{m\,CH_3COOH} = \frac{1000 \times 1.65 \times 10^{-4}}{0.02} = 8.25$$

$$\lambda^{\infty}_{m\,CH_3COOH} = \lambda^{\infty}_{m\,(H^+)} + \lambda^{\infty}_{m\,(CH_3COO^-)}$$
$$= (349.1 + 40.9)\ \text{Scm}^2\ \text{mol}^{-1}$$
$$= 390$$

Degree of dissociation, $\alpha = \frac{\lambda_{m\,CH_3COOH}}{\lambda^{\infty}_{m\,CH_3COOH}}$

$$\alpha = \frac{8.25}{390} = 0.0211$$

111. *(a)* Cristae are folds in the inner membrane of a mitochondrion. Cisternae are flattened membranous disk of the endoplasmic reticulum and Golgi apparatus.

Thylakoids are membranous bound compartments inside the chloroplasts and cyanobacteria.

Centrosome is an organelle which contains two cylindrical structures called centrioles. The central part of the centriole is called the hub. The hub is connected with the tubules of the peripheral triplets by radial spokes made of protein.

112. *(b)* A population with 200 individual has 90 TT, 40 Tt and 70 tt genotypes, i.e. dominant allele (T) is TT+ Tt, i.e. 90 + 90 + 40 = 220

Recessive allele (t) is Tt + tt, i.e. 40 + 70 + 70 = 180

Total allele = 400

Dominant allele (T) frequency = $\frac{220}{400}$ = 0.55

Recessive allele (t) frequency = $\frac{180}{400}$ = 0.45

113. *(d)* The oxygen-haemoglobin dissociation curve, also called the oxyhaemoglobin dissociation curve or oxygen dissociation curve is a sigmoid-shaped curve that plots the proportion of haemoglobin in its saturated form on the vertical axis against the prevailing oxygen tension on the horizontal axis. This curve is an important tool for understanding how our blood carries and releases oxygen.

114. *(c)*

DNA	G_1 →	S →	G_2 →	M
content	C	2C	2C	C
Number	2N	2N	4N	2N

of chromosome

During cell cycle in the G_1-phase, the cellular contents excluding the chromosomes are duplicated. In S-phase, each of the 46 chromosomes is duplicated by the cell but their ploidy is not disturbed.

In G_2-phase, the cell double checks the duplicated chromosomes for error, making any needed repair. In M-phase, the cell division occurs.

115. *(a)* In mammals, BMR is higher per unit of body mass in small animals compared to larger ones. This is because the higher metabolic rate of small animals needs a greater delivery of oxygen to tissues around the body. Also, the smaller animals have a greater surface area to volume ratio, so more heat is lost. Thus, in the graph given animals are distributed throughout the curve with the smaller animals towards the left and progressively bigger animals towards the right.

116. *(a)* Phenylketonuria is caused by mutations in the gene that makes an enzyme phenylalanine hydroxylase. This enzyme is needed to convert the amino acid phenylalanine to tyrosine.

Albinism is a group of inherited disorder characterised by little or no melanin production.

Homocystinuria is a disorder of methionine metabolism, leading to an abnormal accumulation of homocysteine and its metabolites in blood and urine.

Argininemia is an autosomal recessive disorder associated with urea synthesis.

117. *(d)* 3′ TACATGGGTCCG5′

mRNA

5′ AUG UAC CCA GGC 3′

tRNA

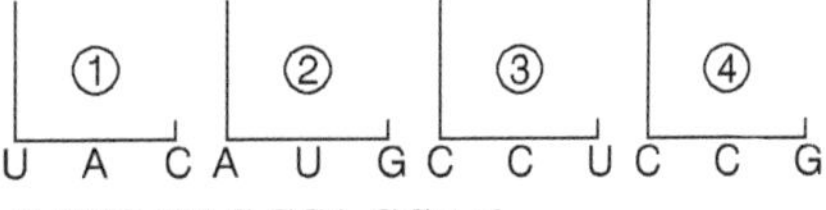

5′ AUG UAC CCA GG5 3′

mRNA → Direction of translation

Option (d) is correct even though the base pairs are not complementary pairing can happen by wobble pairing because the codon (or amino acid) differs only by one base.

118. *(c)* The DNA replication takes place exponentially. Since there are two parent strands, the DNA number at the end of each cycle is increased in powers of 2 thus 2^n. Therefore, the first cycle will use up about 1000 pairs of primers as we have 1000 DNA strands. This will yield 2000 number of products. The next cycle needs 2000 primer pairs and gets used up eventually.

∴ Each DNA multiplies in exponents of 2

Thus for 20 cycles each DNA yields 2^{20} copies

Thus for 1000 DNA templates we will have

$1000 \times 2^{20} = 1000 \times 1048576$
$= 10.5 \times 10^8 = 10^9$ copies approx.

119. *(c)* Black hair (B) is dominant over brown hair (b).

Brown eye (E) is dominant over blue eye (e).

BbEe × bbEE
↓ ↓
(bE)(be)(Be) (bE)

	BE	Be	bE	be
bE	BbEE Black hair brown eye	BbEe Black hair brown eye	bbEE Brown hair brown eye	bbEe Brown hair brown eye

Therefore, we see that the ratio of black hair and brown eye individuals to black haired and brown eyed individuals is 1 : 1.

120. *(b)* SDP (Short Day Plants) or LNP (Long Night Plants) flower only when photoperiod is below critical day length (Critical photoperiod) or they are responsive to night length and flower, when night length is above critical dark period. In this experiment, plant flowers when dark period is above 8 hours. So, it is SDP and actually measures night length to flower.

QUESTION PAPER 2015

Stream : SB/SX

MM : 160

Instructions

1. There are 120 questions in this paper.
2. The question paper contains two parts; Part I (1 Mark Questions) and Part II (2 Marks Questions).
3. There are four sections in each part; Mathematics, Physics, Chemistry and Biology.
4. Out of the four options given with each question, only one is correct.

PART-I (1 Mark Questions)

MATHEMATICS

1. The number of ordered pairs (x, y) of real numbers that satisfy the simultaneous equations $x + y^2 = x^2 + y = 12$ is

(a) 0 (b) 1 (c) 2 (d) 4

2. If z is a complex number satisfying $|z^3 + z^{-3}| \le 2$, then the maximum possible value of $|z + z^{-1}|$ is

(a) 2 (b) $\sqrt[3]{2}$ (c) $2\sqrt{2}$ (d) 1

3. The largest perfect square that divides $2014^3 - 2013^3 + 2012^3 - 2011^3 + \ldots + 2^3 - 1^3$ is

(a) 1^2 (b) 2^2 (c) 1007^2 (d) 2014^2

4. Suppose $BOAC$ is a rectangle in the XY-plane where O is the origin and A, B lie on the parabola $y = x^2$. Then, C must lie on the curve

(a) $y = x^2 + 2$ (b) $y = 2x^2 + 1$
(c) $y = -x^2 + 2$ (d) $y = -2x^2 + 1$

5. Circles C_1 and C_2, of radii r and R respectively, touch each other as shown in the figure. The line l, which is parallel to the line joining the centres of C_1 and C_2, is tangent to C_1 at P and intersects C_2 at A, B. If $R^2 = 2r^2$, then $\angle AOB$ equals

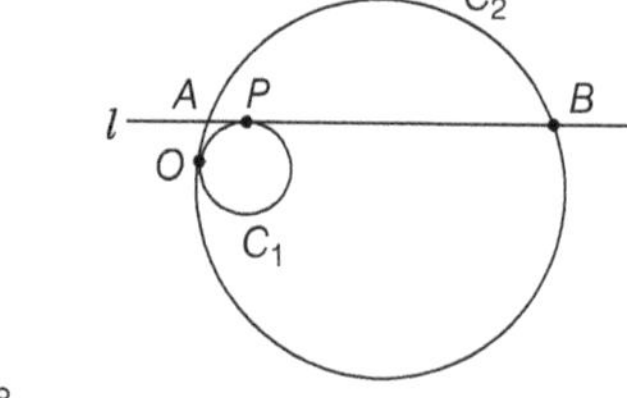

(a) $22\frac{1}{2}^\circ$ (b) 45° (c) 60° (d) $67\frac{1}{2}^\circ$

6. The shortest distance from the origin to a variable point on the sphere $(x-2)^2 + (y-3)^2 + (z-6)^2 = 1$ is

(a) 5 (b) 6 (c) 7 (d) 8

7. The number of real numbers λ for which the equality

$$\frac{\sin(\lambda\alpha)}{\sin\alpha} - \frac{\cos(\lambda\alpha)}{\cos\alpha} = \lambda - 1,$$

holds for all real α which are not integral multiples of $\pi/2$ is

(a) 1 (b) 2
(c) 3 (d) infinite

8. Suppose $ABCDEF$ is a hexagon such that $AB = BC = CD = 1$ and $DE = EF = FA = 2$. If the vertices A, B, C, D, E, F are concyclic, the radius of the circle passing through them is

(a) $\sqrt{\frac{5}{2}}$ (b) $\sqrt{\frac{7}{3}}$ (c) $\sqrt{\frac{11}{5}}$ (d) $\sqrt{2}$

9. Let $p(x)$ be a polynomial such that $p(x) - p'(x) = x^n$, where n is a positive integer. Then, $p(0)$ equals

(a) $n!$ (b) $(n-1)!$ (c) $\frac{1}{n!}$ (d) $\frac{1}{(n-1)!}$

10. The value of the limit $\lim_{x \to 0} \left(\frac{x}{\sin x}\right)^{6/x^2}$ is

(a) e (b) e^{-1} (c) $e^{-1/6}$ (d) e^6

11. Among all sectors of a fixed perimeter, choose the one with maximum area. Then, the angle at the centre of this sector (i.e., the angle between the bounding radii) is

(a) $\frac{\pi}{3}$ (b) $\frac{3}{2}$ (c) $\sqrt{3}$ (d) 2

12. Define a function $f : R \to R$ by $f(x) = \max \{|x|, |x-1|, \ldots, |x-2n|\}$, where n is a fixed natural number. Then, $\int_0^{2n} f(x)\, dx$ is

(a) n (b) n^2 (c) $3n$ (d) $3n^2$

13. If $p(x)$ is a cubic polynomial with $p(1) = 3$, $p(0) = 2$ and $p(-1) = 4$, then $\int_{-1}^{1} p(x)\, dx$ is

(a) 2 (b) 3 (c) 4 (d) 5

14. Let $x > 0$ be a fixed real number. Then, the integral $\int_0^{\infty} e^{-t} |x - t|\, dt$ is equal to

(a) $x + 2e^{-x} - 1$ (b) $x - 2e^{-x} + 1$
(c) $x + 2e^{-x} + 1$ (d) $-x - 2e^{-x} + 1$

15. An urn contains marbles of four colours : red, white, blue and green. When four marbles are drawn without replacement, the following events are equally likely

1. the selection of four red marbles.
2. the selection of one white and three red marbles.
3. the selection of one white, one blue and two red marbles.
4. the selection of one marble of each colour.

The smallest total number of marbles satisfying the given condition is

(a) 19 (b) 21 (c) 46 (d) 69

16. There are 6 boxes labelled $B_1, B_2, \ldots, B_6$. In each trial, two fair dice D_1, D_2 are thrown. If D_1 shows j and D_2 shows k, then j balls are put into the box B_k. After n trials, what is the probability that B_1 contains at most one ball ?

(a) $\left(\frac{5^{n-1}}{6^{n-1}}\right) + \left(\frac{5^n}{6^n}\right)\left(\frac{1}{6}\right)$ (b) $\left(\frac{5^n}{6^n}\right) + \left(\frac{5^{n-1}}{6^{n-1}}\right)\left(\frac{1}{6}\right)$

(c) $\left(\frac{5^n}{6^n}\right) + n\left(\frac{5^{n-1}}{6^{n-1}}\right)\left(\frac{1}{6}\right)$ (d) $\left(\frac{5^n}{6^n}\right) + n\left(\frac{5^{n-1}}{6^{n-1}}\right)\left(\frac{1}{6^2}\right)$

17. Let $\vec{a} = 6\hat{i} - 3\hat{j} - 6\hat{k}$ and $\vec{d} = \hat{i} + \hat{j} + \hat{k}$. Suppose that $\vec{a} = \vec{b} + \vec{c}$, where $\vec{b}$ is parallel to $\vec{d}$ and $\vec{c}$ is perpendicular to $\vec{d}$. Then $\vec{c}$ is

(a) $5\hat{i} - 4\hat{j} - \hat{k}$ (b) $7\hat{i} - 2\hat{j} - 5\hat{k}$
(c) $4\hat{i} - 5\hat{j} + \hat{k}$ (d) $3\hat{i} + 6\hat{j} - 9\hat{k}$

18. If $\log_{(3x-1)}(x-2) = \log_{(9x^2-6x+1)}(2x^2 - 10x - 2)$, then x equals

(a) $9 - \sqrt{15}$ (b) $3 + \sqrt{15}$ (c) $2 + \sqrt{5}$ (d) $6 - \sqrt{5}$

19. Suppose a, b, c are positive integers such that $2^a + 4^b + 8^c = 328$. Then, $\frac{a + 2b + 3c}{abc}$ is equal to

(a) $\frac{1}{2}$ (b) $\frac{5}{8}$ (c) $\frac{17}{24}$ (d) $\frac{5}{6}$

20. The sides of a right angled triangle are integers. The length of one of the sides is 12. The largest possible radius of the incircle of such a triangle is

(a) 2 (b) 3 (c) 4 (d) 5

PHYSICS

21. A small box resting on one edge of the table is struck in such a way that it slides upto the other edge, 1 m away after 2 s. The coefficient of kinetic friction between the box and the table

(a) must be less than 0.05 (b) must be exactly zero
(c) must be more than 0.05 (d) must be exactly 0.05

22. Carbon-11 decays to boron-11 according to the following formula.

$$^{11}_{6}\text{C} \to {}^{11}_{5}\text{B} + e^{+} + \nu_e + 0.96 \text{ MeV}$$

Assume that, positrons (e^+) produced in the decay combine with free electrons in the atmosphere and annihilate each other almost immediately. Also, assume that the neutrinos (ν_e) are massless and do not interact with the environment. At $t = 0$, we have 1 μg of $^{12}_{6}$C. If the half-life of the decay process is t_0, the net energy produced between time $t = 0$ and $t = 2t_0$, will be nearly

(a) 8×10^{18} MeV (b) 8×10^{16} MeV
(c) 4×10^{18} MeV (d) 4×10^{16} MeV

23. Two uniform plates of the same thickness and area but of different materials, one shaped like an isosceles triangle and the other shaped like a rectangle are joined together to form a composite body as shown in the figure alongside.

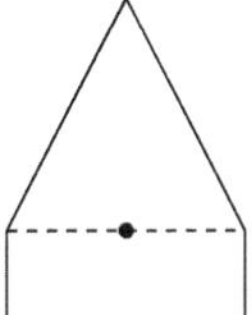

If the centre of mass of the composite body is located at the mid-point of their common side, then the ratio between masses of the triangle to that of the rectangle is

(a) 1 : 1 (b) 4 : 3 (c) 3 : 4 (d) 2 : 1

24. Two spherical objects each of radii R and masses m_1 and m_2 are suspended using two strings of equal length L as shown in the figure ($R << L$). The angle θ which mass m_2 makes with the vertical is approximately

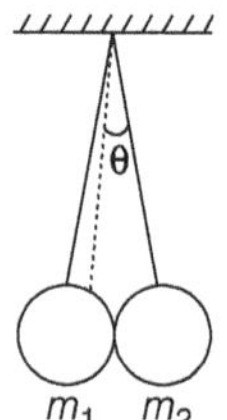

(a) $\frac{m_1 R}{(m_1 + m_2) L}$ (b) $\frac{2m_1 R}{(m_1 + m_2) L}$

(c) $\frac{2m_2 R}{(m_1 + m_2) L}$ (d) $\frac{m_2 R}{(m_1 + m_2) L}$

25. A horizontal disc of moment of inertia 4.25 kg - m^2 with respect to its axis of symmetry is spinning counter clockwise at 15 rps about its axis, as viewed from above. A second disc of moment of inertia 1.80 kg - m^2 with respect to its axis of symmetry is spinning clockwise at 25 rps as viewed from above about the same axis and is dropped on top of the first disc. The two discs stick together and rotate as one about their axis of symmetry. The new angular velocity of the system as viewed from above is close to

(a) 18 rps and clockwise
(b) 18 rps and anti-clockwise
(c) 3 rps and clockwise
(d) 3 rps and anti-clockwise

26. A boy is standing on top of a tower of height 85 m and throws a ball in the vertically upward direction with a certain speed. If 5.25 s later he hears the ball hitting the ground, then the speed with which the boy threw the ball is (take, $g = 10 \text{ m/s}^2$ and speed of sound in air = 340 m/s)

(a) 6 m/s (b) 8 m/s
(c) 10m/s (d) 12 m/s

27. For a diode connected in parallel with a resistor, which is the most likely current I-voltage V characteristic?

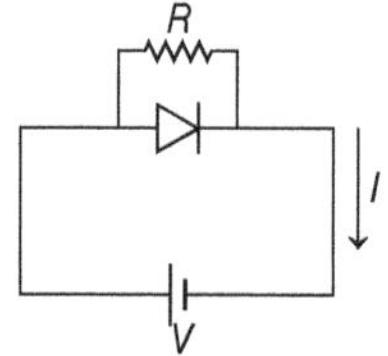

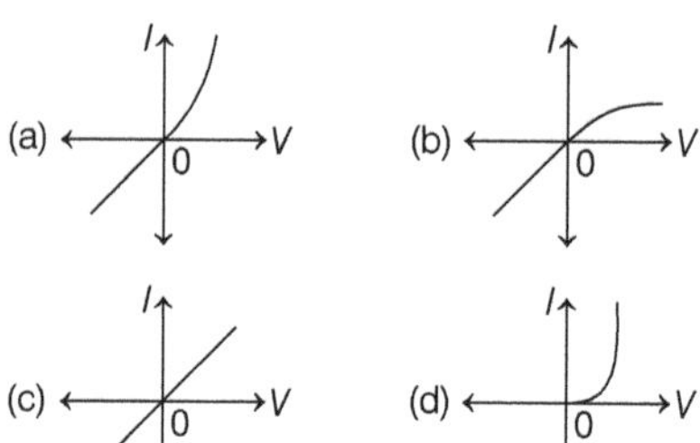

28. A beam of monoenergetic electrons, which have been accelerated from rest by a potential U, is used to form an interference pattern in a Young's double slit experiment. The electrons are now accelerated by potential $4U$. Then, the fringe width

(a) remains same
(b) is half the original fringe width
(c) is twice the original fringe width
(d) is one-fourth the original fringe width

29. A point charge Q $(= 3 \times 10^{-12}$ C) rotates uniformly in a vertical circle of radius R (= 1 mm). The axis of the circle is aligned along the magnetic axis of the earth. At what value of the angular speed ω, the effective magnetic field at the centre of the circle will be reduced to zero? (Horizontal component of earth's magnetic field is 30 μT)

(a) 10^{11} rad/s (b) 10^9 rad/s (c) 10^{13} rad/s (d) 10^7 rad/s

30. A closed bottle containing water at 30°C is open on the surface of the moon. Then,

(a) the water will boil
(b) the water will come out as a spherical ball
(c) the water will freeze
(d) the water will decompose into hydrogen and oxygen

31. A simple pendulum of length l is made to oscillate with an amplitude of 45 degrees. The acceleration due to gravity is g. Let $T_0 = 2\pi \sqrt{l/g}$. The time period of oscillation of this pendulum will be

(a) T_0 irrespective of the amplitude
(b) slightly less than T_0
(c) slightly more than T_0
(d) dependent on whether it swings in a plane aligned with the north-south or east-west directions

32. An AC voltmeter connected between points A and B in the circuit below reads 36 V. If it is connected between A and C, the reading is 39 V. The reading when it is connected between B and D is 25 V. What will the voltmeter read when it is connected between A and D? (Assume that, the voltmeter reads true rms voltage values and that the source generates a pure AC)

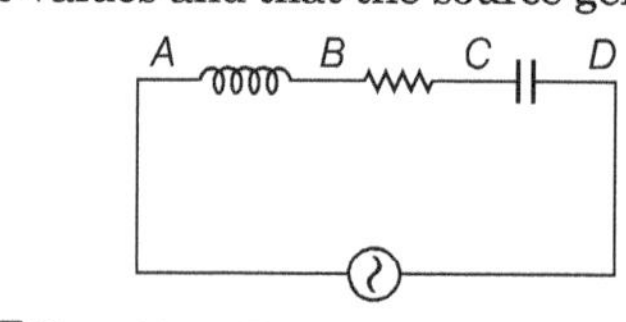

(a) $\sqrt{481}$ V (b) 31 V (c) 61 V (d) $\sqrt{3361}$ V

33. A donor atom in a semiconductor has a loosely bound electron. The orbit of this electron is considerly affected by the semiconductor material but behaves in many ways like an electron orbiting a hydrogen nucleus. Given that the electron has an effective mass of $0.07m_e$, where m_e is mass of the free electron and the space in which it moves has a permittivity $13\varepsilon_0$, then the radius of the electron's lowermost energy orbit will be close to (take, the Bohr radius of the hydrogen atom is 0.53 Å)
(a) 0.53 Å (b) 243 Å (c) 10 Å (d) 100 Å

34. The state of an ideal gas was changed isobarically. The graph depicts three such isobaric lines. Which of the following is true about the pressures of the gas?

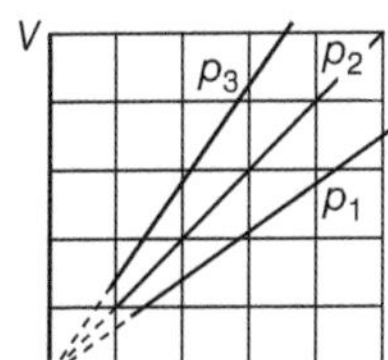

(a) $p_1 = p_2 = p_3$ (b) $p_1 > p_2 > p_3$
(c) $p_1 < p_2 < p_3$ (d) $p_1 / p_2 = p_3 / p_1$

35. A metallic ring of radius a and resistance R is held fixed with its axis along a spatially uniform magnetic field whose magnitude is $B_0 \sin \omega t$. Gravity is neglected. Then,
(a) the current in the ring oscillates with a frequency of 2ω
(b) the joule heating loss in the ring is proportional to a^2
(c) the force per unit length on the ring will be proportional to B_0^2
(d) the net force on the ring is non-zero

36. The dimensions of the area A of a black hole can be written in terms of the universal gravitational constant G, its mass M and the speed of light c as $A = G^{\alpha} M^{\beta} c^{\gamma}$. Here,
(a) $\alpha = -2, \beta = -2$ and $\gamma = 4$
(b) $\alpha = 2, \beta = 2$ and $\gamma = -4$
(c) $\alpha = 3, \beta = 3$ and $\gamma = -2$
(d) $\alpha = -3, \beta = -3$ and $\gamma = 2$

37. A 160 W infrared source is radiating light of wavelength 50000 Å uniformly in all directions. The photon flux at a distance of 1.8 m is the order of
(a) $10 \text{ m}^{-2}\text{s}^{-1}$ (b) $10^{10} \text{ m}^{-2}\text{s}^{-1}$
(c) $10^{15} \text{ m}^{-2}\text{s}^{-1}$ (d) $10^{20} \text{ m}^{-2}\text{s}^{-1}$

38. A wire bent in the shape of a regular n-polygonal loop carries a steady current I. Let l be the perpendicular distance of a given segment and R be the distance of a vertex both from the centre of the loop. The magnitude of the magnetic field at the centre of the loop is given by
(a) $\frac{n\mu_0 I}{2\pi l} \sin(\pi / n)$ (b) $\frac{n\mu_0 I}{2\pi R} \sin(\pi / n)$
(c) $\frac{n\mu_0 I}{2\pi l} \cos(\pi / n)$ (d) $\frac{n\mu_0 I}{2\pi R} \cos(\pi / n)$

39. The intensity of sound during the festival season increased by 100 times. This could imply a decibel (dB) level rise from
(a) 20 dB to 120 dB (b) 70 dB to 72 dB
(c) 100 dB to 10000 dB (d) 80 dB to 100 dB

40. One end of a slack wire (Young's modulus Y, length L and cross-sectional area A) is clamped to a rigid wall and the other end to a block (mass m), which rests on a smooth horizontal plane. The block is set in motion with a speed v. What is the maximum distance, then the block will travel after the wire becomes taut?
(a) $v\sqrt{\frac{mL}{AY}}$ (b) $v\sqrt{\frac{2mL}{AY}}$ (c) $v\sqrt{\frac{mL}{2AY}}$ (d) $L\sqrt{\frac{mv}{AY}}$

CHEMISTRY

41. The Lewis acid strength of BBr_3, BCl_3 and BF_3 is in the order
(a) $BBr_3 < BCl_3 < BF_3$ (b) $BCl_3 < BF_3 < BBr_3$
(c) $BF_3 < BCl_3 < BBr_3$ (d) $BBr_3 < BF_3 < BCl_3$

42. O^{2-} is isoelectronic with
(a) Zn^{2+} (b) Mg^{2+}
(c) K^{+} (d) Ni^{2+}

43. The H—C—H, H—N—H, and H—O—H bond angles (in degrees) in methane, ammonia and water are respectively, closest to
(a) 109.5, 104.5, 107.1
(b) 109.5, 107.1, 104.5
(c) 104.5, 107.1, 109.5
(d) 107.1, 104.5, 109.5

44. In alkaline medium, the reaction of hydrogen peroxide with potassium permanganate produces a compound in which the oxidation state of Mn is
(a) 0 (b) + 2
(c) + 3 (d) + 4

45. The rate constant of a chemical reaction at a very high temperature will approach
(a) Arrhenius frequency factor divided by the ideal gas constant
(b) activation energy
(c) Arrhenius frequency factor
(d) activation energy divided by the ideal gas constant

46. The standard reduction potentials (in V) of a few metal ion/metal electrodes are given below.
$Cr^{3+}/Cr = -0.74$; $Cu^{2+}/Cu = +0.34$;
$Pb^{2+}/Pb = -0.13$; $Ag^{+}/Ag = +0.8$
The reducing strength of the metals follows the order
(a) Ag > Cu > Pb > Cr (b) Cr > Pb > Cu > Ag
(c) Pb > Cr > Ag > Cu (d) Cr > Ag > Cu > Pb

47. Which of the following molecules can exhibit optical activity?
(a) 1-bromopropane (b) 2-bromobutane
(c) 3-bromopentane (d) bromocyclohexane

48. The structure of the polymer obtained by the following reaction is

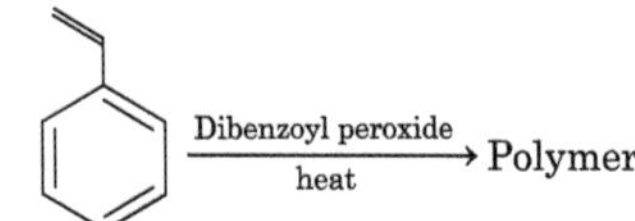

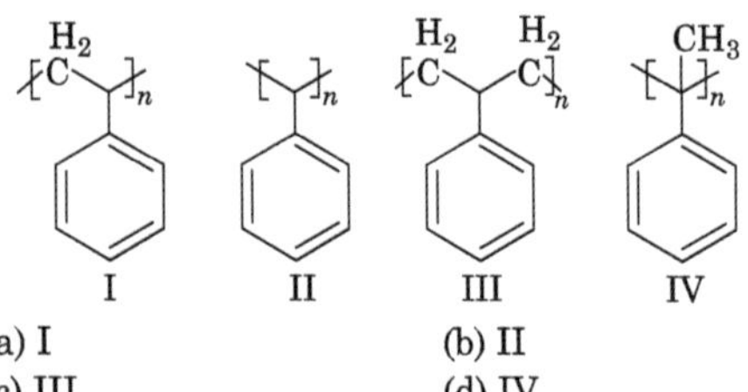

(a) I (b) II
(c) III (d) IV

49. The major product of the reaction between CH_3CH_2ONa and $(CH_3)_3CCl$ in ethanol is
(a) $CH_3CH_2OC(CH_3)_3$
(b) $CH_2{=}C(CH_3)_2$
(c) $CH_3CH_2C(CH_3)_3$
(d) $CH_3CH{=}CHCH_3$

50. When H_2S gas is passed through a hot acidic aqueous solution containing Al^{3+}, Cu^{2+}, Pb^{2+} and Ni^{2+}, a precipitate is formed, which consists of
(a) CuS and Al_2S_3
(b) PbS and NiS
(c) CuS and NiS
(d) PbS and CuS

51. The electronic configuration of an element with the largest difference between the 1st and 2nd ionisation energies is
(a) $1s^2 2s^2 2p^6$ (b) $1s^2 2s^2 2p^6 3s^1$
(c) $1s^2 2s^2 2p^6 3s^2$ (d) $1s^2 2s^2 2p^1$

52. The order of electronegativity of carbon in sp, sp^2 and sp^3 hybridised states follows
(a) $sp > sp^2 > sp^3$ (b) $sp^3 > sp^2 > sp$
(c) $sp > sp^3 > sp^2$ (d) $sp^2 > sp > sp^3$

53. The most abundant transition metal in human body is
(a) copper (b) iron
(c) zinc (d) manganese

54. The molar conductivities of HCl, NaCl, CH_3COOH and CH_3COONa at infinite dilution follow the order
(a) HCl > CH_3COOH > NaCl > CH_3COONa
(b) CH_3COONa > HCl > NaCl > CH_3COOH
(c) HCl > NaCl > CH_3COOH > CH_3COONa
(d) CH_3COOH > CH_3COONa > HCl > NaCl

55. The spin only magnetic moment of $[ZCl_4]^{2-}$ is 3.87 BM, where Z is
(a) Mn (b) Ni
(c) Co (d) Cu

56. If α-D-glucose is dissolved in water and kept for a few hours, the major constituent(s) present in the solution is (are)
(a) α-D-glucose
(b) mixture of β-D-glucose and open chain D-glucose
(c) open chain D-glucose
(d) mixture of α-D-glucose and β-D-glucose

57. The pH of 1 N aqueous solutions of HCl, CH_3COOH and HCOOH follows the order
(a) HCl > HCOOH > CH_3COOH
(b) HCl = HCOOH > CH_3COOH
(c) CH_3COOH > HCOOH > HCl
(d) CH_3COOH = HCOOH > HCl

58. The major product of reaction

$\xrightarrow[H_2O]{H^+}$ Products is

OH
OH
OH HO
I II III IV

(a) I (b) II (c) III (d) IV

59. Reaction of aniline with $NaNO_2$ + dil. HCl at 0°C followed by reaction with CuCN yields

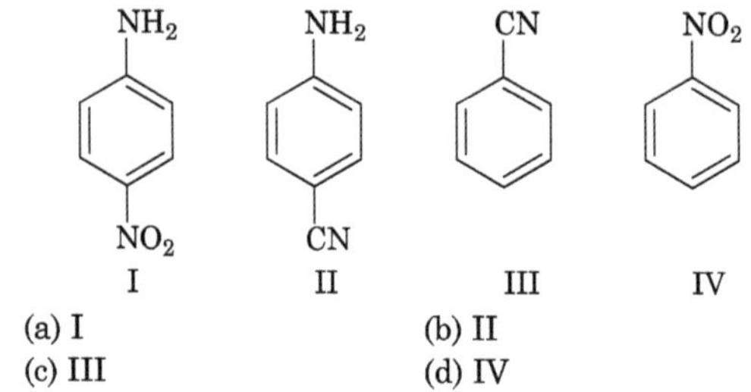

(a) I (b) II
(c) III (d) IV

60. Schottky defect in a crystal arises due to
(a) creation of equal number of cation and anion vacancies
(b) creation of unequal number of cation and anion vacancies
(c) migration of cations to interstitial voids
(d) migration of anions to interstitial voids

BIOLOGY

61. Immunosuppressive drugs like cyclosporine delay the rejection of graft post organ transplantation by
(a) inhibiting T-cell infiltration
(b) killing B-cells
(c) killing macrophages
(d) killing dendritic cells

62. Which one of these substances will repress the *lac* operon?
(a) Arabinose (b) Glucose
(c) Lactose (d) Tryptophan

63. Assume a spherical mammalian cell has a diameter of 27 microns. If a polypeptide chain with alpha helical conformation has to stretch across the cell. How many amino acids should it be comprised of ?

(a) 18000 (b) 1800
(c) 27000 (d) 12000

64. Which one of the following has phosphoric acid anhydride bonds?

(a) Deoxyribonucleic acid (b) Ribonucleic acid
(c) *d*NTPs (d) Phospholipids

65. The two components of autonomous nervous system have antagonistic actions. But in certain cases their effects are mutually helpful. Which of the following statements is correct?

(a) At rest, the control of heartbeat is not by the vagus nerve
(b) During exercise, the sympathetic control decreases
(c) During exercise, the parasympathetic control decreases
(d) Stimulation of sympathetic system results in constriction of the pupil

66. In a random DNA sequence, what is the lowest frequency of encountering a stop codon?

(a) 1 in 20 (b) 1 in 3
(c) 1 in 64 (d) 1 in 10

67. The two alleles that determine the blood group AB of an individual are located on

(a) two different autosomes
(b) the same autosome
(c) two different sex chromosomes
(d) one on sex chromosome and the other on an autosome

68. In biotechnology applications, a selectable marker is incorporated in a plasmid

(a) to increase its copy number
(b) to increase the transformation efficiency
(c) to eliminate the non-transformants
(d) to increase the expression of the gene of interest

69. Spermatids are formed after the second meiotic division from secondary spermatocytes. The ploidy of the secondary spermatocytes is

(a) n (b) $2n$ (c) $3n$ (d) $4n$

70. Phospholipids are formed by the esterification of

(a) three ethanol molecules with three fatty acid molecules
(b) one glycerol and two fatty acid molecules
(c) one glycerol and three fatty acid molecules
(d) one ethylene glycol and two fatty acid molecules

71. Given the fact that histone binds DNA, it should be rich in

(a) arginine, lysine
(b) cysteine, methionine
(c) glutamate, aspartate
(d) isoleucine, leucine

72. If molecular weight of a polypeptide is 15.3 kDa, what would be the minimum number of nucleotides in the *m*RNA that codes for this polypeptide? Assume that molecular weight of each amino acid is 90 Da.

(a) 510 (b) 663
(c) 123 (d) 170

73. Melting temperature for double-stranded DNA is the temperature at which 50% of the double-stranded molecules are converted into single-stranded molecules. Which one of the following DNA will have the highest melting temperature?

(a) DNA with 15% guanine
(b) DNA with 30% cytosine
(c) DNA with 40% thymine
(d) DNA with 50% adenine

74. Following are the types of immunoglobulin and their functions. Which one of the following is incorrectly paired?

(a) IgD – viral pathogen
(b) IgG – phagocytosis
(c) IgE – allergic reaction
(d) IgM – complement fixation

75. Which one of the following can be used to detect amino acids?

(a) Iodine vapour
(b) Ninhydrin
(c) Ethidium bromide
(d) Bromophenol blue

76. Mutation in a single gene can lead to changes in multiple traits. This is an example of

(a) heterotrophy (b) codominance
(c) penetrance (d) pleiotropy

77. Which one of the following is used to treat cancers?

(a) Albumin
(b) Cyclosporin A
(c) Antibodies
(d) Growth hormone

78. Which of the following processes leads to DNA ladder formation?

(a) Necrosis (b) Plasmolysis
(c) Apoptosis (d) Mitosis

79. Coenzymes are the components of an enzyme complex, which are necessary for its function. Which of the following is known as coenzyme?

(a) Zinc (b) Vitamin-B_{12}
(c) Chlorophyll (d) Heme

80. The peptidoglycans of bacteria consist of

(a) sugars, D-amino acids and L-amino acids
(b) sugars and only D-amino acids
(c) sugars and only L-amino acids
(d) sugars and glycine

PART-II (2 Marks Questions)

MATHEMATICS

81. Let $x = (\sqrt{50} + 7)^{1/3} - (\sqrt{50} - 7)^{1/3}$. Then,

(a) $x = 2$
(b) $x = 3$
(c) x is a rational number, but not an integer
(d) x is an irrational number

82. Let $(1 + x + x^2)^{2014} = a_0 + a_1x + a_2x^2 + a_3x^3 + \ldots + a_{4028}x^{4028}$ and let $A = a_0 - a_3 + a_6 - \ldots + a_{4026}$, $B = a_1 - a_4 + a_7 - \ldots - a_{4027}$, $C = a_2 - a_5 + a_8 - \ldots + a_{4028}$.

Then,

(a) $|A| = |B| > |C|$ (b) $|A| = |B| < |C|$
(c) $|A| = |C| > |B|$ (d) $|A| = |C| < |B|$

83. A mirror in the first quadrant is in the shape of a hyperbola whose equation is $xy = 1$. A light source in the second quadrant emits a beam of light that hits the mirror at the point (2, 1/2). If the reflected ray is parallel to the Y-axis, the slope of the incident beam is

(a) $\frac{13}{8}$ (b) $\frac{7}{4}$ (c) $\frac{15}{8}$ (d) 2

84. Let $C(\theta) = \sum_{n=0}^{\infty} \frac{\cos(n\theta)}{n!}$

Which of the following statements is FALSE ?

(a) $C(0) \cdot C(\pi) = 1$ (b) $C(0) + C(\pi) > 2$
(c) $C(\theta) > 0$ for all $\theta \in R$ (d) $C'(\theta) \neq 0$ for all $\theta \in R$

85. Let $a > 0$ be a real number. Then the limit

$$\lim_{x \to 2} \frac{a^x + a^{3-x} - (a^2 + a)}{a^{3-x} - a^{x/2}} \text{ is}$$

(a) $2 \log a$ (b) $-\frac{4}{3}a$ (c) $\frac{a^2 + a}{2}$ (d) $\frac{2}{3}(1 - a)$

86. Let $f(x) = \alpha x^2 - 2 + \frac{1}{x}$, where α is a real constant. The smallest α for which $f(x) \geq 0$ for all $x > 0$ is

(a) $\frac{2^2}{3^3}$ (b) $\frac{2^3}{3^3}$ (c) $\frac{2^4}{3^3}$ (d) $\frac{2^5}{3^3}$

87. Let $f : R \to R$ be a continuous function satisfying

$$f(x) + \int_0^x tf(t)\, dt + x^2 = 0,$$

for all $x \in R$. Then

(a) $\lim_{x \to \infty} f(x) = 2$
(b) $\lim_{x \to -\infty} f(x) = -2$
(c) $f(x)$ has more than one point in common with the X-axis
(d) $f(x)$ is an odd function

88. The figure shows a portion of the graph $y = 2x - 4x^3$. The line $y = c$ is such that the areas of the regions marked I and II are equal. If a, b are the x-coordinates of A, B respectively, then $a + b$ equals

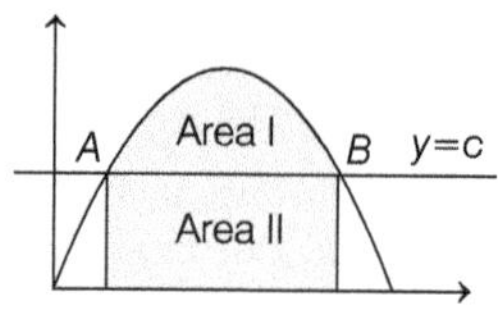

(a) $\frac{2}{\sqrt{7}}$ (b) $\frac{3}{\sqrt{7}}$ (c) $\frac{4}{\sqrt{7}}$ (d) $\frac{5}{\sqrt{7}}$

89. Let $X_n = \{1, 2, 3, \ldots, n\}$ and let a subset A of X_n be chosen so that every pair of elements of A differ by at least 3. (For example, if $n = 5$, A can be ϕ, $\{2\}$ or $\{1, 5\}$ among others). When $n = 10$, let the probability that $1 \in A$ be p and let the probability that $2 \in A$ be q. Then,

(a) $p > q$ and $p - q = \frac{1}{6}$ (b) $p < q$ and $q - p = \frac{1}{6}$
(c) $p > q$ and $p - q = \frac{1}{10}$ (d) $p < q$ and $q - p = \frac{1}{10}$

90. The remainder when the determinant

$$\begin{vmatrix} 2014^{2014} & 2015^{2015} & 2016^{2016} \\ 2017^{2017} & 2018^{2018} & 2019^{2019} \\ 2020^{2020} & 2021^{2021} & 2022^{2022} \end{vmatrix}$$

is divided by 5 is

(a) 1 (b) 2 (c) 3 (d) 4

PHYSICS

91. A cubical vessel has opaque walls. An observer (dark circle in the figure below) is located, such that she can see only the wall CD, but not the bottom. Nearly to what height should water be poured, so that she can see an object placed at the bottom at a distance of 10 cm from the corner C? Refractive index of water is 1.33.

(a) 10 cm (b) 16 cm (c) 27 cm (d) 45 cm

92. The moments of inertia of a non-uniform circular disc (of mass M and radius R) about four mutually perpendicular tangents AB, BC, CD, DA are I_1, I_2, I_3 and I_4, respectively (the square $ABCD$ circumscribes the circle).

The distance of the centre of mass of the disc from its geometrical centre is given by

(a) $\frac{1}{4MR}\sqrt{(I_1 - I_3)^2 + (I_2 - I_4)^2}$

(b) $\frac{1}{12MR}\sqrt{(I_1 - I_3)^2 + (I_2 - I_4)^2}$

(c) $\frac{1}{3MR}\sqrt{(I_1 - I_2)^2 + (I_3 - I_4)^2}$

(d) $\frac{1}{2MR}\sqrt{(I_1 + I_3)^2 + (I_2 + I_4)^2}$

93. A horizontal steel railroad track has a length of 100 m, when the temperature is 25°C. The track is constrained from expanding or bending. The stress on the track on a hot summer day, when the temperature is 40°C is

(**Note** : The linear coefficient of thermal expansion for steel is $1.1 \times 10^{-5}/^\circ C$ and the Young's modulus of steel is 2×10^{11} Pa)

(a) 6.6×10^7 Pa (b) 8.8×10^7 Pa
(c) 3.3×10^7 Pa (d) 5.5×10^7 Pa

94. Electromagnetic waves emanating from a point A (in air) are incident on a rectangular block of material M and emerge from the other side as shown below. The angles i and r are angles of incidence and refraction when the wave travels from air to the medium. Such paths for the rays are possible,

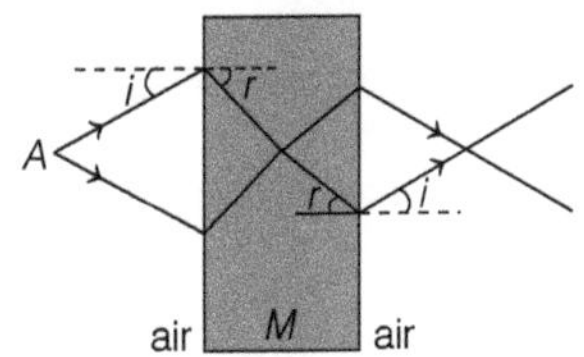

(a) if the material has a refractive index very nearly equal to zero
(b) only with gamma rays with a wavelength smaller than the atomic nuclei of the material
(c) if the material has a refractive index less than zero
(d) only if the wave travels in M with a speed faster than the speed of light in vacuum

95. Two small metal balls of different masses m_1 and m_2 are connected by strings of equal length to a fixed point. When the balls are given equal charges, the angles that the two strings make with the vertical are 30° and 60°, respectively. The ratio m_1/m_2 is close to

(a) 1.7 (b) 3.0 (c) 0.58 (d) 2.0

96. Consider the regular array of vertical identical current carrying wires (with direction of current flow as indicated in the figure alongside) protruding through a horizontal table. If we scatter some diamagnetic particles on the table, then they are likely to accumulate

(a) around regions such as A
(b) around regions such as B
(c) in circular regions around individual wires such as C
(d) uniformly everywhere

97. The distance between the vertex and the centre of mass of a uniform solid planar circular segment of angular size θ and radius R is given by

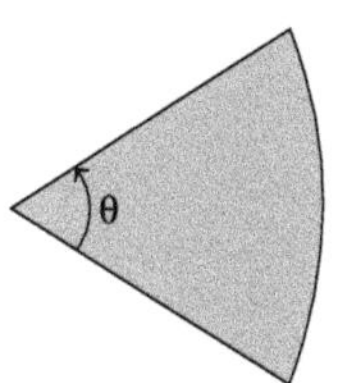

(a) $\frac{4}{3}R\frac{\sin(\theta/2)}{\theta}$ (b) $R\frac{\sin(\theta/2)}{\theta}$
(c) $\frac{4}{3}R\cos\left(\frac{\theta}{2}\right)$ (d) $\frac{2}{3}R\cos\theta$

98. An object is propelled vertically to a maximum height of $4R$ from the surface of a planet of radius R and mass M. The speed of object when it returns to the surface of the planet is

(a) $2\sqrt{\frac{2GM}{5R}}$ (b) $\sqrt{\frac{GM}{2R}}$ (c) $\sqrt{\frac{3GM}{2R}}$ (d) $\sqrt{\frac{GM}{5R}}$

99. In the circuit shown below, all the inductors (assumed ideal) and resistors are identical. The current through the resistance on the right is I after the key K has been switched ON for a long time. The currents through the three resistors (in order, from left to right) immediately after the key is switched OFF are

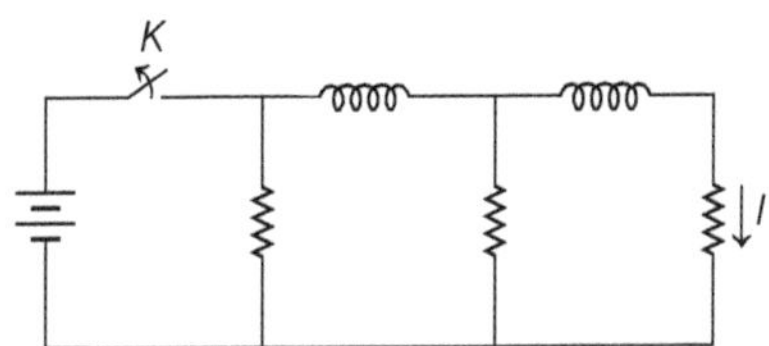

(a) $2I$ upwards, I downwards and I downwards
(b) $2I$ downwards, I downwards and I downwards
(c) I downwards, I downwards and I downwards
(d) 0, I downwards and I downwards

100. An ideal gas undergoes a circular cycle centred at 4 atm, 4 L as shown in the diagram.

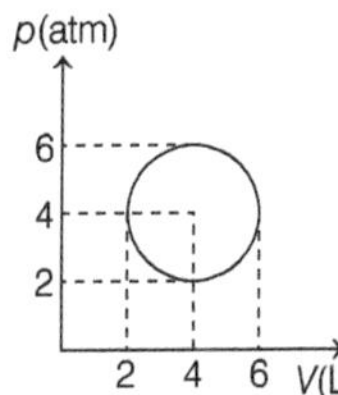

The maximum temperature attained in this process is close to

(a) $\frac{30}{R}$ (b) $\frac{36}{R}$ (c) $\frac{24}{R}$ (d) $\frac{16}{R}$

CHEMISTRY

101. For the reaction $N_2 + 3X_2 \longrightarrow 2NX_3$, where $X = F$, Cl (the average bond energies are F—F = 155 kJ mol^{-1}, N—F = 272 kJ mol^{-1}, Cl—Cl = 242 kJ mol^{-1}, N—Cl = 200 kJ mol^{-1} and N$\equiv$N = 941 kJ mol^{-1}), the heats of formation of NF_3 and NCl_3 in kJ mol^{-1}, respectively, are closest to

(a) -226 and $+467$ (b) $+226$ and -467
(c) -151 and $+311$ (d) $+151$ and -311

102. The equilibrium constants for the reactions $X = 2Y$ and $Z = P + Q$ are K_1 and K_2, respectively. If the initial concentrations and the degree of dissociation of X and Z are the same, the ratio K_1 / K_2 is

(a) 4 (b) 1 (c) 0.5 (d) 2

103. The geometry and the number of unpaired electron(s) of $[MnBr_4]^{2-}$, respectively, are

(a) tetrahedral and 1 (b) square planar and 1
(c) tetrahedral and 5 (d) square planar and 5

104. The standard cell potential for $Zn|Zn^{2+}||Cu^{2+}|Cu$ is 1.10 V. When the cell is completely discharged, $\log [Zn^{2+}] / [Cu^{2+}]$ is closest to

(a) 37.3 (b) 0.026 (c) 18.7 (d) 0.052

105. In the reaction,

Br —(i) x, (ii) y, (iii) z → COOH (benzene rings)

x, y and z are

(a) x = Mg, dry ether; $y = CH_3Cl$; $z = H_2O$
(b) x = Mg, dry methanol; $y = CO_2$; z = dil.HCl
(c) x = Mg, dry ether; $y = CO_2$; z = dil.HCl
(d) x = Mg, dry methanol; $y = CH_3Cl$; $z = H_2O$

106. An organic compound having molecular formula C_2H_6O undergoes oxidation with $K_2Cr_2O_7/H_2SO_4$ to produce X which contains 40% carbon, 6.7% hydrogen and 53.3% oxygen. The molecular formula of the compound X is

(a) CH_2O (b) $C_2H_4O_2$ (c) C_2H_4O (d) $C_2H_6O_2$

107. The maximum number of cyclic isomers (positional and optical) of a compound having molecular formula $C_3H_2Cl_2$ is

(a) 2 (b) 3 (c) 4 (d) 5

108. The volume *versus* temperature graph of 1 mole of an ideal gas is given below

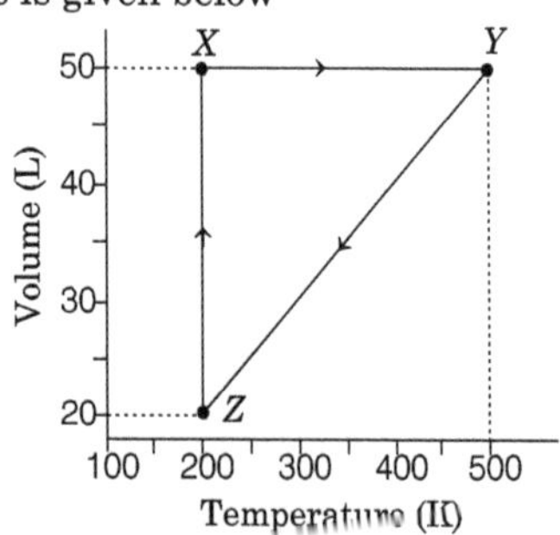

The pressure of the gas (in atm) at X, Y and Z, respectively, are

(a) 0.328, 0.820, 0.820 (b) 3.28, 8.20, 3.28
(c) 0.238, 0.280, 0.280 (d) 32.8, 0.280, 82.0

109. MnO_2 when fused with KOH and oxidised in air gives a dark green compound X. In acidic solution, X undergoes disproportionation to give an intense purple compound Y and MnO_2. The compounds X and Y, respectively, are

(a) K_2MnO_4 and $KMnO_4$ (b) Mn_2O_7 and $KMnO_4$
(c) K_2MnO_4 and Mn_2O_7 (d) $KMnO_4$ and K_2MnO_4

110. A metal (X) dissolves both in dil. HCl and dil. NaOH to liberate H_2. Addition of NH_4Cl and excess NH_4OH to an HCl solution of X produces Y as a precipitate. Y is also produced by adding NH_4Cl to the NaOH solution of X. The species X and Y, respectively, are

(a) Zn and $Zn(OH)_2$ (b) Al and $Al(OH)_3$
(c) Zn and Na_2ZnO_2 (d) Al and $NaAlO_2$

BIOLOGY

111. How many bands are seen when immunoglobulin G molecules are analysed on a Sodium Dodecyl Sulphate-Polyacrylamide Gel Electrophoresis (SDS-PAGE) under reducing conditions?

(a) 6 (b) 1 (c) 2 (d) 4

112. In a mixed culture of slow and fast growing bacteria, penicillin will

(a) kill the fast growing bacteria more than the slow growing
(b) kill slow growing bacteria more than the fast growing
(c) kill both the fast and slow growing bacteria equally
(d) will not kill bacteria at all

113. Consider the following pedigree over four generations and mark the correct answer below about the inheritance of haemophilia.

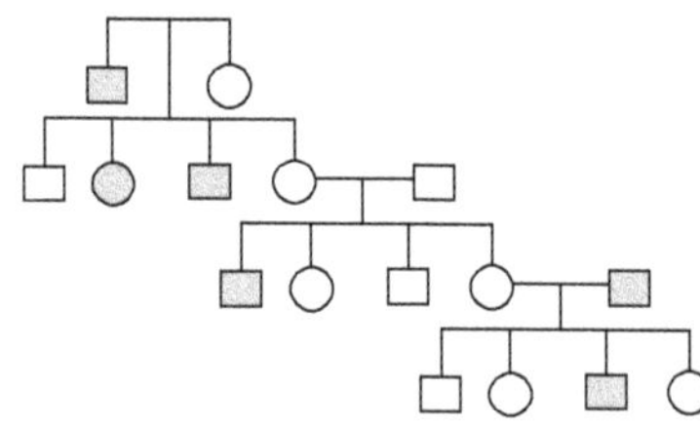

(a) Haemophilia is X-linked dominant
(b) Haemophilia is autosomal dominant
(c) Haemophilia is X-linked recessive
(d) Haemophilia is Y-linked dominant

114. A person has 400 million alveoli per lung with an average radius of 0.1 mm for each alveolus. Considering the alveoli are spherical in shape, the total respiratory surface of that person is closest to

(a) 500 mm^2 (b) 200 mm^2
(c) 100 mm^2 (d) 1000 mm^2

115. A mixture of equal numbers of fast and slow dividing cells is cultured in a medium containing a trace amount of radioactively labelled thymidine for one hour. The cells are then transferred to regular (unlabelled) medium. After 24 hrs of growth in regular media

(a) fast dividing cells will have maximum radioactivity
(b) slow dividing cells will have maximum radioactivity
(c) both will have same amount of radioactivity
(d) there will be no radioactivity in either types of cells

116. If a double-stranded DNA has 15% cytosine, what is the % of adenine in the DNA?

(a) 15% (b) 70% (c) 35% (d) 30%

117. The mitochondrial inner membrane consists of a number of infoldings called cristae. The increased surface area due to cristae helps in

(a) increasing the volume of mitochondria
(b) incorporating more of the protein complexes essential for electron transport chain
(c) changing the pH
(d) increasing diffusion of ions

118. The activity of a certain protein is dependent on its phosphorylation. A mutation in its gene changed a single amino acid, which affected the function of the molecule. Which amino acid change is most likely to account for this observation?

(a) Tyrosine to tryptophan (b) Lysine to valine
(c) Leucine to isoleucine (d) Valine to alanine

119. Consider the linear double-stranded DNA shown below. The restriction enzymes sites and the lengths demarcated are shown. This DNA is completely digested with both *Eco*RI and *Bam*HI restriction enzymes. If the product is analysed by gel electrophoresis, how many distinct bands would be observed?

1 kb | 3 kb | 5 kb | 3 kb

Eco RI *Bam* HI *Bam* HI

(a) 5 (b) 2 (c) 3 (d) 4

120. Enzyme X catalyses hydrolysis of GTP into GDP. The GTP-bound form of X transmits a signal that leads to cell proliferation. The GDP-bound form does not transmit any such signal. Mutations in X are found in many cancers. Which of the following alterations of X are most likely to contribute to cancer?

(a) Mutations that increase the affinity of X for GDP
(b) Mutations that decrease the affinity of X for GTP
(c) Mutations that decrease the rate of GTP hydrolysis
(d) Mutations that prevent expression of enzyme X

Answers

PART-I

1 (d)	2 (a)	3 (c)	4 (a)	5 (b)	6 (b)	7 (b)	8 (b)	9 (a)	10 (a)
11 (d)	12 (d)	13 (d)	14 (a)	15 (b)	16 (d)	17 (b)	18 (b)	19 (c)	20 (d)
21 (a)	22 (d)	23 (c)	24 (b)	25 (d)	26 (b)	27 (a)	28 (b)	29 (a)	30 (a)
31 (c)	32 (a)	33 (d)	34 (b)	35 (c)	36 (b)	37 (d)	38 (a)	39 (a)	40 (a)
41 (c)	42 (b)	43 (b)	44 (d)	45 (c)	46 (b)	47 (b)	48 (a)	49 (b)	50 (d)
51 (a)	52 (a)	53 (b)	54 (a)	55 (c)	56 (d)	57 (c)	58 (a)	59 (c)	60 (a)
61 (a)	62 (b)	63 (a)	64 (c)	65 (c)	66 (a)	67 (b)	68 (c)	69 (a)	70 (b)
71 (a)	72 (a)	73 (b)	74 (a)	75 (b)	76 (d)	77 (c)	78 (c)	79 (b)	80 (a)

PART-II

81 (a)	82 (d)	83 (c)	84 (d)	85 (d)	86 (d)	87 (b)	88 (a)	89 (c)	90 (d)
91 (c)	92 (a)	93 (c)	94 (c)	95 (d)	96 (a)	97 (a)	98 (a)	99 (a)	100 (a)
101 (a)	102 (a)	103 (c)	104 (a)	105 (c)	106 (b)	107 (b)	108 (a)	109 (a)	110 (b)
111 (c)	112 (a)	113 (c)	114 (d)	115 (a)	116 (c)	117 (b)	118 (a)	119 (c)	120 (c)

Solutions

1. *(d)* We have,

$$x + y^2 = x^2 + y = 12$$

$\Rightarrow$ $x + y^2 = 12$ ...(i)

$x^2 + y = 12$ ...(ii)

On subtracting Eq. (i) from Eq. (ii), we get

$(x^2 - x) + (y - y^2) = 0$

$\Rightarrow$ $(x^2 - y^2) - (x - y) = 0$

$\Rightarrow$ $(x - y)(x + y - 1) = 0$

$\Rightarrow$ $x = y, x + y = 1$

When $x = y$

$\therefore$ $x^2 + x = 12 \Rightarrow x^2 + x - 12 = 0$

$\Rightarrow (x + 4)(x - 3) = 0$

$x = -4, 3, y = -4, 3$

When $x + y = 1$

$\therefore$ $x^2 + 1 - x = 12$

$\Rightarrow$ $x^2 - x - 11 = 0$

$\Rightarrow$ $x = \dfrac{1 \pm \sqrt{45}}{2}$

$\therefore$ Total number of ordered pair of (x, y) is 4.

2. *(a)* We have, $|z^3 + z^{-3}| \le 2$

Now, $\left| z^3 + \dfrac{1}{z^3} \right| \le 2$

$\Rightarrow$ $\left| z^3 + \dfrac{1}{z^3} \right| \le |z^3| + \dfrac{1}{|z^3|}$

$\Rightarrow$ $\dfrac{|z|^3 + \dfrac{1}{|z|^3}}{2} \ge \sqrt{|z|^3 \dfrac{1}{|z|^3}}$ [$\because$ AM $\ge$ GM]

$\Rightarrow |z|^3 + \dfrac{1}{|z|^3} \ge 2 \Rightarrow |z|^3 + \dfrac{1}{|z|^3} = 2$

$\therefore$ $|z| = 1$

$\therefore$ Maximum value of

$$\left| z + \frac{1}{z} \right| \le |z| + \frac{1}{|z|} = 2$$

3. *(c)* Let $S = (2014)^3 - (2013)^3 + (2012)^3 - (2011)^3 + ... + 2^3 - 1^3$

$\Rightarrow S = 2(2014^3 + 2012^3 + 2010^3 + ... + 2^3) - (2014^3 + 2013^3 + ... + 2^3 + 1^3)$

$\Rightarrow S = 2 \times 2^3 (1007^3 + 1006^3 + ... + 1^3) - (2014^3 + 2013^3 + ... + 2^3 + 1^3)$

$\Rightarrow S = \dfrac{2 \times 8 (1007)^2 (1008)^2}{4} - \dfrac{2014^2 \times 2015^2}{4}$

$\Rightarrow S = (1007)^2 \left[\dfrac{2 \times 8 \times (1008)^2}{4} - \dfrac{4 \times (2015)^2}{4} \right]$

$\Rightarrow S = (1007)^2 [(2016)^2 - (2015)^2]$

$\Rightarrow S = (1007)^2 (2016 + 2015)(2016 - 2015)$

$\Rightarrow$ $S = (1007)^2 (4031)$

Thus, S is divisible by $(1007)^2$.

4. *(a)* Given,

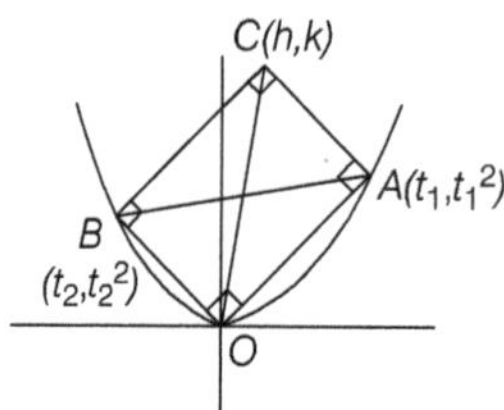

$BOAC$ is a rectangle in the XY-plane where O is the origin and A, B lie on $y = x^2$.

Point $A = (t_1, t_1^2)$

$B = (t_2, t_2^2)$

Slope of $OA = t_1$

Slope of $OB = t_2$

OA is perpendicular to OB.

$\therefore$ $t_1 t_2 = -1$

$h = t_1 + t_2$ and $k = t_1^2 + t_2^2$

$k = (t_1 + t_2)^2 - 2t_1 t_2$

$k = h^2 + 2$

$\therefore$ Locus of C is $y = x^2 + 2$

5. *(b)* Given,

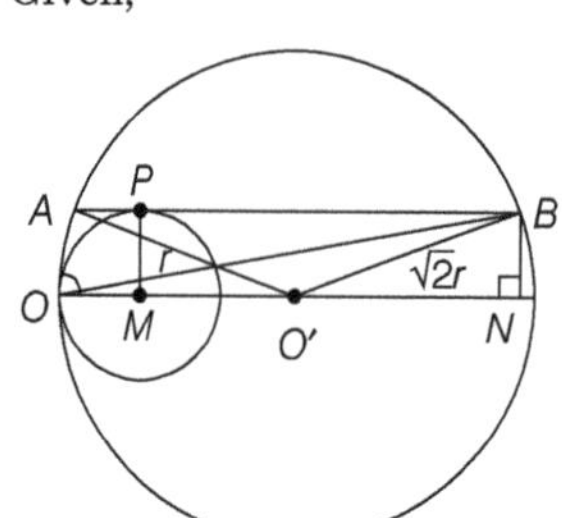

$O'B = R$

$PM = BN = r$

Now,

$R^2 = 2r^2$

$R = \sqrt{2} r$

$O'B = \sqrt{2} r$

$BN = r$

$\therefore$ $O'N = r$

In $\Delta O'BN$,

$O'N = BN = r$

$\therefore$ $\angle BO'N = 45°$

Similarly,

$\angle AO'O = 45°$

$\therefore$ $\angle AO'B = 90°$

$\Rightarrow$ $\angle AOB = \dfrac{1}{2} \angle AO'B$

$\Rightarrow$ $\angle AOB = 45°$

6. *(b)* We have,

$$(x - 2)^2 + (y - 3)^2 + (z - 6)^2 = 1$$

This is the equation of sphere whose centre is (2, 3, 6) and radius = 1

Distance from centre of sphere to origin is $\sqrt{2^2 + 3^2 + 6^2} = 7$

Smallest distance from any point on sphere is $7 - 1 = 6$

7. *(b)* We have,

$$\frac{\sin(\lambda\alpha)}{\sin\alpha} - \frac{\cos(\lambda\alpha)}{\cos\alpha} = \lambda - 1$$

$\Rightarrow \cos\alpha \sin\lambda\alpha - \cos(\lambda\alpha) \sin\alpha = (\lambda - 1)\sin\alpha\cos\alpha$

$\Rightarrow \sin(\lambda - 1)\alpha = \dfrac{\lambda - 1}{2} \sin 2\alpha$

$\therefore$ $\lambda - 1 = 2$ or $\lambda - 1 = 0$

$\therefore$ $\lambda = 3$ or $\lambda = 1$

Hence, λ has two values.

8. *(b)* We have,

$ABCDEF$ is a hexagon such that

$AB = BC = CD = 1$

and $DE = EF = FA = 2$

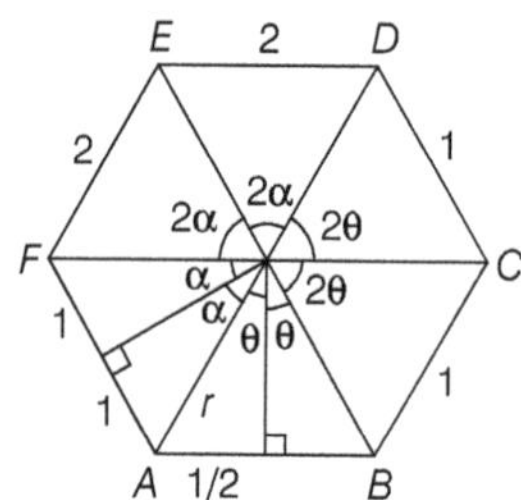

From figure,

$$\sin\theta = \frac{1}{2r} \text{ and } \sin\alpha = \frac{1}{r}$$

$3 \times 2\theta + 3 \times 2\alpha = 360°$

$\Rightarrow$ $\theta + \alpha = 60°$

Now, $\cos(\theta + \alpha) = \dfrac{1}{2}$

$\Rightarrow$ $\cos\theta\cos\alpha - \sin\theta\sin\alpha = \dfrac{1}{2}$

$\Rightarrow$ $\sqrt{\left(1 - \dfrac{1}{4r^2}\right)\left(1 - \dfrac{1}{r^2}\right)} - \dfrac{1}{2r} \times \dfrac{1}{r} = \dfrac{1}{2}$

$\Rightarrow$ $\sqrt{4r^2 - 1}\sqrt{r^2 - 1} - 1 = r^2$

$\Rightarrow$ $(4r^2 - 1)(r^2 - 1) = (r^2 + 1)^2$

$\Rightarrow 4r^4 - 5r^2 + 1 = r^4 + 2r^2 + 1 \Rightarrow 3r^4 = 7r^2$

$\Rightarrow$ $r^2 = \dfrac{7}{3}$

$\Rightarrow$ $r = \sqrt{\dfrac{7}{3}}$

9. *(a)* We have, $p(x) - p'(x) = x^n$

Let $\quad p(x) = a_0x^n + a_1x^{n-1} + a_2x^{n-2} + \dots + a_{n-1}x + a_n$

$p'(x) = na_0x^{n-1} + (n-1)\,a_1x^{n-2} + (n-2)\,a_2x^{n-3} + \dots + a_{n-1}$

$p(x) - p'(x) = a_0x^n + (a_1 - xa_0)\,x^{n-1} + \dots + a_n - a_{n-1}$

But $\quad p(x) - p'(x) = x^n$

$\therefore \quad a_0 = 1$

$a_1 - xa_0 = 0 \Rightarrow a_1 = n$

$a_2 - (n-1)\,a_1 = 0 \Rightarrow a_2 = n(n-1)$

$a_3 - (n-2)\,a_2 = 0 \Rightarrow a_3 = n(n-1)(n-2)$

$\therefore$ Similarly, $a_n = n(n-1)(n-2)\dots 1 = n!$

$\therefore p(0) = a_n = n!$

10. *(a)* Let

$$p = \lim_{x \to 0} \left(\frac{x}{\sin x}\right)^{6/x^2}$$

$$\Rightarrow \quad \log p = \lim_{x \to 0} \frac{6}{x^2} \log\left(\frac{x}{\sin x}\right)$$

$$\Rightarrow \quad \log p = \lim_{x \to 0} \frac{6\log\left(\dfrac{x}{\sin x}\right)}{x^2}$$

Apply L-Hospital rule

$$\log p = \lim_{x \to 0} 6\,\frac{\dfrac{\sin x}{x}\,\dfrac{(\sin x - x\cos x)}{\sin^2 x}}{2x}$$

$$\Rightarrow \log p = \lim_{x \to 0} 6\,\frac{\sin x\,(\sin x - x\cos x)}{x \cdot 2x\sin^2 x}$$

$$\Rightarrow \log p = \lim_{x \to 0} 3\,\frac{\sin x}{x} \times \lim_{x \to 0} \frac{\sin x - x\cos x}{\dfrac{\sin^2 x}{x^2}} \times x^3$$

$$\Rightarrow \quad \log p = 3 \times \lim_{x \to 0} \frac{\sin x - x\cos x}{x^3}$$

Again apply L-Hospital's rule

$$\log p = 3\lim_{x \to 0} \frac{\cos x - \cos x + x\sin x}{3x^2}$$

$$\Rightarrow \log p = \lim_{x \to 0} \frac{\sin x}{x} = 1$$

$\therefore \quad p = e^1 = e$

11. *(d)* Let the radius of circle be r and angle subtended by the radii is θ.

length of an arc $= r\theta$

Perimeter of sector $= 2r + r\theta$

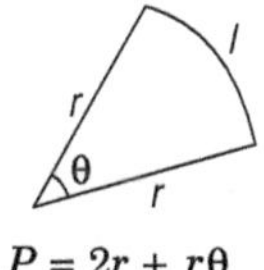

$\Rightarrow \quad P = 2r + r\theta$

Area of sector $= \dfrac{1}{2}r^2\theta$

$$\Rightarrow \quad A = \frac{1}{2}\theta\left(\frac{P}{2+\theta}\right)^2 \quad \left[\because r = \frac{P}{2+\theta}\right]$$

$$\Rightarrow \quad A = \frac{1}{2}\,\frac{\theta P^2}{(2+\theta)^2}$$

$$\Rightarrow \quad \frac{dA}{d\theta} = \frac{P^2}{2}\left[\frac{(2+\theta)^2 - \theta.2\,(2+\theta)}{(2+\theta)^4}\right]$$

For maxima or minima put $\dfrac{dA}{d\theta} = 0$

$$\therefore \quad \frac{(2+\theta)^2 - 2\theta\,(2+\theta)}{(2+\theta)^4} = 0$$

$\Rightarrow \quad (\theta + 2)\,(\theta + 2 - 2\theta) = 0$

$\Rightarrow \quad (\theta + 2)(2 - \theta) = 0 \Rightarrow \theta = 2, -2$

At $\theta = 2$ the arc of sector is maximum.

12. *(d)* We have,

$f(x) = \max\{|x|, |x-1|, \dots, |x-2n|\}$

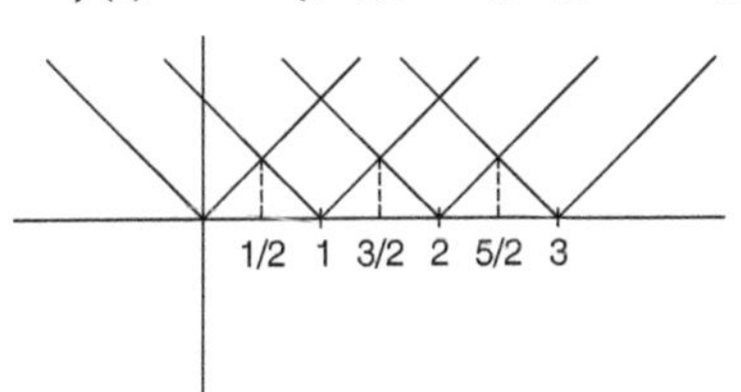

$$\int_0^{2n} f(x)\,dx = \int_0^n f(x)\,dx + \int_n^{2n} f(x)\,dx$$

$$= \int_0^n |x - 2n|\,dx + \int_n^{2n} |x|\,dx$$

$$= \int_0^n (2n - x)\,dx + \int_n^{2n} x\,dx$$

$$= \left[2nx - \frac{x^2}{2}\right]_0^n + \left[\frac{x^2}{2}\right]_n^{2n}$$

$$= \left(2n^2 - \frac{n^2}{2}\right) + \left(\frac{4n^2}{2} - \frac{n^2}{2}\right)$$

$$= 2n^2 - \frac{n^2}{2} + 2n^2 - \frac{n^2}{2} = 3n^2$$

13. *(d)* We have,

$p(x)$ is cubic polynomial.

Let $\quad p(x) = ax^3 + bx^2 + cx + d$

$p(1) = a + b + c + d = 3 \quad \dots(i)$

$p(0) = d = 2 \quad \dots(ii)$

$p(-1) = -a + b - c + d = 4 \quad \dots(iii)$

From Eqs. (i) and (iii), we get

$2\,(b + d) = 7 \Rightarrow b + d = \dfrac{7}{2} \quad \dots(iv)$

On putting the value of d in Eq. (iv), we get

$$b = \frac{3}{2}$$

Now, $\displaystyle\int_{-1}^{1} p(x)\,dx = \int_{-1}^{1} (ax^3 + bx^2 + cx + d)\,dx$

$$= 2\int_{-1}^{1} (bx^2 + d)\,dx$$

[$\because ax^3$ and cx are odd functions]

$$= 2\left[\frac{bx^3}{3} + dx\right]_0^1 = 2\left(\frac{b}{3} + d\right)$$

$$= 2\left[\frac{3}{2 \times 3} + 2\right] = 2\left[\frac{5}{2}\right] = 5$$

14. *(a)* Let $I = \displaystyle\int_0^{\infty} e^{-t}\,|x - t|\,dt, x > 0$

$$\Rightarrow \quad I = \int_0^x e^{-t}\,(x - t)\,dt + \int_x^{\infty} e^{-t}\,(t - x)\,dt$$

$$\Rightarrow \quad I = [-(x - t)\,e^{-t}]_0^x + \int_0^x \frac{d(x-t)}{dx}\,e^{-t}\,dt + \left[\frac{(t - x)\,e^{-t}}{-1}\right]_x^{\infty} + \int_x^{\infty} e^{-t}\,dt$$

$\Rightarrow \quad I = x + [e^{-t}]_0^x + [-e^{-t}]_x^{\infty}$

$\Rightarrow \quad I = x + e^{-x} - 1 + e^{-x}$

$\Rightarrow \quad I = x + 2e^{-x} - 1$

15. *(b)* Let the number of red, white, blue, green balls be r, w, b, g respectively and $r + w + b + g = n$.

Given, $\dfrac{{}^rC_4}{{}^nC_4} = \dfrac{{}^wC_1 \times {}^rC_3}{{}^nC_4}$

$$= \frac{{}^wC_1 \times {}^bC_1 \times {}^rC_2}{{}^nC_4}$$

$$= \frac{{}^rC_1 \times {}^wC_1 \times {}^bC_1 \times {}^gC_1}{{}^nC_4}$$

$\Rightarrow \quad {}^rC_4 = {}^rC_3 \cdot {}^wC_1$

$\Rightarrow \quad r - 3 = 4w \Rightarrow r = 4w + 3$

${}^rC_3\,{}^wC_1 = {}^wC_1\,{}^bC_1\,{}^rC_2 \Rightarrow r = 3b + 2$

${}^rC_2\,{}^wC_1{}^bC_1 = {}^rC_1{}^wC_1{}^bC_1{}^gC_1$

$\Rightarrow \quad r = 2g + 1$

Now, LCM of (4, 3, 2) = 12

$\therefore \quad r_{\min} = 11, \Rightarrow w_{\min} = 2$

$\Rightarrow b_{\min} = 3 \Rightarrow g_{\min} = 5$

$\therefore$ Minimum number of ball

$= r + w + b + g = 11 + 2 + 3 + 5 = 21$

16. *(d)* We have,

6 boxes $B_1, B_2, B_3, B_4, B_5, B_6$ and two dices D_1 and D_2.

Let $P(B_1)$ be the probability that B_1 contains at most one ball.

$\therefore P(B_1) = P(B_1$ contains 0 ball$) + P(B_1$ contains 1 ball$)$ $P(B_1) = (D_1$ never show 1$)$ $+ P(D_2$ shows 1 once when D_1 show 1$)$

$$P(B_1) = \left(\frac{5}{6}\right)^n + n\left(\frac{5^{n-1}}{6^{n-1}}\right)\left(\frac{1}{6^2}\right)$$

17. *(b)* Given,

$\vec{a} = 6\hat{i} - 3\hat{j} - 6\hat{k}, \vec{d} = \hat{i} + \hat{j} + \hat{k}$ and b is parallel to d and c is perpendicular to d.

$$\therefore \quad b = \lambda d \text{ and } \vec{c} \cdot \vec{d} = 0$$
$$b = \lambda(\hat{i} + \hat{j} + \hat{k})$$

Now, $\quad \vec{a} = \vec{b} + \vec{c}$

$$\Rightarrow \vec{c} = \vec{a} - \vec{b} = (6\hat{i} - 3\hat{j} - 6\hat{k}) - \lambda(\hat{i} + \hat{j} + \hat{k})$$
$$\Rightarrow c = (6-\lambda)\hat{i} - (3+\lambda)\hat{j} - (6+\lambda)\hat{k}$$
$$\Rightarrow \vec{c} \cdot \vec{d} = 0$$
$$\therefore ((6-\lambda)\hat{i} - (3+\lambda)\hat{j} - (6+\lambda)\hat{k}) \cdot (\hat{i} + \hat{j} + \hat{k}) = 0$$
$$\Rightarrow 6 - \lambda - 3 - \lambda - 6 - \lambda = 0 \Rightarrow \lambda = -1$$
$$\therefore \quad c = 7\hat{i} - 2\hat{j} - 5\hat{k}$$

18. *(b)* We have,

$$\log_{3x-1}(x-2) = \log_{(9x^2-6x+1)}(2x^2 - 10x - 2)$$
$$\Rightarrow \log_{(3x-1)}(x-2) = \log_{(3x-1)^2}(2x^2 - 10x - 2)$$
$$\Rightarrow \log_{3x-1}(x-2) = \frac{1}{2}\log_{3x-1}(2x^2 - 10x - 2)(x-2)^2$$
$$\Rightarrow (x-2)^2 = 2x^2 - 10x - 2$$
$$\Rightarrow x^2 - 4x + 4 = 2x^2 - 10x - 2$$
$$\Rightarrow x^2 - 6x - 6 = 0$$
$$\Rightarrow x = 3 \pm \sqrt{15}$$
$$\Rightarrow x = 3 - \sqrt{15}, 3 + \sqrt{15}$$

At $\quad x = 3 - \sqrt{15} \Rightarrow x - 2 < 0$

$$\therefore \quad x = 3 + \sqrt{15}$$

19. *(c)* We have,

$$2^a + 4^b + 8^c = 328$$
$$\Rightarrow 2^a(1 + 2^{2b-a} + 2^{3c-a}) = 2^3 \times 41$$
$$\therefore \quad a = 3$$
$$\Rightarrow 2^{2b-3} + 2^{3c-3} = 40$$
$$\Rightarrow 2^{2b-3} + 2^{3c-3} = 2^3 \cdot 5$$
$$\Rightarrow 3c - 3 = 3 \Rightarrow c = 2, b = 4$$
$$\therefore \quad \frac{a + 2b + 3c}{abc} = \frac{3+8+6}{3 \times 2 \times 4} = \frac{17}{24}$$

20. *(d)* In right angle ΔABC.

Let $\quad AB = 12$

$BC = x + r$

$AC = 12 + x - r$

$$AC^2 = AB^2 + BC^2$$
$$\Rightarrow (12 + x - r)^2 = (12)^2 + (x + r)^2$$
$$\Rightarrow 144 + 24(x - r) + (x - r)^2 = 144 + (x + r)^2$$
$$\Rightarrow 24(x - r) = (x + r)^2 - (x - r)^2$$
$$\Rightarrow 24(x - r) = 4xr \Rightarrow 6(x - r) = xr$$
$$\Rightarrow r = \frac{6x}{6 + x}$$

At $\quad x = 6, \; r = 3$

At $\quad x = 12, r = 4$

At $\quad x = 30, r = 5$

$\therefore$ Maximum radius = 5

21. *(a)*

For block, $v = 0, t = 2\,\text{s}, s = 1\,\text{m}$

Now, by equation of motion, we have

$$v = u + at$$
$$0 = u + a \times 2$$
$$\Rightarrow \quad 2a = -u \quad \ldots \text{(i)}$$

Now, $\quad v^2 - u^2 = 2as$

Gives,

$$0 - u^2 = -u \times 1$$
$$\Rightarrow u^2 - u = 0$$
$$\Rightarrow u(u - 1) = 0$$
$$\therefore \quad u = 0 \text{ or } u = 1$$

But $u = 0$ is not a valid condition for given experiment, so

$$u = 1 \quad \ldots \text{(ii)}$$

As total initial energy is lost as work done against friction.

By conservation of energy, we have

$$\frac{1}{2}mu^2 = fs$$
$$\Rightarrow \frac{1}{2}mu^2 = \mu_k mgs$$
$$\Rightarrow \frac{1}{2}\frac{u^2}{gs} = \mu_k$$
$$\Rightarrow \mu_k = \frac{1}{2 \times 10 \times 1} = \frac{1}{20}$$
$$= 0.05$$

In above calculation, we does not consider the air resistance and other dissipative forces. So, μ_k must be slightly less than 0.05 practically.

22. *(d)* We have following reactions,

$${}^{11}_{6}\text{C} \rightarrow {}^{11}_{5}\text{B} + e^+ + \nu_e + 0.96\,\text{MeV}$$

and

$$e^- + e^+ \longrightarrow \gamma + \gamma \text{ (gamma rays photons)}$$

Using $E = mc^2$, energy released in one pair annihilation process is

E = (mass of electron + mass of proton) $\times c^2$

$$= 2 \times 9 \times 10^{-31} \times 9 \times 10^{16}\,\text{J}$$
$$= \frac{2 \times 9 \times 10^{-31} \times 9 \times 10^{16}}{1.6 \times 10^{-19}}\,\text{eV}$$
$$= \frac{1.6 \times 10^{-15}}{1.6 \times 10^{-19}}$$
$$= 1 \times 10^4\,\text{eV}$$
$$= 10^{-2}\,\text{MeV}$$

As, time of reaction is second-half times and initial amount of ${}^{12}_{6}\text{C}$ is 1 μg.

Decay occurs as,

$${}^{12}\text{C}_6 \;(1\,\mu\text{g}) \xrightarrow{\varepsilon_0} {}^{12}\text{C}_6 \;(0.5\,\mu\text{g}) \xrightarrow{\varepsilon_0} {}^{12}\text{C}_6 \;(0.25\,\mu\text{g})$$

So, carbon used in reaction is

$$= (1 - 0.25)\,\mu\text{g} = 0.75\,\mu\text{g}$$
$$= 0.75 \times 10^{-6}\,\text{g}$$

Number of moles of carbon used

$$= \frac{0.75 \times 10^{-6}}{12}$$
$$= 0.0625 \times 10^{-6}\,\text{mol}$$

Number of atoms of carbon ${}^{12}_{6}\text{C}$ participated in reaction

$$= 0.0625 \times 10^{-6} \times 6.023 \times 10^{23}$$
$$= 0.376 \times 10^{17}\,\text{atoms}$$

Total energy obtained from fissions of ${}^{12}_{6}\text{C}$ is

$$E_1 = 0.376 \times 10^{17} \times 0.96\,\text{MeV}$$
$$= 3.6 \times 10^{16}\,\text{MeV}$$

and energy obtained from annihilation is

$$E_2 = 0.376 \times 10^{17} \times 10^{-2}$$
$$= 0.376 \times 10^{15}\,\text{MeV}$$
$$= 0.037 \times 10^{16}\,\text{MeV}$$

Total energy obtained is

$$E_1 + E_2 = 3.6 \times 10^{16} + 0.037 + 10^{16}$$
$$= 3.63 \times 10^{16}$$
$$\approx 4 \times 10^{16}\,\text{MeV}$$

23. *(c)* Let given plate has dimensions, as shown below.

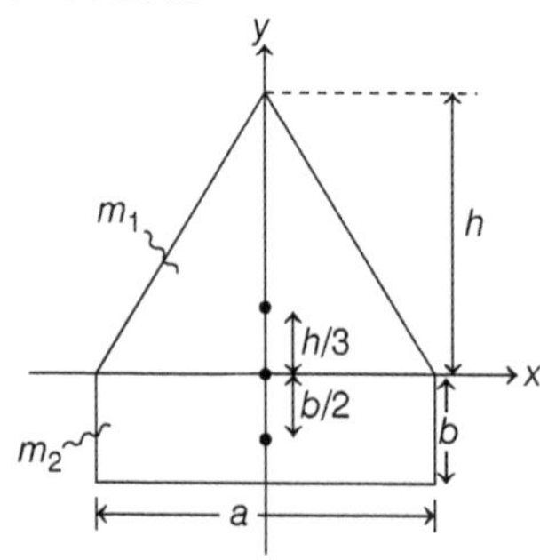

Given, area of triangle = area of rectangle

$\Rightarrow \quad \frac{1}{2} a \times h = ab$

$\Rightarrow \quad \frac{h}{2} = b$

$\Rightarrow \quad \frac{b}{h} = \frac{1}{2}$

From standard results, centre of mass of triangular part is $\frac{h}{3}$ above origin chosen, which is at centre of mass of complete (lamina).

Also, centre of mass of rectangular part is $\frac{b}{2}$ below the origin (shown in figure).

Now, for complete lamina,

$$Y_{CM} = \frac{m_1 y_1 + m_2 y_2}{m_1 + m_2}$$

$\Rightarrow \quad 0 = m_1 y_1 + m_2 y_2$

Here, m_1 = mass of triangular portion,

m_2 = mass of rectangular portion,

y_1 = position of centre of mass of triangular portion $= \frac{h}{3}$

and y_2 = position of centre of mass of rectangular portion $= -\frac{b}{2}$.

$\Rightarrow \quad 0 = m_1 \frac{h}{3} + m_2\left(-\frac{b}{2}\right)$

$\Rightarrow \quad \frac{m_1}{m_2} = \frac{3}{2} \times \frac{b}{h}$

$= \frac{3}{2} \times \frac{1}{2}$

$= \frac{3}{4}$

$\therefore \quad m_1 : m_2 = 3 : 4$

24. *(b)* Given arrangement of spheres is as shown below.

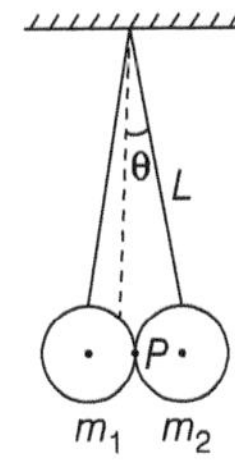

Free body diagram of spheres is

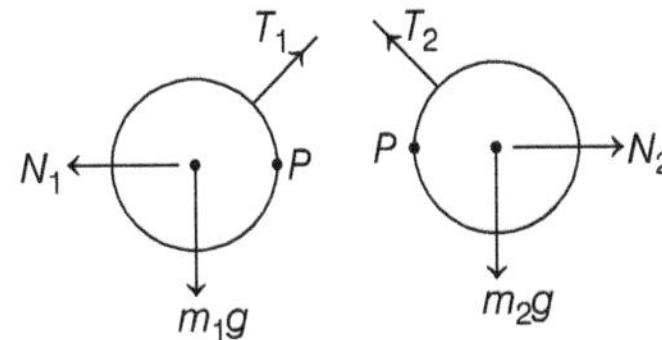

As, there is no rotation about point of contact P.

Equating moments of weights about centre line, we get

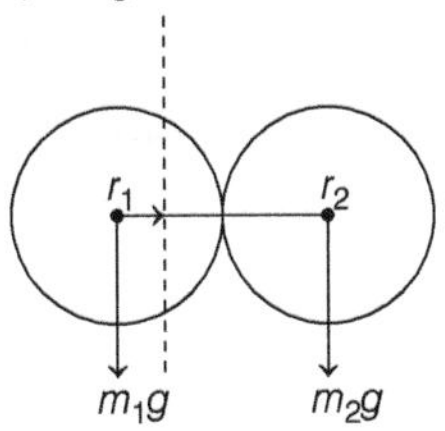

$$m_1 g \times r_1 = m_2 g \times r_2$$

where, $r_1 + r_2 = 2R$

$\Rightarrow \quad \frac{m_2}{m_1} r_2 + r_2 = 2R$

$\Rightarrow \quad r_2\left(\frac{m_2 + m_1}{m_1}\right) = 2R \Rightarrow r_2 = \frac{2m_1 R}{m_1 + m_2}$

Now, if angle made by string of m_2 with vertical line is θ, then

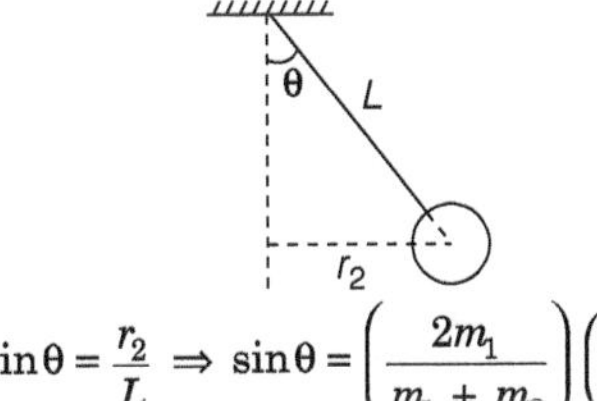

$$\sin\theta = \frac{r_2}{L} \Rightarrow \sin\theta = \left(\frac{2m_1}{m_1 + m_2}\right)\left(\frac{R}{L}\right)$$

As $R << L$, angle θ is small, therefore $\sin\theta \approx \theta$.

$\therefore \quad \theta = \frac{2m_1 R}{(m_1 + m_2)L}$

25. *(d)*

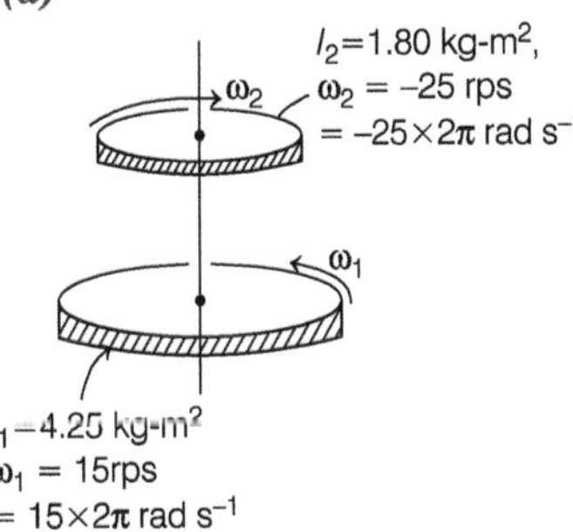

As, there is no external torque on the system of discs, angular momentum of system remains constant.

$\Rightarrow \quad I_1\omega_1 + I_2\omega_2 = (I_1 + I_2)\omega \quad \ldots$(i)

where, I_1 and I_2 are moment of inertias of discs and ω_1 and ω_2 are their angular speeds. Also, ω is the angular speed of combination of the discs.

Here, $I_1 = 4.25\,\text{kg-m}^2, \omega_1 = 15 \times 2\pi \frac{\text{rad}}{\text{s}}$

$I_2 = 1.8\,\text{kg-m}^2, \omega_2 = -25\pi \frac{\text{rad}}{\text{s}}$

where, ω_2 is negative as it is in clockwise sense.

So, from Eq. (i), we have

$4.25 \times 15 \times 2\pi + 1.8 \times (-25 \times 2\pi)$

$= (4.25 + 1.8)\,\omega$

$\Rightarrow \quad \omega = \frac{(63.75 - 45) \times 2\pi}{6.05} \frac{\text{rad}}{\text{s}}$

$= 3 \times 2\pi \frac{\text{rad}}{\text{s}}$

= 3 rps (anti-clockwise)

26. *(b)* Given, sound of ball (hitting the ground) is heard 5.25 s after the ball is thrown.

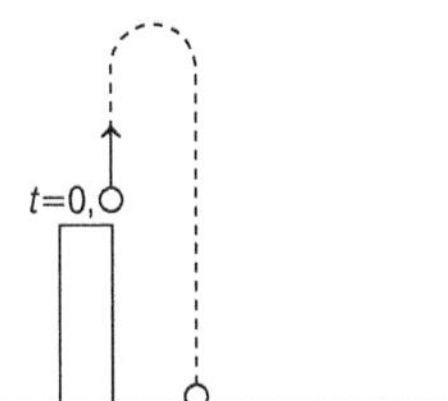

Sound of ball reaches top of tower in time,

$$t_1 = \frac{D}{s} = \frac{85}{340} = 0.25\,\text{s}$$

So, ball reaches ground in time,

$$t_2 = 5.25 - 0.25 = 5\,\text{s}$$

If u = initial speed of ball at $t = 0$, then for the ball,

$s = -85\,\text{m}, a = -g = -10\,\text{ms}^{-2}, t = 5\,\text{s}$

Using, $\quad s = ut + \frac{1}{2}at^2$

$-85 = u(5) + \frac{1}{2}(-10)(5)^2$

$\Rightarrow \quad 5u = 125 - 85 \Rightarrow u = 8\,\text{ms}^{-1}$

27. *(a)* When V is negative, diode is in reverse bias and it does not conducts. So, current-voltage relationship followed from Ohm's law is, $V = IR$

$\therefore V$- I graph is straight line.

As V grows in positive region, diode becomes conductive after knee voltage and after that current is approximately,

$$I = I_s e^{V/nV_T}$$

where, I_s = saturation current in reverse bias, V = voltage across diode n and V_T are diode constants. So, correct graph is

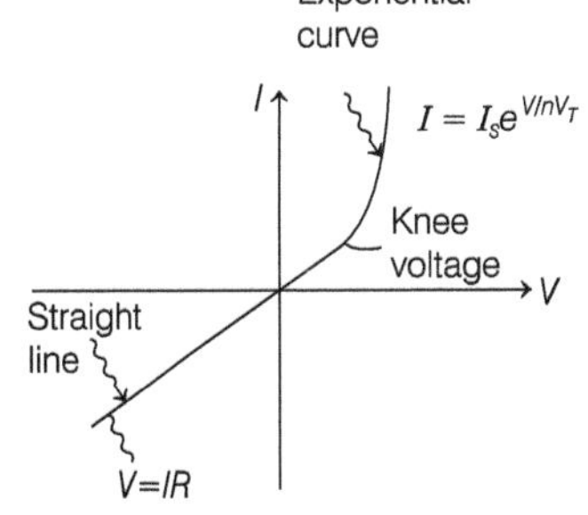

28. *(b)* Accelerated electrons exhibit wave nature and therefore they form interference pattern. Fringe width of pattern is given by

$$\beta = \frac{\lambda D}{d} \quad \text{...(i)}$$

where, λ = wavelength associated with electrons.

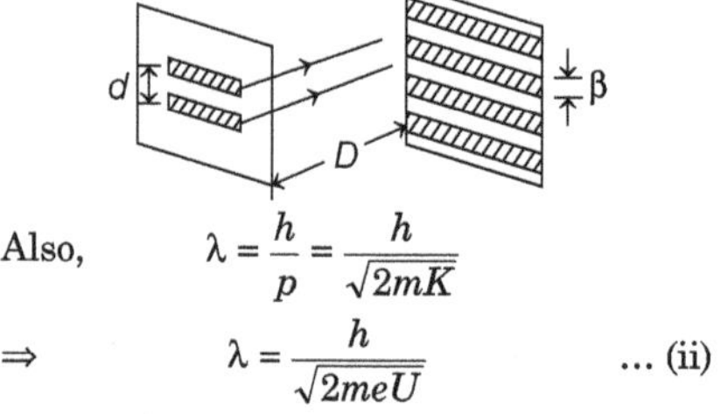

Also, $\lambda = \frac{h}{p} = \frac{h}{\sqrt{2mK}}$

$$\Rightarrow \quad \lambda = \frac{h}{\sqrt{2meU}} \quad \text{... (ii)}$$

So $\beta \propto \frac{1}{\sqrt{U}}$, as other quantities remains constant during the experiment.

So, $\frac{\beta_i}{\beta_f} = \frac{\sqrt{4U}}{\sqrt{U}} = 2$

$$\Rightarrow \quad \beta_f = \frac{1}{2}\beta_i$$

i.e. fringe width becomes half of its initial value.

29. *(a)*

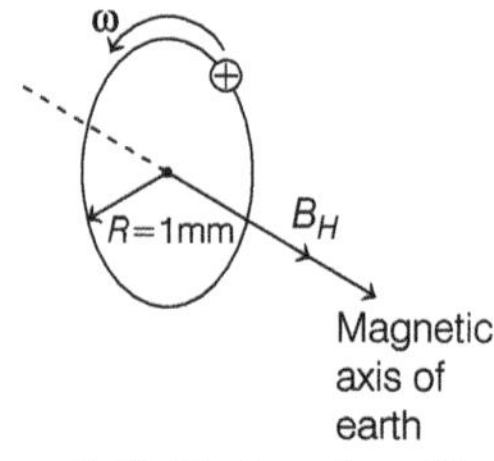

Net magnetic field at centre of loop is zero, when magnetic field due to rotating charge is exactly equal to the horizontal component of earth's magnetic field.

i.e. $\quad B_q = B_H \Rightarrow \frac{\mu_0 i}{2R} = B_H$

$$\Rightarrow \quad \frac{\mu_0\left(\frac{q}{T}\right)}{2R} = B_H \Rightarrow \frac{\mu_0 q\omega}{2R \cdot 2\pi} = B_H$$

$$\Rightarrow \quad \frac{\mu_0 q\omega}{4\pi R} = B_H$$

Here, $\frac{\mu_0}{4\pi} = 10^{-7} \text{TmA}^{-1}$, $q = 3 \times 10^{-12}$C,

$R = 1\text{mm} = 10^{-3}$m, $B_H = 30 \times 10^{-6}$ T

$$\Rightarrow \quad \frac{10^{-7} \times 3 \times 10^{-12} \times \omega}{1 \times 10^{-3}} = 30 \times 10^{-6}$$

$$\Rightarrow \quad \omega = \frac{30 \times 10^{-6} \times 10^{-3}}{3 \times 10^{-12} \times 10^{-7}} = 10^{11} \text{ rad s}^{-1}$$

30. *(a)* When bottle is opened on moon, the atmospheric pressure on water surface is zero. As a liquid boils when vapour pressure on the liquid surface is equal to the atmospheric pressure.

So, this condition met on moon even at 30°C. So, water starts to boil as soon as water bottle is opened.

31. *(c)* For a simple pendulum, equation of tangential motion is

$$F_T = -\ mg\sin\theta$$

$$\Rightarrow \quad ml\alpha = -\ mg\sin\theta$$

$$\Rightarrow \quad \frac{d^2\theta}{dt^2} + g\frac{l}{l}\sin\theta = 0$$

θ, l, F_T=mg sinθ, mg cosθ

Now, $\quad \sin\theta = \theta - \frac{\theta^3}{3!} + ...$

Approximately, $\sin\theta \approx \theta\left(1 - \frac{\theta^2}{6}\right)$

Now, replacing $\theta^2_{av} = \frac{1}{2}\theta_0^2$

We have, $\frac{d^2\theta}{dt^2} + \frac{g}{l}\left(1 - \frac{\theta_0^2}{12}\right)\theta = 0$

$$\therefore \quad \omega = \frac{g}{l}\left(1 - \frac{\theta_0^2}{12}\right)$$

So, time period of oscillation is

$$T = \frac{2\pi}{\omega} = 2\pi\sqrt{\frac{g}{l}}\left[\left(1 - \frac{\theta_0^2}{12}\right)^{-\frac{1}{2}}\right]$$

Now, taking binomial approximately

$$T \approx 2\pi\sqrt{\frac{g}{l}} \cdot \left(1 + \frac{\theta_0^2}{24}\right)$$

Clearly, this is greater than $T_0 = 2\pi\sqrt{\frac{g}{l}}$.

32. *(a)* Given is a series L-C-R circuit.

V_L=36V, V_C+V_R=25V, V_L+V_R=39V (A, B, C, D)

Given, $\quad V_L = 36\text{V} \quad \text{...(i)}$

$$\sqrt{V_L^2 + V_R^2} = 39\text{ V} \quad \text{...(ii)}$$

$$\sqrt{V_C^2 + V_R^2} = 25\text{ V} \quad \text{...(iii)}$$

So, from Eqs. (i) and (ii), we get

$$V_L^2 + V_R^2 = (39)^2$$

$$\Rightarrow \quad V_R^2 = (39)^2 - (36)^2 = (39 - 36)(39 + 36) = 3(75)$$

$$\therefore \quad V_R = 15\text{V} \quad \text{...(iv)}$$

From Eqs. (iii) and (iv), we have

$$V_C^2 + (15)^2 = (25)^2$$

$$\Rightarrow V_C^2 = (25)^2 - (15)^2 = (25 - 15)(25 + 15)$$

$$\therefore \quad V_C^2 = 10 \times 40 \Rightarrow V_C = 20\text{V} \quad \text{...(v)}$$

Now using,

$V_{L\text{-}C\text{-}R} = \sqrt{V_R^2 + (V_L - V_C)^2}$, we have

$$V_{L\text{-}C\text{-}R} = V_{AD} = \sqrt{225 + (36 - 20)^2} = \sqrt{225 + 256} = \sqrt{481}\text{ V}$$

33. *(d)* From Bohr's atomic model, radius of Bohr's orbit is given by

$$r_n = \frac{\varepsilon_0 h^2}{\pi m e^2} \cdot \frac{n^2}{Z}$$

So, in a medium of relative permittivity ε_r, radius of nth orbit of mass 0.07 m_e is

$$r_n = \frac{\varepsilon_0 \varepsilon_r h^2}{0.07\pi m_e e^2} \cdot \frac{n^2}{Z} = \frac{\varepsilon_r h^2}{0.07\, Z}\left(\frac{\varepsilon_0 h^2}{\pi m_e e^2}\right)$$

where, $\frac{\varepsilon_0 h^2}{\pi m_e e^2} = r_0$ = Bohr's radius of H-atom = 0.53 Å.

$$\therefore \quad r_n = \frac{\varepsilon_r h^2}{0.07 Z} \cdot (0.53)\text{ Å}$$

Here, $\quad n = 1, \varepsilon_r = 13$ and $Z = 1$.

∴ Radius of electron's lowermost energy level is

$$r = \frac{13 \times 0.53}{0.07} = 98.42\text{ Å} \approx 100\text{ Å}$$

34. *(b)* Equation of state for ideal gas is

$$pV = nRT \Rightarrow V = \frac{nR}{p} \cdot T$$

This is equation of line with slope $= \frac{nR}{p}$

$$\Rightarrow \quad \text{Slope} \propto \frac{1}{p}$$

So, from graph, $p_3 < p_2 < p_1$.

35. *(c)*

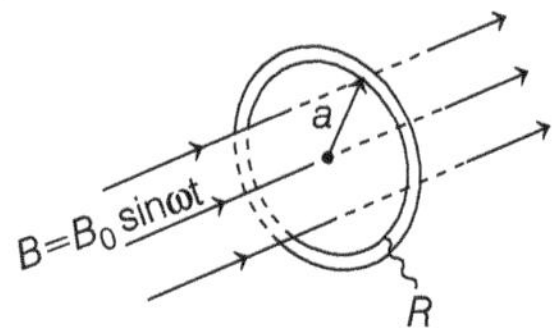

Induced emf in ring is

$$E = \frac{d}{dt}\phi_B = \frac{d}{dt}BA = \frac{d}{dt}B_0 \sin\omega t \cdot 2\pi a^2 = 2B_0\pi a^2 \cdot \omega \cdot \cos\omega t$$

Current in loop, $I = \frac{E}{R} = \frac{2B_0\pi a^2\omega}{R} \cdot \cos\omega t$

So, current oscillates with a frequency ω.

Heat loss per unit time $= I^2R$

$$= \frac{4B_0^2\pi^2a^4\omega^2}{R^2}\cos^2\omega t \cdot R$$

$$= \frac{4B_0^2\pi^2a^4\omega^2}{R}\cdot\cos^2\omega t$$

$\therefore$ Heat loss $\propto a^4$

Force on a small segment dl of ring is

$$F = Bidl = dl \times B_0 \sin\omega t \cdot \frac{2B_0\pi a^2\omega}{R}\cos\omega t$$

$\therefore$ Force per unit length on loop is

$$\frac{F}{dl} = \frac{2B_0^2\pi a^2\omega}{R}\sin\omega t\cos\omega t$$

$\therefore$ Force per unit length $\propto B_0^2$

Also, net force on loop is zero.

36. *(b)* Given, $A = G^\alpha M^\beta c^\gamma$

Substituting dimensions of A, M and c, we have

$$[L]^2 = \left[\frac{MLT^{-2}L^2}{M^2}\right]^\alpha [M]^\beta \left[\frac{L}{T}\right]^\gamma$$

$\Rightarrow \quad [L]^2 = [M^{-1}L^3T^{-2}]^\alpha [M]^\beta [LT^{-1}]^\gamma$

$\Rightarrow \quad [L]^2 = M^{-\alpha+\beta}L^{3\alpha+\gamma}T^{-2\alpha-\gamma}$

Equating dimensions on both sides, we have

$-\alpha + \beta = 0$ ...(i)

$3\alpha + \gamma = 2$ ...(ii)

$-2\alpha - \gamma = 0$ ...(iii)

Adding Eqs. (ii) and (iii), we get

$\Rightarrow \quad 3\alpha + \gamma - 2\alpha - \gamma = 2 + 0 \Rightarrow \alpha = 2$ ...(iv)

Now, putting the value of α in Eq. (i), we get

$\Rightarrow \quad -2 + \beta = 0 \Rightarrow \beta = 2$

Again, putting the value of α in Eq. (iii), we get

$\Rightarrow \quad -2 \times 2 - \gamma = 0 \Rightarrow \gamma = -4$

So, $\quad \alpha = 2, \beta = 2$ and $\gamma = -4$

37. *(d)* Total photons emitted from source is

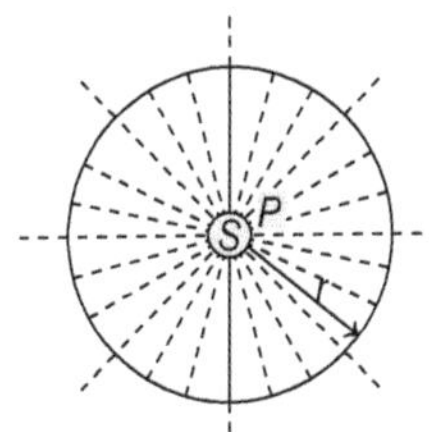

$$N = \frac{\text{Power}}{hc/\lambda} = \frac{160 \times 50000 \times 10^{-10}}{6.6 \times 10^{-34} \times 3 \times 10^8}$$

So, photon flux falling over spherical surface of radius $r = 1.8$ m is

$$n = \frac{N}{A} = \frac{N}{4\pi r^2}$$

$$= \frac{160 \times 50000 \times 10^{-6}}{6.6 \times 10^{-34} \times 3 \times 10^8 \times 4\pi \times (1.8)^2}$$

$$\approx 10^{20}\ \text{m}^{-2}\text{s}^{-1}$$

38. *(a)* In an n-sided polygon magnetic field due to one of the side at

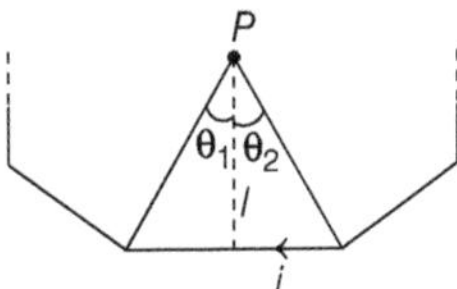

Centre of loop using Biot-Savart's law is

$$B_1 = \frac{\mu_0 I}{4\pi l}(\sin\theta_1 + \sin\theta_2) \quad \text{...(i)}$$

As, there are n sides angle made by one of the side at centre is

$$\alpha = \frac{2\pi}{n}$$

So, $\quad \theta_1 = \theta_2 = \left(\frac{2\pi}{n}\right) \times \frac{1}{2} = \frac{\pi}{n}$

So, from Eq. (i), we have

$$B_1 = \frac{\mu_0 I}{4\pi l}\left(\sin\frac{\pi}{n} + \sin\frac{\pi}{n}\right) = \frac{\mu_0 I}{2\pi l}\left(\sin\frac{\pi}{n}\right)$$

At centre field due to all n segments are added up. So, magnetic field at centre due to complete polygon is

$$B = n \times B_1 = \frac{\mu_0 n I}{2\pi l}\sin\left(\frac{\pi}{n}\right)$$

39. *(a)* Sound level β is generally expressed (in decibel) as,

$$\beta = 10(\text{dB})\cdot\log_{10}\left(\frac{I}{I_0}\right)$$

where, $\quad I_0$ = reference intensity $= 10^{-12}$ W/m^2.

Now, let intensity is initially I, then sound level is

$$\beta_1 = 10\log_{10}\left(\frac{I}{I_0}\right)$$

When intensity increases by 100 times, sound level will be

$$\beta_2 = 10\log_{10}\left(\frac{100I}{I_0}\right)$$

$$= 10\left(\log_{10}100 + \log_{10}\left(\frac{I}{I_0}\right)\right)$$

$\Rightarrow \quad \beta_2 = 10\log_{10}100 + \beta_1$

$\Rightarrow \quad \beta_2 - \beta_1 = 10 \times 2 = 20$ dB

So, sound level increases by 20 dB.

40. *(a)* Kinetic energy of block is converted into potential energy of stretched wire.

$\Rightarrow \quad \frac{1}{2}mv^2 = \frac{1}{2}$ Stress $\times$ Strain $\times$ Volume

$\Rightarrow mv^2 = \frac{\text{Stress}}{\text{Strain}} \times (\text{Strain})^2 \times \text{Volume}$

$\Rightarrow mv^2 = Y \times \frac{\Delta l^2}{L^2} \times A \times L \Rightarrow \Delta l^2 = \frac{mv^2L}{AY}$

$\therefore \quad \Delta l = v\sqrt{\frac{mL}{AY}}$

So, distance travelled by block in stretching the wire is $\Delta l = v\sqrt{\frac{mL}{AY}}$.

41. *(c)* Given compounds of boron are electron deficient molecule, which have tendency to accept a pair of electrons to achieve stable electronic configuration and thus behave as Lewis acids. Lewis acid strength of BBr_3, BCl_3 and BF_3 increases in the order

$$BF_3 < BCl_3 < BBr_3.$$

BF_3 has least acidity due to effective back bonding between $2p$-orbital of B and $2p$-orbital of F. Whereas, no such back bonding is observed in BBr_3 and thus has maximum acidity.

42. *(b)* Isoelectronic species are those species, which have same number of electrons.

Total no. of e^-s in $O^{2-} = 8 + 2 = 10$

Total no. of e^-s in $Zn^{2+} = 30 - 2 = 28$

Total no. of e^-s in $Mg^{2+} = 12 - 2 = 10$

Total no. of e^-s in $K^+ = 19 - 1 = 18$

Total no. of e^-s in $Ni^{2+} = 28 - 2 = 26$

Thus O^{2-} is isoelectronic with Mg^{2+}.

43. *(b)* The bond angle in a molecule gets effected by the electronegativity of the central atom. More is the electronegativity of the central atom, lesser will its bond angle.

Among the given compounds, the electronegativity of central atom increases in the order.

$$C < N < O$$

Thus, the bond angle must decrease in order.

$$CH_2 > NH_2 > H_2O$$

$\therefore$ Bond angles of H—C—H, H—N—H and H—O—H in methane, ammonia and water respectively are 109.5, 107.1, 104.5.

44. *(d)* In alkaline medium, the reaction of hydrogen peroxide with potassium permanganate produces, manganese oxide, MnO_2.

$$KMnO_4 + H_2O_2 \xrightarrow{^-OH} MnO_2 + H_2O$$

Let the oxidation state of Mn in MnO_2 be x.

$$x + 2(-2) = 0$$

$$x = +4$$

45. *(c)* The temperature dependance of rate of a chemical reaction is expressed by Arrhenius equation

$$k = Ae^{-E_a/RT}$$

When T is very high then factor E_a/RT approaches to 0.

$$k = Ae^0$$
$$\therefore \quad k = A$$

Thus, the rate constant of a chemical reaction at very high temperature will approach Arrhenius frequency factor.

46. *(b)* According to electrochemical series, more is negative emf value, stronger is the reducing agent and thus more will be the reducing strength. Therefore, the reducing strength of the metals follows the order

$$Cr > Pb > Cu > Ag$$

47. *(b)* Optical activity can be exhibited only by those molecules, which have chiral centre.

(a) $CH_3—CH_2—CH_2—Br$
1-bromopropane

No chiral centre present thus, it will not exhibit optical activity.

(b) $CH_3—\overset{*}{C}H(Br)—CH_2—CH_3$
2-bromobutane

As there is the presence of one chiral centre (*), so it will exhibit optical activity.

(c) $CH_3—CH_2—CH(Br)—CH_2CH_3$
(3-bromopentane)

It is optically inactive.

(d) Br attached to cyclohexane ring
(Bromocyclohexane)

It is also optically inactive.

48. *(a)*

$$\underset{\text{Styrene}}{C_6H_5CH=CH_2} \xrightarrow[\Delta]{\text{Dibenzoyl peroxide}} \underset{\text{Polystyrene}}{\left[CH(C_6H_5)—CH_2 \right]_n}$$

This polystyrene is clear, hard and rather brittle. It is a rather poor barrier to oxygen and water vapour and has a relatively low melting point. It is a thermoplastic polymer.

49. *(b)* The major product of the reaction between CH_3CH_2ONa and $(CH_3)_3CCl$ in ethanol is 2-methyl propene.

$$CH_3—C(CH_3)_2—Cl \xrightarrow{CH_3CH_2O^-Na^+} CH_3—C(CH_3)=CH_2$$

As $(CH_3)_3CCl$ is tertiary halide and $CH_3CH_2O^-Na^+$ is a strong base, so elimination is favoured over substitution reaction and the resultant product is an alkene. This type of elimination is known as β-elimination, where hydrogen is eliminated from β-carbon and the halogen is lost from α-carbon atom.

50. *(d)* When H_2S gas is passed through a hot acidic aqueous solution containing Al^{3+}, Cu^{2+}, Pb^{2+} and Ni^{2+}, precipitate of only group II cations is formed. Among the given cations, Pb^{2+} and Cu^{2+} lie in group 2 cations.

$$Cu^{2+} + H_2S \longrightarrow \underset{\text{(Black ppt.)}}{CuS} + 2H^+$$

$$Pb^{2+} + H_2S \longrightarrow \underset{\text{(Black ppt.)}}{PbS} + 2H^+$$

51. *(a)* The group number of given electronic configuration are as follows :

Electronic configuration	Group number
$1s^2 2s^2 2p^6$	18
$1s^2 2s^2 2p^6 3s^1$	1
$1s^2 2s^2 2p^6 3s^2$	2
$1s^2 2s^2 2p^1$	13

As electronic configuration, $1s^2 2s^2 2p^6$ corresponds to group 18, i.e. noble gases, thus electron removal from fully-filled stable configuration is very difficult. Therefore, it will have the largest difference between 1 and 2 ionisation energies.

52. *(a)* Electronegativity of carbon is directly proportional to the *s*-character in hybridised state, i.e. more is the *s*-character more will be the electronegativity.

Thus, the order of electronegativity of carbon will be

$$\underset{(50\%\,s\text{-character})}{sp} > \underset{(33.3\%\,s\text{-character})}{sp^2} > \underset{(25\%\,s\text{-character})}{sp^3}$$

53. *(b)* The most abundant transition metal in human body is iron. Iron is present in haemoglobin. It is the iron-containing oxygen transport metalloprotein in the red blood cells. Haemoglobin in the blood carries oxygen from the lungs or gills to the rest of the body.

54. *(a)* Molar conductivity always increases with increase in concentration of electrolyte. Among the given compounds, the molar conductivity depends upon the concentration of H^+ ions. Thus, stronger the acid more will be H^+ ions and more will be the molar conductivity. The order of acidic strength is

$$\underset{\text{Strong acid}}{HCl} > \underset{\text{Weak acid}}{CH_3COOH} > \underset{\text{Neutral salt}}{NaCl} > \underset{\text{Salt of weak acid}}{CH_3COONa}$$

Thus, molar conductivity at infinite dilution follows the order.

$$HCl > CH_3COOH > NaCl > CH_3COONa$$

55. *(c)* Spin magnetic moment of a complex, can be calculated,

$\mu = \sqrt{n(n+2)}$ where, n is the number of unpaired electron.

The oxidation state of Z in $[ZCl_4]^{2-}$ is +2. The spin magnetic moment of given metals in +2 state are as follows:

(a) Mn^{2+}

Electronic configuration = $3d^5 4s^0$

3d: | 1 | 1 | 1 | 1 | 1 |

$\mu = \sqrt{5(5+2)} = 5.9$ BM

(b) Ni^{2+}

Electronic configuration = $3d^8 4s^0$

| 1L | 1L | 1L | 1 | 1 |

$\mu = \sqrt{2(2+2)} = 2.9$ BM

(c) Co^{2+}

Electronic configuration = $3d^7 4s^0$

| 1L | 1L | 1 | 1 | 1 |

$\mu = \sqrt{3(3+2)} = 3.87$ BM

(d) Cu^{2+}

Electronic configuration = $3d^9 4s^0$

| 1L | 1L | 1L | 1L | 1 |

$\mu = \sqrt{1(1+2)} = 1.9$ BM

Therefore, Z in $[ZCl_4]^{2-}$ is Co, where the spin only magnetic moment is 3.87 BM.

56. *(d)* If α-D-glucose is dissolved in water and kept for a few hours, the major constituents present in the solution is a mixture of α-D glucose and β-D glucose. These two cyclic form exist in equilibrium with open chain structure.

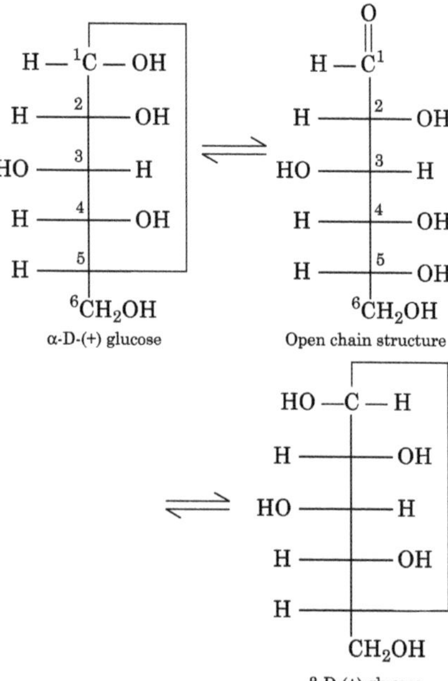

α-D-(+) glucose

Open chain structure

β-D-(+) glucose

57. *(c)* According to Bronsted-Lowry concept of acids and bases, more is the stability of the conjugate base stronger will be the acid. The conjugate bases of the given solutions are :

$$HCl \rightleftharpoons H^+ + \underset{\text{Conjugate base}}{Cl^-}$$

$$HCOOH \rightleftharpoons \underset{\text{Conjugate base}}{HCOO^- + H^+}$$

$$CH_3COOH \rightleftharpoons \underset{\text{Conjugate base}}{CH_3COO^- + H^+}$$

Stability of conjugate base follows the order

$$\underset{\text{Maximum stable as −ve charge is dispersed overlarge size of Cl}}{Cl^-} > H\overset{\overset{O}{\|}}{C}O^- > CH_3\overset{\overset{O}{\|}}{\bar{C}}O \text{ (Least stable because of } -I\text{- effect)}$$

Thus, the acidic strength follows the same order, i.e.

$$HCl > HCOOH > CH_3COOH$$

We also know that pH is inversely proportional to the concentration of H^+ ions (acidic strength)

i.e. $$pH \propto \frac{1}{[H^+]}$$

Thus, the pH of 1N aqueous solution follows the order

$$CH_3COOH > HCOOH > HCl$$

58. *(a)*

[Reaction scheme: alkene $\xrightarrow{H^+}$ carbocation $\xrightarrow[\text{Markownikoff addition}]{H_2O}$ $\overset{+}{O}H_2$ intermediate $\xrightarrow{-H^+}$ HO product (Major product)]

Addition of water to alkene in the presence of an acid gives alcohol, which is in accordance with Markownikoff's rule. According to this rule, negative part (OH^-) of water gets attached to that carbon atom, which possesses lesser number of hydrogen atoms.

59. *(c)* Reaction of aniline with $NaNO_2$ + dil. HCl at 0°C gives benzene diazonium salt, which then reacts with CuCN to give cyanobenzene with the evolution of N_2 gas.

[Scheme: Aniline ($C_6H_5NH_2$) $\xrightarrow[(HNO_2)]{NaNO_2 + \text{dil. HCl}}$ Benzene diazonium chloride ($C_6H_5\overset{+}{N}\equiv N\overset{-}{Cl}$) $\xrightarrow{CuCN}$ Cyanobenzene (C_6H_5CN) + N_2]

60. *(a)* Schottky defect is basically a vacancy defect in ionic solids, where the number of missing cations and anions are equal in order to maintain electrical neutrality. This type of defect decreases the density of the substance and are shown by ionic substances, where the cation and anion are of almost similar sizes, e.g. NaCl, AgBr, etc.

61. *(a)* Cyclosporine works by suppressing the immune system to prevent the white blood cells from trying to get rid of the transplanted organ. Its mode of action is inhibition of the production of cytokines involved in the regulation of T-cell activation. In particular, cyclosporine inhibits the transcription of interleukin 2.

62. *(b)* Glucose represses *lac* operon. This is because the enzymes to metabolise glucose are made constantly by *E. coli*. When both glucose and lactose are available, the genes for lactose metabolism are transcribed at low levels. Maximal transcription of the *lac* operon occurs only when glucose is absent and lactose is present.

63. *(a)* The amino acids in an α-helix are arranged in a right-handed helical structure where each amino acid residue corresponds to a 100° turn in the helix (i.e. the helix has 3.6 residues per turn). Its pitch = 5.4 Å.

∴ The number of amino acids in polypeptide $= \frac{3.6}{5.4 \times 10^{-10}} \times 27 \times 10^{-6}$

$= 1.8 \times 10^4$ amino acids

$= 18000$ amino acids

64. *(c)* *d*NTPs have phophoric acid anhydride bonds. *d*NTPs stand for deoxyribonucleotide triphosphate. Each *d*NTP is made up of a phosphate group, a deoxyribose sugar and a nitrogenous base.

65. *(c)* Sympathetic and parasympathetic nervous system work antagonistic to each other. During exercise, the parasympathetic control decreases and the sympathetic control increases. Vagus nerve is a spinal nerve which controls the heartbeat during rest as well as during exercise. Stimulation of sympathetic system results in the dilation of pupil.

66. *(a)* Genetic codes are triplet. So, the total number of codon = 64. Out of these there is one start (coding for methionine) codon and three stop codons (i.e. do not code for any amino acid).

∴ Functional codon = 64 − 3 = 61

∴ Frequency of encountering stop codon

$$= \frac{3}{61} = \frac{1}{20} \text{ approx.}$$

67. *(b)* The two alleles that determine the blood group AB of an individual are located on the same autosome. Homologous chromosomes have the same genes arranged in the same order but they have slightly different DNA sequences. Different versions of the same genes are called alleles; homologous chromosomes often contain different alleles.

68. *(c)* A selectable marker is a gene introduced into a cell, especially a bacterium or to cells in culture, that confers a trait suitable for artificial selection, i.e. it allows the selection of positively transformed cells and to eliminate the non-transformed cells.

69. *(a)* Primary spermatocytes are diploid (2*n*) cells. After meiosis-I, two secondary spermatocytes are formed. Secondary spermatocytes are haploid (*n*) cells that contain half the number of chromosomes.

70. *(b)* Phospholipids are formed by the esterification of one glycerol and two fatty acid molecules. A phospholipid contains a phosphate group in its molecule. These are a major component of all cell membranes. They can form lipid bilayer, because of their amphiphilic characteristic.

71. *(a)* The histones are rich in the amino acids arginine and lysine. Histones are highly alkaline proteins found in eukaryotic cell nuclei that package and order the DNA into structural units called nucleosomes. They are the chief protein components of chromatin, acting as spools around which DNA winds and playing a role in gene regulation.

72. *(a)* Molecular weight of polypeptide $= 15.3\,\text{kDa}$

Molecular weight of amino acid = 90 Da

$\therefore$ Number of amino acids in polypeptide

$$= \frac{15.3 \times 10^3}{90} = 170$$

We have one amino acid coded by three nitrogen base

$\therefore$ 170 amino acids would be coded by $170 \times 3 = 510$ nitrogen bases.

73. *(b)* The DNA with more cytosine and guanine has high melting temperature, because of the presence of triple bonds between them. There in the question the DNA strand with highest melting temperature is DNA with 30% cytosine.

74. *(a)* The incorrectly matched immunoglobulin is IgD – viral pathogen. Immunoglobulin D (IgD) is an antibody isotype that makes up about 1% of proteins in the plasma membranes of immature B-lymphocytes, where it is usually co-expressed with another cell surface antibody called IgM. In B-cells, the function of IgD is to signal the B-cells to be activated.

75. *(b)* Ninhydrin test is used to detect amino acids. This test is due to a reaction between an amino group of free amino acid and ninhydrin, which gives a coloured product.

76. *(d)* Pleiotropy occurs when one gene influences two or more seemingly unrelated phenotypic traits. Such a gene that exhibits multiple phenotypic expressions is called a pleiotropic gene.

77. *(c)* Antibodies that specifically target a certain antigen, such as one found on cancer cell. They can then make many copies of that antibody in the lab. These are known as monoclonal antibodies (*m*Abs). Monoclonal antibodies are used to treat many diseases, including some types of cancer.

78. *(c)* Apoptosis is programmed cell death. Apoptosis occurring in a cell can be visualised as a ladder pattern of 180-200 bp due to DNA cleavage by the activation of a nuclear endonuclease by standard agarose gel electrophoresis.

79. *(b)* Vitamin-B_{12} is involved in cellular metabolism in two active coenzyme forms-methylcobalamine and 5-deoxyadenosyl cobalamine. Vitamin-B_{12} cooperates with folic acid (folate) in the synthesis of Deoxyribonucleic Acid (DNA).

80. *(a)* Peptidoglycan also known as murein is a polymer consisting of sugars and amino acids that forms a mesh-like layer outside the plasma membrane of most bacteria, forming the cell wall. The glycan component is typically composed of two amino sugars, N-Acetyl-D-Glucosamine (NAG) and N-Acetylmuramic Acid (NAM). The NAM carries a tetrapeptide of alternating D and L amino acids.

81. *(a)* Given,

$$x = (\sqrt{50} + 7)^{1/3} - (\sqrt{50} - 7)^{1/3}$$

On cubing both sides, we get

$$x^3 = (\sqrt{50} + 7) - (\sqrt{50} - 7) - 3 (\sqrt{50} + 7)^{1/3} (\sqrt{50} - 7)^{1/3} [(\sqrt{50} + 7)^{1/3} - (\sqrt{50} - 7)^{1/3}]$$

$$\Rightarrow \quad x^3 = 14 - 3\,(50 - 49)^{1/3}\,(x)$$

$$\Rightarrow \quad x^3 = 14 - 3x$$

$$\Rightarrow \quad x^3 + 3x - 14 = 0$$

$$\Rightarrow (x - 2)\,(x^2 + 2x + 7) = 0$$

$$\Rightarrow \quad x = 2$$

82. *(d)* We have, $(1 + x + x^2)^{2014}$

$$= a_0 + a_1 x + a_2 x^2 + \ldots + a_{4028} x^{4028}$$

Put $x = -1$

$$1 = a_0 - a_1 + a_2 - a_3 + a_4 - \ldots + a_{4028} \quad \ldots\text{(i)}$$

Put $x = -\omega$

$$(1 - \omega + \omega^2)^{2014} = (2\omega)^{2014}$$

$$= a_0 - a_1\omega + a_2\omega^2 - a_3 + \ldots \quad \ldots\text{(ii)}$$

Put $\quad x = -\omega^2$

$$(1 - \omega^2 + \omega)^{2014} = (2\omega^2)^{2014}$$

$$= a_0 - a_1\omega^2 + a_2\omega - a_3 + a_4\omega^2 + \ldots \quad \ldots\text{(iii)}$$

On adding Eqs. (i), (ii) and (iii), we get

$$1 + (2\omega)^{2014} + (2\omega^2)^{2014} = 3(a_0 - a_3 + a_6 + \ldots)$$

$$\Rightarrow a_0 - a_3 + a_6 + \ldots = \frac{1 + 2^{2014}\omega + 2^{2014}\omega}{3}$$

$$\Rightarrow \quad A = \frac{1 - 2^{2014}}{3}$$

$$\Rightarrow \quad |A| = \frac{2^{2014} - 1}{3}$$

and Eq. (i) + Eq. (ii) ω + Eq. (iii) ω^2, we get

$$\frac{1 + 2^{2014}\omega^{2014}\omega + 2^{2014}(\omega^2)^{2014}\omega^2}{3} = a_2 - a_5 + a_8 + \ldots$$

$$\Rightarrow \quad C = \frac{1 + 2^{2014}\omega^{2015} + \omega^{2014}\omega^{4030}}{3}$$

$$\Rightarrow \quad C = \frac{1 - 2^{2014}}{3}$$

$$\Rightarrow \quad |C| = \frac{2^{2014} - 1}{3} = |A| \text{ and Similarly,}$$

Eq. (i) + Eq. (ii) ω^2 + Eq. (iii) ω, we get

$$\frac{1 + 2^{2014}.\,\omega^{2014}\omega^2 + 2^{2014}((\omega)^2)^{2014}.\,\omega}{3} = a_1 - a_4 + a_7 + \ldots$$

$$\Rightarrow \quad B = \frac{1 + 2^{2014}\omega^{2016} + 2^{2014}\omega^{4029}}{3}$$

$$\Rightarrow \quad B = \frac{1 + 2^{2014} + 2^{2014}}{3} = \frac{1 + 2^{2015}}{3}$$

$$\Rightarrow \quad |B| > |A| = |C|$$

83. *(c)* Equation of hyperbola $xy = 1$

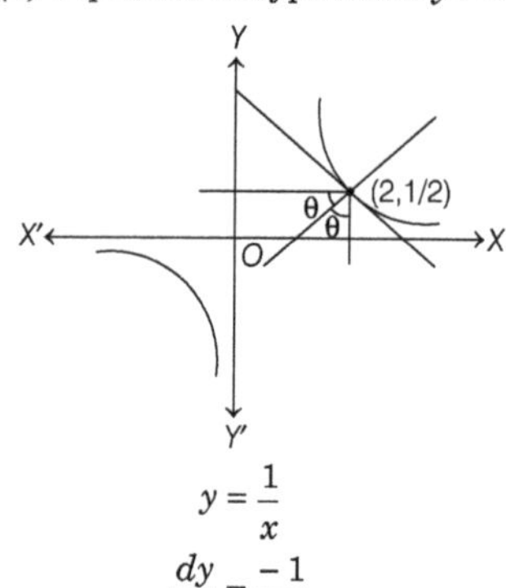

$$y = \frac{1}{x}$$

$$\frac{dy}{dx} = \frac{-1}{x^2}$$

Slope of tangent at $\left(2, \frac{1}{2}\right)$

$$\left(\frac{dy}{dx}\right)_{2, \frac{1}{2}} = \frac{-1}{2^2} = \frac{-1}{4}$$

$$\text{Slope of normal} = \frac{-1}{\frac{-1}{4}} = 4$$

Let the slope of beam of light $= m$

$$\therefore \quad \left|\frac{4 - m}{1 + 4m}\right| = \left|\frac{\infty - 4}{1 - 4 \cdot \infty}\right|$$

$$\Rightarrow \quad \frac{4 - m}{1 + 4m} = \pm\frac{1}{4} \Rightarrow m = \frac{15}{8}$$

84. *(d)* Given, $C(\theta) = \sum_{n=0}^{\infty} \frac{\cos(n\theta)}{n!}$

$$C(0) = \sum_{n=0}^{\infty} \frac{1}{n!} = 1 + \frac{1}{1!} + \frac{1}{2!} + \frac{1}{3!} + \ldots = e$$

$$C(\pi) = \sum_{n=0}^{\infty} (-1)^n \frac{1}{n!} \qquad [\because \cos n\pi = (-1)^n]$$

$$C(\pi) = 1 - \frac{1}{1!} + \frac{1}{2!} - \frac{1}{3!} + \ldots = e^{-1}$$

(A) $C(0)\,.\,C(\pi) = e\,.\,e^{-1} = 1$ True

(B) $C(0) + C(\pi) = e + \frac{1}{e} > 2$ True

(C) $C(\theta) > 0 \;\forall\; \theta \in R$ True

(D) $C'(\theta) = \sum_{n=0}^{\infty} -\frac{n\sin(n\theta)}{n!}$

$\therefore \quad C'(\theta) = 0 \Rightarrow \theta = 0$ False

Hence, option (d) is false.

85. *(d)* Let $L = \lim_{x \to 2} \frac{a^x + a^{3-x} - (a^2 + a)}{a^{3-x} - a^{x/2}}$

Apply L-Hospital rule

$$L = \lim_{x \to 2} \frac{a^x \log a - a^{3-x} \log a}{-a^{3-x} \log a - \frac{1}{2} a^{x/2} \log a}$$

$$L = \lim_{x \to 2} \frac{a^x - a^{3-x}}{-a^{3-x} - \frac{a^{x/2}}{2}}$$

$$L = \frac{a^2 - a}{-a - \frac{a}{2}} = \frac{2}{3}(1 - a)$$

86. *(d)* We have, $f(x) = \alpha x^2 - 2 + \frac{1}{x}$

$$f(x) = \frac{\alpha x^3 - 2x + 1}{x}, x > 0$$

$$f(x) \geq 0 \text{ if } \alpha x^3 - 2x + 1 > 0$$

Let $g(x) = \alpha x^3 - 2x + 1$

$\therefore \quad g'(x) = 3\alpha x^2 - 2 = 0$

$\Rightarrow \quad x = \pm \sqrt{\frac{2}{3\alpha}}$

Clearly, $x = \sqrt{\frac{2}{3\alpha}}$ is point of minima.

$$\therefore g\left(\sqrt{\frac{2}{3\alpha}}\right) \geq 0 \Rightarrow g\left(\sqrt{\frac{2}{3\alpha}}\right) = \alpha\left(\frac{2}{3\alpha}\right)^{3/2} - 2\left(\frac{2}{3\alpha}\right)^{1/2} + 1 \geq 0$$

$$\Rightarrow \left(\frac{2}{3\alpha}\right)^{1/2}\left(\frac{2}{3} - 2\right) + 1 \geq 0$$

$$\Rightarrow \left(\frac{2}{3\alpha}\right)^{1/2} \leq \frac{3}{4} \Rightarrow \frac{2}{3\alpha} \leq \frac{9}{16} \Rightarrow \alpha \geq \frac{32}{27}$$

87. *(b)* We have,

$$f(x) + \int_0^x t\, f(t)\, dt + x^2 = 0$$

On differentiating, we get

$$f'(x) + xf(x) + 2x = 0$$

$$\Rightarrow \quad f'(x) = -x(f(x) + 2)$$

$$\Rightarrow \quad \frac{f'(x)}{f(x) + 2} = -x$$

On integrating, we get

$$\log(f(x) + 2) = -\frac{x^2}{2} + C$$

$$\Rightarrow \quad f(x) = Ae^{-x^2/2} - 2$$

$$\Rightarrow \quad f(0) = 0 = A - 2 \Rightarrow A = 2$$

$$\therefore \quad f(x) = 2e^{-x^2/2} - 2$$

$$\lim_{x \to \infty} f(x) = -2$$

$$\lim_{x \to -\infty} f(x) = -2$$

$f(x)$ is an even function.

$f(x)$ intersect X-axis at one point (0, 0)

$\therefore$ Option (b) is correct.

88. *(a)* We have, $y = 2x - 4x^3$

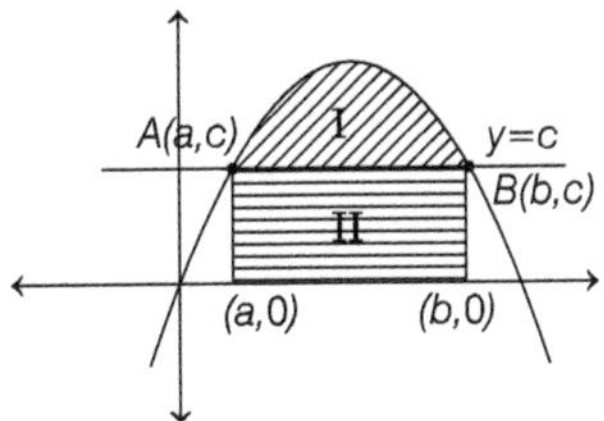

Given, area of I region = area of II region

$$\therefore \quad \int_a^b (2x - 4x^3)\, dx = 2(b - a)c$$

$$[x^2 - x^4]_a^b = 2(b - a)c$$

$$\Rightarrow \quad (b^2 - b^4) - (a^2 - a^4) = 2(b - a)c$$

$$\Rightarrow \quad (b^2 - a^2) - (b^4 - a^4) = 2(b - a)c$$

$$\Rightarrow (b^2 - a^2)(1 - (b^2 + a^2)) = 2(b - a)c$$

$$\Rightarrow \quad (b + a)(1 - b^2 - a^2) = 2c$$

$$\Rightarrow (b + a)(1 - (b + a)^2 + 2ab) = 2c$$

$$[\because c = 2x - 4x^3]$$

$$\Rightarrow \quad c = 2x - 4x^3$$

$$\therefore \quad 4x^3 - 2x + c = 0$$

Let a, b and α are roots.

$$\therefore a + b + \alpha = 0,\ ab + (a + b)\alpha = -\frac{1}{2},$$

$$ab\alpha = -\frac{c}{4}$$

$$a + b = -\alpha,\ ab = \alpha^2 - \frac{1}{2},\ ab = \frac{-c}{4\alpha}$$

$$1 - \alpha^2 + 2\left(\alpha^2 - \frac{1}{2}\right) = -8\alpha\left(\alpha^2 - \frac{1}{2}\right)$$

$$\Rightarrow \quad 1 - \alpha^2 + 2\alpha^2 - 1 = 8\alpha^2 - 4$$

$$\Rightarrow \quad \alpha^2 = 4 \Rightarrow \alpha = \frac{2}{\sqrt{7}}$$

89. *(c)* When $n = 10$, let A_r be number of ways of selecting r numbers.

Number of selection of A is

$$n(A_0) + n(A_1) + n(A_2) + n(A_3) + n(A_4)$$
$$= 1 + 10 + (7 + 6 + 5 + \ldots + 1) + (4 + 3 + 2 + 1) + (3 + 2 + 1) + (2 + 1) + 1$$
$$= 11 + 28 + 10 + 6 + 3 + 1 = 60$$

$N(p) = n$ (Number of ways 1 is selected)

$$= 1 + 7 + 4 + 3 + 2 + 1 + 1 = 19$$

$N(q) = n$ (Number of ways 2 is selected)

$$= 1 + 6 + 3 + 2 + 1 = 13$$

So, $\quad p = \frac{19}{60}, q = \frac{13}{60}$

$$p > q,\ p - q = \frac{19 - 13}{60} = \frac{6}{60} = \frac{1}{10}$$

90. *(d)* Let

$$D = \begin{vmatrix} 2014^{2014} & 2015^{2015} & 2016^{2016} \\ 2017^{2017} & 2018^{2018} & 2019^{2019} \\ 2020^{2020} & 2021^{2021} & 2022^{2022} \end{vmatrix}$$

$$D = \begin{vmatrix} (2015-1)^{2014} & (2015)^{2015} & (2015+1)^{2016} \\ (2015+2)^{2017} & (2020-2)^{2018} & (2020-1)^{2019} \\ (2020)^{2020} & (2020+1)^{2021} & (2020+2)^{2022} \end{vmatrix}$$

Remainder when divided by 5, is

$$D = \begin{vmatrix} 1 & 0 & 1 \\ 2^{2017} & 2^{2018} & -1 \\ 0 & 1 & 2^{2022} \end{vmatrix}$$

$$= 1(2^{4040} + 1) + 2^{2017}$$
$$= (5 - 1)^{2020} + 1 + 2(5 - 1)^{1008}$$
$$= 1 + 1 + 2 = 4$$

91. *(c)* It is given that observer O can view wall CD.

Let h = height of water in vessel at which object P is visible to observer.

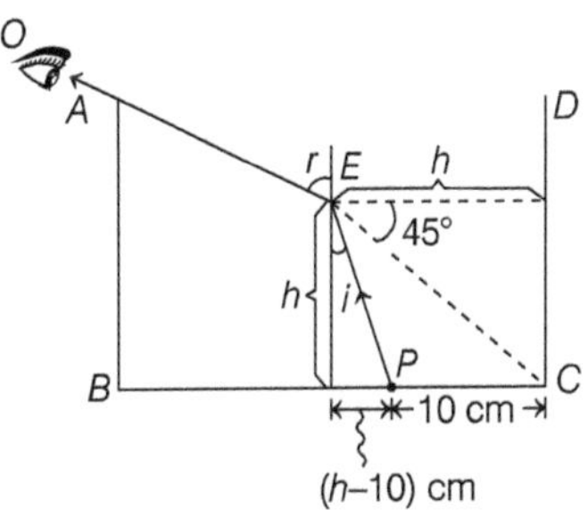

Light from object is refracted and reaches observer.

From geometry of arrangement, we have

$$\sin i \approx \tan i = \frac{h - 10}{h}$$

where, i = angle of incidence for small angles, $\sin i \approx \tan i$.

Also, from geometry of arrangement, angle of refraction $r = 45°$.

Now, μ (air with respect to water) $= \frac{\sin i}{\sin r}$

$$\Rightarrow \quad \frac{3}{4} = \frac{(h - 10)/h}{1/\sqrt{2}} \Rightarrow \frac{3}{4\sqrt{2}} = \frac{h - 10}{h}$$

$$\Rightarrow \quad \frac{10}{h} = 1 - \frac{3}{4\sqrt{2}} \Rightarrow \frac{10}{h} = \frac{26}{57}$$

$$\Rightarrow \quad h \approx 22 \text{ cm}$$

So, if water is filled upto height of 27 cm, object is visible to the observer.

92. *(a)* Let O is centre of disc and C is its centre of mass.

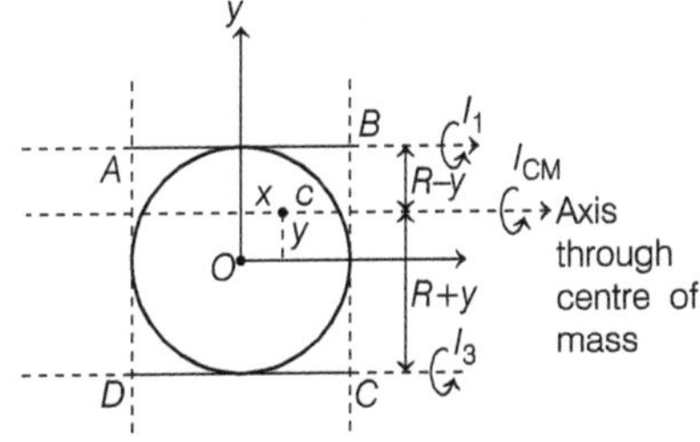

If I_c = moment of inertia about an axis through centre of mass of disc,

I_1 = moment of inertia about an tangential axis AB,

I_3 = moment of inertia about an tangential axis CD,

M = mass of disc and R = radius of disc.

Then, by parallel axes theorem, we have

$$I_1 = I_c + M\cdot(R-y)^2 \quad \ldots\text{(i)}$$

$$I_3 = I_c + M\cdot(R+y)^2 \quad \ldots\text{(ii)}$$

Subtracting Eqs. (i) and (ii), we get

$$\Rightarrow \quad I_1 - I_3 = M((R-y)^2 - (R+y)^2)$$

$$= -4MRy \quad \ldots\text{(iii)}$$

Similarly, from above figure by parallel axes theorem, we have

$$I_2 = I_c + M(R-x)^2$$

and $$I_4 = I_c + M(R+x)^2$$

$$\Rightarrow \quad I_2 - I_4 = -4MRx \quad \ldots\text{(iv)}$$

So, from Eqs. (iii) and (iv), we have

$$(I_1 - I_3)^2 + (I_2 - I_4)^2 = 16M^2R^2(x^2+y^2)$$

$$\Rightarrow \sqrt{x^2+y^2} = \frac{1}{4MR}\sqrt{(I_1-I_3)^2 + (I_2-I_4)^2}$$

But $\sqrt{x^2+y^2}$ = distance of centre of mass from centre of the disc.

93. *(c)* As the steel rail is contrained from expansion, the expansion pressure causes stress in the steel rail.

Thermal stress depends upon coefficient of expansion α and rise of temperature ΔT.

$\therefore$ Thermal stress, $\sigma \propto \alpha\Delta T$

$\Rightarrow \sigma = Y\cdot\alpha\cdot\Delta T$ (where, Y = Young's modulus)

$$\therefore \quad \sigma = 2\times10^{11}\times1.1\times10^{-5}\times(40-25)$$

$$= 3.3\times10^7 \text{ Pa}$$

94. *(c)* In given case, angle of refraction is negative. This is possible only when refractive index of material is negative. Negative Index Metamaterial (NIM) is a material for which refractive index has a negative value for an electromagnetic wave over certain frequency range. These are artificially produced materials, which found their application in optical signal transmission.

95. *(d)* Each ball is in equilibrium under three forces.

(i) Electrostatic repulsion force F_e, equal on both balls along line joining centre to centre of balls.

(ii) Weight m_1g and m_2g, different on each ball vertically downwards through centre of ball.

(iii) Tensions T_1 and T_2, different on each ball along the string.

As, angle between strings is given $30^\circ + 60^\circ = 90^\circ$ and strings are of equal length, strings forms an isosceles triangle as shown below,

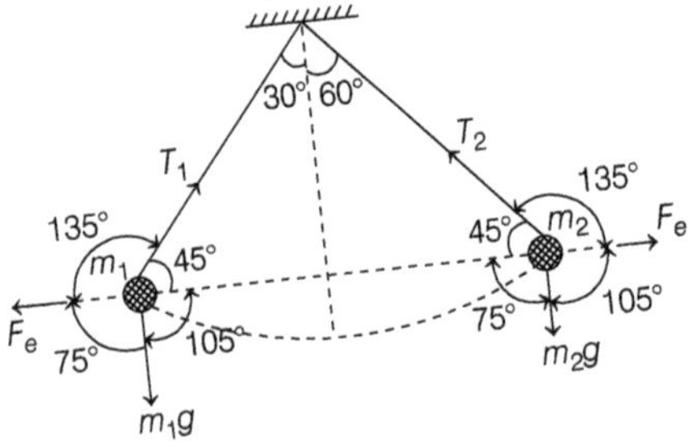

Free body diagrams of balls are

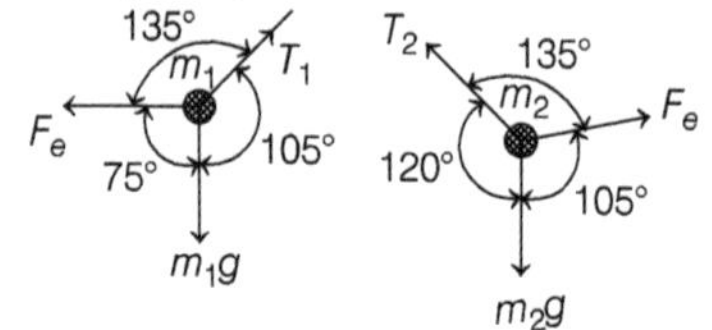

Now, using Lami's theorem for ball 1 and ball 2, we have

$$\frac{F_e}{\sin 150^\circ} = \frac{m_1g}{\sin 135^\circ} \quad \text{(For ball 1)}$$

and $$\frac{F_e}{\sin 120^\circ} = \frac{m_2g}{\sin 135^\circ} \quad \text{(For ball 2)}$$

Dividing these equations, we get

$$\frac{\sin 120^\circ}{\sin 150^\circ} = \frac{m_1}{m_2}$$

$$\Rightarrow \quad \frac{m_1}{m_2} = \frac{\sin(180^\circ - 60^\circ)}{\sin(180^\circ - 30^\circ)}$$

$$= \frac{\sin 60^\circ}{\sin 30^\circ} = \frac{\sqrt{3}/2}{1/2} = \sqrt{3}$$

So, $$\frac{m_1}{m_2} \approx 2.0$$

96. *(a)* A diamagnetic material tries to expel field lines from its volume and therefore experiences a repulsion when placed in region of strong fields. In given arrangement, field is zero in region marked A. So, diamagnetic powder accumulates in region A.

97. *(a)* Refer to figure below,

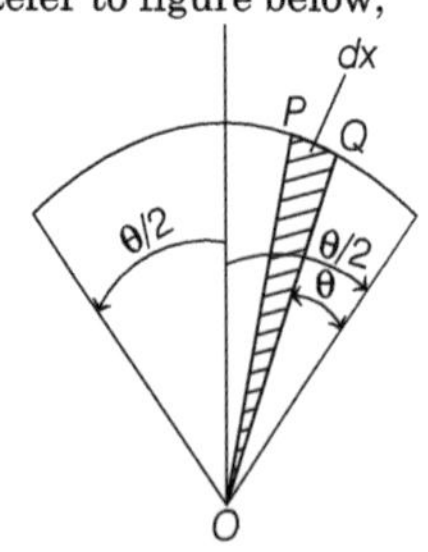

Consider a differentiable are of angle $d\theta$.

$$\text{Area of } \Delta OPQ = \frac{1}{2}\times PQ\times OP$$

$$= \frac{1}{2}R\,d\theta\times R = \frac{1}{2}R^2 d\theta$$

Centre of mass of this ΔOPQ is at a distance of $\frac{2}{3}R\cos\theta$ from O.

So, position of centre of mass of complete segment is

$$\bar{y} = \int y\,dm \Big/ \int dm$$

$$\bar{y} = \frac{\int_0^{\theta/2} \frac{2R}{3}\cos\theta\cdot\rho\frac{r^2}{2}d\theta}{\int_0^{\theta/2}\rho\frac{r^2}{2}d\theta}$$

where, ρ = mass density.

$$\therefore \quad \bar{y} = \frac{\frac{2r}{3}\int_0^{\theta/2}\cos\theta d\theta}{\int_0^{\theta/2}d\theta} = 2\frac{R}{3}\cdot\frac{\sin\theta/2}{\theta/2}$$

$$= \frac{4}{3}R\frac{\sin(\theta/2)}{\theta}$$

98. *(a)* Let v = velocity of object on reaching surface of planet.

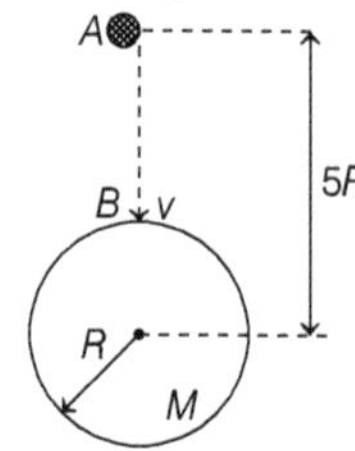

Applying energy conservation between points A and B, we have

Total energy at A = Total energy at B

$\Rightarrow$ Potential energy at separation $(4R + R)$ = Potential energy at separation R + Kinetic energy at surface

$$\Rightarrow \quad \frac{-GMm}{5R} = \frac{-GMm}{R} + \frac{1}{2}mv^2$$

where, m = mass of object.

$$\Rightarrow \quad \frac{GMm}{R}\left(1-\frac{1}{5}\right) = \frac{1}{2}mv^2 \Rightarrow \frac{8}{5}\frac{GM}{R} = v^2$$

$\therefore$ Velocity at surface, $v = 2\sqrt{\frac{2}{5}\frac{GM}{R}}$

99. *(a)* In steady state, after the key remains ON for a long time.

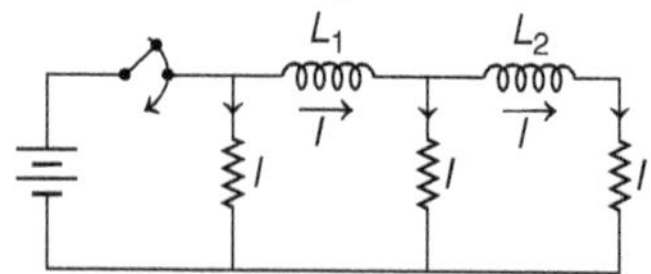

As resistances are identical and parallel (inductors behaves like closed paths), currents in resistors are equal as shown above.

When switch is turned OFF, current in inductors remains same.

So, we have following current distribution in circuit.

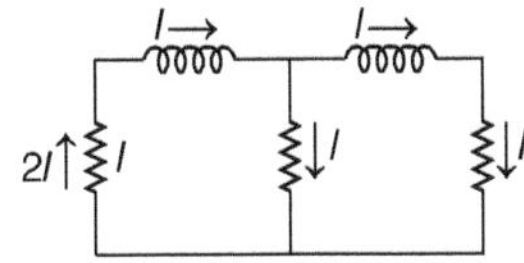

So, currents in resistances are $2I$ upwards, I downwards and I downwards.

100. *(a)* From given cyclic process,

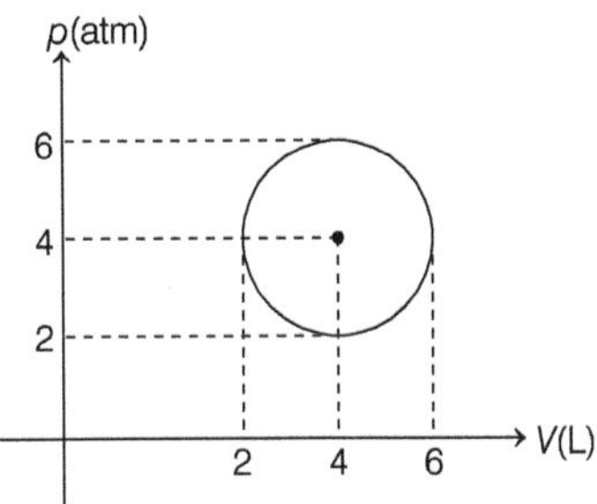

Process equation is

$$(p-4)^2+(V-4)^2=4 \quad \text{...(i)}$$

Now, from $pV = nRT$

We can say that T is maximum when pV is maximum.

Now, for given cyclic process, pV maximum occur when p^2V^2 is maximum.

Now,

$$p^2V^2 = p^2(4-(p-4)^2) \text{ [from Eq. (i)]}$$

Now, p^2V^2 is maximum when $\frac{d}{dp}p^2V^2 = 0$.

$$\Rightarrow \quad \frac{d}{dp}(p^2\cdot(p^2\cdot(4-(p-4)^2)=0$$

$$\Rightarrow \quad p^2-8p+14=0$$

$$\Rightarrow \quad p=\frac{8\pm\sqrt{64-56}}{2}$$

$$= 4\pm\sqrt{2}$$

and from Eq. (i), we get

$$p=4\pm\sqrt{2}, V=4\pm\sqrt{2}$$

Taking positive values, we have

$$(pV)_{max} \Rightarrow p = 4+\sqrt{2}$$

and $V = 4+\sqrt{2}$

So, by gas equation, we have

$$T_{max}=\frac{(pV)_{max}}{R} \quad \text{[for 1 mol of gas]}$$

$$=\frac{(4+\sqrt{2})(4+\sqrt{2})}{R}$$

$$=\frac{16+2+2\times4\times\sqrt{2}}{R}$$

$$=\frac{29.32}{R}\approx\frac{30}{R}$$

101. *(a)* $N_2 + 3F_2 \longrightarrow 2NF_3$

Or

$$N{\equiv}N + 3\,F{-}F \longrightarrow 2\,N(F)_3$$

Given, $BE_{N\equiv N} = 941\,kJ\,mol^{-1}$

$BE_{F-F} = 155\,kJ\,mol^{-1}$

$BE_{N-F} = 272\,kJ\,mol^{-1}$

$$\Delta H_f = \Sigma BE_{reactants} - \Sigma BE_{products}$$

$$= BE_{N\equiv N} + 3BE_{F-F} - 6BE_{N-F}$$

$$= 941 + 3(155) - 6(272)$$

$$= -226\,kJ\,mol^{-1}$$

For the reaction,

$$N_2 + 3Cl_2 \longrightarrow 2NCl_3$$

Or

$$N{\equiv}N + 3\,Cl{-}Cl \longrightarrow 2\,N(Cl)_3$$

$$\Delta H_f = BE_{N\equiv N} + 3BE_{Cl-Cl} - 6(BE_{N-Cl})$$

$$= 941 + 3(242) - 6(200)$$

$$= +467\,kJ\,mol^{-1}$$

102. *(a)* Let the initial concentration of X and Z be 1 and α be the degree of dissociation.

	$X \rightleftharpoons$	$2Y$
Initial conc.	1	0
Equili. conc.	$1-\alpha$	2α

$$K_1=\frac{[Y]^2}{[X]}=\frac{(2\alpha)^2}{1-\alpha}$$

	$Z \rightleftharpoons$	P +	Q
Initial conc.	1	0	0
Equili. conc.	$1-\alpha$	α	α

$$K_2=\frac{[P][Q]}{[Z]}=\frac{\alpha^2}{1-\alpha}$$

$$\therefore \quad \frac{K_1}{K_2}=\frac{\frac{4\alpha^2}{1-\alpha}}{\frac{\alpha^2}{1-\alpha}}=4$$

103. *(c)* The oxidation state of Mn in $[MnBr_4]^{2-}$ is +2. The electronic configuration of Mn^{2+} is $[Ar]\,3d^5\,4s^0$.

Mn^{2+} (Ground state) 3d: [1|1|1|1|1] 4s: [ ] 4p: [| |]

As, Br^- is a weak field ligand pairing of electron will not occur.

$[MnBr_4]^{2-}$ 3d: [1|1|1|1|1] 4s: [ ] 4p: [| |] (sp^3 hybridisation)

As, the hybridisation of Mn in $[MnBr_4]^{2-}$ is sp^3, thus its geometry will be tetrahedral and it contains five unpaired electrons.

104. *(a)* $Zn(s) + Cu^{2+}(aq) \longrightarrow Zn^{2+}(aq) + Cu(s)$

According Nernst equation,

$$E_{cell}=E^\circ_{cell}-\frac{0.0591}{n}\log\frac{[P]}{[R]}$$

When cell is completely discharge then $E_{cell}=0$

$$\therefore \quad 0=1.10-\frac{0.0591}{2}\log\frac{[Zn^{2+}]}{[Cu^{2+}]}$$

$$-1.10=\frac{-0.0591}{2}\log\frac{[Zn^{2+}]}{[Cu^{2+}]}$$

$$\therefore \log\frac{[Zn^{2+}]}{[Cu^{2+}]}=\frac{2.20}{0.0591}=37.3$$

105. *(c)*

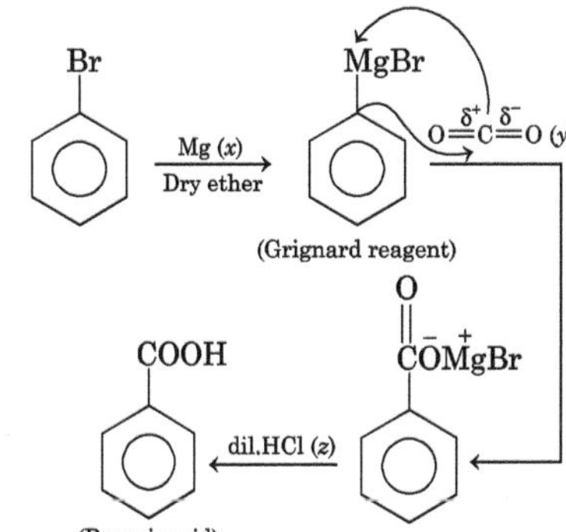

Bromobenzene, when reacts with Mg in the presence of dry ether gives Grignard reagent, which then reacts with CO_2 in the presence of an acid to give benzoic acid.

106. *(b)* An organic compound having molecular formula C_2H_6O can be alcohol or ether, but only alcohol undergoes oxidation to produce carboxylic acid. Thus, alcohol with molecular formula C_2H_6O is ethanol, CH_3CH_2OH which oxidises to acetic acid, CH_3COOH.

$$CH_3CH_2OH \xrightarrow{K_2Cr_2O_7/H_2SO_4} \underset{(X)}{CH_3{-}\overset{O}{\overset{\|}{C}}{-}OH}$$

Confermation of molecular formula of X is as follows:

Symbols	% of element	No. of moles	Simplest molar ratio	Simplest whole no.
C	40	40/12 = 3.33	3.33/3.33 = 1	1
H	6.7	6.7/1 = 6.7	6.7/3.33 = 2.03	2
O	53.3	53.3/16 = 3.33	3.33/3.33 = 1	1

Hence, empirical formula of X is CH_2O.

Empirical formula mass

$= 12 \times 1 + 2 + 16 \times 1$

$= 12 + 2 + 16 = 30$

$$n = \frac{\text{Molecular mass of } CH_3COOH}{\text{Empirical mass}}$$

$$= \frac{60}{30} = 2$$

$\therefore$ Molecular formula will be $(CH_2O)_2$, i.e. $C_2H_4O_2$.

Thus, the above analysis shows that the compound X has molecular formula $C_2H_4O_2$, i.e. CH_3COOH.

107. *(b)* There can be 3 maximum number of cyclic isomers (positional and optical) of a compound having molecular formula C_3H_2Cl. These are as follows

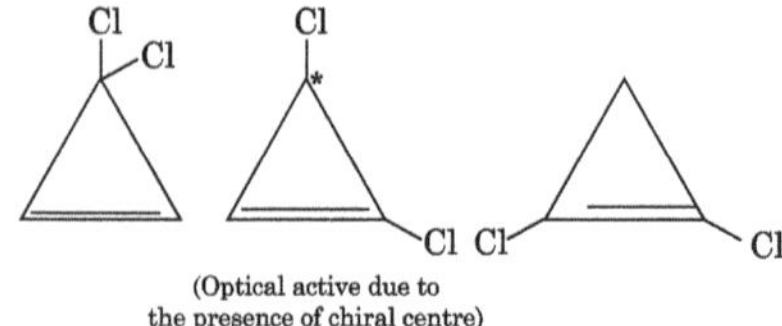

(Optical active due to the presence of chiral centre)

Thus, it will have enantiomeric pair.

108. *(a)* According to ideal gas equation,

$$pV = nRT$$

For 1 mole, $n = 1$

$\therefore \quad pV = RT$

$\Rightarrow \quad p = \frac{RT}{V}$

For gas X,

$V = 50\,\text{L}, T = 200\,\text{K}$

$\therefore$ Pressure of gas $X = \frac{0.0821 \times 200}{50}$

$= 0.328\,\text{atm}$

For gas Y,

$V = 50\,\text{L}, T = 500\,\text{K}$

$\therefore$ Pressure of gas $Y = \frac{0.0821 \times 500}{50}$

$= 0.821\,\text{atm}$

For gas Z,

$V = 20\,\text{L}, T = 200\,\text{K}$

$\therefore$ Pressure of gas $Z = \frac{0.0821 \times 200}{20}$

$= 0.821\,\text{atm}$

109. *(a)* MnO_2, when fused with KOH and oxidised in air gives a dark green compound, X, i.e. potassium manganate (K_2MnO_4). In acidic solution, K_2MnO_4 undergoes diproportionation to give an intense purple solution of potassium permanganate Y and MnO_2.

$$MnO_2 + KOH + O_2 \longrightarrow \underset{(X)}{K_2MnO_4} + 2H_2O$$

$\downarrow$ Disproportionation

$$\underset{(Y)}{KMnO_4} + MnO_2 + H_2O$$

110. *(b)* $Al(X)$ metal dissolves both in dil. HCl and dil. NaOH to liberate H_2. Addition of NH_4Cl and excess NH_4OH to an HCl solution of Al produces $Al(OH)_3$ (Y) as an precipitate. $Al(OH)_3$ is also produced by adding NH_4Cl to the NaOH solution of Al.

$$\underset{(X)}{Al} + HCl \longrightarrow AlCl_3 + H_2$$

$$\underset{(X)}{Al} + NaOH \longrightarrow NaAlO_2 + H_2$$

$\downarrow NH_4Cl$

$$\underset{(Y)}{Al(OH)_3} \longleftarrow AlCl_3 + NH_4Cl + NH_4OH(\text{excess})$$

111. *(c)* On SDS-PAGE of IgG, two (2) principal bands are visible. IgG antibodies is a monomeric structure consisting of heavy chains and two identical light chains. Thus, on SDS-PAGE under reducing conditions shows two bands.

112. *(a)* Penicillin works by bursting the cell wall of bacteria. When a bacterium multiplies, small holes open up in its cell wall as the cells divide. Therefore, in a mixed culture with penicillin, it kills the fast growing bacteria more than the slow growing ones.

113. *(c)* Haemophilia is a sex-linked recessive disorder. These kinds of defects occur more often in men than in women, because men are heterozygous for X-chromosomes and if it carries the gene they are diseased. Whereas, for a woman to be affected she needs to be homozygous for the trait.

114. *(d)* Radius of spherical alveoli

$= 0.1\,\text{mm}$

Total number of alveoli

$= 400 \times 2 = 800$ million

Surface area of sphere $= 4\pi r^2$

Total respiratory surface area

$= 4\pi r^2 \times 800$ million

$= 1000\,\text{mm}^2$ (approx)

115. *(a)* Among slow dividing and fast dividing cells, the radioactivity will be maximum shown by fast dividing cells. This is because, there is more chance that fast dividing cells would have taken up thymidine from the media during that one hour. Radioactivity is mostly incorporated in a rapidly dividing cell during S-phase.

116. *(c)* In a double-stranded DNA, according to Chargaff's rule,

$$A + T = G + C$$

$$A = T \text{ and } G = C$$

Thus, when cytosine = 15%

Guanine = 15%

Both C + G = 30%

Thus, A + T = 70%

and $A = \frac{70}{2} = 35\%$, $T = 35\%$

117. *(b)* The electron transport chain and chemiosmosis take place on this membrane as the part of cellular respiration to create ATP. The cristae increase the surface area of the inner membrane, allowing for faster production of ATP, because there are more places to perform the process. Thus, option (b) is correct.

118. *(a)* Phosphorylation is a biochemical process that involves the addition of phosphate to an organic compound. It is known that eukaryotes rely on the phosphorylation of the hydroxyl group on the side chains of serine, threonine and tyrosine for cell signalling. Thus, any change in these amino acids will affect the function of the molecule. Therefore, a mutation from tyrosine to tryptophan is damaging to cell.

119. *(c)* The gel electrophoresis will show 3 bands because even though 4 fragments are formed but 2 fragments are of 3 kb and they will appear as one single band.

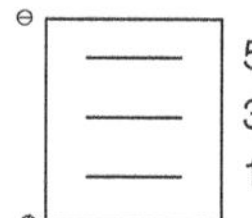

120. *(c)* GTP–X complex → Active cell division

GDP–X-complex → No cell division

and GTP $\xrightarrow[X\text{-enzyme}]{\text{hydrolysis}}$ GDP

Mutation causes the decrease of rate of hydrolysis of GTP. Thus, GTP–X complex remains present for long time and cell division becomes uncontrolled.

QUESTION PAPER 2014

Stream : SB/SX

MM : 160

Instructions

1. There are 120 questions in this paper.
2. The question paper contains two parts; Part I (1 Mark Questions) and Part II (2 Marks Questions).
3. There are four sections in each part; Mathematics, Physics, Chemistry and Biology.
4. Out of the four options given with each question, only one is correct.

PART-I (1 Mark Questions)

MATHEMATICS

1. Let C_0 be a circle of radius 1. For $n \geq 1$, let C_n be a circle whose area equals the area of a square inscribed in C_{n-1}. Then, $\sum_{i=0}^{\infty}$ Area (C_i) equals

(a) π^2 (b) $\frac{\pi-2}{\pi^2}$ (c) $\frac{1}{\pi^2}$ (d) $\frac{\pi^2}{\pi-2}$

2. For a real number r we denote by $[r]$ the largest integer less than or equal to r. If x, y are real numbers with $x, y \geq 1$ then which of the following statements is always true?

(a) $[x+y] \leq [x]+[y]$ (b) $[xy] \leq [x][y]$

(c) $[2^x] \leq 2^{[x]}$ (d) $\left[\frac{x}{y}\right] \leq \frac{[x]}{[y]}$

3. For each positive integer n, let $A_n = \max\left\{\binom{n}{r} \mid 0 \leq r \leq n\right\}$. Then, the number of elements n is $\{1, 2, ..., 20\}$ for which $1.9 \leq \frac{A_n}{A_{n-1}} \leq 2$ is

(a) 9 (b) 10 (c) 11 (d) 12

4. Let $b, d > 0$. The locus of all points $P(r, \theta)$ for which the line OP (where, O is the origin) cuts the line $r \sin\theta = b$ in Q such that $PQ = d$ is

(a) $(r-d)\sin\theta = b$ (b) $(r \pm d)\sin\theta = b$

(c) $(r-d)\cos\theta = b$ (d) $(r \pm d)\cos\theta = b$

5. Let C be the circle $x^2 + y^2 = 1$ in the XY-plane. For each $t \geq 0$, let L_t be the line passing through (0, 1) and $(t, 0)$. Note that L_t intersects C in two points, one of which is (0, 1). Let Q_t be the other point. As t varies between 1 and $1+\sqrt{2}$, the collection of points Q_t sweeps out an arc on C. The angle subtended by this arc at (0, 0) is

(a) $\frac{\pi}{8}$ (b) $\frac{\pi}{4}$

(c) $\frac{\pi}{3}$ (d) $\frac{3\pi}{8}$

6. In an ellipse, its foci and ends of its major axis are equally spaced. If the length of its semi-minor axis is $2\sqrt{2}$, then the length of its semi-major axis is

(a) 4 (b) $2\sqrt{3}$

(c) $\sqrt{10}$ (d) 3

7. Let ABC be a triangle such that $AB = BC$. Let F be the mid-point of AB and X be a point on BC such that FX is perpendicular to AB. If $BX = 3XC$, then the ratio BC / AC equals

(a) $\sqrt{3}$ (b) $\sqrt{2}$ (c) $\sqrt{\frac{3}{2}}$ (d) 1

8. The number of solutions to the equation $\cos^4 x + \frac{1}{\cos^2 x} = \sin^4 x + \frac{1}{\sin^2 x}$ in the interval $[0, 2\pi]$ is

(a) 6 (b) 4 (c) 2 (d) 0

9. Consider the function

$$f(x) = \begin{cases} \frac{x+5}{x-2}, & \text{if } x \neq 2 \\ 1, & \text{if } x = 2 \end{cases}$$

Then, $f(f(x))$ is discontinuous

(a) at all real numbers
(b) at exactly two values of x
(c) at exactly one value of x
(d) at exactly three values of x

10. For a real number x let $[x]$ denote the largest number less than or equal to x. For $x \in R$ let $f(x) = [x] \sin \pi x$. Then,

(a) f is differentiable on R.
(b) f is symmetric about the line $x = 0$.
(c) $\int_{-3}^{3} f(x)\, dx = 0$.
(d) For each real α, the equation $f(x) - \alpha = 0$ has infinitely many roots.

11. Let $f : [0, \pi] \to R$ be defined as

$$f(x) = \begin{cases} \sin x, & \text{if } x \text{ is irrational and } x \in [0, \pi] \\ \tan^2 x, & \text{if } x \text{ is rational and } x \in [0, \pi] \end{cases}$$

The number of points in $[0, \pi]$ at which the function f is continuous is

(a) 6 (b) 4 (c) 2 (d) 0

12. Let $f : [0, 1] \to [0, \infty)$ be a continuous function such that $\int_0^1 f(x)\, dx = 10$. Which of the following statements is NOT necessarily true?

(a) $\int_0^1 e^{-x} f(x)\, dx \leq 10$

(b) $\int_0^1 \frac{f(x)}{(1+x)^2}\, dx \leq 10$

(c) $-10 \leq \int_0^1 \sin(100x) f(x)\, dx \leq 10$

(d) $\int_0^1 f(x)^2\, dx \leq 100$

13. A continuous function $f : R \to R$ satisfies the equation $f(x) = x + \int_0^x f(t)\, dt$. Which of the following options true?

(a) $f(x+y) = f(x) + f(y)$
(b) $f(x+y) = f(x) f(y)$
(c) $f(x+y) = f(x) + f(y) + f(x)(y)$
(d) $f(x+y) = f(xy)$

14. For a real number x let $[x]$ denote the largest integer less than or equal to x and $\{x\} = x - [x]$. Let n be a positive integer. Then, $\int_0^n \cos(2\pi[x]\{x\})\, dx$ is equal to

(a) 0 (b) 1
(c) n (d) $2n - 1$

15. Two persons A and B throw a (fair)die (six-faced cube with faces numbered from 1 to 6) alternately, starting with A. The first person to get an outcome different from the previous one by the opponent wins. The probability that B wins is

(a) $\frac{5}{6}$ (b) $\frac{6}{7}$ (c) $\frac{7}{8}$ (d) $\frac{8}{9}$

16. Let $n \geq 3$. A list of numbers $x_1, x, \ldots, x_n$ has mean μ and standard deviation σ. A new list of numbers $y_1, y_2, \ldots, y_n$ is made as follows

$y_1 = \frac{x_1 + x_2}{2}$, $y_2 = \frac{x_1 + x_2}{2}$ and $y_j = x_j$ for $j = 3, 4, \ldots, n$.

The mean and the standard deviation of the new list are $\hat{\mu}$ and $\hat{\sigma}$. Then, which of the following is necessarily true?

(a) $\mu = \hat{\mu}$ and $\sigma \leq \hat{\sigma}$ (b) $\mu = \hat{\mu}$ and $\sigma \geq \hat{\sigma}$
(c) $\sigma = \hat{\sigma}$ (d) $\mu \neq \hat{\mu}$

17. What is the angle subtended by an edge of a regular tetrahedron at its centre ?

(a) $\cos^{-1}\left(\frac{-1}{2}\right)$ (b) $\cos^{-1}\left(\frac{-1}{\sqrt{2}}\right)$
(c) $\cos^{-1}\left(\frac{-1}{3}\right)$ (d) $\cos^{-1}\left(\frac{-1}{\sqrt{3}}\right)$

18. Let $S = \{(a, b) : a, b, \in Z, 0 \leq a, b \leq 18\}$. The number of elements (x, y) in S such that $3x + 4y + 5$ is divisible by 19, is

(a) 38 (b) 19
(c) 18 (d) 1

19. For a real number r let $[r]$ denote the largest integer less than or equal to r. Let $a > 1$ be a real number which is not an integer, and let k be the smallest positive integer such that $[a^k] > [a]^k$. Then, which of the following statements is always true?

(a) $k \leq 2([a]+1)^2$ (b) $k \leq ([a]+1)^4$
(c) $k \leq 2^{[a]+1}$ (d) $k \leq \frac{1}{a - [a]} + 1$

20. Let X be set of 5 elements. The number d of ordered pairs (A, B) of subsets of X such that $A \neq \phi, B \neq \phi, A \cap B \neq \phi$ satisfies

(a) $50 \leq d \leq 100$ (b) $101 \leq d \leq 150$
(c) $151 \leq d \leq 200$ (d) $201 \leq d$

PHYSICS

21. An uniform thin rod of length $2L$ and mass m lies on a horizontal table. A horizontal impulse J is given to the rod at one end. There is no friction. The total kinetic energy of the rod just after the impulse will be

(a) $\frac{J^2}{2m}$ (b) $\frac{J^2}{m}$ (c) $\frac{2J^2}{m}$ (d) $\frac{6J^2}{m}$

22. A solid cylinder P rolls without slipping from rest down an inclined plane attaining a speed v_p at the bottom. Another smooth solid cylinder Q of same mass and dimensions slides without friction from rest down the inclined plane attaining a speed v_q at the bottom. The ratio of the speeds $\frac{v_q}{v_p}$ is

(a) $\sqrt{\frac{3}{4}}$ (b) $\sqrt{\frac{3}{2}}$ (c) $\sqrt{\frac{2}{3}}$ (d) $\sqrt{\frac{4}{3}}$

23. A body moves in a circular orbit of radius R under the action of a central force. Potential due to the central force is given by $V(r) = kr$ (k is a positive constant). Period of revolution of the body is proportional to

(a) $R^{1/2}$ (b) $R^{-1/2}$ (c) $R^{-3/2}$ (d) $R^{-5/2}$

24. A simple pendulum is attached to a block which slides without friction down an inclined plane ABC having an angle of inclination α as shown below.

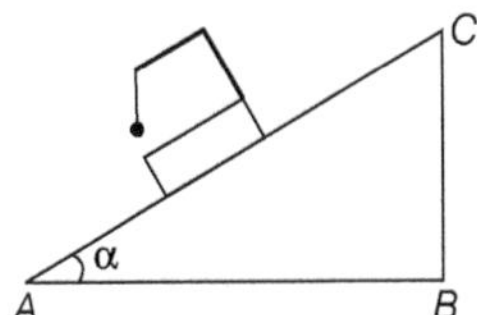

While the block is sliding down the pendulum oscillates in such a way that at its mean position the direction of the string is

(a) at angle α to the perpendicular to the inclined plane AC
(b) parallel to the inclined plane AC
(c) vertically downwards
(d) perpendicular to the inclined plane AC

25. Water containing air bubbles flows without turbulence through a horizontal pipe which has a region of narrow cross-section. In this region, the bubbles

(a) move with greater speed and are smaller in size than in the rest of the pipe
(b) move with greater speed and are larger in size than in the rest of the pipe
(c) move with lesser speed and are smaller in size than in the rest of the pipe
(d) move with lesser speed and are of the same size as in the rest of the pipe

26. A solid expands upon heating because

(a) the potential energy of interaction between atoms in the solid is asymmetric about the equilibrium positions of atoms
(b) the frequency of vibration of the atoms increases
(c) the heating generates a thermal gradient between opposite sides
(d) a fluid called the caloric flows into the interatomic spacing of the solid during heating there by expanding it

27. Consider two thermometers T_1 and T_2 of equal length, which can be used to measure temperature over the range θ_1 to θ_2. T_1 contains mercury as the thermometric liquid, while T_2 contains bromine. The volumes of the two liquids are the same at the temperature θ_1. The volumetric coefficients of expansion of mercury and bromine are $18 \times 10^{-5}\ K^{-1}$ and $108 \times 10^{-5}\ K^{-1}$, respectively. The increase in length of each liquid is the same for the same increase in temperature. If the diameters of the capillary tubes of the two thermometers are d_1 and d_2, respectively. Then, the ratio of $d_1 : d_2$ would be closest to

(a) 6.0 (b) 2.5 (c) 0.6 (d) 0.4

28. An ideal gas follows a process described by $pV^2 = C$ from (p_1, V_1, T_1) to (p_2, V_2, T_2) and C is a constant. Then,

(a) if $p_1 > p_2$ then $T_2 > T_1$ (b) if $V_2 > V_1$ then $T_2 < T_1$
(c) if $V_2 > V_1$ then $T_2 > T_1$ (d) if $p_1 > p_2$ then $V_1 > V_2$

29. A whistle emitting a loud sound of frequency 540 Hz is whirled in a horizontal circle of radius 2 m and at a constant angular speed of 15 rad/s. The speed of sound is 330 m/s. The ratio of the highest to the lowest frequency heard by a listener standing at rest at a large distance from the centre of the circle is

(a) 1.0 (b) 1.1
(c) 1.2 (d) 1.4

30. Monochromatic light passes through a prism. Compared to that in air, inside the prism the light's

(a) speed and wavelength are different, but frequency remains same
(b) speed and frequency are different, but wavelength remains same
(c) wavelength and frequency are different, but speed remains same
(d) speed, wavelength and frequency are all different

31. The flat face of a plano-convex lens of focal length 10 cm is silvered. A point source placed 30 cm in front of the curved surface will produce a

(a) real image 15 cm away from the lens
(b) real image 6 cm away from the lens
(c) virtual image 15 cm away from the lens
(d) virtual image 6 cm away from the lens

32. Two identical metallic square loops L_1 and L_2 are placed next to each other with their sides parallel on a smooth horizontal table. Loop L_1 is fixed and a current which increases as a function of time is passed through it. Then, loop L_2 is

(a) rotates about its centre of mass
(b) moves towards L_1
(c) remains stationary
(d) moves away from L_1

33. An electron enters a parallel plate capacitor with horizontal speed u and is found to deflect by angle θ on leaving the capacitor as shown below. It is found that $\tan\theta = 0.4$ and gravity is negligible.

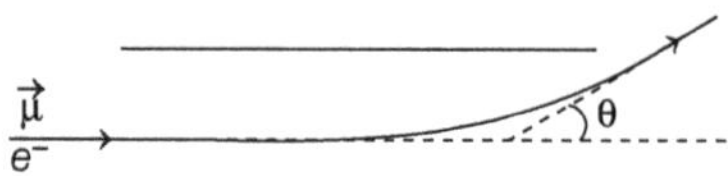

If the initial horizontal speed is doubled, then the value of $\tan\theta$ will be

(a) 0.1 (b) 0.2 (c) 0.8 (d) 1.6

34. Consider a spherical shell of radius R with a total charge $+Q$ uniformly spread on its surface (centre of the shell lies at the origin $x = 0$). Two point charges $+q$ and $-q$ are brought, one after the other from far away and placed at $x = -a/2$ and $x = +a/2$ $(a < R)$, respectively. Magnitude of the work done in this process is

(a) $(Q+q)^2/4\pi\varepsilon_0 a$ (b) zero
(c) $q^2/4\pi\varepsilon_0 a$ (d) $Qq/4\pi\varepsilon_0 a$

35. Two identical parallel plate capacitors of capacitance C each are connected in series with a battery of emf, E as shown below. If one of the capacitors is now filled with a dielectric of dielectric constant k, then the amount of charge which will flow through the battery is (neglect internal resistance of the battery)

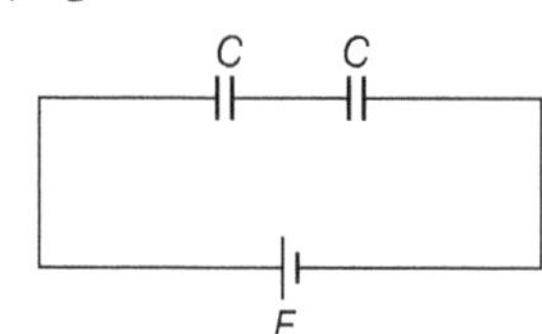

(a) $\frac{k+1}{2(k-1)} \cdot CE$ (b) $\frac{k-1}{2(k+1)} \cdot CE$
(c) $\frac{k-2}{k+2} \cdot CE$ (d) $\frac{k+2}{k-2} \cdot CE$

36. A certain p-n junction, having a depletion region of width 20 μm, was found to have a breakdown voltage of 100 V. If the width of the depletion region is reduced to 1 μm during its production, then it can be used as a zener diode for voltage regulation of

(a) 5 V (b) 10 V
(c) 7.5 V (d) 2000 V

37. The half-life of a particle of mass 1.6×10^{-26} kg is 6.9 s and a stream of such particles is travelling with the kinetic energy of a particle being 0.05 eV. The fraction of particles which will decay, when they travel a distance of 1 m is

(a) 0.1 (b) 0.01
(c) 0.001 (d) 0.0001

38. A 160 W light source is radiating light of wavelength 6200 Å uniformly in all directions. The photon flux at a distance of 1.8 m is of the order of (Planck's constant $= 6.63 \times 10^{-34}$ J-s)

(a) 10^2 m^{-2}s^{-1} (b) 10^{12} m^{-2}s^{-1}
(c) 10^{19} m^{-2}s^{-1} (d) 10^{25} m^{-2}s^{-1}

39. The wavelength of the first Balmer line caused by a transition from the $n = 3$ level to the $n = 2$ level in hydrogen is λ_1. The wavelength of the line caused by an electronic transition from $n = 5$ to $n = 3$ is

(a) $\frac{375}{128}\lambda_1$ (b) $\frac{125}{64}\lambda_1$
(c) $\frac{64}{125}\lambda_1$ (d) $\frac{128}{375}\lambda_1$

40. The binding energy per nucleon of ${}_5B^{10}$ is 8.0 MeV and that of ${}_5B^{11}$ is 7.5 MeV. The energy required to remove a neutron from ${}_5B^{11}$ is (take, mass of electron and proton are 9.11×10^{-31} kg and 1.67×10^{-27} kg, respectively)

(a) 2.5 MeV (b) 8.0 MeV
(c) 0.5 MeV (d) 7.5 MeV

CHEMISTRY

41. When 1.88 g of AgBr(s) is added to a 10^{-3} M aqueous solution of KBr, the concentration of Ag^+ is 5×10^{-10}M. If the same amount of AgBr(s) is added to a 10^{-2}M aqueous solution of $AgNO_3$, the concentration of Br^- is

(a) 9.4×10^{-9} M (b) 5×10^{-10}M
(c) 1×10^{-11}M (d) 5×10^{-11}M

42. Aniline reacts with excess Br_2/H_2O to give the major product.

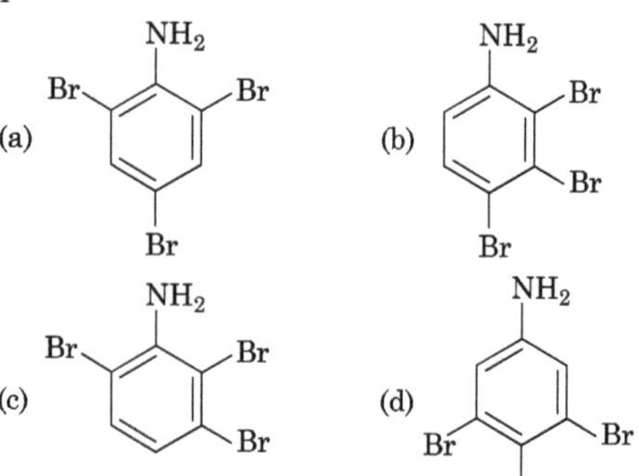

43. The metal with the highest oxidation state present in K_2CrO_4, $NbCl_5$ and MnO_2 is

(a) Nb (b) Mn
(c) K (d) Cr

44. The number of geometrical isomers of $[CrCl_2(en)(NH_3)_2]$, where en = ethylenediamine, is

(a) 2 (b) 3
(c) 4 (d) 1

45. The element that combines with oxygen to give an amphoteric oxide is

(a) N (b) P (c) Al (d) Na

46. The Arrhenius plots of two reactions, I and II are shown graphically

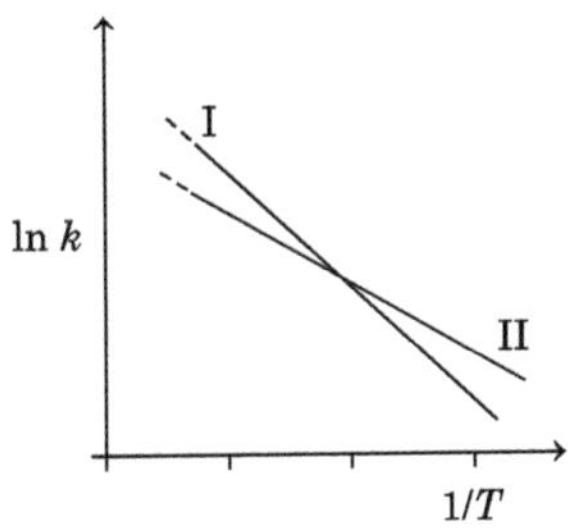

The graph suggests that

(a) $E_I > E_{II}$ and $A_I > A_{II}$
(b) $E_{II} > E_I$ and $A_{II} > A_I$
(c) $E_I > E_{II}$ and $A_{II} > A_I$
(d) $E_{II} > E_I$ and $A_I > A_{II}$

47. $Ni(CO)_4$ is

(a) tetrahedral and paramagnetic
(b) square planar and diamagnetic
(c) tetrahedral and diamagnetic
(d) square planar and paramagnetic

48. In the following reaction,

1,2-dimethylcyclohexene $\xrightarrow[\text{(ii) } \overset{\ominus}{O}H]{\text{(i) Ozonolysis}} X$

the major product X is

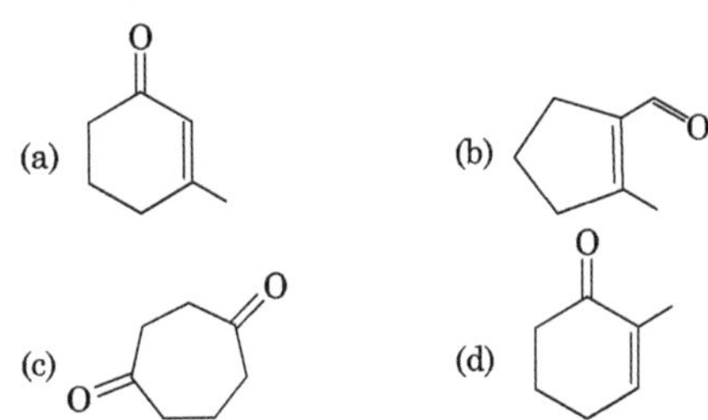

49. Given, the structure of D - (+)-glucose as

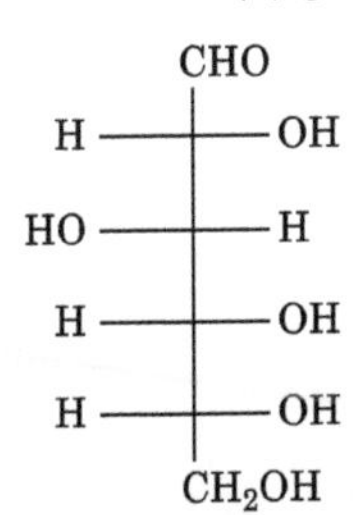

The structure of L- (–)-glucose is

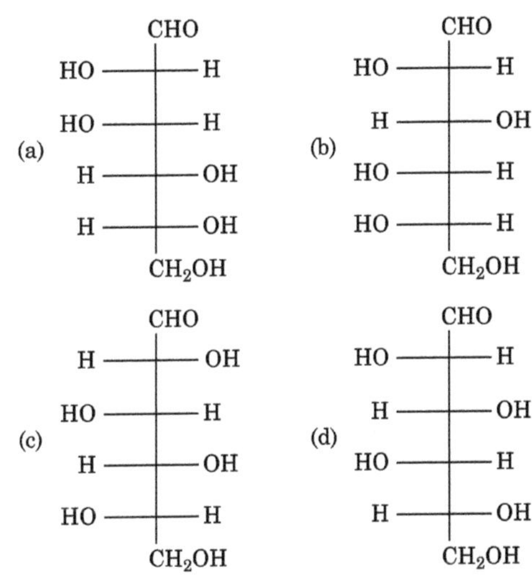

50. In a cubic close packed structure, fractional contributions of an atom at the corner and at the face in the unit cell are, respectively

(a) 1/8 and 1/2
(b) 1/2 and 1/4
(c) 1/4 and 1/2
(d) 1/4 and 1/8

51. The equilibrium constant K_C of the reaction, $2A \rightleftharpoons B + C$ is 0.5 at 25°C and 1 atm. The reaction will proceed in the backward direction, when concentrations $[A]$, $[B]$ and $[C]$ are, respectively

(a) 10^{-3}, 10^{-2} and 10^{-2} M
(b) 10^{-1}, 10^{-2} and 10^{-2} M
(c) 10^{-2}, 10^{-2} and 10^{-3} M
(d) 10^{-2}, 10^{-3} and 10^{-3} M

52. Major products formed in the reaction of *t*-butyl methyl ether with HI are

(a) $H_3C—I$ and $(CH_3)_3C—OH$
(b) $(CH_3)_2C=CH_2$ and $H_3C—OH$
(c) $H_3C—OH$ and $(CH_3)_3C—I$
(d) $(CH_3)_2CHCH_2I$ and $H_3C—OH$

53. If the molar conductivities (in $S\ cm^2\ mol^{-1}$) of NaCl, KCl and NaOH at infinite dilution are 126, 150 and 250, respectively, the molar conductivity of KOH (in $S\ cm^2\ mol^{-1}$) is

(a) 526 (b) 226
(c) 26 (d) 274

54. 4-formyl benzoic acid on treatment with one equivalent of hydrazine followed by heating with alc. KOH gives the major product

(a) O H, O NHNH$_2$

(b)

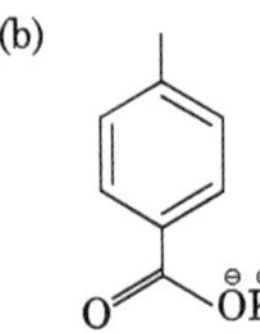

(c)

(d)

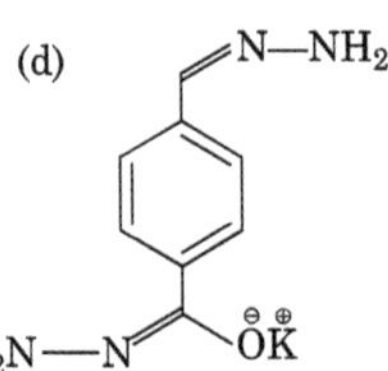

55. Two elements, X and Y, have atomic numbers 33 and 17, respectively. The molecular formula of a stable compound formed between them is

(a) XY (b) XY_2 (c) XY_3 (d) XY_4

56. The number of moles of $KMnO_4$ required to oxidise one equivalent of KI in the presence of sulphuric acid is

(a) 5 (b) 2 (c) 1/2 (d) 1/5

57. Three successive measurements in an experiment gave the values 10.9, 11.4042 and 11.42. The correct way of reporting the average value is

(a) 11.2080 (b) 11.21
(c) 11.2 (d) 11

58. The latent heat of melting of ice at 0°C is 6 kJ mol^{-1}. The entropy change during the melting in J K^{-1} mol^{-1} is closest to

(a) 22 (b) 11 (c) – 11 (d) – 22

59. The major product of the following reaction,

I $\xrightarrow{\text{Cu, }\Delta}$

is

(a) (b)

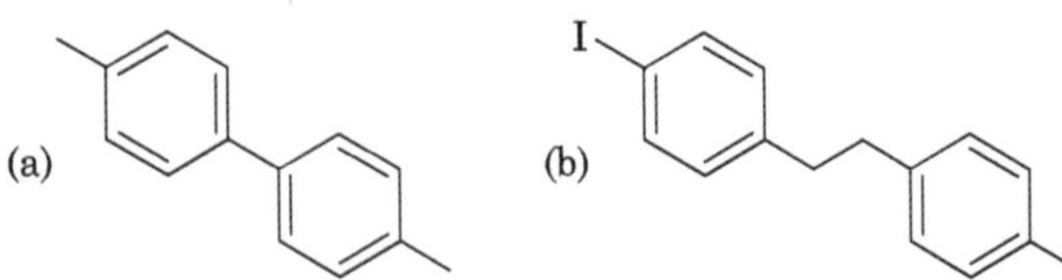

(c) (d)

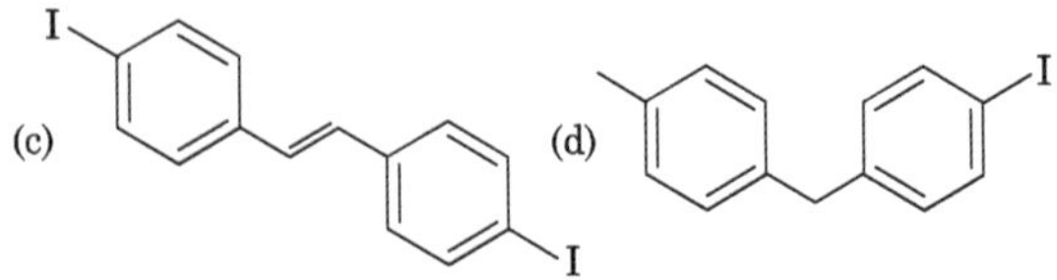

60. The energies of d_{xy} and d_z^2-orbitals in octahedral and tetrahedral transition metal complexes are such that

(a) $E(d_{xy}) > E(d_z^2)$ in both tetrahedral and octahedral complexes
(b) $E(d_{xy}) < E(d_z^2)$ in both tetrahedral and octahedral complexes
(c) $E(d_{xy}) < E(d_z^2)$ in tetrahedral but $E(d_{xy}) < E(d_z^2)$ in octahedral complexes
(d) $E(d_{xy}) < E(d_z^2)$ in tetrahedral but $E(d_{xy}) > E(d_z^2)$ in octahedral complexes

BIOLOGY

61. In which of the following types of glands is the secretion collected inside the cell and discharged by disintegration of the entire gland?

(a) Apocrine (b) Merocrine
(c) Holocrine (d) Epicrine

62. Which one of the following interactions does not promote coevolution?

(a) Commensalism
(b) Mutualism
(c) Parasitism
(d) Interspecific competition

63. Stratification is more common in which of the following?

(a) Deciduous forest (b) Tropical rainforest
(c) Temperate forest (d) Tropical savannah

64. Where is the third ventricle of the brain located?

(a) Cerebrum (b) Cerebellum
(c) Pons Varolii (d) Diencephalon

65. Which of the following is the final product of a gene?

(a) A polypeptide only
(b) An RNA only
(c) Either polypeptide or RNA
(d) A nucleotide only

66. Forelimbs of whales, bats, humans and cheetah are examples of which of the following processes?

(a) Divergent evolution (b) Convergent evolution
(c) Adaptation (d) Saltation

67. Which of the following results from conjugation in *Paramecium*?

(a) Cell death (b) Cell division
(c) Budding (d) Recombination

68. In an experiment investigating photoperiodic response, the leaves of a plant are removed. What is the most likely outcome?

(a) Photoperiodism is not affected
(b) Photoperiodic response does not occur
(c) The plant starts flowering
(d) The plant starts to grow taller

69. Testosterone is secreted by which endocrine part of testis?

(a) Leydig cells (b) Seminiferous tubules
(c) Tunica albugenia (d) Sertoli cells

70. The mutation of a purine to a pyrimidine is known as
(a) transition (b) frame shift
(c) non-sense (d) transversion

71. Which of the following is secreted at the ends of an axon?
(a) Ascorbic acid (b) Acetic acid
(c) Acetylcholine (d) Acetyl Co-A

72. A bacterial colony is produced from
(a) a single bacterium by its repetitive division
(b) multiple bacteria without replication
(c) clumping of two to three bacteria
(d) a single bacterium without cell division

73. Rhinoviruses are the causative agents of
(a) diarrhoea (b) AIDS
(c) dengue (d) common cold

74. What is the genetic material of ebola virus?
(a) Single-stranded DNA (b) Double-stranded RNA
(c) Single-stranded RNA (d) Double-stranded DNA

75. Name the terminal acceptor of electrons in the mitochondrial electron transport chain.
(a) Nitrate (b) Fumarate
(c) Succinate (d) Oxygen

76. Two tubes labelled 'P' and 'Q' contain food stuff. Tube 'P' gave positive test with Benedict's solution while tube 'Q' gave positive test with nitric acid. Which of the following is correct?
(a) Tube 'P' contains sugar, tube 'Q' contains protein
(b) Tube 'P' contains protein, tube 'Q' contains sugar
(c) Both, tube 'P' and tube 'Q' contain sugar
(d) Both, tube 'P' and tube 'Q' contain protein

77. How many linear DNA fragments will be produced when a circular plasmid is digested with a restriction enzyme having 3 sites?
(a) 4 (b) 5
(c) 3 (d) 2

78. If the humidity of the atmosphere suddenly increases substantially, the water flow in the xylem will
(a) increase
(b) decrease
(c) remain unaltered
(d) increase sharply and then reduce slowly to the pre-existing level

79. Which one of the following is the complementary sequence for the DNA with 5′-CGTACTA-3′ ?
(a) 5′-TAGTACG-3′ (b) 5′-ATCATGC-3′
(c) 5′-UTCUTGC-3′ (d) 5′-GCUAGCA-3′

80. A diploid plant has 14 chromosomes, but its egg cell has 6 chromosomes. Which one of the following is the most likely explanation of this?
(a) Non-disjunction in meiosis-I and II
(b) Non-disjunction in meiosis-I
(c) Non-disjunction in mitosis
(d) Normal meiosis

PART-II (2 Marks Questions)

MATHEMATICS

81. Let $n \geq 3$ be an integer. For a permutation $\sigma = (a_1, a_2, \dots, a_n)$ of (1, 2, ... , n) we let $f_\sigma(x) = a_n x^{n-1} + a_{n-1}x^{n-2} + \dots a_2 x + a_1$. Let S_σ be the sum of the roots of $f_\sigma(x) = 0$ and let S denote the sum over all permutations σ of $(1, 2, \dots, n)$ of the numbers S_σ. Then,
(a) $S < -n!$ (b) $-n! < S < 0$
(c) $0 < S < n!$ (d) $n! < S$

82. If n is a positive integer and $\omega \neq 1$ is a cube root of unity, the number of possible values of $\left| e^{\sum_{k=0}^{n} {}^nC_k \omega^k} \right|$ is
(a) 2 (b) 3 (c) 4 (d) 6

83. Suppose a parabola $y = ax^2 + bx + c$ has two x intercepts, one positive and one negative, and its vertex is $(2, -2)$. Then, which of the following is true?
(a) $ab > 0$ (b) $bc > 0$
(c) $ca > 0$ (d) $a + b + c > 0$

84. Let $n \geq 3$ and let $C_1, C_2, \dots, C_n$, be circles with radii $r_1, r_2, \dots, r_n$, respectively. Assume that C_i and C_{i+1} touch externally for $1 \leq i \leq n-1$. It is also given that the X-axis and the line $y = 2\sqrt{2}\,x + 10$ are tangential to each of the circles. Then, $r_1, r_2, \dots, r_n$ are in
(a) an arithmetic progression with common difference $3 + \sqrt{2}$
(b) a geometric progression with common ratio $3 + \sqrt{2}$
(c) an arithmetic progression with common difference $2 + \sqrt{3}$
(d) a geometric progression with common ratio $2 + \sqrt{3}$

85. The number of integers n for which $3x^3 - 25x + n = 0$ has three real roots is
(a) 1 (b) 25 (c) 55 (d) infinite

86. An ellipse inscribed in a semi-circle touches the circular arc at two distinct points and also touches the bounding diameter. Its major axis is parallel to the bounding diameter. When the ellipse has the maximum possible area, its eccentricity is
(a) $\frac{1}{\sqrt{2}}$ (b) $\frac{1}{2}$ (c) $\frac{1}{\sqrt{3}}$ (d) $\sqrt{\frac{2}{3}}$

87. Let $I_n = \int_0^{\pi/2} x^n \cos x dx$, where n is a non-negative integer. Then, $\sum_{n=2}^{\infty} \left(\frac{I_n}{n!} + \frac{I_{n-2}}{(n-2)!} \right)$ equals

(a) $e^{\pi/2} - 1 - \frac{\pi}{2}$ (b) $e^{\pi/2} - 1$

(c) $e^{\pi/2} - \frac{\pi}{2}$ (d) $e^{\pi/2}$

88. For a real number x let $[x]$ denote the largest integer less than or equal to x. The smallest positive integer n for which the integral $\int_1^n [x][\sqrt{x}]\, dx$ exceeds 60 is

(a) 8 (b) 9 (c) 10 (d) $[60^{2/3}]$

89. Choose a number n uniformly at random from the set $\{1, 2, \ldots, 100\}$. Choose one of the first seven days of the year 2014 at random and consider n consecutive days starting from the chosen day. What is the probability that among the chosen n days, the number of Sundays is different from the number of Mondays?

(a) $\frac{1}{2}$ (b) $\frac{2}{7}$ (c) $\frac{12}{49}$ (d) $\frac{43}{175}$

90. Let $S = \{(a, b)\ a, b \in Z, 0 \le a, b \le 18\}$. The number of lines in R^2 passing through (0, 0) and exactly one other point in S is

(a) 16 (b) 22 (c) 28 (d) 32

PHYSICS

91. A solid sphere spinning about a horizontal axis with an angular velocity ω is placed on a horizontal surface. Subsequently it rolls without slipping with an angular velocity of

(a) $2\omega/5$ (b) $7\omega/5$ (c) $2\omega/7$ (d) ω

92. Consider the system shown below.

A horizontal force F is applied to a block X of mass 8 kg, such that the block Y of mass 2 kg adjacent to it does not slip downwards under gravity. There is no friction between the horizontal plane and the base of the block X. The coefficient of friction between the surfaces of blocks X and Y is 0.5. The minimum value of F is (take, acceleration due to gravity to be 10 ms^{-2})

(a) 200 N (b) 160 N (c) 40 N (d) 240 N

93. The maximum value attained by the tension in the string of a swinging pendulum is four times the minimum value it attains. There is no slack in the string. The angular amplitude of the pendulum is

(a) 90° (b) 60° (c) 45° (d) 30°

94. One mole of a monoatomic ideal gas is expanded by a process described by $pV^3 = C$, where C is a constant. The heat capacity of the gas during the process is given by (R is the gas constant)

(a) $2R$ (b) $\frac{5}{2}R$

(c) $\frac{3}{2}R$ (d) R

95. A concave mirror of radius of curvature R has a circular outline of radius r. A circular disc is to be placed normal to the axis at the focus, so that it collects all the light that is reflected from the mirror from a beam parallel to the axis. For $r << R$, the area of this disc has to be at least

(a) $\frac{\pi r^6}{4R^4}$ (b) $\frac{\pi r^4}{4R^2}$ (c) $\frac{\pi r^5}{4R^3}$ (d) $\frac{\pi r^4}{R^2}$

96. The angles of incidence and refraction of a monochromatic ray of light of wavelength λ at an air-glass interface are i and r, respectively. A parallel beam of light with a small spread $\delta\lambda$ in wavelength about a mean wavelength λ is refracted at the same air-glass interface. The refractive index μ of glass depends on the wavelength λ as $\mu(\lambda) = a + b/\lambda^2$, where a and b are constants. Then, the angular spread in the angle of refraction of the beam is

(a) $\left| \frac{\sin i}{\lambda^3 \cos r} \delta\lambda \right|$ (b) $\left| \frac{2b}{\lambda^3} \delta\lambda \right|$

(c) $\left| \frac{2b \tan r}{a\lambda^3 + b\lambda} \delta\lambda \right|$ (d) $\left| \frac{2b(a + b/\lambda^2) \sin i}{\lambda^3} \delta\lambda \right|$

97. What are the charges stored in the 1 μF and 2 μF capacitors in the circuit below, once the currents become steady?

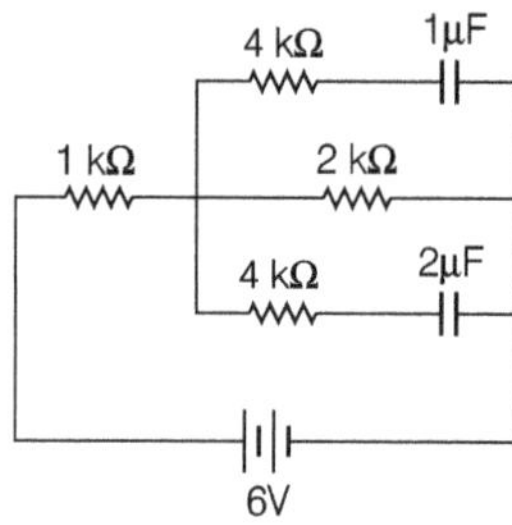

(a) 8 μC and 4 μC, respectively
(b) 4 μC and 8 μC, respectively
(c) 3 μC and 6 μC, respectively
(d) 6 μC and 3 μC, respectively

98. A 1.5 kW laser beam of wavelength 6400 Å is used to levitate a thin aluminium disc of same area as the cross-section of the beam. The laser light is reflected by the aluminium disc without any absorption. The mass of the foil is close to

(a) 10^{-9} kg (b) 10^{-3} kg
(c) 10^{-4} kg (d) 10^{-6} kg

99. When Ultraviolet radiation of a certain frequency falls on a potassium target, the photo electrons released can be stopped completely by a retarding potential of 0.6 V. If the frequency of the radiation is increased by 10%, this stopping potential rises to 0.9 V. The work function of potassium is

(a) 2.0 eV (b) 2.4 eV
(c) 3.0 eV (d) 2.8 eV

100. The dimensions of Stefan-Boltzmann's constant σ can be written in terms of Planck's constant h, Boltzmann's constant k_B and the speed of light c as $\sigma = h^{\alpha} k_B^{\beta} c^{\gamma}$. Here,

(a) $\alpha = 3, \beta = 4$ and $\gamma = -3$
(b) $\alpha = 3, \beta = -4$ and $\gamma = 2$
(c) $\alpha = -3, \beta = 4$ and $\gamma = -2$
(d) $\alpha = 2, \beta = -3$ and $\gamma = -1$

CHEMISTRY

101. In the reaction sequence,

NH_2 (benzene ring) $\xrightarrow[\text{(ii) } Br_2/CH_3CO_2H]{\text{(i) } (CH_3CO)_2O, \text{ pyridine}} X \xrightarrow{Aq. \text{ conc. NaOH}} Y$

X and Y are, respectively

(a) O, NH, Br ; NH_2, Br

(b) O, NH, Br, Br, Br ; NH_2, Br, Br, Br

(c) NH_2 O, Br ; NH_2 O, OH

(d) O, NH, Br, Br ; O, NH, OH, OH

102. The density of acetic acid vapor at 300 K and 1 atm is 5 mg cm^{-3}. The number of acetic acid molecules in the cluster that is formed in the gas phase is closest to

(a) 5 (b) 2 (c) 3 (d) 4

103. The molar enthalpy change for $H_2O\,(l) \rightleftharpoons H_2O\,(g)$ at 373 K and 1 atm is 41 kJ/mol. Assuming ideal behaviour, the internal energy change for vaporisation of 1 mol of water at 373 K and 1 atm in kJ mol^{-1} is

(a) 30.2 (b) 41.0 (c) 48.1 (d) 37.9

104. The equilibrium constants (K_C) of two reactions $H_2 + I_2 \rightleftharpoons 2HI$ and $N_2 + 3H_2 \rightleftharpoons 2NH_3$ are 50 and 1000, respectively. The equilibrium constant of the reaction $N_2 + 6HI \rightleftharpoons 2NH_3 + 3I_2$ is closest to

(a) 50000 (b) 20 (c) 0.008 (d) 0.005

105. Given, that the bond energies of N$\equiv$N is 946 kJ mol^{-1}, H—H is 435 kJ mol^{-1}, N—N is 159 kJ mol^{-1}, and N—H is 389 kJ mol^{-1}, the heat of formation of hydrazine in the gas phase in kJ mol^{-1} is

(a) 833 (b) 101
(c) 334 (d) 1268

106. The radius of K^+ is 133 pm and that of Cl^- is 181 pm. The volume of the unit cell of KCl expressed in 10^{-22} cm^3 is

(a) 0.31 (b) 1.21
(c) 2.48 (d) 6.28

107. The reaction, $K_2Cr_2O_7 + mFeSO_4 + nH_2SO_4 \longrightarrow Cr_2(SO_4)_3 + pFe_2(SO_4)_3 + K_2SO_4 + qH_2O$ when balanced, m, n, p and q are, respectively

(a) 6, 14, 3, 14 (b) 6, 7, 3, 7
(c) 3, 7, 2, 7 (d) 4, 14, 2, 14

108. The standard free energy change (in J) for the reaction $3Fe^{2+}(aq) + 2Cr(s) = 2Cr^{3+}(aq) + 3Fe(s)$ given $E^{\circ}_{Fe^{2+}/Fe} = -0.44$ V and $E^{\circ}_{Cr^{3+}/Cr} = -0.74$ V is ($F = 96500$ C)

(a) 57,900 (b) – 57,900
(c) – 173,700 (d) 173,700

109. Calcium butanoate on heating followed by treatment with 1, 2-ethanediol in the presence of catalytic amount of an acid, produces a major product which is

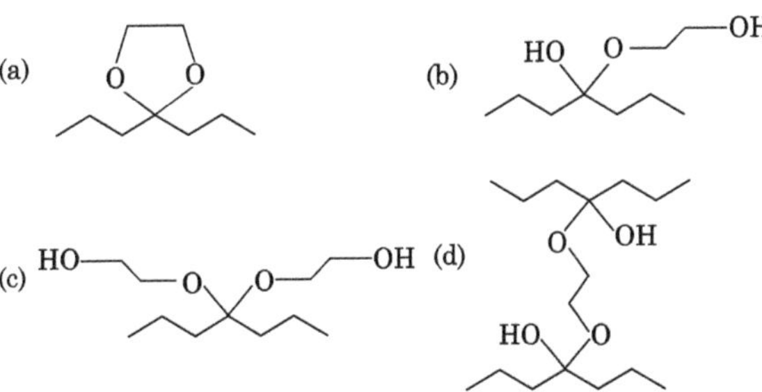

110. XeF_6 on complete hydrolysis yields 'X'. The molecular formula of X and its geometry, respectively are
(a) XeO_2 and linear (b) XeO_3 and trigonal planar
(c) XeO_3 and pyramidal (d) XeO_4 and tetrahedral

BIOLOGY

111. Following the cell cycle scheme given below, what is the probability that a cell would be in M-phase at any given time?

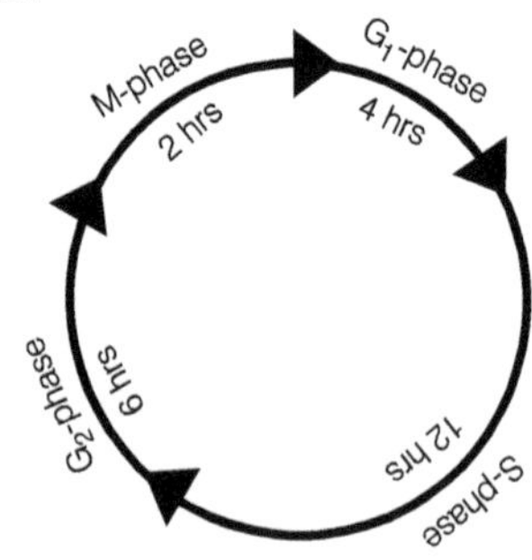

(a) 1/24 (b) 1/12 (c) 1/6 (d) 1/2

112. A flower with Tt genotype is cross-pollinated by TT pollens. What will the genotypes of the resulting endosperm and embryo, respectively, be?
(a) TTT, (TT + Tt) (b) (TTT + TTt), TT
(c) TTt, Tt (d) TTt, (TT + Tt)

113. A new life form discovered on a distant planet has a genetic code consisting of five unique nucleotides and only one stop codon. If each codon has four bases, what is the maximum number of unique amino acids this life form can use?
(a) 624 (b) 20 (c) 124 (d) 3124

114. A spontaneous mutation results in a couple having only female progeny. When the daughter marries and has children, none of them is male. However, in the third generation, there is few male offsprings. What is the most likely explanation of this observation?
(a) The mutation reverses spontaneously in the third generation
(b) The mutation occurs on the X-chromosome and is both recessive and lethal
(c) The mutation occurs on the X-chromosome and is both recessive and dominant
(d) The mutation occurs on an autosome and is dominant

115. A circular plasmid of 10,000 base pairs (bp) is digested with two restriction enzymes, A and B, to produce a 3,000 bp and a 2,000 bp bands when visualised on an agarose gel. When digested with one enzyme at a time, only one band is visible at 5,000 bp. If the first site for enzyme A (A1) is present at the 100 th base, the order in which the remaining sites (A2, B1 and B2) are present is
(a) 3,100, 5,100, 8,100 (b) 8,100, 3,100, 5,100
(c) 5,100, 3,100, 8,100 (d) 8,100, 5,100, 3,100

116. After meiosis-II, daughter cells differ from the parent cells and each other in their genotypes. This can occur because of which one of the following mechanism(s)?
(a) Only synaptic crossing over
(b) Only crossing over and independent assortment of chromosomes
(c) Only crossing over and chromosomal segregation
(d) Crossing over, independent assortment and segregation of chromosomes

117. A desert lizard (an ectotherm) and a mouse (an endotherm) are placed inside a chamber at 15°C and their body temperatures [T(L) for the lizard and T(M) for the mouse] and metabolic rates [M(L) for the lizard and M(M) for the mouse] are monitored. Which one of the following is correct?
(a) T(L) and M(L) will fall while T(M) and M(M) will increase
(b) T(L) and M(L) will increase while T(M) and M(M) will fall
(c) T(L) and M(L) will fall, T(M) will remain same and M(M) will increase
(d) T(L) and M(L) will remain same and T(M) and M(M) will decrease

118. In Griffith's experiments, mice died when injected with
(a) heat-killed S-strain
(b) heat-killed S-strain combined with R-strain
(c) heat-killed R-strain
(d) live R-strain

119. Human height is a multigenic character. If the heights of all the individuals living in a metropolis are measured and the percentages of the population belonging to a specific height are plotted as shown below, which of the plots would represent the most realistic distribution?

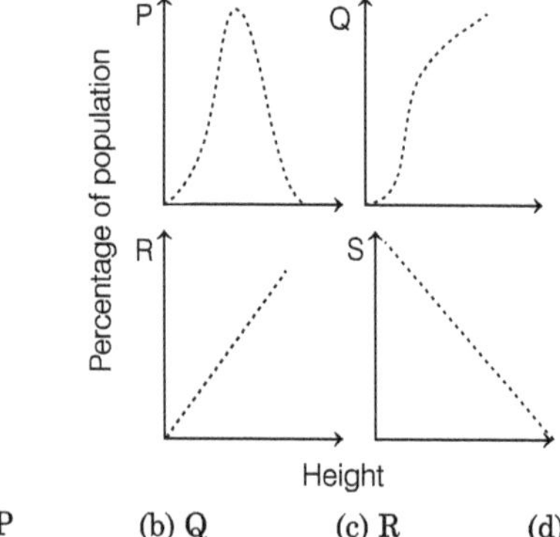

(a) P (b) Q (c) R (d) S

120. If mitochondria isolated from a cell are first placed without carbon source in a buffer at pH 8.0 and then transferred to a buffer at pH 4, it will lead to
(a) an increase in intra-mitochondrial acidity
(b) a decrease in intra-mitochondrial acidity
(c) blockage of ATP synthesis
(d) synthesis of ATP

Answers

PART-I

1 (d)	2 (d)	3 (c)	4 (b)	5 (b)	6 (d)	7 (c)	8 (b)	9 (b)	10 (d)
11 (b)	12 (d)	13 (c)	14 (b)	15 (b)	16 (b)	17 (c)	18 (b)	19 (d)	20 (c)
21 (c)	22 (b)	23 (a)	24 (d)	25 (b)	26 (a)	27 (d)	28 (b)	29 (c)	30 (a)
31 (b)	32 (d)	33 (a)	34 (c)	35 (b)	36 (a)	37 (d)	38 (c)	39 (b)	40 (a)
41 (d)	42 (a)	43 (d)	44 (b)	45 (c)	46 (a)	47 (c)	48 (a)	49 (b)	50 (a)
51 (a)	52 (c)	53 (d)	54 (b)	55 (c)	56 (d)	57 (c)	58 (a)	59 (a)	60 (c)
61 (c)	62 (a)	63 (b)	64 (d)	65 (c)	66 (a)	67 (d)	68 (b)	69 (a)	70 (d)
71 (c)	72 (a)	73 (d)	74 (c)	75 (d)	76 (a)	77 (c)	78 (b)	79 (a)	80 (b)

PART-II

81 (a)	82 (c)	83 (b)	84 (d)	85 (c)	86 (d)	87 (a)	88 (b)	89 (b)	90 (*)
91 (c)	92 (a)	93 (b)	94 (d)	95 (a)	96 (c)	97 (b)	98 (d)	99 (b)	100 (c)
101 (a)	102 (b)	103 (d)	104 (c)	105 (b)	106 (c)	107 (b)	108 (c)	109 (a)	110 (c)
111 (b)	112 (d)	113 (a)	114 (b)	115 (c)	116 (d)	117 (c)	118 (b)	119 (a)	120 (d)

** No options are correct.*

Solutions

1. *(d)* We have, C_0 be a circle of radius 1. C_n be a circle whose area equals the area of a square inscribed in C_{n-1}.

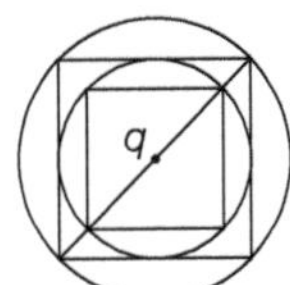

Let $a_0, a_1, a_2, a_3, \ldots, a_n$ be the length of sides of square inscribed in circle $C_0, C_1, C_2, \ldots, C_n$ and $r_0, r_1, r_2, \ldots, r_n$ be radius of circle.

$$2a_0^2 = 4$$
$$a_0^2 = 2$$
$$\pi r_1^2 = a_0^2$$
$$r_1^2 = \frac{2}{\pi}$$
$$2a_1^2 = (2r_1)^2 = 4r_1^2$$
$$a_1^2 = \frac{4}{\pi}$$
$$\pi r_2^2 = a_1^2 \Rightarrow r_2^2 = \frac{4}{\pi^2}$$

Similarly, $r_n^2 = \dfrac{2^n}{\pi^n}$

Now, $\sum_{i=0}^{\infty} \text{Area } (C_{ij})$

$$= \pi\left(-1 + \frac{2}{\pi} + \frac{2^2}{\pi^2} + \frac{2^3}{\pi^3} + \ldots\right)$$

$$= \pi\left(\frac{1}{1 - \frac{2}{\pi}}\right)$$

$$\left[S_\infty = 1 + r + r^2 + \ldots = \frac{1}{1-r}\right]$$

$$= \frac{\pi^2}{\pi - 2}$$

2. *(d)* (a) $[x + y] \leq [x] + [y]$

Let $x = 0.6,\ y = 0.5$

$$[0.6 + 0.5] \leq [0.6] + [0.5]$$
$$1 \leq 0 + 0 \quad \text{False}$$

(b) $[xy] \leq [x]\,[y]$

Let $x = 1.5,\ y = 1.6$

$$[(1.5)\ (1.6)] \leq [1.5]\,[1.6]$$
$$2 \leq 1 \times 1 \quad \text{False}$$

(c) $[2^x] \leq 2^{[x]}$

Let $x = \dfrac{5}{2}$

$$[2^{5/2}] \leq 2^{[5/2]}$$
$$[4\sqrt{2}] \leq 2^2$$
$$5 \leq 4 \ \text{False}$$

(d) $\left[\dfrac{x}{y}\right] \leq \dfrac{[x]}{[y]}$

Let $x, y \geq 1$

If $x < y$, then $\left[\dfrac{x}{y}\right] = 0 \Rightarrow 0 \leq \dfrac{[x]}{[y]}$ True

If $x \geq y$, then $\left[\dfrac{x}{y}\right] \leq \dfrac{[x]}{[y]}$ always true

Hence, option (d) is correct.

3. *(c)* We have,

$$A_n = \max\{{}^nC_r \mid 0 \leq r \leq n\}$$
$$n \in \{1, 2, 3, \ldots, 20\}$$

***Case* I** When n is even

$$A_n = {}^nC_{n/2}$$

$$\therefore \quad \frac{A_n}{A_{n-1}} = \frac{{}^nC_{n/2}}{{}^{n-1}C_{\frac{n-1-1}{2}}} = 2$$

So for all n even given relation is true.

***Case* II** When n is odd

$$A_n = {}^nC_{\frac{n-1}{2}}$$

$$\therefore \quad \frac{A_n}{A_{n-1}} = \frac{{}^nC_{\frac{n-1}{2}}}{{}^{n-1}C_{\frac{n-1}{2}}} = \frac{2n}{n+1}$$

$$1.9 \leq \frac{2n}{n+1} \leq n \text{ if } n = 19$$

$\therefore$ Total number of elements are 10 even number and $19 = 10 + 1 = 11$

4. *(b)* We have, $P\cdot(r,\theta)=(x,y)$

Equation of line OP is

$$y = x\tan\theta$$

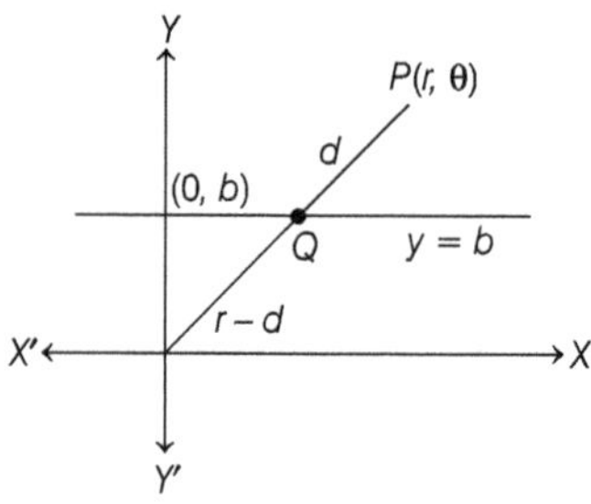

OP cut the line $r\sin\theta = b$

i.e. $y = b$

Given $PQ = d$

$\therefore$ point P is $y = b \pm d\sin\theta$

$\Rightarrow r\sin\theta = b \pm d\sin\theta \Rightarrow (r \pm d)\sin\theta = b$

Hence, locus of $P(r,\theta)$ is $(r \pm d)\sin\theta = b$

5. *(b)* Given, equation of circle

$$x^2 + y^2 = 1 \qquad \text{...(i)}$$

Equation of line L_t passing through (0, 1) and $(t, 0)$ is

$$\frac{x}{t} + \frac{y}{1} = 1,\ t \ge 0$$

$$y = 1 - \frac{x}{t} \qquad \text{...(ii)}$$

Putting the value of y in Eq. (i), we get

$$x^2 + \left(1 - \frac{x}{t}\right)^2 = 1$$

$$x^2 + 1 - \frac{2x}{t} + \frac{x^2}{t^2} = 1$$

$$x^2\left(1 + \frac{1}{t^2}\right) = \frac{2x}{t}$$

$$\Rightarrow \quad x = 0,\ x = \frac{2t}{1+t^2}$$

When, $x = 0$, $y = 1$

$$x = \frac{2t}{1+t^2},\ y = 1 - \frac{2t}{t(1+t^2)} = -\frac{(1-t^2)}{1+t^2}$$

$$Q_t = \left(\frac{2t}{1+t^2}, -\frac{(1-t^2)}{1+t^2}\right),$$

$Q_t = (\sin 2\theta, -\cos\theta)$ when $t = \tan\theta$

$t \in [1, 1+\sqrt{2}]$

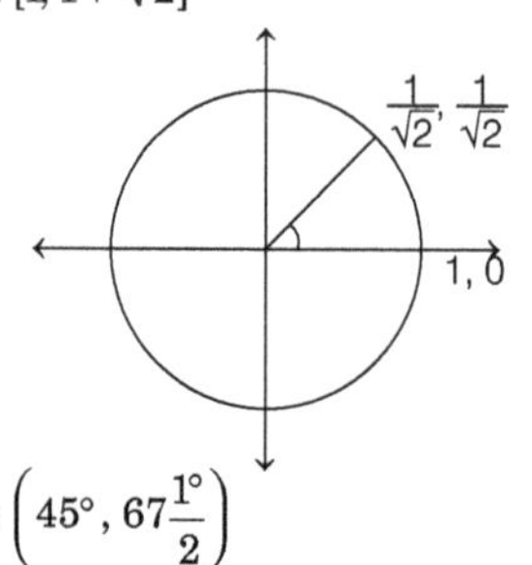

$$\theta \in \left(45°, 67\frac{1°}{2}\right)$$

$\therefore$ Angle subtended by this arc is $\frac{\pi}{4}$.

6. *(d)* Given, in ellipse

$$A'S' = SS' = SA$$

$$\therefore \quad SS' = \frac{1}{3}AA'$$

$$\Rightarrow \quad 2ae = \frac{1}{3}(2a)$$

$$\Rightarrow \quad e = \frac{1}{3}$$

Also given, $b = 2\sqrt{2}$

$$e^2 = 1 - \frac{b^2}{a^2}$$

$$\Rightarrow \quad \frac{1}{9} = 1 - \frac{8}{a^2}$$

$$\Rightarrow \quad \frac{8}{a^2} = 1 - \frac{1}{9} = \frac{8}{9}$$

$$\Rightarrow \quad a^2 = 9$$

$$\therefore \quad a = 3$$

7. *(c)* In ΔABC, $AB = BC$

F be the mid-point of AB and X be a point on BC such that FX is perpendicular to AB

$$BX = 3XC$$

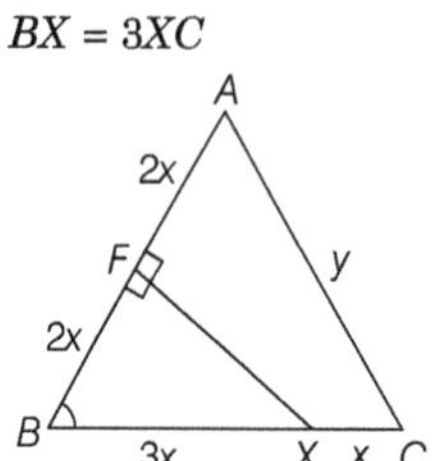

In ΔBFX,

$$\cos B = \frac{BF}{BX} = \frac{2x}{3x}$$

$$\Rightarrow \quad \cos B = \frac{2}{3}$$

In ΔABC,

$$\cos B = \frac{(AB)^2 + (BC)^2 - AC^2}{2(AB)(BC)}$$

$$\Rightarrow \quad \frac{2}{3} = \frac{(4x)^2 + (4x)^2 - y^2}{2(4x)(4x)}$$

$$\Rightarrow \quad 4(4x)^2 = 6(4x)^2 - 3y^2$$

$$\Rightarrow \quad 3y^2 = 2(4x)^2$$

$$\Rightarrow \quad \left(\frac{4x}{y}\right)^2 = \frac{3}{2} \Rightarrow \frac{4x}{y} = \sqrt{\frac{3}{2}}$$

$$\Rightarrow \quad \frac{BC}{AC} = \sqrt{\frac{3}{2}} \qquad [\because 4x = BC,\ y = AC]$$

8. *(b)* We have,

$$\cos^4 x + \frac{1}{\cos^2 x} = \sin^4 x + \frac{1}{\sin^2 x}$$

$$\Rightarrow \quad \cos^4 x - \sin^4 x = \frac{1}{\sin^2 x} - \frac{1}{\cos^2 x}$$

$$\Rightarrow \quad (\cos^2 x - \sin^2 x)(\cos^2 x + \sin^2 x) = \frac{\cos^2 x - \sin^2 x}{\sin^2 x\cos^2 x}$$

$$\Rightarrow \quad \cos^2 x - \sin^2 x = \frac{\cos^2 x - \sin^2 x}{\sin^2 x\cos^2 x}$$

$$\Rightarrow \quad \cos 2x\left(1 - \frac{4}{\sin^2 2x}\right) = 0$$

$$\Rightarrow \quad \cos 2x = 0 \text{ or } \sin^2 2x = 4$$

$$\therefore \quad \cos 2x = 0 \text{ or } \sin^2 2x \ne 4$$

$$\Rightarrow \quad 2x = (2n+1)\frac{\pi}{2}$$

$$\Rightarrow \quad x = (2n+1)\frac{\pi}{4}$$

In $x \in (0, 2\pi)$

$$x = \frac{\pi}{4}, \frac{3\pi}{4}, \frac{5\pi}{4}, \frac{7\pi}{4}$$

$\therefore$ Total number of solution = 4

9. *(b)* We have,

$$f(x) = \begin{cases} \dfrac{x+5}{x-2}, & x \ne 2 \\ 1, & x = 2 \end{cases}$$

$f(x)$ is discontinuous at $x = 2$

$\therefore f(f(x))$ is also discontinuous at $x = 2$

$$\text{Now,} \quad f(f(x)) = \frac{f(x)+5}{f(x)-2} = \frac{\frac{x+5}{x-2} + 5}{\frac{x+5}{x-2} - 2}$$

$$\Rightarrow \quad f(f(x)) = \frac{6x-5}{9-x}$$

Clearly $f(f(x))$ is discontinuous at $x = 9$

$\therefore f(f(x))$ is discontinuous at $x = 2$ and 9

Hence, $f(f(x))$ is discontinuous at exactly two values of x.

10. *(d)* We have, $f(x) = [x]\sin \pi x$

Graph of $f(x)$ are

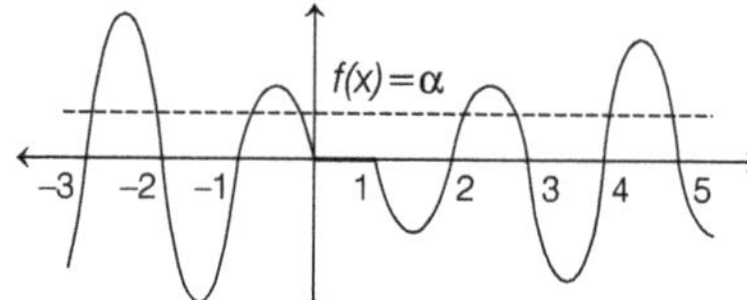

Clearly, $f(x)$ is not differentiable at $x = 1$

$f(x)$ is not symmetric about line $x = 0$

$$\int_{-3}^{3} f(x)\,dx \ne 0$$

$f(x) = \alpha$ will have infinite solutions.

11. *(b)* We have,

$$f(x)=\begin{cases}\sin x, & \text{if } x \text{ is irrational and } x\in[0,\pi]\\ \tan^2 x, & \text{if } x \text{ is rational and } x\in[0,\pi]\end{cases}$$

$f(x)$ is continuous at $x = 0$ and π

For other point, then

$$\sin x = \tan^2 x$$

$$\sin x = \frac{\sin^2 x}{\cos^2 x}$$

$$\Rightarrow \sin x(1-\sin^2 x) = \sin^2 x$$

$$\Rightarrow 1-\sin^2 x = \sin x$$

$$\Rightarrow \sin^2 x + \sin x - 1 = 0$$

$$\Rightarrow \sin x = \frac{-1\pm\sqrt{5}}{2}$$

$$\sin x = \frac{\sqrt{5}-1}{2},\ \sin x \neq \frac{-1-\sqrt{5}}{2}$$

$$x = \sin^{-1}\left(\frac{\sqrt{5}-1}{2}\right)$$

$$x = \pi - \sin^{-1}\left(\frac{\sqrt{5}-1}{2}\right)$$

$\therefore$ $f(x)$ is continuous at $x = 0, \pi$, $\sin^{-1}\frac{\sqrt{5}-1}{2}$ and $\pi - \sin^{-1}\frac{\sqrt{5}-1}{2}$

Hence, $f(x)$ is continuous at 4 points.

12. *(d)* We have, $\int_0^1 f(x)\,dx = 10$

(a) $\int_0^1 e^{-x} f(x)\,dx \le 10$

$$0 < e^{-x} < 1\ x\in(0,1)$$

$$\therefore\ 0 \le e^{-x} f(x) \le 10 \qquad x\in(0,10)$$

$$\therefore\ \int_0^1 e^{-x} f(x) \le 10 \text{ it is true}$$

(b) $\int_0^1 \frac{f(x)}{(1+x)^2}\,dx \le 10$

Now, $\frac{1}{(1+x)^2}\in\left[\frac{1}{4},1\right], x\in(0,1)$

$\therefore$ It is also true.

(c) $-10 \le \int_0^1 \sin(100x) f(x)\,dx \le 10$

$$\sin(100x)\in(-1,1)\ x\in R$$

$\therefore$ It is also true.

(d) $\int_0^1 f(x)^2\,dx \le 100$ not necessarily true

because $f(x)^2$ can take very high value area bounded by $f(x)^2$, X-axis, $x = 0$ to 1 may be greater than 100.

13. *(c)* We have, $f(x) = x + \int_0^x f(t)\,dt$

$$f'(x) = 1 + f(x)$$

On integrating, we get

$$\log(1+f(x)) = x + c$$

$$1 + f(x) = Ae^x$$

$$f(x) = Ae^x - 1$$

$$f(0) = A - 1$$

$$0 = A - 1 \qquad [\because f(0) = 0]$$

$$\Rightarrow A = 1$$

$$\therefore f(x) = e^x - 1$$

$$f(y) = e^y - 1$$

$$f(x+y) = e^{x+y} - 1 = e^x\cdot e^y - 1$$

$$f(x)+f(y)+f(x)f(y)$$

$$= e^x - 1 + e^y - 1 + (e^x-1)(e^y-1)$$

$$= e^x - 1 + e^y - 1 + e^{x+y} - e^x - e^y + 1$$

$$= e^{x+y} - 1$$

Hence, $f(x+y) = f(x) + f(y) + f(x)f(y)$

14. *(b)* Let $I = \int_0^n \cos(2\pi[x]\{x\})\,dx$

$$\{x\} = x - [x]$$

$$\therefore I = \int_0^n \cos[2\pi[x](x-[x])]\,dx$$

$$\Rightarrow I = \int_0^1 \cos 0\,dx + \int_1^2 \cos 2\pi(x-1)\,dx + \int_2^3 \cos 2\pi(x-2)\,dx + \dots + \int_{n-1}^n \cos 2\pi(x-(n-1))\,dx$$

$$\Rightarrow I = 1 + \int_1^2 \cos 2\pi x\,dx + \int_2^3 \cos 4\pi x\,dx + \dots + \int_{n-1}^n \cos 2\pi(n-1)x\,dx$$

$$\Rightarrow I = 1 + \left[\frac{\sin 2\pi x}{2\pi}\right]_1^2 + \left(\frac{\sin 4\pi x}{4\pi}\right)_2^3 + \dots + \left[\frac{\sin 2\pi(n-1)x}{2\pi(n-1)}\right]_{n-1}^n$$

$$\Rightarrow I = 1 + 0 + 0 + \dots + 0 \Rightarrow I = 1$$

15. *(b)* Given, Probability of win $= \frac{5}{6}$

Probability of loss $= \frac{1}{6}$

Probability of B wins =

$$P(B) + P(\overline{B}\overline{A}B) + P(\overline{B}\ \overline{A}\ \overline{B}\ \overline{A}\ B) + \dots$$

$$= \frac{5}{6} + \frac{1}{6}\times\frac{1}{6}\times\frac{5}{6} + \frac{1}{6}\times\frac{1}{6}\times\frac{1}{6}\times\frac{1}{6}\times\frac{5}{6} + \dots$$

$$= \frac{5}{6}\left(1 + \frac{1}{6^2} + \frac{1}{6^4} + \frac{1}{6^6} + \dots\right)$$

$$= \frac{5}{6}\left(\frac{1}{1-\frac{1}{36}}\right) = \frac{30}{35} = \frac{6}{7}$$

16. *(b)* Given,

$$\mu = \frac{\Sigma x_i}{n}$$

$$\sigma = \sqrt{\frac{\Sigma x_1^2}{n} - (\mu)^2}$$

$$\hat{\mu} = \frac{\Sigma y_i}{n}$$

$$= \frac{\frac{x_1+x_2}{2} + \frac{x_1+x_2}{2} + x_3 + x_4 + \dots + x_n}{n}$$

$$\hat{\mu} = \frac{x_1 + x_2 + x_3 \dots + x_n}{n} = \frac{\Sigma x_i}{n} = \mu$$

$$\sigma = \sqrt{\frac{\Sigma y_1^2}{n} - (\mu')^2} = \sqrt{\frac{\Sigma y_1^2}{n} - \mu^2}$$

$$\Sigma x_1^2 = x_1^2 + x_2^2 + x_3^2 + \dots + x_n^2$$

$$\Sigma y_1^2 = \frac{(x_1+x_2)^2}{4} + \frac{(x_1+x_2)^2}{4} + x_3^2 + x_4^2 + \dots + x_n^2$$

$$\Sigma x_1^y - \Sigma y_1^2 = x_1^2 + x_2^2 - 2x_1x_2 = (x_1 - x_2)^2 \ge 0$$

$$\Sigma x_1^2 \ge \Sigma y_1^2$$

Hence, $\sigma \ge \hat{\sigma}$

17. *(c)* G be the centroid of tetrahedron centre and centroid of tetrahedron is same

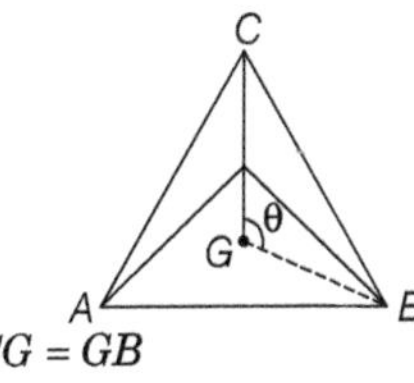

$$CG = GB$$

We know in regular tetrahedron

$CG = \frac{3}{4}$ Height regular tetrahedron from base

$$CG = \frac{3}{4}\left(\frac{\sqrt{6}}{3}\right)BC$$

$$\left[\because H = \frac{\sqrt{6}}{3} \text{ side of regular tetrahedron}\right]$$

$$CG = \frac{\sqrt{6}}{4}BC = GB$$

In ΔGCB,

$$\cos\theta = \frac{GB^2 + GC^2 - BC^2}{2\cdot GB \times GC}$$

$$\cos\theta = \frac{2CG^2 - BC^2}{2CG^2}$$

$$\cos\theta = \frac{\frac{2\times 6}{16}BC^2 - BC^2}{\frac{2\times 6}{16}BC^2}$$

$$= -\frac{4BC^2}{12BC^2} = -\frac{1}{3}$$

$$\therefore \theta = \cos^{-1}\left(\frac{-1}{3}\right)$$

18. *(b)* We have,

$$S = \{(a, b) : a, b, \in Z, 0 \le a, b \le 18\}$$

$$3x + 4y + 5 \ (x, y) \in S$$

$$\therefore \quad S \le 3x + 4y + 5 \le 131$$

$[\because \min (x, y) = (0, 0), \max (x, y) = (15, 18)]$

Given, $3x + 4y + 5$ is divisible by 19

$\therefore \ 3x + 4y + 5 = 19, 38, 57, 76, 95, 114$

Case I

$$3x + 4y + 5 = 19$$
$$3x + 4y = 14$$

Only (2, 2) satisfies

Case II

$$3x + 4y + 5 = 38$$
$$3x + 4y = 33$$

Possible values of (x, y) is (3, 6), (7, 3), (11, 0)

Case III

$$3x + 4y + 5 = 57$$
$$3x + 4y = 52$$

Possible values of (x, y) is (0, 13), (4, 10), (8, 7), (12, 4), (16, 1)

Case IV

$$3x + 4y + 5 = 76$$

Possible values of (x, y) is (1, 17), (5, 14), (9, 11), (13, 8), (17, 5)

Case V

$$3x + 4y + 5 = 95$$

Possible values of (x, y) is (6, 18), (10, 15), (14, 12), (18, 9)

Case VI

$$3x + 4y + 5 = 114$$
$$(x, y) = (15, 15)$$

$\therefore$ Total solution = 19

19. *(d)* We have,

$$[a^k] > [a]^k$$

It is true only $k \le \dfrac{1}{a - [a]} + 1$

$$\frac{1}{a - [a]} + 1 \to \infty$$

$\therefore$ k is smaller than ∞.

20. *(c)* We have,

(A, B) is subset of X, where X has 5 elements $A \ne \phi, B \ne \phi, A \cap B = \phi$

The order pairs if (A, B) is

$$^5C_2 \times 2! + {}^5C_3 \ ({}^3C_2 \times 2!) + {}^5C_4 \ ({}^4C_3 \times 2 + {}^4C_2) + {}^5C_5 \ ({}^5C_4 \times 2! + {}^5C_3 \times 2!)$$
$$= 20 + 60 + 70 + 30 = 180$$

$\therefore \ 151 \le d \le 200$

Hence, option (c) is correct.

21. *(c)* Impulsive forces provides both linear momentum and angular momentum to the rod.

Let v = linear velocity and ω = angular velocity of rod after impulse is given.

Then, conservation of linear momentum gives,

$$J = mv_{CM} \Rightarrow v_{CM} = \frac{J}{m}$$

Angular momentum is also provided by the impulse, so

$$J \times L = I\omega$$

$$\Rightarrow \quad JL = \frac{m(2L)^2}{12}\omega$$

$$\Rightarrow \quad \omega = \frac{3J}{mL}$$

Final kinetic energy of rod is sum of rotational kinetic energy and translational kinetic energy.

$\because$ Kinetic energy (KE)

$$= (KE)_{translation} + (KE)_{rotation}$$
$$= \frac{1}{2}mv^2 + \frac{1}{2}I\omega^2$$
$$= \frac{1}{2}m\left(\frac{J^2}{m^2}\right) + \frac{1}{2}\frac{m(2L)^2}{12} \times \left(\frac{3J}{mL}\right)^2$$
$$= \frac{J^2}{2m} + \frac{36J^2}{24m} = \frac{4J^2}{2m} = \frac{2J^2}{m}$$

22. *(b)* In presence of friction, cylinder rolls and by energy conservation, we have

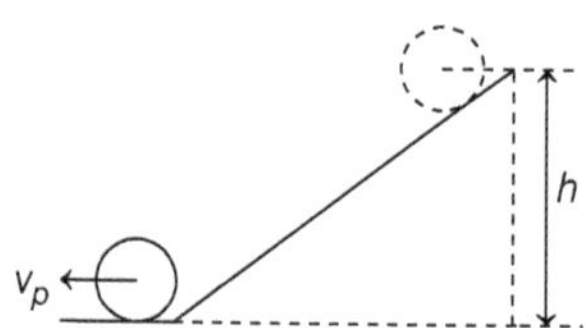

Initial potential energy = Kinetic energy at bottom

$$mgh = (KE)_{translation} + (KE)_{rotation}$$
$$= \frac{1}{2}mv^2 + \frac{1}{2}I\omega^2$$
$$= \frac{1}{2}mv^2 + \frac{1}{2}\left(\frac{1}{2}mr^2\right)\omega^2$$
$$= \frac{3}{4}mv^2 \quad [\because v = r\omega]$$

So, $\quad v_p = \sqrt{\dfrac{4}{3}gh}$

In absence of friction, cylinder slips without rolling.

Hence, $\dfrac{1}{2}mv^2 = mgh$

$$\Rightarrow \quad v_q = \sqrt{2gh}$$

So, ratio of $v_q / v_p = \dfrac{\sqrt{2gh}}{\sqrt{\dfrac{4}{3}gh}} = \sqrt{\dfrac{3}{2}}$

23. *(a)* Force on the particle is

$$F = -\frac{d}{dr}(V) = -\frac{d}{dr}kr = -k$$

Now, this force must acts like centripetal force for rotation of body.

$\therefore \ F = mR\omega^2 = k$ (equating only magnitude of force)

$$\Rightarrow \quad mR\left(\frac{2\pi}{T}\right)^2 = k \Rightarrow T^2 = \frac{4\pi^2 m}{k} \cdot R$$

$$\Rightarrow \quad T^2 \propto R \ \text{ or } \ T \propto R^{\frac{1}{2}}$$

24. *(d)* As block slides down the plane, pendulum bob is subjected to two forces.

(i) Weight mg perpendicularly downwards to the horizontal.

(ii) Pesudo force $mg\sin\theta$ acting upwards and parallel to the inclined plane.

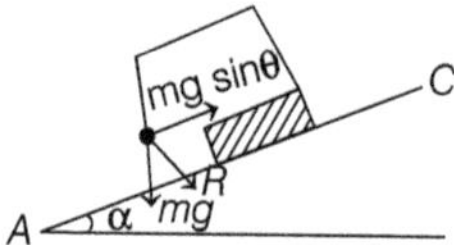

Resultant of these forces is perpendicular to inclined plane AC. So, mean position of pendulum is perpendicular to AC.

25. *(b)* In narrow region, velocity of flow is more.

$$v \propto \frac{1}{\text{Area}}$$

As a result, pressure in narrow region is less.

As, $p + \dfrac{1}{2}\rho v^2$ = constant for a horizontal tube.

So, bubbles grow in size and they move fast through narrow region.

26. *(a)* Potential energy *versus* separation plot for two atoms in a solid is as shown below.

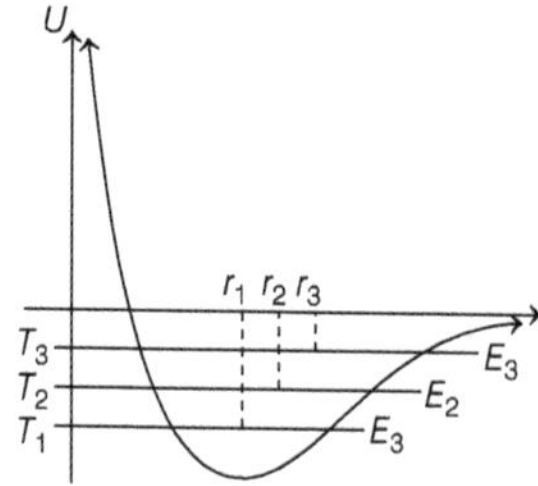

From above graph, we see that as temperature rises ($T_3 > T_2 > T_1$), total energy of atoms also increases ($E_3 > E_2 > E_1$).

As a result, mean separation between atoms also increases ($r_3 > r_2 > r_1$).

Hence, crystalline solids in general expands on heating.

27. *(d)* Ratio of increase in volume per unit original volume per degree rise of temperature of mercury and bromine is

$$\frac{\left[\frac{(V_\theta - V_{\theta_1})_{Hg}}{V_{\theta_1}\cdot\Delta\theta}\right]}{\left[\frac{(V_\theta - V_{\theta_1})_{Br}}{V_{\theta_1}\cdot\Delta\theta}\right]} = \frac{\gamma_{Hg}}{\gamma_{Br}}$$

$$\Rightarrow \quad \frac{(V_\theta - V_{\theta_1})_{Hg}}{(V_\theta - V_{\theta_1})_{Br}} = \frac{\gamma_{Hg}}{\gamma_{Br}}$$

$$[\because (V_{\theta_1})_{Hg} = (V_{\theta_1})_{Br}]$$

$$\Rightarrow \quad \frac{\left(\Delta l \cdot \frac{\pi d_1^2}{4}\right)_{Hg}}{\left(\Delta l \cdot \frac{\pi d_2^2}{4}\right)_{Br}} = \frac{\gamma_{Hg}}{\gamma_{Br}}$$

$$\Rightarrow \quad \frac{d_1^2}{d_2^2} = \frac{\gamma_{Hg}}{\gamma_{Br}} \qquad [\because \Delta l_{Hg} = \Delta l_{Br}]$$

$$\Rightarrow \quad \frac{d_1}{d_2} = \sqrt{\frac{\gamma_{Hg}}{\gamma_{Br}}}$$

Substituting values in above equation, we have

$$\Rightarrow \quad \frac{d_1}{d_2} = \sqrt{\frac{18\times10^{-5}}{108\times10^{-5}}} = \sqrt{\frac{1}{6}} \approx 0.4$$

28. *(b)* $pV^2 = C \Rightarrow \left(\frac{nRT}{V}\right)\cdot V^2 = C$

$[\because$ for an ideal gas, $pV = nRT]$

$$\Rightarrow \quad TV = \frac{C}{nR} = \text{a constant}$$

$$\therefore \quad T \propto \frac{1}{V}$$

So, if $V_2 > V_1$, then $T_2 < T_1$.

29. *(c)* Given situation is as shown below.

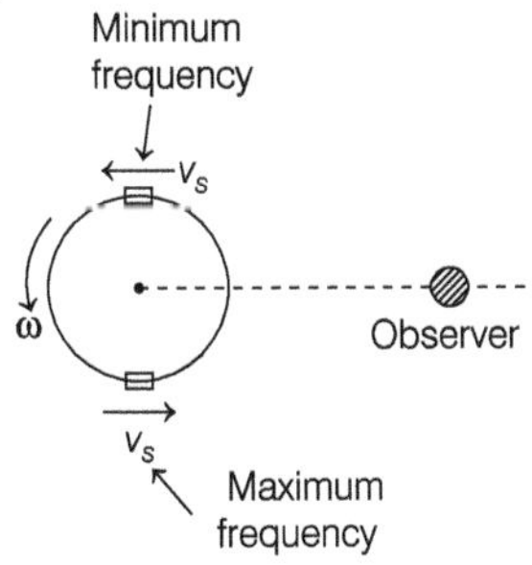

Speed of approach of source

$$= v_s = r\omega = 30\ \text{ms}^{-1}$$

Using formula for Doppler's effect,

Maximum frequency,

$$f_{max} = \left(\frac{v}{v - v_s}\right)\cdot f$$

Minimum frequency,

$$f_{min} = \left(\frac{v}{v + v_s}\right)\cdot f$$

So, ratio of

$$\frac{f_{max}}{f_{min}} = \frac{v + v_s}{v - v_s} = \frac{330 + 30}{330 - 30} \Rightarrow \frac{f_{max}}{f_{min}} = 1.2$$

30. *(a)* Frequency is a characteristics of source it does not changes during refraction process.

Inside the prism wavelength and speed are reduced than that in air.

31. *(b)* When a lens is silvered, it acts like a mirror with focal length f.

It is given by $\frac{-1}{f} = \frac{2}{f_l} + \frac{1}{f_m}$

where, f_l = focal length of lens and f_m = focal length of mirror .

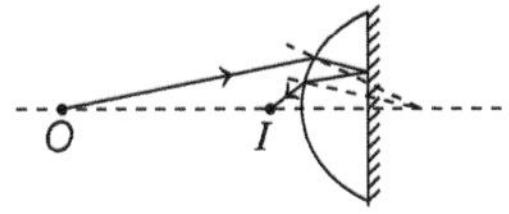

$$\Rightarrow \quad \frac{-1}{f} = \frac{2}{10} + \frac{1}{\infty}$$

$$\Rightarrow \quad f = -5 \text{ cm}$$

Here, note that negative sign appears in formula because mirror obtained is concave in nature.

Here, $u = -30$ cm

So, by mirror formula, we have

$$\frac{1}{f} = \frac{1}{v} + \frac{1}{u}$$

$$\Rightarrow \quad \frac{1}{-5} = \frac{1}{v} + \frac{1}{-30}$$

$$\Rightarrow \quad \frac{1}{v} = \frac{1}{-5} + \frac{1}{30} = \frac{-6+1}{30}$$

$$\Rightarrow \quad v = -6\,\text{cm}$$

So, image is real and 6 cm in front.

32. *(d)* As current increases in loop

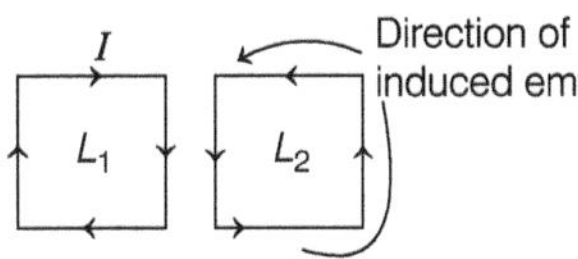

L_1 flux linked with L_2 (due to field produced by L_1) also increases.

Now, induced emf, current due to in loop L_2 is such that, in nearby arms of L_1 and L_2 currents are in same direction.

So, there is repulision between parallel arms of L_1 and L_2.

As a result, L_2 moves away from L_1.

33. *(a)* Electron is subjected to electric force due to field of capacitor plates.

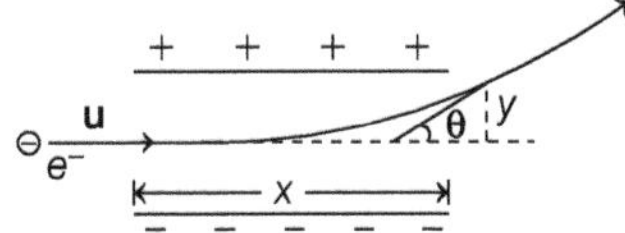

Time taken by e^- to cross region between plates $= t = \frac{x}{u}$

In this time acceleration of e^- in y direction is $a_y = \frac{F}{m} = \frac{eE}{m}$

Deflection y is then,

$$y = u_y t + \frac{1}{2}a_y t^2$$

$$\Rightarrow \quad y = \frac{eEx^2}{2mu^2}$$

Now, $\tan\theta \propto y \propto \frac{1}{u^2}$

$$\text{So,} \quad \frac{\tan\theta_2}{\tan\theta_1} = \frac{u_1^2}{u_2^2}$$

Given that,

$$\tan\theta_1 = 0.4,\ u_2 = 2u_1$$

$$\therefore \quad \tan\theta_2 = 0.4\left(\frac{u_1^2}{(2u_1)^2}\right) = \frac{0.4}{4} = 0.1$$

34. *(c)* Work done in the process

= Potential energy of the system

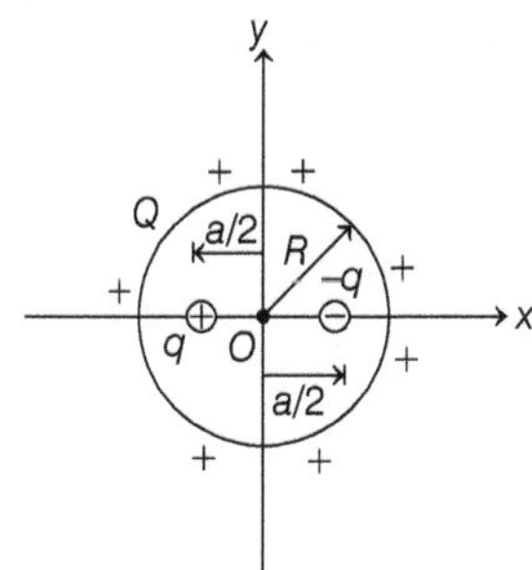

Also from shell theorem, charge Q on shell behaves as a point charge at centre.

So, magnitude of work done is

$$|U_{system}| = |U_{12} + U_{23} + U_{31}|$$

$$= \left|\frac{kQq}{a/2} + \frac{kq(-q)}{a} + \frac{kQ(-q)}{a/2}\right|$$

$$= \left|\frac{k}{a}(2Qq - q^2 - 2Qq)\right|$$

$$= \left|\frac{-q^2}{4\pi\varepsilon_0 a}\right| = \frac{q^2}{4\pi\varepsilon_0 a}$$

35. *(b)*

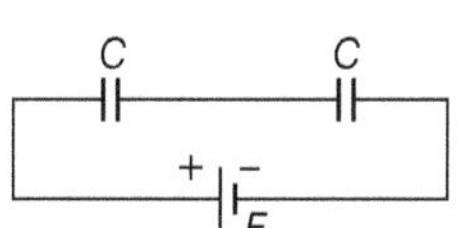

$$\text{Initially,} \quad C_{eq} = \frac{C}{2}, V = E$$

So, charge that is delivered by cell is

$$Q_1 = C_{eq}E = \frac{CE}{2}$$

In series charge remains same for both capacitors.

When one of capacitors is filled with a dielectric then,

$$C_{eq} = \frac{kC^2}{(C + kC)} = \left(\frac{k}{1+k}\right)C$$

As battery remains connected, $V = E$.

So, charge of combination

$$Q_2 = C_{eq}V = \left(\frac{k}{1+k}\right)CE$$

So, extra charge given by cell after insertion of dielectric is

$$\Delta Q = Q_2 - Q_1 = \left(\frac{k}{1+k} - \frac{1}{2}\right)CE$$

$$= \frac{k-1}{2(k+1)} \cdot CE$$

36. *(a)* Breakdown occurs for a semiconductor at a particular value of field which is independent of amount of doping concentration. Hence, breakdown field is same for both cases.

Now using, $E = \frac{dV}{dr}$, we have

$$\frac{V_1}{d_1} = \frac{V_2}{d_2} \Rightarrow \frac{100}{20} = \frac{V_2}{1}$$

$$\Rightarrow \quad V_2 = 5\ \text{V}$$

37. *(d)* Kinetic energy of particle is

$$K = \frac{1}{2}mv^2$$

$$\Rightarrow 0.05 \times 1.6 \times 10^{-19} = \frac{1}{2} \times 1.6 \times 10^{-26} \times v^2$$

$$\Rightarrow \quad v^2 = \frac{0.05 \times 1.6 \times 10^{-19} \times 2}{1.6 \times 10^{-26}}$$

$$\Rightarrow \quad v = 10^3\ \text{ms}^{-1}$$

Time taken by particle beam to travel through $1\text{m} = t = \frac{D}{v} = \frac{1}{10^3} \Rightarrow t = 10^{-3}$ s

Number of half lives occured in 10^{-3} s

$$= n = \frac{t}{T_{1/2}} = \frac{10^{-3}\ \text{s}}{6.9\text{s}} \approx 0.0001$$

$$\text{So, fraction decayed} = 1 - \left(\frac{1}{2}\right)^{0.0001}$$

$$\approx 0.0001$$

38. *(c)* Photon flux = Number of photons per unit area per unit time

$$= \frac{I}{\left(\frac{hc}{\lambda}\right)} = \frac{I\lambda}{hc} = \frac{P\lambda}{4\pi r^2 hc}$$

$$= \frac{160 \times 6200 \times 10^{-10}}{4 \times \pi \times (1.8)^2 \times 6.63 \times 10^{-34} \times 3 \times 10^8}$$

$$= 1.22 \times 10^{19}\ \text{m}^{-2}\text{s}^{-1}$$

39. *(b)* Using Balmer's formula, we have

For transition $n = 3$ to $n = 2$,

$$\frac{1}{\lambda_1} = R\left(\frac{1}{2^2} - \frac{1}{3^2}\right) = R \cdot \frac{5}{9 \times 4}$$

and for transition $n = 5$ to $n = 3$,

$$\frac{1}{\lambda_2} = R\left(\frac{1}{3^2} - \frac{1}{5^2}\right) = R \cdot \frac{16}{9 \times 25}$$

$$\text{So,} \quad \frac{\lambda_2}{\lambda_1} = \frac{5}{9 \times 4} \times \frac{9 \times 25}{16}$$

$$\text{or} \quad \lambda_2 = \frac{125}{64}\lambda_1$$

40. *(a)* Consider following reaction,

$${}_5\text{B}^{11} \longrightarrow {}_5\text{B}^{10} + {}_0n^1$$

So, energy required to remove a neutron from ${}_5\text{B}^{11}$ is

ΔE = Binding energy of ${}_5\text{B}^{11}$ – Binding energy of ${}_5\text{B}^{10}$.

$$= 7.5 \times 11 - 8.0 \times 10 = 2.5\ \text{MeV}$$

41. *(d)* $\underset{}{\text{AgBr}} \rightleftharpoons \underset{5 \times 10^{-10}\ \text{M}}{\text{Ag}^+} + \underset{10^{-3}\ \text{M}}{\text{Br}^-}$

$$K_{sp} = [\text{Ag}^+][\text{Br}^-]$$

$$= (5 \times 10^{-10})(10^{-3})$$

$$= 5 \times 10^{-13}$$

If same amount of AgBr(*s*) is added to a 10^{-2} M aqueous solution of $AgNO_3$, then the concentration of $Ag^+ = 10^{-2}$ M

$$\therefore \quad 5 \times 10^{-13} = [10^{-2}][\text{Br}^-]$$

$$[\text{Br}^-] = 5 \times 10^{-11}\text{M}$$

42. *(a)* Aniline reacts with excess Br_2/H_2O to give 2, 4, 6-tribromoaniline as a major product. The NH_2 group attached to benzene ring makes it more activating and occupy all the *ortho* and *para*-positions.

$$\text{C}_6\text{H}_5\text{NH}_2 + 3\text{Br}_2 \xrightarrow{\text{H}_2\text{O}} \text{C}_6\text{H}_2\text{Br}_3\text{NH}_2 + 3\text{HBr}$$

2, 4, 6-tribromoaniline

43. *(d)* Let the oxidation state of metals in given compounds be *x*.

(i) K_2CrO_4

$$2(+1) + x + 4(-2) = 0$$

$$x = 8 - 2 = +6$$

The oxidation state of K is K_2CrO_4 is + 6.

(ii) $NbCl_5$

$$x + 5(-1) = 0$$

$$x = +5$$

Oxidation state of Nb in $NbCl_5$ is + 5.

(iii) MnO_2

$$x + 2(-2) = 0$$

$$x = +4$$

Oxidation state of Mn is MnO_2 is + 4.

Thus, Cr has the highest oxidation state (+ 6) among the given metals.

44. *(b)* $[CrCl_2(en)(NH_3)_2]$ is a type of $M(AA)X_2Y_2$ complex. It can have 3 geometrical isomers (2 *trans* and 1 *cis* form) as shown below.

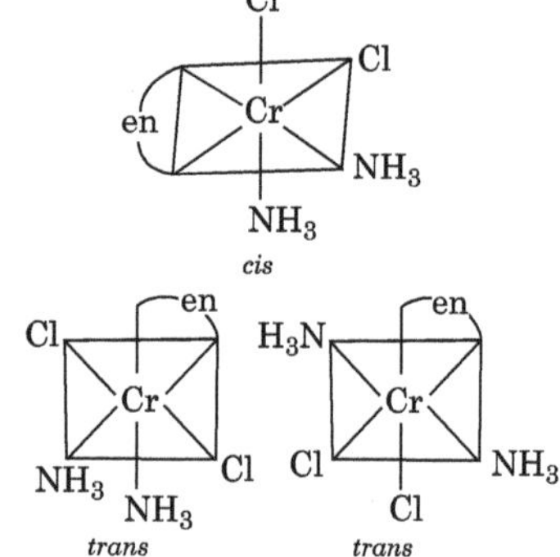

45. *(c)* Amphoteric oxide are those oxides, which shows both the properties of acidic and basic oxides. Among the given elements, Al combine with oxygen to give Al_2O_3, which is an amphoteric oxide.

$$\text{N}(s) + \text{O}_2(g) \longrightarrow \underset{\text{(Acidic oxide)}}{\text{NO}_2(s)}$$

$$\text{P}_4(s) + 5\text{O}_2(g) \longrightarrow \underset{\text{(Acidic oxide)}}{\text{P}_4\text{O}_{10}(s)}$$

$$2\text{Al}(s) + \frac{3}{2}\text{O}_2(g) \longrightarrow \underset{\text{(Amphoteric oxide)}}{\text{Al}_2\text{O}_3(s)}$$

$$2\text{Na}(s) + \text{O}_2(g) \longrightarrow \underset{\text{(Basic oxide)}}{\text{Na}_2\text{O}_2(s)}$$

46. *(a)* The temperature dependance of rate of a chemical reaction is expressed by Arrhenius equation,

$$k = Ae^{-E_a/RT}$$

$$\ln k = \ln A - \frac{E_a}{RT}$$

If we compare the above equation with the equation of a straight line, $y = mx + c$

$\therefore$ For the plot between log *k versus* $1/T$,

slope = $-E_a/R$

and intercept = $\ln A$

$\therefore$ $E_{\text{I}} > E_{\text{II}}$ and $A_{\text{I}} > A_{\text{II}}$

47. *(c)* The oxidation state of Ni in Ni $(CO)_4$ is 0. Thus, the electronic configuration of Ni(O) is $[Ar]3d^8 4s^0$.

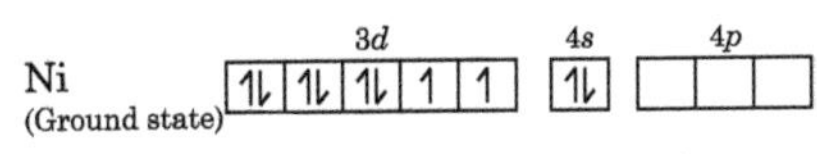

As CO is a strong ligand pairing of electrons occur.

$Ni(CO)_4$ 3d 4s 4p

sp^3-hybridisation

$Ni(CO)_4$ has tetrahedral geometry (sp^3 hybridisation) and is diamagnetic due to absence of unpaired electrons.

48. *(a)*

O_3 (Step 1) → Ozonide; Zn+H_2O; OH^- (Step 2) Intramolecular aldol condensation; $-H_2O$

(β-hydroxy ketone) (X) (α,β-unsaturated carbonyl)

In step 1 Ozonolysis of alkene involves the addition of ozone molecule to alkene to form ozonide and then cleavage of ozomide by Zn - H_2O to smaller carboxyl groups.

In step 2 This step involves intramolecular aldol condensation to form β-hydroxy ketone, which readily loses water to form α, β-unsaturated carbonyl compound.

49. *(b)* D and L before the name of any compound commonly indicates the relative configuration. For assigning the configuration of monosaccharides it is the lowest asymmetric carbon atom, which is compared to glyceraldehyde. For glucose, if —OH on the lowest asymmetric carbon is on the right side. Then, it is assigned D-configuration and if OH is on the left side then it is assigned L-configuration. Both D and L configuration are mirror images of each other.

D-(+)-glucose		L-(–)-glucose	
CHO		CHO	
H	OH	HO	H
HO	H	H	OH
H	OH	HO	H
H	OH	HO	H
CH_2OH		CH_2OH	

50. *(a)* The cubic lattice possesses 8 atoms on the corner.

∴ Fractional contribution by of atoms at the corner $= \frac{1}{8}$.

Each atom located at the face centered of ccp structure is shared between two adjacent unit cell and only $\frac{1}{2}$ of each atom belongs to a unit cell.

Thus, fractional contribution of each atom at face $= \frac{1}{2}$.

51. *(a)* For the reaction,

$$2A \rightleftharpoons B + C$$

$$Q_C = \frac{[B][C]}{[A]^2} \text{(Given, } K_c = 0.5)$$

If $Q_C > K_C$ then the reaction will proceed in backward direction. The value of Q_C in the given options can be calculated as follows:

(i) $Q_C = \frac{[10^{-2}][10^{-2}]}{[10^{-3}]^2} = \frac{10^{-4}}{10^{-6}} = 10^2$

As $Q_C > K_C$ the reaction will proceed in backward direction.

(ii) $Q_C = \frac{[10^{-2}][10^{-2}]}{[10^{-1}]^2} = \frac{10^{-4}}{10^{-2}} = 10^{-2}$

As $Q_C < K_C$, the reaction will proceed in forward direction.

(iii) $Q_C = \frac{[10^{-2}][10^{-3}]}{[10^{-2}]^2} = \frac{10^{-5}}{10^{-4}} = 10^{-1}$

Here, $Q_C < K_C$, the reaction will proceed in forward direction.

(iv) $Q_C = \frac{[10^{-3}][10^{-3}]}{[10^{-2}]^2} = \frac{10^{-6}}{10^{-4}} = 10^{-2}$

The reaction will proceed in forward direction as $Q_C < K_C$.

52. *(c)* Major products formed when *t*-butyl methyl ether reacts with HI are *tert*-butyl iodide and methanol.

$$CH_3-C(CH_3)_2-OCH_3 \xrightarrow{HI} CH_3-C(CH_3)_2-I + CH_3OH$$

The cleavage of C—O bond in ethers takes place under drastic conditions with excess of hydrogen halide, HX. This reaction follows S_N1 mechanism.

53. *(d)* Given,

$$\lambda^\infty_{NaCl} = 126 \text{ S cm}^2\text{mol}^{-1}$$
$$\lambda^\infty_{KCl} = 150 \text{ S cm}^2\text{mol}^{-1}$$
$$\lambda^\infty_{NaOH} = 250 \text{ S cm}^2\text{mol}^{-1}$$
$$\lambda^\infty_{KOH} = \lambda^\infty_{KCl} + \lambda^\infty_{NaOH} - \lambda^\infty_{NaCl}$$
$$= (150 + 250) - 126$$
$$= 400 - 126$$
$$= 274 \text{ S cm}^2\text{mol}^{-1}$$

Thus, the molar conductivity of KOH is 274 S cm^2mol^{-1}.

54. *(b)* 4-formyl benzoic acid on treatment with one equivalent of hydrazine followed by heating with alc. KOH gives 4-methyl benzoic acid the major product.

CHO / COOH $\xrightarrow[\text{alc. KOH}]{NH_2-NH_2}$ CH_3 / COO^-K^+

This reaction is known as Wolf-Kishner reduction, where the aldehyde or ketone when treated with hydrazine and KOH, get reduced to methylene group to form a hydrocarbon. This reagent doesn't affect COOH group.

55. *(c)* X with atomic number 33 and Y with atomic number 17 belong to 15th and 17th group respectively. Thus, the valency of X and Y are 3 and 1 respectively.

X^+ Y^- (3, 1) $= XY_3$

∴ The molecular formula of a stable compound formed between X and Y is XY_3. Here, X could be As (Z = 33) and Y could be Cl(Z = 17) which gives compound $AsCl_3$.

56. *(d)* $K\overset{+7}{Mn}O_4 + K\overset{-1}{I} + H_2SO_4 \longrightarrow \overset{+2}{Mn}SO_4 + \overset{0}{I_2} + K_2SO_4 + H_2O$

Balance factor for Mn = 5

Balance factor for I = 1

Equivalent of $KMnO_4$ = equivalent of KI

Number of equivalents = $y \times$ Number of moles

where, y is the balance factor.

$$= y_{Mn} \times n_{Mn} = y_{KI} \times n_{KI}$$
$$= 5 \times x_{Mn} = 1 \times 1$$
$$n_{Mn} = \frac{1}{5}$$

57. *(c)* The correct way of reporting the average value is that it must have exactly same number of digits after decimal, which has least digit after decimal.

$$\text{Average value} = \frac{10.9 + 11.4042 + 11.42}{3}$$
$$= 11.2414 \approx 11.2$$

As among the given data least digit after decimal is 1 digit value, so the average value must also be have one digit after decimal, i.e. 11.2.

58. *(a)* Given, latent heat of melting of ice, $\Delta H = 6\,\text{kJ mol}^{-1}$

$$\Delta S = \frac{\Delta H}{T_{\text{freezing}}}$$
$$= \frac{6\,\text{kJ}}{273\,\text{K}} = \frac{6 \times 10^3\,\text{J}}{273\,\text{K}}$$
$$= 21.9 \approx 22\,\text{J/K}$$

59. *(a)*

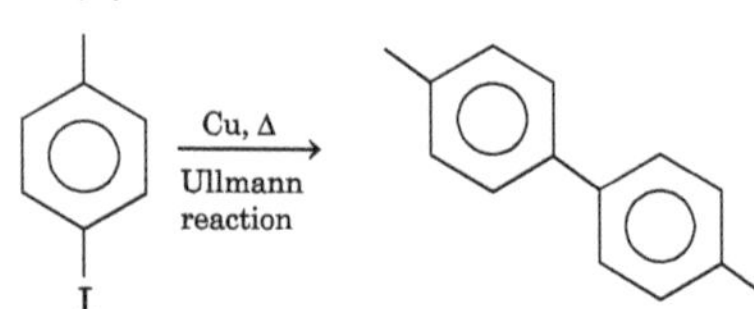

This reaction is known as Ullmann reaction. It is a coupling reaction between aryl halides and copper.

60. *(c)* The five degenerate *d*-orbitals of the metal ion split into different sets of orbital having different energies in the presence of electrical field of ligand.

In octahedral metal complexes $d_{x^2-y^2}$ and d_z^2-orbitals point towards the axes along the direction of the ligand will experience more repulsion and will be raised in energy and the d_{xy}, d_{zx} and d_{yz}-orbitals are directed between the axes will be lowered in energy.

In tetrahedral metal complexes d_{xy}, d_{zx} and d_{yz} experience more repulsion and have more energy than $d_{x^2-y^2}$ and d_z^2.

61. *(c)* Holocrine is a term used to classify the mode of secretion in exocrine glands. Holocrine secretions are produced in the cytoplasm of the cell and released by the rupture of the plasma membrane, which destroys the cells and results in the secretion of the product into the lumen.

62. *(a)* Commensalism does not promote coevolution. Coevolution occurs when two or more species reciprocally affect each other's evolution. Each party in coevolutionary relationship exerts selective pressure on the other, thereby affecting each other's evolution. Coevolution includes many forms of mutualism, host parasite and predator-prey relationship between species, as well as interspecific competition, (i.e. competition within or between species).

63. *(b)* The process of arrangement of plants into defined vertical layers depending on their height is called stratification. The tropical rainforest is a hot and moist biome found near the equator of the earth. It has a dense growth of flora arranged at different strata of the forest.

Thus, the correct answer is option (b).

64. *(d)* The third ventricle is one of the four connected fluid-filled cavities comprising the ventricular system within the mammalian brain. It is a median cleft in the diencephalon between the two thalami and is filled with Cerebrospinal Fluid (CSF).

65. *(c)* Gene expression is the process by which information from a gene is used in the synthesis of a functional gene product. These products are often protein, but in non-protein coding genes such as transfer RNA (*t*RNA) or small nuclear RNA (*sn*RNA) genes, the product is a functional RNA. Thus, option (c) is correct.

66. *(a)* Divergent evolution occurs when two different species share a common ancestor but have different characteristics from one another. An example of divergent evolution is human's arm, the whale's fin, cheetah's legs and bat's wing. All four forelimbs perform different purposes, but all contain the same bones in different sizes and shapes because all four animals (human, cheetah, whale, bat) evolved from a common ancestor whose legs contained those bones.

67. *(d)* Conjugation is a type of sexual reproduction in *Paramecium*. It is a temporary union of two individuals of same species for mutual exchange of genetic materials.

Conjugation results in rejuvenation and transference of hereditary material (i.e. recombination) in different strains.

68. *(b)* The chemicals which perceive the photoperiodic stimulus are called phytochromes, which are present on leaves. Thus, if leaves of a plant are removed, photoperiodic response will not occur in it.

69. *(a)* Cells in the testes called Leydig cells, produce testosterone in response to the production of Luteinising Hormone (LH). Testosterone is the primary sex hormone produced in males. It maintains typical male characteristics of the body, e.g. facial hair.

70. *(d)* Transversion refers to a point mutation in deoxyribonucleic acid, where a single purine is changed for a pyrimidine or *vice-versa*. A transversion can be spontaneous or can be caused by ionising radiation or alkylating agent. It can only be reversed by spontaneous reversion.

71. *(c)* Acetylcholine is the neurotransmitter used at the neuromuscular junction by the ends of axon–in other words, it is the chemical that motor neurons of the nervous system release in order to activate muscles. The synaptic bulbs of axon have vesicles which are filled with acetylcholine.

72. *(a)* If a single bacterium is placed in a nutrient medium, it will consume the nutrients and begin to grow and multiply by repetitive division. This generates thousands to millions to billions of cells that begin to pile up, becoming visible to the naked eye. This pile of cells originates from one cell and is called bacterial colony.

Thus, option (a) is correct.

73. *(d)* Rhinoviruses are the causative agents of common cold. The common cold is viral infectious disease that infects the upper respiratory system. It is also known as acute viral rhinopharyngitis and acute coryza. Common cold symptoms include dry or sore throat, blocked or runny nose and sneezing.

74. *(c)* Ebola virus is a member of the Filovirus family and causes a severe febrile disease in humans with high case fatality rates. The non-segmented negative-sense single-stranded RNA genome of ebola virus is 19 kb in length and contains seven genes.

75. *(d)* The main function of electron transport chain is to produce ATP, during which water is produced as a byproduct. At the end of the chain, the electrons are taken up by the oxygen molecules to make water. This is why, oxygen is called the final electron acceptor.

76. *(a)* The Benedict's test is used to test for simple sugars. The nitric acid reacts with proteins to form yellow nitrated products, thus used to detect proteins in sample. Therefore, the tube 'P', positive for Benedict's test and tube 'Q' positive for nitric acid contains sugar and protein, respectively.

77. *(c)* For a circular DNA, 'N' represents the number of fragments of DNA obtained if the plasmid has 'N' recognition sequences for a given restriction endonucleus. There, if a restriction enzyme have 3 sites of action, the number of fragments produced will be 3 for a circular DNA.

78. *(b)* Xylem moves water from roots to the leaves by the process of transpiration. As the relative humidity of the air surrounding the plant rises, the transpiration rate falls. It is easier for water to evaporate into dryer air than into more saturated air.

Thus, option (b) is correct.

79. *(a)* For writing complementary sequence just use the base pairing rules, C bonds with G and A bonds with T and for each base in the original sequence.

Suppose we have

5′-CGTACTA-3′

Its complementary sequence will be

3′-GCATGAT-5′

But since the two strands run in opposite directions. So, technically more correct answer is 5′-TAGTACG-3′.

80. *(b)* If non-disjunction occurs during meiosis-I, this means that atleast one pair of homologous chromosomes did not separate. The end result is two cells that have an extra copy of one chromosome and two cells that are missing that chromosome.

Therefore, if the egg cell has 6 chromosomes produced from $2n = 14$ chromosome plant, it would be because of non-disjunction in meiosis-I.

81. *(a)* We have,

$$f_\sigma(x) = a_n x^{n-1} + a_{n-1}x^{n-2} + \dots + a_2 x + a_1$$

$\sigma = (a_1, a_2, a_3, \dots, a_n)$ of $(1, 2, 3, \dots, n)$

S_σ = Sum of roots of $f_\sigma(x) = 0$

$S = \Sigma S_\sigma$

$$\therefore \quad S = -\left[\frac{\lambda - a_n}{a_n} + \frac{\lambda - a_{n-1}}{a_{n-1}} + \dots + \frac{\lambda - a_1}{a_1}\right]$$

$\forall\, \lambda = a_1 + a_2 + a_3 + \dots + a_n$

$$S = -\left[(a_1 + a_2 + a_3 + \dots + a_n)\left(\frac{1}{a_1} + \frac{1}{a_2} + \dots + \frac{1}{a_n}\right) - n\right]$$

$$S = n - \left[(a_1 + a_2 + a_3 + \dots + a_n)\left(\frac{1}{a_1} + \frac{1}{a_2} + \dots + \frac{1}{a_n}\right)\right]$$

From AM ≥ HM

$$\frac{a_1 + a_2 + a_3 + \dots + a_n}{n} \geq \frac{n}{\frac{1}{a_1} + \frac{1}{a_2} + \dots + \frac{1}{a_n}}$$

$$\Rightarrow (a_1 + a_2 + a_3 + \dots + a_n)\left(\frac{1}{a_1} + \frac{1}{a_2} + \dots + \frac{1}{a_n}\right) \geq n^2$$

$S \leq -n(n-1)$

$\therefore \quad S \leq -n!$

82. *(c)* We have, $\left| e^{\sum_{k=0}^{n} {}^nC_k \omega^k} \right|$, where ω is cube root of unity.

$$\sum_{k=0}^{n} {}^nC_k\omega^k = {}^nC_0 + {}^nC_1\omega + {}^nC_2\omega^2 + \dots + {}^nC_n\omega^n$$

$= (1+\omega)^n$

$= (-\omega^2)^n \quad [\because 1 + w + w^2 = 0]$

$= (-1)^n \omega^{2n}$

$$\therefore \quad \left| e^{\sum_{k=0}^{n} {}^nC_k\omega^k} \right| = \left| e^{(-1)^n \omega^{2n}} \right| = \left| e^{(-\omega^2)^n} \right|$$

$$= \left| e^{\left(-\cos\frac{4\pi}{3} - i\sin\frac{4\pi}{3}\right)^n} \right|$$

$$= \left| e^{\left(\cos\frac{\pi}{3} + i\sin\frac{\pi}{3}\right)^n} \right|$$

$$= \left| e^{\cos\frac{n\pi}{3} + i\sin\frac{n\pi}{3}} \right|$$

$$= \left| e^{\cos\frac{n\pi}{3}} \right| \cdot \left| e^{i\sin\frac{n\pi}{3}} \right|$$

$$= \left| e^{\cos\frac{n\pi}{3}} \right| \quad [\because |e^{i\theta}| = 1]$$

$= e^1, e^{1/2}, e^{-1/2}, e^{-1}$

only four values possible.

83. *(b)* We have, $y = ax^2 + bx + c$

Parabola has two roots one is positive and one is negative.

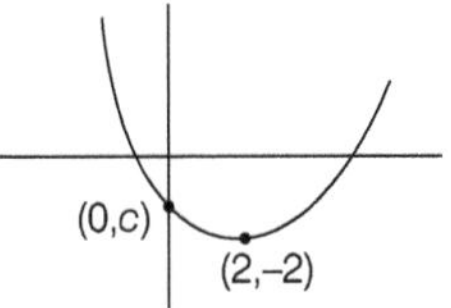

Clearly, c is negative

$a > 0$

$\frac{-b}{a} = 2$

$\therefore \quad -\frac{b}{a} > 0$

$b < 0 \quad [\because a > 0]$

$\therefore \quad bc > 0$

Hence, option (b) is correct.

84. *(d)* We have,

$C_1, C_2, C_3, \dots, C_n$ be circle with radii $r_1, r_2, \dots, r_n$ respectively. C_i and C_{i+1} touch externally X-axis and $y = 2\sqrt{2}x + 10$ are tangent of each circle.

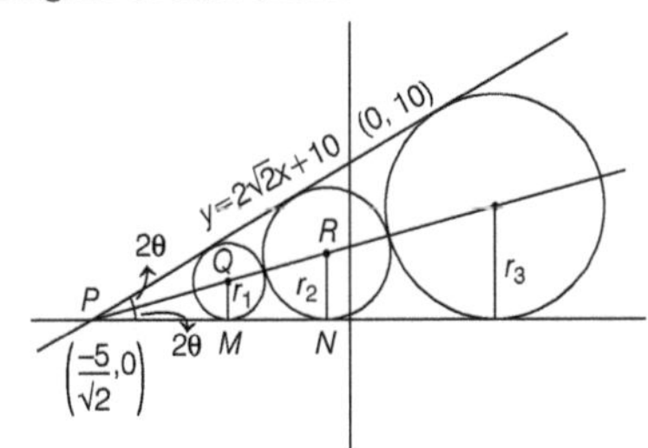

Slope of line $y = 2\sqrt{2}x + 10$ is $2\sqrt{2}$

$\therefore \quad \tan 2\theta = 2\sqrt{2}$

$$\frac{2\tan\theta}{1 - \tan^2\theta} = 2\sqrt{2}$$

$\sqrt{2}\tan^2\theta + \tan\theta - \sqrt{2} = 0$

$(\sqrt{2}\tan\theta - 1)(\tan\theta + \sqrt{2}) = 0$

$\tan\theta = \frac{1}{\sqrt{2}} \quad \tan\theta \neq -\sqrt{2}$

$\sin\theta = \frac{1}{\sqrt{3}}$

In ΔPQM, $\sin\theta = \frac{QM}{PQ}$

$\Rightarrow \quad PQ = \sqrt{3}\,QM \Rightarrow PQ = \sqrt{3}r_1$

In ΔPRN,

$$\sin\theta = \frac{RN}{PR} = \frac{r_2}{PQ + r_1 + r_2} = \frac{r_2}{\sqrt{3}r_1 + r_1 + r_2}$$

$$\Rightarrow \quad \frac{1}{\sqrt{3}} = \frac{r_2}{(\sqrt{3}+1)r_1 + r_2}$$

$\Rightarrow \quad r_2 + (\sqrt{3}+1)r_1 = \sqrt{3} - r_2$

$$\Rightarrow \quad \frac{r_2}{r_1} = \frac{\sqrt{3}+1}{\sqrt{3}-1} = 2 + \sqrt{3}$$

r_1, r_2, r_3 a geometric progression with common ratio $2 + \sqrt{3}$.

85. *(c)* We have, $3x^3 - 25x + n = 0$

Let $f(x) = 3x^3 - 25x + n$

$f'(x) = 9x^2 - 25$

Put $f'(x) = 0$

$9x^2 - 25 = 0$

$$x = \pm\frac{5}{3} \Rightarrow x_1 = \frac{-5}{3}, x_2 = \frac{5}{3}$$

$f(x)$ has three real roots.

$\therefore \quad f(x_1)f(x_2) < 0$

$$\therefore \left(3\left(\frac{-5}{3}\right)^3 - 25\left(\frac{-5}{3}\right) + n\right)\left(3\left(\frac{5}{3}\right)^3 - 25\left(\frac{5}{3}\right) + n\right) < 0$$

$$\left(\frac{-125}{9} + \frac{125}{3} + n\right)\left(\frac{125}{9} - \frac{125}{3} + n\right) < 0$$

$$\left(n + \frac{250}{9}\right)\left(n - \frac{250}{9}\right) < 0$$

$$n \in \left(\frac{-250}{9}, \frac{250}{9}\right]$$

Hence, $n \in I$

$\therefore$ Total integer is 55.

86. *(d)* Let equation of ellipse is

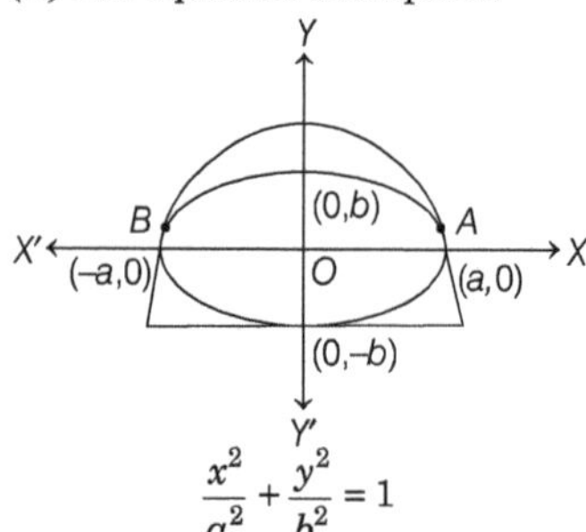

$$\frac{x^2}{a^2} + \frac{y^2}{b^2} = 1$$

$\therefore$ Equation of circle is

$$x^2 + (y + b)^2 = r^2$$

Put $x^2 = a^2 - \frac{a^2y^2}{b^2}$ in circle

$$a^2 - \frac{a^2y^2}{b^2} + (y + b)^2 = r^2$$

$$\Rightarrow \left(1 - \frac{a^2}{b^2}\right)y^2 + 2by + (a^2 + b^2 - r^2) = 0$$

$$D = 0 \Rightarrow r^2 = \frac{a^4}{a^2 - b^2} \Rightarrow b = a\sqrt{1 - \frac{a^2}{r^2}}$$

Area of ellipse $= \pi ab$

$$A = \pi a^2\sqrt{1 - \frac{a^2}{r^2}}$$

$$\frac{dA}{da} = 0 \Rightarrow a^2 = \frac{2r^2}{3} \Rightarrow a = \sqrt{\frac{2}{3}}r$$

$$\therefore \quad b = a\sqrt{1 - \frac{2}{3}} = \frac{a}{\sqrt{3}}$$

$$\Rightarrow \quad e = \sqrt{1 - \frac{b^2}{a^2}} = \sqrt{1 - \frac{1}{3}} = \sqrt{\frac{2}{3}}$$

87. *(a)* We have,

$$I_n = \int_0^{\pi/2} x^n \cos x\, dx$$

$$\Rightarrow I_n = [x^n \sin x]_0^{\pi/2} - \int_0^{\pi/2} nx^{n-1}\sin x\, dx$$

$$\Rightarrow I_n = \left(\frac{\pi}{2}\right)^n - [n\, x^{n-1}(-\cos x)]_0^{\pi/2} - \int_0^{\pi/2} n(n-1)\, x^{n-2}\cos x\, dx$$

$$\Rightarrow I_n = \left(\frac{\pi}{2}\right)^n - n(n-1)\, I_{n-2}$$

$$\Rightarrow I_n + n\,(n-1)\, I_{n-2} = \left(\frac{\pi}{2}\right)^n$$

Now, $\sum_{n=2}^{\infty}\left[\frac{I_n}{n!} + \frac{I_{n-2}}{(n-2)!}\right]$

$$\Rightarrow \sum_{n=2}^{\infty}\left[\frac{I_n}{n!} + \frac{n\,(n-1)I_{n-2}}{n!}\right]$$

$$\sum_{n=2}^{\infty}\frac{\left(\frac{\pi}{2}\right)^n}{n!} = e^{\pi/2} - 1 - \frac{\pi}{2}$$

88. *(b)* Let $I = \int_1^n [x][\sqrt{x}]\, dx$

$$\Rightarrow I = \int_1^2 dx + \int_2^3 dx + \int_3^4 3\, dx + \int_4^5 8\, dx + \int_5^6 10\, dx + \int_6^7 12\, dx + \int_7^8 14\, dx + \int_8^9 16\, dx + \int_9^{10} 27\, dx + \dots$$

But $I > 60$

$$I = 1 + 2 + 3 + 8 + 10 + 12 + 14 + 16 = 66$$

So, least value of $n = 9$

89. *(b)* We have, $n \in \{1, 2, 3, \dots, 100\}$

There are fourteen weeks in 1 to 100

In fourteen weeks number of Sunday is equal to number of Monday.

Only 2 days are different number of Sundays out of 7 days.

$\therefore$ Required probability $= \frac{2}{7}$

90. *(*)* We have,

$S = \{(a, b) : a, b \in Z, 0 \le a, b \le 18\}$

Number of line in R^2 passing through (0, 0) and one other point in S is 34.

(1, 17), (2, 17), (3, 17),...,(18, 17) = 17 points

(17, 1), (17, 2), (17, 3),..., (17, 18) = 17 points

$\therefore$ Total number of lines = 34

91. *(c)* Initially slipping occurs at point of contact and friction acts in forward direction.

Friction produces acceleration of centre of mass.

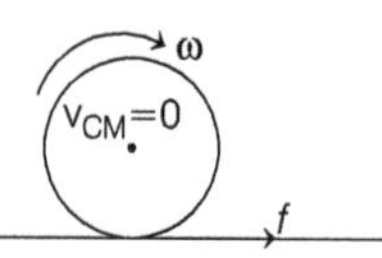

Acceleration of centre of mass caused by friction is

$$a_{CM} = \frac{F}{m} = \frac{\mu_k mg}{m} = \mu_k g$$

Angular deacceleration caused by friction is

$$\alpha = \frac{\tau}{I} = \frac{fR}{I} = \frac{-\mu_k mgR}{\frac{2}{5}mR^2} = -\frac{5\mu_k g}{2R}$$

After time t, velocity of centre of mass is

$$v_{CM} = u_{CM} + a_{CM}t = 0 + \mu_k gt = \mu_k gt$$

and angular velocity is

$$\omega = \omega_0 + \alpha t = \omega_0 - \frac{5\mu_k g}{2R}.t$$

Pure rolling begins

when $\omega = \frac{v_{CM}}{R} = \frac{\mu_k gt}{R}$

So, $\frac{\mu_k gt}{R} = \omega_0 - \frac{5}{2}\frac{\mu_k gt}{R}$

$$\Rightarrow \frac{7}{2}\frac{\mu_k gt}{R} = \omega \Rightarrow t = \frac{2R\omega_0}{7\mu_k g}$$

So, angular speed when pure rolling occurs is

$$\omega_f = \omega + \alpha t = \omega - \frac{5\mu_k g}{2R}.\frac{2R\omega}{7\mu_k g}$$

$$= \omega\left(1 - \frac{5}{7}\right) = \frac{2}{7}\omega$$

92. *(a)* Acceleration of system of blocks

$$= \frac{F}{8+2} = \frac{F}{10}\ \text{ms}^{-2}.$$

Now, from free body diagram below, we see that

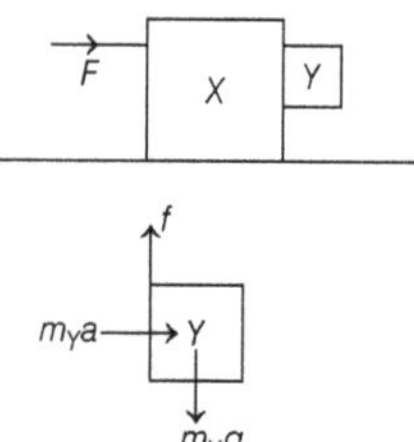

Reaction of X and Y is

$$R = m_Y a = 2 \times \frac{F}{10} \Rightarrow R = \frac{F}{5}$$

Block Y does not slips, if

Force of friction = Weight of Y

$$\Rightarrow \quad \mu R = m_Y g \Rightarrow 0.5 \times \frac{F}{5} = 2 \times 10$$

So, $\quad F = 200\,\text{N}$

93. *(b)* For a pendulum with angular amplitude θ,

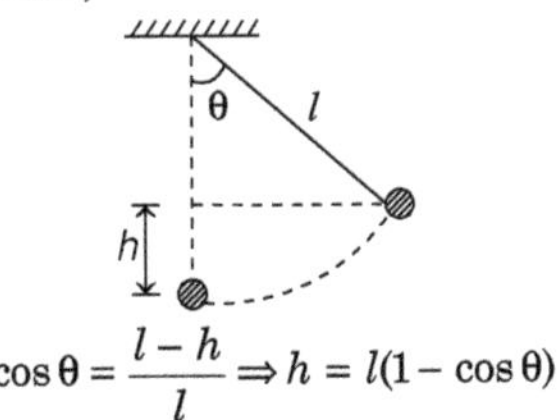

$$\cos\theta = \frac{l-h}{l} \Rightarrow h = l(1-\cos\theta)$$

When pendulum is released at mean position. If its speed is v, then by energy conservation, we have

$$\frac{1}{2}mv^2 = mgh \Rightarrow v^2 = 2gl(1-\cos\theta)$$

Now, maximum tension occurs at mean position and it is given by

$$T_{max} = mg + \frac{mv^2}{l} = mg + m\cdot 2g(1-\cos\theta)$$

And minimum tension occurs at extreme position, its value is

$$T_{min} = mg\cos\theta$$

Given, $T_{max} = 4\,T_{min}$

$$\Rightarrow mg + 2mg(1-\cos\theta) = 4\,mg\cos\theta$$

$$\Rightarrow 3mg = 6\,mg\cos\theta$$

$$\Rightarrow \cos\theta = \frac{1}{2} \Rightarrow \theta = 60°$$

94. *(d)* For an ideal gas in process,

$pV^3 = C$ (a constant)

Work done in the process is

$$W = \int p\,dV = \int \frac{C}{V^3}dV = \frac{CV^{-2}}{-2} = \frac{pV^3\cdot V^{-2}}{-2} = \frac{pV}{-2}$$

$$\Rightarrow \Delta W = -\frac{nR\Delta T}{2}$$

Also, for ideal gas,

$$\Delta U = nC_V\Delta T$$

So, by first law of thermodynamics, we have

$$\Delta Q = \Delta U + \Delta W$$

$$\Rightarrow nC\Delta T = nC_V\Delta T - \frac{nR\Delta T}{2}$$

$$\Rightarrow C = C_V - \frac{R}{2} \Rightarrow C = \frac{3}{2}R - \frac{R}{2} = R$$

[∴ for monoatomic ideal gas, $C_V = \frac{3}{2}R$]

∴ Heat capacity for given process is

$$C = R$$

95. *(a)* For a small mirror, rays parallel to axis converges at focus. But if apperture of mirror is large, rays reflected from mirror forms a circular (not point) image.

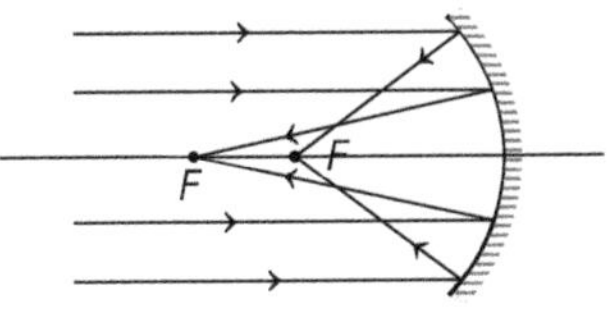

Now, consider a ray reflected from periphery of mirror as shown below.

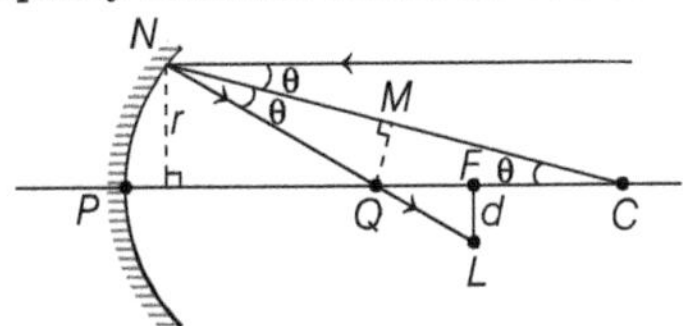

In above figure,

C = centre of curvature,
CN = normal,
F = focus of mirror,
P = pole of mirror,
d = radius of disc

and r = radius of circular section of mirror.

In ΔQNC, perpendicular QM is drawn.

From ΔQMC,

$$QC = \frac{CM}{\cos\theta} = \frac{R}{2\cos\theta}$$

So, $$PQ = R - \frac{R}{2\cos\theta}$$

∴ $$QF = PF - PQ = \frac{R}{2} - \left(R - \frac{R}{2\cos\theta}\right) = \frac{R}{2\cos\theta} - \frac{R}{2}$$

Now, from similar triangles ΔPQN and ΔQFL, we have

$$\frac{r}{d} = \frac{PQ}{QF} \Rightarrow \frac{r}{d} = \frac{\left(R - \frac{R}{2\cos\theta}\right)}{\frac{R}{2\cos\theta} - \frac{R}{2}} = \frac{2\cos\theta - 1}{1-\cos\theta} \quad \text{...(i)}$$

Now, $$\cos\theta = \frac{\sqrt{R^2 - r^2}}{R} = \left(1 - \frac{r^2}{R^2}\right)^{\frac{1}{2}} \approx 1 - \frac{r^2}{2R^2}$$

Substituting in Eq. (i), we get

$$\frac{r}{d} = \frac{2 - \frac{r^2}{R^2} - 1}{\frac{r^2}{2R^2}} = \frac{\left(1 - \frac{r^2}{R^2}\right)}{\left(\frac{r^2}{2R^2}\right)}$$

$$\Rightarrow d = \frac{r^3}{2R^2} \Big/ \left(1 - \frac{r^2}{R^2}\right)$$

$$\Rightarrow d = \frac{r^3}{2R^2}\cdot\left(1 - \frac{r^2}{R^2}\right)^{-1}$$

∴ $$d \approx \frac{r^3}{2R^2}$$

So, area of disc required is

$$A = \pi d^2 = \frac{\pi r^6}{4R^4}$$

96. *(c)* From Snell's law, we have

$$\sin i = \mu\sin r$$

$$\Rightarrow \sin i = \left(a + \frac{b}{\lambda^2}\right)\sin r$$

Differentiating with respect to λ, we get

$$\Rightarrow \frac{d}{d\lambda}\sin i = \frac{d}{d\lambda}\left(a + \frac{b}{\lambda^2}\right)\cdot\sin r$$

$$\Rightarrow 0 = \left(a + \frac{b}{\lambda^2}\right)\cdot\cos r\cdot\frac{dr}{d\lambda} + \sin r\cdot\left(\frac{-b}{\lambda^3}\right)$$

$$\Rightarrow \frac{b}{\lambda^3}\sin r = \left(a + \frac{b}{\lambda^2}\right)\cos r\cdot\frac{dr}{d\lambda}$$

$$\Rightarrow dr = \frac{2bd\lambda}{\lambda}\cdot\frac{\tan r}{(a\lambda^2 + b)}$$

Angular spread of angle of refraction is

$$\delta r = \left|\frac{2b\tan r}{a\lambda^3 + b\lambda}\cdot\delta\lambda\right|$$

97. *(b)* In steady state (after the capacitors are fully charged), no current flows through the branches containing capacitors. So, in steady state, given circuit is equivalent to circuit shown below.

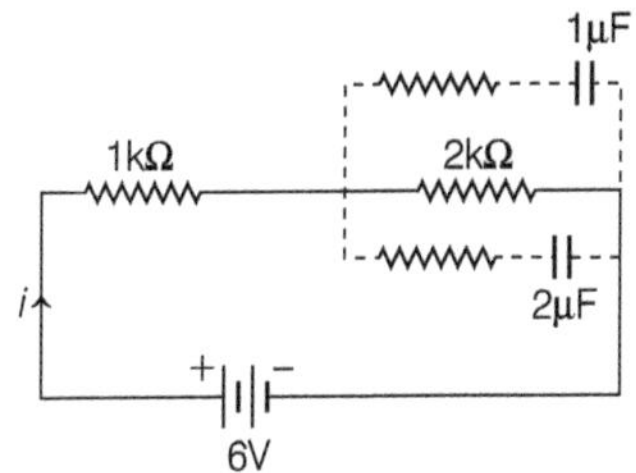

Now, current in circuit is

$$i = \frac{E}{R_{total}} = \frac{6}{3\times10^3} = 2\times10^{-3}\text{ A}$$

Now, potential drop across 1 μF and 2 μF capacitors = potential drop across 2 kΩ resistance

$$= iR = 2\times10^{-3}\times2\times10^3 = 4\text{ V}$$

So, charges stored in capacitors are

$$Q_1 = C_1V = 1\times10^{-6}\times4 = 4\mu\text{C}$$

and $Q_2 = 8\mu$C

98. *(d)* Force due to radiation pressure = Weight of disc

$$\Rightarrow \frac{2p}{c} = mg$$

$$\Rightarrow \frac{2\times1.5\times10^3}{3\times10^8} = m\times10$$

$$\Rightarrow m = 10^{-6}\text{ kg}$$

99. *(b)* Initially let frequency is f.

Then, $K_{max} = eV_0 = hf - \phi_0$

$\Rightarrow \quad e(0.6) = hf - \phi_0 \quad \ldots(i)$

When frequency is increased by 10%, frequency will be $f' = 1.1f$ and stopping potential $= V_0' = 0.9V$

$\Rightarrow \quad e(0.9) = h(1.1f) - \phi_0 \quad \ldots(ii)$

Substituting hf from Eq. (i) in Eq. (ii), we get

$$e(0.9) = 1.1(e(0.6) + \phi_0) - \phi_0$$

$\Rightarrow \quad e(0.9) = e(0.66) + 0.1\phi_0$

$\Rightarrow \quad e(0.9 - 0.66) = 0.1\phi_0$

$\Rightarrow \quad 0.24 \text{ eV} = 0.1\phi_0$

So, work function is $\phi_0 = 2.4 \text{ eV}$

100. *(c)* Let $\sigma = h^\alpha k^\beta c^\gamma$, then equating dimensions of both sides, we have

$[MT^{-3}K^{-4}] = [ML^2T^{-1}]^\alpha [ML^2T^{-2}K^{-1}]^\beta [LT^{-1}]^\gamma$

$\alpha + \beta = 1 \quad \ldots(i)$

$2\alpha + 2\beta + \gamma = 0 \quad \ldots(ii)$

$-\alpha - 2\beta - \gamma = -3 \quad \ldots(iii)$

In Eq. (i), we multiplying with 2, we get

$2\alpha + 2\beta = 2 \quad \ldots(iv)$

Now, subtracting Eq. (ii) from Eq. (iv), we get

$2\alpha - 2\alpha + 2\beta - 2\beta + \gamma = 0 - 2$

$\gamma = -2$

Now, putting the value of γ in Eq. (iii), we get

$\alpha + 2\beta = 5 \ldots(v)$

And solve the Eq. (i) and (v), we get

$\alpha = -3, \beta = 4$

So, $\alpha = -3, \beta = 4$ and $\gamma = -2$.

101. *(a)*

Aniline (NH_2) $\xrightarrow[\text{Pyridine}]{(CH_3CO)_2O}$ Acetanilide $\xrightarrow{Br_2}$ p-bromoacetanilide (X) $\xrightarrow{aq.\ \text{conc. NaOH}}$ p-bromoaniline (Y)

Note Aniline is a strong activating group. During electrophilic substitution reaction of aniline, reactivity towards substitution at *ortho* and *para*-positions are very high. In order to get a monohalogenated product, NH_2 is protected by acetylation.

102. *(b)* Given,

density of acetic acid = 5 mg cm^{-3}

Pressure = 1 atm

Temperature = 300 K

We know that

$pV = RT$, (when, $n = 1$)

Also, $V = \frac{m}{\rho}$

$\therefore \quad pM = \rho RT$

$$M = \frac{5 \times 0.0821 \times 300}{1} = 123.15 \text{ g}$$

Molar mass of acetic acid, CH_3COOH
$= 12 \times 1 + 1 \times 3 + 12 \times 1 + 16 \times 2 + 1 = 60 \text{ g}$

$\therefore$ Number of molecules of acetic acid

$$= \frac{123.15}{60} \approx 2$$

103. *(d)* For the reaction,

$H_2O(l) \rightleftharpoons H_2O(g)$;

$\Delta H = 41 \text{ kJ}$

$W = -p\Delta V = -\Delta n_g RT$

$= -(1 \times 8.314 \times 10^{-3} \times 373) \text{ kJ}$

$= -3.10 \text{ kJ}$

At constant pressure,

$q = \Delta H = 41 \text{ kJ}$

From first law of thermodynamics

$\Delta U = q + w$

Substituting the value of q and w in above equation we get,

$\Delta U = 41 - 3.1 = 37.9 \text{ kJ}$

104. *(c)* For the reactions,

$H_2 + I_2 \rightleftharpoons 2HI, K_{C_1} = 50 \quad \ldots(i)$

$$K_{C_1} = \frac{[HI]^2}{[H_2][I_2]}$$

$N_2 + 3H_2 \rightleftharpoons 2NH_3, K_{C_2} = 1000 \quad \ldots(ii)$

$$K_{C_2} = \frac{[NH_3]^2}{[N_2][H_2]^3}$$

$N_2 + 6HI \rightleftharpoons 2NH_3 + 3I_2 \quad \ldots(iii)$

In order to get Eq. (iii), we multiply Eq. (i) by (ii) and then reverse it and then add Eq. (ii) to it.

$$\therefore \quad K_{C_3} = \frac{[NH_3]^2[I_2]^3}{[N_2][HI]^6} = K_{C_2} \times \frac{1}{(K_{C_1})^3}$$

$$K_{C_3} = \frac{1000}{(50)^3} = 0.008$$

105. *(b)* Given, bond energy of

$N \equiv N, BE_{N \equiv N} = 946 \text{ kJ mol}^{-1}$

Bond energy of

$H—H, BE_{H—H} = 435 \text{ kJ mol}^{-1}$

Bond energy of

$N—N, BE_{N—N} = 159 \text{ kJ mol}^{-1}$

Bond energy of

$N—H, BE_{N—H} = 389 \text{ kJ mol}^{-1}$

$\Delta H_f = \Sigma BE_{\text{reactants}} - \Sigma BE_{\text{products}}$

For the reaction,

$N \equiv N + 2H—H \longrightarrow H_2N—NH_2$

$\Delta H_f = 1 \times BE_{N \equiv N} + 2BE_{H—H} - 4BE_{N—H^{-1}} BE_{N—N}$

$= [(1 \times 946) + (2 \times 435) - 4(389) - 1 \times (159)]$

$= 101 \text{ kJ/mol}$

106. *(c)* Given, radius of $K^+ = 133$ pm

radius of $Cl^- = 181$ pm

$$r_{K^+} + r_{Cl^-} = \frac{a}{2}$$

$a = 2(r_{K^+} + r_{Cl^-})$

$= 2(133 + 181) = 628 \text{ pm}$

$= 628 \times 10^{-10}$ cm or

$= 6.28 \times 10^{-8}$ cm

Volume of a unit cell $= a^3 = (6.28 \times 10^{-8})^3$

$= 2.48 \times 10^{-22} \text{ cm}^3$

107. *(b)* $K_2Cr_2O_7 + mFeSO_4 + nH_2SO_4 \longrightarrow Cr_2(SO_4)_3 + pFe_2(SO_4)_3 + K_2SO_4 + qH_2O$

We can balance the equation by using half reaction method. In this method, the two half equations are balanced seperately and then added together to give balanced equation.

Thus, we get balanced reaction,

$K_2Cr_2O_7 + 6FeSO_4 + 7H_2SO_4 \longrightarrow Cr_2(SO_4)_3 + 3Fe_2(SO_4)_3 + K_2SO_4 + 7H_2O$

$\therefore m = 6, n = 7, p = 3$ and $q = 7$.

108. *(c)* For the reaction

$3Fe^{2+}(aq) + 2Cr(s) \rightleftharpoons 2Cr^{3+}(aq) + 3Fe(s)$

Given, $E^\circ_{Fe^{2+}/Fe} = -0.44V$

$E^\circ_{Cr^{3+}/Cr} = -0.74V$

$E^\circ_{Cell} = E_{Fe^{2+}/Fe} - E_{Cr^{3+}/Cr}$

$= -0.44 + 0.74 = 0.3 \text{ V}$

We know that, $\Delta G = -nFE^\circ_{cell}$

For this reaction, $n = 6$

$\therefore \Delta G = -6 \times 96500 \times 0.3 = -173700 \text{ J}$

109. *(a)* Calcium butanoate on heating gives heptan-4-one, which on further reaction with 1, 2-ethanediol in presence of catalytic amount of an acid produces cyclic ketal [product given in option (a)] as a major product.

$(CH_3CH_2CH_2COO)_2Ca \xrightarrow{\Delta}$

$CH_3—CH_2—CH_2\overset{O}{\overset{\|}{C}}CH_2CH_2CH_3$

$\Big\downarrow -H_2O \;\; \begin{matrix}CH_2OH\\|\\CH_2OH\end{matrix}, H^+$

$CH_3—CH_2—CH_2—C(—O—CH_2—CH_2—O—)—CH_2CH_2CH_3$ (cyclic ketal, 2,2-dipropyl-1,3-dioxolane)

110. *(c)* XeF_6 on complete hydrolysis yield XeO_3 (*X*).

$$XeF_6 + 3H_2O \longrightarrow XeO_3 + 6HF$$

Xe has eight valence electrons. It forms 3 bond pairs with oxygen and has one lone pair. This its geometry will be pyramidal.

Pyramidal XeO_3

111. *(b)* Total time for cell cycle = 24 hrs

Time for M-phase = 2 hrs

So, probability of cell in M-phase at any given time is $\frac{2}{24} = \frac{1}{12}$

The cell cycle is the series of events that take place in a cell leading to duplication of its DNA and division of cytoplasm and organelles to produce two daughter cells.

112. *(d)* A flower with Tt genotype undergoes meiosis during gamete formation. It produces two types, T and t. Since, the nuclei of embryo sac are haploid they might either be T or t. So, when the two nuclei fuse to form secondary nucleus the genotype of it might be TT or Tt or tt. After the fusion with the male gamete (which produces a single type of gamete T) the genotype of endosperm will be TTT or TTt or Ttt.

Same in the case is with embryo. As there are two types of gametes formed from female parent, i.e. T and t. The genotype of the embryo will be TT + Tt.

Thus, option (d) is correct.

113. *(a)* If a genetic code of a certain life form consists of five unique nucleotides and each codon consists of four bases, then by using permutation combination, i.e. $5^4 (5 \times 5 \times 5 \times 5)$, it would generate 625 codons. Since, one codon is stop codon, then maximum number of unique amino acids this life form can use are 624.

114. *(b)* If a spontaneous mutation leads to the production of female progeny only, it means there is recessive lethal gene in X-chromosome, which when occurs in male child causes death.

i.e. $X^1X \longrightarrow$ carrier female (survive)

$X^1Y \longrightarrow$ male (died)

When these carrier daughters produce offspring there will be same results, as before.

While in third generation, the possibility of male child occurs as in the second generation, some daughters are normal (XX). Thus, option (b) is correct.

115. *(c)* Lets have a look at the single enzyme digests first. The digest with enzyme A and B only leads to products which are 5kb (5,000 bp) away from each other. Since, they are of the same size, both equally sized restriction fragments appear as one band. So, each enzyme cuts the plasmid exactly in half.

The double digest will form two products from it, with 2 kb and 3 kb. This means that enzyme B cuts between the restriction sites for enzyme A, resulting in these two fragments. See the sketch below

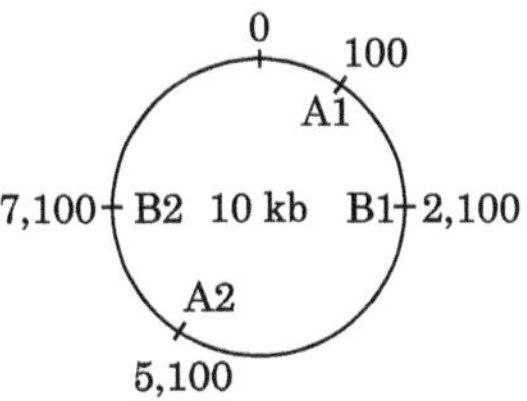

Digest

Band size	Enzyme A	Enzyme B	Enzyme A+B
8,000	▬	▬	
3,000			▬
2,000			▬

A1 and A2 are the cutting sites for enzyme A, B1 and B1 are the sites for enzyme B. The numbers on the outside are the positions on the DNA sequence. You can see that a single digest leads to two fragments of 5kb and that both A and both B sites are located 5 kb from each other.

A1 is at position 100, A2 has to be at position 5,100, B1 is located 2kb behind A1 (and therefore at position 2,100) and B2 2kb behind position A2 (and therefore at position 7,100). To get the same restriction pattern it is also possible that B1 and B2 are located 3kb behind their respective enzyme A position (so at 3,100 and 8,100) and you still get the same pattern of 2kb and 3kb on the gel. Thus option (c) is correct.

116. *(d)* Meiosis, involves two divisions that produce a total of four daughter cells. During both meiosis-I and meiosis-II, cells undergo the same phases found in mitosis but the processes and results are different. Within meiosis-I, homologous chromosomes become paired and crossing over occurs. Homologous pairs are separated (i.e. independent assortment occurs here) and the two resulting daughter cells have half as many chromosomes per cell (i.e. segregation of chromosomes). The two daughter cells produced from meiosis-I, enter meiosis-II where they each are divided again to produce a total of four haploid daughter cells.

117. *(c)* Endotherms (animals that use body heat to maintain a constant internal temperature) increases metabolic heat production to keep their body temperature constant, when its cold outside. For ectotherms, on the other hand, body temperature mainly depends on external heat sources. That is why ectotherm body temperature and metabolic rate rises and falls along with the temperature of the surrounding environment. Thus, option (c) is correct.

118. *(b)* Through a series of experiments, Griffith established that the virulence of the S-strain was destroyed by heating the bacteria. Thus, he was surprised to find that mice died when they were injected with a mixture of heat-killed S-bacteria and living R-bacteria, neither of which caused mice to die when they were injected alone.

119. *(a)* In polygenic inheritance, a trait is controlled by more than one gene. Human height is a multigenic or polygenic trait, controlled by at least three genes with six alleles. Polygenic inheritance often results in a bell-shaped curve when you analyse the population. That means that most people fall in the middle of phenotypic range, such as average height, while very few people are at the extremes, such as very tall or very short. At one end of the curve will be individuals who are recessive for all the alleles (for example aabbcc) at the other end will be individuals, who are dominant for all the alleles (for example AABBCC). Through the middle of the curve will be individuals who have a combination of dominant and recessive alleles (for example AaBbCc or AaBBcc). Thus, curve 'P' is the correct representation of human height in a metropolis.

120. *(d)* Low pH increases the concentration of base causing mitochondria to pump out H^+ to the intermembrane space leading to ATP production. The high external acid concentration causes an increase in H^+ ions in the intermembrane space leading to increased ATP production by ATP synthetase. Thus, option (d) is correct.

QUESTION PAPER 2013

Stream : SB/SX

MM : 160

Instructions

1. There are 120 questions in this paper.
2. The question paper contains two parts; Part I (1 Mark Questions) and Part II (2 Marks Questions).
3. There are four sections in each part; Mathematics, Physics, Chemistry and Biology.
4. Out of the four options given with each question, only one is correct.

PART-I (1 Mark Questions)

MATHEMATICS

1. The sum of non-real roots of the polynomial equation $x^3 + 3x^2 + 3x + 3 = 0$ is

(a) equals 0
(b) lies between 0 and 1
(c) lies between -1 and 0
(d) has absolute value bigger than 1

2. Let n be a positive integer such that $\log_2 \log_2 \log_2 \log_2 \log_2 (n) < 0 < \log_2 \log_2 \log_2 \log_2 (n)$. Let l be the number of digits in the binary expansion of n. Then the minimum and the maximum possible values of l are

(a) 5 and 16 (b) 5 and 17
(c) 4 and 16 (d) 4 and 17

3. Let ω be a cube root of unity not equal to 1. Then, the maximum possible value of $|a + b\omega + c\omega^2|$, where $a, b, c \in \{+1, -1\}$ is

(a) 0 (b) 2
(c) $\sqrt{3}$ (d) $1+\sqrt{3}$

4. If a, b are positive real numbers such that the lines $ax + 9y = 5$ and $4x + by = 3$ are parallel, then the least possible value of $a + b$ is

(a) 13 (b) 12
(c) 8 (d) 6

5. Two line segments AB and CD are constrained to move along the X and Y-axes, respectively, in such a way that the points A, B, C, D are concyclic. If $AB = a$ and $CD = b$, then the locus of the centre of the circle passing through A, B, C, D in polar coordinates is

(a) $r^2 = \dfrac{a^2 + b^2}{4}$ (b) $r^2 \cos 2\theta = \dfrac{a^2 - b^2}{4}$
(c) $r^2 = 4(a^2 + b^2)$ (d) $r^2 \cos 2\theta = 4(a^2 - b^2)$

6. Consider a ΔABC in the XY-plane with vertices $A = (0, 0)$, $B = (1, 1)$ and $C = (9, 1)$. If the line $x = a$ divides the triangle into two parts of equal area, then a equals

(a) 3 (b) 3.5
(c) 4 (d) 4.5

7. Let ABC be an acute-angled triangle and let D be the mid-point of BC. If $AB = AD$, then $\tan(B) / \tan(C)$ equals

(a) $\sqrt{2}$ (b) $\sqrt{3}$
(c) 2 (d) 3

8. The angles α, β, γ of a triangle satisfy the equations $2\sin\alpha + 3\cos\beta = 3\sqrt{2}$ and $3\sin\beta + 2\cos\alpha = 1$. Then, angle γ equals

(a) 150° (b) 120°
(c) 60° (d) 30°

9. Let $f: R \to R$ be a function such that $\lim_{x\to\infty} f(x) = M > 0$.

Then which of the following is false ?

(a) $\lim_{x\to\infty} x\sin(1/x) f(x) = M$ (b) $\lim_{x\to\infty} \sin(f(x)) = \sin M$
(c) $\lim_{x\to\infty} x\sin(e^{-x}) f(x) = M$ (d) $\lim_{x\to\infty} \frac{\sin x}{x} \cdot f(x) = 0$

10. For $x, t \in R$ let $p_t(x) = (\sin t)x^2 - (2\cos t)x + \sin t$ be a family of quadratic polynomials in x with variable coefficients. Let $A(t) = \int_0^1 p_t(x)\,dx$. Which of the following statements are true?

I. $A(t) < 0$ for all t.

II. $A(t)$ has infinitely many critical points.

III. $A(t) = 0$ for infinitely many t.

IV. $A'(t) < 0$ for all t.

(a) I and II only (b) II and III only
(c) III and IV only (d) IV and I only

11. Let $f(x) = \sqrt{2 - x - x^2}$ and $g(x) = \cos x$. Which of the following statements are true?

I. Domain of $f((g(x))^2)$ = Domain of $f(g(x))$

II. Domain of $f(g(x)) + g(f(x))$ = Domain of $g(f(x))$

III. Domain of $f(g(x))$ = Domain of $g(f(x))$

IV. Domain of $g((f(x))^3)$ = Domain of $f(g(x))$

(a) Only I (b) Only I and II
(c) Only III and IV (d) Only I and IV

12. For real x with $-10 \le x \le 10$ define $f(x) = \int_{-10}^{x} 2^{[t]}\,dt$, where for a real number r we denote by $[r]$ the largest integer less than or equal to r. The number of points of discontinuity of f in the interval $(-10, 10)$ is

(a) 0 (b) 10
(c) 18 (d) 19

13. For a real number x, let $[x]$ denote the largest integer less than or equal to x and $\{x\} = x - [x]$. The smallest possible integer value of n for which $\int_1^n [x]\{x\}\,dx$ exceeds 2013 is

(a) 63 (b) 64
(c) 90 (d) 91

14. The area bounded by the curve $y = \cos x$, the line joining $(-\pi/4, \cos(-\pi/4))$ and $(0, 2)$ and the line joining $(\pi/4, \cos(\pi/4))$ and $(0, 2)$ is

(a) $\left(\frac{4+\sqrt{2}}{8}\right)\pi - \sqrt{2}$ (b) $\left(\frac{4+\sqrt{2}}{8}\right)\pi + \sqrt{2}$
(c) $\left(\frac{4+\sqrt{2}}{4}\right)\pi - \sqrt{2}$ (d) $\left(\frac{4+\sqrt{2}}{4}\right)\pi + \sqrt{2}$

15. A box contains coupons labelled 1, 2, 3,... , n. A coupon is picked at random and the number x is noted. The coupon is put back into the box and a new coupon is picked at random. The new number is y. Then, the probability that one of the numbers x, y divides the other is (in the options below $[r]$ denotes the largest integer less than or equal to r)

(a) $\frac{1}{2}$ (b) $\frac{1}{n^2}\sum_{k=1}^{n}\left[\frac{n}{k}\right]$
(c) $-\frac{1}{n} + \frac{1}{n^2}\sum_{k=1}^{n}\left[\frac{n}{k}\right]$ (d) $-\frac{1}{n} + \frac{2}{n^2}\sum_{k=1}^{n}\left[\frac{n}{k}\right]$

16. Let $n \ge 3$. A list of numbers $0 < x_1 < x_2 < \ldots < x_n$ has mean μ and standard deviation σ. A new list of numbers is made as follows: $y_1 = 0$, $y_2 = x_2, \ldots, x_{n-1} = x_{n-1}$, $y_n = x_1 + x_n$. The mean and the standard deviation of the new list are $\hat{\mu}$ and $\hat{\sigma}$. Which of the following is necessarily true?

(a) $\mu = \hat{\mu}, \sigma \le \hat{\sigma}$
(b) $\mu = \hat{\mu}, \sigma \ge \hat{\sigma}$
(c) $\sigma = \hat{\sigma}$
(d) μ may or may not be equal to $\hat{\mu}$

17. Let $\mathbf{v}_1, \mathbf{v}_2, \mathbf{v}_3, \mathbf{v}_4$ be unit vectors in the XY-plane, one each in the interior of the four quadrants. Which of the following statements is necessarily true?

(a) $\mathbf{v}_1 + \mathbf{v}_2 + \mathbf{v}_3 + \mathbf{v}_4 = 0$
(b) There exist $\mathbf{i}, \mathbf{j}$ with $1 \le \mathbf{i} < \mathbf{j} \le 4$ such that $\mathbf{v}_i + \mathbf{v}_j$ is in the first quadrant
(c) There exist $\mathbf{i}, \mathbf{j}$ with $1 \le \mathbf{i} < \mathbf{j} \le 4$ such that $\mathbf{v}_i \cdot \mathbf{v}_j < 0$
(d) There exist $\mathbf{i}, \mathbf{j}$ with $1 \le \mathbf{i} < \mathbf{j} \le 4$ such that $\mathbf{v}_i \cdot \mathbf{v}_j > 0$

18. The number of integers n with $100 \le n \le 999$ and containing at most two distinct digits is

(a) 252 (b) 280
(c) 324 (d) 360

19. For an integer n let $S_n = \{n+1, n+2, \ldots\ldots, n+18\}$. Which of the following is true for all $n \ge 10$?

(a) S_n has a multiple of 19
(b) S_n has a prime
(c) S_n has at least four multiples of 5
(d) S_n has at most six primes

20. Let P be a closed polygon with 10 sides and 10 vertices (assume that the sides do not intersect except at the vertices). Let k be the number of interior angles of P that are greater than 180°. The maximum possible value of k is

(a) 3 (b) 5 (c) 7 (d) 9

PHYSICS

21. Consider an initially neutral hollow conducting spherical shell with inner radius r and outer radius $2r$. A point charge $+Q$ is now placed inside the shell at a distance $r/2$ from the centre. The shell is then grounded by connecting the outer surface to the earth. P is an external point at a distance $2r$ from the point charge $+Q$ on the line passing through the centre and the point charge $+Q$ as shown in the figure.

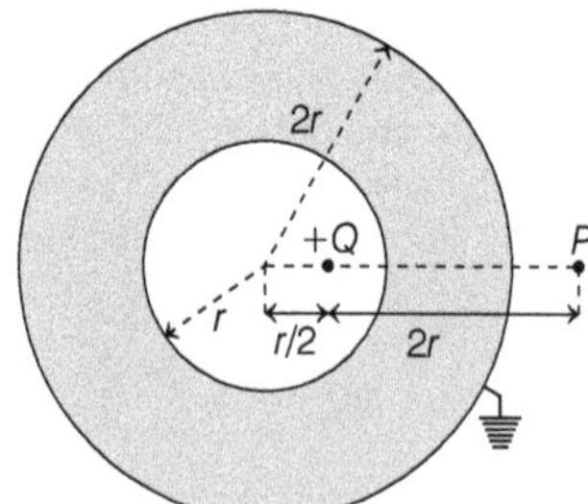

The magnitude of the force on a test charge $+q$ placed at P will be

(a) $\frac{1}{4\pi\varepsilon_0}\frac{qQ}{4r^2}$ (b) $\frac{1}{4\pi\varepsilon_0}\frac{9qQ}{100r^2}$ (c) $\frac{1}{4\pi\varepsilon_0}\frac{4qQ}{25r^2}$ (d) 0

22. Consider the circuit shown in the figure below.

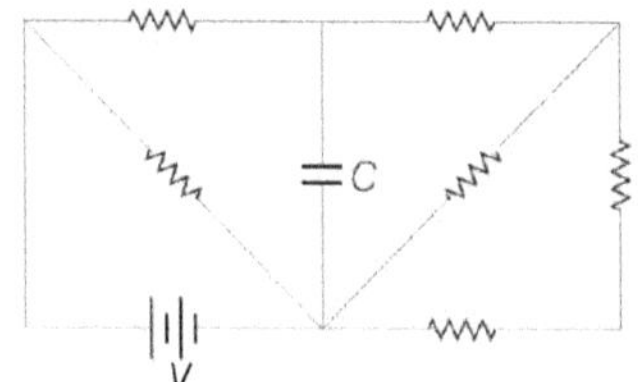

All the resistors are identical. The charge stored in the capacitor, once it is fully charged is

(a) 0 (b) $\frac{5}{13}CV$ (c) $\frac{2}{3}CV$ (d) $\frac{5}{8}CV$

23. A nuclear decay is possible if the mass of the parent nucleus exceeds the total mass of the decay particles. If $M(A, Z)$ denotes the mass of a single neutral atom of an element with mass number A and atomic number Z, then the minimal condition that the β decay $X_Z^A \to Y_{Z+1}^A + \beta^- + \bar{\nu}_e$ will occur is (m_e denotes the mass of the β particle and the neutrino mass m_ν can be neglected)

(a) $M(A, Z) > M(A, Z+1) + m_e$
(b) $M(A, Z) > M(A, Z+1)$
(c) $M(A, Z) > M(A, Z+1) + Zm_e$
(d) $M(A, Z) > M(A, Z+1) - m_e$

24. The equation of state of n moles of a non-ideal gas can be approximated by the equation

$$\left(p + \frac{n^2a}{V^2}\right)(V - nb) = nRT,$$

where a and b are constant characteristics of the gas. Which of the following can represent the equation of a quasistatic adiabat for this gas (assume that, C_V is the molar heat capacity at constant volume is independent of temperature)?

(a) $T(V - nb)^{R/C_V} = \text{constant}$
(b) $T(V - nb)^{C_V/R} = \text{constant}$
(c) $\left(T + \frac{ab}{V^2R}\right)(V - nb)^{R/C_V} = \text{constant}$
(d) $\left(T + \frac{n^2ab}{V^2R}\right)(V - nb)^{C_V/R} = \text{constant}$

25. A blackbox (BB) which may contain a combination of electrical circuit elements (resistor, capacitor or inductor) is connected with other external circuit elements as shown below in the figure (A). After the switch S is closed at time $t = 0$, the current I as a function of time t is shown in the figure (B).

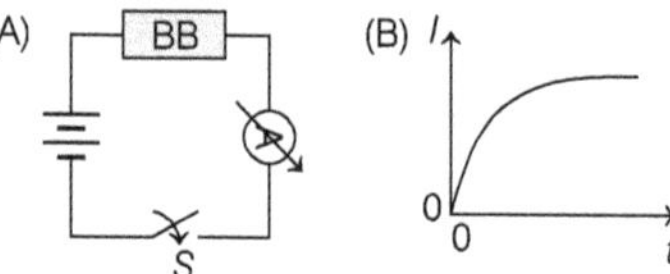

From this we can infer that the blackbox contains

(a) a resistor and a capacitor in series
(b) a resistor and a capacitor in parallel
(c) a resistor and an inductor in series
(d) a resistor and an inductor in parallel

26. In a photocell circuit the stopping potential V_0 is a measure of the maximum kinetic energy of the photoelectrons. The following graph shows experimentally measured values of stopping potential *versus* frequency ν of incident light.

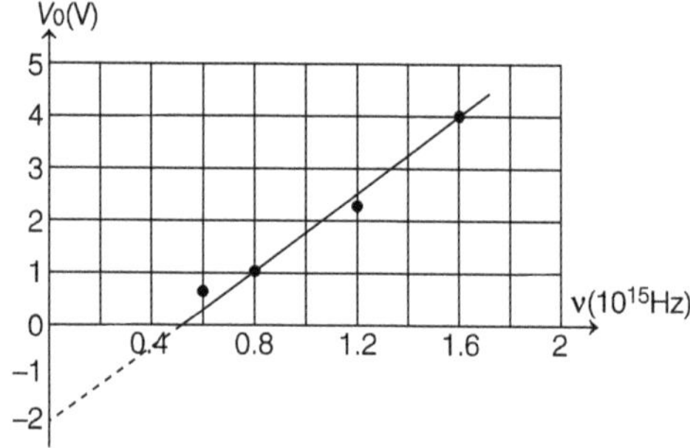

The values of Planck's constant and the work function as determined from the graph are (taking, the magnitude of electronic charge to be $e = 1.6 \times 10^{-19}$ C)

(a) 6.4×10^{-34} Js, 2.0 eV (b) 6.0×10^{-34} Js, 2.0 eV
(c) 6.4×10^{-34} Js, 3.2 eV (d) 6.0×10^{-34} Js, 3.2 eV

27. An engine moving away from a vertical cliff blows a horn at a frequency f. Its speed is 0.5% of the speed of sound in air. The frequency of the reflected sound received at the engine is

(a) $0.990f$ (b) $0.995f$
(c) $1.005f$ (d) $1.010f$

28. An arrangement with a pair of quarter circular coils of radii r and R with a common centre C and carrying a current I is shown in the figure.

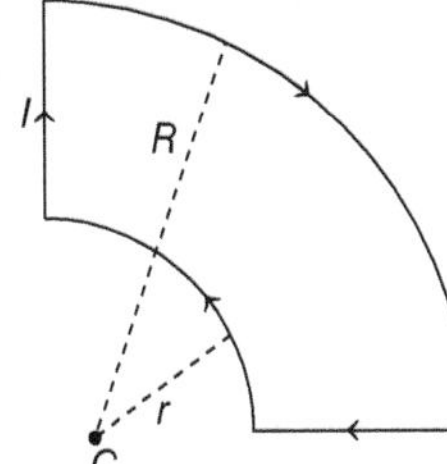

The permeability of free space is μ_0. The magnetic field at C is

(a) $\mu_0 I\ (1/r - 1/R)/8$ into the page
(b) $\mu_0 I\ (1/r - 1/R)/8$ out of the page
(c) $\mu_0 I\ (1/r + 1/R)/8$ out of the page
(d) $\mu_0 I\ (1/r + 1/R)/8$ into the page

29. The circuit shown has been connected for a long time. The voltage across the capacitor is

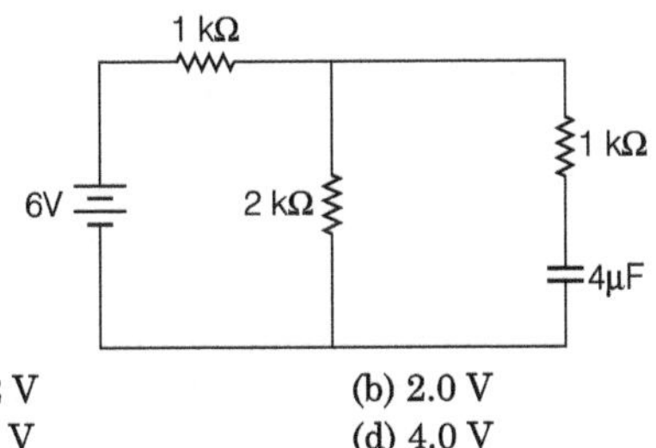

(a) 1.2 V (b) 2.0 V
(c) 2.4 V (d) 4.0 V

30. A wheel of radius R with an axle of radius $R/2$ is as shown in the figure and is free to rotate about a frictionless axis through its centre and perpendicular to the page. Three forces $(F, F, 2F)$ are exerted tangentially to the respective rims as shown in the figure.

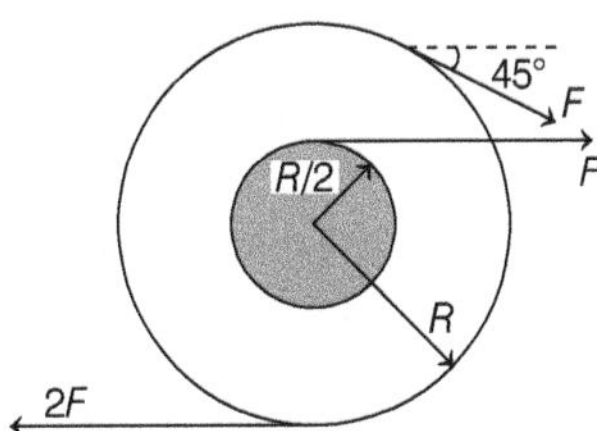

The magnitude of the net torque acting on the system is nearly

(a) $3.5FR$ (b) $3.2FR$
(c) $2.5FR$ (d) $1.5FR$

31. Two species of radioactive atoms are mixed in equal number. The disintegration constant of the first species is λ and of the second is $\lambda/3$. After a long time the mixture will behave as a species with mean life of approximately

(a) $0.70/\lambda$ (b) $2.10/\lambda$ (c) $1.00/\lambda$ (d) $0.52/\lambda$

32. The bulk modulus of a gas is defined as $B = -V dp/dV$. For an adiabatic process the variation of B is proportional to p^n. For an ideal gas n is

(a) zero (b) 1
(c) 5/3 (d) 2

33. Photons of energy 7 eV are incident on two metals A and B with work functions 6 eV and 3 eV, respectively. The minimum de-Broglie wavelengths of the emitted photoelectrons with maximum energies are λ_A and λ_B respectively, where λ_A/λ_B is nearly

(a) 0.5 (b) 1.4
(c) 4.0 (d) 2.0

34. An electron enters a chamber in which an uniform magnetic field is present as shown in figure. Ignore gravity.

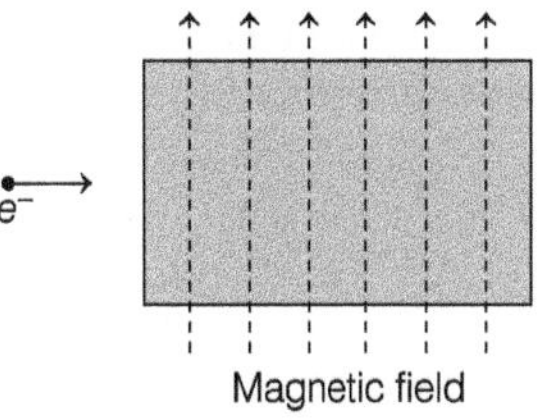

During its motion inside the chamber

(a) the force on the electron remains constant
(b) the kinetic energy of the electron remains constant
(c) the momentum of the electron remains constant
(d) the speed of the electron increases at a uniform rate

35. A ray of light incident on a glass sphere (refractive index $\sqrt{3}$) suffers total internal reflection before emerging out exactly parallel to the incident ray. The angle of incidence was

(a) 75° (b) 30°
(c) 45° (d) 60°

36. Young-Laplace law states that the excess pressure inside a soap bubble of radius R is given by $\Delta P = 4\sigma/R$, where σ is the coefficient of surface tension of the soap. The EÖTVÖS number E_0 is a dimensionless number that is used to describe the shape of bubbles rising through a surrounding fluid. It is a combination of g, the acceleration due to gravity ρ the density of the surrounding fluid σ and a characteristic length scale L which could be the radius of the bubble. A possible expression for E_0 is

(a) $\frac{\rho g}{\sigma L^3}$ (b) $\frac{\rho L^2}{\sigma g}$ (c) $\frac{\rho g L^2}{\sigma}$ (d) $\frac{g L^2}{\sigma \rho}$

37. A plank is resting on a horizontal ground in the northern hemisphere of the earth at a 45° latitude. Let the angular speed of the earth be ω and its radius r_e. The magnitude of the frictional force on the plank will be

(a) $mr_e\omega^2$ (b) $\frac{mr_e\omega^2}{\sqrt{2}}$ (c) $\frac{mr_e\omega^2}{2}$ (d) zero

38. The average distance between molecules of an ideal gas at STP is approximately of the order of

(a) 1 nm (b) 100 nm (c) 100 cm (d) 1 μm

39. A point particle of mass 0.5 kg is moving along the *X*-axis under a force described by the potential energy *V* shown below. It is projected towards the right from the origin with a speed *v*.

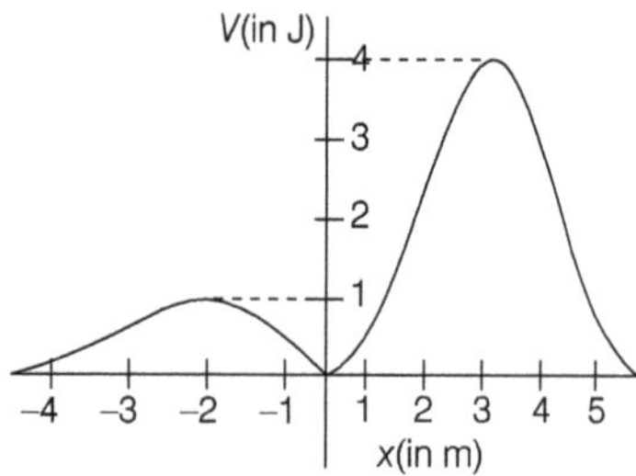

What is the minimum value of *v* for which the particle will escape infinitely far away from the origin?

(a) $2\sqrt{2}$ ms^{-1}

(b) 2 ms^{-1}

(c) 4 ms^{-1}

(d) The particle will never escape

40. The figure below shows pressure variation in two different sound waves in air with time at a given position. Both the figures are drawn to the same scale.

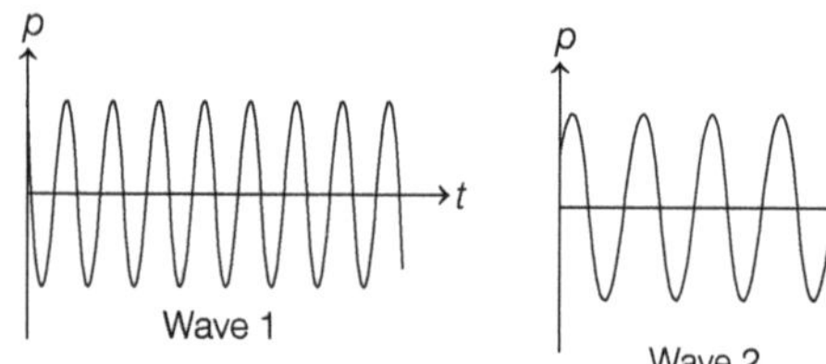

Which of the following statements is true?

(a) Wave 1 has lower frequency and smaller amplitude compared to wave 2

(b) Wave 1 has higher frequency and greater amplitude compared to wave 2

(c) Wave 1 has shorter wavelength and greater amplitude compared to wave 2

(d) Wave 1 has shorter wavelength and smaller amplitude compared to wave 2

CHEMISTRY

41. Among the following the set of isoelectronic ions is

(a) Na^+, Mg^{2+}, F^-, Cl^- (b) Na^+, Ca^{2+}, F^-, O^-

(c) Na^+, Mg^{2+}, F^-, O^{2-} (d) Na^+, K^+, S^{2-}, Cl^-

42. For a zero-order reaction with rate constant *k*, the slope of the plot of reactant concentration against time is

(a) $k/2.303$ (b) k

(c) $-k/2.303$ (d) $-k$

43. The compound which reacts with excess bromine to produce 2, 4, 6-tribromophenol, is

(a) 1,3-cyclohexadiene (b) 1,3-cyclohexanedione

(c) salicyclic acid (d) cyclohexanone

44. Ethyl acetate reacts with $NH_2NHCONH_2$ to form

(a) $CH_3CONHCONHNH_2$

(b) $CH_3CON(NH_2)CONH_2$

(c) $CH_3CONHNHCONH_2$

(d) $CH_3CH_2NHNHCONH_2$

45. The variation of solubility of four different gases (G1, G2, G3, G4 etc.) in a given solvent with pressure at a constant temperature is shown in the plot.

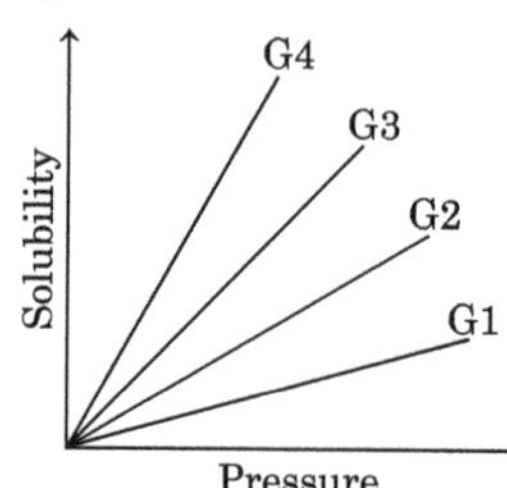

The gas with the highest value of Henry's law constant is

(a) G4 (b) G2 (c) G3 (d) G1

46. For the reaction, $A \rightleftharpoons nB$, the concentration of *A* decreases from 0.06 to 0.03 mol L^{-1} and that of *B* rises from 0 to 0.06 mol L^{-1} at equilibrium. The values of *n* and the equilibrium constant for the reaction, respectively, are

(a) 2 and 0.12 (b) 2 and 1.2

(c) 3 and 0.12 (d) 3 and 1.2

47. The reaction of ethyl methyl ketone with Cl_2/excess OH^- gives the following major product

(a) $ClCH_2CH_2COCH_3$ (b) $CH_3CH_2COCCl_3$

(c) $ClCH_2CH_2COCH_2Cl$ (d) $CH_3CCl_2COCH_2Cl$

48. The compound that readily tautomerises is

(a) $CH_3COCH_2CO_2C_2H_5$ (b) $CH_3COCH_2CHCH_3$

(c) $CH_3COCH_2CH_2CH_3$ (d) $(CH_3)_3CCOC(CH_3)_3$

49. Hydrolysis of BCl_3 gives *X*, which on treatment with sodium carbonate produces *Y*. *X* and *Y*, respectively, are

(a) H_3BO_3 and $NaBO_2$ (b) H_3BO_3 and $Na_2B_4O_7$

(c) B_2O_3 and $NaBO_2$ (d) B_2O_3 and $Na_2B_4O_7$

50. The numbers of lone pair (s) on Xe in XeF_2 and XeF_4 are, respectively

(a) 2 and 3 (b) 4 and 1

(c) 3 and 2 (d) 4 and 2

51. The entropy change in the isothermal reversible expansion of 2 moles of an ideal gas from 10 to 100 L at 300 K is

(a) 42.3 J K^{-1} (b) 35.8 J K^{-1}

(c) 38.3 J K^{-1} (d) 32.3 J K^{-1}

52. D-glucose upon treatment with bromine-water gives

(a)
```
      COOH
   H──┼──OH
  HO──┼──H
   H──┼──OH
   H──┼──OH
      CH2OH
```

(b)
```
      CHO
   H──┼──OH
  HO──┼──H
   H──┼──OH
   H──┼──OH
      CHO
```

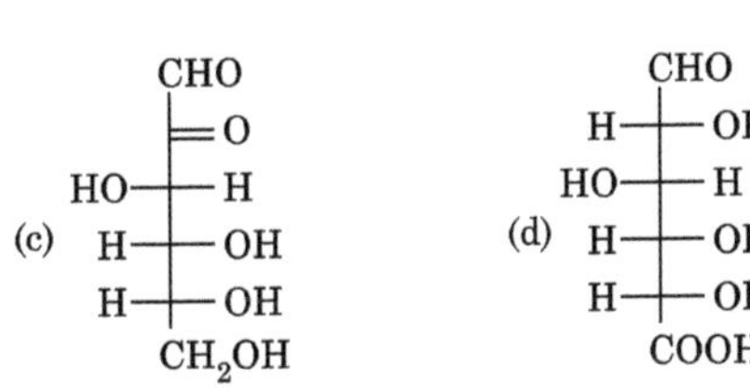

53. In the structure of borax, the numbers of boron atoms and B—O—B units, respectively are

(a) 4 and 5 (b) 4 and 3
(c) 5 and 4 (d) 5 and 3

54. The number of peptide bonds in the compound,

$H_3C-CO-NH-CH(CH_3)-CO-NH-CH(CH(CH_3)_2)-CO-NHNH_2$ is

(a) 1 (b) 2
(c) 3 (d) 4

55. For the isothermal reversible expansion of an ideal gas

(a) $\Delta H > 0$ and $\Delta U = 0$ (b) $\Delta H > 0$ and $\Delta U < 0$
(c) $\Delta H = 0$ and $\Delta U = 0$ (d) $\Delta H = 0$ and $\Delta U > 0$

56. If the angle of incidence of X-ray of wavelength 3Å which produces a second order diffracted beam from the (100) planes in a simple cubic lattice with interlayer spacing $a = 6$ Å is 30°, the angle of incidence that produces a first order diffracted beam from the (200) planes is

(a) 15° (b) 45° (c) 30° (d) 60°

57. The number of ions produced in water by dissolution of the complex having the empirical formula, $CoCl_3 \cdot 4NH_3$, is

(a) 1 (b) 2 (c) 4 (d) 3

58. The spin-only magnetic moments of $[Fe(NH_3)_6]^{3+}$ and $[FeF_6]^{3-}$ in BM are, respectively

(a) 1.73 and 1.73 (b) 5.92 and 1.73
(c) 1.73 and 5.92 (d) 5.92 and 5.92

59. The order of S_N1 reactivity in aqueous acetic acid solution for the compounds

1. $H_3C-\overset{O}{\overset{||}{C}}-CH_2-Cl$
2. $H_3C-CH_2-CH_2-Cl$
3. $(H_3C)_3C-Cl$

is

(a) 1 > 2 > 3 (b) 1 > 3 > 2
(c) 3 > 2 > 1 (d) 3 > 1 > 2

60. An ionic compound is formed between a metal M and a non-metal Y. If M occupies half the octahedral voids in the cubic close-packed arrangement formed by Y, the chemical formula of the ionic compound is

(a) MY (b) MY_2 (c) M_2Y (d) MY_3

BIOLOGY

61. Human foetal haemoglobin differs from the adult haemoglobin in that it has

(a) higher affinity for oxygen
(b) lower affinity for oxygen
(c) two subunits only
(d) glycosylated

62. Nucleolus is an organelle responsible for the production of

(a) carbohydrates (b) messenger RNA
(c) lipids (d) ribosomal RNA

63. The sequences of four DNA molecules are given below

(i) TATATATATATATA
ATATATATATATAT
(ii) TTTCCCGGGAAA
AAAGGGCCCTTT
(iii) TTGCGTTGCGCC
AACGCAACGCGG
(iv) GCCGGATCCGGC
CGGCCTAGGCCG

Which one of these DNA molecules will have the highest melting temperature (T_m)?

(a) i (b) ii (c) iii (d) iv

64. If DNA codons are ATG GAA, insertion of thymine after the first codon results in

(a) non-sense mutation (b) mis-sense mutation
(c) frameshift mutation (d) silent mutation

65. Genetic content of a cell reduces to half during

(a) meiotic prophase-I (b) mitotic prophase
(c) meiotic prophase-II (d) meiotic telophase

66. Which one of the following techniques is used for the detection of proteins?

(a) Northern blotting (b) Western blotting
(c) Southern blotting (d) *In situ* hybridisation

67. Fission yeasts are

(a) archaebacteria (b) eubacteria
(c) prokaryotes (d) eukaryotes

68. In green leaves, the light and dark reactions occur in

(a) stroma and grana respectively
(b) grana and stroma respectively
(c) cristae and matrix respectively
(d) both occur in cytoplasm

69. According to Mendel, ... segregate and .... assort independently.
(a) alleles of a gene, alleles of different genes
(b) alleles of different genes, alleles of a gene
(c) dominant traits, recessive traits
(d) recessive traits, recessive traits

70. The two enzymatic activities associated with RuBisCO are
(a) oxidase and oxygenase
(b) oxygenase and carboxylase
(c) oxidase and carboxylase
(d) oxygenase and carbamylation

71. Chlorofluorocarbons (CFCs) are believed to be associated with cancers because
(a) CFCs react with DNA and cause mutations
(b) CFCs react with proteins involved in DNA repair
(c) CFCs destroy the ozone layer and permit harmful UV rays to reach on the earth
(d) CFCs react with DNA polymerase and reduce fidelity of DNA replication

72. Morphogenetic movements take place predominantly during the following embryonic stage
(a) blastula (b) morula
(c) gastrula (d) fertilised eggs

73. The only organ which is capable of producing fructose in human is
(a) liver (b) pancreas
(c) seminal vesicles (d) muscle

74. Stroke could be prevented/treated with
(a) balanced diet (b) clotting factors
(c) insulin (d) blood thinners

75. In orange and lemon, the edible part of the fruit is
(a) placenta (b) thalamus
(c) hairs of the ovary wall (d) succulent mesocarp

76. Which one of the following statements about nitrogenase is correct?
(a) It is sensitive to CO_2 and therefore present in isolated nodules
(b) It requires O_2 and therefore functional during the day
(c) It is sensitive to O_2 and therefore is functional in anaerobic environments
(d) It is sensitive to light and therefore functions only in dark

77. Part of epidermis that keeps out unwanted particles is called
(a) columnar epithelium (b) squamous epithelium
(c) ciliated epithelium (d) cuboidal epithelium

78. Species that are most effective at colonising new habitats show
(a) low reproductive ability
(b) high dispersal ability
(c) slow growth and maturation
(d) high competitive ability

79. In a large isolated population, alleles p and q at a locus are at Hardy-Weinberg equilibrium. The frequencies are $p = 0.6$ and $q = 0.4$. The proportion of the heterozygous genotype in the population is
(a) 0.24 (b) 1
(c) 0.48 (d) 0.12

80. In vertebrates, glycogen is stored chiefly in
(a) heart and blood (b) spleen and stomach
(c) bones and lymph (d) liver and muscles

PART-II (2 Marks Questions)

MATHEMATICS

81. Let $f(x)$ be a non-constant polynomial with real coefficients such that $f\left(\frac{1}{2}\right) = 100$ and $f(x) \le 100$ for all real x. Which of the following statements is NOT necessarily true?
(a) The coefficient of the highest degree term in $f(x)$ is negative.
(b) $f(x)$ has at least two real roots.
(c) If $x \ne 1/2$ then $f(x) < 100$.
(d) At least one of the coefficients of $f(x)$ is bigger than 50.

82. Let a, b, c, d be real numbers such that
$$\sum_{k=1}^{n} (ak^3 + bk^2 + ck + d) = n^4, \text{ for every natural number } n.$$
Then, $|a| + |b| + |c| + |d|$ is equal to
(a) 15 (b) 16 (c) 31 (d) 32

83. The vertices of the base of an isosceles triangle lie on a parabola $y^2 = 4x$ and the base is a part of the line $y = 2x - 4$. If the third vertex of the triangle lies on the X-axis, its coordinates are
(a) $\left(\frac{5}{2}, 0\right)$ (b) $\left(\frac{7}{2}, 0\right)$ (c) $\left(\frac{9}{2}, 0\right)$ (d) $\left(\frac{11}{2}, 0\right)$

84. In a ΔABC, let G denote its centroid and let M, N be points in the interiors of the segments AB, AC, respectively, such that M, G, N are collinear. If r denotes the ratio of the area of ΔAMN to the area of ABC, then
(a) $r = 1/2$ (b) $r > 1/2$
(c) $4/9 \le r < 1/2$ (d) $4/9 < r$

85. Let XY be the diameter of a semi-circle with centre O. Let A be a variable point on the semi-circle and B another point on the semi-circle such that AB is parallel to XY.

The value of $\angle BOY$ for which the inradius of ΔAOB is maximum, is

(a) $\cos^{-1}\left(\frac{\sqrt{5}-1}{2}\right)$ (b) $\sin^{-1}\left(\frac{\sqrt{5}-1}{2}\right)$

(c) $\frac{\pi}{3}$ (d) $\frac{\pi}{5}$

86. Let $f(x) = 1 + \frac{x}{1!} + \frac{x^2}{2!} + \frac{x^3}{3!} + \frac{x^4}{4!}$. The number of real roots of $f(x) = 0$ is

(a) 0 (b) 1 (c) 2 (d) 4

87. Suppose that the earth is a sphere of radius 6400 km. The height from the earth's surface from where exactly a fourth of the earth's surface is visible, is

(a) 3200 km (b) $3200\sqrt{2}$ km

(c) $3200\sqrt{3}$ km (d) 6400 km

88. Let n be a positive integer. For a real number x, let $[x]$ denote the largest integer not exceeding x and $\{x\} = x - [x]$. Then, $\int_1^{n+1} \frac{(\{x\})^{[x]}}{[x]}\, dx$ is equal to

(a) $\log_e(n)$ (b) $\frac{1}{n+1}$

(c) $\frac{n}{n+1}$ (d) $1 + \frac{1}{2} + \ldots.. + \frac{1}{n}$

89. A box contains coupons labelled 1, 2, ..., 100. Five coupons are picked at random one after another without replacement. Let the numbers on the coupons be $x_1, x_2, \ldots, x_5$. What is the probability that $x_1 > x_2 > x_3$ and $x_3 < x_4 < x_5$?

(a) 1/120 (b) 1/60 (c) 1/20 (d) 1/10

90. In a tournament with five teams, each team plays against every other team exactly once. Each game is won by one of the playing teams and the winning team scores one point, while the losing team scores zero. Which of the following is NOT necessarily true?

(a) There are at least two teams which have at most two points each.

(b) There are at least two teams which have at least two points each.

(c) There are at most three teams which have at least three points each.

(d) There are at most four teams which have at most two points each.

PHYSICS

91. A bullet of mass m is fired horizontally into a large sphere of mass M and radius R resting on a smooth horizontal table.

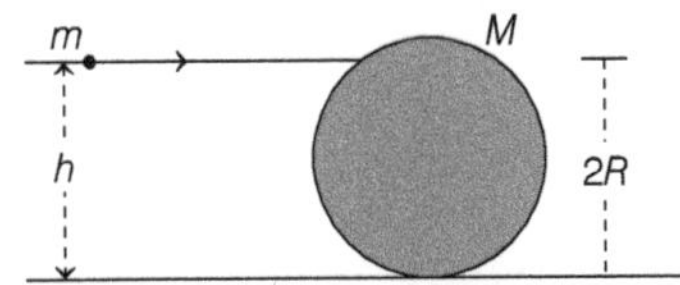

The bullet hits the sphere at a height h from the table and sticks to its surface. If the sphere starts rolling without slipping immediately on impact, then

(a) $\frac{h}{R} = \frac{4m + 3M}{2(m + M)}$ (b) $\frac{h}{R} = \frac{m + M}{m + 2M}$

(c) $\frac{h}{R} = \frac{10m + 7M}{5(m + M)}$ (d) $\frac{h}{R} = \frac{4m + 3M}{m + M}$

92. A small boy is throwing a ball towards a wall 6 m in front of him. He releases the ball at a height of 1.4 m from the ground. The ball bounces from the wall at a height of 3 m, rebounds from the ground and reaches the boy's hand exactly at the point of release. Assuming the two bounces (one from the wall and the other from the ground) to be perfectly elastic, how far ahead of the boy did the ball bounce from the ground?

(a) 1.5 m (b) 2.5 m

(c) 3.5 m (d) 4.5 m

93. In the p-V diagram below, the dashed curved line is an adiabat.

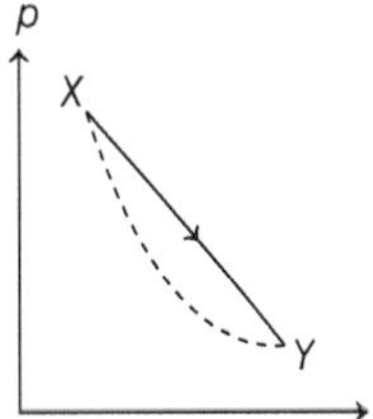

For a process that is described by a straight line joining two points X and Y on the adiabat (solid line in the diagram) heat is

(**Hint** consider the variation in temperature from X to Y along the straight line)

(a) absorbed throughout from X to Y

(b) released throughout from X to Y

(c) absorbed from X upto an intermediate point Z (not shown in the figure) and then released from Z to Y

(d) released from X upto an intermediate point Z (not shown in the figure) and then absorbed from Z to Y

94. A singly ionised helium atom in an excited state ($n = 4$) emits a photon of energy 2.6 eV. Given that the ground state energy of hydrogen atom is −13.6 eV, the energy E_f and quantum number n of the resulting state are respectively,

(a) $E_f = -13.6$ eV, $n = 1$ (b) $E_f = -6.0$ eV, $n = 3$

(c) $E_f = -6.0$ eV, $n = 2$ (d) $E_f = -13.6$ eV, $n = 2$

95. The figure below shows a circuit and its input voltage V_i as function of time t.

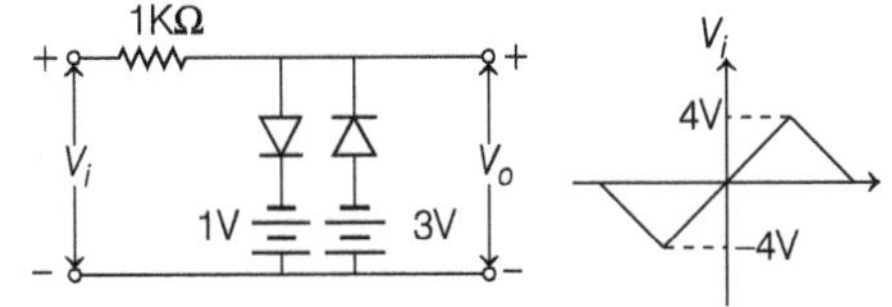

Assuming the diodes to be ideal, which of the following graphs depicts the output voltage V_o as function of time t ?

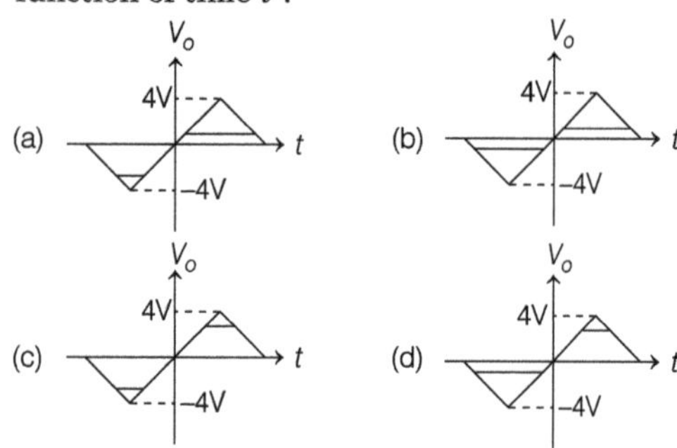

96. A ball is rolling without slipping in a spherical shallow bowl (radius R) as shown in the figure and is executing simple harmonic motion. If the radius of the ball is doubled, then the time period of oscillation

(a) increases slightly
(b) is reduced by a factor of 1/2
(c) is increased by a factor of 2
(d) decreases slightly

97. A solid sphere rolls without slipping, first horizontally and then up to a point X at height h on an inclined plane before rolling down, as shown in the figure below.

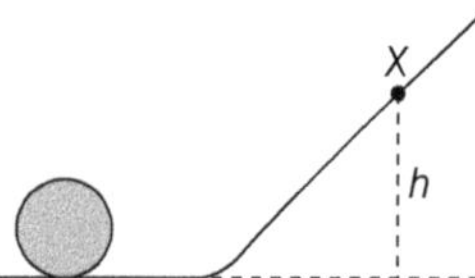

The initial horizontal speed of the sphere is
(a) $\sqrt{10gh/7}$ (b) $\sqrt{7gh/5}$ (c) $\sqrt{5gh/7}$ (d) $\sqrt{2gh}$

98. The three processes in a thermodynamic cycle shown in the figure are : Process $1 \rightarrow 2$ is isothermal; Process $2 \rightarrow 3$ is isochoric (volume remains constant); Process $3 \rightarrow 1$ is adiabatic.

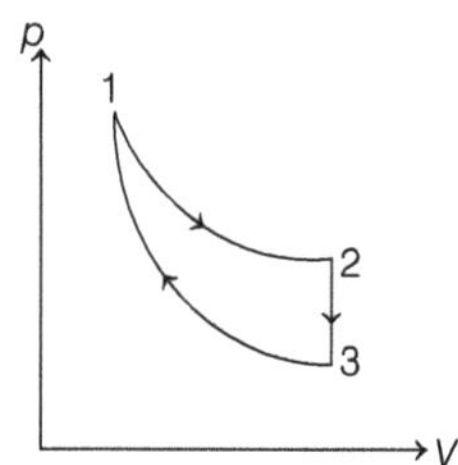

The total work done by the ideal gas in this cycle is 10 J. The internal energy decreases by 20 J in the isochoric process. The work done by the gas in the adiabatic process is –20 J. The heat added to the system in the isothermal process is
(a) 0 J (b) 10 J (c) 20 J (d) 30 J

99. A block of mass m slides from rest at a height H on a frictionless inclined plane as shown in the figure. It travels a distance d across a rough horizontal surface with coefficient of kinetic friction μ and compresses a spring of spring constant k by a distance x before coming to rest momentarily. Then the spring extends and the block travels back attaining a final height of h.

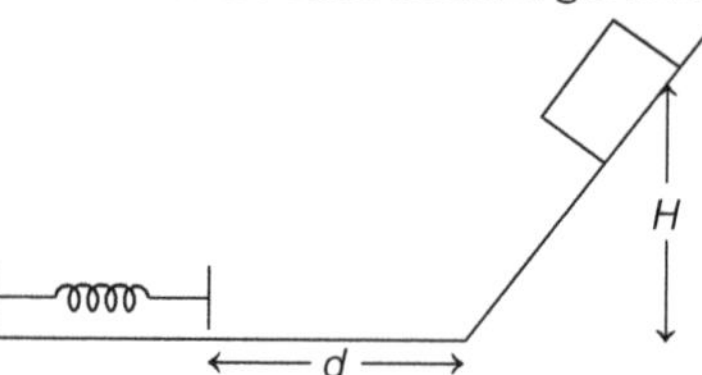

Then,
(a) $h = H - 2\mu\ (d + x)$
(b) $h = H + 2\mu\ (d - x)$
(c) $h = H - 2\mu d + kx^2 / mg$
(d) $h = H - 2\mu\ (d + x) + kx^2 / 2mg$

100. A metallic prong consists of 4 rods made of the same material, cross-sections and same lengths as shown below. The three forked ends are kept at 100°C and the handle end is at 0°C. The temperature of the junction is

T = 0°C T = 100°C T = 100°C T = 100°C

(a) 25°C (b) 50°C (c) 60°C (d) 75°C

CHEMISTRY

101. The major product obtained in the reaction of aniline with acetic anhydride is

(a) C_6H_5–$NHCOCH_3$

(b) H_3C–CO–C_6H_4–$NHCOCH_3$

(c) C_6H_5–N($COCH_3$)$_2$

(d) H_3C–CO–C_6H_4–NH_2

102. The maximum number of isomers that can result from monobromination of 2-methyl-2-pentene with N-bromosuccinimide in boiling CCl_4 is
(a) 1 (b) 2 (c) 3 (d) 4

103. The compound $X(C_7H_9N)$ reacts with benzene sulphonyl chloride to give $Y(C_{13}H_{13}NO_2S)$ which is insoluble in alkali. The compound X is

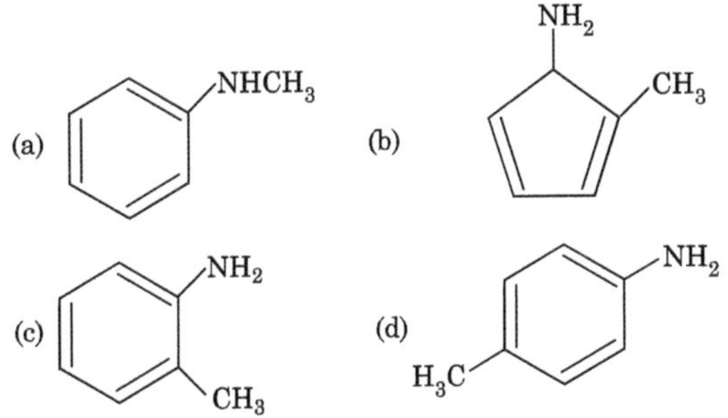

104. In 108 g of water, 18 g of a non-volatile compound is dissolved. At 100°C, the vapour pressure of the solution is 750 mm Hg. Assuming that the compound does not undergo association or dissociation, the molar mass of the compound in g mol^{-1} is
(a) 128 (b) 182
(c) 152 (d) 228

105. The standard electrode potential of Zn^{2+}/Zn is -0.76 V and that of Cu^{2+}/Cu is 0.34 V. The emf (V) and the free energy change (kJ mol^{-1}), respectively for a daniell cell will be
(a) -0.42 and 81 (b) 1.1 and -213
(c) -1.1 and 213 (d) 0.42 and -81

106. Consider the equilibria (1) and (2) with equilibrium constants K_1 and K_2, respectively.

$$SO_2(g) + 1/2\,O_2(g) \rightleftharpoons SO_3(g) \quad \ldots(i)$$
$$2SO_3(g) \rightleftharpoons 2SO_2(g) + O_2(g) \quad \ldots(ii)$$

K_1 and K_2 are related as
(a) $2K_1 = K_2^2$ (b) $K_1^2 = \frac{1}{K_2}$
(c) $K_2^2 = \frac{1}{K_1}$ (d) $K_2 = \frac{2}{K_1^2}$

107. Aqueous solution of a metallic nitrate X reacts with NH_4OH to form Y, which dissolves in excess NH_4OH. The resulting complex is reduced by acetaldehyde to deposit the metal. X and Y, respectively, are
(a) $Cs(NO_3)$ and $CsOH$ (b) $Zn(NO_3)_2$ and ZnO
(c) $AgNO_3$ and Ag_2O (d) $Mg(NO_3)_2$ and $Mg(OH)_2$

108. The density and equivalent weight of a metal are 10.5 g cm^{-3} and 100, respectively. The time required for a current of 3 amp to deposit a 0.005 mm thick layer of the same metal on an area of 80 cm^2 is closest to
(a) 120 s (b) 135 s
(c) 67.5 s (d) 270 s

109. The amount of $Na_2S_2O_3 \cdot 5H_2O$ required to completely reduce 100 mL of 0.25 N iodine solution, is
(a) 6.20 g (b) 9.30 g (c) 3.10 g (d) 7.75 g

110. In aqueous solution, $[Co(H_2O)_6]^{2+}$ (X), reacts with molecular oxygen in the presence of excess liquor NH_3 to give a new complex Y. The number of unpaired electrons in X and Y are, respectively
(a) 3, 1 (b) 3, 0 (c) 3, 3 (d) 7, 0

BIOLOGY

111. 10^9 bacteria were spread on an agar plate containing penicillin. After incubation overnight at 37°C, 10 bacterial colonies were observed on the plate. That the colonies are likely to be resistant to penicillin can be tested by
(a) measuring their growth rate
(b) observing the colour of the colonies
(c) checking their ability to grow on another plate containing penicillin
(d) checking their ability to cause disease

112. Watson and Crick model of DNA is a
(a) B-form DNA with a spiral length of 34 Å and a diameter of 20 Å
(b) A-form DNA with a spiral length of 15 Å and a diameter of 20 Å
(c) Z-form DNA with a spiral length of 34 Å and a diameter of 20 Å
(d) B-form DNA with a spiral length of 28 Å and a diameter of 14 Å

113. *Eco* RI and *Rsa* I restriction endonucleases require 6 and 4 bp sequences, respectively for cleavage. In a 10 kb DNA fragment how many probable cleavage sites are present for these enzymes?
(a) 0 *Eco* RI and 10 *Rsa* I (b) 1 *Eco* RI and 29 *Rsa* I
(c) 4 *Eco* RI and 69 *Rsa* I (d) 2 *Eco* RI and 39 *Rsa* I

114. From an early amphibian embryo, the cells that would give rise to skin in adults were transplanted into the developing brain region of another embryo. The transplanted cells developed into brain tissue in the recipient embryo. What do you infer from this experiment?
(a) Cell fate is permanently determined during early embryonic development
(b) Development fate of donor cells is influenced by the surrounding cells
(c) Developmental fate of donor cells is not influenced by recipient cells
(d) Any cell which is transplanted into another embryo always develops into a brain

115. The presence of plastids in *Plasmodium* suggests
(a) it is a plant species
(b) it is a parasite with a cyanobacterium as an endosymbiont
(c) it is a parasite with an archaebacterium as an endosymbiont.
(d) it is a plant species with an archaebacterium as an endosymbiont.

116. The figure below demonstrates the growth curves of two organisms A and B growing in the same area. What kind of relation exists between A and B?

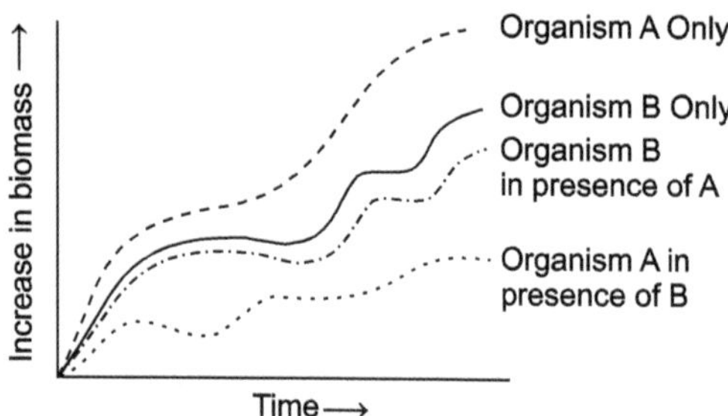

(a) Competition (b) Symbiosis
(c) Commensalism (d) Mutualism

117. A scientist has cloned an 8 kb fragment of a mouse gene into the *Eco* RI site of a vector of 6 kb size. The cloned DNA has no other *Eco* RI site within. Digestions of the cloned DNA is shown below.

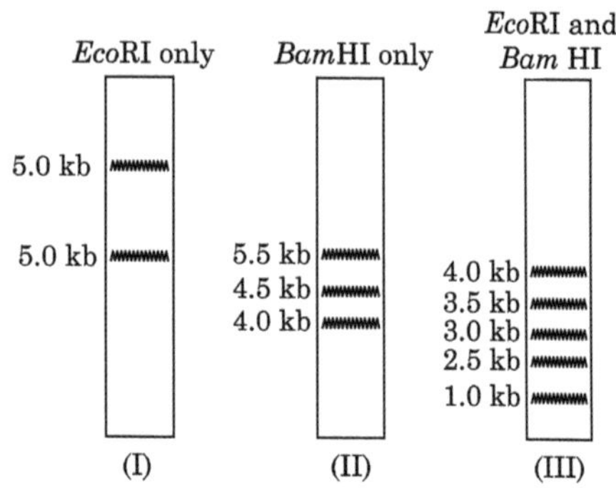

Which one of the following sets of DNA fragments generated by digestion with both *Eco* RI and *Bam* HI as shown in (III) is from the gene?

(a) 1 kb and 4 kb (b) 1 kb and 2.5 kb
(c) 1 kb and 3 kb (d) 1 kb and 3.5 kb

118. Brown fat is a specialised adipose tissue with abundant mitochondria and rich blood supply. Brown fat

(a) insulates animals that are acclimatised to cold
(b) is the major source of heat production in birds
(c) provides energy to muscles
(d) produces heat without producing ATP

119. In some species, individuals forego reproduction and help bring up another individual's offspring. Such altruistic behaviour cannot be explained by which of the following?

(a) An individual helps relatives only and gets indirect genetic benefits
(b) The individual benefits because it can later inherit the breeding position
(c) The individual benefits because it gets access to resources, such as food and security from predators, in return
(d) The species benefits from a reduction in competition among offspring

120. Lions in India are currently restricted to Gir, Gujarat. Efforts are being made to move them to other parts of the country. This is because they are most susceptible to extinction due to infectious diseases under the following conditions when present as

(a) several small, isolated populations
(b) one large population
(c) several large, connected populations
(d) several large, isolated populations

Answers

PART-I

1 *(c)*	2 *(a)*	3 *(b)*	4 *(b)*	5 *(b)*	6 *(a)*	7 *(d)*	8 *(d)*	9 *(c)*	10 *(b)*
11 *(a)*	12 *(a)*	13 *(d)*	14 *(a)*	15 *(d)*	16 *(a)*	17 *(c)*	18 *(a)*	19 *(d)*	20 *(c)*
21 *(d)*	22 *(d)*	23 *(b)*	24 *(a)*	25 *(c)*	26 *(b)*	27 *(a)*	28 *(b)*	29 *(d)*	30 *(a)*
31 *(b)*	32 *(b)*	33 *(d)*	34 *(b)*	35 *(*)*	36 *(c)*	37 *(c)*	38 *(a)*	39 *(b)*	40 *(d)*
41 *(c)*	42 *(d)*	43 *(c)*	44 *(c)*	45 *(d)*	46 *(a)*	47 *(b)*	48 *(a)*	49 *(b)*	50 *(c)*
51 *(c)*	52 *(a)*	53 *(a)*	54 *(a)*	55 *(c)*	56 *(c)*	57 *(b)*	58 *(c)*	59 *(c)*	60 *(b)*
61 *(a)*	62 *(d)*	63 *(d)*	64 *(a)*	65 *(c)*	66 *(b)*	67 *(d)*	68 *(b)*	69 *(a)*	70 *(b)*
71 *(c)*	72 *(c)*	73 *(c)*	74 *(d)*	75 *(c)*	76 *(c)*	77 *(c)*	78 *(b)*	79 *(c)*	80 *(d)*

PART-II

81 *(c)*	82 *(a)*	83 *(c)*	84 *(c)*	85 *(a)*	86 *(a)*	87 *(d)*	88 *(c)*	89 *(c)*	90 *(d)*
91 *(c)*	92 *(a)*	93 *(c)*	94 *(b)*	95 *(a)*	96 *(d)*	97 *(a)*	98 *(d)*	99 *(a)*	100 *(d)*
101 *(a)*	102 *(d)*	103 *(a)*	104 *(d)*	105 *(b)*	106 *(b)*	107 *(c)*	108 *(b)*	109 *(a)*	110 *(b)*
111 *(c)*	112 *(a)*	113 *(d)*	114 *(b)*	115 *(b)*	116 *(a)*	117 *(c)*	118 *(d)*	119 *(d)*	120 *(b)*

** No options are correct.*

Solutions

1. *(c)* We have,

$$x^3 + 3x^2 + 3x + 3 = 0$$

Let $\quad f(x) = x^3 + 3x^2 + 3x + 3$

$$f'(x) = 3x^2 + 6x + 3$$

$$f'(x) = 3(x+1)^2$$

$$f'(x) > 0, \forall x \in R$$

$\therefore f(x)$ is monotonic increasing function.

Hence, $x^3 + 3x^2 + 3x + 3 = 0$ has only one real root.

Now, $\quad f(-3) = (-3)^3 + 3(-3)^2 + 3(-3) + 3 = -6 < 0$

$f(-2) = (-2)^3 + 3(-2) + 3(-2) + 3 = 1 > 0$

$\therefore$ Roots lie between -3 and -2

$$\alpha + \beta + \gamma = -3$$

Let $\alpha \in (-3, -2)$, i.e. $-3 < \alpha < -2$

$\therefore \quad \alpha = -3 - \beta - \gamma$

$\therefore \quad -3 < -3 - \beta - \gamma < -2$

$\Rightarrow \quad 0 < -(\beta + \gamma) < 1$

$\Rightarrow \quad -1 < (\beta + \gamma) < 0$

$\therefore$ Sum of non-real roots lie between -1 and 0.

2. *(a)* We have,

$$\log_2\log_2\log_2\log_2\log_2(n) < 0 < \log_2\log_2\log_2\log_2(n)$$

$\Rightarrow \quad \log_2\log_2\log_2\log_2\log_2(n) < 0$

$\Rightarrow \quad \log_2\log_2\log_2\log_2(n) < 2^0$

$\Rightarrow \quad \log_2\log_2\log_2\log_2(n) < 1$

$\Rightarrow \quad \log_2\log_2\log_2(n) < 2$

$\Rightarrow \quad \log_2\log_2(n) < 2^2$

$\Rightarrow \quad \log_2(n) < 2^4$

$\Rightarrow \quad n < 2^{16}$

Similarly, for

$$\log_2\log_2\log_2\log_2(n) > 0 \Rightarrow n > 2^4$$

Hence, $\quad 2^4 < n < 2^{16}$

$\therefore$ The minimum number of digits in binary expansion of n is 5 and maximum numbers of digits in binary expansion of n is 16.

3. *(b)* We have,

$$|a + b\omega + c\omega^2|, a, b, c \in \{-1, 1\}$$

For maximum put $a = 1, b = -1, c = -1$

$|1 - \omega - \omega^2|$

$$|1 - (\omega + \omega^2)| = |1 + 1| = 2$$

$[\because \omega + \omega^2 = -1]$

4. *(b)* We have,

$ax + 9y = 5$ and $4x + by = 3$ are parallel.

$\therefore \quad \frac{a}{4} = \frac{9}{b} \Rightarrow ab = 36$

AM $\geq$ GM

$\therefore \quad \frac{a+b}{2} \geq \sqrt{ab}$

$$a + b \geq 2\sqrt{36}$$

$$a + b \geq 12$$

Hence, least possible value of $a + b = 12$.

5. *(b)* Let the equation of circle passing through $ABCD$ is

$$x^2 + y^2 + 2gx + 2fy + c = 0$$

Polar coordinates of centre of circle is

$$-g = r\cos\theta$$

$$-f = r\sin\theta$$

AB = length of intercept of X-axis

$\Rightarrow \quad a = 2\sqrt{g^2 - c}$

$\Rightarrow \quad \frac{a^2}{4} = g^2 - c \quad \ldots(\text{i})$

CD = length of intercept in Y-axis

$\Rightarrow \quad b = 2\sqrt{f^2 - c}$

$\Rightarrow \quad \frac{b^2}{4} = f^2 - c \quad \ldots(\text{ii})$

From Eqs. (i) and (ii), we get

$$g^2 - f^2 = \frac{a^2 - b^2}{4}$$

$\Rightarrow \quad r^2\cos^2\theta - r^2\sin^2\theta = \frac{a^2 - b^2}{4}$

$\Rightarrow \quad r^2(\cos^2\theta - \sin^2\theta) = \frac{a^2 - b^2}{4}$

$\Rightarrow \quad r^2\cos 2\theta = \frac{a^2 - b^2}{4}$

Hence, locus of centre of circle passing through $ABCD$ in polar coordinates is

$$r^2\cos 2\theta = \frac{a^2 - b^2}{4}.$$

6. *(a)* Given, in ΔABC

$A(0, 0), B(1, 1)\, C(9, 1)$

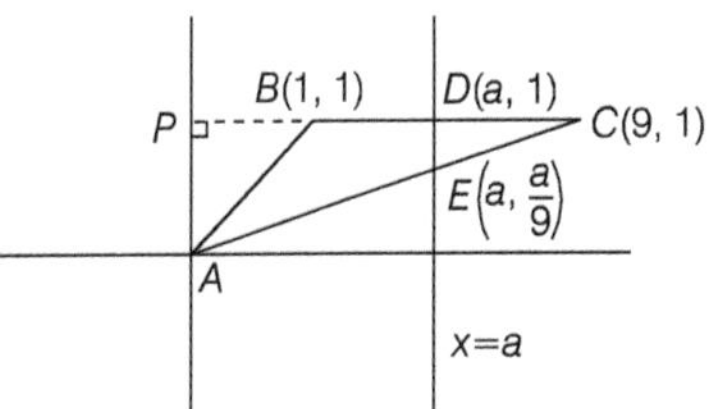

Area of $\Delta CDE = \frac{1}{2}$ Area of ΔABC

$\Rightarrow \quad \frac{1}{2} \times CD \times DE = \frac{1}{4} \times BC \times AP$

$\Rightarrow \quad \frac{1}{2}(9 - a) \times \left(1 - \frac{a}{9}\right) = \frac{1}{4} \times 8 \times 1$

$\Rightarrow \quad (9 - a)(9 - a) = 36$

$\Rightarrow \quad 9 - a = 6 \Rightarrow a = 3$

7. *(d)* Given, in ΔABC, D is mid-point of BC and $AB = AD$

$\therefore \quad \angle B = \angle ADB$

$$\theta = \pi - \angle ADB = \pi - B$$

$$BD = DC = 1:1$$

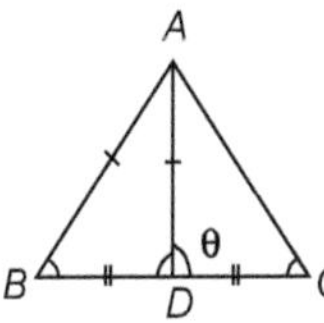

Apply $(m - n)$ theorem,

$$(m + n)\cot\theta = n\cot B - m\cot C$$

$\Rightarrow (1 + 1)\cot(\pi - B) = \cot B - \cot C$

$[\because m = n = 1]$

$\Rightarrow \quad -2\cot B = \cot B - \cot C$

$[\because \cot(\pi - B) = -\cot B]$

$\Rightarrow \quad 3\cot B = \cot C \Rightarrow \frac{\tan B}{\tan C} = 3$

8. *(d)* We have,

α, β, γ are angle of triangle.

$\therefore \quad \alpha + \beta + \gamma = 180°$

Given, $\quad 2\sin\alpha + 3\cos\beta = 3\sqrt{2} \quad \ldots(\text{i})$

and $\quad 3\sin\beta + 2\cos\alpha = 1 \quad \ldots(\text{ii})$

On squaring and adding Eqs. (i) and (ii), we get

$$4\sin^2\alpha + 9\cos^2\beta + 12\sin\alpha\cos\beta + 9\sin^2\beta + 4\cos^2\alpha + 12\cos\alpha\sin\beta = 18 + 1$$

$\Rightarrow 4(\sin^2\alpha + \cos^2\alpha) + 9(\cos^2\beta + \sin^2\beta) + 12(\sin\alpha\cos\beta + \cos\alpha\sin\beta) = 19$

$\Rightarrow \quad 4 + 9 + 12\sin(\alpha + \beta) = 19$

$\Rightarrow \quad 12\sin(\alpha + \beta) = 19 - 9 - 4$

$\Rightarrow \quad \sin(\alpha+\beta) = \frac{1}{2}$

$\Rightarrow \quad \alpha + \beta = 150°$

$\therefore \quad \gamma = 30°$

9. *(c)* We have,

$$\lim_{x\to\infty} f(x) = M > 0$$

(a) $\lim_{x\to\infty} x\sin\left(\frac{1}{x}\right) f(x) = \lim_{x\to\infty} \frac{\sin\left(\frac{1}{x}\right)}{\frac{1}{x}} f(x)$

$$= \lim_{x\to\infty} \frac{\sin\left(\frac{1}{x}\right)}{\frac{1}{x}} \times \lim_{x\to\infty} f(x) = 1\times M = M$$

Hence, option (a) is true.

(b) $\lim_{x\to\infty} \sin(f(x))$

$\sin(\lim_{x\to\infty} f(x)) = \sin M$

Option (b) is also true.

(c) $\lim_{x\to\infty} x\sin(e^{-x}) f(x)$

$\lim_{x\to\infty} xe^{-x} \frac{\sin(e^{-x})}{e^{-x}} f(x)$

$= 0\times 1\times M = 0$

Hence, option (c) is false.

(d) $\lim_{x\to\infty} \frac{\sin x}{x} f(x) = 0\times M = 0$

Option (d) is also true.

10. *(b)* We have,

$p_t(x) = (\sin t)x^2 - (2\cos t)x + \sin t$

$A(t) = \int_0^1 p_t(x)\,dx$

$A(t) = \int_0^1 ((\sin t)x^2 - (2\cos t)x + \sin t)\,dx$

$A(t) = \left[\frac{x^3}{3}\sin t - x^2\cos t + x\sin t\right]_0^1$

$A(t) = \frac{1}{3}\sin t - \cos t + \sin t$

$A(t) = \frac{4}{3}\sin t - \cos t$

$A'(t) = \frac{4}{3}\cos t + \sin t$

$A(t)$ has infinitely critical points.

$A(t) = 0$ for infinitely many value of t.

Hence, II and III statements are true.

11. *(a)* We have, $f(x) = \sqrt{2 - x - x^2}$

and $g(x) = \cos x$

$(g(x))^2 = \cos^2 x$

$f((g(x))^2) = \sqrt{2 - (g(x))^2 - (g(x))^4}$

$= \sqrt{2 - \cos^2 x - \cos^4 x}$

$f((g(x))^2)$ is defined if

$2 - \cos^2 x - \cos^4 x \ge 0$

$\Rightarrow \quad \cos^4 x + \cos^2 x - 2 \le 0$

$\Rightarrow \cos^4 x + 2\cos^2 x - \cos^2 x - 2 \le 0$

$\Rightarrow \quad (\cos^2 x + 2)(\cos^2 x - 1) \le 0$

$\Rightarrow \quad \cos^2 x \in [-2, 1]$

$\therefore \quad x \in R$

$f(g(x)) = \sqrt{2 - \cos x - \cos^2 x}$

Similarly, for domain of $f(g(x))$ is $x \in R$

Hence, domain of $f((g(x)^2)) =$ Domain of $f(g(x))$.

12. *(a)* We have,

$$f(x) = \int_{-10}^{x} 2^{[t]}\,dt, x \in [-10, 10]$$

Let r be an integer $r \in [-10, 10]$

For LHL $\lim_{x\to r^-} \int_{-10}^{x} 2^{[t]}\,dt$

$\lim_{x\to r-h} \int_{-10}^{r-h} 2^{[t]}\,dt$

$\lim_{h\to 0}\left[\int_{-10}^{-9} 2^{[t]}\,dt + \int_{-9}^{-8} 2^{[t]}\,dt + \ldots + \int_{r-1}^{r-h} 2^{[t]}\,dt\right]$

$\lim_{h\to 0}[2^{-10} + 2^{-9} + 2^{-8} + \ldots + 2^{r-1}(1-h)]$

$2^{-10} + 2^{-9} + 2^{-8} + \ldots + 2^{r-1}$ ...(i)

RHL $\lim_{x\to r^+} \int_{-10}^{x} 2^{[t]}\,dt \Rightarrow \lim_{x\to r+h} \int_{-10}^{r+h} 2^{[t]}\,dt$

$\lim_{h\to 0}\left[\int_{-10}^{-9} 2^{[t]}\,dt + \int_{-9}^{-8} 2^{[t]}\,dt + \ldots + \int_{r}^{r+h} 2^{[t]}\,dt\right]$

$= 2^{-10} + 2^{-9} + 2^{-8} + \ldots + 2^{r-1}$ ...(ii)

$f(r) = \int_{-10}^{r} 2^{[t]}dt = 2^{-10} + 2^{-9} + \ldots + 2^{r-1}$ ...(iii)

$\lim_{x\to r^-} f(x) = \lim_{x\to r^+} f(x) = f(x)$

Hence, $f(x)$ is continuous for all $r \in (-10, 10)$.

13. *(d)* Let $I = \int_1^n [x]\{x\}\,dx$

$I = \int_1^2 \{x\}\,dx + \int_2^3 2\{x\}\,dx + \ldots + \int_{n-1}^{n} (n-1)\{x\}\,dx$

$I = (1 + 2 + 3 + 4 + \ldots + (n-1))\int_0^1 \{x\}\,dx$

$I = \frac{n(n-1)}{2}\int_0^1 x\,dx$

$I = \frac{n(n-1)}{2}\left[\frac{x^2}{2}\right]_0^1 = \frac{n(n-1)}{4}$

Given, $I \ge 2013$

$\frac{n(n-1)}{4} \ge 2013$

$\Rightarrow \quad n^2 - n \ge 4052$

$\Rightarrow \quad n^2 - n + \frac{1}{4} \ge 4052 + \frac{1}{4}$

$\Rightarrow \quad \left(n - \frac{1}{2}\right)^2 \ge \frac{32209}{4}$

$\Rightarrow \quad n \ge \frac{\sqrt{32209}}{2} + \frac{1}{2}$

$\Rightarrow \quad n \ge \frac{179.46 + 1}{2}$

$= \frac{180.46}{2}$

$\Rightarrow \quad n \ge 90.23$

$\therefore$ Least value of x is 91.

14. *(a)* Given, $y = \cos x$

Equation of line joining $\left(\frac{-\pi}{4}, \cos\left(\frac{-\pi}{4}\right)\right)$ and $(0, 2)$ is

$$y - 2 = \frac{2 - 1/\sqrt{2}}{0 + \pi/4}(x - 0)$$

$\Rightarrow \quad y - 2 = (8 - 2\sqrt{2})x$

$\Rightarrow \quad y = \frac{(8 - 2\sqrt{2})}{\pi}x + 2$

Equation of line joining $\left(\frac{\pi}{4}, \cos\frac{\pi}{4}\right)$ and $(0, 2)$ is

$$y - 2 = \frac{2 - \frac{1}{\sqrt{2}}}{0 - \frac{\pi}{4}}(x - 0)$$

$\Rightarrow \quad y = \left(\frac{-8 + 2\sqrt{2}}{\pi}\right)x + 2$

Graph of given curve, $y = \cos x$, and line are

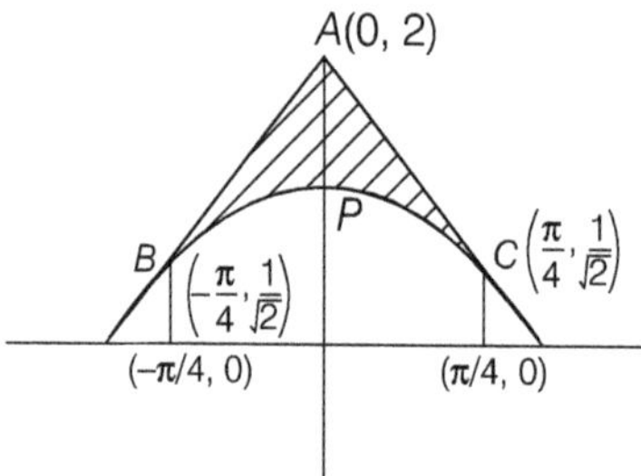

Area of shaded region

$= 2$ area of curve $APCA$

$$= 2\int_0^{\pi/4}\left[\frac{(-8 + 2\sqrt{2})}{\pi}x + 2 - \cos x\right]dx$$

$$= 2\left[\frac{-8x^2}{2\pi} + \frac{2\sqrt{2}x^2}{2\pi} + 2x - \sin x\right]_0^{\pi/4}$$

$$= 2\left[\frac{-4}{\pi}\left(\frac{\pi}{4}\right)^2 + \frac{\sqrt{2}}{\pi}\left(\frac{\pi}{4}\right)^2 + \frac{2\pi}{4} - \frac{1}{\sqrt{2}}\right]$$

$$= 2\left[\frac{\pi}{4}\left(\frac{-4}{4} + \frac{\sqrt{2}}{4} + 2\right) - \frac{1}{\sqrt{2}}\right]$$

$$= \left(\frac{4+\sqrt{2}}{8}\right)\pi - \sqrt{2}$$

15. *(d)* We have,

A box contains coupons $\{1, 2, 3, \ldots, n\}$

Let a number $x = 1$ is selected.

Then, favourable outcomes $(1, 1)\ (1, 2), \ldots,$.

$(1, n) = \left[\frac{n}{1}\right]$

Number of favourable outcomes, when $x = 1,\ y = 1$ is $2\left[\frac{n}{1}\right] - 1$

Number of favourable outcomes, when $x = 2,\ y = 2,\ x \neq 1,\ y \neq 1 = 2\left[\frac{n}{2}\right] - 1$

Similarly, for number of favourable outcomes, when

$x = k,\ y = k$ but $x, y \neq \{1, 2, \ldots, k-1\}$

$$= 2\left[\frac{n}{k}\right] - 1$$

So, required probability

$$= \frac{2\sum_{k=1}^{n}\left[\frac{n}{k}\right] - (1 + 1 + \ldots n \text{ times})}{n^2}$$

$$= \frac{2}{n^2}\sum_{k=1}^{n}\left[\frac{n}{k}\right] - \frac{n}{n^2}$$

$$= \frac{-1}{n} + \frac{2}{n^2}\sum_{k=1}^{n}\left[\frac{n}{k}\right]$$

16. *(a)* We have,

Mean $(\mu) = \dfrac{x_1 + x_2 + x_3 + \ldots + x_n}{n}$

$$\mu = \frac{\Sigma x_i}{n}$$

Standard deviation $\sigma = \sqrt{\dfrac{\Sigma x_i^2}{n} - (\mu)^2}$

Mean of other observations

Mean $(\mu') = \dfrac{y_1 + y_2 + y_3 + \ldots + y_{n-1} + y_n}{n}$

$$= \frac{0 + x_2 + x_3 + \ldots + x_{n-1} + x_1 + x_n}{n}$$

$$= \frac{\Sigma x_i}{n} = \mu$$

$\therefore \quad \mu' = \mu$

$$\sigma' = \sqrt{\frac{\sum y_i^2}{n} - (\mu')^2}$$

$$\sigma' = \sqrt{\frac{0 + x_2^2 + x_3^2 + \ldots + x_{n-1}^2 + (x_1 + x_n)^2}{n} - \mu}$$

$$\Sigma x_i^2 = x_1^2 + x_2^2 + \ldots + x_n^2$$

$$\Sigma y_i^2 = 0 + x_2^2 + \ldots + x_{n-1}^2 + x_1^2 + x_n^2 + 2x_1x_n$$

Clearly, $\Sigma y_i^2 \geq \Sigma x_i^2$

$\therefore \quad \sigma' \geq \sigma$

Hence, option (a) is correct.

17. *(c)* We have, $\mathbf{v}_1, \mathbf{v}_2, \mathbf{v}_3, \mathbf{v}_4$ be unit vectors lie in XY-plane.

Let

$\mathbf{v}_1 = \cos\alpha\mathbf{i} + \sin\alpha\mathbf{j}$	$\alpha \in (0, 90°)$
$\mathbf{v}_2 = \cos\beta\mathbf{i} + \sin\beta\mathbf{j}$	$\beta \in (90°, 180°)$
$\mathbf{v}_3 = \cos\gamma\mathbf{i} + \sin\gamma\mathbf{j}$	$\gamma \in (180°, 270°)$
$\mathbf{v}_1 = \cos\delta\mathbf{i} + \sin\delta\mathbf{j}$	$\delta \in (270°, 360°)$

(a) $\mathbf{v}_1 + \mathbf{v}_2 + \mathbf{v}_3 + \mathbf{v}_4 = 0$ not necessarily true for all $\mathbf{v}_1, \mathbf{v}_2, \mathbf{v}_3$ and $\mathbf{v}_4$

(b) $\mathbf{v}_i + \mathbf{v}_j\ (1 \leq \mathbf{i} < \mathbf{j} \leq 4)$

$\mathbf{v}_1 + \mathbf{v}_2 = (\cos\alpha + \cos\beta)\,\mathbf{i} + (\sin\alpha + \sin\beta)\mathbf{j}$

If y coordinate is positive i.e.

α in 1st quadrant β in 2nd quadrant.

But in this case x-coordinate is not necessarily positive.

(c) $\quad \mathbf{v}_i = \mathbf{v}_j\ (1 \leq \mathbf{i} < \mathbf{j} \leq 4)$

Let α in 1st quadrant, γ in 3rd quadrant

$\therefore \quad \mathbf{v}_i \cdot \mathbf{v}_j < 0$

(d) $\mathbf{v}_i \cdot \mathbf{v}_j = (\cos\alpha\ \cos\beta) + \sin\alpha\ \sin\beta$

$= \cos(\alpha - \beta)$ is positive $0 < |\alpha - \beta| < \dfrac{\pi}{2}$

$= \dfrac{3\pi}{2} < |\alpha - \beta| < 2\pi$

Not positive for all values of $\mathbf{v}_1, \mathbf{v}_2, \mathbf{v}_3, \mathbf{v}_4$.

18. *(a)* We have,

Integer n with $100 \leq n \leq 999$

$\therefore \quad n = 999 - 99 = 900$

n containing three distinct digits are

$9 \times 9 \times 8 = 648$

n containing at most two distinct digit is

Total number $- n$ containing three distinct digits $= 900 - 648 = 252$

19. *(d)* We have,

$S_n = \{n + 1, n + 2, n + 3, \ldots, n + 18\},$

$n \geq 10$

(a) $n = 19$, then S_n has not a multiple of 19.

Hence, option (a) is false.

(b) S_n has more than one prime for $n \geq 10$.

(c) $n = 10$

11, 28

Only 15, 20, 25 are multiple of 5.

(d) Number of odd integer in $S_n = 9$

So, every third integer is multiple of 3.

$\therefore$ S_n has at most six primes.

20. *(c)* We have,

A polygon of 10 sides.

$\therefore$ Sum of total interior angle of 10 sided polygon $= (10 - 2) \times \pi = 8\pi$

Let k be the number of interior angle of P that are greater than 180°.

$\therefore \quad k \times 180°$

$\therefore \quad k\pi + x = 8\pi$

$\therefore$ Maximum value of k is 7, x is the sum of three angles can be less than π.

21. *(d)* When a charge $+Q$ is placed inside the shell, it induces a charge $-Q$ on the inner surface of shell and a charge of $+Q$ appears on outer surface of shell.

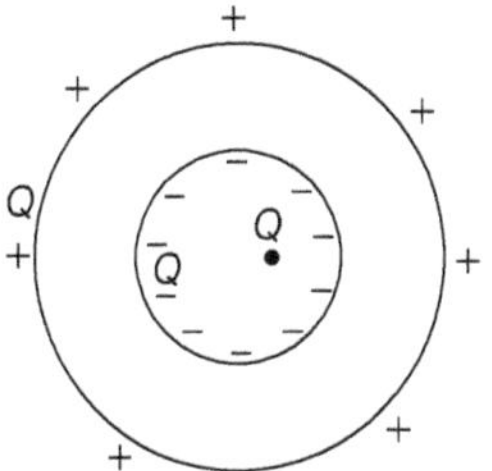

When outer surface is earthed, charge on outer surface is neutralised.

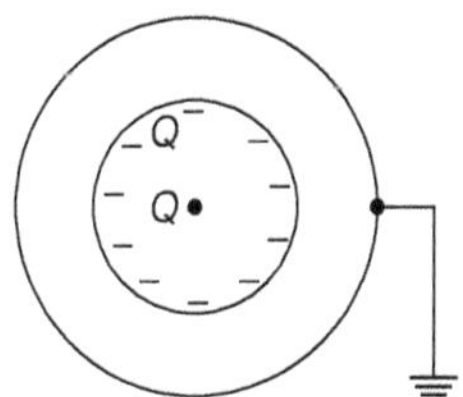

As, now net charge inside the shell is zero.

Hence, net field outside the shell is zero.

$\therefore$ Force on charge at an external point P is zero.

22. *(d)* When capacitor is fully charged, it acts like an open switch or path. So, we have following equivalent circuit

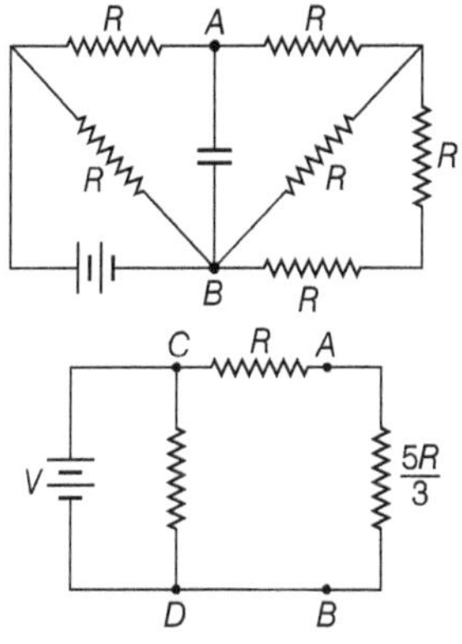

Clearly, $\quad V_{CD} = V$

So, $V_{AB} = I_{AB} \cdot R_{AB}$

$$= \frac{V}{\left(R + \frac{5}{3}R\right)} \cdot \left(\frac{5}{3}R\right) = \frac{5}{8}V$$

Charge stored on the capacitor is

$$Q = C_{AB} \cdot V_{AB} = \frac{5}{8}CV$$

23. *(b)* For a spontaneous reaction to occur is that Q value of reaction must be positive.

This happens when mass of reactants is greater than mass of products.

i.e. $\underbrace{M(A, Z)}_{\text{Mass of reactants}} > \underbrace{M(A, Z+1)}_{\text{Mass of products}}$

$\therefore$ Mass of β and neutrino is neglected, as these are very small.

24. *(a)* Given a van der Waal's gas with state equation, $\left(p + \frac{n^2 a}{V^2}\right)(V - nb) = nRT$

For adiabatic process, $dW = 0$

So, from first law of thermodynamics, we have

$$dU = dQ + dW$$

$$\Rightarrow \quad dU - dW = 0$$

$$\Rightarrow \quad dU + pdV = 0$$

$$\Rightarrow \quad \left(\frac{\partial U}{\partial T}\right)_V dT + \left(\frac{\partial U}{\partial V}\right)_T dV + \left(p + \frac{n^2 a}{V^2}\right) dV = 0$$

$$\Rightarrow \quad C_V dT + \left(\frac{nkT}{V - nb}\right) dV = 0$$

$$\Rightarrow \quad C_V \frac{dT}{T} + \left(\frac{nk}{V - nb}\right) dV = 0$$

Integrating above equation, we get

$$\Rightarrow \quad C_V \int \frac{dT}{T} + nk \int \frac{dV}{V - nb} = \text{constant}$$

$$\Rightarrow \quad \log T^{C_V} + \log (V - nb)^{nk} = C$$

$$\Rightarrow \quad (V - nb)^R \cdot T^{C_V} = e^C$$

$$\Rightarrow \quad (V - nb)^{R/C_V} \cdot T = e^{C/C_V} = \text{constant}$$

$\therefore$ Equation for quasistatic adiabatic process is $(V - nb)^{R/C_V} \cdot T = \text{constant}$.

25. *(c)* As current increases exponentially and reaches its maximum value and after that it is constant. So, box contains a resistor and an inductor in series combination.

26. *(b)* For a photoelectron,

$$eV_0 = h\nu - \phi_0 \Rightarrow V_0 = \frac{h}{e} \cdot \nu - \frac{\phi_0}{e}$$

So, in a V_0 *versus* ν graph, slope $= \frac{h}{e}$ and intercept $= \frac{\phi_0}{e}$.

Now from given graph,

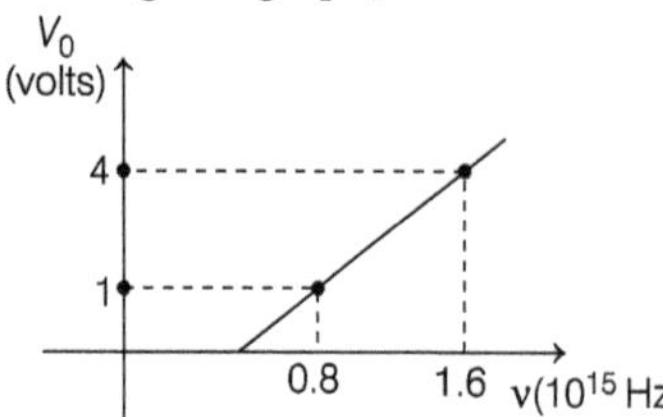

$$\text{Slope} = \frac{h}{e} = \frac{V_{0_2} - V_{0_1}}{\nu_2 - \nu_1} = \frac{4 - 1}{(1.6 - 0.8) \times 10^{15}}$$

$$\Rightarrow \quad h = \left(\frac{3 \times 1.6 \times 10^{-19}}{0.8 \times 10^{15}}\right) = 6 \times 10^{-34} \text{ Js}$$

Also, from graph intercept is -2 V.

Hence, work function is $\phi = 2$ eV.

27. *(a)* Frequency of sound received at cliff is let f_1.

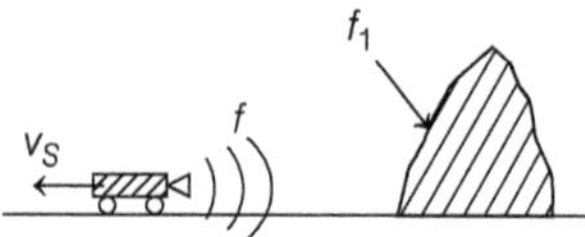

By Doppler's formula, we have

$$f_1 = \left(\frac{v}{v - v_s}\right) \cdot f$$

$$= \frac{v}{v - 0.5v} f = 2f \quad [\because v_s = 0.5\,v]$$

Now, reflected sound is observed on a moving engine.

Let observed frequency is f_2, then

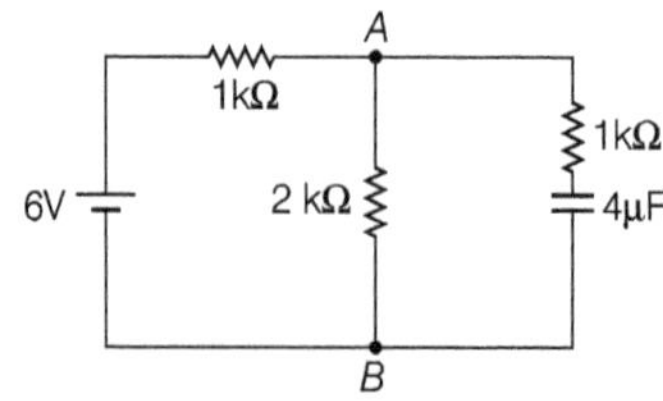

$$f_2 = \left(\frac{v - v_0}{v}\right) f_1 = \frac{1}{2} f_1 = \frac{1}{2} \times 2f$$

$$\therefore \quad f_2 \approx 0.990\,f$$

28. *(b)* Magnetic field at C is resultant field due to arcs of radius R and r.

$$\therefore \quad B = \frac{1}{4}\left(\frac{\mu_0 I}{2r}\right) \odot - \frac{1}{4}\left(\frac{\mu_0 I}{2R}\right) \otimes$$

$$= \frac{1}{8} \cdot \mu_0 I \left(\frac{1}{r} - \frac{1}{R}\right) \odot$$

Net field is out of the plane of paper.

29. *(d)* After a long time capacitor is charged and it acts like an open circuit.

So, equivalent circuit is

So, current in circuit is

$$I = \frac{6}{(1 + 2) \times 10^3} = 2 \times 10^{-3} \text{ A}$$

Potential drop across 2 kΩ resistor is

$$V_{AB} = I \cdot R_{AB}$$

$$= 2 \times 10^{-3} \times 2 \times 10^3 = 4 \text{ V}$$

Clearly, potential drop across capacitor is also 4 V.

30. *(a)* Net torque is resultant of all three torques over the wheel.

So, $\tau_{net} = \tau_F + \tau_F + \tau_{2F}$ (as all torques produces clockwise rotation).

$$= R \times F + \frac{R}{2} \times F + R \times 2F$$

$$= \left(1 + \frac{1}{2} + 2\right) R \times F$$

$$= \frac{7}{2} RF = 3.5\,RF$$

31. *(b)* Mean life of first species $= \tau_1 = \frac{1}{\lambda}$

Mean life of second species $= \tau_2 = \frac{1}{\lambda/3} = \frac{3}{\lambda}$

So, mean life of sample consisting of equal number of atoms of both species is

$$\tau_{mean} = \frac{\tau_1 + \tau_2}{2}$$

$$= \frac{\frac{1}{\lambda} + \frac{3}{\lambda}}{2} = \frac{2}{\lambda} \approx \frac{2.10}{\lambda}$$

32. *(b)* Adiabatic bulk modulus for an ideal gas is

$$K = \frac{dp}{-dV/V} = -V\left(\frac{dp}{dV}\right)$$

$$= \gamma P \qquad \ldots\text{(i)}$$

Now given,

$$K \propto p^n \qquad \ldots\text{(ii)}$$

$\therefore$ From Eqs. (i) and (ii), we have

$$\Rightarrow \quad n = 1$$

33. *(d)* Given situation is

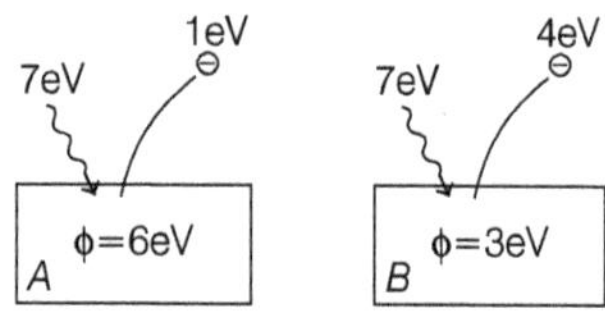

Energy associated with most energised electron is

for metal A, $E_A = 7 - 6 = 1$eV

and for metal B, $E_B = 7 - 3 = 4$ eV

Now, de-Broglie wavelength associated with these electrons is

$$\lambda = \frac{h}{p} = \frac{h}{\sqrt{2mE}}$$

$$\therefore \quad \frac{\lambda_A}{\lambda_B} = \frac{\sqrt{E_B}}{\sqrt{E_A}} = \sqrt{\frac{4}{1}} \Rightarrow \frac{\lambda_A}{\lambda_B} = 2:1 = 2.0$$

34. *(b)* Magnetic force **F** acting on the electron is perpendicular to **v** and **B**.

$$\mathbf{F} = q\,.\,(\mathbf{v} \times \mathbf{B})$$

As **F** is perpendicular to **v** so, $\mathbf{F} \cdot \mathbf{v} = 0$.

Hence, work done by force is zero.

As force is not doing any work so magnitude of velocity or kinetic energy of electron remains constant.

35. *(*)* From geometry of figure,

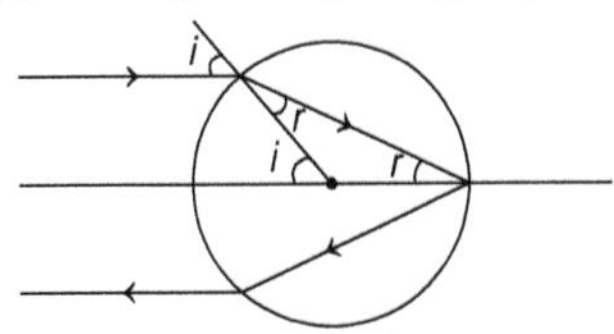

$$\angle i = \angle r + \angle r = 2\angle r$$

So, $$\frac{\sin i}{\sin r} = \sqrt{3}$$

$\Rightarrow$ $$\frac{\sin 2r}{\sin r} = \sqrt{3}$$

$\Rightarrow$ $$\frac{2 \sin r \cdot \cos r}{\sin r} = \sqrt{3}$$

$\Rightarrow$ $$2 \cos r = \sqrt{3} \Rightarrow \cos r = \frac{\sqrt{3}}{2}$$

So, $$r = 30^\circ$$

$\therefore$ Angle of incidence is

$$i = 2\angle r = 2 \times 30^\circ = 60^\circ$$

But for $r = 30^\circ$, TIR is not possible. Hence, value of refractive index is incorrect.

36. *(c)* As EÖTVÖS number E_0 is dimensionless, we check dimensions of options given to the choose correct answer.

Now, $$\left[\frac{\rho g L^2}{\sigma}\right] = \frac{[\rho] \cdot [g] \cdot [L]^2}{[\sigma]}$$

$$= \frac{[\mathrm{ML^{-3}}] \cdot [\mathrm{LT^{-2}}] \cdot [\mathrm{L^2}]}{[\mathrm{MT^{-2}}]}$$

$$= [\mathrm{M^0 L^0 T^0}] = \text{Dimensionless}$$

37. *(c)* As plank is resting, force of friction is just equal to centripetal force due to rotation of earth.

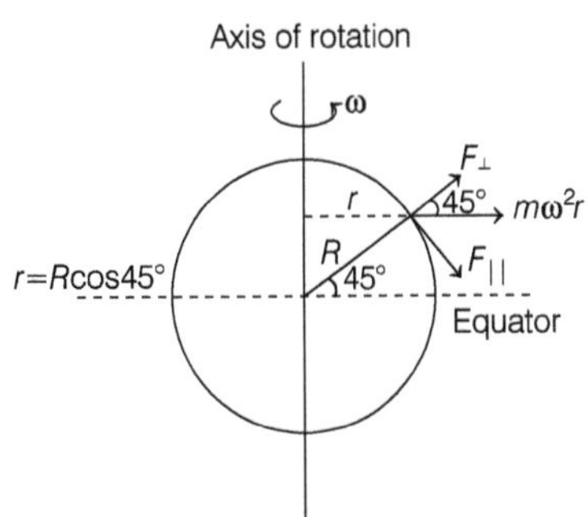

Force trying to slip plank on the ground is parallel component of centripetal force.

$\therefore$ friction, $f = F_{||}$

$\Rightarrow$ $$f = m\omega^2 r_e \sin 45^\circ$$

$$= m\omega^2 r_e \cos 45^\circ \sin 45^\circ = \frac{m\omega^2 r_e}{2}$$

38. *(a)* 1 mole of an ideal gas has $\sim 10^{24}$ molecules and it occupies 22 L or $\sim 22 \times 10^{-3}$ m^3 of space.

So, average spacing between molecules is

nearly $$d = \left(\frac{V}{N}\right)^{\frac{1}{3}} = \left(\frac{22 \times 10^{-3}}{10^{24}}\right)^{\frac{1}{3}}$$

$$\approx 3 \times 10^{-9} \text{ m}$$

or $$d = 3 \text{ nm}$$

So, order will be 1 nm.

39. *(b)*

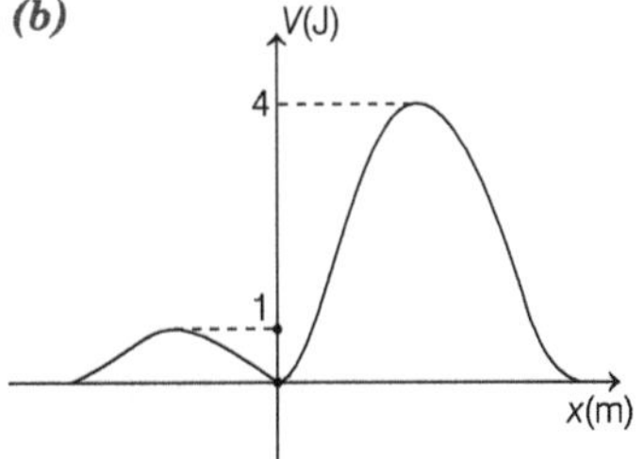

When particle is projected towards right side of origin with velocity v, it must cross a potential barrier of 4 J to escape to infinity.

i.e. $$\frac{1}{2} mv^2 \geq 4 \Rightarrow \frac{1}{2} \times \frac{1}{2} \times v^2 \geq 4$$

or $$v^2 \geq 16 \Rightarrow v \geq 4 \text{ ms}^{-1}$$

But it is possible that kinetic energy of particle is less than 4 J, then it is reflected back towards left with same velocity v. Here, we are taking repulsion from field as elastic collision.

Then particle can escape to infinity from left side of origin when,

$$\frac{1}{2} mv^2 \geq 1 \Rightarrow \frac{1}{2} \times \frac{1}{2} v^2 \geq 1 \text{ or } v \geq 2 \text{ ms}^{-1}$$

Hence, $v_{\min}$ for escape is 2 ms^{-1}.

40. *(d)*

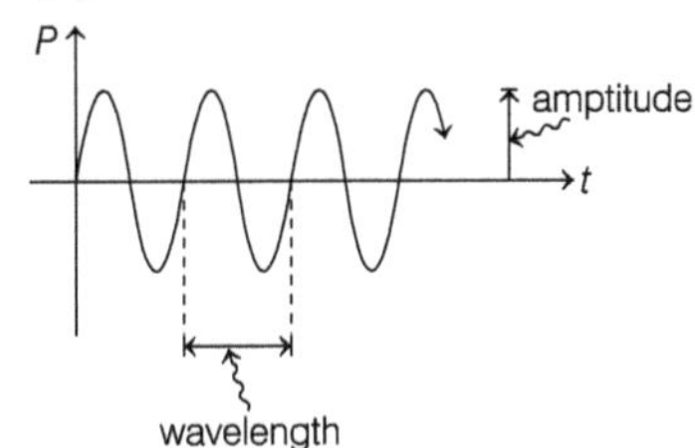

As,

Amplitude $\propto$ Maximum pressure and
Wavelength $\propto$ Time duration for 1 complete variation

So, wave 1 has shorter wavelength and smaller amplitude compared to wave 2.

41. *(c)* Isoelectronic species are those species which have same number of electrons.

Number of electrons in $Na^+ = 10$

$Mg^{2+} = 10$

$F^- = 10$

$Cl^- = 18$

$O^{2-} = 10$

$S^{2-} = 18$

Thus, Na^+, Mg^{2+}, F^- and O^{2-} are isoelectronic with each other. So,option (c) is correct.

42. *(d)* For a zero order reaction,

$$[R] = -kt + [R]_0$$

Comparing with equation of a straight line $y = mx + c$.

The slope of plot of reactant concentration against time is $-k$.

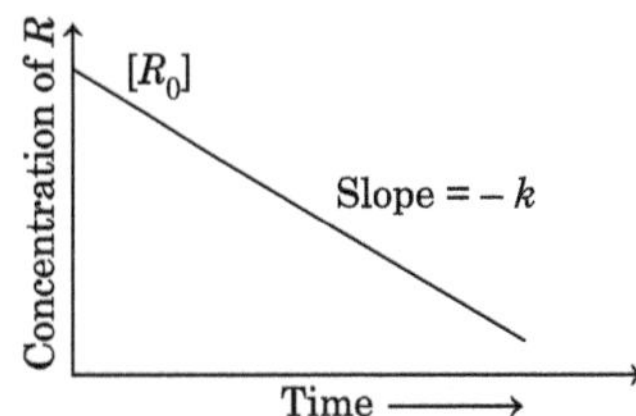

43. *(c)* Salicyclic acid reacts with excess bromine to produce 2, 4, 6-tribromo phenol.

OH, COOH (salicylic acid) $+ Br_2 + H_2O \longrightarrow$ OH, Br, Br, Br (2,4,6-tribromophenol)

Firstly bromine gets substituted in place of COOH group. Phenol being highly activating group undergo bromination at all three available positions.

44. *(c)* $NH_2NHCONH_2$ being more nucleophilic attack at the electrophilic carbonyl carbon of ethyl acetate group. Thus, ethyl acetate reacts with $NH_2NHCONH_2$ to form $CH_3CONHNHCONH_2$, which is a peptide linkage.

$$CH_3\underset{\underset{O}{||}}{C}OC_2H_5 + H_2NNH\underset{\underset{O}{||}}{C}NH_2 \longrightarrow$$

$$CH_3-\underset{\underset{O}{||}}{C}-NHNH\,\underset{\underset{O}{||}}{C}-NH_2$$

45. *(d)* According to Henry's law,

$$p = K_H \chi$$

where, K_H is Henry's constant

$$\Rightarrow \quad \chi = \frac{p}{K_H}$$

Thus, higher the value of solubility at given pressure, lower will be the value of Henry's law constant of the gas. As the solubility of G_1 is least, thus it will have the highest value of Henry's law constant.

46. *(a)* For the reaction, $A \rightleftharpoons nB$

	A	nB
$t = 0$	0.06	0
$t = t$	0.03	0.06

$$\alpha c = 0.03$$
$$n\,\alpha\,c = 0.06$$
$$n = 2$$

Thus, the reaction becomes $A \rightleftharpoons 2B$

$$k_c = \frac{[B]^2}{[A]} = \frac{(0.06)^2}{0.03} = 0.12$$

47. *(b)* As α-hydrogen to carbonyl group is more acidic gets attacked by strong OH^- base to form enol which then attack by Cl_2 molecule to give chlorinated product.

$$CH_3CH_2\overset{O}{\overset{\|}{C}}CH_2{-}H \xrightarrow{OH^\ominus} CH_3CH_2\overset{O}{\overset{\|}{C}}\overset{\ominus}{C}H_2$$

$$\longrightarrow CH_3CH_2{-}\underset{O^-}{\underset{|}{C}}{=}CH_2 \xrightarrow{Cl-Cl}$$

$$CH_3CH_2{-}\overset{O}{\overset{\|}{C}}{-}\overset{\ominus}{C}HCl \xleftarrow{OH^\ominus} CH_3CH_2{-}\overset{O}{\overset{\|}{C}}{-}CH_2Cl$$

$$\downarrow$$

$$CH_3CH_2{-}\underset{O^-}{\underset{|}{C}}{=}CHCl \xrightarrow{Cl-Cl} CH_3CH_2{-}\overset{O}{\overset{\|}{C}}{-}CHCl_2$$

$$\downarrow OH^-/Cl_2$$

$$CH_3CH_2{-}\overset{O}{\overset{\|}{C}}{-}CCl_3$$

48. *(a)* Aceto acetic ester has active methylene $[\overset{O}{\overset{\|}{C}}{-}CH_2{-}\overset{O}{\overset{\|}{C}}]$ group, which readily tautomerises to give enol.

$$CH_3{-}\overset{O}{\overset{\|}{C}}{-}\overset{H}{\overset{|}{C}}{-}\overset{O}{\overset{\|}{C}}{-}OC_2H_5 \rightleftharpoons$$

$$CH_3{-}\underset{OH}{\underset{|}{C}}{=}CH{-}\underset{O}{\underset{\|}{C}}{-}OC_2H_5$$

49. *(b)* Hydrolysis of BCl_3 gives boric acid (X), which on treatment with sodium carbonate produces sodium metaborate (Y), carbon dioxide and water.

$$BCl_3 + 3H_2O \longrightarrow \underset{X}{H_3BO_3} + 3HCl$$

$$4H_3BO_3 + Na_2CO_3 \longrightarrow \underset{Y}{Na_2B_4O_7} + CO_2 + 6H_2O$$

50. *(c)* As Xe has eight valence electrons. In XeF_2, Xe forms 2 bonds with fluorine thus 6 electrons remains unbonded. So, it has 3 lone pairs. While in XeF_4, Xe form 4 bonds with fluorine, thus 4 electrons remain unbonded. Thus, it has 2 lone pairs.

51. *(c)* For isothermal reversible expansion

$$\Delta S = nR\ln\frac{V_2}{V_1} \text{ or } 2.303nR\log\frac{V_2}{V_1}$$

Given, $n = 2, V_2 = 100, V_1 = 10$

$$= 2.303 \times 2 \times 8.314 \times \log\frac{100}{10}$$
$$= 2.303 \times 2 \times 8.314 \times 1$$
$$= 38.3 \text{ J/K}$$

52. *(a)* D-glucose upon treatment with bromine water gives gluconic acid.

D-glucose		Gluconic acid
CHO		COOH
H—C—OH		H—C—OH
HO—C—H	$\xrightarrow{Br_2, H_2O}$	HO—C—H
H—C—OH		H—C—OH
H—C—OH		H—C—OH
CH_2OH		CH_2OH

53. *(a)* The structure of borax $H_4B_4O_9$ is shown below

[Structure: ring of four boron atoms — two tetrahedral $B^\ominus$ (top and bottom, each bearing OH) and two trigonal B (left and right, each bearing OH: HO—B and B—OH), linked through O bridges, with a central O bridging the two $B^\ominus$ atoms.]

Thus, number of boron atoms are 4 and B—O—B units are 5.

54. *(a)* A peptide bond is an amide bond formed between two molecules, when the carboxyl group of one molecule reacts with the amino group of the other molecule.

[Structure: $H_3C{-}\overset{O}{\overset{\|}{C}}{-}\underset{H}{N}{-}CH(CH_3){-}\overset{O}{\overset{\|}{C}}{-}\underset{H}{N}{-}CH(CH(CH_3)_2){-}\overset{O}{\overset{\|}{C}}{-}NHNH_2$, with the C(=O)—NH link circled and labelled "Peptide bond"]

Thus, the given structure has only one peptide bond.

55. *(c)* $\Delta U = nC_VRT$

For isothermal reversible expansion

$$\Delta T = 0$$
$$\therefore \quad \Delta U = 0$$
$$\Delta H = \Delta U + nR\Delta T$$
$$\text{As,} \quad \Delta T = 0$$
$$\text{Also,} \quad \Delta U = 0$$
$$\therefore \quad \Delta H = 0$$

56. *(c)* According to Bragg's equation

$$n\lambda = 2d\sin\theta \quad \text{...(i)}$$

In case (1),

Given, $\lambda = 3\text{Å}, \theta = 30°, n = 2$

For (100) plane $d_{hkl} = \dfrac{a}{\sqrt{(1)^2 + (0)^2 + (0)^2}}$

$$\Rightarrow \quad d = \frac{6}{\sqrt{1}}$$

Substituting the above values in Eq. (i)

$$2 \times 3 = 2 \times \frac{6}{\sqrt{1}}\sin 30 \quad \text{....(ii)}$$

In case (2),

Given, $\lambda = 3\text{Å}, n = 1, d = 6$

For (200) plane

$$d_{hkl} = \frac{a}{\sqrt{(2)^2 + (0)^2 + (0)^2}}$$

$$\Rightarrow \quad d = \frac{6}{2} = 3$$

Substituting the above values in (1)

$$1 \times 3 = 2 \times 3\sin\theta$$
$$\theta = 30°$$

57. *(b)* The structure of complex having empirical formula

$CoCl_3.4NH_3$ can $[Co(NH_3)_4Cl_2]$ Cl.

$$[Co(NH_3)_4Cl_2]Cl \longrightarrow \underbrace{[Co(NH_3)_4Cl_2]^+ + Cl^-}_{\text{(2 ions)}}$$

Thus, the number of ions produced in water of the complex having empirical formula $CoCl_3 \cdot 4NH_3$ is 2.

58. *(c)* The magnetic moment of coordination compound is gives as

$$\mu = \sqrt{n(n+2)}$$

In case of $[Fe(NH_3)_6]^{3+}$, Fe is in +3 oxidation state thus the electronic configuration of Fe^{3+} is $3d^5\,4s^0\,(t_{2g}^3\,e_g^2)$, but NH_3 is a strong ligand pairing of electrons occur and thus configuration becomes $t_{2g}^5\,e_g^0$

$$\Rightarrow \quad n = 1$$

$$\mu = \sqrt{1(1+2)} = \sqrt{3} = 1.73 \text{ BM}$$

In case of $[FeF_6]^{3-}$, Fe is in +3 oxidation state, as Fe is a weak ligand, pairing will not occur thus its configuration will be $3d^5\,4s^0(t_{2g}^3\,e_g^0) \Rightarrow n = 5$

$$\mu = \sqrt{5(5+2)} = \sqrt{35} = 5.92 \text{ BM}$$

59. *(c)* The S_N1 reaction depends upon the stability of carbocations. Higher is the stability of carbocation more will be the reactivity of compounds.

The carbocation of given compounds are as follows :

$$CH_3-\overset{\overset{O}{||}}{C}-CH_2-Cl \longrightarrow CH_3\overset{\overset{O}{||}}{C}-CH_2^+ + Cl^-$$

$$CH_3-CH_2-CH_2-Cl \longrightarrow CH_3CH_2CH_2^+ + Cl^-$$

$$(CH_3)_3CCl \longrightarrow (CH_3)_3C^+ + Cl^-$$

The order of stability of carbocation is

$$\underset{3^\circ}{(CH_3)_3C^+} > \underset{1^\circ}{CH_3CH_2CH_2^+} > CH_3-\overset{\overset{O}{||}}{C}-\overset{+}{C}H_2$$

[Stability is least due to presence of electron withdrawing group]

Thus, the order of reactivity of given compounds in S_N1 reaction is 3 > 2 > 1.

60. *(b)* The number of atoms present in closed cubic packed arrangement is 4.

∴ Number of Y-atoms = 4

Number of octahedral voids = N

∴ Number of M-atoms $= \frac{1}{2} \times N$

$$\frac{1}{2} \times 4 = 2$$

∴ The formula of ionic compound becomes MY_2.

61. *(a)* Foetal haemoglobin has greater affinity for oxygen as compared to the adult haemoglobin mainly to facilitate more oxygen supply to the developing foetus. Foetal haemoglobin lifespan is 80 days as compared to that of adult haemoglobin is 120 days.

62. *(d)* The nucleolus is a distinct structure in the nucleus of the cell composed of filamentous and granular material. It is the site of synthesis and processing of ribosomal RNA and the assembly of this RNA with ribosomal proteins into ribosomal subunits.

63. *(d)* The melting temperature of DNA is directly proportional to the number of hydrogen bonds in the DNA.

We know, that there are two H-bonds between adenine and thymine and three H-bonds between guanine and cytosine. Thus, the DNA molecule, 'iv', i.e.

GCCGGATCCGGC

CGGCCTAGGCCG has the highest melting temperature.

64. *(a)* If thymine is inserted in between ATG GAA. The new sequence will be ATG TGAA. Since the RNA transcript of new sequence will be

AUG UGAA

Start codon, Stop codon

It is a non-sense mutation. A non-sense mutation is the one in which a sense codon that corresponds to one of the 20 amino acids specified by the genetic code is changed to a chain terminating codon.

65. *(c)* Meiosis is a special type of cell division that reduces the chromosome number by half, creating four haploid cells, each genetically distinct from the parent cell.

The genetic content of cell reduces to half during meiotic prophase-II.

66. *(b)* Western blotting (protein blotting) is a rapid and sensitive assay for detection and characterisation of proteins. It is based on the principle of immunochromatography where proteins are separated into polyacrylamide gel according to their molecular weight.

67. *(d)* *Schizosaccharomyces pombe*, also called 'fission yeast' is a species of yeast used in traditional brewing and as a model organism in molecular and cell biology. It is a unicellular eukaryote, whose cells are rod-shaped.

68. *(b)* Light reaction takes place in the grana thylakoids of the chloroplast. Dark reaction takes place in the stroma of the chloroplast. In light reaction, H^+ ions are utilised by NADP to form NADPH. In dark reaction, the hydrogen of NADPH is used to combine with CO_2.

69. *(a)* According to Mendel, alleles of a gene segregate and alleles of different genes assort independently. Mendel's law of segregation states that allele pairs separate or segregate during gamete formation. Mendel's law of independent assortment states that the alleles of two (or more) different genes get sorted into gametes independently of one another.

70. *(b)* RuBisCO has both oxygenase and carboxylase activities. Ribulose-1, 5-Bisphosphate Carboxylase/Oxygenase, commonly known by the abbreviations RuBisCO, is an enzyme involved in the first major step of carbon fixation, a process by which atmospheric carbon dioxide is converted by plants and other photosynthetic organisms to energy rich molecules such as glucose.

71. *(c)* Dermal contact with Chlorofluorocarbons (CFCs) can cause skin irritation and dermatitis. CFCs are involved in the destruction of the stratospheric ozone layer resulting in increased exposure to UV-radiation which is known to cause skin cancer.

72. *(c)* Gastrulation occurring during gastrula stage is one of the most important morphogenetic movement in the formation of the basic tubular structure of animals. Morphogenetic movement in the early embryo takes place mainly through the movement and deformation of the epithelial cells, which constitute the embryo and include invagination, extension and locomotive movements.

73. *(c)* The only organ which is capable of producing fructose in human is seminal vesicles. The function of seminal vesicles is to produce and store fluid that will eventually become semen. Fructose is a sugar that provides sperm with energy in males.

74. *(d)* Blood thinners or anticoagulants, decrease the chances of blood clot formation in the heart, thereby reducing the risk of stroke. Stroke is a medical emergency in which damage to the brain occurs due to interruption of its blood supply.

75. *(c)* In a fruit, the endocarp is the inner layer surrounding the seed. In hesperidium type of fruit, the epicarp forms the cover, the mesocarp is fibrous and the endocarp bends inwards giving rise to a chamber from which hair-like structures are formed. These are juicy and form the edible part in orange and lemon. Thus, the correct answer is option (c).

76. *(c)* The enzyme, nitrogenase which is capable of nitrogen reduction is present exclusively in prokaryotes (e.g. *Rhizobium*). It is highly sensitive to O_2 and gets inactivated when exposed to it, thus does not require oxygen (i.e. anaerobic condition) for its functioning.

77. *(c)* Ciliated epithelium is a thin tissue that has hair-like structures on it. These hairs, called cilia, move back and forth to help move particles out of our body.

78. *(b)* Dispersal is the process by which individuals move from the immediate environment of their parents to establish in an area more or less distant from them. Because dispersal can enable escape from low quality environments and access to higher quality resources, many species that specialise in colonising disturbed areas tend to have greater dispersal abilities than species that live in relatively stable habitats.

79. *(c)* The equation for Hardy-Weinberg equilibrium is–$p^2 + 2pq + q^2 = 1$

where p^2 represents the frequency of the homozygous genotype A A, q^2 represents the frequency of the homozygous genotype aa and $2pq$ represents the frequency of the heterozygous genotype A a.

Since $p = 0.6$ and $q = 0.4$

$2pq = 2 \times 0.6 \times 0.4 = 0.48$

80. *(d)* Glycogen is a large multi-branched polymer of glucose which is accumulated in response to insulin and broken down into glucose in response to glycogen. Glycogen is mainly stored in the liver and muscles and provides the body with a readily available source of energy if blood glucose levels decrease.

81. *(c)* We have, $f\left(\frac{1}{2}\right) = 100$

$$f(x) \le 100, \forall x \in R$$

$$\therefore \quad f(x) = a\left(x - \frac{1}{2}\right)$$

$[a_0x^{n-1} + a_1 x^{n-2} + ... + a_{n-1}] + 100$

If $f(x) \le 100, \forall x \in R$

$\therefore a < 0$ and $f(x)$ must be even degree polynomial.

Since, there may be more value of x at which $f(x)$ attains maximum.

$\therefore$ If $x \ne \frac{1}{2}$, then $f(x) < 100$ may be false.

82. *(a)* We have,

$$\sum_{k=1}^{n} (ak^3 + bk^2 + ck + d) = n^4, n \in N$$

$$a\left(\frac{n(n+1)}{2}\right)^2 + b\,\frac{n(n+1)(2n+1)}{6} + c\,\frac{n(n+1)}{2} + dn = n^4$$

$$\Rightarrow \frac{a}{4}n^4 + \left(\frac{2a}{4} + \frac{2b}{6}\right)n^3 + \left(\frac{a}{4} + \frac{3b}{6} + \frac{c}{2}\right)n^2 + \left(\frac{b}{6} + \frac{c}{2} + d\right)n = n^4$$

$$\therefore \frac{a}{4} = 1, \frac{2a}{4} + \frac{2b}{6} = 0, \frac{a}{4} + \frac{3b}{6} + \frac{c}{2} = 0, \frac{b}{6} + \frac{c}{2} + d = 0$$

On solving these equations, we get

$$a = 4, b = -6, c = 4, d = -1$$

$\therefore |a| + |b| + |c| + |d| = 4 + 6 + 4 + 1 = 15$

83. *(c)* We have,

Equation of parabola

$$y^2 = 4x \quad ...(i)$$

Equation of line $y = 2x - 4$ ...(ii)

On solving Eqs. (i) and (ii), we get (1, – 2) and (4, 4)

Now, ABC is an isosceles triangle

$AB = AC$

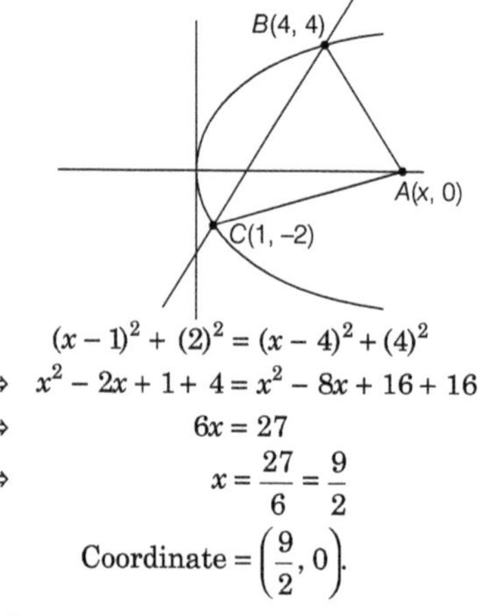

$$\therefore \quad (x-1)^2 + (2)^2 = (x-4)^2 + (4)^2$$

$$\Rightarrow x^2 - 2x + 1 + 4 = x^2 - 8x + 16 + 16$$

$$\Rightarrow \quad 6x = 27$$

$$\Rightarrow \quad x = \frac{27}{6} = \frac{9}{2}$$

$$\therefore \quad \text{Coordinate} = \left(\frac{9}{2}, 0\right).$$

84. *(c)* We have,

G is the centroid of ΔABC, M and N are interior point on sides AB and AC respectively.

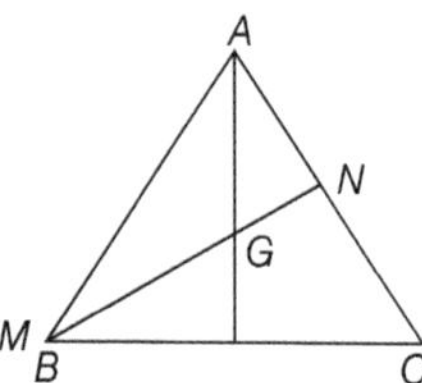

MGN are collinear.

If M and B are coincide.

$\therefore MGN$ are median of ΔABC.

$\therefore$ Area of $\Delta AMN = \frac{1}{2}$ area of ΔABC

Given, $\frac{\text{area of } \Delta AMN}{\text{area of } \Delta ABC} = r$

$$\therefore \quad r_{\max} = \frac{1}{2} \quad ...(i)$$

***Case* I** When MGN are parallel to BC

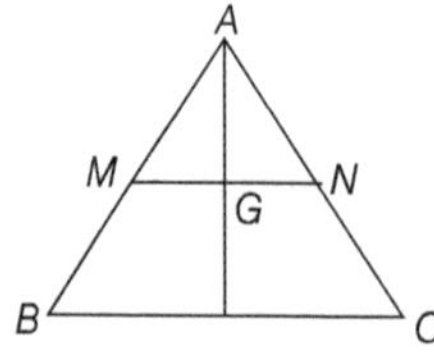

$$AG : AD = 2 : 3$$

$$\Delta AMN \sim \Delta ABC$$

$$\frac{ar(\Delta AMN)}{ar(\Delta ABC)} = \left(\frac{2}{3}\right)^2 = \frac{4}{9}$$

$$r \ge \frac{4}{9} \quad ...(ii)$$

From Eqs. (i) and (ii), we get

$$\frac{4}{9} \le r < \frac{1}{2}$$

85. *(a)* XY be the diameter of a semi-circle with centre O. A, B point on a semi-circle such that AB is parallel to XY.

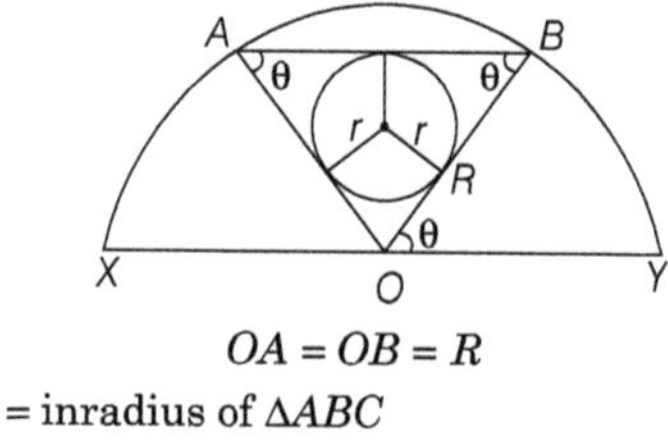

$$OA = OB = R$$

r = inradius of ΔABC

$$r = \frac{\text{Area of } \Delta ABC}{\frac{OA + OB + AB}{2}}$$

$$r = \frac{\frac{1}{2}R^2 \sin(\pi - 2\theta)}{\frac{2R + 2R\cos\theta}{2}}$$

$[\because AB = 2R\cos\theta]$

$$r = \frac{R^2 \sin 2\theta}{2R(1 + \cos\theta)} = \frac{R\sin\theta\cos\theta}{1 + \cos\theta}$$

$$\frac{dr}{d\theta} = R\,\frac{(1+\cos\theta)(\cos 2\theta) + \sin^2\theta\cos\theta}{(1+\cos\theta)^2}$$

For maxima or minima $\frac{dr}{d\theta} = 0$

$\therefore (1 + \cos\theta)(\cos^2\theta \sin^2\theta) + \sin^2\theta\cos\theta = 0$

$\Rightarrow \cos^2\theta - \sin^2\theta + \cos^3\theta - \cos\theta\sin^2\theta + \sin^2\theta\cos\theta = 0$

$$\Rightarrow \quad 2\cos^2\theta - 1 + \cos^3\theta = 0$$

$$\Rightarrow \quad \cos^3\theta + 2\cos^2\theta - 1 = 0$$

$$\Rightarrow (\cos\theta + 1)(\cos^2\theta + \cos\theta - 1) = 0$$

$$\Rightarrow \cos\theta = -1 \text{ or } \cos^2\theta + \cos\theta - 1 = 0$$

$$\cos\theta = -\frac{1 \pm \sqrt{5}}{2}$$

r is maximum, when $\cos\theta = \frac{\sqrt{5} - 1}{2}$

$$\therefore \quad \angle BOY = \cos^{-1}\left(\frac{\sqrt{5} - 1}{2}\right)$$

86. *(a)* We have,

$$f(x) = 1 + \frac{x}{1!} + \frac{x^2}{2!} + \frac{x^3}{3!} + \frac{x^4}{4!}$$

$$f(x) = 1 + x + \frac{x^2}{2} + \frac{x^3}{6} + \frac{x^4}{24}$$

$$f'(x) = 1 + x + \frac{x^2}{2} + \frac{x^3}{6}$$

$$f''(x) = 1 + x + \frac{x^2}{2}$$

$$f''(x) > 0, \forall x \in R$$

$\therefore f'(x)$ is increasing.

Let $f'(x) = 0$ at $x = x_0$

$$\therefore \quad 1 + x_0 + \frac{x_0^2}{2} + \frac{x_0^3}{6} = 0$$

$$\Rightarrow \quad f'(-2) = 1 - 2 + 2 - \frac{8}{6} < 0$$

$$\Rightarrow \quad f'(-1) = 1 - 1 + \frac{1}{2} - \frac{1}{6} > 0$$

$$\Rightarrow \quad f'(-2)' f(-1) < 0$$

$$\Rightarrow \quad x_0 \in (-2, -1)$$

$$f(x_0) = 1 + \frac{x_0}{1!} + \frac{x_0^2}{2!} + \frac{x_0^3}{3!} + \frac{x_0^4}{4!} > 0$$

$\therefore f(x)$ has no solution.

87. *(d)* Radius of earth = 6400 km

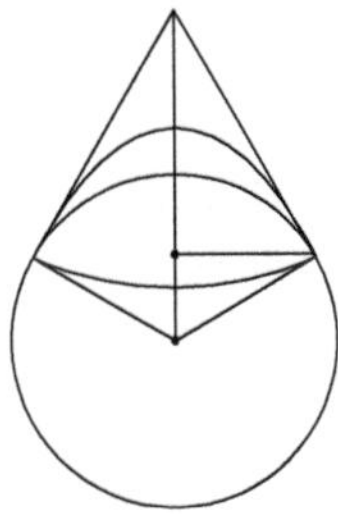

For a cone solid angles.

Total solid angle full round = 4π

for $\pi r^2 \to \pi$

[$\because$ one-fourth of area is covered]

$$w = \pi = \text{ solid angle}$$

$$w = 2\pi (1 - \cos\theta)$$

$$\pi = 2\pi (1 - \cos\theta)$$

$$\cos\theta = \frac{1}{2}$$

$$\theta = 60^\circ$$

In ΔOPB,

$$\cos 60^\circ = \frac{OB}{OP}$$

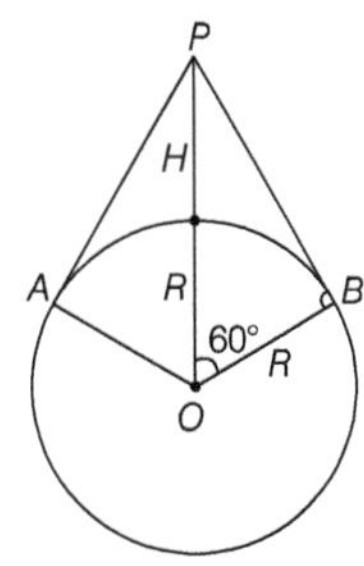

$$\frac{1}{2} = \frac{R}{OP}$$

$$2R = OP$$

$$2R = H + R \qquad [\because OP = R + H]$$

$$R = H$$

$$\therefore \quad H = 6400 \text{ km}$$

88. *(c)* Let

$$I = \int_1^{n+1} \frac{(\{x\})^{[x]}}{[x]} dx$$

$$\Rightarrow \quad I = \int_1^2 \frac{\{x\}}{1} dx + \int_2^3 \frac{\{x\}^2}{2} dx + \int_3^4 \frac{\{x\}^3}{3} dx + \dots + \int_n^{n+1} \frac{\{x\}^n}{n} dx$$

$$\Rightarrow \quad I = \int_0^1 \left(\frac{x}{1} + \frac{x^2}{2} + \frac{x^3}{3} + \dots + \frac{x^n}{n} \right) dx$$

$$\Rightarrow I = \left[\frac{x^2}{1 \times 2} + \frac{x^3}{2 \times 3} + \frac{x^4}{3 \times 4} + \dots + \frac{x^{n+1}}{n(n+1)} \right]_0^1$$

$$\Rightarrow I = \frac{1}{1 \times 2} + \frac{1}{2 \times 3} + \frac{1}{3 \times 4} + \dots + \frac{1}{n(n+1)}$$

$$\Rightarrow I = \left(1 - \frac{1}{2}\right) + \left(\frac{1}{2} - \frac{1}{3}\right) + \left(\frac{1}{3} - \frac{1}{4}\right) + \dots + \left(\frac{1}{n} - \frac{1}{n+1}\right)$$

$$\Rightarrow I = 1 - \frac{1}{n+1} = \frac{n+1-1}{n+1} = \frac{n}{n+1}$$

89. *(c)* We have, 100 coupons labelled 1, 2, 3, ... 100

Five coupons are random selected and arranged.

$\therefore$ Total numbers of outcomes = ${}^{100}C_5 \times 5!$

Five coupons x_1, x_2, x_3, x_4, x_5 are arranged such that $x_1 > x_2 > x_3$ and $x_3 < x_4 < x_5$

Favourable outcomes = ${}^{100}C_5 \times \frac{4!}{2!2!}$

$$\therefore \text{Required probability} = \frac{{}^{100}C_5 \times \frac{4!}{2!2!}}{{}^{100}C_5 \times 5!} = \frac{1}{20}$$

90. *(d)* We have,

Five teams in a tournament and each team plays against every team.

$\therefore$ Total number of one team = 4

Possible number of distribution

T_1	T_2	T_3	T_4	T_5
4	4	2	0	0
4	4	1	1	0
4	3	2	1	0
4	3	3	0	0
4	3	1	1	1
4	2	2	2	0
4	2	2	1	1
3	3	2	1	1
3	2	2	2	1
3	3	3	1	0
3	3	2	2	0
2	2	2	2	2

Clearly, option (d) is correct.

91. *(c)* Let v_0 = linear velocity of centre of mass after collision and ω_0 = angular velocity of rotation of composite body about centre of rotation.

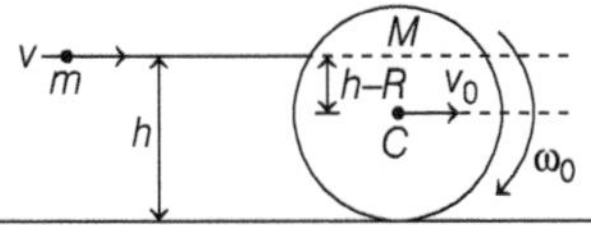

Linear momentum is conserved

$$mv = (m + M) v_0 \qquad \dots(i)$$

Angular momentum about axis of rotation is conserved, $mv (R - h) = I\omega_0$

Substituting for mv from Eq. (i), we get

$$(m + M) v_0 (R - h) = \left(\frac{2}{5} MR^2 + mR^2\right) \omega_0$$

As pure rolling occurs, $v_0 = R\omega_0$

So, $(m + M) \cdot R\omega_0 \cdot (h - R)$

$$= \left(\frac{2}{5} M + m\right) R^2 \omega_0$$

$$\Rightarrow (m + M) R^2 \omega_0 \left(\frac{h}{R} - 1\right) = \left(\frac{2}{5} M + m\right) R^2 \omega_0$$

$$\Rightarrow \frac{h}{R} = \frac{\left(\frac{2}{5} M + m\right)}{(m + M)} + 1 \Rightarrow \frac{h}{R} = \frac{2M + 5m}{5(m + M)} + 1$$

$$\Rightarrow \quad \frac{h}{R} = \frac{7M + 10m}{5(m + M)}$$

92. *(a)* Motion of the ball is as shown below.

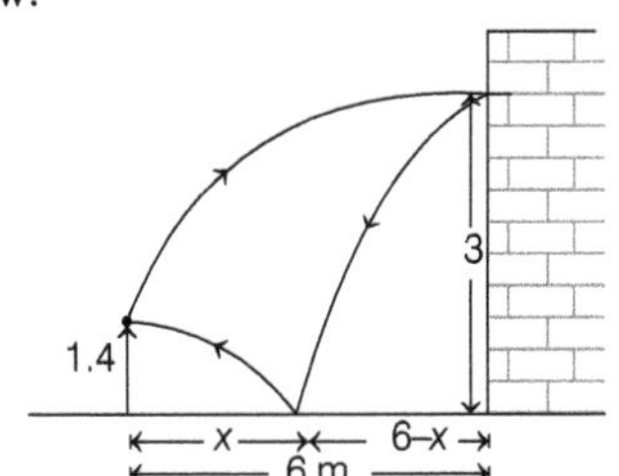

We can combine all three portions one after other to get a perfect parabola obtained by throwing ball from a point on the ground. It is as shown below.

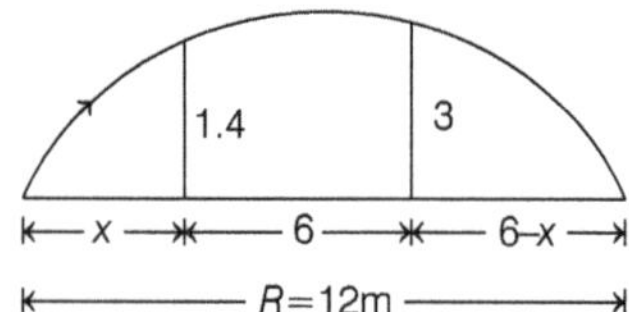

Now, for the standard parabolic motion, equation of trajectory can be written as

$$y = x\tan\theta - \frac{gx^2}{2u^2\cos^2\theta}$$

Substituting $R = \frac{u^2\sin 2\theta}{g}$ in above equation, we have

$$Y = X\tan\theta\cdot\left(1-\frac{X}{R}\right)$$

When $y = 14$ m, $X = x$ and $R = 12$ m, so

We have $14 = x\tan\theta\cdot\left(1-\frac{x}{12}\right)$ ...(i)

Again, when $y = 3$ m and $X = 6 + x$, so

We have

$$3 = (6+x)\tan\theta\cdot\left(1-\frac{6+x}{12}\right) \quad ...(i)$$

Dividing Eq. (i) by Eq. (ii), we get

$$\frac{14}{3} = \frac{x(12-x)}{(6+x)(6-x)}$$

$\Rightarrow 7(36 - x^2) = 15(12x - x^2)$

$\Rightarrow \quad 2x^2 - 45x + 63 = 0$

$\Rightarrow \quad x^2 - \left(\frac{3}{2} + 21\right)x + \left(\frac{3}{2}\times 21\right) = 0$

$\Rightarrow \quad x^2 - \frac{3}{2}x - 21x + \frac{3}{2}\times 21 = 0$

$\Rightarrow \quad x\left(x-\frac{3}{2}\right) - 21\left(x-\frac{3}{2}\right) = 0$

$\Rightarrow \quad \left(x-\frac{3}{2}\right)(x-21) = 0$

So, $\quad x = \frac{3}{2}$ or $x = 21$

As, x must be less than 6m.

So, $x = \frac{3}{2} = 1.5$ m.

93. *(c)*

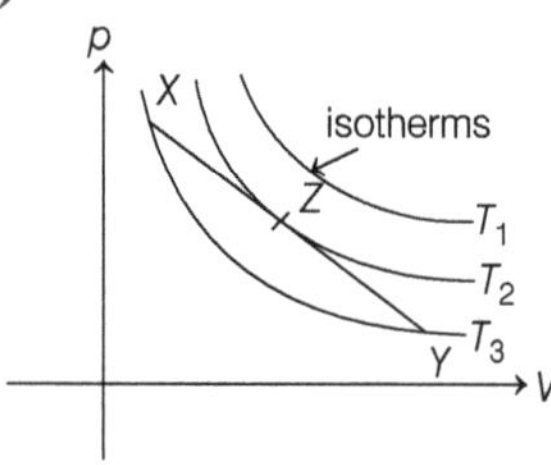

As we move along line XY, it cuts a isotherm at higher temperature $(T_2 > T_3)$ but then from Z to Y line moves to a low temperature isotherm. So, heat is absorbed upto Z and then it is released from Z to Y.

94. *(b)* Energy of a singly ionised helium atom in nth state is $\frac{-13.6\,Z^2}{n^2}$, $Z = 2$

So, energy levels are

–3.4 ↓ ⇝ 2.6 eV — $n = 4$
–6.04 — $n = 3$
–13.6 — $n = 2$
–54.4 eV — $n = 1$

In $n = 4$ state,

Energy, $E_4 = -\frac{13.6\times 4}{16} = -3.4$ eV

When a photon of 2.6 eV is emitted, energy of atom will be

$$E_f = -3.4 - 2.6 = -6 \text{ eV}$$

Clearly, $n = 3$.

95. *(a)*

D_1 D_2 V_i 1V 3V V_o

As input voltage V_i falls from 0 to –4 V, diode D_1 is in reverse bias and it does not conducts.

As input voltage V_i falls from 0 to – 3 V, diode D_2 is in reverse bias and it does not conducts. So, potential drop $V_0 = V_i$.

As input voltage drops below – 3 V, diode D_2 is in forward bias and it conducts and bypass the excess current.

So, $V_0 = -3$ V when $-4 \le V_i \le -3$ V.

As V_i increases from 0 to 1 V, diode D_1 remains in reverse bias and D_2 is also in reverse bias. So, $V_0 = V_i$ as $0 \le V_i \le 1$.

When V_i increases from 0 to 4 V, diode D_1 conducts while diode D_2 is in reverse bias.

So, $\quad V_0 = 1$ V as $V_i > 1$ V

Hence, correct graph is (a).

96. *(d)* Time period of oscillation of ball (radius r) in a bowl (radius R) rolling without slipping is

$$T = 2\pi\sqrt{\frac{R-r}{g}}$$

So, if r increases, then time period is decreases.

97. *(a)* Let initial horizontal speed of sphere is v.

Then, total kinetic energy of sphere on horizontal part

$= (\text{KE})_{\text{translation}} + (\text{KE})_{\text{rotation}}$

$= \frac{1}{2}mv^2 + \frac{1}{2}I\omega^2$

$= \frac{1}{2}mv^2 + \frac{1}{2}\times\frac{2}{5}mR^2\times\frac{v^2}{R^2} = \frac{7}{10}mv^2$

If sphere rises upto height h then, by conservation of energy, we have

$$mgh = \frac{7}{10}mv^2$$

or $\quad v = \sqrt{\frac{10}{7}gh}$

98. *(d)* Work done in complete cycle,

$$W = W_{12} + W_{23} + W_{31} \quad ...(i)$$

p 1 2 3 V

Given $\quad W = 10$ J, $W_{23} = 0$

and $\quad W_{31} = -20$ J

So, from Eq. (i), we have $W_{12} = 30$ J

As in isothermal process,

Heat absorbed = Work done

$\therefore$ Heat absorbed in process $1 \to 2 = 30$ J

99. *(a)*

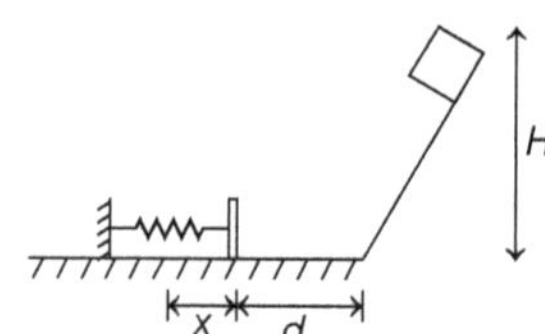

As spring is ideal, it gives energy stored back to the block.

Applying energy conservation, we have

Initial potential energy

= Work done against friction

+ Final potential energy

$\Rightarrow \quad mgH = 2\mu mg(d + x) + mgh$

$\Rightarrow \quad h = H - 2\mu(d + x)$

100. *(d)* Let junction temperature $= T°C$.

As, heat inflow through three prongs = heat outflow through fourth prong.

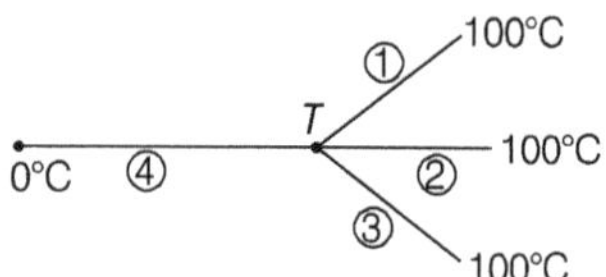

So, $\quad 3\frac{kA}{l}(100 - T) = \frac{kA}{l}(T - 0)$

$\Rightarrow \quad 300 - 3T = T$

or $\quad T = 75°$ C

101. *(a)* The major product obtained in the reaction of aniline with acetic anhydride $[(CH_3CO)_2O]$ is N-phenylethanamide (acetanilide).

$$C_6H_5NH_2 \xrightarrow{(CH_3CO)_2O} C_6H_5NH-CO-CH_3$$

Acetanilide

102. *(d)* When 2-methyl-2 pentene reacts with N-bromosuccinimide (NBS) in boiling CCl_4 the bromination occurs at allylic position and the product obtained is as follows.

$$CH_3-\underset{CH_3}{\underset{|}{C}}=CH-CH_2-CH_3 \xrightarrow{NBS}$$

$$CH_3-\underset{CH_2Br}{\underset{|}{C}}=CH-CH_2CH_3 \quad (I)$$

$$+\ CH_3-\underset{CH_3}{\underset{|}{C}}=CH-\underset{CH_3}{\underset{|}{\overset{Br}{\overset{|}{CH}}}} \quad (II)$$

As the products obtained are alkenes, thus alkene (I) and (II) can have 2-geometrical isomers *E* and *Z* form each.

Thus, maximum number of isomers that can be obtained from this reaction are 4.

103. *(a)* Benzene sulphonyl chloride is also known as Hinsberg's reagent, reacts with 1° and 2° amines to form sulphonamides. The sulphonamide of 1° amine is soluble in alkali whereas that of 2° amine is insoluble in alkali.

Thus, the given compound $X(C_7H_9N)$ is 2° amine, among the options (a)

$$C_6H_5NHCH_3$$

is a 2° amine, which reacts with benzene sulphonyl chloride to give $Y(C_3H_{13}NO_2S)$, which is insoluble in alkali.

$$\underset{(X)}{C_6H_5NHCH_3} + C_6H_5-\overset{O}{\overset{||}{\underset{O}{\underset{||}{S}}}}-Cl \longrightarrow \underset{(Y)}{C_6H_5-\overset{O}{\overset{||}{\underset{O}{\underset{||}{S}}}}-\underset{C_6H_5}{\underset{|}{N}}-CH_3}$$

104. *(d)* The relative lowering of vapour pressure is related to mole fraction by the following formula.

$$\frac{p_0 - p_s}{p_0} = \chi_2$$

$$\frac{p_0 - p_s}{p_0} = \frac{n_2}{n_1 + n_2}$$

As the solution is diluted one

$$n_2 << n_1 \Rightarrow \frac{p_0 - p_s}{p_0} = \frac{n_2}{n_1}$$

For given non-volatile compound

$$n_2 = \frac{18}{M}$$

$$n_1 = \frac{108}{18}$$

$$\therefore \frac{760 - 750}{760} = \frac{18}{M} \times \frac{18}{108}$$

$$\frac{18}{6M} = \frac{10}{760}$$

$$6M = 18 \times 76$$

$$M = 228 \text{ g mol}^{-1}$$

105. *(b)* Given, $E^\circ_{Zn^{2+}/Zn} = -0.76\text{V}$

$$E^\circ_{Cu^{2+}/Cu} = 0.34\text{V}$$

$$E^\circ_{cell} = 0.34 - (-0.76) = 1.1\text{V}$$

$$\Delta G^\circ = -nFE^\circ_{cell} = -2 \times 96500 \times 1.1 = -213\text{V}$$

106. *(b)* If the equation is multiplied by the factor 2, then equilibrium constant will be the square of the equilibrium constant of that equation. If the reaction is reversed, the value of equilibrium constant is inversed.

$$SO_2(g) + \frac{1}{2}O_2(g) \rightleftharpoons SO_3(g) \quad \text{...(i)}$$

$$2SO_3(g) \rightleftharpoons 2SO_2(g) + O_2(g) \quad \text{...(ii)}$$

Eq. (ii) can be obtained by multiplying Eq. (i) by 2 and the reversing it.

Thus, $K_2 = \frac{1}{K_1^2} \Rightarrow K_1^2 = \frac{1}{K_2}$

107. *(c)* Aqueous solution of silver nitrate (*X*) reacts with NH_4OH to form silver oxide (*Y*), which dissolves in excess NH_4OH to give ammoniacal silver nitrate solution (Tollen's reagent). This resulting complex is reduced by acetaldehyde to deposit the metal.

$$\underset{X}{AgNO_3} + NH_4OH \longrightarrow \underset{Y}{Ag_2O} \xrightarrow[\text{Excess}]{NH_4OH} \underset{\text{Tollen's reagent}}{[Ag(NH_3)_2]^+}$$

$$[Ag(NH_3)_2]^+ + RCHO \longrightarrow R\text{—}COOH + 2Ag + 2H_2O + 4NH_3$$

108. *(b)* Given,

density of metal = 10.5g cm^{-3}

Equivalent weight of metal = 100

Current, $I = 3\text{A}$

Area of cross section = 80cm^2, length = $0.005\text{mm} = 5 \times 10^{-4}\text{cm}$

According to Faraday's second law,

$$W = \frac{\text{eq. wt.} \times I \times t}{n \times 96500}$$

$$W = \frac{100 \times 3 \times t}{1 \times 96500} \quad \text{...(i)}$$

Also, W = density × volume

= density × area × length

$$= 10.5 \times 80 \times 5 \times 10^{-4} = 42 \times 10^{-2} \quad \text{...(ii)}$$

Substituting the value of W in Eq. (i),

$$\frac{100 \times 3 \times t}{1 \times 96500} = 42 \times 10^{-2}$$

$$t = \frac{42 \times 10^{-2} \times 96500}{100 \times 3} = 135\text{ s}$$

109. *(a)*

$$I_2 + Na_2S_2O_3 \cdot 5H_2O \rightarrow Na_2S_4O_6 + I^-$$

Eq. wt of $I_2 = 0.25 \times \frac{100}{1000} = 0.025$

Milli Eq. wt of $Na_2S_2O_3 \cdot 5H_2O = 0.25 \times 100$

$$\therefore \text{Eq. wt of } Na_2S_2O_3 \cdot 5H_2O = \frac{0.25 \times 100}{100} = 0.025$$

Mole of $Na_2S_2O_3 \cdot 5H_2O = 0.025 \times 1$

Weight of $Na_2S_2O_3 \cdot 5H_2O$

$= 0.025 \times 248 = 6.2\text{ g}$

110. *(b)* In aqueous solution $[Co(H_2O)_6]^{2+} X$ reacts with molecular oxygen in the presence of excess liquor NH_3 to give a new complex $[Co(NH_3)_6]$

$$\underset{(X)}{[Co(H_2O)_6]_2^+} + 6NH_3 \longrightarrow$$

$$[Co(NH_3)_6]_{2+} \xrightarrow{O_2} \underset{(Y)}{[Co(NH_3)_6]^{3+}}$$

In $[Co(H_2O)_6]^{2+}$, Co is in +2 oxidation H_2O being weak ligand will not pair up the electrons, thus its electronic configuration will be $t_{2g}^5\, e_g^2$. So, it will have three unpaired electrons.

In $[Co(NH_3)_6]^{3+}$ Co is in +3 oxidation, NH_3 being strong ligand will pair up the electrons, thus the electronic configuration will be $t_{2g}^6\, e_g^0$. So, it will have no unpaired electron.

111. *(c)* Penicillin is an antibiotic, i.e. a chemical which stops bacterial growth. Sometimes bacterial colonies develop resistance to certain antibiotics, i.e. bacteria change and become resistant to the antibiotics used to treat the infections they cause. In order to check bacterial colonies which are likely to be resistant to penicillin in the experiment. The bacterial colonies should be taken and then allowed to grow on another plate containing penicillin.

112. *(a)* The Watson and Crick model proposed that DNA is 20 nm or with a 20 Å in diameter. Adjacent bases are separated with a distance of 0.34 nm or 3.4 Å along the axis. The length of a complete turn of helix is 3.4 nm or 34 Å i.e. there are 10 bp per turn. Also that DNA molecule consists of two polynucleotide chain coiled around a common axis to form a right-handed double helix, i.e. B-DNA.

113. *(d)* The frequency by which a restriction enzyme will cut DNA is mainly a function of the length of the sequence, it is sensitive to, like *Rsa* I recognition sequence has 4 base pairs. Out of sheer probability, we see that with 4 positions and each position having potentially 4 different values there are $4^4 = 256$ different possibilities for given 4-base long strand. Therefore, theoretically, this enzyme will cut 1 in 256 4-base pair long sites.

On the other hand, *Eco* RI recognises a sequence of 6 base pairs $4^6 = 4096$ possible combinations with this length.

Total DNA length = 10 kb = 10,000 bp

So, probable cleavage would be

$$Eco\ \text{RI} = \frac{10000}{4096} = 2.44$$

$$Rsa\ \text{I} = \frac{10000}{256} = 39.06$$

114. *(b)* The fate of cells during embryonic life depends on their surrounding cells. This process is called cell differentiation.

Generic embryonic cells become specialised cells by gene expression which regulates cell differentiation. The surrounding cells at the place of transplant have genes that are turned on or off (expressed or repressed) and this is what decides cells fate.

115. *(b)* The plastid in *Plasmodium* is apicoplast. The apicoplast is a non-photosynthetic plasmid that was believed to be endosymbiotic with the cyanobacterium. These organelles are necessary for the survival of the parasite as they are involved in biochemical processes. Thus the correct answer is option 'b'.

116. *(a)* Graph represents competition as when the organism A and B are kept along there biomass is higher to that when they are kept together. Competition is an interaction between organisms or species in which both the organisms or species are harmed.

117. *(c)* When 8kb fragment of a mouse gene is cloned into the *Eco*RI site of a vector of 6 kb size, the total length of DNA becomes 14 kb. When digested with *Eco*RI only two fragments of 8 kb and 6kb are produced, hence two bands are obtained. When digested with *Bam* HI only three fragments of 5.5 kb, 4.5 kb and 4 kb are produced, hence three bands are obtained. When digested with both *Eco* RI and *Bam*HI five fragments of 4 kb, 3.5 kb, 3 kb, 2.5 kb and 1 kb are produced hence five bands are obtained. The 1 kb, 3 kb and 4 kb add upto make 8 kb mouse gene while 3.5 kb and 2.5 kb add up to make 6 kb vector DNA. Hence 1 kb and 3 kb are fragments of mouse gene.

118. *(d)* Brown adipose tissue is a unique mammalian tissue that protects neonatal body temperature around the time of birth by combusting triglyceride in the numerous mitochondria to emit heat. This heat generation is called thermogenesis. It does not produce ATP.

119. *(d)* Altruism in animals describes a range of behaviours performed by animals that may be to their own disadvantage but which benefits others. The costs and benefits are measured in terms of reproductive fitness or expected number of offsprings. Competition among species decrease the number of offsprings. Therefore, altruistic behaviour help species in its survival but the species showing it does not benefit from reduction in competition among offspring.

120. *(b)* There are chances that the lions may get infected in a large interconnected population, i.e. if one get affected the infection spreads in an area easily. This is the reason for movement of Gir lions to other parts of the country.

QUESTION PAPER 2012

Stream : SB/SX

MM : 160

Instructions

1. There are 120 questions in this paper.
2. The question paper contains two parts; Part I (1 Mark Questions) and Part II (2 Marks Questions).
3. There are four sections in each part; Mathematics, Physics, Chemistry and Biology.
4. Out of the four options given with each question, only one is correct.

PART-I (1 Mark Questions)

MATHEMATICS

1. Three children, each accompanied by a guardian, seek admission in a school. The principal wants to interview all the 6 persons one after the other subject to the condition that no child is interviewed before its guardian. In how many ways can this be done?

(a) 60 (b) 90 (c) 120 (d) 180

2. In the real number system, the equation $\sqrt{x+3-4\sqrt{x-1}}+\sqrt{x+8-6\sqrt{x-1}}=1$ has

(a) no solution
(b) exactly two distinct solutions
(c) exactly four distinct solutions
(d) infinitely many solutions

3. The maximum value M of $3^x+5^x-9^x+15^x-25^x$, as x varies over reals, satisfies

(a) $3<M<5$ (b) $0<M<2$
(c) $9<M<25$ (d) $5<M<9$

4. Suppose two perpendicular tangents can be drawn from the origin to the circle $x^2+y^2-6x-2py+17=0$, for some real p. Then, $|p|$ is equal to

(a) 0 (b) 3 (c) 5 (d) 17

5. Let a, b, c, d be numbers in the set $\{1, 2, 3, 4, 5, 6\}$ such that the curves $y=2x^3+ax+b$ and $y=2x^3+cx+d$ have no point in common. The maximum possible value of $(a-c)^2+b-d$ is

(a) 0 (b) 5
(c) 30 (d) 36

6. Consider the conic $ex^2+\pi y^2-2e^2x-2\pi^2y+e^3+\pi^3=\pi e$. Suppose P is any point on the conic and S_1, S_2 are the foci of the conic, then the maximum value of (PS_1+PS_2) is

(a) πe (b) $\sqrt{\pi e}$
(c) $2\sqrt{\pi}$ (d) $2\sqrt{e}$

7. Let $f(x)=\dfrac{\sin(x-a)+\sin(x+a)}{\cos(x-a)-\cos(x+a)}$, then

(a) $f(x+2\pi)=f(x)$ but $f(x+\alpha)\neq f(x)$ for any $0<\alpha<2\pi$
(b) f is a strictly increasing function
(c) f is a strictly decreasing function
(d) f is a constant function

8. The value of $\tan 81°-\tan 63°-\tan 27°+\tan 9°$ is

(a) 0 (b) 2 (c) 3 (d) 4

9. The mid-point of the domain of the function $f(x)=\sqrt{4-\sqrt{2x+5}}$ for real x is

(a) $\frac{1}{4}$ (b) $\frac{3}{2}$ (c) $\frac{2}{3}$ (d) $-\frac{2}{5}$

10. Let n be a natural number and let a be a real number. The number of zeroes of $x^{2n+1}-(2n+1)x+a=0$ in the interval $[-1,1]$ is

(a) 2 if $a>0$
(b) 2 if $a<0$
(c) atmost one for every value of a
(d) atleast three for every value of a

11. Let $f:R\to R$ be the function $f(x)=(x-a_1)(x-a_2)+(x-a_2)(x-a_3)+(x-a_3)(x-a_1)$ with $a_1, a_2, a_3 \in R$. Then, $f(x)\geq 0$ if and only if

(a) at least two of a_1, a_2, a_3 are equal
(b) $a_1=a_2=a_3$
(c) a_1, a_2, a_3 are all distinct
(d) a_1, a_2, a_3 are all positive and distinct

12. The value of $\dfrac{\int_0^{\pi/2}(\sin x)^{\sqrt{2}+1}\,dx}{\int_0^{\pi/2}(\sin x)^{\sqrt{2}-1}\,dx}$ is

(a) $\frac{\sqrt{2}+1}{\sqrt{2}-1}$ (b) $\frac{\sqrt{2}-1}{\sqrt{2}+1}$ (c) $\frac{\sqrt{2}+1}{\sqrt{2}}$ (d) $2-\sqrt{2}$

13. The value of $\int_{-2012}^{2012}(\sin(x^3)+x^5+1)\,dx$ is

(a) 2012 (b) 2013 (c) 0 (d) 4024

14. Let $[x]$ and $\{x\}$ be the integer part and fractional part of a real number x respectively. The value of the integral $\int_0^5 [x]\{x\}\,dx$ is

(a) $\frac{5}{2}$ (b) 5 (c) 34.5 (d) 35.5

15. Let $S_n=\sum_{k=1}^{n} k$, denotes the sum of the first n positive integers. The numbers $S_1, S_2, S_3, \ldots, S_{99}$ are written on 99 cards. The probability of drawing a card with an even number written on it is

(a) $\frac{1}{2}$ (b) $\frac{49}{100}$ (c) $\frac{49}{99}$ (d) $\frac{48}{99}$

16. A purse contains 4 copper coins and 3 silver coins. A second purse contains 6 copper coins and 4 silver coins. A purse is chosen randomly and a coin is taken out of it. What is the probability that it is a copper coin?

(a) $\frac{41}{70}$ (b) $\frac{31}{70}$ (c) $\frac{27}{70}$ (d) $\frac{1}{3}$

17. Let H be the orthocentre of an acute angled ΔABC and O be its circumcenter. Then, **HA**+ **HB**+ **HC**

(a) is equal to **HO**
(b) is equal to 3**HO**
(c) is equal to 2**HO**
(d) is not a scalar multiple of **HO** in general

18. The number of ordered pairs (m, n), where $m, n \in \{1, 2, 3, \ldots, 50\}$, such that 6^m+9^n is a multiple of 5 is

(a) 1250 (b) 2500
(c) 625 (d) 500

19. Suppose $a_1, a_2, a_3, \ldots, a_{2012}$ are integers arranged on a circle. Each number is equal to the average of its two adjacent numbers. If the sum of all even indexed numbers is 3018, what is the sum of all numbers?

(a) 0 (b) 1509
(c) 3018 (d) 6036

20. Let $S=\{1,2,3,\ldots\ldots,n\}$ and $A=\{(a,b)\mid 1\leq a,b\leq n\}=S\times S$. A subset B of A is said to be a good subset if $(x,x)\in B$ for every $x\in S$. Then, the number of good subsets of A is

(a) 1 (b) 2^n
(c) $2^{n(n-1)}$ (d) 2^{n^2}

PHYSICS

21. An ideal monoatomic gas expands to twice its volume. If the process is isothermal, the magnitude of work done by the gas is W_i. If the process is adiabatic, the magnitude of work done by the gas is W_a. Which of the following is true?

(a) $W_i=W_a>0$ (b) $W_i>W_a>0$
(c) $W_i>W_a=0$ (d) $W_a>W_i=0$

22. The capacitor of capacitance C in the circuit shown is fully charged initially resistance is R.

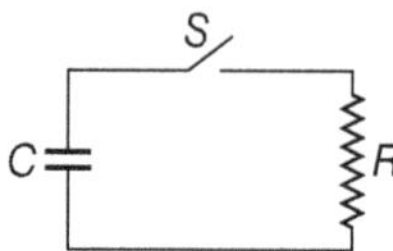

After the switch S is closed, the time taken to reduce the stored energy in the capacitor to half its initial value is

(a) $\frac{RC}{2}$ (b) $RC\ln 2$
(c) $2RC\ln 2$ (d) $\frac{RC\ln 2}{2}$

23. A liquid drop placed on a horizontal plane has a near spherical shape (slightly flattened due to gravity). Let R be the radius of its largest horizontal section. A small disturbance causes the drop to vibrate with frequency ν about its equilibrium shape. By dimensional analysis, the ratio $\dfrac{\nu}{\sqrt{\sigma/\rho R^3}}$ can be (Here, σ is surface tension, ρ is density, g is acceleration due to gravity and k is an arbitrary dimensionless constant)

(a) $k\rho g R^2/\sigma$ (b) $k\rho R^3/g\sigma$
(c) $k\rho R^2/g\sigma$ (d) $k\rho/g\sigma$

24. Seven identical coins are rigidly arranged on a flat table in the pattern shown below, so that each coin touches its neighbours. Each coin is a thin disc of mass m and radius r.

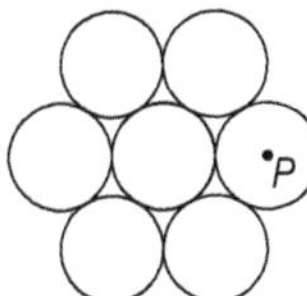

Here note, that the moment of inertia of an individual coin about an axis passing through centre and perpendicular to the plane of the coin is $\frac{mr^2}{2}$.

The moment of inertia of the system of seven coins about an axis that passes through the point P (the centre of the coin positioned directly to the right of the central coin) and perpendicular to the plane of the coins is

(a) $\frac{55}{2}mr^2$ (b) $\frac{127}{2}mr^2$ (c) $\frac{111}{2}mr^2$ (d) $55\,mr^2$

25. A planet orbits in an elliptical path of eccentricity e around a massive star considered fixed at one of the foci. The point in space, where it is closest to the star is denoted by P and the point, where it is farthest is denoted by A. Let v_P and v_A be the respective speeds at P and A, then

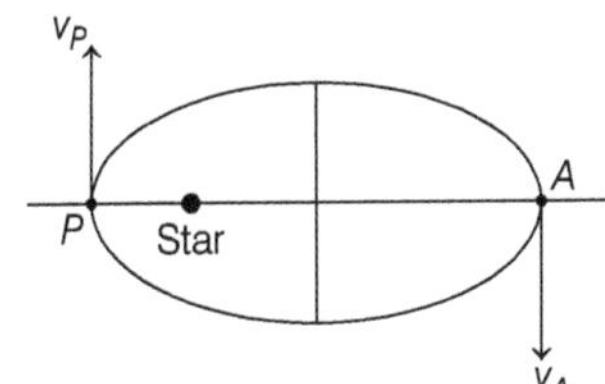

(a) $\frac{v_P}{v_A} = \frac{1+e}{1-e}$ (b) $\frac{v_P}{v_A} = 1$

(c) $\frac{v_P}{v_A} = \frac{1+e^2}{1-e}$ (d) $\frac{v_P}{v_A} = \frac{1+e^2}{1-e^2}$

26. In a Young's double slit experiment, the intensity of light at each slit is I_0. Interference pattern is observed along a direction parallel to the line S_1S_2 on screen S.

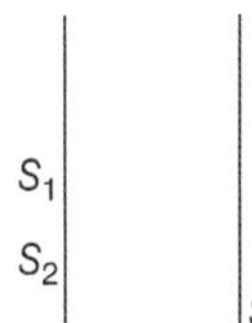

The minimum, maximum and the intensity averaged over the entire screen are respectively.

(a) $0, 4I_0, 2I_0$ (b) $I_0, 2I_0, 3I_0/2$

(c) $0, 4I_0, I_0$ (d) $0, 2I_0, I_0$

27. A loop carrying current I has the shape of a regular polygon on n sides. If R is the distance from the centre to any vertex, then the magnitude of the magnetic induction vector **B** at the centre of the loop is

(a) $n\frac{\mu_0 I}{2\pi R}\tan\frac{\pi}{n}$ (b) $n\frac{\mu_0 I}{2\pi R}\tan\frac{2\pi}{n}$

(c) $\frac{\mu_0 I}{2R}$ (d) $\frac{\mu_0 I}{\pi R}\tan\frac{\pi}{n}$

28. A conducting rod of mass m and length l is free to move without friction on two parallel long conducting rails, as shown below. There is a resistance R across the rails. In the entire space around, there is a uniform magnetic field B normal to the plane of the rod and rails. The rod is given an impulsive velocity v_0.

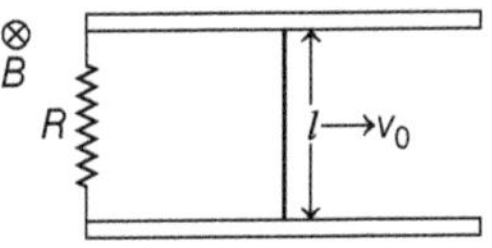

Finally, the initial energy $\frac{1}{2}mv_0^2$

(a) will be converted fully into heat energy in the resistor
(b) will enable rod to continue to move with velocity v_0, since the rails are frictionless
(c) will be converted fully into magnetic energy due to induced current
(d) will be converted into the work done against the magnetic field

29. A steady current I flows through a wire of radius r, length L and resistivity ρ. The current produces heat in the wire. The rate of heat loss in a wire is proportional to its surface area. The steady temperature of the wire is independent of

(a) L (b) r (c) I (d) ρ

30. The ratio of the speed of sound to the average speed of an air molecule at 300 K and 1 atm pressure is close to

(a) 1 (b) $\sqrt{300}$ (c) $\sqrt{1/300}$ (d) 300

31. In one model of the electron, the electron of mass m_e is thought to be a uniformly charged shell of radius R and total charge e, whose electrostatic energy E is equivalent to its mass m_e *via* Einstein's mass energy relation $E = m_ec^2$. In this model, R is approximately ($m_e = 9.1\times10^{-31}$ kg, $c = 3\times10^8$ ms^{-1}, $1/4\pi\varepsilon_0 = 9\times10^9$ Fm^{-1}, magnitude of the electron charge $= 1.6\times10^{-19}$ C)

(a) 1.4×10^{-15} m (b) 2×10^{-13} m

(c) 5.3×10^{-11} m (d) 2.8×10^{-35} m

32. A body is executing simple harmonic motion of amplitude a and period T about the equilibrium position $x = 0$. Large numbers of snapshots are taken at random of this body in motion. The probability of the body being found in a very small interval x to $x + |dx|$ is highest at

(a) $x = \pm a$ (b) $x = 0$

(c) $x = \pm\frac{a}{2}$ (d) $x = \pm\frac{a}{\sqrt{2}}$

33. Two identical bodies are made of a material for which the heat capacity increases with temperature. One of these is held at a temperature of 100°C, while the other one is kept at 0°C. If the two are brought into contact, then assuming no heat loss to the environment, the final temperature that they will reach is

(a) 50°C (b) more than 50°C
(c) less than 50°C (d) 0°C

34. A particle is acted upon by a force given by $F = -\alpha x^3 - \beta x^4$, where α and β are positive constants. At the point $x = 0$, the particle is

(a) in stable equilibrium (b) in unstable equilibrium
(c) in neutral equilibrium (d) not in equilibrium

35. The potential energy of a point particle is given by the expression $V(x) = -\alpha x + \beta \sin(x/\gamma)$. A dimensionless combination of the constants α, β and γ is

(a) $\frac{\alpha}{\beta\gamma}$ (b) $\frac{\alpha^2}{\beta\gamma}$ (c) $\frac{\gamma}{\alpha\beta}$ (d) $\frac{\alpha\gamma}{\beta}$

36. A ball of mass m suspended from a rigid support by an inextensible massless string is released from a height h above its lowest point. At its lowest point, it collides elastically with a block of mass $2m$ at rest on a frictionless surface. Neglect the dimensions of the ball and the block. After the collision, the ball rises to a maximum height of

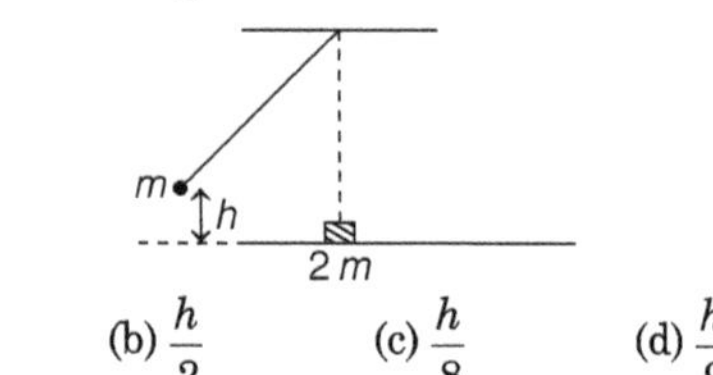

(a) $\frac{h}{3}$ (b) $\frac{h}{2}$ (c) $\frac{h}{8}$ (d) $\frac{h}{9}$

37. A particle released from rest is falling through a thick fluid under gravity. The fluid exerts a resistive force on the particle proportional to the square of its speed. Which one of the following graphs best depicts the variation of its speed v with time t?

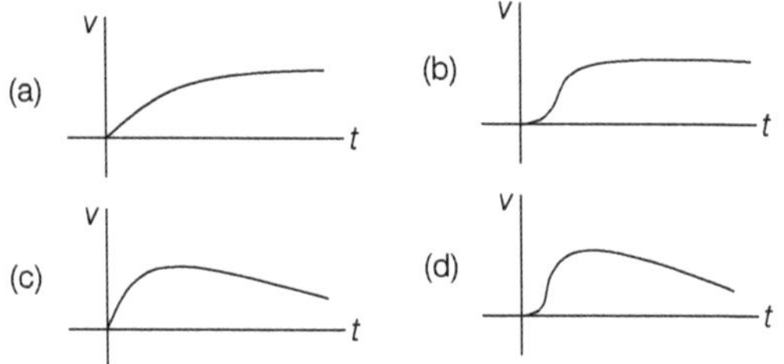

38. A cylindrical steel rod of length 0.10 m and thermal conductivity 50 $Wm^{-1}K^{-1}$ is welded end to end to copper rod of thermal conductivity 400 $Wm^{-1}K^{-1}$ and of the same area of cross-section but 0.20 m long. The free end of the steel rod is maintained at 100°C and that of the copper rod at 0°C. Assuming that the rods are perfectly insulated from the surrounding, the temperature at the junction of the two rods is

(a) 20°C (b) 30°C (c) 40°C (d) 50°C

39. A parent nucleus X is decaying into daughter nucleus Y which in turn decays to Z. The half lives of X and Y are 40000 yr and 20 yr, respectively. In a certain sample, it is found that the number of Y nuclei hardly changes with time. If the number of X nuclei in the sample is 4×10^{20}, the number of Y nuclei present in it is

(a) 2×10^{17} (b) 2×10^{20} (c) 4×10^{23} (d) 4×10^{20}

40. An unpolarised beam of light of intensity I_0 passes through two linear polarisers making an angle of 30° with respect to each other. The emergent beam will have an intensity

(a) $\frac{3I_0}{4}$ (b) $\frac{\sqrt{3}I_0}{4}$ (c) $\frac{3I_0}{8}$ (d) $\frac{I_0}{8}$

CHEMISTRY

41. Among the following, the species with the highest bond order is

(a) O_2 (b) F_2 (c) O_2^+ (d) F_2^-

42. The molecule with non-zero dipole moment is

(a) BCl_3 (b) $BeCl_2$ (c) CCl_4 (d) NCl_3

43. For a one-electron atom, the set of allowed quantum numbers is

(a) $n = 1, l = 0, m_l = 0, m_s = +1/2$
(b) $n = 1, l = 1, m_l = 0, m_s = +1/2$
(c) $n = 1, l = 0, m_l = -1, m_s = -1/2$
(d) $n = 1, l = 1, m_l = 1, m_s = -1/2$

44. In the reaction of benzene with an electrophile E^+, the structure of the intermediate σ-complex can be represented as

(a) 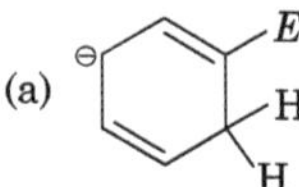

(b) 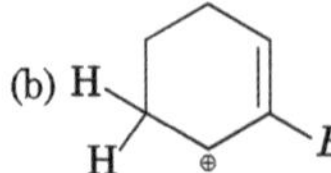

(c) 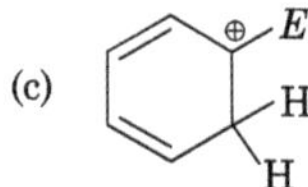

(d) 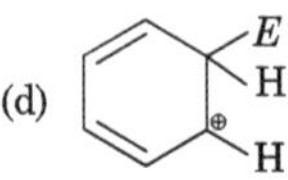

45. The most stable conformation of 2, 3-dibromobutane is

(a) 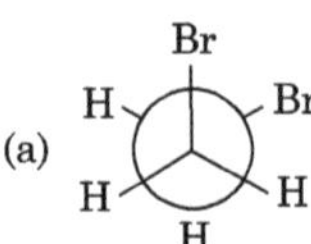

(b) 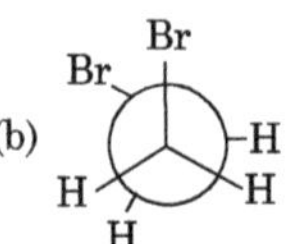

(c) 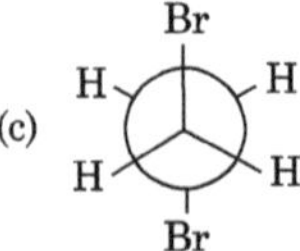

(d) 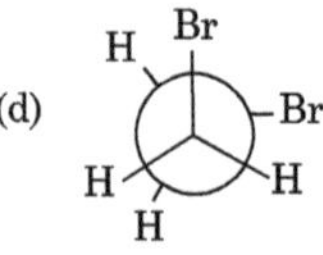

46. Typical electronic energy gaps in molecules are about 1.0 eV. In terms of temperature, the gap is closest to

(a) 10^2 K (b) 10^4 K (c) 10^3 K (d) 10^5 K

47. The major final product in the following reaction is

$$CH_3CH_2CN \xrightarrow[\text{(ii) } H_3O^+]{\text{(i) } CH_3MgBr}$$

(a) H_3C–CH_2–C(=NH)–CH_3 (b) H_3C–CH_2–CH=N–CH_3

(c) H_3C–CH_2–C(=O)–CH_3 (d) H_3C–CH_2–C(=O)–NH–CH_3

48. A zero-order reaction, $A \rightarrow$ Product, with an initial concentration $[A]_0$ has a half-life of 0.2 s. If one starts with the concentration $2[A]_0$, then the half-life is

(a) 0.1 s (b) 0.4 s
(c) 0.2 s (d) 0.8 s

49. The isoelectronic pair of ions is

(a) Sc^{2+} and V^{3+} (b) Mn^{2+} and Fe^{3+}
(c) Mn^{3+} and Fe^{2+} (d) Ni^{3+} and Fe^{2+}

50. The major product in the following reaction is

$$H_2C{=}CHBr \xrightarrow{NaNH_2}$$

(a) H—C≡C—H (b) $H_2C{=}CH$–NH_2

(c) H_2N–H_2C—CH_2–NH_2 (d) H_3C–CH_2–NH_2

51. The major product of the following reaction is

$$HO\text{–}CH_2\text{–}CH{=}CH_2 \xrightarrow{\text{Conc. HBr}}$$

(a) HO–CH_2–CH(Br)–CH_3 (b) Br–CH_2–CH=CH_2

(c) Br–CH_2–CH=CH–CH_2–Br (d) H_3C–CH(Br)–CH_2–Br

52. The oxidation state of cobalt in the following molecule is

$(OC)_3Co(\mu\text{-}CO)_2Co(CO)_3$ (Co—Co bonded, two bridging C=O)

(a) 3 (b) 1 (c) 2 (d) 0

53. The pK_a of a weak acid is 5.85. The concentrations of the acid and its conjugate base are equal at a pH of

(a) 6.85 (b) 5.85 (c) 4.85 (d) 7.85

54. For a tetrahedral complex $[MCl_4]^{2-}$, the spin-only magnetic moment is 3.83 BM. The element M is

(a) Co (b) Cu (c) Mn (d) Fe

55. Among the following graphs showing variation of rate (k) with temperature (T) for a reaction, the one that exhibits Arrhenius behaviour over the entire temperature range is

(a)
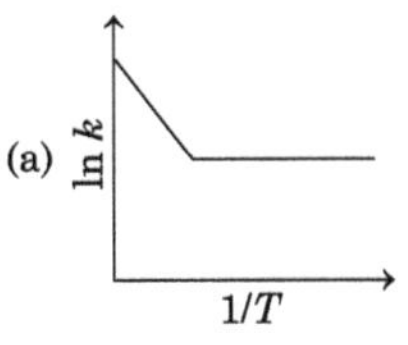

(b)
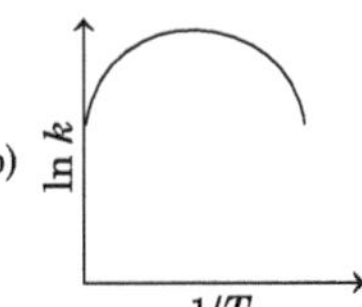

(c)
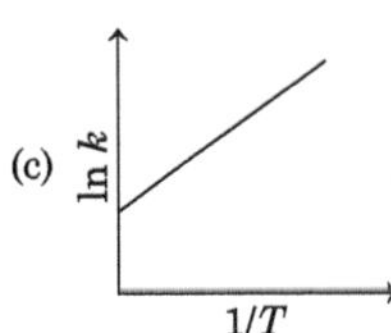

(d)
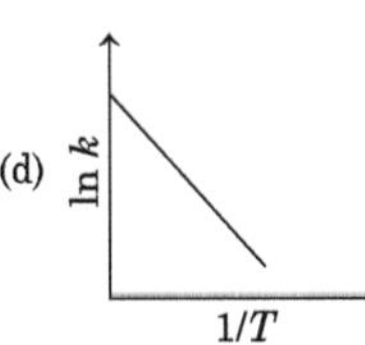

56. The reaction that gives the following molecule as the major product is

$(H_3C)_3C$–O–CH_3

(a) $(H_3C)_3C$–Br + CH_3ONa

(b) $(H_3C)_3C$–ONa + CH_3Br

(c) $(H_3C)_3C$–OH + CH_3ONa

(d) $(H_3C)_2C{=}CH_2$ + CH_3ONa

57. The C—O bond length in CO, CO_2 and CO_3^{2-} follows the order.

(a) $CO < CO_2 < CO_3^{2-}$ (b) $CO_2 < CO_3^{2-} < CO$
(c) $CO > CO_2 > CO_3^{2-}$ (d) $CO_3^{2-} < CO_2 < CO$

58. The equilibrium constant for the following reactions are K_1 and K_2, respectively.

$$2P(g) + 3Cl_2(g) \rightleftharpoons 2PCl_3(g)$$
$$PCl_3(g) + Cl_2(g) \rightleftharpoons PCl_5(g)$$

Then, the equilibrium constant for the reaction,

$$2P(g) + 5Cl_2(g) \rightleftharpoons 2PCl_5(g) \text{ is}$$

(a) K_1K_2 (b) $K_1K_2^2$ (c) $K_1^2K_2^2$ (d) $K_1^2K_2$

59. The major product of the following reaction is

C_6H_5–CH_2CH_3 + $(CH_3C)_2CHCH_2Cl$ $\xrightarrow{AlCl_3}$

(a) $(H_3C)_3C$—[benzene ring]—CH_2CH_3

(b) $(H_3C)_2HCH_2C$—[benzene ring]—CH_2CH_3

(c) [benzene ring] CH_2CH_3, $CH_2CH(CH_3)_2$

(d) [benzene ring] CH_2CH_3, $C(CH_3)_3$

60. Doping silicon with boron produces a
(a) n-type semiconductor
(b) p-type semiconductor
(c) metallic conductor
(d) insulator

BIOLOGY

61. The disorders that arise when the immune system destroys 'self' cells are called autoimmune disorders. Which of the following would be classified under this?
(a) Rheumatoid arthritis
(b) Rhinitis
(c) Asthma
(d) Eczema

62. Which of the following classes of immunoglobulins can trigger the complement cascade?
(a) IgA (b) IgM (c) IgD (d) IgE

63. Diabetes insipidus is due to
(a) hypersecretion of vasopressin
(b) hyposecretion of insulin
(c) hypersecretion of insulin
(d) hyposecretion of vasopressin

64. Fossils are most often found in which kind of rocks?
(a) Meteorites
(b) Igneous rocks
(c) Sedimentary rocks
(d) Metamorphic rocks

65. Peptic ulcers are caused by
(a) a fungus, *Candida albicans*
(b) a virus, *Cytomegalovirus*
(c) a parasite, *Trypanosoma brucei*
(d) a bacterium, *Helicobacter pylori*

66. Transfer RNA (*t*RNA)
(a) is present in the ribosomes and provides structural integrity
(b) usually has clover leaf-like structure
(c) carries genetic information from DNA to ribosomes
(d) codes for proteins

67. Some animals excrete uric acid in urine (uricotelic) as it requires very little water. This is an adaptation to conserve water loss. Which animals among the following are most likely to be uricotelic?
(a) Fishes (b) Birds
(c) Amphibians (d) Mammals

68. A ripe mango, kept with unripe mangoes causes their ripening. This is due to the release of a gaseous plant hormone
(a) auxin (b) cytokinin
(c) gibberellin (d) ethylene

69. Human chromosomes undergo structural changes during the cell cycle. Chromosomal structure can be best visualised if a chromosome is isolated from a cell at
(a) G_1-phase (b) G_2-phase
(c) S-phase (d) M-phase

70. By which of the following mechanisms is glucose reabsorbed from the glomerular filtrate by the kidney tubule?
(a) Osmosis (b) Active transport
(c) Diffusion (d) Passive transport

71. In mammals, the hormones secreted by the pituitary, the master gland, is itself regulated by
(a) hypothalamus (b) pineal gland
(c) median cortex (d) cerebrum

72. Which of the following is true for TCA cycle in eukaryotes?
(a) Takes place in mitochondrion
(b) Produces no ATP
(c) Takes place in Golgi complex
(d) Independent of electron transport chain

73. A hormone molecule binds to a specific protein on the plasma membrane inducing a signal. The protein it binds to is called
(a) ligand (b) receptor
(c) antibody (d) histone

74. DNA mutations that do not cause any functional change in the protein product are known as
(a) non-sense mutations
(b) deletion mutations
(c) mis-sense mutations
(d) silent mutations

75. Plant roots are usually devoid of chlorophyll and cannot perform photosynthesis. However, there are exceptions. Which of the following plant's root can perform photosynthesis?
(a) *Arabidopsis* (b) Rice
(c) *Tinospora* (d) *Hibiscus*

76. Vitamin-A deficiency leads to night blindness. Which of the following is the reason for the disease?
(a) Rod cells are not converted to cone cells
(b) Rhodopsin pigment of rod cells is defective
(c) Melanin pigment is not synthesised in cone cells
(d) Cornea of eye gets dried

77. In dengue virus infection, patients often develop haemorrhagic fever due to internal bleeding. This happens due to the reduction of
(a) platelets (b) WBCs
(c) RBCs (d) lymphocytes

78. If the sequence of base in sense strand of DNA is 5′-GTTCATCG-3′, then the sequence of bases in its RNA transcript would be
(a) 5′-GTTCATCG-3′
(b) 5′-GUUCAUCG-3′
(c) 5′-CAAGTAGC-3′
(d) 5′-CAAGUAGC-3′

79. A reflex action is a quick involuntary response to stimulus. Which of the following is an example of both, unconditioned and conditioned reflex?
(a) Knee-jerk reflex
(b) Secretion of saliva in response to the aroma of food
(c) Sneezing reflex
(d) Contraction of the pupil in response to bright light

80. In a food chain such as grass → deer → lion, the energy cost of respiration as a proportion of total assimilated energy at each level would be
(a) 60%–30%–20% (b) 20%–30%–60%
(c) 20%–60%–30% (d) 30%–30%–30%

PART-II (2 Marks Questions)

MATHEMATICS

81. Suppose a, b, c are real numbers, and each of the equations $x^2 + 2ax + b^2 = 0$ and $x^2 + 2bx + c^2 = 0$ has two distinct real roots. Then, the equation $x^2 + 2cx + a^2 = 0$ has
(a) two distinct positive real roots
(b) two equal roots
(c) one positive and one negative root
(d) no real roots

82. The coefficient of x^{2012} in $\dfrac{1+x}{(1+x^2)(1-x)}$ is
(a) 2010 (b) 2011 (c) 2012 (d) 2013

83. Let (x, y) be a variable point on the curve $4x^2 + 9y^2 - 8x - 36y + 15 = 0$. Then, min $(x^2 - 2x + y^2 - 4y + 5) +$ max $(x^2 - 2x + y^2 - 4y + 5)$ is
(a) $\dfrac{325}{36}$ (b) $\dfrac{36}{325}$ (c) $\dfrac{13}{25}$ (d) $\dfrac{25}{13}$

84. The sum of all $x \in [0, \pi]$ which satisfy the equation
$\sin x + \dfrac{1}{2}\cos x = \sin^2\left(x + \dfrac{\pi}{4}\right)$ is
(a) $\dfrac{\pi}{6}$ (b) $\dfrac{5\pi}{6}$ (c) π (d) 2π

85. A polynomial $P(x)$ with real coefficients has the property that $P''(x) \neq 0$ for all x. Suppose $P(0) = 1$ and $P'(0) = -1$. What can you say about $P(1)$?
(a) $P(1) \geq 0$ (b) $P(1) \neq 0$
(c) $P(1) \leq 0$ (d) $-\dfrac{1}{2} < P(1) < \dfrac{1}{2}$

86. Define a sequence $\langle a_n \rangle$ by $a_1 = 5$, $a_n = a_1 a_2 \ldots a_{n-1} + 4$ for $n > 1$. Then, $\lim\limits_{n \to \infty} \dfrac{\sqrt{a_n}}{a_{n-1}}$
(a) equals $\dfrac{1}{2}$ (b) equals 1
(c) equals $\dfrac{2}{5}$ (d) does not exist

87. The value of the integral $\displaystyle\int_{-\pi}^{\pi} \frac{\cos^2 x}{1+a^x}\,dx$, where $a > 0$, is
(a) π (b) $a\pi$ (c) $\dfrac{\pi}{2}$ (d) 2π

88. Consider, $L = \sqrt[3]{2012} + \sqrt[3]{2013} + \ldots + \sqrt[3]{3011}$,
$R = \sqrt[3]{2013} + \sqrt[3]{2014} + \ldots + \sqrt[3]{3012}$ and $I = \displaystyle\int_{2012}^{3012} \sqrt[3]{x}\,dx$.
Then,
(a) $L + R < 2I$ (b) $L + R = 2I$ (c) $L + R > 2I$ (d) $\sqrt{LR} = I$

89. A man tosses a coin 10 times, scoring 1 point for each head and 2 points for each tail. Let $P(K)$ be the probability of scoring at least K points. The largest value of K such that $P(K) > \dfrac{1}{2}$ is
(a) 14 (b) 15 (c) 16 (d) 17

90. Let $f(x) = \dfrac{x+1}{x-1}$ for all $x \neq 1$. Let $f^1(x) = f(x)$, $f^2(x) = f(f(x))$ and generally $f^n(x) = f(f^{n-1}(x))$ for $n > 1$. Let $P = f^1(2) f^2(3) f^3(4) f^4(5)$
Which of the following is a multiple of P?
(a) 125 (b) 375 (c) 250 (d) 147

PHYSICS

91. The total energy of a black body radiation source is collected for five minutes and used to heat water. The temperature of the water increases from 10.0°C to 11.0°C. The absolute temperature of the black body is doubled and its surface area halved and the experiment repeated for the same time. Which of the following statements would be most nearly correct?
(a) The temperature of the water would increase from 10.0°C to a final temperature of 12°C
(b) The temperature of the water would increase from 10.0°C to a final temperature of 18°C
(c) The temperature of the water would increase from 10.0°C to a final temperature of 14°C
(d) The temperature of the water would increase from 10.0°C to a final temperature of 11°C

92. A small asteroid is orbiting around the sun in a circular orbit of radius r_0 with speed v_0. A rocket is launched from the asteroid with speed $v = \alpha v_0$, where v is the speed relative to the sun. The highest value of α for which the rocket will remain bound to the solar system is (ignoring gravity due to the asteroid and effects of other planets)

(a) $\sqrt{2}$ (b) 2 (c) $\sqrt{3}$ (d) 1

93. A radioactive nucleus A has a single decay mode with half-life τ_A. Another radioactive nucleus B has two decay modes 1 and 2. If decay mode 2 were absent, the half-life of B would have been $\tau_A / 2$. If decay mode 1 were absent, the half-life of B would have been $3\tau_A$. If the actual half life of B is τ_B, then the ratio τ_B / τ_A is

(a) 3/7 (b) 7/2 (c) 7/3 (d) 1

94. A stream of photons having energy 3 eV each impinges on a potassium surface. The work function of potassium is 2.3 eV. The emerging photo-electrons are slowed down by a copper plate placed 5 mm away. If the potential difference between the two metal plates is 1 V, the maximum distance the electrons can move away from the potassium surface before being turned back is

(a) 3.5 mm (b) 1.5 mm (c) 2.5 mm (d) 5.0 mm

95. Consider three concentric metallic spheres A, B and C of radii a, b, c, respectively where $a < b < c$. A and B are connected, whereas C is grounded. The potential of the middle sphere B is raised to V, then the charge on the sphere C is

(a) $-4\pi\varepsilon_0 V \dfrac{bc}{c-b}$ (b) $+4\pi\varepsilon_0 V \dfrac{bc}{c-b}$

(c) $-4\pi\varepsilon_0 V \dfrac{ac}{c-a}$ (d) zero

96. On a bright sunny day a diver of height h stands at the bottom of a lake of depth H. Looking upward, he can see objects outside the lake in a circular region of radius R. Beyond this circle he sees the images of objects lying on the floor of the lake. If refractive index of water is 4/3, then the value of R is

(a) $\dfrac{3(H-h)}{\sqrt{7}}$ (b) $3h\sqrt{7}$ (c) $\dfrac{(H-h)}{\sqrt{\frac{7}{3}}}$ (d) $\dfrac{(H-h)}{\sqrt{\frac{5}{3}}}$

97. A cube is formed with ten identical resistances R (thick lines) and two shorting wires (dotted lines) along the arms AC and BD as shown in the figure below.

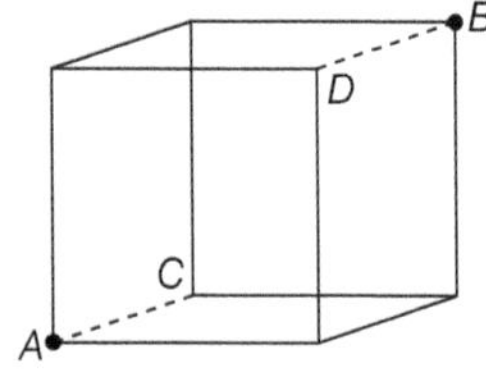

Resistance between point A and B is

(a) $\dfrac{R}{2}$ (b) $\dfrac{5R}{6}$ (c) $\dfrac{3R}{4}$ (d) R

98. A standing wave in a pipe with a length $L = 1.2$ m is described by

$y(x, t) = y_0 \sin[(2\pi / L)x] \sin[(2\pi / L)x + \pi / 4]$

based on above information, which one of the following statement is incorrect?

(Speed of sound in air is 300 ms^{-1})

(a) The pipe is closed at both ends
(b) The wavelength of the wave could be 1.2 m
(c) There could be a node at $x = 0$ and antinode at $x = L/2$
(d) The frequency of the fundamental mode of vibrations is 137.5 Hz

99. Two blocks 1 and 2 of equal mass m are connected by an ideal string (see figure below) over a frictionless pulley. The blocks are attached to the ground by springs having spring constants k_1 and k_2 such that $k_1 > k_2$. Initially, both springs are unstretched. Block 1 is slowly pulled down a distance x and released. Just after the release the possible values of the magnitudes of the accelerations of the blocks a_1 and a_2 can be

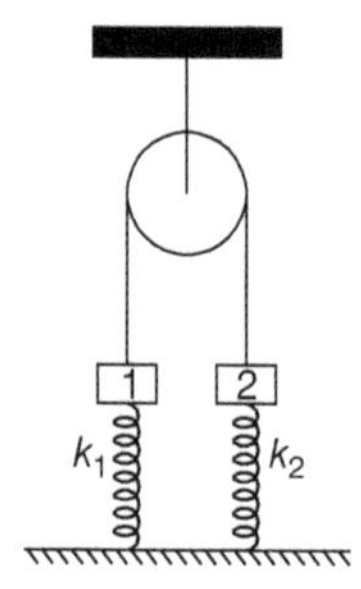

(a) either $\left(a_1 = a_2 = \dfrac{(k_1 + k_2)x}{2m}\right)$ or $\left(a_1 = \dfrac{k_1 x}{m} - g \text{ and } a_2 = \dfrac{k_2 x}{m} + g\right)$

(b) $\left(a_1 = a_2 = \dfrac{(k_1 + k_2)x}{2m}\right)$ only

(c) $\left(a_1 = a_2 = \dfrac{(k_1 - k_2)x}{2m}\right)$ only

(d) either $\left(a_1 = a_2 = \dfrac{(k_1 - k_2)x}{2m}\right)$ or $\left(a_1 = a_2 = \dfrac{(k_1 k_2)x}{(k_1 + k_2)m} - g\right)$

100. A simple pendulum is released from rest at the horizontally stretched position. When the string makes an angle θ with the vertical, the angle ϕ which the acceleration vector of the bob makes with the string is given by

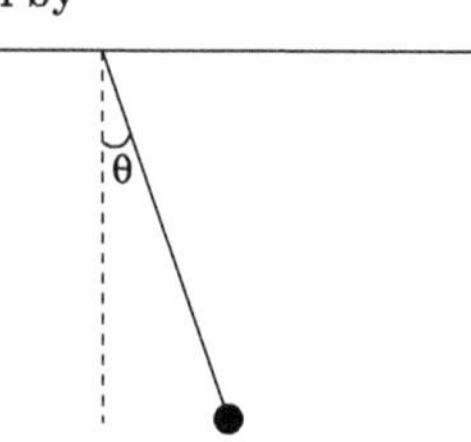

(a) $\phi = 0$

(b) $\phi = \tan^{-1}\left(\dfrac{\tan\theta}{2}\right)$

(c) $\phi = \tan^{-1}(2\tan\theta)$

(d) $\phi = \dfrac{\pi}{2}$

CHEMISTRY

101. The final major product obtained in the following sequence of reaction is

Ph—≡ $\xrightarrow[\text{(ii) } CH_3I \text{ (iii) } H_2Pd/C]{\text{(i) } NaNH_2NH_3}$

(a) Ph, H / H, CH_3 (b) Ph, CH_3 / H, H

(c) Ph, H / H_3C, H (d) Ph—CH$_2$CH$_2$—CH_3

102. In the DNA of *E. coli* the mole ratio of adenine to cytosine is 0.7. If the number of moles of adenine in the DNA is 350000, the number of moles of guanine is equal to

(a) 350000 (b) 500000 (c) 225000 (d) 700000

103. (*R*)-2-bromobutane upon treatment with *aq.* NaOH gives

(a) OH, H, CH_3, H_3C

(b) OH, H, CH_3 + H_3C, OH, H, CH_3 (1 : 1 mixture)

(c) H_3C, OH, H, CH_3 (d) H_3C, OH, CH_3

104. Phenol on treatment with dil. HNO_3 gives two products *P* and *Q*. *P* is steam volatile but *Q* is not. *P* and *Q* are, respectively.

(a) OH, NO_2 and OH, NO_2

(b) OH, NO_2 and OH, NO_2

(c) OH, OH, NO_2 and OH, NO_2, OH

(d) OH, NO_2 and OH, OH

105. A metal is irradiated with light of wavelength 660 nm. Given that the work function of the metal is 1.0 eV, the de Broglie wavelength of the ejected electron is close to

(a) 6.6×10^{-7} m (b) 8.9×10^{-11} m
(c) 1.3×10^{-9} m (d) 6.6×10^{-13} m

106. The inter-planar spacing between the (2 2 1) planes of a cubic lattice of length 450 pm is

(a) 50 pm (b) 150 pm
(c) 300 pm (d) 450 pm

107. The ΔH for vaporisation of a liquid is 20 kJ/mol. Assuming ideal behaviour, the change in internal energy for the vaporisation of 1 mole of the liquid at 60°C and 1 bar is close to

(a) 13.2 kJ/mol (b) 17.2 kJ/mol
(c) 19.5 kJ/mol (d) 20.0 kJ/mol

108. Among the following the species that is both tetrahedral and diamagnetic is

(a) $[NiCl_4]^{2-}$ (b) $[Ni(CN)_4]^{2-}$
(c) $Ni(CO)_4$ (d) $[Ni(H_2O)_6]^{2+}$

109. Three moles of an ideal gas expand reversibly under isothermal condition from 2 L to 20 L at 300 K. The amount of heat-change (in kJ/mol) in the process is

(a) 0 (b) 7.2 (c) 10.2 (d) 17.2

110. The following data are obtained for a reaction, $X + Y \rightarrow$ Products.

Expt.	$[X_0]$/mol	$[Y_0]$/mol	rate/mol L^{-1} s^{-1}
1	0.25	0.25	1.0×10^{-6}
2	0.50	0.25	4.0×10^{-6}
3	0.25	0.50	8.0×10^{-6}

The overall order of the reaction is

(a) 2 (b) 4 (c) 3 (d) 5

BIOLOGY

111. When hydrogen peroxide is applied on the wound as a disinfectant, there is frothing at the site of injury, which is due to the presence of an enzyme in the skin that uses hydrogen peroxide as a substrate to produce

(a) hydrogen (b) water (c) carbon dioxide (d) oxygen

112. Persons suffering from hypertension (high blood pressure) are advised a low-salt diet because

(a) more salt is absorbed in the body of a patient with hypertension
(b) high salt leads to water retention in the blood that further increases the blood pressure
(c) high salt increases nerve conduction and increases blood pressure
(d) high salt causes adrenaline release that increases blood pressure

113. Insectivorous plants that mostly grow on swampy soil use insects as a source of

(a) carbon (b) phosphorus
(c) nitrogen (d) magnesium

114. In cattle, the coat colour red and white are two dominant traits, which express equally in F_1 to produce roan (red and white colour in equal proportion). If F_1 progeny are self-bred, the resulting progeny in F_2 will have phenotypic ratio (red : roan : white) is

(a) 1 : 1 : 1 (b) 3 : 9 : 3 (c) 1 : 2 : 1 (d) 3 : 9 : 4

115. The restriction endonuclease *Eco*RI recognises and cleaves DNA sequence as shown below

5′-G A A T T C-3′ 3′-C T T A A G-5′

What is the probable number of cleavage sites that can occur in a 10 kb long random DNA sequence?

(a) 10 (b) 2 (c) 100 (d) 50

116. Which one of the following is true about enzyme catalysis?

(a) The enzyme changes at the end of the reaction
(b) The activation barrier of the process is lower in the presence of an enzyme
(c) The rate of the reaction is retarded in the presence of an enzyme
(d) The rate of the reaction is independent of substrate concentration

117. *Vibrio cholerae* causes cholera in humans. Ganga water was once used successfully to combat the infection. The possible reason could be

(a) high salt content of Ganga water
(b) low salt content of Ganga water
(c) presence of bacteriophages in Ganga water
(d) presence of antibiotics in Ganga water

118. When a person begins to fast, after some time glycogen stored in the liver is mobilised as a source of glucose. Which of the following graphs best represents the change of glucose level (y-axis) in his blood, starting from the time (x-axis) when he begins to fast?

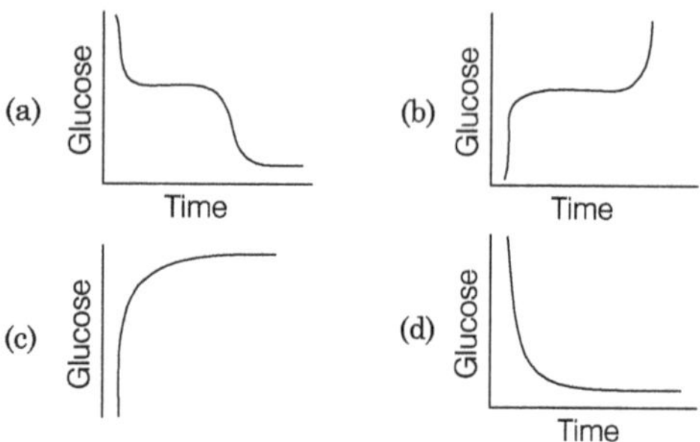

119. The following sequence contains the open reading frame of a polypeptide. How many amino acids will the polypeptide consists of ?

5′-AGCATATGATCGTTTCTCTGCTTTGAACT-3′

(a) 4 (b) 2 (c) 10 (d) 7

120. Insects constitute the largest animal group on the earth. About 25-30% of the insect species are known to be herbivores. In spite of such huge herbivore pressure, globally, green plants have persisted. One possible reason for this persistence is

(a) food preference of insects has tended to change with time
(b) herbivore insects have become inefficient feeders of green plants
(c) herbivore population has been kept in control by predators
(d) decline in reproduction of herbivores with time

Answers

PART-I

1 (b)	2 (d)	3 (b)	4 (c)	5 (b)	6 (c)	7 (d)	8 (d)	9 (b)	10 (c)
11 (b)	12 (d)	13 (d)	14 (b)	15 (c)	16 (a)	17 (c)	18 (a)	19 (d)	20 (b)
21 (b)	22 (d)	23 (a)	24 (c)	25 (a)	26 (a)	27 (a)	28 (a)	29 (a)	30 (*)
31 (a)	32 (a)	33 (b)	34 (a)	35 (d)	36 (d)	37 (a)	38 (a)	39 (a)	40 (c)
41 (c)	42 (d)	43 (a)	44 (*)	45 (c)	46 (d)	47 (c)	48 (b)	49 (b)	50 (a)
51 (b)	52 (d)	53 (b)	54 (a)	55 (d)	56 (b)	57 (a)	58 (b)	59 (a)	60 (a)
61 (a)	62 (b)	63 (d)	64 (c)	65 (d)	66 (b)	67 (c)	68 (d)	69 (d)	70 (b)
71 (a)	72 (a)	73 (c)	74 (d)	75 (c)	76 (b)	77 (a)	78 (b)	79 (b)	80 (b)

PART-II

81 (d)	82 (*)	83 (a)	84 (c)	85 (b)	86 (b)	87 (c)	88 (a)	89 (b)	90 (b)
91 (b)	92 (d)	93 (a)	94 (a)	95 (a)	96 (a)	97 (a)	98 (d)	99 (b)	100 (b)
101 (b)	102 (b)	103 (b)	104 (a)	105 (c)	106 (b)	107 (b)	108 (c)	109 (d)	110 (d)
111 (d)	112 (b)	113 (c)	114 (c)	115 (b)	116 (b)	117 (d)	118 (a)	119 (d)	120 (c)

** No options are correct.*

Solutions

1. *(b)* We have,

Three children with their guardians

$\therefore$ Total number of persons = 6

Total number of ways when principal want to interview all the 6 persons such that no child interviewed before their guardians is $\frac{6!}{2!2!2!} = \frac{720}{8} = 90$ ways

2. *(d)* We have,

$$\sqrt{x+3-4\sqrt{x-1}} + \sqrt{x+8-6\sqrt{x-1}} = 1$$

$$\Rightarrow \sqrt{(\sqrt{x-1})^2 - 2(2)\sqrt{x-1} + (2)^2} + \sqrt{(\sqrt{x-1})^2 - 2\times 3\sqrt{x-1} + (3)^2} = 1$$

$$\Rightarrow \sqrt{(\sqrt{x-1}-2)^2} + \sqrt{(\sqrt{x-1}-3)^2} = 1$$

$$\Rightarrow |\sqrt{x-1}-2| + |\sqrt{x-1}-3| = 1$$

$$x \in [5, 10]$$

$$\sqrt{x-1} - 2 - \sqrt{x-1} + 3 = 1$$

$$\therefore \quad 1 = 1$$

Hence, x has infinite solutions in $x \in [5, 10]$.

3. *(b)* Let

$$f(x) = 3^x + 5^x - 9^x + 15^x - 25^x$$

$$f(x) = 3^x + 5^x - 3^{2x} + 3^x \cdot 5^x - 5^{2x}$$

Maximum value of $f(x)$ is M.

$$\therefore \quad M = 3^x + 5^x - 3^{2x} - 3^x \cdot 5^x - 5^{2x}$$

$$M = a + b - a^2 + ab - b^2$$

$$[\because 3^x = a, 5^x = b]$$

$$\Rightarrow \quad a + b + ab - (a^2 + b^2)$$

$$\Rightarrow \quad M \le a + b + ab - 2ab$$

$$[\because -(a^2+b^2) \le -2ab]$$

$$\Rightarrow \quad M \le a + b - ab$$

$$\Rightarrow \quad M \le 1 - (a-1)(b-1)$$

So, maximum value of M is 1 at $x = 0$.

$$\therefore \quad 0 < M < 2.$$

4. *(c)* We have,

Equation of circle

$$x^2 + y^2 - 6x - 2py + 17 = 0$$

$$\Rightarrow \quad (x-3)^2 + (y-p)^2 = p^2 - 8$$

Given two perpendicular tangents drawn from origin to the circle.

$\therefore$ Locus is director circle of the given circle.

$\therefore$ Equation of director circle is

$$(x-3)^2 + (y-p)^2 = 2(p^2 - 8)$$

This passes through (0, 0).

$$\therefore \quad 9 + p^2 = 2p^2 - 16$$

$$\Rightarrow \quad p^2 = 25 \Rightarrow p = \pm 5$$

$$\therefore \quad |p| = |\pm 5| = 5$$

5. *(b)* We have,

$$y = 2x^3 + ax + b \quad \ldots\text{(i)}$$

$$y = 2x^3 + cx + d \quad \ldots\text{(ii)}$$

From Eqs. (i) and (ii), we get

$$2x^3 + ax + b = 2x^3 + cx + d$$

$$\Rightarrow \quad (a-c)x = (d-b)$$

Since, Eqs. (i) and (ii) have no common point.

$$\therefore \quad (a-c)x \ne (d-b)$$

If $(a-c) = 0$ and $(d-b) \ne 0$

$\therefore$ Maximum value of $(a-c)^2 + (b-d)$

$$= 0 + (b-d)$$

$$= 6 - 1 = 5 \quad [\because b, d \in \{1, 2, 3, 4, 5, 6\}]$$

6. *(c)* We have,

$$ex^2 + \pi y^2 - 2e^2x - 2\pi^2 y + e^3 + \pi^3 = \pi e$$

$$\Rightarrow ex^2 - 2e^2x + \pi y^2 - 2\pi^2 y = \pi e - e^3 - \pi^3$$

$$\Rightarrow e(x^2 - 2ex + e^2) + \pi(y^2 - 2\pi y + \pi^2) = \pi e - e^3 - \pi^3 + e^3 + \pi^3$$

$$\Rightarrow e(x-e)^2 + \pi(y-\pi)^2 = \pi e$$

$$\Rightarrow \frac{(x-e)^2}{(\sqrt{\pi})^2} + \frac{(y-\pi)^2}{(\sqrt{e})^2} = 1$$

$$\therefore \quad \sqrt{\pi} > \sqrt{e}$$

$$\therefore \quad PS_1 + PS_2 = 2a$$

$$\therefore \quad PS_1 + PS_2 = 2\sqrt{\pi}$$

7. *(d)* We have,

$$f(x) = \frac{\sin(x-a) + \sin(x+a)}{\cos(x-a) - \cos(x+a)}$$

$$\Rightarrow \quad f(x) = \frac{2\sin x \cos a}{2\sin x \sin a}$$

$$\Rightarrow \quad f(x) = \cot a$$

$\therefore f$ is constant function.

8. *(d)* We have,

$$\tan 81^\circ - \tan 63^\circ - \tan 27^\circ + \tan 9^\circ$$

$$= (\tan 81^\circ + \tan 9^\circ) - (\tan 63^\circ + \tan 27^\circ)$$

$$= (\cot 9^\circ + \tan 9^\circ) - (\cot 27^\circ + \tan 27^\circ)$$

$$= \left(\frac{\cos 9^\circ}{\sin 9^\circ} + \frac{\sin 9^\circ}{\cos 9^\circ}\right) - \left(\frac{\cos 27^\circ}{\sin 27^\circ} + \frac{\sin 27^\circ}{\cos 27^\circ}\right)$$

$$= \left(\frac{\cos^2 9^\circ + \sin^2 9^\circ}{\sin 9^\circ \cos 9^\circ}\right) - \left(\frac{\cos^2 27^\circ + \sin^2 27^\circ}{\sin 27^\circ \cos 27^\circ}\right)$$

$$= \left(\frac{2}{2\sin 9^\circ \cos 9^\circ}\right) - \left(\frac{2}{2\sin 27^\circ \cos 27^\circ}\right)$$

$$= \left[\frac{2}{\sin 18^\circ} - \frac{2}{\sin 54^\circ}\right]$$

$$= 2\left[\frac{1}{\sin 18^\circ} - \frac{1}{\cos 36^\circ}\right] = 2\left[\frac{4}{\sqrt{5}-1} - \frac{4}{\sqrt{5}+1}\right]$$

$$\left[\because \sin 18^\circ = \frac{\sqrt{5}-1}{4}, \cos 36^\circ = \frac{\sqrt{5}+1}{4}\right]$$

$$= 8\left[\frac{\sqrt{5}+1-\sqrt{5}+1}{5-1}\right] = 4$$

9. *(b)* We have, $f(x) = \sqrt{4 - \sqrt{2x+5}}$

$f(x)$ is defined if

$$4 - \sqrt{2x+5} \ge 0 \text{ and } 2x + 5 \ge 0$$

$$\Rightarrow 4 \ge \sqrt{2x+5} \text{ and } x \ge -5/2$$

$$\Rightarrow 16 \ge 2x + 5 \text{ and } x \ge -5/2$$

$$\Rightarrow x \le \frac{11}{2} \text{ and } x \ge -5/2$$

$\therefore$ Domain of $f(x)\ x \in \left[\frac{-5}{2}, \frac{11}{2}\right]$

Mid-point of domain $\dfrac{\frac{-5}{2} + \frac{11}{2}}{2} = \dfrac{3}{2}$

10. *(c)* We have,

$$x^{2n+1} - (2n+1)x + a = 0$$

Let $f(x) = x^{2n+1} - (2n+1)x + a$

$$f'(x) = (2n+1)x^{2n} - (2n+1)$$

$$f'(x) = (2n+1)(x^{2n} - 1)$$

$$f'(x) \le 0 \quad x \in [-1, 1]$$

$f(x)$ is strictly decreasing in $x \in [-1, 1]$.

$\therefore f(x)$ cut X-axis at most one point in the given interval.

11. *(b)* We have,

$$f(x) = (x-a_1)(x-a_2) + (x-a_2)(x-a_3) + (x-a_3)(x-a_1)$$

$$f(x) = 3x^2 - 2(a_1 + a_2 + a_3)x + a_1a_2 + a_2a_3 + a_1a_3$$

$$f(x) \ge 0 \text{ iff } D \le 0$$

$$\therefore 4(a_1 + a_2 + a_3)^2 - 4 \times 3(a_1a_2 + a_2a_3 + a_1a_3) \le 0$$

$$= a_1^2 + a_2^2 + a_3^2 - (a_1a_2 + a_2a_3 + a_1a_3) \le 0$$

$$= \frac{1}{2}[(a_1 - a_2)^2 + (a_2 - a_3)^2 + (a_3 - a_1)^2] \le 0$$

It is possible only

$$a_1 = a_2, a_2 = a_3, a_3 = a_1$$

$$\therefore \quad a_1 = a_2 = a_3$$

12. *(d)* Let $I = \dfrac{\int_0^{\pi/2} (\sin x)^{\sqrt{2}+1}\,dx}{\int_0^{\pi/2} (\sin x)^{\sqrt{2}-1}\,dx}$

$$\Rightarrow \quad I = \frac{\int_0^{\pi/2} (\sin x)^{\sqrt{2}} \sin x\,dx}{\int_0^{\pi/2} (\sin x)^{\sqrt{2}-1}\,dx}$$

$$[-(\sin x)^{\sqrt{2}}\cos x]_0^{\pi/2}+\sqrt{2}$$

$$\Rightarrow I=\frac{\int_0^{\pi/2}(\sin x)^{\sqrt{2}-1}\cos^2 x\,dx}{\int_0^{\pi/2}(\sin x)^{\sqrt{2}-1}\,dx}$$

$$\Rightarrow I=\frac{\sqrt{2}\int_0^{\pi/2}(\sin x)^{\sqrt{2}-1}(1-\sin^2 x)\,dx}{\int_0^{\pi/2}(\sin x)^{\sqrt{2}-1}\,dx}$$

$$\Rightarrow I=\frac{\sqrt{2}\left[\int_0^{\pi/2}(\sin x)^{\sqrt{2}-1}-\int_0^{\pi/2}(\sin x)^{\sqrt{2}+1}\,dx\right]}{\int_0^{\pi/2}(\sin x)^{\sqrt{2}-1}\,dx}$$

$$\Rightarrow I=\sqrt{2}\left[1-\frac{\int_0^{\pi/2}(\sin x)^{\sqrt{2}+1}dx}{\int_0^{\pi/2}(\sin x)^{\sqrt{2}-1}\,dx}\right]$$

$$\Rightarrow I=\sqrt{2}\,[1-I]$$

$$\Rightarrow I=\sqrt{2}-\sqrt{2}\,I \Rightarrow I+\sqrt{2}\,I=\sqrt{2}$$

$$\Rightarrow I=\frac{\sqrt{2}}{\sqrt{2}+1}\Rightarrow I=\frac{\sqrt{2}\,(\sqrt{2}-1)}{2-1}$$

$$\Rightarrow I=2-\sqrt{2}$$

13. *(d)* Let

$$I=\int_{-2012}^{2012}(\sin(x^3)+x^5+1)\,dx$$

$$\Rightarrow I=\int_{-2012}^{2012}\sin x^3\,dx+\int_{-2012}^{2012}x^5\,dx+\int_{-2012}^{2012}dx$$

$$\Rightarrow I=0+0+2\int_0^{2012}dx$$

[$\because \sin x^3$ and x^5 are odd function]

$$\Rightarrow I=2\,[x]_0^{2012}=2\times(2012)=4024$$

14. *(b)* Let $I=\int_0^5 [x]\,\{x\}\,dx$

$$I=\int_0^1 0\,\{x\}\,dx+\int_1^2 1\cdot\{x\}\,dx+\int_2^3 2\{x\}\,dx+\int_3^4 3\,\{x\}\,dx+\int_4^5 4\,\{x\}\,dx$$

$$I=0+1\int_0^1 x\,dx+2\int_0^1 x\,dx+3\int_0^1 x\,dx+4\int_0^1 x\,dx$$

$$I=(1+2+3+4)\int_0^1 x\,dx$$

$$I=10\left[\frac{x^2}{2}\right]_0^1\Rightarrow I=10\left(\frac{1}{2}-0\right)=5$$

15. *(c)* We have, $S_n=\sum_{k=1}^{n}k$

$$S_1=1$$
$$S_2=1+2=3$$
$$S_3=1+2+3=6$$
$$S_4=1+2+3+4=10$$
$$S_5=1+2+3+4+5=15$$
$$S_6=1+2+3+4+5+6=21$$

Clearly, from two terms are even two term are odd.

$\therefore S_1, S_2, S_3, \dots, S_{98}$

Among these 48 terms are even and 48 terms are odd

$$S_{99}=\frac{99\times100}{2}=99\times50=\text{even}$$

$\therefore$ Total even terms $=48+1=49$

$\therefore$ Required probability $=\frac{49}{99}$

16. *(a)* Consider the event

E_1 = Purse contains 4 copper coins and 3 silver coins.

E_2 = Purse contains 6 copper coins and 4 silver coins

A = draw copper coins

$$\therefore \quad P(E_1)=\frac{1}{2},\ P(A/E_1)=4/7$$

$$P(E_2)=\frac{1}{2},\ P(A/E_2)=6/10$$

$\therefore$ Required probability

$$P(A)=P(E_1)\times P(A/E_1)+P(E_2)\times P(A/E_2)$$

$$=\frac{1}{2}\times\frac{4}{7}+\frac{1}{2}\times\frac{6}{10}$$

$$=\frac{1}{2}\left(\frac{4}{7}+\frac{6}{10}\right)=\frac{1}{2}\left(\frac{82}{70}\right)=\frac{41}{70}$$

17. *(c)* Given, H is orthocentre of ΔABC and O is circumcentre of ΔABC.

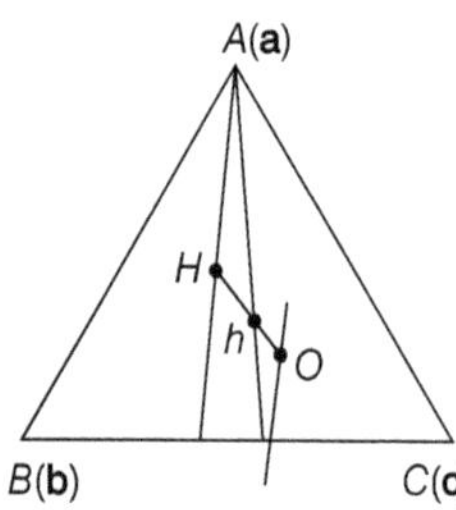

Let centroid of ΔABC is G.

In acute angle ΔABC,

We know, $HG:GO=2:1$

$$\mathbf{G}=\frac{\mathbf{A}+\mathbf{B}+\mathbf{C}}{3}$$

Now, $\quad \mathbf{G}=\frac{2\mathbf{O}+\mathbf{H}}{3}$

$$\Rightarrow \quad 3\mathbf{G}=2\mathbf{O}+\mathbf{H}$$

$$\therefore\ \mathbf{HA}+\mathbf{HB}+\mathbf{HC}$$
$$=\mathbf{A}-\mathbf{H}+\mathbf{B}-\mathbf{H}+\mathbf{C}-\mathbf{H}$$
$$=(\mathbf{A}+\mathbf{B}+\mathbf{C})-3\mathbf{H}$$
$$=3\mathbf{G}-3\mathbf{H}$$
$$=2\mathbf{O}+\mathbf{H}-3\mathbf{H}$$
$$=2\mathbf{O}-2\mathbf{H}$$
$$=2\mathbf{HO}$$

18. *(a)* We have,

$$6^m+9^n=(5+1)^m+(10-1)^n$$
$$=5k+(1)^m+10\lambda-(1)^n$$
$$=5k+10\lambda+(1)^m-(1)^n$$

It is multiple of 5 if $m\in\{1,2,3,\dots,50\}$ and $n\in\{1,3,5,\dots,49\}$ [$\because n$ must be odd]

$\therefore$ Total order pairs of $(m,n)=50\times25$
$=1250$

19. *(d)* Given,

$$a_2=\frac{a_1+a_3}{2}$$
$$a_3=\frac{a_2+a_4}{2}$$
$$a_4=\frac{a_3+a_5}{2}$$
$$\vdots$$
$$a_{2012}=\frac{a_{2011}+a_1}{2}$$

$$a_2+a_4+a_6+\dots+a_{2012}=3018 \quad \dots\text{(i)}$$
$$2a_2+2a_4+2a_6+\dots+2a_{2012}=6036$$
$$(a_1+a_3)+(a_3+a_5)+(a_5+a_7)+\dots+(a_{2011}+a_1)=6036$$
$$2(a_1+a_3+a_5+\dots+a_{2011})=6036$$
$$a_1+a_3+a_5+\dots+a_{2011}=3018 \quad \dots\text{(ii)}$$

On adding Eqs. (i) and (ii),

$$a_1+a_2+a_3+a_4+\dots+a_{2011}+a_{2012}=6036$$

20. *(b)* We have,

$$S=\{1,2,3,4,\dots,n\}$$
$$A=\{(a,b):1\le a, b\le n\}=S\times S$$
$$B=\{(x,x):x\in S\}$$
$$\therefore\quad B=\{(1,1),(2,2),(3,3),\dots,(n,n)\}$$

Number of elements in $B=n$

Total number of subset of B is 2^n.

$\therefore$ Total number of good subset of A is 2^n.

21. *(b)* Indicator diagrams for isothermal and adiabatic processes are as shown below.

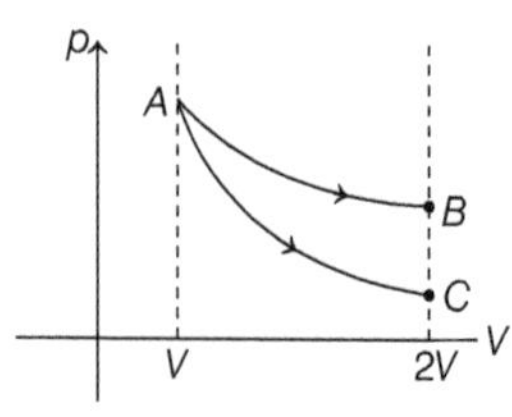

In above diagram, AB is isothermal process and AC is adiabatic process.

Work done = Area under p-V graph

So, $W_{AB} > W_{AC}$

Also, gas expands, so work done is positive.

$\Rightarrow \quad W_i > W_a > 0$

22. *(d)* Stored energy in capacitor is

$$U = \frac{q^2}{2C}$$

and charge of capacitor in an RC circuit during discharging is

$$q = q_0 e^{-t/RC}$$

Combining both, we have

$$U = \frac{q_0^2}{2C} \cdot e^{-2t/RC} = U_0 e^{-2t/RC}$$

Given, $U = \frac{U_0}{2}$

So, $\frac{U_0}{2} = U_0 e^{\frac{-2t}{RC}} \Rightarrow e^{\frac{-2t}{RC}} = \frac{1}{2}$

$$\Rightarrow \quad \frac{-2t}{RC} = \ln \frac{1}{2}$$

$$\Rightarrow \quad t = \frac{1}{2} \cdot RC \ln 2$$

23. *(a)* Dimensions of

$$\frac{\nu}{\sqrt{\frac{\sigma}{\rho R^3}}} = \frac{[\nu]}{\sqrt{\frac{[\sigma]}{[\rho R^3]}}} = \frac{[T^{-1}]}{\sqrt{\frac{[MT^{-2}]}{\left[\frac{M}{L^3} \cdot L^3\right]}}} = [M^0L^0T^0]$$

Now, from option (a),

$$\text{Dimensions of } \frac{k\rho g R^2}{\sigma} = \frac{[k\rho g R^2]}{[\sigma]} = \frac{\left[\frac{M}{L^3}\right] \cdot [LT^{-2}] \cdot [L^2]}{[MT^{-2}]} = [M^0L^0T^0]$$

So, by dimensional analysis, option (a) is correct.

24. *(c)* Moment of inertia of given configuration about an axis through point A (centre of middle disc), using parallel axes theorem is

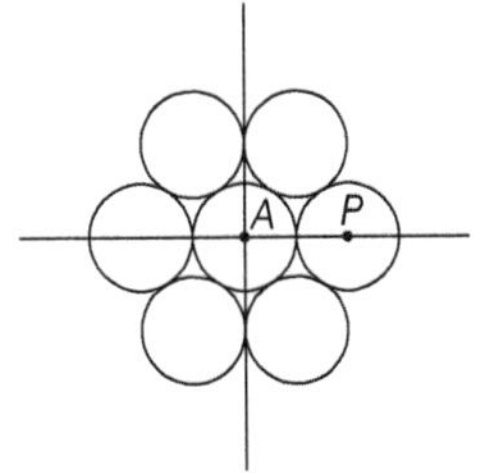

$$I_A = \frac{mr^2}{2} + 6\left(\frac{mr^2}{2} + m(2r)^2\right)$$

$$= mr^2\left(\frac{1}{2} + \frac{6}{2} + 24\right) = \frac{55}{2} mr^2$$

Now, again using parallel axes theorem, moment of inertia of configuration about P is

$$I_P = I_A + M_{\text{total}} \times d^2$$

$$= \frac{55}{2} mr^2 + 7m\,(2r)^2 = \frac{111}{2} mr^2$$

25. *(a)* By conservation of angular momentum,

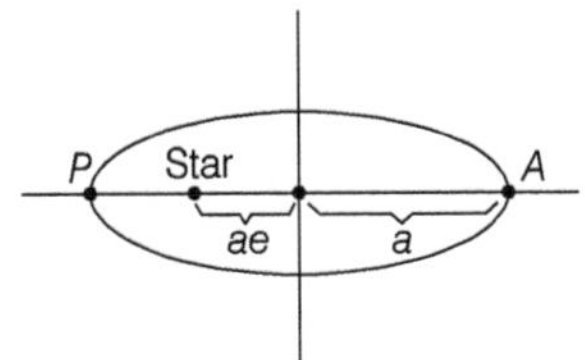

We have,

$$L_P = L_A$$

$$\Rightarrow \quad m v_P r_P = m v_a r_a$$

So, ratio of speeds at points P and A is

$$\Rightarrow \quad \frac{v_P}{v_a} = \frac{r_a}{r_P} = \frac{a + ae}{a - ae} = \frac{a(1+e)}{a(1-e)} = \frac{1+e}{1-e}$$

26. *(a)* In an interference pattern of Young's double slit experiment,

Minimum intensity = $I_{\min} = 0$

Maximum intensity = $I_{\max} = 4I_0$

Average intensity = $I_{\text{av}} = \frac{0 + 4I_0}{2} = 2I_0$

27. *(a)*

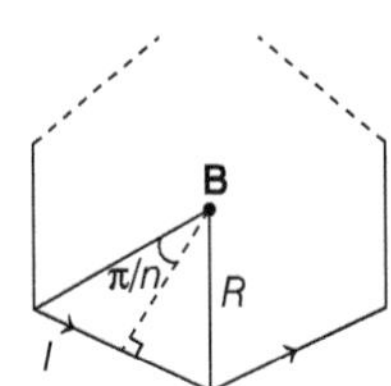

From geometry of figure, distance of side from centre is

$$d = R \cos \frac{\pi}{n}$$

So, magnetic field due to a single side length at centre is

$$B_1 = \frac{\mu_0 I}{4\pi\left(R\cos\frac{\pi}{n}\right)} \cdot 2\sin\left(\frac{\pi}{n}\right)$$

$$= \frac{2\mu_0 I}{4\pi R} \cdot \tan\frac{\pi}{n}$$

Hence, total magnetic field due to complete polygon is

$$B = n \times B_1$$

$$= n \frac{\mu_0 I}{2\pi R} \tan \frac{\pi}{n}$$

28. *(a)* Due to motion of conductor in region of magnetic field, an emf is generated in the conducting loop.

Induced current produces heat when it flows through resistor.

29. *(a)* Given, heat loss per second Q_1 through wire is proportional to surface area of wire.

$$\Rightarrow \quad Q_1 = k\pi r^2 L(\Delta T)$$

where, k = proportionality constant

and ΔT = temperature difference of wire and surroundings.

and heat generated per second is

$$Q_2 = I^2 R = \frac{I^2 \rho L}{\pi r^2}$$

In steady state, $Q_1 = Q_2$

$$\Rightarrow \quad k\pi r^2 L\,(\Delta T) = \frac{I^2 \rho L}{\pi r^2} \Rightarrow \Delta T = \frac{I^2 \rho}{k\pi^2 r^4}$$

$\therefore$ Steady state temperature is independent of length of wire.

30. *(*)* Speed of sound in air is

$$v_s = \sqrt{\frac{\gamma P}{\rho}} = \sqrt{\frac{\gamma RT}{M}}$$

Average speed of air molecule is

$$v_{\text{av}} = \sqrt{\frac{8}{\pi} \frac{RT}{M}}$$

So, ratio is $\frac{v_s}{v_{\text{av}}} = \frac{\sqrt{\gamma}}{\sqrt{\frac{8}{\pi}}} = \sqrt{\frac{\gamma\pi}{8}}$

For air $\gamma \approx 14$,

So, $\frac{v_s}{v_{\text{av}}} \approx \frac{1}{2}$

Hence, no option is matching.

Note $v_s \approx 340 \text{ ms}^{-1}$ and $v_{\text{av}} \approx 500 \text{ ms}^{-1}$ practically.

31. *(a)* Electrostatic energy of charged shell is

$$E = \frac{e^2}{8\pi\varepsilon_0 R}$$

and from Einstein's mass - energy equivalence, $E = mc^2$

Equating both, we get

$$\frac{e^2}{8\pi\varepsilon_0 R} = mc^2$$

$$\Rightarrow \quad R = \frac{e^2}{8\pi\varepsilon_0 \cdot mc^2}$$

Substituting values, we have

$$\Rightarrow \quad R = \frac{(1.6 \times 10^{-19})^2 \times 9 \times 10^9}{2 \times 9.1 \times 10^{-31} \times 9 \times 10^{16}}$$

$$= 1.4 \times 10^{-15} \text{ m}$$

32. *(a)* In simple harmonic motion, at positions near extreme positions speed of body is less.

So, more time is spend by the body in extreme positions and less near mean position.

Hence, probability of finding the body at extremes is much higher.

So, correct option is (a).

33. *(b)* Let specific heats of body at 100°C is S_1 and that of body at 0°C is S_2.

Let T = common steady state temperature, then as the bodies are kept in contact.

Heat lost = Heat gained

$$mS_1(100 - T) = mS_2(T - 0)$$

$$\Rightarrow \quad 100\, S_1 = T\,(S_1 + S_2)$$

$$\Rightarrow \quad T = \frac{100 \cdot S_1}{S_1 + S_2}$$

If $S_1 = S_2$, then $T = 50°$ C.

But given $S_1 > S_2$ and $S_1 \approx S_2$

Under these conditions,

$$\frac{S_1}{S_1 + S_2} > \frac{1}{2}$$

So, T is more than 50°C.

34. *(a)* Force on the particle is

$$F = -\alpha x^3 - \beta x^4$$

As, $$F = -\frac{dU}{dx}$$

$(\because U = \text{potential energy})$

$$\Rightarrow \quad \frac{dU}{dx} = \alpha x^3 + \beta x^4$$

Now, $$\frac{dU}{dx} = 0 \text{ at } x = 0$$

and $$\frac{d^2U}{dx^2} = \frac{d^3U}{dx^3} = 0 \text{ at } x = 0$$

But $$\frac{d^4U}{dx^4} = 6\alpha \text{ at } x = 0$$

As $\frac{d^4U}{dx^4}$ is positive, so there is a minima for U at $x = 0$.

Hence, it is position of a stable equilibrium for the particle.

35. *(d)* Potential energy of the particle is

$$V(x) = -\alpha x + \beta \sin\left(\frac{x}{\gamma}\right)$$

Clearly, dimensions of α, β and γ are

$$[\alpha] = \frac{[V]}{[x]} = \frac{[ML^2T^{-2}]}{[L]} = [MLT^{-2}]$$

$$[\beta] = [V] = [ML^2T^{-2}]$$

and $[\gamma] = [x] = [L]$

$$\text{So, } \left[\frac{\alpha\gamma}{\beta}\right] = \frac{[MLT^{-2}][L]}{[ML^2T^{-2}]} = [M^0L^0T^0]$$

36. *(d)* In collision, both kinetic energy and linear momentum are conserved.

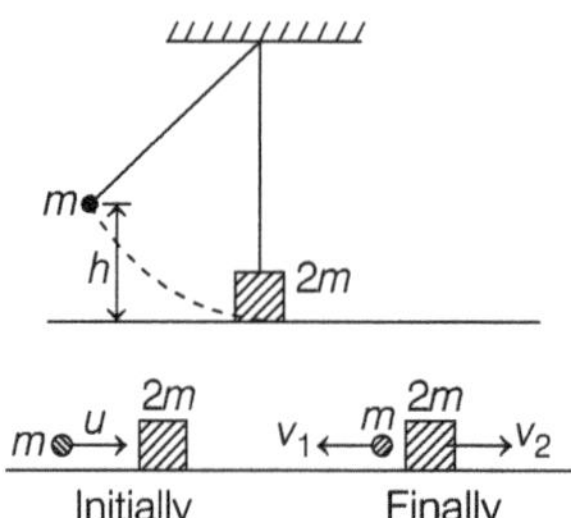

So, we have

$$\frac{1}{2}mu^2 = \frac{1}{2}mv_1^2 + \frac{1}{2}(2m)v_2^2$$

and $$mu = 2mv_2 - mv_1$$

$$\Rightarrow \quad u^2 = v_1^2 + 2v_2^2 \qquad \text{...(i)}$$

and $$u = 2v_2 - v_1 \qquad \text{...(ii)}$$

Substituting for v_2 from Eq. (ii) in Eq. (i), we get

$$u^2 = v_1^2 + 2\left(\frac{u + v_1}{2}\right)^2$$

$$\Rightarrow \quad 2u^2 = 2v_1^2 + u^2 + v_1^2 + 2uv_1$$

$$\Rightarrow \quad u^2 - v_1^2 = 2v_1(u + v_1)$$

$$\Rightarrow \quad u - v_1 = 2v_1$$

$$\Rightarrow \quad v_1 = \frac{u}{3}$$

Let ball rises to height h' after collision, then

$$mgh' = \frac{1}{2}mv_1^2$$

$$\Rightarrow \quad h' = \frac{v_1^2}{2g} = \frac{u^2}{9(2g)} = \frac{h}{9}$$

37. *(a)* Velocity initially increases and this results in increase in resistive stokes force $(= 6\pi\eta rv)$.

After some time, resistive stokes force and force of buoyancy are equal to gravitational pull on the particle.

After this instant velocity of particle becomes constant.

Hence, correct graph is (a).

38. *(a)* In steady state, rate of heat flow through both rods is equal.

$$\text{So, } \left(\frac{kA\,(T_1 - T)}{l}\right)_{\text{steel}} = \left(\frac{kA\,(T - T_2)}{l}\right)_{\text{copper}}$$

T

100°C | Steel | Copper | 0°C

Substituting given values, we have

$$\frac{50 \cdot A \cdot (100 - T)}{0.1} = \frac{400 \cdot A \cdot (T - 0)}{0.2}$$

$$\Rightarrow \quad 100\,(100 - T) = 400\,(T)$$

$$\Rightarrow \quad T = 20°\text{C}$$

39. *(a)* Decay occurs as

$$X \xrightarrow{40000 \text{ yr}} Y \xrightarrow{20 \text{ yr}} Z$$

As number of Y nuclei does not changes with time, this means decay rate of X = decay rate of Y.

$$\Rightarrow \quad \lambda_X N_X = \lambda_Y N_Y$$

$$\Rightarrow \quad \frac{N_X}{T_X} = \frac{N_Y}{T_Y}$$

$$\Rightarrow \quad N_Y = \frac{T_Y}{T_X} N_X$$

$$= \frac{20}{40000} \times 4 \times 10^{20}$$

$$= 2 \times 10^{17} \text{ nuclei}$$

40. *(c)*

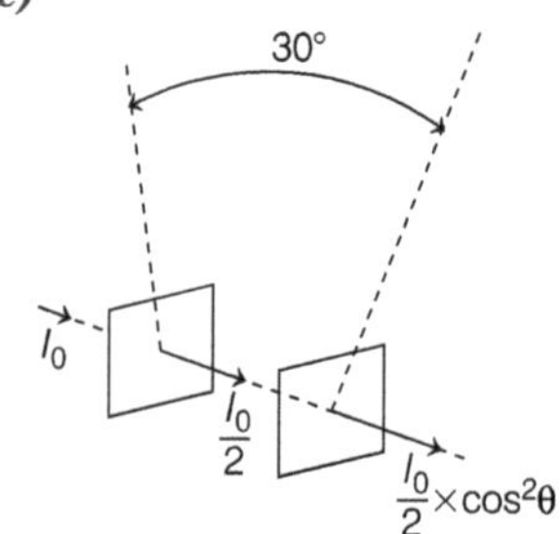

Intensity after first polaroid $= \frac{I_0}{2}$

and intensity after second polaroid from Malus' law is

$$I' = \frac{I_0}{2} \times \cos^2\theta$$

$$= \frac{I_0}{2} \times \cos^2 30°$$

$$= \frac{I_0}{2} \times \frac{3}{4} = \frac{3}{8}I_0$$

41. *(c)* The bond order of a species can be calculated as

$$\text{BO} = \frac{N_b - N_a}{2}$$

where, N_b is the number of electrons in bonding orbitals.

N_a is the number of electrons in antibonding orbitals.

(a) O_2

The electronic configuration of O_2 is

$\sigma 1s^2 \sigma * 1s^2 \sigma 2s^2 \sigma * 2s^2$

$\sigma 2p_z^2 \pi 2p_x^2 \pi 2p_y^2 \pi * 2p_x^1 \pi * 2p_y^1$

$$\text{BO} = \frac{10 - 6}{2} = 2$$

(b) F_2

The electronic configuration of F_2

$\sigma 1s^2 \sigma^* 1s^2 \sigma 2s^2 \sigma^* 2s^2$

$\sigma 2p_z^2\, \pi 2p_x^2\, \pi 2p_y^2\, \pi^* 2p_x^2\, \pi^* 2p_y^2$

$$BO = \frac{10-8}{2} = 1$$

(c) O_2^+

The electronic configuration is

$\sigma 1s^2 \sigma^* 1s^2 \sigma 2s^2 \sigma^* 2s^2 \sigma 2p_z^2$

$\pi 2p_x^2\, \pi 2p_y^2\, \pi^* 2p_x^1$

$$BO = \frac{10-5}{2} = 2.5$$

(d) F_2^+

The electronic configuration

$\sigma 1s^2 \sigma^* 1s^2 \sigma 2s^2 \sigma^* 2s^2 \sigma^* 2p_z^2$

$\pi 2p_x^2\, \pi 2p_y^2\, \pi^* 2p_x^2\, \pi^* 2p_y^1$

$$BO = \frac{10-7}{2} = 1.5$$

Thus O_2^+ has the highest BO among the given options.

42. *(d)* The symmetrical molecule has a zero dipole moment whereas unsymmetrical molecule has non-zero dipole moment.

BCl_3: Symmetrical molecule $\mu = 0$

Cl—Be—Cl: Symmetrical molecule $\mu = 0$

CCl_4: Symmetrical molecule $\mu = 0$

NCl_3: Unsymmetrical molecule $\mu \neq 0$

43. *(a)* The quantum numbers of an electron which are allowed must follow the given conditions

(i) l should range from 0 to $n-1$.

(ii) m_l should range from $-l$ to l

(iii) m_s should be equal to $+\frac{1}{2}, -\frac{1}{2}$, for a given value of m.

Thus for an electron atom, the set of allowed quantum number is

$$n = 1,\ l = 0,\ m_l = 0,\ m_s = +\frac{1}{2}$$

44. (*) In the reaction of benzene with an electrophile E^+, the structure of the intermediate σ-complex can be represented as

Benzene $+ E^+ \longrightarrow$ σ-complex (Arenium ion)

The arenium ion is resonance stabilised, which can shown as

[resonance structures of arenium ion]

Option (d) would be correct only, if there would be a positive charge.

45. *(c)* The most stable conformation of 2, 3-dibromobutane is staggered form which is shown in option (c), where 2 bulkier bromine atoms are far apart.

(Staggered form)

46. *(d)* Electronic energy of a molecule can be calculated as

$$E = \frac{3}{2}kT$$

Given, $E = 1.0\,\text{eV} = 1.6 \times 10^{-19}$ C

$K = 1.38 \times 10^{-23}$

$\therefore \quad T = E \times \frac{2}{3k}$

$$= 1.6 \times 10^{-19} \times \frac{2}{3 \times 1.38 \times 10^{-23}}$$

$$T = 10^5 \text{ K}$$

47. *(c)*

$$CH_3CH_2CN \xrightarrow{CH_3MgBr} CH_3CH_2-\underset{\text{Imine salt}}{\overset{NMgBr}{\overset{||}{C}}}-CH_3$$

$$\xrightarrow{H_3O^+} CH_3CH_2-\overset{O}{\overset{||}{C}}-CH_3$$

Nitriles being moderate electrophiles, get attacked by nucleophilic Grignard reagent to give imine salt, which then gets converted to ketone by acidic hydrolysis.

48. *(b)* For zero order reaction,

$$A \longrightarrow \text{Product}$$

$$t_{1/2} = \frac{A_0}{2k}$$

$$\frac{(t_{1/2})_1}{(t_{1/2})_2} = \frac{(A_0)_1}{(A_0)_2}$$

$$\frac{0.2}{(t_{1/2})_2} = \frac{[A_0]}{2[A_0]}$$

$$(t_{1/2})_2 = 0.4 \text{ s.}$$

49. *(b)* The isoelectronic species are those species which have same number of electrons.

(a) Sc^{2+} and V^{3+}

Number of electrons in Sc^{2+} = 19

Number of electrons in V^{3+} = 21

(b) Mn^{2+} and Fe^{3+}

Number of electrons in Mn^{2+} = 23

Number of electrons in Fe^{3+} = 23

(c) Mn^{3+} and Fe^{2+}

Number of electrons in Mn^{3+} = 22

Number of electrons in Fe^{2+} = 24

(d) Ni^{3-} and Fe^{2-}

Number of electrons in Ni^{3-} = 31

Number of electrons in Fe^{2-} = 28

Thus, Mn^{2+} and Fe^{3+} are isoelectronic pair and hence the correct option is (b).

50. *(a)*

Vinylbroxide $\xrightarrow{NaNH_2}$ H—C≡C—H (Acetyline) $+ NaBr + NH_3$

This reaction is known as dehydro halogenation. A good yield of acetylene is obtained from vinyl bromide using stronger base such as $NaNH_2$.

51. *(b)*

HO—CH₂CH=CH₂ ⇌ Allylic carbocation $\xrightarrow{\text{Conc. HBr}}$ Br—CH₂CH=CH₂

The above reaction is nucleophilic substitution reaction, where Br gets substituted over OH and takes place *via* S_N1 mechanism as allylic carbocation is formed which gets resonance stabilised.

52. *(d)*

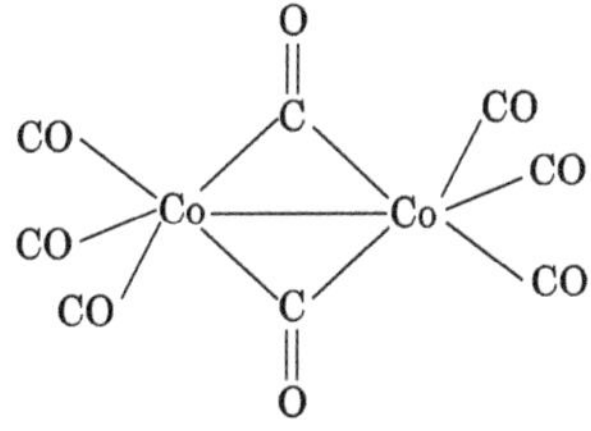

As Co is a neutral ligand. Thus, the oxidation state of cobalt in the above molecule will be 0.

53. *(b)* Given, pK_a of a weak acid = 5.85

Given, [salt] = [acid]

According to Derson's-Hassel Balch equation,

$$pH = pK_a + \log\frac{[salt]}{[acid]}$$
$$pH = 5.85 + \log 1$$
$$pH = 5.85$$

54. *(a)* Given, $\mu_{[MCl_4]^{2-}} = 3.83$

The spin magnetic moment can be calculated as

$$\mu = \sqrt{n(n+2)}$$
$$3.83 = \sqrt{n(n+2)} \Rightarrow n = 3$$

Thus, the number of unpaired electrons in $[MCl_4]^{2-}$ complex is 3.

The electronic configuration of given metals are

(a) Co = [Ar] $3d^7 4s^2$

3d: ⇅ ⇅ ↑ ↑ ↑ | 4s: □ | 4p: □ □ □

It has 3 unpaired electrons.

(b) Cu = [Ar] $3d^{10} 4s^1$

3d: ⇅ ⇅ ⇅ ⇅ ⇅ | 4s: ↑ | 4p: □ □ □

It has 1 unpaired electron.

(c) Mn = [Ar] $3d^5 4s^2$

3d: ↑ ↑ ↑ ↑ ↑ | 4s: □ | 4p: □ □ □

It has 5 unpaired electrons.

(d) Fe = [Ar] $3d^6 4s^2$

3d: ⇅ ↑ ↑ ↑ ↑ | 4s: □ | 4p: □ □ □

It has 4 unpaired electrons.

Thus, the element *M* is Co.

55. *(d)* According to Arrhenius equation,

$$\ln k = \ln A - \frac{E_a}{RT}$$

On comparing the above equation with the straight line, $y = mx + c$, we found that the plot of ln *k versus* $\frac{1}{T}$ gives a straight line with slope $= \frac{-E_a}{R}$ and intercept = ln *A*.

ln *k* vs 1/*T*

Thus, among the given options, the one that exhibits Arrhenius behaviour over the entire temperature range is shown in option (d).

56. *(b)* The reaction that gives the

$(CH_3)_3C{-}O{-}CH_3$

as the major product is

$(CH_3)_3C{-}ONa + CH_3Br \longrightarrow (CH_3)_3C{-}O{-}CH_3$

This reaction is known as Williamson synthesis. It is an important laboratory method for the preparation of symmetrical and unsymmetrical ethers.

57. *(a)* The bond length is inversely proportional to bond order.

Bond order can be calculated as

$$= \frac{N_b - N_a}{2}$$

(i) Total number of electrons in CO = 6 + 8 = 14.

The electronic configuration will be

$\sigma 1s^2 \sigma^* 1s^2 \sigma 2s^2 \sigma^* 2s^2 \sigma 2p_z^2 \pi 2p_x^2 \pi 2p_y^2$

$$BO(CO) = \frac{10-4}{2} = \frac{6}{2} = 3$$

(ii) Total number of electrons in

$CO_2 = 6 + 16 = 22$

The electronic configuration will be

$\sigma 1s^2 \sigma^* 1s^2 \sigma 2s^2 \sigma^* 2s^2$

$\sigma 2p_z^2 \pi 2p_x^2 \pi 2p_y^2 \pi^* 2p_x^2 \pi^* 2p_y^2 \sigma^* 2p_z^2 \sigma 3s^2$

$$BO(CO_2) = \frac{12-8}{2} = \frac{4}{2} = 2$$

(iii) The calculated BO for CO_3^{2-} is 1.33.

The order of bond order will be

$$CO_3^{2-} < CO_2 < CO$$

Thus, the order of bond length will be

$$CO < CO_2 < CO_3^{2-}$$

58. *(b)* If the equation is multiplied by 2 the equilibrium constant for the new equation is the square of *K*, i.e. K^2.

If we add two equation, then the equilibrium constant of new equations is the product of the equilibrium constant of those equations.

For the reactions,

$2P(g) + 3Cl_2(g) \rightleftharpoons 2PCl_3(g)$ ...(i)

$PCl_3(g) + Cl_2(g) \rightleftharpoons PCl_5(g)$ ...(ii)

$2P(g) + 5Cl_2(g) \rightleftharpoons 2PCl_5(g)$ ...(iii)

To obtain (iii) we multiply Eq. (ii) by 2 and then add Eq. (i) to it.

Thus, equilibrium constant for Eq. (iii), $K = K_1 K_2^2$

59. *(a)*

Ethylbenzene ($C_6H_5CH_2CH_3$) + $(CH_3)_2CHCH_2Cl$ $\xrightarrow[\text{AlCl}_3]{\text{Rearrangement}}$ p-$CH_3CH_2C_6H_4C(CH_3)_3$

In the above reaction, first Cl^- is lost giving $(CH_3)_2CHCH_2^+$, which is a primary carbocation rearranges to give tertiary carbocation $(CH_3)_2C^+CH_3$ and then ethyl benzene attack at 3° carbocation to give required product.

60. *(a)* When silicon is doped with boron a *n*-type semi-conductor is formed. This is because B has 3 valence electrons and Bi has 4 valence electrons, out of which 3 electrons from bonds with boron. So, there is a presence of free electron. This extra electron helps in conduction of electricity.

61. *(a)* Rheumatoid Arthritis (RA) is an autoimmune disease in which the body's immune system which normally protects its health by attacking foreign substance like bacteria and viruses mistakenly attacks the joints.

62. *(b)* The complement system has four major functions including lysis of infectious organisms, activation of inflammation, opsonisation and immune clearance. The complement system is a part of the innate immune system. It can be activated through bound IgM and IgG molecules but also by DNA, collagen and C-reactive protein.

63. *(c)* Diabetes insipidus occurs when the pituitary gland fails to secrete the hormone vasopressin, which regulates bodily fluids.

64. *(c)* Fossils are formed when a layer of sediment covers dead plants, animals and microorganisms. This makes sedimentary rocks like limestone and shale the best for the formation of fossils.

65. *(d)* Peptic ulcers are open sores that develop on the inside lining of your stomach and the upper portion of your small intestine. The most common causes of peptic ulcers are infection with bacterium *Helicobacter pylori* and long term use of aspirin and non-steroidal anti-inflammatory drugs.

66. *(b)* Transfer RNA(*t*RNA), small molecule in cells that carries amino acids to organelles called ribosomes where they are linked into proteins. Molecules of *t*RNA typically contain fewer than 100 nucleotide units and fold into a characteristic clover leaf structure.

67. *(c)* Birds and reptiles convert ammonia to uric acid, which is even less toxic than urea. In fact, uric acid does not need any water at all, they can excrete it as a solid and conserve as much water as possible. Thus, they are uricotelic. Mammals, amphibians and fishes are ureotelic animals.

68. *(d)* Ethylene is a gaseous plant hormone that speeds up the ripening process. Therefore, a ripe mango will release ethylene which will cause the other unripe mangoes to ripe if kept together.

69. *(d)* During metaphase stage of M-phase of cell cycle, all the chromosomes assume their most compacted state, as the centromeres of all the cell's chromosomes line up at the equator of the spindle. Thus, metaphase is particularly useful in cytogenetics, because chromosomes can be most easily visualised at this stage.

70. *(b)* Glucose that enters the nephron along with the filtrate after passing through the glomerulus, passes from the tubule of nephron where it is selectively reabsorbed by active transport and sent back to blood.

71. *(a)* The hypothalamus sends signals to the pituitary to release or inhibits pituitary hormone production. Thus, the hypothalamus links the nervous and endocrine systems by way of the pituitary gland.

72. *(a)* In eukaryotic cells, the citric acid cycle or TCA cycle occurs in the matrix of the mitochondrion. It yields 2 molecules of ATP for every molecule of glucose. The high energy electrons released during TCA are used in the electron transport chain to generate ATP.

73. *(c)* A receptor is a chemical group or a protein molecule on the cell surface or in the cell interior that has an affinity for a specific chemical group, molecule or virus.

74. *(d)* A silent mutation is a change in the sequence of nucleotide bases which constitutes DNA, without a subsequent change in the amino acid or the function of the overall protein.

75. *(c)* *Tinospora* is an example of plant with assimilatory roots. These are green, aerial, adventitious roots which prepare food material by photosynthesis.

76. *(b)* Vitamin-A is a group of unsaturated nutritional organic compounds that includes retinol, retinal, retinoic acid and several provitamin-A carotenoids. As the retinal component of rhodopsin is derived from vitamin-A, a deficiency of vitamin-A causes a deficit in the pigment needed by rod cells thus causing night blindness.

77. *(a)* Dengue fever is a mosquito borne tropical disease caused by the dengue virus. This may include a high fever, headache, vomiting, muscle and joint pains and skin rashes. In severe dengue, there occurs dengue haemorrhagic fever, resulting in internal bleeding due to low levels of blood platelets and blood plasma leakage.

78. *(b)* The sequence of bases in a RNA transcript of the sense strand of DNA will be exactly same except that in place of thymine there will be uracil. This is because RNA does not have thymine.

Therefore, 5′-GTTCATCG-3′ DNA sense strand 5′-GUUCAUCG-3′ RNA transcript.

79. *(b)* Secretion of saliva in human can be conditioned by giving food with sounding a buzzer and switching on a light.

Also, secretion of saliva is a unconditioned reflex as when food enters the mouth saliva is secreted to digest it.

80. *(b)* According to 10% law at each trophic level 90% of assimilated energy is lost in respiration and only 10% is passed on to the next trophic level, so respiration energy loss would be constant at each trophic level. Thus, in a food chain such as grass → deer → lion, the energy cost of respiration as a proportional of total assimilated energy at each level would be 20% – 30% – 60%.

81. *(d)* We have, $x^2 + 2ax + b^2 = 0$

It has two distinct real roots.

$\therefore \quad 4a^2 - 4b^2 > 0 \Rightarrow a^2 > b^2 \quad \ldots(i)$

$x^2 + 2bx + c^2 = 0$ has also two distinct real roots.

$\therefore \quad 4b^2 - 4c^2 > 0$

$\Rightarrow \quad b^2 > c^2 \quad \ldots(ii)$

From Eqs. (i) and (ii),

$$a^2 > b^2 > c^2$$

Now, $\quad x^2 + 2cx + a^2 = 0$

$$D = 4c^2 - 4a^2$$

$$D < 0 \qquad [\because a^2 > c^2]$$

$\therefore x^2 + 2cx + a^2 = 0$ has no real roots.

82. (*) We have, $\dfrac{1+x}{(1+x^2)(1-x)}$

$$\frac{1+x}{(1+x^2)(1-x)} = \frac{x}{1+x^2} + \frac{1}{2(1+x^2)} + \frac{1}{2(1-x)}$$

$$= x(1+x^2)^{-1} + \frac{1}{2}(1+x^2)^{-1} + \frac{1}{2}(1-x)^{-1}$$

Coefficient of x^{2012} in

$\left[x(1+x^2)^{-1} + \frac{1}{2}(1+x^2)^{-1} + \frac{1}{2}(1-x)^{-1}\right]$ is

$$= 0 + \frac{1}{2} + \frac{1}{2} = 1$$

No option is correct.

83. *(a)* We have,

$$4x^2 + 9y^2 - 8x - 36y + 15 = 0$$

$$\Rightarrow 4(x^2 - 2x + 1) + 9(y^2 - 4y + 4) = -15 + 4 + 36$$

$$\Rightarrow \quad 4(x-1)^2 + 9(y-2)^2 = 25$$

$$\Rightarrow \quad \frac{(x-1)^2}{25/4} + \frac{(y-2)^2}{25/9} = 1$$

Now, $\min(x^2 - 2x + y^2 - 4y + 5) + \max(x^2 - 2x + y^2 - 4y + 5)$

$$= \min[(x-1)^2 + (y-2)^2] + \max[(x-1)^2 + (y-2)^2]$$

$$= \frac{25}{9} + \frac{25}{4}$$

$$= 25\left(\frac{4+9}{36}\right) = \frac{25 \times 13}{36} = \frac{325}{36}$$

84. *(c)* We have,

$$\sin x + \frac{1}{2}\cos x = \sin^2\left(x + \frac{\pi}{4}\right)$$

$$\Rightarrow \quad 2\sin x + \cos x = 2\left[\sin\left(x + \frac{\pi}{4}\right)\right]^2$$

$$\Rightarrow 2\sin x + \cos x = 2\left[\sin x \sin\frac{\pi}{4} + \cos x \cos\frac{\pi}{4}\right]^2$$

$$\Rightarrow 2\sin x + \cos x = 2 \times \frac{1}{2}(\sin x + \cos x)^2$$

$$\Rightarrow 2\sin x + \cos x = \sin^2 x + \cos^2 x + 2\sin x \cos x$$

$$\Rightarrow \quad 2\sin x + \cos x = 1 + 2\sin x \cos x$$

$$\Rightarrow \quad 2\sin x - 2\sin x \cos x + \cos x - 1 = 0$$

$$\Rightarrow \quad 2\sin x(1 - \cos x) - 1(1 - \cos x) = 0$$

$$\Rightarrow \quad (2\sin x - 1)(1 - \cos x) = 0$$

$\Rightarrow \quad \sin x = \frac{1}{2}$ and $\cos x = 1$

$= x = \frac{\pi}{6}, \frac{5\pi}{6}$ and $x = 0$ [$\because x \in [0, \pi]$]

Sum of all x is $\frac{\pi}{6} + \frac{5\pi}{6} + 0 = \pi$

85. *(b)* Let

$P(x) = ax^2 + bx + c, a \neq 0$

$P'(x) = 2ax + b$

$P''(x) = 2a$

$P(0) = c = 1$

$P'(0) = b = -1$

$\therefore \quad P(x) = ax^2 - x + 1$

$P(1) = a - 1 + 1 = a$

$a \neq 0$

$\therefore \quad P(1) \neq 0$

86. *(b)* We have,

$a_n = a_1 . a_2 . a_3 \dots a_{n-1} + 4, a_1 = 5$

$a_2 = a_1 + 4 = 5 + 4 = 9$

$a_3 = a_1 a_2 + 4 = 5 \times 9 + 4 = 49$

$a_4 = a_1 a_2 a_3 = 5 \times 9 \times 49 + 4 = 2209$

$a_4 = (a_3 - 2)^2 = (49 - 2)^2 = 47^2 = 2209$

$a_3 = (a_2 - 2)^2 = (9 - 2)^2 = 49$

$\therefore \quad a_n = (a_{n-1} - 2)^2$

$$\therefore \lim_{n \to \infty} \frac{\sqrt{a_n}}{a_{n-1}} = \lim_{n \to \infty} \frac{\sqrt{(a_{n-1} - 2)^2}}{a_{n-1}}$$

$$= \lim_{n \to \infty} \left(\frac{a_{n-1} - 2}{a_{n-1}} \right) = 1$$

87. *(c)* Let $I = \int_{-\pi}^{\pi} \frac{\cos^2 x}{1 + a^x} dx$

$$\Rightarrow \quad I = \int_{-\pi}^{\pi} \frac{\cos^2 (\pi - \pi - x)}{1 + a^{\pi - \pi - x}} dx$$

$$\left[\because \int_a^b f(x)\, dx = \int_a^b f(a + b - x)\, dx \right]$$

$$\Rightarrow \quad I = \int_{-\pi}^{\pi} \frac{\cos^2 x}{1 + a^{-x}} dx$$

$$\Rightarrow \quad 2I = \int_{-\pi}^{\pi} \cos^2 x\, dx$$

$$\Rightarrow \quad 2I = 2\int_0^{\pi} \cos^2 x\, dx$$

$$\Rightarrow \quad I = \int_0^{\pi} \cos^2 x\, dx$$

$$\Rightarrow \quad I = 2 \int_0^{\pi/2} \cos^2 x\, dx$$

$$\Rightarrow \quad I = 2 \int_0^{\pi/2} \cos^2 \left(\frac{\pi}{2} - x \right) dx$$

$$\Rightarrow \quad I = 2 \int_0^{\pi/2} \sin^2 x\, dx$$

$$\Rightarrow \quad 2I = 2 \int_0^{\pi/2} (\cos^2 x + \sin^2 x)\, dx$$

$$\Rightarrow \quad 2I = 2 \int_0^{\pi/2} x\, dx \Rightarrow I = [x]_0^{\pi/2} = \frac{\pi}{2}$$

88. *(a)* We have,

$L = (2012)^{1/3} + (2013)^{1/3} + \dots + (3011)^{1/3}$

$R = (2013)^{1/3} + (2014)^{1/3} + \dots + (3012)^{1/3}$

and $I = \int_{2012}^{3012} (x)^{1/3}\, dx$

Let $\quad f(x) = x^{1/3}$

$nh = b - a$

$nh = 3012 - 2012 = 1000$

$\Rightarrow \quad n = 1000 \qquad [\because h = 1]$

$I = \frac{b - a}{h} [f(a) + f(a + h) + f(a + 2h) + \dots + f(a + (n - 1)h]$

$I = 1000[(2012)^{1/3} + (2011)^{1/3} + \dots + (3011)^{1/3}]$

$2I = 1000 \times 2 [2012]^{1/3} + (2013)^{1/3} + \dots + (3011)^{1/3}]$

$2I = 1000 [(2012)^{1/3} + L + R - (3012)^{1/3}]$

$\therefore L + R < 2I$

89. *(b)* A man tosses a coin 10 times, scoring 1 point for head and 2 points for tails.

$P(k)$ be the probability of scoring at least k points.

$\therefore$ Required probability

$$= \frac{{}^{10}C_0 + {}^{10}C_1 + {}^{10}C_2 + \dots + {}^{10}C_{k-10}}{2^{10}}$$

Given, $P(k) \geq \frac{1}{2}$

$\therefore {}^{10}C_0 + {}^{10}C_1 + {}^{10}C_2 + \dots + {}^{10}C_{k-10} \geq 2^9 = 512$

$1 + 10 + 45 + 120 + 210 + 252 + \dots > 512$

$\therefore \quad k - 10 \geq 5 \Rightarrow k \geq 15$

$\therefore$ Largest value of k is 15.

90. *(b)* We have, $f^1(x) = \frac{x + 1}{x - 1}$

$$f^2(x) = f(f(x)) = \frac{f(x) + 1}{f(x) - 1} = \frac{\frac{x + 1}{x - 1} + 1}{\frac{x + 1}{x - 1} - 1} = x$$

$f^3(x) = f(x), f^4(x) = x$

$P = f^1(2) \cdot f^2(3) \cdot f^3(4) \cdot f^4(5)$

$P = 3 \times 3 \times \frac{5}{3} \times 5 = 75$

$\therefore$ Multiple of P is 375.

91. *(b)* Energy radiated from a black body is given by

$$U = \sigma \cdot A \cdot T^4 \cdot t$$

where, σ = Stefan's constant, A = area, T = absolute temperature and t = time.

Now, ratio of energy collected in two given cases is

$$\frac{U_2}{U_1} = \frac{(A/2)\, 2T^4}{AT^4} = 8$$

Hence, ratio of temperature rise of water is

$$\frac{ms\Delta T_2}{ms\Delta T_1} = \frac{U_2}{U_1} = 8 \; .$$

$\Rightarrow \quad \Delta T_2 = 8\, \Delta T_1$

As, $\Delta T_1 = 1°$ C and $\Delta T_2 = 8°$C.

So, temperature of water increases from 10°C to 18° C.

92. *(d)* Total energy of rocket at instant of launch = Potential energy due to its position on asteroid + Kinetic energy due to motion of asteroid + Kinetic energy given for launch

$$= -\frac{GMm}{r_0} + \frac{1}{2} m V_0^2 + \frac{1}{2} m (\alpha V_0)^2$$

Rocket remains bounded to the solar system, if this energy is less than or equal to zero.

i.e. $\frac{-GMm}{r_0} + \frac{1}{2} m V_0^2 (1 + \alpha^2) \leq 0$

$$\Rightarrow \quad \frac{1}{2} m V_0^2 (1 + \alpha^2) = \frac{GMm}{r_0}$$

$$\Rightarrow \quad \frac{1}{2} m \frac{GM}{r_0} (1 + \alpha^2) = \frac{GMm}{r_0}$$

$\Rightarrow \quad \alpha^2 = 1 \Rightarrow \alpha = 1$

93. *(a)* For material B, total disintegration constant.

$$\lambda = \lambda_1 + \lambda_2$$

$$\Rightarrow \quad \frac{1}{\tau_B} = \frac{1}{T_1} + \frac{1}{T_2}$$

where, λ = decay constant

and τ_B = half-life time of sample.

$$\Rightarrow \quad \frac{1}{\tau_B} = \frac{1}{(\tau_A/2)} + \frac{1}{3\tau_A}$$

$$\Rightarrow \quad \tau_B = \frac{\left(\frac{\tau_A}{2}\right) 3\tau_A}{\left(\frac{\tau_A}{2}\right) + 3\tau_A} = \left(\frac{\frac{3}{2}}{\frac{7}{2}}\right) \tau_A$$

or $\quad \tau_B = \frac{3}{7} \tau_A \Rightarrow \frac{\tau_B}{\tau_A} = \frac{3}{7}$

94. *(a)* Energy of emergent photo electron is E = Incident photon energy – Work function

$\Rightarrow \quad E = 3 - 2.3 = 0.7$ eV

As energy of photoelectron is 0.7 eV this implies that an electron turns back when it crosses a potential difference of 0.7 V. Now, in 5 mm a potential difference of 1 V is occured.

So, a potential difference of 0.7 V occurs in $0.7 \times 5 = 3.5$ mm.

So, photoelectron turns back after 3.5 mm.

95. *(a)*

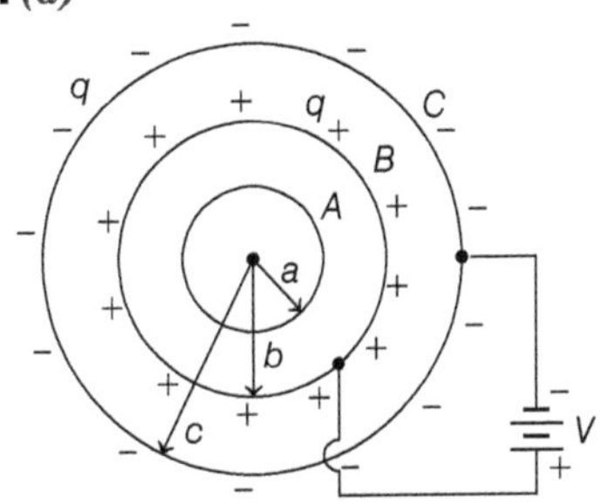

Let potential source (cell) of V volts is connected across B and C. Then potential on surface of $B = V_B = \frac{kq}{b} + \frac{k(-q)}{c}$

As given, $V_B = V$

So, $V = kq\left(\frac{1}{b} - \frac{1}{c}\right)$

$= kq\left(\frac{c-b}{bc}\right)$

$\Rightarrow \quad q = \frac{Vbc}{k(c-b)}$

So, charge on C is $-q = \frac{-Vbc}{k(c-b)}$

where, $k = \frac{1}{4\pi\varepsilon_0}$ = constant.

Hence, the charge on the sphere is $-4\pi\varepsilon_0 \frac{Vbc}{(c-b)}$.

96. *(a)* Let R = radius of circular region through which outside objects are visible.

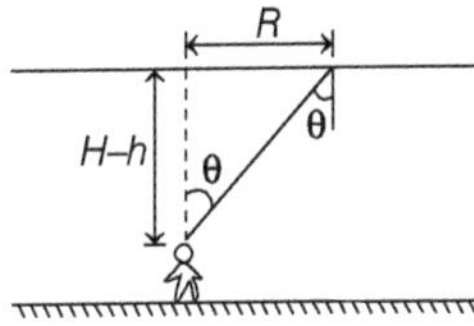

Then, θ = angle of critical incidence.

$\therefore \quad \sin\theta = \frac{1}{\mu} = \frac{1}{4/3} = \frac{3}{4}$

Now, $\tan\theta = \frac{\sin\theta}{\cos\theta}$

$= \frac{3/4}{\sqrt{1-(9/16)}}$

$\Rightarrow \quad \tan\theta = \frac{3}{\sqrt{7}}$

From above diagram, $\tan\theta = \frac{R}{H-h}$

So, $\frac{R}{H-h} = \frac{3}{\sqrt{7}}$

or $\quad R = 3(H-h)/\sqrt{7}$

97. *(a)* First we connect points of equal potentials and redraw equivalent circuit.

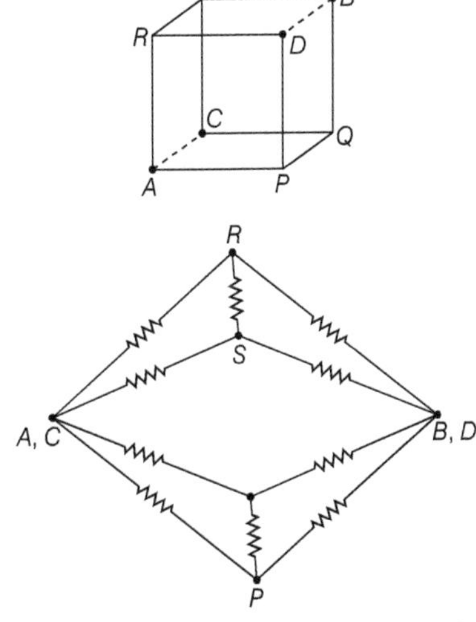

Above equivalent circuit is a combination of two balanced Wheat stones bridge. Hence, simplified circuit is

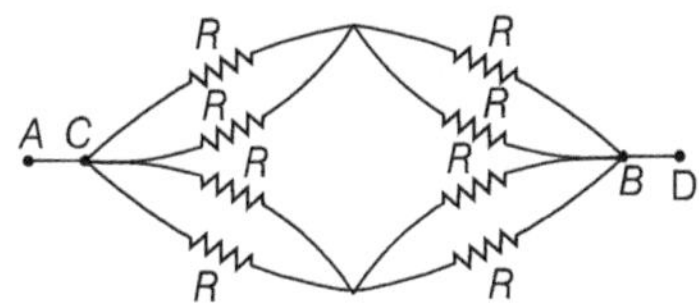

In above circuit, resistance combination gives,

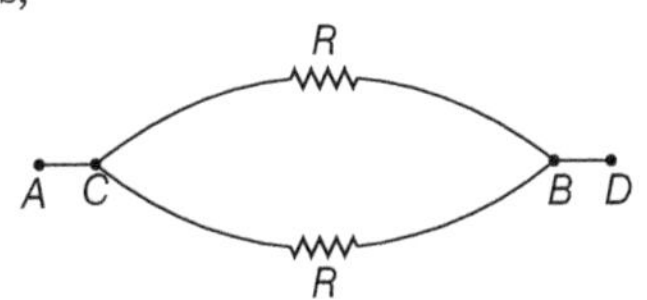

So, equivalent resistance between A and B is $R_{AB} = \frac{R}{2}\Omega$.

98. *(d)* Given speed of sound,

$$v = 300\,\text{ms}^{-1}$$

And wave equation is

$$y = y_0 \sin\left(\frac{2\pi}{L}x\right) \cdot \sin\left(\frac{2\pi}{L}x + \frac{\pi}{4}\right)$$

So, angular wave number,

$$k = \frac{2\pi}{\lambda} = \frac{2\pi}{L}$$

$\therefore \quad \lambda = L = 1.2$ m

Frequency of fundamental vibration is

$$\nu = \frac{v}{\lambda} = \frac{300}{1.2} = 250\,\text{Hz}$$

So, option (d) is incorrect.

99. *(b)* Free body diagram for block 1 and 2 are as shown below.

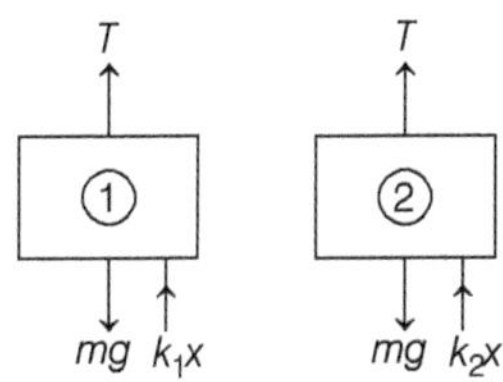

So, equations of force balance are

$$T + k_1x - mg = ma_1$$

and $\quad k_2x + mg - T = ma_2$

As string is in extensible,

$$a_1 = a_2 = a \text{ (let)}$$

Then adding both equations, we have

$$(k_1 + k_2)\,x = 2ma$$

$\Rightarrow \quad a = \left(\frac{k_1 + k_2}{2m}\right) \cdot x$

100. *(b)* Let v is the velocity of bob at position θ after being released from horizontal position.

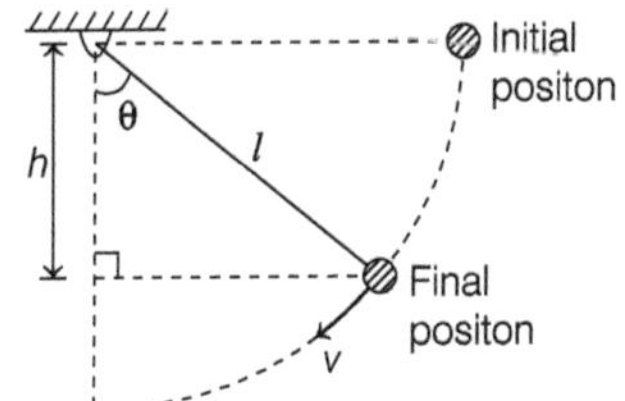

From energy conservation, we have

$$mgh = \frac{1}{2}mv^2$$

$\Rightarrow \quad mg\,(l\cos\theta) = \frac{1}{2}mv^2$

$\Rightarrow \quad \frac{v^2}{l} = 2g\cos\theta \quad \ldots\text{(i)}$

Restoring force on bob is component of weight.

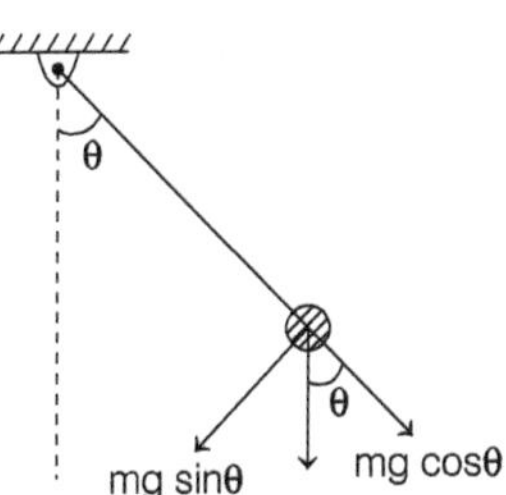

So, tangential component of acceleration is

$$a_t = g\sin\theta \quad \ldots\text{(ii)}$$

If total acceleration **a** makes angle ϕ with string. Then,

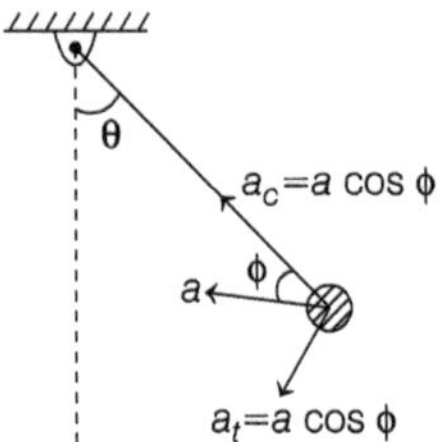

Tangential component,

$$a_t = a \sin \phi$$

and radial component, $a_c = a \cos \phi$

So, $$\frac{a_t}{a_c} = \frac{a \sin \phi}{a \cos \phi} = \tan \phi$$

$\Rightarrow$ $$\tan \phi = \frac{a_t}{a_c} \quad ...(iii)$$

Substituting values of a_t and a_c from Eqs. (i) and (ii) in Eq. (iii), we get

$$\tan \phi = \frac{g \sin \theta}{2g \cos \theta}$$

$\Rightarrow$ $$\tan \phi = \frac{\tan \theta}{2}$$

So, $$\phi = \tan^{-1}\left(\frac{\tan \theta}{2}\right)$$

101. *(b)*

$$Ph—C\equiv CH \xrightarrow[\text{Liq. } NH_3]{\text{(i) } NaNH_2} Ph—C\equiv C^- \; Na^+ \xrightarrow{\text{(ii) } 2CH_3I}$$

Ph(H)C=C(CH₃)(H) *cis*-form $\xrightarrow{\text{(iii) } 3.H_2 \text{ Pd/C}}$ Ph(H)C=C(CH₃)(H) *cis*

In reaction (i), the Ph—C ≡ C—H (alkyne) hydrogen is very acidic and thus get attacked by strong base $NaNH_2$ to give Ph—C ≡ C^-Na^+. This compound when reacts with CH_3I, CH_3 replaces Na from it. In reaction (iii), reduction of alkyne takes place with Pd/C which gives a *cis* alkene.

102. *(b)* Given,

molar ratio of adenine to cytosine, i.e. A : C = 0.7

Number of moles of adenine = 350000

Molar ratio of A : C

$$= \frac{\text{Number of moles of adenine}}{\text{Number of moles of cytosine}}$$

$$0.7 = \frac{350000}{\text{Number of moles of cytosine}}$$

$$\text{Number of moles of cytosine} = \frac{350000}{0.7}$$

$$= 500000$$

In DNA, number of moles of cytosine is equal to number of moles of guanine, which is equal to 500000.

103. *(b)*

Br (*R*-2-bromobutane) + NaOH (*aq*) $\longrightarrow$ H_3C–CH(OH)(H)–CH_3 + CH_3–CH(OH)(H)–CH_3

(1 : 1 mixture)

As (*R*) 2-bromobutane is a 3° halide, so it will undergo S_N1 mechanism. As it is optically active alkyl halide and follows S_N1 mechanism, it is accompanied by racemisation. Thus, it will give both *R* and *S* products.

104. *(a)* Phenol on treatment with dil. HNO_3 gives two products *ortho* and *para*. *Ortho* product (*P*) is steam volatile where as *para* product (*Q*) is not.

Phenol (OH) $\xrightarrow{\text{Dil. } HNO_3}$ (OH, NO_2) *P* *o*-isomer (Steam volatile) + (OH, NO_2) *Q* *p*-isomer

105. *(c)* Given,

Wavelength of light = 660 nm

Work function of metal

$$= 1.0 \text{eV} = 1.6 \times 10^{-19} \text{ J}$$

Energy of photon is given as

$$E = \frac{hc}{\lambda}$$

$$= \frac{6.6 \times 10^{-34} \times 3 \times 10^8}{660 \times 10^{-9}}$$

$$= 3 \times 10^{-9} \text{ J}$$

According to Einstein equation, kinetic energy of the ejected electron is given by

$$\text{KE} = hv - hv_0$$

$$= 3 \times 10^{-19} - 1.6 \times 10^{-19}$$

$$= 1.4 \times 10^{-19} \text{ J}$$

$$\text{Also, } \lambda = \frac{h}{\sqrt{2m\text{KE}}}$$

$$= \frac{6.6 \times 10^{-34}}{\sqrt{2 \times 9.1 \times 10^{-31} \times 1.4 \times 10^{-19}}}$$

$$= 1.32 \times 10^{-9} \text{ m}$$

106. *(b)* Interplanar distance,

$$d = \frac{a}{\sqrt{h^2 + k^2 + l^2}}$$

Given, length of cubic lattice = 450 pm

For (221) plane, $h = 2$, $k = 2$, $l = 1$

$$\therefore \quad d = \frac{450}{\sqrt{(2)^2 + (2)^2 + (1)^2}}$$

$$= \frac{450}{\sqrt{9}} = \frac{450}{3} = 150 \text{ pm}$$

107. *(b)* Given,

ΔH for vaporisation of a liquid

$$= 20 \text{ kJ/mol}$$

Number of moles of gas = 1

Temperature = 60° C = 60 + 273 = 333 K

$$\Delta H = \Delta U = \Delta n_g RT$$

$$\text{or } \Delta U = \Delta H - \Delta n_g RT$$

$$= 20 - 1 \times 8.314 \times 10^{-3} \times 333$$

$$= 17.2 \text{ kJ/mol}$$

108. *(c)* In $Ni(CO)_4$, Ni, is in 0 oxidation number thus, its electronic configuration is $3d^8 4s^2$.

Ni (Ground state) 3d: ⇅ ⇅ ⇅ ↑ ↑ 4s: ⇅ 4p: □ □ □

As CO is a strong field ligand and pairing of electrons occurs thus its configuration becomes $3d^{10} 4s^0$

$Ni(CO)_4$ 3d: ⇅ ⇅ ⇅ ⇅ ⇅ 4s: □ 4p: □ □ □ (sp^3 hybridisation)

As $Ni(CO)_4$ has sp^3-hybridisation, so its geometry will be tetrahedral. Also, all the electrons are paired, so it is diamagnetic in nature.

109. *(d)* Given,

Number of moles of ideal gas, $n = 3$

$V_1 = 2\text{L}$, $V_2 = 20\text{L}$, $T = 300 \text{K}$

From 1st law of thermodynamic

$$w = u + q$$

For isothermal reaction, $u = 0$

$$\therefore \quad w = q$$

Work done in isothermal reversible is given by

$$w = -2.303nRT \log \frac{V_2}{V_1}$$

$$= -2.303 \times 3 \times 8.314 \times 300 \times \log \frac{20}{2}$$

$$= -17200 \text{ J/mol}$$

$$W = q = -17.2 \text{ kJ/mol}$$

110. *(d)* According to rate law

$$R = k[X]^x [Y]^y$$

$$\frac{\text{Rate}_1}{\text{Rate}_2} = \frac{k[0.25]^x [0.25]^y}{k[0.50]^x [0.25]^y} = \frac{1 \times 10^{-6}}{4 \times 10^{-6}}$$

On solving we get , $\frac{1}{2^x} = \frac{1}{4}$

$$\frac{1}{2^x} = \frac{1}{2^2}$$

$\Rightarrow \quad x = 2$

Also, $\frac{\text{rate}_1}{\text{rate}_3} = \frac{k[0.25]^x[0.25]^y}{k[0.25]^x[0.50]^y} = \frac{1\times10^{-6}}{8\times10^{-6}}$

On solving we get,

$$\frac{1}{2y} = \frac{1}{8} \Rightarrow \frac{1}{2y} = \frac{1}{2^3}$$

$\Rightarrow \quad y = 3$

The rate of reaction $= k[X]^2 [Y]^3$

Thus, the order of reaction

$$= x + y = 2 + 3 = 5$$

111. *(d)* Peroxidase enzyme uses hydrogen peroxidase as a substrate to produce oxygen that causes frothing at the site of injury.

112. *(b)* High salt leads to water retention in the blood that increases blood volume and further increases the blood pressure.

When the concentration of salts in the blood increases, the volume of the blood increases due to excessive absorption of water and hence the blood pressure increases in the blood vessels.

113. *(c)* Insectivorous plants use insects as a source of nitrogen.

Since, insectivorous plants grow in places like swampy soil or where the soil is thin or poor in nutrients. They trap insects to fulfil their nitrogen requirement.

114. *(c)* Phenotypic ratio (Red : Roan : White) is 1 : 2 : 1.

As red and white colour traits are codominant on selfing,

F_2-generation–RW

	R	W
R	RR (Red)	RW (Roan)
W	RW (Roan)	WW (White)

115. *(b)* The restriction site of *Eco*RI contains 6 bases. The probability of this configuration is $\frac{1}{4^6}$. Hence, for a given length of genome, 10 kb in this case, the expected number of cuts is

$10{,}000 \times \frac{1}{4^6} = 2.44$ or 2 approximately.

116. *(b)* During enzyme catalysis, enzymes speed up chemical reactions by lowering activation energy without being consumed.

117. *(d)* Ganga water may possibly contain antibiotic that may kill *Vibrio cholerae* bacteria certainly. *Vibrio cholerae* are cholera causing bacteria and antibiotics are chemicals that kill bacteria by destroying their cell membrane.

118. *(a)* On starting fast, glucose level decreases and then remains constant as glucose begins to mobilise from stored glycogen in liver and then further decreases. Thus, option (a) is correct.

119. *(d)* The given polypeptide will give 7 amino acids.

As,

5′-AGCATATGATCGTTTCTCTGCTTTG AACT-3′ is an open reading frame (i.e. contains both initiating and stop codons), thus will possibly encode 7 amino acids.

120. *(c)* Herbivore population has been kept in control by predators. When there are more predators than prey, the predators do not have enough to eat and they die. When there are more prey than predators, the prey eat everything until there is nothing left to eat and they die. So, nature fell into a delicate balance where, under normal conditions, there are just enough predators to keep the herbivore population in check and just enough herbivores to maintain the predator population while keeping this own population stable.

QUESTION PAPER 2011

Stream : SB/SX

MM : 160

Instructions

1. There are 120 questions in this paper.
2. The question paper contains two parts; Part I (1 Mark Questions) and Part II (2 Marks Questions).
3. There are four sections in each part; Mathematics, Physics, Chemistry and Biology.
4. Out of the four options given with each question, only one is correct.

PART-I (1 Mark Questions)

MATHEMATICS

1. Suppose $\log_a b + \log_b a = c$. The smallest possible integer value of c for all $a, b > 1$ is

(a) 4 (b) 3 (c) 2 (d) 1

2. Suppose n is a natural number such that $|i + 2i^2 + 3i^3 + \ldots ni^n| = 18\sqrt{2}$, where i is the square root of -1. Then, n is

(a) 9 (b) 18 (c) 36 (d) 72

3. Let P be an $m \times m$ matrix such that $P^2 = P$. Then, $(I + P)^n$ equals

(a) $I + P$ (b) $I + nP$ (c) $I + 2^n P$ (d) $I + (2^n - 1) P$

4. Consider the cubic equation $x^3 + ax^2 + bx + c = 0$, where a, b, c are real numbers. Which of the following statements is correct?

(a) If $a^2 - 2b < 0$, then the equation has one real and two imaginary roots

(b) If $a^2 - 2b \geq 0$, then the equation has all real roots.

(c) If $a^2 - 2b > 0$, then the equation has all real and distinct roots.

(d) If $4a^3 - 27b^2 > 0$, then the equation has real and distinct roots.

5. All the points (x, y) in the plane satisfying the equation $x^2 + 2x \sin (xy) + 1 = 0$ lie on

(a) a pair of straight lines (b) a family of hyperbolas

(c) a parabola (d) an ellipse

6. Let $A = (4, 0), B = (0, 12)$ be two points in the plane. The locus of a point C such that the area of ΔABC is 18 sq units is

(a) $(y + 3x + 12)^2 = 81$ (b) $(y + 3x + 81)^2 = 12$

(c) $(y + 3x - 12)^2 = 81$ (d) $(y + 3x - 81)^2 = 12$

7. In a rectangle $ABCD$, the coordinates of A and B are $(1, 2)$ and $(3, 6)$ respectively and some diameter of the circumscribing circle of $ABCD$ has equation $2x - y + 4 = 0$. Then, the area of the rectangle is

(a) 16 (b) $2\sqrt{10}$

(c) $2\sqrt{5}$ (d) 20

8. In the XY-plane, three distinct lines l_1, l_2, l_3 concur at a point $(\lambda, 0)$. Further the lines l_1, l_2, l_3 are normals to the parabola $y^2 = 6x$ at the points $A = (x_1, y_1)$, $B = (x_2, y_2)$, $C = (x_3, y_3)$ respectively. Then, we have

(a) $\lambda < -5$ (b) $\lambda > 3$

(c) $-5 < \lambda < -3$ (d) $0 < \lambda < 3$

9. Let $f(x) = \cos 5x + A \cos 4x + B \cos 3x + C \cos 2x + D \cos x + E$, and

$$T = f(0) - f\left(\frac{\pi}{5}\right) + f\left(\frac{2\pi}{5}\right) - f\left(\frac{3\pi}{5}\right) + \dots + f\left(\frac{8\pi}{5}\right) - f\left(\frac{9\pi}{5}\right).$$

Then, T

(a) depends on A, B, C, D, E
(b) depends on A, C, E, but independent of B and D
(c) depends on B, D, but independent of A, C, E
(d) is independent of A, B, C, D, E

10. In ΔABC, we are given that $3 \sin A + 4 \cos B = 6$ and $4 \sin B + 3 \cos A = 1$. Then, the measure of the $\angle C$ is

(a) 30° (b) 150° (c) 60° (d) 75°

11. Which of the following intervals is a possible domain of the function $f(x) = \log_{\{x\}} [x] + \log_{[x]} \{x\}$, where $[x]$ is the greatest integer not exceeding x and $\{x\} = x - [x]$?

(a) (0, 1) (b) (1, 2) (c) (2, 3) (d) (3, 5)

12. If $f(x) = (2011 + x)^n$, where x is a real variable and n is a positive integer, then the value of

$$f(0) + f'(0) + \frac{f''(0)}{2!} + \dots + \frac{f^{(n-1)}(0)}{(n-1)!}$$ is

(a) $(2011)^n$ (b) $(2012)^n$
(c) $(2012)^n - 1$ (d) $n\ (2011)^n$

13. The minimum distance between a point on the curve $y = e^x$ and a point on the curve $y = \log_e x$ is

(a) $\frac{1}{\sqrt{2}}$ (b) $\sqrt{2}$ (c) $\sqrt{3}$ (d) $2\sqrt{2}$

14. Let $f : (2, \infty) \to N$ be defined by $f(x) =$ the largest prime factor of $[x]$. Then, $\int_2^8 f(x)\, dx$ is equal to

(a) 17 (b) 22 (c) 23 (d) 25

15. Let $[x]$ denote the largest integer not exceeding x and $\{x\} = x - [x]$. Then, $\int_0^{2012} \frac{e^{\cos(\pi\{x\})}}{e^{\cos(\pi\{x\})} + e^{-\cos(\pi\{x\})}}\, dx$ is equal to

(a) 0 (b) 1006 (c) 2012 (d) 2012π

16. The value of $\lim_{n \to \infty} \left(\frac{1}{\sqrt{4n^2 - 1}} + \frac{1}{\sqrt{4n^2 - 4}} + \dots + \frac{1}{\sqrt{4n^2 - n^2}} \right)$ is

(a) $\frac{1}{4}$ (b) $\frac{\pi}{12}$ (c) $\frac{\pi}{4}$ (d) $\frac{\pi}{6}$

17. Two players play the following game : A writes 3, 5, 6 on three different cards: B writes 8, 9, 10 on three different cards. Both draw randomly two cards from their collections. Then, A computes the product of two numbers he/she has drawn, and B computes the sum of two numbers he/she has drawn. The player getting the larger number wins. What is the probability that A wins?

(a) $\frac{1}{3}$ (b) $\frac{5}{9}$ (c) $\frac{4}{9}$ (d) $\frac{1}{9}$

18. Let $\vec{a}, \vec{b}, \vec{c}$ be three vectors in the xyz -space such that $\vec{a} \times \vec{b} = \vec{b} \times \vec{c} = \vec{c} \times \vec{a} \neq 0$?

If A, B, C are points with position vectors $\vec{a}, \vec{b}, \vec{c}$ respectively, then the number of possible positions of the centroid of ΔABC is

(a) 1 (b) 2 (c) 3 (d) 6

19. The sum of $(1^2 - 1 + 1)(1!) + (2^2 - 2 + 1)(2!) + \dots + (n^2 - n + 1)(n!)$ is

(a) $(n + 2)!$ (b) $(n - 1)((n + 1)!) + 1$
(c) $(n + 2)! - 1$ (d) $n((n + 1)!) - 1$

20. Let X be a non-empty set and let $P(X)$ denote the collection of all subsets of X. Define

$$f : X \times P(X) \to R \text{ by } f(x, A) = \begin{cases} 1, & \text{if } x \in A \\ 0, & \text{if } x \notin A \end{cases}.$$

Then, $f(x, A \cup B)$ equals

(a) $f(x, A) + f(x, B)$
(b) $f(x, A) + f(x, B) - 1$
(c) $f(x, A) + f(x, B) - f(x, A)\, f(x, B)$
(d) $f(x, A) + |\, f(x, A) - f(x, B)\,|$

PHYSICS

21. A narrow but tall cabin is falling freely near the earth's surface. Inside the cabin, two small stones A and B are released from rest (relative to the cabin). Initially A is much above the centre of mass and B much below the centre of mass of the cabin. A close observation of the motion of A and B will reveal that

(a) both A and B continue to be exactly at rest relative to the cabin
(b) A moves slowly upward and B moves slowly downward relative to the cabin
(c) both A and B fall to the bottom of the cabin with constant acceleration due to gravity
(d) A and B move slightly towards each other vertically

22. Two plates each of mass m are connected by a massless spring as shown below.

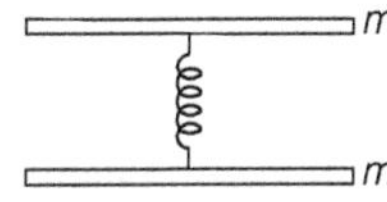

A weight w is put on the upper plate which compresses the spring further. When w is removed, the entire assembly jumps up. The minimum weight w needed for the assembly to jump up when the weight is removed is just more than

(a) m (b) $2m$ (c) $3m$ (d) $4m$

23. If the speed v of the bob in a simple pendulum is plotted against the tangential acceleration a, the correct graph will be represented by

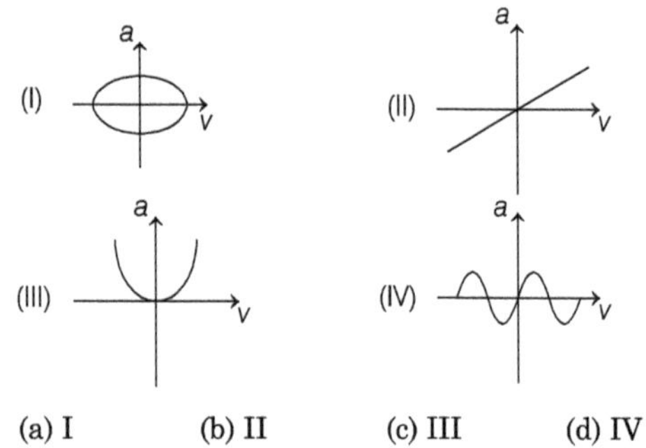

(a) I (b) II (c) III (d) IV

24. A container with rigid walls is covered with perfectly insulating material. The container is divided into two parts by a partition. One part contains a gas while the other is fully evacuated (vacuum). The partition is suddenly removed. The gas rushes to fill the entire volume and comes to equilibrium after a little time. If the gas is not ideal, then

(a) the initial internal energy of the gas equals its final internal energy
(b) the initial temperature of the gas equals its final temperature
(c) the initial pressure of the gas equals its final pressure
(d) the initial entropy of the gas equals its final entropy

25. Two bulbs of identical volumes connected by a small capillary are initially filled with an ideal gas at temperature T. Bulb 2 is heated to maintain a temperature $2T$, while bulb 1 remains at temperature T. Assume throughout that the heat conduction by the capillary is negligible. Then, the ratio of final mass of the gas in bulb 2 to the initial mass of the gas in the same bulb is close to

(a) 1/2 (b) 2/3 (c) 1/3 (d) 1

26. Two rods, one made of copper and the other steel of the same length and cross-sectional area are joined together. The thermal conductivity of copper is 385 $Js^{-1}m^{-1}K^{-1}$ and steel is 50 $Js^{-1}m^{-1}K^{-1}$. If the copper end is held at 100°C and the steel end is held at 0°C, what is the junction temperature? (Assuming no other heat losses)

(a) 12°C (b) 50°C (c) 73°C (d) 88°C

27. Jet aircrafts fly at altitudes above 30000 ft, where the air is very cold at −40° C and the pressure is 0.28 atm. The cabin is maintained at 1 atm pressure by means of a compressor which exchanges air from outside adiabatically. In order to have a comfortable cabin temperature of 25°C, we will require in addition

(a) a heater to warm the air injected into the cabin
(b) an air-conditioner to cool the air injected into the cabin
(c) neither a heater nor an air-conditioner, the compressor is sufficient
(d) alternatively heating and cooling in the two halves of the compressor cycle

28. A speaker emits a sound wave of frequency f_0. When it moves towards a stationary observer with speed u, the observer measures a frequency f_1. If the speaker is stationary and the observer moves towards it with speed u, the measured frequency is f_2. Then,

(a) $f_1 = f_2 < f_0$ (b) $f_1 > f_2$ (c) $f_1 < f_2$ (d) $f_1 = f_2 > f_0$

29. A plane polarised light passed through successive polarisers which are rotated by 30° with respect to each other in the clockwise direction. Neglecting absorption by the polarisers and given that the first polariser's axis is parallel to the plane of polarisation of the incident light, the intensity of light at the exit of the fifth polariser is closest to

(a) same as that of the incident light
(b) 17.5% of the incident light
(c) 30% of the incident light
(d) zero

30. At 23°C, a pipe open at both ends resonates at a frequency of 450 Hz. At what frequency does the same pipe resonate on a hot day when the speed of sound is 4 % higher than it would be at 23°C?

(a) 446 Hz (b) 454 Hz (c) 468 Hz (d) 459 Hz

31. In a Young's double slit experiment, light from a laser source falls on a pair of very narrow slits separated by 1.0 μm and bright fringes separated by 1.0 mm are observed on a distant screen. If the frequency of the laser light is doubled, what will be the separation of the bright fringes?

(a) 0.25 mm (b) 0.5 mm (c) 1.0 mm (d) 2.0 mm

32. For a domestic AC supply of 220 V at 50 cps, the potential difference between the terminals of a two-pin electric outlet in a room is (in volt) given by

(a) $V(t) = 220\sqrt{2}\cos 100\pi t$ (b) $V(t) = 220\cos 50t$
(c) $V(t) = 220\cos 100\pi t$ (d) $V(t) = 220\sqrt{2}\cos 50t$

33. In the circuit shown below, the resistances are given (in Ω) and the battery is assumed ideal with emf equal to 3.0 V. The resistor that dissipates the most power is

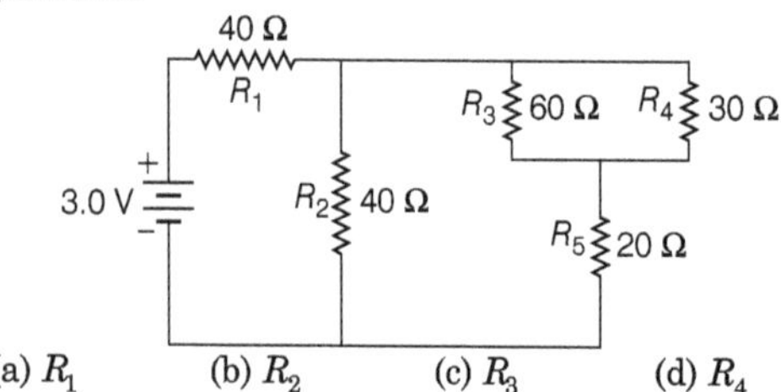

(a) R_1 (b) R_2 (c) R_3 (d) R_4

34. An electron collides with a free molecule initially in its ground state. The collision leaves the molecule in an excited state that is metastable and does not decay to the ground state by radiation. Let K be the sum of the initial kinetic energies of the electron and the molecule and $\mathbf{p}$ be the sum of their initial momenta. Let K' and $\mathbf{p}'$ represent the same physical quantities after the collision. Then,

(a) $K = K'$, $\mathbf{p} = \mathbf{p}'$ (b) $K' < K$, $\mathbf{p} = \mathbf{p}'$
(c) $K = K'$, $\mathbf{p} \neq \mathbf{p}'$ (d) $K' < K$, $\mathbf{p} \neq \mathbf{p}'$

35. In the circuit shown below, the switch is closed at time $t = 0$.

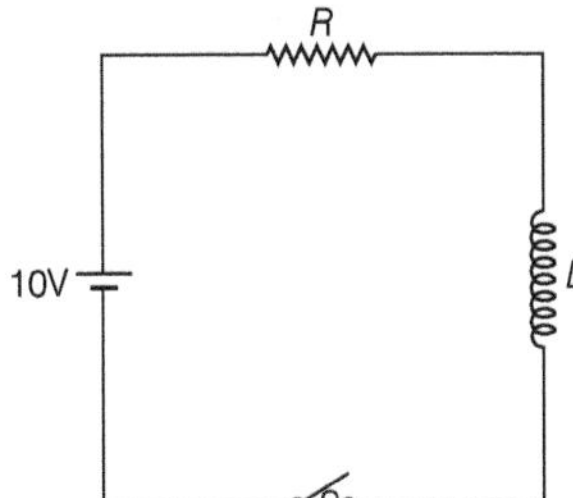

Which of the graphs shown below best represents the voltage across the inductor, as seen on an oscilloscope?

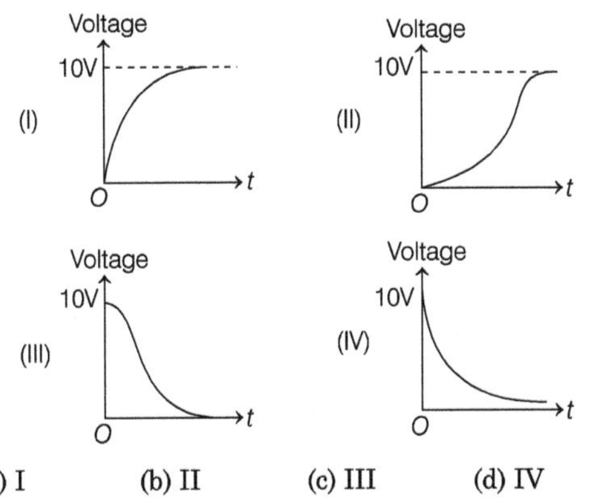

(a) I (b) II (c) III (d) IV

36. Given below are three schematic graphs of potential energy $V(r)$ *versus* distance r for three atomic particles : electron (e^-), proton (p^+) and neutron (n), in the presence of a nucleus at the origin O. The radius of the nucleus is r_0. The scale on the V-axis may not be the same for all figures. The correct pairing of each graph with the corresponding atomic particle is

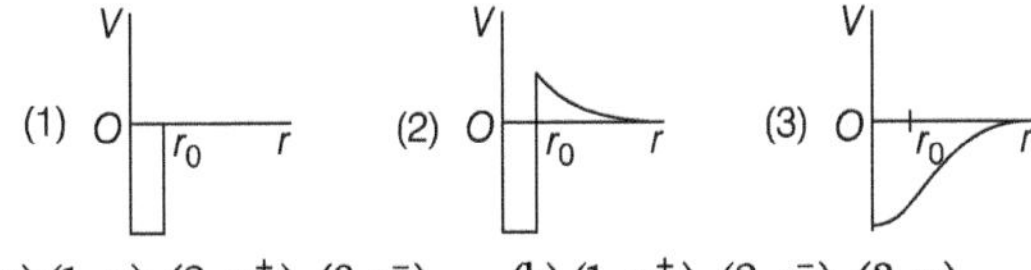

(a) $(1, n), (2, p^+), (3, e^-)$ (b) $(1, p^+), (2, e^-), (3, n)$
(c) $(1, e^-), (2, p^+), (3, n)$ (d) $(1, p^+), (2, n), (3, e^-)$

37. Due to transitions among its first three energy levels, hydrogenic atom emits radiation at three discrete wavelengths λ_1, λ_2 and λ_3 $(\lambda_1 < \lambda_2 < \lambda_3)$. Then,

(a) $\lambda_1 = \lambda_2 + \lambda_3$ (b) $\lambda_1 + \lambda_2 = \lambda_3$
(c) $\frac{1}{\lambda_1} + \frac{1}{\lambda_2} = \frac{1}{\lambda_3}$ (d) $\frac{1}{\lambda_1} = \frac{1}{\lambda_2} + \frac{1}{\lambda_3}$

38. The total radiative power emitted by spherical black body with radius R and temperature T is P. If the radius is doubled and the temperature is halved, then the radiative power will be

(a) $\frac{P}{4}$ (b) $\frac{P}{2}$ (c) $2P$ (d) $4P$

39. The quantum hall resistance R_H is a fundamental constant with dimensions of resistance. If h is Planck's constant and e is the electron charge, then the dimension of R_H is the same as

(a) $\frac{e^2}{h}$ (b) $\frac{h}{e^2}$ (c) $\frac{h^2}{e}$ (d) $\frac{e}{h^2}$

40. Four students measure the height of a tower. Each student uses a different method and each measures the height many different times. The data for each are plotted below. The measurement with highest precision is

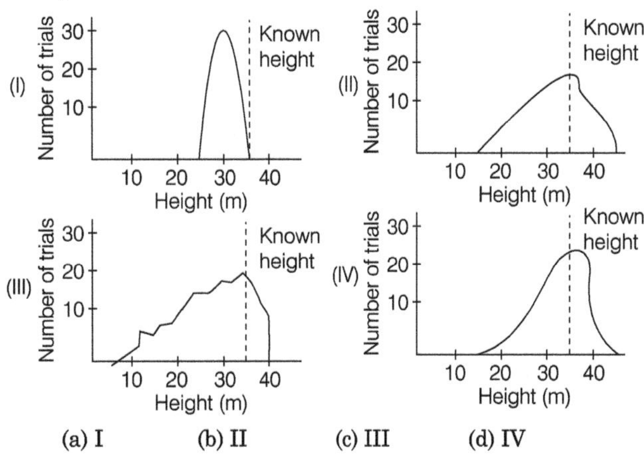

(a) I (b) II (c) III (d) IV

CHEMISTRY

41. The hybridisations of $Ni(CO)_4$ and $[Cr(H_2O)_6^2]^+$ respectively, are

(a) sp^3 and d^3sp^2 (b) dsp^2 and d^2sp^3
(c) sp^3 and d^2sp^3 (d) dsp^2 and sp^3d^2

42. Extraction of silver is achieved by initial complexation of the ore (argentite) with X followed by reduction with Y. X and Y, respectively are

(a) CN^- and Zn (b) CN^- and Cu
(c) Cl^- and Zn (d) Br^- and Zn

43. Assuming ideal behaviour, the enthalpy and volume of mixing of two liquids respectively, are

(a) zero and zero (b) + ve and zero
(c) – ve and zero (d) – ve and – ve

44. At 298 K, the ratio of osmotic pressures of two solutions of a substance with concentrations of 0.01 M and 0.001 M respectively, is

(a) 1 (b) 100 (c) 10 (d) 1000

45. The rate of gas phase chemical reactions generally increases rapidly with rise in temperature. This is mainly because

(a) the collision frequency increases with temperature
(b) the fraction of molecules having energy in excess of the activation energy increases with temperature
(c) the activation energy decreases with temperature
(d) the average kinetic energy of molecule increases with temperature

46. Among I-IV

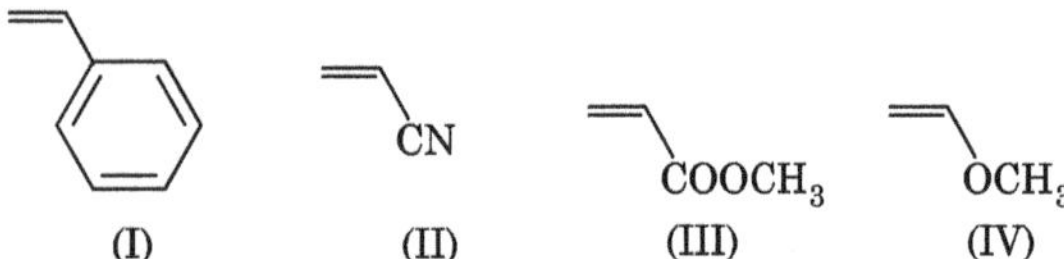

the compound that does not undergo polymerisation under radical initiation, is

(a) I (b) II (c) III (d) IV

47. Two possible stereoisomers for

H, HO, COOH, H

are

(a) enantiomers (b) diastereomers
(c) conformers (d) rotamers

48. For a process to occur spontaneously

(a) only the entropy of the system must increase
(b) only the entropy of the surroundings must increase
(c) either the entropy of the system or that of the surroundings must increase
(d) the total entropy of the system and the surroundings must increase

49. When the size of a spherical nanoparticle decreases from 30 nm to 10 nm, the ratio surface area/volume becomes

(a) 1/3 of the original (b) 3 times the original
(c) 1/9 of the original (d) 9 times the original

50. The major product of the following reaction is

Me, OH, Me, OH $\xrightarrow{H^+}$

(a) 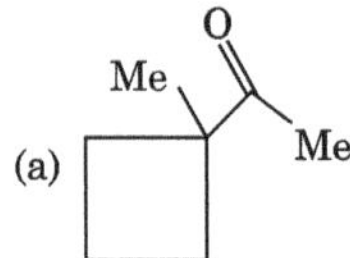

(b) 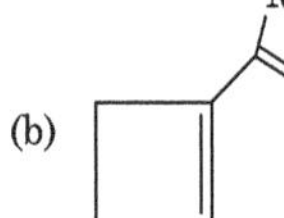

(c) 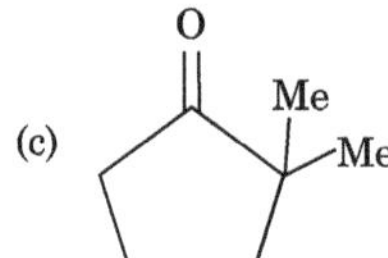

(d) Me, Me

51. For the transformation,

Br, $\overset{\oplus}{N_2}\overset{\ominus}{Cl}$ $\longrightarrow$ Br

the reagent used is

(a) $LiAlH_4$ (b) H_3PO_2 (c) H_3O^+ (d) H_2/Pt

52. The values of the limiting molar conductivity (Λ°) for NaCl, HCl and NaOAc are 126.4, 425.9 and 91.0 S cm^{2}mol^{-1}, respectively. For HOAc, Λ° in S cm^2 mol^{-1} is

(a) 390.5 (b) 299.5 (c) 208.5 (d) 217.4

53. To obtain a diffraction peak, for a crystalline solid with interplane distance equal to the wavelength of incident X-ray radiation. The angle of incidence should be

(a) 90° (b) 0° (c) 30° (d) 60°

54. The standard Gibbs free energy change (ΔG° in kJ mol^{-1}), in a Daniell cell ($E^\circ_{cell} = 1.1$ V), when 2 moles of Zn(s) is oxidised at 298 K, is closest to

(a) – 212.3 (b) – 106.2 (c) – 424.6 (d) – 53.1

55. All the products formed in the oxidation of $NaBH_4$ by I_2 are

(a) B_2H_6 and NaI (b) B_2H_6, H_2 and NaI
(c) BI_3 and NaH (d) $NaBI_4$ and HI

56. The spin-only magnetic moments of $[Mg(CN)_6]^{2-}$ and $[MnBr_4]^{2-}$ in Bohr magnetons, respectively are

(a) 5.92 and 5.92 (b) 4.89 and 1.73
(c) 1.73 and 5.92 (d) 1.73 and 1.73

57. In a zero-order reaction, if the initial concentration of the reaction is doubled, the time required for half the reactant to be consumed

(a) increases two-fold (b) increases four-fold
(c) decrease by half (d) does not change

58. The adsorption isotherm for a gas is given by the relation $x = ap(1 + bp)$, where x is moles of gas adsorbed per gram of the adsorbent, p is the pressure of the gas, and a and b are constants. Then x

(a) increases with p
(b) remains unchanged with p
(c) decreases with p
(d) increases with p at low pressures and then remains the same at high pressures

59. The reaction,

OH + $CHCl_3$ $\xrightarrow[H^+]{NaOH/Heat}$ OH, CHO

is known as

(a) Perkin reaction (b) Sandmeyer reaction
(c) Reimer-Tiemann reaction (d) Cannizzaro reaction

60. Among I-III

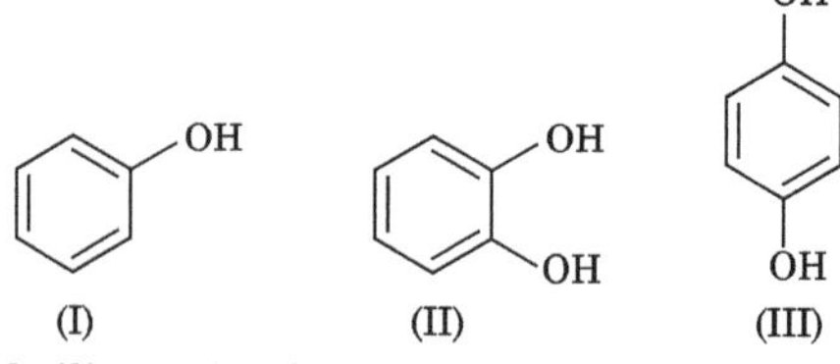

the boiling point follows the order

(a) II< I < III (b) III < II < I
(c) I < II < III (d) II < III < I

BIOLOGY

61. The major constituents of neurofilaments are
(a) microtubules (b) intermediate filaments
(c) actin filaments (d) protofilaments

62. In which phase of the cell cycle are sister chromatids available as template for repair?
(a) G_1-phase (b) G_2-phase (c) S-phase (d) M-phase

63. A person has difficulty in breathing at higher altitudes because
(a) oxygen is likely to diffuse from lungs to blood
(b) oxygen is likely to diffuse from blood to lungs
(c) partial pressure of O_2 is lower than partial pressure of CO_2
(d) overall intake of O_2 by the blood becomes low

64. In humans, the composition of a zygote that will develop into a female is
(a) 44A+ XX (b) 44A+ XY (c) 22+ X (d) 23A

65. If you fractionate all the organelles from the cytoplasm of a plant cell, in which one of the following sets of fractions will you find nucleic acids?
(a) Nucleus, mitochondria, chloroplast, cytoplasm
(b) Nucleus, mitochondria, chloroplast, glyoxysome
(c) Nucleus, chloroplast, cytoplasm, peroxisome
(d) Nucleus, mitochondria, chloroplast, Golgi bodies

66. A protein with 100 amino acid residues has been translated based on triplet genetic code, had the genetic code been quadruplet the gene that codes for the protein would have been
(a) same in size (b) longer in size by 25%
(c) longer in size by 100% (d) shorter in size

67. If the sequence of bases in DNA is 5′-ATGTATCTCAAT-3′, then the sequence of bases in its transcript will be
(a) 5′-TACATAGAGTTA-3′ (b) 5′-UACAUAGAGUUA-3′
(c) 5′-AUGUAUCUCAAU-3′ (d) 5′-AUUGAGAUACAU-3′

68. The Na^+ / K^+ pump is present in the plasma membrane of mammalian cells where it
(a) expels potassium from the cell
(b) expels sodium and potassium from the cell
(c) pumps sodium into the cell
(d) expels sodium from the cell

69. The CO_2 in the blood is mostly carried
(a) by haemoglobin in RBCs
(b) in the cytoplasm of WBCs
(c) in the plasma as bicarbonate ions
(d) by plasma proteins

70. Patients who have undergone organ transplants are given anti-rejection medications to
(a) minimise infection
(b) stimulate B-macrophage cell interaction
(c) prevent T-lymphocyte proliferation
(d) adopt the HLA of donor

71. Saline drip is given to a cholera patient because
(a) NaCl kills *Vibrio cholerae*
(b) NaCl generates ATP
(c) Na^+ ions stop nerve impulse and hence sensation of pain
(d) Na^+ ions help in retention of water in body tissue

72. A water molecule can form a maximum of .... hydrogen bond(s).
(a) 1 (b) 2 (c) 3 (d) 4

73. Circadian rhythm is an endogenously driven cycle for biochemical, physiological and behavioural processes. In human, the approximate duration of this 'biological clock' is
(a) 1 h (b) 6 h
(c) 12 h (d) 24 h

74. Modern evolutionary theory consists of the concepts of Darwin modified by knowledge concerning
(a) population statistics
(b) Mendel's laws
(c) the idea of the survival of the fittest
(d) competition

75. Soon after the three germ layers are formed in a developing embryo, the process of organogenesis starts. The human brain is formed from the
(a) ectoderm
(b) endoderm
(c) mesoderm
(d) partly endoderm and partly mesoderm

76. Puffs in the polytene chromosomes of *Drosophila melanogaster* salivary glands represent
(a) transcriptionally active genes
(b) transcriptionally inactive genes
(c) heterochromatin
(d) housekeeping genes

77. The process of cell death involving DNA cleavage in cells is known as
(a) necrosis (b) apoptosis
(c) cytokinesis (d) endocytosis

78. According to the original model of DNA, as proposed by Watson and Crick in 1953, DNA is a
(a) left-handed helix
(b) helix that makes full turn every 70 mm
(c) helix where one turn of DNA contains 20 basepairs
(d) two-stranded helix where each strand has opposite polarity

79. At which stage of meiosis-I does crossing over occur?
(a) Leptotene (b) Zygotene
(c) Pachytene (d) Diplotene

80. An electrode is placed in the axioplasm of a mammalian axon and another electrode is placed just outside the axon. The potential difference measured will be
(a) 0 (b) – 70 mV
(c) – 70 μV (d) + 70 μV

PART-II (2 Marks Questions)

MATHEMATICS

81. Let A and B be any two $n \times n$ matrices such that the following conditions hold : $AB = BA$ and there exist positive integers k and l such that $A^k = I$ (the identity matrix) and $B^l = 0$ (the zero matrix). Then,

(a) $A + B = I$
(b) $\det(AB) = 0$
(c) $\det(A + B) \neq 0$
(d) $(A + B)^m = 0$ for some integer m

82. The minimum value of n for which
$$\frac{2^2 + 4^2 + 6^2 + \ldots + (2n)^2}{1^2 + 3^2 + 5^2 + \ldots + (2n-1)^2} < 1.01 \text{ is}$$

(a) 101 (b) 121 (c) 151 (d) does not exist

83. The locus of the point $P = (a, b)$ where a, b are real numbers such that the roots of $x^3 + ax^2 + bx + a = 0$ are in arithmetic progression is

(a) an ellipse
(b) a circle
(c) a parabola whose vertex is on the Y-axis
(d) a parabola whose vertex is on the X-axis

84. The smallest possible positive slope of a line whose y-intercept is 5 and which has a common point with the ellipse $9x^2 + 16y^2 = 144$ is

(a) $\frac{3}{4}$ (b) 1 (c) $\frac{4}{3}$ (d) $\frac{9}{16}$

85. Let $A = \{\theta \in R \mid \cos^2(\sin\theta) + \sin^2(\cos\theta) = 1\}$ and $B = \{\theta \in R \mid \cos(\sin\theta)\sin(\cos\theta) = 0\}$. Then, $A \cap B$

(a) is the empty set
(b) has exactly one element
(c) has more than one but finitely many elements
(d) has infinitely many elements

86. Let $f(x) = x^3 + ax^2 + bx + c$, where a, b, c are real numbers. If $f(x)$ has a local minimum at $x = 1$ and a local maximum at $x = -\frac{1}{3}$ and $f(2) = 0$, then $\int_{-1}^{1} f(x)\, dx$ equals

(a) $\frac{14}{3}$ (b) $\frac{-14}{3}$ (c) $\frac{7}{3}$ (d) $\frac{-7}{3}$

87. Let $f(x) = x^{12} - x^9 + x^4 - x + 1$. Which of the following is true?

(a) f is one-one (b) f has a real root
(c) f' never vanishes (d) f takes only positive values

88. For each positive integer n, defined $f_n(x) = \text{minimum}\left(\frac{x^n}{n!}, \frac{(1-x)^n}{n!}\right)$, for $0 \le x \le 1$. Let $I_n = \int_0^1 f_n(x)\, dx$, $n \ge 1$. Then, $\sum_{n=1}^{\infty} I_n$ is equal to

(a) $2\sqrt{e} - 3$ (b) $2\sqrt{e} - 2$ (c) $2\sqrt{e} - 1$ (d) $2\sqrt{e}$

89. The maximum possible value of $x^2 + y^2 - 4x - 6y$, x, y real, subject to the condition $|x + y| + |x - y| = 4$ is

(a) 12 (b) 28 (c) 72 (d) does not exist

90. The arithmetic mean and the geometric mean of two distinct 2-digit numbers x and y are two integers one of which can be obtained by reversing the digits of the other (in base 10 representation). Then, $x + y$ equals

(a) 82 (b) 116 (c) 130 (d) 148

PHYSICS

91. An isolated sphere of radius R contains uniform volume distribution of positive charge. Which of the curve shown below, correctly illustrates the dependence of the magnitude of the electric field of the sphere as a function of the distance r from its centre?

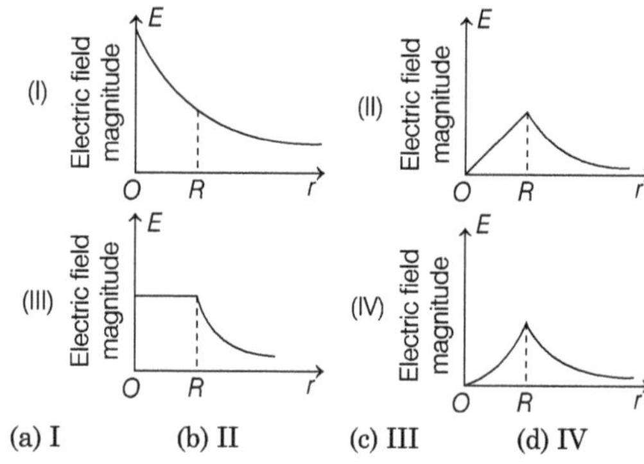

(a) I (b) II (c) III (d) IV

92. The surface of a planet is found to be uniformly charged. When a particle of mass m and no charge is thrown at an angle from the surface of the planet, it has a parabolic trajectory as in projectile motion with horizontal range L. A particle of mass m and charge q, with the same initial conditions has a range $L/2$. The range of particle of mass m and charge $2q$, with the same initial conditions is

(a) L (b) $\frac{L}{2}$ (c) $\frac{L}{3}$ (d) $\frac{L}{4}$

93. Figure below shows a small mass connected to a string, which is attached to a vertical post. If the ball is released, when the string is horizontal as shown below. The magnitude of the total acceleration (including radial and tangential) of the mass as a function of the angle θ is

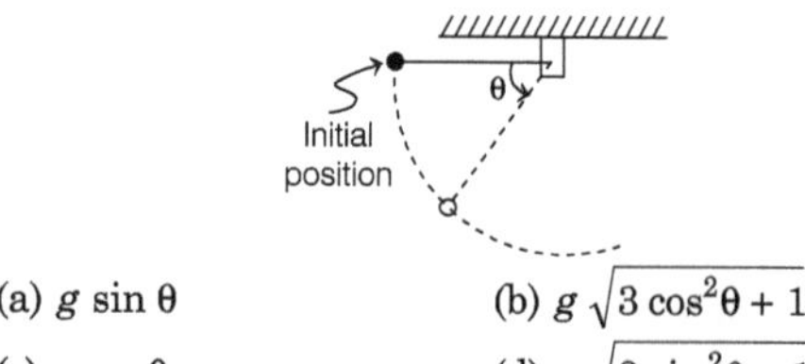

(a) $g\sin\theta$ (b) $g\sqrt{3\cos^2\theta + 1}$
(c) $g\cos\theta$ (d) $g\sqrt{3\sin^2\theta + 1}$

94. One mole of an ideal gas at initial temperature T, undergoes a quasi-static process during which the volume V is doubled. During the process, the internal energy U obeys the equation $U = aV^3$, where a is a constant. The work done during this process is

(a) $\frac{3RT}{2}$ (b) $\frac{5RT}{2}$

(c) $\frac{5RT}{3}$ (d) $\frac{7RT}{3}$

95. A constant amount of an ideal gas undergoes the cyclic process $ABCA$ in the p-V graph shown below.

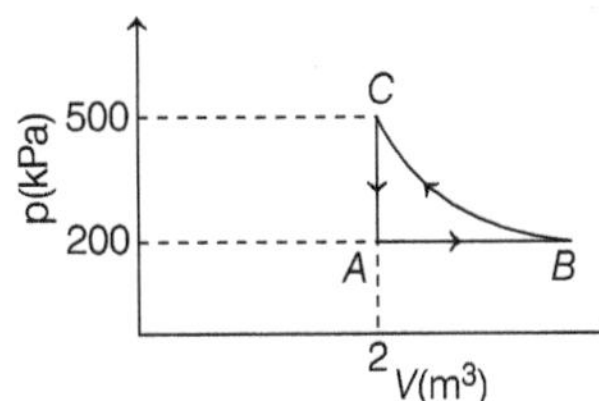

The path BC is an isothermal. The work done by the gas during one complete cycle, beginning and ending at A is nearly

(a) 600 kJ (b) 300 kJ (c) – 300 kJ (d) – 600 kJ

96. A material is embedded between two glass plates. Refractive index n of the material varies with thickness as shown below. The maximum incident angle (in degrees) on the material for which beam will pass through the material is

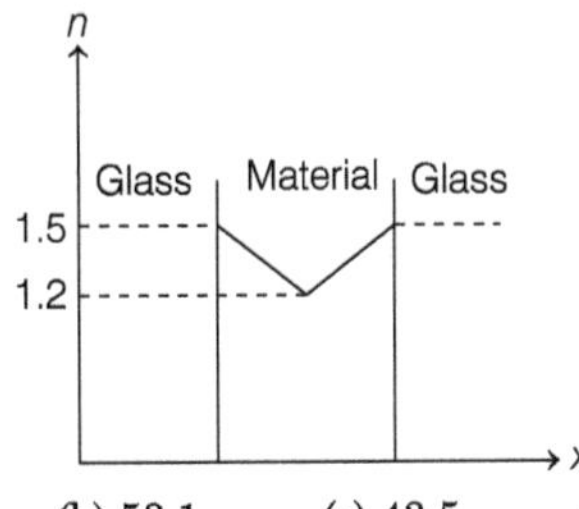

(a) 60.0 (b) 53.1 (c) 43.5 (d) 32.3

97. At a distance l from a uniformly charged long wire, a charged particle is thrown radially outward with a velocity u in the direction perpendicular to the wire. When the particle reaches a distance $2l$ from the wire, its speed is found to be $\sqrt{2}u$. The magnitude of the velocity, when it is a distance $4l$ away from the wire is (ignore gravity)

(a) $\sqrt{3}u$ (b) $2u$ (c) $2\sqrt{2}u$ (d) $4u$

98. A rectangular loop of wire shown below is coplanar with a long wire carrying current I.

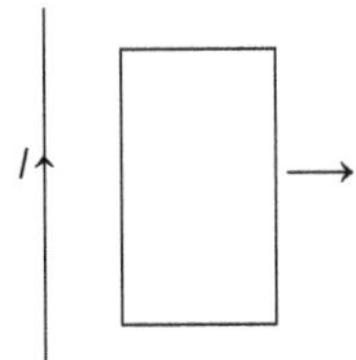

The loop is pulled to the right as indicated. What are the directions of the induced current in the loop and the magnetic forces on the left and the right sides of the loop?

	Induced current	Force on left side	Force on right side
a.	Counter clockwise	To the left	To the right
b.	Clockwise	To the left	To the right
c.	Counter clockwise	To the right	To the left
d.	Clockwise	To the right	To the left

99. Two batteries V_1 and V_2 are connected to three resistors as shown below.

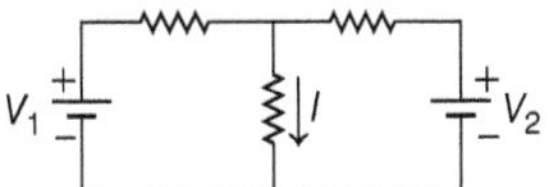

If $V_1 = 2$ V and $V_2 = 0$ V, then the current $I = 3$ mA.
If $V_1 = 0$ V and $V_2 = 4$ V, then the current $I = 4$ mA.
Now, if $V_1 = 10$ V and $V_2 = 10$ V, then the current I will be

(a) 7 mA (b) 15 mA (c) 20 mA (d) 25 mA

100. A particle moves in a plane along an elliptic path given by $\frac{x^2}{a^2} + \frac{y^2}{b^2} = 1$. At point (0, b), the x-component of velocity is u. The y-component of acceleration at this point is

(a) $-bu^2/a^2$ (b) $-u^2/b$

(c) $-au^2/b^2$ (d) $-u^2/a$

CHEMISTRY

101. XeF_6 hydrolyses to give an oxide. The structures of XeF_6 and the oxide, respectively are

(a) octahedral and tetrahedral
(b) distorted octahedral and pyramidal
(c) octahedral and pyramidal
(d) distorted octahedral and tetrahedral

102. MnO_4 oxidises (i) oxalate ion in acidic medium at 333 K and (ii) HCl. For balanced chemical equations, the ratios $[MnO_4^- : C_2O_4^{2-}]$ in (i) and $[MnO_4^- : HCl]$ in (ii) respectively are

(a) 1 : 5 and 2 : 5 (b) 2 : 5 and 1 : 8

(c) 2 : 5 and 1 : 5 (d) 5 : 2 and 1 : 8

103. If $E^\circ_{Fe^{2+}/Fe} = -0.440$ V and $E^\circ_{Fe^{3+}/Fe^{2+}} = 0.770$ V, then $E^\circ_{Fe^{3+}/Fe}$ is

(a) 0.330 V (b) – 0.037 V

(c) – 0.330 V (d) – 1.210 V

104. The electron in hydrogen atom is, in the first Bohr orbit ($n = 1$). The ratio of transition energies, E ($n = 1 \rightarrow n = 3$) to E ($n = 1 \rightarrow n = 2$), is

(a) 32/27 (b) 16/27

(c) 32/9 (d) 8/9

105. In the following conversion,

$$C_6H_5CN \xrightarrow[\text{(ii) } H_3O^+]{\text{(i) MeMgBr}} X \xrightarrow[H_3O^+]{NaOH/I_2} Y$$

the major products X and Y, respectively are

I. $C_6H_5CONHCH_3$ and $C_6H_5NHCH_3$

II. $C_6H_5CONH_2$ and $C_6H_5NH_2$

III. $C_6H_5COCH_3$ and C_6H_5COOH

IV. $C_6H_5CON(CH_3)_2$ and C_6H_5COOH

(a) I (b) II
(c) III (d) IV

106. In the reaction sequence,

$$C_6H_5OH \xrightarrow{HNO_2} X \xrightarrow[H_2SO_4]{C_6H_5OH} Y$$

the major products X and Y, respectively are

I. (cyclohexadienone with N–OH at para position) and (cyclohexadienone with =N– linked to 4-hydroxyphenyl)

II. (cyclohexadienone with =N–OH at ortho position) and (cyclohexadienone with =N– linked to 2-hydroxyphenyl)

III. (cyclohexadienone with =N–OH at para position) and (cyclohexadienone with =N– linked to 2-hydroxyphenyl)

IV. (cyclohexadienone with =N–OH at ortho position) and (cyclohexadienone with =N– linked to 4-hydroxyphenyl)

(a) I (b) II (c) III (d) IV

107. Optically active (S)-α-methoxyacetaldehyde on reaction with MeMgX gave a mixture of alcohols. The major diastereomer P on treatment with MeI / K_2CO_3 gave an optically inactive compound. P is

I. H_3C–CH(OCH_3)–CH(OH)–CH_3 (both hashed)
II. H_3C–CH(OCH_3)–CH(OH)–CH_3 (OCH_3 hashed, OH wedged)
III. H_3C–CH(OCH_3)–CH(OH)–CH_3 (both wedged)
IV. H_3C–CH(OCH_3)–CH(OH)–CH_3 (OCH_3 wedged, OH hashed)

(a) I (b) II (c) III (d) IV

108. At 300 K the vapour pressures of two pure liquids. A and B are 100 and 500 mm Hg, respectively. If in a mixture of A and B, the vapour pressure is 300 mm Hg, the mole fractions of A in the liquid and in the vapour phase, respectively are

(a) 1/2 and 1/10 (b) 1/4 and 1/6
(c) 1/4 and 1/10 (d) 1/2 and 1/6

109. The Crystal Field Stabilisation Energies (CFSE) of high spin and low spin d^6 metal complexes in terms of Δ_o, respectively are

(a) -0.4 and -2.4 (b) -2.4 and -0.4
(c) -0.4 and 0.0 (d) -2.4 and 0.0

110. Emulsification of 10 mL of oil in water produces 2.4×10^{18} droplets. If the surface tension at the oil-water interface is 0.03 Jm^{-2} and the area of each droplet is $12.5 \times 10^{-16} m^2$, the energy spent in the formation of oil droplets is

(a) 90 J (b) 30 J (c) 900 J (d) 10 J

BIOLOGY

111. Which sequence of events gives rise to flaccid guard cells and stomatal closure at night?

(a) Low [Glucose] $\Rightarrow$ low osmotic pressure $\Rightarrow$ low pH $\Rightarrow$ high pCO_2
(b) Low pH $\Rightarrow$ high pCO_2 $\Rightarrow$ low [Glucose] $\Rightarrow$ low osmotic pressure

(c) Low osmotic pressure ⇒ high pCO_2 ⇒ low pH ⇒ low [Glucose]

(d) High pCO_2 ⇒ low pH ⇒ low [Glucose] ⇒ low osmotic pressure

112. Rice has a diploid genome with $2n = 24$. If crossing-over is stopped in a rice plant and then selfed seeds are collected, will all the offsprings be genetically identical to the parent plant?

(a) Yes, because crossing-over is the only source of genetic variation

(b) No, because stopping of crossing-over automatically increases rate of point mutation

(c) Yes, only if the parent plant was a completely inbred line

(d) Yes, only if the parent plant was a hybrid between two pure-bred lines

113. Rodents can distinguish between many different types of odours. The basis for odour discrimination is that

(a) they have a small number of odorant receptors that bind to many different odorant molecules

(b) the mechanoreceptors in the nasal cavity are activated by different odorant molecules found in the air passing through the nostrils

(c) the part of the brain that processes the sense of smell has many different receptors for odorant molecules

(d) a large number of different chemoreceptors are present in the nasal cavity that binds a variety of odorant molecules

114. Although blood flows through large arteries at high pressure, when the blood reaches small capillaries the pressure decreases because

(a) the valves in the arteries regulate the rate of blood flow into the capillaries

(b) the volume of blood in the capillaries is much lesser than that in the arteries

(c) the total cross-sectional area of capillaries arising from an artery is much greater than that of the artery

(d) elastin fibres in the capillaries help to reduce the arterial pressure

115. *E.coli* about to replicate was pulsed with tritiated thymidine for 5 min and then transferred to normal medium. After one cell division which one of the following observations would be correct?

(a) Both the strands of DNA will be radioactive

(b) One strand of DNA will be radioactive

(c) None of the strands will be radioactive

(d) Half of one strand of DNA will be radioactive

116. Selection of lysine auxotroph (bacteria which requires lysine for growth) from a mixed population of bacteria can be done by growing the bacterial population in the presence of

(a) lysine (b) penicillin

(c) lysine and penicillin (d) glucose

117. Increasing the number of measurements of an experimental variable will

(a) increase the standard error of the sample

(b) increase the mean of the sample

(c) decrease the standard error of the sample

(d) result in all of the above

118. For a human male what is the probability that all the maternal chromosomes will end up in the same gamete?

(a) 1/23 (b) 2^{23} (c) 2^{46} (d) $(1/2)^{23}$

119. Nocturnal animals have retinas that contain

(a) a high percentage of rods to increase sensitivity to low light conditions

(b) a high percentage of cones so that nocturnal colour vision can be improved in low light conditions

(c) an equal number of rods and cones so that vision can be optimised

(d) retinas with the photoreceptor layer present in the eye to increase light sensitivity

120. The length of one complete turn of a DNA double helix is

(a) 34 Å (b) 34 nm (c) 3.4 Å (d) 3.4 μm

Answers

PART-I

1	(c)	2	(c)	3	(d)	4	(a)	5	(a)	6	(c)	7	(a)	8	(b)	9	(c)	10	(a)
11	(c)	12	(c)	13	(b)	14	(b)	15	(b)	16	(d)	17	(c)	18	(a)	19	(b)	20	(c)
21	(b)	22	(b)	23	(a)	24	(a)	25	(b)	26	(d)	27	(b)	28	(b)	29	(c)	30	(c)
31	(b)	32	(a)	33	(a)	34	(b)	35	(d)	36	(a)	37	(d)	38	(a)	39	(b)	40	(a)
41	(c)	42	(a)	43	(a)	44	(c)	45	(b)	46	(d)	47	(a)	48	(d)	49	(b)	50	(c)
51	(b)	52	(a)	53	(c)	54	(c)	55	(b)	56	(c)	57	(a)	58	(d)	59	(c)	60	(c)
61	(b)	62	(b)	63	(d)	64	(a)	65	(a)	66	(b)	67	(b)	68	(d)	69	(c)	70	(c)
71	(d)	72	(d)	73	(d)	74	(b)	75	(a)	76	(a)	77	(b)	78	(d)	79	(c)	80	(b)

PART-II

81	(b)	82	(c)	83	(c)	84	(b)	85	(a)	86	(b)	87	(d)	88	(a)	89	(b)	90	(c)
91	(b)	92	(c)	93	(d)	94	(d)	95	(c)	96	(b)	97	(a)	98	(b)	99	(d)	100	(a)
101	(b)	102	(b)	103	(b)	104	(a)	105	(c)	106	(b)	107	(*)	108	(d)	109	(a)	110	(a)
111	(d)	112	(c)	113	(d)	114	(c)	115	(b)	116	(b)	117	(c)	118	(d)	119	(a)	120	(a)

* No options are correct.

Solutions

1. *(c)* We have,

$$\log_a b + \log_b a = c$$

$$AM \geq GM$$

$$\therefore \quad \frac{\log_a b + \log_b a}{2} \geq \sqrt{\log_a b \log_b a}$$

$$\Rightarrow \frac{c}{2} \geq \sqrt{\frac{\log b}{\log a} \times \frac{\log a}{\log b}} \Rightarrow \frac{c}{2} \geq 1 \Rightarrow c \geq 2$$

$\therefore$ Smallest positive integer value of $c = 2$.

2. *(c)* Let $S_n = i + 2i^2 + 3i^3 + \dots + ni^n \dots$(i)

$iS_n = i^2 + 2i^2 + \dots (n-1)\, i^n + ni^{n+1} \dots$(ii)

On subtracting Eq. (ii) from Eq. (i), we get

$$S_n(1-i) = i + i^2 + i^3 + \dots + i^n - ni^{n+1}$$

$$\Rightarrow S_n(1-i) = \frac{i(1-i^n)}{1-i} - ni^{n+1}$$

$$\Rightarrow \quad S_n = \frac{i(1-i^n)}{(1-i)^2} - \frac{ni^{n+1}}{(1-i)}$$

$$\Rightarrow \quad S_n = \frac{i(1-i^n)}{-2i} - \frac{ni^{n+1}}{1-i}$$

$$\Rightarrow \quad S_n = \frac{1-i^n}{-2} - \frac{ni^{n+i}(1+i)}{2}.$$

Let $\quad Z_1 = \frac{1-i^n}{-2}$ and $Z_2 = \frac{ni^{n+1}(1+i)}{2}$

$$\therefore \quad |Z_1| = \frac{1}{\sqrt{2}} \text{ or } 0 \text{ and } |Z_2| = \frac{n}{2}\sqrt{2}$$

$$\therefore \quad \frac{n}{2}\sqrt{2} = 18\sqrt{2} \Rightarrow n = 36$$

3. *(d)* Given, $\quad P^2 = P$

$$\because \quad (I+P)^n = (I+I)^n$$

$$[\because P^2 = P \Rightarrow P^{-1}P^2 = P^{-1}P = P = I]$$

$$\Rightarrow \quad (I+P)^n = (2I)^n$$

$$= 2^n I$$

$$= (2^n - 1 + 1)\, I$$

$$= I + (2^n - 1)I$$

$$= I + (2^n - 1)P \quad [\because I = P]$$

4. *(a)* We have, $x^3 + ax^2 + bx + c = 0$

Let $\quad f(x) = x^3 + ax^2 + bx + c$

$$P'(x) = 3x^2 + 2ax + b$$

$$D = (2a)^2 - 4(3)(b)$$

$$D = 4a^2 - 12b$$

$$D = 4(a^2 - 3b)$$

For three roots $a^2 - 3b > 0$

but given $a^2 - 2b < 0$

$$\therefore \quad D < 0$$

Hence, $f(x)$ has one real roots and two imaginary roots.

5. *(a)* We have, $x^2 + 2x \sin xy + 1 = 0$

$$\Rightarrow x^2 + 2x \sin xy + \sin^2 xy + 1 - \sin^2 xy = 0$$

$$\Rightarrow \quad (x + \sin xy)^2 + \cos^2 xy = 0$$

$\therefore$ $x + \sin xy = 0$ and $\cos^2 xy = 0$

$$\cos^2 xy = 0$$

$$\Rightarrow \quad xy = (2n+1)\frac{\pi}{2}$$

$$\Rightarrow \quad x + 1 = 0$$

$$[\because \cos^2 xy = 0 \Rightarrow \sin xy = 1]$$

$$\Rightarrow \quad x = -1$$

$$\therefore \quad y = -(2n+1)\frac{\pi}{2}$$

which represent the pair of straight line.

6. *(c)* Given, $A(4, 0)$ and $B(0,12)$.

Let $C\ (x, y)$

$$\text{Area of } \Delta ABC = \frac{1}{2}\begin{vmatrix} x & y & 1 \\ 4 & 0 & 1 \\ 0 & 12 & 1 \end{vmatrix}$$

$$\Rightarrow 18 = \frac{1}{2}\left|\{x(0-12) - y(4-0) + 1(48-0)\}\right|$$

$$\Rightarrow 18 = \frac{1}{2}\left|(-12x - 4y + 48)\right|$$

$$\Rightarrow 18 = \frac{-4}{2}\left|(3x + y - 12)\right|$$

$$\Rightarrow \quad (3x + y - 12)^2 = 81$$

$$\Rightarrow \quad (y + 3x - 12)^2 = 81$$

7. *(a)* Given,

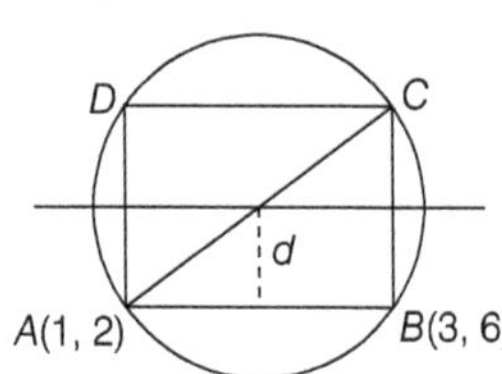

$ABCD$ is a rectangle $A(1, 2)$ and $B(3, 6)$, equation of one of the diameter of circle circumscribing the rectangle is

$$2x - y + 4 = 0.$$

Slope of diameter $= 2$

and slope of line $AB = \frac{6-2}{3-1} = \frac{4}{2} = 2$

$\therefore$ Side AB is parallel to diameter of circle equation of line AB

$$y - 2 = 2(x-1) \Rightarrow 2x - y = 0$$

Distance between diameter of circle and line AB,

$$d = \left|\frac{4}{\sqrt{2^2+1}}\right| = \frac{4}{\sqrt{5}}$$

$$\therefore \quad BC = 2d = \frac{8}{\sqrt{5}}$$

Now, $\quad AB = \sqrt{(3-1)^2 + (6-2)^2}$

$$= \sqrt{4+16} = 2\sqrt{5}$$

Area of rectangle $= AB \times BC$

$$= 2\sqrt{5} \times \frac{8}{\sqrt{5}} = 16$$

8. *(b)* We have, $y^2 = 6x$

Equation of normal of parabola $y^2 = 4ax$ is $y = mx - 2am - am^3$

$\therefore$ Equation of normal of parabola $y^2 = 6x$ is

$$y = mx - 3m - \frac{3}{2}m^3$$

Normal is passes through $(\lambda, 0)$.

$$\therefore \quad 0 = \lambda m - 3m - \frac{3}{2}m^3$$

$$\Rightarrow \quad \lambda = 3 + \frac{3}{2}m^2 \Rightarrow m^2 = \frac{2}{3}(\lambda - 3)$$

$$\Rightarrow \quad m = \sqrt{\frac{2}{3}(\lambda - 3)}$$

$\therefore$ m is real.

$$\therefore \quad \lambda - 3 > 0 \Rightarrow \lambda > 3$$

9. *(c)* We have,

$$f(x) = \cos 5x + A\cos 4x + B\cos 3x + C\cos 2x + D\cos x + E$$

We know that,

$$\cos x = \cos(2\pi - x)$$

$$\therefore \quad f(x) = f(2\pi - x)$$

[$\because f(x)$ contains only cosines terms]

$$\because \quad f\left(\frac{\pi}{5}\right) = f\left(2\pi - \frac{\pi}{5}\right) = f\left(\frac{9\pi}{5}\right)$$

Similarly, $f\left(\frac{2\pi}{5}\right) = f\left(\frac{8\pi}{5}\right), f\left(\frac{3\pi}{5}\right) = f\left(\frac{7\pi}{5}\right)$

$$f\left(\frac{4\pi}{5}\right) = f\left(\frac{6\pi}{5}\right)$$

Let $T = f(0) - f\left(\frac{\pi}{5}\right) + f\left(\frac{2\pi}{5}\right) - f\left(\frac{3\pi}{5}\right) + \dots f\left(\frac{9\pi}{5}\right)$

$$T = f(0) - 2\left[f\left(\frac{\pi}{5}\right) + f\left(\frac{3\pi}{5}\right)\right] + 2\left[f\left(\frac{2\pi}{5}\right) + f\left(\frac{4\pi}{5}\right)\right] - f(\pi)$$

Now, $f(0) = 1 + A + B + C + D + E$

$$f(\pi) = -1 + A - B + C - D + E$$

$$\because f(0) - f(\pi) = 2(1 + B + D)$$

$$f\left(\frac{\pi}{5}\right)+f\left(\frac{3\pi}{5}\right)$$
$$=2\left(1+B\cos\frac{3\pi}{5}+D\cos\frac{\pi}{5}\right)$$
$$f\left(\frac{2\pi}{5}\right)+f\left(\frac{4\pi}{5}\right)$$
$$=2\left(1+B\cos\frac{6\pi}{5}+D\cos\frac{2\pi}{5}\right)$$

Clearly, T contains only B and D terms.

10. *(a)* We have,

In ΔABC $\quad 3\sin A+4\cos B=6 \quad$ ...(i)

and $\quad 4\sin B+3\cos A=1 \quad$ ...(ii)

On squaring and adding Eqs. (i) and (ii), we get

$$9(\sin^2 A+\cos^2 A)+16(\sin^2 B+\cos^2 B)$$
$$+24(\sin A\cos B+\cos A\sin B)=36+1$$
$$\Rightarrow \quad 9+16+24\sin(A+B)=37$$
$$\Rightarrow \quad 24\sin(A+B)=37-25$$
$$\Rightarrow \quad \sin(A+B)=\frac{1}{2}$$
$$\Rightarrow \quad \sin(\pi-C)=\frac{1}{2} \quad [\because A+B+C=\pi]$$
$$\Rightarrow \quad \sin C=\frac{1}{2}$$
$$\therefore \quad C=30^\circ$$

11. *(c)* We have, $\log_{\{x\}}[x]+\log_{[x]}\{x\}$

$\log_{\{x\}}[x]$ is defined if $[x]>0,\ x\notin$ integer

$\therefore \quad x>1 \quad$ ...(i)

$\log_{[x]}\{x\}$ is defined if $[x]>1,\ x\notin$ integer

$\because \quad x>2 \quad$ ...(ii)

From Eqs. (i) and (ii), $x>2$

$\therefore$ Option (c) $x\in(2,3)$ satisfied and in option (d) $x\in(3,5)$ not satisfied because 4 is an integer.

12. *(c)* We have,

$$f(x)=(2011+x)^n$$
$$f(0)=(2011)^n$$
$$f'(x)=n(2011+x)^{n-1}$$

and $\quad f'(0)=n\cdot(2011)^{n-1}$

$$f''(x)=n(n-1)(2011+x)^{n-2}$$

and $\quad f''(0)=n(n-1)(2011)^{n-2}$

Similarly,

$$f^{n-1}(x)=n(n-1)=3\times2\times1$$

and $\quad f^{n-1}(0)=n(n-1)\ldots\times3\times2\times1$

$$\therefore f(0)+f'(0)+\frac{f''0}{2!}+\ldots+\frac{f^{(n-1)}(0)}{(n-1)!}$$
$$=(2011)^n+n(2011)^{n-1}$$
$$+\frac{n(n-1)}{2!}(2011)^{n-2}+\ldots+\frac{n!}{(n-1)!}$$
$$=(2011)^n+{}^nC_1(2011)^{n-1}$$
$$+{}^nC_2(2011)^{n-2}+\ldots+{}^nC_{n-1}$$
$$=(2011+1)^n-1=(2012)^n-1$$

13. *(b)* We have, $y=e^x$ and $y=\log_e x$

$y=e^x$ and $y=\log_e x$ are inverse of each other they are symmetric about line $y=x$

Hence, $AB=BC$

$$AB=\left|\frac{h-e^h}{\sqrt{2}}\right|$$

Let $AB=D=\left|\dfrac{h-e^h}{\sqrt{2}}\right|$

$$\frac{dD}{dh}=1-e^h$$

For maximum or minimum $\dfrac{dD}{dh}=0$

$$\Rightarrow \quad e^h=1\Rightarrow h=0$$
$$\Rightarrow \quad D=\left|\frac{0-1}{\sqrt{2}}\right|=\frac{1}{\sqrt{2}}$$
$$\Rightarrow \quad AC=2AB=\frac{2}{\sqrt{2}}=\sqrt{2}$$

$\therefore$ Minimum distance $=\sqrt{2}$

14. *(b)* We have, $f:(2,\infty)\to N$

$f(x)=$ The largest prime factor of $[x]$.

Let $I=\int_2^8 f(x)\,dx$

$$\Rightarrow \quad I=\int_2^3 2\,dx+\int_3^4 3\,dx+\int_4^5 2\,dx+\int_5^6 5\,dx$$
$$+\int_6^7 3\,dx+\int_7^8 7\,dx$$
$$\Rightarrow \quad I=2+3+2+5+3+7=22$$

15. *(b)* Let $I=\displaystyle\int_0^{2012}\frac{e^{\cos(\pi\{x\})}}{e^{\cos(\pi\{x\})}+e^{-\cos(\pi\{x\})}}\,dx$

$f(x)=\{x\}$ is periodic function of period 1

$$\therefore I=2012\int_0^1\frac{e^{\cos\pi x}}{e^{\cos\pi x}+e^{-\cos\pi x}}\,dx$$
$$I=2012\int_0^1\frac{e^{\cos\pi(1-x)}}{e^{\cos\pi(1-x)}+e^{-\cos\pi(1-x)}}\,dx$$
$$\left[\because\int_0^a f(x)\,dx=\int_0^a f(a-x)\,dx\right]$$
$$I=2012\int_0^I\frac{e^{-\cos\pi x}}{e^{-\cos\pi x}+e^{\cos\pi x}}\,dx$$
$$2I=2012\int_0^1 dx=2012\Rightarrow I=\frac{2012}{2}=1006$$

16. *(d)* Let

$$I=\lim_{n\to\infty}\left(\frac{1}{\sqrt{4n^2-1}}+\frac{1}{\sqrt{4n^2-4}}+\ldots+\frac{1}{\sqrt{4n^2-n^2}}\right)$$
$$=I=\lim_{n\to\infty}\frac{1}{n}\sum_{r=1}^{n}\frac{1}{\sqrt{4-\left(\frac{r}{n}\right)^2}}$$
$$I=\int_0^1\frac{1}{\sqrt{4-x^2}}\,dx$$
$$I=\left[\sin^{-1}\frac{x}{2}\right]_0^1=\sin^{-1}\frac{1}{2}-\sin^{-1}0=\frac{\pi}{6}$$

17. *(c)* Total outcomes of A is $\{(3,5)\,(3,6)\,(5,6)\}$.

Total outcomes of B is $\{(8,9)\,(8,10)\,(9,10)\}$

***Case* I**

A wins A get $(3,6)$ and B gets $(8,9)$.

$\therefore$ Probability $=\dfrac{1}{3}\times\dfrac{1}{3}=\dfrac{1}{9}$

***Case* II** A wins A get $(5,6)$ and B gets any outcomes

$$\text{Probability}=\frac{1}{3}\times1=\frac{1}{3}$$

$$\text{Total probability}=\frac{1}{9}+\frac{1}{3}=\frac{1+3}{9}=\frac{4}{9}$$

18. *(a)* We have,

The position vector of the vertices of ΔABC, $A(\vec{a})$, $B(\vec{b})$ and $C(\vec{c})$ respectively.

Given,

$$\vec{a}\times\vec{b}=\vec{b}\times\vec{c}=\vec{c}\times\vec{a}\neq0$$
$$\vec{a}\times\vec{b}=\vec{b}\times\vec{c}$$
$$\Rightarrow \quad \vec{a}\times\vec{b}+\vec{c}\times\vec{b}=0$$
$$\Rightarrow \quad (\vec{a}+\vec{c})\times\vec{b}=0$$
$$\therefore \quad \vec{a}+\vec{c}=\lambda\vec{b} \quad \text{...(i)}$$
$$[\because \vec{a}+\vec{c}\text{ is parallel to }\vec{b}]$$

Similarly, $\quad \vec{a}+\vec{b}=\mu\,\vec{c} \quad$ ...(ii)

and $\quad \vec{b}+\vec{c}=n\,\vec{a} \quad$ ...(iii)

From Eqs. (i), (ii) and (iii), we get

$$\vec{a}+\vec{b}+\vec{c}=0$$

$\therefore$ Centroid of $\Delta ABC=0$ only one position.

19. *(b)* Let

$$S_n=(1^2-1+1)(1!)+(2^2-2+1)(2!)$$
$$+\ldots+(n^2-n+1)(n!)$$

Here, $T_r=(r^2-r+1)(r!)$

$$T_r=(r^2-1-r+2)(r!)$$
$$T_r=(r^2-1)\,r!-(r-2)\,r!$$
$$T_r=(r-1)(r+1)!-(r-2)\,r!$$

$S_n = T_1 + T_2 + T_3 + ... + T_n$

$S_n = (1-0) + (3!-0) + (2\cdot 4! - 1\cdot 3!) + (3\cdot 5! - 2\cdot 4!) + (4\cdot 6!) - 3\cdot 5! ... + [(n-1)(n+1)! - (n-2)(n!)]$

$S_n = 1 + (n-1)(n+1)!$

$S_n = (n-1)(n+1)! + 1$

20. *(c)* We have,

$f : X \times P(X) \to R$

$$f(x, A) = \begin{cases} 1, & \text{if } x \in A \\ 0, & \text{if } x \notin A \end{cases}$$

$$f(x, A \cup B) = \begin{cases} 1, & \text{if } x \in A \cup B \\ 0, & \text{if } x \notin A \cup B \end{cases}$$

If $x \in A, x \in B \Rightarrow f(x, A \cup B) = 1$

If $x \in A, x \notin B \Rightarrow f(x, A \cup B) = 1$

If $x \notin A, x \in B \Rightarrow f(x, A \cup B) = 1$

If $x \notin A, x \notin B \Rightarrow f(x, A \cup B) = 0$

$\therefore f(x, A \cup B) = f(x,A) + f(x, B) - f(x, A) \cdot f(x, B)$

$[\because n(A \cup B) = n(A) + n(B) - n(A \cap B)]$

21. *(b)* Acceleration due to gravity decreases as we move up from the earth's surface.

Let A and B are stones and C is centre of mass of cabin. Their positions given are as shown below.

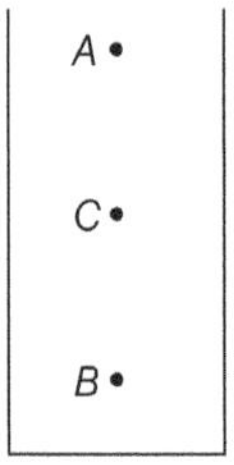

A •

C •

B •

Clearly, $a_B > a_C > a_A$.

Hence, B is accelerated more than C and A is accelerated less than C.

So, with time A move upwards and B move downwards relative to the cabin. So, option (b) is correct.

22. *(b)*

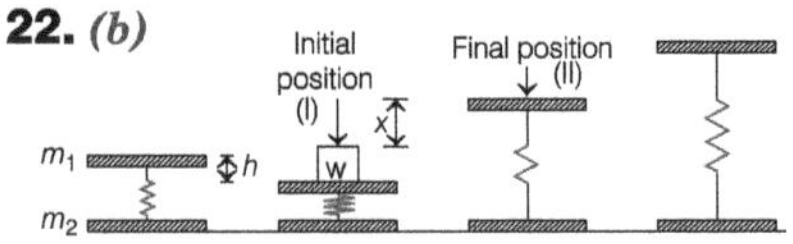

Initial position (I) Final position (II) m_1 m_2 h x w

For block m_2 to lift off,

Spring force = Weight of block

$\Rightarrow \quad kx = mg$ or $x = \dfrac{mg}{k}$

Now, from energy conservation (position (I and II marked in the diagram),

Initial stored energy (due to h compression) = Energy of extended spring + Potential energy of mass of upper block

$\Rightarrow \quad \frac{1}{2}kh^2 = mg(h + x) + \frac{1}{2}kx^2$

$\Rightarrow \quad \frac{1}{2}kh^2 = mg\left(h + \frac{mg}{k}\right) + \frac{1}{2}k\left(\frac{mg}{k}\right)^2$

$\Rightarrow \quad \frac{1}{2}kh^2 = mgh + \frac{m^2g^2}{k} + \frac{m^2g^2}{2k}$

$\Rightarrow \quad \frac{kh^2}{2} = mgh + \frac{3m^2g^2}{2k}$

$\Rightarrow \quad k^2h^2 - (2mgk)h - 3m^2g^2 = 0$

$\Rightarrow h = \dfrac{2mgk \pm \sqrt{4m^2g^2k^2 - 4(k^2)(-3m^2g^2)}}{2k^2}$

$= \dfrac{2mgk \pm 4mgk}{2k^2} = \dfrac{3mg}{k}$

or $hk = 3mg$

Now, considering equilibrium in position 1,

Spring force = Weight of upper block + Weight of mass M

$\Rightarrow \quad kh = mg + Mg$

$\Rightarrow \quad 3mg = mg + Mg$

or $\quad M = 2m$

23. *(a)* For an oscillating simple pendulum,

Acceleration, $a = -\omega^2 x$

and velocity, $v = \omega\sqrt{A^2 - x^2}$

$\Rightarrow \quad v^2 = \omega^2(A^2 - x^2)$

$\Rightarrow \quad v^2 + \omega^2x^2 = \omega^2A^2$

$\Rightarrow \quad v^2 + \frac{\omega^4x^2}{\omega^2} = \omega^2A^2$

$\Rightarrow \quad v^2 + \frac{a^2}{\omega^2} = \omega^2A^2$

$\Rightarrow \quad \frac{v^2}{\omega^2A^2} + \frac{a^2}{\omega^4A^2} = 1$

This is an ellipse $\left(\frac{x^2}{a^2} + \frac{y^2}{b^2} = 1\right)$ with centre at origin.

24. *(a)* Expansion is in vacuum, so work done by expanding gas is $\Delta W = 0$.

Container is insulated, so heat exchanged with surroundings is $\Delta Q = 0$.

Now, from first law of thermodynamics, we have

$$\Delta Q = \Delta U + \Delta W$$

$\Rightarrow \quad \Delta U = 0$

$\Rightarrow \quad U_i = U_f$

Hence, initial and final internal energies of gas are equal.

As, the gas is not ideal expansion results in reduction of pressure, so intermolecular potential energy increases. This results in reduction of kinetic energy, so temperature reduces.

25. *(b)* Let n = number of moles of gas in both of the bulbs and capillary.

When bulbs are maintained at different temperatures, let number of moles are n_1 and n_2 in bulbs at temperatures T and $2T$, respectively.

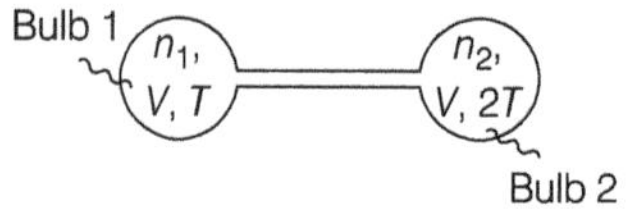

Bulb 1 n_1, V, T — n_2, $V, 2T$ Bulb 2

As gas is ideal,

$$n_1 = \frac{pV}{RT} \text{ and } n_2 = \frac{pV}{R(2T)} \Rightarrow \frac{n_1}{n_2} = \frac{2}{1}$$

Also, $\quad n_1 + n_2 = n$

$\Rightarrow \quad n_1 = \frac{2n}{3}$ and $n_2 = \frac{n}{3}$

Now, ratio of final mass of gas in bulb 2 to the initial mass of gas in same bulb is

$= \dfrac{\text{Mass of gas finally in bulb 2}}{\text{Mass of gas initially in bulb 2}}$

$= \dfrac{\text{Number of moles finally present in bulb 2}}{\text{Number of moles initially present in bulb 2}}$

$= \dfrac{n/3}{n/2} = \dfrac{2}{3}$

26. *(d)* Let junction temperature is T°C.

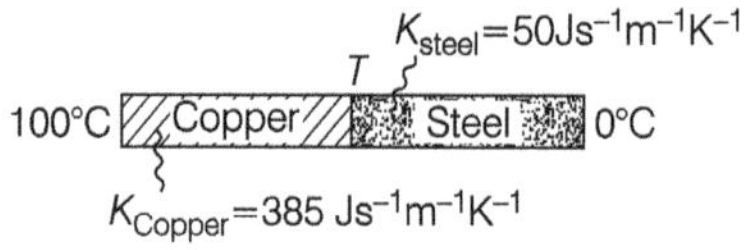

$K_{steel} = 50 \text{ Js}^{-1}\text{m}^{-1}\text{K}^{-1}$, T, 100°C Copper Steel 0°C, $K_{Copper} = 385 \text{ Js}^{-1}\text{m}^{-1}\text{K}^{-1}$

Then, assuming no heat loss and steady state,

Heat flow through copper rod per second = Heat flow through steel rod per second

$\Rightarrow \dfrac{K_{Copper}A(100 - T)}{l} = \dfrac{K_{steel}A(T - 0)}{l}$

$\Rightarrow \quad K_{Copper}(100 - T) = K_{steel} \cdot T$

$\therefore$ Area and length of both rods are equal.

Substituting values, we get

$385(100 - T) = 50T$

$\Rightarrow \quad 385 \times 100 = (385 + 50)T$

$\Rightarrow \quad T = \dfrac{385 \times 100}{435} \approx 88°\text{ C}$

27. *(b)* Compression of a gas in a compressor is nearly an adiabatic process.

So, by using $p_{in}^{1-\gamma} \cdot T_{in}^{\gamma} = p_{out}^{1-\gamma} \cdot T_{out}^{\gamma}$

We get, $(0.28)^{1-\gamma}(233)^{\gamma} = (1)^{1-\gamma}(T)^{\gamma}$

Here, for air, $\gamma = 1.4 = \frac{7}{5}$

Hence, $T = (233)(0.28)^{-2/7} = \dfrac{233}{(0.28)^{2/7}}$

This temperature is much higher than 298 K or 25°C.

So, an air-conditioner is needed to cool the air coming out of compressor.

28. *(b)* When observer is stationary and source is moving, observed frequency is

$$f_1 = f_0\left(\frac{v}{v-u}\right)$$

where, v = speed of sound and u = speed of source.

When source is stationary and the observer is moving towards source, then measured frequency is

$$f_2 = f_0\left(\frac{v+u}{v}\right)$$

So, difference in frequencies observed is

$$f_2 - f_1 = f_0\left(\frac{v+u}{v} - \frac{v}{v-u}\right)$$

$$= -\frac{u^2 f_0}{v(v-u)}$$

$\Rightarrow \quad f_2 - f_1 < 0$ or $f_2 < f_1$

29. *(c)* Let intensity of incident light is I_0. From Malus' law, intensity of light after first polaroid is $I_1 = I_0$.

Here note that, light is plane polarised and first polariser's axis is parallel to plane of polarisation of incident light.

Intensity of light after second polaroid is

$$I_2 = I_0 \cos^2 30^\circ = \frac{3}{4} I_0$$

Similarly, intensity of light after fifth polaroid will be

$$I_5 = \left(\frac{3}{4}\right)^4 I_0 = 0.31\, I_0$$

So, intensity of light at exit is nearly 30% of the incident light.

30. *(c)* For an open end pipe,

Frequency, $f = \frac{v}{\lambda} = \frac{nv}{2L}$

$\Rightarrow \quad L = \frac{nv}{2f}$

As, length of tube remains same in both cases.

So, $\quad L = \frac{nv_1}{2f_1} = \frac{nv_2}{2f_2}$

$\Rightarrow \quad f_2 = \frac{v_2}{v_1} \cdot f_1$

Here, $v_2 = 1.04\, v_1$ and $f_1 = 450$ Hz

So, $\quad f_2 = \frac{1.04 v_1}{v_1} \times 450$

$= 1.04 \times 450$

$= 468$ Hz

31. *(b)* In Young's double slit experiment, fringe width or fringe separation is

$$\beta = \frac{\lambda D}{d} = \frac{cD}{fd}$$

As, $\quad \lambda = c/f$

$\therefore\ \beta \propto \frac{1}{f}$, keeping other parameters constant.

So, when frequency f is doubled, fringe separation β is halved.

$\therefore \quad \beta = \frac{1}{2}$ mm $= 0.5$ mm

32. *(a)* For a domestic AC supply of 220 V at 50 cps,

Peak voltage $= V_{\text{rms}} \times \sqrt{2}$

$= 220\sqrt{2}$ V

Angular frequency, $\omega = 2\pi f = 2\pi \times 50$

$= 100\,\pi \frac{\text{rad}}{\text{s}}$

So, instantaneous potential difference between terminals of a two pin outlet is

$$V = V_{\max} \cos \omega t$$

$$= 220\sqrt{2} \cos 100\,\pi t \text{ volt}$$

33. *(a)* Given circuit is

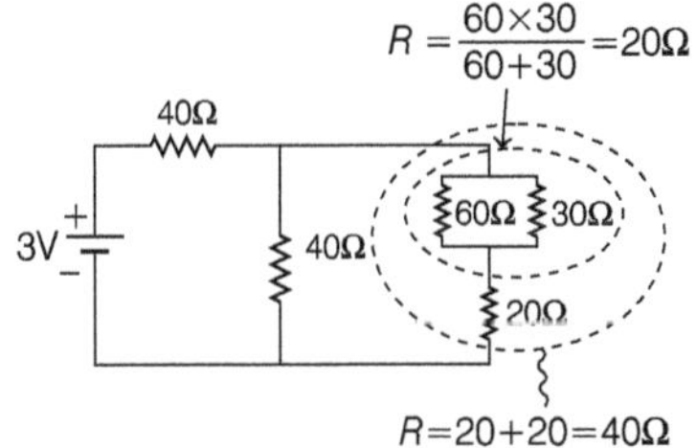

Above circuit can be reduced to following equivalent circuit.

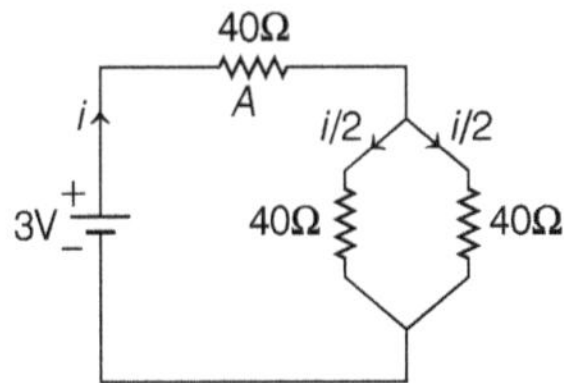

So, power dissipation is maximum in 40 Ω resistor marked A as current through it is maximum.

34. *(b)* Part of kinetic energy of incoming electron is absorbed by molecule and it gets into excited state.

So, kinetic energy after collision is reduced (collision is inelastic).

Linear momentum is conserved as momentum conservation holds correct for elastic as well as inelastic collision.

So, $\quad K' < K$ and $\mathbf{p}' = \mathbf{p}$

35. *(d)* At $t = 0$, as the switch is closed, current in circuit is zero and potential drop across resistor (= iR) is zero. So, potential drop across inductor is 10 V.

With time, current begins to grow in circuit and so potential drop across resistor increases, while potential drop across inductor reduces.

When current reaches a maximum steady value, potential drop across inductor $\left(= L\frac{di}{dt}\right)$ tends to zero.

So, correct variation is

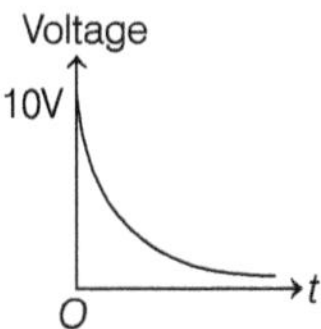

36. *(a)* For an electron and nucleus pair,

Potential energy $= \frac{K(-e)(+Ze)}{r}$

$= \frac{-KZe^2}{r}$

So, potential energy of electron is negative and it tends to zero as separation r increases.

Hence, correct variation of potential energy with r is as shown in graph (3).

For a neutron, force outside nucleus is zero.

Hence, potential energy of neutron is zero as $r > r_0$.

So, correct variation of potential energy with r for a neutron is as shown in graph (1).

For a proton, as $r > r_0$, force is repulsive. Hence, potential energy is positive.

So, correct variation of potential energy with r is as shown in graph (2).

37. *(d)* Given transitions are

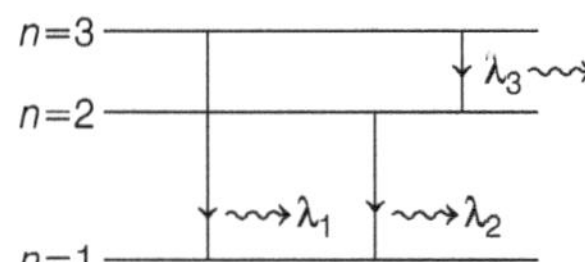

As given, $\lambda_3 > \lambda_2 > \lambda_1$

$\Rightarrow \quad f_3 < f_2 < f_1$

Hence, λ_1 corresponds to $n = 3$ to $n = 1$ transition, λ_2 corresponds to $n = 2$ to $n = 1$ transition and λ_3 corresponds to $n = 3$ to $n = 2$ transition.

Clearly, $\quad E_{3\to1} = E_{3\to2} + E_{2\to1}$

$$\frac{hc}{\lambda_1} = \frac{hc}{\lambda_3} + \frac{hc}{\lambda_2}$$

$\Rightarrow \quad \frac{1}{\lambda_1} = \frac{1}{\lambda_3} + \frac{1}{\lambda_2}$

38. *(a)* Radiative power of a spherical body is given by

$$P = \sigma AT^4$$
$$= \sigma 4\pi R^2 T^4$$
$$\Rightarrow \quad P \propto R^2 T^4$$

When temperature is halved and radius is doubled, then power becomes

$$P' = P \times (2)^2 \times \left(\frac{1}{2}\right)^4$$
$$= P\left(\frac{4}{16}\right) = \frac{P}{4}$$

So, power is reduced to $\frac{1}{4}$th.

39. *(b)* Let $R_H = kh^a e^b$ ...(i)

As, $R = \frac{V}{I}$

$\therefore \quad [R_H] = [ML^2T^{-3}A^{-2}]$

$h = E \cdot t$

$\Rightarrow \quad [h] = [ML^2T^{-1}]$

$e = I \cdot t \Rightarrow [e] = [A \cdot T]$

Substituting above values in Eq. (i), we have

$$[ML^2T^{-3}A^{-2}] = k\,[ML^2T^{-1}]^a\,[AT]^b$$
$$= k\,[M^aL^{2a}T^{-a+b}A^b]$$

Equating dimensions, we get

$$a = 1 \text{ and } b = -2$$

Hence, $R_H = k\left(\frac{h}{e^2}\right)$

So, dimensions of hall resistance are same as that of $\frac{h}{e^2}$.

40. *(a)* Precision refers to the limit of resolution. Precision is determined by the least count of measuring instrument.

A smaller least count means greater precision.

So, more precise readings have least spread of readings as in case I.

41. *(c)* The oxidation state of Ni in $Ni(CO)_4$ is O. It's electronic configuration will be $[Ar]3d^8 4s^2$.

Ni Ground state: 3d [⇅|⇅|⇅|↑|↑] 4s [⇅] 4p [| |]

As CO is a strong field ligand, pairing of electrons will occur.

$Ni(CO)_4$: 3d [⇅|⇅|⇅|⇅|⇅] 4s [ ] 4p [| |] — sp^3-hybridised (4s, 4p)

The oxidation state of Cr in $[Cr(H_2O)_6]^{2+}$ is +2. It's electronic configuration will $[Ar]3d^4 4s^0$.

Cr^{2+}: 3d [↑|↑|↑|↑|] 4s [ ] 4p [| |]

As H_2O is also a strong field ligand, pairing of electrons will occur.

$[Cr(H_2O)_6]^{2+}$: 3d [⇅|↑|↑| |] 4s [ ] 4p [| |] — d^2sp^3-hybridised

42. *(a)* Extraction of silver is achieved by initial complexation of the ore (argentite) with cyanide ion (X) followed by reduction with Zn (Y). This process is known as cyanide process.

$$\underset{}{Ag_2S} + \underset{(X)}{CN^-} \longrightarrow [Ag(CN)_2]^- \xrightarrow{Zn(Y)} [Zn(CN_4)^{2-}] + Ag$$

43. *(a)* Assuming ideal behaviour,

$$\Delta H_{mix} = 0, \Delta V_{mix} = 0$$

It means that no heat is absorbed or evolved when the two liquids are mixed. Also, the volume of solution would be equal to the sum of the volumes of two components.

44. *(c)* The osmotic pressure of a solution is given as, $\pi = CRT$

At constant temperature,

$$\pi \propto C$$
$$\Rightarrow \quad \frac{\pi_1}{\pi_2} = \frac{C_1}{C_2} = \frac{0.01M}{0.001M} = 10$$

45. *(b)* The rate of gas phase chemical reaction generally, increases rapidly with rise in temperature. This is because on increasing the temperature, the fraction of molecules also increases, having the energies greater than activation energy.

This is according to Arrhenius equation $K = Ae^{-E_a/RT}$, where the factor $e^{-E_a/RT}$ corresponds to the fraction of molecules having kinetic energy greater than E_a.

46. *(d)* In free radical polymerisation, initiation is the first step. During initiation, an active centre is created from which a polymer chain is generated. In case of given options, compound (d), 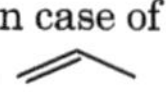 OCH_3 does not undergo polymerisation under initiation because of the presence of electronegative O-atom.

47. *(a)*

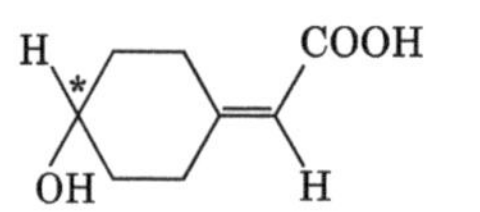

As the compound is optically active, because it contains chiral centre (*), thus it will have two mirror images, which are non-superimposable on each other. Thus, the two possible stereoisomers (for given compounds) are enantiomers (*d* and *l* form).

48. *(d)* According to second law of thermodynamics, whenever a spontaneous process takes place, it is always accompanied by an increase in total entropy of the universe (system and surrounding). Thus, the correct option is (d).

49. *(b)* $\frac{\text{Surface area}}{\text{Volume}} = \frac{4\pi r^2}{\frac{4}{3}\pi r^3}$

or $\frac{\pi d^2}{\frac{\pi d^3}{6}} = \frac{6}{d}$

When $d_1 = 30\,\text{nm}$ and $d_2 = 10\,\text{nm}$

$$\therefore \quad \frac{\left(\frac{SA}{V}\right)_2}{\left(\frac{SA}{V}\right)_1} = \frac{\frac{6}{10}}{\frac{6}{30}} = 3$$

50. *(c)*

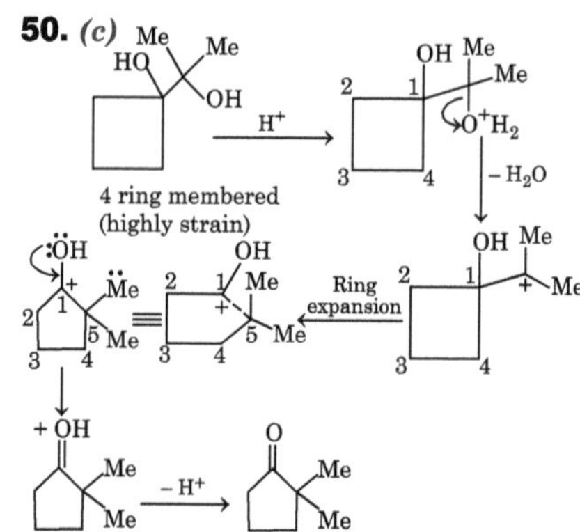

This reaction is known as Pinacol-pinacolone reaction, where diol is converted to a carbonyl compound. Also, the given compound gives a four membered ring carbocation, which changes to five-memered ring carbocation, so as to minimise the strain.

51. *(b)*

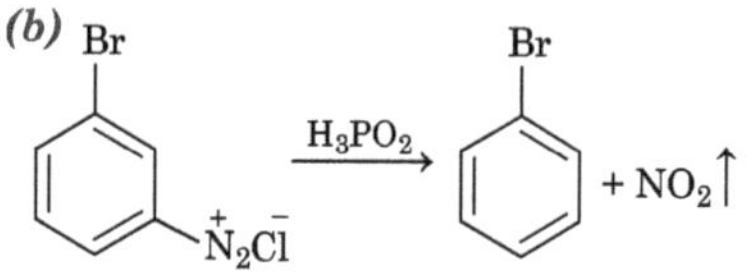

The best reagent is H_3PO_2 because on reduction, $N_2^+Cl^-$ is reduced to H and it is an effective means of removing amino (nitro) group and unlike nucleophillic substitution reactions, this reaction probably proceeds by free radical mechanism.

52. *(a)* Given,

$\lambda^\circ_{NaCl} = 126.4\ S\ cm^2\ mol^{-1}$

$\lambda^\circ_{HCl} = 425.9\ S\ cm^2\ mol^{-1}$

$\lambda^\circ_{NaOAc} = 91.0\ S\ cm^2\ mol^{-1}$

$$\lambda_{HOAc} = \lambda^\circ_{HCl} + \lambda^\circ_{NaOAC} - \lambda^\circ_{NaCl}$$
$$= 425.9 + 91 - 126.4$$
$$= 390.5\ S\ cm^2\ mol^{-1}$$

53. *(c)* According to Bragg's equation,

$$n\lambda = 2d\sin\theta$$

As, $\lambda = d$ [Given]

$n = 1$

$\therefore \quad \lambda = 2\lambda\sin\theta$

$$\sin\theta = \frac{1}{2}$$

$$\theta = 30^\circ$$

54. *(c)* Given, $E^\circ_{cell} = 1.1V$

1 mole of Zn gives 2 electrons

$\therefore$ 2 moles of Zn gives 4 electrons

$\Rightarrow \quad n = 4$

$T = 298\,K$

$$\Delta G^\circ = -nFE^\circ$$
$$= -4\times 96500\times 1.1$$
$$= -424600\,J = -424.6\,kJ$$

55. *(b)* The products formed in the oxidation of $NaBH_4$ by I_2 are NaI, B_2H_6 and H_2.

$$NaBH_4 + I_2 \longrightarrow NaI + B_2H_6 + H_2$$

56. *(c)* The oxidation state for Mn in $Mn[(CN)_6]^{2-}$ is + 4. The electronic configuration for Mn^{4+} is $[Ar]3d^3$.

	3d	4s	4p
Mn^{2+}	1 1 1 _ _	_	_ _ _

As CN^- is a strong field ligand, pairing of electrons occur

	3d	4s	4p
$[Mn(CN)_6]^{2-}$	1⇂ 1 _ _ _	_	_ _ _

$$\mu = \sqrt{n(n+2)}$$

In case of $[Mn(CN)_6]^{2-}$,

$n = 1$

$$\therefore \quad \mu = \sqrt{1(1+2)} = \sqrt{3} = 1.73\,BM$$

The oxidation state of Mn in $[Mn\,Br_4]^{2-}$ is +2. The electronic configuration for Mn^{2+} is $[Ar]3d^5\,4s^0$.

	3d	4s	4p
Mn^{2+}	1 1 1 1 1	_	_ _ _

As Br^- is a weak field ligand, therefore pairing of electrons will not occur.

In this case,

$n = 5$

$$\therefore \quad \mu = \sqrt{5(5+2)} = \sqrt{35} = 5.92\,BM$$

57. *(a)* Time required to reduce the concentration of the reactant to half of its initial value is known as half-life.

For zero order,

$$t_{1/2} = \frac{A_0}{2k}$$

If $\quad A_0 = 2A_0$

$$\therefore \quad t_{1/2} = 2\left(\frac{A_0}{2k}\right)$$

So, if the initial concentration of reactant is double, the time required for half the reactant to be consumed increases by two fold.

58. *(d)* According to Langmuir adsorption isotherm

$$x = \frac{ap}{1+bp}$$

If pressure p is very high, then bp >>1

$$\therefore \quad x = \frac{ap}{bp} = \frac{a}{b}$$

If pressure is very low, then bp<<1

$$\therefore \quad x = \frac{ap}{1} \Rightarrow x \propto p$$

Thus, x increases with p at low pressure and then remains the same at high pressures.

59. *(c)*

Phenol + $CHCl_3$ (Chloroform) $\xrightarrow[H^+]{NaOH/\Delta}$ Salicylaldehyde

This reaction is known as Reimer-Tiemann reaction. The above reaction takes place through following mechanism.

Mechanism

$$CHCl_3 + OH^- \rightleftharpoons \ddot{C}Cl_3^- + H_2O$$

$$:CCl_3^- \longrightarrow :CCl_2 + Cl^-$$

Phenol (OH) $\rightleftharpoons$ Phenoxide (O^-) $+ H^+$

Phenoxide $+ :CCl_2 \longrightarrow$ intermediate $\longrightarrow$ (O^-, $CHCl_2$) $\xrightarrow{OH}$ (O^-, CHO) $\xrightarrow{H^+}$ (OH, CHO)

60. *(c)* The boiling point of a compound depends upon the H-bonding present in it. More is the H-bonding more will be the boiling point. In case of phenol (I), there is only H-bonding, so it's boiling point will be least.

In compound (III) there is intermolecular H-bonding and compound (II), there is intramolecular H-bonding. Thus, compound (III) has maximum boiling point.

Thus, the correct order is I < II < III.

61. *(b)* Neurofilaments are intermediate filaments found in the cytoplasm of neurons. They are protein polymers measuring approximately 10 nm in diameter and many micrometres in length.

62. *(b)* In G_2-phase of the cell cycle, the sister chromatids are available as template for repair. Since, whole DNA is replicated during the S-phase of cell cycle, and it leads to a multiple Double Strand Break (DSB) which needs to be repaired.

So, a cell must properly choose the specific pathway to repair broken DNA molecule which occurs in G_2-phase.

63. *(d)* A person has difficulty in breathing at higher altitudes because overall intake of O_2 by the blood becomes low. Although the percentage of oxygen in inspired air is constant at different altitudes, the fall in atmospheric pressure at higher altitudes decreases the partial pressure of inspired oxygen and hence the driving pressure for gas exchange in the lungs.

64. *(a)* A zygote is formed by the fusion of male and female gametes thus it is diploid. In human, the composition of a zygote that will develop a female will be 44A+ XX and for a male is 44A+ XY.

65. *(a)* On isolation of different cell organelles by differential centrifugation the set of fraction we will find is nucleus, mitochondria, chloroplast, cytoplasm.

Differential centrifugation is a procedure to separate organelles and other sub-cellular particles on the basis of sedimentation rate.

The sequence in which a cell fraction is generated is

Nuclei $\rightarrow$ mitochondria, chloroplast, lysosomes and peroxisomes $\rightarrow$ Plasma membrane, microsomal fraction and large polyribosomes $\rightarrow$ Ribosomal subunits, small polyribosomes $\rightarrow$ Soluble part of cytoplasm.

66. *(b)* Since, the genetic code is a triplet, 64 triplets of – nucleotides will be involved ($4^3 = 64$). If the genetic code is quadruplet it consists of 256 triplets of nucleotides ($4^4 = 256$). 64 triplets coded for 100 amino acid residues. So, when 256 triplets are involved in coding the genes that code for the proteins would have been longer in size by 25%.

Calculation

$(100 \times 256)/64 = 400$ amino acids.
Hence, the percentage change will be $(100/400) \times 100 = 25\%$.
Thus, the correct answer is option (b).

67. *(b)* DNA is made up of two polynucleotide chains having antiparallel polarity. Therefore, when one chain has $5' \to 3'$ polarity the other chain will have $3' \to 5'$ polarity. Thus, if the sequence of the bases in the coding strand of a double-stranded DNA is 5′-ATGTATCTCAAT-3′, its complementary strand will have 3′-TACATAGAGTTA-5′. But in the case of bases in the transcript (i.e. when *m*RNA is formed).

The sequence will be 5′-UACAUAGAGUUA-3′, i.e. exactly similar to the coding strand excepting at the place of thymine (T) which will be replaced by uracil (U).

68. *(d)* The sodium-potassium pump, an important pump in animal cells, expends energy to move potassium ions into the cell and a different number of sodium ions out of the cell. The action of this pump results in a concentration and charge difference across the membrane.

69. *(c)* Carbon dioxide in gaseous form diffuse out of the cell into the capillaries, where it is transported in three ways, i.e in dissolved state (7% of all CO_2 transported), in the form of bicarbonate (70% of CO_2) and as carboxyhaemoglobin (23% of CO_2).

70. *(c)* Patients undergone organ transplants are given anti-rejection medications to prevent T-lymphocytes proliferation. Immunosuppressive are drugs or immunosuppressive agents or anti-rejection medications are drugs that inhibit or prevent activity of the immune system.

71. *(d)* Severe diarrhoea, vomiting, watery stools are the chief symptoms of cholera. All these lead to dehydration. Actually, cholera toxin secreted by *Vibrio cholerae* causes, through a series of metabolic reactions, the continuous activation of adenylate cyclase of intestinal epithelial cells. The resultant high concentration of cAMP triggers continual secretion of Cl^-, HCO_3^- and water in the lumen of the intestine. Administration of saline not only supports the sodium-potassium pump through which water in cell is restored, but glucose is also symported along with sodium.

72. *(d)* A single water molecule can participate in a maximum of four hydrogen bonds because it can accept two bonds using the lone pairs on oxygen and denote two hydrogen atoms.

73. *(d)* Sleep-wake and other daily patterns are part of our circadian rhythms which are governed by the body's interval or biological clock, housed deep with in the brain. The average length of a person's intrinsic circadian rhythm has been shown to be around 24 hours and 11 minutes.

74. *(b)* Modern evolutionary theory consists of the concepts of Darwin modified by knowledge concerning Mendel's laws. Mendel's work provided a way for Darwin's beneficial traits to be preserved. Instead of mixtures that were blended, Mendel proposed particles that could be recombined. In simple terms, Mendel's theory says that individual traits are 'coded' by pairs of particles.

75. *(a)* The ectoderm differentiates to form the nervous system (spine, peripheral nerves and brain), tooth enamel and the epidermis (the outer part of integument). It also forms the lining of mouth, anus, nostrils, sweat glands, hair and nails.

76. *(a)* Chromosome puff is a localised separation of polytene chromosomes (e.g. chromosomes of *Drosophila melanogaster*) that represent a site of active RNA synthesis, i.e. transcription which, when very large is known as Balbiani ring.

77. *(b)* Apoptosis is a form of programmed cell death that occurs in multicellular organisms. Biochemical events lead to characteristic cell changes and death. These changes include blebbing, cell shrinkage, nuclear fragmentation, chromatin condensation, DNA cleavage and global *m*RNA decay.

78. *(d)* JD Watson and FHC Crick (1953) proposed a double helix model for DNA molecule. According to this model, the DNA molecule consists of two strands which are connected together by hydrogen bonds and helically twisted.

79. *(c)* During pachytene stage of meiosis-I, a tetrad of the chromosomes form known as a bivalent. This is the stage when homologous recombination, including chromosomal crossover (crossing over) occurs.

80. *(b)* The resting membrane potential of a neuron is about – 70 mV. This means that the inside of the neuron is 70 mV less than the outside. Therefore, if an electrode is placed in the axioplasm of a mammalian axon and another electrode is placed just outside the axon then the potential difference measured will be – 70 mV. This is because the neuron is at resting stage.

81. *(b)* We have, $AB = BA$

$$A^k = I,\ B^l = 0$$
$$|A^k| = 1,\ |B^l| = 0$$
$$\Rightarrow \quad |B| = 0$$
$$\therefore \quad \det(AB) = |A|\,|B| = 0$$

82. *(c)* We have,

$$\frac{2^2 + 4^2 + 6^2 + \ldots + (2n)^2}{1^2 + 3^2 + 5^2 + \ldots + (2n-1)^2} < 1.01$$

$$= \frac{\Sigma 4n^2}{\Sigma(4n^2 - 4n + 1)} < 1.01$$

$$= \frac{4\frac{n(n+1)(2n+1)}{6}}{4\frac{n(n+1)(2n+1)}{6} - 4\frac{n(n)(n+1)}{2} + n} < 1.01$$

$$= \frac{4\frac{n(n+1)(2n+1)}{6}}{\frac{n(2n+1)(2n-1)}{3}} < 1.01 = \frac{2(n+1)}{2n-1} < \frac{101}{100}$$

$$= 200n + 200 < 202n - 101$$

$$\Rightarrow 2n > 301 \Rightarrow n > \frac{301}{2} = 150.5$$

$$\therefore n > 151$$

83. *(c)* We have, $x^3 + ax^2 + bx + a = 0$

Let α, β, γ are roots of equation and α, β, γ are in AP.

$$\therefore \quad \alpha + \beta + \gamma = -a \qquad \ldots(i)$$
$$\alpha\beta + \beta\gamma + \gamma\alpha = b \qquad \ldots(ii)$$
$$\alpha\beta\gamma = -a \qquad \ldots(iii)$$

From Eq. (i) $3\beta = -a$ $\quad [\because \alpha + \gamma = 2\beta]$

$$\Rightarrow \quad \beta = -\frac{a}{3}$$

and $\quad \alpha\gamma \frac{(-a)}{3} = -a \Rightarrow \alpha\gamma = 3$

Put the values of β and $\alpha\gamma$ in Eq. (ii), we get

$$\beta(\alpha + \gamma) + \gamma\alpha = b \Rightarrow 2\beta^2 + 3 = b$$

$$\Rightarrow \quad 2\left(\frac{a^2}{9}\right) + 3 = b \Rightarrow 2a^2 + 27 = 9b$$

$$\Rightarrow \quad 2a^2 = 9b - 27$$

$\therefore$ Locus of $P(a, b)$ is $2x^2 = 9y - 27$ which represent the parabola whose vertex on Y-axis.

84. *(b)* We have equation of ellipse

$$9x^2 + 16y^2 = 144$$

$$\Rightarrow \quad \frac{x^2}{16} + \frac{y^2}{9} = 1$$

Equation of tangent of ellipse is

$$y = mx \pm \sqrt{a^2m^2 + b^2}$$

$$\therefore \quad y = mx \pm \sqrt{16m^2 + 9}$$

Now, given y-intercept = 5

$$\therefore \sqrt{16m^2 + 9} = 5 \Rightarrow 16m^2 + 9 = 25$$

$$\Rightarrow \quad 16m^2 = 16 \Rightarrow m = \pm 1$$

$\therefore$ Positive slope = 1

85. *(a)* We have,

$A = \{\theta \in R \mid \cos^2(\sin\theta) + \sin^2(\cos\theta) = 1\}$

$\therefore \ \cos^2(\sin\theta) + \sin^2(\cos\theta) = 1$

$\therefore \ \sin\theta = \cos\theta \Rightarrow \tan\theta = 1$

$$\Rightarrow \quad \theta = n\pi + \frac{\pi}{4}$$

$$B = \{\theta \in R \mid \cos(\sin\theta)\sin(\cos\theta) = 0\}$$

$$\cos(\sin\theta)\sin(\cos\theta) = 0$$

$$\cos(\sin\theta) = 0 \text{ or } \sin(\cos\theta) = 0$$

$$\sin\theta = (2n+1)\frac{\pi}{2} \text{ or } \cos\theta = 2n\pi$$

$\therefore$ Clearly $A \cap B = \phi$.

$\therefore$ $A \cap B$ is an empty set.

86. *(b)* We have,

$$f(x) = x^3 + ax^2 + bx + c$$

$$f(2) = 8 + 4a + 2b + c = 0 \quad \text{...(i)}$$

$$[\because f(2) = 0]$$

$$f'(x) = 3x^2 + 2ax + b$$

$$f'(1) = 3 + 2a + b = 0 \quad \text{...(ii)}$$

$$[\because f'(1) = 0]$$

$$f'\left(\frac{-1}{3}\right) = \frac{1}{3} - \frac{2a}{3} + b = 0 \quad \text{...(iii)}$$

$$\left[\because f'\left(\frac{-1}{3}\right) = 0\right]$$

From Eqs. (ii) and (iii), we get

$$a = -1, b = -1$$

Putting the values of a and b in Eq. (i), we get $c = -2$

$$\therefore \quad f(x) = x^3 - x^2 - x - 2$$

$$\int_{-1}^{1} f(x)\,dx = \int_{-1}^{1} (x^3 - x^2 - x - 2)\,dx$$

$$= -2\int_0^1 (x^2 + 2)\,dx$$

$[\because x^3$ and $-x$ are odd functions$]$

$$= -2\left[\frac{x^3}{3} + 2x\right]_0^1$$

$$= -2\left[\frac{1}{3} + 2\right] = -\frac{14}{3}$$

87. *(d)* We have,

$$f(x) = x^{12} - x^9 + x^4 - x + 1$$

$$f(0) = f(1) = 1$$

$\therefore$ f is not one-one.

$$f'(x) = 12x^{11} - 9x^8 + 4x^3 - 1$$

$f'(x)$ is never vanishes and $f(x)$ has not a real root

$$x > 1 \text{ and } x \le 0$$

$$x^{12} - x^9 + x^4 - x + 1 > 0$$

$$[\because x^{12} > x^9, x^4 > x]$$

When $x \in (0, 1)$

$$1 - x + x^4 - x^9 + x^{12} > 0$$

$$[\because 1 - x > 0, x^4 - x^9 > 0]$$

$\therefore$ $f(x)$ has only positive value.

88. *(a)* We have,

$$f_n(x) = \text{minimum}\left(\frac{x^n}{n!}, \frac{(1-x)^n}{n!}\right), x \in [0, 1]$$

$$I_n = \int_0^1 f_n(x)\,dx$$

$$I_n = \left|\int_0^{1/2} \frac{x^n}{n!}\,dx\right| + \left|\int_{1/2}^1 \frac{(1-x)^n}{n!}\,dx\right|$$

$$I_n = \left|\left[\frac{x^{n+1}}{(n+1)n!}\right]_0^{1/2}\right| + \left|\left[\frac{(1-x)^{n+1}}{(n+1)\,n!}\right]_{1/2}^1\right|$$

$$I_n = \frac{2}{(n+1)!}\left[\left(\frac{1}{2}\right)^{n+1}\right]$$

$$\sum_{n=1}^{\infty} I_n = 2\left[\frac{1}{2!}\left(\frac{1}{2}\right)^2 + \frac{1}{3!}\left(\frac{1}{2}\right)^3 + \dots + \frac{1}{(n+1)!}\left(\frac{1}{2}\right)^{n+1}\right]$$

$$\sum_{n=1}^{\infty} I_n = 2\left[e^{1/2} - \left(1 + \frac{1}{2}\right)\right]$$

$$= 2\sqrt{e} - 3$$

89. *(b)* We have,

$$|x + y| + |x - y| = 4$$

$$\Rightarrow \quad x + y + x - y = 4$$

$$[\because x + y \ge 0, x - y \ge 0]$$

$$x = 2 \Rightarrow x + y - x + y = 4$$

$$[\because x + y \ge 0, x - y < 0]$$

$$y = 2 \Rightarrow -x - y + x - y = 4$$

$$[\because x + y < 0, x - y \ge 0]$$

$$y = -2 \Rightarrow -x - y - x + y = 4$$

$$[\because x + y < 0, x - y < 0]$$

$$x = -2$$

The maximum possible value of

$$x^2 + y^2 - 4x - 6y \text{ at } (-2, -2)$$

$$\therefore \quad (-2)^2 + (-2)^2 - 4(-2) - 6(-2)$$

$$\Rightarrow \quad 4 + 4 + 8 + 12 = 28$$

90. *(c)* We have,

Let two-digits numbers are $10a + b$.

Given, $10a + b$ is AM of x and y and $10b + a$ is GM of x and y.

$$\therefore \quad \frac{x+y}{2} = 10a + b$$

$$\Rightarrow \quad \sqrt{xy} = 10b + a$$

$$\Rightarrow \quad xy = (10b + a)^2$$

$$\Rightarrow (x + y)^2 - (x - y)^2 = 4xy$$

$$\therefore \quad (x - y)^2 = (x + y)^2 - 4xy$$

$$\Rightarrow \quad (x - y)^2 = 4(10a + b)^2 - 4(10b + a)^2$$

$$\Rightarrow \quad (x - y)^2 = 4(10a + b + 10b + a)(10a + b - 10b - a)$$

$$\Rightarrow \quad (x - y)^2 = 4(11)(a + b)\cdot 9(a - b)$$

$$\Rightarrow \quad (x - y)^2 = 4 \times 11\cdot(a + b)\cdot 9(a - b)$$

$4 \times 11(a + b) \times 9(a - b)$ must be a perfect square.

$$\therefore \quad a + b = 11, a - b = 1$$

On solving these equations, we get

$$a = 6, b = 5$$

$$\therefore \quad x + y = 2(10a + b)$$

$$\Rightarrow \quad x + y = 2(60 + 5)$$

$$\Rightarrow \quad x + y = 130$$

91. *(b)* Electric field E of a solid sphere varies with distance from centre r as

$$E = \begin{cases} C_1 \cdot r, & r < 0 \\ \dfrac{C_2}{r^2}, & r \ge 0 \end{cases}$$

where, C_1 and C_2 are constants.

So, correct graph of E *versus* r is

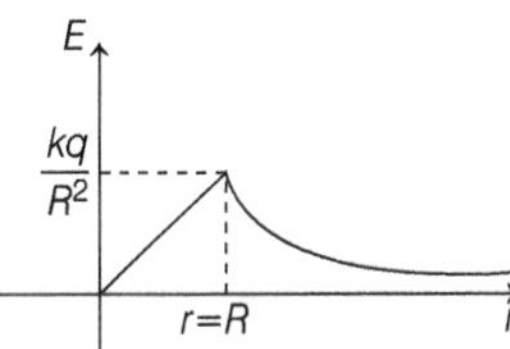

92. *(c)* For uncharged particle in projectile motion, range is

$$L = \frac{u^2\sin 2\theta}{g} \quad \text{...(i)}$$

For a charged particle, acceleration is $\left(g + \dfrac{qE}{m}\right)$

So, its range will be $\dfrac{L}{2} = \dfrac{u^2\sin 2\theta}{\left(g + \dfrac{qE}{m}\right)}$

Equating for L from above equations, we get

$$\frac{1}{g} = \frac{2}{g + \dfrac{qE}{m}}$$

$$\Rightarrow \quad g = qE/m \quad \text{...(ii)}$$

Now, for a particle of mass m and charge $2q$, range will be

$$R = \frac{u^2 \sin 2\theta}{g + \frac{2qE}{m}}$$

Here, $q\frac{E}{m} = g$ [From Eq. (ii)]

$$= \frac{u^2 \sin 2\theta}{3g}$$

Again, $\frac{u^2 \sin 2\theta}{g} = L,$ [From Eq. (i)]

$$= \frac{L}{3}$$

93. *(d)* As ball falls through height h and string turns by angle θ,

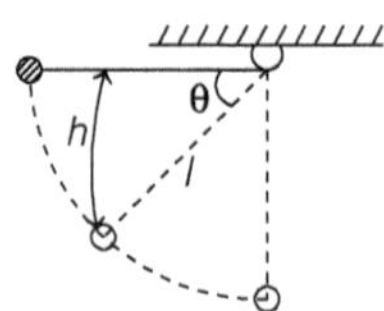

Velocity of ball is obtained by equating kinetic and potential energies

$$\Rightarrow \frac{1}{2} mv^2 = mgh$$

$$\Rightarrow v^2 = 2gh$$

$$\Rightarrow v^2 = 2gl \sin \theta$$

So, radial acceleration is

$$a_r = \frac{v^2}{l} = 2g \sin\theta$$

Tangential acceleration is produced by component of weight along the string.

$\therefore$ Tangential acceleration is

$$a_t = g \cos\theta$$

Hence, total acceleration is

$$a = \sqrt{a_t^2 + a_r^2}$$

$$= \sqrt{4g^2 \sin^2 \theta + g^2 \cos^2\theta}$$

$$= g\sqrt{1 + 3 \sin^2\theta}$$

94. *(d)* In the process given, internal energy is

$$U = aV^3$$

$$\Rightarrow \frac{fnRT}{2} = aV^3$$

where, f = degree of freedom = 3 (for ideal gas) and n = number of moles = 1.

$$\Rightarrow \frac{3}{2}RT = aV^3$$

$$\Rightarrow \frac{3}{2}pV = aV^3 \quad [\because pV = RT]$$

So, pressure in this process is given by

$$p = \frac{2a}{3} . V^2$$

Now, work done during the process is

$$W = \int_V^{2V} pdV = \int_V^{2V} \frac{2a}{3} V^2 dV$$

$$= \frac{2a}{3} \times \frac{1}{3} ((2V)^3 - V^3)$$

$$= \frac{2a}{9} (7V^3)$$

$$= \frac{2}{9} \times 7 \times \frac{3}{2} RT \quad \left[\because aV^3 = \frac{3}{2} RT\right]$$

$$= \frac{7}{3} RT$$

95. *(c)* In isothermal process BC,

$$pV = \text{constant}$$

$$\Rightarrow p_C V_C = p_B V_B$$

$$\Rightarrow V_B = \frac{p_C V_C}{p_B} = \frac{500 \times 2}{200} = 5 \text{m}^3$$

Now, work done in the complete cycle $CABC$ is

$$W_{CABC} = W_{CA} + W_{AB} + W_{BC}$$

$$= 0 + p\Delta V + \int pdV$$

$$= 200 (5 - 2) + \int_5^2 k \frac{dV}{V}$$

$$= 600 + k\int_5^2 \frac{dV}{V}$$

$$= 600 + 1000 (\log 2 - \log 5)$$

$$= 600 + 1000 (0.69 - 1.60)$$

$$= 600 - 910 \approx - 300 \text{ kJ}$$

96. *(b)* Light beam passes through material, if no total internal reflection occurs at surface of material.

So, $\sin i \leq \frac{n_2}{n_1}$

$$\Rightarrow \sin i \leq \frac{1.2}{1.5}$$

$$\Rightarrow \sin i \leq 0.8$$

or $\sin(i_{\text{maximum}}) = 0.8$

$$\Rightarrow i_{\text{maximum}} = \sin^{-1} (0.8)$$

$$\Rightarrow i_{\max} = 53.1°$$

97. *(a)* Distances and velocities of charged particle given are as below.

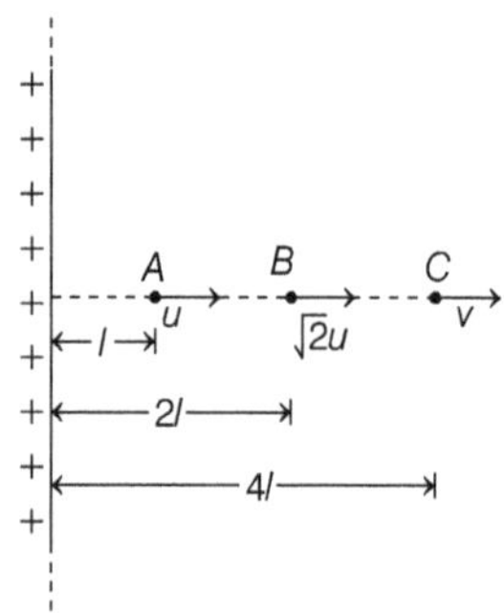

Energy conservation between A and B gives,

$$qV_A + \frac{1}{2} mu^2 = qV_B + \frac{1}{2} m (\sqrt{2}u)^2$$

$$\Rightarrow q(V_A - V_B) = \frac{1}{2} mu^2$$

$$\Rightarrow q \frac{\lambda}{2\pi\varepsilon_0} . \ln 2 = \frac{1}{2} mu^2 \quad \text{...(i)}$$

Now, energy conservation between A and C gives,

$$qV_A + \frac{1}{2}mu^2 = qV_C + \frac{1}{2} mv^2$$

$$\Rightarrow \frac{1}{2} mv^2 = q(V_A - V_C) + \frac{1}{2}mu^2$$

$$= q \frac{\lambda}{2\pi\varepsilon_0} . \ln 4 + \frac{1}{2} mu^2$$

$$= \frac{2q\lambda}{2\pi\varepsilon_0} . \ln 2 + \frac{1}{2} mu^2$$

Substituting value of $\left(\frac{q\lambda}{2\pi\varepsilon_0} \ln 2\right)$ from Eq. (i) in above equation, we have

$$\frac{1}{2} mv^2 = 2\left(\frac{1}{2} mu^2\right) + \frac{1}{2} mu^2$$

$$\Rightarrow v = \sqrt{3}u$$

98. *(b)* The loop $ABCD$ is moving from region of high flux to low flux in region of a downward magnetic field.

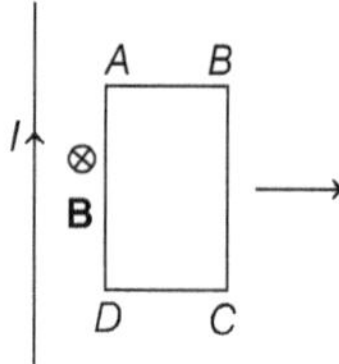

Following Lenz's law induced emf must be directed to increase the flux. Hence, direction of induced emf is clockwise in the loop.

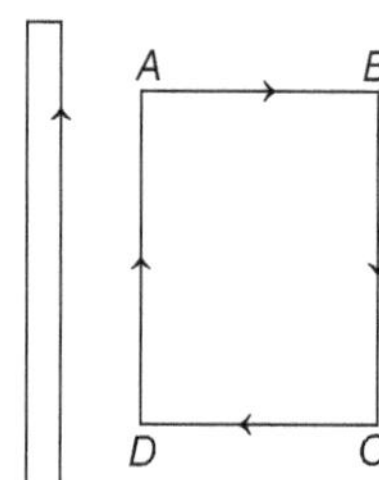

Hence, current in wire and AD length are parallel. So, there is attraction in wire and AD length.

Similarly, there is repulsion in BC and wire.

99. *(d)* Given circuit is a parallel combination of two cells connected across a resistance R.

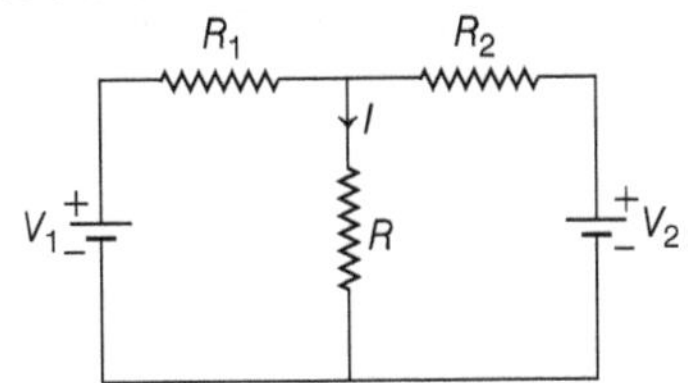

Now, for parallel combination of cells

$$V_{eq} = \frac{\frac{V_1}{R_1} + \frac{V_2}{R_2}}{\frac{1}{R_1} + \frac{1}{R_2}}$$

$$\Rightarrow \quad V_{eq} = \frac{V_1R_2 + V_2R_1}{R_1 + R_2} \quad \ldots(i)$$

$$\text{and } R_{eq} = \frac{R_1R_2}{R_1 + R_2} \quad \ldots(ii)$$

So, current through resistance R is

$$I = \frac{V_{eq}}{R + R_{eq}} \quad \ldots(iii)$$

In case A

$V_1 = 2\,V, V_2 = 0\,V$ and $I = 3\,mA$

So, from Eqs. (i), (ii) and (iii), we get

$$V_{eq} = \frac{2 \cdot R_2 + 0 \cdot R_1}{R_1 + R_2} = \frac{2R_2}{R_1 + R_2}$$

$$R_{eq} = \frac{R_1R_2}{R_1 + R_2} \text{ and } I = \frac{2R_2 / R_1 + R_2}{R + \frac{R_1R_2}{R_1 + R_2}}$$

$$\Rightarrow \quad 3\,mA = \frac{2R_2}{R(R_1 + R_2) + R_1R_2} \quad \ldots(iv)$$

In case B

$V_1 = 0\,V, V_2 = 4\,V$ and $I = 4\,mA$

$$\Rightarrow \quad 4\,mA = \frac{4R_1}{R(R_1 + R_2) + R_1R_2} \quad \ldots(v)$$

From Eqs. (iv) and (v), we have

$$\frac{3}{4} = \frac{2R_2}{4R_1} \Rightarrow \frac{3}{4} = \frac{R_2}{2R_1} \Rightarrow \frac{R_2}{R_1} = \frac{3}{2}$$

In case C

$V_1 = 10\,V$ and $V_2 = 10\,V$

So, if current through resistance R is I, then

$$V_{eq} = \frac{10R_1 + 10R_2}{R_1 + R_2}$$

$$\text{and so } \quad I = \frac{10R_1 + 10R_2}{R(R_1 + R_2) + R_1R_2} \quad \ldots(vi)$$

Now, from Eqs. (iv) and (vi), we have

$$\frac{3}{I} = \frac{\left(\frac{2R_2}{R(R_1 + R_2) + R_1R_2}\right)}{\left(\frac{10R_1 + 10R_2}{R(R_1 + R_2) + R_1R_2}\right)}$$

$$\Rightarrow \quad \frac{3}{I} = \frac{2R_2}{10R_1 + 10R_2}$$

$$\Rightarrow \quad \frac{I}{3} = 5\frac{R_1}{R_2} + 5$$

$$\Rightarrow \quad \frac{I}{3} = 5\left(\frac{2}{3}\right) + 5$$

$$\Rightarrow \quad I = 10 + 15 = 25\,mA$$

100. *(a)* Path of particle is

$$\frac{x^2}{a^2} + \frac{y^2}{b^2} = 1$$

Differentiating with time t, we get

$$\frac{x}{a^2} \cdot \frac{dx}{dt} + \frac{y}{b^2} \cdot \frac{dy}{dt} = 0 \quad \ldots(i)$$

Again, differentiating with time t, we have

$$\frac{1}{a^2}\left(x, \frac{d^2x}{dt^2} + \left(\frac{dx}{dt}\right)^2\right) + \frac{1}{b^2}\left(y\frac{d^2y}{dt^2} + \left(\frac{dy}{dt}\right)^2\right) = 0 \ldots(ii)$$

Now, given $u_x = u$ at $(0, b)$

So, from Eq. (i), we have

$$\frac{1}{b}\frac{dy}{dt} = 0 \Rightarrow \frac{dy}{dt} = u_y = 0$$

Now at $(0, b)$ from Eq. (ii), we have

$$\frac{1}{a^2}\left(\frac{dx}{dt}\right)^2 + \frac{1}{b^2}\left(b\frac{d^2y}{dt^2} + \left(\frac{dy}{dt}\right)^2\right) = 0$$

$$\text{or} \quad \frac{1}{a^2} \cdot u^2 + \frac{1}{b^2} \cdot b\,(a_y) = 0 \quad [\because u_y = 0]$$

$$\Rightarrow \quad a_y = -\frac{b}{a^2}u^2$$

101. *(b)* XeF_6 hydrolyses to give XeO_3 and HF.

$$XeF_6 + 3H_2O \longrightarrow XeO_3 + 6HF$$

Structure of XeF_6

Xe has eight valence electrons XeF_6 has 6 bond pairs and 1 lone pair with sp^3d^3-hybridisation. Thus, its structure will distorted octahedral.

(Distorted octahedral)

Structure of XeO_3

XeO_3 has 3 bond pairs and 1 lone pair with sp^3-hybridisation. Thus, its structure will be pyramidal.

102. *(b)* $2MnO_4^- + 5C_2O_4^{2-} \xrightarrow{16H^+} 2Mn^{2+} + 10\,CO_2 + 8H_2O$

As 2 moles of MnO_4^- oxidises 5 moles of $C_2O_4^{2-}$.

$\therefore MnO_4^- : C_2O_4^{2-} = 2 : 5$

$$2MnO_4^- + 16HCl \longrightarrow 2MnCl_2 + 5Cl_2 + 2Cl^- + 8H_2O$$

As 2 moles of MnO_4^- oxidises 16 moles of HCl.

$\therefore MnO_4^- : HCl = 2 : 16 = 1 : 8$

103. *(b)* For the reactions,

$Fe^{2+} + 2e^- \longrightarrow Fe\,;\ E_1^\circ = -0.440V \quad \ldots(i)$

$Fe^{3+} + e^- \longrightarrow Fe^{2+};\ E_2^\circ = 0.770\ V \quad \ldots(ii)$

To obtain reaction,

$Fe^{3+} + 3e^- \longrightarrow Fe;\ E_3^\circ = ?$

We add Eq. (i) + (ii)

$$\therefore \Delta G_3^\circ = \Delta G_1^\circ + \Delta G_2^\circ$$

$$= -n_3FE_3^\circ = -n_1FE_1^\circ + (-n_2FE_2^\circ)$$

$$\Rightarrow E_3^\circ = \frac{-n_1E_1^\circ - n_2E_2^\circ}{n_3E_3^\circ}$$

$$= \frac{2(-0.44) + 1 \times 0.770}{3}$$

$$= \frac{-0.88 + 0.770}{3} = -0.037\,V$$

104. *(a)* According to Bohr's energy,

$$\Delta E = 13.6\left(\frac{1}{n_1^2} - \frac{1}{n_2^2}\right) \text{eV/atom}$$

For transition from $n = 1$ to $n = 3$

$$\Delta E_{1\to3} = 13.6\left[\frac{1}{1^2} - \frac{1}{3^2}\right] = 13.6\left[\frac{8}{9}\right]$$

For transition from $n = 1$ to $n = 2$

$$\Delta E_{1\to2} = 13.6\left[\frac{1}{1^2} - \frac{1}{2^2}\right] = 13.6\left[\frac{3}{4}\right]$$

$$\therefore \frac{\Delta E_{1\to3}}{\Delta E_{1\to2}} = \frac{13.6\left[\frac{8}{9}\right]}{13.6\left[\frac{3}{4}\right]} = \frac{32}{27}$$

105. *(c)*

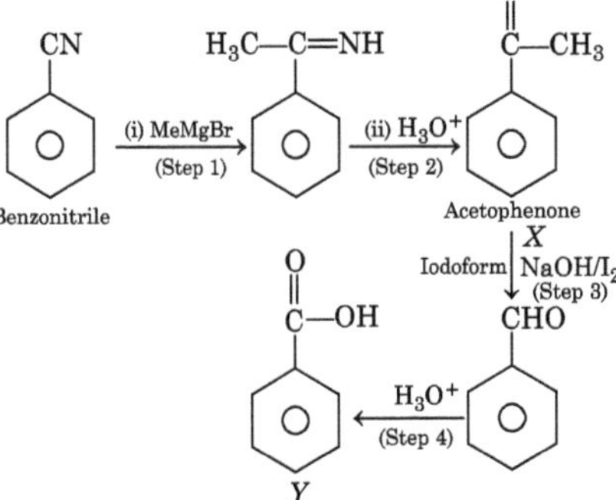

In step 1, benzonitrile reacts with Grignard reagent to give imine compound, which undergoes hydrolysis to give acetophenone (X).

In step 3, acetophenone undergoes iodoform reaction and then hydrolysis to give benzoic acid (Y).

106. *(b)*

OH → (HNO_2) → I (OH, NO) → (Tautomerisc) → X (O, NOH); X + OH (phenol), H_2SO_4 → Y (O, N, OH)

In given series of reaction phenol reacts with HNO_2 to give compound I, which tautomerises to form oxime (X). As the phenol is a nucleophile, it will attack at the oxime [electrophile] at the most electrophilic site to give compound Y.

107. *(*)* Incorrect question as S-α-methoxyacetaldehyde is optically inactive.

$$H—\overset{CHO}{\underset{H}{C}}—OCH_3$$

S-α-methoxy acetaldehyde

[optically inactive due to presence of 2Hs]

108. *(d)* In liquid phase, according to Dalton's law of partial pressure

$$p_T = p_A + p_B$$
$$= x_A p_A^\circ + x_B p_B^\circ$$
$$= x_A p_A^\circ + (1 - x_A) p_B^\circ$$
$$p_T = (p_A^\circ - p_B^\circ)x_A + p_B^\circ$$
$$300 = (-400)x_A + 500$$
$$x_A = \frac{1}{2}$$

In vapour phase, $x_A = \frac{p_A}{p_T} = \frac{p^\circ_A x_A}{p_A^\circ x_A + p_B^\circ x_B}$

$$= \frac{100 \times \frac{1}{2}}{100 \times \frac{1}{2} + 500 \times \frac{1}{2}} = \frac{50}{300} = \frac{1}{6}$$

109. *(a)* For high spin d^6 metal complex

1⇂1111 → e_g: 1 1 ; t_{2g}: ⇅ 1 1

$$CFSE = -0.4 \times x + 0.6 \times y$$
$$= -0.4 \times 4 + 0.6 \times 2$$
$$= -1.6 + 1.2 = -0.4\Delta_o$$

For low spin d^6 metal complex

1⇂1111 → e_g: — ; t_{2g}: ⇅ ⇅ ⇅

$$CFSE = -0.4 \times 6 + 0.6 \times 0$$
$$= -2.4 + 0$$
$$= -2.4\Delta_o$$

110. *(a)* Given,

Surface tension = $0.03\ Jm^{-2}$

Total droplets = 2.4×10^{18}

Area of each droplet = $12.5 \times 10^{-6} m^2$

⇒ Total area = Total droplet × Area of one droplet covered by droplets

$$= 2.4 \times 10^{18} \times 12.5 \times 10^{-16}$$
$$= 12.5 \times 2.4 \times 10^2 m^2$$

Energy spent = Total area covered by droplets × Surface tension

$= 12.5 \times 2.4 \times 10^2 \times 0.03 = 90\ J$

111. *(d)* Stomata are open during the day and close during night. There are certain factors affecting stomatal movement.

Stomatal opening during day	Stomatal closure during night
Rise in pH	Low pH
Low pCO_2	High pCO_2
High humidity	Dry environment
High osmotic pressure	Low osmotic pressure
High glucose	Low glucose

112. *(c)* Crossing-over leads to mixing of genes in sexually reproducing organisms. Therefore, if crossing-over is stopped in a rice plant ($2n = 24$) and then selfed seeds are collected, all the offsprings will be genetically identical to the parent if the parent plant was a completely inbred line.

113. *(d)* Olfaction is a chemoreceptor that forms the sense of smell. Olfaction occurs when odorants bind to specific sites on olfactory receptors located in the nasal cavity. Thus the basis for odour discrimination in rodent is the presence of a large number of different chemoreceptors in the nasal cavity.

114. *(c)* As the heart pumps, blood is pushed through the body through the entire circulatory system. From the aorta, blood flows into the arteries and arterioles and ultimately to the capillary beds. As it reaches the capillary beds, the rate of flow is dramatically (one thousand times) slower than the rate of flow in the aorta while the diameter of each individual arteriole and capillary is far narrower than the diameter of the aorta, the rate is actually slower due to the over all diameter of all the combined capillaries being far greater than the diameter of the individual aorta.

115. *(b)* Radiolabelled thymidine (TdR), such as tritiated thymidine is commonly used in cell proliferation assays. A DNA consists of four bases adenine, guanine, cytosine and thymine. Thus, *E.coli* treated with tritiated thymidine will have one strand of radio labelled DNA after its replication.

116. *(b)* Isolation of lysine auxotroph from a mixture of auxotrophs is done in the following steps :

(i) Grow cells in media containing penicillin (which will kill dividing cells) and all amino acids except lysine.

(ii) Remove the penicillin laced media and replace with media containing lysine.

(iii) Screen for lysine auxotrophs.

117. *(c)* As you increase your sample size the standard error of the mean will become smaller. With bigger sample sizes, the sample mean becomes a more accurate estimate of the parametric mean, so the standard error of the mean becomes smaller.

118. *(d)* During meiosis and due to independent assortment at metaphase-I, the homologous chromosome pairs line up together. The first time the cell divides, one of each chromosome will go into each new cell, so they will be separated from each other. Each pair lines up independently of every other pair. So, each cell is a random combination of female and male chromosomes. There are 2 chromosomes in each pair and 23 pairs of chromosomes.

So, for a human male the probability that all the maternal chromosomes will end up in the same gamete is $\left(\frac{1}{2}\right)^{23}$

119. *(a)* The retina of nocturnal animals is almost entirely composed of rods. The other type of vision cells, cones are absent or almost absent, leaving nocturnal animals with virtually no colour vision. The photosensitive pigments inside the rods, rhodopsin is particularly sensitive low levels of light.

120. *(a)* The DNA molecule has the shape of a double helix. The radius of each helix is about 10 Å. Each helix rises about 34 Å during each complete turn and there are about 2.9×10^8 complete turns.

QUESTION PAPER 2010

Stream : SB/SX

MM : 160

Instructions

1. There are 120 questions in this papor.
2. The question paper contains two parts; Part I (1 Mark Questions) and Part II (2 Marks Questions).
3. There are four sections in each part; Mathematics, Physics, Chemistry and Biology.
4. Out of the four options given with each question, only one is correct.

PART-I (1 Mark Questions)

MATHEMATICS

1. Let A denote the matrix $\begin{bmatrix} 0 & i \\ i & 0 \end{bmatrix}$, where $i^2 = -1$, and let I denote the identity matrix $\begin{bmatrix} 1 & 0 \\ 0 & 1 \end{bmatrix}$. Then, $I + A + A^2 + \ldots + A^{2010}$ is

(a) $\begin{bmatrix} 0 & 0 \\ 0 & 0 \end{bmatrix}$ (b) $\begin{bmatrix} 0 & i \\ i & 0 \end{bmatrix}$ (c) $\begin{bmatrix} 1 & i \\ i & 1 \end{bmatrix}$ (d) $\begin{bmatrix} -1 & 0 \\ 0 & -1 \end{bmatrix}$

2. Suppose the sides of a triangle form a geometric progression with common ratio r. Then, r lies in the interval

(a) $\left(0, \frac{-1+\sqrt{5}}{2}\right]$ (b) $\left(\frac{1+\sqrt{5}}{2}, \frac{2+\sqrt{5}}{2}\right]$

(c) $\left(\frac{-1+\sqrt{5}}{2}, \frac{1+\sqrt{5}}{2}\right]$ (d) $\left(\frac{2+\sqrt{5}}{2}, \infty\right)$

3. The number of rectangles that can be obtained by joining four of the twelve vertices of a 12-sided regular polygon is

(a) 66 (b) 30 (c) 24 (d) 15

4. Let 1, ω and ω^2 be the cube roots of unity. The least possible degree of a polynomial with real coefficients, having $2\omega^2, 3 + 4\omega, 3 + 4\omega^2$ and $5 - \omega - \omega^2$ as roots is

(a) 4 (b) 5 (c) 6 (d) 8

5. A circle touches the parabola $y^2 = 4x$ at (1, 2) and also touches its directrix. The y-coordinate of the point of contact of the circle and the directrix is

(a) $\sqrt{2}$ (b) 2 (c) $2\sqrt{2}$ (d) 4

6. Let ABC be an equilateral triangle, let $KLMN$ be a rectangle with K, L on BC, M on AC and N on AB. Suppose $AN / NB = 2$ and the area of ΔBKN is 6.The area of the ΔABC is

(a) 54 (b) 108 (c) 48
(d) not determinable with the above data

7. Let P be an arbitrary point on the ellipse $\frac{x^2}{a^2} + \frac{y^2}{b^2} = 1$, $a > b > 0$. Suppose F_1 and F_2 are the foci of the ellipse. The locus of the centroid of the ΔPF_1F_2 as P moves on the ellipse is

(a) a circle (b) a parabola
(c) an ellipse (d) a hyperbola

8. The number of roots of the equation $\cos^7\theta - \sin^4\theta = 1$ that lie in the interval $[0, 2\pi]$ is
(a) 2 (b) 3 (c) 4 (d) 8

9. The product $(1+\tan 1^\circ)(1+\tan 2^\circ)(1+\tan 3^\circ)$ $\dots(1+\tan 45^\circ)$ equals
(a) 2^{21} (b) 2^{22} (c) 2^{23} (d) 2^{25}

10. Let $f: R \to R$ be a differentiable function such that $f(a) = 0 = f(b)$ and $f'(a)f'(b) > 0$ for some $a < b$. Then, the minimum number of roots of $f'(x) = 0$ in the interval (a, b) is
(a) 3 (b) 2 (c) 1 (d) 0

11. The roots of $(x-41)^{49} + (x-49)^{41} + (x-2009)^{2009} = 0$ are
(a) all necessarily real
(b) non-real except one positive real root
(c) non-real except three positive real roots
(d) non-real except for three real roots of which exactly one is positive

12. The figure shown below is the graph of the derivative of some function $y = f(x)$.

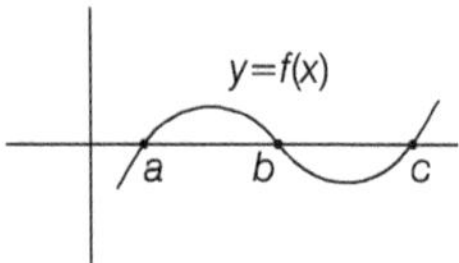

Then,
(a) f has local minima at $x = a, b$ and a local maximum at $x = c$
(b) f has local minima at $x = b, c$ and a local maximum at $x = a$
(c) f has local minima at $x = c, a$ and a local maximum at $x = b$
(d) the given figure is insufficient to conclude any thing about the local minima and local maxima of f

13. The following figure shows the graph of a continuous function $y = f(x)$ on the interval [1, 3]. The points A, B, C have coordinates (1, 1), (3, 2), (2, 3) respectively, and the lines l_1 and l_2 are parallel, with l_1 being tangent to the curve at C. If the area under the graph of $y = f(x)$ from $x = 1$ to $x = 3$ is 4 sq units, then the area of the shaded region is

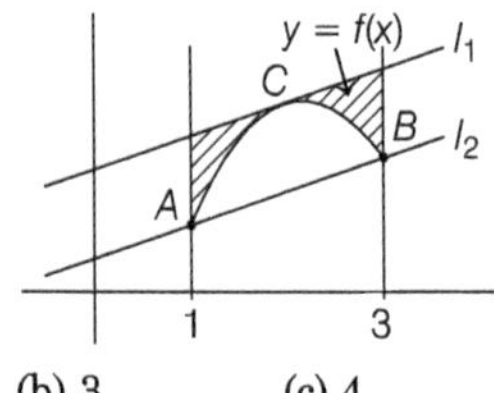

(a) 2 (b) 3 (c) 4 (d) 5

14. Let $I_n = \int_1^0 (\log x)^n \, dx$, where n is a non-negative integer. Then, $I_{2011} + 2011\, I_{2010}$ is equal to
(a) $I_{1000} + 999\, I_{998}$ (b) $I_{889} + 890\, I_{891}$
(c) $I_{100} + 100 I_{99}$ (d) $I_{53} + 54 I_{52}$

15. Consider the regions $A = \{(x, y) \mid x^2 + y^2 \le 100\}$ and $R = \{(x, y) \mid \sin(x + y) > 0\}$ in the plane. Then, the area of the region $A \cap B$ is
(a) 10π (b) 100 (c) 100π (d) 50π

16. Three vertices are chosen randomly from the seven vertices of a regular 7-sided polygon. The probability that they form the vertices of an isosceles triangle is
(a) $\frac{1}{7}$ (b) $\frac{1}{3}$ (c) $\frac{3}{7}$ (d) $\frac{3}{5}$

17. Let $\vec{u} = 2\hat{i} - \hat{j} + \hat{k}$, $\vec{v} = -3\hat{j} + 2\hat{k}$ be vectors in R^3 and $\vec{w}$ be a unit vector in the XY-plane. Then, the maximum possible value of $|(\vec{u} \times \vec{v}) \cdot \vec{w}|$ is
(a) $\sqrt{5}$ (b) $\sqrt{12}$ (c) $\sqrt{13}$ (d) $\sqrt{17}$

18. How many six-digit numbers are there in which no digit is repeated, even digits appear at even places, odd digits appear in odd places and the number is divisible by 4?
(a) 3600 (b) 2700 (c) 2160 (d) 1440

19. The number of natural number n in the interval [1005, 2010] for which the polynomial. $1 + x + x^2 + x^3 + \dots + x^{n-1}$ divides the polynomial $1 + x^2 + x^4 + x^6 + \dots + x^{2010}$ is
(a) 0 (b) 100 (c) 503 (d) 1006

20. Let $a_0 = 0$ and $a_n = 3a_{n-1} + 1$ for $n \ge 1$. Then, the remainder obtained dividing a_{2010} by 11 is
(a) 0 (b) 7 (c) 3 (d) 4

PHYSICS

21. A pen of mass m is lying on a piece of paper of mass M placed on a rough table. If the coefficients of friction between the pen and paper and the paper and the table are μ_1 and μ_2, respectively. Then, the minimum horizontal force with which the paper has to be pulled for the pen to start slipping is given by
(a) $(m + M)(\mu_1 + \mu_2)g$ (b) $(m\mu_1 + M\mu_2)g$
(c) $(m\mu_1 + (m + M)\mu_2)g$ (d) $m(\mu_1 - \mu_2)g$

22. Two masses m_1 and m_2 connected by a spring of spring constant k rest on a frictionless surface. If the masses are pulled apart and let go, the time period of oscillation is
(a) $T = 2\pi\sqrt{\frac{1}{k}\left(\frac{m_1 m_2}{m_1 + m_2}\right)}$ (b) $T = 2\pi\sqrt{k\left(\frac{m_1 + m_2}{m_1 m_2}\right)}$
(c) $T = 2\pi\sqrt{\frac{m_1}{k}}$ (d) $T = 2\pi\sqrt{\frac{m_2}{k}}$

23. A bead of mass m is attached to the mid-point of a tant, weightless string of length l and placed on a frictionless horizontal table.

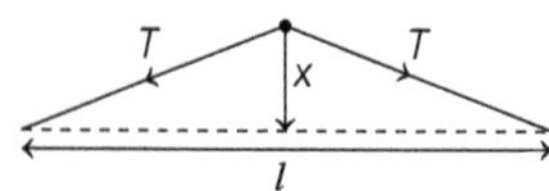

Under a small transverse displacement x, as shown in above figure. If the tension in the string is T, then the frequency of oscillation is

(a) $\frac{1}{2\pi}\sqrt{\frac{2T}{ml}}$ (b) $\frac{1}{2\pi}\sqrt{\frac{4T}{ml}}$ (c) $\frac{1}{2\pi}\sqrt{\frac{4T}{m}}$ (d) $\frac{1}{2\pi}\sqrt{\frac{2T}{m}}$

24. A camet (assumed to be in an elliptical orbit around the sun) is at a distance of 0.4 AU from the sun at the perihelion. If the time period of the camet is 125 yr. What is the aphelion distance? (AU : Astronomical Unit)

(a) 50 AU (b) 25 AU
(c) 49.6 AU (d) 24.6 AU

25. The circuit shown consists of a switch S, a battery B of emf E, a resistance R and an inductor L.

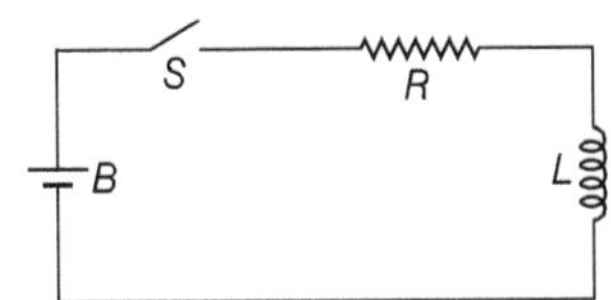

The current in the circuit at the instant. The switch is closed to

(a) E/R (b) $E/R(1-e)$
(c) ∞ (d) 0

26. Consider a uniform spherical volume charge distribution of radius R. Which of the following graphs correctly represents the magnitude of the electric field E at a distance r from the centre of the sphere?

(a) 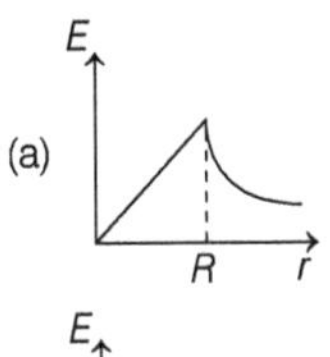

(b) 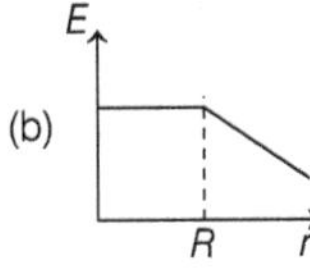

(c) 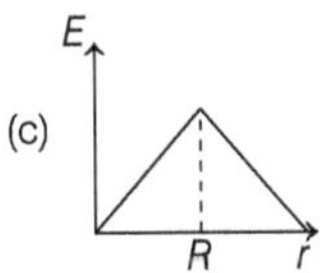

(d) 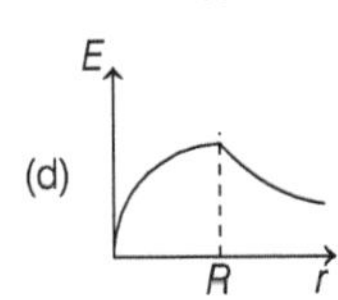

27. A charge $+q$ is placed somewhere inside the cavity of a thick conducting spherical shell of inner radius R_1 and outer radius R_2. A charge $+Q$ is placed at a distance $r > R_2$ from the centre of the shell. Then the electric field in the hollow cavity

(a) depends on both $+q$ and $+Q$
(b) is zero
(c) is only that due to $+Q$
(d) is only that due to $+q$

28. The following travelling electromagnetic wave $E_x = 0$, $E_y = E_0 \sin(kx + \omega t)$, $E_z = -2E_0 \sin(kx - \omega t)$ is

(a) elliptically polarised
(b) circularly polarised
(c) linearly polarised
(d) unpolarised

29. A point source of light is placed at the bottom of a vessel, which is filled with water of refractive index μ to a height h. If a floating opaque disc has to be placed exactly above it, so that the source is invisible from above. The radius of the disc should be

(a) $\frac{h}{\sqrt{\mu-1}}$ (b) $\frac{h}{\sqrt{\mu^2-1}}$ (c) $\frac{h}{\mu^2-1}$ (d) $\frac{\mu h}{\sqrt{\mu^2-1}}$

30. Three transparent media of refractive indices μ_1, μ_2, μ_3, respectively, are stacked as shown below. A ray of light follows the path shown. No light enters the third medium.

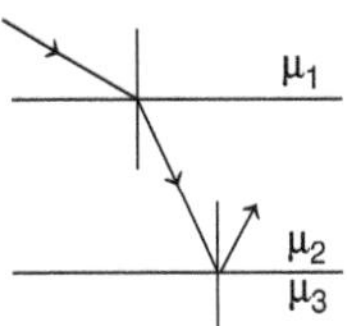

Then,

(a) $\mu_1 < \mu_2 < \mu_3$ (b) $\mu_2 < \mu_1 < \mu_3$
(c) $\mu_1 < \mu_3 < \mu_2$ (d) $\mu_3 < \mu_1 < \mu_2$

31. A nucleus has a half-life of 30 min. At 3 PM its decay rate was measured as 120000 cps. What will be the decay rate at 5 PM?

(a) 120000 cps (b) 30000 cps
(c) 60000 cps (d) 7500 cps

32. A book is resting on a shelf that is undergoing vertical simple harmonic oscillations with an amplitude of 2.5 cm. What is the minimum frequency of oscillation of the shelf for which the book will lose contact with the shelf ? (Assume that, $g = 10\ \text{ms}^{-2}$)

(a) 20 Hz (b) 3.18 Hz (c) 125.6 Hz (d) 10 Hz

33. A van der Waal's gas obeys the equation of state $\left(p + \frac{n^2a}{V^2}\right)(V - nb) = nRT$. Its internal energy is given by $U = CT - \frac{n^2a}{V}$. The equation of a quasistatic adiabat for this gas is given by

(a) $T^{C/nR} \cdot V =$ constant
(b) $T^{(C+nR)/nR} \cdot V =$ constant
(c) $T^{C/nR} \cdot (V - nb) =$ constant
(d) $p^{(C+nR)/nR} \cdot (V - nb) =$ constant

34. An ideal gas is made to undergo a cycle depicted by the p-V diagram given below. The curved line from A to B is an adiabat.

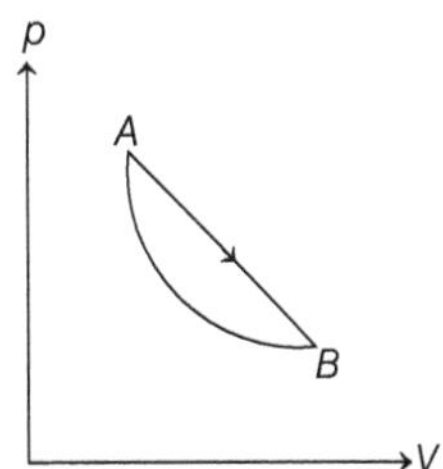

Then,

(a) the efficiency of this cycle is given by unity as on heat is released during the cycle

(b) heat is absorbed in the upper part of the straight line path and released in the lower part

(c) if T_1 and T_2 are the maximum and minimum temperatures reached during the cycle, then the efficiency is given by $1-\frac{T_2}{T_1}$

(d) the cycle can only be carried out in the reverse of the direction shown in above figure

35. A bus driving along at 39.6 km/h is approaching a person who is standing at the bus stop, while honking repeatedly at an interval of 30 s. If the speed of sound is 330 ms^{-1}, at what interval will the person hear the horn?

(a) 31 s (b) 29 s (c) 30 s

(d) The interval will depend on the distance of the bus from the passenger

36. Velocity of sound measured at a given temperature in oxygen and hydrogen is in the ratio

(a) 1 : 4 (b) 4 : 1

(c) 1 : 1 (d) 32 : 1

37. In Young's double slit experiment, the distance between the two slits is 0.1 mm, the distance between the slits and the screen is 1 m and the wavelength of the light used is 600 nm. The intensity at a point on the screen is 75% of the maximum intensity. What is the smallest distance of this point from the central fringe?

(a) 1.0 mm (b) 2.0 mm

(c) 0.5 mm (d) 1.5 mm

38. Two masses m_1 and m_2 are connected by a massless spring of spring constant k and unstretched length l. The masses are placed on a frictionless straight channel, which we consider our X-axis. They are initially at rest at $x=0$ and $x=l$, respectively. At $t=0$, a velocity of v_0 is suddenly imparted to the first particle. At a later time t, the centre of mass of the two masses is at

(a) $x=\frac{m_2 l}{m_1+m_2}$ (b) $x=\frac{m_1 l}{m_1+m_2}+\frac{m_2 v_0 t}{m_1+m_2}$

(c) $x=\frac{m_2 l}{m_1+m_2}+\frac{m_2 v_0 t}{m_1+m_2}$ (d) $x=\frac{m_2 l}{m_1+m_2}+\frac{m_1 v_0 t}{m_1+m_2}$

39. A charged particle of charge q and mass m, gets deflected through an angle θ upon passing through a square region of side a, which contains a uniform magnetic field B normal to its plane. Assuming that the particle entered the square at right angles to one side, what is the speed of the particle?

(a) $\frac{qB}{m}a\cot\theta$ (b) $\frac{qB}{m}a\tan\theta$

(c) $\frac{qB}{m}a\cot^2\theta$ (d) $\frac{qB}{m}a\tan^2\theta$

40. A piece of hot copper at 100°C is plunged into a pond at 30°C. The copper cools down to 30°C, while the pond being huge stays at its initial temperature. Then,

(a) copper loses some entropy, the pond stays at the same entropy

(b) copper loses some entropy and the pond gains exactly the same amount of entropy

(c) copper loses entropy and the pond gains more than this amount of entropy

(d) both copper and the pond gain in entropy

CHEMISTRY

41. The number of isomers of Co (diethylene triamine) Cl_3 is

(a) 2 (b) 3 (c) 4 (d) 5

42. Among the following, the π-acid ligand is

(a) F^- (b) NH_3 (c) CN^- (d) I^-

43. The bond order in O_2^{2-} is

(a) 2 (b) 3 (c) 1.5 (d) 1

44. The energy of a photon of wavelength $(\lambda)=1$ m is (Planck's constant $=6.626\times10^{-34}$ Js, speed of light $=3\times10^8 ms^{-2}$)

(a) 1.988×10^{-23} J (b) 1.988×10^{-28} J

(c) 1.988×10^{-30} J (d) 1.988×10^{-25} J

45. The concentration of a substance undergoing a chemical reaction becomes one-half of its original value after time, regardless of the initial concentration. The reaction is an example of a

(a) zero order reaction (b) first order reaction

(c) second order reaction (d) third order reaction

46. The shape of the molecule ClF_3 is

(a) trigonal planar (b) pyramidal

(c) T - shaped (d) Y - shaped

47. Friedel-Craft's acylation is

(a) α - acylation of a carbonly compound

(b) acylation of phenols to genarate esters

(c) acylation of aliphatic olefins

(d) acylation of aromatic nucleus

48. The order of acidity of compounds I-IV, is

I: $C_6H_5-CH_2OH$ II: $C_6H_5-CO_2H$

III: $H_3C-C_6H_4-OH$ IV: $C_6H_5-SO_3H$

(a) I < III < II < IV (b) III < I < II < IV

(c) IV < I < II < III (d) II < IV < III < I

49. The most stable conformation for n-butane is

(a) 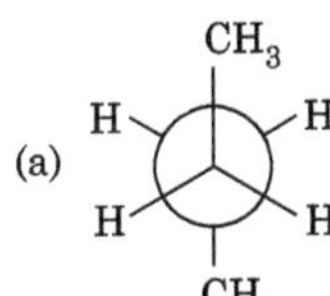

(b) CH_3 (front, top), H, CH_3, H, H, H

(c) 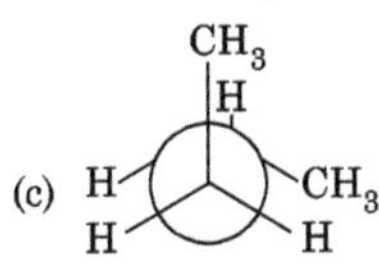

(d) CH_3, CH_3, H, H, H, H

50. In the nuclear reaction,

$$^{234}_{94}Th \longrightarrow {}^{234}_{91}Pa + X,$$

X is

(a) $_{-1}^{0}e$ (b) $_{1}^{0}e$
(c) H (d) $_{1}^{2}H$

51. A concentrated solution of copper sulphate, which is dark blue in colour, is mixed at room temperature with a dilute solution of copper sulphate, which is light blue. For this process

(a) entropy change is positive, but enthalpy change is negative
(b) entropy and enthalpy changes are both positive
(c) entropy change is positive and enthalpy does not change
(d) entropy change is negative and enthalpy change is positive

52. Increasing the temperature increases the rate of reaction, but does not increase the

(a) number of collisions
(b) activation energy
(c) average energy of collisions
(d) average velocity of the reactant molecules

53. In metallic solids, the number of atoms for the face centered and the body centered cubic unit cells, are respectively.

(a) 2, 4 (b) 2, 2 (c) 4, 2 (d) 4, 4

54. From equations 1 and 2,

$$CO_2 \rightleftharpoons CO + \frac{1}{2}O_2 [K_{C_1} = 9.1 \times 10^{-12} \text{at } 1000^1C] \text{ (Eq. (i))}$$

$$H_2O \rightleftharpoons H_2 + \frac{1}{2}O_2 \ [K_{C_2} = 7.1 \times 10^{-12} \text{at } 1000^0C] \text{ (Eq. (ii))}$$

The equilibrium constant for the reaction, $CO_2 + H_2 \rightleftharpoons CO + H_2O$ at the same temperature, is

(a) 0.78 (b) 2.0
(c) 16.2 (d) 1.28

55. For a first order reaction $R \longrightarrow P$, the rate constant is k, if the initial concentration of R is $[R_n]$. The concentration of R at any time t is given by the expression.

(a) $[R_0]e^{kt}$ (b) $[R_0]e^{-kt}$
(c) $[R_0](1-e^{-kt})$ (d) $[R_0](1-e^{-kt})$

56. The correct structure of PCl_3F_2 is

(a) 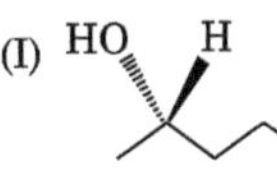

(b) 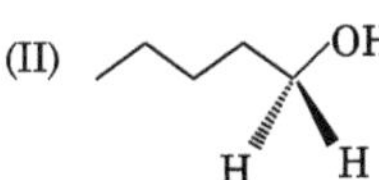

(c) 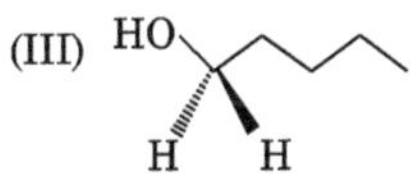

(d) 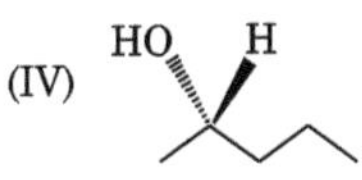

57. The enantiomeric pair among the following four structures is

(I) (II) (III) (IV)

(a) I and II (b) I and IV
(c) II and III (d) II and IV

58. Consider the reaction, $2NO_2(g) \rightarrow 2NO(g) + O_2(g)$. In the figure below, identify the curves X, Y and Z associated with the three species in the reaction,

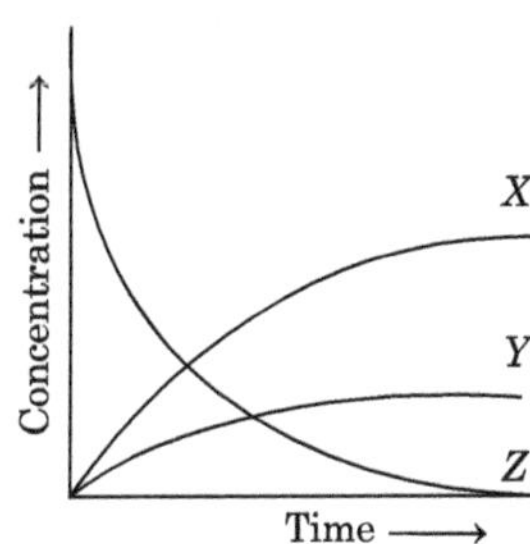

(a) $X = NO, Y = O_2, Z = NO_2$
(b) $X = O_2, Y - NO, Z = NO_2$
(c) $X = NO_2, Y = NO, Z = O_2$
(d) $X = O_2, Y = NO_2, Z = NO$

59. The aromatic carbocation among the following is

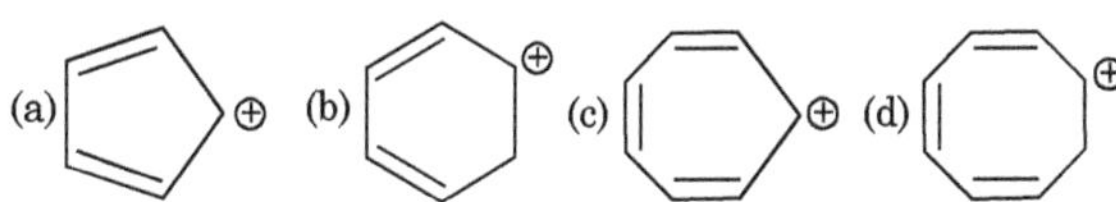

60. Cyclohexene is reacted with bromine in CCl_4 in the dark. The product of the reaction is,

(a) 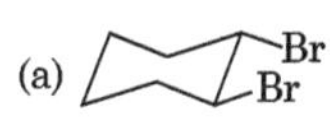

(b) 

(c) 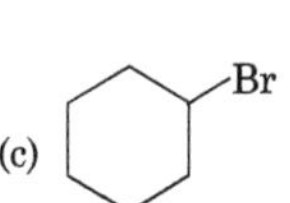

(d) 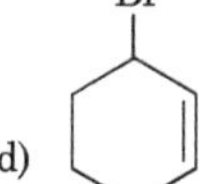

BIOLOGY

61. Ribonucleic Acids (RNA) that catalyse enzymatic reactions are called ribozymes. Which one of the following acts as a ribozyme?

(a) Ribosome
(b) Amylase
(c) *t*RNA
(d) Riboflavin

62. In 1670, Robert Boyle conducted an experiment wherein he placed a viper (a poisonous snake) in a chamber and rapidly reduced the pressure in that chamber. Which of the following would be true?

(a) Gas bubbles developed in the tissues of the snake
(b) The basal metabolic rate of the snake increased tremendously
(c) The venom of the snake was found to decrease in potency
(d) The venom of the snake was found to increase in potency

63. Bacteria can survive by absorbing soluble nutrients *via* their outer body surface but animals cannot, because

(a) bacteria cannot ingest particles but animals can
(b) bacteria have cell walls and animals do not
(c) animals have too small a surface area per unit volume as compared to bacteria
(d) animals cannot metabolise soluble nutrients

64. A horse has 64 chromosomes and a donkey has 62. Mules result from crossing a horse and a donkey. State which of the following is incorrect?

(a) Mules can have either 64, 63 or 62 chromosomes
(b) Mules are infertile
(c) Mules have well-defined gender (male/female)
(d) Mules have 63 chromosomes

65. If the total number of photons falling per unit area of a leaf per minute is kept constant then which of the following will result in maximum photosynthesis?

(a) Shining green light
(b) Shining sunlight
(c) Shining blue light
(d) Shining ultraviolet light

66. Path-finding by ants is by means of

(a) visually observing landmarks
(b) visually observing other ants
(c) chemical signals between ants
(d) using the earth's magnetic field

67. Sometimes urea is fed to ruminants to improve their health. It works by

(a) helping growth of gut microbes that breakdown cellulose
(b) killing harmful microorganisms in their gut
(c) increasing salt content in the gut
(d) directly stimulating blood cell proliferation

68. If you compare adults of two herbivore species of different sizes, but from the same geographical area, the amount of faeces producted per kg body weight would be

(a) more in the smaller one than the larger one
(b) more in the larger one than the smaller one
(c) roughly the same amount in both
(d) not possible to predict which would be more

69. Fruit wrapped in paper ripens faster than when kept in open air because

(a) heat of respiration is retained better
(b) a chemical in the paper helps fruit ripening
(c) a volatile substance produced by the fruit is retained better and helps in ripening
(d) the fruit is cut off from the ambient oxygen which is an inhibitor to fruit ripening

70. When a person is suffering from high fever, it is sometimes observed that the skin has a reddish tinge. Why does this happen?

(a) Red colour of the skin radiates more heat
(b) Fever causes the release of a red pigment in the skin
(c) There is more blood circulation to the skin to keep the body warm
(d) There is more blood circulation to the skin to release heat from the body

71. Bacteriochlorophylls are photosynthetic pigments found in phototrophic bacteria. Their function is distinct from the plant chlorophylls in that they

(a) do not produce oxygen
(b) do not conduct photosynthesis
(c) absorb only blue light
(d) function without a light source

72. Athletes often experience muscle cramps. Which of the following statements is true about muscle cramps?

(a) Muscle cramp is caused due to conversion of pyruvic acid into lactic acid in the cytoplasm
(b) Muscle cramp is caused due to conversion of pyruvic acid into lactic acid in the mitochondria
(c) Muscle cramp is caused due to non-conversion of glucose to pyruvate in the cytoplasm
(d) Muscle cramp is caused due to conversion of pyruvic acid into ethanol in the cytoplasm

73. A couple went to a doctor and reported that both of them are carriers for a particular disorder, their first child is suffering from that disorder and that they are expecting their second child. What is the probability that the new child would be affected by the same disorder?

(a) 100% (b) 50% (c) 25% (d) 75%

74. Of the following combinations of cell biological processes, which one is associated with embryogenesis?

(a) Mitosis and meiosis
(b) Mitosis and differentiation
(c) Meiosis and differentiation
(d) Differentiation and reprogramming

75. Conversion of the *Bt* protoxin produced by *Bacillus thuringiensis* to its active form in the gut of the insects is mediated by

(a) acidic pH of the gut
(b) alkaline pH of the gut
(c) lipid modification of the protein
(d) cleavage by chymotrypsin

76. If you dip a sack full of paddy seeds in water overnight and then keep it out for a couple of days, it feels warm. What generates this heat?

(a) Imbibition
(b) Exothermic reaction between water and seed coats
(c) Friction among seeds due to swelling
(d) Respiration

77. Restriction endonucleases are enzymes that cleave DNA molecules into smaller fragments. Which type of bond do they act on?

(a) N-glycosidic bond
(b) Hydrogen bond
(c) Phosphodiester bond
(d) Disulphide bond

78. The fluid part of blood flows in and out of capillaries in tissues to exchange nutrients and waste materials. Under which of the following conditions will fluid flow out from the capillaries into the surrounding tissue?

(a) When arterial blood pressure exceeds blood osmotic pressure
(b) When arterial blood pressure is less than blood osmotic pressure
(c) When arterial blood pressure is equal to blood osmotic pressure
(d) Arterial blood pressure and blood osmotic pressure have nothing to do with the outflow of fluid from capillaries

79. The distance between two consecutive DNA base pairs is 0.34 nm. If the length of a chromosome is 1 mm, the number of base pairs in the chromosome is approximately

(a) 3 million (b) 30 million
(c) 1.5 million (d) 6 million

80. Estimate the order of the speed of propagation of an action potential or nerve impluse.

(a) nm/s (b) micron/s (c) cm/s (d) m/s

PART-II (2 Marks Questions)

MATHEMATICS

81. Arrange the expansion of $\left(x^{1/2}+\frac{1}{2x^{1/4}}\right)^n$ in decreasing powers of x. Suppose the coefficients of the first three terms form an arithmetic progression. Then, the number of terms in the expansion having integer power of x is

(a) 1 (b) 2
(c) 3 (d) more than 3

82. Let r be a real number and $n \in N$ be such that the polynomial $2x^2+2x+1$ divides the polynomial $(x+1)^n - r$. Then, (n, r) can be

(a) $(4000, 4^{1000})$ (b) $\left(4000, \frac{1}{4^{1000}}\right)$
(c) $\left(4^{1000}, \frac{1}{4^{1000}}\right)$ (d) $\left(4000, \frac{1}{4000}\right)$

83. Suppose a, b are real numbers such that $ab \neq 0$. Which of the following four figures represent the curve $(y-ax-b)(bx^2+ay^2-ab)=0$?

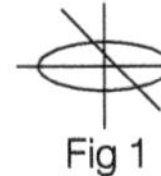

Fig 1 Fig 2 Fig 3 Fig 4

(a) Fig 1 (b) Fig 2 (c) Fig 3 (d) Fig 4

84. Among all cyclic quadrilaterals inscribed in a circle of radius R with one of its angles equal to 120°. Consider the one with maximum possible area. Its area is

(a) $\sqrt{2}\,R^2$ (b) $\sqrt{3}\,R^2$ (c) $2R^2$ (d) $2\sqrt{3}\,R^2$

85. The following figure shows the graph of a differentiable function $y = f(x)$ on the interval $[a, b]$ (not containing 0).

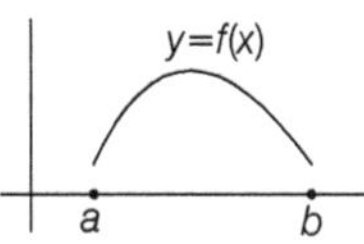

Let $g(x)=\frac{f(x)}{x}$. Which of the following is a possible graph of $y = g(x)$?

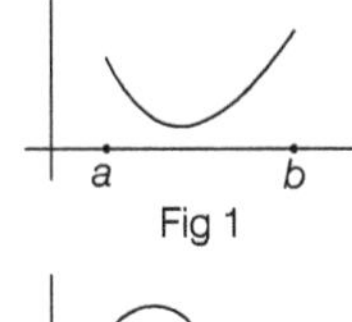

Fig 1

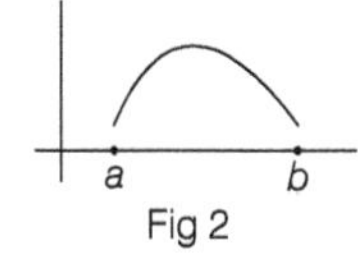

Fig 2

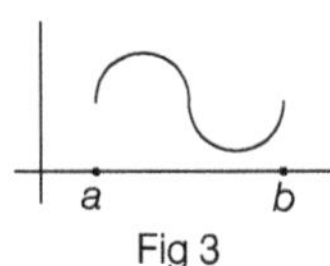

Fig 3

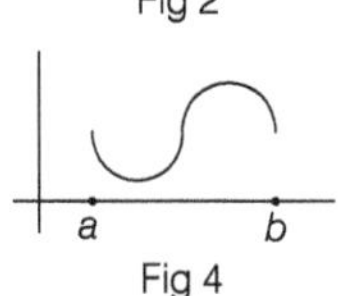

Fig 4

(a) Fig 1 (b) Fig 2
(c) Fig 3 (d) Fig 4

86. Let V_1 be the volume of a given right circular cone with O as the centre of the base and A as its apex. Let V_2 be the maximum volume of the right circular cone inscribed in the given cone whose apex is O and whose base is parallel to the base of the given cone. Then, the ratio V_2 / V_1 is

(a) $\frac{3}{25}$ (b) $\frac{4}{9}$ (c) $\frac{4}{27}$ (d) $\frac{8}{27}$

87. Let $f: R \to R$ be a continuous function satisfying $f(x) = x + \int_0^x f(t)\,dt$, for all $x \in R$. Then, the number of elements in the set $S = \{x \in R : f(x) = 0\}$ is

(a) 1 (b) 2 (c) 3 (d) 4

88. The value of $\int_0^{2\pi} \min\{|x-\pi|, \cos^{-1}(\cos x)\}\,dx$ is

(a) $\frac{\pi^2}{4}$ (b) $\frac{\pi^2}{2}$ (c) $\frac{\pi^2}{8}$ (d) π^2

89. Let ABC be a triangle and P be a point inside ABC such that $\overrightarrow{PA} + 2\,\overrightarrow{PB} + 3\,\overrightarrow{PC} = 0$. The ratio of the area of ΔABC to that of ΔAPC is

(a) 2 (b) $\frac{3}{2}$ (c) $\frac{5}{3}$ (d) 3

90. Suppose m, n are positive integers such that $6^m + 2^{m+n} \cdot 3^w + 2^n = 332$. The value of the expression $m^2 + mn + n^2$ is

(a) 7 (b) 13 (c) 19 (d) 21

PHYSICS

91. A ball is dropped vertically from a height of h onto a hard surface. If the ball rebounds from the surface with a fraction r of the speed with which it strikes the latter on each impact, what is the net distance travelled by the ball up to the 10th impact?

(a) $2h\left(\frac{1-r^{10}}{1-r}\right)$ (b) $h\left(\frac{1-r^{20}}{1-r^2}\right)$

(c) $2h\left(\frac{1-r^{20}}{1-r^3}\right) - h$ (d) $2h\left(\frac{1-r^{20}}{1-r^2}\right) - h$

92. A certain planet completes one rotation about its axis in time T. The weight of an object placed at the equator on the planet's surface is a fraction f (f is close to unity) of its weight recorded at a latitude of 60°. The density of the planet (assumed to be a uniform perfect sphere) is given by

(a) $\left(\frac{4-f}{1-f}\right) \cdot \frac{3\pi}{4GT^2}$ (b) $\left(\frac{4-f}{1+f}\right) \cdot \frac{3\pi}{4GT^2}$

(c) $\left(\frac{4-3f}{1-f}\right) \cdot \frac{3\pi}{4GT^2}$ (d) $\left(\frac{4-2f}{1-f}\right) \cdot \frac{3\pi}{4GT^2}$

93. Three equal charges $+q$ are placed at the three vertices of an equilateral triangle centred at the origin. They are held in equilibrium by a restoring force of magnitude $F(r) = kr$ directed towards the origin, where k is a constant. What is the distance of the three charges from the origin?

(a) $\left[\frac{1}{6\pi\varepsilon_0}\frac{q^2}{k}\right]^{1/2}$ (b) $\left[\frac{\sqrt{3}}{12\pi\varepsilon_0}\frac{q^2}{k}\right]^{1/3}$

(c) $\left[\frac{1}{6\pi\varepsilon_0}\frac{q^2}{k}\right]^{2/3}$ (d) $\left[\frac{\sqrt{3}}{4\pi\varepsilon_0}\frac{q^2}{k}\right]^{2/3}$

94. Consider the infinite ladder circuit shown below.

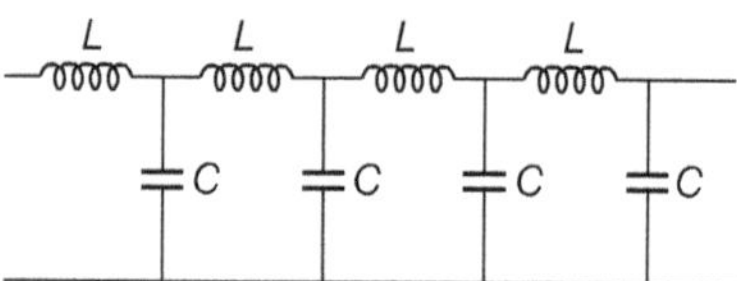

For which angular frequency ω will the circuit behave like a pure inductance?

(a) $\frac{LC}{\sqrt{2}}$ (b) $\frac{1}{\sqrt{LC}}$ (c) $\frac{2}{\sqrt{LC}}$ (d) $\frac{2L}{\sqrt{C}}$

95. A narrow parallel beam of light falls on a glass sphere of radius R and refractive index μ at normal incidence. The distance of the image from the outer edge is given by

(a) $\frac{R(2-\mu)}{2(\mu-1)}$ (b) $\frac{R(2+\mu)}{2(\mu-1)}$ (c) $\frac{R(2-\mu)}{2(\mu+1)}$ (d) $\frac{R(2+\mu)}{2(\mu+1)}$

96. A particle of mass m undergoes oscillations about $x = 0$ in a potential given by $V(x) - \frac{1}{2}kx^2 - V_0 \cos\left(\frac{x}{a}\right)$, where V_0, k, a are constants. If the amplitude of oscillation is much smaller than a, the time period is given by

(a) $2\pi\sqrt{\frac{ma^2}{ka^2 + V_0}}$ (b) $2\pi\sqrt{\frac{m}{k}}$

(c) $2\pi\sqrt{\frac{ma^2}{V_0}}$ (d) $2\pi\sqrt{\frac{ma^2}{ka^2 - V_0}}$

97. An ideal gas with heat capacity at constant volume C_V undergoes a quasistatic process described by pV^α in a p-V diagram, where α is a constant. The heat capacity of the gas during this process is given by

(a) C_V (b) $C_V + nR$ (c) $C_V + \frac{nR}{1-\alpha}$ (d) $C_V + \frac{nR}{1-\alpha^2}$

98. An ideal gas with constant heat capacity $C_V = \frac{3}{2}nR$ is made to carry out a cycle that is depicted by a triangle in the figure given below.

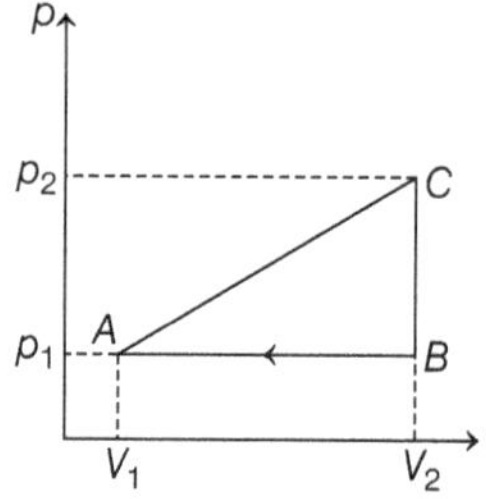

The following statement is true about the cycle.

(a) The efficiency is given by $1 - \frac{p_1V_1}{p_2V_2}$

(b) The efficiency is given by $1 - \frac{1}{2}\frac{p_1V_1}{p_2V_2}$

(c) Net heat absorbed in the cycle is $(p_2 - p_1)(V_2 - V_1)$

(d) Heat absorbed in part AC is given by $2(p_2V_2 - p_1V_1) + \frac{1}{2}(p_1V_2 - p_2V_1)$

99. Two identical particles of mass m and charge q are shot at each other from a very great distance with an initial speed v. The distance of closest approach of these charges is

(a) $\frac{q^2}{8\pi\varepsilon_0 mv^2}$ (b) $\frac{q^2}{4\pi\varepsilon_0 mv^2}$ (c) $\frac{q^2}{2\pi\varepsilon_0 mv^2}$ (d) 0

100. At time $t = 0$, a container has N_0 radioactive atoms with a decay constant λ. In addition, c numbers of atoms of the same type are being added to the container per unit time. How many atoms of this type are there at $t = T$?

(a) $\frac{c}{\lambda}\exp(-\lambda T) - N_0 \exp(-\lambda T)$

(b) $\frac{c}{\lambda}\exp(-\lambda T) + N_0 \exp(-\lambda T)$

(c) $\frac{c}{\lambda}(1-\exp(-\lambda T)) + N_0 \exp(-\lambda T)$

(d) $\frac{c}{\lambda}(1+\exp(-\lambda T)) + N_0 \exp(-\lambda T)$

CHEMISTRY

101. 2.52 g of oxalic acid dihydrate was dissolved in 100 mL of water. 10 mL of this solution was diluted to 500 mL. The normality of the final solution and the amount of oxalic acid (mg/mL) in the solution are respectively

(a) 0.16 N, 5.04 (b) 0.08 N, 3.60
(c) 0.04 N, 3.60 (d) 0.02 N, 10.08

102. Two isomeric compounds I and II are heated with HBr.

I: OH, CH_2OH II: OH, OCH_3

The products obtained are

(a) OH, CH_2Br , OH, OH

(b) Br, CH_2OH , Br, OCH_3

(c) Br, CH_2Br , Br, Br

(d) OH, Br, CH_2OH , OH, Br, OCH_3

103. The number of possible enantiomeric pair(s) produced from the bromination of I and II, respectively, are

I: H_3C, CH_3, H_3C, CH_3 II: H_3C, H_3C, $=CH_2$

(a) 0, 1 (b) 1, 0 (c) 0, 2 (d) 1, 1

104. For the reaction $A \longrightarrow B$,
$\Delta H° = 7.5$ kJ mol^{-1} and $\Delta S° = 2.5$ J mol^{-1}. The value of $\Delta G°$ and the temperature, at which the reaction reaches equilibrium are, respectively

(a) 0 kJ mol^{-1} and 400 K (b) -2.5 kJ mol^{-1} and 400 K
(c) 2.5 kJ mol^{-1} and 200 K (d) 0 kJ mol^{-1} and 300 K

105. The solubility product of $Mg(OH)_2$ is 1.0×10^{-12}. Concentrated aqueous NaOH solution is added to a 0.01 M aqueous solution of $MgCl_2$. The pH at which precpitation occurs is

(a) 7.2 (b) 7.8 (c) 8.0 (d) 9.0

106. A metal with an atomic radius of 141.4 pm crystallises in the face centrod cubic structure. The volume of the unit cell in pm^3 is

(a) 2.74×10^7 (b) 2.19×10^7 (c) 6.40×10^7 (d) 9.20×10^7

107. Identify the cyclic silicate ion given in the figure below,

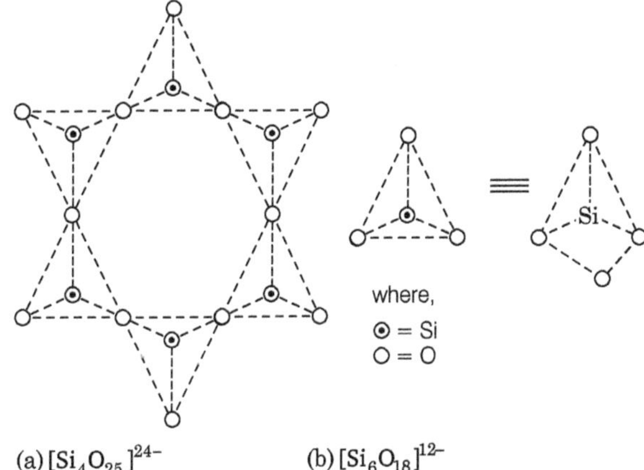

(a) $[Si_4O_{25}]^{24-}$ (b) $[Si_6O_{18}]^{12-}$
(c) $[Si_4O_{18}]^{12}$ (d) $[Si_6O_{24}]^{12-}$

108. Diborane is formed from the elements as shown in equation (i)

$$2B(s) + 3H_2(g) \longrightarrow B_2H_6(g) \quad \ldots(i)$$

Given, that $H_2O(l) \longrightarrow H_2O(g), \Delta H_1^\circ = 44\text{kJ}$

$$2B + \frac{3}{2}O_2(g) \longrightarrow B_2O_3(s), \Delta H_2^\circ = -1273\text{kJ}$$

$$B_2H_6(g) + 3O_2(g) \longrightarrow B_2O_3(s) + 3H_2O(g), \Delta H_3^\circ = -2035\text{kJ}$$

$$H_2(g) + \frac{1}{2}O_2(g) \longrightarrow H_2O(l), \Delta H_4^\circ = -286\text{kJ}$$

The $\Delta H°$ for the reaction (i) is

(a) 36 kJ (b) 520 kJ (c) 509 kJ (d) -3550 kJ

109. The crystal field stabilisation energy (CFSE) and the spin-only magnetic moment in Bohr Magneton (BM) for the complex $K_3[Fe(CN)_6]$ are, respectively

(a) $0.0\Delta_o$ and $\sqrt{35}$ BM
(b) $-2.0\Delta_o$ and $\sqrt{3}$ BM
(c) $-0.4\Delta_o$ and $\sqrt{24}$ BM
(d) $-2.4\Delta_o$ and 0 BM

110. A solution containing 8.0 g of nicotine in 92 g of water freezes 0.925 degrees below the normal freezing point of water. If the moles freezing point depression constant, $k_f = 1.85°C\ mol^{-1}$, then the molar mass of nicotine is

(a) 16 (b) 80 (c) 320 (d) 160

BIOLOGY

111. A host cell has intracellular bacterial symbionts. If the growth rate of the bacterial symbionts is always 10% higher than that of the host cell, after 10 generations of the host cell the density of bacteria in host cells will increase

(a) by 10% (b) ten-fold
(c) two-fold (d) hundred-fold

112. In a diploid organism, there are three different alleles for a particular gene. Of these three alleles one is recessive and the other two alleles exhibit codominance. How many phenotypes are possible with this set of alleles?

(a) 3 (b) 6 (c) 4 (d) 2

113. Two students are given two different double-stranded DNA molecules of equal length. They are asked to denature the DNA molecules by heating. The DNA given to student A has the following composition of bases (A : G : T : C :: 35 : 15 : 35 : 15) while that given to student B is (A : G : T : C :: 12 : 38 : 12 : 38). Which of the following statements is true?

(a) Both the DNA molecules would denature at the same rate
(b) The information given is insufficient to draw any conclusion
(c) DNA molecule given to student B would denature faster than that of student A
(d) DNA molecule given to student A would denature faster than that given to student B

114. The amino acid sequences of a bacterial protein and a human protein carrying out similar function are found to be 60% identical. However, the DNA seqeunces of the genes coding for these proteins are only 45% identical. This is possible because

(a) protein sequence does not depend on DNA sequence
(b) DNA codons having different nucleotides in the third position can code for the same amino acids
(c) DNA codons having different nucleotides in the second position can code for the same amino acids
(d) same DNA codons can code for multiple amino acids

115. The following DNA sequence ($5' \rightarrow 3'$) specifies part of a protein coding sequence, starting from position 1. Which of the following mutations will give rise to a protein that is shorter than the full-length protein?

1	2	3	4	5	6	7	8	9	10	11	12	13	14	15
A	T	G	C	A	A	G	A	T	A	T	A	G	C	T

(a) Deletion of nucleotide 13
(b) Deletion of nucleotide 5
(c) Insertion of a single nucleotide between 3 and 4
(d) Insertion of a single nucleotide between 10 and 11

116. Which of the following correctly represents the results of an enzymatic reaction? Enzyme is E_3 substrate is S and products are P1 and P2.

(a) P1 + S ⇔ P2 + E (b) E + S ⇔ P1 + P2
(c) P1 + P2 + E ⇔ S (d) E + S ⇔ P1 + P2 + E

117. Four species of birds have different egg colours, [1] white with no markings, [2] pale brown with no markings, [3] grey-brown with dark streaks and spots, [4] pale blue with dark blue-green spots. Based on egg colour, which species is most likely to nest in a deep tree hole?

(a) 1 (b) 2 (c) 3 (d) 4

118. Consider a locus with two alleles, *A* and *a*. If the frequency of *AA* is 0.25, what is the frequency of *A* under Hardy-Weinberg equilibrium?

(a) 1 (b) 0.25 (c) 0.5 (d) 0

119. Which of the following graphs accurately represents the insulin levels (*Y*-axis) in the body as a function of time (*X*-axis) after eating sugar and bread/roti ?

(a)
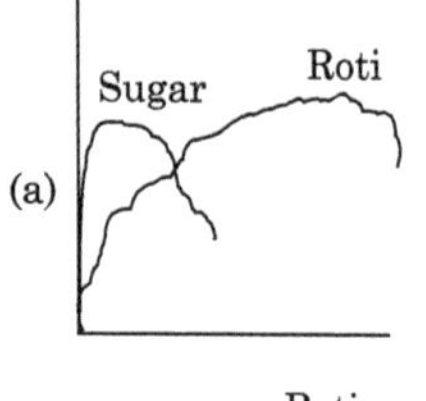

(b)
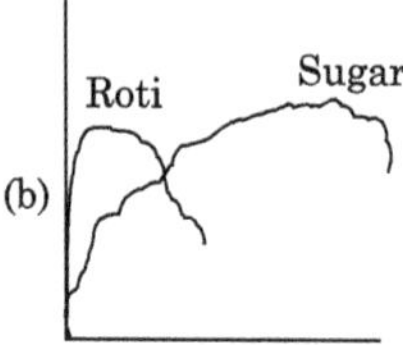

(c)
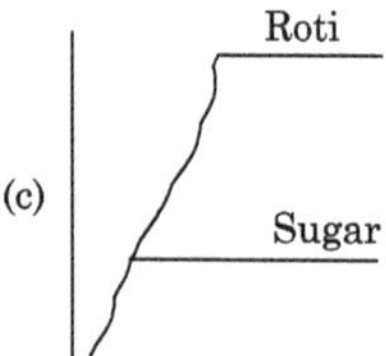

(d)
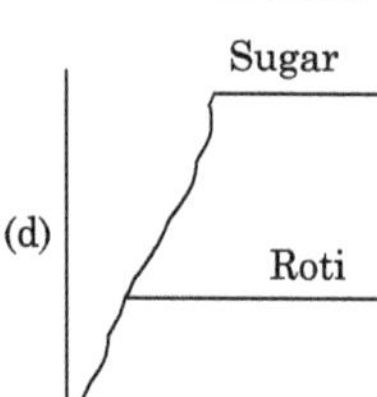

120. You marked two ink-spots along the height at the base of a coconut tree and also at the top of the tree. When you examine the spots next year when the tree has grown taller, you will see

(a) the two spots at the top have grown more apart than the two spots at the bottom
(b) the top two spots have grown less apart than the bottom two spots
(c) both sets of spots have grown apart to the same extent
(d) both sets of spots remain un-altered

Answers

PART-I

1 (b)	2 (c)	3 (d)	4 (b)	5 (c)	6 (b)	7 (c)	8 (a)	9 (c)	10 (b)
11 (b)	12 (c)	13 (a)	14 (c)	15 (d)	16 (d)	17 (d)	18 (d)	19 (c)	20 (a)
21 (a)	22 (a)	23 (b)	24 (c)	25 (d)	26 (a)	27 (d)	28 (c)	29 (b)	30 (d)
31 (d)	32 (b)	33 (c)	34 (b)	35 (b)	36 (a)	37 (a)	38 (d)	39 (a)	40 (c)
41 (a)	42 (c)	43 (d)	44 (a)	45 (b)	46 (c)	47 (d)	48 (a)	49 (a)	50 (a)
51 (c)	52 (b)	53 (c)	54 (d)	55 (b)	56 (a)	57 (b)	58 (a)	59 (c)	60 (a)
61 (a)	62 (a)	63 (c)	64 (a)	65 (c)	66 (c)	67 (a)	68 (a)	69 (c)	70 (d)
71 (a)	72 (a)	73 (c)	74 (b)	75 (b)	76 (d)	77 (c)	78 (a)	79 (a)	80 (d)

PART-II

81 (c)	82 (b)	83 (b)	84 (b)	85 (b)	86 (c)	87 (a)	88 (b)	89 (d)	90 (c)
91 (d)	92 (a)	93 (b)	94 (*)	95 (a)	96 (a)	97 (c)	98 (d)	99 (b)	100 (c)
101 (c)	102 (a)	103 (a)	104 (d)	105 (d)	106 (c)	107 (b)	108 (a)	109 (a)	110 (d)
111 (b)	112 (c)	113 (d)	114 (b)	115 (b)	116 (d)	117 (a)	118 (c)	119 (a)	120 (a)

* No options are correct.

Solutions

1. *(b)* We have,

$$A=\begin{bmatrix}0 & i\\ i & 0\end{bmatrix},$$

$$A^2=\begin{bmatrix}0 & i\\ i & 0\end{bmatrix}\begin{bmatrix}0 & i\\ i & 0\end{bmatrix}$$

$$=\begin{bmatrix}i^2 & 0\\ 0 & i^2\end{bmatrix}=\begin{bmatrix}-1 & 0\\ 0 & -1\end{bmatrix}$$

$$=-\begin{bmatrix}1 & 0\\ 0 & 1\end{bmatrix}=-I$$

$$A^3=A^2\cdot A=\begin{bmatrix}-1 & 0\\ 0 & -1\end{bmatrix}\begin{bmatrix}0 & i\\ i & 0\end{bmatrix}$$

$$=\begin{bmatrix}0 & -i\\ -i & 0\end{bmatrix}$$

$$=-\begin{bmatrix}0 & i\\ i & 0\end{bmatrix}=-A$$

and $A^4=A^2\cdot A^2=(-I)(-I)=I=\begin{bmatrix}1 & 0\\ 0 & 1\end{bmatrix}$

$\therefore\ I+A+A^2+A^3+A^4+A^5+\dots+A^{2008}+A^{2009}+A^{2010}$

$=I+A+A^2+A^3+A^4\,[I+A+A^2+A^3]+\dots+A^{2008}[I+A+A^2]$

$=0+0+\dots+[I+A+A^2]$ $\quad[\because I+A+A^2+A^3=0]$

$$\therefore I+A+A^2=\begin{bmatrix}1 & 0\\ 0 & 1\end{bmatrix}+\begin{bmatrix}0 & i\\ i & 0\end{bmatrix}+\begin{bmatrix}-1 & 0\\ 0 & -1\end{bmatrix}$$

$$=\begin{bmatrix}0 & i\\ i & 0\end{bmatrix}$$

2. *(c)* Let the sides of triangle are a, ar, ar^2. $\quad[\because$ sides of triangle in GP$]$

***Case* I** $r>1$

We know sum of two sides is greater than third side.

$\therefore\ a+ar>ar^2 \Rightarrow r^2-r-1<0$

$\Rightarrow\quad r=\dfrac{1\pm\sqrt{5}}{2}$

$\Rightarrow\quad 1<r<\dfrac{\sqrt{5}+1}{2},\ r>1$

***Case* II** $0<r<1$

$\therefore\ ar^2+ar>a \Rightarrow r^2+r-1>0$

$\Rightarrow\quad r=\dfrac{-1\pm\sqrt{5}}{2}$

$\Rightarrow\quad 1>r>\dfrac{\sqrt{5}-1}{2},\ 0<r<1$

$\therefore\quad r\in\left(\dfrac{\sqrt{5}-1}{2},\dfrac{\sqrt{5}+1}{2}\right)$

3. *(d)* We have,

12-sided regular polygon the diagonal of rectangle are equal.

$\therefore$ Diagonals of rectangle passes through the centre of 12-sided regular polygon = 6

$\therefore$ Number of rectangles $={}^6C_2=\dfrac{6!}{2!4!}=15$

4. *(b)* Given, $1, \omega, \omega^2$ are cube roots of unity.

Let the roots of polynomials are

$\alpha=2\omega^2$

$\beta=3+4\omega$

$\gamma=3+4\omega^2$

$\delta=5-\omega+\omega^2$

$\quad=5-(\omega+\omega^2)=5+1=6$

$[\because 1+\omega+\omega^2=0]$

If $\alpha=2\omega^2$ is a root the 2ω has to be also root.

$\therefore$ Total roots of polynomial = 5.

Hence, minimum degree of polynomial = 5

5. *(c)* We have,

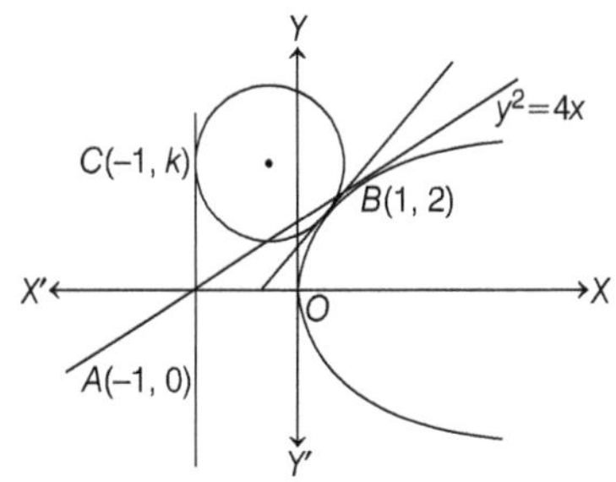

Equation of parabola

$y^2=4x$

Equation of tangent of parabola at (1, 2), is

$$2y=4\frac{(x+1)}{2}\ \Rightarrow\ y=x+1$$

The tangent of parabola $y = x + 1$ intersect the directrix of parabola at $A(-1, 0)$.

Now, AB and AC are tangent of circle

$\because \quad AB = AC$

$\Rightarrow \sqrt{(1+1)^2 + (2-0)^2}$

$= \sqrt{(-1+1)^2 + (0-k)^2}$

$\Rightarrow \quad 4 + 4 = k^2 \Rightarrow k^2 = 8 \Rightarrow k = 2\sqrt{2}$

6. *(b)* Given,

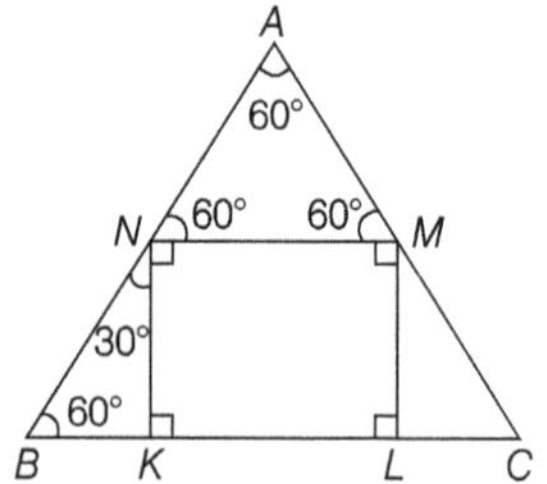

ABC is an equilateral triangle.

$\therefore \quad AB = BC = AC$

$KLMN$ be a rectangle.

$\therefore \quad KL = MN$

and $NK = LM$

$$\frac{AN}{NB} = 2$$

$\therefore \quad AN = 2NB = AM = MN$

$AB = AN + NB = 3NB.$

Area of $\Delta BKN = 6$

In ΔBKN,

$$\sin 60° = \frac{NK}{BN} \Rightarrow NK = \frac{\sqrt{3}}{2} BN.$$

Area of $\Delta BKN = \frac{1}{2} \cdot BN \cdot NK \sin 30°$

$$\Rightarrow \quad 6 = \frac{1}{2}.BN.\frac{\sqrt{3}}{2}.BN.\frac{1}{2} \Rightarrow BN^2 = \frac{48}{\sqrt{3}}$$

$\therefore$ Area of $\Delta ABC = \frac{\sqrt{3}}{4} \times AB^2$

$$= \frac{\sqrt{3}}{4} \times 9BN^2 = \frac{\sqrt{3}}{4} \times 9 \times \frac{48}{\sqrt{3}} = 108$$

7. *(c)* We have, equation of ellipse

$$\frac{x^2}{a^2} + \frac{y^2}{b^2} = 1$$

Point $P(a\cos\theta, b\sin\theta)$ lie on ellipse

Foci of ellipse $(F_1)(ae, 0)$ and $F_2(-ae, 0)$

$\therefore$ Centroid of $\Delta PF_1F_2 = \left(\frac{a\cos\theta}{3}, \frac{b\sin\theta}{3}\right)$

$$\therefore \quad h = \frac{a\cos\theta}{3}, K = \frac{b\sin\theta}{3}$$

$$\Rightarrow \left(\frac{3h}{a}\right)^2 + \left(\frac{3k}{b}\right)^2 = \cos^2\theta + \sin^2\theta$$

$$\Rightarrow \quad \frac{9x^2}{a^2} + \frac{9y^2}{b^2} = 1$$

which represent the locus of ellipse.

8. *(a)* We have, $\cos^7\theta - \sin^4\theta = 1$

$\Rightarrow \quad \cos^7\theta = 1 + \sin^4\theta$

LHS $\quad \cos^7\theta \in [-1, 1]$

RHS ≥ 1.

Hence, $\cos^7\theta = 1$ and $\sin^4\theta = 0$

$\therefore \theta = 0, 2\pi$ in $[0, 2\pi]$

9. *(c)* We have,

$(1 + \tan 1°)(1 + \tan 2°)(1 + \tan 3°) \ldots (1 + \tan 45°)$

We know that,

$(1 + \tan\theta)(1 + \tan(45° - \theta)) = 2$

$\therefore (1 + \tan 1°)(1 + \tan 44°)(1 + \tan 2°)(1 + \tan 43°) \ldots (1 + \tan 22°)(1 + \tan 23°)(1 + \tan 45°)$

$\Rightarrow \quad 2^{22} \cdot 2 = 2^{23}$

10. *(b)* Given,

$f : R \to R$ be a differentiable function in (a, b)

and $\quad f(a) = f(b) = 0, a < b$

and $f'(a)f'(b) > 0$

Since, $f'(a)f'(b) > 0$.

$\therefore f'(a)$ and $f'(b)$ both are positive or negative when both are positive then at least one root. $f'(x) = 0$ lie in interval, also when both are negative then at least one root lie in the interval.

[by Rolle's theorem]

$\therefore$ Minimum number of roots are 2.

11. *(b)* We have,

$(x - 41)^{49} + (x - 49)^{41} + (x - 2009)^{2009} = 0$

Let $f(x) = (x - 41)^{49} + (x - 49)^{41} + (x - 2009)^{2009}$

$\Rightarrow f'(x) = 49(x - 41)^{48} + 41(x - 49)^{40} + 2009(x - 2009)^{2008}$

Here, $f'(x) > 0, \forall x \in R$

$\therefore f(x)$ is monotonic increasing function.

Hence, $f(x)$ has only one real root.

12. *(c)* Clearly from graph

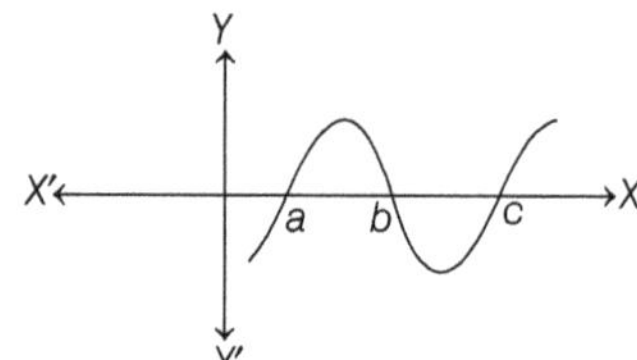

$f'(a) = f'(b) = f'(c) = 0$

$f'(a^-) < 0$ and $f'(a) > 0$

and $f'(c^-) < 0$ and $f'(c) > 0$

$\therefore$ Minimum at a and c.

$f'(b^-) > 0$ and $f'(b^+) < 0$

$\therefore$ Maximum at b.

$\therefore f(x)$ has local minima at $x = a, c$ and local maxima at $x = b$.

13. *(a)* Given,

$A(1, 1), B(3, 2), C(2, 3)$

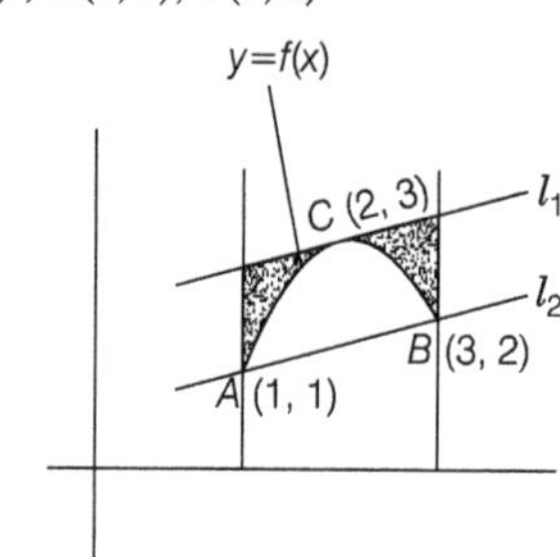

Slope of line $l_2 = \frac{2-1}{3-1} = \frac{1}{2}$

Line l_1 is parallel to l_2.

$\therefore$ Equation of line l_1 is

$$\Rightarrow \quad y - 3 = \frac{1}{2}(x - 2)$$

$$\Rightarrow \quad y = \frac{x}{2} + 2$$

Area of shaded region

$$= \int_1^3 (\text{line } l_1 - \text{curve } f(x))\, dx$$

$$= \int_1^3 \left(\frac{x}{2} + 2\right) dx - 4$$

[$\because$ area of curve = 4]

$$= \left[\frac{x^2}{4} + 2x\right]_1^3 - 4$$

$$= \left(\frac{9}{4} + 6\right) - \left(\frac{1}{4} + 2\right) - 4$$

$$= \frac{9}{4} + 6 - \frac{1}{4} - 2 - 4 = 2$$

14. *(c)* We have,

$$I_n = \int_1^0 (\log x)^n\, dx$$

Put $\log x = t \Rightarrow x = e^t$

$\Rightarrow \quad dx = e^t\, dt$

When $x = 1, t = 0$

and $\quad x = e \Rightarrow t = 1$

$$\therefore \quad I_n = \int_0^1 t^n e^t dt$$

$$\Rightarrow \quad I_n = [t^n e^t]_0^1 - n\int_0^1 t^{n-1} e^t\, dt$$

[$\because$ by parts]

$\Rightarrow \quad I_n = e - nI_{n-1}$

$\Rightarrow \quad I_n + n\, I_{n-1} = e$

$\Rightarrow \quad I_{2011} + 2011 I_{2010} = e$

Similarly, $I_{100} + 100 I_{99} = e$

Hence, option (c) is correct.

15. *(d)* We have,

$$A = \{(x, y) : x^2 + y^2 \le 100\}$$
$$B = \{(x, y) : \sin (x + y) > 0\}$$

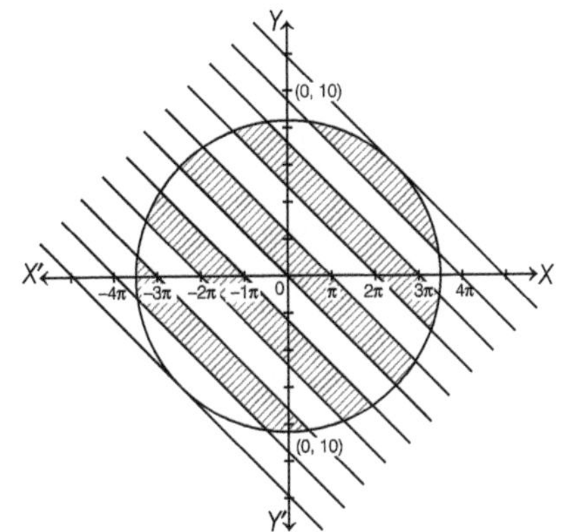

$\sin (x + y) > 0$

$\therefore \quad x + y \in (0, \pi) \cup (2\pi, 3\pi)$

$x + y = c$ equation of line

Required area = shaded region

$\frac{1}{2}$ area of circle $= \frac{1}{2} \pi (10)^2 = 50\pi$

16. *(d)* We have,

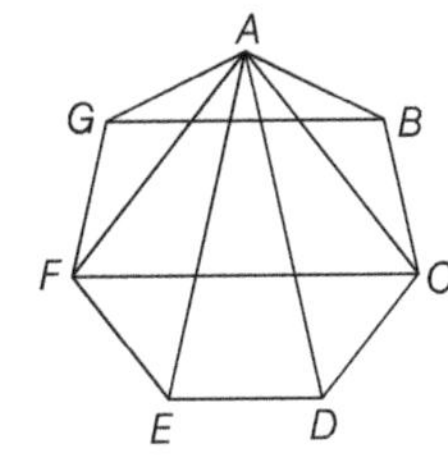

Regular Heptagon from vertices 'A'.

There are three isosceles triangle formed

AED, AFC, AGB.

Similarly, from other vertices also formed three isosceles triangle.

$\therefore$ Required probability $= \frac{7 \times 3}{{}^7C_3}$

$$= \frac{7 \times 3 \times 1 \times 2 \times 3}{7 \times 6 \times 5} = \frac{3}{5}$$

17. *(d)* We have,

$\vec{u} = 2\hat{i} - \hat{j} + \hat{k}$ and $\vec{v} = -3\hat{j} + 2\hat{k}$

$$\vec{u} \times \vec{v} = \begin{vmatrix} \hat{i} & \hat{j} & \hat{k} \\ 2 & -1 & 1 \\ 0 & -3 & 2 \end{vmatrix}$$

$$= (-2 + 3)\hat{i} - (4 - 0)\hat{j} + (-6 - 0)\hat{k}$$

$$\vec{u} \times \vec{v} = \hat{i} - 4\hat{j} - 6\hat{k}$$

$\vec{w}$ is a unit vector in XY-plane.

Let $\vec{w} = \cos\theta\,\hat{i} + \sin\theta\hat{j}$

$$\therefore |(\vec{u} \times \vec{v}) \cdot \vec{w}| = |(\hat{i} - 4\hat{j} - 6\hat{k}) . (\cos\theta\hat{i} + \sin\theta\hat{j})|$$

$= \cos\theta - 4\sin\theta$

$\therefore$ Maximum value of

$\cos\theta - 4\sin\theta = \sqrt{1 + 4^2} = \sqrt{17}$

18. *(d)* 6-digit number are there in which no digits is repeated and even digits appear on even places and odd digits appear in odd place. Such 6-digit number which is divisible by 4. If last two digits are divisible by 4.

i.e. 12, 16, 32, 36, 52, 56, 72, 76, 92, 96

$\therefore$ Last digits appears only 2 and 6

3 way	3 ways	4 ways	4 way	5 way	2/6

$\{1, 3, 5, 7, 9\}$

$\therefore$ Total numbers $= 3 \times 3 \times 4 \times 4 \times 5 \times 2$
$= 1440$

19. *(c)* Let $P(x) = 1 + x^2 + x^4 + \ldots + x^{2010}$

$$P(x) = \frac{1 - x^{2012}}{1 - x^2}$$

$$P(x) = \frac{(1 - x^{1006})(1 + x^{1006})}{(1 - x)(1 + x)}$$

$$P(x) = (1 + x^{1006})\left(\frac{1 - x^{503}}{1 - x}\right)\left(\frac{1 + x^{503}}{1 + x}\right)$$

$$P(x) = (1 + x^{1006})\,1 + x + x^2 + x^3 + \ldots + x^{502})\,(1 - x + x^2 - x^3 + \ldots + x^{502})$$

Thus, $P(x)$ is divisible by $1 + x + x^2 + \ldots x^{n-1}$.

If $n - 1 = 502 \Rightarrow n = 503$

20. *(a)* Given,

$$a_0 = 0$$
$$a_n = 3a_{n-1} + 1, \forall\, n \ge 1$$
$$a_1 = 3a_0 + 1 = 1$$
$$a_2 = 3a_1 + 1 = 3 + 1$$
$$a_3 = 3a_2 + 1 = 3(3 + 1) + 1 = 3^2 + 3 + 1$$

and $\quad a_4 = 3a_3 + 1 = 3(3^2 + 3 + 1) + 1 = 3^3 + 3^2 + 3 + 1$

Similarly,

$$a_n = 1 + 3 + 3^2 + \ldots + 3^{n-1} = \frac{3^n - 1}{3 - 1} = \frac{3^n - 1}{2}$$

$$a_{2010} = \frac{3^{2010} - 1}{2}$$

a_{2010} is divided by 11

$$\therefore \quad \frac{3^{2010} - 1}{2} \div 11 = \frac{3^{2010} - 1}{22}$$

$$\Rightarrow \quad 3^{2010} = (3^5)^{402}$$
$$= (243)^{402} = (242 + 1)^{402}$$
$$= 402(242)k + 1$$

[$\because k$ is positive integer]

$$= 402 \times 22 \times 11 \times k + 1$$
$$= 22m + 1$$

$$\therefore \frac{3^{2010} - 1}{22} = \frac{22m + 1 - 1}{22} = \frac{22m}{22}$$

$\therefore$ Remainder $= 0$

21. *(a)*

Table Pen Paper
F
μ_1
μ_2

Limiting value of friction on pen is

$$f_1 = \mu_1\, mg$$

So, for pen to start slipping, force acting over pen is

$$F_1 \ge f_1 = \mu_1\, mg$$

So, minimum acceleration of pen is

$$a_{\min} = \frac{F_1}{m} = \mu_1 g$$

Hence, minimum acceleration of paper is $a_{\min} = \mu_1 g$.

Now, consider free body diagram of paper

$f_1 = \mu_1 mg$
a
M
F
$f_2 = \mu_2 mg$

We have,

$$F_{\text{net}} = Ma_{\min} = F - f_1 - f_2$$

$$\Rightarrow \quad F = Ma_{\min} + f_1 + f_2$$
$$= M\mu_1 g + \mu_1 mg + \mu_2(m + M)g$$
$$= (M + m)(\mu_1 + \mu_2)g$$

22. *(a)* Let displacements of masses m_1 and m_2 are x_1 and x_2, respectively.

x_1 x_2
m_1 m_2

Total elongation of spring is $x = x_1 + x_2$.

So, when spring snaps back, pull on each of mass is

$$F = -kx$$

Hence, by second law equation for m_1 and m_2 are

$$m_1 a_1 = -kx \Rightarrow m_1 \frac{d^2x_1}{dt^2} = -kx$$

and $\quad m_2 a_2 = -kx \Rightarrow m_2 \frac{d^2x_2}{dt^2} = -kx$

Now, from $x = x_1 + x_2$, we have

$$\frac{d^2x}{dt^2} = \frac{d^2x_1}{dt^2} + \frac{d^2x_2}{dt^2}$$

$$\Rightarrow \quad \frac{d^2x}{dt^2} = \frac{-k}{m_1}x + \frac{-k}{m_2}x$$

$$\Rightarrow \quad \frac{d^2x}{dt^2} = -k\left(\frac{1}{m_1} + \frac{1}{m_2}\right)x$$

So, $\quad \omega^2 = k\left(\frac{1}{m_1} + \frac{1}{m_2}\right) = k\left(\frac{m_1 + m_2}{m_1 m_2}\right)$

Hence, time period of oscillation is

$$T = \frac{2\pi}{\omega} = 2\pi\sqrt{\frac{1}{k}\left(\frac{m_1 m_2}{m_1 + m_2}\right)}$$

23. *(b)* Resolving tension T, we have

Restoring force on bead $= 2T\cos\theta$

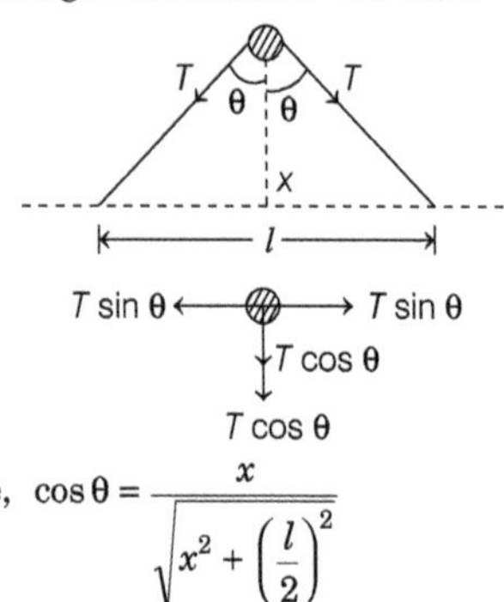

Here, $\cos\theta = \dfrac{x}{\sqrt{x^2 + \left(\dfrac{l}{2}\right)^2}}$

As $l >> x$, $\dfrac{l^2}{4} + x^2 \approx \dfrac{l^2}{4}$

So, $\cos\theta = \dfrac{2x}{l}$

Now, if a = acceleration of bead. Then,

Restoring force = Mass × Acceleration

$\Rightarrow \quad 2T\cos\theta = -ma$

$\Rightarrow \quad \dfrac{2T\,(2x)}{l} = -ma$

$\Rightarrow \quad a = -\left(\dfrac{4T}{ml}\right)\cdot x$

Hence, $\omega^2 = \dfrac{4T}{ml}$.

∴ Time period of oscillation is

$$T = \frac{2\pi}{\omega} = 2\pi\sqrt{\frac{ml}{4T}}$$

Hence, frequency of oscillation is

$$f = \frac{1}{T} = \frac{1}{2\pi}\sqrt{\frac{4T}{ml}}$$

24. *(c)*

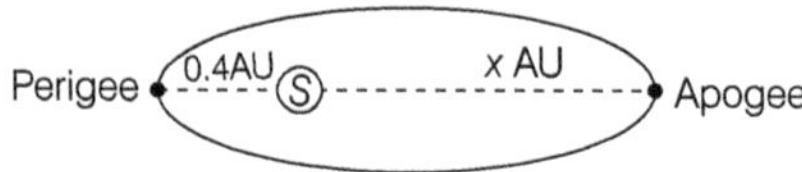

For camet,

Mean orbital radius,

$$r_1 = \frac{0.4 + x}{2}\,\text{AU}$$

Now, using $T^2 \propto r^3$, we get

$$\left(\frac{T_1}{T_2}\right)^2 = \left(\frac{r_1}{r_2}\right)^3 \quad \ldots(\text{i})$$

Let T_2 and r_2 are time period and mean orbital radius for earth.

Then, $T_2 = 1\,\text{yr}$ and $r_2 = 1\,\text{AU}$

Also given,

$$T_1 = 125\,\text{yr}$$

Substituting values in Eq. (i), we get

$$\left(\frac{125}{1}\right)^2 = \left(\frac{0.4 + x}{2\,(1)}\right)^3$$

$\Rightarrow \quad (0.4 + x)^3 = (125)^2(2^3)$

$\Rightarrow \quad (0.4 + x)^3 = (5^6)\,(2^3)$

$\Rightarrow \quad 0.4 + x = (5^2)\,(2)$

$\Rightarrow \quad x = 50 - 0.4 = 49.6\,\text{AU}$

25. *(d)* Just when switch S is closed, inductor offers a very large opposing emf in the circuit. Hence, current is zero at $t = 0$.

26. *(a)* Electric field due to a volume charge distribution varies with distance r as,

$$E = \begin{cases} \dfrac{kQr}{R^3}, & r < R \\ \dfrac{kQ}{r^2}, & r \ge R \end{cases}$$

or
$$E \propto \begin{cases} r, & r < R \text{ (straight line)} \\ \dfrac{1}{r^2}, & r \ge R \text{ (curve)} \end{cases}$$

So, correct graph is option (a).

27. *(d)* Following Gauss' law, electric field inside cavity is proportional to enclosed charge $+q$ only.

28. *(c)* $E_y = E_0 \sin(kx + \omega t)$

and $\quad E_z = -2E_0 \sin(kx - \omega t)$

As, sine function is variable for both components. So, in plane the tip of **E**-vector goes back and forth along a straight line.

So, correct option is (c).

29. *(b)* Radius r of disc must be such that rays beyond it are totally internally reflected.

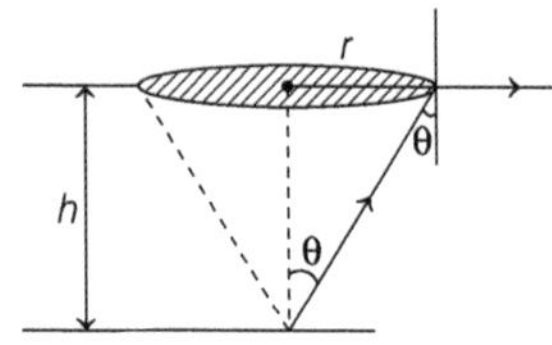

For $\theta \ge i_c$,

$\Rightarrow \quad \sin\theta \ge \sin i_c$

$\Rightarrow \quad \dfrac{r}{\sqrt{r^2 + h^2}} \ge \dfrac{1}{\mu} \qquad \left[\because \mu = \dfrac{1}{\sin i_c}\right]$

$\Rightarrow \quad r = \dfrac{h}{\sqrt{\mu^2 - 1}}$

30. *(d)* In first incidence, light ray is bends towards normal so, it must be going into a denser medium.

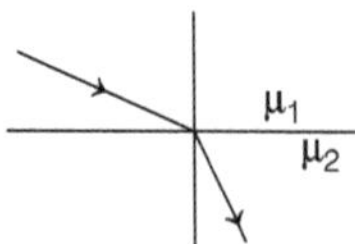

i.e. $\quad \mu_2 > \mu_1$

As, in second incidence, no light goes into third medium so, total internal reflection takes place at second incidence.

Hence, light is going from a denser to a rare medium.

i.e. $\quad \mu_2 > \mu_3$

Also $\mu_1 \ne \mu_3$ and $\mu_1 > \mu_3$, because there is no emergent light. Hence, $\mu_3 < \mu_1 < \mu_2$ is correct order.

31. *(d)* Half-life is 30 min so decay rate varies as shown below.

120000 cps (3 PM) ⟶ 60000 cps (3:30 PM) ⟶ 30000 cps (4:00 PM) ⟶ 15000 cps (4:30 PM) ⟶ 7500 cps (5:00 PM)

32. *(b)* Book started to loose contact with shelf when maximum acceleration a of shelf is such that,

$$a_{\max} \ge g$$

$\Rightarrow \quad \omega^2 A \ge g$

$\Rightarrow \quad \omega^2 = \dfrac{g}{A}$

So, frequency of oscillations is

$$f = \frac{\omega}{2\pi} = \frac{1}{2\pi}\sqrt{\frac{g}{A}}$$

$\Rightarrow \quad f = \dfrac{1}{2\pi}\cdot\sqrt{\dfrac{10}{2.5 \times 10^{-2}}} = \dfrac{2 \times 10}{2\pi}$

$\Rightarrow \quad f \approx 3.18\,\text{Hz}$

33. *(c)* Process is quasistatic adiabatic.

So, $\quad dQ = 0$

$\Rightarrow \quad dW = -dU$

$\Rightarrow \quad p\Delta V = -nC_V dT \quad \ldots(\text{i})$

$\quad U = CT - \dfrac{n^2 a}{V} \quad \ldots(\text{ii})$

and $\quad p = \dfrac{nRT}{V - nb} - \dfrac{n^2 a}{V^2} \quad \ldots(\text{iii})$

From Eq. (ii), $dU = CdT + \dfrac{n^2 a}{V^2}\,dV$

So, from Eq. (i), we have

$$dW = -dU$$

$\Rightarrow \quad pdV = -\left(CdT + \dfrac{n^2 a}{V^2}\,dV\right)$

Substituting for p from Eq. (iii) in above equation, we have

$\Rightarrow \left(\dfrac{nRT}{V - nb} - \dfrac{n^2 a}{V^2}\right)dV$

$\quad = -CdT - \dfrac{n^2 a}{V^2}\cdot dV$

$\Rightarrow \left(\dfrac{nRT}{V - nb}\right)dV = -CdT$

$\Rightarrow \quad \dfrac{dV}{V-nb} = \dfrac{-C}{nR} \cdot \dfrac{dT}{T}$

$\Rightarrow \quad \int \dfrac{dV}{V-nb} = -\dfrac{C}{nR}\int \dfrac{dT}{T}$

$\Rightarrow \log (V-nb) + k = -\log T^{C/nR}$

$\Rightarrow \log \left(T^{\frac{C}{nR}} \cdot (V-nb) \right) = \text{constant}$

$\Rightarrow \quad T^{\frac{C}{nR}} \cdot (V-nb) = \text{constant}$

34. *(b)*

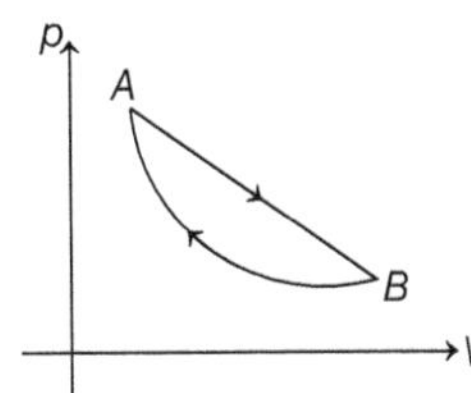

A to B is an expansion process and B to A is adiabatic compression.

So, heat is absorbed in process A to B and is released in process B to A.

35. *(b)* Here, source is moving towards a stationary observer.

Given,

Velocity of source, $v_s = 39.6$ km/h

$$= 39.6 \times \frac{5}{18} = 11 \text{ms}^{-1}$$

Velocity of sound, $v = 330 \text{ ms}^{-1}$

Frequency of source, $f = \dfrac{1}{30} \text{ s}^{-1}$

So, apparent frequency of sound as heard by the observer is

$$f' = \left(\frac{v}{v - v_s}\right) \cdot f = \left(\frac{330}{330-11}\right) \cdot \frac{1}{30}$$

$$\Rightarrow \quad f' = \frac{11}{319}$$

$$\Rightarrow \quad f' = \frac{1}{29}$$

So, time interval between two honks of bus horn observed by observer (person at bus stop) is 29 s.

36. *(a)* Velocity of sound in a gaseous medium is

$$v = \sqrt{\frac{\gamma RT}{M}}$$

$$\Rightarrow \quad v \propto \frac{1}{\sqrt{M}}$$

So, $\dfrac{v_{\text{Oxygen}}}{v_{\text{Hydrogen}}} = \sqrt{\dfrac{M_H}{M_o}} = \sqrt{\dfrac{2}{32}}$

$$\Rightarrow \quad \frac{v_{\text{Oxygen}}}{v_{\text{Hydrogen}}} = \frac{1}{4}$$

37. *(a)* Let intensity of source is I_0.

Then, intensity of maxima is $4I_0$.

So, intensity at given point on screen = 75% of maximum intensity = $3I_0$

Now, using $I = 4I_0 \cos^2 \dfrac{\phi}{2}$, we get

$$3I_0 = 4I_0 \cos^2 \frac{\phi}{2}$$

$$\Rightarrow \quad \cos\frac{\phi}{2} = \frac{\sqrt{3}}{2}$$

$$\phi = \frac{\pi}{3}$$

So, phase difference between rays reaching given point from upper and lower slit is $\dfrac{\pi}{3}$.

Hence, by using formula

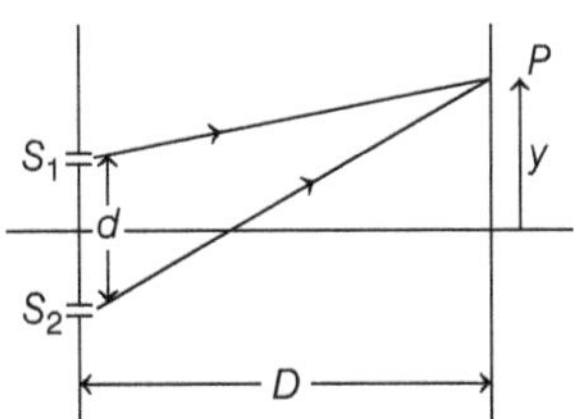

For path difference in Young's experiment, we get

$$\Delta\phi = \frac{\pi}{3} = \frac{yd}{D} \times \frac{2\pi}{\lambda}$$

$$\Rightarrow \quad y = \frac{\lambda D}{6d} = \frac{600 \times 10^{-9} \times 1}{6 \times 0.1 \times 10^{-3}}$$

$$\Rightarrow \quad y = 1 \times 10^{-3} \text{ m} \Rightarrow y = 1 \text{mm}$$

38. *(d)* Initial position of masses is given as,

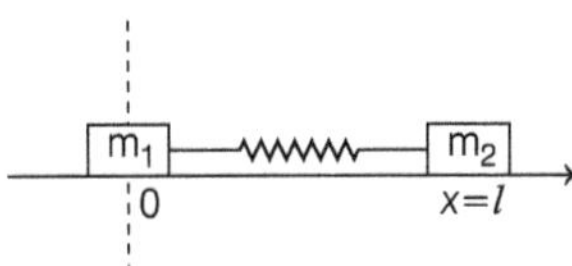

So, initial position and velocity of centre of mass at $t = 0$ are

$$x_{CM} = \frac{m_1x_1 + m_2x_2}{m_1 + m_2}$$

$$= \frac{m_1 \times 0 + m_2 \times l}{m_1 + m_2} = \frac{m_2 l}{m_1 + m_2}$$

$$v_{CM} = \frac{m_1 v_1 + m_2 v_2}{m_1 + m_2}$$

$$= \frac{m_1(v_0) + m_2(0)}{m_1 + m_2} = \frac{m_1 v_0}{m_1 + m_2}$$

At instant t, position of centre of mass is

$$x'_{CM} = x_{CM} + (v_{CM} \times t)$$

$$= \frac{m_2 l}{m_1 + m_2} + \frac{m_1 v_0}{m_1 + m_2} \cdot t$$

39. *(a)*

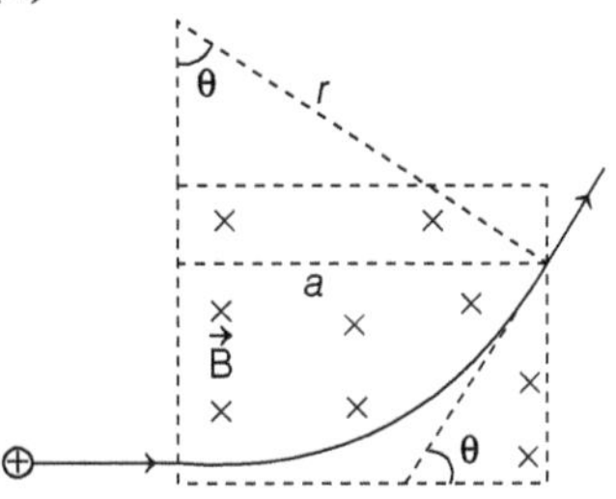

In given condition, if r is radius of curved path travelled by changed particle.

Then,

$$\sin\theta \approx \tan\theta = \frac{a}{r}$$

or $\quad r = a \operatorname{cosec}\theta \approx a \cot\theta \qquad \ldots\text{(i)}$

Also, $\quad r = \dfrac{mv}{Bq} \qquad \ldots\text{(ii)}$

Combining Eqs. (i) and (ii), we get

$$v = \frac{qB}{m} a \cot\theta$$

40. *(c)* Copper loses entropy and cools. Pond gains same amount of heat and there is some agitation in pond's water caused by plunging of copper.

Hence, pond gain more entropy than copper.

41. *(a)* Co (diethylene triamine) is of Ma_3b_3 type complex, therefore it shows two isomers *cis* and *trans*-form

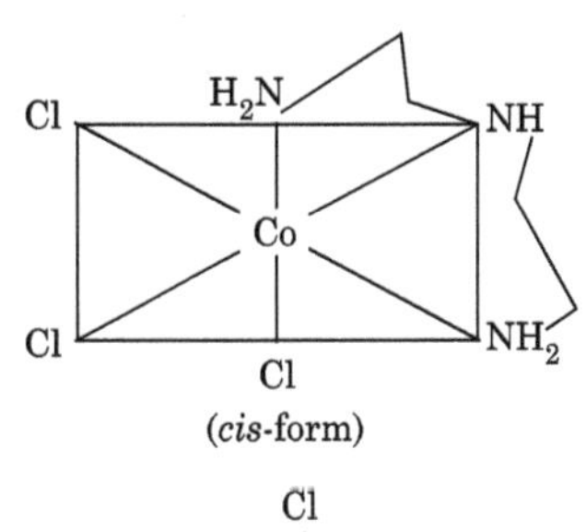

(*cis*-form)

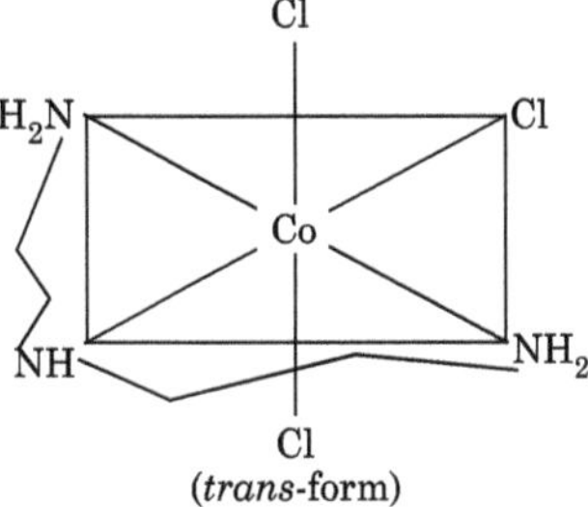

(*trans*-form)

42. *(c)* The π-acid ligands are those ligands, which have filled orbitals that can donate electrons to metals and also have empty anti-bonding orbitals, which can also accept electrons from the central metals. Among the given options CN is a π-acid ligand.

43. *(d)* Total number of electrons in

$O_2^{2-} = 18$

The electronic configuration of O_2^{2-} will be $\sigma 1s^2 \sigma * 1s^2 \sigma 2s^2 \sigma * 2s^2 \sigma 2p_z^2$
$\pi 2p_x^2 \pi 2p_y^2 \pi * 2p_x^2 \pi * 2p_y^2$

Bond order of a diatomic molecule is given by the formula, $BO = \frac{N_b - N_a}{2}$

Bond order of O_2^{2-} will be $= \frac{10-8}{2} = 1$

44. *(a)* Given, wavelength of photon $= 1$ m

Speed of light $= 3 \times 10^8 \text{ ms}^{-1}$

Energy of photon $= \frac{hc}{\lambda}$

Substituting the values in above equation we get, $E = \frac{6.626 \times 10^{-34} \text{Js} \times 3 \times 10^8 \text{ms}^{-1}}{1\text{m}}$

$= 1.988 \times 10^{-23}$ J

45. *(b)* As given in the question, the concentration of a substance becomes one half of its original value, that means it is talking about half-life. Half-life period is the time required to reduce the initial concentration to half of its value.

$t_{1/2}$ (for zero order) $= \frac{A_0}{2k}$

$t_{1/2}$ (for first order) $= \frac{0.693}{k}$

$t_{1/2}$ (for second order) $= \frac{1}{k[A_0]}$

$t_{1/2}$ (for third order) $= \frac{1}{k[A_0]^2}$

where, A_0 is the initial concentration. Thus, the correct option is (b). This is because, the half-life of first order does not depend upon the initial concentration.

46. *(c)* Central atom Cl has seven valence electrons, out of which it forms 3 bond pairs with fluorine and 2 pairs of electrons remain non-bonded (lone pair). Thus, its shape will be T-shaped, where lone pairs occupy equatorial positions, so as to minimise the repulsion.

(T-shaped)

47. *(d)* Friedel-Craft's acylation is an electrophilic aromatic substitution, which allows the synthesis of monoacylated products from the reaction between aromatic compounds and acyl chloride or anhydrides.

$C_6H_6 + CH_3-CO-Cl \xrightarrow{AlCl_3} C_6H_5-CO-CH_3$

48. *(a)* The strength of acidic character of a compound depends upon the stability of its conjugate base more is the stability of the conjugate base more will be the acidic character.

The conjugate bases of given compounds are as follows:

I. $C_6H_5-CH_2OH \longrightarrow C_6H_5-CH_2O^- + H^+$
(No resonance)

II. $C_6H_5-COOH \longrightarrow C_6H_5-COO^- + H^+$
(Resonance stabilised)

III. $H_3C-C_6H_4-OH \longrightarrow H_3C-C_6H_4-O^- + H^+$
(Resonance is possible, but-*I*-effect of CH_3 makes it less stable)

IV. $C_6H_5-SO_3H \longrightarrow C_6H_5-SO_3^- + H^+$
(The negative charge is delocalised over 3 electronegative oxygen atom)

Thus, the order of stability of conjugate base will be

$C_6H_5-CH_2O^- < CH_3-C_6H_4-O^- < C_6H_5-COO^- < C_6H_5-SO_3^-$

$\therefore$ The order of acidity will be,

I < III < II < IV

49. *(a)* The most stable conformation for *n*-butane is the staggered form, where the methyl group and H-atoms of one carbon are far apart from the groups of other carbon atom and, hence have minimum repulsion between them. Such, type of conformation is shown in option (a).

(Staggered form)

50. *(a)* In the nuclear reaction,

$^{234}_{90}Th \longrightarrow ^{234}_{91}Pa + X$

The mass number remain same, but the atomic number increases by 1. Thus, an electron is ejected. Hence, X will be, $^{0}_{-1}e$.

51. *(c)* A concentrated solution of copper sulphate, which is dark blue in colour is mixed at room temperature with a dilute solution of copper sulphate, which is light blue. For this process entropy change will be positive and enthalpy does not change. As the solution is getting diluted, the randomness of molecules increases thus the entropy of the process increases, but the enthalpy remains constant.

52. *(b)* Rise in temperature increases the fraction of molecules that collides with energies more than the activation energy. Thus, number of collisions, average energy of collisions and average velocity of the reactants, increases with increase in temperature which increases the rate of reaction but does not increase activation energy.

53. *(c)* A face centered cubic unit cell contains atoms at all the corners and at the centre of all faces of the cube.

Thus, number of atoms in fcc

$= \frac{1}{8} \times 8$ (present at the corners

$+ \frac{1}{2} \times 6$ (present at face)

$= 4$

A body centered cubic unit cell has an atom at each of its corner and also one atom at its body centre.

Thus, number of atoms in bcc

$= \frac{1}{8} \times 8$ (present at the corners)

+1 (present at the body)

$= 2$

54. *(d)* $CO_2 \rightleftharpoons CO + \frac{1}{2}O_2$,

$K_{C_1} = 9.1 \times 10^{-12}$ ...(i)

$H_2O \rightleftharpoons H_2 + \frac{1}{2}O_2$, $K_{C_2} = 7.1 \times 10^{-12}$ ...(ii)

$CO_2 + H_2 \rightleftharpoons CO + H_2O$ ...(iii)

We can obtain Eq. (iii) by reversing Eq. (ii) and then adding Eq. (i) to it.

Thus, $K = \frac{1}{K_2} \times K_1 = \frac{1}{7.1 \times 10^{-12}} \times 9.1 \times 10^{-12}$

$K = 1.28$

55. *(b)* For the first order reaction,

$$R \longrightarrow P$$

$$\text{Rate} = -\frac{d[R]}{dt} = k[R]$$

or $\frac{d[R]}{[R]} = -kdt$

On integration, we get

$$\ln R = -kt + I \quad \text{...(i)}$$

When $t = 0, R = R_0$

$\therefore \quad \ln[R]_0 = -k \times 0 + I$

$$\ln[R]_0 = I \quad \text{...(ii)}$$

Substituting (ii) in (i) we get

$$\ln R = -kt + \ln[R_0]$$

$$\frac{\ln R}{[R_0]} = -kt$$

Taking exponential on both sides, we get

$$\frac{R}{R_0} = e^{-kt}$$

$\Rightarrow \quad R = R_0 e^{-kt}$.

56. *(a)* The correct structure of PCl_3F_2 is trigonal bipyramidal, where the F atoms occupies the axial positions and Cl atoms due to large size occupies the equitorial positions, so as to minimise the repulsion between the atoms.

57. *(b)* Optically active compounds (having chiral centres), which are non-superimposable mirror images of each other are called enantiomers.

Molecules II and III does not have chiral centres thus they cannot be enantiomers of any of the given compounds. Thus, option (b) is correct.

Enantiomeric pair (Non-superimposable mirror images and have chiral carbons)

58. *(a)* For the reaction,

$$2NO_2(g) \longrightarrow 2NO(g) + O_2(g)$$

$$r = -\frac{1}{2}\frac{d[NO_2]}{dt} = +\frac{1}{2}\frac{d[NO]}{dt} = +\frac{d[O]}{dt}$$

From the above rate equation, it is clear that rate of disappearance of NO_2 is equal to the rate of formation of NO.

Thus, the curve X is associated with NO, Y with O_2 and Z with NO_2.

59. *(c)* The aromatic compound is one which obeys Huckel's rule $[(4n + 2)\pi]$, is planar and have conjugated system.

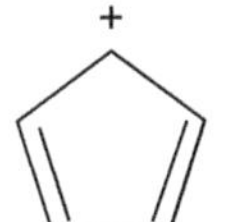

4π electrons, planar, non-conjugated system does not follow Huckel's rule (anti-aromatic)

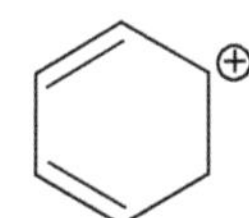

4π electrons, non planar, non-conjugated system does not follow Huckel's rule (non-aromatic)

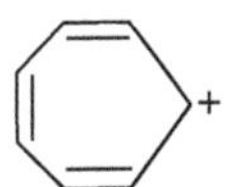

6π electrons, planar, conjugated system follow Huckel's rule (aromatic)

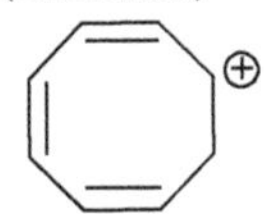

6π electrons, non-planar, non-conjugated system (non-aromatic)

Thus, the aromatic compound is given in option (c).

60. *(a)* When cyclohexene reacts with bromine in dark in presence of CCl_4 gives dibromo substituted product.

Its chair form is the most stable conformation, where both the bromine atoms occupy equitorial positions to minimise the repulsion.

(Chair form)

61. *(a)* Ribosome acts as a ribozyme. A ribozyme is a Ribonucleic Acid (RNA) enzyme that catalyses a chemical reaction. The ribozyme catalyses specific reactions in a similar way to that of protein enzymes. Also called catalytic RNA, ribozymes are found in the ribosome where they join amino acids together to form protein chains.

62. *(a)* In 1670, Sir Robert Boyle, performed an experiment with a viper in a vacuum. A gas bubble was observed in the eye of the snake when the pressure in the chamber was rapidly reduced. It displayed signs of extreme discomfort. This was the first recorded description of decompression sickness.

63. *(c)* Bacteria accumulate nutrients from the environment by diffusion. Bacterial cells are very small about 10 times smaller than most plant and animal cells. Smaller cells have a large surface area compared to cell volume, while larger cells have a reduced ratio of surface area to volume. One reason why bacterial cells are so small is that they need a large surface area to cell volume to take in nutrients.

64. *(a)* Mules are a hybrid of two species, i.e. a female horse and a male donkey. So, they end up with an odd number of chromosomes. A horse has 64 chromosomes and a donkey has 62. A mule inherits 63 chromosomes. An even number of chromosome is needed to divide into pairs and reproduce. Thus option (a) is incorrect.

65. *(c)* The rate of photosynthesis will be maximum in shining blue light if the total number of photons falling per unit area of a leaf per minute is kept constant. This is because, blue light is absorbed not only by chlorophyll, but also by carotenoids, further the blue wavelength of light have exactly the right amount of energy to energies or excite, chlorophyll electrons and boost them out of their orbits to a higher energy level.

66. *(c)* Path-finding by ants is by means of chemical signals between ants. These chemicals are called pheromones, which help them to communicate with each other.

67. *(a)* Urea is mixed with fed of ruminants to improve their health as it helps grow gut microbes that breakdown cellulose. Cattle and other ruminants convert urea to protein through the production of ammonia and carbon dioxide. It is vitally important that the right level of urea is fed and that there will be sufficient bacterial action to produce protein.

68. *(a)* On comparing adults of two herbivore species of different sizes from the same geographical area, the amount of faeces produced per kg body weight would be more in the smaller one than the larger one. This is because, the quantity of faeces is proportional to the amount of food they eat and smaller animals have a higher metabolic rate.

69. *(c)* The paper serves to trap the ethylene (a volatile gas) given off by the fruit in a layer around it rather than let it dissipate. This gas causes the plant tissue to chemically relax and begins breaking down faster, hence the effect of ripening quicker.

70. *(d)* There occurs more blood circulation to the skin to release heat from the body during high fever, which gives skin a reddish tinge. As, during high body temperature, the hypothalamus gives signal to cold down the body, which led to increase blood circulation, thereby release of heat occurs through body surface.

71. *(a)* Bacteria that contain bacteriochlorophyll do not use water as an electron donor during photosynthesis and therefore do not produce oxygen. This is known as anoxygenic photosynthesis.

72. *(a)* A muscle cramp is an involuntarily and forcibly contracted muscle that does not relax. Muscle cramp is caused due to conversion of pyruvic acid into lactic acid in the cytoplasm of muscle cell.

73. *(c)* A person can be carrier to only autosomal recessive disorder. Thus, a couple (both carriers to a disease) will have 25% chances that their child be a diseased individual.

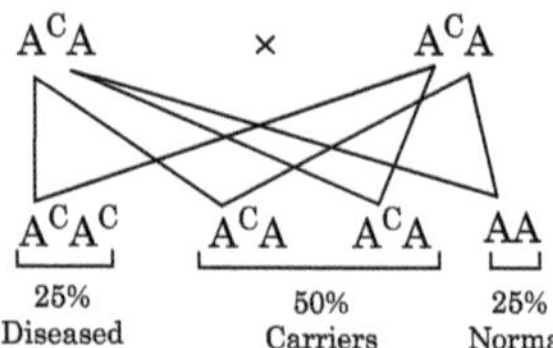

74. *(b)* Embryonic development also embryogenesis is the process by which the embryo forms and develops. The zygote undergoes mitotic divisions with no significant growth (a process known as cleavage) and cellular differentiation, leading to development of a multicellular embryo.

75. *(b)* The *Bt* toxin kills insects by converting the *Bt* protoxin into its active form in the insect gut due to the alkaline pH of the gut. The toxins then attack the gut cells of the insects, punching holes in the lining. The *Bt* spores spill out of the gut and germinate in the insect causing death within a couple of days.

76. *(d)* Paddy seeds soacked in water overnight and then kept out for a couple of days could feel warm due to the heat released during respiration. As the seed respires it uses oxygen and produces carbon dioxide gas into the surrounding. Thereby making the surrounding warmer.

77. *(c)* Restriction enzymes are classified as endonucleases. Their biochemical activity in the hydrolysis (digestion) of the phosphodiester backbone at specific sites is a DNA sequence.

78. *(a)* When arterial blood pressure exceeds blood osmotic pressure, the fluid part of blood, (i.e. plasma) flows out of the capillaries into the surrounding tissue. Colloid osmotic pressure is a form of osmotic pressure exerted by proteins on the blood plasma or interstitial fluid.

79. *(a)* The distance between two consecutive DNA bp is 0.34 nm

The length of a chromosome is 1 mm $= 10^6$ nm

$\therefore$ The number of bases in 1 mm

chromosome $= \dfrac{10^6 \text{nm}}{0.34\text{bp nm}} = 2{,}941{,}176.47$

$= 3$ million base pairs

80. *(d)* The order of the speed of propagation of an action potential of nerve impulse is m/s. The nerve impulses in unmyelinated neurons have a maximum speed of around 1m/s, in myelinated neurons they travel at 100 m/s.

81. *(c)* We have, $\left(x^{1/2} + \dfrac{1}{2x^{1/4}}\right)^n$

$$T_{r+1} = {}^nC_r\,(x^{1/2})^{n-r}\,\frac{1}{(2x^{1/4})^r}$$

$$T_{r+1} = {}^nC_r\,x^{\frac{n-r}{2}-\frac{r}{4}}\,.\,2^{-r}$$

$$T_{r+1} = {}^nC_r\,\frac{2n-3r}{4}\cdot 2^{-r} = \frac{{}^nC_r}{2^r}\,x^{\frac{2n-3r}{4}}$$

Given T_1, T_2, T_3 are in AP.

$\therefore \quad 2T_2 = T_1 + T_3$

$$\frac{2\,{}^nC_1}{2} = {}^nC_0 + \frac{{}^nC_2}{2^2}$$

$$\Rightarrow \quad {}^nC_1 = {}^nC_0 + \frac{n(n-1)}{2\cdot 2^2}$$

$$\Rightarrow \quad n = 1 + \frac{n(n-1)}{8}$$

$$\Rightarrow \quad n - 1 = \frac{n(n-1)}{8}$$

$\Rightarrow n - 1 = 0$ or $n = 8$

When $n = 8$,

$\dfrac{2n-3r}{4} = \dfrac{16-3r}{4}$ is integer

If $\quad r = 0, 4, 8 \qquad [r \le n]$

82. *(b)* We have,

$2x^2 + 2x + 1 = 0$

$$\Rightarrow x = -\frac{2 \pm \sqrt{4-8}}{4}$$

$$x = -\frac{2 \pm 2i}{4} = \frac{-1 \pm i}{2}$$

x satisfies the equation $(x+1)^n - r = 0$

$$\therefore \quad \left(\frac{-1 \pm i}{2} + 1\right)^n - r = 0 \qquad [r \in R]$$

$$\Rightarrow \quad \left(\frac{-1 \pm i + 2}{2}\right)^n = r$$

$$\Rightarrow \quad \left(\frac{1 \pm i}{2}\right)^n = r$$

$$\Rightarrow \quad \frac{1}{2^n}\,[(1 \pm i)^2]^{n/2} = r$$

$$\Rightarrow \quad \frac{1}{2^n}\,(\pm 2i)^{n/2} = r$$

$$\Rightarrow \quad \frac{1}{2^{n/2}}\,(\pm i)^{n/2} = r$$

r is real

$\therefore \quad n \in 4\text{m}$

$\therefore \quad n = 4000$

and $\quad r = \dfrac{1}{2^{\frac{4000}{2}}} = \dfrac{1}{4^{1000}}$

83. *(b)* We have,

$(y - ax - b)\,(bx^2 + ay^2 - ab) = 0$

$\Rightarrow y - ax - b = 0$ or $bx^2 + ay^2 - ab = 0$

$\Rightarrow y = ax + b$ or $bx^2 + ay^2 = ab$

$\Rightarrow y = ax + b$ or $\dfrac{x^2}{a} + \dfrac{y^2}{b} = 1$

***Case* I** When $a, b > 0$

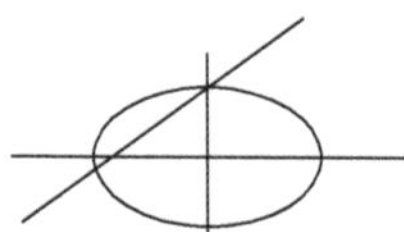

***Case* II** When $a < 0, b > 0$

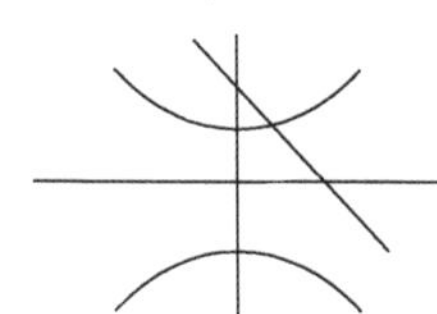

***Case* III** When $a > 0, b < 0$

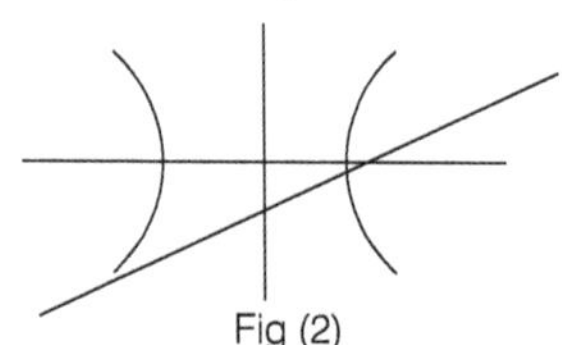

Fig (2)

$\therefore$ Figure (2) represent the curve.

84. *(b)* $ABCD$ is a cyclic quadrilateral $\angle A = 120°$.

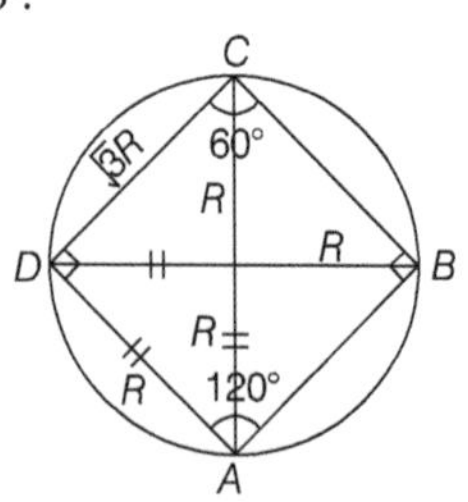

$ABCD$ has maximum area possible.

When $\angle B = \angle D = 90°$

$\therefore$ Area of quadrilateral

$$= 2 \times \frac{1}{2} \times AD \times DC$$

$$= 2 \times \frac{1}{2} \times R \times \sqrt{3}R \quad [\because AD = R, DC = \sqrt{3}R]$$

$$= \sqrt{3}R^2$$

85. *(b)* Let $f'(x) = 0$ at c

$\therefore \quad f'(c) = 0$

$f'(c^-) > 0 \Rightarrow f'(c^+) < 0$

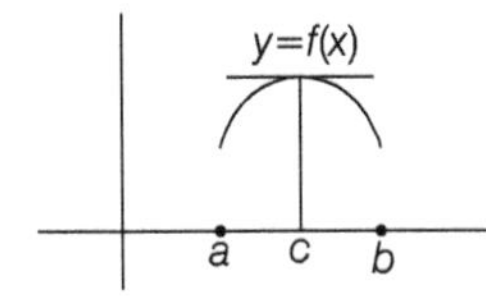

$\Rightarrow \quad g(x) = \dfrac{f(x)}{x}$

$\Rightarrow \quad g'(x) = \dfrac{xf'(x) - f(x)}{x^2}$

$$g'(c^+) = \lim_{h \to 0} \frac{(c+h)f'(c+h) - f(c)}{h} \quad [f'(c+h) < 0]$$

$$g'(c^+) < 0$$

Hence, option (b) is correct.

86. *(c)* V_1 = Volume of cone ABC

H = Height of cone V_1

R = Radius of cone V_1

V_2 = Volume of cone inscribed in V_1

$(H - h)$ = Height of cone V_2

r = Radius of cone V_2

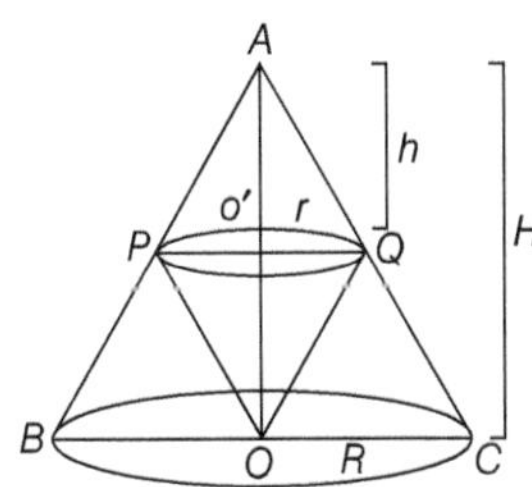

$\Delta AO'Q$ and ΔAOC are similar.

$\therefore \quad \dfrac{AO'}{AO} = \dfrac{O'Q}{OC}$

$\Rightarrow \dfrac{h}{H} = \dfrac{r}{R} \Rightarrow h = \dfrac{rH}{R}$

$\Rightarrow V_2 = \dfrac{1}{3}\pi r^2(H - h)$

$\Rightarrow V_2 = \dfrac{1}{3}\pi r^2\left(H - \dfrac{rH}{R}\right)$

$\Rightarrow \dfrac{dV_2}{dr} = \dfrac{1}{3}\pi H\left(2r - \dfrac{3r^2}{R}\right)$

For maxima or minima,

$$\frac{dV_2}{dr} = 0$$

$\Rightarrow \quad 2r - \dfrac{3r^2}{R} = 0$

$\Rightarrow \quad r = \dfrac{2R}{3}$

$\Rightarrow \quad V_2 = \dfrac{1}{3}\pi\left(\dfrac{2}{3}R\right)^2\left(H - \dfrac{2H}{3}\right)$

$$= \frac{4\pi R^2 H}{81}$$

$$V_1 = \frac{1}{3}\pi R^2 H$$

$\therefore \quad \dfrac{V_2}{V_1} = \dfrac{4\pi R^2 H}{81} \div \dfrac{1}{3}\pi R^2 H$

$$= 4 : 27 = \frac{4}{27}$$

87. *(a)* We have,

$$f(x) = x + \int_0^x f(t)\,dt$$

$f'(x) = 1 + f(x)$ [$\because$ by Leibnitz rule]

Let $\quad f'(x) = \dfrac{dy}{dx}\,f(x) = y$

$\therefore \quad \dfrac{dy}{dx} = 1 + y$

$\Rightarrow \quad \dfrac{dy}{1+y} = dx$

On integrating, we get

$$\log(1+y) = x + c$$

$\Rightarrow \quad 1 + y = Ae^x$

$\Rightarrow \quad y = Ae^x - 1$

$\Rightarrow \quad f(x) = Ae^x - 1$

$\Rightarrow \quad f(x) = e^x - 1 \quad [\because f(0) = 0, A = 1]$

Put $f(x) = 0$

$\therefore \quad 0 = e^x - 1 \Rightarrow e^x = 1$

Hence, only one solution.

88. *(b)* We have,

$$\int_0^{2\pi} \min\{|x - \pi|, \cos^{-1}(\cos x)\}\,dx$$

Graph of $y = |x - \pi|$ and $y = \cos^{-1}(\cos x)$

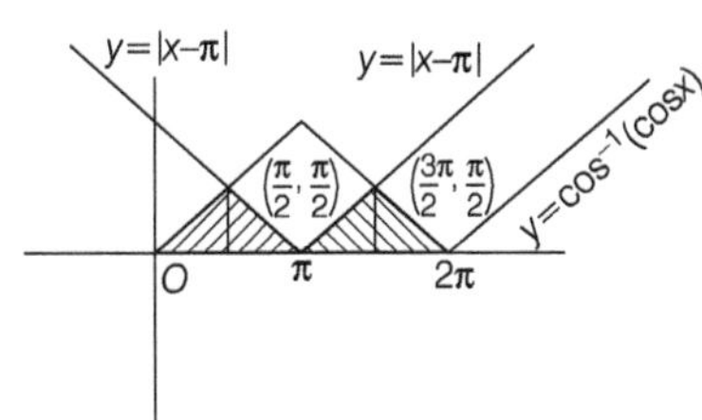

$\int_0^{2\pi} \min\{|x - \pi|, \cos^{-1}(\cos x)\,dx$ = Area of shaded region.

$$= \frac{1}{2} \times \pi \times \frac{\pi}{2} + \frac{1}{2} \times \pi \times \frac{\pi}{2}$$

$$= \frac{\pi^2}{4} + \frac{\pi^2}{4} = \frac{\pi^2}{2}$$

89. *(d)* Given,

Point P inside the ΔABC such that

$$\mathbf{PA} + 2\mathbf{PB} + 3\mathbf{PC} = 0$$

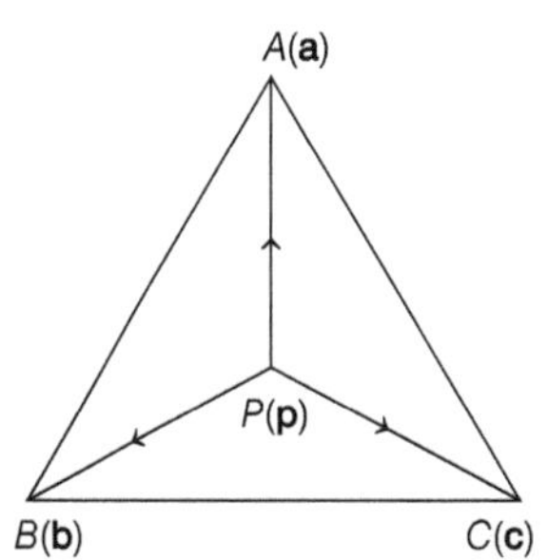

Let $\quad \mathbf{PA} = \mathbf{a} - \mathbf{P}$

$\mathbf{PB} = \mathbf{b} - \mathbf{P}$

$\mathbf{PC} = \mathbf{c} - \mathbf{P}$

$\therefore (\mathbf{a} - \mathbf{P}) + 2(\mathbf{b} - \mathbf{P}) + 3(\mathbf{c} - \mathbf{P}) = 0$

$\Rightarrow \quad \mathbf{a} + 2\mathbf{b} + 3\mathbf{c} = 3\mathbf{P}$

$\Rightarrow \quad \mathbf{P} = \dfrac{\mathbf{a} + 2\mathbf{b} + 3\mathbf{c}}{6}$

Area of $\Delta APC = \dfrac{1}{2}|\mathbf{a} \times \mathbf{P} + \mathbf{P} \times \mathbf{c} + \mathbf{c} \times \mathbf{a}|$

$$= \frac{1}{2}\left|\mathbf{a} \times \frac{(\mathbf{a} + 2\mathbf{b} + 3\mathbf{c})}{6} + \frac{(\mathbf{a} + 2\mathbf{b} + 3\mathbf{c})}{6} \times \mathbf{c} + \mathbf{c} \times \mathbf{a}\right|$$

$$= \frac{1}{2 \times 6}|\,2(\mathbf{a} \times \mathbf{b}) + 3(\mathbf{a} \times \mathbf{c}) + (\mathbf{a} \times \mathbf{c}) + 2(\mathbf{b} \times \mathbf{c}) + 6\,\mathbf{c} \times \mathbf{a}\,|$$

$$= \frac{1}{6}|\mathbf{a} \times \mathbf{b} + \mathbf{b} \times \mathbf{c} + \mathbf{c} \times \mathbf{a}|$$

$\therefore$

$$\frac{\text{Area of } \Delta ABC}{\text{Area of } \Delta APC} = \frac{\frac{1}{2}|\mathbf{a} \times \mathbf{b} + \mathbf{b} \times \mathbf{c} + \mathbf{c} \times \mathbf{a}|}{\frac{1}{6}|\mathbf{a} \times \mathbf{b} + \mathbf{b} \times \mathbf{c} + \mathbf{c} \times \mathbf{a}|}$$

$$= 3$$

$\therefore$ Ratio $= 3 : 1$

90. *(c)* We have,

$$6^m + 2^{m+n} \cdot 3^w + 2^n = 332$$

When $m = 4$, LHS > RHS

$\therefore$ Maximum value of $m = 3$

When $m = 3$,

$6^3 + 2^3 \cdot 2^n \cdot 3^w + 2^n = 332$

$2^n(8 \cdot 3^w + 1) = 332 - 216$

$2^n(8 \cdot 3^w + 1) = 116$

$2^n(8 \times 3^w + 1) = 4 \times 29$

$\therefore n = 2$

$8 \cdot 3^w + 1 = 29 \Rightarrow 3^w = \frac{7}{2}$

[not possible $\because w \in I$]

Put $m = 2$,

$\therefore 6^2 + 2^2 \cdot 2^n \cdot 3^w + 2^n = 332$

$2^n (4 \cdot 3^w + 1) = 332 - 36$

$2^n (4 \cdot 3^w + 1) = 296$

$2^n (4 \cdot 3^w + 1) = 2^3 \times 37$

$\therefore 2^n = 2^3$ and $4 \cdot 3^w + 1 = 37$

$n = 3$ and $3^w = 9 \Rightarrow w = 2$

Hence, m, n, w are positive integer.

$\therefore m^2 + mn + n^2 = (2)^2 + (2)(3) + (3)^2$

$= 4 + 6 + 9 = 19$

91. *(d)* Net distance travelled by ball is

$$d = 2h_1 + 2h_2 + 2h_3 + \dots$$

or, $$d = \left(\frac{v_0^2}{g} + \frac{(rv_0)^2}{g} + \frac{(r^2v_0)^2}{g} + \dots + \text{upto 10 terms}\right) - h$$

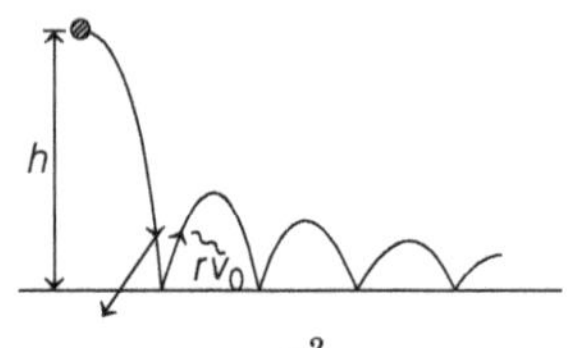

$$v_0 = \sqrt{2gh} \Rightarrow \frac{v_0^2}{g} = 2h$$

$$\Rightarrow d = \frac{v_0^2}{g}(1 + r^2 + r^4 + \dots + 10 \text{ terms}) - h$$

$$= 2h\left(\frac{1 - (r^2)^{10}}{1 - r^2}\right) - h$$

$$= 2h\left(\frac{1 - r^{20}}{1 - r^2}\right) - h$$

92. *(a)* Due to rotation of earth, acceleration due to gravity at latitude λ is

$$g' = g - \omega^2 R \cos^2 \lambda$$

At equator $\lambda = 0°$,

$$\Rightarrow g_e = g - \omega^2 R$$

At latitude of 60°, $\lambda = 60°$

$$\Rightarrow g_\lambda = g - \omega^2 R\left(\frac{1}{2}\right)^2$$

Given,

Weight at equator $= f \times$ Weight at latitude λ

$$\Rightarrow mg_e = f \times mg_\lambda \Rightarrow g_e = fg_\lambda$$

$$\Rightarrow g - \omega^2 R = f\left(g - \frac{\omega^2 R}{4}\right)$$

$$\Rightarrow g(1 - f) = \frac{\omega^2 R}{4}(4 - f) \quad \dots(i)$$

Now, as $g = \frac{GM}{R^2}$ and $\omega = \frac{2\pi}{T}$

Also, $M = \frac{4}{3}\pi R^3 \cdot \rho$

Substituting these in Eq. (i), we get

$$\frac{4}{3}\pi\rho G R(1 - f) = \frac{4\pi^2 R}{4T^2}(4 - f)$$

$$\Rightarrow \rho = \left(\frac{4 - f}{1 - f}\right) \cdot \frac{3\pi}{4GT^2}$$

93. *(b)* In given charge configuration,

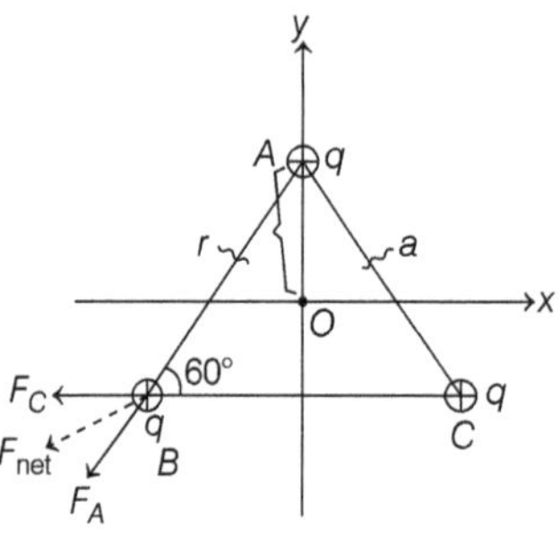

Net force on any of charge is

$$F_{net} = \sqrt{F_A^2 + F_B^2 + 2F_A F_B \cos 60°}$$

$$= \frac{\sqrt{3} \cdot q^2}{4\pi\varepsilon_0 a^2}$$

where, a = side length of equilateral triangle.

So, radius r is $r = \frac{2}{3}\left(\frac{\sqrt{3}}{2}a\right)$

$$\Rightarrow a = \sqrt{3}r$$

Hence, $$F_{net} = \frac{\sqrt{3} \cdot q^2}{(4\pi\varepsilon_0)3r^2}$$

or $$F_{net} = \frac{q^2}{(4\pi\varepsilon_0)\sqrt{3}r^2}.$$

Now, given that this force is balanced by a force $F(r) = kr$

$$\therefore kr = \frac{q^2}{(4\pi\varepsilon_0)\sqrt{3}r^2}$$

$$\Rightarrow r^3 = \frac{\sqrt{3}q^2}{12\pi\varepsilon_0 k}$$

So, $$r = \left(\frac{\sqrt{3}q^2}{12\pi\varepsilon_0 k}\right)^{\frac{1}{3}}$$

94. *(*)*

In an infinite network, adding one more loop does not affects total impedence of circuit.

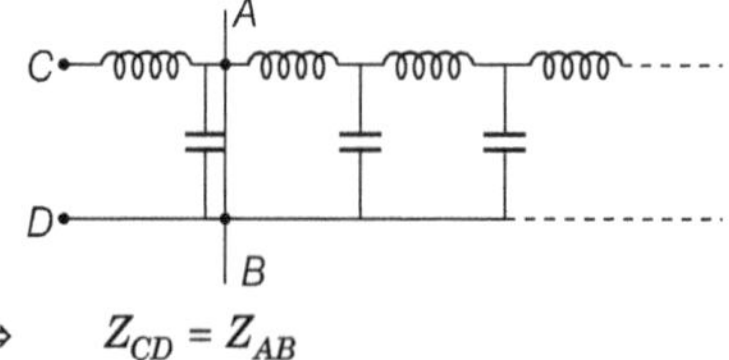

$$\Rightarrow Z_{CD} = Z_{AB}$$

$\Rightarrow (Z \parallel C)$ series $L = Z$

$$\Rightarrow \frac{ZX_C}{Z + X_C} + X_L = Z$$

$$\Rightarrow \frac{Z/\omega C}{Z + \frac{1}{\omega C}} = Z - \omega L$$

$$\Rightarrow \frac{\frac{Z}{\omega C}}{Z + \frac{1}{\omega C}} = Z - \omega L$$

$$\Rightarrow Z^2 - \omega L Z - \frac{L}{C} = 0$$

$$\Rightarrow Z = \omega L \pm \sqrt{\omega^2 L^2 + \frac{4L}{C}}$$

Now, $Z = \omega L$ or circuit is purely inductive when,

$$\omega^2 L^2 + \frac{4L}{C} = 0 \Rightarrow \omega^2 = -\frac{4}{LC}$$

But this is not possible or ω is imaginary. Hence, circuit does not behaves like an inductor.

95. *(a)* Refraction through glass sphere is as shown below.

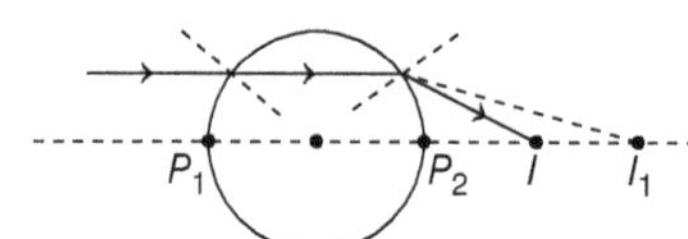

For refraction at first surface,

$$\frac{\mu}{v} - \frac{1}{u} = \frac{\mu - 1}{R} \quad \dots(i)$$

Here, $u = \infty$

So, from Eq. (i), we have

$$\frac{1}{v} = \frac{\mu - 1}{\mu R} \Rightarrow v = P_1 I_1 = \frac{\mu R}{\mu - 1}$$

Now, image of first surface acts like object for second.

For refraction at second surface,

$$\frac{1/\mu}{v} - \frac{1}{u} = \frac{\left(\frac{1}{\mu} - 1\right)}{-R} \quad \dots(ii)$$

Here, $$u = P_1 I_1 - P_1 P_2$$

$$= \frac{\mu R}{\mu - 1} - 2R = \frac{(2 - \mu)R}{\mu - 1}$$

So, from Eq. (ii), we have

$$\frac{1}{\mu v} - \frac{(\mu - 1)}{(2 - \mu)R} = \frac{\mu - 1}{\mu R}$$

$$\Rightarrow \frac{1}{\mu v} = \frac{\mu - 1}{\mu R} + \frac{\mu - 1}{(2 - \mu)R}$$

$$\Rightarrow \frac{R}{\mu(\mu - 1)v} = \frac{1}{\mu} + \frac{1}{2 - \mu}$$

$$\Rightarrow \frac{R}{\mu(\mu - 1)v} = \frac{2 - \mu + \mu}{\mu(2 - \mu)}$$

$$\Rightarrow \quad \frac{R}{(\mu-1)v} = \frac{2}{2-\mu}$$

$$\Rightarrow \quad \frac{1}{v} = \frac{2(\mu-1)}{(2-\mu)R} \text{ or } v = \frac{(2-\mu)R}{2(\mu-1)}$$

Hence, distance of final image from second surface is $P_2I = \frac{R(2-\mu)}{2(\mu-1)}$.

96. *(a)* Force on the particle is $F = -\frac{dV}{dx}$

$$\Rightarrow \quad F = -\frac{d}{dx}\left(\frac{1}{2}kx^2 - V_0\cos\left(\frac{x}{a}\right)\right)$$

$$= -\left(kx + \frac{V_o}{a}\sin\left(\frac{x}{a}\right)\right)$$

As $x << a, \Rightarrow \frac{x}{a} << 1$

So, $\sin\frac{x}{a} \approx \frac{x}{a}$

Hence, $F = -\left(kx + \frac{V_0}{a}.\frac{x}{a}\right) = -\left(k + \frac{V_0}{a^2}\right)\cdot x$

Acceleration A of the particle is

$$A = \frac{F}{m} = -\frac{1}{m}\left(k + \frac{V_0}{a^2}\right)\cdot x$$

As, acceleration $A = -\omega^2 x$, we have

$$\omega = \sqrt{\frac{1}{m}\left(k + \frac{V_0}{a^2}\right)} = \sqrt{\left(\frac{ka^2 + V_0}{ma^2}\right)}$$

$\therefore$ Time period of oscillation is

$$T = \frac{2\pi}{\omega} = 2\pi\sqrt{\left(\frac{ma^2}{ka^2 + V_0}\right)}$$

97. *(c)* Process equation is

$$pV^\alpha = \text{constant } (k) \Rightarrow p = \frac{k}{V^\alpha}$$

Work done by the gas in given process is

$$\Delta W = \int_{V_i}^{V_f} pdV$$

$$= \int_{V_i}^{V_f} \frac{kdV}{V^\alpha} = \left[\frac{kV^{1-\alpha}}{1-\alpha}\right]_{V_i}^{V_f}$$

$$= \left[\frac{pV}{1-\alpha}\right]_{V_i}^{V_f} = \frac{p(V_f - V_i)}{1-\alpha}$$

$$= \frac{p\Delta V}{1-\alpha} = \frac{nR\Delta T}{1-\alpha}$$

The change of internal energy of gas in this process will be

$$\Delta U = C_V \Delta T$$

And if ΔQ is heat supplied to the gas then,

$$\Delta Q = C\Delta T$$

Now, by first law of thermodynamics, we have

$$\Delta Q = \Delta U + \Delta W$$

$$\Rightarrow \quad C\Delta T = C_V\Delta T + \frac{nR\Delta T}{1-\alpha}$$

Heat capacity of the gas is

$$\Rightarrow \quad C = C_V + \frac{nR}{1-\alpha}$$

98. *(d)* As $B \to A \to C$ is a closed cyclic process, we have

ΔU (complete cycle) = 0

So, by first law of thermodynamics, we have $\Delta Q = \Delta W$

or $\Delta Q_{AB} + \Delta Q_{BC} + \Delta Q_{AC} = \Delta W$

$= \Delta W_{AB} + \Delta W_{BC} + \Delta W_{AC}$

= Area enclosed under p-V graph

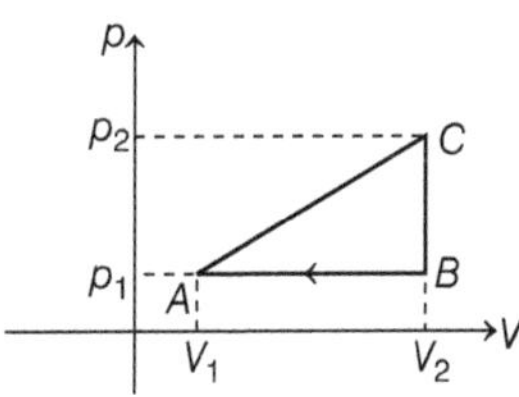

$$\Rightarrow \Delta Q_{AC} = -(\Delta Q_{AB} + \Delta Q_{BC}) + \frac{1}{2}(V_2 - V_1)(p_2 - p_1)$$

$$= -\left(\frac{3}{2}nR\Delta T - p_1(V_2 - V_1) + \frac{3}{2}nR\Delta T - V_2(p_2 - p_1)\right) + \frac{1}{2}(V_2 - V_1)(p_2 - p_1)$$

$$\therefore \Delta Q_{AC} = 2(p_2V_2 - p_1V_1) + \frac{1}{2}(V_2 - V_1)(p_2 - p_1)$$

99. *(b)* At distance of closest approach,

Total initial KE = Total final PE

$$\therefore \quad \frac{1}{2}mv^2 + \frac{1}{2}mv^2 = \frac{kq^2}{r}$$

$$\Rightarrow \quad mv^2 = \frac{q^2}{4\pi\varepsilon_0 r}$$

$$\Rightarrow \quad r = \frac{q^2}{4\pi\varepsilon_0 mv^2}$$

100. *(c)* Net decay rate of active nuclii is

$$\frac{dN}{dt} = c - \lambda N$$

$$\Rightarrow \quad \frac{dN}{c - \lambda N} = dt$$

Integrating both sides, we get

$$\int \frac{dN}{c - \lambda N} = \int dt \quad \text{...(i)}$$

Now, $c - \lambda N = u$

Differentiating,

$$-\lambda dN = du$$

above equation, we have

$$dN = \frac{dx}{-\lambda}$$

Substituting is Eq. (i), we get

$$\frac{-1}{\lambda}\int \frac{du}{u} = \int dt$$

$$\Rightarrow \quad -\frac{1}{\lambda}\log u = t + k$$

where, k is constant of integration.

$$\Rightarrow \quad -\frac{1}{\lambda}\log(c - \lambda N) = t + k \quad \text{...(ii)}$$

Now, at $t = 0$ and $N = N_0$,

So, $\frac{-1}{\lambda}\log(c - \lambda N_0) = k$.

Substituting for k in Eq. (ii), we get

$$-\frac{1}{\lambda}\log(c - \lambda N) = t - \frac{1}{\lambda}\log(c - \lambda N_0)$$

$$\log\left(\frac{c - \lambda N}{c - \lambda N_0}\right) = -\lambda t$$

Taking antilog we get,

$$\frac{c - \lambda N}{c - \lambda N_0} = e^{-\lambda t}$$

$$\Rightarrow \quad c - \lambda N = (c - \lambda N_0)e^{-\lambda t}$$

Solving for N, we get

$$\Rightarrow \quad N = \frac{c}{\lambda}(1 - e^{-\lambda t}) + N_0 e^{-\lambda t}$$

101. *(c)* Normality

$$= \frac{\text{Gram equivalent of solute}}{\text{Volume of solution}} \times 1000$$

Gram equivalent of oxalic acid

$$= \frac{\text{Weight}}{\text{Equivalent weight}}$$

$$= \frac{2.52}{\frac{126}{2}}$$

$$\therefore \quad N = \frac{2.52 \times 1000}{\frac{126}{2} \times 100} = 0.4$$

Now, $N_1V_1 = N_1V_2$

$$0.4 \times 10 = N_2 \times 500$$

$$N_2 = 0.08\,\text{N}$$

Now, amount of oxalic acid is calculated as

$$N_2 = \frac{\text{wt.}_{\text{oxalic acid}} \times 1000}{\text{eq. wt.}_{\text{oxalic acid}} \times V_2}$$

$$0.08 = \frac{\text{wt}}{\frac{126}{2}} \times \frac{1000}{500}$$

wt = 3.60 mg/mL

102. *(a)*

m-(hydroxymethyl)phenol ($C_6H_4(OH)CH_2OH$) + HBr ⟶ 3-(bromomethyl)phenol ($C_6H_4(OH)CH_2Br$) + H_2O ...(I)

m-methoxyphenol ($C_6H_4(OH)OCH_3$) + HBr ⟶ benzene-1,3-diol ($C_6H_4(OH)_2$) + H_2O + CH_3Br ...(II)

Reaction (I) takes place *via* S_N1 mechanism OH of CH_2OH group gets substituted by Br because it form stable carbocation and yield 3-(bromo methyl) phenol.

In reaction (II) protonation of ether takes place. The bond between O—CH_3 is weaker than the bond between O—C_6H_5 because the carbon of phenyl group is sp^2-hybridised and there is a partical double bond character.Thus, the reaction yields benzene1, 3-diol.

103. *(a)*

(i) $(H_3C)_2C{=}C(CH_3)_2$ (I) $+ Br_2$ ⟶ $Br-C(CH_3)_2-C(CH_3)_2-Br$ (*Meso* compound)

(ii) $(H_3C)_2C{=}CH_2$ (II) + Br ⟶ $Br-C^*(CH_3)(C_2H_5)-CH_2Br$ (Chiral compound)

Bromination of compound I will give a *meso* product. So, will not have any enantiomer product, but bromination of compound II gives a chiral compound, hence its two form *d* and *l* are possible, which are mirror images of each other and are non-superimposable. It's other enantiomeric form will be

$CH_2Br-C(CH_3)(C_2H_5)-Br$

So, it will have one enantiomeric pair. Thus, correct option is (a).

104. *(d)* For the reaction,

$$A \longrightarrow B$$
$$\Delta H^\circ = 7.5 \text{ kJ mol}^{-1}$$
$$\Delta S^\circ = 25 \text{ kJ mol}^{-1}$$

As the reaction reaches equilibrium,

$\therefore \quad \Delta G^\circ = 0$

Also, $\quad \Delta G^\circ = \Delta H^\circ - T\Delta S^\circ$

$$0 = \Delta H^\circ - T\Delta S^\circ$$
$$\Delta H^\circ = T\Delta S^\circ$$
$$T = \frac{\Delta H^\circ}{\Delta S^\circ} = \frac{7.5\times 1000}{25} = 300 \text{ K}$$

105. *(d)* For the reaction,

$$Mg(OH)_2 \rightleftharpoons Mg^{2+} + 2OH^-$$
$$K_{sp} = [Mg^{2+}][OH^-]^2$$
$$10^{-12} = (0.01)[OH^-]^2$$
$$\frac{10^{-12}}{10^{-2}} = [OH^-]^2$$

$\therefore \quad [OH^-] = 10^{-5}$

$\Rightarrow \quad pOH = +5$

Also, pH + pOH = 14

$$pH = 14 - pOH$$
$$14 - 5 = 9$$

106. *(c)* Given,

atomic radius of metal, $r = 141.4$ pm

For fcc

$$r = \frac{a}{2\sqrt{2}}$$

$\Rightarrow \quad a = 2\sqrt{2}r$

$$= 2\sqrt{2}\times 1414$$
$$= 400 \text{ pm}$$

Volume of unit cell = a^3

$$= (400)^3 = 64\times 10^6$$

or $\quad 6.4\times 10^7 \text{ pm}^3$

107. *(b)* In cyclic silicate, two oxygen atoms of each tetrahedron are shared to form a ring with general formula $[SiO_3]_n^{2n-}$. Thus, in the given figure cyclic silicate is $[Si_6O_{18}]^{12-}$.

108. *(a)* Given,

$H_2O(l) \longrightarrow H_2O(g)$; $\Delta H_1^\circ = 44$ kJ ...(i)

$2B(s) + \frac{3}{2}O_2(g) \longrightarrow B_2O_3(s)$;

$\Delta H_2^\circ = -1273$ kJ ...(ii)

$B_2H_6(g) + 3O_2(g) \longrightarrow B_2O_3(s) + 3H_2O(g)$;

$\Delta H_3^\circ = -2035$ kJ...(iii)

$H_2(g) + \frac{1}{2}O_2(g) \longrightarrow H_2O(l)$;

$\Delta H_4^\circ = -286$ kJ ...(iv)

$2B(s) + 3H_2(g) \longrightarrow B_2H_6(g)$ (1)

We can obtain the above Eq. (i) by Eq.(ii) + Eq.(iv) × 3 + Eq.(i) × 3 − Eq.(iii)

$\therefore \Delta H_r^\circ = -1273 + (-286)\times 3 + 44\times 3 - (-2035)$

$= 36$ kJ.

109. *(a)* The oxidation state of Fe in $K_3[Fe(CN)_6]$ is +3. Thus, the electronic configuration for Fe^{3+} will be $[Ar]3d^5 4s^0$

(Fe^{3+}) [1|1|1|1|1] → e_g: ↿ ↿ ; t_{2g}: ↿ ↿ ↿

CFSE = $-(0.4\times t_{2g}\ e^-) + 0.6\times e_g e^-$

$= -0.4\times 3 + 0.6\times 2$

$= -1.2 + 1.2 = 0.0\Delta_o$

Magnetic moment, $\mu = \sqrt{n(n+2)}$

In this case, $n = 5$

$$\mu = \sqrt{5(5+2)} = \sqrt{35} \text{ BM}$$

110. *(d)* Given,

Weight of nicotine, $w_2 = 8$ g

Weight of water $w_1 = 92$ g

$$\Delta T_f = 0.925^\circ C$$
$$k_f = 1.85^\circ C \text{ mol}^{-1}$$

The relation between molar mass and freezing point is given by

$$M = \frac{k_f \times w_2 \times 1000}{\Delta T_f \times w_1}$$
$$= \frac{1.82\times 8\times 1000}{0.925\times 92} = 171 \text{g [Closest to 160g]}$$

111. *(b)* A bacterium doubles itself after each division, i.e. if we start from a single bacteria, after 10 generations/cell divisions, it will be 1024 in number.

An intercellular bacterial symbiont has 10% higher growth rate than the host bacteria. Thus, after 10 generations the density of bacteria in host cell will increase by 11 times. If the host cell number is taken in the right side column and bacterial population on the right side, then

1-1

2-2.2

3-3.3
4-4.4
5-5.5
6-6.6
7-7.7
8-8.8
9-9.9
10-11 i.e. 10 folds.

112. *(c)* The given question can be related to human ABO blood group, where alleles I^A and I^B show codominance and allele 'i' is recessive.

Genotype	I^AI^A	I^BI^B	I^AI^B	I^Ai	I^Bi	ii
Phenotype	Blood group -A	Blood group -B	Blood group -AB	Blood group -A	Blood group -B	Blood group -O

Number of genotypes = 6

Number of phenotypes = 4

113. *(d)* In a DNA, adenine is paired with thymine with double bond and guanine is paired with cytosine with triple bond. We know that double bonds are easier to break than triple bonds.

Student A has more A = T than G ≡ C, while student B has more G ≡ C than A = T.

Therefore, the DNA molecule given to student A would denature faster than that given to student B.

114. *(b)* The genetic code is degenerate. Some amino acids are encoded by more than one codon, in as much as there are 64 possible base triplets and only 20 amino acids. Therefore, the reason why inspite of having 60% identical proteins only 45% identical genes are present in bacteria and human is that DNA codons having different nucleotides in the third position can code for the same amino acids.

115. *(b)* The DNA sequence given is 5′ ATGCAAGATATAGCT 3′.

Its complementary DNA strand would be 3′ TACGTTCTATATCGA 5′. But for *m*RNA transcription from DNA only the strand with polarity from 3′→ 5′ acts as template and is referred to as template strand thus the *m*RNA strand will be

5′ AUG CAA GAU AUA GCU 3′
↓ ↓ ↓ ↓ ↓
Methionine Gln Asp Ile Ala

So, there are 5 amino acids formed from this sequence, but if we delete the nucleotide 5, the new *m*RNA would be

5′ AUG CAG UAA UAG CU 3′
↓ ↓ ↓ ↓
Methionine Gln Stop codon Stop codon

Only 2 amino acids will form. Therefore, option (b) is correct.

116. *(d)* The Michaelis-Menten equation for an enzymatic reaction is that

$E + S \Leftrightarrow ES \Leftrightarrow P1 + P2 + E$

(where there are two products formed)

This shows that enzymes are biological catalyst in a reaction which lowers the activation energy of a reaction without being consumed.

117. *(a)* From all the four species of birds given in the question, the species 1 which gives white egg with no markings would most likely to nest in a deep tree hole in order to protect its egg from harsh environment and predators. Since, the white eggs are easily detected by predators while the other three types of eggs have certain colour or marking on them, they would not be easily identified by predators.

118. *(c)* Hardy-Weinberg equation

$p^2 + 2pq + q^2 = 1$

where, $p^2 = AA = 0.25$

$q^2 = aa$

$2pq = 2Aa$

We also know that, $p + q = 1$

$\therefore p = \sqrt{0.25} = 0.5$

$q = 1 - 0.5 = 0.5$

$q = a = 0.5$

119. *(a)* The sugar we consume is sucrose, i.e. glucose + fructose. Bread contains fibre and glucose.

Insulin is a blood sugar lowering hormone. When we consume sugar and bread, the insulin level would shoot up to lower down the glucose level of the blood, present in table sugar (sucrose) as it will cause an abundant increase in blood sugar level. But in case of eating bread/roti, these food are rich in fibre, they are converted into glucose at a much more slower rate leading to a much smaller rise and the fall in blood sugar levels. Therefore, graph (a) is correct.

120. *(a)* In a plant, meristematic activity is seen at the apical region. Thus, the growth of top portion of a tree is more than the bottom of the tree. Therefore, if we have marked two spots at the top and bottom of a tree, the two spots at the top will grow more apart than the two spots at the bottom.

QUESTION PAPER 2009

Stream : SB/SX

MM : 160

Instructions

1. There are 120 questions in this paper.
2. The question paper contains two parts; Part I (1 Mark Questions) and Part II (2 Marks Questions).
3. There are four sections in each part; Mathematics, Physics, Chemistry and Biology.
4. Out of the four options given with each question, only one is correct.

PART-I (1 Mark Questions)

MATHEMATICS

1. Suppose the sequence $a_1, a_2, a_3, \ldots$ is an arithmetic progression of distinct numbers such that the sequence $a_1, a_2, a_4, a_8, \ldots$ is a geometric progression. The common ratio of the geometric progression is

(a) 2 (b) 4
(c) a_1 (d) not determinable

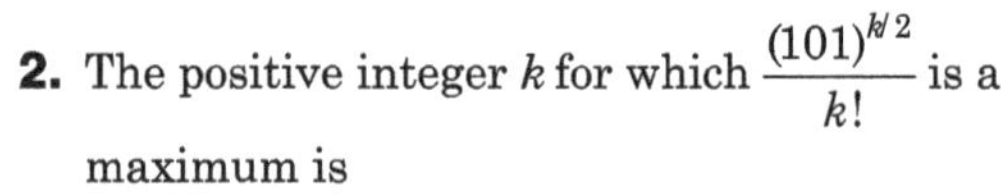

2. The positive integer k for which $\dfrac{(101)^{k/2}}{k!}$ is a maximum is

(a) 9 (b) 10 (c) 11 (d) 101

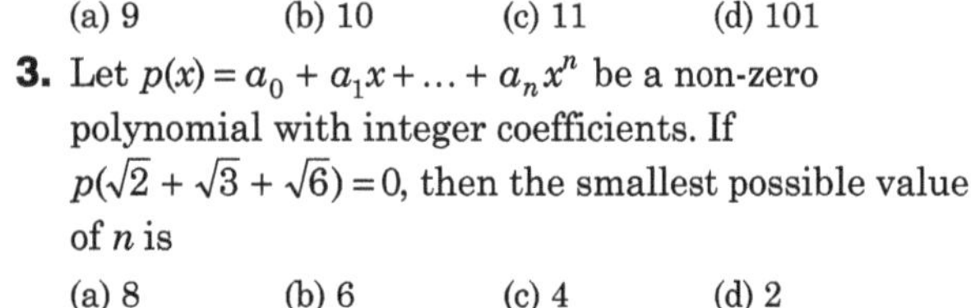

3. Let $p(x) = a_0 + a_1x + \ldots + a_nx^n$ be a non-zero polynomial with integer coefficients. If $p(\sqrt{2} + \sqrt{3} + \sqrt{6}) = 0$, then the smallest possible value of n is

(a) 8 (b) 6 (c) 4 (d) 2

4. Three players play a total of 9 games. In each game, one person wins and the other two lose; the winner gets 2 points and the losers get -1 each. The number of ways in which they can play all the 9 games and finish each with a zero score is

(a) 84 (b) 1680 (c) 7056 (d) 0

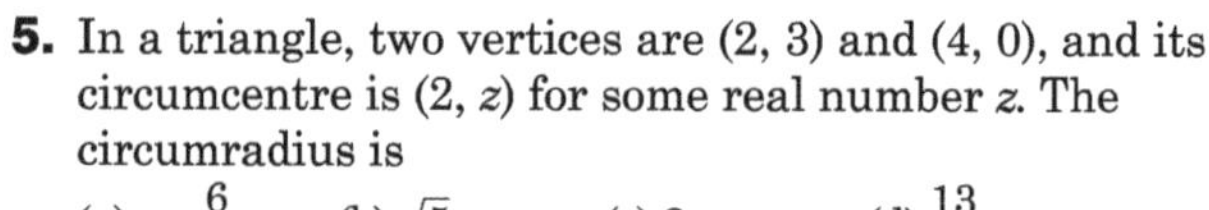

5. In a triangle, two vertices are (2, 3) and (4, 0), and its circumcentre is (2, z) for some real number z. The circumradius is

(a) $\dfrac{6}{2+\sqrt{13}}$ (b) $\sqrt{5}$ (c) 2 (d) $\dfrac{13}{6}$

6. Consider an ellipse with foci at (5, 15) and (21, 15). If the X-axis is a tangent to the ellipse, then the length of its major axis equals

(a) 17 (b) 34
(c) 13 (d) $\sqrt{416}$

7. Let the line $2x + 3y = 18$ intersect the Y-axis at B. Suppose C ($\neq B$), with coordinates (a, b), is a point on the line such that $PB = PC$, where $P = (10, 10)$. Then, $8a + 2b$ equals

(a) 60 (b) 62
(c) 66 (d) 78

8. If $\operatorname{cosec}^2(\alpha + \beta) - \sin^2(\beta - \alpha) + \sin^2(2\alpha - \beta) = \cos^2(\alpha - \beta)$, where $\alpha, \beta \in \left(0, \dfrac{\pi}{2}\right)$, then $\sin(\alpha - \beta)$ is equal to

(a) $-\dfrac{1}{2}$ (b) $\dfrac{1}{2}$
(c) $\dfrac{-\sqrt{3}}{2}$ (d) $\dfrac{\sqrt{3}}{2}$

9. If $\sin x + \sin y = \frac{7}{5}$ and $\cos x + \cos y = \frac{1}{5}$, then $\sin(x+y)$ equals

(a) $\frac{7}{25}$ (b) $\frac{24}{25}$

(c) $\frac{-7}{25}$ (d) $\frac{-24}{25}$

10. The number of solutions to $\sin x = \frac{6}{x}$ with $0 \le x \le 12\pi$ is

(a) 1 (b) 6

(c) 10 (d) 12

11. Define a function $f : R \to R$ by

$$f(x) = \begin{cases} \frac{\sin x^2}{x}, & \text{for } x < 0 \\ x^2 + ax + b, & \text{for } x \ge 0 \end{cases}$$

Suppose $f(x)$ is differentiable of R. Then,

(a) $a = 0, b = 0$ (b) $a = 1, b = 0$

(c) $a = 0, b = 1$ (d) $a = 1, b = 1$

12. The shortest distance from (0, 3) to the parabola $y^2 = 4x$ is

(a) 2 (b) $\sqrt{2}$

(c) 5 (d) $\sqrt{5}$

13. Ten trucks, numbered 1 to 10, are carrying packets of sugar. Each packet weights either 999 g or 1000 g and each truck carries only the packets equal weights. The combined weight of 1 packet selected from the first truck, 2 packets from the second, 4 packets from the third, and so on, and 2^9 packet from the tenth truck is 1022870 g. The trucks that have the lighter bags are

(a) 1, 3, 5 (b) 2, 4, 5

(c) 1, 9 (d) 2, 8

14. What is the value of $\int_0^1 \cos(\pi x)\cos([2x]\pi)dx$?

(Here $[t]$ denotes the integral part of the real number t.)

(a) 1 (b) −1

(c) $\frac{2}{\pi}$ (d) $\frac{-2}{\pi}$

15. The value of the limit $\lim_{n\to\infty} \int_0^1 x^{10} \sin(nx)dx$ equals

(a) 0 (b) $\frac{1}{10!}$ (c) $\frac{\pi}{2}$ (d) 1

16. The area bounded by the parabolas $y = x^2$ and $y = 1 - x^2$ equals

(a) $\frac{\sqrt{2}}{3}$ (b) $\frac{2\sqrt{2}}{3}$ (c) $\frac{1}{3}$ (d) $\frac{2}{3}$

17. A vector which bisects the angle between $\mathbf{a} = 3\mathbf{i} - 4\mathbf{k}$ and $\mathbf{b} = 5\mathbf{j} + 12\mathbf{k}$ is

(a) $39\mathbf{i} - 25\mathbf{j} + 8\mathbf{k}$ (b) $39\mathbf{i} + 25\mathbf{j} + 8\mathbf{k}$

(c) $3\mathbf{i} - 5\mathbf{j} + \frac{8}{5}\mathbf{k}$ (d) $3\mathbf{i} + 5\mathbf{j} + \frac{8}{5}\mathbf{k}$

18. An envelope has space for at most 3 stamps. If you are given three stamps of denomination 1, and three stamps of denomination a, $(a > 1)$, the least positive integer for which there is no stamp value is

(a) 7 (b) 8 (c) 9 (d) 10

19. If m, n are positive integers such that $m < n$ and

$\sum_{d|m} d = \sum_{d|n} d$ (here $d|k$ means d is a positive divisor of k), then

(a) $\sum_{d|m} \frac{1}{d} < \sum_{d|n} \frac{1}{d}$

(b) $\sum_{d|m} \frac{1}{d} = \sum_{d|n} \frac{1}{d}$

(c) $\sum_{d|m} \frac{1}{d} > \sum_{d|n} \frac{1}{d}$

(d) no relationship can be determined

20. The number of relations R from an m-element set A to an n-element set B satisfying the condition

$(a, b_1) \in R, (a, b_2) \in R \Rightarrow b_1 = b_2$

for $a \in A$, $b_1, b_2 \in B$ is

(a) n^m (b) $2^{m+n} - 2^m - 2^n$

(c) mn (d) $(n+1)^m$

PHYSICS

21. The relation $C_p - C_V = R$ (C_p and C_V are the molar specific heats at constant pressure and volume) is exactly true for

(a) an ideal monoatomic gas

(b) any ideal gas, whether monoatomic, diatomic or polyatomic

(c) any real gas above its critical temperature

(d) All real gases

22. The molecules of air in the room that you are sitting are all experiencing the force of gravity tending to bring them down. The molecules are also frequently and randomly undergoing collisions, which tend to oppose the effect of fall under gravity. The density of air is nearly uniform throughout the room because

(a) the mass of the molecules is very small

(b) the gravitational potential energy mgh is much lesser than the average thermal energy kT

(c) the gravitational potential energy mgh is much greater than the average thermal energy kT

(d) mgh is nearly of the same magnitude as kT, which results in the cancellation of the two opposing factors

23. A parallel plate capacitor is charged fully by using a battery. Then, without disconnecting the battery, the plates are moved further apart. Then,

(a) the charge on the capacitor increases

(b) the voltage difference between the plates decreases

(c) the capacitance increases

(d) the electrostatic energy stored in the capacitor decreases

24. The five sides of a regular pentagon are represented by vectors $\mathbf{A}_1, \mathbf{A}_2, \mathbf{A}_3, \mathbf{A}_4$ and $\mathbf{A}_5$, in cyclic order as shown below.

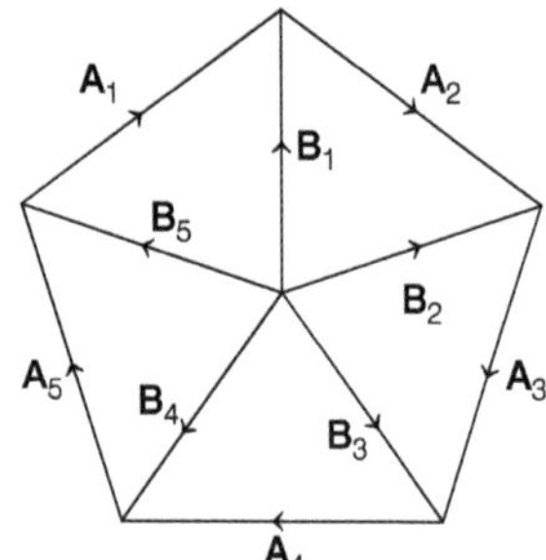

Corresponding vertices are represented by $\mathbf{B}_1, \mathbf{B}_2, \mathbf{B}_3, \mathbf{B}_4$ and $\mathbf{B}_5$, drawn from the centre of the pentagon.

Then, $\mathbf{B}_2 + \mathbf{B}_3 + \mathbf{B}_4 + \mathbf{B}_5$ is equal to

(a) $\mathbf{A}_1$ (b) $-\mathbf{A}_1$ (c) $\mathbf{B}_1$ (d) $-\mathbf{B}_1$

25. Four metallic plates each of surface area (of one side) A, are placed at a distance d apart from each other. The two outer plates are connected to a point P and the two inner plates are connected to another point Q as shown in figure below.

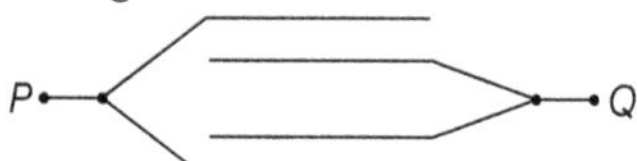

Then, the capacitance of the system is

(a) $\varepsilon_0 \frac{A}{2d}$ (b) $\varepsilon_0 \frac{A}{d}$ (c) $2\varepsilon_0 \frac{A}{d}$ (d) $3\varepsilon_0 \frac{A}{d}$

26. A progressive wave travelling in positive x-direction given by $y = a\cos(kx - \omega t)$ meets a denser surface at $x = 0, t = 0$. The reflected wave is then given by

(a) $y = -a\sin(kx - \omega t)$ (b) $y = a\sin(\omega t - kx)$
(c) $y = -a\cos(kx + \omega t)$ (d) $y = a\cos(kx - \omega t)$

27. A charge Q is spread non-uniformly on the surface of a hollow sphere of radius R, such that the charge density is given by $\sigma = \sigma_0 (1 - \sin\theta)$, where θ is the usual polar angle. The potential at the centre of the sphere is

(a) $\frac{Q}{2\pi\varepsilon_0 R}$ (b) $\frac{Q}{\pi\varepsilon_0 R}$ (c) $\frac{Q}{8\pi\varepsilon_0 R}$ (d) $\frac{Q}{4\pi\varepsilon_0 R}$

28. An ideal diatomic gas is heated at constant pressure. The ratio of the work done to the heat supplied is

(a) $\frac{3}{5}$ (b) $\frac{2}{5}$ (c) $\frac{2}{7}$ (d) $\frac{4}{7}$

29. In the hydrogen spectrum, the ratio of the wavelengths for Lyman-alpha radiation to Balmer-alpha radiation is

(a) 5/27 (b) 5/48 (c) 27/5 (d) 1/3

30. Two identical conducting spheres carry identical charges. If the spheres are set at a certain distance apart, they repel each other with a force F. A third conducting sphere identical to the other two, but initially uncharged is touched to one sphere and then to the other before being removed. The force between the original two spheres is now

(a) $\frac{F}{2}$ (b) $\frac{F}{4}$ (c) $\frac{3F}{4}$ (d) $\frac{3F}{8}$

31. A small rectangular loop of wire in the plane of the paper is moved with uniform speed across a limited region of uniform magnetic field perpendicular to the plane of the paper as shown below.

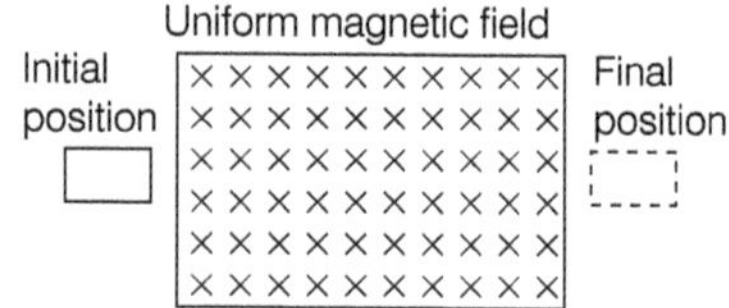

Which graph would best represent the variation of the electric current I in the wire with time t?

(a) 
(b) 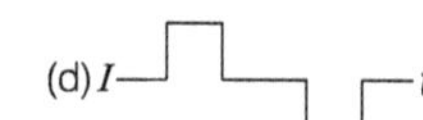
(c) 
(d)

32. The moment of inertia of a solid disc made of thin metal of radius R and mass M about one of its diameters is given by $\frac{MR^2}{4}$. What will be the moment of inertia about this axis, if the disc is folded in half about this diameter?

(a) $\frac{MR^2}{8}$ (b) $\frac{MR^2}{2}$ (c) $\frac{MR^2}{4}$ (d) MR^2

33. A plane electromagnetic wave propagating in the direction of the unit vector $\hat{\mathbf{n}}$ with a speed c is described by electric and magnetic field vectors $\mathbf{E}$ and $\mathbf{B}$, respectively. Which of the following relations (in SI units) between $\mathbf{E}$ and $\mathbf{B}$ can be ruled out on dimensional grounds alone?

(a) $\mathbf{E} = \frac{\hat{\mathbf{n}} \times \mathbf{B}}{c}$ (b) $\mathbf{E} = -c\hat{\mathbf{n}} \times \mathbf{B}$
(c) $\mathbf{B} = \frac{\hat{\mathbf{n}} \times \mathbf{E}}{c}$ (d) $\hat{\mathbf{n}} \times \mathbf{E} \times \mathbf{B} = 0$

34. A point electric dipole placed at the origin has a potential given by $V(r, \theta) = \frac{p\cos\theta}{4\pi\varepsilon_0 r^2}$, where θ is the angle made by the position vector with the direction of the dipole. Then,

(a) since the potential vanishes at $\theta = \frac{\pi}{2}$, the electric field is zero everywhere on the $\theta = \frac{\pi}{2}$ plane
(b) the electric field everywhere on the $\theta = \frac{\pi}{2}$, plane is normal to the plane
(c) the electric field everywhere on the $\theta = \frac{\pi}{2}$, plane is along the plane
(d) the electric field vanishes on the $\theta = 0$ line

35. A uniform non-deformable cylinder of mass m and radius R is rolling without slipping on a horizontal rough surface. The force of friction is

(a) $\mu\ mg$, where μ is the coefficient of sliding friction
(b) zero
(c) increases with time
(d) decreases with time

36. Consider a one-dimensional potential $V(x)$ as shown in the figure below.

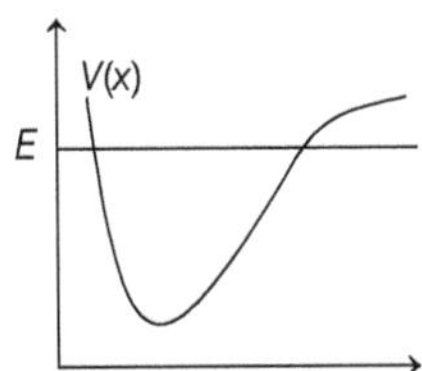

A classical particle of mass m moves under its influence and has total energy E as shown below. The motion is

(a) non-periodic
(b) stationary
(c) periodic but not a simple harmonic
(d) simple harmonic

37. A source of frequency f is emitting sound waves. If temperature of the medium increases, then

(a) wavelength of the sound wave increases
(b) speed of the sound wave decreases
(c) wavelength of the sound wave decreases
(d) amplitude of the sound wave increases

38. A block of mass m is stationary on a rough plane of mass M inclined at an angle θ to the horizontal, while the whole set up is accelerating upwards at an acceleration a. If the coefficient of friction between the block and the plane is μ, then the force that the plane exerts on the block is

(a) $m(g + a)$ upwards
(b) $mg \cos\theta$ normal to the plane
(c) resultant of $mg \cos\theta$ normal to the plane and $\mu mg \cos\theta$ along the plane
(d) resultant of $m(g + a) \cos\theta$ normal to the plane and $\mu mg \cos\theta$ along the plane

39. A stream of charged particles enter into a region with crossed electric and magnetic fields as shown in the figure below. On the other side is a screen with a hole that is right on the original path of the particles.

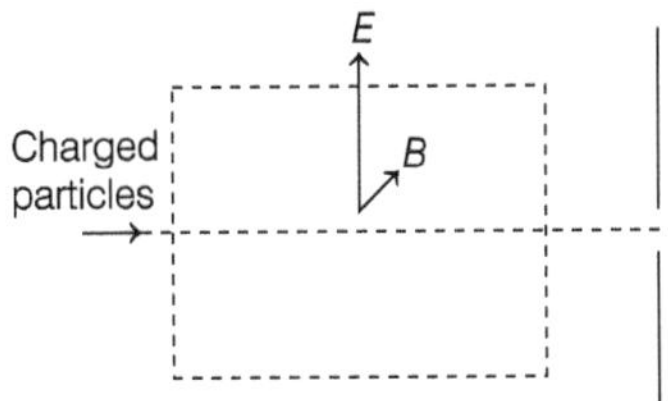

Then,

(a) no particle can get through the hole
(b) all particles can get through the hole
(c) only positively charged particles with speed $\frac{E}{B}$ can get through the hole
(d) all particles with speed $\frac{E}{B}$ can get through the hole

40. A small body is released from a height H of an inclined plane. At the bottom of the plane is a loop of radius R as shown in the figure below.

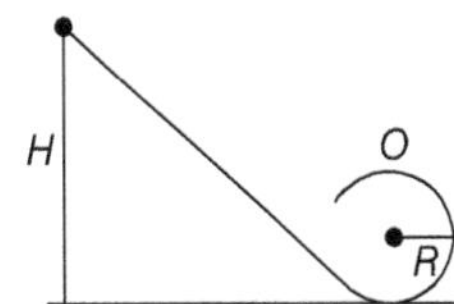

Ignoring friction, the minimum H required for the body to just complete the loop (that is, reach the point O) is

(a) $2R$ (b) $\frac{5R}{2}$ (c) $3R$ (d) $\frac{7R}{2}$

CHEMISTRY

41. The gas that has the slowest rate of diffusion among O_2, H_2, CO_2 and CH_4 is

(a) O_2 (b) H_2 (c) CO_2 (d) CH_4

42. Assuming ideal behaviour, the ratio of kinetic energies of 3 g of H_2 and 4 g of O_2 at any temperature is

(a) 3 : 4 (b) 1 : 16
(c) 4 : 3 (d) 12 : 1

43. The IUPAC name for the compound

O
Me Cl
Me

is

(a) 1-chloro-3-methyl-4-pentanone
(b) 1-chloro-2-methyl-4-pentanone
(c) 5-chloro-3-methyl-2-pentanone
(d) 5-chloro-2-methyl-3-pentanone

44. The shape of the molecule ClF_3 is

(a) triangular (b) pyramidal
(c) T-shape (d) linear

45. Among CO_3^{2}, OH^-, NH_3 and HCO_3^-, the species that acts as a Bronsted acid as well as a Bronsted base, is

(a) Na_2CO_3 (b) OH (c) NH_3 (d) HCO_3^-

46. The ratio of the heat capacities C_p/C_V for one mole of a gas is 1.67. The gas is

(a) He (b) H_2 (c) CO_2 (d) CH_4

47. The ion that is isoelectronic with CO is

(a) O_2^+ (b) O_2^- (c) CN^- (d) N_2

48. Among CH_4, CO_2, H_2O and SO_2 the bond angle is the highest in

(a) CH_4 (b) CO_2 (c) H_2O (d) SO_2

49. The solvent of choice for carrying out a Grignard reaction is
(a) diethyl ether (b) chloroform
(c) ethyl acetate (d) ethanol

50. The reaction of butanal with n-propylmagnesium bromide gives a
(a) chiral secondary alcohol
(b) achiral secondary alcohol
(c) chiral tertiary alcohol
(d) achiral tertiary alcohol

51. The hybridisation of Ni centre in $Ni[(PPh_3)_2Cl_2]$ and $[NiCl_4]^2$, respectively are
(a) dsp^2 and sp^3 (b) dsp^2 and sp^2d
(c) sp^3 and sp^3 (d) sp^3 and dsp^2

52. Oxalic acid when treated with potassium permanganate in the presence of an acid, produces
(a) O_2 (b) C
(c) CO (d) CO_2

53. The equilibrium constant for the reaction,

$$N_2 + 3H_2 \rightleftharpoons 2NH_3$$

at 400 K is 41. The equilibrium constant for the reaction,

$$\frac{1}{2}N_2 + \frac{3}{2}H_2 \rightleftharpoons NH_3$$

at the same temperature will be closest to
(a) 41 (b) 20.5
(c) 6.4 (d) 1681

54. In a one component second order reaction, if the concentration of the reactant is reduced to half, the rate
(a) increases two times (b) increases four times
(c) decreases to one half (d) decreases to one fourth

55. The conjugate bases for HCO_3^- and NH_3, respectively, are
(a) H_2CO_3 and NH_4^+ (b) CO_3^{2-} and NH_2^-
(c) H_2CO_3 and NH_2^- (d) CO_3^{2-} and NH_4^+

56. Among the following

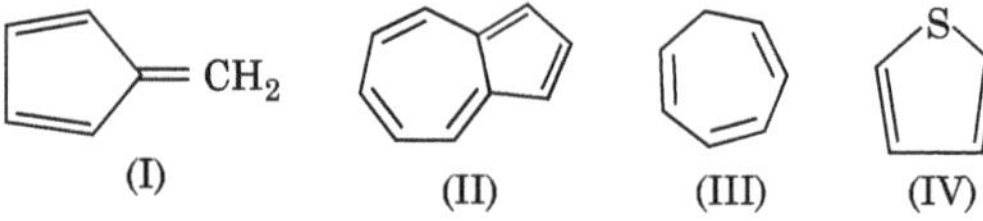

The aromatic compounds are
(a) I and II (b) I and III
(c) II and III (d) II and IV

57. Among the compounds

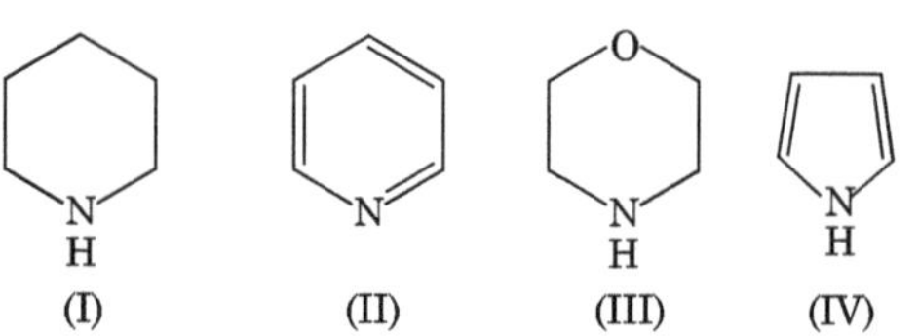

The order of basicity is
(a) I > III > II > IV (b) III > I > IV > II
(c) II > IV > I > III (d) II > I > III > IV

58. The Newmann projection of

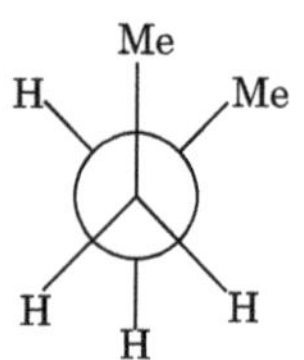

is known as the
(a) eclipsed conformer (b) staggered conformer
(c) skewed conformer (d) gauche conformer

59. The half-life of a first order reaction is 30 min. The time required for 75% completion of the same reaction will be
(a) 45 min (b) 60 min
(c) 75 min (d) 90 min

60. The hydrogen ion concentration in a mixture of 10 mL of 0.1 M H_2SO_4 and 10 mL of 0.1 M KOH solution in water, is
(a) 0.1 M (b) 0.05 M (c) 0.2 M (d) 0.02 M

BIOLOGY

61. During photosynthesis, the chemical conversion of water is termed as
(a) photolysis (b) hydrolysis
(c) hydration (d) condensation

62. In the organism mussel, oxygen is carried by
(a) albumin (b) myosin
(c) myoglobin (d) haemoglobin

63. Enzymes do the following
(a) Make products and reactants of equal energy
(b) Help the chemical processes by lowering the energy of products
(c) Reduce the activation barrier and speed up chemical processes
(d) Hydrolyse all the biopolymers indiscriminately

64. Glycolysis is
(a) biosynthesis of glucose
(b) biosynthesis of glycine
(c) degradation of glucose
(d) reaction of glucose with proteins

65. Plants are attracted to light through the hormonal action of
(a) gibberellic acid (b) auxin
(c) chlorophyll (d) thiamine

66. During development, unspecified cells become cells having unique functions. This process is called
(a) evolution (b) differentiation
(c) translation (d) replication

67. The chromosomal attachment site of the spindle microtubule is
(a) centrosome (b) liposome
(c) centromere (d) telomere

68. Which of the following diseases is not sexually transmitted?
(a) Syphilis (b) AIDS
(c) Gonorrhoea (d) Tuberculosis

69. This cell organelle consists of two granule-like centrioles and is found in animal cells only. It helps in cell division. What is it called?
(a) Centrosome (b) Chromosome
(c) Centromere (d) Chromatids

70. Nucleotides are monomers of DNA. What does each nucleotide consist of ?
(a) A nitrogenous base and a pentose sugar
(b) A nitrogenous base and a phasphate group
(c) A pentose sugar and a phosphate group
(d) A nitrogenous base, a pentose sugar and a phosphate group

71. Fertilisation in humans usually takes place in
(a) uterus (b) Graafian follicle
(c) ovary (d) Fallopian tube

72. ELISA, the standard screening test for HIV, detects which of the following?
(a) HIV DNA (b) HIV RNA
(c) HIV proteins (d) Antibodies to HIV proteins

73. Sickle-cell anaemia is caused by
(a) complete absence of the haemoglobin gene
(b) point mutation of the haemoglobin gene
(c) increased affinity of haemoglobin for oxygen
(d) truncation of the haemoglobin protein

74. The natural source of Ti-plasmid is
(a) bacteria (b) virus
(c) plants (d) animals

75. Earthworms are bisexual but still cross-fertilisation is common. This is because
(a) spermatozoa of different earthworms are different
(b) spermatozoa and ova mature at different times in the same earthworm
(c) ova from other earthworms may be larger
(d) sperm and ova from the same earthworm cannot fertilise

76. One difference between blood and lymph is that
(a) blood contains WBC and lymph contains RBC
(b) blood contains RBC and WBC and lymph contains only WBC
(c) blood contains RBC and lymph contains WBC
(d) blood is liquid while lymph is solid

77. The abnormal development of which of the following lymphoid organs would result in the most severe immunodeficiency?
(a) Spleen (b) Tonsil
(c) Thymus (d) Lymph node

78. Mitochondria are associated with all of the following functions, except
(a) ATP synthesis
(b) DNA synthesis
(c) protein synthesis
(d) protein glycosylation

79. The probability of having a girl child with blood group O when the parents have blood group A and B is
(a) 0%
(b) at least 50%
(c) at most 25%
(d) exactly 75%

80. Wooden doors and windows swell up in the rainy season by
(a) a special type of diffusion called imbibition
(b) evaporation of stored water in wood
(c) conduction of water from walls
(d) transpiration

PART-II (2 Marks Questions)

MATHEMATICS

81. Let $p(x) = a_0 + a_1x + \dots + a_nx^n$. If $p(-2) = -15$, $p(-1) = 1$, $p(0) = 7$, $p(1) = 9$, $p(2) = 13$ and $p(3) = 25$, then the smallest possible value of n is
(a) 5 (b) 4
(c) 3 (d) 2

82. Let a, b, c be the sides of a triangle. If t denotes the expression $\frac{(a^2 + b^2 + c^2)}{(ab + bc + ca)}$, the set of all possible values of t is
(a) $\{x \in R | x > 1\}$ (b) $\{x \in R | 1 < x < 2\}$
(c) $\{x \in R | 1 \le x < 2\}$ (d) $\{x \in R | 1 \le x \le 2\}$

83. The area of the region bounded by $y = ||x-3|-4|-5$ and the X-axis is
(a) 24.5 (b) 37
(c) 49 (d) $35\sqrt{2}$

84. The lengths of the sides and the diagonal of an isosceles trapezium form a two-element set $\{a, b\}$. If $a > b$, then a / b equals
(a) $\frac{1}{2}(\sqrt{6} + \sqrt{2})$
(b) $\frac{1}{2}(\sqrt{5} + 1)$
(c) $\sqrt{3}$
(d) $\sqrt{2}$

85. Define a sequence $\{a_n\}_{n \geq 0}$ by

$$a_n = \sqrt{\frac{1 + a_{n-1}}{2}} \text{ for } n \geq 1, a_0 = \cos\theta \neq \pm 1$$

Then, $\lim_{n \to \infty} 4^n (1 - a_n)$ equals

(a) θ^2 (b) $\frac{\theta^2}{2}$ (c) $\frac{\theta}{2}$ (d) θ

86. The range of the function $f(x) = (\sin x)^{\sin x}$ defined on $(0, \pi)$ is

(a) $(0, 1)$ (b) $(e^{-1/e}, 1)$
(c) $[e^{-1/e}, 1)$ (d) $[e^{-1/e}, 1]$

87. Let A denote the area bounded by the curve $y = \frac{1}{x}$ and the lines $y = 0$, $x = 1$, $x = 10$, let $B = 1 + \frac{1}{2} + \ldots + \frac{1}{9}$, and let $C = \frac{1}{2} + \frac{1}{3} + \ldots + \frac{1}{10}$. Then,

(a) $C < B < A$
(b) $A < C < B$
(c) $C < A < B$ and $A - C < B - A$
(d) $C < A < B$ and $B - A < A - C$

88. Two points are randomly chosen on the circumference of a circle of radius r. The probability that the distance between the two points is at least r is equal to

(a) $\frac{2}{\pi}$ (b) $\sin r$ (c) $\frac{1}{2}$ (d) $\frac{2}{3}$

89. Consider all natural number whose decimal expansion has only the even digits 0, 2, 4, 6, 8. Suppose these are arranged in increasing order. If a_n denotes the nth number in this sequence, then $\frac{\lim_{n \to \infty} \log a_n}{\log n}$ equals

(a) 0 (b) $\log_5 10$
(c) $\log_2 10$ (d) 2

90. The sum of all absolute values of the difference of the numbers 1, 2, 3, ..., n, taken two at a time, i.e. $\sum_{1 \leq i < j \leq n} |i - j|$ equals

(a) $\binom{n-1}{3}$ (b) $\binom{n}{3}$ (c) $\binom{n+1}{3}$ (d) $\binom{n+2}{3}$

PHYSICS

91. A spherical cavity of radius r is curved out of a uniform solid sphere of radius R as shown in the figure below.

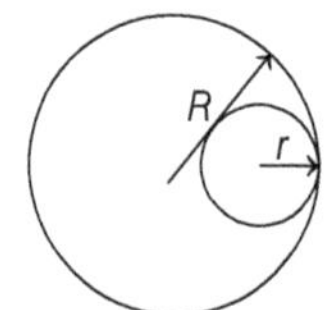

The distance of the centre of mass of the resulting body from that of the solid sphere is given by

(a) $\frac{R - r}{2}$ (b) $\frac{R + r}{2}$
(c) 0 (d) $\frac{r^3}{R^2 + Rr + r^2}$

92. A plano-convex lens made of material of refractive index μ with radius of curvature R is silvered on the curved side. How far away from the lens-mirror must you place a point object, so that the image coincides with the object?

(a) $\frac{R}{\mu}$ (b) R (c) $\frac{R}{\mu - 1}$ (d) μR

93. n moles of a van der Waals' gas obeying the equation of state $\left(p + \frac{n^2 a}{V^2}\right)(V - nb) = nRT$, where a and b are gas dependent constants, is made to undergo a cyclic process that is depicted by a rectangle in the p-V diagram as shown below. What is the heat absorbed by the gas in one cycle?

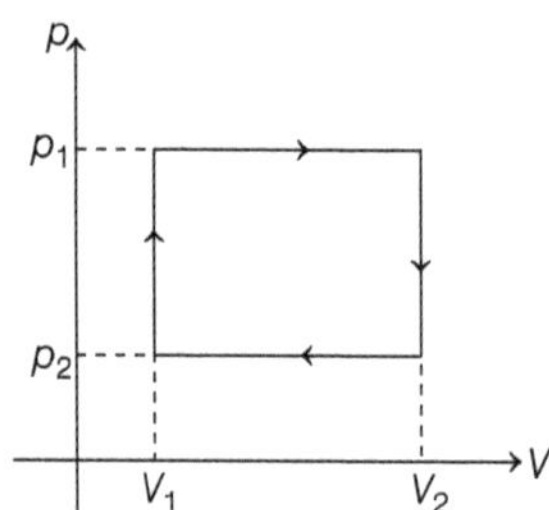

(a) $n(p_1 - p_2)(V_2 - V_1)$
(b) $(p_1 - p_2)(V_2 - V_1)$
(c) $\left(p_1 + \frac{n^2 a}{V_1^2} - p_2 - \frac{n^2 a}{V_2^2}\right)(V_1 - V_2)$
(d) $\left(p_1 + \frac{n^2 a}{V_1^2} - p_2 - \frac{n^2 a}{V_2^2}\right)(V_2 - V_1)$

94. For what value of the resistor X will the equivalent resistance of the two circuits shown be the same?

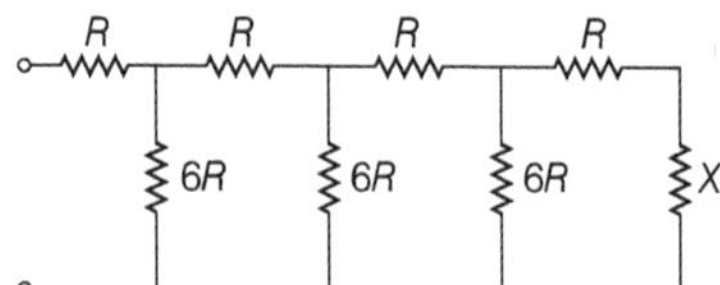

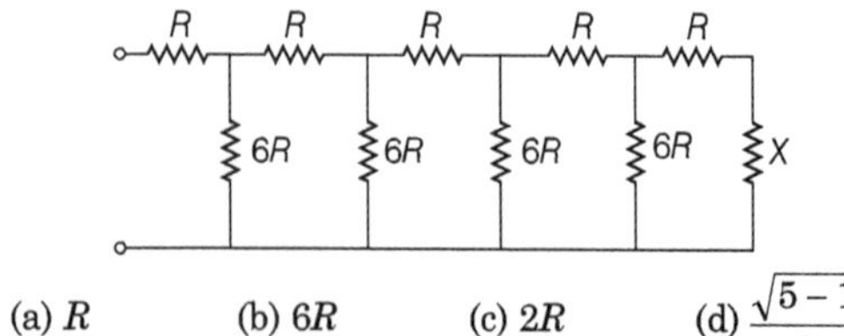

(a) R (b) $6R$ (c) $2R$ (d) $\frac{\sqrt{5-1}}{2} R$

95. A solid uniform sphere having a mass M, radius R and moment of inertia of $\frac{2}{5}MR^2$ rolls down a plane inclined at an angle θ to the horizontal starting from rest. The coefficient of static friction between the sphere and the plane is μ_s. Then,

(a) the sphere will always roll without slipping
(b) the sphere will always slide
(c) the sphere will roll without slipping only, if $\theta \le \sin\frac{7\mu_s}{2}$
(d) the sphere will roll without slipping only, if $\theta \le \tan^{-1}\frac{7\mu_s}{2}$

96. A cubical box of side a sitting on a rough table-top is pushed horizontally with a gradually increasing force until the box moves. If the force is applied at a height from the table top which is greater than a critical height H, the box topples first. If it is applied at a height less than H, the box starts sliding first. Then, the coefficient of friction between the box and the table top is

(a) $\frac{a}{2H}$ (b) $\frac{2H}{a}$ (c) $\frac{a}{H}$ (d) $\frac{H}{a}$

97. A vehicle is moving with speed v on a curved road of radius r. The coefficient of friction between the vehicle and the road is μ. The angle θ of banking needed is given by

(a) $\tan\theta = \frac{v^2 - \mu rg}{v^2 - rg}$ (b) $\tan\theta = \frac{v^2 - \mu rg}{v^2 + \mu rg}$
(c) $\tan\theta = \frac{v^2 - \mu rg}{rg + \mu v^2}$ (d) $\tan\theta = \frac{\mu rg - v^2}{rg + \mu v^2}$

98. Two small identical speakers are connected in phase to the same source. The speakers are 3 m apart and at ear level. An observer stands at P, 4 m in front of one speaker as shown alongside.

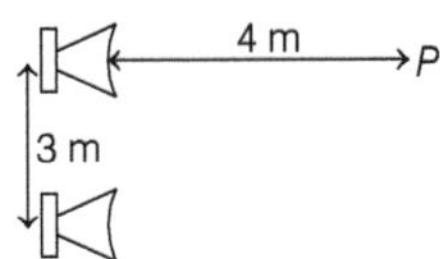

The sound she hears is least intense when the wavelength is λ_1 and most intense when the wavelength is λ_2. Then, the possible values of λ_1 and λ_2 are

(a) $\lambda_1 = 1$ m and $\lambda_2 = 2$ m
(b) $\lambda_1 = 4$ m and $\lambda_2 = 3$ m
(c) $\lambda_1 = 2$ m and $\lambda_2 = 1$ m
(d) $\lambda_1 = 0.5$ m and $\lambda_2 = 0.25$ m

99. Two small blocks slide without losing contact with the surface along two frictionless tracks 1 and 2, starting at the same time with same initial speed v. Track 1 is perfectly horizontal, while track 2 has a dip in the middle, as shown in the figure.

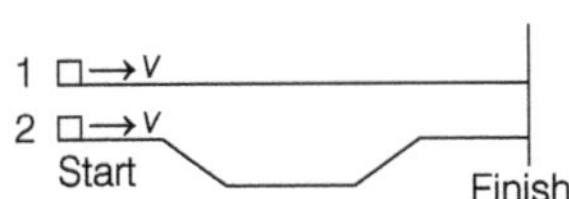

Which block reaches the finish line first?
[**Hint** : Use velocity-time graph to solve]

(a) Block on track 1 reaches the finish line first
(b) Block on track 2 reaches the finish line first
(c) Both blocks reach the finish line at the same time
(d) It depends on the length of the dip in the second track, relative to the total length of the tracks

100. Consider 1 kg of liquid water undergoing change in phase to water vapour at 100°C. At 100°C, the vapour pressure is 1.01×10^5 N-m^2 and the latent heat of vaporization is 22.6×10^5 Jkg^{-1}. The density of liquid water is 10^3 kg m^{-3} and that of vapour is $\frac{1}{1.8}$ kg m^{-3}. The change in internal energy in this phase change is nearly

(a) 1.8×10^5 J kg^{-1} (b) 20.8×10^5 J kg^{-1}
(c) 22.6×10^5 J kg^{-1} (d) 11.3×10^5 J kg^{-1}

CHEMISTRY

101. If the pH of a mixture of 10 mL of 0.1 M NH_4OH and 10 mL of 1 M NH_4Cl solution is 8, the pK_b value of NH_4OH is then closest to

(a) 3 (b) 5 (c) 7 (d) 9

102. A cylinder of cooking gas in a household contains 11.6 kg of butane. The thermochemical reaction for the combustion of butane is,

$$2C_4H_{10}(g) + 13O_2(g) \longrightarrow 8CO_2(g) + 10H_2O(l), \quad \Delta H = -2658 \text{ kJ/mol}$$

If the household needs 15000 kJ of energy per day, the cooking gas cylinder will last for about

(a) 64 days (b) 45 days
(c) 20 days (d) 35 days

103. The addition of 0.643 g of a compound to 50 mL of benzene (density = 0.879 g mL^{-1}) lowers the freezing point from 5.51° C to 5.03°C. If the freezing point constant, K_f for benzene is 5.12 K kg mol^{-1}, the molar mass of the compound is approximately

(a) 156 g mol^{-1} (b) 88 g mol^{-1}
(c) 60 g mol^{-1} (d) 312 g mol^{-1}

104. Consider the following electrochemical cell,

$$Zn(s) + 2Ag^+(0.04M) \longrightarrow Zn^{2+}(0.28M) + 2Ag(s)$$

If $E^\circ_{cell} = 2.57$ V, then emf of the cell at 298 K, is

(a) 2.5 V (b) 1.5 V
(c) 0.5 V (d) −0.5 V

105. When Co (II) chloride is dissolved in concentrated HCl a blue solution is obtained. Upon dilution with water, the colour changes to pink because

(a) $[CoCl_6]^{4-}$ is converted to $[CoCl_6]^{3-}$
(b) $[CoCl_4]^{2-}$ is converted to $[Co(OH_2)_6]^{2+}$
(c) $[Co(OH_2)_6]^{2+}$ is converted to $[Co(OH_2)_6]^{3+}$
(d) $[CoCl_4]^{2-}$ is converted to $[Co(OH_2)_6]^{3+}$

106. The rate constant for the reaction, $COCl_2(g) \longrightarrow CO(g) + Cl_2(g)$ is given by $\ln [k/(\text{min}^{-1})] = -11067/T\text{ K} + 31.33$. The temperature at which the rate of this reaction will be doubled from that at 25°C is

(a) 75°C (b) 100°C (c) 31°C (d) 50°C

107. For the reactions and their equilibrium constants given below,

$CuCl_4^{2-} + Br^- \rightleftharpoons CuCl_3Br^{2-} + Cl^-; K_1$

$CuCl_3Br^{2-} + Br^- \rightleftharpoons CuCl_2Br_2^{2-} + Cl^-; K_2$

$CuCl_2Br_2^{2-} + Br^- \rightleftharpoons CuClBr_3^{2-} + Cl^-; K_3$

$CuClBr_3^{2-} + Br^- \rightleftharpoons CuBr_4^{2-} + Cl^-; K_4$

The equilibrium constant, K for the reaction $CuCl_4^{2-} + 3Br^- \rightleftharpoons CuClBr_3^{2-} + 3Cl^-$, is

(a) $K_1K_2K_3$ (b) $K_1K_2K_3K_4$ (c) $K_1 + K_2 + K_3$ (d) $\frac{1}{(K_1K_2K_3)}$

108.

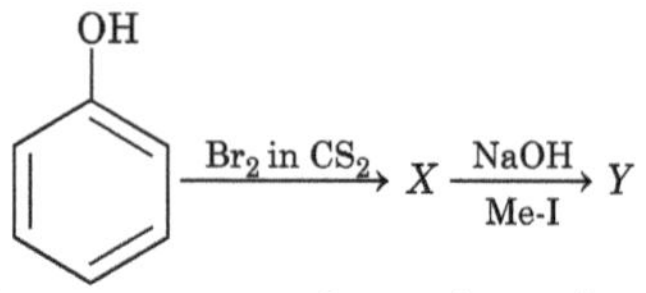

Phenol (OH on benzene) $\xrightarrow{Br_2 \text{ in } CS_2}$ X $\xrightarrow[\text{Me-I}]{\text{NaOH}}$ Y

In the above sequence of reactions, the major products X and Y are

(a) X = 2-bromophenol (OH, Br ortho), Y = 2-bromoanisole (OMe, Br ortho)

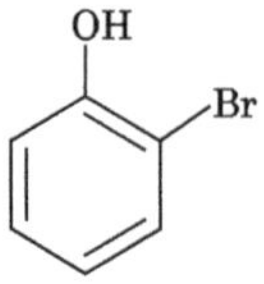

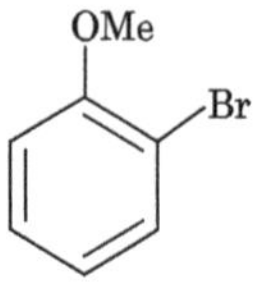

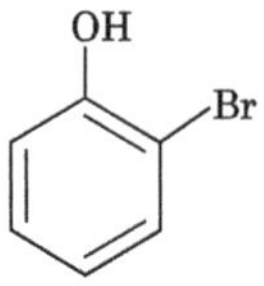

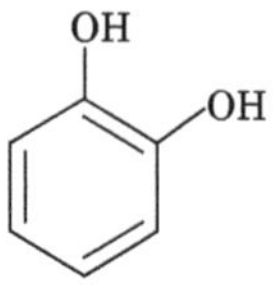

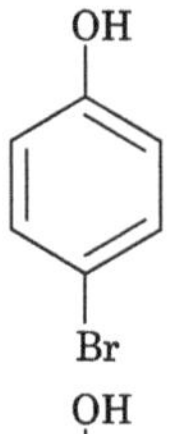

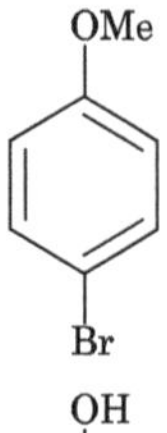

(b) X = 2-bromophenol (OH, Br ortho), Y = benzene-1,2-diol (OH, OH ortho)

(c) X = 4-bromophenol (OH, Br para), Y = 4-bromoanisole (OMe, Br para)

(d) X = 4-bromophenol (OH, Br para), Y = benzene-1,4-diol (OH, OH para)

109. $^{234}Th_{90}$ gets converted to $^{206}Pb_{82}$ through a series of radioactive decay processes. The number of alpha and beta particles lost in this transformation respectively, are

(a) 6 and 6 (b) 4 and 2
(c) 6 and 7 (d) 7 and 6

110. In the following transformation

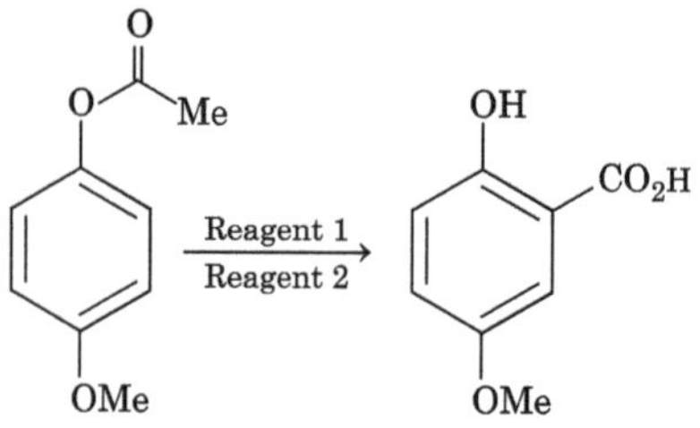

4-methoxyphenyl acetate (O–C(=O)Me, OMe para) $\xrightarrow[\text{Reagent 2}]{\text{Reagent 1}}$ 2-hydroxy-5-methoxybenzoic acid (OH, CO_2H ortho, OMe para to OH)

Reagents 1 and 2 are

(a) H_2SO_4 ; alk. $KMnO_4$ (b) $AlCl_3; I_2$ / NaOH
(c) $H_3PO_4; CHCl_3$ / KOH (d) KOH; $CHCl_3$ / KOH

BIOLOGY

111. The mode of action of penicillin is as follows

(a) it inhibits viral replication
(b) it enhances immunity
(c) it inhibits bacterial cell wall synthesis
(d) it inhibits transcription

112. Which of the following statements is true for meiosis?

(a) One round of chromosome duplication and one round of cell division
(b) One round of chromosome duplication and two rounds of cell division
(c) Two rounds of chromosome duplication and one round of cell division
(d) Two rounds of chromosome duplication and two rounds of cell division

113. Instead of 3, if 2 bases code for an amino acid, the degeneracy of codons coding for the same amino acid would have

(a) increased (b) remained the same
(c) decreased (d) been uncertain

114. Gregor Mendel showed that unit factors exist in pairs and exhibit a dominant-recessive relationship. These unit factors in modern terminology are called

(a) genes (b) loci
(c) alleles (d) determinants

115. *E. coli* has optimal growth temperature of 37°C. Which of the following is an incorrect explanation for this?

(a) The membrane is most permeable at this temperature
(b) DNA synthesis makes the least mistakes at this temperature
(c) Most enzymes in the cell have the highest activity at this temperature
(d) Protein synthesis is most efficient at this temperature

116. Male offsprings of which of the following couples have the highest chance of haemophilia?

(a) Haemophilic father and normal, non-carrier mother
(b) Haemophilic father and normal, carrier mother
(c) Normal father and normal, carrier mother
(d) Normal father and haemophilic mother

117. The effect of consumption of excess protein by normal individuals would result in

(a) excretion of excess protein in urine
(b) increase in the amount of adipose tissue
(c) increase in the synthesis of muscle protein
(d) increase in the circulatory plasma proteins

118. The condition varicose veins is swelling of veins, that occurs due to

(a) loss of elasticity of the muscular layer
(b) condition of high blood pressure
(c) condition of low blood pressure
(d) condition of anoxia

119. Greatest proportion of photosynthesis in the world is carried out by

(a) trees in the rainforests of the world
(b) trees in the temperate forests of the world
(c) algae in oceans
(d) irrigated crop fields

120. Energetically unfavourable reactions occur in human cells through

(a) heat energy supplied by the body
(b) heat energy released through exercise
(c) coupling of energetically favourable reactions with unfavourable ones
(d) photosynthesis

Answers

PART-I

1 (a)	2 (b)	3 (c)	4 (b)	5 (d)	6 (a)	7 (*)	8 (a)	9 (a)	10 (c)
11 (b)	12 (b)	13 (d)	14 (c)	15 (a)	16 (b)	17 (b)	18 (a)	19 (c)	20 (a)
21 (b)	22 (b)	23 (d)	24 (d)	25 (c)	26 (c)	27 (d)	28 (c)	29 (c)	30 (d)
31 (d)	32 (c)	33 (a)	34 (c)	35 (b)	36 (c)	37 (a)	38 (a)	39 (a)	40 (a)
41 (c)	42 (d)	43 (c)	44 (c)	45 (c)	46 (a)	47 (c)	48 (b)	49 (a)	50 (b)
51 (a)	52 (d)	53 (c)	54 (d)	55 (b)	56 (d)	57 (a)	58 (d)	59 (b)	60 (a)
61 (a)	62 (d)	63 (c)	64 (c)	65 (b)	66 (b)	67 (c)	68 (d)	69 (a)	70 (d)
71 (d)	72 (d)	73 (b)	74 (c)	75 (b)	76 (b)	77 (b)	78 (d)	79 (c)	80 (a)

PART-II

81 (c)	82 (c)	83 (c)	84 (a)	85 (b)	86 (d)	87 (d)	88 (d)	89 (b)	90 (c)
91 (d)	92 (a)	93 (b)	94 (c)	95 (d)	96 (a)	97 (c)	98 (c)	99 (b)	100 (b)
101 (c)	102 (d)	103 (a)	104 (a)	105 (b)	106 (c)	107 (a)	108 (c)	109 (d)	110 (b)
111 (c)	112 (b)	113 (c)	114 (b)	115 (c)	116 (d)	117 (c)	118 (b)	119 (c)	120 (c)

* *No options are correct.*

Solutions

1. *(a)* We have, $a_1, a_2, a_3, \ldots, a_n$ in an AP and a_1, a_2, a_4, a_8 in GP.

Let $a_1 = a$, $a_2 = a + d$, $a_3 = a + 2d$ and a_1, a_2, a_4, a_8 are in GP.

Let common ratio is r.

$\therefore \quad a_1 = a, a_2 = ar, a_4 = ar^2, a_8 = ar^3$

$\therefore a + d = ar, a + 3d = ar^2, a + 7d = ar^3$

$\Rightarrow \quad 2d = ar(r-1), 4d = ar^2(r-1)$

$\Rightarrow \quad r = 2$

2. *(b)* We have, $\dfrac{(101)^{k/2}}{k!}$

$$\sqrt{101} > 10$$

$$\frac{(\sqrt{101})^9}{9!} - \frac{(\sqrt{101})^{10}}{10!}$$

$$\frac{(\sqrt{101})^9}{9!}\left[1 - \frac{\sqrt{101}}{10}\right] < 0 \quad \left[\because \frac{\sqrt{101}}{10} > 1\right]$$

$$\therefore \quad \frac{(\sqrt{101})^9}{9!} < \frac{(\sqrt{101})^{10}}{10!}$$

$$\frac{(\sqrt{101})^{10}}{10!} - \frac{(\sqrt{101})^{11}}{11!}$$

$$= \frac{(\sqrt{101})^{10}}{10!}\left[1 - \frac{\sqrt{101}}{11}\right] > 0 \quad \left[\because \frac{\sqrt{101}}{11} < 1\right]$$

$$\therefore \quad \frac{(\sqrt{101})^{10}}{10!} > \frac{(\sqrt{101})^{11}}{11!}$$

$$\frac{(\sqrt{101})^{10}}{10!} - \frac{(\sqrt{101})^{101}}{(101)!}$$

$$= \frac{\sqrt{101}}{10!}\left[1 - \frac{(\sqrt{101})^{91}}{11 \times 12 \ldots 101}\right] > 0$$

$\therefore$ For $k = 10$, Maximum value of $\dfrac{(101)^{k/2}}{k!}$.

3. *(c)* We have,

$$p(x) = a_0 + a_1x + a_2x^2 + \ldots + a_nx^n$$

$$p(\sqrt{2} + \sqrt{3} + \sqrt{6}) = 0$$

$\therefore \quad x = \sqrt{2} + \sqrt{3} + \sqrt{6}$

$\Rightarrow \quad (x - \sqrt{6})^2 = (\sqrt{2} + \sqrt{3})^2$

$\Rightarrow \quad x^2 - 2\sqrt{6}x + 6 = 2 + 3 + 2\sqrt{6}$

$\Rightarrow \quad (x^2 + 1)^2 = 2\sqrt{6}(x + 1)^2$

$\Rightarrow \quad x^4 + 2x^2 + 1 = 24(x^2 + 2x + 1)$

$\Rightarrow \quad x^4 - 22x^2 - 48x - 23 = 0$

$\therefore \quad p(x) = x^4 - 22x^2 - 48x - 23 = 0$

Minimum value of $x = 4$

4. *(b)* Total number of game = 9

Total number of player = 3

Total score be zero

If every team wins 3 games and lose their 6 games, then final score be zero.

$\therefore$ Total number of ways

$${}^9C_3 \times {}^6C_2 \times {}^3C_3$$

$$\Rightarrow \quad \frac{9!}{3!6!} \times \frac{6!}{3!3!} \times 1 = \frac{9!}{3!3!3!} = 1680$$

5. *(d)* $\because O(2, z)$ is circumcentre of ΔABC.

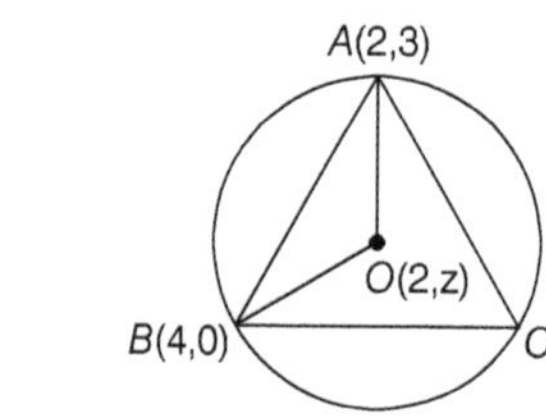

$\therefore \quad OA^2 = OB^2$

$\Rightarrow \quad (2 - z)^2 + (z - 3)^2 = (4 - 2)^2 + (z)^2$

$\Rightarrow \quad z^2 - 6z + 9 = 4 + z^2$

$\Rightarrow \quad z = \dfrac{5}{6}$

$$\Rightarrow \quad OA = r = \sqrt{\left(\frac{5}{6} - 3\right)^2} = \frac{13}{6}$$

6. *(b)*

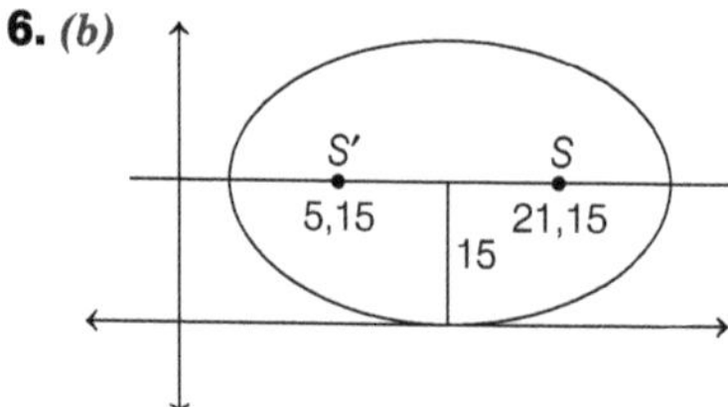

Given, (5, 15) and (21, 15) are foci of parabola and X-axis is tangent of ellipse.

$\therefore \quad 2ae = 16$ and $b = 15$

$\Rightarrow \quad ae = 8$ and $b = 15$

$\Rightarrow \quad ae^2 = a^2 - b^2$

$\Rightarrow \quad 16 = a^2 - 225$

$\Rightarrow \quad a^2 = 289$

$\Rightarrow \quad a = 17$

$\therefore$ Length of major axis = 34

7. *(*)* The line $2x + 3y = 18$

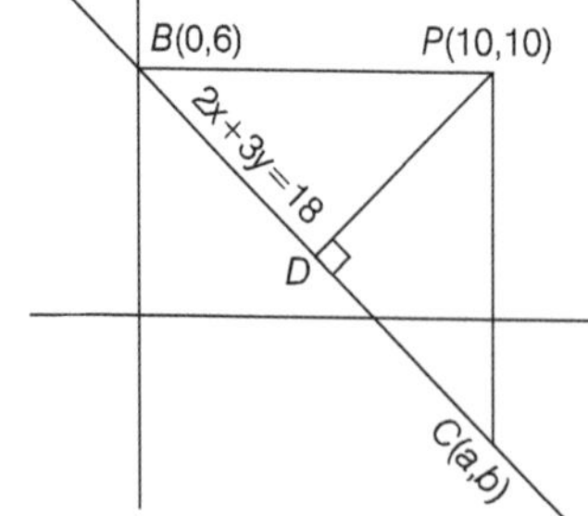

Cut the Y-axis at B.

$\therefore \quad B = (0, 6), C = (a, b)$

C lie on line $2x + 3y = 18$

$\therefore \quad 2a + 3b = 18$

Now, $\quad PB = PC$

$\because PD$ is perpendicular bisector of line segment BC

Equation of line PD is perpendicular to BC.

$\therefore \quad 3x - 2y = \lambda$

Since, PD is passing through $P(10, 10)$.

$\therefore \quad 30 - 20 = \lambda \Rightarrow \lambda = 10$

Equation of line PD is $3x - 2y = 10$

Solving the equation $2x + 3y = 18$ and $3x - 2y = 10$,

we get $D\left(\dfrac{66}{13}, \dfrac{34}{13}\right)$.

D is mid-point of BC

$\therefore \quad \dfrac{a + 0}{2} = \dfrac{66}{13}, \dfrac{b + 6}{2} = \dfrac{34}{13}$

$\Rightarrow \quad a = \dfrac{132}{13}, b = \dfrac{-10}{13}$

$$\therefore \quad 8a + 2b = \frac{8 \times 132}{13} - \frac{20}{13}$$

$$= \frac{1056 - 20}{13} = \frac{1036}{13}$$

8. *(a)* $\text{cosec}^2(\alpha + \beta) - \sin^2(\beta - \alpha) + \sin^2(2\alpha - \beta) = \cos^2(\alpha - \beta)$

$\Rightarrow \text{cosec}^2(\alpha + \beta) + \sin^2(2\alpha - \beta) = \cos^2(\alpha - \beta) + \sin^2(\alpha - \beta)$

$\Rightarrow \text{cosec}^2(\alpha + \beta) + \sin^2(2\alpha - \beta) = 1$

It is possible only $\text{cosec}^2(\alpha + \beta) = 1$ and $\sin^2(2\alpha - \beta) = 0$

$\therefore \quad \alpha + \beta = \dfrac{\pi}{2}$ and $2\alpha - \beta = 0$

On solving these equations, we get

$$\alpha = \frac{\pi}{6} \text{ and } \beta = \frac{\pi}{3}$$

$$\therefore \sin(\alpha - \beta) = \sin\left(\frac{\pi}{6} - \frac{\pi}{3}\right) = \sin\left(-\frac{\pi}{6}\right) = -\frac{1}{2}$$

9. *(a)* We have,

$$\sin x + \sin y = \frac{7}{5}$$

and $\quad \cos x + \cos y = \dfrac{1}{5}$

$$\Rightarrow \quad 2\sin\left(\frac{x + y}{2}\right)\cos\left(\frac{x - y}{2}\right) = \frac{7}{5} \quad \ldots\text{(i)}$$

$$\text{and} \quad 2\cos\left(\frac{x + y}{2}\right)\cos\left(\frac{x - y}{2}\right) = \frac{1}{5} \quad \ldots\text{(ii)}$$

On dividing Eq. (ii) by Eq. (i), we get

$$\tan\left(\frac{x+y}{2}\right) = 7$$

$$\Rightarrow \quad \sin(x+y) = \frac{2\tan\left(\frac{x+y}{2}\right)}{1+\tan^2\left(\frac{x+y}{2}\right)}$$

$$\Rightarrow \quad \sin(x+y) = \frac{14}{1+49} = \frac{14}{50}$$

$$\Rightarrow \quad \sin(x+y) = \frac{7}{25}$$

10. *(c)* We have, $\sin x = \frac{6}{x}, x \in [0, 12\pi]$

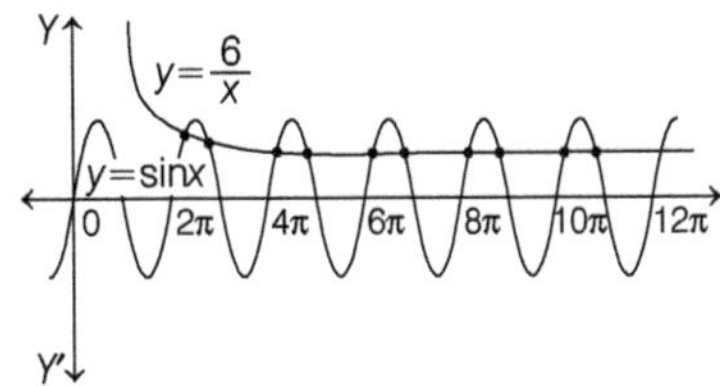

Clearly, from graph total number of solution = 10

11. *(b)* We have, $f : R \to R$ by

$$f(x) = \begin{cases} \frac{\sin x^2}{x}, & x < 0 \\ x^2 + ax + b, & x \geq 0 \end{cases}$$

$f(x)$ is differentiable on R.

$\therefore f(x)$ is also continuous on R.

$$\lim_{x\to 0^-} f(x) = \lim_{x\to 0^+} f(x)$$

$$\lim_{x\to 0^-} \frac{\sin x^2}{x} = \lim_{x\to 0^+} x^2 + ax + b$$

$$\Rightarrow \quad 0 = b \Rightarrow b = 0$$

Now, $f(x)$ is differentiable at $x = 0$.

$$\therefore \lim_{h\to 0} \frac{f(0-h) - f(0)}{h} = \lim_{h\to 0} \frac{f(0+h) - f(0)}{h}$$

$$\Rightarrow \lim_{h\to 0} \frac{\sin^2 h}{h^2} = \lim_{h\to 0} \frac{h^2 + ah}{h} \Rightarrow 1 = a$$

$$\therefore \quad a = 1, b = 0$$

12. *(b)* Let a point B on parabola is $B\left(\frac{K^2}{4}, K\right)$ and $A(0, 3)$.

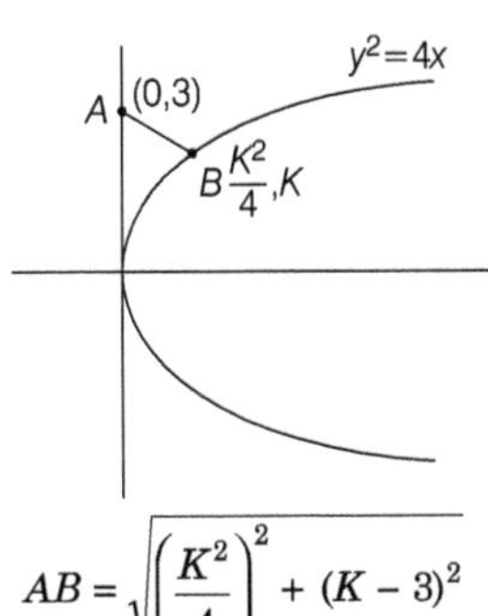

$$\therefore \quad AB = \sqrt{\left(\frac{K^2}{4}\right)^2 + (K-3)^2}$$

$$\Rightarrow \quad AB^2 = \frac{K^4}{16} + K^2 - 6K + 9$$

AB is shortest.

$\therefore AB^2$ is also shortest.

Let $\quad AB = Z$

$$\therefore \quad Z = \frac{K^4}{16} + K^2 - 6K + 9$$

$$\Rightarrow \quad \frac{dZ}{dK} = \frac{K^3}{4} + 2K - 6$$

For maxima or minima

put $\quad \frac{dZ}{dK} = 0 \Rightarrow \frac{K^2}{4} + 2K - 6 = 0$

$$\Rightarrow \quad K^2 + 8K - 24 = 0$$

$$\Rightarrow \quad (K-2)(K^2 + 2K + 12) = 0$$

$$K = 2 \Rightarrow \frac{d^2z}{dK^2} = \frac{3K^2}{4} + 2$$

$$\Rightarrow \quad \left(\frac{d^2Z}{dK^2}\right)_{K=2} > 0$$

$\because Z$ is minimum at $K = 2$

$$\therefore \quad AB = \sqrt{\frac{16}{16} + (1)^2} = \sqrt{2}$$

13. *(d)* If all trucks had packets of 1000 g, then total weight is

$$1000(1 + 2 + 2^2 + \dots + 2^9)$$

$$= 1000(2^{10} - 1)$$

$$= 1000(1024 - 1) = 1023000$$

We have given total weight is 1022870.

$\therefore$ 1022870 < 1023000

Extra amount of weight

$$= 1023000 - 1022870$$

$$= 130$$

$$130 = 2^7 + 2^1 = 128 + 2$$

$\therefore$ The trucks have the lighter bags are 2, 8.

14. *(c)* Let

$$I = \int_0^1 \cos(\pi x)\cos[2x]\pi dx$$

$$\Rightarrow I = \int_0^{1/2} \cos(\pi x)\cos 0 dx + \int_{1/2}^1 \cos \pi x \cos \pi dx$$

$$\Rightarrow I = \int_0^{1/2} \cos \pi x dx - \int_{1/2}^1 \cos \pi x dx$$

$$\Rightarrow I = \left[\frac{\sin \pi x}{\pi}\right]_0^{1/2} - \left[\frac{\sin \pi x}{\pi}\right]_{1/2}^1$$

$$\Rightarrow I = \frac{1}{\pi}\left[\sin\frac{\pi}{2} - \sin 0\right] - \frac{1}{\pi}\left[\sin \pi - \sin\frac{\pi}{2}\right]$$

$$\Rightarrow I = \frac{1}{\pi}[1 - 0] - \frac{1}{\pi}[0 - 1] = \frac{2}{\pi}$$

15. *(a)* Let $I = \int_0^1 x^{10} \sin nx dx$

$$\Rightarrow \quad I = \left[\frac{-x^{10}\cos nx}{n}\right]_0^1 + \frac{10}{n}\int_0^1 x^9 \cos nx dx$$

$$\Rightarrow \quad I = \frac{-\cos n}{n} + \frac{10}{n}\left[\frac{x^9 \sin nx}{n^2}\right]_0^1 - \frac{90}{n}\int_0^1 \frac{x^8 \sin nx}{n^2} dx$$

$$\Rightarrow I = -\frac{\cos n}{n} + \frac{10 \sin n}{n^2} - \frac{90}{n}\int_0^1 x^8 \sin nx dx$$

$\lim_{n\to\infty} I = 0 \qquad n \to \infty \; I = 0$

16. *(b)* We have, $y = x^2$ and $y = 1 - x^2$

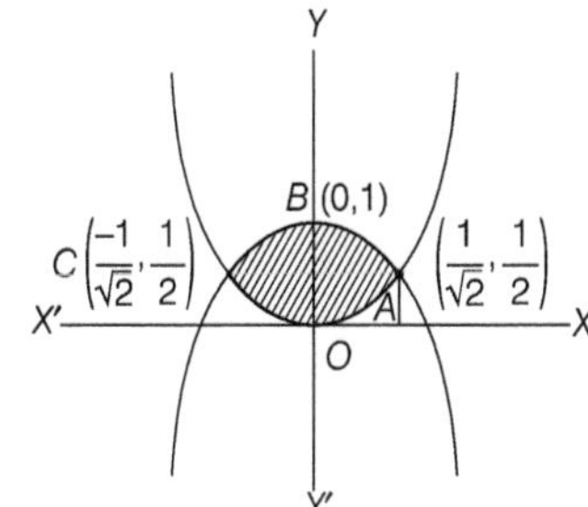

Intersection point of $y = x^2$ and $y = 1 - x^2$ is $A\left(\frac{1}{\sqrt{2}}, \frac{1}{2}\right)$ and $C\left(\frac{-1}{\sqrt{2}}, \frac{1}{2}\right)$

Area of shaded region

$$= 2\int_0^{1/\sqrt{2}} [(1 - x^2) - (x^2)]dx$$

$$= 2\int_0^{1/\sqrt{2}} (1 - 2x^2)dx$$

$$= 2\left[x - \frac{2x^3}{3}\right]_0^{1/\sqrt{2}}$$

$$= 2\left[\frac{1}{\sqrt{2}} - \frac{1}{3\sqrt{2}}\right]$$

$$= \frac{2}{\sqrt{2}}\left[\frac{3-1}{3}\right] = \frac{2\sqrt{2}}{3} \text{ sq units}$$

17. *(b)* We have,

$\mathbf{a} = 3\mathbf{i} - 4\mathbf{k}$ and $\mathbf{b} = 5\mathbf{j} + 12\mathbf{k}$

We know that,

angle bisector of vector **a** and **b**

$$= \lambda\left[\frac{\mathbf{a}}{|\mathbf{a}|} + \frac{\mathbf{b}}{|\mathbf{b}|}\right]$$

$$\therefore \lambda\left[\frac{3\hat{\mathbf{i}} - 4\hat{\mathbf{k}}}{5} + \frac{5\hat{\mathbf{j}} + 12\hat{\mathbf{k}}}{13}\right]$$

$$= \lambda\left[\frac{39\hat{\mathbf{i}} - 52\hat{\mathbf{k}} + 25\hat{\mathbf{j}} + 60\hat{\mathbf{k}}}{65}\right]$$

$$= \lambda[39\hat{\mathbf{i}} + 25\hat{\mathbf{j}} + 8\hat{\mathbf{k}}]$$

18. *(a)* An envelope has space for at most 3 stamps. Three stamps of denomination is 1 and three stamps of denomination is a.

The value of stamps of envelope must be

$1+1+1=3, a+a+a=3a,$

$a+a+1=2a+1, a+1+1=a+2$

$a>1$, If $a=2$,

then maximum value of stamps is

$$3\times 2=6$$

$\therefore$ Least positive integer has not stamp value is 7.

19. *(c)* We have,

$\left(\frac{d}{m}\right)=d$ is positive division of m.

Let $d_1, d_2, d_3, \ldots, d_k$ are divisor of m and $D_1, D_2, D_3, \ldots, D_r$ are divisor of n

$\sum d_i = \sum D_i$ (given)

Now, $\sum \frac{1}{d_i} = \frac{1}{d_1}+\frac{1}{d_2}+\ldots+\frac{1}{d_k} = \frac{\sum d_i}{m}$

$\sum \frac{1}{D_i} = \frac{1}{D_1}+\frac{1}{D_2}+\ldots+\frac{1}{D_r} = \frac{\sum D_i}{n}$

$\therefore \quad \frac{1}{m} > \frac{1}{n} \quad [\because m<n]$

$$\frac{\sum d_i}{m} > \frac{\sum d_i}{n}$$

$\Rightarrow \quad \sum_{d|m} \frac{1}{d} > \sum_{d|n} \frac{1}{d}$

20. *(a)* Set A have m-elements,

Set B have n-elements

$(a, b_1)\in R, (a, b_2)\in R \Rightarrow (b_1 = b_2)$

By condition relation is a function.

$\therefore$ Total number of function (Relation) $= n^m$

21. *(b)* Mayer's relation,

$C_p - C_V = R$ is true only for ideal gases whether they are monoatomic, diatomic or polyatomic as for real gases.

$$C_p - C_V = TV\beta^2 k$$

where, k = isothermal compressibility and β = isobaric thermal expansion coefficient.

So, correct option is (b).

22. *(b)* Gas molecules does not settle in room because they have too much kinetic energy ($k_B T > mgh$).

If heat energy is dissipated or extracted from that room, gas molecules does settle down. For example, earth's atmosphere is more denser near the surface. So, correct option is (b).

23. *(d)* As plates are moved far away further, capacity of capacitor $\left(C=\frac{\varepsilon_0 A}{d}\right)$ decreases.

As battery remains connected during this activity, potential difference between plates remains same.

So, the charge on plates ($Q=CV$) decreases.

Also, energy of capacitor $\left(U=\frac{1}{2}CV^2\right)$ decreases as capacitance decreases.

So, correct option is (d).

24. *(d)* From triangle law, in given vector pentagon,

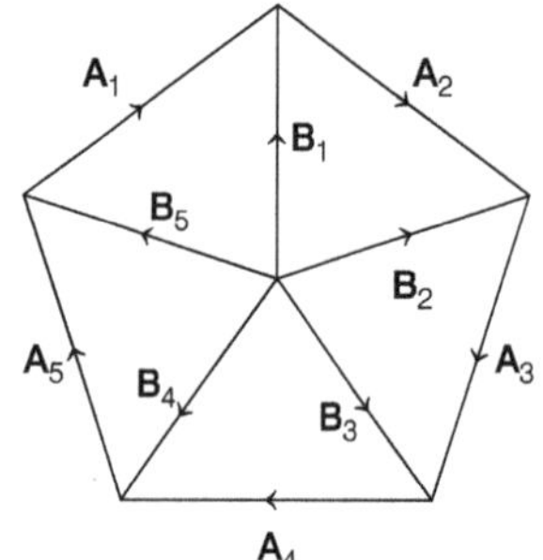

We have,

$$\mathbf{B}_2+\mathbf{A}_3=\mathbf{B}_3 \quad \ldots(i)$$
$$\mathbf{B}_3+\mathbf{A}_4=\mathbf{B}_4 \quad \ldots(ii)$$
$$\mathbf{B}_4+\mathbf{A}_5=\mathbf{B}_5 \quad \ldots(iii)$$
$$\mathbf{B}_5+\mathbf{A}_1=\mathbf{B}_1 \quad \ldots(iv)$$

Adding these equations, we have

$\Rightarrow (\mathbf{B}_2+\mathbf{B}_3+\mathbf{B}_4+\mathbf{B}_5) + (\mathbf{A}_3+\mathbf{A}_4+\mathbf{A}_5+\mathbf{A}_1)$

$= (\mathbf{B}_3+\mathbf{B}_4+\mathbf{B}_5+\mathbf{B}_1)$

$\Rightarrow \mathbf{B}_2+\mathbf{B}_3+\mathbf{B}_4+\mathbf{B}_5 + (-\mathbf{A}_2) = (-\mathbf{B}_2) \ldots(v)$

As from polygon law of vector addition,

$$\mathbf{A}_3+\mathbf{A}_4+\mathbf{A}_5+\mathbf{A}_1 = -\mathbf{A}_2$$

and $\mathbf{B}_3+\mathbf{B}_4+\mathbf{B}_5+\mathbf{B}_1 = -\mathbf{B}_2$

So, from Eq. (v), we have

$$\mathbf{B}_2+\mathbf{B}_3+\mathbf{B}_4+\mathbf{B}_5 = \mathbf{A}_2-\mathbf{B}_2 = -\mathbf{B}_1$$

25. *(c)*

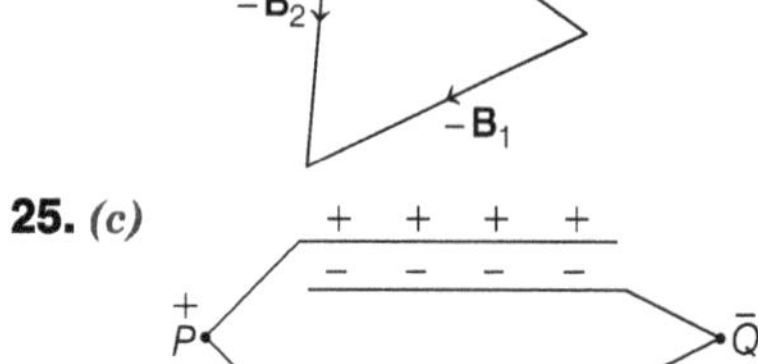

Above system is equivalent to two capacitors in parallel,

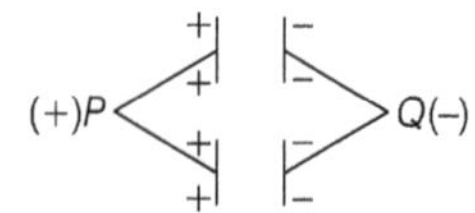

So, capacitance of system is

$$C_{eq} = C_1 + C_2 = \frac{\varepsilon_0 A}{d}+\frac{\varepsilon_0 A}{d} = \frac{2\varepsilon_0 A}{d}$$

26. *(c)* Due to hard boundary reflection, phase of wave changes by π radians and its direction of travel is reversed.

So, resultant wave is

$$y = a\cos(kx+\omega t+\pi)$$

$\Rightarrow \quad y = -a\cos(kx+\omega t)$

27. *(d)* Potential due to an elemental charge dQ at centre of sphere is

$$dV = \frac{kdQ}{R}$$

Potential due to complete charge distribution at centre is

$$V = \oint dV = \oint \frac{kdQ}{R}$$
$$= \frac{k}{R}\oint dQ = \frac{kQ}{R}$$

or $\quad V = \frac{kQ}{R} = \frac{Q}{4\pi\varepsilon_0 R}$

28. *(c)* Heat supplied at constant pressure is $\Delta Q = nC_p\Delta T$.

Work done by the gas,

$$\Delta W = p\Delta V = nR\Delta T$$

Ratio of work done to heat supplied is

$$\frac{\Delta W}{\Delta Q} = \frac{nR\Delta T}{nC_p\Delta T}$$
$$= \frac{R}{C_p} = \frac{C_p - C_V}{C_p}$$
$$= 1-\frac{1}{\gamma}$$
$$= 1-\frac{1}{(7/5)} = 1-\frac{5}{7} = \frac{2}{7}$$

29. *(c)* Lyman-alpha line sometimes denoted as Ly-α line is obtained when an electron falls from $n=2$ to $n=1$ orbital in an one electron ion or atom.

Similarly, Balmer-alpha radiation is emitted when an electron makes a transition from $n=3$ to $n=2$ quantum state.

Now, using $\frac{1}{\lambda} = R\left(\frac{1}{n_1^2}-\frac{1}{n_2^2}\right)$

We have,

$$\frac{1}{\lambda_{\text{Ly-}\alpha}} = R\left(\frac{1}{1^2}-\frac{1}{2^2}\right) = \frac{3R}{4}$$

and $\quad \frac{1}{\lambda_{\text{Ba-}\alpha}} = R\left(\frac{1}{4}-\frac{1}{9}\right) = \frac{5}{36}R$

Hence, $\quad \frac{\lambda_{\text{Ly-}\alpha}}{\lambda_{\text{Ba-}\alpha}} = \frac{\frac{3}{4}R}{\frac{5}{36}R} = \frac{27}{5}$

30. *(d)* Initial condition is

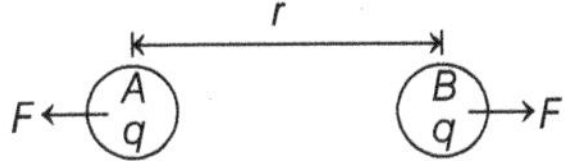

So, initial value of force between spheres is $F = kq^2 / r^2$.

When another identical and uncharged sphere C is touched to A, charge on each of A and C will be $\frac{q}{2}$.

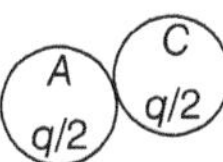

When C is touched with B, then charge on each of them will be mean value of total charge.

So, charge on B = charge on C

$$= \frac{q + \frac{q}{2}}{2} = \frac{3q}{4}$$

Now, the final condition is

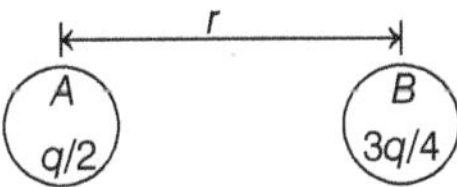

So, final value of force between spheres is

$$F' = \frac{kq_1 \cdot q_2}{r^2}$$

$$= \frac{k\frac{q}{2} \times \frac{3q}{4}}{r^2} = \frac{3}{8}\frac{kq^2}{r^2}$$

or $\quad F' = \frac{3}{8}F$

31. *(d)* As the loop is moving in the region of magnetic flux, induced emf is

$$E = Blv \text{ (volts)}$$

Value of emf remains same till the loop is going in or out of loop.

Direction of induced emf is (in accordance with Lenz's law) anti-clockwise (taken positive) when loop is going in the region of field and it becomes clockwise as the loop is moving out of region of field. So, correct graph is (d).

32. *(c)* As mass distribution in rotation about the diameter is same in both cases, moment of inertia also remains same in both cases.

$\therefore$ Moment of inertia (half folded disc)

$$= \frac{MR^2}{4}$$

33. *(a)* As $\frac{E}{B} = c$, for an electromagnetic wave, relation $\mathbf{E} = \frac{\hat{\mathbf{n}} \times \mathbf{B}}{c}$ is dimensionally incorrect.

34. *(c)*

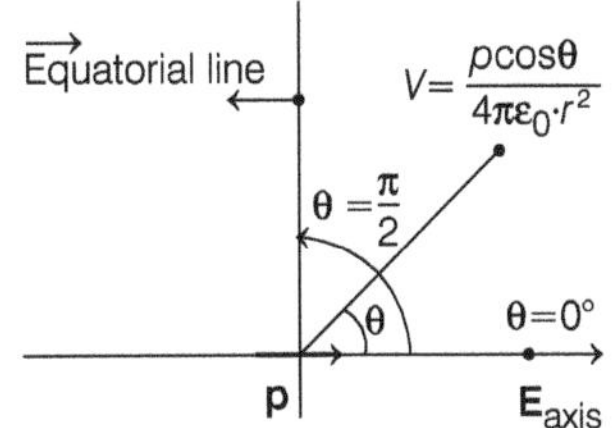

A dipole and its field an axis and equatorial line are shown in above figure.

Clearly at $\theta = \frac{\pi}{2}$ plane, electric field is along the plane. So, option (c) is correct.

35. *(b)* In pure rolling, there is no relative motion between body and surface at the point of contact, so dissipation of energy and friction is zero in ideal condition.

36. *(c)* Given, potential energy curve closely represents potential energy of a simple harmonic oscillator $\left(U = \frac{1}{2}kx^2, \text{shown in dotted line}\right)$.

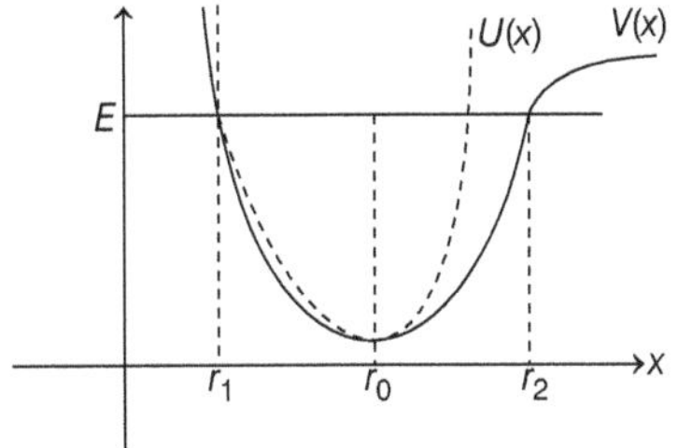

In region $r_1 < x < r_2$, the potential energy is less than total energy, so motion of particle in this region is oscillatory.

Also, the potential energy curve is not symmetric about $x = r_0$.

$\therefore$ Motion is not a simple harmonic.

37. *(a)* Velocity of sound in an isentropic gaseous medium is given by

$$v = \sqrt{\frac{\gamma RT}{M}} \Rightarrow v \propto \sqrt{T}$$

Also, $\quad v = f\lambda \Rightarrow \lambda \propto \sqrt{T}$

Because frequency is a characteristics of source and it remains constant with change in temperature of medium.

So, correct option is (a).

38. *(a)* As the inclined plane is accelerating upwards with acceleration a, net acceleration of block is $(a + g)$.

We have following free body diagram,

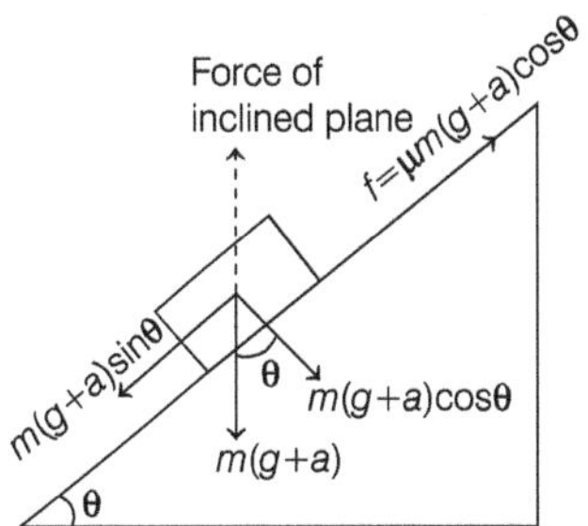

Clearly, force on block by inclined plane is $m(g + a)$ in upward direction.

39. *(a)* Direction of force on positive charged particle by Fleming's left hand rule is towards direction of electric field.

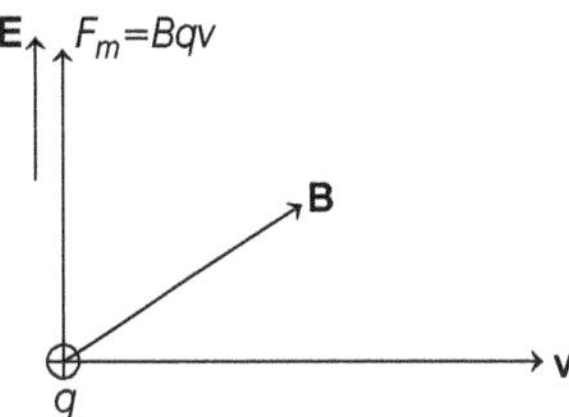

As, electric force on positively charged particle is also in same direction. So, a positively charged particle will be deflected upwards.

Hence, no particle can get through the hole in slit.

So, option (a) is correct.

40. *(a)* From conservation of energy, energy of particle at height H over inclined plane must be equal to energy of particle at point O.

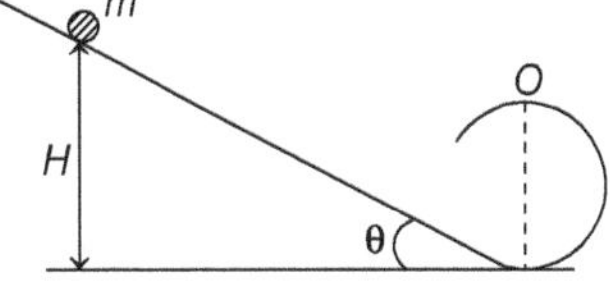

$\Rightarrow \quad mgH = 2mgR \Rightarrow H = 2R$

41. *(c)* The rate of diffusion of gases under similar conditions of temperature and pressure is inversely proportional to the square roots of their molecular masses. i.e.

$$\text{Rate of diffusion} \propto \frac{1}{\sqrt{M}}$$

The correct increasing order of molecular masses of given gases is,

$$H_2 < CH_4 < O_2 < CO_2$$

$\therefore$ Increasing order of rate of diffusion will be

$$CO_2 < O_2 < CH_4 < H_2$$

Thus, CO_2 has the slowest rate of diffusion.

42. *(d)* The formula for kinetic energy for diatomic gas is given as

$$KE = \frac{5}{2}nRT$$

Number of moles of $H_2 \Rightarrow n_{H_2} = \frac{3}{2}$

$\therefore KE_{H_2} = \frac{5}{2} \times \frac{3}{2} RT$

Number of moles of $O_2 \Rightarrow n_{O_2} = \frac{4}{32} = \frac{1}{8}$

$\therefore K_{O_2} = \frac{5}{2} \times \frac{1}{8} RT$

Thus, $\frac{KE_{H_2}}{KE_{O_2}} = \frac{\frac{5}{2} \times \frac{3}{2} RT}{\frac{5}{2} \times \frac{1}{8} RT} = \frac{\frac{3}{2}}{\frac{1}{8}} = \frac{12}{1}$

The ratio of kinetic energies of 3 g of H_2 and 4g of O_2 at any temperature is 12 : 1.

43 *(c)*

$$\underset{1}{Me}-\overset{\overset{O}{\|}}{\underset{2}{C}}-\underset{\underset{Me}{|}}{\overset{3}{C}H}-\overset{4}{C}H_2-\overset{5}{C}H_2-Cl$$

The correct IUPAC name of the above compound will be, 5-chloro-3-methyl-2-pentanone.

44. *(c)* Central chlorine atom has seven valence electrons, out of which it forms three bond pairs with fluorine atoms and has two lone pair. Thus, it's shape will be T-shaped, where, the lone pair will occupy the equitorial positions, so as to minimise the repulsion.

$$F-\underset{\underset{F}{|}}{\overset{\overset{F}{|}}{Cl}}<\!:$$

ClF_3 (T-shaped)

45. *(c)*Bronsted acid is that species that can donate H^+ ions and Bronsted base is one that can accept H^+ ions. Thus, the species that acts as a Bronsted acid as well as a Bronsted base is NH_3.

$$NH_3 + NH_3 \rightleftharpoons NH_4^+ + NH_2^-$$

Such, species are known as amphoteric species.

46. *(a)* The ratio of molar heat capacities at constant pressure to that at constant volume is represented by γ. The value of γ gives information about the atomicity of gases.

If $\frac{C_p}{C_V} = 1.67$, gas will be monoatomic.

If $\frac{C_p}{C_V} = 1.40$, gas will be diatomic.

If $\frac{C_p}{C_V} = 1.33$, gas will be triatomic.

As given in question, C_p/C_V for one mole of a gas is 1.67. i.e.

$$\frac{C_p}{C_V} = \gamma = 1.67$$

$\therefore$ It will be a monoatomic gas. Thus, from the given options, He is a monoatomic gas.

47. *(c)* Isoelectronic species are those species, which have same number of electrons.

Total number of electrons in CO is
$= 6 + 8 = 14$

(a) Total number of electrons in O_2^+ is
$= 8 + 8 - 1 = 15$

(b) Total number of electrons in CN^- is
$= 6 + 7 + 1 = 14$

(c) Total number of electrons in O_2^- is
$= 8 + 8 + 1 = 17$

(d) Total number of electrons in N_2^+ is
$= 7 + 7 - 1 = 13$

Thus, CN^- is isoelectronic with CO.

48. *(b)* As the number of lone pairs of electrons increases, bond angle decreases. The number of lone pair in CH_4, CO_2, H_2O and SO_2 respectively are 0, 0, 2 and 1. So, CH_4 and CO_2 will have the maximum bond angle. Also, multiple bond orbitals repel other orbitals more strongly than single bond orbital. Thus, the bond angle will increase.

As CH_4 has single bonds whereas CO_2 has double bonds. Therefore, CO_2 has the highest bond angle.

49. *(a)* Anhydrous diethyl ether is the solvent for carrying out Grignard reactions. This is because it is highly volatile solvent which helps in preventing oxygen from reaching the reaction solution. Also, ether molecules coordinate with and help in stabilising the Grignard reagent.

$$\begin{array}{c} Et\ddot{O}Et \\ \downarrow \\ R-Mg-Br \\ \uparrow \\ Et\ddot{O}Et \end{array}$$

50. *(b)* The reaction of butanal with n-propyl magnesium bromide (Grignard reagent) gives heptan-4-ol, which is a secondary alcohol as the 2 functional groups ($-CH_2CH_2CH_3$) are same, so the formed alcohol is a chiral. So, correct option is (b).

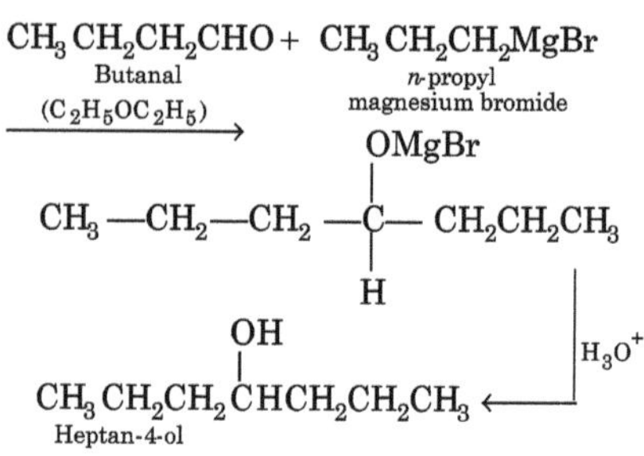

51. *(a)* The oxidation number of Ni in both $[Ni(PPh_3)_2Cl_2]$ and $[NiCl_4]^{2-}$ is +2.

The electronic configuration of Ni (II) is [Ar] $3d^8 4s^0$.

Ni (II): 3d [⥮ ⥮ ⥮ 1 1], 4s [ ], 4p [ ][ ][ ]

As PPh_3 is a strong ligand, so it forces the electrons to get pair up.

Ni $(PPh_3)_2Cl_2$: [⥮ ⥮ ⥮ ⥮ ××] [××] [×× ××] [ ] — dsp^2-hybridised

As Cl^- is not a strong ligand therefore pairing will not occurs.

$NiCl_4$: [⥮ ⥮ ⥮ 1 1] [××] [×× ×× ××] — sp^3-hybridised

Thus, the hybridisation of Ni central atom in $Ni[(PPh_3)_2Cl_2]$ and $[NiCl_4]^{2-}$ respectively are dsp^2 and sp^3.

52. *(d)* Oxalic acid when treated with potassium permanganate in the presence of an acid produces CO_2 gas.

$$2KMnO_4 + 5H_2C_2O_4 + 16H^+ \longrightarrow 2Mn^{2+} + 2K^+ + 10CO_2 + 8H_2O$$

53. *(c)* $N_2 + 3H_2 \rightleftharpoons 2NH_3 \; K_C = 4.1$ ...(i)

$$\frac{1}{2}N_2 + \frac{3}{2}H_2 \rightleftharpoons NH_3 \quad \text{...(ii)}$$

We divide equation (i) by 2 to get equation (ii) .

$\therefore$ The new equilibrium constant, K'_C for Eq. (ii) will be the square root of K'_C Eq. (i), i.e. $K'_C = \sqrt{K_C}$

$$K'_C = \sqrt{41} = 6.4$$

54. *(d)* For second order reaction,

$$2A \longrightarrow \text{Product}$$

Rate of reaction, $r = k[A]^2$.

If the concentration is reduced to half, i.e. $\frac{A}{2}$.

$$\therefore \quad r' = k\left[\frac{A}{2}\right]^2$$

$$r' = \frac{1}{4}k[A]^2 \text{ or } r' = \frac{1}{4}r$$

55. *(b)* The conjugate base differ by a proton from its respective acid. The conjugate bases for HCO_3^- and NH_3 respectively are CO_3^{2-} and NH_2^-.

$$\underset{\text{Acid}}{HCO_3^-} \rightleftharpoons \underset{\text{Conjugate base}}{CO_3^{2-}} + H^+$$

$$\underset{\text{Acid}}{NH_3} \rightleftharpoons \underset{\text{Conjugate base}}{NH_2^-} + H^+$$

56. *(d)* The aromatic compounds are those compounds, which follow three given conditions

(i) the compound has to be planar and cyclic.

(ii) delocalisation of electrons should occur.

(iii) follows Huckel's rule $[(4n+2)\pi]$.

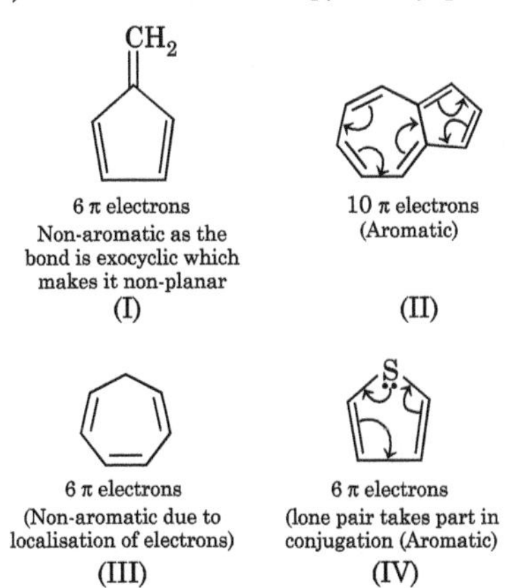

Thus, correct option is (d).

57. *(a)* In compound I and compound III the central atom N is sp^3-hybridised, so their basicity will be more than compound II and IV. Compound III will be less basic as compared to I because of the presence of more electronegative atom, O. Between II and IV, II will be more basic because the lone pair on N-atom will not take part in resonance and will be available for donation while the lone pair on N-atom of compound IV will take part in resonance and will be least available.Thus, the correct order of basicity will be I > III > II > IV.

58. *(d)*

Me
H Me
H H
H

As in the given Newmann projection the two methyl groups are adjacent to each other, thus they are known as gauche conformers.This can be obtained by rotating one of C_2 or C_3 carbon atoms roughly at an angle of 60°.

59. *(b)* Given,

$t_{1/2}$ of first order reaction = 30 min

For first order, $k = \frac{0.693}{t_{1/2}}$

$$k = \frac{0.693}{30} = 0.0231$$

Also, if the reaction is 75% completed.

$$t = \frac{2.303}{k} \log \frac{a}{a-x}$$

$$t = \frac{2.303}{0.0231} \log \frac{100}{25}$$

$$= 99.69 \times \log 4$$

$$= 99.69 \times 0.602$$

$$= 60 \text{ min}$$

60. *(a)* Number of moles of H_2SO_4

$$= 0.1 \times 10 = 1$$

Number of moles of KOH $= 0.1 \times 10 = 1$

As number of moles of KOH and H_2SO_4 are same, i.e. KOH is completely neutralising H_2SO_4.

$\therefore$ The resulting molarity of a mixture will give the hydrogen ion concentration.

$$MV = M_1V_1 + M_2V_2$$

$$M(20) = 1 + 1$$

$$M = \frac{2}{20} = 0.1$$

61. *(a)* The chemical conversion of water during photosynthesis is termed as photolysis. It is a part of photosynthesis which occurs in the granum of a chloroplast where light is absorbed by chlorophyll.

Chlorophyll reacts with water and splits the oxygen and hydrogen molecules apart.

62. *(d)* In the organism mussel (shell-fish), oxygen is carried by haemoglobin. Haemoglobin is the iron containing O_2-transport metalloprotein in the RBCs of almost all vertebrates as well as the tissues of some invertebrates. Haemoglobin in blood carries oxygen from the lungs or gills to the rest of the body.

63. *(c)* Enzymes are proteins that lower the activation energy / barrier of a reaction by binding to one of the reactants, called a substrate. Likewise, an enzyme holds its substrate in such a way that the reaction is much more likely to occur, i.e. speed up the chemical processes.

64. *(c)* Glycolysis is the degradation of glucose, which breaks down glucose to pyruvate. The main purpose of glycolysis is the generation of energy (ATP).

65. *(b)* Auxin is a plant hormone which is sensitive to light. Auxin stimulates the growth of cells on the shady side of the plant which causes bending of the plant to the other side. This gives the appearance that the stem of the plant bends in the direction of light.

66. *(b)* Differentiation is the process by which cells, tissues and organs acquire specialised features, especially during embryonic development.

67. *(c)* The chromosomal attachment site of the spindle microtubule is centromere. The microtubules during cell division extend from the centrioles and begin to attach to the centomeres of chromosomes.

68. *(d)* Tuberculosis is a potentially serious infectious bacterial disease that mainly affects the lungs. It is not sexually transmitted like AIDS, syphilis, gonorrhoea, etc.

Tuberculosis spreads by airborne respiratory droplets and by saliva.

69. *(a)* Centrosomes are structure found inside of cells. They are made from two centrioles. Centrioles are microtubule rings. The main purpose of a centrosome is to organise microtubules and provide structure for the cell, as well as work to pull chromatids apart during cell division.

70. *(d)* Nucleotides are the building blocks of nucleic acids, they are composed of three subunit molecules, i.e. a nitrogenous base (also known as nucleobase), a pentose sugar (ribose or deoxyribose) and at least one phosphate group.

71. *(d)* Fertilisation in humans usually takes place in the Fallopian tube. The steps of fertilisation involve the joining of an egg and a sperm.

72. *(d)* ELISA is (Enzyme Linked Immuno Sorbent Assay) is a standard test for finding out if someone has antibodies to a particular antigen/protein. During HIV test, it looks for HIV antibodies in the blood.

73. *(b)* Sickle-cell anaemia is caused by point mutation in the β-globin chain of haemoglobin causing the hydrophilic amino acid glutamic acid to be replaced with the hydrophobic amino acid valine at the sixth position. The β-globin gene is found on the short arm of chromosome 11.

74. *(c)* The natural source of Ti-plasmid is bacteria, i.e. *Agrobacterium tumefaciens*. It is a Gram negative soil bacteria, which has natural genetic engineering capability by which it integrates its plasmid genes into plant genome.

75. *(b)* Earthworms are bisexual / harmaphrodites but they cannot fertilise their own eggs because of their relative position of male and female genital aperture and they are protrandous (i.e. male sex matures earlier than female gametes). So, cross-fertilisation takes place.

76. *(b)* One difference between blood and lymph is that blood contains both RBC and WBC while lymph contains only WBC. Blood is the red liquid which circulates inside arteries and veins, carrying respiratory gases and nutrients throughout the body. Lymph is a colourless fluid which bathes tissues and drains through the lymphatic system.

77. *(b)* Abnormal development of thymus can cause most severe immunodeficiency. Because thymus gland is responsible for producing hormones which activate the immune system by activating the T-cells.

78. *(d)* Mitochondria are not associated with protein glycosylation. Protein glycosylation helps in proper folding of proteins, stability and in cell to cell adhesion commonly needed by cells of the immune system. The major sites of this are ER, Golgi body, nucleus and the cell fluid.

79. *(c)* A child with blood group O is born by parents with genotypes I^AI^O and I^BI^O.

$I^AI^O \times I^BI^O$

$I^AI^B \quad I^AI^O \quad I^BI^O \quad I^aI^O$

$\therefore$ There is 25% probability of having a O blood group here. We know that there is 50% probability of a girl child to be born each time a couple reproduces.

$\therefore$ The probability of having a girl child with blood group O when the parents have blood group A and B is at most 25%.

80. *(a)* Wooden doors swell up and get struck during the rainy season, this phenomenon is due to imbibition. Imbibition is defined as the displacement of one fluid by another immiscible fluid.

81. *(c)* We have,

$p(x) = a_0 + a_1x + a_2x^2 + \ldots + a_nx^n$

$p(-2) = -15, p(-1) = 1, p(0) = 7\ p(1) = 9, p(2) = 13, p(3) = 25$

Clearly, $p(x)$ is an increasing function.

$\therefore n$ is an odd number.

Here, $n = 3$ or 5

put $n = 3$,

$$p(x) = a_0 + a_1x + a_2x^2 + a_3x^3$$

$$p(0) = a_0 = 7$$

$$p(-1) = 1 = a_0 - a_1 + a_2 - a_3 \quad \ldots(\text{i})$$

$$p(1) = 9 = a_0 + a_1 + a_2 + a_3 \quad \ldots(\text{ii})$$

On adding Eqs. (i) and (ii), we get

$$a_2 = -2$$

$$\therefore \quad 1 = 7 - a_1 - 2 - a_3$$

$$\Rightarrow \quad a_1 + a_3 = 7 - 2 - 1 = 4 \quad \ldots(\text{iii})$$

$$\Rightarrow \quad p(2) = 13 = a_0 + 2a_1 + 4a_2 + 8a_3$$

$$\Rightarrow \quad 13 = 7 + 2a_1 - 8 + 8a_3$$

$$\Rightarrow \quad a_1 + 4a_3 = 7 \quad \ldots(\text{iv})$$

On solving Eqs. (iii) and (iv), we get

$$a_1 = 3, a_3 = 1$$

$$\therefore \quad p(x) = 7 + 3x - 2x^2 + x^3$$

which is satisfies $p(-2)$ and $p(3)$

$$\therefore \quad n = 3$$

82. *(c)* Let a, b, c be the sides of a triangle.

$$\therefore \quad a^2 + b^2 \geq 2ab \quad [\because \text{AM} \geq \text{GM}]$$

Similarly, $b^2 + c^2 \geq 2bc$

$$c^2 + a^2 \geq 2ac$$

$$\Rightarrow 2(a^2 + b^2 + c^2) \geq 2(ab + bc + ca)$$

$$\Rightarrow \frac{a^2 + b^2 + c^2}{ab + bc + ac} \geq 1$$

$$\therefore \quad t \geq 1 \quad \left[\because \frac{a^2 + b^2 + c^2}{ab + bc + ac} = t\right]$$

$\because a, b, c$ be the side of a triangle.

$$\therefore \quad a + b > c$$

$$|a| > |c - b|$$

$$\Rightarrow \quad a^2 > (c - b)^2$$

$$\Rightarrow \quad a^2 > c^2 + b^2 - 2bc$$

$$\Rightarrow \quad b^2 + c^2 - a^2 < 2bc \quad \ldots(\text{i})$$

Similarly, $a^2 + b^2 - c^2 < 2ab \quad \ldots(\text{ii})$

and $c^2 + a^2 - b^2 < 2ac \quad \ldots(\text{iii})$

On adding Eqs. (i), (ii) and (iii), we get

$$a^2 + b^2 + c^2 < 2(ab + bc + ca)$$

$$\Rightarrow \frac{a^2 + b^2 + c^2}{ab + bc + ca} < 2$$

$$\therefore \quad t < 2$$

$$\therefore \quad \{x \in R \mid 1 \leq x < 2\}$$

83. *(c)* We have, $y = ||x - 3| - 4| - 5$

$$y = \begin{cases} -x - 6, & x < -1 \\ x - 4, & -1 \leq x < 3 \\ -x + 2, & 3 \leq x < 7 \\ x - 12, & x \geq 7 \end{cases}$$

The graph of function are

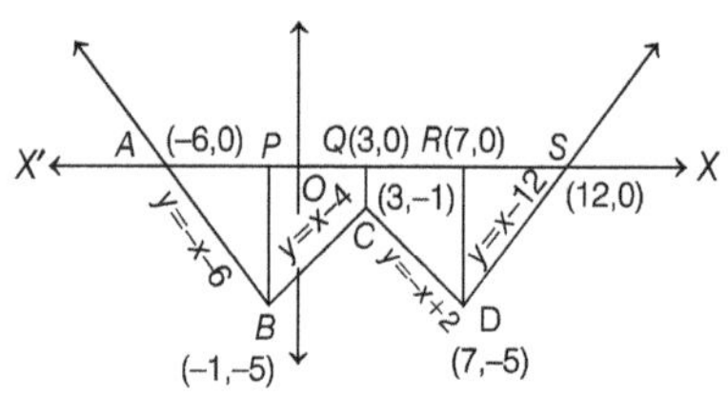

Area of region bounded by $y = ||x - 3| - 4| - 5$ and X-axis

= Area of ΔAPB + Area of trapezium $BPQC$ + Area of trapezium $QCDR$ + Area of ΔRSD

$$= \frac{1}{2} \times 5 \times 5 + \frac{1}{2} \times 4 \times (5 + 1) + \frac{1}{2} \times 4 \times (5 + 1) + \frac{1}{2} \times 5 \times 5$$

$$= 49$$

85. *(b)* We have,

$$a_n = \sqrt{\frac{1 + a_{n-1}}{2}} \text{ for } n \geq 1, a_0 = \cos\theta$$

$$a_1 = \sqrt{\frac{1 + a_0}{2}} = \sqrt{\frac{1 + \cos\theta}{2}}$$

$$a_1 = \sqrt{\frac{2\cos^2\frac{\theta}{2}}{2}} = \cos\frac{\theta}{2}$$

$$a_2 = \sqrt{\frac{1 + \cos\frac{\theta}{2}}{2}} = \cos\frac{\theta}{2^2}$$

$$a_n = \cos\frac{\theta}{2^n}$$

$$\lim_{n \to \infty} 4^n(1 - a_n)$$

$$= \lim_{n \to \infty} 4^n\left(1 - \cos\frac{\theta}{2^n}\right)$$

$$= \lim_{n \to \infty} 4^n\left(2\sin^2\frac{\theta}{2^{n+1}}\right)$$

$$= \lim_{n \to \infty} 4^n \cdot 2\left(\frac{\sin\frac{\theta}{2^{n+1}}}{\frac{\theta}{2^{n+1}}}\right)^2 \times \frac{\theta^2}{2^{2n+2}}$$

$$= \lim_{n \to \infty} \frac{4^n}{4^n \cdot 2}\left(\frac{\sin\frac{\theta}{2^{n+1}}}{\frac{\theta}{2^{n+1}}}\right)^2 \times \theta^2 = \frac{\theta^2}{2}$$

86. *(d)* We have,

$$f(x) = (\sin x)^{\sin x}, x \in (0, \pi)$$

$$x = \frac{\pi}{2}$$

$$\Rightarrow \quad f\left(\frac{\pi}{2}\right) = 1$$

Hence, maximum value of $f(x)$ is 1.

Taking log both sides

$\log f(x) = \sin x \log \sin x$

On differentiating w.r.t. x, we get

$$f'(x) = f(x)[\cos x + \cos x \log \sin x]$$

put $f'(x) = 0$,

$\cos x(1 + \log \sin x) = 0$

$\cos x = 0, \log \sin x = -1$

$$x = \frac{\pi}{2}, \sin x = e^{-1}$$

$\therefore$ Range of $f(x)$ is $\left[\left(\frac{1}{e}\right)^{1/e}, 1\right] = [e^{-1/e}, 1]$

87. *(d)* Area bounded by curve

$$y = \frac{1}{x}, y = 0, x = 1, x = 10 \text{ is}$$

$$\int_1^{10} \frac{1}{x} dx = [\log_e x]_1^{10} = \log_e 10$$

$$A = \log_e 10 = 2.303$$

$$B = 1 + \frac{1}{2} + \frac{1}{3} + \frac{1}{4} + \ldots + \frac{1}{9}$$

$$C = \frac{1}{2} + \frac{1}{3} + \frac{1}{4} \ldots + \frac{1}{10}$$

Clearly, $B > C$

Also, $B > A$

$$\left[\because \log_e (1 + n) < 1 + \frac{1}{2} + \frac{1}{3} + \ldots + \frac{1}{n}\right]$$

$\therefore$ $B > A > C$

$2A > B + C$

$A - C > B - A$

88. *(d)* Two point on a circumference of circle of radius r.

Distance between two points is atleast r.

$\therefore$ Angle subtends to chord of length atleast r is greater than or equal to 60°.

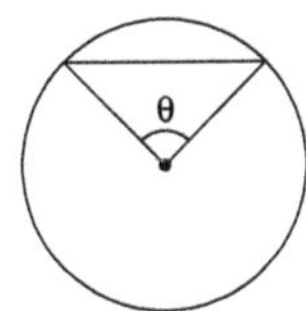

$$\therefore \text{Required probability} = 1 - \frac{60}{180}$$

$$= 1 - \frac{1}{3} = \frac{2}{3}$$

89. *(b)* We have all natural numbers whose decimal expansion has only even digits 0, 2, 4, 6, 8.

a_n = nth number of sequence if these are arranged in increasing orders.

Let $n = 5^K$, then 5 based representation of n is a 1 followed by K zeroes,

$\therefore$ Decimal representation and doubling its value, we obtain a 2 followed by K zeroes that is 2.10^K.

$\therefore$ $\log a_n = K + \log_{10} 2$

$\Rightarrow$ $\log n = K \log_{10} 5$

$$\Rightarrow \lim_{n \to \infty} \frac{\log a_n}{\log n} = \frac{K + \log_{10} 2}{K \log_{10} 5}$$

K goes to infinity

$$\therefore \lim_{n \to \infty} \frac{\log a_n}{\log n} = \frac{1}{\log_{10} 5} = \log_5 10$$

90. *(c)* $\sum_{1 \le i < j \le n} |i - j|$

If we fix $j = 1$, then i range, we get

$1 + 2 + 3 + \ldots + (n - 1)$,

if we fix $j = 2$, then i range, we get

$1 + 2 + 3 + \ldots + (n - 2)$

Similarly, we fix $j = n - 1$, then i range, we get

if we put them all together we get a total of $(n - 1)$ numbers 1's, $(n - 2)$ number of 2's ...... 1 number of $(n - 1)$'s then sum equals

$$\sum_{k=1}^{n-1} k(n - k) = n \sum_{k=1}^{n-1} k - \sum_{k=1}^{n-1} k^2$$

$$= \frac{n(n)(n - 1)}{2} - \frac{n(n - 1)(2n - 1)}{6}$$

$$= \frac{n(n - 1)}{2}\left[n - \frac{2n - 1}{3}\right]$$

$$= \frac{n(n - 1)(n + 1)}{2 \cdot 3} = {}^{n+1}C_3$$

91. *(d)* Centre of mass of solid sphere is at origin (centre of sphere).

Now, centre of mass of sphere with cavity is calculated as follows.

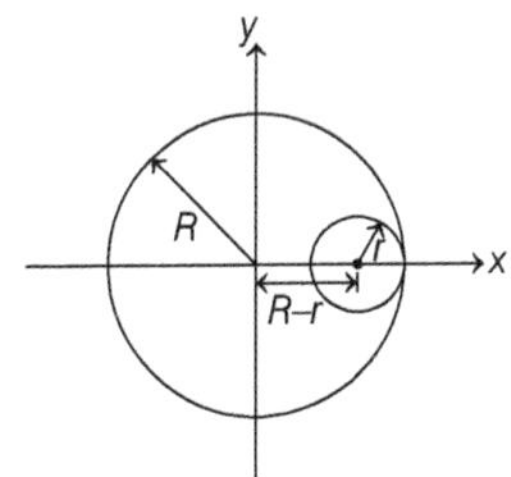

$$x_{CM} = \frac{m_1 x_1 - m_2 x_2}{m_1 - m_2}$$

$$= \frac{\left(\frac{4}{3}\pi R^3 \cdot \rho\right)(0) - \left(\frac{4}{3}\pi r^3 \cdot \rho\right)(R - r)}{\frac{4}{3}\pi R^3 \rho - \frac{4}{3}\pi r^3 \rho}$$

where, ρ = density of material of sphere.

$$\Rightarrow x_{CM} = -\frac{r^3 (R - r)}{(R^3 - r^3)}$$

$$= \frac{-r^3 (R - r)}{(R - r)(R^2 + Rr + r^2)}$$

$$= \frac{-r^3}{R^2 + Rr + r^2}$$

So, distance between two mass centres is

$$d = \frac{r^3}{R^2 + Rr + r^2}$$

92. *(a)* Due to silvering on curved surface,

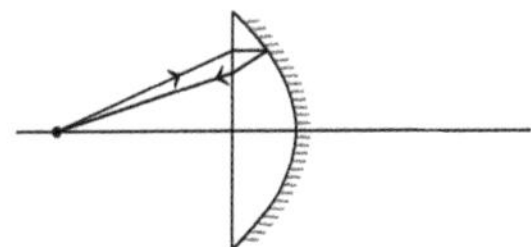

lens acts like a concave mirror and its focal length f is given by

$$\frac{1}{f} = \frac{2}{f_e} + \frac{1}{f_m}$$

$$= \frac{2}{R/(\mu - 1)} + \frac{1}{R/2}$$

$$\Rightarrow f = \frac{R}{2\mu}$$

Now, by mirror formula, we have

$$\frac{1}{f} = \frac{1}{v} + \frac{1}{u}$$

Here, $u = v = -x$

$$\therefore \frac{2\mu}{R} = \frac{(-1)}{x} + \frac{(-1)}{x}$$

$$\Rightarrow \frac{2\mu}{R} = \frac{-2}{x} \text{ or } x = \frac{-R}{\mu}$$

So, distance of object is $\left(\frac{R}{\mu}\right)$ from the pole of silvered plano-convex lens.

93. *(b)* Work done = Area enclosed under p-V diagram = $(p_1 - p_2)(V_2 - V_1)$

94. *(c)* Given circuits are

Circuit I,

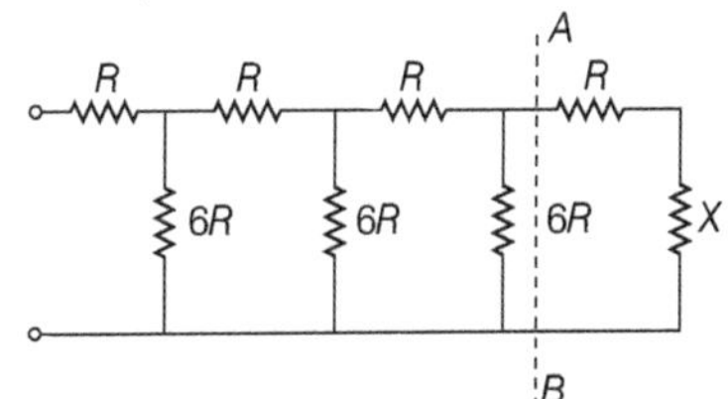

Circuit II,

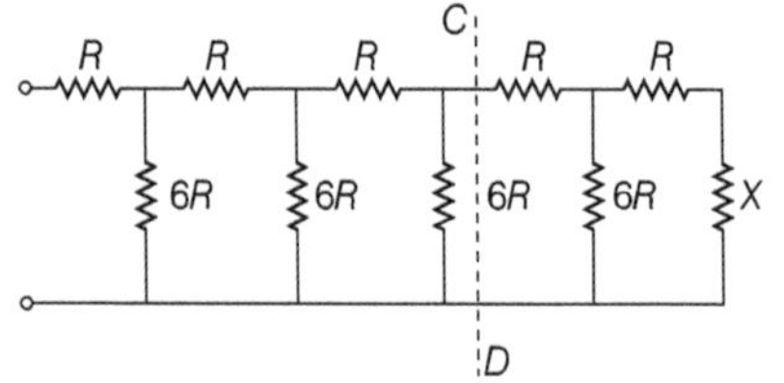

Clearly, both circuits are different after sections AB and CD.

So, resistances of two circuits are same, if $R_{AB} = R_{CD}$.

So, $\quad R + X = R + \dfrac{6R\,(R + X)}{6R + R + X}$

$\Rightarrow \quad X = \dfrac{6R\,(R + X)}{7R + X}$

$\Rightarrow \; 7RX + X^2 = 6R^2 + 6RX$

$\Rightarrow \; X^2 + RX - 6R^2 = 0$

From sridharacharya formula, we have

$$\Rightarrow \quad X = \frac{-R \pm \sqrt{R^2 - 4(1)\,(-6R^2)}}{2(1)}$$

$$= \frac{-R \pm \sqrt{25R^2}}{2}$$

or $\quad X = \dfrac{-R \pm 5R}{2}$

$\therefore \quad X = 2R$

95. *(d)*

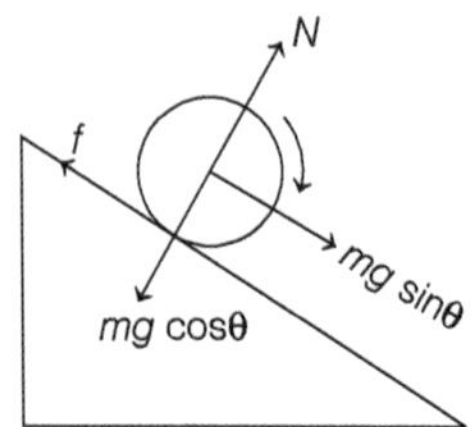

For a cylinder in pure rolling, we have following equations,

$$N - Mg\cos\theta = 0 \qquad \text{...(i)}$$

$$Ma_{CM} = Mg\sin\theta - f \qquad \text{...(ii)}$$

$$\tau = f \times R = I\alpha = I\frac{a_{CM}}{R}$$

or $\quad f = \dfrac{MK^2}{R^2} \cdot a_{CM} \qquad \text{...(iii)}$

From Eqs. (ii) and (iii), we have

$$a_{CM} = \frac{g\sin\theta}{(1 + K^2/R^2)}$$

So, from Eq. (ii), we have

Friction, $f = Mg\sin\theta - Ma_{CM}$

$$= M\left(g\sin\theta - \frac{g\sin\theta}{1 + K^2/R^2}\right)$$

$$= M\left(\frac{K^2/R^2}{1 + K^2/R^2}\right) g\sin\theta$$

For a sphere,

$$I = \frac{2}{5}MR^2 = MK^2$$

$$\Rightarrow \quad \frac{K^2}{R^2} = \frac{2}{5}$$

So, $\quad f = Mg\sin\theta\left(\dfrac{\frac{2}{5}}{1 + \frac{2}{5}}\right)$

$$\Rightarrow \quad f = \frac{2}{7}\,Mg\sin\theta$$

For no skidding,

$$\frac{f}{N} \geq \mu_s$$

or $\quad \dfrac{\frac{2}{7}Mg\sin\theta}{Mg\cos\theta} \geq \mu_s$

$$\Rightarrow \quad \tan\theta \geq \frac{7}{2}\mu_s$$

or $\quad \theta \leq \tan^{-1}\left(\dfrac{7}{2}\mu_s\right)$

If inclination is more than above value, then sphere will started to skid.

96. *(a)* When the block started to topple, normal reaction passes through the edge along which block tends to turn, as shown below.

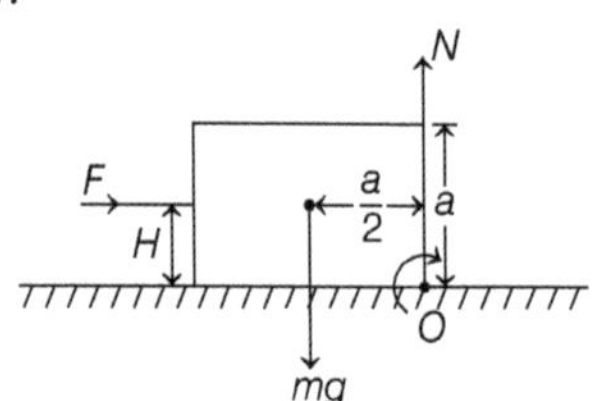

For toppling, torque of applied force F must be greater than torque of weight about O.

$$\Rightarrow \quad F \times H \geq mg\,\frac{a}{2} \qquad \text{...(i)}$$

and block begins to slide when

$$F \geq \text{friction}$$

$$\Rightarrow \quad F \geq \mu mg \qquad \text{...(ii)}$$

Equating limiting values from Eqs. (i) and (ii), we have

$$\mu mgH = mg\frac{a}{2}$$

or $\quad \mu = \dfrac{a}{2H}$

97. *(c)* On a banked road,

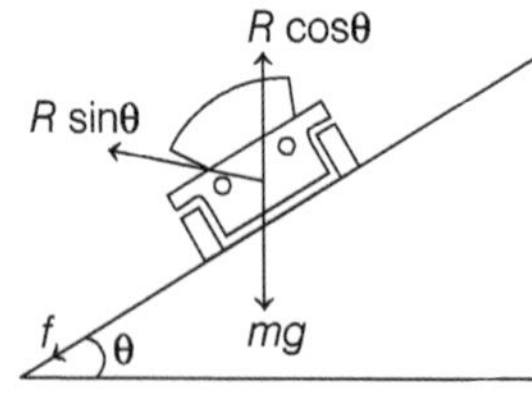

Resolving forces in horizontal and vertical directions, we have

$$R\sin\theta + f\cos\theta = \frac{mv^2}{r}$$

and $\quad R\cos\theta - f\sin\theta = mg$

$$\Rightarrow \quad \frac{R\sin\theta + f\cos\theta}{R\cos\theta - f\sin\theta} = \frac{v^2}{rg}$$

$$\Rightarrow \quad \frac{\tan\theta + f/R}{1 - f/R\,\tan\theta} = \frac{v^2}{rg}$$

$$\Rightarrow \quad \frac{\tan\theta + \mu}{1 - \mu\tan\theta} = \frac{v^2}{rg}$$

$$\Rightarrow \quad (\mu v^2 + rg)\tan\theta = v^2 - \mu rg$$

$$\Rightarrow \quad \tan\theta = \frac{v^2 - \mu rg}{\mu v^2 + rg}$$

98. *(c)*

4m, P, 3m, 5m

Path difference of sounds from speakers A and B is 1 m (= 5m − 4m).

For least intense sound, 1 m can cause a destructive interference. So, path difference is an odd multiple of half wavelength.

$$\Rightarrow \quad \Delta L = (2n + 1)\frac{\lambda_1}{2}$$

$$\Rightarrow \quad 1\text{m} = \frac{\lambda_1}{2}, \frac{3\lambda_1}{2}, \frac{5\lambda_1}{2}, \ldots, \text{etc.}$$

Above condition is satisfied, when $\lambda_1 = 2$ m.

And for most intense sound, path difference of 1 m can cause a constructive interference. So, path difference must be a whole number multiple of one wavelength.

$$\Rightarrow \quad \Delta L = n\lambda_2$$

$$\Rightarrow \quad 1\text{m} = \lambda_2, 2\lambda_2, 3\lambda_2, \ldots, \text{etc.}$$

Above condition is satisfied with $\lambda_2 = 1$ m.

So, correct option is (c).

99. *(b)* Block 2 moves for a certain period of time with higher velocity (compared to that of block 1).

So, block 2 finishes the race earlier.

100. *(b)* Change in volume of 1 kg of water due to phase change is

$$\Delta V = V_{\text{steam}} - V_{\text{water}}$$

$$= \frac{m_{\text{steam}}}{\rho_{\text{steam}}} - \frac{m_{\text{water}}}{\rho_{\text{water}}}$$

$$= \frac{1}{(1/1.8)} - \frac{1}{1000}$$

$$= 1.800 - 0.001 = 1.799\ \text{m}^3$$

Work done against atmospheric pressure during phase change is

$$\Delta W = p\Delta V$$
$$= 101\times 10^5 \times 1.799$$
$$= 181699\,J$$

Heat absorbed during phase change is

$$\Delta Q = mL$$
$$= 1\times 22.6\times 10^5$$
$$= 2260000\,J$$

So, change in internal energy is

$$\Delta U = \Delta Q - \Delta W$$
$$= 2260000 - 181699$$
$$= 2078301\,J\,kg^{-1}$$
$$= 20.8\times 10^5\,J\,kg^{-1}$$

101. *(c)* Given,

pH of a mixture = 8

Concentration of $NH_4OH = 0.1M$

Concentration of $NH_4Cl = 1M$

pOH = 14 − pH

pOH = 14 − 8 = 6

$$\therefore pK_b = pOH + \log\frac{[NH_4Cl]}{[NH_4OH]}$$
$$= 6 + \log\frac{[1]}{[0.1]}$$
$$= 6 + \log 10$$
$$= 6 + 1 = 7$$

102. *(d)* Molar mass of butane,

$$C_4H_{10} = 12\times 4 + 1\times 10 = 58\,g$$

Heat of combustion of butane,

$$\Delta H = -2658\,kJ/mol$$

58 g of butane gives 2658 kJ/mol of heat

$\therefore$ 11.6 kg of butane will give heat

$$= \frac{2658}{58}\times 11.6\times 10^3$$
$$= 531600\,kJ$$

Daily requirement for cooking = 15000 kJ

$\therefore$ Number of days cylinder will last

$$= \frac{531600}{15000}$$
$$= 35.44 \approx 35\text{ days}$$

103. *(a)* Given, weight of compound,

$$w_2 = 0.643\,g$$
$$k_f = 5.12\,K\,kg\,mol^{-1}$$
$$\Delta T_f = T_f^{\circ} - T_f$$
$$278.51 - 278.03 = 0.48\,K$$

Volume of benzene = 50 mL

Density of benzene = 0.879 g mL^{-1}

$\therefore$ Weight of benzene,

$$w_1 = 50\times 0.879 = 43.95\,g$$

Molar mass of a compound can be calculated by using the following formula

$$M_2 = \frac{k_f\times w_2\times 1000}{\Delta T_f\times w_1}$$
$$= \frac{5.12\times 0.643\times 1000}{0.48\times 43.95}$$
$$= 156.05\,g\,mol^{-1}$$

104. *(a)* Given, $E^{\circ}_{cell} = 2.57\,V$

$$[Ag^+] = 0.04\,M$$
$$[Zn^{2+}] = 0.28\,M$$

For the reaction,

$$Zn(s) + 2Ag^+ \longrightarrow Zn^{2+} + 2Ag(s)$$

Using Nernst equation,

$$E = E^{\circ} - \frac{2.303}{nF}\log\frac{[Zn^{2+}]}{[Ag^+]^2}$$
$$E = 2.57 - \frac{2.303}{2\times 96500}\log\frac{[0.28]}{[0.04]^2}$$
$$2.57 - \frac{2.303}{2\times 96500}\times 2.24 = 2.5\,V$$

105. *(b)* When Co (II) chloride is dissolved in conc. HCl gives $[CoCl_4]^{2-}$, which is blue is colour upon dilution with water this $[CoCl_4]^{2-}$ complex gets converted to $[Co(H_2O)_6]^{2+}$, which is pink in colour.

$$CoCl_2 + 2HCl \longrightarrow \underset{\text{Blue colour}}{[CoCl_4]^{2-}} + 2H^+$$

$$[CoCl_4]^{2-}(aq) + 6H_2O(l) \rightleftharpoons \underset{\text{Pink colour}}{[Co(H_2O)_6]^{2+}(aq)}$$

106. *(c)* Given,

$$\ln k = -\frac{11067}{T}k + 31.33$$

Comparing with Arrhenius equation,

$$\ln k = \ln A - \frac{E_a}{RT}$$
$$\frac{-E_a}{RT} = -\frac{11067}{T}$$
$$\Rightarrow \quad \frac{-E_a}{R} = -11067$$

Also, $\ln A = 31.33$

If the rate constant gets doubled, then

$$\log\left(\frac{2k}{k}\right) = \frac{E_a}{2.303R}\left[\frac{1}{T_1} - \frac{1}{T_2}\right]$$

When $T_1 = 25^{\circ}C = 298K$

$$\therefore \quad \log\frac{2k}{k} = \frac{11067}{2.303}\left[\frac{1}{298} - \frac{1}{T_2}\right]$$
$$0.301 = 4805.47\left[\frac{1}{298} - \frac{1}{T_2}\right]$$
$$6.62\times 10^{-5} = \frac{1}{298} - \frac{1}{T_2}$$
$$\frac{1}{T_2} = \frac{1}{298} - 6.62\times 10^{-5}$$
$$\frac{1}{T_2} = 3.289\times 10^{-3}$$
$$T_2 = 304.1\,K \text{ or } 31°C$$

107. *(a)* For the reactions,

$$CuCl_4^{2-} + Br^- \rightleftharpoons CuCl_3Br^{2-} + Cl^-; K_1 \quad ...(i)$$
$$CuCl_3Br^{2-} + Br^- \rightleftharpoons CuCl_2Br_2^{2-} + Cl^-; K_2 \quad ...(ii)$$
$$CuCl_2Br_2^{2-} + Br^- \rightleftharpoons CuClBr_3^{2-} + Cl^-\ K_3 \quad ...(iii)$$
$$CuClBr_3^{2-} + Br^- \rightleftharpoons CuBr_4^{2-} + Cl^-; K_4 \quad ...(iv)$$

To obtain reaction,

$$CuCl_4^{2-} + 3Br^- \rightleftharpoons CuClBr_3^{2-} + 3Cl^-$$

Add equations (i), (ii) and (iii)

$$\therefore \quad K = K_1\times K_2\times K_3$$

108. *(c)*.

Phenol (OH) $\xrightarrow{Br_2 \text{ in } CS_2}$ 4-bromophenol (OH, Br) *X* $\xrightarrow[\text{Me-I}]{\text{NaOH}}$ 4-bromoanisole (OMe, Br) *Y*

In first reaction phenol undergoes halogenation in presence of CS_2 solvent where 4-bromo phenol (*X*) is obtained as major product. In the second reaction 1-bromo-4-methoxy benzene (*Y*) is obtained when *X* reacts with NaOH in presence of MeI. This reaction takes place *via* S_N2 1-mechanism and is known as Williamson's synthesis.

109. *(d)* For the decay process,

$$^{234}Th_{90} \longrightarrow {}^{206}Pb_{82}$$

An alpha particle has 2 neutrons and 2 protons.

Thus, combined mass is 4. Thus, mass number increases by 4. Let the number of alpha particles be *x*.

$$\therefore \quad 234 = 4(x) + 206$$
$$28 = 4x \Rightarrow x = 7$$

A beta particle consist of an electron and neutron, where this neutron suddenly changes to electron. Thus, atomic number is increased by 2.

Let the number of beta particle be *y*.

$$90 = 82 + 2(y)$$
$$12 = 2y \Rightarrow y = 6$$

110. *(b)*

O Me OMe $\xrightarrow[\text{Reagent 1}]{AlCl_3 / H^+}$ OH C Me O OMe

$\xrightarrow[\text{Reagent 2}]{I_2/NaOH}$ OH COOH OMe

In first reaction, $AlCl_3$ act as a reagent and this reaction is known as fries rearrangement . It involves the migration of acyl group of phenol ester to aryl ring.

In 2nd reaction, I_2/NaOH act as a reagent. This reaction is haloform reaction, where methyl group linked to carboxyl carbon gets oxidised to carboxylic acid having one carbon atom less than of carbonyl carbon.

111. *(c)* The mode of action of penicillin is that it inhibits bacterial cell wall synthesis. Penicillin binds at the active site of the transpeptidase enzyme that cross links the peptidoglycan strands. Penicillin irreversibly inhibits the enzyme transpeptidase by reacting with a serine residue in the transpeptidase. This reaction is irreversible and, so the growth of the bacterial cell wall is inhibited.

112. *(b)* Meiosis is one round of chromosome duplication and two rounds of cell division. During meiosis, the parent cell undergoes one round of DNA replication followed by two separate cycles of nuclear division. The process results in four daughter cells that are haploid, which means they contain half the number of chromosomes of the diploid parent cell.

113. *(c)* The genetic code is degenerate because there are many instances in which different codons specify the same amino acid. This happens as there are 64 codons and only 20 amino acids. Thus, we can say that if instead of 3, if 2 bases code for an amino acid, the degeneracy of codons coding for the same amino acid would have remained the same.

114. *(b)* The unit factors (as called by Mendel) in modern terminology are called alleles. Mendel called the dominant allele, the expressed unit factor, the recessive allele was referred to as the latent unit factor. Alleles are pairs or series of genes on a chromosome that determine the hereditary characteristics.

115. *(c)* Most of the enzymes in the cell have the highest activity at a certain temperature called optimal growth temperature which is 37°C in *E.coli.* All bacteria have their own optimum environmental surrounding and temperature in which they thrive the most.

116. *(d)* The options given are

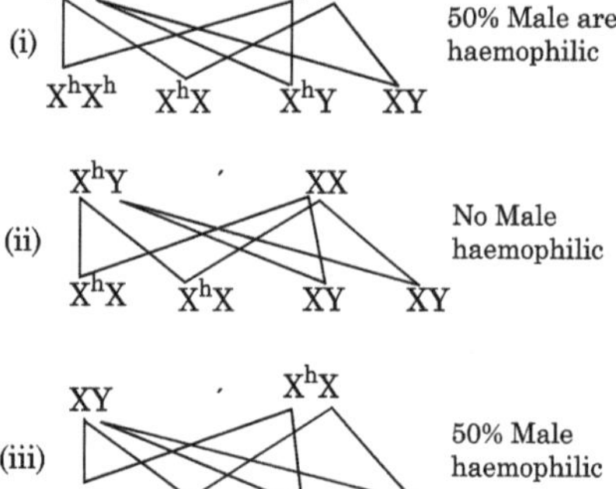

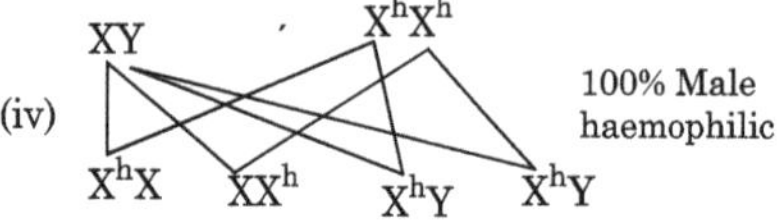

Therefore, 100% male are haemophilic in case of normal father and haemophilic mother.

117. *(c)* Protein is an important component of every cell in the body. Hair and nails are mostly made up of protein. Dietary protein intake helps in building, maintaining and repairing body tissues. The body's structural components, such as skin, muscles, bones and organs are made up of in large part by protein. Therefore, excess of protein by normal individuals would result in increase in the synthesis of muscle protein.

118. *(b)* Varicose veins are enlarged, swollen and twisting veins, often appearing blue or dark purple. They happen when faulty valves in the veins allow blood to flow in the wrong direction. The cause of varicose veins is high blood pressure, pregnancy, obesity and standing for long periods.

119. *(c)* Greatest proportion of photosynthesis in the world is carried out by algae in oceans.

It is estimated that marine plants produce between 70% to 80% of the oxygen in the atmosphere. Nearly all marine plants are single-celled photosynthetic algae.

120. *(c)* Energetically unfavourable reactions occur in human cells through coupling of energetically favourable reactions with unfavourable ones. Many energetically unfavourable reactions in cells are coupled to the hydrolysis of the molecule ATP. The breakdown of ATP to ADP is a highly energetically favourable reaction that releases energy. This energy is then used to carry out the unfavourable reaction.

PRACTICE SETS (1-5)

PRACTICE SET 1

Stream : SB/SX

MM : 160

Instructions

1. There are 120 questions in this paper.
2. The question paper contains two parts; Part I (1 Mark Questions) and Part II (2 Marks Questions).
3. There are four sections in each part; Mathematics, Physics, Chemistry and Biology.
4. Out of the four options given with each question, only one is correct.

PART-I (1 Mark Questions)

MATHEMATICS

1. If $|z-4+3i| \le 1$ and α and β are the least and greatest value of $|z|$ and K be the least value of $\dfrac{x^4+x^2+4}{x}$ on the interval $(0, \infty)$, then K is equal to

(a) α (b) β
(c) $\alpha\beta$ (d) None of these

2. The angle α, β, γ of a triangle satisfy the equation $2\sin\alpha + 3\cos\beta = 3\sqrt{2}$ and $3\sin\beta + 2\cos\alpha = 1$. Then, γ equal

(a) 150° (b) 120° (c) 60° (d) 30°

3. Let E' denote the complement of an event E. Let E, F, G be pairwise independent events with $P(G) > 0$ and $P(E \cap F \cap G) = 0$. Then, $\dfrac{P(E' \cap F')}{G}$ equals

(a) $P(E') + P(F')$ (b) $P(E') - P(F')$
(c) $P(E') - P(F)$ (d) $(E) - P(F')$

4. Let $f(x) = \dfrac{x}{(1+x^n)^{1/n}}$ for $n \ge 2$ and $g(x) = \underbrace{(f_0 f_0 \cdots f)}_{n \text{ times}} f(x)$.

Then, $\int x^{n-2} g(x)dx$ is equal to

(a) $\dfrac{1}{n(n-1)}(1+nx^n)^{1-\frac{1}{n}} + C$

(b) $\dfrac{1}{n-1}(1+nx^n)^{1-\frac{1}{n}} + C$

(c) $\dfrac{1}{n(n+1)}(1+nx^n)^{1+\frac{1}{n}} + C$

(d) $\dfrac{1}{n+1}(1+nx^n)^{1+\frac{1}{n}} + C$

5. Suppose two perpendicular tangents can be drawn from the origin to the circle $x^2 + y^2 - 6x - 2py + 17 = 0$, for some real p. Then, $|p|$ is equal to

(a) 0 (b) 3
(c) 5 (d) 17

6. A hyperbola, having the transverse axis of length $2\sin\theta$ is confocal with the ellipse $3x^2+4y^2=12$. Then, its equation is

(a) $x^2\text{cosec}^2\theta - y^2\sec^2\theta = 1$
(b) $x^2\sec^2\theta - y^2\text{cosec}^2\theta = 1$
(c) $x^2\sin^2\theta - y^2\cos^2\theta = 1$
(d) $x^2\cos^2\theta - y^2\sin^2\theta = 1$

7. One Indian and four American men and their wives are to be seated randomly around a circular table. Then, the conditional probability that the Indian man is seated adjacent to his wife given that each American man is seated to his wife is

(a) $\frac{1}{2}$ (b) $\frac{1}{3}$ (c) $\frac{2}{5}$ (d) $\frac{1}{5}$

8. The sides of a right angled triangle are in arithmetic progression. If the triangle has area 24. Then, the length of smallest side is

(a) 6 (b) 8 (c) 10 (d) 12

9. For every $x \in R$ the value of expression $y = \frac{x^2}{8} + x\cos x + \cos 2x$ is never less than

(a) −1 (b) 0 (c) 1 (d) 2

10. The function $f(x)=0$ has eight distinct real solutions and f also satisfy $f(4+x)=f(4-x)$. The sum of all eight solution of $f(x)=0$ is

(a) 12 (b) 32 (c) 16 (d) 15

11. The total number of distinct $x \in [0, 1]$ for which $\int_0^x \frac{t^2}{1+t^4}\,dt = 2x-1$ is

(a) 1 (b) 2
(c) 3 (d) 4

12. Let the curve C be the mirror image of the parabola $y^2=4x$ with respect to the line $x+y+4=0$. If A and B are the points of intersection of C with the line $y=-5$, then the distance between A and B is

(a) 3 (b) 4
(c) 5 (d) 6

13. The value of a for which the points A, B, C with position vector $2\hat{\mathbf{i}} - \hat{\mathbf{j}} + \hat{\mathbf{k}}$, $\hat{\mathbf{i}} - 3\hat{\mathbf{j}} - 5\hat{\mathbf{k}}$ and $a\hat{\mathbf{i}} - 3\hat{\mathbf{j}} + \hat{\mathbf{k}}$ respectively are the vertices of a right angle triangle with $C = \frac{\pi}{2}$ are

(a) 2 and 1 (b) −2 and −1
(c) −2 and 1 (d) 2 and −1

14. The value of $\int_1^a [x]f'(x)dx$, $a>1$ where $[x]$ denotes the greatest integer not exceeding x is

(a) $af(a) - \{f(1)+f(2)+\ldots f([a])\}$
(b) $[a]f(a) - \{f(1)+f(2)+\ldots+f([a])\}$
(c) $[a]f(a) - \{f(1)+f(2)+\ldots+f(a)\}$
(d) $af(a) - \{f(1)+f(2)+\ldots+f(a)\}$

15. If $\log_x 8 = z$, $\log_y x = -1$ and $\log_{\frac{1}{4}} y = -1$, then $\left(\frac{1}{x}+1\right)^{\log\sqrt{5}(y^2+4z^2)}$ is equal to

(a) $\sqrt{5}$ (b) 5 (c) 25 (d) 625

16. If $0<x<\pi$ and $\cos x + \sin x = \frac{1}{2}$, then $\tan x$ is

(a) $\frac{1-\sqrt{7}}{4}$ (b) $\frac{4-\sqrt{7}}{3}$ (c) $-\left(\frac{4+\sqrt{7}}{3}\right)$ (d) $\frac{1+\sqrt{7}}{4}$

17. One ticket is selected at random from 50 tickets numbered 00, 01, 02, ..., 49. Then, the probability that the sum of the digits on the selected ticket is 8 given that the product of these digits is zero, equals

(a) $\frac{1}{14}$ (b) $\frac{1}{7}$ (c) $\frac{5}{14}$ (d) $\frac{1}{50}$

18. The number of ordered pair (a, b), where $a, b \in \{1, 2, 3, \ldots, 50\}$ such that $6^a + 9^b$ is multiple of 5 is

(a) 1250 (b) 2500 (c) 625 (d) 500

19. In a binomial distribution $B\left(n, p=\frac{1}{4}\right)$. If the probability of at least one success is greater than or equal to $\frac{9}{10}$, then n is greater than

(a) $\frac{1}{\log_{10}4 - \log_{10}3}$ (b) $\frac{1}{\log_{10}4 + \log_{10}3}$
(c) $\frac{9}{\log_{10}4 - \log_{10}3}$ (d) $\frac{4}{\log_{10}4 - \log_{10}3}$

20. If $\left|z - \frac{4}{z}\right| = 2$, then maximum value of $|z|$ is equal to

(a) $\sqrt{3}+1$ (b) $\sqrt{5}+1$ (c) 3 (d) $2+\sqrt{2}$

PHYSICS

21. A pulley system with frictionless massless pulleys and ideal string is used to lift a weight of 100 N. F is the force that is sufficient to lift the load.

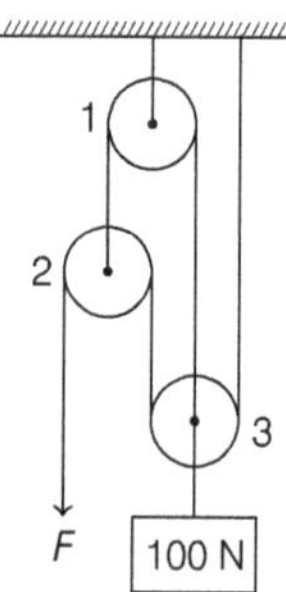

Now, choose the correct option.

(a) F is atleast 25 N
(b) F is atmost 25 N
(c) F is exactly 25 N
(d) Load cannot be lifted by given arrangement

22. Water flows out from a beaker through a capillary of length 200 mm and diameter 2 mm.

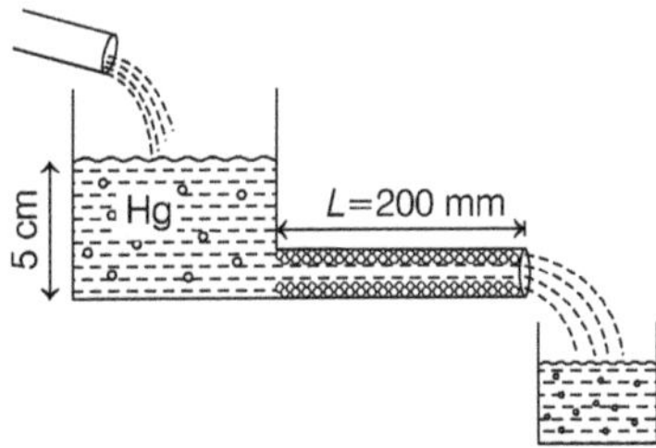

Density of water is 1000 kgm^{-3} and its viscosity is 0.8 centipoise. If a level of 5 cm is continuously maintained in the beaker, then flow rate through capillary is nearly

(a) 5 mLs^{-1} (b) 1 mLs^{-1}
(c) 15 mLs^{-1} (d) 13 mLs^{-1}

23. Partial pressures of components of a gaseous mixture are as given below.

Hydrogen	:	200 mm Hg
Carbon-dioxide	:	150 mm Hg
Methane	:	320 mm Hg
Ethylene	:	105 mm Hg

Percentage of mass of hydrogen in mixture is around

(a) 2.5% (b) 0.025%
(c) 2.3% (d) 0.23%

24. Let's assume that our atmosphere consists only of nitrogen gas with molecular mass is nearly 4.5×10^{-26} kg. If mean temperature of earth is taken as 27°C, then height of our atmosphere is nearly

(a) 15 km (b) 28 km
(c) 40 km (d) 12 km

25. A container is separated into two chambers A and B of equal volumes by a conducting wall.

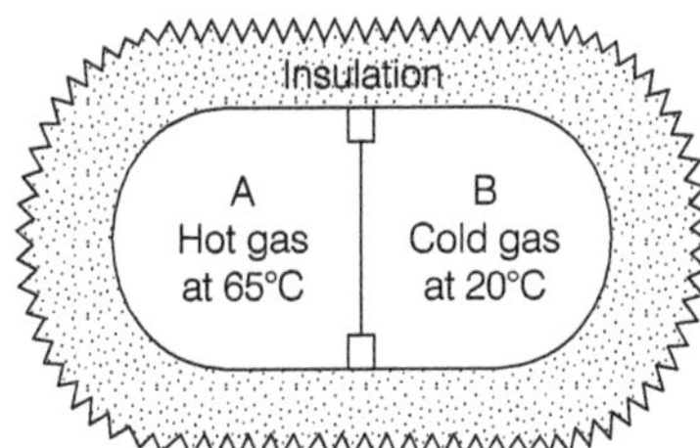

Heat cannot leave or enters the chambers A and B from outer boundary except through partition AB. Compartment A is filled with hot gas and compartment B is filled with same cold gas in equal amount. Hot gas cool down from 65°C to 63° C, choose correct,

(a) entropy change of A is more
(b) entropy change of B is more
(c) entropy change of A and B is same
(d) total entropy change of A and B is negative

26. An aerofoil shaped plane wing is being tested in an wind tunnel.

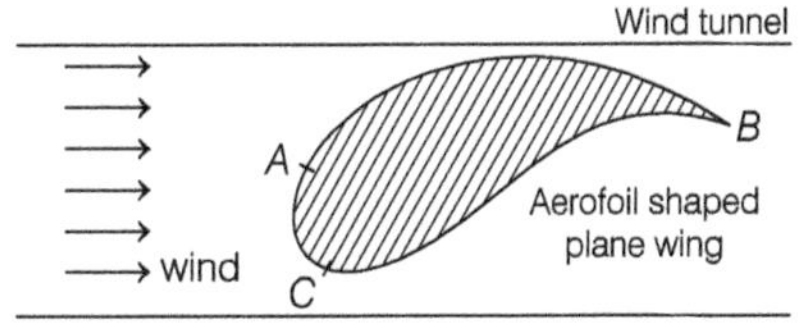

Choose the most appropriate depiction of pressure over aerofoil.

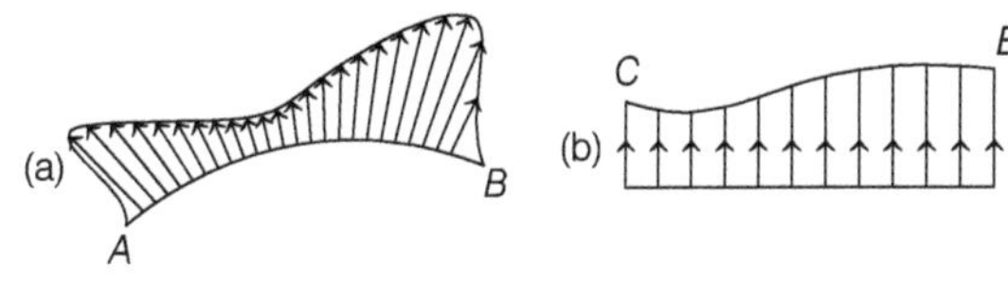

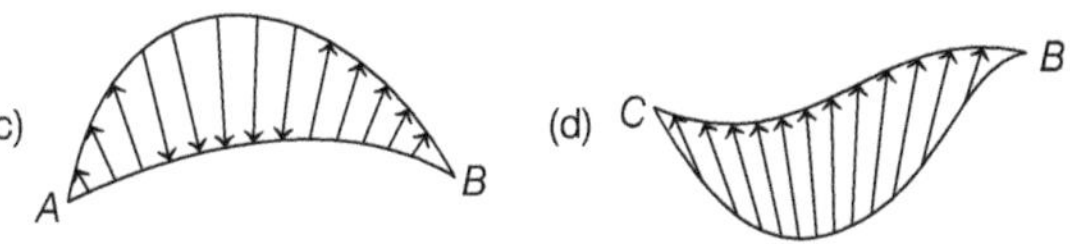

27. Consider a block held stationary over a rough incline by applying a force F parallel to the incline as shown below.

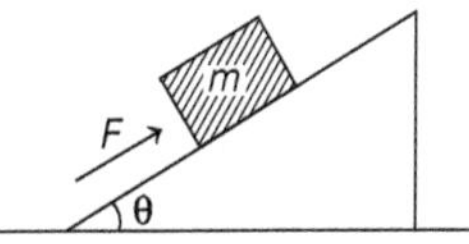

Force applied is initially, $F_i = mg(\sin\theta - \mu_s \cos\theta)$ and is increased finally to $F_f = mg(\sin\theta + \mu_K \cos\theta)$, where μ_s and μ_K are coefficient of friction and θ is angle of inclination.

Now, correct graph of F (Force) *versus* f (friction) is

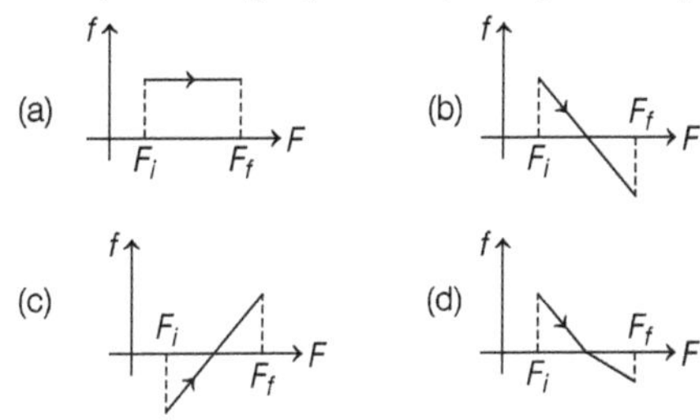

28. A cubical block of side l is given a kinetic energy K. It moves over a smooth flat surface and strikes a ridge at some point P as shown below.

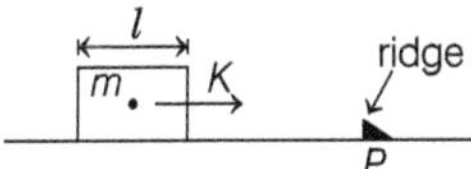

Moment of inertia of cube about its centre is $\frac{ml^2}{6}$.

Angular speed of block just after it strikes P will be

(a) $\sqrt{2}\frac{Kl}{m}$ (b) $\frac{\sqrt{2K}\cdot m}{l^2}$

(c) $\frac{3}{2\sqrt{2}}\cdot\sqrt{\frac{K}{ml^2}}$ (d) $\frac{2}{\sqrt{3}}\cdot\sqrt{\frac{Km}{l^2}}$

29. A typical atomic nucleus is about 5×10^{-15} m in radius. Using uncertainty principle, the least value of energy of an electron so that it resides in nucleus is around (Use $\frac{h}{2\pi}=1.054\times10^{-34}$ Js)

(a) 2.6×10^{-12} J (b) 2.6×10^{12} J
(c) 2.6×10^{-12} eV (d) 2.6×10^{12} eV

30. A long copper wire carries a steady current. Force on an electron moving with drift speed on surface of wire will be take current in wire $I=10$ A, radius of wire $a=0.5$ mm and number of conduction electrons in wire $n=5\times10^{28}\text{m}^{-3}$.

(a) $\pi\times10^{-25}$ N (b) $2\pi\times10^{-26}$ N
(c) $4\pi\times10^{-25}$ N (d) $8\pi\times10^{-26}$ N

31. A thin metallic disc of radius R is rotating with constant angular velocity ω about a vertical axis that is perpendicular to plane of disc and passes through its centre.

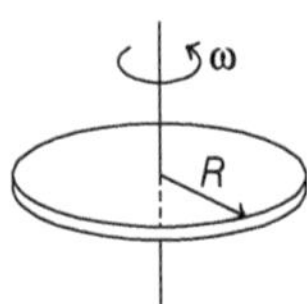

Rotation causes free electrons in the disc to redistribute. If there is no external electric or magnetic field is present, then potential difference between centre and rim of disc is

(a) 0 (b) $\frac{m\omega^2R^2}{e}$ (c) $\frac{m\omega^2R^2}{2e}$ (d) $\frac{2m\omega^2R^2}{e}$

32. An ideal gas is adiabatically compressed to one tenth of its initial volume. Ratio of final temperatures T_A and T_B will be (where, in case A a monoatomic gas is taken and in case B a diatomic gas is taken)

(a) $\frac{T_A}{T_B}=10^{\frac{7}{5}}$ (b) $\frac{T_A}{T_B}=10^{\frac{2}{5}}$ (c) $\frac{T_A}{T_B}=10^{\frac{4}{15}}$ (d) $\frac{T_A}{T_B}=10^{\frac{7}{15}}$

33. In given cycle, heat is dissipated in the process.

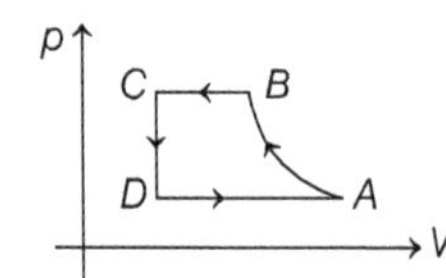

(a) Process AB (b) Process BC
(c) Process CD (d) Process DA

34. If we consider air resistance, then in case of a small ball thrown up from the ground correct is

(a) its initial velocity from ground is more than its final velocity on reaching ground
(b) its final velocity on reaching ground is more than its initial velocity
(c) its final velocity on ground is equal to its initial velocity
(d) time to reach maximum height is more than its time of fall

35. A light ray strikes centre of one end of a cylinder rod of refractive index n, at an angle α with its axis.

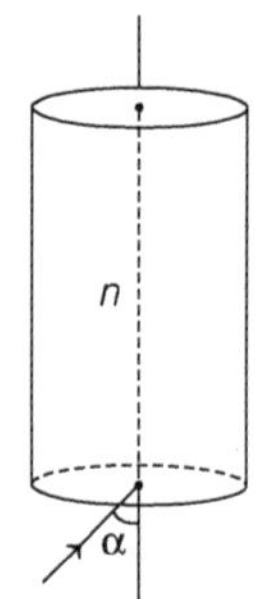

Least value of n such that light ray does not emerge from curved surface of cylinder is

(a) $\sqrt{3}$ (b) $\sqrt{2}$ (c) 2 (d) $\sqrt{5}$

36. Consider following nuclear reactions and their reaction energies Q.

$$^{14}\text{N}+{}^{2}\text{H}\longrightarrow{}^{15}\text{N}+{}^{1}\text{H}+8.53\text{ MeV}$$
$$^{15}\text{N}+{}^{2}\text{H}\longrightarrow{}^{13}\text{C}+{}^{4}\text{He}+7.58\text{ MeV}$$
$$^{13}\text{C}+{}^{2}\text{H}\longrightarrow{}^{11}\text{B}+{}^{4}\text{He}+5.16\text{ MeV}$$

It is given that, masses of involved nuclii are

${}^{4}_{2}\text{He}=4.0026$ u, ${}^{2}_{1}\text{H}=2.0140$ u,

${}^{1}_{1}\text{H}=1.0078$ u, ${}^{1}_{0}n=1.0087$ u

$1\text{ u}=931\frac{\text{MeV}}{c^2}$

Using above data, Q value of reaction

$$^{11}\text{B}+{}^{4}\text{He}\longrightarrow{}^{14}\text{N}+{}^{1}_{0}n \text{ is}$$

(a) 0.5 MeV (b) 0.005 MeV
(c) 0.05 MeV (d) 0.5 eV

37. A long metal rod of radius R carries a DC current I. Magnetic field associated with rod varies with distance from its centre r as,

(a) 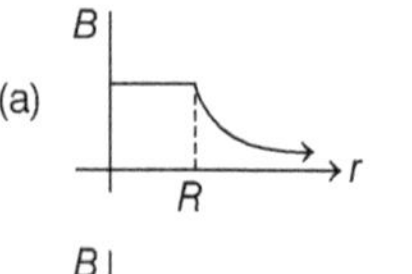

(b) 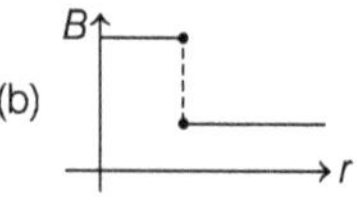

(c) 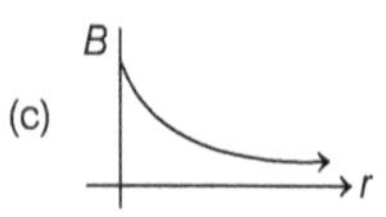

(d) 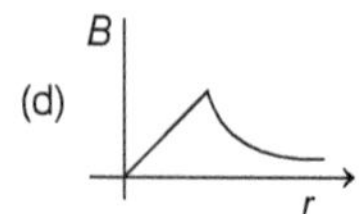

38. Temperature at which probability of having an atom in $n = 2$ state is 0.001 will be

(a) -273 K (b) 300 K
(c) 1430 K (d) 14300 K

39. In a wind tunnel used to test drag force on car using their models a 20 cm high car model is used for testing a car 550 cm high.

Let you want to calculate drag force on car, when it is moving at 15 ms^{-1} in still air. Wind speed in the tunnel must be kept around

(a) 4 kms^{-1} (b) 4 kmh^{-1}
(c) 0.4 kms^{-1} (d) 0.4 kmh^{-1}

40. A thin isentropic dielectric rod is placed between two unlike charged spheres having charges $+q_1$ and $-q_2$ as shown below.

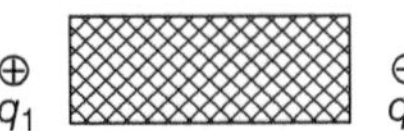

Now, choose the correct option.

(a) Force on q_1 will be more
(b) Force on q_1 will be less
(c) Force on q_1 remains same
(d) Magnitude of force depends on dielectric constant of rod

CHEMISTRY

41. Total number of stereoisomers possible for this molecule are $CH_3CH = CHCH_2CHBrCH_3$

(a) 8 (b) 2
(c) 4 (d) 6

42. Equivalent conductance of NaCl, HCl and C_2H_5COONa, at infinite dilution are 126.45, 426.16 and 91 Ω^{-1} cm^2 respectively. The equivalent conductance of C_2H_5COOH is

(a) 201.28 Ω^{-1} cm^{-2} (b) 390.71 Ω^{-1} cm^{-2}
(c) 698.28 Ω^{-1} cm^{-2} (d) 540.48 Ω^{-1} cm^{-2}

43. For the following reaction,

$$H_2(g) + Cl_2(g) \rightleftharpoons 2HCl(g)$$

$\Delta G°$ is -262 kJ. The equilibrium constant k for the reaction at 298 K is

(a) 1.27×10^{23} (b) 3.63×10^{40}
(c) 8.279×10^{45} (d) 0.03×10^{23}

44. Which of the following sets of ions are colourless ?

(a) $Zn^{2+}, Cu^{2+}, Ti^{3+}, Co^{2+}$ (b) $Zn^{2+}, Cr^{+}, Ti^{4+}, V^{2+}$
(c) $Cr^{3+}, Mn^{2+}, Zn^{2+}, Ti^{4+}$ (d) $Mn^{7+}, Cr^{6+}, Cu^{+}, V^{2+}$

45. Which of the following solutions will have the minimum freezing points?

(a) 0.01 M NaCl (b) 0.005 M C_2H_5OH
(c) 0.005 M MgI_2 (d) 0.005 M $MgSO_4$

46. Number of α and β-particles emitted, when an atom of $_{90}Th^{232}$ undergoes disintegration to produce an atom of $_{82}Pb^{208}$, are

(a) 6α and 4β (b) 2α and 4β
(c) 6α and 2β (d) 4α and 6β

47. The four quantum numbers of the 19th electron in Ti($Z = 22$) are

(a) $n = 4, l = 1, m = +2, s = +\frac{1}{2}$
(b) $n = 4, l = 0, m = 0, s = +\frac{1}{2}$
(c) $n = 3, l = 2, m = +2, s = +\frac{1}{2}$
(d) $n = 4, l = 2, m = -1, s = +\frac{1}{2}$

48. Synthetic polymer that resembles natural rubber is

(a) chloroprene (b) isoprene
(c) neoprene (d) glyptal

49. Among the following ions, the $p\pi$-$d\pi$ overlapping could be present in

(a) PO_4^{3-} (b) NO_2^- (c) NO_3^- (d) CO_3^{2-}

50. A compound forms hexagonal close-packed structure. What is the total number of voids in 0.5 mole of it?

(a) 3.011×10^{23} (b) 6.022×10^{23}
(c) 9.033×10^{23} (d) 1.802×10^{24}

51. 90% of a first order reaction was completed in 100 min. The half-life of the reaction is

(a) 63.3 min (b) 53.3 min
(c) 43.3 min (d) 30 min

52. Copper metal on treatment with dilute HNO_3 produces a gas (X). X, when passed through acidic solution of stannous chloride, a nitrogen containing (Y) is obtained. Which on reaction with nitrous acid produces a gas (Z) is

(a) NO (b) N_2 (c) NO_2 (d) N_2O

53. The major structure of the following reaction,

HO–C_6H_4–CH_2OH + HCl $\xrightarrow{\text{Heat}}$

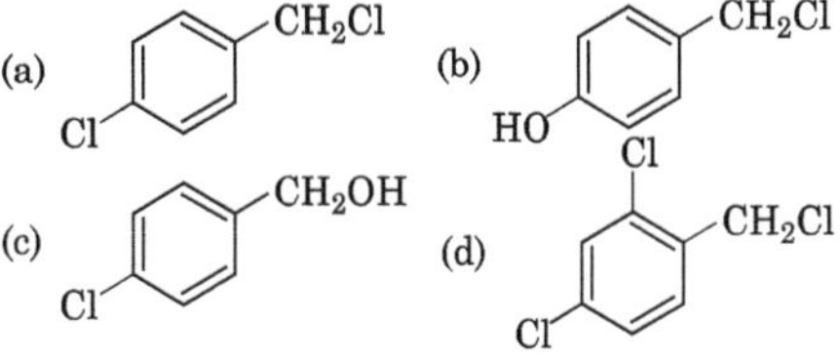

54. Which amino acid have more than one chiral centre?

(a) Aspartic acid (b) Arginine
(c) Lysine (d) Histidine

55. A 0.02 M solution of pyridinium hydrochloride has pH = 3.44. The ionisation constant of pyridine is
(a) 1.84×10^{-7} (b) 2.36×10^{-9}
(c) 1.5×10^{-9} (d) 3.62×10^{-7}

56. The correct order of increasing basicity of given compound is

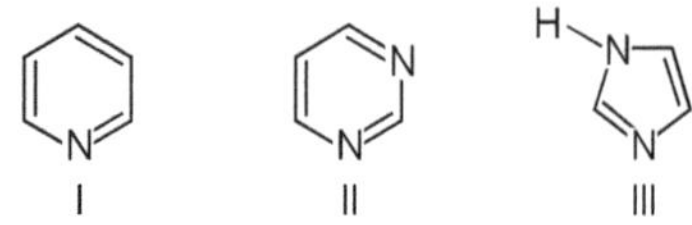

(a) II < III < I (b) I < II < III
(c) II < I < III (d) III < II < I

57. A mixture of benzaldehyde and formaldehyde on heating with aqueous NaOH solution gives
(a) benzyl alcohol and sodium formate
(b) sodium benzoate and methyl alcohol
(c) sodium benzoate and sodium formate
(d) benzyl alcohol and methyl alcohol

58. Standard enthalpy of formation of $C_3H_7NO_2$ (*g*), CO_2 (*g*) and $H_2O(l)$ are −133.57, −94.05 and −68.32 kcal/mol respectively. Standard enthalpy of combustion of CH_4 at 25°C is −212.8 kcal/mol. Calculate the enthalpy for combustion in kcal/mol of $C_3H_7NO_2(s)$.
(a) −372.69 (b) −387.70
(c) −370.89 (d) −372.16

59. The rate constant, the activation energy and the Arrhenius parameter of a chemical reaction at 25°C are $3 \times 10^{-4} s^{-1}$, 104.4 kJ mol^{-1} and $6.0 \times 10^{14} s^{-1}$ respectively. The value of rate constant at $T \to \infty$ is closest to
(a) $2.0 \times 10^{18} s^{-1}$ (b) $6 \times 10^{14} s^{-1}$
(c) $3.6 \times 10^{30} s^{-1}$ (d) infinity

60. The B—F bond length in BF_3 is shorter than that in BF_4^-. This is because of
(a) resonance in BF_3 but not in BF_4^-
(b) $p\pi$ - $p\pi$ back bonding in BF_4^- but not in BF_3
(c) $p\pi$ - $p\pi$ back bonding in BF_3 but not in BF_4^-
(d) $p\pi$ - $p\pi$ back bonding in BF_3 but not in BF_4^-

BIOLOGY

61. Certain nerve gases developed for military purposes work by producing convulsive muscular contractions upon the slightest stimulaiton. This suggests that their function is to inhibit the action of
(a) acetylcholine (b) atropine
(c) cholinesterase (d) eserine

62. Male gametes in angiosperms are formed by the division of
(a) microspore (b) generative cell
(c) vegetative cell (d) microspore mother cell

63. What happens when the pacemaker is non-functional?
(a) Only the auricles will contract rhythmically
(b) The cardiac muscles do not contract in a coordinated manner rhythmically
(c) Only ventricles will contract rhythmically
(d) Cardiac muscles will contract in a coordinated manner

64. Throughout which phase of the mitotic cycle would DNA be present in the least amount in a nucleus?
(a) Anaphase (b) Interphase
(c) Metaphase (d) Telophase

65. Which of the following is a complex of globular protein with non-proteinaceous material?
(a) Collagen (b) Egg albumen
(c) Haemoglobin (d) Fibrinogen

66. Bacteria containing recombinant plasmids are often identified by which process?
(a) By examining the cells with an electron microscope
(b) By using radioactive tracers to locate the plasmids
(c) By exposing the bacteria to an antibiotic that kills the cells lacking the plasmid
(d) By producing antibodies specific for each bacterium containing a recombinant plasmid

67. Which of the following causes prostate cancer?
(a) Chromium (b) Cadmium oxide
(c) Vinyl chloride (d) Aflatoxins

68. Beadle and Tatum showed that each kind of mutant bread mould they studied lacked a specific enzyme. Their experiments demonstrated that
(a) cells need specific enzymes in order to function
(b) genes are made of DNA
(c) genes carry information for making proteins
(d) enzymes are required to repair damaged DNA information

69. Identify the incorrect statement about DNA.
(a) The length of DNA is defined as the number of base pairs present in it
(b) Cytosine is common to both DNA and RNA
(c) In a nucleotide, the nitrogenous base is linked to a phosphate group
(d) Thymine is chemically 5-methyl uracil

70. Which one of the following cannot be explained on the basis of Mendel's law of dominance?
(a) The discrete unit controlling a particular character is called a factor
(b) Out of one pair of factors one is dominant and the other is recessive
(c) Alleles do not show any blending and both the characters recover as such in F_2-generation
(d) Factors occur in pairs

71. Which of the following increases variation within a gene pool?
(a) Chromosome inversion
(b) Crossing over
(c) Gene mutation
(d) Random fusion of gametes

72. Axenic culture is

(a) pure culture of a microbe without any nutrient
(b) pure culture without any contamination
(c) culture of tissue
(d) culture of gene

73. For every 100 units of sunlight falling on a chloroplast of a green plant, 50 units are not used for photosynthesis. This is because

(a) the wavelength are inappropriate
(b) they are converted into heat energy
(c) they are used to evaporate water vapour
(d) they fall on non-photosynthetic structures

74. How many moles of carbon dioxide are produced by the complete oxidation of 1 mole of pyruvate?

(a) 1 (b) 2 (c) 3 (d) 4

75. Match the following types of chloroplast with the respective algae.

Column I	Column II
A. Cup-shaped	1. *Ulothrix*
B. Girdle-shaped	2. *Oedogonium*
C. Stellate	3. *Chlamydomonas*
D. Reticulate	4. *Zygnema*

	A	B	C	D
(a)	2	4	3	1
(b)	3	1	4	2
(c)	3	4	2	1
(d)	4	3	1	2

76. Two cells A and B are contiguous. Cell A has osmotic pressure 10 atm, turgor pressure 7 atm and diffusion pressure deficit 3 atm. Cell B has osmotic pressure 8 atm, turgor pressure 3 atm and diffusion pressure deficit 5 atm. The result will be

(a) movement of water from cell B to A
(b) no movement of water
(c) equilibrium between the two cells
(d) movement of water from cell A to B

77. A mutation in one of the replication enzyme causes DNA polymerase to be unable to add nucleotides at the origin of replication, hence no daughter strands of DNA can be synthesised. Which of the following is the defective enzyme?

(a) Helicase (b) Primase
(c) DNA ligase (d) Topoisomerase

78. At which stage of HIV infection does one usually show symptoms of AIDS?

(a) Within 15 days of sexual contact with an infected person
(b) When the infected retrovirus enters host cells
(c) When HIV damages large number of helper T-lymphocytes
(d) When the viral DNA is produced by reverse transcriptase

79. The oxygen toxicity is related with

(a) blood poisoning (b) collapse of alveolar walls
(c) failure of ventilation (d) Both (a) and (b)

80. Which property of embryonic stem cells makes them particularly useful in medical research?

(a) They can be fused together to form zygote
(b) They can divide and are pluripotent
(c) They can divide and are multipotent
(d) They will continue to divide indefinitely

PART-II (2 Marks Questions)

MATHEMATICS

81. A red ball and a green ball are randomly and independently tossed into bins numbered with the positive integers so that for each ball, the probability that it is tossed into bin 2^{-k} for $k = 1, 2, 3, \ldots$ what is the probability that the red ball is tossed into a higher numbered bin than green ball?

(a) $\frac{1}{4}$ (b) $\frac{2}{7}$ (c) $\frac{1}{3}$ (d) $\frac{3}{8}$

82. In a circle with centre '*O*', *PA* and *PB* are two chords *PC* is the chord that bisects the $\angle APB$ the tangent to the circle at *C* is drawn meeting *PA* and *PB* extended at *Q* and *R* respectively. If $QC = 3$, $QA = 2$ and $RC = 4$, then length of *RB* equals

(a) 2 (b) $\frac{8}{3}$ (c) $\frac{10}{3}$ (d) $\frac{11}{3}$

83. If $\alpha = \int_0^1 (e^{9x + 3\tan^{-1} x})\left(\frac{12 + 9x^2}{1 + x^2}\right) dx$, where $\tan^{-1} x$ takes only principal values, then the value of $\left(\log_e |1 + \alpha| - \frac{3\pi}{4}\right)$ is

(a) 6 (b) 7 (c) 8 (d) 9

84. Let $n_1 < n_2 < n_3 < n_4 < n_5$ be positive integers such that $n_1 + n_2 + n_3 + n_4 + n_5 = 20$. Then, the number of such distinct arrangement $(n_1, n_2, n_3, n_4, n_5)$ is

(a) 7 (b) 6
(c) 8 (d) 5

85. The number of points in $(-\infty, \infty)$ for which $x^2 - x\sin x - \cos x = 0$ is

(a) 6 (b) 4
(c) 2 (d) 0

86. Let $f:\left[\frac{1}{2},1\right]\to R$ (the set of all real numbers) be a positive non-constant and differentiable function such that $f'(x)<2f(x)$ and $f\left(\frac{1}{2}\right)=1$. Then, the value of $\int_{1/2}^{1} f(x)dx$ lies in the interval

(a) $(2e-1, 2e)$ (b) $(e-1, 2e-1)$
(c) $\left(\frac{e-1}{2}, e-1\right)$ (d) $\left(0, \frac{e-1}{2}\right)$

87. Let $z=\cos\theta+i\sin\theta$. Then, the value of $\sum_{m=1}^{15} I_m(z^{2m-1})$ at $\theta=2°$ is

(a) $\frac{1}{\sin 2°}$ (b) $\frac{1}{2\sin 2°}$
(c) $\frac{1}{3\sin 2°}$ (d) $\frac{1}{4\sin 2°}$

88. Tangent drawn from the point $(1, 8)$ to the circle $x^2+y^2-6x-4y-11=0$ touch the circle at the points A and B. The equation of the circumcircle of ΔPAB is

(a) $x^2+y^2+4x-6y+19=0$
(b) $x^2+y^2-4x-10y+19=0$
(c) $x^2+y^2-2x+6y-29=0$
(d) $x^2+y^2-6x-4y+19=0$

89. Let $n_1=\sin 7+\cos 7$, $n_2=\sqrt{\sin 7}+\sqrt{\cos 7}$, $n_3=\sqrt{1+\sin 14}$ and $n_4=1$. Then,

(a) $n_2>n_3>n_4>n_1$ (b) $n_3>n_4>n_2>n_1$
(c) $n_3=n_1>n_4>n_2$ (d) $n_2>n_1=n_3>n_4$

90. If $|f(x)|+\sqrt{1+\cos 2\pi x}=\tan^2\left(\frac{\pi x}{9}\right)f(x)$, then $f(3)$ is equal to

(a) $\frac{1}{\sqrt{2}}$ (b) $\frac{1}{2\sqrt{2}}$ (c) $\frac{-1}{\sqrt{2}}$ (d) $-\frac{1}{2\sqrt{2}}$

PHYSICS

91. Which of these transitions are not frequently possible?

(a) $3p\longrightarrow 1s$ (b) $2p\longrightarrow 1s$
(c) $3p\longrightarrow 2p$ (d) $4p\longrightarrow 3p$

92. A gas is contained in a small container which is connected to a vaccum chamber via a stop value.

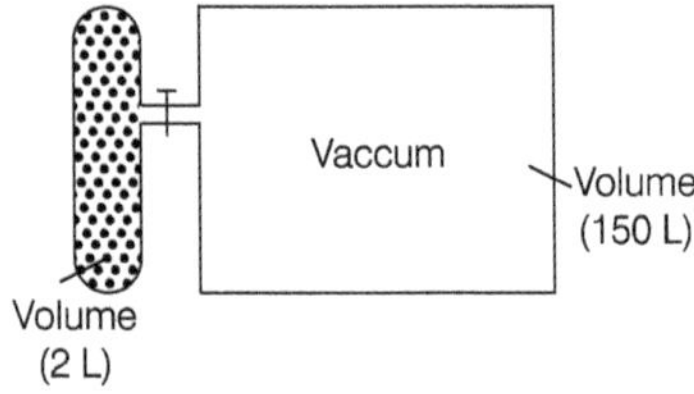

The 2L container contains 5 moles of gas at 300 K. Given, ln 76 = 4.3. At time $t=0$, stop value is opened and gas expands in 150 L vaccum chamber. Now, Choose the correct option.

(a) Due to expansion, temperature of gas falls
(b) Due to expansion, 54 kJ of energy becomes unavailable
(c) Due to expansion, gas does 54 kJ of work
(d) Internal energy changes by 54 kJ due to expansion process

93. Consider the following statements regarding emission of heat radiation by a substance.

I. Radiation occurs in solids, liquids and gases.
II. In solids and liquids radiation is a surface phenomena.
III. In liquids radiation is a surface phenomena.
IV. In all fluids radiation is a volumetric phenomena.

Which of the above statements are correct?

(a) Statements I, II and III are correct
(b) Statements I, II and IV are correct
(c) Statements II and III are correct
(d) Only statement IV is correct

94. Power dissipated in given circuit is

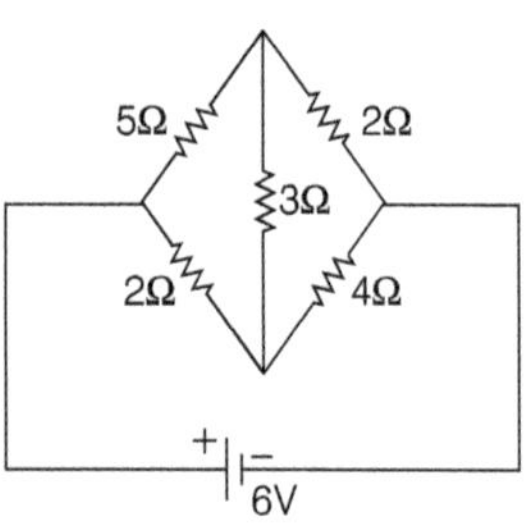

(a) 12 W (b) 24 W
(c) 36 W (d) 72 W

95. Moment of inertia of a rod of mass M and length L that rotated about its centre along an axis, which makes angle of 30° with the length of rod will be

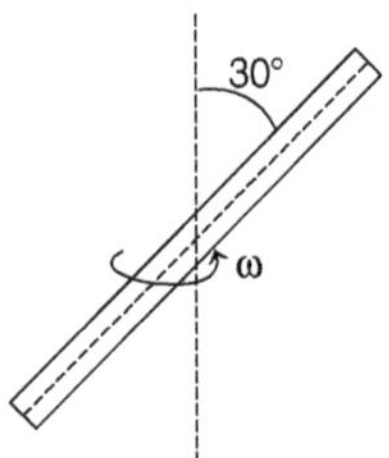

(a) $\frac{ML^2}{12}$ (b) $\frac{ML^2}{3}$ (c) $\frac{ML^2}{48}$ (d) $\frac{ML^2}{16}$

96. Two nucleus are projected in a region of magnetic field **B**, such that their velocity vectors **v** are parallel as shown below.

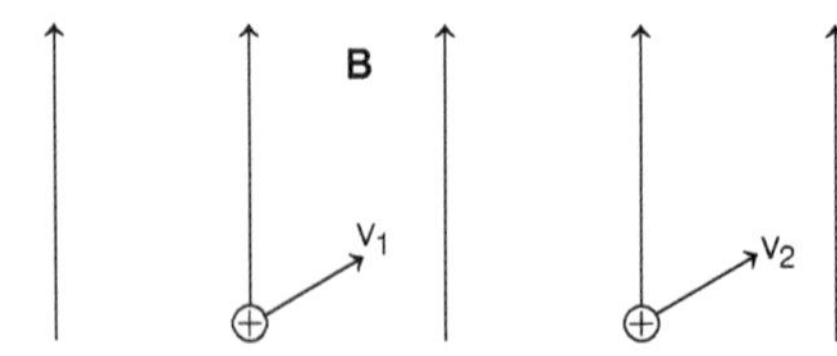

Now, choose the correct option.
(a) Trajectories of isobars are similar circles
(b) Trajectories are always similar helix
(c) Trajectories of isobars are similar helix
(d) Heavier isotope traces a helix of larger radius

97. A uniform rod of length L is placed over a horizontal frictionless surface. It is pulled over the surface by applying a force F on one of its end. Area of rod is A.

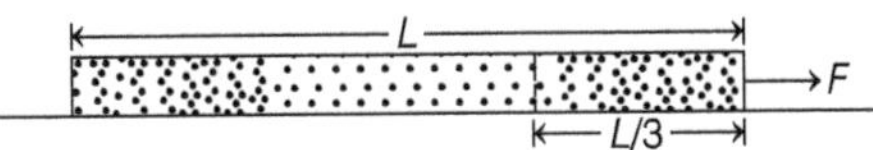

Stress in the rod at a distance $L/3$ from the end, where the force is applied is
(a) $\frac{F}{A}$ (b) $\frac{1}{3}\frac{F}{A}$ (c) $\frac{2}{3}\frac{F}{A}$ (d) $\frac{3}{4}\frac{F}{A}$

98. A rigid massless rod of length L connects two particles each of mass m. The rod lies over a horizontal frictionless table. Another particle of mass m and velocity v_0 struck one of two particles attached with rod.

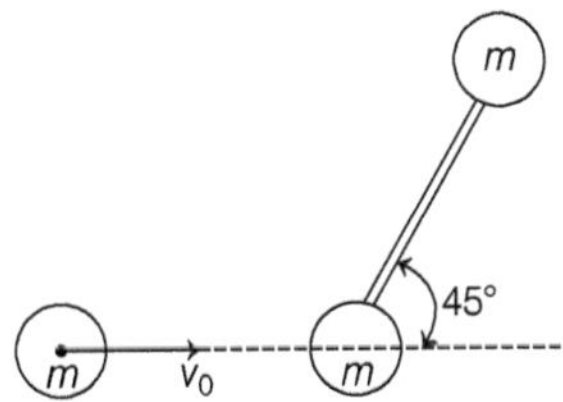

Angular velocity of rod about its centre of mass just after collision will be (take, collision elastic)
(a) $\frac{\sqrt{2}\,v_0}{7L}$ (b) $\frac{2\sqrt{2}\,v_0}{7L}$ (c) $\frac{3\sqrt{2}\,v_0}{7L}$ (d) $\frac{4\sqrt{2}\,v_0}{7L}$

99. A square loop of area 100 cm^2 with sides parallel to X and Y-axes is moved with velocity 8 cms^{-1} in positive direction of X-axis. A magnetic field exists in region which points in z-direction. Magnetic field decreases along negative x-direction at rate of 0.1 Tm^{-1} and also decreases in time at a rate of 10^{-3} Ts^{-1}. Emf induced in the loop is
(a) 10 μV (b) 70 μV (c) 90 μV (d) 100 μV

100. Heating curve for water is shown below.

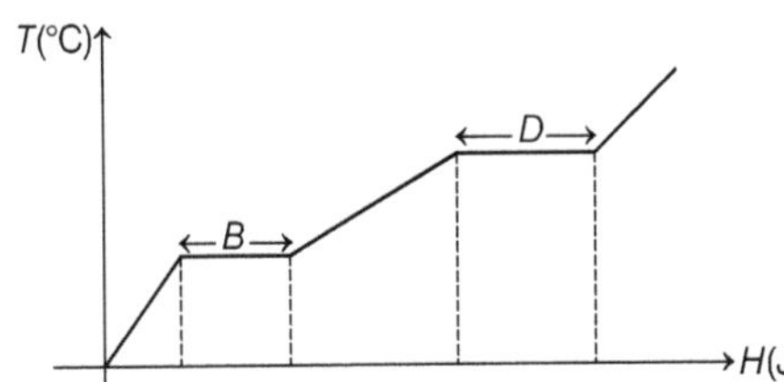

Now, choose the correct option.
(a) Length B is larger than length D as average distance between two liquid molecules is much less than distance between two gas molecules
(b) Length B is equal to length D as average distance between two liquid molecules is much less than distance between two gas molecules
(c) Length B is less than length D as average distance between two liquid molecules is much less than distance between two gas molecules
(d) Length B can be less or more than D depending on temperature

CHEMISTRY

101. Which of the following molecules is least resonance stabilised?

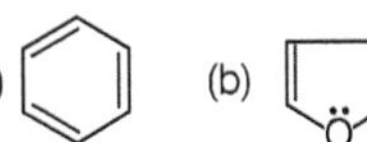
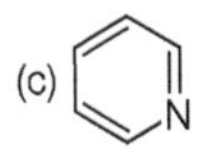
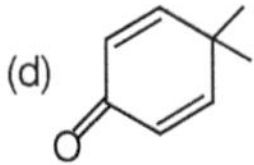

102. The pair having the same magnetic moment is [at. no. Cr = 24, Mn = 25, Fe = 26 and Co = 27]
(a) $[Cr(H_2O)_6]^{2+}$ and $[Fe(H_2O)_6]^{2+}$
(b) $[Mn(H_2O)_6]^{2+}$ and $[Cr(H_2O)_6]^{2+}$
(c) $[CoCl_4]^{2-}$ and $[Fe(H_2O)_6]^{2+}$
(d) $[Cr(H_2O)_6]^{2+}$ and $[CoCl_4]^{2-}$

103. The hybridisation and geometry of XeF_4 is
(a) dsp^2, square planar (b) sp^3d^2, octahedral
(c) sp^3, tetrahedral (d) sp^3d^2, square planar

104. Which of the following cannot be made by using Williamson's synthesis?
(a) Di-*tert* butyl ether (b) Methoxy benzene
(c) Benzyl *p*-nitro phenyl ether
(d) Methyl tertiary butyl ether

105. Which of the following complex ion will exhibit geometrical isomerism?
(a) $[Cr(H_2O)_4Cl_2]^+$ (b) $[Pt(NH_3)_3Cl]$
(c) $[Co(NH_3)_6]^{3+}$ (d) $[Co(CN)_5Cl]^{3-}$

106. An organic compound (A) with molecular formula $C_9H_{10}O$ forms an orange-red precipitate with 2, 4-DNP reagent and gives yellow precipitate on heating with iodine and NaOH. It does not reduces Tollen's reagent or Fehling's solution nor it decolourises bromine water as Baeyer's reagent. On drastic oxidation with chromic acid, it gives a carboxylic acid having molecular formula $C_7H_6O_2$. Identify the compound (A).

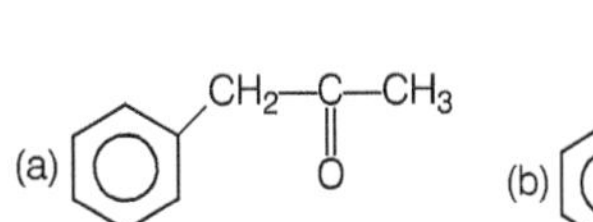

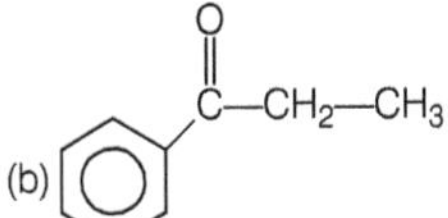

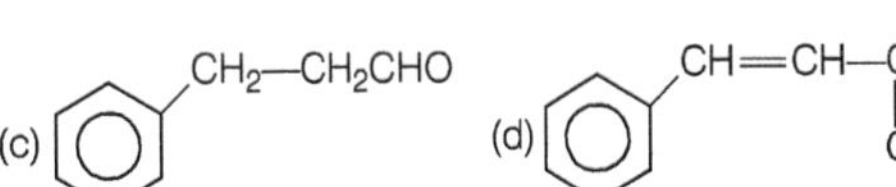

107. The emf of the cell, $Pb|PbCl_2||AgCl||Ag$ at 300 K is 0.50 V. If temperature coefficient of emf is -2×10^{-4} volt deg^{-1}. Then calculate the enthalpy change for the cell reaction.

(a) -108.08 kJ (b) 108.08 kJ
(c) 121.31 kJ (d) -121.31 kJ

108. Reagents x, y and z for the following conversion will be

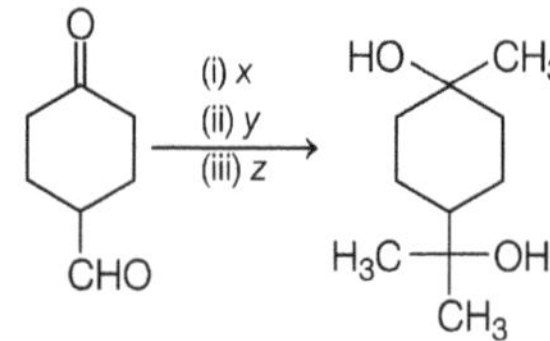

(a) $[Ag(NH_3)_2]^+ OH^-$, H^+, CH_3OH, CH_3MgBr
(b) $CH_3MgBr, H^+]^+$ CH_3OH, $[Ag(NH_3)_2^+ OH^-$
(c) CH_3MgBr, $[Ag(NH_3)_2]^+ OH^-$, H^+/CH_3OH
(d) $Ag(NH_3)_2]^+ OH^-$, CH_3MgBr, $H^+|CH_3OH$

109. The graph given below expresses the various steps of the system containing 1 mole of gas. Which type of process, system has when it moves from C to A?

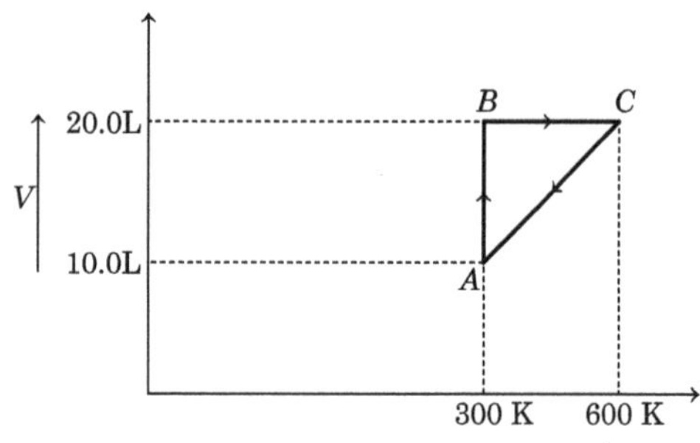

(a) Isochoric (b) Isobaric
(c) Isothermal (d) Cyclic

110. Caesium broxide has CsCl structure (bcc lattice). Its density is 4.49 g cm^{-3}. The side of the unit cell is closest to

(a) 5.584 Å (b) 4.285 Å
(c) 47.392 Å (d) 1.121 Å

BIOLOGY

111. A man has normal red-green colour vision. His blood group is rhesus negative (homozygous recessive). His wife also has normal colour vision but is rhesus positive. She is heterozygous at both the red-green colour vision locus and the blood group locus. What is the probability that their first child will be rhesus negative, red green colourblind boy?

(a) 0 (b) 0.0625 (c) 0.125 (d) 0.25

112. A biologist studied the population of rats in a barn. He found that the average natality was 250, average mortality 240, immigration 20 and emigration 30. The net increase in population is

(a) 10 (b) 15
(c) 5 (d) 0

113. The gene map of a plasmid, which has been used for genetic engineering is shown below. It contains two antibiotic resistance genes. The positions of several restriction endonuclease binding sites are also shown below

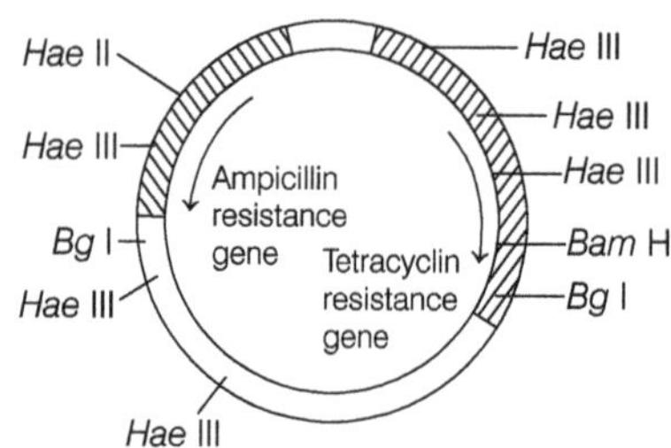

Which enzyme(s) would be the most suitable for the cleavage of the plasmid and the DNA containing the gene of interest?

(a) *Bam* HI (b) *Bg*I
(c) *Hae* III (d) *Bam* HI and *Hae* III

114. If there were 34 amino acids and DNA only contained two types of nitrogenous bases, what would be the minimum number of bases per codon that could code for proteins?

(a) 3 (b) 4
(c) 6 (d) 8

115. Which of the following statements about productivity is correct?

(a) Primary productivity of all ecosystems is a constant
(b) Net primary productivity is the amount of biomass available for consumption by carnivores
(c) Secondary productivity is defined as the rate of formation of new organic matter by decomposers
(d) Primary productivity depends on the plant species inhabiting a particular area

116. How many different polypeptides, each consisting of r amino acids, can be made if the number of different amino acids available is n?

(a) n^r (b) r^n
(c) nr (d) n

117. Read the following statements and choose the correct option.

I. RNA polymerase associates transiently with 'Rho' factor to initiate transcription.
II. In bacteria, transcription and translation take place in the same compartment.

III. RNA polymerase I is responsible for transcription of *t*RNA.
IV. When *hn*RNA undergoes capping process, adenylate residues are added at 3′-end in a template independent manner.
V. *hn*RNA is the precursor of *m*RNA.

(a) Only I (b) II, III and IV
(c) III and IV (d) I and V

118. Each graph shows the rate of reaction of an uninhibited enzyme and that of the same enzyme in the presence of a constant amount of either a competitive or a non-competitive inhibitor.

Which graph is correctly labelled?

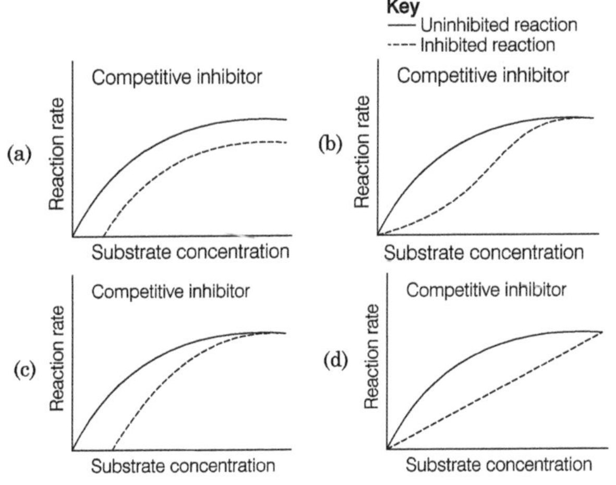

119. Match the following and choose the correct combinations from the given options.

Column I (Mineral)	Column II (Characteristic feature)
A. Potassium	1. Constituent of ferredoxin
B. Sulphur	2. Involved in stomatal movement
C. Molybdenum	3. Needed in the synthesis of auxin
D. Zinc	4. Component of nitrogenase

Codes

	A	B	C	D		A	B	C	D
(a)	2	1	4	3	(b)	1	2	3	4
(c)	4	3	2	1	(d)	1	3	4	2

120. The family tree shows the inheritance of a condition caused by the recessive allele 'r'.

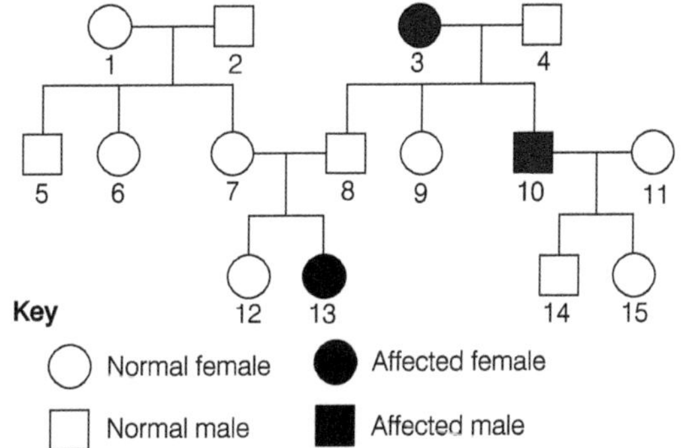

Which of the females are certain to have the genotype Rr?

(a) 1, 6 and 7 (b) 1, 7 and 12
(c) 7, 9 and 15 (d) 9, 12 and 15

Answers

PART-I

1 (b)	2 (d)	3 (c)	4 (a)	5 (c)	6 (a)	7 (c)	8 (a)	9 (a)	10 (b)
11 (a)	12 (b)	13 (a)	14 (b)	15 (d)	16 (c)	17 (a)	18 (a)	19 (a)	20 (b)
21 (a)	22 (b)	23 (a)	24 (b)	25 (b)	26 (d)	27 (d)	28 (c)	29 (a)	30 (d)
31 (c)	32 (c)	33 (b)	34 (a)	35 (b)	36 (c)	37 (d)	38 (d)	39 (c)	40 (a)
41 (c)	42 (b)	43 (c)	44 (b)	45 (a)	46 (a)	47 (b)	48 (c)	49 (a)	50 (c)
51 (d)	52 (d)	53 (b)	54 (b)	55 (c)	56 (c)	57 (a)	58 (b)	59 (b)	60 (c)
61 (a)	62 (b)	63 (b)	64 (d)	65 (c)	66 (c)	67 (b)	68 (c)	69 (c)	70 (c)
71 (c)	72 (b)	73 (a)	74 (c)	75 (b)	76 (a)	77 (b)	78 (c)	79 (c)	80 (c)

PART-II

81 (c)	82 (b)	83 (d)	84 (a)	85 (c)	86 (d)	87 (d)	88 (b)	89 (d)	90 (a)
91 (c)	92 (b)	93 (a)	94 (a)	95 (c)	96 (d)	97 (c)	98 (d)	99 (c)	100 (c)
101 (d)	102 (a)	103 (d)	104 (a)	105 (a)	106 (a)	107 (a)	108 (a)	109 (a)	110 (b)
111 (c)	112 (d)	113 (a)	114 (c)	115 (d)	116 (a)	117 (d)	118 (d)	119 (a)	120 (c)

Solutions

1. *(b)* We have,

$$|z-4+3i|\le 1$$
$$|z-(4-3i)|\le 1$$
$$-1+|4-3i|\le |z|\le 1+|4-3i|$$
$$-1+5\le |z|\le 1+5$$
$$4\le |z|\le 6$$
$$\because \quad \alpha=4, \beta=6$$

Let $y=\dfrac{x^4+x^2+4}{x}$

$\Rightarrow \quad y=x^3+x+\dfrac{4}{x}$

$\Rightarrow \quad y=x^3+x+\dfrac{1}{x}+\dfrac{1}{x}+\dfrac{1}{x}+\dfrac{1}{x}$

Since, $y\in(0,\infty)$, therefore

$x^3, x, \dfrac{1}{x}$ are positive.

Sum will be least of $x^3=x=\dfrac{1}{x}$

$\Rightarrow \quad x=1$

$\therefore$ Least value of y is 6

$$K=6\Rightarrow K=\beta$$

2. *(d)* Given,

$$2\sin\alpha+3\cos\beta=3\sqrt{2} \quad \text{...(i)}$$
$$3\sin\beta+2\cos\alpha=1 \quad \text{...(ii)}$$

On squaring and adding Eqs. (i) and (ii), we get

$$4+9+12(\sin\alpha\cos\beta+\sin\beta\cos\alpha)=18+1$$

$\Rightarrow \quad 12\sin(\alpha+\beta)=6$

$\Rightarrow \quad \sin(\alpha+\beta)=\dfrac{1}{2}$

$\Rightarrow \quad \alpha+\beta=150^\circ$

$\Rightarrow \quad \alpha+\beta\ne 30^\circ$

$\therefore \quad \gamma=180-(\alpha+\beta)$

$\Rightarrow \quad \gamma=30^\circ$

3. *(c)* $P\left(\dfrac{E'\cap F'}{G}\right)=\dfrac{P(E'\cap F'\cap G)}{P(G)}$

$$=\frac{P(G)-P(E\cap G)-P(G\cap F)}{P(G)}$$
$$=\frac{P(G)-P(E)\cdot P(G)-P(F)\cdot P(G)}{P(G)}$$
$$=\frac{P(G)(1-P(E)-P(F))}{P(G)}$$
$$=1-P(E)-P(F)$$
$$=P(E')-P(F)$$

4. *(a)* We have,

$$f(x)=\frac{x}{(1+x^n)^{\frac{1}{n}}}$$

$$f(f(x))=\frac{f(x)}{(1+f(x)^n)^{\frac{1}{n}}}=\frac{x}{(1+2x^n)^{\frac{1}{n}}}$$

$$f(f(f(x)))=\frac{x}{(1+3x^n)^{1/n}}$$

$$f(f(f(x)))=\underbrace{(f_0f_0f_0\ldots f)}_{n\text{ times}}x=\frac{x}{(1+nx^n)^{\frac{1}{n}}}$$

$\Rightarrow \quad I=\displaystyle\int x^{n-2}g(x)dx$

$$=\int x^{n-2}\frac{x}{(1+nx^n)^{\frac{1}{n}}}dx$$
$$=\int\frac{x^{n-1}}{(1+nx^n)^{\frac{1}{n}}}dx$$

Put $1+nx^n=t\Rightarrow x^{n-1}dx=\dfrac{dt}{n^2}$

$\therefore \quad I=\dfrac{1}{n^2}\displaystyle\int\frac{dt}{t^{1/n}}$

$\Rightarrow \quad I=\dfrac{1}{n^2}\dfrac{t^{-\frac{1}{n}+1}}{-\frac{1}{n}+1}+C$

$\Rightarrow \quad I=\dfrac{1}{n(n-1)}t^{1-\frac{1}{n}}+C$

$\Rightarrow \quad I=\dfrac{1}{n(n-1)}(1+nx^n)^{1-\frac{1}{n}}+C$

5. *(c)* We have,

$$x^2+y^2-6x-2py+17=0$$
$$(x-3)^2+(y-p)^2=9-17+p^2$$

Director circle of the given circle is

$(x-3)^2+(y-p)^2=2(p^2-8)$

It is passes through (0, 0).

$\therefore \quad 9+p^2=2p^2-16$

$\Rightarrow \quad p^2=25$

$\Rightarrow \quad |p|=5$

6. *(a)* Equation of ellipse

$$3x^2+4y^2=12$$

$\Rightarrow \quad \dfrac{x^2}{4}+\dfrac{y^2}{3}=1$

Focus of ellipse $(\pm1, 0)$.

Focus of ellipse is also focus of hyperbola.

$\therefore \quad ae=1$

Length of transverse axis $(2a)=2\sin\theta$

$\therefore \quad a=\sin\theta$

In hyperbola $\quad (ae)^2=a^2+b^2$

$$1=\sin^2\theta+b^2$$

$\Rightarrow \quad b^2=\cos^2\theta$

Equation of hyperbola

$$\frac{x^2}{\sin^2\theta}-\frac{y^2}{\cos^2\theta}=1$$

$\Rightarrow \quad x^2\text{cosec}^2\theta-y^2\sec^2\theta=1$

7. *(c)* Let event A = Each American man is seated adjacent to his wife.

B = Indian men is seated adjacent to his wife

Now, $n(A\cap B)=4!\times(2!)^5$

Event when each American man is seated to his wife

$\therefore \quad n(A)=(5!)\times(2!)^4$

$\Rightarrow \quad P\left(\dfrac{B}{A}\right)=\dfrac{n(A\cap B)}{P(B)}$

$$=\frac{4!\times(2!)^5}{5!\times(2!)^4}=\frac{2}{5}$$

8. *(a)* Let the side of right angled triangle are $a-d, a, a+d$.

$\therefore \quad (a-d)^2+a^2=(a+d)^2$

$\Rightarrow \quad a^2=(a+d)^2-(a-d)^2$

$\Rightarrow \quad a^2=4ad$

$\Rightarrow \quad a=4d \quad \text{...(i)}$

Area of triangle $=\dfrac{1}{2}a(a-d)=24$

$\Rightarrow \quad a(a-d)=48 \quad \text{...(ii)}$

From Eqs. (i) and (ii) we get

$$a=8, d=2$$

$\because$ Smallest side is $a-d=8-2=6$

9. *(a)* We have,

$$y=\frac{x^2}{8}+x\cos x+\cos 2x$$

$\Rightarrow \quad y=\dfrac{1}{8}[x^2+8x\cos x+8\cos 2x]$

$\Rightarrow \quad y=\dfrac{1}{8}[x^2+8x\cos x+4\cos^2x+8(2\cos^2x-1)-4\cos^2x]$

$\Rightarrow \quad y=\dfrac{1}{8}[(x+2\cos x)^2+12\cos^2x-8]$

$\Rightarrow \quad y=\dfrac{1}{8}(x+2\cos x)^2+\dfrac{1}{8}(12\cos^2x-8)$

$\therefore$ Minimum value of y is $\dfrac{1}{8}(0-8)=-1$

10. *(b)* We have,

$$f(4+x)=f(4-x)$$

$\therefore f(x)$ is symmetric about $x=4$.

Sum of its solution

$$2\left[\left(\frac{x_1+x_2}{2}\right)+\left(\frac{x_3+x_4}{2}\right)+\left(\frac{x_5+x_6}{2}\right)+\left(\frac{x_7+x_8}{2}\right)\right]$$

$$= 2[4+4+4+4]$$
$$= 2\times 16 = 32$$

11. *(a)* Let $f(x)=\int_0^x \frac{t^2}{1+t^4}dt - 2x + 1$

$\Rightarrow \quad f'(x) = \frac{x^2}{1+x^4} - 2$ as $\frac{1+x^4}{x^2} \ge 2$

$\therefore \quad \frac{x^2}{1+x^4} \le \frac{1}{2}$

$\Rightarrow \quad f'(x) \le -\frac{3}{2}$

$\Rightarrow f(x)$ is continuous and decreasing.

$\therefore f(0)=1$ and $f(1)=\int_0^1 \frac{t^2}{1+t^4}dt - 2 \le -\frac{3}{2}$

By intermediate value theorem $f(x)=0$ possesses exactly one solution in [0, 1].

12. *(b)* Let point $P(t^2, 2t)$ lie on parabola $y^2 = 4x$. Image of $P(t^2, 2t)$ with respect to line $x+y+4=0$ is $8(h,k)$.

$\therefore \quad \frac{h-t^2}{1} = \frac{k-2t}{1} = \frac{-2(t^2+2t+4)}{2}$

$h = -2t-4,\ k = -t^2-4$

Hence curve C becomes

$$\frac{(x+4)^2}{4} + 4 = -y$$

Since it intersect with line $y=-5$

$\therefore \quad (x+4)^2 + 16 = 20$

$\Rightarrow \quad (x+4)^2 = 4$

$\Rightarrow \quad x = -2$ or $x = 6$

$\therefore \quad A = (-2, -5)$ and $B = (-6, -5)$

$\Rightarrow \quad AB = \sqrt{(-2+6)^2 + (-5+5)^2} = 4$

13. *(a)* We have,

$A = 2\hat{\mathbf{i}} - \hat{\mathbf{j}} + \hat{\mathbf{k}},\ B = \hat{\mathbf{i}} - 3\hat{\mathbf{j}} - 5\hat{\mathbf{k}}$

$C = a\hat{\mathbf{i}} - 3\hat{\mathbf{j}} + \hat{\mathbf{k}}$

$A(2\hat{i} - \hat{j} + \hat{k})$

$(\hat{i} - 3\hat{j} - 5\hat{k})\ B$ $\quad C(a\hat{i} - 3\hat{j} + \hat{k})$

$\mathbf{CA} = (2-a)\hat{\mathbf{i}} + 2\hat{\mathbf{j}}$

$\mathbf{CB} = (1-a)\hat{\mathbf{i}} - 6\hat{\mathbf{k}}$

$\mathbf{CA}\cdot\mathbf{CB} = 0$

$\therefore (2-a)(1-a) = 0$

$a = 2, 1$

14. *(b)* We have,

$$I = \int_1^a [x]f'(x)dx$$

$$I = \int_1^2 f'(x)dx + 2\int_2^3 f'(x)dx + 3\int_3^4 f'(x)dx + \dots + [a]\int_{[a]}^a f'(x)dx$$

$$I = [f(x)]_1^2 + 2[f(x)]_2^3 + \dots + [a][f(x)]_{a-1}^a$$

$$I = f(2) - f(1) + 2f(3) - 2f(2) + 3f(3) - 3f(2) + \dots [a][f(a) - f[a]]$$

$$I = [a]f(a) - \{f(1) + f(2) + \dots f([a])\}$$

15. *(d)* We have,

$\log_x 8 = z \Rightarrow x^z = 8$

$\log_y x = -1 \Rightarrow x = \frac{1}{y} \Rightarrow xy = 1$

$\log_{1/4} y = -1,\ y = 4,\ x = \frac{1}{4}$

$\Rightarrow \quad \left(\frac{1}{4}\right)^z = 8 \Rightarrow 2^{-2z} = 2^3$

$\Rightarrow \quad z = -\frac{3}{2}$

Now, $\left(\frac{1}{x}+1\right)^{\log\sqrt{5}(y^2+4z^2)}$

$(5)^{\log\sqrt{5}\left(16+4\times\frac{9}{4}\right)}$

$= (\sqrt{5})^{2\log\sqrt{5}(16+9)} = \sqrt{5}^{\log\sqrt{5}(25)^2}$

$= 25^2 = 625$

16. *(c)* Given,

$\sin x + \cos x = \frac{1}{2},\ x \in (0, \pi)$

On squaring both sides, we get

$1 + \sin 2x = \frac{1}{4}$

$\Rightarrow \quad \sin 2x = \frac{1}{4} - 1 = -\frac{3}{4} < 0$

$\therefore \quad x \in \left(\frac{\pi}{2}, \pi\right)$

$\cos 2x = -\sqrt{1-\sin^2 2x}$

$= -\sqrt{1-\frac{9}{16}} = -\frac{\sqrt{7}}{4}$

$\tan x = \frac{1-\cos 2x}{\sin 2x}$

$= \frac{1+\frac{\sqrt{7}}{4}}{\frac{-3}{4}} = -\left(\frac{4+\sqrt{7}}{3}\right)$

17. *(a)* A = Events that sum of digits on selected ticket is 8.

$A = \{08, 17, 26, 35, 44\}$

$n(A) = 5\ P(A) = \frac{5}{50}$

B = Events that product of digits is zero

$B = \{00, 01, \dots, 10, 20, 30, 40\}$

$n(B) = 14,\ P(B) = \frac{14}{50}$

$A \cap B = \{08\}$

$n(A\cap B) = 1\ P(A \cap B) = \frac{1}{50}$

Required probability $P\left(\frac{A}{B}\right) = \frac{P(A\cap B)}{P(B)}$

$= \frac{1/50}{14/50} = \frac{1}{14}$

18. *(a)* $6^a + 9^b = (5+1)^a + (10-1)^b$

$= 5\lambda + (1)^a + 10\mu + (-1)^b$

$= 5(\lambda + 24) + (1)^a + (-1)^b$

is divisible by 5 if b is odd.

$\therefore$ Total number of ordered pair (a, b)

$= 50 \times 25 = 1250$

19. *(a)* Here $p = \frac{1}{4},\ q = \frac{3}{4}$

$p(X \ge 1) = 1 - p(X = 0)$

$= 1 - {}^nC_0 p^0 q^n$

$1 - \left(\frac{3}{4}\right)^n \ge \frac{9}{10} \quad \left[\because p(x \ge 1) \ge \frac{9}{10}\right]$

$\Rightarrow \quad \frac{1}{10} \ge \left(\frac{3}{4}\right)^n \Rightarrow 10 \le \left(\frac{4}{3}\right)^n$

$\Rightarrow \quad n \ge \frac{1}{\log_{10} 4/3} = \frac{1}{\log_{10} 4 - \log_{10} 3}$

20. *(b)* Given,

$\left|z - \frac{4}{z}\right| = 2$

Now, $|z| = \left|z - \frac{4}{z} + \frac{4}{z}\right| \le \left|z - \frac{4}{z}\right| + \left|\frac{4}{z}\right|$

$\Rightarrow \quad |z| \le 2 + \frac{4}{|z|}$

$\Rightarrow \quad |z|^2 - 2|z| - 4 \le 0$

$\Rightarrow \quad 0 \le |z| \le \sqrt{5} + 1$

21. *(a)* Tensions involved are as shown below.

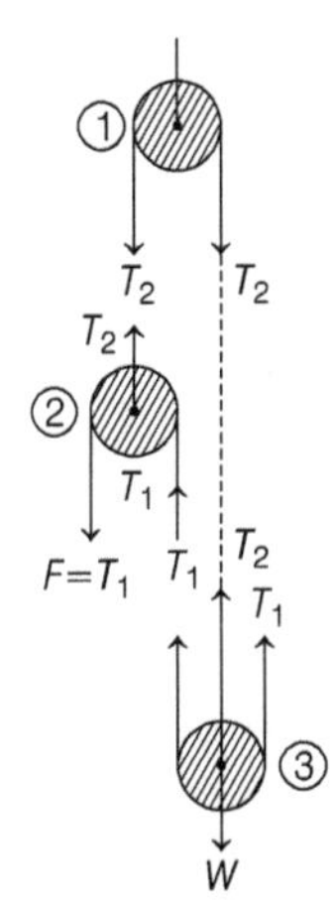

For equilibrium of pulley 2,

$\Rightarrow \quad F = T_1$

$\Rightarrow \quad T_2 - T_1 - F = 0$

$\Rightarrow \quad T_2 = 2F$

For equilibrium of pulley 3,

$\Rightarrow \; T_1 + T_1 + T_2 - W = 0$

$\Rightarrow \quad 4F = W, F = 25\text{ N}$

To lift the load F is atleast 25 N is required.

22. *(b)* Pressure differential across capillary is $p_i - p_o = \rho g h$

$$= 1000 \times 10 \times 0.05$$
$$= 5000\text{ Nm}^{-2}$$

Viscosity of water is

$$\eta = \left(0.8\text{cp} \times \frac{10^{-3}}{\text{cp}}\text{ Kgm}^{-1}\text{s}^{-1}\right)$$
$$= 8 \times 10^{-4}\text{kgm}^{-1}\text{s}^{-1}$$

Flow rate using Poiseuille's law is

$$j = \frac{\pi r^4 (p_i - p_o)}{8\eta L}$$
$$= \frac{\pi \times 1 \times 10^{-4} \times 5000}{8 \times 8 \times 10^{-4} \times 0.2}$$
$$= \frac{5\pi}{128} \times 10^{-5}\text{ m}^3\text{s}^{-1} \approx 1.0\text{ mLs}^{-1}$$

23. *(a)* From gas laws, we have

$$m = M \cdot \left(\frac{pV}{RT}\right)$$

Mass of hydrogen gas present is

$$m_H = M_H \cdot p_H \left(\frac{V}{RT}\right)$$

Percentage of mass of hydrogen is

$$m_H = \frac{M_H p_H \left(\frac{V}{RT}\right) \times 100}{(M_H p_H + M_{CO_2} p_{CO_2} + M_{CH_4} p_{CH_4} + M_{C_2H_6} p_{C_2H_6})\left(\frac{V}{RT}\right)}$$
$$= \frac{2 \times 200 \times 100}{2 \times 200 + 44 \times 150 + 16 \times 320 + 30 \times 105}$$
$$= 0.026 \times 100 = 2.6\%$$

24. *(b)* Kinetic energy of a molecule at temperature T is

$$\text{KE} = \frac{1}{2} m v_{\text{rms}}^2 = \frac{3}{2} K_B T$$

The molecule can rises until whole of its kinetic energy is changed to potential energy.

Let this height is h. Then,

$$mgh = \frac{3}{2} K_B T$$
$$\Rightarrow \quad h = \frac{3K_B T}{2mg} = \frac{3 \times 1.38 \times 10^{-23} \times 300}{2 \times 4.5 \times 10^{-26} \times 10}$$
$$= 27.6\text{ km} \approx 28\text{ km}$$

25. *(b)* Heat lost by hot gas in A

$$= -\Delta Q = -mC_V \Delta T$$

Heat gained by cold gas in B

$$= +\Delta Q = mC_V \Delta T$$

Entropy change for hot gas in A is

$$S_A = \frac{-\Delta Q}{T_A} = \frac{-\Delta Q}{273 + 65}$$

Entropy change for cold gas in B is

$$S_B = \frac{+\Delta Q}{273 + 20}$$

Clearly, $|S_B| > |S_A|$.

So, more entropy gain occurs in B than entropy loss in A. Total entropy of universe increases due to this process.

26. *(d)* As air passes around plane wing,

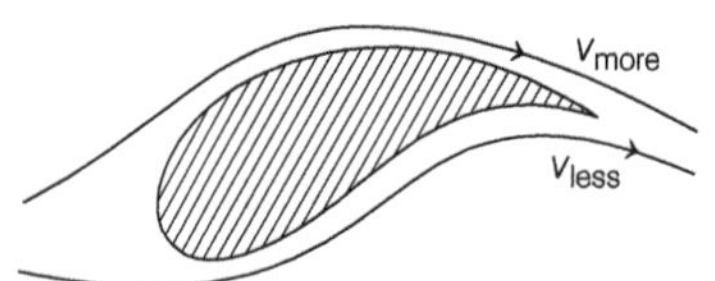

Due to shape of aerofoil, velocity of air layers over above surface of wing is more than lower surface of wing.

As, $\frac{1}{2}\rho v^2 + p = \text{constant}$

When v velocity increases, pressure will fall. So, a pressure difference acts on lower surface of wing causing lift of plane.

27. *(d)*

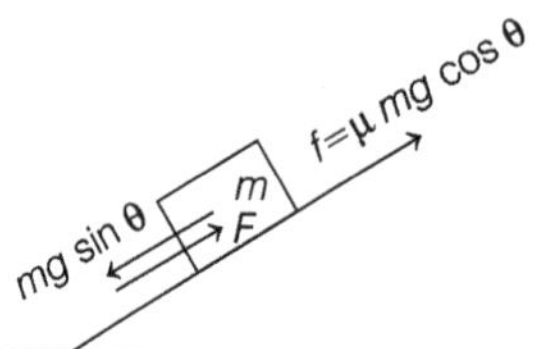

If applied force is just sufficient to hold the block over incline, friction acts in upwards direction (so treated positive).

$$F = F_i = mg(\sin\theta - \mu_s \cos\theta)$$

$\Rightarrow \; F_i = mg\sin\theta - f \;$ or $f = mg\sin\theta - F$

When $F = mg\sin\theta$, friction f is zero.

If applied force exceeds $mg\sin\theta$, then block moves upwards and friction now acts downwards.

$\therefore \; F_f = mg(\sin\theta + \mu_K \cos\theta)$

Hence, friction in now negative.

As $\mu_s > \mu_K$, graph must be (d).

28. *(c)* Moment of inertia of cube about an edge is (by parallel axis theorem)

$$I = I_{CM} + m\left(\frac{l}{\sqrt{2}}\right)^2$$
$$= \frac{ml^2}{6} + \frac{ml^2}{2} = \frac{2}{3}ml^2$$

As there is no external torque, angular momentum about P is conserved.

i.e. $\quad L_i = L_f$

$$\therefore \quad mv\left(\frac{l}{2}\right) = I\omega$$
$$\Rightarrow \quad mv\frac{l}{2} = \frac{2}{3}ml^2 \cdot \omega$$
$$\Rightarrow \quad \omega = \frac{3mv}{4lm}$$
$$= \frac{3\sqrt{2mK}}{4lm} = \frac{3}{2\sqrt{2}} \cdot \sqrt{\frac{K}{ml^2}}$$

29. *(a)* With $\Delta x = 5 \times 10^{-15}$ m,

By uncertainty principle, we have

$$\Delta p \cdot \Delta x \geq \frac{h}{4\pi}$$
$$\Rightarrow \quad \Delta p \geq \frac{\left(\frac{h}{2\pi}\right)}{2\Delta x} \geq \frac{1.054 \times 10^{-34}}{(2 \times 5 \times 10^{-15})}$$
$$\Rightarrow \quad \Delta p \geq 1.1 \times 10^{-20}\text{ kg ms}^{-1}$$

Kinetic energy, $K = \frac{1}{2}mc^2 = \frac{1}{2}pc$

$$= \frac{1}{2} \times 1.1 \times 10^{-20} \times 3 \times 10^8$$
$$= 2.6 \times 10^{-12}\text{ J}$$

30. *(d)*

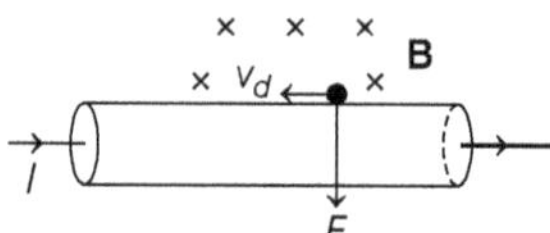

Direction of force on an electron using Flemming's right hand rule is towards centre of wire.

Force on electron is

$F = Bqv_d$

where, $v_d = \frac{I}{nqA}$.

$$\therefore \quad F = \frac{\mu_0 I}{2\pi a} \cdot q \cdot \frac{I}{nq\pi a^2}$$
$$F = \frac{\mu_0 I^2}{2\pi^2 n a^3}$$

Substituting values in above equation, we have

$$F = \frac{4 \times \pi \times 10^{-7} \times 10^2}{2 \times \pi^2 \times 5 \times 10^{28} \times (0.5 \times 10^{-3})^3}$$
$$\approx 8\pi \times 10^{-26}\text{ N}$$

31. *(c)* In rotating disc, let a potential difference dV occurs across radial distance dr.

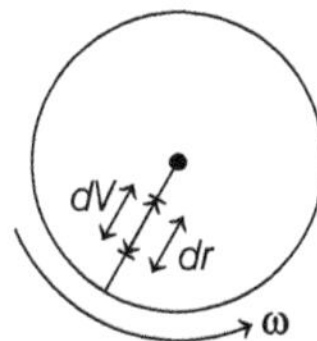

In equilibrium, electric field balances centrifugal force on an electron.

$$\therefore \quad e\frac{dV}{dr} = m\omega^2 r \Rightarrow dV = \frac{m}{e}\omega^2 r dr$$

$\therefore$ Integrating between limits, we have

$$V = \int_0^{V(R)} dV = \int_0^R \frac{m}{e}\omega^2 r dr$$

$$V(R) = \frac{-m\omega^2 R^2}{e \cdot 2}$$

Note *That outer rim will be negatively charged.*

In practice the potential difference is very-very small.

32. *(c)* For adiabatic process,

$$T_i V_i^{\gamma-1} = T_f V_f^{\gamma-1} \Rightarrow \frac{T_f}{T_i} = \left(\frac{V_i}{V_f}\right)^{\gamma-1}$$

Here, $\frac{V_i}{V_f} = 10$

So, $T_f = T_i (10)^{\gamma-1}$

For monoatomic gas, $\gamma = \frac{5}{3}$ and for diatomic gas, $\gamma = \frac{7}{5}$.

Hence, $\frac{T_A}{T_B} = \frac{10^{\frac{5}{3}-1}}{10^{\frac{7}{5}-1}} = 10^{\frac{4}{15}}$

33. *(b)* Given is a refrigeration cycle.

In process BC, pressure is decreasing at constant volume. i.e. heat is being extracted from the working medium (gas).

34. *(a)* We can throw a ball up with any velocity but it cannot have velocity greater than its terminal velocity during free fall. So, option (a) is correct.

35. *(b)*

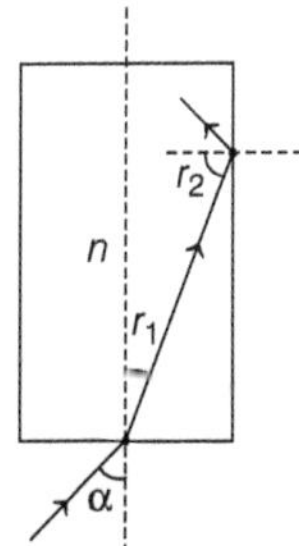

From figure, we have

$$r_2 = 90 - r_1$$
$$(r_2)_{\min} = 90 - (r_1)_{\max}$$

Also, for TIR,

$$(r_2)_{\min} \geq \theta_c \text{ (critical angle)}$$
$$\sin(r_2)_{\min} \geq \sin\theta_c$$
$$\Rightarrow \quad \sin(90 - (r_1)_{\max}) \geq \sin\theta_c$$
$$\Rightarrow \quad \cos(r_1)_{\max} \geq \frac{1}{n} \quad \ldots\text{(i)}$$

Also, $\frac{\sin\alpha}{\sin r_1} = n \Rightarrow \frac{\sin(\alpha)_{\max}}{\sin(r_1)_{\max}} \leq n$

But $\alpha_{\max} = 90°$

$$\therefore \quad \frac{\sin 90}{\sin(r_1)_{\max}} \leq n \Rightarrow \frac{1}{\sin(r_1)_{\max}} \leq n$$

$$\Rightarrow \sin(r_1)_{\max} \geq \frac{1}{n} \quad \ldots\text{(ii)}$$

From Eqs. (i) and (ii), we have

$$1 - \cos^2(r_1)_{\max} = \sin^2(r_1)_{\max}$$
$$1 - \frac{1}{n^2} = \frac{1}{n^2}$$
$$\Rightarrow \quad \frac{2}{n^2} = 1$$
$$\Rightarrow \quad n = \sqrt{2}$$
$$\therefore \quad n_{\min} = \sqrt{2}$$

36. *(c)* First reaction is

$$^{14}N + ^{2}H \longrightarrow ^{15}N + ^{1}H + 8.53\,\text{MeV}$$

We can rearrange above equation as,

$$\Rightarrow \quad ^{14}N - ^{15}N \longrightarrow p - d + 8.53$$

where, $p = {}^{1}H$, $d = {}^{2}H$

So, rewriting given reactions, we have

$$^{14}N - ^{15}N = p - d + 8.53 \quad \ldots\text{(i)}$$
$$^{15}N - ^{13}C = \alpha - d + 7.58 \quad \ldots\text{(ii)}$$
$$^{13}C - ^{11}B = \alpha - d + 5.16 \quad \ldots\text{(iii)}$$

Adding above equations, we have

$$^{14}N - ^{11}B = p + 2\alpha - 3d + 21.27$$
$$\Rightarrow \quad ^{11}B - ^{14}N + \alpha - n = 3d - \alpha - p - n - 21.27$$

Now, $3d - \alpha - p - n$
$= 3 \times 2.0140 - 4.0026 - 1.0078 - 1.0087$
$= 0.0229\,\text{u}$

So, Q value for reaction is

$$^{11}B(\alpha, n)^{14}N = 0.229 \times 931 - 21.27$$
$$= 0.05\,\text{MeV}$$

37. *(d)* Using Ampere's law, magnetic field of a solid conducting rod is

$$B = \begin{cases} \frac{\mu_0 I}{2\pi R^2} \cdot r, & r \leq R \\ \frac{\mu_0 I}{2\pi r}, & r > R \end{cases}$$

So, correct graph is (d).

38. *(d)* Given ratio of number of atoms in state 2 to number of atoms in state 1 is 0.001.

So, $\frac{n(E_2)}{n(E_1)} = \frac{1}{1000}$

$$\Rightarrow \quad \frac{1}{1000} = \frac{N_2}{N_1} \cdot \left(e^{-(E_2 - E_1)/kT}\right) \quad \ldots\text{(i)}$$

where, N_2 and N_1 are number of possible electron states in $n = 2$ and $n = 1$ quantum states.

Also, E_2 and E_1 are energies in $n = 2$ and $n = 1$ quantum states.

Here, $N_2 = 8$, $N_1 = 2$, $E_2 = \frac{E_1}{4}$ and

$$E_1 = -13.6\,\text{eV}$$
$$k = 8.62 \times 10^{-5}\,\frac{\text{eV}}{\text{K}}$$

Taking log in Eq. (i), we have

$$T = \frac{1}{k} \cdot \frac{\frac{3}{4}(-E_1)}{\log(4000)}$$
$$= \frac{\frac{3}{4} \times 13.6}{8.62 \times 10^{-5} \times \log 4000}$$
$$\approx 1.43 \times 10^4\,\text{K}$$
$$= 14300\,\text{K}$$

39. *(c)* To keep situations similar, we must keep reynolds number of flow same in both cases.

$$\Rightarrow \quad N_R = \left(\frac{\rho v D}{\eta}\right)_{\text{model}} = \left(\frac{\rho v D}{\eta}\right)_{\text{car}}$$
$$\Rightarrow \quad v_m D_m = v_c D_c$$
$$\Rightarrow \quad v_m = v_c \cdot \frac{D_c}{D_m} = 15 \times \left(\frac{550}{20}\right) = 0.4\frac{\text{km}}{\text{s}}$$

40. *(a)* Due to induced charge produced by polarisation of dielectric, charge q_1 will be acted upon by an additional charge.

$$\oplus_{q_1} \quad \boxed{\begin{matrix} - & + \\ - & + \\ - & + \end{matrix}} \quad \ominus_{q_2}$$

As negative induced charge is now more near to q_1 than induced positive charge, attraction is more prominent.

Hence, net force on q_1 will be more.

41. *(c)* Total number of stereoisomers = number of geometrical + number of optical isomers

$CH_3CH{=}CHCH_2\overset{*}{C}HBrCH_3$ is an alkene, so it can have two geometrical forms *cis* and *trans*. Since, the given compound has a chiral centre, therefore each geometrical isomer has a pair of enantiomer. Thus, in all there are $2 \times 2 = 4$ stereoisomers possible for given compound.

42. *(b)* Given,

$$\lambda^{\infty}_{NaCl} = 126.45\ \Omega^{-1}\ cm^2$$
$$\lambda^{\infty}_{HCl} = 426.16\ \Omega^{-1}\ cm^2$$
$$\lambda^{\infty}_{C_2H_5COONa} = 91\ \Omega^{-1}\ cm^2$$
$$\lambda_{C_2H_5COOH} = \lambda^{\infty}_{C_2H_5COONa} + \lambda^{\infty}_{HCl} - \lambda^{\infty}_{NaCl}$$
$$= 91 + 426.16 - 126.45$$
$$= 390.71\Omega^{-1}\ cm^{-2}$$

43. *(c)* $\Delta G° = -262\ kJ = -262000J$

$$\Delta G = -2.303RT \log k$$
$$\log k = \frac{\Delta G}{-2.30RT}$$
$$= \frac{26200}{2.303 \times 8.314 \times 298}$$
$$\log k = 45.918$$
$$k = 8.279 \times 10^{45}$$

44. *(b)* Ions of elements which have paired electrons or no electrons in *d*-orbitals are colourless.

The electronic configuration of the elements are as follows:

$Zn^{2+} = [Ar]\ 3d^{10}\ 4s^0 \Rightarrow$ Paired *d*-electrons.

$Cr^{+} = [Ar]\ 3d^{10}\ 4s^0 \Rightarrow$ Paired *d*-electrons.

$Ti^{4+} = [Ar]\ 3d^0 \Rightarrow$ No *d*-electrons.

$V^{2+} = [Ar]\ 3d^0 \Rightarrow$ No *d*-electrons.

Thus, the set of above ions are colourless.

45. *(a)* The depression in freezing point is a colligative property. Thus, it depends upon number of particles or ions. More is the number of ions, minimum will be the freezing point.

For 0.01 M NaCl,
Number of ions = $0.01 \times 2 = 0.02$
For 0.005 M C_2H_5OH,
Number of ions $0.005 \times 1 = 0.005$
For 0.005 M MgI_2,
Number of ions = $0.005 \times 3 = 0.015$
For $MgSO_4$,
Number of ions = $0.005 \times 2 = 0.01$

As, 0.01 M NaCl has highest number of ions, so it will have minimum freezing point.

46. *(a)* ${}_{90}Th^{232} \longrightarrow {}_{82}Pb^{208} + x{}_2^4\alpha + y{}_1\beta$

Let the number of α-particles be *x*.

Let the number of β-particles be *y*.

In radioactive disintegration atomic number as well as mass number must be equal to two sides of equation.

$$\therefore\ 208 + 4x + 0y = 232$$
$$4x = 24$$
$$x = 6$$
and $82 + 2x - y = 90$
$$2x - y = 8 \Rightarrow y = 4$$

$\therefore$ Number of α-particles = 6
Number of β-particles = 4.

47. *(b)* Given, conditions should be followed for an electron in an orbital.

(i) The value of *l* should ranges between 0 to $n - 1$.

(ii) The value of *m* should ranges between $-l$ to $+l$.

(iii) The value of $s = +\frac{1}{2}$ or $-\frac{1}{2}$.

The electronic configuration of Ti is [Ar] $3d^2, 4s^2$.

Therefore, $n = 4$, $l = 0$, $m = 0$, $s = +\frac{1}{2}$.

48. *(c)* Natural rubber is made from monomer isoprene (2-methyl, 1, 3-butadiene). It resembles synthetic rubber, neoprene, which is produced by the monomer chloroprene.

$$\underset{\text{Neoprene}}{-\!\!\left[CH_2—CH{=}C(Cl)—CH_2\right]\!\!-_n}\quad \text{Polymerisation}$$
$$\underset{\text{Natural rubber}}{-\!\!\left[CH_2—CH{=}C(CH_3)—CH_2\right]\!\!-_n}$$

49. *(a)* $p\pi$ - $d\pi$ overlapping is formed when *p*-orbital of one atom and *d*-orbital of another atom overlap laterally.

In case of PO_4^{3-}, P is sp^3-hybridised, it forms four P—O σ bonds by overlap of sp^3-hybridised orbitals of P with *p*-orbitals of O. The bond is however formed by overlap of *p*-orbitals of O and *d*-orbitals of P. N and C on the other hand cannot form $p\pi$ - $d\pi$ bond, since they do not have *d*-orbitals.

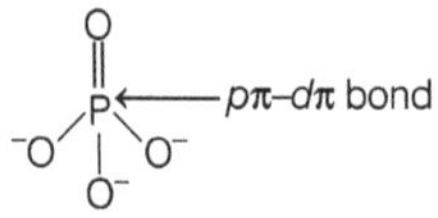

O
‖
$^-$O—P—O$^-$ ← $p\pi$–$d\pi$ bond
|
O$^-$

50. *(c)* Total number of voids = Number of octahedral voids + Number of tetrahedral voids.

Number of octahedral voids = Total number of atoms (*N*) is a closed structure (0.5 mol)

$$= 0.5 \times 6.022 \times 10^{23}$$
$$= 3.011 \times 10^{23}$$

Number of tetrahedral voids = 2*N*

$$= 2 \times 3.011 \times 10^{23} = 6.022 \times 10^{23}$$

Total number of voids

$$= 3.011 \times 10^{23} + 6.022 \times 10^{23}$$
$$= 9.033 \times 10^{23}$$

51. *(d)* For first order reaction

$$t = \frac{2.303}{k} \log \frac{a}{a - x}$$

For 90% reaction,

$$100 = \frac{2.303}{k} \log \frac{100}{100 - 90}$$
$$k = \frac{2.303}{100} \log 10$$
$$k = 0.023$$

Also, $t_{1/2} = \frac{0.693}{k} = \frac{0.693}{0.023} = 30.1$ min
≈ 30 min

52. *(d)* Cu produces NO (*X*) gas when treated with dil. HNO_3. This gas is reduced to $NH_2OH \cdot HCl$ (*Y*) by $SnCl_2$/HCl and then oxidised by HNO_2 to give N_2O (*Z*). The reaction is as follows:

$$Cu + Dil.\,HNO_3 \longrightarrow \underset{(X)}{NO} \xrightarrow{SnCl_2/HCl}$$
$$\underset{(Y)}{NH_2OH \cdot HCl} \xrightarrow{HNO_2} \underset{(Z)}{N_2O}$$

53. *(b)*

4-hydroxymethyl phenol (HO–C₆H₄–CH₂OH) + HCl $\xrightarrow{\text{Heat}}$ HO–C₆H₄–CH₂Cl + H_2O

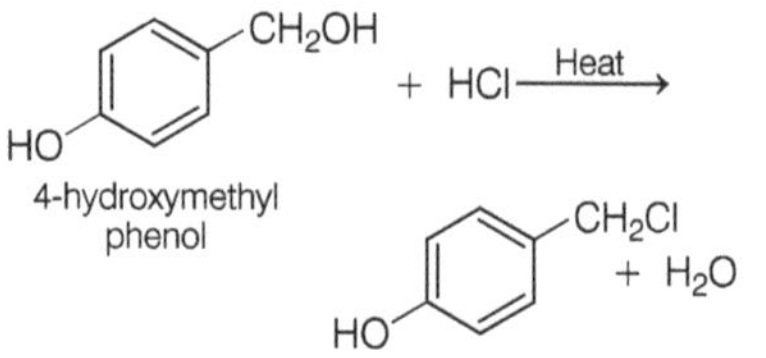

This reaction undergoes S_N1 reaction. Where, the OH of benzyl group gets substituted by the chloride group through carbocation formation. Here, the phenolic OH group does not get substituted by Cl group.

54. *(b)* The structure of given acids are as follows

(a) **Aspartic acid**

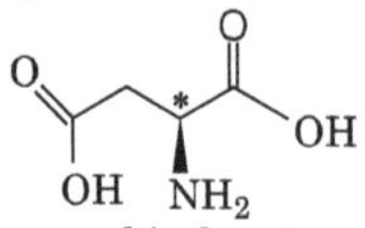

It has one chiral centre.

(b) **Arginine**

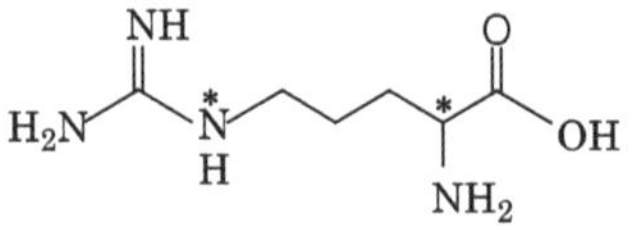

It has two chiral centres.

(c) **Lysine**

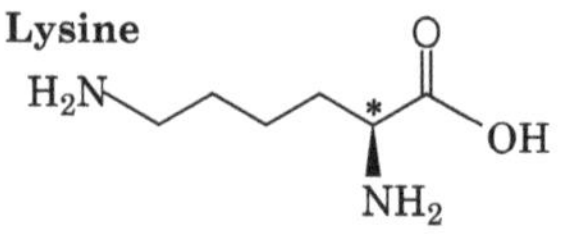

It has one chiral centre.

(d) **Histidine**

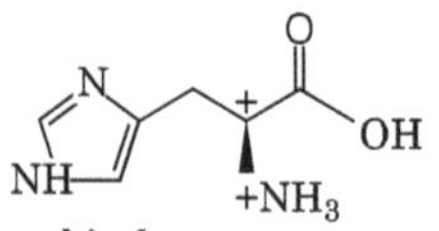

It has one chiral centre.

Thus, arginine has more than chiral centre.

55. *(c)* Pyridinium hydrochloride is a salt of pyridine (weak base) and hydrochloric acid (strong acid). pH of a solution of strong acids and weak base can be calculated as

$$pH = 7 - \frac{1}{2}(pK_b + \log c)$$

$$\therefore \quad 3.44 = 7 - \frac{pK_b}{2} - \frac{\log 0.02}{2}$$

$$pK_b = 8.82$$

$$pK_b = 8.82 = -\log K_b$$

$$K_b = 1.5 \times 10^{-9}$$

56. *(c)* Among the given compounds, III is most basic as one of the nitrogen atom is sp^3-hybridised. Between I and II, I is more basic as the lone pair on N atom is relatively more available for donation as compared to II. Also, in compound II, the other N-atom shows $-I$-effect which makes it a weaker base.

Thus, the correct increasing order of basicity is II < I < III.

57. *(a)* A mixture of benzaldehyde and formaldehyde on heating with aqueous NaOH solution gives benzyl alcohol and sodium formate. This reaction is known as Cannizzaro reaction, where the formaldehyde gets oxidised to sodium formate and benzaldehyde gets reduced to benzyl alcohol.

$$C_6H_5\text{—}CHO + H\text{—}\underset{\underset{O}{\|}}{C}\text{—}H \xrightarrow[H_2O]{NaOH} C_6H_5\text{—}CH_2OH + HCOONa$$

Benzaldehyde, Formaldehyde → Benzyl alcohol, Sodium formate

58. *(b)* For the reaction,

$$C_3H_7NO_2(s) + \frac{11}{2}O_2(g) \longrightarrow 3CO_2(g) + \frac{1}{2}N_2(g) + \frac{7}{2}H_2O(g)$$

$$\Delta H_{comb} = 3\Delta H_f CO_2 + 0 + \frac{7}{2}\Delta H_f (H_2O) - \Delta H_f (C_3H_7NO_2) + 0$$

$$= 3(-94.05) + \frac{7}{2}(-68.32) - (-133.57)$$

$$= -387.70 \text{ kcal/mol.}$$

59. *(b)* Given, $T_1 = 25°C = 298K$ $T_2 = T$

$E_a = 104.4 \text{ kJ mol}^{-1} = 104.4 \times 10^3 \text{ J mol}^{-1}$

$k_1 = 3 \times 10^{-4} \text{ s}^{-1}$, $k_2 = ?$

According to Arrhenius equation.

$$\log\frac{k_2}{k_1} = \frac{E_a}{2.303R}\left[\frac{1}{T_1} - \frac{1}{T_2}\right]$$

$$\log\frac{k_2}{3 \times 10^{-4}} = \frac{104.4 \times 10^3}{2.303 \times 8.314} \times \frac{1}{298}$$

$$\left[\text{as } T \to \infty, \frac{1}{T} \to 0\right]$$

$$\log\frac{k_2}{3 \times 10^{-4}} = 18.297$$

$$\frac{k_2}{3 \times 10^{-4}} = 1.98 \times 10^{18}$$

$$k_2 = 1.98 \times 10^{18} \times 3 \times 10^{-4}$$

$$= 5.94 \times 10^4 \approx 6 \times 10^{14} \text{ s}^{-1}$$

60. *(c)* In BF_3 boron is sp^2 hybridised. It has a vacant $2p$-orbital, each fluorine atom in BF_3 has completely filled unutilised $2p$-orbitals. So, $p\pi$ - $p\pi$ back bonding occurs, where a lone pair of electrons of F-atoms gets transferred into vacant $2p$-orbital. Therefore, B—F, bond has some double bond character. That is why all the 3B — F bonds are shorter than the usual single B—F bond. Whereas, in $[BF_4]$, boron is sp^3-hybridised and thus does not have empty $2p$-orbital, so no such $p\pi$ - $p\pi$ back bonding can occur. BF_4^- has pure B—F bonds. Hence, B—F bond length in BF_3 are shorter than BF_4^-.

F
B
F
F

Back bonding in BF_3

61. *(a)* Acetylcholine released from an excited terminal axon, depolarises the membrane and is then very rapidly inactivated by the enzyme cholinesterase, which is present in great concentrations in the neuromuscular junction. The inhibition of cholinesterase will result in the muscle being unable to restore the state of polarisation.

62. *(b)* Before pollination, the pollen grain cytoplasm divides between the generative cell and vegetative cell. The generative cell divides to form two male gametes.

63. *(b)* In a case, when SA-node or the pacemaker becomes non-functional, then there will be no origin of heartbeat and will be no transmission of impulses of atria.

Also, the ventricles will fail to receive atrial impulse by obstruction in AV conduction. Thus, overall conducting system of heart is disrupted. Hence, the complete cardiac muscles fail to contract rhythmically in a well-coordinated manner.

64. *(d)* Nuclear division would have been completed by telophase, resulting in the amount of DNA in the nucleus being halved, compared to the amount present at interphase, prophase and metaphase of cell divison cycle.

65. *(c)* Haemoglobin is a globular protein that binds to a non-proteinaceous substance, haem, which actually carries the oxygen. This material is known as a prosthetic group and influences the shape of the polypeptide.

66. *(c)* Only transformed bacteria containing plasmids coding for resistances to antibiotics will be able to survive in a medium containing antibiotics. Bacteria lacking the plasmids will be destroyed by the antibiotic.

67. *(b)* Any agent or factor that induces cancer is called carcinogen or carcinogenic factor, e.g. cadmium oxide, which induces prostate cancer. Chromium and nickel compounds induce lung cancer, whereas vinyl chloride and aflatoxins induce liver cancer.

68. *(c)* The experiments demonstrated that genes carry information for making proteins. Beadle and Tatum proposed one gene one enzyme hypothesis on the basis of biochemical genetics in *Neurospora crassa*. Beadle and Tatum showed that each kind of mutant bread mould lacks a specific enzyme and were unable to form the protein.

69. *(c)* In a nucleotide, the 5′-OH of a nucleoside (rather than a nitrogen base) is linked to a phosphate group. DNA acts as a genetic material in most organisms. It is a long polymer of deoxyribonucleotides. The number of nucleotides present in the DNA determines its length. A nucleotide is composed of a nitrogenous base, pentose sugar and a phosphate group. Rest all other statements are correct for DNA.

70. *(c)* Option (c) cannot be explained on the basis of Mendel's law of dominance. When F_1 hybrids exhibit a mixture or blending of characters of two parents, the case is considered as that of incomplete blending inheritance. It simply means that the two genes of an allelomorphic pair are not related as dominant or recessive, but each of them can express itself partially.

71. *(c)* Gene mutations are spontaneous events which result in the alternation of existing genes and the occurrence of new sequences. These mutations are thus responsible for variations occurring in the gene pool of a population.

72. *(b)* Pure culture without any contamination is called axenic culture. It is not associated with any living organisms and is raised under sterile conditions.

73. *(a)* Photosynthesis occurs most efficiently at wavelengths of red and blue light, because the chlorophyll pigments absorb light maximally at these wavelengths. Thus, for every 100 units of sunlight falling on a chloroplast of a green plant, so units are not used for photosynthesis as the wavelength are inappropriate.

74. *(c)* Pyruvate is a three carbon compound ($CH_3C[O]C[O]OH$), which on complete oxidation releases three moles of CO_2.

75. *(b)* The correct matches are as follows

Cup-shaped — *Chlamydomonas*

Girdle-shaped — *Ulothrix*

Stellate — *Zygnema*

Reticulate — *Oedogonium*

76. *(a)* The water moves from lower DPD to higher DPD. Cell A possesses diffusion pressure deficit of –3 atm and cell B –5 atm DPD. Hence, it will move from cell B to A.

77. *(b)* DNA polymerase binds to the single-stranded RNA primer and catalyses the formation of a phosphodiester bond between the free nucleotide and the preceding nucleotide, thus extending the new strand of DNA. In the event that the primase is defective, RNA primer will be absent and DNA polymerase will be unable to add nucleotides at the origin of replication.

78. *(c)* AIDS symptoms appear in the most advanced stage of HIV disease. Some people show HIV symptom shortly after incubation period but it usually takes more than 10 years. HIV attacks T-helper cells which regulate both hormonal and cellular immunity. This causes their depletion due to which patients immune system becomes very weak and susceptible to even minor infections.

79. *(c)* Oxygen toxicity develops when 100% pure oxygen is breathed for a prolonged period, it causes failure of ventilation of lungs.

80. *(c)* Embryonic stem cells are not totipotent but are multipotent cells. They have the ability to differentiate into almost any cell type to form any organ or type of cell.

81. *(c)* Suppose the green ball goes to bin i, for some $i \geq 1$.

The probability of this occurring is $\frac{1}{2^i}$

Given that this occurs, the probability that the red ball goes in a higher numbered bin is

$$\frac{1}{2^{i+1}} + \frac{1}{2^{i+2}} + \ldots = \frac{1}{2^i}$$

Probability that the green ball goes to bin i and the red ball goes in a bin greater than i is $\left(\frac{1}{2^i}\right)^2 - \frac{1}{2^{2i}} = \frac{1}{4^i}$

$$\text{Required probability} = \sum_{i=1}^{\infty} \frac{1}{4^i}$$

$$= \frac{1}{4} + \frac{1}{4^2} + \frac{1}{4^3} \ldots = \frac{\frac{1}{4}}{1 - \frac{1}{4}} = \frac{1}{3}$$

82. *(b)* $QC^2 = PQ \times AQ$

$$PQ = \frac{QC^2}{AQ} = \frac{9}{2}$$

PC is angle bisector of APQ

$$\therefore \quad \frac{PQ}{PR} = \frac{QC}{RC}$$

$$PR = \frac{PQ \times RC}{QC} = \frac{\frac{9}{2} \times 4}{3} = 6$$

$$RC^2 = PR \times BR$$

$$\Rightarrow \quad 16 = 6 \times BR \Rightarrow BR = \frac{16}{6} = \frac{8}{3}$$

83. *(d)* We have,

$$\alpha = \int_0^1 (e^{9x + 3\tan^{-1} x}) \left(\frac{12 + 9x^2}{1 + x^2} \right) dx$$

$$\alpha = \int_0^1 (e^{9x + 3\tan^{-1} x}) \left(9 + \frac{3}{1 + x^2} \right) dx$$

Put $9x + 3\tan^{-1} x = t$

$$\Rightarrow \quad \left(9 + \frac{3}{1 + x^2} \right) dx = dt$$

$$\therefore \quad \alpha = \int_0^{9 + \frac{3\pi}{4}} e^t \, dt = [e^t]_0^{9 + \frac{3\pi}{4}}$$

$$\Rightarrow \quad \alpha = e^{9 + \frac{3\pi}{4}} - 1$$

$$\Rightarrow \quad (\alpha + 1) = e^{9 + \frac{3\pi}{4}}$$

$$\Rightarrow \quad \log_e(\alpha + 1) - \frac{3\pi}{4} = 9$$

84. *(a)* We have, n_1, n_2, n_3, n_4, n_5 are positive integer and $n_5 > n_4 > n_3 > n_2 > n_1$

$\therefore \; n_1 \geq 1, n_2 \geq 2, n_3 \geq 3, n_4 \geq 4, n_5 \geq 5$

Let $n_1 - 1 = x_1 \geq 0$, $n_2 - 2 = x_2 \geq 0$

$x_5 - 5 = x_5; x_5 \geq 0$

$\therefore \; x_1 + 1 + x_2 + 2 + \ldots x_5 + 5 = 20$

$\Rightarrow \; x_1 + x_2 + x_3 + x_4 + x_5 = 5$

Now, $x_1 \leq x_2 < x_3 \leq x_4 \leq x_5$

x_1	x_2	x_3	x_4	x_5
0	0	0	0	5
0	0	0	1	4
0	0	0	2	3
0	0	1	1	3
0	0	1	2	2
0	1	1	1	2
1	1	1	1	1

So, 7 possible cases will be there.

85. *(c)* We have,

$$x^2 - x\sin x - \cos x = 0$$

Let $f(x) = x^2 - x\sin x - \cos x$

$$f'(x) = 2x - x\cos x$$

$$\lim_{x \to \infty} f(x) = \infty$$

$$\lim_{x \to -\infty} f(x) = \infty$$

$$f(0) = -1$$

Hence, 2 solutions exists.

86. *(d)* Given,

$$f'(x) < 2f(x)$$

$$\Rightarrow \quad \frac{f'(x)}{f(x)} < 2$$

On integrating, we get

$$f(x) < Ce^{2x}$$

put $x = \frac{1}{2} \Rightarrow C > \frac{1}{e}$

Hence, $f(x) < \frac{e^{2x}}{e}$

$$\Rightarrow \quad f(x) < e^{2x - 1}$$

$$\Rightarrow \quad 0 < \int_{1/2}^{1} f(x) dx < \int_{1/2}^{1} e^{2x - 1} dx$$

$$0 < \int_{1/2}^{1} f(x) < \frac{e - 1}{2}$$

87. *(d)* We have,

$$z = \cos\theta + i\sin\theta$$
$$z^{2m-1} = (\cos\theta + i\sin\theta)^{2m-1}$$
$$z^{2m-1} = \cos(2m-1)\theta + i\sin(2m-1)\theta$$
$$I_m(z^{2m-1}) = \sin(2m-1)\theta$$
$$\sum_{m=1}^{15} I_m(z^{2m-1}) = \sum_{m=1}^{15}\sin(2m-1)\theta$$
$$= \frac{1}{2\sin\theta}\sum_{m=1}^{15} 2\sin\theta\sin(2m-1)\theta$$
$$= \frac{1}{2\sin\theta}\sum_{m=1}^{15}(\cos(2m-2)\theta - \cos 2m\theta)$$
$$= \frac{1-\cos 30°\cdot\theta}{2\sin\theta}$$
$$= \frac{1-\cos 60°}{2\sin 2°} \quad [\theta = 2°]$$
$$= \frac{1}{4\sin 2°}$$

88. *(b)* Equation of circle

$$x^2 + y^2 - 6x - 4y - 11 = 0$$

Centre $O(3, 2)$.

Since, OA and OB is perpendicular to PA and PB

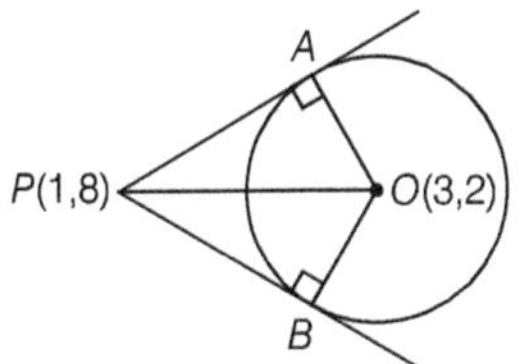

OP is diameter of the circumcircle of ΔPAB equation of circle is

$$(x-3)(x-1) + (y-2)(y-8) = 0$$
$$\Rightarrow \quad x^2 - 4x + 3 + y^2 - 10y + 16 = 0$$
$$\Rightarrow \quad x^2 + y^2 - 4x - 10y + 19 = 0$$

89. *(d)* Given, $n_1 = \sin 7 + \cos 7$

$$n_2 = \sqrt{\sin 7} + \sqrt{\cos 7}$$
$$n_3 = \sqrt{1+\sin 14}$$
$$n_4 = 1$$
$$n_1^2 = \sin^2 7 + \cos^2 7 + 2\sin 7\cos 7$$
$$n_1^2 = 1 + \sin 14$$
$$n_1 = \sqrt{1+\sin 14} = n_3$$
$$\sqrt{1+\sin 14} > 1$$
$$\therefore \quad n_3 > n_4$$

Hence, $n_2 > n_1 = n_3 > n_4$

90. *(a)* We have,

$$|f(x)| + \sqrt{1+\cos 2\pi x} = \tan^2\left(\frac{\pi x}{9}\right)f(x)$$

At $x = 3$

$$|f(3)| + \sqrt{1+\cos 6\pi} = \tan^2\left(\frac{\pi}{3}\right)f(3)$$
$$|f(3)| + \sqrt{2} = 3f(3)$$

If $f(3) \geq 0 \Rightarrow f(3) = \frac{1}{\sqrt{2}}$

If $f(3) < 0 \Rightarrow f(3) = \frac{\sqrt{2}}{4}$ not possible.

Hence, $f(3) = \frac{1}{\sqrt{2}}$

91. *(c)* For conservation of angular momentum, in a transition the angular momentum change (in final and in initial state) must be exactly of one unit.

i.e. $\quad \Delta l = |l_f - l_i| = \pm 1$

This selection rule must followed in all transitions. In given cases, for $3p \longrightarrow 2p$ transition $\Delta l = 0$.

So, this transition is forbidden.

Hence, it occurs very rarely.

92. *(b)* As gas expands in vaccum, work done by gas is zero.

So, $\Delta U = 0$ and temperature of gas remains constant.

Unavailable energy is

$$E = T\Delta S = nRT\ln\left(\frac{V_2}{V_1}\right)$$
$$= 5 \times 8.314 \times 300 \times \ln\left(\frac{152}{5}\right)$$
$$= 5 \times 8.314 \times 300 \times 4.3 \approx 54\,\text{kJ}$$

93. *(a)* All substances emits radiation. In general radiation is a volumetric phenomena but in solids and liquids emitted radiation by a molecule deep inside matter is absorbed by surrounding molecules and dissipation of energy occurs only from surface.

94. *(a)* Let current distribution is

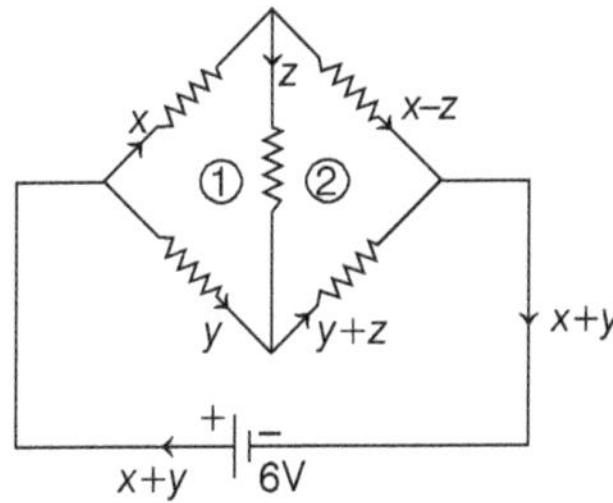

By loop rule in loops 1 and 2, we have

$$5x - 2y + 3z = 0 \quad \text{...(i)}$$
$$2x - 4y - 9z = 0 \quad \text{...(ii)}$$
$$\Rightarrow \quad \frac{x}{30} = \frac{y}{51} = \frac{z}{-16} = k \text{ (let)}$$
$$\Rightarrow \quad x = 30k,\ y = 51k,\ z = -16k$$

If R = equivalent resistance of circuit, then

$$V = R(x+y)$$
$$\Rightarrow \quad 7x - 2z = R(x+y)$$
$$\Rightarrow R = \frac{7x-2z}{x+y} = \frac{7\times 30k - 2(-16k)}{30k + 51k}$$
$$\Rightarrow R = \frac{242}{81} = 3\Omega$$

So, power dissipated,

$$P = \frac{V^2}{R} = \frac{36}{3} = 12\,\text{W}$$

95. *(c)* Consider an element of length dx at a distance x from centre.

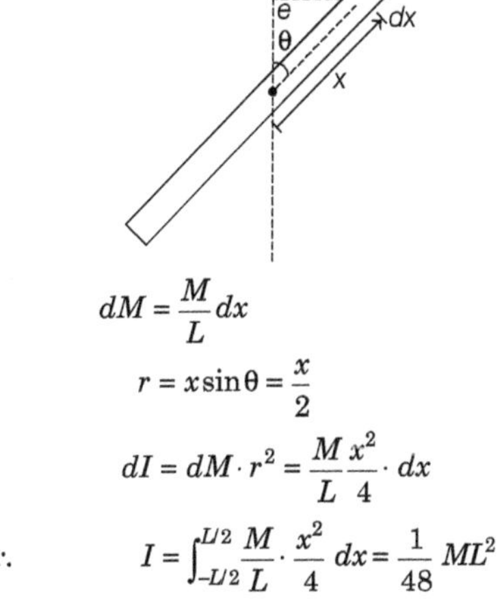

$$dM = \frac{M}{L}dx$$
$$r = x\sin\theta = \frac{x}{2}$$
$$dI = dM\cdot r^2 = \frac{M}{L}\frac{x^2}{4}\cdot dx$$
$$\therefore \quad I = \int_{-L/2}^{L/2}\frac{M}{L}\cdot\frac{x^2}{4}\,dx = \frac{1}{48}ML^2$$

96. *(d)* Velocity of nucleus can be resolved into parallel and perpendicular components.

Due to parallel component nucleus moves in direction of **B** and due to perpendicular component, it rotates in circular path.

Radius of circular path, $r = \frac{mv_\perp}{Bq} \Rightarrow r \propto m$

and pitch of helix, $p = \frac{2\pi m}{Bq}\cdot v_{||} \Rightarrow p \propto m$

As, both isotopes have same charge q, heavier traces a path with larger radius and larger pitch.

97. *(c)*

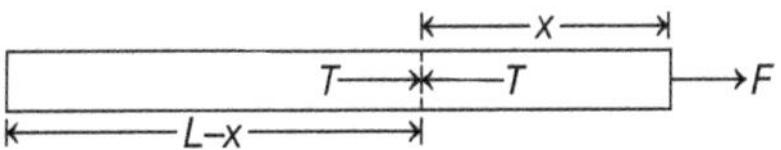

Acceleration of rod $= \frac{F}{M}$

Let T = tension in the rope at a distance x from the end where force F is applied.

For $L - x$ length, T is only force, so

$$T = \text{Mass} \times \text{Acceleration}$$

$$= \frac{M}{L} \times (L - x) \times \frac{F}{M}$$

= Mass per unit length × Length of remaining part × Acceleration

$$= \left(\frac{L - x}{L}\right) F = \left(\frac{L - \frac{L}{3}}{L}\right) F = \frac{2}{3} F$$

$\therefore$ Stress, $\sigma = \frac{2F}{3A}$

98. *(d)* Linear momentum is conserved,

$$mv_0 = 2mv_1 - mv_2$$

$$\Rightarrow \quad 2v_1 - v_2 = v_0$$

$$\Rightarrow \quad v_2 = 2v_1 - v_0 \qquad \ldots(i)$$

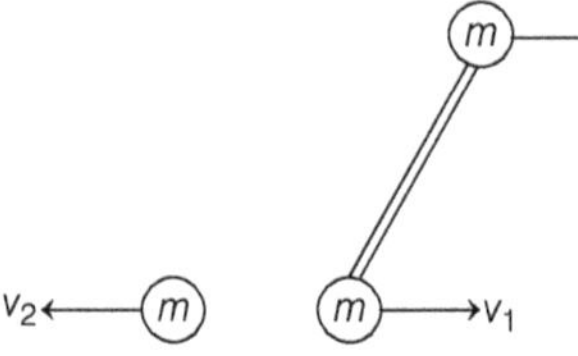

Angular momentum is also conserved, so

$$mv_0\left(\frac{L}{2}\sin 45^\circ\right) = 2m\left(\frac{L}{2}\right)^2 \omega - mv_2\left(\frac{L}{2}\sin 45^\circ\right)$$

$$\Rightarrow \quad v_0 = \sqrt{2}L\omega - v_2 \qquad \ldots(ii)$$

As, collision is elastic.

Relative velocity of separation

= Relative velocity of approach

$$\Rightarrow \quad v_2 + v_1 + \frac{L}{2}\omega \times \frac{1}{\sqrt{2}} = v_0 \qquad \ldots(iii)$$

Putting the value of v_2 from Eq. (i) in Eq. (iii), we get

$$2v_1 - v_0 + v_1 + \frac{L\omega}{2\sqrt{2}} = v_0$$

$$3v_1 - 2v_0 + \frac{L\omega}{2\sqrt{2}} = 0$$

Value of v_1 from Eqs. (i) and (ii), we get

$$\Rightarrow \quad \frac{3\sqrt{2}}{2}L\omega - 2v_0 + \frac{L\omega}{2\sqrt{2}} = 0$$

$$\left[\begin{array}{l} \Rightarrow v_0 = \sqrt{2}L\omega + v_0 - 2v_1 \\ \Rightarrow 2v_1 = \sqrt{2}L\omega \\ \Rightarrow \quad v_1 = \frac{\sqrt{2}}{2}L\omega \end{array}\right]$$

$$\Rightarrow \quad L\omega\left(\frac{3\sqrt{2}}{2} + \frac{1}{2\sqrt{2}}\right) = 2v_0$$

$$\Rightarrow \quad L\omega\left(\frac{6+1}{2\sqrt{2}}\right) = 2v_0$$

$$\Rightarrow \quad \frac{7L\omega}{2\sqrt{2}} = 2v_0$$

$$\Rightarrow \quad \omega = \frac{4\sqrt{2}\,v_0}{7L}$$

99. *(c)* Induced emf, $E = \frac{d}{dt}\phi_B$

$$= a\int_0^a \left(v\frac{\partial B}{\partial x} + \frac{\partial B}{\partial T}\right)dx$$

$$= a^2\left(v\frac{\partial B}{\partial x} + \frac{\partial B}{\partial T}\right)$$

$$= (0.1)^2\,(0.08 \times 0.1 \times 10^{-3})$$

$$= 9 \times 10^{-5}\ \text{V}$$

$$= 90\,\mu\text{V}$$

100. *(c)* Latent heat is used in increasing molecular separation. Length D shows heat absorbed during liquid to vapour change. So, length D is large.

101. *(d)* Aromatic compounds are stable than non-aromatics are due to resonance. Aromaticity can be determined by Huckel's rule according to which (or $4n + 2$ rule), "a planar, cyclic compound is aromatic, when its π cloud must contain $(4n + 2)\pi$ electrons, where n is any whole number.

(6πe^- system), (4πe^- + 2e^- system) and (6πe^- system)

is non-aromatic, hence, least stabilised by resonance.

102. *(a)* The complexes that has same number of unpaired electrons will have same magnetic moment. The number of unpaired electrons in complexes are as follows:

Complex ion	Electronic configuration of metal ion	Number of unpaired electrons (*n*)
$[Cr(H_2O)_6]^{2+}$	Cr^{2+} ; $[Ar]3d^4$	1 1 1 1 ;4
$[Fe(H_2O)_6]^{2+}$	Fe^{2+} ; $[Ar]3d^6$	⥮ 1 1 1 1 ;4
$[Mn(H_2O)_6]^{2+}$	Mn^{2+} ; $[Ar]3d^5$	1 1 1 1 1 ;5
$[CoCl]^{2-}$	Co^{2+} ; $[Ar]3d^7$	⥮ ⥮ 1 1 1 ;3

Thus, $[Cr(H_2O)_6]^{2+}$, $[Fe(H_2O)_6]^{2+}$ have same magnetic moment.

103. *(d)* The hybridisation of a compound can be calculated using formula.

$$H = \frac{1}{2}[V + MO \pm \text{anion / cation}]$$

where, H = Hybridised orbitals

V = Valence

MO = Number of monoatomic atoms

For XeF_4

$$H = \frac{1}{2}[8 + 4] = 6$$

$$\Rightarrow \quad sp^3d^2$$

It has 4 bond pairs and 2 lone pair, where 2 lone pair occupies axial positions and four F occupies equitorial positions. So, its geometry is square planar and not octahedral.

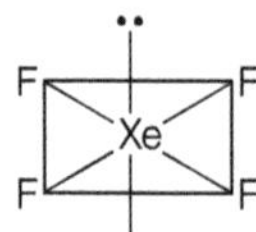

104. *(a)* Williamson's synthesis is an organic reaction used to convert an alcohol and an alkyl halide to an ether using a base such as NaOH. This reaction proceed *via* S_N1 mechanism.

Among the given compounds, Di-*tert* butyl ether cannot be made by Williamson's synthesis, since *tert*-alkyl (3°) halides prefer to undergo elimination rather than substitution. i.e.

$$CH_3-C(CH_3)_2-Br + CH_3-C(CH_3)_2-O^-Na^+ \xrightarrow[-NaBr]{\Delta} CH_3-C(CH_3)=CH_2 + CH_3-C(CH_3)_2-OH$$

105. *(a)* Octahedral complexes of the type $[MA_4B_2]$ exhibit geometrical isomers. The type of complexes in the given options are as follows:

	Complexes	Types of complexes
(a)	$[Cr(H_2O)_4Cl_2]^+$	$[MA_4B_2]$
(b)	$[Pt(NH_3)_3Cl]$	$[MA_3B]$
(c)	$[Co(NH_3)_6]^{3+}$	$[MA_6]$
(d)	$[Co(CN)_5Cl]^{3-}$	$[MA_5B]$

Hence, the given options only $[Cr(H_2O)_4Cl_2]^+$ exhibit geometrical isomerism, *cis* and *trans* form.

106. *(a)* The organic compound (*A*) with molecular formula $C_9H_{10}O$ forms an orange-red ppt. with 2, 4-DNP reagent and gives yellow ppt. on heating with I_2 and NaOH. This confirms the presence of carboxyl group in compound (*A*). Also, it does not reduces Tollen's reagent or Fehling's solution, this shows it is a ketonic group and not an aldehyde. As it does not decolourises bromine water it indicates the absence of unsaturated bonds. The reactions are as follows:

C_6H_5COOH $(C_7H_6O_2)$ $\xleftarrow{H_2Cr_2O_7}$ $C_6H_5CH_2—\overset{O}{\overset{\|}{C}}—CH_3$ (*A*) 1-phenyl propanone

$C_6H_5CH_2—\overset{O}{\overset{\|}{C}}Na + CHI_3$ (Yellow ppt.) $\xleftarrow{I_2/NaOH}$

$\downarrow$ $H_2NHN—C_6H_3(O_2N)—NO_2$ (2, 4-DNP)

$C_6H_5—CH_2—\overset{CH_3}{\overset{|}{C}}=NH—N—C_6H_3(O_2N)—NO_2$

(Red-orange ppt.)

107. *(a)* $Pb|PbCl_2||AgCl||Ag$

The cell reaction is

$$Pb + 2AgCl \longrightarrow PbCl_2 + 2Ag$$

$$\Delta G = -nFE$$
$$= -2 \times 96500 \times 0.5 \text{ J}$$
$$= -96500 \text{ J or } -96.5 \text{ kJ}$$
$$\Delta S = nF\left(\frac{\partial E}{\partial T}\right)_p$$
$$= 2 \times 96500 \times -2 \times 10^{-4}$$
$$= -38.6 \text{ JK}^{-1}$$

Also,

$$\Delta G = \Delta H - T\Delta S$$
$$= -96500$$
$$= \Delta H - 300(-38.6)$$
$$\Delta H = -96500 - 300 \times 38.6$$
$$= -108080 \text{ J}$$
$$= -108.08 \text{ kJ}$$

108. *(a)*

4-formylcyclohexanone (CHO) $\xrightarrow[\text{Tollen's reagent } (x)]{[Ag(NH_3)_2]^+OH^-}$ 4-oxocyclohexane (COOH) $\xrightarrow[\text{Esterification } (y)]{CH_3OH, H^+}$ (COOH$_3$)

$\downarrow$ Grignard reagent CH_3MgBr Excess (*z*)

HO, CH_3 ... OH

Reagent $x[Ag(NH_3)_2]^+ OH^-$ is a Tollen's reagent . It will only oxidises aldehyde group and not ketone group. Reagent *y*, CH_3OH is an alcohol, which then reacts with formed carboxylic acid to give ester. In presence of reagent *z*, CH_3MgBr (excess) it will reduce both ketone and ester group to alcoholic group.

109. *(a)* At *A*

Temp. = 300 K, volume = 10 L

Let pressure at this point = p_1 At *C*

Temp. = 600 K, volume = 20 L

Let the pressure at this point = p_2

According to ideal gas,

$$pV = RT \text{ (For 1 mole)}$$
$$R = \frac{pV}{T}$$

$\therefore$ For system *A*, $R = \dfrac{p_1 \times 10}{300}$

For system *C*, $R = \dfrac{p_2 \times 20}{600}$

$$\frac{p_1 \times 10}{300} = \frac{p_2 \times 20}{600}$$
$$p_1 = p_2$$

Thus, it is an isochoric process, when system moves from $A \to C$.

110. *(b)* Molar mass of CsBr

$$= 212.8 \text{ g mol}^{-1}$$

$\therefore$ Molar volume $= \dfrac{\text{molar mass}}{\text{density}}$

$$= \frac{212.8}{4.49} = 47.39 \text{ cm}^3$$

6.023×10^{23} molecules are there in 47.39 cm^3

1 molecule is there in $\dfrac{47.39}{6.023 \times 10^{23}}$

$$= 7.868 \times 10^{-23} \text{ cm}^3$$

$\therefore$ Volume of the unit cell

$$= a^3 = 7.868 \times 10^{-23} \text{ cm}^3$$

Side of the unit cell

$$= a = (7.868 \times 10^{-23})^{1/3}$$
$$= 4.285 \times 10^{-8} \text{ cm}$$
$$= 4.285 \text{ Å}$$

111. *(c)* A normal red-green colour vision man with Rh^- blood group is XY Rh^-Rh^-. A normal woman with Rh^+ blood group with heterozygocity at both red-green colour vision and blood group is $XX^C Rh^+Rh^-$.

The inheritance of colour vision in the progeny is

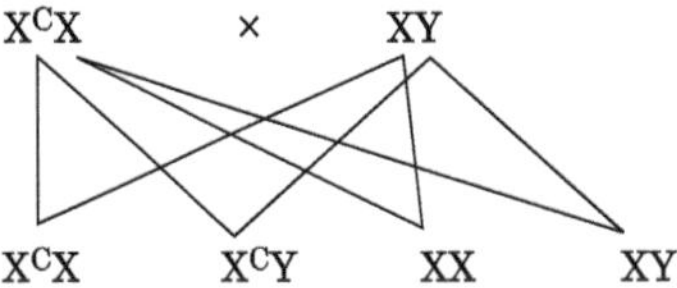

i.e. 50% boys are colourblind also 50% of offspring are boys. Thus, 0.25 is the probability the first boy is colourblind.

The inheritance of Rhesus blood group in the progeny is

Rh^-Rh^- Rh^+Rh^-

Rh^+Rh^-	Rh^-Rh^-	Rh^+Rh^-	Rh^-Rh^-
Rhesus positive	Rhesus negative	Rhesus positive	Rhesus negative

i.e. 50% are Rhesus positive and 50% Rhesus negative or 0.5 is the probability of the progeny being Rhesus negative.

Probability of their first child being a rhesus negative = 0.5.

Probability of their first child being a red green colourblind boy = 0.25.

Therefore, probability of their first child being both = 0.5× 0.25= 0.125.

112. *(d)* The net increase in population will be

Birth rate (B) + Immigration (I) – Death rate (D) + Emigration (E), i.e.

250 – 20 – (240 + 30) = 0

113. *(a)* A suitable restriction endonuclease should cut the plasmid at only one position in order to allow the DNA containing the gene of interest to be inserted into the circular plasmid without loss of any part of the plasmid. Thus, in the map of the plasmid given, the most suitable enzyme will be *Bam* HI.

114. *(c)* As there are 2 types of base, 2^n where n = number of bases per codon. Thus, 2^6 would provide 64 combination, which is more than sufficient. Thus, 6 are the minimum number of bases per codon that could code for proteins.

115. *(d)* Option (d) is correct. Other statements can be corrected as primary productivity varies for all ecosystems.

Net primary productivity is the amount of biomass available for consumption by heterotrophs. Secondary productivity is defined as the rate of formation of new organic matter by consumers.

116. *(a)* Since, there are different amino acids, the number of different polypeptides that can be formed, each with r amino acid is n^r.

117. *(d)* Out of the statements given, I and V are the correct options, remaining three are wrong. These can be corrected as in bacteria, transcription and translation do not take place in same compartments they are instead coupled in the cytosol itself as there is no separation of nucleus and cytosol in bacteria therefore, translation starts much before the *m*RNA is fully transcribed.

RNA polymerase-I is responsible for the transcription of *hn*RNA (not *t*RNA).

When *hn*RNA undergoes capping process, methyl guanosine triphosphate are added at 5′ end in a template independent manner.

118. *(d)* With increasing substrate concentration in the presence of a competitive inhibitor, the rate of reaction only remains constant at a later stage. This is because the additional substrates can compete with the inhibitor for the enzyme's active site. Thus, graph (d) is correct.

119. *(a)* The correct matching is

Potassium – Involved in stomatal movement

Sulphur – Constituent of ferredoxin

Molybdenum – Component of nitrogenase

Zinc – Needed in the synthesis of auxin

120. *(c)* Parent 3 is certain to have the genotype rr. Since, the male 10 is also having the genotype rr, this means that parent 4 must have the genotype Rr. Therefore, female 9 must have the genotype Rr. As female 13 has the genotype rr, both her parents (7 and 8) must have the genotypes Rr. Female 15 must have the genotype Rr because her father (male 10) has the genotype rr and her mother (female 11) is normal.

PRACTICE SET 2

Stream : SB/SX

MM : 160

Instructions

1. There are 120 questions in this paper.
2. The question paper contains two parts; Part I (1 Mark Questions) and Part II (2 Marks Questions).
3. There are four sections in each part; Mathematics, Physics, Chemistry and Biology.
4. Out of the four options given with each question, only one is correct.

PART-I (1 Mark Questions)

MATHEMATICS

1. If equation $x^n - nx^{n-1} + a_2x^{n-2} + a_3x^{n-3} + \dots + a_{n-1}x + (-1)^n = 0$ has n positive roots, then the least value of n for which $a_2 + a_3$ is negative is

(a) 2 (b) 6 (c) 4 (d) 1

2. If in an isosceles triangle with base 'a' vertical angle 20° and lateral side each of length 'b' is given, then the value of $a^3 + b^3$ equals

(a) $3ab$ (b) $3ab^2$ (c) $3a^2b$ (d) 3

3. The number of triangles with each side having integral length and the longest side is of 11 units is

(a) 30 (b) 34 (c) 35 (d) 36

4. If $\log_{245} 175 = a$ and $\log_{1715} 875 = b$, then $\frac{1-ab}{a-b}$ is equal to

(a) 2 (b) 3 (c) 5 (d) 7

5. Given $\int_0^{\frac{\pi}{2}} \frac{\sin x}{1+\sin x+\cos x}\,dx = K$, then the value of the definite integral $\int_0^{\frac{\pi}{2}} \frac{dx}{1+\sin x+\cos x}$ is equal to

(a) $\frac{K}{2}$ (b) $\frac{\pi}{2} - K$

(c) $\frac{\pi}{2} + K$ (d) $\frac{\pi}{2} - 2K$

6. If the line $y = \sqrt{3}x$ cuts the curve $x^4 + ax^2y + bxy + cx + dy + 6 = 0$ at A, B, C and D, then the value of $\frac{1}{12} OA \cdot OB \cdot OC \cdot OD$ (where O is origin) is equal to

(a) 6 (b) 8 (c) 12 (d) 16

7. Let $f(x) = \int \frac{x^2 dx}{(1+x^2)(1+\sqrt{1+x^2})}$ and $f(0) = 0$, then $f(1)$ is

(a) $\log(\sqrt{2}+1)$ (b) $\log(\sqrt{2}+1) - \frac{\pi}{4}$

(c) $\log(\sqrt{2}+1) + \frac{\pi}{4}$ (d) None of these

8. The number of ways in which the letters of the word '$ABBCABBC$' can be arranged such that the word $ABBC$ does not appear in any word is

(a) 420 (b) 392 (c) 361 (d) 360

9. If $f(x)$ is a function satisfying the relation $x^2 f(x) - 2f\left(\frac{1}{x}\right) = g(x)$, where $g(x)$ is odd function, then the value of $f(2)$ is

(a) $-f(2)$ (b) $-f(-2)$

(c) $f\left(\frac{1}{2}\right)$ (d) $-f\left(\frac{1}{2}\right)$

10. If A, B and C are $n \times n$ matrices and $\det(A) = 2$, $\det(B) = 3$ and $\det(C) = 5$, then the value of $\det(A^2BC^{-1})$ is

(a) $\frac{6}{5}$ (b) $\frac{12}{5}$ (c) $\frac{18}{5}$ (d) $\frac{24}{5}$

11. The sum of all the roots of the equation $\sin\left(\pi \log_3\left(\frac{1}{x}\right)\right) = 0$ in $(0, 2\pi)$ is

(a) $\frac{3}{2}$ (b) 4 (c) $\frac{9}{2}$ (d) $\frac{13}{3}$

12. If the chords of contact of tangents from two points $(-4, 2)$ and $(2, 1)$ to the hyperbola $\frac{x^2}{a^2} - \frac{y^2}{b^2} = 1$ are at right angle, then the eccentricity of the hyperbola is

(a) $\frac{\sqrt{7}}{2}$ (b) $\sqrt{\frac{5}{3}}$ (c) $\sqrt{\frac{3}{2}}$ (d) $\sqrt{2}$

13. If Z is a complex number satisfying $|Z^3 + Z^{-3}| \le 2$, then the maximum possible value of $|Z + Z^{-1}|$ is

(a) 2 (b) $3\sqrt{2}$ (c) $2\sqrt{2}$ (d) 1

14. If $f(x) = x^2 + \alpha x^2 + \beta x + \gamma$, where α, β, γ, are rational numbers and two roots of $f(x) = 0$ are eccentricities of a parabola and a rectangular hyperbola, then $\alpha + \beta + \gamma$ is equal to

(a) -1 (b) 0 (c) 1 (d) 2

15. $\int_{-\pi}^{\pi} (\cos \alpha x - \sin \beta x)^2 dx$ is equal to

(a) 0 (b) $\frac{\pi}{2}$ (c) π (d) 2π

16. If a, b, c and d are four positive real numbers such that $abcd = 1$, then the minimum value of $(1 + a)(1 + b)(1 + c)(1 + d)$ is

(a) 4 (b) 1 (c) 16 (d) 18

17. There are three coplaner parallel lines. If any point p taken on each the lines, the maximum number of triangles with vertices at these points is

(a) $3p^2(p - 1) + 1$ (b) $3p^2(p - 1)$

(c) $p^2(4p - 3)$ (d) $p^2(2p - 3)$

18. The value of $\lim_{n \to \infty} n^2 \int_{-1/n}^{1/n} (2018 \sin x + 2019 \cos x) |x| \, dx$ is

(a) 0 (b) 1 (c) 2018 (d) 2019

19. The value of $\tan^6 40° - 33 \tan^4 40° + 27 \tan^2 40°$ is equal to

(a) 2 (b) 3

(c) 4 (d) None of these

20. A ten-digit number is formed without repeating any digit the probability that the difference of digits at equal distances from the beginning and the end is always 1, is

(a) $\frac{17}{1944}$ (b) $\frac{4}{27}$ (c) $\frac{1}{945}$ (d) $\frac{34}{243}$

PHYSICS

21. A number of particles start simultaneously from same point in all possible directions with same speed in a vertical plane.

Now, choose the correct option.

(a) After time t they all lie on a parabola

(b) After time t they all lie on a circle

(c) After time t parabola described by particles has focal distance ut

(d) After time t circle described by particles has radius $2ut$

22. Two particles each of mass m are attached to ends of a light string of length a and placed on a horizontal turntable at distances a and $2a$ from centre of rotation. Let turntable rotates with a maximum angular speed ω, such that particle remains over it without any slip. Necessary friction coefficient must be

(a) $\frac{a\omega^2}{2g}$ (b) $\frac{2a\omega^2}{g}$ (c) $\frac{3a\omega^2}{2g}$ (d) $\frac{a\omega^2}{3g}$

23. A chord of length 90 m is used to tie a 200 kg astronaut (with his space suit and instruments) with his spaceship. Tension in the chord is nearly

(a) 0.0084 N (b) 0.084 N (c) 0.84 N (d) 8.40 N

24. A non-uniform rod of length l has a linear mass density $\lambda = \lambda_0 + kx$, is placed over X-axis with its one end at origin. The x-coordinate of centre of mass of rod is

(a) $\frac{\lambda_0 l + 2kl^2}{3(2\lambda_0 + kl)}$ (b) $\frac{3\lambda_0 l + 2kl^2}{(2\lambda_0 + kl)}$

(c) $\frac{4\lambda_0 l + kl^2}{3(2\lambda_0 + kl)}$ (d) $\frac{3\lambda_0 l + 2kl^2}{3(2\lambda_0 + kl)}$

25. Two masses m_1 and m_2 $(m_1 > m_2)$ are connected by a light spring. This system is placed on a horizontal frictionless table. At $t = 0$, blocks are pushed closer by compressing the spring and then released.

The system oscillates with a time period T. Time period of oscillation of centre of mass of system is

(a) $\frac{m_1 m_2}{m_1 + m_2} \cdot T$ (b) $\frac{m_1}{m_1 + m_2} \cdot T$

(c) $\frac{m_2}{m_1 + m_2} \cdot T$ (d) zero

26. A particle moves around a circular path of radius r with a constant speed v. If particle describes angle θ in 4 s, then average acceleration of the particle is

(a) $\frac{v^2}{r} \cdot \sin\left(\frac{2v}{R}\right)$ (b) $\frac{2v^2}{R} \cdot \sin\left(\frac{\theta}{2}\right)$

(c) $\frac{2v^2}{R\theta} \cdot \cos\left(\frac{\theta}{2}\right)$ (d) $\frac{v}{2} \cdot \sin\left(\frac{2v}{R}\right)$

27. A block of mass m_1 is placed over a plank of mass m_2. Block and plank are given initial velocities v_1 and v_2 relative to ground $(v_1 > v_2)$.

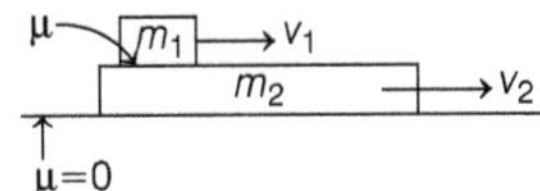

There is friction between block and plank, but there is no friction between plank and ground.

Correct variation of velocity of block v_b and velocity of plank v_p is

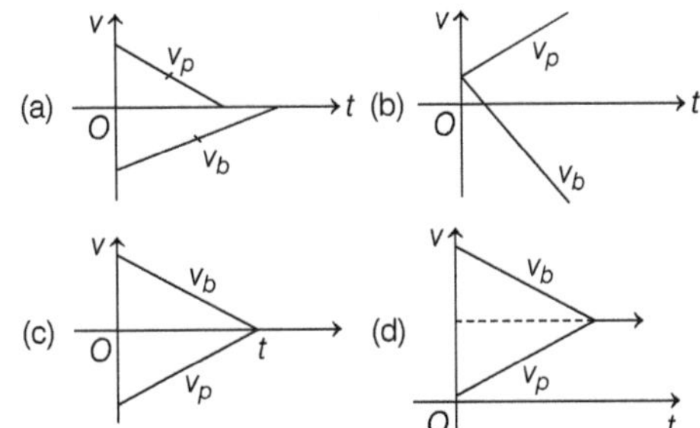

28. If you are pulling one end of an ideal spring of mass m with velocity v, while keeping other end fixed to a wall, then kinetic energy of the spring at this instant is

(a) $\frac{1}{2}mv^2$ (b) $\frac{1}{4}mv^2$ (c) $\frac{1}{6}mv^2$ (d) $\frac{1}{8}mv^2$

29. Consider given pulley and mass system consisting of ideal pulley, ideal springs and ideal identical strings.

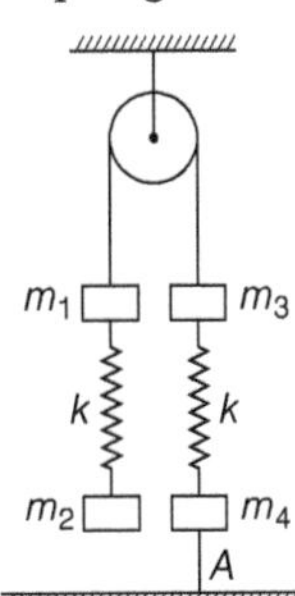

Let accelerations of m_1, m_2, m_3 and m_4 are a_1, a_2, a_3 and a_4, respectively immediately after cutting thread A which keeps the system stable.

Now, choose the correct option.

(a) $a_4 = 0$ (b) $a_4 = \left(\frac{m_3 + m_4 - m_1 - m_2}{m_4}\right) \cdot g$

(c) $a_4 = \frac{(m_1 + m_2)g}{m_4}$ (d) $a_4 = \left(\frac{m_3 - m_1 - m_2}{m_4}\right) \cdot g$

30. Colours of moist objects seems deeper and richer than those of dry ones. This happens because

(a) water oxidises the dye of colour
(b) interference occurs from reflections from top and bottom layers of water film
(c) thin water covered surface reflects light in one direction only
(d) water layer acts like a converging lens

31. XOY-plane is a boundary between two transparent media. Medium 1 with $z \geq 0$, has a refractive index $\sqrt{2}$ and medium 2 with $z \leq 0$, has a refractive index $\sqrt{3}$. A light ray $\mathbf{AO} = 6\sqrt{3}\hat{\mathbf{i}} + 8\sqrt{3}\hat{\mathbf{j}} - 10\hat{\mathbf{k}}$ is incident over point O in medium 1. If refracted ray in medium 2 is $\mathbf{r} = a\hat{\mathbf{i}} + b\hat{\mathbf{j}} + c\hat{\mathbf{k}}$. Then, correct relation is

(a) $a = \frac{4b}{3}$ (b) $c = \frac{5a}{3}$

(c) $c = -\frac{5a}{3}$ (d) $a = \frac{-4b}{3}$

32. When a spherical ball is suspended in an air or water jet as shown alongside.

Vertical elevation Z depends on

d = sphere diameter,

g = gravitational constant,

ρ_1 = fermi density,

ρ_2 = mass density of ball

and F = force of fluid on ball.

Then, which of these graphs is correct experimental observation of

$\frac{Z}{d}$ *versus* x, where $x = \frac{F}{d^3 g(\rho_2 - \rho_1)}$ is

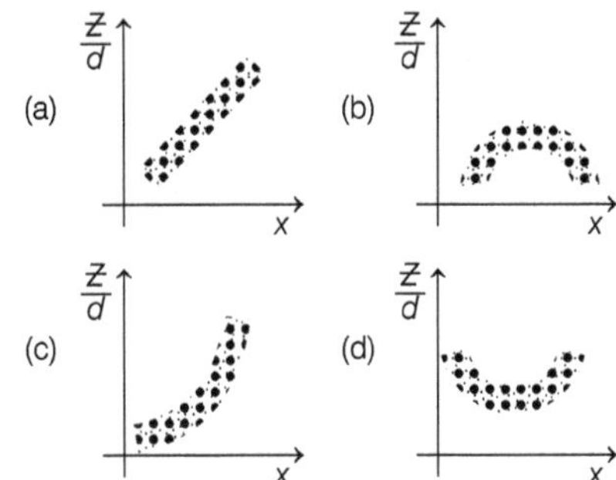

33. An unstable radioactive substance decays into a daughter product, but this in turn will probably itself decay until, a stable non-decaying isotope is formed. The list of all members is called a radioactive series. An example of a series is

$${}^{232}_{90}\text{Th} \xrightarrow{\alpha} {}^{228}_{88}\text{Ra} \xrightarrow{\beta} {}^{228}_{89}\text{Ac} \xrightarrow{\beta}$$

$${}^{228}_{90}\text{Th} \xrightarrow{\alpha} {}^{224}_{88}\text{Ra} \xrightarrow{\alpha} {}^{220}_{80}\text{Ra} \xrightarrow{\alpha} {}^{216}_{84}\text{Po}$$

Now, choose the correct option.

(a) ^{232}Th and ^{228}Th are isobars
(b) ^{228}Ra and ^{228}Ac are isotopes
(c) ^{232}Th and ^{220}Ra are isotones
(d) After some time all the members of series are in equilibrium

34. Which region of binding energy per nucleon *versus* mass number curve indicates that the nucleus also have some kind of its own structure.

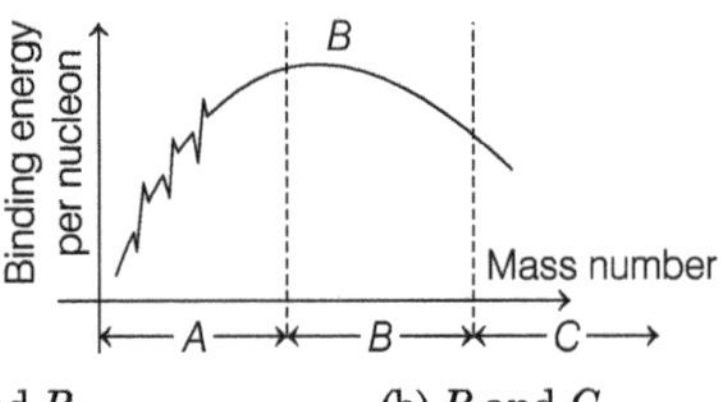

(a) A and B (b) B and C
(c) Only A (d) Only C

35. Consider two heat transfer processes A and B occuring as shown below.

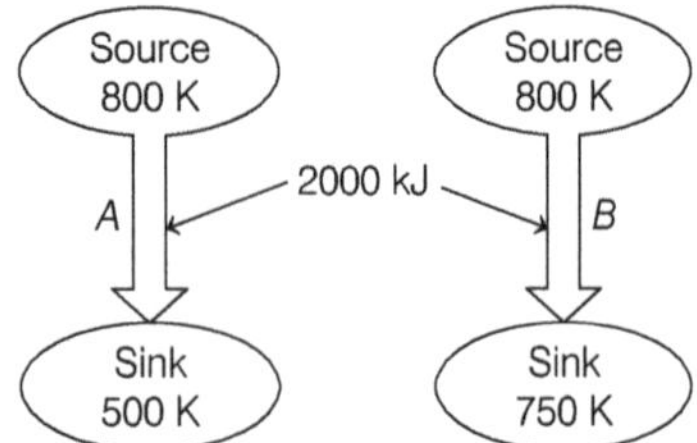

In both cases, 2000 kJ of heat is taken from source and transferred to sink. Source temperature is 800 K and sink temperatures in A and B are 500 K and 750 K, respectively. Which of the processes is easily reversed?

(a) Both A and B (b) A
(c) B (d) None of A and B

36. $\frac{pV}{RT}$ *versus* p graph for methane gas at several temperatures T_1, T_2, T_3 and T_4 are as shown below.

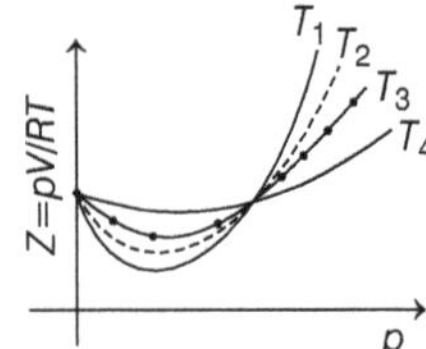

Now, choose the correct option.

(a) $T_1 = T_2 = T_3 = T_4$ (b) $T_1 < T_2 < T_3 < T_4$
(c) $T_1 > T_2 > T_3 > T_4$ (d) Data insufficient

37. A ring of N turns with radius a and resistance R is pulled into region of magnetic field B at an angle of 45° with the horizontal as shown below.

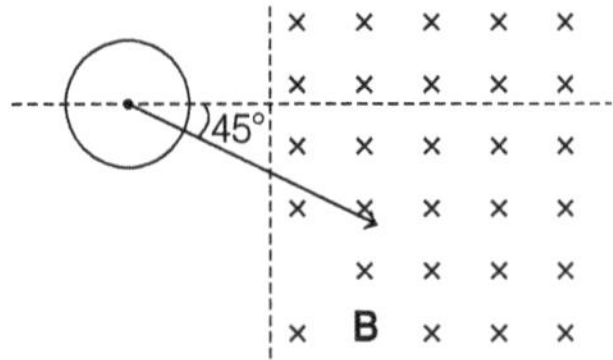

Charge passes through coil in time duration in which coil is completely immersed in field will be

(a) $\frac{\pi a^2 NB}{R}$ (b) $\frac{\pi a^2 NB}{\sqrt{2}R}$
(c) $\frac{\sqrt{2}\pi a^2 NB}{R}$ (d) $\frac{2\pi a^2 NB}{R}$

38. In the circuit given below.

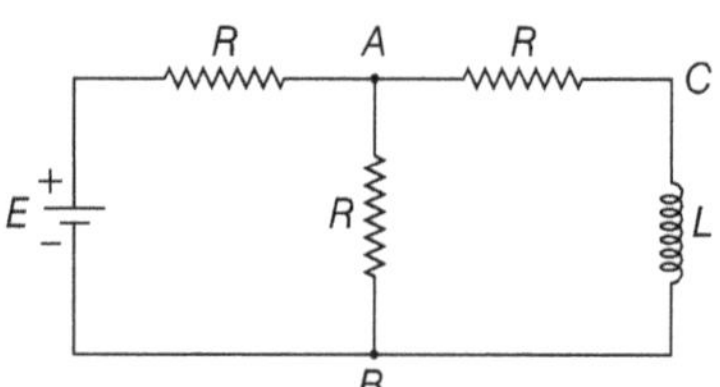

Potential difference of points A and B at time t is

(a) $\frac{E}{3}\left(1+\frac{1}{2}e^{-t\left(\frac{3R}{2L}\right)}\right)$ (b) $\frac{E}{3}\left(1-\frac{1}{2}e^{-t\left(\frac{3R}{2L}\right)}\right)$
(c) $\frac{E}{2}\left(1+e^{-t\left(\frac{3R}{2L}\right)}\right)$ (d) $\frac{E}{2}\left(1-e^{-t\left(\frac{3R}{2L}\right)}\right)$

39. Energy of a hydrogen atom in its ground state is (r = atomic radius of hydrogen atom)

(a) $-\frac{e^2}{4\pi\varepsilon_0 r}$ (b) $\frac{e^2}{4\pi\varepsilon_0 r}$
(c) $\frac{e^2}{8\pi\varepsilon_0 r}$ (d) $-\frac{e^2}{8\pi\varepsilon_0 r}$

40. Consider the given statements.

I. An electron can absorbs a photon.
II. An electron can absorbs two photons (one after other).
III. A free electron cannot absorbs a photon.
IV. A bound electron can absorbs two photons (one after other).

Which of the above statement is correct?

(a) Only I (b) Only III
(c) Only IV (d) Only II

CHEMISTRY

41. In the compound $CH_2{=}CH{-}CH_2{-}CH_2{-}C{\equiv}CH$, the $C_2{-}C_3$ bond is of

(a) $sp-sp^2$ (b) sp^3-sp^3 (c) $sp-sp^3$ (d) sp^2-sp^3

42. When hydrochloric acid gas is treated with propene in the presence of benzoyl peroxide, it gives

(a) 2-chloropropane (b) allyl chloride
(c) n-propyl chloride (d) isopropyl chloride

43. Plots showing the variation of the rate constant (k) with temperature (T) are given below. The plot that follows Arrhenius equation is

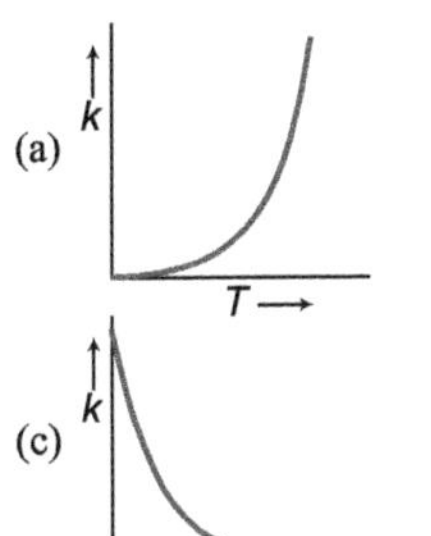

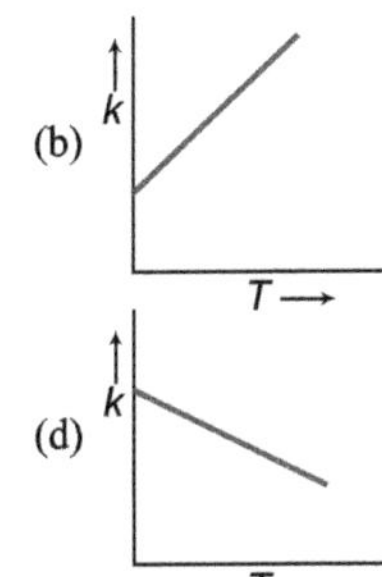

44. For a linear plot of log (x/m) *versus* log p in a Freundlich adsorption isotherm, which of the following statements is correct? (k and n are constants)

(a) $1/n$ appears as the intercept

(b) Only $1/n$ appears as the slope

(c) $\log\left(\frac{1}{n}\right)$ appears as the intercept

(d) Both k and $1/n$ appear in the slope term

45. A gaseous substance dissolves in water giving a pale blue solution which decolorises $KMnO_4$ and oxidises KI to I_2. Gaseous substance is

(a) N_2O_5 (b) NH_3 (c) N_2O_3 (d) HNO_3

46. The geometry of $Ni(CO)_4$ and $[Ni(PPh_3)_2]Cl_2$ are

(a) both square planar

(b) tetrahedral and square planar respectively

(c) both tetrahedral

(d) square planar and tetrahedral respectively

47. Given, $E^\circ_{Cr^{3+}/Cr} = -0.74$ V; $E^\circ_{MnO_4^-/Mn^{2+}} = 1.51$ V

$E^\circ_{Cr_2O_7^{2-}/Cr^{3+}} = 1.33$ V; $E^\circ_{Cl/Cl^-} = 1.36$ V

Based on the data given above strongest oxidising agent will be

(a) Cl (b) Cr^{3+} (c) Mn^{2+} (d) MnO_4^-

48. Consider the following compounds:

$$\underset{\text{I}}{CH_3CH_2NH_2},\quad \underset{\text{II}}{CH_3\overset{\overset{O}{\|}}{C}NH_2},\quad \underset{\text{III}}{Ph\overset{\overset{O}{\|}}{C}NH_2}$$

Basic nature of above compounds is in the order as:

(a) I > II > III (b) I > III > II

(c) II > III > I (d) III > II > I

49. Element A burns in nitrogen to give an ionic compound B. Compound B reacts with water to give C and D. A solution of C becomes 'milky' on bubbling carbon dioxide. The element A is

(*a*) Na (*b*) Mg

(*c*) Ca (*d*) Be

50. Iron oxide crystallises in a hexagonal close-packed array of oxide ions with two out of every three octahedral holes occupied by iron ions. Derive the formula of the iron oxide.

(a) FeO (b) Fe_2O_3

(c) Fe_3O_4 (d) All are possible

51. If at 298 K the bond energies of C—H, C—C, C=C and H — H bonds are respectively 414, 347, 615 and 435 kJ mol^{-1}, the value of enthalpy change for the reaction,

$$H_2C{=}CH_2(g) + H_2(g) \longrightarrow H_3C{-}CH_3(g)$$

at 298 K will be

(a) + 250 kJ (b) –250kJ

(c) + 125 kJ (d) –125 kJ

52. The oxidation state of chromium in the final product formed by the reaction between KI and acidified potassium dichromate solution is

(a) + 3 (b) + 2 (c) + 6 (d) + 4

53. Phenol is treated with excess bromine water and shaken well. The white precipitate formed during the process is

(a) 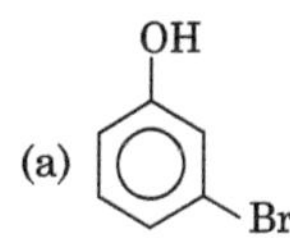

(b) 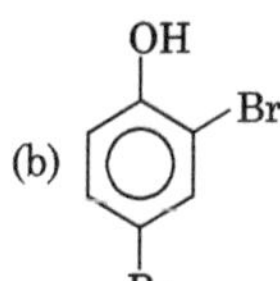

(c) 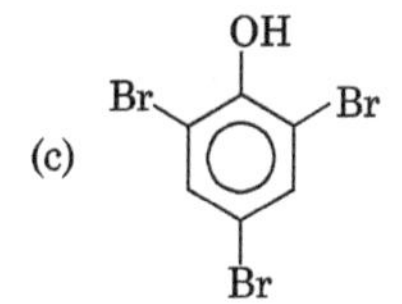

(d) OH, Br (para) + OH, Br (ortho)

54. The correct representation of $p\pi$ - $d\pi$ bonding is

(a) 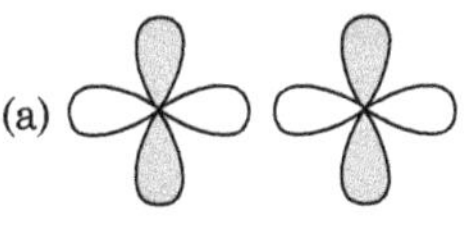

(b) 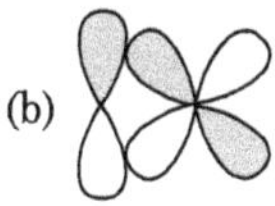

(c) 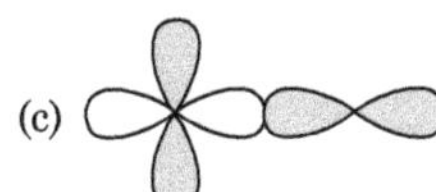

(d) 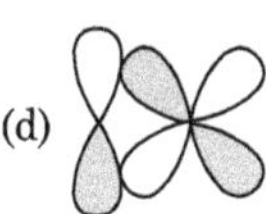

55. 2-hexyne gives *trans*-2-hexene on treatment with

(a) Pt / H_2 (b) Li / NH_3

(c) Pd / $BaSO_4$ (d) $LiAlH_4$

56. In which of the following polymers ethylene glycol is one of the monomer units?

(a) 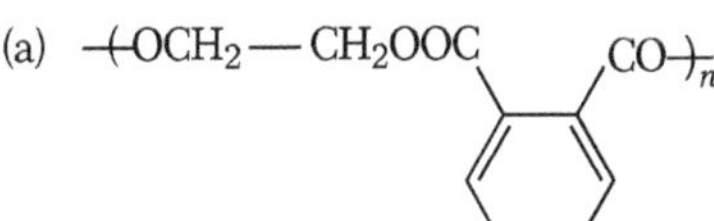

(b) $+CH_2—CH_2+_n$

(c) 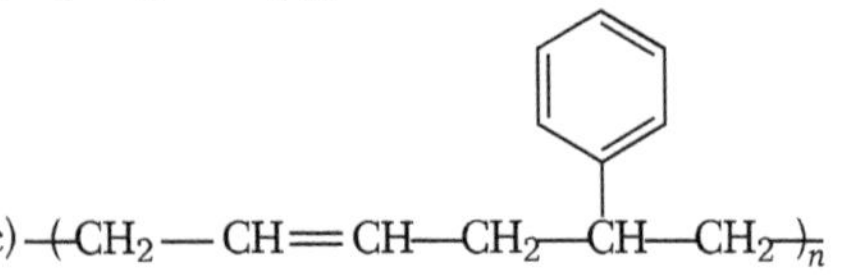 $+CH_2—CH=CH—CH_2—CH(C_6H_5)—CH_2+_n$

(d) 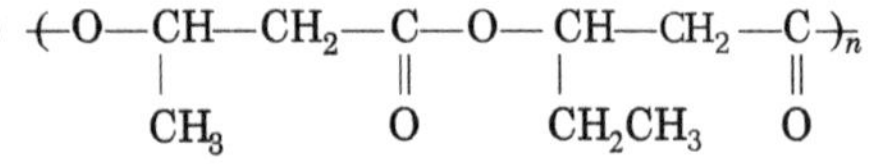 $+O—CH(CH_3)—CH_2—C(=O)—O—CH(CH_2CH_3)—CH_2—C(=O)+_n$

57. ^{23}Na is the more stable isotope of Na. Find out the process by which $^{24}_{11}Na$ can undergo radioactive decay

(a) β^--emission (b) α-emission
(c) β^+-emission (d) K-electron capture

58. The reduction,

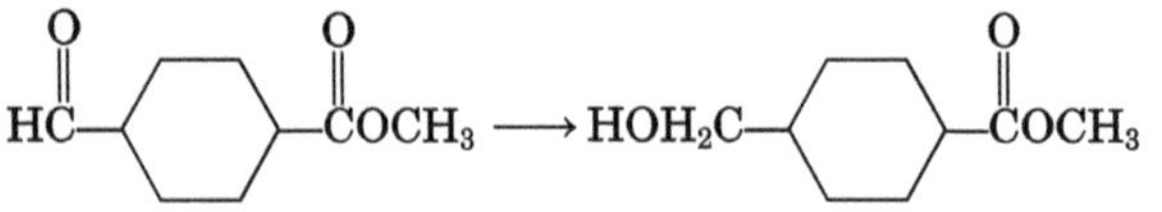

$$HC(=O)—C_6H_{10}—C(=O)OCH_3 \longrightarrow HOH_2C—C_6H_{10}—C(=O)OCH_3$$

can be achieved by using

(a) $NaBH_4$ (b) $LiAlH_4$
(c) $CuO \cdot CuCN_2O_4$ (d) None of these

59. Which of the following compounds is formed when glucose react with Br_2 water at pH 5-6?

(a) Glucaric acid (b) Saccharic acid
(c) Gluconic acid (d) Acetic acid

60. Which of the following are sets of diamagnetic species?

(a) $TiCl_4$, $[Ni(H_2O)_4]^{2+}$ (b) $TiCl_4$, O_2
(c) $TiCl_4$, $[Ni(CN)_4]^{2-}$ (d) $[Ni(CN)_4]^{2-}$ $[Ni(H_2O)_4]^{2+}$

BIOLOGY

61. Sliding filament theory can be best explained as

(a) when myofilaments slide pass each other actin filaments shorten while myosin filaments do not shorten
(b) actin and myosin filaments shorten and slide pass each other
(c) actin and myosin filaments do not shorten but rather slide pass each other
(d) when myofilaments slide pass each other myosin filament shorten while actin filaments do not shorten

62. Which of the following cells during gametogenesis is normally diploid?

(a) Primary polar body (b) Spermatid
(c) Secondary polar body (d) Spermatogonia

63. An artificial pacemaker is implanted subcutaneously and connected to the heart in patients

(a) having 90% blockage of the three main coronary arteries
(b) having a very high blood pressure
(c) with irregularity in the heart rhythm
(d) suffering from arteriosclerosis

64. Which one of the following would exocrine cells be expected to contain as a result of their function?

(a) Increased amounts of DNA
(b) Increased amounts of rough endoplasmic reticulum
(c) Increased numbers of lysosomes
(d) Large mitochondria

65. How many fatty acid molecule(s) is/are normally present in a fat or oil molecule?

(a) 3 (b) 2 (c) 4 (d) 1

66. The specificity of enzymes is due to

(a) their high molecular weight
(b) their hydrogen bonding
(c) their pH sensitivity
(d) their surface configuration

67. What is the role of centrioles during meiosis in animal cells?

(a) Breaking down the nuclear membrane during prophase
(b) Helping homologous chromosomes to pair and form bivalents
(c) Holding the two chromatids of a chromosome together
(d) Organising microtubules to form spindle fibres

68. How does DNA synthesis along the lagging strand differ from that on the leading strand?

(a) Okazaki fragments, synthesised $5' \rightarrow 3'$ are linked by DNA ligase
(b) An RNA primer is needed on the lagging strand, but not on the leading strand
(c) Deoxyribonucleotides are added to the 5′ end instead of the 3′ end
(d) Helicase synthesises Okazaki fragments, which are then linked together

69. Insulin is a protein containing 51 amino acids. These include 17 of the 20 different amino acids commonly occurring in proteins.

What is the minimum number of different kinds of *t*RNA molecules involved in the synthesis of insulin?

(a) 3 (b) 51 (c) 17 (d) 20

70. During transcription of the DNA fragment shown, a single base is paired incorrectly.

– C – T – A – A – C – T Sense strand
– G – A – T – T – G – A Anti-sense strand

Which *m*RNA strand results?

(a) C – A – T – T – G – A
(b) G – A – U – U – C – A
(c) G – A – U – U – G – A
(d) G – T – U – U – G – U

71. Which one of the following statements about gene mutation is incorrect?

(a) It can occur in both somatic and sex cells
(b) It can cause Down's syndrome in humans
(c) It can change a dominant allele into a recessive one
(d) It can be brought about by exposure to ionising radiation

72. Promoters and control elements work together to regulate transcription. What shows the possible locations of these in relation to a transcription start site on the DNA molecule?

	Promoters	Control elements
(a)	Downstream	Distal or proximal
(b)	Upstream	Distal or proximal
(c)	Distal or proximal	Downstream
(d)	Distal or proximal	Upstream

73. Which of the following would cause phenotypic variation among organisms of the same genotype?
(a) Continuous variation within the species
(b) Different varieties of the same species
(c) Different sexes
(d) Exposure to different environments

74. During the light phase of photosynthesis, the photoactivated pigments remove an electron from the hydroxylation derived from the water molecule.
The fate of the free hydroxyl radical is that it
(a) is broken down into oxygen and a free radical of hydrogen
(b) is used to raise the activation level of chlorophyll by donating a positive charge
(c) is used to produce adenosine triphosphate from adenosine diphosphate
(d) reduces carbon dioxide to sugar

75. When mitochondria are extracted from cells for biochemical study, they are usually kept in a 0.25 mol dm^{-3} sucrose solution.
Why is the sucrose solution used?
(a) To act as a solvent
(b) To provide a source of food
(c) To assist in the extraction of enzymes
(d) To prevent the mitochondria from changing in structure

76. Four identical samples of plant tissue, each with a water potential (ψ) of sign – 700 kPa, are placed in four different solutions. Which solution induces full plasmolysis within the tissue?

Solution	(ψ)
(a) A	–700 kPa
(b) B	–1000 kPa
(c) C	–400 kPa
(d) D	–200 kPa

77. Protein was the only food available to a mammal. If this supply of protein was only half of the minimum required to supply its energy needs, the mammal would show an increase in the
(a) concentration of amino acids in the blood
(b) concentration of insulin in the blood
(c) synthesis of glycogen
(d) hydrolysis of glycogen

78. The diagram shows an action potential

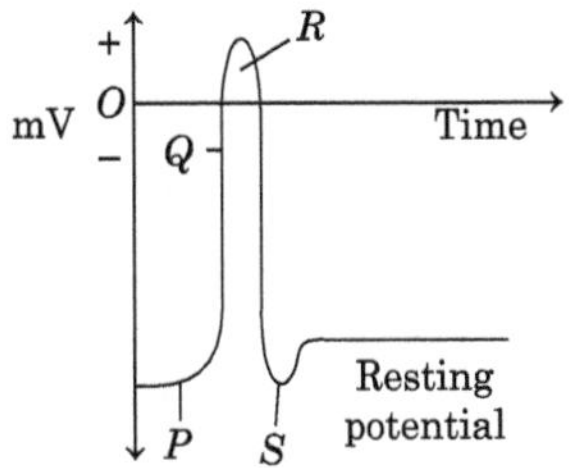

At which point on the graph is the membrane most permeable to sodium ions?
(a) *P* (b) *Q* (c) *R* (d) *S*

79. The frequency of a mutant gene in a population would be likely to increase if
(a) the genes were selectively advantageous
(b) the genes were dominant
(c) the genes were sex-linked
(d) the population increased

80. Which genetic modification could increase the yield of crop measured as mass of crop produced per unit area per year?
(a) Herbicide resistant plants
(b) Delayed ripening in the fruits
(c) More essential amino acids in the seeds
(d) More vitamin-A in the grain

PART-II (2 Marks Questions)

MATHEMATICS

81. A triangle has base 10 cm long and the base angles of 50° and 70°. If the perimeter of the triangle is $x + y\cos z°$, where $z \in (0, 90°)$, then the value of $x + y + z$ equals to
(a) 60 (b) 55 (c) 50 (d) 40

82. If $g(x)$ is a differential real valued function satisfying $g''(x) - 3g'(x) > 3, x \geq 0$ and $g'(0) = -1$, then $g(x) + x$ for $x > 0$ is
(a) an increasing function
(b) a decreasing function
(c) a constant function
(d) None of these

83. The number of integers between 1 and 10000 with at least one 8 and at least one 9 as digits is
(a) 972 (b) 974
(c) 970 (d) 980

84. If the expression $z^5 - 32$ can be factorized into linear and quadratic factors over real coefficient as $(z^5 - 32) = (z - 2)(z^2 - pz + 4)(z^2 - qz + 4)$, then $(p^2 + 2q)$ is equal to
(a) 2 (b) 3 (c) 4 (d) 8

85. If the line $x+y=1$ is a tangent to a parabola with focus (1, 2) at A and intersects the directrix at B and tangent at vertex at C respectively, then $AC \cdot BC$ is equal to

(a) 2 (b) 1 (c) $\frac{1}{2}$ (d) $\frac{1}{4}$

86. The line $2x-y+1=0$ is a tangent to the circle at the point (2, 5) and the centre of the circles lies on $x-2y=4$. The radius of the circle is

(a) $3\sqrt{5}$ (b) $5\sqrt{3}$ (c) $2\sqrt{5}$ (d) $5\sqrt{2}$

87. $X=\{x \in R: \cos(\sin x)=\sin(\cos x)\}$. The number of elements in X is

(a) 0 (b) 2
(c) 4 (d) Not finite

88. There are 6 boxes labelled $B_1, B_2, B_3, \dots, B_6$. In each trial two fair dice D_1, D_2 are thrown. If D_1 shows j and D_2 shows k, then j balls are put into the box B_K, After n trials, what is the probability that B_1 contains at most one ball?

(a) $\left(\frac{5^{n-1}}{6^{n-1}}\right)+\left(\frac{5^n}{6^n}\right)\left(\frac{1}{6}\right)$ (b) $\left(\frac{5^n}{6^n}\right)+\left(\frac{5^{n-1}}{6^{n-1}}\right)\left(\frac{1}{6}\right)$

(c) $\left(\frac{5^n}{6^n}\right)+n\left(\frac{5^{n-1}}{6^{n-1}}\right)\left(\frac{1}{6}\right)$ (d) $\left(\frac{5^n}{6^n}\right)+n\left(\frac{5^{n-1}}{6^{n-1}}\right)\left(\frac{1}{6^2}\right)$

89. The minimum distance between a point on the curve $y=e^x$ and a point on the curve $y=\log_e x$ is

(a) $\frac{1}{\sqrt{2}}$ (b) $\sqrt{2}$ (c) $\sqrt{3}$ (d) $2\sqrt{2}$

90. Let $[x]$ and $\{x\}$ be the integral part and fractional part of a real number x respectively. Then the value of the integral $\int_0^5 [x]\{x\}\,dx$ is

(a) $\frac{5}{2}$ (b) 5 (c) $\frac{69}{2}$ (d) $\frac{71}{2}$

PHYSICS

91. A light beam is incident over a spherical drop of water at an angle of 60°. Total deviation of light ray after a single reflection from internal surface of drop is let δ. Then, for $\mu_{\text{water}}=\frac{4}{3}$, value of δ will be

(a) 180° (b) 145°
(c) 138° (d) 176°

92. A cart of mass m_0 starts moving right due to a constant horizontal force F. At same time, sand starts falling into cart from a stationary hopper. Rate of loading is constant and equal to μ kg s^{-1}.

Velocity of cart at time t is

(a) $\frac{Ft}{m_0+\mu t}$ (b) $\frac{Ft}{2(m_0+\mu t)}$

(c) $\frac{Ft}{m_0-\mu t}$ (d) $\frac{Ft}{2(m_0-\mu t)}$

93. A ladder PQ of length 5 m is inclined to a vertical wall is slipping over horizontal floor with velocity of 2 ms^{-1}.

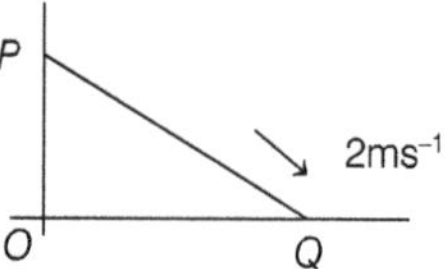

Velocity of centre of mass of ladder, when Q is 3 m from the wall is

(a) 2 ms^{-1} (b) 1.25 ms^{-1}
(c) 175 ms^{-1} (d) 2.75 ms^{-1}

94. A voltmeter with resistance 500 Ω is used to measure the emf of a cell of internal resistance 4 Ω.

Percentage error in reading of voltmeter is around below

(a) 1% (b) 0.8% (c) 0.1% (d) 8%

95. A gas undergoes two processes as shown below.

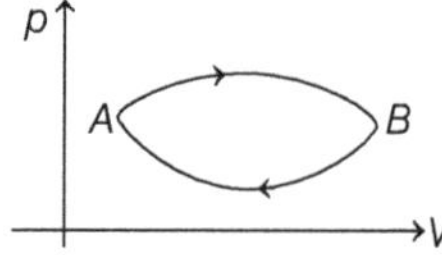

Ratio of molar heat capacities for process AB and BA is

(a) less than 1 (b) greater than 1
(c) equal to 1 (d) Data insufficient

96. A regular hexagonal mesh using equal resistances each of 2 Ω is made as shown below.

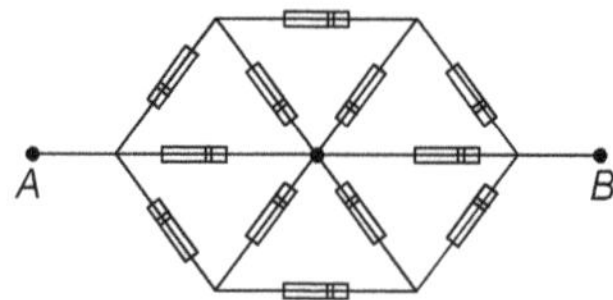

Then, equivalent resistance between terminals A and B is

(a) $\frac{8}{5}\Omega$ (b) $\frac{5}{8}\Omega$ (c) $\frac{3}{8}\Omega$ (d) 2Ω

97. A ^{238}U nucleus emits a 4.2 MeV α-particle. If we take account of recoil energy of ^{234}Th daughter nucleus formed in the process, then Q value (disintegration energy) of the reaction will be

(a) $4.2\left(1+\frac{4}{238}\right)$ MeV (b) $4.2\left(\frac{234}{238}\right)$ MeV

(c) $4.2\left(\frac{238}{234}\right)$ MeV (d) $4.2\left(1-\frac{4}{234}\right)$ MeV

98. Two towers AB and CD are situated d distance apart.

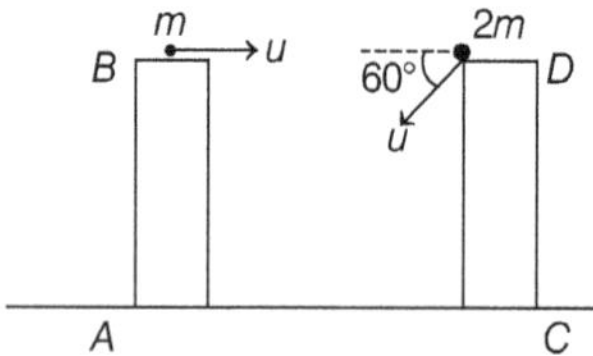

AB is 20 m high and CD is 30 m high. From top of AB a particle of mass m is thrown towards CD with speed $u=10$ m/s.

At same instant, another particle of mass 2 m is thrown from CD towards AB at an angle of 60° with the horizontal in same vertical plane with same speed of 10 ms^{-1}. Distance between towers is

(a) $5\sqrt{3}$ m (b) $6\sqrt{3}$ m (c) $8\sqrt{3}$ m (d) $10\sqrt{3}$ m

99. An ideal gas undergoes an adiabatic process in which internal energy U is $U = a + b(pV)$, where a and b are constants. Value of γ for this process is

(a) $\frac{b}{1+b}$ (b) b (c) $\frac{b+1}{b+2}$ (d) $1+\frac{1}{b}$

100. A measurement establishes the position of a proton with an accuracy of $\pm 1.00 \times 10^{-11}$ m. Uncertainly in the proton's position 1 s later is (take, $v << c$)

(a) 3.15 mm (b) 3.15 cm (c) 3.15 m (d) 3.15 km

CHEMISTRY

101. Calculate the mass of a non-volatile solute (molar mass 40 g mol^{-1}), which should be dissolved in 114 g octane to reduce its vapour pressure to 80%.

(a) 20 g (b) 10 g (c) 30 g (d) 45 g

102. Percentage of free space in cubic close packed structure and in body centred packed structure are respectively

(a) 30% and 26% (b) 26% and 32%
(c) 32% and 48% (d) 48% and 26%

103. (1-methylcyclohexene, CH_3) $\xrightarrow[\text{Alkaline } KMnO_4]{\text{Cold}} A \xrightarrow[\text{AcOH}]{CrO_3} B$

A and B are

(a) (CH_3, OH, OH) , (CH_3, OH, O)

(b) (CH_3, OH, OH) , (CH_3, =O, O)

(c) (CH_3, OH, OH) , (CH_3, OH)

(d) no formation of A and B takes place

104. Which of the following will produce a buffer solution when mixed in equal volumes?

(a) 0.1 mol dm^{-3} NH_4OH and 0.1 mol dm^{-3} HCl
(b) 0.05 mol dm^{-3} NH_4OH and 0.1 mol dm^{-3} HCl
(c) 0.1 mol dm^{-3} NH_4OH and 0.05 mol dm^{-3} HCl
(d) 0.1 mol dm^{-3} CH_3COONa and 0.1 mol dm^{-3} NaOH

105. The bond dissociation energy of B — F in BF_3 is 646 kJ mol^{-1} whereas that of C — F in CF_4 is 515 kJ mol^{-1}. The correct reason for higher B — F bond dissociation energy as compared to that of C — F is

(a) smaller size of B-atom as compared to that of C-atom.
(b) stronger σ-bond between B and F in BF_3 as compared to that between C and F in CF_4.
(c) significant $p\pi$ - $p\pi$ interaction between B and F in BF_3 whereas there is no possibility of such interaction between C and F in CF_4.
(d) lower degree of $p\pi$ - $p\pi$ interaction between B and F in BF_4 than that between C and F in CF_4.

106. Electrode potential for Mg electrode varies according to the equation

$E_{Mg^{2+}/Mg} = E^\circ_{Mg^{2+}/Mg} - \frac{0.059}{2} \log \frac{1}{[Mg^{2+}]}$. The graph of $E_{Mg^{2+}/Mg}$ vs log $[Mg^{2+}]$ is

(a)

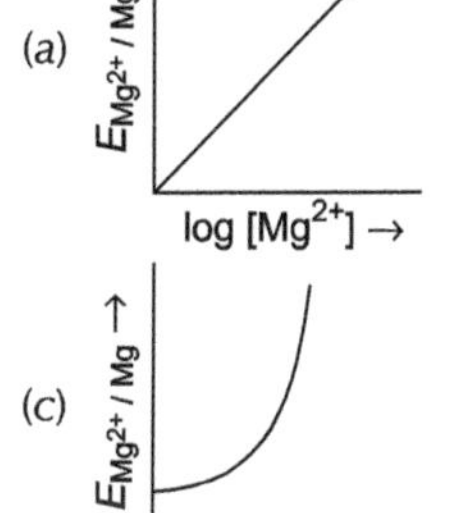

(b)

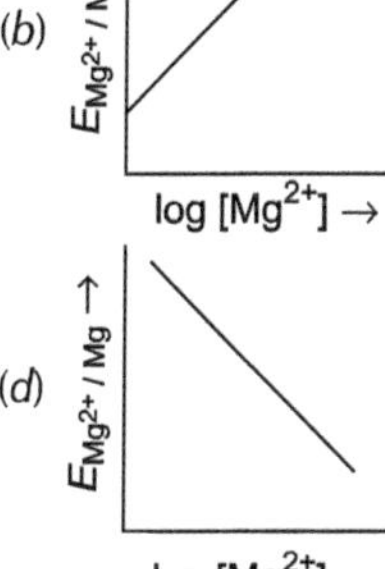

(c) $E_{Mg^{2+}/Mg}$ → ; log $[Mg^{2+}]$ →

(d) $E_{Mg^{2+}/Mg}$ → ; log $[Mg^{2+}]$ →

107.

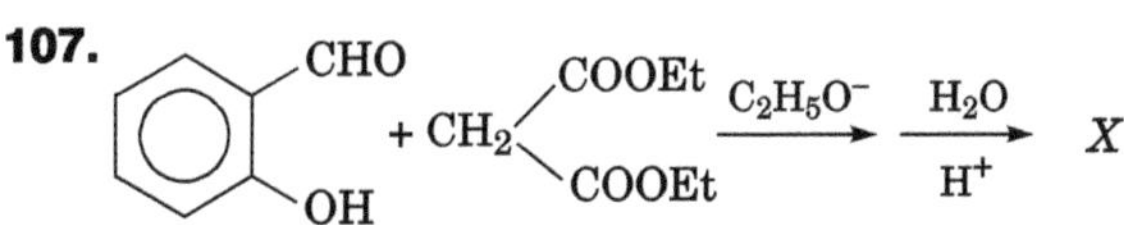

Identify the final product X.

(a) $COOC_2H_5$, O, O

(b) $COOC_2H_5$, O, O

(c) O, O

(d) $COOC_2H_5$, O

108. A metal complex having composition $Cr(NH_3)_4Cl_2Br$ has been isolated in two forms A and B. A reacts with $AgNO_3$ producing a white precipitate which was soluble in dilute ammonia solution. B reacts with $AgNO_3$ producing a pale yellow precipitate soluble in concentrated ammonia solution. The formulae of A and B are respectively

(a) $[Cr(NH_3)_4Br]Cl_2; [Cr(NH_3)Cl_2]Br$
(b) $[Cr(NH_3)_4BrCl]Cl; [Cr(NH_3)_3Cl_2Br]NH_3$
(c) $[Cr(NH_3)_4Cl_2]Br; [Cr(NH_3)_4BrCl]Cl$
(d) $[Cr(NH_3)_4BrCl]Cl; [Cr(NH_3)_4Cl_2]Br$

109. On reaction with Cl_2, phosphorus forms two types of halides A and B. Halide A is yellowish white powder, but halide B is colourless oily liquid. Which of the following is/are their hydrolysis products?

(I) H_3PO_2 (II) H_3PO_3

(III) H_3PO_4 (IV) H_3PO_4

(a) I and II (b) II and III

(c) III and IV (d) I and IV

110. The major product of the following reaction is

Phenol (OH on benzene) $\xrightarrow[\text{2. HCHO,OH}^-]{\text{1. CHCl}_3\text{+NaOH/H}^+}$

(a) OH, CH_2OH (b) OH, CHO

(c) HCOONa (d) OH, COOH

BIOLOGY

111. In an experiment to determine the number of rats in a field, 80 rats were initially captured, marked and released. After one month, 100 rats were captured in the same field, of which 20 were previously marked ones. Based on the above observation, estimated population size of rats in the field will be

(a) 160 (b) 200 (c) 400 (d) 1600

112. To understand prey-predator relationship, *Didinium* (predator) and *Paramecium* (prey) were used. *Paramecium* population was grown with sand sediment as hiding place or refuge. To this population, *Didinium* was introduced only once. What could happen to the prey population in the course of time?

(a) The population will steadily decrease and vanish
(b) The population will initially increase and then stabilise
(c) The population will initially decrease, then increase and stabilise
(d) The population will steadily increase

113. A recessive inherited disease is expressed only in individuals of blood group O and not expressed in blood groups A, B and AB. Alleles controlling the disease and blood group are independently inherited. A normal woman with blood group A and her normal husband with blood group B already has one child with the disease. The woman is pregnant for second time. What is the probability that the second child will also have the disease?

(a) $\frac{1}{2}$ (b) $\frac{1}{4}$ (c) $\frac{1}{16}$ (d) $\frac{1}{64}$

114. The region of the genome containing the RFLP used in this analysis is shown below.

Allele Q —— *Hind* III —— *Hind* III* —— *Hind* III

Hind III indicates the restriction sites for this enzyme and * indicates the polymorphic site which is missing in the recessive allele q.

The black bar indicates the position of the probe used to detect the RFLP. DNA fragments from three different individuals, X, Y and Z, were subjected to restriction digestion by *Hind* III and separate by gel electrophoresis. The following results were obtained.

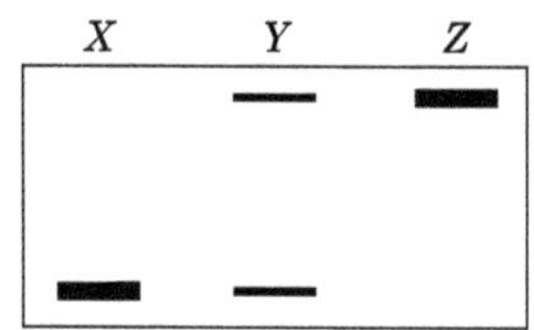

Deduce the genotype of individual Z.

(a) QQ (b) Qq

(c) qq (d) X^QY

115. A virus has a base ratio of $\frac{(A+G)}{(U+C)} = 1$.

What type of virus is this?

(a) A single-stranded DNA virus
(b) A double-stranded DNA virus
(c) A single-stranded RNA virus
(d) A double-stranded RNA virus

116. In an investigation to determine the effect of temperature on the activity of an enzyme, the time for all the substrates to disappear from a standard solution was recorded. Which graph shows the result of this investigation?

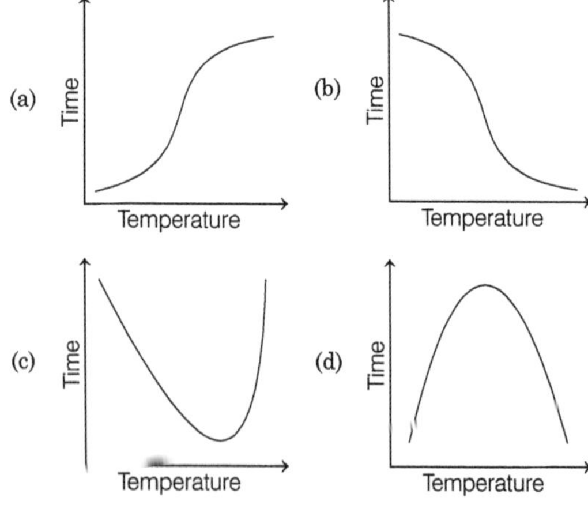

117. The correct order of periods from Palaeozoic to Mesozoic era is

(a) Triassic → Jurassic → Cretaceous → Cambrian → Ordovician → Silurian → Devonian → Carboniferous → Permian
(b) Palaeocene → Eocene → Oligocene → Miocene → Pliocene → Pleistocene → Holocene
(c) Cambrian → Ordovician → Silurian → Devonian → Carboniferous → Permian → Triassic → Jurassic → Cretaceous
(d) Pliocene → Eocene → Oligocene → Silurian → Devonian → Carboniferous → Triassic → Jurassic → Cretaceous

118. In animals, four separate families of cell-cell adhesion proteins are listed in Column A and their functional characteristics are given in Column B.

	Column A		Column B
A.	Integrin	1.	Lectins that mediate a variety of transient, cell-cell adhesion interactions in the bloodstream
B.	Cadherin	2.	Contains extracellular Ig-like domains and are mainly involved in the fine tuning of cell-cell adhesive interaction during development and regeneration
C.	Ig super family	3.	Mediates Ca^{2+} dependent strong homophilic cell-cell adhesion
D.	Selectin	4.	Transmembrane cell adhesion protein that acts as extracellular matrix receptors.

Which one of the following is the correct combination?

	A	B	C	D		A	B	C	D
(a)	1	2	3	4	(b)	2	3	4	1
(c)	3	4	1	2	(d)	4	3	2	1

119. The diagram shows the inheritance of haemophilia in a family.

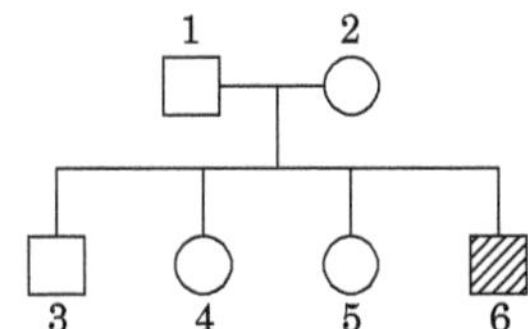

If daughter 4 marries a normal male, what is the probability that their first child would suffer from haemophilia?

(a) 0 (b) 0.125 (c) 0.25 (d) 0.5

120. The complete oxidation of one mole of glucose yields 2880 kJ of energy. The addition of one phosphate molecule to ADP requires 30.6 kJ of energy per mole. In aerobic respiration, 38 molecules of ATP are formed as a result of the breakdown of each glucose molecule.

Which figure best represents the efficiency of aerobic respiration in trapping the energy released by the glucose molecule?

(a) 23% (b) 36%
(c) 45% (d) 40%

Answers

PART-I

1	(b)	2	(b)	3	(d)	4	(c)	5	(d)	6	(b)	7	(b)	8	(c)	9	(b)	10	(b)
11	(c)	12	(c)	13	(a)	14	(a)	15	(d)	16	(c)	17	(c)	18	(d)	19	(b)	20	(c)
21	(b)	22	(c)	23	(b)	24	(d)	25	(d)	26	(d)	27	(d)	28	(c)	29	(b)	30	(c)
31	(c)	32	(a)	33	(d)	34	(c)	35	(c)	36	(b)	37	(a)	38	(a)	39	(d)	40	(b)
41	(d)	42	(a)	43	(a)	44	(b)	45	(c)	46	(c)	47	(d)	48	(a)	49	(c)	50	(b)
51	(d)	52	(a)	53	(c)	54	(b)	55	(b)	56	(a)	57	(a)	58	(a)	59	(c)	60	(c)
61	(c)	62	(d)	63	(c)	64	(b)	65	(a)	66	(d)	67	(d)	68	(a)	69	(b)	70	(b)
71	(c)	72	(b)	73	(d)	74	(a)	75	(d)	76	(b)	77	(d)	78	(b)	79	(a)	80	(a)

PART-II

81	(d)	82	(a)	83	(b)	84	(c)	85	(a)	86	(a)	87	(a)	88	(d)	89	(b)	90	(b)
91	(c)	92	(a)	93	(b)	94	(b)	95	(b)	96	(a)	97	(c)	98	(d)	99	(d)	100	(d)
101	(b)	102	(b)	103	(a)	104	(c)	105	(c)	106	(b)	107	(a)	108	(d)	109	(b)	110	(a)
111	(c)	112	(c)	113	(c)	114	(c)	115	(d)	116	(c)	117	(c)	118	(d)	119	(b)	120	(d)

Solutions

1. *(b)* We have,

$$x^n - nx^{n-1} + a_2x^{n-2} + a_3x^{n-3} + \dots + a_{n-1}x + (-1)^n = 0$$

has n positive roots.

$\therefore$ AM of all roots is equal to GM of roots, so all roots are equal and equal to 1.

$\therefore \quad a_2 = {}^nC_2$ and $a_3 = -{}^nC_3$

$\Rightarrow \quad a_2 + a_3 = {}^nC_2 - {}^nC_3 < 0$

$\Rightarrow \quad n > 2 + 3$

$\Rightarrow \quad n > 5$

2. *(b)* In ΔABD,

$$\sin 10° = \frac{BD}{AB} = \frac{9a}{2b}$$

$$\sin 30° = 3\sin 10° - 4\sin^3 10°$$

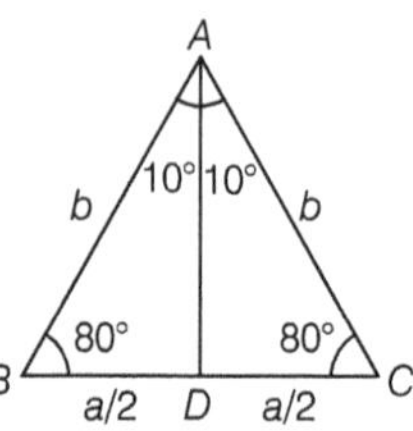

$\Rightarrow \quad \frac{1}{2} = \frac{3a}{2b} - \frac{4a^3}{8b^3}$

$\Rightarrow \quad b^3 = 3ab^2 - a^3$

$\Rightarrow \quad a^3 + b^3 = 3ab^2$

3. *(d)* Let the three sides of triangle are a, b and c.

$$a \le b \le c = 11$$

$\therefore 6 \le b \le 11$ and $c - b < a \le b$.

As b decreased by 1 unit the range of a decreased by 2 when $b = 11$, we have

$$1 \le a \le 11$$

Hence, the total number of triangle is

$$11 + 9 + 7 + 5 + 3 + 1 = 36$$

4. *(c)* We have,

$$\log_{245} 175 = a$$

$\Rightarrow \quad \frac{\log 175}{\log 245} = a$

$\Rightarrow \quad \frac{2\log 5 + \log 7}{\log 5 + 2\log 7} = a$

$\Rightarrow \quad \frac{2\log 5 + \log 7 + \log 5 + 2\log 7}{2\log 5 + \log 7 - \log 5 - 2\log 7} = \frac{a+1}{a-1}$

$\Rightarrow \quad 3\left(\frac{\log 5 + \log 7}{\log 5 - \log 7}\right) = \frac{a+1}{a-1} \quad \dots(i)$

and $\quad \log_{1715} 875 = b$

$\Rightarrow \quad \frac{\log 875}{\log 1715} = b \Rightarrow \frac{3\log 5 + \log 7}{\log 5 + 3\log 7} = b$

Apply componendo and dividendo, we get

$$2\left(\frac{\log 5 + \log 7}{\log 5 - \log 7}\right) = \frac{b+1}{b-1} \quad \dots(ii)$$

From Eqs. (i) and (ii), we get

$$\frac{1}{3}\left(\frac{a+1}{a-1}\right) = \frac{1}{2}\left(\frac{b+1}{b-1}\right)$$

$\Rightarrow \quad 2(a+1)(b-1) = 3(b+1)(a-1)$

$\Rightarrow \quad 2ab + 2b - 2a - 2 = 3ab + 3a - 3b - 3$

$\Rightarrow \quad 1 - ab = 5(a - b)$

$\Rightarrow \quad \frac{1-ab}{a-b} = 5$

5. *(d)* We have, $K = \int_0^{\pi/2} \frac{\sin x}{1 + \sin x + \cos x} dx$

$\Rightarrow \quad K = \int_0^{\pi/2} \frac{\sin\left(\frac{\pi}{2} - x\right)}{1 + \sin\left(\frac{\pi}{2} - x\right) + \cos\left(\frac{\pi}{2} - x\right)} dx$

$\Rightarrow \quad K = \int_0^{\pi/2} \frac{\cos x}{1 + \cos x + \sin x} dx$

$\Rightarrow \quad 2K = \int_0^{\pi/2} \frac{\sin x + \cos x}{1 + \cos x + \sin x} dx$

$\Rightarrow \quad 2K = \int_0^{\pi/2} \frac{1 + \sin x + \cos x}{1 + \cos x + \sin x} dx - \int_0^{\pi/2} \frac{dx}{1 + \sin x + \cos x}$

$\Rightarrow \quad \int_0^{\pi/2} \frac{dx}{1 + \sin x + \cos x} = \frac{\pi}{2} - 2K$

6. *(b)* We have, $y = \sqrt{3}x$.

Let $\quad x = \frac{r}{2}, \; y = \frac{\sqrt{3}r}{2}$

Putting the value of x and y in

$x^4 + ax^2y + bxy + cx + dy + 6 = 0$, we get

$$\frac{r^4}{16} + \frac{\sqrt{3}ar^3}{8} + \frac{\sqrt{3}br^2}{4} + \frac{cr}{2} + \frac{\sqrt{3}dr}{2} + 6 = 0$$

$\Rightarrow \quad \frac{1}{12}(OA \cdot OB \cdot OC \cdot OD) = \frac{1}{12}|r_1r_2r_3r_4|$

$$= \frac{1}{12}(16 \times 6) = \frac{96}{12} = 8$$

7. *(b)* We have,

$$f(x) = \int \frac{x^2}{(1+x^2)(1+\sqrt{1+x^2})} dx$$

$\Rightarrow \quad f(x) = \int \frac{x^2(\sqrt{1+x^2} - 1)}{(1+x^2)(1+x^2-1)} dx$

$\Rightarrow \quad f(x) = \int \frac{\sqrt{1+x^2}}{1+x^2} dx - \int \frac{1}{1+x^2} dx$

$\Rightarrow \quad f(x) = \int \frac{1}{\sqrt{1+x^2}} dx - \int \frac{1}{1+x^2} dx$

$\Rightarrow \quad f(x) = \log(x + \sqrt{1+x^2}) - \tan^{-1} x + C$

$\Rightarrow \quad f(0) = 0 = C$

$\therefore \quad f(x) = \log(x + \sqrt{1+x^2}) - \tan^{-1} x$

$\Rightarrow \quad f(1) = \log(1 + \sqrt{2}) - \tan^{-1} 1$

$$f(1) = \log(\sqrt{2} + 1) - \frac{\pi}{4}$$

8. *(c)* Total number of ways that can be formed $ABBCABBC$ is $\frac{8!}{4!2!2!} = 420$

Let $ABBC$ be a unit, then this unit and other $ABBC$ can be arranged in $\frac{5!}{2!} = 60$

But this includes two times

$\therefore$ Actual number = 59

$\therefore$ Required number of ways

$$= 420 - 59 = 361$$

9. *(b)* We have, $x^2 f(x) - 2f\left(\frac{1}{x}\right) = g(x)$,

where $g(x)$ is odd function.

$\therefore \quad g(x) + g(-x) = 0$

$$x^2 [f(x) + f(-x)] - 2\left[f\left(\frac{1}{x}\right) + f\left(-\frac{1}{x}\right)\right] = 0 \quad \dots(i)$$

$x \to \frac{1}{x}$, we get

$$\frac{1}{x^2}\left[f\left(\frac{1}{x}\right) + f\left(-\frac{1}{x}\right) - 2f(x) + f(-x)\right] = 0 \quad \dots(ii)$$

From Eqs. (i) and (ii), we get

$$f(x) = -f(-x)$$

$$f(2) = -f(-2)$$

10. *(b)* We have,

det $(A) = 2$ det $(B) = 3$, det $(C) = 5$

$\Rightarrow \quad \det(A^2BC^{-1}) = \det(A) \times \det(A) \times \det(B) \times \frac{1}{\det C}$

$$= 2 \times 2 \times 3 \times \frac{1}{5} = \frac{12}{5}$$

11. *(c)* Given, $\sin\left[\pi \log_3\left(\frac{1}{x}\right)\right] = 0$

$\Rightarrow \quad \pi \log_3\left(\frac{1}{x}\right) = n\pi$

$\Rightarrow \quad \log_3\left(\frac{1}{x}\right) = n$

$\Rightarrow \quad x = 3^{-n}$

Sum of roots $= 3+1+\frac{1}{3}+\frac{1}{3^2}+\frac{1}{3^3}+...+\infty$

$[\because x \in (0, 2\pi)]$

$$= 4+\frac{1}{3}\left(\frac{1}{1-\frac{1}{3}}\right) = 4+\frac{1}{2} = \frac{9}{2}$$

12. *(c)* Chord of contact of tangent from (–4, 2) to the hyperbola $\frac{x^2}{a^2}-\frac{y^2}{b^2}=1$ is

$$\frac{-4x}{a^2}-\frac{2y}{b^2}=1 \quad ...(i)$$

and from (2, 1) is $\frac{2x}{a^2}-\frac{y}{b^2}=1 \quad ...(ii)$

Since, chord of contact are perpendicular.

$$\therefore \left(\frac{-4}{a^2}\right)\left(\frac{2}{a^2}\right)+\left(\frac{2}{b^2}\right)\left(\frac{1}{b^2}\right)=0$$

$$\frac{a^4}{b^4}=4$$

$$\Rightarrow a^2=2b^2$$

$$\Rightarrow \frac{b^2}{a^2}=\frac{1}{2}$$

$$\Rightarrow e=\sqrt{1+\frac{b^2}{a^2}}$$

$$=\sqrt{1+\frac{1}{2}}=\frac{\sqrt{3}}{\sqrt{2}}=\sqrt{\frac{3}{2}}$$

13. *(a)* Given, $\left|Z^3+\frac{1}{Z^3}\right|\le 2$

$$\left|Z^3+\frac{1}{Z^3}\right|\le |Z^3|+\frac{1}{|Z^3|}\le 2$$

By AM ≥ GM

$$|Z^3|+\frac{1}{|Z^3|}\ge 2$$

$$\therefore |Z^3|+\frac{1}{|Z^3|}=2 \; |Z|=1$$

$$\therefore \left|Z+\frac{1}{Z}\right|\le |Z|+\frac{1}{|Z|}=2$$

$\therefore$ Maximum value of $\left|Z+\frac{1}{Z}\right|=2$

14. *(a)* We have,

$$f(x)=x^3+\alpha x^2+\beta x+\gamma$$

Roots of $f(x)$ are eccentricity of parabola and rectangular hyperbola.

Eccentricity of parabola = 1

Eccentricity of rectangular hyperbola $=\sqrt{2}$

$\therefore x^3+\alpha x^2+\beta x+\gamma=(x-1)(x-\sqrt{2})(x+\sqrt{2})$

$\Rightarrow x^3+\alpha x^2+\beta x+\gamma$

$$=(x-1)(x^2-2)=x^3-x^2-2x+2$$

$$\therefore \alpha=-1, \beta=-2, \gamma=2$$

$$\therefore \alpha+\beta+\gamma=-1-2+2=-1$$

15. *(d)* Let $I=\int_{-\pi}^{\pi}(\cos\alpha x-\sin\beta x)^2 dx$

$\Rightarrow$

$$I=\int_{-\pi}^{\pi}(\cos^2\alpha x+\sin^2\beta x-2\sin\beta x\cos\alpha x)dx$$

$$\Rightarrow I=2\int_0^{\pi}(\cos^2\alpha x+\sin^2\beta x)dx$$

$[\because \sin\beta x\cos\alpha x$ is an odd function]

$$\Rightarrow I=2\int_0^{\pi}\left[\frac{1+\cos 2\alpha x}{2}+\frac{1-\cos 2\beta x}{2}\right]dx$$

$$\Rightarrow I=2\left[x+\frac{\sin 2\alpha x}{2\alpha}-\frac{\sin 2\beta x}{2\beta}\right]_0^{\pi}$$

$$\Rightarrow I=2[\pi]=2\pi$$

16. *(c)* Given, $abcd=1$

$$\frac{1+a}{2}\ge\sqrt{a}$$

AM ≥ GM

$$1+a\ge 2\sqrt{a}$$

Similarly, $1+b\ge 2\sqrt{b}$

$$1+c\ge 2\sqrt{c}$$

$$\Rightarrow 1+d\ge 2\sqrt{d}$$

$$\therefore (1+a)(1+b)(1+c)(1+d)\ge 16\sqrt{abcd}$$

$$(1+a)(1+b)(1+c)(1+d)\ge 16$$

17. *(c)* Maximum number of triangles

$$={}^{3P}C_3-3({}^{P}C_3)$$

$$=\frac{3p(3p-1)(3p-2)}{6}-\frac{3p(p-1)(p-2)}{6}$$

$$=\frac{p}{2}[(3p-1)(3p-2)-(p-1)(p-2)]$$

$$=\frac{p}{2}[9p^2-9p+2-p^2+3p-2]$$

$$=\frac{p}{2}[8p^2-6p]=p^2(4p-3)$$

18. *(d)* Let

$$L=\lim_{n\to\infty}n^2\int_{-1/n}^{1/n}(2018\sin x+2019\cos x)|x|dx$$

$$\Rightarrow L=\lim_{n\to\infty}\frac{2\int_0^{1/n}2019\cos x|x|\,dx}{\frac{1}{n^2}}$$

$[\because 2018\sin x|x|$ is odd function]

$$\Rightarrow L=\lim_{n\to\infty}\frac{2\cdot 2019\cos\left(\frac{1}{n}\right)\left(\frac{1}{n}\right)\times\left(\frac{-1}{n^2}\right)}{\frac{-2}{n^3}}$$

[$\because$ by Leibnitz rule]

$$\Rightarrow L=\lim_{n\to\infty}2019\cos\left(\frac{1}{n}\right)=2019$$

19. *(b)* We have,

$\tan^6 40°-33\tan^4 40°+27\tan^2 40°$

$$\Rightarrow \tan 3(40°)=\frac{3\tan 40°-\tan^3 40°}{1-3\tan^2 40°}$$

$$\Rightarrow \tan 120°=\frac{3\tan 40°-\tan^3 40°}{1-3\tan^2 40°}$$

$$\Rightarrow -\sqrt{3}(1-3\tan^2 40°)=3\tan 40°-\tan^3 40°$$

$$\Rightarrow 3(1-3\tan^2 40°)^2$$

$$=\tan^2 40°(3-\tan^2 40°)^2$$

$$\Rightarrow \tan^6 40°-33\tan^4 40°+27\tan^2 60°=3$$

20. *(c)* Ten digits are split the digits into pairs (0, 1), (1, 2), (2, 3), ..., (8, 9)

Disjoints pairs out of these are

(0, 1)(2, 3)(4,5)(6, 7)(8, 9)

Now, two cases are

(a) When the pair (0, 1) is not used for first and last place, the number of ways

$$=4\times 4!\times 2^5$$

(b) When the pair (0, 1) is used for first and last place, the number of ways

$$=1\times 4!\times 2^4$$

$\therefore$ Total number of favourable cases

$$=9\times 4!\times 2^4$$

Required probability $=\frac{9\times 4!\times 2^4}{9\times 9!}=\frac{1}{945}$

21. *(b)* For any projected particle,

$$x=u\cos\theta\cdot t \quad ...(i)$$

$$y=u\sin\theta\cdot t-\frac{1}{2}gt^2 \quad ...(ii)$$

Squaring both Eqs. (i) and (ii), we get

$$x^2=u^2\cos^2\theta\cdot t^2 \quad ...(iii)$$

$$y^2=u^2\sin^2\theta\cdot t^2-\left(\frac{1}{2}gt^2\right)^2 \quad ...(iv)$$

Now, adding Eqs. (iii) and (iv), we get

$$u^2(\cos^2\theta+\sin^2\theta)t^2=x^2+\left(y+\frac{1}{2}gt^2\right)^2$$

$$\Rightarrow u^2t^2=x^2+\left(y+\frac{1}{2}gt^2\right)^2$$

This is a circle with centre $\left(0,-\frac{1}{2}gt^2\right)$ and radius ut.

22. *(c)* Let O is centre of rotation, then

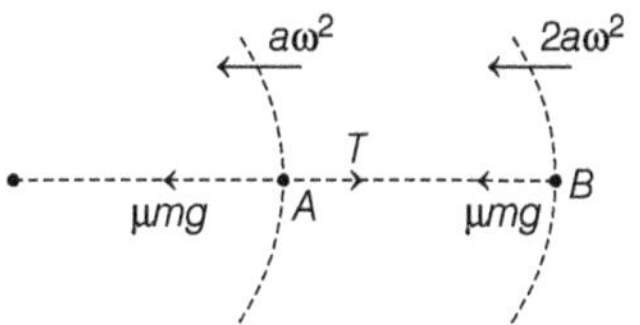

In above figure,

T = tension in string when masses about to slip.

Accelerations of A and B are

$$a_A=a\omega^2$$

$$a_B=2a\omega^2$$

Frictions of A and B are

$$f_A=\mu mg$$

$$f_B=\mu mg$$

Net force on A is

$$F_A = \mu mg - T = ma\omega^2$$

Net force on B is

$$F_B = \mu mg + T = m(2a)\omega^2$$

For no slip; $F_A + F_B$

$$f_A + f_B = 3am\omega^2 \Rightarrow \mu = \frac{3a\omega^2}{2g}$$

23. *(b)* Let angular speed of spaceship and astronaut is ω.

Forces on astronaut provides necessary centripetal pull.

$$\Rightarrow \frac{GM_e m}{(r+L)^2} + T = m(r+L)\omega^2$$

$$\Rightarrow mg\left(1 - \frac{L}{r}\right)^2 + T = \frac{mg}{r}(r+L)$$

Taking $r \approx R$ (radius of earth)

$$\Rightarrow mg - \frac{2mgL}{R} + T = mg + \frac{mgL}{R}$$

$$\Rightarrow T = \frac{3mgL}{R} = \frac{3 \times 200 \times 10 \times 90}{6400 \times 10^3} = 0.084 \text{ N}$$

24. *(d)*

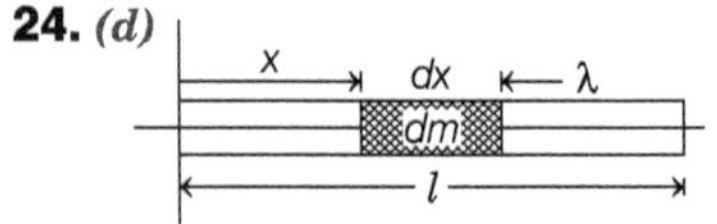

We have, $dm = \lambda\, dx = (\lambda_0 + kx)dx$

$$\therefore X_{CM} = \frac{\int x\, dm}{\int dm} = \frac{\int_0^l x(\lambda_0 + kx)\, dx}{\int_0^l (\lambda_0 + kx)\, dx}$$

$$= \frac{\lambda_0 \frac{l^2}{2} + \frac{kl^3}{3}}{\lambda_0 l + \frac{kl^2}{2}} = \frac{3\lambda_0 l + 2kl^2}{3(2\lambda_0 + kl)}$$

25. *(d)* As system is released at $t = 0$, momentum of system at any instant t is

$$p(t) = M_{\text{system}} \times v_{CM} = p_{\text{system}}\ (t = 0) = 0$$

$\therefore v_{CM} = 0$

So, centre of mass of system is always at rest and is not oscillating.

26. *(d)* As particle moves by an angle θ.

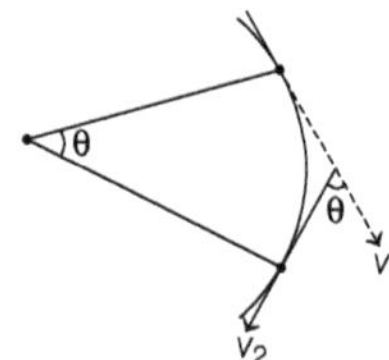

Then, change in velocity is $\Delta \mathbf{v} = \mathbf{v}_2 - \mathbf{v}_1$.

Magnitude of change in velocity is

$$|\Delta \mathbf{v}| = \sqrt{(\mathbf{v}_2 - \mathbf{v}_1)^2}$$

$$\Rightarrow |\Delta \mathbf{v}| = \sqrt{v_1^2 + v_2^2 - 2v_1 v_2 \cos\theta}$$

Here, $v_1 = v_2 = v$

$$\Rightarrow |\Delta \mathbf{v}| = \sqrt{v^2 + v^2 - 2v^2 \cos\theta}$$

$$= \sqrt{2v^2(1 - \cos\theta)}$$

$$= \sqrt{2v^2\left[1 - \left(1 - 2\sin^2\frac{\theta}{2}\right)\right]}$$

$$= \sqrt{2v^2\left(1 - 1 + 2\sin^2\frac{\theta}{2}\right)}$$

$$= \sqrt{2v^2\left(2\sin^2\frac{\theta}{2}\right)} = \sqrt{4v^2 \sin^2\frac{\theta}{2}}$$

$$\therefore |\Delta \mathbf{v}| = 2v \sin\frac{\theta}{2}$$

Time duration of motion is given 4 s

So, $\Delta t = 4 = \frac{R\theta}{v} \Rightarrow \theta = \frac{4v}{R}$

So, average acceleration of the particle is

$$|\mathbf{a}_{\text{avg}}| = \frac{|\Delta \mathbf{v}|}{\Delta t} = \frac{2v \cdot \sin\frac{\theta}{2}}{4} = \frac{v}{2}\sin\left(\frac{2v}{R}\right)$$

27. *(d)* For $v_1 > v_2$, m_1 slows down and m_2 speed up till their velocities are equal at some time t_0.

where, $t_0 = \dfrac{|v_1 - v_2|}{\left(1 + \dfrac{m_1}{m_2}\right)\mu_k g}$

28. *(c)* Mass dm of spring at a distance x moves with velocity is

$$v_x = \frac{x}{l} \cdot v$$

Kinetic energy of mass dm is

$$dK = \frac{1}{2}(dm)v_x^2 \Rightarrow dK = \frac{1}{2} \cdot \frac{v^2 x^2}{l^2} \cdot dm$$

Kinetic energy of complete spring is

$$K = \int dK = \frac{1}{2} \cdot \frac{v^2}{l^2} \cdot \int_0^l x^2\, dm$$

Here, $dm = \dfrac{m}{l} dx$

$$\therefore K = \frac{1}{2} \cdot \frac{v^2}{l^2} \times \frac{m}{l} \int_0^l x^2 dx = \frac{1}{6} mv^2$$

29. *(b)* Here, $m_1 + m_2 > m_3 + m_4$ otherwise equilibrium is not possible.

Force balance for mass m_3 is

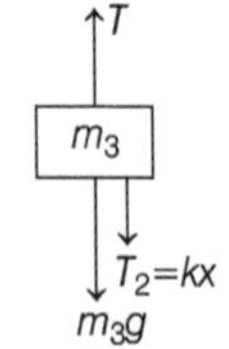

Also, force balance for mass m_1 is

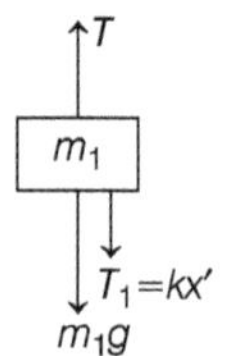

Here, $T_1 = m_2 g$

So, $T = (m_1 + m_2)g$

$$m_3 g + T_2 - T = 0$$

$\Rightarrow T_2 = T - m_3 g$ or $T_2 = (m_1 + m_2 - m_3)g$

When lower thread A is cut equations of motion for all masses can be written as

$$m_1 a_1 = m_1 g + T_1 - T$$
$$m_2 a_2 = m_2 g - T_1$$
$$m_3 a_3 = T_2 + m_3 g - T$$
$$m_4 a_4 = m_4 g - T_2$$

Solving these equations, we have

$$a_1 = a_2 = a_3 = 0$$

and $a_4 = \left(\dfrac{m_3 + m_4 - m_1 - m_2}{m_4}\right) g$

30. *(c)* A thin water film covering a moist object reflects the incident light in one definite direction.The surface of object no longer diffuses white light in all directions and its own colour becomes prominent. The diffused light is not superimposed on the light reflected from the object and for this reason the colours seems more richer.

31. *(c)* Let refracted ray is

$$\mathbf{r} = a\hat{\mathbf{i}} + b\hat{\mathbf{j}} + c\hat{\mathbf{k}}$$

Then, normal to plane of incidence is

$$\mathbf{N} = \begin{vmatrix} \hat{\mathbf{i}} & \hat{\mathbf{j}} & \hat{\mathbf{k}} \\ 6\sqrt{3} & 8\sqrt{3} & -10 \\ 0 & 0 & 1 \end{vmatrix} = 8\sqrt{3}\hat{\mathbf{i}} - 6\sqrt{3}\hat{\mathbf{j}}$$

This must be normal to refracted ray

$$\therefore \quad \mathbf{r} \cdot \mathbf{N} = 0$$

$$\Rightarrow \quad 8\sqrt{3}a - 6\sqrt{3}b = 0 \Rightarrow b = \frac{4}{3}a$$

Also, $\cos(\pi - \hat{\mathbf{i}}) = \dfrac{(6\sqrt{3}\hat{\mathbf{i}} + 8\sqrt{3}\hat{\mathbf{j}} - 10\hat{\mathbf{k}}) \cdot \hat{\mathbf{k}}}{|6\sqrt{3}\hat{\mathbf{i}} + 8\sqrt{3}\hat{\mathbf{j}} - 10\hat{\mathbf{k}}|\,|\hat{\mathbf{k}}|}$

$$= -\frac{1}{2}$$

$\Rightarrow \quad \cos(\pi - i) = \cos 120° \quad \Rightarrow i = 60°$

As, $\sqrt{3}\sin r = \sqrt{2}\sin i$

$\Rightarrow \quad \sin r = \dfrac{1}{\sqrt{2}} \quad \Rightarrow r = 45°$

So, angle between refracted ray and normal = 45°

$$\cos 45° = \frac{(a\hat{\mathbf{i}} + b\hat{\mathbf{j}} + c\hat{\mathbf{k}}) \cdot \hat{\mathbf{k}}}{\sqrt{a^2 + b^2 + c^2}}$$

$$\Rightarrow \quad \frac{1}{\sqrt{2}} = \frac{c}{\sqrt{a^2 + b^2 + c^2}}$$

$$\Rightarrow \quad \sqrt{2}c = \sqrt{a^2 + b^2 + c^2}$$

$$\Rightarrow \quad c = \pm \frac{5a}{3} \Rightarrow c = -\frac{5a}{3}$$

As, positive value is not possible in given case.

32. *(a)* As $\frac{Z}{d}$ is dimensionless and $\frac{F}{d^3 g\Delta\rho}$ is also dimensionless.

We can say that, $\frac{Z}{d}$ *versus* $\frac{F}{d^3 g \Delta\rho}$ is a linear relation.

Hence, correct graph of $\frac{Z}{d}$ *versus* $\frac{F}{d^3 g \Delta\rho}$ must be a straight line.

33. *(d)* As, $\lambda_1 N_1 = \lambda_2 N_2 = \lambda_3 N_3$ .... etc., production rates of daughter nuclii are equal to decay rate of parent.

34. *(c)* In region A, binding energy increases and then decreases, it again increases and then decreases.

This indicates that binding energy reaches a local maximum and then reduces as nucleon number increases.

This is possible, if one of nucleus is stable and then next nucleus with more nucleons is unstable because now nucleus may form a half filled shell.

35. *(c)* In process A,

$$\Delta S_{\text{source}} = \frac{Q}{T} = \frac{-2000}{800} = -2.5\,\frac{\text{kJ}}{\text{K}}$$

$$\Delta S_{\text{sink}} = \frac{Q}{T} = \frac{2000}{500} = 4\,\frac{\text{kJ}}{\text{K}}$$

$$\Delta S_{\text{Total}} = \Delta S_{\text{source}} + \Delta S_{\text{sink}} = 1.5\ \text{kJ/K}$$

In process B,

$$\Delta S_{\text{source}} = \frac{-2000}{800} = -2.5\,\frac{\text{kJ}}{\text{K}}$$

$$\Delta S_{\text{sink}} = \frac{2000}{750} = 2.7\,\frac{\text{kJ}}{\text{K}}$$

$$\Delta S_{\text{Total}} = -2.5 + 2.7 = 0.2\,\frac{\text{kJ}}{\text{K}}$$

As, ΔS_{Total} is less for process B. So, it has less irreversibility or B can be more easily reversed.

36. *(b)* At low pressure and at high temperature the compressibility coefficient $Z = \frac{pV}{RT}$, is always equal to one for even real gases. But at high pressures and low temperatures gases begun to deviate from ideal gas behaviour. So, T_4 is greatest and T_1 is least temperature.

37. *(a)* Current induced in coil,

$$i = \frac{E}{R} = \frac{1}{R}\left(\frac{d\phi_B}{dt}\right)$$

$$\Rightarrow \quad \frac{dq}{dt} = \frac{1}{R}\frac{d\phi_B}{dt}$$

$$\Rightarrow \quad \Delta Q = \frac{1}{R} \times \Delta\phi_B = \frac{\pi a^2 NB}{R}$$

Total charge passing through coil is independent of time.

Also, $\frac{d\phi}{dt}$ can be determined only when velocity is given.

38. *(a)* Effective emf of circuit is

$$E_{\text{eff}} = \frac{E}{2}$$

Effective resistance of circuit is

$$R_{\text{eff}} = \frac{3}{2}R$$

So, time constant of circuit is

$$\tau = \frac{L_{\text{eff}}}{R_{\text{eff}}} = \frac{2L}{3R}$$

Current through inductor at time t is

$$i = \frac{E}{3R}\left(1 - e^{-t\left(\frac{3R}{2L}\right)}\right)$$

$$iR = \frac{E}{3}(1 - e^{-t/a}) \qquad [\because iR = V]$$

$$V = \frac{E}{3}(1 - e^{-t/a})$$

Potential drop across inductor is

$$V_L = L\frac{di}{dt} = \frac{E}{2}e^{-t\left(\frac{3R}{2L}\right)}$$

So, $V_{AB} = V_{AC} + V_L$

$$= \frac{E}{3}(1 - e^{-t/\tau}) + \frac{E}{2}e^{-t/\tau}$$

$$= \frac{E}{3}\left(1 + \frac{1}{2}e^{-t\left(\frac{3R}{2L}\right)}\right)$$

39. *(d)* Centripetal force on electron is

$$\frac{mv^2}{r} = \frac{e^2}{4\pi\varepsilon_0 r^2}$$

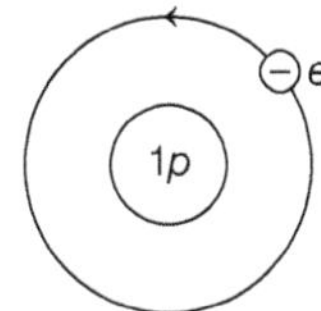

So, $\quad mv^2 = \frac{e^2}{4\pi\varepsilon_0 r}$

$$\text{Kinetic energy} = \frac{1}{2}mv^2 = \frac{e^2}{8\pi\varepsilon_0 r}$$

$$\text{Potential energy} = \frac{1}{4\pi\varepsilon_0}\cdot\frac{e}{r}(-e) = -\frac{e^2}{4\pi\varepsilon_0 r}$$

Total energy = Kinetic energy + Potential energy

$$\frac{e^2}{8\pi\varepsilon_0 r} - \frac{e^2}{\pi\varepsilon_0 r} = \frac{e^2 - 2e^2}{8\pi\varepsilon_0 r} = \frac{-e^2}{8\pi\varepsilon_0 r}$$

40. *(b)* A free electron cannot absorbs a photon. In an electron-photon collision, when electron is bound. i.e. with in a metal surface, electron absorbs the photon and it cannot further absorbs photon.

41. *(d)*

$$\overset{1}{C}H_2{=}\overset{2}{C}H{-}\overset{3}{C}H_2{-}\overset{4}{C}H_2{-}\overset{5}{C}{\equiv}\overset{6}{C}H$$

Hybridisation at $C_2 = sp^2$

Hybridisation at $C_3 = sp^3$

Thus, the C_2—C_3 bond is of sp^2- sp^3 type.

42. *(a)* Peroxide effect is observed only in case of HBr. Therefore, addition of HCl to propene even in the presence of benzoyl peroxide occurs according to Markownikoff's rule.

$$\underset{\text{Propene}}{CH_3CH{=}CH_2} \xrightarrow[(C_6H_5COO)_2]{HCl} \underset{\text{2-chloropropane}}{CH_3{-}\underset{|}{\overset{Cl}{C}}H{-}CH_3}$$

43. *(a)* According to Arrhenius equation, rate constant increases exponentially with temperature :

$$k = Ae^{-E_a/RT}$$

Thus, correct plot is given in option (a).

44. *(b)* According to Freundlich adsorption isotherm, $\frac{x}{m} = kp^{1/n}$

On taking logarithm of both sides, we get

$$\log\frac{x}{m} = \log k + \log p^{1/n}$$

$$\text{or} \quad \log\frac{x}{m} = \log k + \frac{1}{n}\log p$$

log x/m → ; Slope $= \frac{1}{n}$; θ ; log k ; log p →

On comparing the above equation, with equation of straight line, i.e. $y = c + mx$, we get

$$y = \log\frac{x}{m},$$

$$c = \text{intercept} = \log k$$

$$m = \text{slope} = \frac{1}{n} \text{ and } x = \log p$$

Thus, option (b) is correct.

45. *(c)* N_2O_3 dissolves in water giving a pale blue solution of HNO_2 which decolorises $KMnO_4$ and oxidises KI to I_2.

$$N_2O_3 + H_2O \longrightarrow \underset{\text{Pale blue}}{HNO_2}$$

HNO_2 is a reducing agent

$$MnO_4^- \xrightarrow{HNO_2} Mn^{2+}$$

HNO_2 is also an oxidising agent

$$I^- \xrightarrow{HNO_2} I_2$$

46. *(c)*
$Ni(CO)_4$; Oxidation state of Ni = 0

$$Ni = 1s^2, 2s^2 2p^6, 3s^2 3p^6, 4s^2, 3d^8$$

As CO is a strong ligand and hence pairing of electrons occurs.

Ni = [3d: ⥮ ⥮ ⥮ ⥮ ⥮] [4s: □] [4p: □□□]
sp^3-hybridisation

Oxidation state of Ni in $[Ni(PPh_3)_2]Cl_2$ is

$$x + 0 \times 2 + (-2) = 0$$
$$x = +2$$

Ni^{2+} = [3d: ⥮ ⥮ ⥮ ⥮ ⥮] [4s: □] [4p: □□□]
sp^3-hybridisation

As both the complexes has sp^3-hybridisation, thus their geometry is tetrahedral.

47. *(d)* Higher the standard reduction potential, better is the oxidising agent. Among the given options, $E^\circ_{MnO_4^-/Mn^{2+}}$ is highest, hence MnO_4^- is the strongest oxidising agent.

48. *(a)* Basic nature of a compound depends upon the ease with which it can donate its unshared pair of electrons.

$CH_3CH_2 \rightarrow NH_2$ (I), $CH_3—\overset{O}{\overset{\|}{C}} \leftarrow \ddot{N}H_2$ (II)

$C_6H_5—\overset{O}{\overset{\|}{C}} \leftarrow \ddot{N}H_2$ (III)

$CH_3—\underset{O}{\underset{\|}{C}}$ is an

I. electron donating group. (Maximum basic)

II. electron withdrawing. The lone pair of N is involved in resonance with CH_3CO.

III. Lone-pair of N-atom is also used in delocalisation of π-electrons, resonance of carbonyl benzene nucleus (least basic)

Thus, the correct order of basic nature of given compounds are as follows:

$$I > II > III.$$

49. *(c)* $\underset{A}{3M} + N_2 \longrightarrow \underset{B}{M_3N_2}$

$$\underset{B}{M_3N_2} + 6H_2O \longrightarrow \underset{C}{3M(OH)_2} + \underset{D}{2NH_3}$$

$$\underset{C}{M(OH)_2} + CO_2 \longrightarrow \underset{D}{MCO_3} + H_2O$$

M may be either Ca or Ba. It is not magnesium because $Mg(OH)_2$ has very low solubility in water. Thus, the correct option is (c).

50. *(b)* There is one octahedral hole for each atom in hexagonal close packed arrangement. If the number of oxide ions (O^{2-}) per unit cell is 1, then the number of Fe^{3+} ions = $2/3 \times$ octahedral holes = $2/3 \times 1 = 2/3$. Thus, the formula of the compound = $Fe_{2/3}\ O_1$ or Fe_2O_3.

51. *(d)*

$$CH_2 = CH_2 + H_2 \longrightarrow CH_3 — CH_3$$

$$\Delta H = (BE)_{reactants} - (BE)_{products}$$
$$= 4(BE)_{C—H} + (BE)_{C=C} + (BE)_{H—H} - [6(BE)_{C—H} + (BE)_{C—C}]$$
$$= [4 \times 414 + 615 + 435] - [6 \times 414 + 347]$$
$$= -125 \text{ kJ}$$

52. *(a)* $Cr_2O_7^{2-} + 14H^+ + 6I^- \longrightarrow 2Cr^{3+} + 7H_2O + 3I_2$

Since, $Cr_2O_7^{2-}$ is reduced to Cr^{3+}, thus final state of Cr is +3, which is formed by the reaction between KI and acidified potassium dichromate solution.

53. *(c)* Phenol on treatment with bromine water produces a white precipitate of 2, 4, 6-tribromophenol.

Phenol $\xrightarrow[Br_2, \text{water}]{\text{Excess}}$ 2,4,6-tribromophenol (white precipitate)

54. *(b)* This problem includes basic concept of bonding. It can be solved by using the concept of molecular orbital theory.

← + ve phase
← − ve phase

When two same phase overlap with each other, it forms bonding molecular orbital otherwise antibonding.

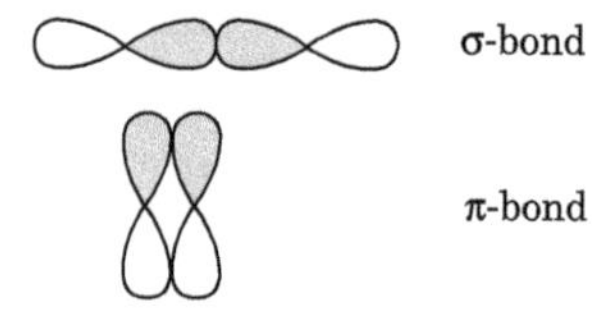

The type of molecular orbital shown in given representations are as follows :

(a) → *d-d* σ bonding

(b) → *p-d* π bonding

(c) → *d-d* σ antibonding

(d) → *p-d* π antibonding

55. *(b)* 2 hexyne gives *trans*-2-hexene on treatment with Li/NH_3.

$$\underset{\text{2-hexyne}}{CH_3CH_2CH_2 — C \equiv C — CH_3} \xrightarrow{(b)\ Li/NH_3}$$

$(H)(CH_3CH_2CH_2)C = C(CH_3)(H)$
trans-2-hexene

The products obtained by the other given reagents on treatment with 2-hexyne are as follows:

$$\text{2-hexyne} \xrightarrow{(a)\ Pt/H_2} n\text{-hexane}$$
$$\xrightarrow{(c)\ Pd/BaSO_4} cis\text{-2-hexene}$$
$$\xrightarrow{(d)\ LiAlH_4} \text{no reaction}$$

56. *(a)* Given polymer can be obtained by condensation polymerisation of ethylene glycol and phthalic acid with the elimination of water molecule.

nHOCH$_2$—CH$_2$OH (Glycol) + n HOOC–C_6H_4–COOH (Phthalic acid) $\xrightarrow[-nH_2O]{420\text{-}460\ K,\ Zn(OCOCH_3)_2 + Sb_2O_3}$

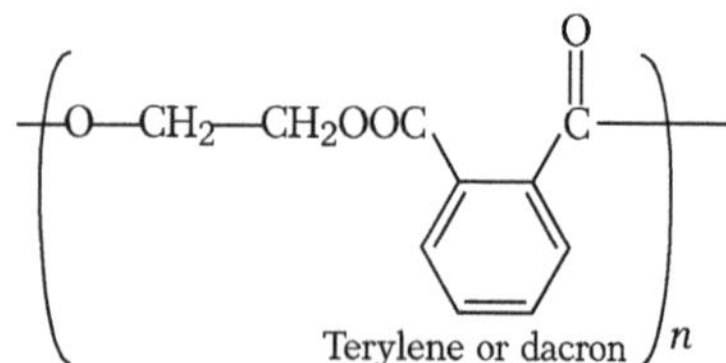

Terylene or dacron

57. *(a)* In stable isotope of Na, there are 11 protons and 12 neutrons. In the given radioactive isotope of sodium (Na^{24}), there are 13 neutrons, one neutron is more than that required for stability. A neutron rich isotope always decay by β^--emission as

$$_0n^1 \longrightarrow {}_{-1}\beta^0 + {}_1H^1$$

58. *(a)*

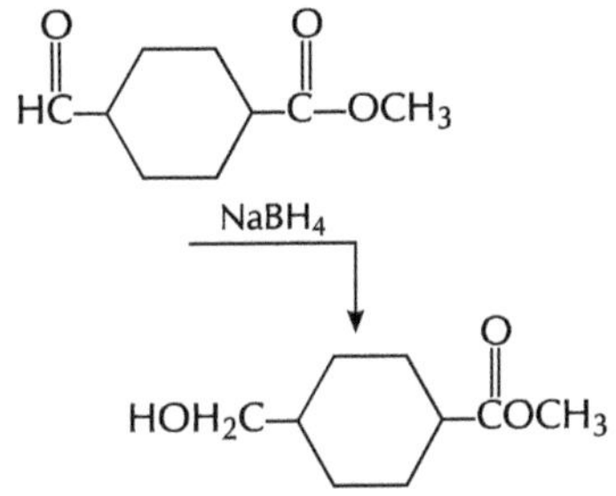

$$OHC\text{–}C_6H_{10}\text{–}COOCH_3 \xrightarrow{NaBH_4} HOH_2C\text{–}C_6H_{10}\text{–}COOCH_3$$

$NaBH_4$ is a selective reducing agent, it only reduces aldehyde and ketone group. Groups like ester, cyanide remains unaffected.

59. *(c)* When glucose reacts with Br_2 water at pH 5-6, then gluconic acid is formed.

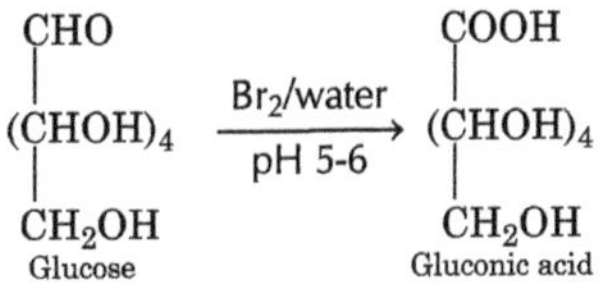

$$\underset{\text{Glucose}}{CHO(CHOH)_4CH_2OH} \xrightarrow[\text{pH 5-6}]{Br_2/\text{water}} \underset{\text{Gluconic acid}}{COOH(CHOH)_4CH_2OH}$$

60. *(c)* $TiCl_4$ Oxidation state of Ti = + 4, the electronic configuration is [Ar] $3d^0$.

Thus, all electrons are paired and hence, diamagnetic.

O_2 No. of electron in $O_2 = 16$.

Electronic configuration :

$\sigma 1s^2 \sigma * 1s^2 \sigma 2s^2 \sigma * 2s^2 \sigma 2p_z{}^2$

$\pi 2p_x{}^2 = \pi 2p_y{}^2 \pi * 2p_x{}^1 = \pi * 2p_y{}^1$

Two electrons are unpaired (by MO therory).

Thus, paramagnetic.

$[Ni(CN)_4]^{2-}$ Electronic configuration of $Ni^{2+} = 3d^8 4s^0$

As CN^- is a strong ligand, pairing of electrons occurs.

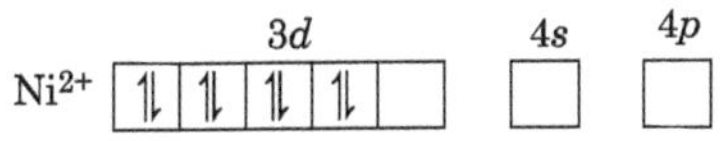

	3d					4s	4p
Ni^{2+}	⥮	⥮	⥮	⥮			

Hybridisation of $[Ni(CN)_4]^{2-}$ is d_{sp^2} and shape is square plannar.

As all electrons paired thus, diamagnetic in nature.

Thus, $[Ni(H_2O)_4]$ No. pairing occurs as H_2O is a weak ligand.

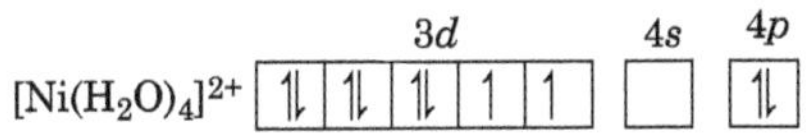

	3d					4s	4p
$[Ni(H_2O)_4]^{2+}$	⥮	⥮	⥮	1	1		⥮

sp^3-hybridisation as two electrons are unpaired. Thus, paramagnetic in nature.

Therefore, option (c) is correct.

61. *(c)* Sliding filament theory is explained as actin and myosin filaments do not shorten but rather slide pass each other.

62. *(d)* Spermatogonia are produced from the undifferentiated primordial germ cells which divide continuously by mitosis to form diploid cells, i.e. primary spermatocyte. These are the precursors of sperms (the haploid cells). Primary polar body, secondary polar body (in oogenesis) are produced after first meiotic division as haploid cells. Spermatids are haploid cells produced by second meiotic division of secondary spermatocytes (n).

63. *(c)* SA node (Sino-Atrial node) sets the basic pace of the heartbeat, hence it is called pacemaker which is a bundle of modified cardiac muscles. It is generally implanted subcutaneously and connected to heart in patients who have irregularity in their heart rhythm.

64. *(b)* Secretory cells always have abundant RER and Golgi apparatus. Newly made secretory proteins are localised to the lumen of the rough ER. The proteins then migrate to the Golgi complex in the membrane bound vesicles.

65. *(a)* The typical storage forms of fatty acids in cells are triacylglycerols, which is formed from one molecule of glycerol by esterification of fatty acids to each of the 3 hydroxyl groups.

$HOCH_2 — CH(OH) — CH_2OH$

66. *(d)* The specificity of the enzymes is due to their surface configuration. It means that the active site of enzyme consists of certain amino acid side chains whose linear arrangement and the appropriate folding of the peptide chain give the enzyme its specificity. This site recognises and binds the substrate(s) and catalyses the reaction once the substrate(s) have been bound.

67. *(d)* Centrioles migrate to the poles at mid prophase-I and start to form the spindle fibres at late prophase-I.

68. *(a)* Option (a) is correct as the lagging strand is synthesised discontinuously, the newly synthesised DNAs exist as small fragments (called Okazaki fragments), which are linked together by DNA ligase.

Other statements can be corrected as: RNA primer is needed on both the lagging and leading strands.

Deoxyribonucleotides are added to the 3'-end.

DNA polymerase synthesises Okazaki fragments.

69. *(b)* Since, there are 17 different amino acids that constitute insulin and each amino acid is coded by a triplet codon attached to a *t*RNA molecule, the minimum number of different *t*RNA molecules is $17 \times 3 = 51$.

70. *(b)* The correct *m*RNA strand should be G – A – U – U – G – A, in order to be complementary to the sense strand of the DNA fragments. However, with a single cytosine base incorrectly paired the *m*RNA could end up as G – A – U – U – C – A.

71. *(c)* A gene mutation cannot cause a dominant allele to become recessive, but can disallow it from being expressed phenotypically.

72. *(b)* The promoter is made up of a basal promoter which is usually located within 40 base pairs from the transcription start site and an upstream promoter (proximal control element) which may be located as far as 200 base pairs from the transcription start site.

Control elements can be divided into two kinds, depending on the distance from the basal promoters, proximal control elements and distal control elements. Enhancers and silencers are distal control elements.

73. *(d)* Some phenotypes are influenced by the different environment even if they are individuals of the same species, e.g. height and skin colour in humans.

74. *(a)* The water molecule is split as shown in the equation

$$H_2O \rightarrow \frac{1}{2}O_2 + 2H^+ + 2e^-$$

forming oxygen and hydrogen radicals.

75. *(d)* The osmotic pressure of 0.25 mol dm^{-3} sucrose solution approximates that of the internal environment of the mitochondria thus this solution is applied to prevent the mitochondria from undergoing structural changes.

76. *(b)* The tissue has a water potential of – 700 kPa. 'A' is isotonic to the tissue. Addition of solute tends to lower the water potential. Therefore, the most negative (hypertonic) will have more solutes in it, causing water to move out of the tissue into the solute. This would be the answer (b) (– 1000 kPa). (c) and (d) are both hypotonic to the tissue, causing water to move in, making it more turgid instead.

77. *(d)* Glycogen is stored in the liver and muscles. When the external nutrition source is inadequate, glycogen is hydrolysed to provide the energy required in its metabolic processes.

78. *(b)* The opening of sodium ion gates increases the permeability of the axon membrane to sodium ions. As more sodium ions enter the axon, the membrane potential increases until it reaches a positive value.

79. *(a)* Each gene probably has its own characteristic mutational behaviour. Some genes undergo mutations more frequently than others in the same organisation and are called unstable or mutable. This gene is likely to mutate if its mutant form will allow the population to adapt successfully in the environment.

Thus, frequency of a mutant gene in a population would be likely to increase if the genes were selectively advantageous.

80. *(a)* Genetic modification of plants into herbicide resistant plants can improve crop yield. This is because herbicides are used to get rid of weeds which compete with crops for soil nutrients. Transgenic plants are not inhibited by herbicide. Thus, farmers may apply a specific herbicide to control weeds population, without damaging their herbicide tolerant crops. This leads to an increase in crop yield because of less competition from weeds.

81. *(d)* In ΔABC,

$$\frac{10}{\sin 60^\circ} = \frac{b}{\sin 50^\circ} = \frac{c}{\sin 70^\circ} = K$$

$$\frac{20}{\sqrt{3}} = \frac{b}{\sin 50^\circ} = \frac{c}{\sin 70^\circ} = K$$

$$b = \frac{20}{\sqrt{3}}\sin 50^\circ$$

$$c = \frac{20}{\sqrt{3}}\sin 70^\circ$$

$$\Rightarrow \quad b + c = \frac{20}{\sqrt{3}}(\sin 50^\circ + \sin 70^\circ)$$

$$= \frac{40}{\sqrt{3}}\sin 60^\circ \cos 10^\circ$$

$$= \frac{40}{\sqrt{3}} \times \frac{\sqrt{3}}{2}\cos 10^\circ$$

$$b + c = 20\cos 10^\circ$$

Perimeter of ΔABC

$$= a + b + c = x + y\cos z^\circ$$

$$= 10 + 20\cos 10^\circ = x + y\cos z^\circ$$

$$\therefore \quad x = 10,\ y = 20,\ z = 10$$

$$\therefore x + y + z = 10 + 20 + 10 = 40$$

82. *(a)* We have,

$$g''(x) - 3g'(x) > 3$$

$$\Rightarrow \quad \frac{g''(x)}{g'(x) + 1} > 3$$

On integrating, we get

$$\log(g'(x) + 1) > 3x$$

$$\Rightarrow \quad g'(x) + 1 > e^{3x}$$

$$g'(x) + 1 > 0 \ \forall\, x > 0$$

$\therefore g(x) + x$ is an increasing function.

83. *(b)*

9's	8's	other	Number of ways
1	1	2	${}^4C_2 \times 2 \times 8 \times 8 = 768$
1	2	1	$\frac{4!}{2!} \times 8 = 96$
1	3	0	$\frac{4!}{3!} = 4$
2	1	1	$\frac{4!}{2!} \times 8 = 96$
2	2	0	$\frac{4!}{2!2!} = 6$
3	1	0	$\frac{4!}{3!} = 4$

Total number of ways = 974

84. *(c)* We have

$$z^5 - 32 = (z-2)(z^2 - pz + 4)(z^2 - qz + 4)$$

$$z^5 - 32 = (z - z_0)(z - z_1)(z - z_2)(z - z_3)(z - z_4)$$

$$\Rightarrow \quad z_i = 2\left(\cos\frac{2k\pi}{5} + i\sin\frac{2k\pi}{5}\right)$$

where $z_i = 0, 1, 2, 3, 4$

$$\Rightarrow \quad z_0 = 2,\ z_1 = 2\left(\cos\frac{2\pi}{5} + i\sin\frac{2\pi}{5}\right),$$

$$z_2 = 2\left(\cos\frac{4\pi}{5} + i\sin\frac{4\pi}{5}\right)$$

$$z_3 = 2\left(\cos\frac{6\pi}{5} + i\sin\frac{6\pi}{5}\right),$$

$$z_4 = 2\left(\cos\frac{8\pi}{5} + i\sin\frac{8\pi}{5}\right)$$

$$z^5 - 32 = (z-2)\left(z^2 - 4\cos\frac{2\pi}{5} + 4\right)\left(z^2 - 4\cos\frac{2\pi}{5} + 4\right)$$

$$[\because (z - \alpha)(z - \bar{\alpha}) = (z^2 - (\alpha + \bar{\alpha})z + \alpha\bar{\alpha})]$$

$$\therefore \quad p = q = 4\cos\frac{2\pi}{5} = 4\cos 72^\circ = 4\sin 18^\circ$$

$$p^2 + 2q = 16\sin^2 18^\circ + 8\sin 18^\circ$$

$$= 8(1 - \cos 36^\circ + \sin 18^\circ)$$

$$= 8\left(1 + \frac{\sqrt{5}-1}{4} - \frac{\sqrt{5}+1}{4}\right) = 4$$

85. *(a)* Using power of C

$$BC \cdot AC = CS^2$$

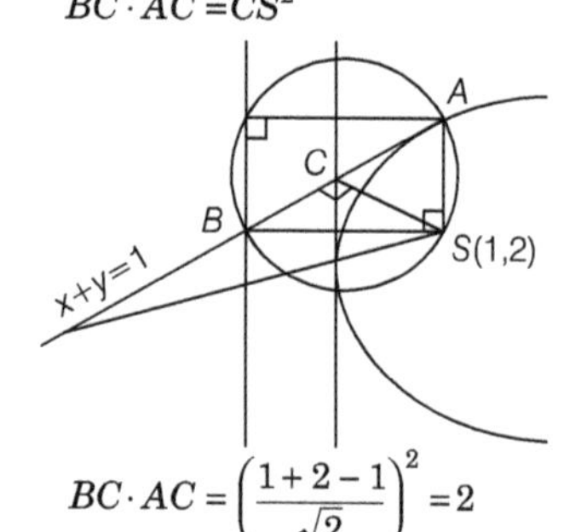

$$\therefore \quad BC \cdot AC = \left(\frac{1+2-1}{\sqrt{2}}\right)^2 = 2$$

86. *(a)* Given, $2x - y + 1 = 0$ is tangent of circle at (2, 5).

Centre of circle lies on $x - 2y = 4$

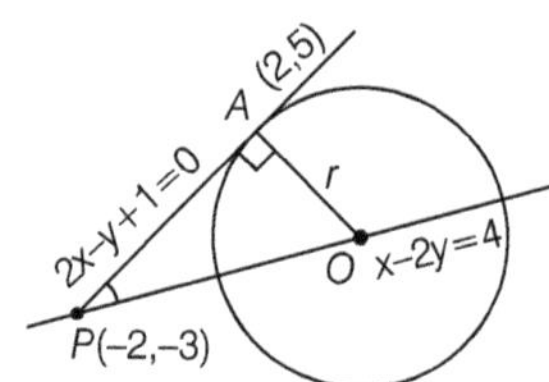

Solving $2x - y + 1 = 0$ and $x - 2y = 4$, we get intersecting point $P(-2, -3)$

$$PA = \sqrt{(2+2)^2 + (5+3)^2}$$

$$= \sqrt{16 + 64} = \sqrt{80} = 4\sqrt{5}$$

Angle between lines are

$$\tan\theta = \frac{2 - \frac{1}{2}}{1 + 2\left(\frac{1}{2}\right)} = \frac{3}{4}$$

In ΔPAO,

$$\tan\theta = \frac{r}{PA}$$

$$\Rightarrow \quad \frac{3}{4} = \frac{r}{4\sqrt{5}}$$

$$\Rightarrow \quad r = 3\sqrt{5}$$

87. *(a)* We have,

$$\cos(\sin x) = \sin(\cos x)$$

$$\Rightarrow \quad \sin\left(\frac{\pi}{2} \pm \sin x\right) = \sin(\cos x)$$

$$\Rightarrow \quad \cos x = n\pi + (-1)^n\left(\frac{\pi}{2} + \sin x\right), n \in I$$

$$\Rightarrow \quad \cos x \pm \sin x = n\pi + (-1)^n \frac{\pi}{2}, n \in I$$

As LHS $\in [-\sqrt{2}, \sqrt{2}]$, and it does not satisfies RHS.

Hence, no solution exists.

88. *(d)* $B_1B_2, \ldots, B_n$

Case I D_1 never show 1

$$\text{Probability} = \left(\frac{5}{6}\right)^n$$

Case II D_2 shows 1 (one time) then D_1 shows 1

$$\text{Probability} = \left\{{}^nC_1\left(\frac{5}{6}\right)^{n-1}\left(\frac{1}{6}\right)\right\}\left(\frac{1}{6}\right)$$

$$\text{Total probability} = \left(\frac{5}{6}\right)^n + n\left(\frac{5^{n-1}}{6^{n-1}}\right)\left(\frac{1}{6^2}\right)$$

89. *(b)* $y = e^x$ and $y = \log_e x$ are inverse of each other

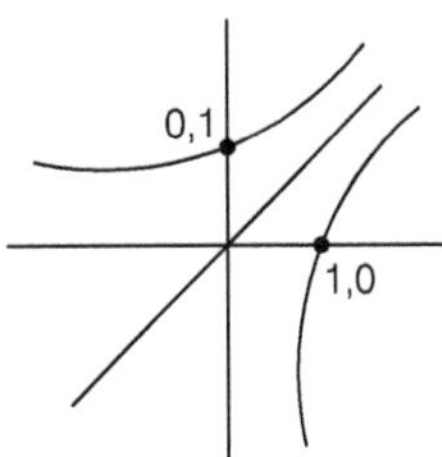

Minimum distance of the curve at (0, 1) and (1, 0).

$$\sqrt{1+1} = \sqrt{2}$$

90. *(b)* Let

$$I = \int_0^5 [x]\{x\}\, dx$$

$$\Rightarrow \quad I = \int_0^1 \{0\}\, dx + \int_1^2 \{x\}\, dx + 2\int_2^3 \{x\}\, dx + 3\int_3^4 \{x\}\, dx + \int_4^5 4\{x\} dx$$

$$\Rightarrow \quad I = (1+2+3+4)\int_0^1 x\, dx$$

$[\because \{x\}$ is periodic with 1$]$

$$\Rightarrow \quad I = 10\left[\frac{x^2}{2}\right]_0^1$$

$$I = \frac{10}{2} = 5$$

91. *(c)* Refraction and reflection occurs as shown below

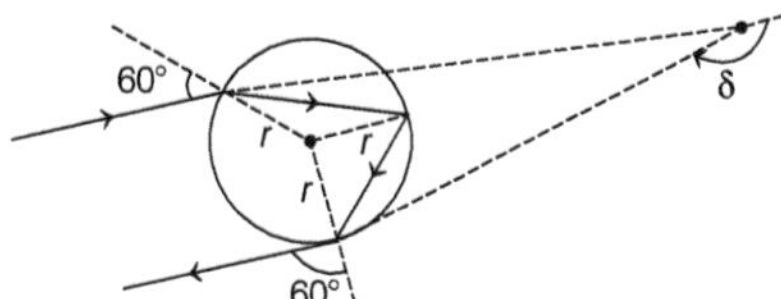

We have, $\dfrac{4}{3} = \dfrac{\sin 60°}{\sin r}$

$$\therefore \quad \sin r = \frac{3}{4} \times \sin 60° = 0.65$$

$$\therefore \quad r = 40°\ 30'$$

So, deviation, $\delta = 180° + 2 \times 60° - 40°30' \times 4 = 138°$

92. *(a)* Let velocity of cart in v at time t. When sand particle falls onto cart, its' relative velocity with respect to cart is $-v$ (backwards), so force of sand on cart is backwards.

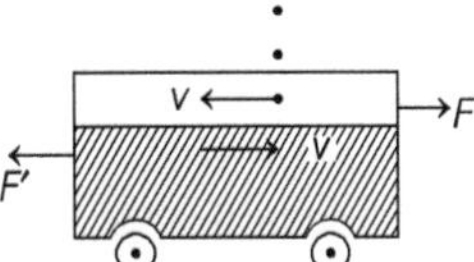

Net force on cart is

$$F_{\text{net}} = F - F' = F - \mu v$$

$$\Rightarrow \quad m\frac{dv}{dt} = F - \mu v$$

$$\Rightarrow \quad (m_0 + \mu t)\frac{dv}{dt} = F - \mu v$$

$$\Rightarrow \quad \frac{1}{F - \mu v} \cdot dv = \frac{1}{m_0 + \mu t} \cdot dt$$

$$\Rightarrow \quad \int_0^v \frac{1}{F - \mu v} \cdot dv = \int_0^t \frac{1}{m_0 + \mu t} \cdot dt$$

$$\Rightarrow \quad \ln\left(\frac{F}{F - \mu v}\right) = \ln\left(\frac{m_0 + \mu t}{m_0}\right)$$

$$\Rightarrow \quad v = \frac{Ft}{m_0 + \mu t}$$

93. *(b)*

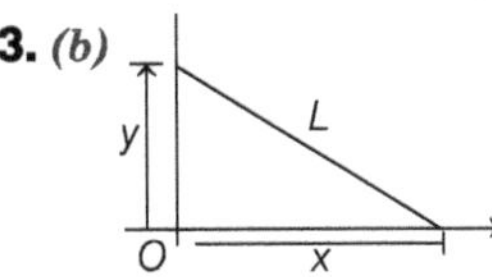

At any instant, $y = \sqrt{L^2 - x^2}$

$$\Rightarrow \quad \frac{dy}{dt} = \frac{-x}{\sqrt{L^2 - x^2}} \cdot \frac{dx}{dt} = \frac{-3}{\sqrt{5^2 - 3^2}} \times 2$$

or $\dfrac{dy}{dt} = -\dfrac{3}{2}\ \text{ms}^{-1}$

At any instant, position of centre of mass is at $\left(\dfrac{x}{2}, \dfrac{y}{2}\right)$.

$$(v_{\text{CM}})_x = v_x = \frac{d}{dt}\left(\frac{x}{2}\right) \text{ or } v_x = \frac{1}{2}\frac{dx}{dt} = 1\ \text{ms}^{-1}$$

Similarly, $v_y = \dfrac{1}{2}\left(\dfrac{3}{2}\right) = \dfrac{3}{4}\ \text{ms}^{-1}$

So, velocity of centre of mass of ladder is

$$v = \sqrt{v_x^2 + v_y^2} = \sqrt{1 + \frac{9}{16}} = \frac{5}{4} = 1.25\ \text{ms}^{-1}$$

94. *(b)* Let E = emf of cell.

Current in the circuit is

$$i = \frac{E}{R + r} = \frac{E}{500 + 4}$$

Potential drop across terminals of voltmeter is

$$V = iR = 500 \times \frac{E}{500 + 4} = \frac{500E}{504}$$

So, voltmeter reads $\dfrac{500E}{504}$ instead of E.

Hence, per cent error in reading is

$$\text{Per cent error} = \frac{\left(E - \frac{500}{504}E\right)}{\left(\frac{500}{504}E\right)} \times 100 \approx 0.8\%$$

95. *(b)* We have, $C = \dfrac{\Delta Q}{\Delta T} = \dfrac{\Delta U + \Delta W}{\Delta T}$

For paths A and B,

$$\Delta U_1 = \Delta U_2 \text{ and } \Delta W_2 > \Delta W_1$$

So, $\quad C_2 > C_1 \Rightarrow \dfrac{C_2}{C_1} > 1$

96. *(a)* We take current distribution as follows

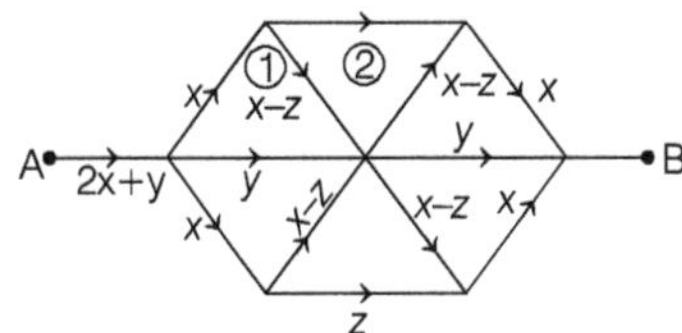

Now, from loops 1 and 2, we have

$$2x - y - z = 0 \quad \text{...(i)}$$

$$2x - 3z = 0 \quad \text{...(ii)}$$

So, $\dfrac{x}{3} = \dfrac{z}{2} = k$ (let) $\Rightarrow x = 3k, z = 2k$

Substituting these in Eq. (i), we get

$$y = 4k$$

Let R = Equivalent resistance between A and B, then $V_{AB} = R\,(2x + y)$.

$$\Rightarrow \quad 2ry = R(2x + y) \text{ or } R = \frac{2ry}{2x + y} = \frac{4}{5}r$$

Here, $r = 2\Omega$, so $R = \dfrac{8}{5}\ \Omega$.

97. *(c)* Uranium nucleus is initially at rest. Conservation of momentum gives

$$0 = p_\alpha + p_{\text{Th}} \Rightarrow p_{\text{Th}} = -p_\alpha$$

Also, $Q = K_\alpha + K_{\text{Th}}$

$$\Rightarrow \quad Q = K_\alpha + p_{\text{Th}}^2 / 2m_{\text{Th}}$$

$$= K_\alpha + \frac{p_\alpha^2}{2m_{\text{Th}}} = K_\alpha + \frac{2m_\alpha \cdot K_\alpha}{2m_{\text{Th}}}$$

$$= K_\alpha\left(1+\frac{m_\alpha}{m_{Th}}\right) = 4.2\left(1+\frac{4}{234}\right)$$
$$= 4.2\left(\frac{238}{234}\right)\text{MeV}$$

98. *(d)* Let particles collide in mid air at sometime instant t after $t = 0$.

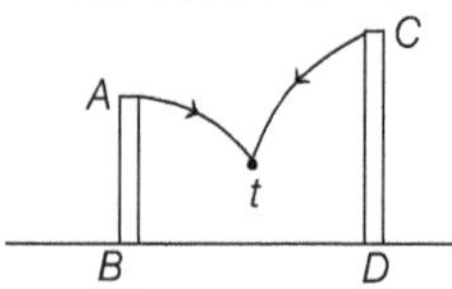

Then, $20 - y = \frac{1}{2}gt^2$

and $30 - y = 10\sin 60° t + \frac{1}{2}gt^2$

[taking, B as origin]

Solving, we get $t = \frac{2}{\sqrt{3}}$ s

Now, x_1 = distance covered horizontally by first particle $= 10t = \frac{20}{\sqrt{3}}$ m

x_2 = horizontal distance of second particle $= 10\cos 60° \times t = \frac{10}{\sqrt{3}}$ m

$\therefore$ Distance between towers,

$$d = x_1 + x_2 = 10\sqrt{3}\text{ m}$$

99. *(d)* As process is adiabatic, $\Delta Q = 0$.

$\Rightarrow \quad dU + pdV = 0$

$\Rightarrow \quad d(a + bpV) + pdV = 0$

$\Rightarrow \quad (b+1)\frac{dV}{V} = -b\frac{dp}{p}$

Integrating and rearranging, we get

$\Rightarrow \quad (b+1)\ln V + b\ln p = \text{constant}$

$\Rightarrow \quad pV^{\left(\frac{b+1}{b}\right)} = \text{a constant}$

$\Rightarrow \quad \gamma = \frac{b+1}{b} = 1 + \frac{1}{b}$

100. *(d)* Let Δx_0 = uncertainly in position at time $t = 0$ s, so uncertainly in momentum is $\Delta p \geq \frac{h}{4\pi\Delta x_0}$

As $v << c$, momentum uncertainly is

$$\Delta p = \Delta(mv) = m_0\Delta v;$$

Also, the uncertainly in proton's velocity is

$$\Delta v = \frac{\Delta p}{m_0} = \frac{h}{4\pi m_0 \Delta x_0}$$

So, the distance x that proton covers in time t cannot be known more accurately than

$$\Delta x = t\Delta v \geq \frac{ht}{4\pi m_0 \Delta x_0}$$

At $t = 1$ s, value of Δx is

$$\Delta x \geq \frac{1.054 \times 10^{-34} \times 1}{2(1.672 \times 10^{-27}) \times (1 \times 10^{-11})}$$
$$\geq 3.15 \times 10^3 \text{ m}$$

or $\quad \Delta x \geq 3.15$ km

101. *(b)* According to Raoult's law, relative lowering of vapour pressure,

$$\frac{p_A^\circ - p_S}{p_A^\circ} = \chi_B \quad \ldots\text{(i)}$$

$$\chi_B = \frac{n_B}{n_B + n_A} = \frac{W_B/M_B}{\frac{W_B}{M_B} + \frac{W_A}{M_A}} \quad \ldots\text{(ii)}$$

Given, vapour pressure is reduced to 80% when non-volatile solute is dissolved in octane. It means

If $p_A^\circ = 1$ atm then $p_S = 0.8$ atm;

$p_A^\circ - p_S = 0.2$ atm;

M_A (C_8H_{18}) = 114 g mol^{-1}; $W_A = 114$ g;

$M_B = 40$ g mol^{-1}; W_B = ?

On applying Eq. (i) and (ii), we get,

$$\frac{0.2}{1} = \frac{W_B/40}{\frac{W_B}{40} + \frac{114}{114}} = \frac{W_B/40}{\frac{W_B}{40} + 1}$$

$$0.2 = \frac{W_B}{W_B + 40}$$

$$0.2W_B + 8 = W_B$$

$$W_B = 10\text{ g}$$

102. *(b)* Packing % in lattice

$$= \frac{\text{Total volume of spheres}}{\text{Volume of the unit cell}}$$

Packing fraction of ccp $= \frac{\pi}{3\sqrt{2}} = 0.74 \Rightarrow 74\%$

% free space = 100 − packing fraction %

% free space in ccp = 26%

Packing fraction of bcc $= \frac{\pi\sqrt{3}}{8} = 0.68$

$= 68\%$

% free space in bcc = 32%

103. *(a)*

[1-methylcyclohexene] —(Cold, MnO_4^-/OH^-)→ [1-methylcyclohexane-1,2-diol: CH_3, OH, OH] —(CrO_3, AcOH)→ [2-hydroxy-2-methylcyclohexanone: CH_3, OH, O]

Only 2° alcohol is oxidised

Alkenes on treatment with cold alkaline $KMnO_4$ gives diol. CrO_3 is a selective oxidising agent which only oxidises 2° alcohol.

104. *(c)* Buffer solutions are aqueous solutions consisting of a mixture of a weak acid and its conjugate base or *vice-versa*.

When the concentration of NH_4OH (weak base) is higher than the strong acid (HCl), a mixture of weak base and its conjugate acid is obtained, which acts as basic buffer.

	NH_4OH	+ HCl ⟶	NH_4Cl	+ H_2O
Initially	0.1 M	0.05 M	0	0
After reaction	0.05 M	0	0.05	0

105. *(c)* In BF_3 there is significant $p\pi$-$p\pi$ interaction (back bonding) between unshared p-orbital (having no electron) of boron and the lone pair of electron over fluorine in $2p$-orbital. This provides extra stability and thus bond dissociation energy of B — F in BF_3 is more than C — F in CF_4, where no such back bonding exist.

[Resonance structures of BF_3: $F_2B—F \longleftrightarrow F_2B{=}\overset{+}{F} \longleftrightarrow \overset{+}{F}{=}B(F)—F \longleftrightarrow$ F, $\overset{+}{F}{=}$B—F]

106. *(b)* $E_{Mg^{2+}/Mg} = E^\circ_{Mg^{2+}/Mg} - \frac{0.059}{2}\log\frac{1}{Mg^{2+}}$

Or $E_{Mg^{2+}/Mg} = E^\circ_{Mg^{2+}/Mg} + \frac{0.0591}{2}\log Mg^{2+}$

The given equation is of $y = mx + c$ type, which is equation of straight line.

So, graph of $E_{Mg^{2+}|Mg}(y)$ *vs* log $[Mg^{2+}](x)$ is a straight line with slope $\frac{0.059}{2}$ and intercept $E^\circ_{Mg^{2+}|Mg}$.

107. *(a)* Formation of α, β-unsaturated carboxylic acid by the action of acetic anhydride and sodium acetate on aromatic aldehyde is known as **Perkin reaction**. The other Perkin like condensation involves condensation of aromatic aldehyde and α-hydrogen containing compound. The final product X of the given reaction is

$CH_2(COOEt)_2 \xrightarrow{C_2H_5O^-} \bar{C}H(COOEt)_2$ + [salicylaldehyde: O=C(H)–C₆H₄–HO] ⟶ [$C_6H_4(OH)$–$CH(O^-)$—$CH(COOEt)_2$] $\xleftarrow[H^+]{-H_2O}$

$-H_2O$ … COOEt … O H C … HO O … COOEt … (X)

108. *(d)* Since, the compound *A* gives white precipitate with $AgNO_3$, which is soluble in dilute ammonia solution, it contains at least one Cl atom outside the coordination sphere. Thus, its formula (*A*) is $[Cr(NH_3)_4BrCl]Cl$.

$$[Cr(NH_3)_4BrCl]Cl \longrightarrow [Cr(NH_3)_4BrCl]^+ + Cl^-$$

$$AgNO_3 + Cl^- \longrightarrow \underset{\text{White ppt.}}{AgCl} + NO_3^-$$

Further, the compound *B* on reaction with $AgNO_3$ produces a pale yellow precipitate soluble in conc. NH_3 solution.

Thus, '*B*' is $[Cr(NH_3)_4Cl_2]Br$

$$[Cr(NH_3)_4Cl_2]Br \longrightarrow [Cr(NH_3)_4Cl_2]^+ + Br^-$$

$$AgNO_3 + Br^- \longrightarrow \underset{\text{Pale yellow ppt.}}{Ag\,Br} + NO_3^-$$

109. *(b)* On reaction with Cl_2, phosphorus forms two halides PCl_5 (*A*) which is yellowish powder and PCl_3 (*B*) is colourless oily liquid.

'*A*' is PCl_5 and '*B*' is PCl_3.

$$P_4 + 10Cl_2 \longrightarrow 4PCl_5$$

$$P_4 + 6Cl_2 \longrightarrow 4PCl_3$$

When '*A*' and '*B*' are hydrolysed.

(a) $$\underset{\substack{[A]\\ \text{Phosphorus pentachloride}}}{PCl_5} + 4H_2O \longrightarrow \underset{\text{Phosphoric acid}}{H_3PO_4} + 5HCl$$

(b) $$\underset{\substack{[B]\\ \text{Phosphorus trichloride}}}{PCl_5} + 3H_2O \longrightarrow \underset{\text{Phosphoric acid}}{H_3PO_3} + 3HCl$$

110. *(a)* In step 1, when phenol reacts with chloroform in presence of NaOH then salicyaldehyde is formed. This reaction is known as **Reimer-Tiemann reaction**

In step 2, the formed salicylaldehyde then undergoes **Cannizzro reaction** with formaldehyde in presence of strong base to give an oxidised product (minor) and reduced product (major).

OH … + $CHCl_3$+NaOH $\xrightarrow[\text{Reimer-Tiemann reaction}]{H^+}$ OH, CHO + HCHO $\xrightarrow{\text{Cannizzaro reaction}}$ OH (Major) + CH_2OH + HCOONa (Minor)

111. *(c)* $N = \dfrac{n_1 n_2}{m_2}$

(N = population size of rats in the field)

Here, $n_1 = 80$, $m_2 = 20$, $n_2 = 100$

Hence, $N = \dfrac{80 \times 100}{20} = \dfrac{8000}{20} = 400$

112. *(c)* First prey were eliminated from the clear fluid medium by the predators and then afterwards the predators starved to death in the absence of prey. The sand sediment served as a refuge for the prey from where they emerged later, increased its number *via* reproduction and finally reached the stable equilibrium.

113. *(c)* It is already mentioned that the couple has a child with blood group O and the child is diseased. Thus, the parents are heterozygotes for the diseased allele (as parents are normal and the disease is recessively inherited) and for the blood group. The probability of O blood group in progeny is

$$I^Ai \times I^Bi$$
$$\downarrow$$
$$I^AI^B I^Ai I^Bi\ ii$$

i.e. $\frac{1}{4}$ individual is blood group O.

Similarly, for the disease to be inherited, the individuals should have both the recessive alleles, thus $Dd \times Dd = dd$ (probability to get the disease is $\frac{1}{4}$). Therefore, probability that the second child has the disease = P (the child is blood group O) × P (the child has both the recessive alleles). Thus, the P(diseased child) $= \frac{1}{4} \times \frac{1}{4} = \frac{1}{16}$.

114. *(c)* Upon digestion with *Hind* III:

For allele Q, there is the presence of 3 restriction sites, 2 restriction fragments will be formed.

For allele q, there is the presence of 2 restriction sites, 1 large restriction fragment will be formed, resulting in a band that is nearest to the well. From the analysis, there is only 1 band (nearest to the well) observed. This indicates that individual Z carries allele q. The band is more intense as there is twice as much DNA in it proving that individual Z is a homozygote carrying two copies of the allele (*qq*).

115. *(d)* The presence of U indicates a RNA virus. A and G are purines while U and C are pyrimidines. When ratio of purines: pyrimidines = 1, this suggests that there is complementary base pairing between the bases. Thus, this must be double-stranded RNA virus.

116. *(c)* Since, the *y*-axis is time, the graph obtained would be a vertically inverted reaction rate *vs* temperature graph. As the temperature increases, the time needed decreases at first until a minimum because an increased number of molecules have sufficient energy to pass over the energy barrier and react to form products. A further increase in temperature results in a more rapid increase of time required because of temperature induced denaturation of the enzyme.

117. *(c)* Palaeozoic era is divided into the periods, Cambrian, Ordovician, Silurian, Devonian, Carboniferous and Permian from past to recent, respectively. The' mesozoic era is divided into three periods, Triassic, Jurassic and Cretaceous from past to recent, respectively.

118. *(d)* **Integrin** Transmembrane cell adhesion proteins that act as extracellular matrix receptors.

Cadherin Mediates Ca^{2+} dependent strong homophilic cell-cell adhesion.

Ig super family Contains extracellular Ig like domains and are mainly involved in the fine tuning of cell-cell adhesive interaction during development and regeneration.

Selectin Lectins that mediate a variety of transient cell-cell adhesion interactions in the bloodstream.

119. *(b)* Daughter 4 :

XX^h, XX where X^h is the gene for haemophilia.

Probability of her being a carrier = 0.5

Normal male XY

X^hX — XY

XX | X^hX Carrier | XY | X^hY Haemophilia

Probability of a haemophilic = 0.25

Therefore, probability of daughter 4 having a haemophilic child

= 0.5 × 0.25 = 0.125

120. *(d)* The total amount of energy used in forming 38 ATP = 38 × 30.6 = 1162. 8 kJ

Thus, the efficiency of aerobic respiration is

$$\frac{1162.8}{2880} \times 100\% = 40\%$$

PRACTICE SET 3

Stream : SB/SX

MM : 160

Instructions

1. There are 120 questions in this paper.
2. The question paper contains two parts; Part I (1 Mark Questions) and Part II (2 Marks Questions).
3. There are four sections in each part; Mathematics, Physics, Chemistry and Biology.
4. Out of the four options given with each question, only one is correct.

PART-I (1 Mark Questions)

MATHEMATICS

1. Let $a, b, p, q \in Q$ and suppose that $f(x) = x^2 + ax + b = 0$ and $g(x) = x^3 + px + q = 0$ have a common irrational roots. Then,

(a) $f(x)$ divides $g(x)$ (b) $g(x) = xf(x)$
(c) $g(x) = (x - b - q) f(x)$ (d) None of these

2. If $0 < \theta, \phi < \pi/2$ and $x = \sum_{n=0}^{\infty} \sin^{2n}\theta$, $y = \sum_{n=0}^{\infty} \cos^{2n}\phi$ and

$z = \sum_{n=0}^{\infty} \cos^n(\theta + \phi)\cos^n(\theta - \phi)$, then

(a) $xyz + 1 = yz - zx$ (b) $xyz - 1 = yz + zx$
(c) $xyz - xy = yz - zx$ (d) $xyz + 1 = yz + zx$

3. Solution set of the inequality $\log_3(x+2)(x+4) + \log_{1/3}(x+2) < \frac{1}{2}\log_{\sqrt{3}} 7$ is

(a) $(-2, -1)$ (b) $(-2, 3)$
(c) $(-1, 3)$ (d) $(3, \infty)$

4. The number of ways in which a mixed double game can be arranged from 9 married couples if no husband and wife play in the same game is

(a) 756 (b) 1512
(c) 3024 (d) None of these

5. Let $f(x) : \dfrac{\tan[e^2]x^3 - \tan[-e^2]x^3}{\sin^3 x}$, $x \neq 0$ ([·] represents greatest integer function). The value of $f(0)$ for which $f(x)$ is continuous, is

(a) 15 (b) 12 (c) -12 (d) 14

6. If $I = \int_0^{\pi/2} \cos^n x \sin^n x\, dx = \lambda \int_0^{\pi/2} \sin^n x\, dx$, then λ equals

(a) 2^{-n+1} (b) 2^{n+1} (c) 2^{-n-1} (d) 2^{-n}

7. If $x = \tan 27\theta - \tan\theta$ and $y = \dfrac{\sin\theta}{\cos 3\theta} + \dfrac{\sin 3\theta}{\cos 9\theta} + \dfrac{\sin 9\theta}{\cos 27\theta}$, then

(a) $x = y$ (b) $x + y = 0$ (c) $x = 2y$ (d) $y = 2x$

8. C is the centre of ellipse $\frac{x^2}{16}+\frac{y^2}{9}=1$ and A and B are two points on the ellipse such that $\angle ACB=90°$, then $\frac{1}{CA^2}+\frac{1}{CB^2}$ is equal to

(a) $\frac{25}{144}$ (b) $\frac{144}{25}$
(c) $\frac{7}{12}$ (d) $\frac{12}{7}$

9. The shortest distance from the line $3x+4y=25$ to the circle $x^2+y^2=6x-8y$ is equal to

(a) 7/5 (b) 9/5
(c) 11/5 (d) 32/5

10. Let $g(x)=\cos x^2$, $f(x)=\sqrt{x}$ and α, β $(\alpha<\beta)$ be the roots of the quadratic equation $18x^2-9\pi x+\pi^2=0$. Then, the area (in sq units) bounded by the curve $y=(gof)(x)$ and the lines $x=\alpha$, $x=\beta$ and $y=0$ is

(a) $\frac{\sqrt{3}+1}{2}$ (b) $\frac{\sqrt{3}-\sqrt{2}}{2}$ (c) $\frac{\sqrt{2}-1}{2}$ (d) $\frac{\sqrt{3}-1}{2}$

11. Let O be an interior point of ΔABC such that $\mathbf{OA}+2\mathbf{OB}+3\mathbf{OC}=\mathbf{O}$. Then, the ratio of area of ΔABC to area of ΔAOC is

(a) 1 : 3 (b) 3 : 1 (c) 2 : 3 (d) 3 : 2

12. Let $f(x)=\int_0^x \sin(t^2-t+x)\,dt$, then the value of $f''\left(\frac{-1}{2}\right)+f\left(\frac{-1}{2}\right)$ is

(a) 0 (b) 1 (c) $\frac{1}{2}$ (d) $-\frac{1}{2}$

13. The positive integer k for which $\frac{k^2}{3k^3+500}$ is a maximum is

(a) 5 (b) 6 (c) 7 (d) 8

14. The radius of an arc given by locus of z if $\arg\left(\frac{z-4i}{z-3}\right)=\frac{\pi}{3}$ is

(a) $5\sqrt{3}$ (b) $3\sqrt{5}$ (c) $\frac{5}{\sqrt{3}}$ (d) $\frac{5}{2}$

15. In a ΔABC, bisector of $\angle C$ meets the side AB at D and circumcentre at E. The maximum value of $CD\cdot DE$ is equal to

(a) $\frac{a^2}{4}$ (b) $\frac{b^2}{4}$ (c) $\frac{c^2}{4}$ (d) $\frac{(a+b)^2}{4}$

16. The number of integral values of k for which the equation $7\cos x+5\sin x=2k+1$ has a solution is

(a) 4 (b) 8 (c) 10 (d) 12

17. A pair of unbiased dice is rolled together till a sum of either 5 or 7 is obtained. The probability that 5 comes before 7 is

(a) 2/5 (b) 3/5 (c) 4/5 (d) $\frac{1}{5}$

18. The number of continuous function $f:[0,1]\to[0,1]$, such that $f(x)<x^2$ for all x and $\int_0^1 f(x)\,dx=\frac{1}{3}$ is

(a) 0 (b) 1
(c) 2 (d) infinite

19. Let X be a set of 5 elements. The number d of ordered pairs (A, B) of subsets of X such that $A\neq\phi$, $B\neq\phi$, $A\cap B=\phi$ satisfies

(a) $50\leq d\leq 100$ (b) $101\leq d\leq 150$
(c) $151\leq d\leq 200$ (d) $201\leq d$

20. If $f(x)=(x+2019)^n$, where x is a real variable and n is a positive integer, then the value of $f(0)+f'(0)+\frac{f''(0)}{2!}+\frac{f'''(0)}{3!}+\ldots+\frac{f^{n-1}(0)}{(n-1)!}$ is

(a) $(2019)^n$ (b) $(2020)^n$
(c) $(2020)^n-1$ (d) $n(2019)^n$

PHYSICS

21. A rod mass m and length R is held very close to surface of earth (radius R) radially.
Weight of the rod is

(a) $\frac{mg}{3}$ (b) $\frac{2}{3}mg$ (c) $\frac{mg}{2}$ (d) $\frac{mg}{4}$

22. A particle starts from origin with speed u and angle α with horizontal. Its velocity is perpendicular to initial velocity after time

(a) $\frac{u}{g}\sin\alpha$ (b) $\frac{u}{g}\sec\alpha$
(c) $\frac{u}{g}\cos\alpha$ (d) $\frac{u}{g}\operatorname{cosec}\alpha$

23. A homogeneous circular plate of radius r has a circular hole as shown below.

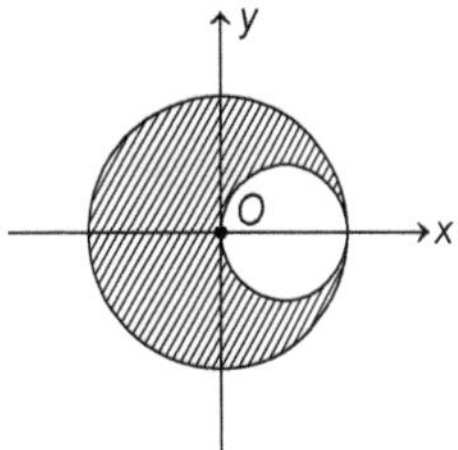

Coordinates of centre of mass of plate with respect to given origin is

(a) (0, 0) (b) $\left(-\frac{r}{2}, 0\right)$ (c) $\left(-\frac{r}{6}, 0\right)$ (d) $\left(-\frac{r}{6}, \frac{r}{6}\right)$

24. A system with two 3 kg mass particles having velocities $2\hat{\mathbf{i}}+3\hat{\mathbf{j}}\ \text{ms}^{-1}$ and $4\hat{\mathbf{i}}-6\hat{\mathbf{j}}\ \text{ms}^{-1}$ is subjected to a constant force of $24\hat{\mathbf{i}}$ N for 5 s.
Final velocity of centre of mass is

(a) $3\hat{\mathbf{i}}-1.5\hat{\mathbf{j}}\ \text{ms}^{-1}$ (b) $60\hat{\mathbf{i}}-30\hat{\mathbf{j}}\ \text{ms}^{-1}$
(c) $20\hat{\mathbf{i}}-3\hat{\mathbf{j}}\ \text{ms}^{-1}$ (d) $23\hat{\mathbf{i}}-1.5\hat{\mathbf{j}}\ \text{ms}^{-1}$

25. A block of mass 0.5 kg is placed over top of a wedge of mass 50 kg.

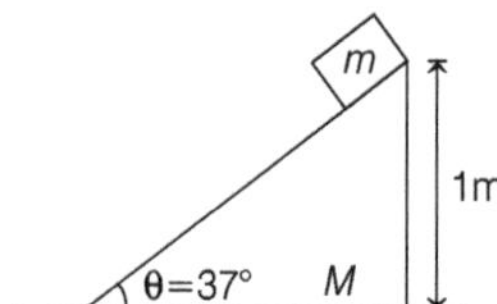

Angle of inclination θ is 37° with horizontal. Block is released at $t = 0$ and it slide down the incline and reaches bottom with a speed of $1.2625\ ms^{-1}$.

There is no friction. Speed of the wedge with respect to ground, when block reaches at bottom is

(a) $0.01\ ms^{-1}$ (b) $0.02\ ms^{-1}$
(c) $0.03\ ms^{-1}$ (d) $0.05\ ms^{-1}$

26. A small block of mass 1 kg is pushed with a velocity $2\hat{\mathbf{i}}\ ms^{-1}$ over a horizontal plane which itself is moving with a velocity of $-2\hat{\mathbf{j}}\ ms^{-1}$. Coefficient of friction between block and plane is 0.25. Force of friction acting on the block (in newton) nearly is

(a) $1.7(\hat{\mathbf{i}} + \hat{\mathbf{j}})$ (b) $1.7(-\hat{\mathbf{i}} - \hat{\mathbf{j}})$
(c) $1.7(\hat{\mathbf{i}} - \hat{\mathbf{j}})$ (d) $1.7(-\hat{\mathbf{i}} + \hat{\mathbf{j}})$

27. A pendulum bob is released from angular position θ, such that the magnitude of its initial acceleration and acceleration at lowest position are equal.

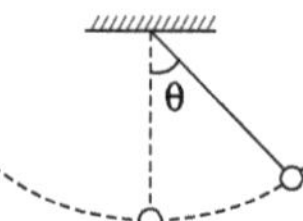

The angular position θ is

(a) $\theta = \cos^{-1}\left(\frac{3}{5}\right)$ (b) $\theta = \sin^{-1}\left(\frac{3}{5}\right)$
(c) $\theta = \tan^{-1}\left(\frac{3}{5}\right)$ (d) $\theta = \cos^{-1}\left(\frac{2}{5}\right)$

28. A metal sphere having radius r, charged to a potential V is enveloped by a thin walled conducting spherical shell of radius $4r$ asymetrically.

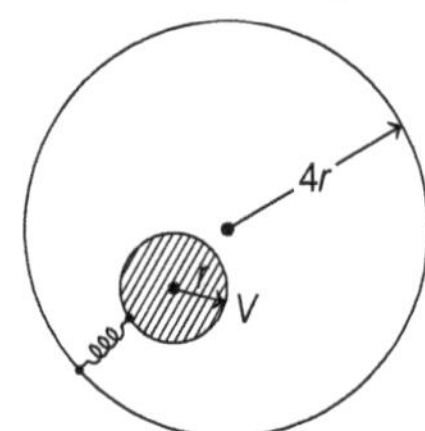

Sphere is connected with shell by a conducting wire for a short time, then connection is removed.

Potential of sphere after removing connecting wire is

(a) V (b) $\frac{V}{2}$ (c) $4V$ (d) $\frac{V}{4}$

29. In steady state, heat conduction equations governing heat flow are identical to the electric current flowing through a conductor. What are the dimensions of quantity analogous to electrical resistance?

(a) $[ML^{-2}TK]$ (b) $[M^{-1}L^{-2}TK]$
(c) $[ML^{-2}T^{-3}K]$ (d) $[M^{-1}L^{-2}T^{3}K]$

30. A rectangular glass wedge is partially dipped in water $\left(n_{glass} = \frac{3}{2}\right)$. A beam of light entring face AB normally reaches entirely to AC.

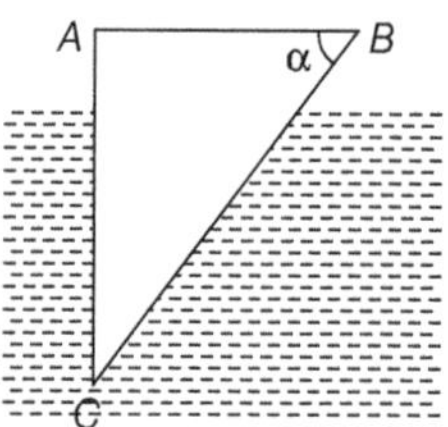

Now, value of angle α must be

(a) less than 60° (b) less than 30°
(c) more than 60° (d) between 30° to 60°

31. A thick transparent slab is made such that its refractive index changes from n_1 to n_2 nearly linearly. This slab is used by a student to measure lateral displacement in lab. She draw following diagram using four pins. Angle β must be

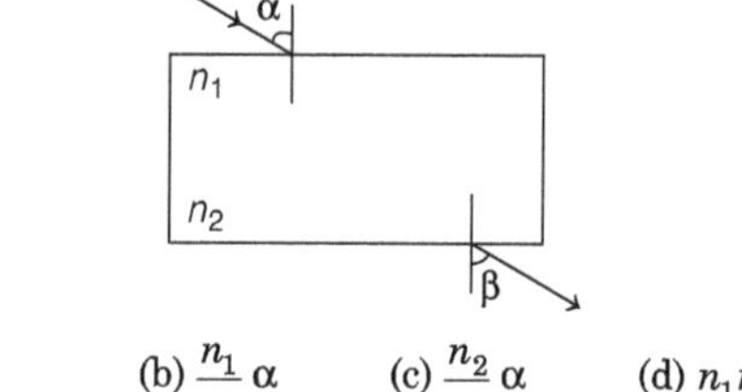

(a) α (b) $\frac{n_1}{n_2}\alpha$ (c) $\frac{n_2}{n_1}\alpha$ (d) $n_1 n_2 \alpha$

32. The word "excellent" is written on a sheet of white paper with a red pencil and the word "good" is written with a green pencil, on a school notice board. A student look at this notice board through a green and red piece of glass.

(a) He can see "excellent" through red glass
(b) He can see "excellent" through green glass
(c) He can see "excellent" through both glasses
(d) He cannot see "excellent" through any of the glasses

33. In an experiment of suspending spherical plastic balls in vertical turbulent jets of air and water it is found that if sphere is displaced slightly to left or right (from its suspended position), the air or water stream again pushes it back to its original position.

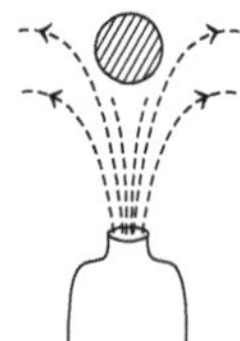

This can be explained by using

(a) Archemedes' principle (b) Bernoulli's principle
(c) Pascal's law (d) Stoke's theorem

34. If 8×10^{10} atoms of Radon are separated from Radium, how many disintegrations will occur in 11.46 days?

(a) 8×10^{10} (b) 8×10^{6} (c) 7×10^{10} (d) 7×10^{9}

35. From examining binding energy per nucleon *versus* mass number curve shown below.

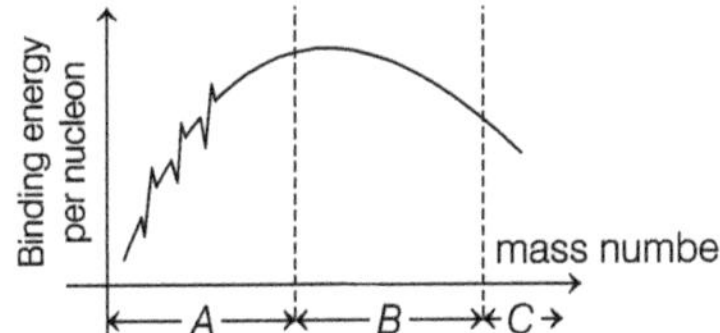

Saturation property of nuclear force is concluded from region.

(a) *A* only (b) *A* and *C* only
(c) *B* only (d) *B* and *C* only

36. A piston moves upwards by 5 cm, when 200 J of heat is added.

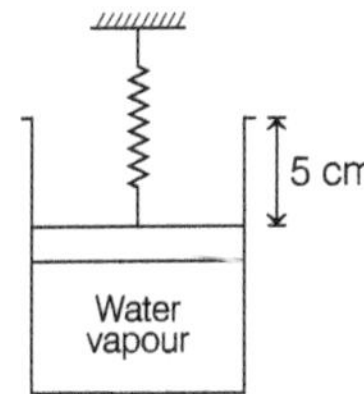

Spring has a spring constant of 50 kNm^{-1} and is initially unstretched. Mass of piston is 60 kg and its diameter is 20 cm. Change of internal energy of vapour is

(a) 200 J (b) 280 J (c) 50 J (d) −50 J

37. A piston cylinder contains a liquid vapour mixture of water at 300 K.

During a constant vapour process, 750 kJ of heat is transferred to water.

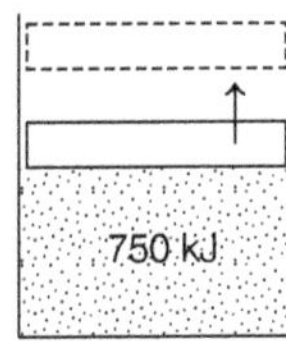

Now, choose the correct option.

(a) Entropy change of water in process is 2.5 kJ/K
(b) Entropy change of water in process is −2.5 kJ / K
(c) Entropy change of water is zero
(d) Entropy change of system is 2.5 kJ / K

38. Charge q is distributed uniformly through out a non-conducting spherical volume of radius R. Energy of charge distribution is (k = dielectric constant)

(a) $\frac{8q^2}{4\pi \epsilon_0 kR}$ (b) $\frac{q^2}{8\pi \epsilon_0 kR}$
(c) $\frac{30q^2}{\pi \epsilon_0 kR}$ (d) $\frac{3q^2}{20\pi \epsilon_0 kR}$

39. Let λ_1 is the wavelength of an emitted photon in deexcitation of a Bohr's atom without considering recoil of atom. Also, λ_2 is the wavelength of an emitted photon in deexcitation of a Bohr's atom considering recoil of atom.

Then, correct statement is

(a) $\frac{\lambda_1}{\lambda_2} > 1$ (b) $\frac{\lambda_1}{\lambda_2} = 1$ (c) $\frac{\lambda_1}{\lambda_2} > \sqrt{2}$ (d) $\frac{\lambda_1}{\lambda_2} < 1$

40. Consider decay process.

$$^{232}_{92}U \longrightarrow ^{228}_{90}Th + ^{4}_{2}He$$

masses involved are

$$m(^{232}U) = 232.1095 \text{ u}$$
$$m(^{228}Th) = 228.0998 \text{ u}$$
$$m(^{4}He) = 4.0039 \text{ u}$$

Kinetic energy of emitted α-particle is

(a) always 5.30 MeV
(b) always less than 5.30 MeV
(c) always greater than 5.30 MeV
(d) either 5.30 MeV or less

CHEMISTRY

41. Which of the following oxides is amphoteric in character?

(a) SnO_2 (b) SiO_2
(c) CO_2 (d) CaO_2

42. When salicylic acid is distilled with Zn-dust, the product obtained is

(a) zinc salicylate (b) salicyl aldehyde
(c) phenol (d) benzoic acid

43. The rate constant of the reaction $A \rightarrow B$ is 0.6×10^{-3} $mole^{-1}$ per second. If the concentration of A is 5 M then concentration of B after 20 min is

(a) 1.08 M (b) 3.60 M
(c) 0.36 M (d) 0.72 M

44. The elevation in boiling point, when 13.44 g of freshly prepared $CuCl_2$ are added to one kilogram of water, is [Some useful data, $K_b = 0.52$ K kg mol^{-1}, molecular weight of $CuCl_2$ = 134.4 g].

(a) 0.05 (b) 0.1 (c) 0.16 (d) 0.21

45. Which of the following reaction is an example of a redox reactions?

(a) $XeF_4 + O_2F_2 \longrightarrow XeF_6 + O_2$
(b) $XeF_2 + PF_5 \longrightarrow [XeF]^+ PF_6^-$
(c) $XeF_6 + PF_5 \longrightarrow XeOF_4 + 2HF$
(d) $XeF_6 + 2H_2O \longrightarrow XeO_2F_2 + 4HF$

46. A metal crystallises into two cubic phases, face centred cubic (fcc) and body centred cubic (bcc), whose unit cell lengths are 3.5 and 3.0 Å, respectively. Calculate the ratio of densities of fcc and bcc.

(a) 1.26 (b) 3.25 (c) 7.8 (d) 5.35

47. Concentrated nitric acid upon long standing, turns yellow-brown due to the formation of

(a) NO (b) NO_2
(c) N_2O (d) N_2O_4

48. The standard reduction potential values of three metallic cations, *X, Y, Z* are 0.52, – 3.03 and – 1.18 V respectively. The order of reducing power of the corresponding metals is

(a) $Y > Z > X$ (b) $X > Y > Z$
(c) $Z > Y > X$ (d) $Z > X > Y$

49. The rate of a first order reaction is 0.04 mol L^{-1} s^{-1} at 10 min and 0.03 mol L^{-1} s^{-1} at 20 min. Thus, half-life of the reaction is

(a) 24.1 min (b) 34.8 min
(c) 50.1 min (d) 48.2 min

50. When phenol is treated with D_2SO_4/D_2O, some of the hydrogens get exchanged. The final product in this exchange reaction is

(a) 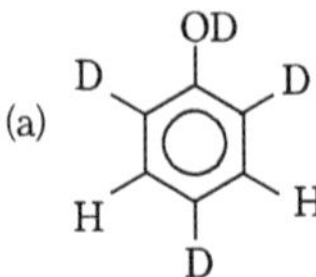

(b) OD, D, H, H, H, D

(c) OD, D, H, H, D, H

(d) OD, H, H, D, D, H

51. The enthalpy changes for the following processes are listed below :

$$Cl_2(g) \longrightarrow 2Cl(g), 242.3 \text{ kJ mol}^{-1}$$
$$I_2(g) \longrightarrow 2I(g), 151.0 \text{ kJ mol}^{-1}$$
$$ICl(g) \longrightarrow I(g) + Cl(g), 211.3 \text{ kJ mol}^{-1}$$
$$I_2(s) \longrightarrow I_2(g), 62.76 \text{ kJ mol}^{-1}$$

Given that the standard states for iodine and chlorine are $I_2(s)$ and $Cl_2(g)$, the standard enthalpy for the formation of $ICl(g)$ is

(a) –14.6 kJ mol^{-1} (b) –16.8 kJ mol^{-1}
(c) +16.8 kJ mol^{-1} (d) + 244.8 kJ mol^{-1}

52. How many chiral compounds are possible on monochlorination of 2-methyl butane?

(a) 8 (b) 2 (c) 4 (d) 6

53. 2-bromopentane is heated with potassium ethoxide in ethanol. The major product obtained is

(a) pent-1-ene (b) 2-ethoxypentane
(c) *trans*-pent-2-ene (d) *cis*-pent-2-ene

54. Among the following metal carbonyls, the C — O bond order is lowest in

(a) $[Mn(CO)_6]^+$ (b) $[Fe(CO)_5]$
(b) $[Cr(CO)_6]$ (d) $[V(CO)_6]^-$

55. Reaction by which benzaldehyde cannot be prepared is

(a) $C_6H_5CH_3$ + CrO_2Cl_2 in CS_2 followed by H_3O^+

(b) C_6H_5COCl + H_2 in the presence of Pd-$BaSO_4$

(c) C_6H_6 + CO + HCl in the presence of anhy. $AlCl_3$

(d) C_6H_5COOH + Zn/Hg and conc. HCl

56. The polymer obtained by the copolymerisation of styrene and butadiene is

(a) $\left[CH_2—CH{=}CH—CH_2—CH_2—\underset{\underset{CH}{|}}{CH} \right]_n$

(b) $\left[CH_2—CH{=}CH—CH_2—CH_2—\underset{\underset{Ph}{|}}{CH} \right]_n$

(c) $\left[CH_2—\overset{\overset{Cl}{|}}{C}{=}CH—CH_2 \right]_n$

(d) $\left[OC\,(CH_2)_5NH \right]_n$

57. The solubility of $Pb(OH)_2$ in water is 6.7×10^{-6} M. Calculate the solubility of $Pb(OH)_2$ in a buffer solution of pH = 8.

(a) 1.2×10^{-3} M (b) 1.8×10^{-5} M
(c) 3.6×10^{-3} M (d) 4.1×10^{-4} M

58. The hyperconjugative stabilities of *tert*-butyl cation and but-2-ene, respectively, are due to

(a) $\sigma \to p$ (empty) and $\sigma \to \pi^*$ electron delocalisations
(b) $\sigma \to \sigma^*$ and $\sigma \to \pi$ electron delocalisations
(c) $\sigma \to p$ (filled) and $\sigma \to \pi$ electron delocalisations
(d) p (filled) $\to \sigma^*$ and $\sigma \to \pi^*$ electrons delocalisations

59. Which of the following species is not paramagnetic?

(a) NO (b) CO
(c) O_2 (d) B_2

60. The number of spherical nodes in 3*p*-orbital is

(a) 0 (b) 1 (c) 2 (d) 3

BIOLOGY

61. During some surgical operations the drug curare, which has a similar shape to acetylcholine is injected into the muscles to relax them. Why do the muscles remain relaxed?

(a) Calcium ions cannot be taken up by membrane vesicles
(b) Cholinesterase cannot remove acetylcholine
(c) Post-synaptic membrane receptors are blocked
(d) Sodium channels remain open

62. Viroids have
(a) *ss*RNA not enclosed by protein coat
(b) *ss*DNA not enclosed by protein coat
(c) *ds*DNA enclosed by protein coat
(d) *ds*RNA enclosed by protein coat

63. A gymnospermic leaf carries 16 chromosomes. The number of chromosomes in its endosperm is
(a) 24 (b) 16 (c) 12 (d) 8

64. The difference between rough endoplasmic reticulum and smooth endoplasmic reticulum is that rough endoplasmic reticulum
(a) does not contain ribosomes
(b) contains ribosomes
(c) does not transport proteins
(d) transports proteins

65. Identify the polysaccharide with β-glycosidic bond.
(a) Starch (b) Glycogen
(c) Sucrose (d) Cellulose

66. Interphase nucleus is enclosed by
(a) non-porous nuclear membrane
(b) porous double nuclear membrane
(c) non-porous double discontinuous nuclear membrane
(d) a single porous unit membrane

67. Which of the following in guard cell is responsible for opening of stomata?
(a) Decrease in CO_2 concentration and more H^+ ion concentration
(b) Decrease in CO_2 concentration and less H^+ ion concentration
(c) Increase in CO_2 concentration and more H^+ ion concentration
(d) More free H^+ ions and less Cl^- ions

68. In which one of the following reactions, oxidative decarboxylation does not occur?
(a) Malic acid → Pyruvic acid
(b) Pyruvic acid → Acetyl Co-A
(c) Glyceraldehyde → 1,3-bisphosphoglyceric acids-3-phosphate
(d) α-ketoglutaric acid → Succinyl Co-A

69. In 24 h, total glomerular filtrate formed in human kidney is
(a) 1.7 L (b) 7 L
(c) 17 L (d) 170 L

70. Damage to thymus in a child may lead to
(a) a reduction in haemoglobin content of blood
(b) a reduction in stem cell production
(c) loss of antibody-mediated immunity
(d) loss of cell-mediated immunity

71. If mammalian ovum fails to get fertilised, which one of the following is unlikely?
(a) Corpus luteum will disintegrate
(b) Oestrogen secretion further decreases
(c) Primary follicle starts developing
(d) Progesterone secretion rapidly declines

72. Genotypic and phenotypic ratio in monohybrid cross remain same in the case of
(a) sex-linked genes
(b) pseudoallelic genes
(c) intermediate inheritance
(d) dominant and recessive genes

73. In *lac* operon, if mutation occurs in the middle gene of the structural gene, then
(a) β-galactosidase will not be synthesised
(b) permease will not be synthesised
(c) transacetylase will not be synthesised
(d) lactose digestion will be rapid

74. Which option is correct for the amino acid and the total number of their genetic code?
(a) Val = 6, Pro = 6 (b) Arg = 6, His = 6
(c) Pro = 4, Thr = 4 (d) Thr = 4, Arg = 4

75. Origin of life as a result of chemical evolution has been properly explained by or the most logical biochemical theory of origin of life has been given by
(a) Stanley Miller (b) Darwin
(c) AI Oparin (d) S Fox

76. Which cell secretes antibody?
(a) Lymphocytes (b) Monocytes
(c) Eosinophils (d) Neutrophils

77. Dipsomania is
(a) a type of depression
(b) insanity due to psychosis
(c) severe addiction to alcohol
(d) tendency to take much water

78. Which of the following is incorrectly matched?
(a) *Indigofera* — Dye (b) *Sesbania* — Fodder
(c) *Petunia* — Fumigatory (d) *Aloe* — Medicine

79. If 20J of energy is trapped at producer level, then how much energy will be available to peacock as food in the following chain?
Plant → Mice → Snake → Peacock
(a) 0.02 J (b) 0.002 J
(c) 0.2 J (d) 0.0002 J

80. Which of the following metals is a water pollutant and causes sterility in human being?
(a) As (b) Mn (c) Mg (d) Hg

PART-II (2 Marks Questions)

MATHEMATICS

81. If $a^2 + b^2 = 7$ and $a^3 + b^3 = 10$, then the maximum value of $|a + b|$ is

(a) 4 (b) 5 (c) 6 (d) 3

82. The area of the portion common to $y = \sin^{-1}(\sin x)$ and $y = [\sin^{-1}(\sin x)]$ in $[0, \pi]$, where [·] denotes the greatest integer function is

(a) $(\pi - 2)^2$ (b) $\frac{(\pi - 2)^2}{2}$

(c) $\frac{(\pi - 2)^2}{4}$ (d) None of these

83. If $I_1 = \int_0^1 \frac{e^x}{x+3} dx$ and $I_2 = \int_0^1 \frac{x^3}{e^{x^4}(4 - x^4)} dx$, then $\frac{I_1}{I_2}$ is

(a) 4 (b) $4e$ (c) $3e$ (d) $2e$

84. If $\sin\alpha + \cos\beta = \frac{1}{\sqrt{2}}$ and $\cos\alpha + \sin\beta = \sqrt{\frac{2}{3}}$, then the value of $\tan\left(\frac{\alpha - \beta}{2}\right)$ is

(a) $3\sqrt{3} - 6$ (b) $\sqrt{3} - 2$

(c) $4\sqrt{3} - 7$ (d) $4\sqrt{3} + 7$

85. The digit in the unit place of $15^{981} + 9^{789} - 6^{50}$ is

(a) 8 (b) 6 (c) 5 (d) 9

86. A die is so loaded that the probability of throwing a number is proportional to K^2. Then the probability that the number 3 appears given that when the die is rolled the number turned up is not even is equal to

(a) $\frac{9}{70}$ (b) $\frac{35}{91}$ (c) $\frac{9}{91}$ (d) $\frac{9}{35}$

87. The circle $x^2 + y^2 - 8x = 0$ and hyperbola $\frac{x^2}{9} - \frac{y^2}{4} = 1$.

The equation of a common tangent with positive slope to the circle as well as to the hyperbola is

(a) $2x - \sqrt{5}y - 20 = 0$ (b) $2x - \sqrt{5}y + 4 = 0$

(c) $3x - 4y + 8 = 0$ (d) $4x - 3y + 4 = 0$

88. Two vertices of a triangle are $(5, -1)$ and $(-2, 3)$ if orthocenter of the triangle is origin, then the coordinate of third vertex is

(a) (4, 7) (b) (3, 7)

(c) (– 4, – 7) (d) (4, – 7)

89. Which of the following intervals is a possible domain of the function $f(x) = \log_{\{x\}}[x] + \log_{[x]}\{x\}$? Where $[x]$ is the greatest integer function and $\{x\}$ is the fractional part of x.

(a) (0, 1) (b) (1, 2)

(c) (2, 3) (d) (3, 5)

90. If $f: R \to R$ is a differentiable function $f(1) = 4$. Then, the value of $\lim_{x \to 1} \int_4^{f(x)} \frac{2t}{x-1} dt$ is

(a) $8f'(1)$ (b) $4f'(1)$ (c) $2f'(1)$ (d) $f'(1)$

PHYSICS

91. Consider the process of pair production.

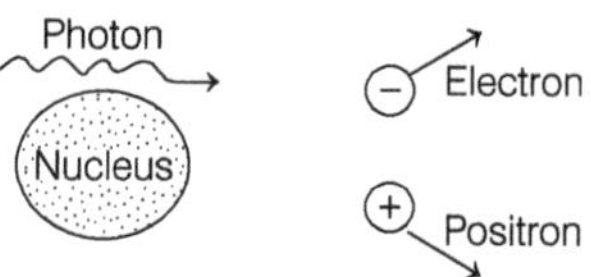

It is possible for a photon to materialize into an electron and a positron. In this way, electromagnetic energy is converted into matter. Both energy and linear momentum are conserved when an electron-positron pair is created near an atomic nucleus.

In absence of nucleus, pair production cannot occur in empty space because

(a) it is impossible to conserve momentum without presence of nucleus

(b) it is impossible to conserve energy without presence of nucleus

(c) Both a and b are correct

(d) pair production is possible even in empty space

92. Particle deaccelerates on a straight line whose magnitude is $|\mathbf{a}| = \alpha\sqrt{v}$, where α = a positive constant. If initial velocity of the particle is v_0, then the distance travelled by particle before it stops is

(a) $\frac{2v_0^{3/2}}{3\alpha}$ (b) $\frac{3v_0^{3/2}}{2\alpha}$ (c) $\frac{v_0^{3/2}}{\alpha}$ (d) $\frac{2\alpha^3}{3v_0}$

93. Two candles of equal heights, 15 cm, each are placed in between vertical screens at a distance of 10 cm from each other and also from the nearer screen.

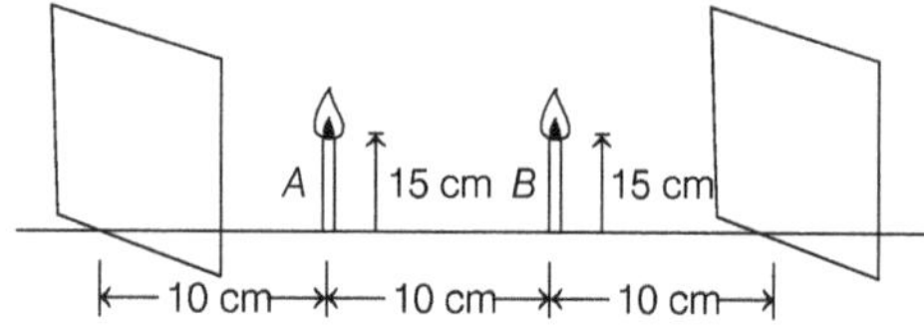

Candle A burns completely in 1.5 h and candle B burns completely in 1 h. Shadows of A and B moves with speeds

(a) $\frac{1}{4}$ cm s^{-1}, $\frac{1}{8}$ cm s^{-1} (b) $\frac{1}{12}$ cm s^{-1}, $\frac{1}{12}$ cm s^{-1}

(c) $\frac{1}{3}$ cm s^{-1}, $\frac{1}{12}$ cm s^{-1} (d) $\frac{1}{3}$ cm s^{-1}, $\frac{1}{4}$ cm s^{-1}

94. $v-t$ graph of a particle is

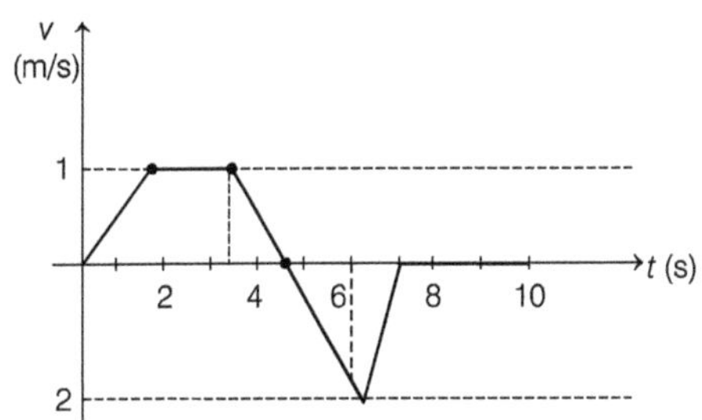

Its distance-time graph is

(a) 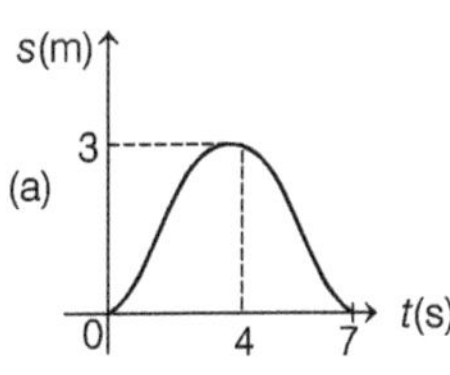

(b) 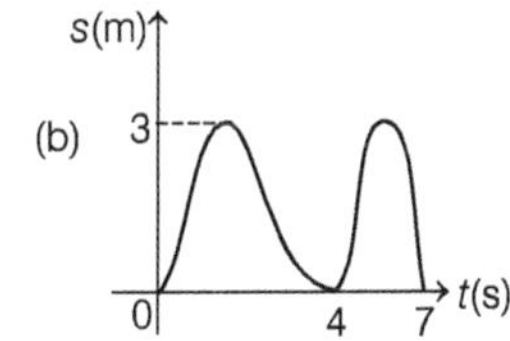

(c) 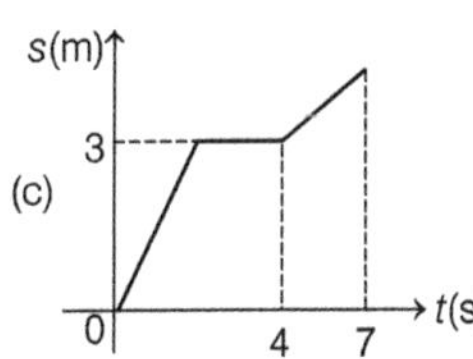

(d) 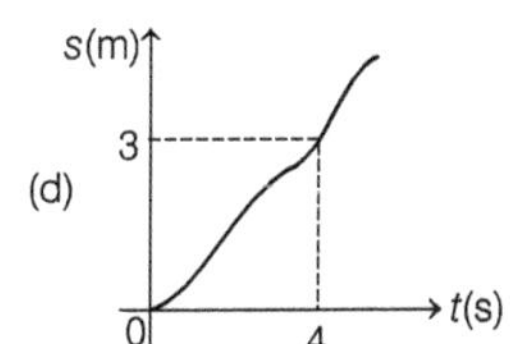

95. Force of repulsion between two charges of $+2\,\mu C$ and $+3\mu C$, separated by a distance of 10 cms^{-1} is F_1. If a dielectric slab of dielectric constant $k=4$ is placed, such that it fills 5 cm distance (as shown), then force between charges will be

(a) $\frac{4}{9}F_1$ (b) $\frac{9}{4}F_1$ (c) $\frac{7}{5}F_1$ (d) $\frac{5}{7}F_1$

96. A point source of light is placed at depth h below the surface of a lake. Fraction of light energy that escapes from the lake depends on

(a) h (b) h^2 (c) h^{-2} (d) h^0

97. A diatomic ideal gas undergoes following two steps process.

***Step* 1** Constant volume heating to triple pressure of gas.

***Step* 2** Constant pressure heating to double volume.

Molar heat capacity of the gas for whole process is

(a) $\frac{5}{4}R$ (b) $\frac{31}{30}R$ (c) $\frac{27}{32}R$ (d) $\frac{30}{25}R$

98. Instantaneous position of a particle is given by

$$x=t^3-12t^2+36t$$

Motion of the particle for first 8 s is best illustrated by

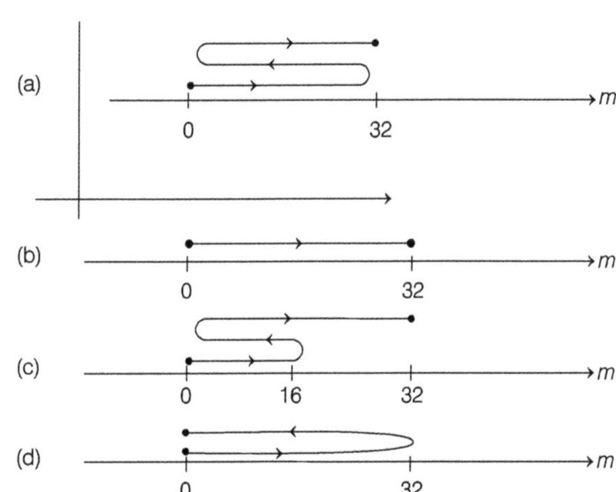

99. A student measures magnitude of electric field due to a charged sphere S by two different ways.

I. He places a very small similarly charged sphere near S and then measures force per unit charge.

II. He places a large similarly charged sphere near S and then measures force per unit charge.

Let his recordings are E_1 and E_2.

(a) $E_2 > E_1$, when charge on S is negative
(b) $E_1 = E_2$
(c) $E_1 > E_2$
(d) $E_2 > E_1$, when charge on S is positive

100. An automatic rotating water sprinkler is mounted over top of a 5 m high pillar in a field.

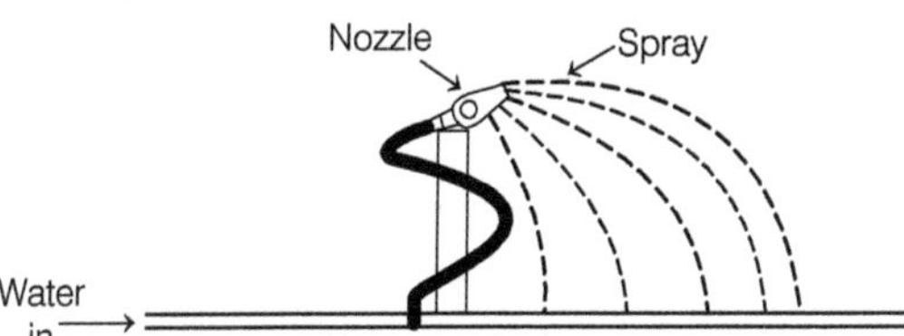

Water is flowing at a rate of $5\times10^{-4}\text{ m}^3$ in 1 s through a nozzle of area 1 cm^2. Maximum possible area of field that can be irrigated by the nozzle is (take, $g=10\text{ ms}^{-2}$)

(a) $15\pi\text{ m}^2$ (b) $25\pi\text{ m}^2$
(c) $100\pi\text{ m}^2$ (d) $75\pi\text{ m}^2$

CHEMISTRY

101. Which of the following statement(s) is/are incorrect?

(a) The coordination number of each type of ion in CsCl crystal is 8.
(b) A metal that crystallises in bcc structure has a coordination number of 12.
(c) A unit cell of an ionic crystal shares some of its ions with other unit cells.
(d) The length of the unit cell in NaCl is 276 pm. ($r_{Na^+}=95\text{ pm}$; $r_{Cl^-}=181\text{ pm}$).

102. Which one of the following complexes shows optical isomerism?

(a) *cis*-$[Co(en)_2Cl_2]Cl$ (b) *trans*-$[Co(en)_2Cl_2]Cl$
(c) *cis*-$[Co(NH_3)_4Cl_2]Cl$ (d) *trans*-$[Co(NH_3)_4Cl_3]$

103. The major product of the following reaction is

NH_2 (benzene ring) $\xrightarrow[\text{2. } Br_2 \quad \text{3. } KOH/C_2H_5OH]{\text{1. } CH_3COCl}$

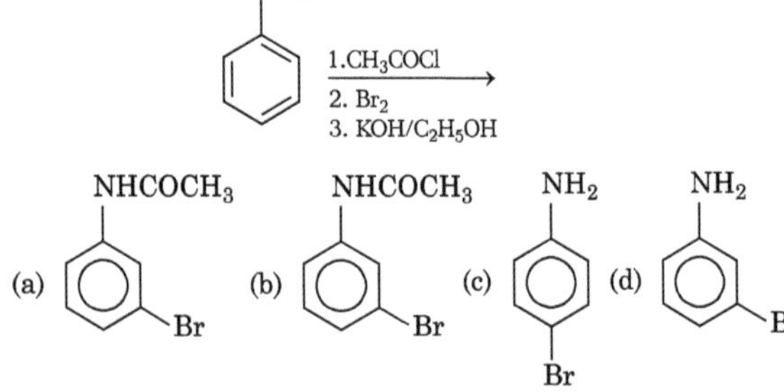

104. On treatment of 100 mL of 0.1 M solution of $CoCl_3.6H_2O$ with excess of $AgNO_3$; 1.2×10^{22} ions are precipitated. The complex is

(a) $[Co(H_2O)_4Cl_2]\,Cl{\cdot}2H_2O$ (b) $[Co(H_2O)_3Cl_3]{\cdot}3H_2O$
(c) $[Co(H_2O)_6]Cl_3$ (d) $[Co(H_2O)_5Cl]\,Cl_2\cdot H_2O$

105. Identify the set of reagents/reaction conditions X and Y in the following set of transformations

$$CH_3—CH_2—CH_2Br \xrightarrow{X} \text{Product} \xrightarrow{Y} CH_3—\underset{\underset{Br}{|}}{CH}—CH_3$$

(a) X = dilute aqueous NaOH
Y = HBr/acetic acid
(b) X = concentrated alcoholic NaOH
Y = HBr/acetic acid
(c) X = dilute aqueous NaOH, $Y = Br_2/CHCl_3$
(d) X = concentrated aqueous NaOH
$Y = Br_2/CHCl_3$

106. For one mole of a van der Waals' gas when $b = 0$ and $T = 300$ K, the pV *vs* $1/V$ plot is shown below. The value of the van der Waals' constant a (atm L mol^{-2})

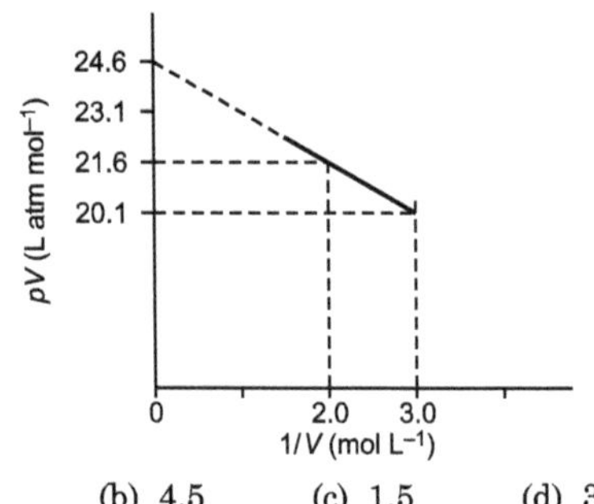

(a) 1.0 (b) 4.5 (c) 1.5 (d) 3.0

107. The molar conductivity of 0.25 mol L^{-1} formic acid is 40.1 S cm^2 mol^{-1}. Calculate its dissociation constant. [Assuming, $\Lambda^\circ_{m(H^+)} = 349.6$ S cm^2 mol^{-1};

$\Lambda^\circ_{(m)\,(HCOO^-)} = 54.6$ S cm^2 mol^{-1}]

(a) 2.73×10^{-3} mol L^{-1} (b) 3.74×10^{-3} mol L^{-1}
(c) 4.75×10^{-3} mol L^{-1} (d) 5.27×10^{-3} mol L^{-1}

108. A hexapeptide has the composition Ala, Gly, Phe, Val. Both the N-terminal and C-terminal units are Val. Cleavage of the hexapeptide by chemotrypsin gives two different tripeptides, both having Val as the N-terminal group. Among the products of random hydrolysis is a Ala-Val dipeptide fragment. What is the primary structure of the hexapeptide?

(a) Val-Galy-Phe-Val-Ala-Val
(b) Val-Ala-Phe-Val-Gly-Val
(c) Val-Gly-Ala-Val-Phe-Val
(d) Val-Phe-Val-Ala-Gly-Val

109. One mole of an organic compound A with the formula C_3H_8O reacts completely with two moles of HI to from X and Y. When Y is boiled with aqueous alkali it forms Z. Z answers the iodoform test. The compound A is

(a) propan-2-ol (b) propan-1-ol
(c) ethoxyethane (d) methoxyethane

110. Consider the following conversion:

$$C_6H_{11}—\overset{\overset{O}{\|}}{C}CH_3 \xrightarrow{?} C_6H_{11}—\overset{\overset{CH_3}{|}}{C}H\underset{\underset{OH}{|}}{C}HCH_2CH_3$$

Which of the following sets of reagents, would successfully accomplish the conversion shown?

(a) $CH_3CH_2CH_2MgBr; H_3O^+; PCC, CH_2Cl_2$
(b) $CH_3CH_2CH_2MgBr; H_3O^+; H_2SO_4$, heat; PCC, CH_2Cl_2
(c) $(C_6H_5)_3\overset{+}{P}—\overset{\bullet\bullet}{\overset{-}{C}}HCH_2CH_3; B_2H_6; H_2O_2, HO^-$
(d) $(C_6H_5)_3\overset{+}{P}—\overset{\bullet\bullet}{\overset{-}{C}}HCH_2CH_3; H_2SO_4, H_2O$

BIOLOGY

111. Match the following columns and choose the correct option from the codes given below.

	Column I		Column II
A.	Sap vacuole	1.	Contains digestive enzyme
B.	Contractile vacuole	2.	Stores metabolic gases
C.	Food vacuole	3.	Osmoregulation
D.	Air vacuole	4.	Stores lipids
E.	Spherosome	5.	Stores and concentrates mineral, salts and nutrients

Codes

	A	B	C	D	E
(a)	5	3	1	2	4
(b)	2	3	4	5	1
(c)	5	2	3	1	4
(d)	5	3	2	4	1

112. The restriction fragment shown below contains a gene whose recessive allele is lethal. The normal allele has restriction sites for restriction enzyme pst I at sites I and II. The recessive allele lacks restriction sites I. An individual who had a sister with the lethal trait is being tested to determine if he is a carrier of that lethal trait.

Which of the band patterns would be produced on a gel if he is a carrier?

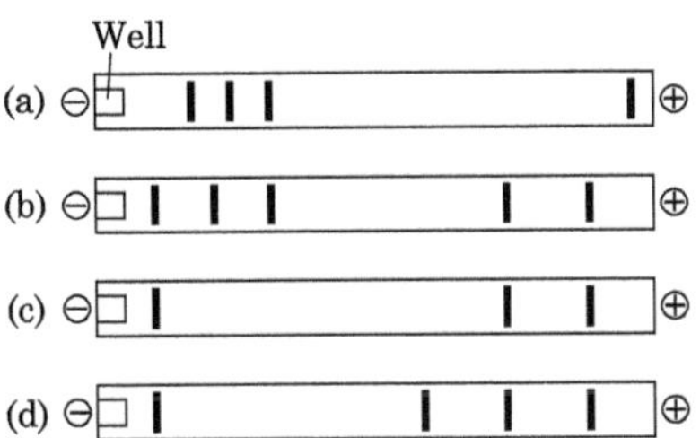

113. The Galapagos islands are a group of volcanic islands in the Eastern Pacific Ocean about 1000 km from South America. Thirteen species of finch are found on the islands, they resemble each other closely but differ in their feeding habit and in the shape of their beaks.

Assuming that an ancestral stock of finches came from the mainland, what is the most likely explanation for the existence of similar but distinct species of Galapagos finches?

(a) Finches developed different kinds of beak in order to feed on different kinds of food
(b) Finches evolved separately according to the habitat in which they settled
(c) Finches from the mainland bred with a resident population of a related species and produced new genotypes
(d) Finches underwent convergent evolution to produce very similar species

114. The diagram below represents the proportions of a population of newborn deer calves falling into various birth weight classes. The graph superimposed on the diagram represents mortality in relation to birth weight.

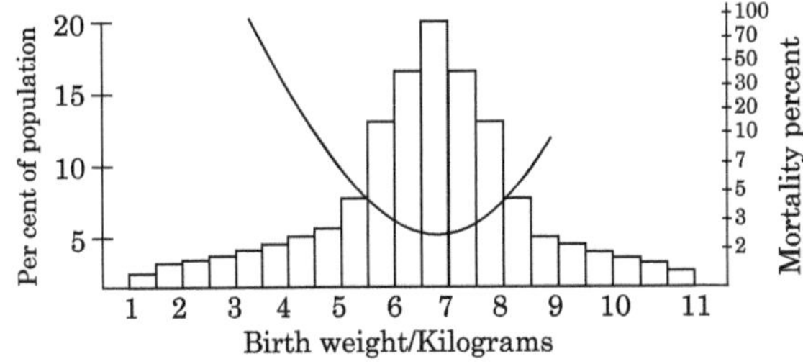

From the information given which one of the following interpretations is correct?

(a) Birth weight is undergoing stabilising selection
(b) Birth weight is an example of discontinuous variation
(c) Birth weight is inversely proportional to mortality
(d) Birth weight is genetically linked to mortality

115. ATP can be formed by oxidative phosphorylation in the electron transport system and by glycolysis.
In the complete oxidation of one molecule of glucose, approximately what percentage of ATP is formed by oxidative phosphorylation?

(a) 10% (b) 25%
(c) 75% (d) 90%

116. The diagram shows the inheritance of a form of breast cancer associated with the presence of just one allele of the autosomal gene BRCA 1.

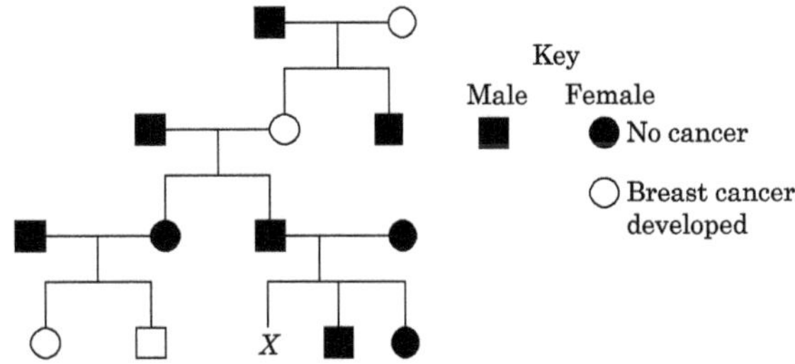

What is the probability that the woman X inherits the BRCA 1 allele associated with breast cancer?

(a) 0.00 (b) 0.25 (c) 0.50 (d) 1.00

117. The diagram represents a sequence of reactions taking place in a bacterium in which amino acids are produced from one another by the action of specific enzymes. Number 1 to 6 represent different amino acids, letters V to Z represent different enzymes. All the amino acids are essential for survival.

$$1 \xrightarrow{V} 2 \xrightarrow{W} 4 \xrightarrow{Y} 6$$
$$2 \xrightarrow{X} 3, \quad 4 \xrightarrow{Z} 5$$

The original strain of the bacterium required only amino acid 1. A mutant strain of this bacterium could only survive when produced with amino acids 1, 2 and 5 in its culture medium.

Which enzymes were missing in the mutant strain?

(a) V and Z (b) W and Z
(c) X and Y (d) V, W and Z

118. The table shows the result of an analysis of percentage concentration of three bases in nucleic acids from four sources. Three of the sources are DNA and one is RNA. Which source is RNA?

	Source	Adenine	Cytosine	Guanine
(a)	A	19.7	30.4	30.2
(b)	B	25.5	24.6	23.8
(c)	C	26.7	28.8	22.3
(d)	D	31.1	18.3	18.7

119. The graph shows the amount of product formed by a standard concentration of enzyme and a standard concentration of substrate at a temperature of 20°C.

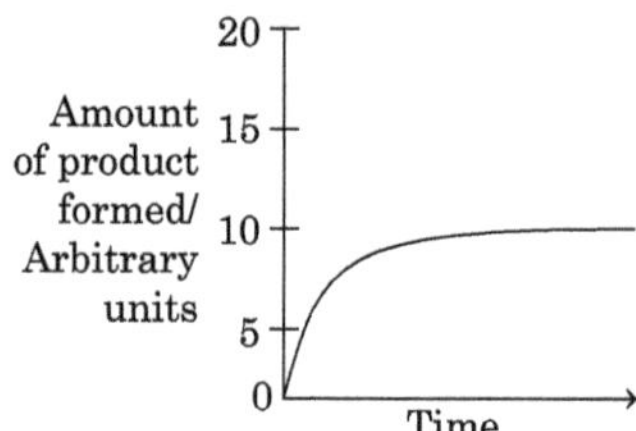

Which graph shows the effect on the activity of the enzyme of decreasing the temperature to 15°C?

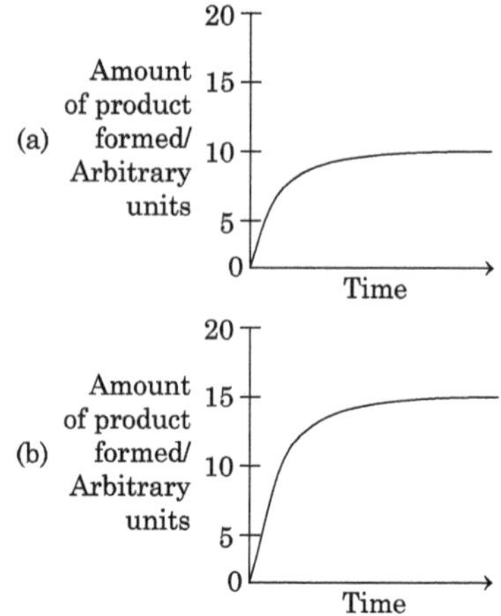

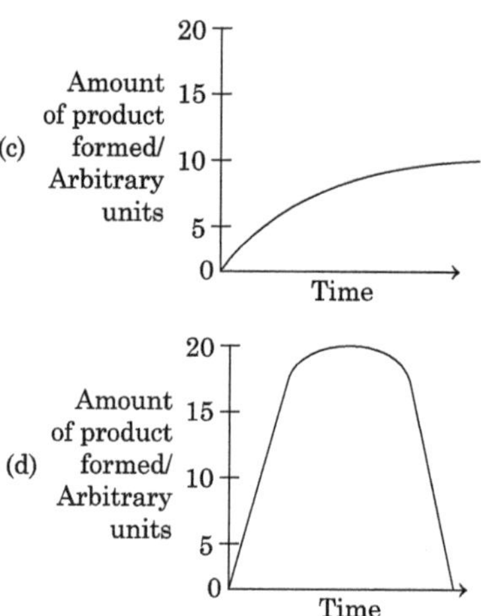

120. Match the five (A– E) group of organisms with their correct taxonomic rank (1-5) and choose the correct option from the codes given below.

Group	Taxonomic rank
A. Crustacea	1. Order
B. Hominidae	2. Domain
C. Dermaptera	3. Class
D. Ctenophora	4. Phylum
E. Archaea	5. Family

Codes

	A	B	C	D	E
(a)	3	1	5	4	2
(b)	1	2	3	4	5
(c)	4	3	2	1	5
(d)	3	5	1	4	2

Answers

PART-I

1 *(a)*	2 *(c)*	3 *(b)*	4 *(b)*	5 *(a)*	6 *(d)*	7 *(c)*	8 *(a)*	9 *(a)*	10 *(d)*
11 *(b)*	12 *(a)*	13 *(c)*	14 *(c)*	15 *(c)*	16 *(b)*	17 *(a)*	18 *(a)*	19 *(c)*	20 *(c)*
21 *(c)*	22 *(d)*	23 *(c)*	24 *(d)*	25 *(a)*	26 *(c)*	27 *(a)*	28 *(d)*	29 *(d)*	30 *(c)*
31 *(a)*	32 *(b)*	33 *(b)*	34 *(c)*	35 *(c)*	36 *(d)*	37 *(a)*	38 *(d)*	39 *(d)*	40 *(d)*
41 *(a)*	42 *(d)*	43 *(d)*	44 *(c)*	45 *(a)*	46 *(a)*	47 *(b)*	48 *(a)*	49 *(a)*	50 *(b)*
51 *(c)*	52 *(b)*	53 *(c)*	54 *(b)*	55 *(d)*	56 *(b)*	57 *(a)*	58 *(a)*	59 *(b)*	60 *(b)*
61 *(c)*	62 *(a)*	63 *(d)*	64 *(b)*	65 *(d)*	66 *(b)*	67 *(b)*	68 *(c)*	69 *(d)*	70 *(d)*
71 *(c)*	72 *(c)*	73 *(b)*	74 *(c)*	75 *(c)*	76 *(a)*	77 *(c)*	78 *(c)*	79 *(a)*	80 *(b)*

PART-II

81 *(b)*	82 *(c)*	83 *(b)*	84 *(c)*	85 *(a)*	86 *(d)*	87 *(b)*	88 *(c)*	89 *(c)*	90 *(a)*
91 *(c)*	92 *(a)*	93 *(c)*	94 *(d)*	95 *(a)*	96 *(d)*	97 *(b)*	98 *(a)*	99 *(c)*	100 *(b)*
101 *(b,d)*	102 *(a)*	103 *(c)*	104 *(d)*	105 *(b)*	106 *(c)*	107 *(a)*	108 *(a)*	109 *(d)*	110 *(c)*
111 *(a)*	112 *(d)*	113 *(a)*	114 *(a)*	115 *(d)*	116 *(a)*	117 *(a)*	118 *(c)*	119 *(c)*	120 *(d)*

Solutions

1. *(a)* Let $\alpha \in R - Q$ be a common root of $f(x) = 0$ and $g(x) = 0$, then,

$$\alpha^2 + a\alpha + b = 0 \Rightarrow \alpha^2 = -a\alpha - b$$

On putting this in $\alpha^3 + p\alpha + q = 0$, we get

$$(a^2 - b + p)\alpha + ab + q = 0$$

As α is irrational and $a, b, p, q \in Q$,

$$p = b - a^2, q = -ab$$

This give, $g(x) = (x - a) f(x)$

$\therefore f(x)$ divides $g(x)$.

2. *(c)* Given,

$$x = \sum_{n=0}^{\infty} \sin^{2n}\theta$$

$$= 1 + \sin^2\theta + \sin^4\theta + \ldots$$

$$\Rightarrow x = \frac{1}{1 - \sin^2\theta} = \frac{1}{\cos^2\theta}$$

$$\Rightarrow y = \sum_{n=0}^{\infty} \cos^{2n}\phi$$

$$= 1 + \cos^2\phi + \cos^4\phi + \ldots$$

$$\Rightarrow y = \frac{1}{1 - \cos^2\phi} = \frac{1}{\sin^2\phi}$$

$$z = \sum_{n=0}^{\infty} \cos^n(\theta + \phi)\cos^n(\theta - \phi)$$

$$\Rightarrow z = 1 + \cos(\theta + \phi)\cos(\theta - \phi) + \cos^2(\theta + \phi)\cos^2(\theta - \phi) + \ldots$$

$$\Rightarrow z = \frac{1}{1 - \cos(\theta + \phi)\cos(\theta - \phi)}$$

$$\Rightarrow z = \frac{1}{1 - (\cos^2\theta - \sin^2\phi)}$$

$$\Rightarrow z = \frac{1}{1 - \left(\frac{1}{x} - \frac{1}{y}\right)} = \frac{xy}{xy - (y - x)}$$

$$\Rightarrow xyz - (zy - xz) = xy$$

$$\Rightarrow xyz - xy = yz - xz$$

3. *(b)* We have,

$$\log_3(x + 2)(x + 4) + \log_{1/3}(x + 2) < \frac{1}{2}\log_{\sqrt{3}} 7 \quad \ldots\text{(i)}$$

$$(x + 2)(x + 4) > 0, (x + 2) > 0$$

$$\therefore \quad x > -2$$

Now, Eq. (i) becomes

$$\log_3(x + 2)(x + 4) - \log_3(x + 2) < \log_3 7$$

$$\Rightarrow \log_3(x + 4) < \log_3 7$$

$$\Rightarrow x + 4 < 7$$

$$\Rightarrow x < 3$$

$\therefore\ x \in (-2, 3)$

4. *(b)* We have, 9 married couples. We can select two men out of 9 in ${}^9C_2$ ways. Since no husband and wife are to play in the same game, two women out of the remaining 7 can be chosen in ${}^7C_2$ ways. If M_1, M_2, W_1 W_2 are chosen, then a team consist of M_1W_1 and M_2W_2. Thus, the number of ways of arrangement is ${}^9C_2 \times {}^7C_2 \times 2 = 1512$.

5. *(a)* We have,

$$f(x) = \frac{\tan[e^2]x^3 - \tan[-e^2]x^3}{\sin^3 x}$$

$7 < e^2 < 8$, so $[e^2] = 7$ and $[-e^2] = -8$

$$\text{So } f(0) = \lim_{x \to 0} \frac{\tan 7x^3 - \tan(-8)x^3}{\sin^3 x}$$

$$= \lim_{x \to 0}\left[\frac{\tan 7x^3}{\sin^3 x} + \frac{\tan 8x^3}{\sin^3 x}\right]$$

$$= 7 + 8 = 15$$

6. *(d)* We have,

$$I = \int_0^{\pi/2} \cos^n x \sin^n dx$$

$$\Rightarrow I = \int_0^{\pi/2} \frac{(2\sin x\cos x)^n}{2^n} dx$$

$$\Rightarrow I = \frac{1}{2^n}\int_0^{\pi/2} (\sin 2x)^n dx$$

$$\left[\text{put } 2x = t \Rightarrow dx = \frac{dt}{2}\right]$$

$$\Rightarrow I = \frac{1}{2^n}\int_0^{\pi} (\sin t)^n \frac{dt}{2}$$

$$\Rightarrow I = \frac{1}{2^{n+1}}\int_0^{\pi} (\sin t)^n dt$$

$$\Rightarrow I = \frac{2}{2^{n+1}}\int_0^{\pi/2} \sin^n t dt$$

$$\left[\because \int_0^{2a} f(x)\,dx = 2\int_0^a f(2a - x)\,dx\right]$$

$$\Rightarrow I = 2^{-n}\int_0^{\pi/2} \sin^n x\,dx$$

$$\therefore \quad \lambda = 2^{-n}$$

7. *(c)* We have, $x = \tan 27\theta - \tan\theta$

$$\text{and} \quad y = \frac{\sin\theta}{\cos 3\theta} + \frac{\sin 3\theta}{\cos 9\theta} + \frac{\sin 9\theta}{\cos 27\theta}$$

$$\because \tan A - \tan B = \frac{\sin(A - B)}{\cos A\cos B}$$

$$\therefore \tan 3\theta - \tan\theta = \frac{\sin 2\theta}{\cos 3\theta\cos\theta}$$

$$= \frac{2\sin\theta\cos\theta}{\cos 3\theta\cos\theta} = \frac{2\sin\theta}{\cos 3\theta}$$

$$\therefore \quad \frac{\sin\theta}{\cos 3\theta} = \frac{1}{2}(\tan 3\theta - \tan\theta)$$

$$\therefore \quad y = \frac{1}{2}[\tan 3\theta - \tan\theta + \tan 9\theta - \tan 3\theta + \tan 27\theta - \tan 9\theta]$$

$$\Rightarrow \quad y = \frac{1}{2}(\tan 27 - \tan\theta)$$

$$\Rightarrow \quad 2y = x \Rightarrow x = 2y$$

8. *(a)* Equation of ellipse

$$\frac{x^2}{16} + \frac{y^2}{9} = 1$$

$$\angle ACB = 90^\circ$$

$$CA = 4, CB = 3$$

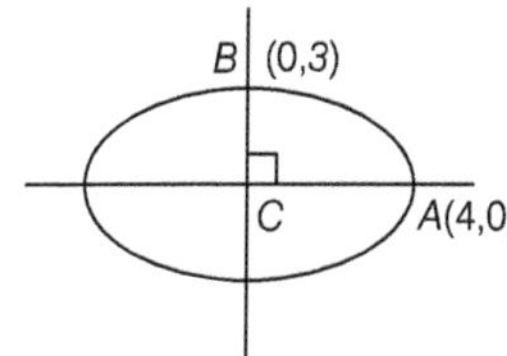

$$\Rightarrow \quad \frac{1}{CA^2} + \frac{1}{CB^2} = \frac{1}{16} + \frac{1}{9} = \frac{25}{144}$$

9. *(a)* Equation of circle

$$x^2 + y^2 - 6x + 8y = 0$$

$$(x - 3)^2 + (y + 4)^2 = 25$$

Centre $(3, -4)$, radius $= 5$

Distance from $(3, -4)$ to the line $3x + 4y - 25 = 0$ is $d = \left|\frac{9 - 16 - 25}{5}\right| = \frac{32}{5}$

Shortest distance $= d - r = \frac{32}{5} - 5 = \frac{7}{5}$

10. *(d)* $18x^2 - 9\pi x + \pi^2 = 0$

$$(3x - \pi)(6x - \pi) = 0$$

$$\Rightarrow \quad \alpha = \frac{\pi}{6}, \beta = \frac{\pi}{3}$$

$$f(x) = \sqrt{x},\ g(x) = \cos x^2$$

$$(gof)(x) = \cos x$$

Area between $\cos x$, $x = \frac{\pi}{6}$, $x = \frac{\pi}{3}$ and $y = 0$ is $A = \int_{\pi/6}^{\pi/3} \cos x\,dx = [\sin x]_{\pi/6}^{\pi/3}$

$$A = \sin\frac{\pi}{3} - \sin\frac{\pi}{6} = \frac{\sqrt{3} - 1}{2}$$

11. *(b)* $\dfrac{\text{Area of } \Delta ABC}{\text{Area of } \Delta OAC}$

$$= \frac{\frac{1}{2}[|\mathbf{OA} \times \mathbf{OB}| + |\mathbf{OB} \times \mathbf{OC}| + |\mathbf{OC} \times \mathbf{OA}|}{\frac{1}{2}|\mathbf{OA} \times \mathbf{OC}|}$$

$$= \frac{|\mathbf{OA} \times \mathbf{OB}| + |\mathbf{OB} \times \mathbf{OC}|}{|\mathbf{OA} \times \mathbf{OC}|} + 1$$

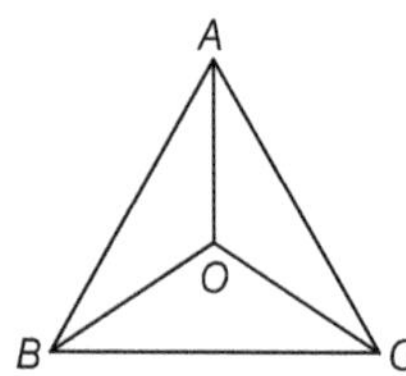

Given, $\mathbf{OA} + 2\mathbf{OB} + 3\mathbf{OC} = 0$

$$2\mathbf{OB} \times \mathbf{OA} + 3\mathbf{OC} \times \mathbf{OA} = 0$$

$$\Rightarrow \frac{|\mathbf{OB} \times \mathbf{OA}|}{3} = \frac{|\mathbf{OC} \times \mathbf{OA}|}{2} = \lambda$$

$$\mathbf{OB} \times \mathbf{OA} + 3\mathbf{OB} \times \mathbf{OC} = 0$$

$$\Rightarrow \frac{|\mathbf{OB} \times \mathbf{OA}|}{3} = \frac{|\mathbf{OB} \times \mathbf{OC}|}{1} = \lambda$$

$$\frac{\text{Area of } \Delta ABC}{\text{Area of } \Delta AOC} = \frac{3\lambda + \lambda}{2\lambda} + 1 = 3:1$$

12. *(a)* We have,

$$f(x) = \int_0^x \sin(t^2 - t + x)\, dt$$

$$\Rightarrow f'(x) = \sin(x^2 - x + x) + \int_0^x \cos(t^2 - t + x)\, dt$$

$$\Rightarrow f''(x) = 2x\cos x^2 + \cos(x^2) - \int_0^x \sin(t^2 - t + x)\, dt$$

$$\Rightarrow f''(x) + f(x) = (2x + 1)\cos x^2$$

$$\Rightarrow f''\left(-\frac{1}{2}\right) + f\left(\frac{-1}{2}\right) = \left(2\left(\frac{-1}{2}\right) + 1\right)\cos\left(\frac{1}{4}\right) = 0$$

13. *(c)* Let $f(x) = \dfrac{x^2}{3x^3 + 500}$

and $g(x) = 3x + \dfrac{500}{x^2}$

If $g(x)$ is minimum, then $f(x)$ is maximum

$$g'(x) = 3 - \frac{1000}{x^3} \Rightarrow g''(x) = \frac{3000}{x^4} > 0$$

Minimum value of $g(x)$ at $x = \left(\dfrac{1000}{3}\right)^{1/3}$

$$= (333.33)^{1/3}$$

where $6 < x < 7$

So greatest term of the sequence $\left(\dfrac{k^2}{3k^3 + 500}\right)$ is either $k = 6$ or $k = 7$

$$\therefore \quad f(6) = \frac{36}{1148}$$

$$f(7) = \frac{49}{1529}$$

$$\therefore \quad f(7) > f(6)$$

$\therefore$ At $k = 7$ is maximum.

14. *(c)* We have,

$$\arg\left(\frac{z - 4i}{z - 3}\right) = \frac{\pi}{3}$$

In ΔOAB,

$$\cos\frac{2\pi}{3} = \frac{OA^2 + OB^2 - AB^2}{2OA \cdot OB}$$

$$\Rightarrow \frac{-1}{2} = \frac{2r^2 - 25}{2r^2} \Rightarrow 3r^2 = 25$$

$$r = \frac{5}{\sqrt{3}}$$

15. *(c)* $AB = c$

$$BC = a$$

$$AC = b$$

$$AD + DB = AB = c$$

$$CD \times DE = AD \times BD$$

$$\frac{AD + DB}{2} \geq \sqrt{AD \times BD}$$

$[\because AB \geq 4M]$

$$\frac{c}{2} \geq \sqrt{CD \times DE}$$

$$\Rightarrow \quad CD \times DE \leq \frac{c^2}{4}$$

16. *(b)* We have,

$$7\cos x + 5\sin x = 2k + 1$$

The minimum and maximum value of $7\cos x + 5\sin x$

is $-\sqrt{7^2 + 5^2}, \sqrt{7^2 + 5^2}$

$$-\sqrt{74}, \sqrt{74}$$

$$\therefore \quad -\sqrt{74} < 2k + 1 < \sqrt{74}$$

$$\Rightarrow \quad -8 < 2k + 1 < 8$$

$$\Rightarrow \quad -4 \leq k \leq 3$$

$$k = -4, -3, -2, -1, 0, 1, 2, 3$$

Total 8 integral value.

17. *(a)* Consider the events

A = Sum of 5 occurs

B = Sum of 7 occurs

C = Neither sum of 5 nor sum of 7 occurs

$$P(A) = \frac{1}{9}, P(B) = \frac{1}{6}, P(C) = \frac{13}{18}$$

$P(A$ occurs before $B)$

$$= P(A \text{ or } (C \cap A) \text{ or } (C \cap C \cap A) \text{ or } \ldots)$$

$$= P(A) + P(CA) + P(CCA) + \ldots$$

$$= \frac{1}{9} + \frac{13}{18} \times \frac{1}{9} + \left(\frac{13}{18}\right)^2 \times \frac{1}{9} + \ldots$$

$$= \frac{1}{9}\left(1 + \frac{13}{18} + \left(\frac{13}{18}\right)^2 + \ldots\right)$$

$$= \frac{1}{9}\left(\frac{1}{1 - \left(\frac{13}{18}\right)}\right) = \frac{2}{5}$$

18. *(a)* Given,

$f(x) < x^2 f(x)$ is always positive for $x \in [0, 1]$

$$\int_0^1 f(x)\, dx < \int_0^1 x^2 dx$$

$$\frac{1}{3} < \left[\frac{x^3}{3}\right]_0^1 \Rightarrow \frac{1}{3} < \frac{1}{3}$$

It is not possible.

19. *(c)* We have,

$$n(X) = 5$$

A and B are subset of X

Such that $A \neq \phi$, $B \neq \phi$ $A \cap B = \phi$

Total number of order pair (A, B) is

$${}^5C_2 \times 2! + {}^5C_3\left(\frac{3!}{2!} \times 2!\right) + {}^5C_4\left(\frac{4!}{3!}2! + \frac{4!}{2!2!}\right) + {}^5C_5\left(\frac{5!}{4!}2! + \frac{5!}{2!3!} \times 2!\right)$$

$$= 10(2) + 10(6) + 5(8 + 6) + (10 + 20)$$

$$= 20 + 60 + 70 + 30 = 180$$

$$\therefore \quad d \in [151, 200]$$

$$\Rightarrow 151 \leq d \leq 200$$

20. *(c)* $f(x) = (x + 2019)^n$

$$f(0) = (2019)^n$$

$$f'(x) = n(x + 2019)^{n-1},$$

$$f'(0) = n(2019)^{n-1}$$

$$f''(x) = n(x - 1)(x + 2019)^{n-2},$$

$$f''(0) = n(n - 1)(2019)^{x-2}$$

$$f^{n-1}(n) = n(n - 1) \ldots 1$$

$$f^{n-1}(0) = n(n - 1) \ldots \times 1$$

$$f(0) + f'(0) + \frac{f''0}{2!} + \ldots + \frac{f^{n-1}(0)}{(n-1)!}$$

$$(2019)^n + {}^nC_1(2019)^{n-1} + {}^nC_2(2019)^{n-2} + ... + {}^nC_{n-1}\,2019 + {}^nC_{n-1}$$
$$= (2019+1)^n - 1$$
$$= (2020)^n - 1$$

21. *(c)*

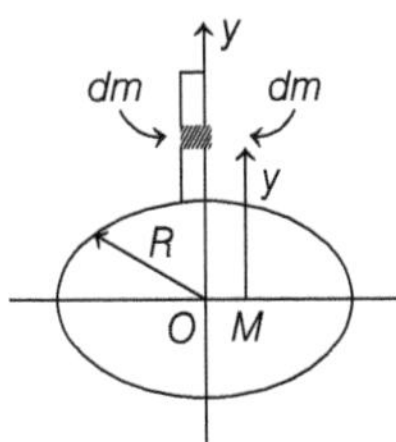

Force on an elemental length dy of rod due to earth is

$$dF = dm \cdot g$$
$$= \frac{m}{R}\cdot dy \cdot gy = \frac{m}{R}\cdot dy \cdot g\left(\frac{R^2}{(R+y)^2}\right)$$
$$\Rightarrow \quad dF = \frac{mg}{R}\left(\frac{1}{\left(1+\frac{y}{R}\right)^2}\right)dy$$

Total force on complete length of rod is

$$F = \int_0^R dF = \frac{mg}{R}\left(\int_0^R \frac{dy}{\left(1+\frac{y}{R}\right)^2}\right) = \frac{mg}{2}$$

22. *(d)* Let particle is moving perpendicular to its initial direction after time t.

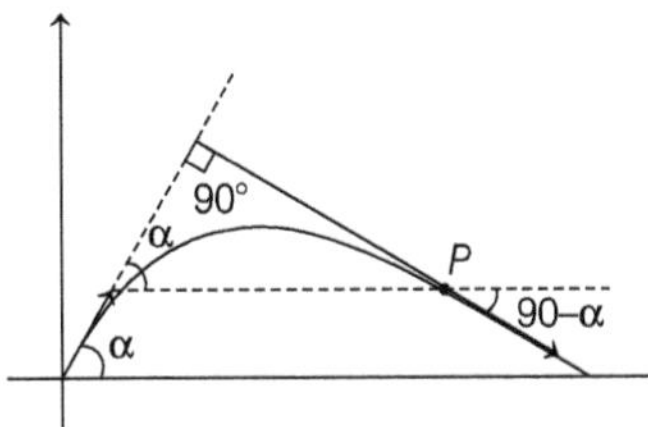

After time t velocity is let v, then velocity components of particle at time t are

$$v_{x_f} = v\sin(90-\alpha) = v\cos\alpha$$

As there is no change in magnitude of horizontal component, $v_{x_f} = v_{x_i}$

$$v\cos\alpha = u\cos\alpha$$

And vertical component at instant t is

$$-v\sin(90-\alpha) = u\sin\alpha - gt$$

From these equations, we have

$$-\cot\alpha = \frac{u\sin\alpha - gt}{u\cos\alpha}$$
$$\Rightarrow \quad \frac{gt}{u\cos\alpha} = \frac{1}{\cos\alpha\sin\alpha}$$
$$\Rightarrow \quad t = \frac{u}{g\sin\alpha} \text{ or } t = \frac{u}{g}\operatorname{cosec}\alpha$$

23. *(c)* The x-coordinate of complete disc is $\quad X_{CM} = \dfrac{m_1x_1 + m_2x_2}{m_1 + m_2}$

Treating removed mass m_2 as negative mass, we have

$$0 = \frac{\frac{3}{4}m\cdot x_0 + \frac{m}{4}\cdot\frac{r}{2}}{\frac{3}{4}m + \frac{m}{4}} \Rightarrow x_0 = \frac{-r}{6}$$

24. *(d)* Initial velocity of centre of mass is

$$u = \frac{m_1u_1 + m_2u_2}{m_1 + m_2}$$
$$= \frac{3(2\hat{\mathbf{i}} + 3\hat{\mathbf{j}}) + 3(4\hat{\mathbf{i}} - 6\hat{\mathbf{j}})}{3+3} = 3\hat{\mathbf{i}} - 1.5\hat{\mathbf{j}}\ \text{ms}^{-1}$$

Acceleration of centre of mass is

$$a = \frac{F}{m_{\text{system}}} = \frac{24\hat{\mathbf{i}}}{3+3} = 4\hat{\mathbf{i}}\ \text{ms}^{-2}$$

Final velocity of centre of mass is

$$v = u + at$$
$$= (3\hat{\mathbf{i}} - 1.5\hat{\mathbf{j}}) + (4\hat{\mathbf{i}} \times 5) = 23\hat{\mathbf{i}} - 1.5\hat{\mathbf{j}}\ \text{ms}^{-1}$$

25. *(a)* Let block reaches bottom with speed u and at that instant block is moving with a velocity v, then by conservation of momentum, we have

$$Mv + m(v - u\cos\theta) = 0$$

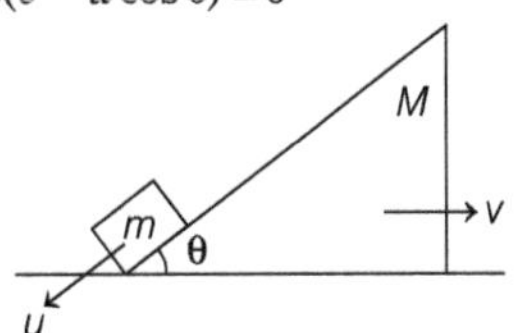

Substituting values in above equation, we get

$$5.0v + 0.5v - 0.5\times\frac{4}{5} = 0$$
$$\Rightarrow \quad v(50.5) = 0.5\times 1.2625\times\frac{4}{5}$$
$$\Rightarrow \quad v = \frac{0.5\times 1.2625\times 4}{5\times 50.5} = 0.01\ \text{ms}^{-1}$$

26. *(c)* Velocity of block with respect to plane is $\mathbf{v}_{12} = \mathbf{v}_1 - \mathbf{v}_2$

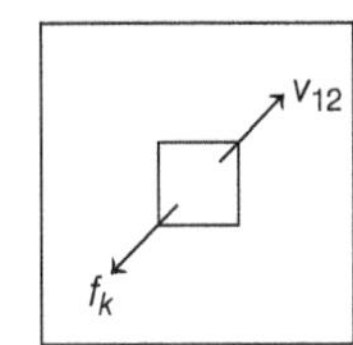

$$\therefore\ v_{12} = -2\hat{\mathbf{i}} - (-2\hat{\mathbf{j}}) = -2(\hat{\mathbf{i}} - \hat{\mathbf{j}})\ \text{ms}^{-1}$$

Since, friction opposes this velocity.
So, friction is $f_k = \mu mg(\hat{\mathbf{n}})$
where, $\hat{\mathbf{n}}$ = unit vector opposite to $\mathbf{v}_{12}$.

$$\therefore\ f_k = \frac{0.25\times 1\times 10\times 2(\hat{\mathbf{i}} - \hat{\mathbf{j}})}{\sqrt{2^2 + 2^2}} = 1.78(\hat{\mathbf{i}} - \hat{\mathbf{j}})\ \text{N}$$

27. *(a)*

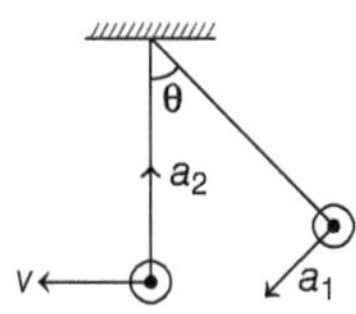

Let a_1 and a_2 are the accelerations, at given positions. Then,

$$a_2 = \frac{v^2}{l} = \frac{2gl(1-\cos\theta)}{l}$$
$$\Rightarrow \quad a_2 = 2g(1-\cos\theta)$$

Also, $a_1 = g\sin\theta$

As, $a_1 = a_2$

$$\Rightarrow\ g\sin\theta = 2g(1-\cos\theta)$$
$$\Rightarrow\ \sin\theta = 2(1-\cos\theta)$$
$$\Rightarrow 5\cos^2\theta - 8\cos\theta + 3 = 0$$
$$\Rightarrow \cos\theta = \frac{3}{5} \text{ or } \theta = \cos^{-1}\left(\frac{3}{5}\right)$$

28. *(d)* Charge on inner sphere is

$$q = 4\pi\varepsilon_0\cdot V\cdot r$$

When sphere is connected to shell, whole of the charge is transferred to outer shell.

So, potential of sphere inside shell will be potential of any inner point of shell.

As, potential at any interior point of shell is equal to its surface potential, final potential of sphere is

$$V_2 = \frac{q}{4\pi\varepsilon_0(4r)} = \frac{4\pi\varepsilon_0 Vr}{4\pi\varepsilon_0 4r} = \frac{V}{4}$$

29. *(d)* Electrical resistance, $R = \dfrac{V}{I}$

So, heat resistance or thermal resistance is

$$R_{Th} = \frac{\Delta T}{\Delta Q}$$

where, ΔT = temperature difference and ΔQ = heat flow rate = $\dfrac{H}{t}$.

$$\therefore\ R_{Th} = \frac{\Delta T}{\left(\frac{KA\Delta T}{l}\right)} \Rightarrow R_{Th} = \frac{l}{KA}$$

So, its dimensions are

$$[R_{Th}] = \frac{[L]}{[MLT^{-3}K^{-1}L^2]} = [M^{-1}L^{-2}T^3K]$$

30. *(c)* Beam reaches AC, if TIR occurs at BC.

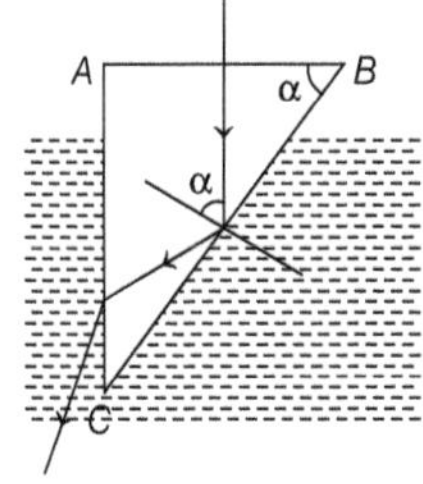

From geometry of figure, angle of incidence over face $BC = \alpha$.

For beam to reach AC,

$$\sin\alpha > \frac{n_2}{n_1}$$

where, n_2 = refractive index of water $= \frac{4}{3}$

and n_1 = refractive index of glass $= \frac{3}{2}$.

$$\therefore \sin\alpha > \frac{\frac{4}{3}}{\frac{3}{2}} \Rightarrow \sin\alpha > \frac{8}{9} = 0.89$$

$\therefore \quad \alpha > 60°$

$\therefore \quad \sin 60° \approx 0.86$

31. *(a)* We divide slab into a series of parallel slabs with different refractive indices.

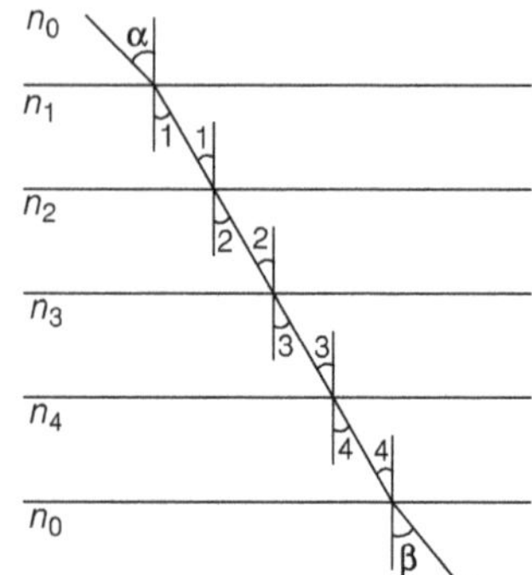

We have, $\frac{\sin\alpha}{\sin 1} = \frac{n_1}{n_0}$

$$\frac{\sin 1}{\sin 2} = \frac{n_2}{n_1}, \frac{\sin 2}{\sin 3} = \frac{n_3}{n_2}$$

$$\frac{\sin 3}{\sin 4} = \frac{n_4}{n_3}, \frac{\sin 4}{\sin\beta} = \frac{n_0}{n_4}$$

So, we have

$$\frac{\sin\alpha}{\sin 1} \times \frac{\sin 1}{\sin 2} \times \frac{\sin 2}{\sin 3} \times \frac{\sin 3}{\sin 4} \times \frac{\sin 4}{\sin\beta} = 1$$

$\Rightarrow \quad \sin\alpha = \sin\beta \;$ or $\; \alpha = \beta$

32. *(b)* When student look through green glass "excellent" appears as black against the green background of the paper.

Red colour of "excellent" does not pass through green glass.

33. *(b)*

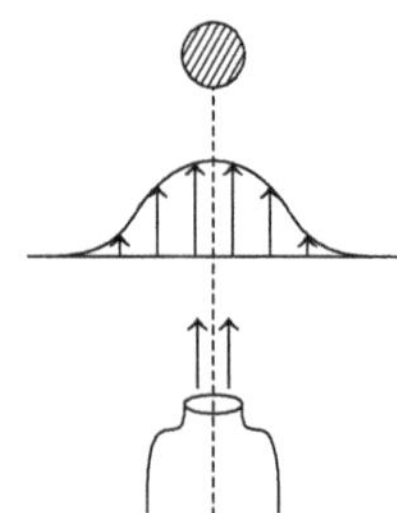

The sphere shows stable equilibrium for small horizontal displacement because of velocity distribution of approaching fluid.

Near edge, velocity is low, so pressure is high. Hence, when ball is slightly displaced, it is pushed back to centre where velocity is more. As, by Bernoulli's principle where v is high, p is low.

34. *(c)* As, 11.46 days is three times the half-life of Radon. So, decay scheme is as follows.

8×10^{10} Nucleus $\xrightarrow{\text{3.82 days}}$ 4×10^{10} Nucleus $\xrightarrow{\text{3.82 days}}$ 2×10^{10} Nucleus $\xrightarrow{\text{3.82 days}}$ 1×10^{10} Nucleus

So, only 1×10^{10} active nuclii remains after 3-half lives.

So, 7×10^{10} disintegrations would occur in 11.46 days.

35. *(c)* In region B, we observe that binding energy is not increasing with mass number.

This indicates a nucleon possibly interacts with other nucleons which lies near to it but not with those which are far off from it.

This indicates about saturation property of nuclear forces.

36. *(d)* Work needed to raise the piston and compress the spring is

$$W = mgh + \frac{1}{2}kx^2 + p_{atm} A \cdot h$$

$$= 60 \times 9.8 \times 0.05 + \frac{1}{2} \times 50000 \times (0.05)^2 + \frac{100000 \times \pi \times (0.2)^2}{4} \times 0.05$$

$$= 250 \text{ J}$$

By first law of thermodynamics, we have

$$\Delta U = \Delta Q - \Delta W = 200 - 250 = -50 \text{ J}$$

37. *(a)* System undergoes an internally reversible isothermal process.

So, entropy change of system is

$$\Delta S = \frac{Q}{T_{\text{system}}} = \frac{750}{300} = 2.5 \frac{\text{kJ}}{\text{K}}$$

38. *(d)* Consider a sphere of radius z charge $\rho \cdot V$, where ρ = charge density (cm^{-3}) and V = volume (m^3).

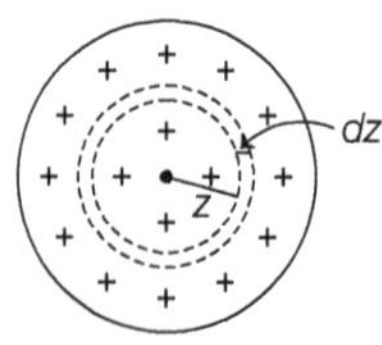

Potential of sphere $= \frac{kq}{r} = \frac{\frac{4}{3}\pi z^3 \cdot \rho}{4\pi \in z} = \frac{\rho z^2}{3 \in}$

To increase charge layer by thickness dz,

Work done, $dW = qV = \frac{z^2 \rho}{3 \in} 4\pi z^2 dz \cdot \rho$

or $\quad dW = \frac{4\pi\rho^2}{3 \in} \cdot z^4 \cdot dz$

Work done in assembling complete sphere

$$\int_0^R dW = \int_0^R \frac{4\pi\rho^2}{3 \in} z^4 \cdot dz$$

$$= \frac{4\pi\sigma^2}{3 \times 5 \times \in} \cdot (z^5)/_0^R = \frac{4\pi\sigma^2 R^5}{3 \times 5 \in}$$

$$= \frac{4\pi \times \left(\frac{3q}{4\pi R^3}\right)^2 \times R^5}{3 \times 5 \times \in}$$

$$= \frac{3q^2}{20\,\pi \in R} = \frac{3q^2}{20\pi \in_0 kR}$$

39. *(d)* When an excited atom at rest in lab frame emits a photon of energy hf, the photon also carries a momentum $p = \frac{hf}{c}$. Conservation of momentum requires that atom also recoils with same momentum. So, energy carried by atom is $\frac{p^2}{2m}$. Hence, in emission we can write,

$\Delta E = hf$ + recoil energy of atom.

So, emission energy of photon is smaller in case of recoil of atom.

As $E \propto \frac{1}{\lambda}$, emitted wavelength is larger.

So, $\lambda_2 > \lambda_1$ and $\frac{\lambda_1}{\lambda_2} < 1$.

40. *(d)* Using, $Q = 931.5\,(m_x - m_y - m_\alpha)$, we get

$$Q = 5.40 \text{ MeV}$$

This energy is distributed between α-particle and daughter nucleus.

Energy is shared in inverse proportion of masses. So, ratio of kinetic energies is

$$\frac{E_k(\text{Th})}{E_k(\alpha)} = \frac{4}{228}$$

So, $E_k(\text{Th}) = 0.10$ MeV

and $E_k(\alpha) = 5.30$ MeV

But if ^{228}Th is in an excited state, then $E_k(\alpha)$ is less.

41. *(a)* A species is amphoteric, if it is soluble in acid (behaves as a base) as well as in base (behaves as an acid).

$$\underset{\text{Basic}}{SnO_2} + \underset{\text{Acid}}{4HCl} \longrightarrow SnCl_4 + 2H_2O$$

$$\underset{\text{Acid}}{SnO_2} + \underset{\text{Base}}{2NaOH} \longrightarrow Na_2SnO_3 + H_2O$$

42. *(d)* When salicylic acid is treated, with zinc dust, benzoic acid is obtained as a product.

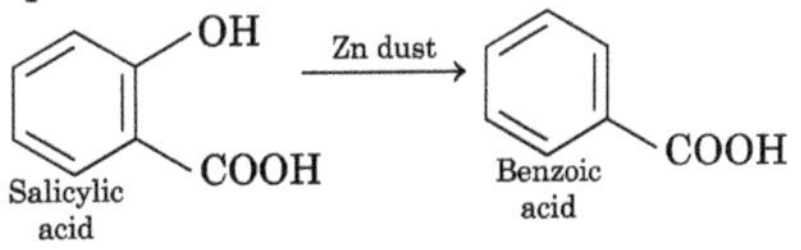

Salicylic acid (OH, COOH) $\xrightarrow{\text{Zn dust}}$ Benzoic acid (COOH)

43 *(d)* For a zero order reaction, unit of rate constant is mole $L^{-1}s^{-1}$. Hence, we can easily calculate concentration of B after 20 minutes by the following formula

$$x = kt$$

$$x = kt = 0.6\times10^{-3}\times20\times60 = 0.72\ \text{M}$$

44. *(c)* Given, weight of $CuCl_2 = 13.44$ g

Molecular weight of $CuCl_2 = 134.4$ g

$$\text{Molality} = \frac{\text{number of moles of solute}}{\text{mass of solvent (kg)}}$$

$\because$ No. of moles of solute, $CuCl_2 = \frac{13.44}{134.1}$

$$\therefore \quad \text{Molality} = \frac{\frac{13.44}{134.1}}{1} = 0.1$$

Now, $CuCl_2 \rightleftharpoons \underbrace{Cu^{2+} + 2Cl^-}_{\text{3 particles}}$

$\therefore \quad i = 3$

$\Rightarrow \quad \Delta T_b = iK_b \cdot m = 3\times0.52\times0.1 = 0.156 \approx 0.16$

45. *(a)* The reaction in which oxidation and reduction occurs simultaneously are termed as redox reaction or where there is change in oxidation number on both reactants and product sides.

(a) $\overset{+4}{(Xe)}F_4 + \overset{+1}{O_2}(F_2) \longrightarrow \overset{+6}{Xe}F_6 + \overset{0}{O_2}$

Since, Xe undergoes oxidation while O undergoes reduction. So, it is an example of redox reaction.

(b) $\overset{+4}{Xe}F_2 + \overset{+5}{P}F_5 \longrightarrow \overset{+6}{[XeF]^+} + \overset{+5}{P}F_6$

As the oxidation state of P is same on both sides. Thus, it is not a redox reaction.

(c) $\overset{+6}{Xe}F_6 + H_2O \longrightarrow \overset{+6}{Xe}F_4$

Not a redox reaction.

(d) $\overset{+6}{Xe}F_6 + 2H_2O \longrightarrow \overset{+6}{Xe}O_2F_2 + 4HF$

Not a redox reaction.

46. *(a)* Density $= \frac{M.Z}{N_A\times a^3}$ Density $\propto \frac{Z}{a^3}$

In case of fcc, $Z = 4, a = 3.5\text{Å}$

$$d_1 = \frac{4}{(3.5)^3}$$

In case of bcc $Z = 2, a = 3\text{Å}$

$$d_1 = \frac{2}{(3)^3}$$

$$\Rightarrow \quad \frac{d_1}{d_2} = \frac{N_1}{N_2}\left(\frac{a_2}{a_1}\right)^3 = \frac{4}{2}\left(\frac{3}{3.5}\right)^3 = 1.26$$

47. *(b)* NO_2 is a brown coloured gas and imparts this colour to concentrated HNO_3 during long standing.

$$4HNO_3 \longrightarrow 2H_2O + 2NO_2 + 3O_2$$

48. *(a)* According to spectrochemical series, lower the value of $E°$ more easily, it can loose electron. Thus, the order of decreasing value of reduction potential of corresponding metals is

$$Y\ (E° = -3.03\ \text{V}) > Z(E° = -1.18\ \text{V}) > X\ (E° = 0.52\ \text{V}).$$

Thus, the order of reducing power of the corresponding metal is

$$Y > Z > X$$

49. *(a)* For first order reaction

$$k = \frac{2.303}{t}\log\left(\frac{a}{a-x}\right)$$

$$\frac{kt}{2.303} = \log a - \log(a-x)$$

$$\frac{kt_1}{2.303} = \log a - \log(a-x_1) \quad \text{at time } t_1$$

$$\frac{kt_2}{2.303} = \log a - \log(a-x_2) \quad \text{at time } t_2$$

$$\therefore \quad \frac{k}{2.303}(t_2-t_1) = \log\left(\frac{a-x_1}{a-x_2}\right)$$

Rate, $r_1 = k(a-x_1)$

$r_2 = k(a-x_2)$

$$\therefore \quad \frac{(a-x_1)}{(a-x_2)} = \frac{r_1}{r_2} \Rightarrow \frac{0.04}{0.03} = \frac{4}{3}$$

$$\frac{k(20-10)}{2.303} = \log\frac{4}{3}$$

$$\therefore \quad k = \frac{2.303}{10}\log\frac{4}{3} = \frac{2.303}{10}\log 1.33$$

$$= \frac{2.303\times0.125}{10} = 0.02878\ \text{min}^{-1}$$

Also, for 1st order $t_{1/2} = \frac{0.693}{k}$

$$T_{50} = \frac{0.693}{0.02878} = 24.1\ \text{min}$$

50. *(b)* In acidic medium, phenol exists in the following resonaning structures.

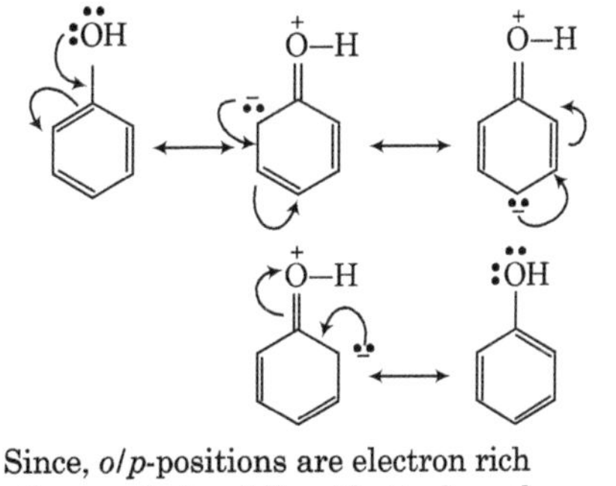

Since, o/p-positions are electron rich sites, so electrophile will attack on these sites, i.e. hydrogen of these sites get exchanged by D (deuterium). Hence, the final product of the reaction is

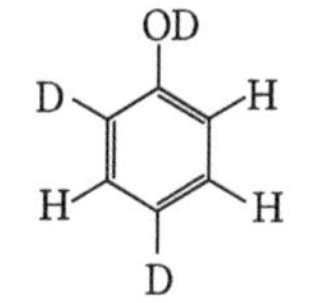

which is formed as:

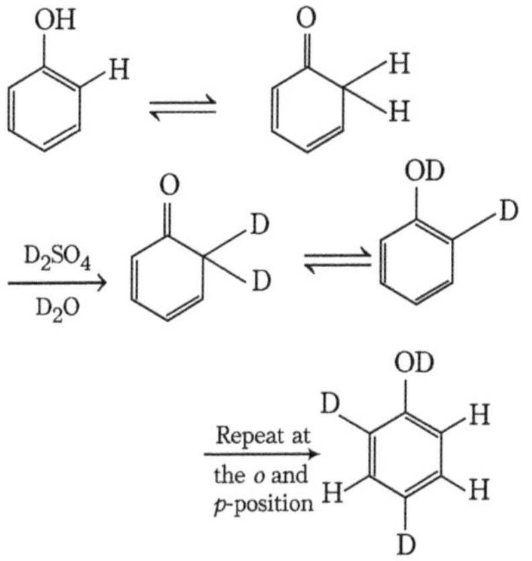

$\xrightarrow[D_2O]{D_2SO_4}$ $\xrightarrow[\text{the } o \text{ and } p\text{-position}]{\text{Repeat at}}$

51. *(c)* $\frac{1}{2}I_2(s) + \frac{1}{2}Cl_2(g) \longrightarrow ICl\ (g)$

$$\Delta H = \left[\frac{1}{2}\Delta H_{s\to g} + \frac{1}{2}\Delta H_{\text{diss}}(Cl_2) + \frac{1}{2}\Delta H_{\text{diss}}(I_2)\right] - \Delta H_{ICl}$$

$$= \left(\frac{1}{2}\times62.76 + \frac{1}{2}\times242.3 + \frac{1}{2}\times151.0\right) - 211.3$$

$$= 228.03 - 211.3$$

$$\Delta H = 16.73 \approx 16.8\ \text{kJ mol}^{-1}$$

52. *(b)* The compound 2-methyl butane on monochlorination gives following isomers:

$CH_3-CH_2-CH(CH_3)-CH_3 \xrightarrow{Cl_2}$

2-methylbutane

$CH_3-CH_2-\overset{*}{C}H(CH_3)-CH_2Cl$ (I)

$+ CH_3-CH_2-C(Cl)(CH_3)-CH_3$ (II)

$+ CH_3-\underset{*}{C}H(Cl)-CH(CH_3)-CH_3$ (III)

$+ CH_2(Cl)-CH_2-CH(CH_3)-CH_3$ (IV)

I and III are chiral compounds, which are possible on monochlorination of 2-methyl butane.

53. *(c)* Secondary alkyl halides on treatment with alcoholic KOH preferably undergo elimination rather than substitution to give alkenes in accordance with Saytzeff rule. Usually the more stable *trans*-alkenes predominate in this reaction. Thus, 2-bromopentane on heating potassium ethoxide in ethanol gives *trans*-pent-2-ene.

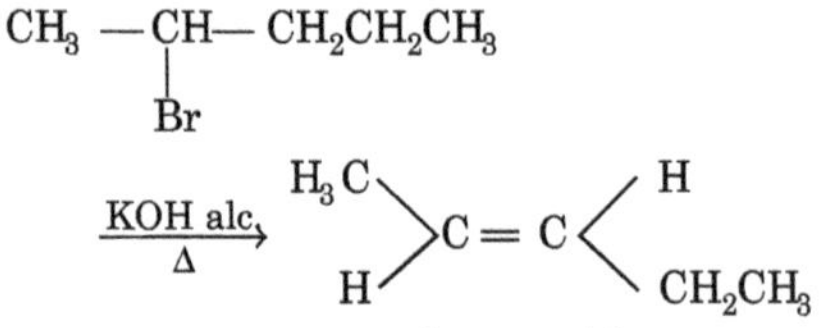

$CH_3—CH(Br)—CH_2CH_2CH_3 \xrightarrow[\Delta]{\text{KOH alc.}} (H_3C)(H)C=C(H)(CH_2CH_3)$

trans-pent-2-ene

54. *(b)* Greater the extent of $d\pi$ - $p\pi$ back bonding, smaller will be the bond order of CO bond in metal carbonyls. More are the number of valence shell electrons (d - electrons), maximum is the chance of $p\pi$ - $d\pi$ back bonding and thus lowest will be the bond order of CO bond.

The number of d-etectrons of metal in given complexes are as follows:

(a) $[Mn(CO)_6]^+$

Mn is present as Mn^+, thus the electronic configuration is $3d^5 4s^1$.

(b) $[Fe(CO)_5]$

Fe is present in zero oxidation state, thus the electronic configuration will be $3d^6 4s^0$.

(c) $[Cr(CO)_6]$

Cr is present in zero oxidation state, thus the electronic configuration is $3d^5 4s^1$.

(d) $[V(CO)_6]^-$

V is present as V^+ in given complex. Thus, the electronic configuration is $3d^3 4s^1$.

Thus, $[Fe(CO)_5]$ has lowest bond order of CO bond.

55. *(d)* The product obtained from the reactions given in options are as follows:

(a) $C_6H_5CH_3 \xrightarrow[CS_2]{2CrO_2Cl_2} C_6H_5CH_3\cdot 2CrO_2Cl_2$ Brown (addition product) $\xrightarrow{H_2O} C_6H_5CHO$ Benzaldehyde

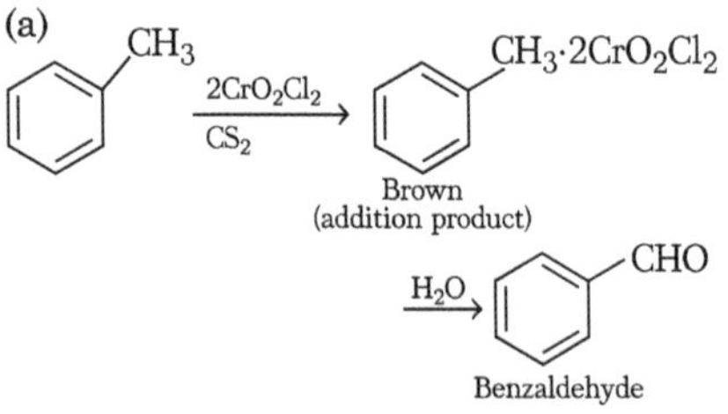

This reaction is known as **Etard reaction.**

(b) $C_6H_5COCl \xrightarrow[\text{(Boiling xylene)}]{Pd/BaSO_4/H_2} C_6H_5CHO + HCl$

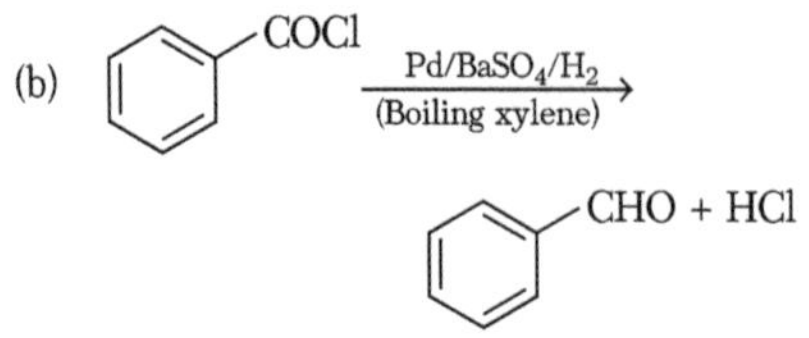

This reaction is known as **Rosenmund reduction.**

(c) $C_6H_6 + CO + HCl \xrightarrow{\text{Anhy. } AlCl_3} C_6H_5CHO + HCl$

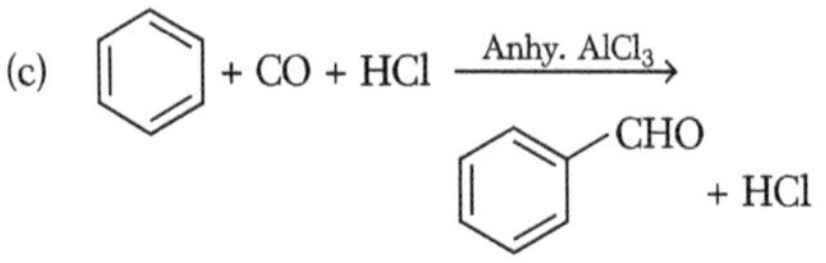

This reaction is known as **Gattermann-Koch** aldehyde synthesis.

(d) $C_6H_5COOH + Zn/Hg \xrightarrow{\text{Conc. HCl}}$ No reaction

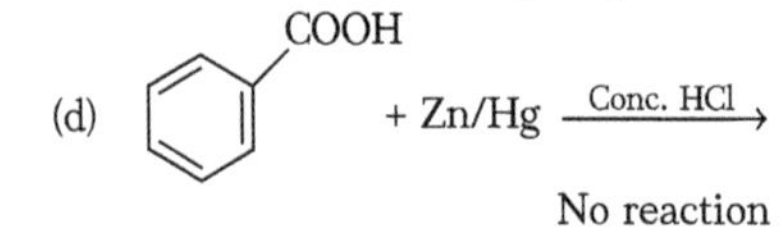

Thus, benzaldehyde cannot be prepared from the reagents given in option (d).

56. *(b)* The polymer obtained by the copolymerisation of styrene and butadiene is buna-S. It is an elastomer.

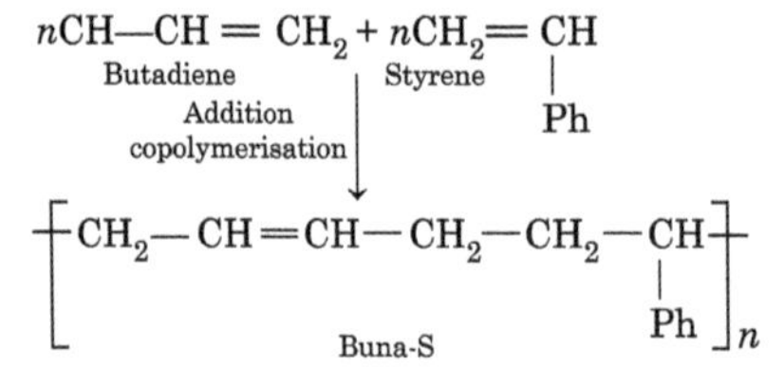

$nCH—CH=CH_2 + nCH_2=CH(Ph)$ (Butadiene, Styrene) $\xrightarrow{\text{Addition copolymerisation}}$ $[—CH_2—CH=CH—CH_2—CH_2—CH(Ph)—]_n$ Buna-S

57. *(a)* In water,

$Pb(OH)_2 \rightleftharpoons \underset{S}{Pb^{2}} + \underset{2S}{2OH^-}$

$K_{sp} = 4S^3 = 4(6.7 \times 10^{-6})^3 = 1.2 \times 10^{-15}$

In buffer of pH = 8

pOH = 6, $[OH^-] = 10^{-6}$

$K_{sp} = S[OH^-]^2$

$$S = \frac{1.2 \times 10^{-15}}{10^{-12}} = 1.2 \times 10^{-3}\ M$$

58. *(a)* Spreading out charge by the overlap of an empty p-orbital with an adjacent σ-bond is called hyperconjugation. This overlap (the hyperconjugation) delocalises the positive charge on the carbocation, spreading it over a larger volume and this stabilises the carbocation.

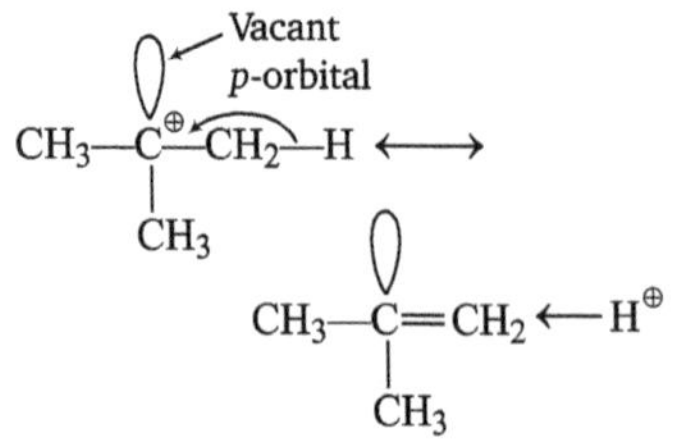

$CH_3—\overset{\oplus}{C}(CH_3)—CH_2—H \longleftrightarrow CH_3—C(CH_3)=CH_2 \longleftarrow H^{\oplus}$ (Vacant p-orbital)

Tertiary butyl carbocation has one vacant p-orbital, hence, it is stabilised by σ-p (empty) hyperconjugation.

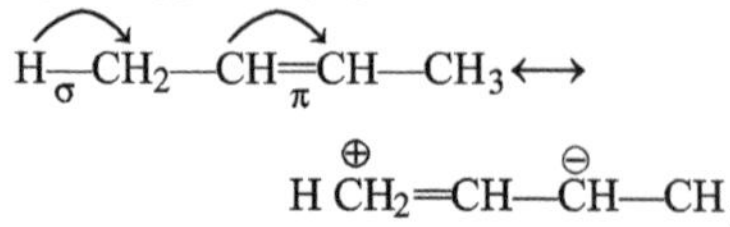

$H \underset{\sigma}{—} CH_2—CH\underset{\pi}{=}CH—CH_3 \longleftrightarrow H\ \overset{\oplus}{C}H_2=CH—\overset{\ominus}{C}H—CH_3$

In but-2-ene, stabilisation is due to hyperconjugation between σ - π^* electron delocalisation.

59. *(b)* Those species, which have one or more than one unpaired electron, are paramagnetic in nature.

The electronic configuration of given species are as follows:

(a) NO = 15

$= \sigma 1s^2 \sigma^* 1s^2 \sigma 2s^2 \sigma^* 2s^2 \sigma 2p_z^2\ \pi 2p_x^2 = \pi 2p_y^2$
$\pi^* 2p_x^1$

(b) CO = 14

$1s^2\sigma^* 1s^2 \sigma 2s^2 \sigma^* 2s^2 \sigma 2p_z^2\ \pi 2p_x^2 = \pi 2p_y^2$

(c) $O_2 = 16 = \sigma 1s^2 \sigma^* 1s^2 \sigma 2s^2 \sigma^* 2s^2 \sigma 2p_z^2$

$\pi 2p_x^2 = \pi 2p_y^2\ \pi^* 2p_x^1 = \pi^* 2p_y^1$

(d) $B_2 = 10 = \sigma 1s^2 \sigma^* 1s^2 \sigma 2s^2 \sigma^* 2s^2$

$\pi 2p_x^1 = \pi 2p_y^1$

As all the electrons in CO are paired, so it is not a paramagnetic species.

60 *(b)* Number of spherical nodes in oribital $= n - l - 1$

For, $3p$ - orbital $n = 3$, $l = 1$

∴ No. of spherical nodes $= 3 - 1 - 1 = 1$

61. *(c)* Acetylcholine is a neurotransmitter. It transmits message across the synapse (excitatory synapse). The drug curare blocks the post-synaptic membrane receptors, so that the muscles remain relaxed during surgical operations.

62. *(a)* Viroids are small, single-stranded, circular RNA molecules not enclosed by protein coat. They were discovered by TO Diener in 1971. Viroid replication requires host encoded RNA polymerase.

63. *(d)* Endosperm in a gymnospermic plant is haploid structure, which is formed without fusion of gametes and represents female gametophyte.

$2n = 16, n = 8$

So, the number of chromosomes in endosperm of gymnosperm will be 8.

64. *(b)* The difference between rough and smooth Endoplasmic Reticulum (ER) is that rough endoplasmic reticulum contains ribosomes. Endoplasmic reticulum is a network of interconnected cisternae, tubules and vesicles present in cytoplasm.

Depending on the presence or absence of ribosomes it is of two types

(i) **Rough ER** It has ribosomes attached on its surface by ribophorin.

(ii) **Smooth ER** The ribosomes are not attached to its surface.

65. *(d)* Cellulose is the most important structural component of the cell wall of plants. It is a linear polymer of β-D-glucose units connected through β-1,4-glycosidic linkage.

66. *(b)* The interphase though called the resting phase is the time during which the cell prepares itself for division by undergoing both cell growth and DNA replication in an orderly manner. In interphase, nucleus is enclosed by a porous double nuclear membrane which does not disappear till the time M-phase does not occur in which actual division of cell takes place.

67. *(b)* Decrease in CO_2 concentration and less H^+ ion concentration is responsible for the opening of stomata. According to active K^+ transport theory (Levitt), rise in pH of the guard cells (due to increased concentration of H^+ present) causes hydrolysis of starch into phosphoenol pyruvate, which further combines with CO_2 present in guard cells to form oxalic acid which gets changed to malic acid.

H^+ ions pass out of the guard cells in exchange of K^+ ions that react with malate inside guard cells. This leads to absorption of water from adjacent cells, which increases the turgor of guard cells and causes opening of stomata.

68. *(c)* The reaction given, i.e. glyceraldehyde-3-phosphate → 1, 3 bisphosphoglyceric acid is the step of glycolysis and it does not undergo decarboxylation.

69. *(d)* About 120 mL glomerular filtrate is produced per minute,

i.e.$120 \times 1440\,\text{min} = 172800\,\text{ml}$

approximately 170 L filtrate is produced in 24 h. The blood pressure in glomerular capillaries becomes very high, so that there is continuous process of ultrafiltration. Glomerular filtrate contains a large amount of water and other dissolved substances such as urea, uric acid, creatinine, amino acids, glucose, sodium, potassium, vitamins, etc.

70. *(d)* Thymus gland secretes thymosine hormone which plays a major role in differentiation of T-lymphocytes which are responsible for cell-mediated immunity. It also promotes production of antibodies.

71. *(c)* If mammalian ovum fails to get fertilised, the corpus luteum degenerates. This causes disintegration of endometrium leading to menstruation and start of a new cycle. Progesterone secretion declines and the primary follicle starts to develop again.

72. *(c)* In incomplete dominance, genotypic and phenotypic ratio remain same and is 1 : 2 : 1. Intermediate inheritance is incomplete dominance in which dominant factor of a heterozygote does not completely mask the expression of a recessive allele.

73. *(b)* If the mutation occurs in the middle gene of structural gene, i.e. '*y*' gene permease will not be synthesised. *Lac* operon is a transcriptionally regulated system where a polycistronic structural gene is regulated by a common promoter and regulatory gene. It has three structural genes, i.e. *z,y,a,* out of which '*z*' gene codes for β-galactosidase, '*y*' gene, i.e. middle gene codes for permease and '*a*' gene codes for transacetylase.

74. *(c)* Out of the given options, the right option is that proline (Pro) which is encoded by four genetic codes, i.e. CCU, CCC, CCA and CCG and threonine (Thr) which is also encoded by four codons, i.e. ACU, ACC, ACA and ACG. The genetic code consists of 64 triplets of nucleotides. These triplets are called codons. With three exceptions each codon encodes for one of the 20 amino acids used in the synthesis of proteins. But most of the amino acids are encoded by more than one codon.

75. *(c)* Modern theory of origin of life was proposed by AI Oparin and JBS Haldane. As per this theory, origin of life is the result of long series of physicochemical changes which was bought about first by chemical evolution and then by biological evolution.

76. *(a)* Lymphocytes are the agranular leukocytes. They manufacture the serum, globin and secrete antibodies. They play a key role in immunological reactions.

77. *(c)* Dipsomania is a historical term, describing a medical condition which involves severe addiction to alcohol. The term was coined by a German physician CW Hufeland (1819).

78. *(c)* *Petunia* belongs to family–Solanaceae and is an ornamental plant.

79. *(a)* According to 10% law of energy flow by Lindemann,

Plant → 20 J

Mice → $20 \times 10\% = 2\,\text{J}$

Snake → $2 \times 10\% = 0.2\,\text{J}$

Peacock → $0.2 \times 10\% = 0.02\,\text{J}$

80. *(b)* Prolonged exposure to Mn can cause sterility in humans. The effect of manganese occurs, mainly in the respiratory tract and in brain. It can also cause Parkinson's disease, lung embolism and bronchitis.

81. *(b)* We have,

$$a^2 + b^2 = 7$$

and $$a^3 + b^3 = 10$$

$\Rightarrow a^3 + b^3 = (a+b)(a^2+b^2-ab)$

$\Rightarrow 10 = (a+b)(7-ab)$

$\Rightarrow 10 = (a+b)\left(7 - \dfrac{(a+b)^2-7}{2}\right)$

$[\because a^2+b^2 = (a+b)^2 - 2ab]$

$\Rightarrow 20 = (a+b)(21-(a+b)^2)$

$\Rightarrow (a+b)^3 - 21(a+b) + 20 = 0$

Let $a + b = x$

$\therefore x^3 - 21x + 20 = 0$

$(x-1)(x-4)(x+5) = 0$

$\therefore x = 1, 4, -5$

$a + b = 1, 4, -5$

$|a+b| = 1, 4, 5$

$\therefore$ Maximum value of $|a+b| = 5$

82. *(c)* We have,

$y = \sin^{-1}(\sin x) = x$ in $[0, \pi/2]$

$= \pi - x$ in $[\pi/2, \pi]$

$y = [\sin^{-1}(\sin x)] = 0$ in $[0, 1)$

$= 1$ in $[1, \pi - 1)$

$= 0$ in $[\pi - 1, \pi]$

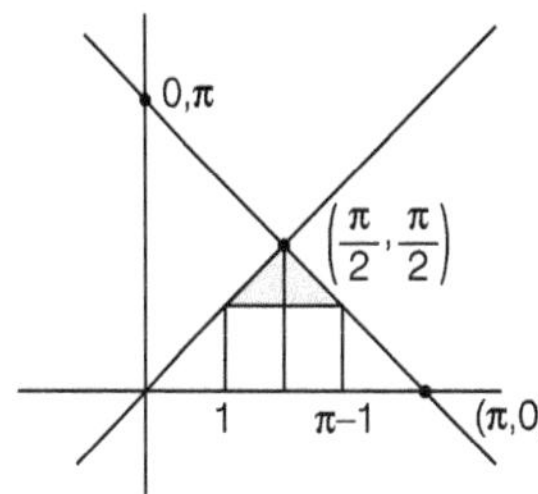

Required area $= \dfrac{1}{2}(\pi - 2)\left(\dfrac{\pi}{2} - 1\right)$

$= \dfrac{(\pi-2)^2}{4}$

83. *(b)* We have,

$$I_1 = \int_0^1 \frac{e^x}{x+3}\,dx$$

$$I_2 = \int_0^1 \frac{x^3}{e^{x^4}(4-x^4)}\,dx$$

In I_2 put $x^4 = t$

$\Rightarrow \quad 4x^3\,dx = dt$

$$\therefore \quad I_2 = \frac{1}{4}\int_0^1 \frac{dt}{e^t(4-t)}$$

$$I_2 = \frac{1}{4}\int_0^1 \frac{dt}{e^{1-t}(4-(1-t))}$$

$$\left[\because \int_0^a f(x)\,dx = \int_0^a f(a-x)\,dx\right]$$

$$I_2 = \frac{1}{4}\int_0^1 \frac{e^t}{e(3+t)}\,dt$$

$$I_2 = \frac{1}{4e}I_1 \Rightarrow \frac{I_1}{I_2} = 4e$$

84. *(c)* Given,

$$\sin\alpha + \cos\beta = \frac{1}{\sqrt{2}} \quad \text{...(i)}$$

$$\cos\alpha + \sin\beta = \frac{\sqrt{2}}{\sqrt{3}} \quad \text{...(ii)}$$

On subtracting Eq. (ii) from Eq. (i), we get

$$(\sin\alpha - \sin\beta) + (\cos\beta - \cos\alpha) = \frac{1}{\sqrt{2}} - \frac{\sqrt{2}}{\sqrt{3}}$$

$$2\cos\frac{\alpha+\beta}{2}\sin\left(\frac{\alpha-\beta}{2}\right) + 2\sin\frac{\alpha+\beta}{2}$$

$$\sin\frac{\alpha-\beta}{2} = \frac{\sqrt{3}-2}{\sqrt{6}}$$

$$2\sin\left(\frac{\alpha+\beta}{2}\right)\left[\cos\frac{\alpha+\beta}{2} + \sin\frac{\alpha+\beta}{2}\right] = \frac{\sqrt{3}-2}{\sqrt{6}} \quad \text{...(iii)}$$

On adding Eqs. (i) and (ii), we get

$$2\sin\frac{\alpha+\beta}{2}\cos\frac{\alpha-\beta}{2} + 2\cos\frac{\alpha+\beta}{2}\cos\frac{\alpha-\beta}{2} = \frac{\sqrt{3}+2}{\sqrt{6}}$$

$$2\cos\left(\frac{\alpha-\beta}{2}\right)\left[\sin\frac{\alpha+\beta}{2} + \cos\frac{\alpha+\beta}{2}\right] = \frac{\sqrt{3}+2}{\sqrt{6}} \quad \text{...(iv)}$$

On dividing Eq. (iii) by Eq. (iv), we get

$$\tan\left(\frac{\alpha-\beta}{2}\right) = \frac{\sqrt{3}-2}{\sqrt{3}+2} = 4\sqrt{3}-7$$

85. *(a)* The unit digit in $(15)^{981}$ is 5.

The unit digit in $(9)^{789}$ is 9.

The unit digit in $(6)^{50}$ is 6

$\therefore$ The unit digit in $(15)^{981} + (9)^{789} - (6)^{50}$ is $5 + 9 - 6 = 8$.

86. *(d)* Given,

Probability of throwing a number is proportional to K^2.

$$\therefore \quad \sum_{K=1}^{6} \lambda K^2 = 1$$

$$\Rightarrow \quad \frac{\lambda(6)(6+1)(12+1)}{6} = 1 \Rightarrow \lambda = \frac{1}{91}$$

$P(A)$ = Probability of not getting an even number

$$= \lambda(1^2 + 3^2 + 5^2)$$

$$= \lambda(35) = \frac{35}{91}$$

$P(B)$ = Probability of getting 3

$$= \lambda(3)^2 = \frac{9}{91}$$

$$P(B/A) = \frac{(A\cap B)}{P(A)} = \frac{P(B)}{P(A)}$$

$$= \frac{9/91}{35/91} = 9/35$$

87. *(b)* Equation of tangent $\frac{x^2}{9} - \frac{y^2}{4} = 1$ is

$$y = mx + \sqrt{9m^2-4},\ m > 0$$

Equation of circle $(x-4)^2 + y^2 = 16$ is

$$y = mx - 4m + 4\sqrt{1+m^2}$$

Equating the constant terms, we get

$$\sqrt{9m^2-4} = -4m + 4\sqrt{1+m^2}$$

On squaring, we get

$$9m^2 - 4 = 16m^2 - 32m\sqrt{1+m^2} + 16 + 16m^2$$

$$= 23m^2 + 20 = 32m\sqrt{1+m^2}$$

On squaring again, we get

$$529m^4 + 920m^2 + 400 = 1024m^2 + 1024m^4$$

$$\Rightarrow 495m^4 + 104m^2 - 400 = 0$$

$$\Rightarrow (5m^2-4)(99m^2+100) = 0$$

$$\Rightarrow \quad m^2 = \frac{4}{5}$$

$$\Rightarrow \quad m = \frac{2}{\sqrt{5}}$$

Equation of tangent $= y = \frac{2}{\sqrt{5}}x + \frac{4}{\sqrt{5}}$

$$\Rightarrow \quad 2x - \sqrt{5}y + 4 = 0$$

88. *(c)* Slope of $AH = \frac{k}{h}$

Slope of $BC = \frac{4}{-7}$

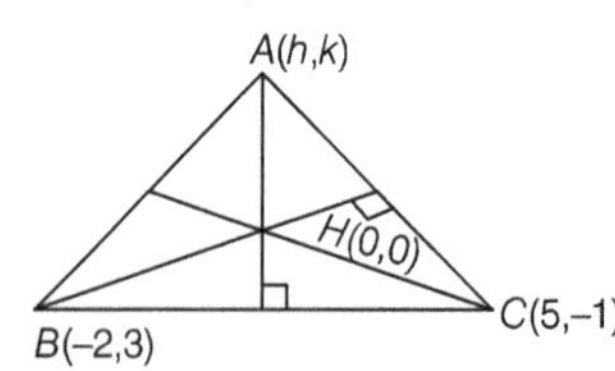

Slope of $AH \times$ slope of $BC = -1$

$$\therefore \quad \frac{k}{h} \times \frac{4}{-7} = -1$$

$$\frac{k}{h} = \frac{7}{4} \quad \text{...(i)}$$

Slope of $AC = \frac{k+1}{h-5}$

Slope of $BH = \frac{-3}{2}$

Slope of $AC \times$ Slope of $BH = -1$

$$\therefore \quad \left(\frac{k+1}{h-5}\right)\left(\frac{-3}{2}\right) = -1$$

$$\Rightarrow \quad \frac{k+1}{h-5} = \frac{2}{3} \quad \text{...(ii)}$$

On solving Eqs. (i) and (ii), we get

$h = -4,\ k = -7$

$\therefore$ Third vertex of triangle is $(-4, -7)$.

89. *(c)* We have,

$f(x) = \log_{\{x\}}[x] + \log_{[x]}\{x\}$

$[x] > 1\, x \notin$ integer

$\therefore x \in (2, 3)$ only satisfy the option.

90. *(a)* We have,

$$\lim_{x\to 1}\int_4^{f(x)} \frac{2t}{x-1}\,dt$$

$$\lim_{x\to 1}\frac{\int_4^{f(x)} 2t\,dt}{x-1}$$

Apply Leibnitz's rule, we get

$$\lim_{x\to 1}\frac{2f(x)\cdot f'(x)}{1}$$

$$= 2f(1)\,f'(1)$$

$$= 2\cdot(4)\cdot f'(1) \quad [\because f(1) = 4]$$

$$= 8f'(1)$$

91. *(c)* Following is the vector diagram of the momenta involved if a photon were to materialise into an electron-positron pair in empty space.

For momentum to conserve,

$$\frac{hf}{c} = 2p\cos\theta$$

$$\Rightarrow \quad hf = 2pc\cdot\cos\theta$$

(here, f = frequency of incident photon)

As, $p = mv$

$$hf = 2mc^2\left(\frac{v}{c}\right)\cos\theta$$

As, $\frac{v}{c} < 1$ and $\cos\theta \le 1$

We can say that,

$$hf < 2mc^2 \quad \text{... (i)}$$

But energy conservation required

$$hf = 2mc^2 \quad \text{... (ii)}$$

Hence, it is impossible to satisfy both Eqs. (i) and (ii) simultaneously.

But if there is some other object is present to carry away part of initial momentum, then both condition (i) and (ii) can be satisfied.

92. *(a)* $a = \frac{dv}{dt} = -\alpha\sqrt{v}$

$$\Rightarrow \int_{v_0}^{v} \frac{dv}{\sqrt{v}} = -\alpha \int_{t_0=0}^{t} dt$$

(here, to is initial time)

$$\Rightarrow \quad v = \left(\sqrt{v_0} - \frac{\alpha t}{2}\right)^2 = v_0 - \alpha\sqrt{v_0}.t + \frac{\alpha^2 t^2}{4}$$

$$\Rightarrow \frac{dx}{dt} = v_0 - \alpha\sqrt{v_0}t + \frac{\alpha^2 t^2}{4}$$

Integrating

$$\Rightarrow x = v_0 t - \frac{\alpha\sqrt{v_0}t^2}{2} + \frac{\alpha^2 t^3}{12}$$

Particle stops, when $v = 0$

$$\Rightarrow \quad \sqrt{v_0} - \frac{\alpha t_0}{2} = 0$$

$$\Rightarrow \quad t_0 = \frac{2\sqrt{v_0}}{\alpha}$$

Substituting in Eq. (i), we get

$$x = v_0 \times \frac{2\sqrt{v_0}}{\alpha} - \frac{\alpha\sqrt{v_0}}{2} \times \frac{4v_0}{\alpha^2} + \frac{\alpha^2}{12} \times \frac{8v_0^{3/2}}{\alpha^3}$$

$$\Rightarrow x = \frac{2v_0^{3/2}}{3\alpha}$$

93. *(c)* Let candles burn down by l_1 and l_2 and the shadows shorten by x_1 and x_2.

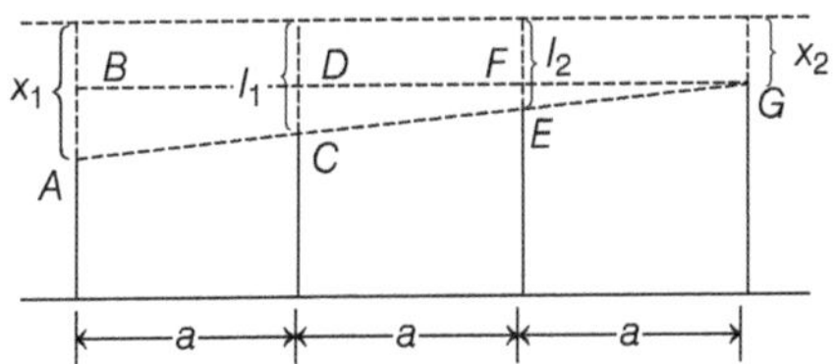

From similar triangles ΔABG, ΔDCG and ΔFEG, we have

$$\frac{x_1 - x_2}{3a} = \frac{l_1 - x_2}{2a} = \frac{l_2 - x_2}{a}$$

$$\Rightarrow 2(x_1 - x_2) = 3(l_1 - x_2) = 6(l_2 - x_2)$$

where, $\dot{l}_1 = \frac{h}{t_1} = \frac{15}{15 \times 60} = \frac{1}{6}\ \text{cms}^{-1}$

and $\dot{l}_2 = \frac{h}{t_2} = \frac{15}{1 \times 60} = \frac{1}{4}\ \text{cm s}^{-1}$

Let $\dot{x}_1 = v_A$ and $\dot{x}_2 = v_B$. Then,

$$2(v_A - v_B) = 3\left(\frac{1}{6} - v_B\right) = 6\left(\frac{1}{4} - v_B\right)$$

$$\Rightarrow \quad v_A = \frac{1}{3}\ \text{cm s}^{-1}$$

$$v_B = \frac{1}{12}\ \text{cm s}^{-1}$$

94. *(d)* $s(t) = \int_0^t |v_x|\, dt = \frac{t^2}{2},\ 0 \le t \le 1$

$$t - \frac{1}{2},\ 1 < t \le 3$$

$$4t - \frac{t^2}{2} - 5,\ 3 < t \le 4$$

$$\frac{t^2}{2} - 4t + 11,\ 4 < t \le 6$$

$$14t - t^2 - 43,\ 6 < t \le 7$$

95. *(a)* Effective air separation between charges $= r - t + t\sqrt{k}$

$$= 10 - 5 + 5\sqrt{4} = 20 - 5 = 15\,\text{cm}$$

$$\therefore \ F_1 \propto \frac{q_1 q_2}{(10)^2} \text{ and } F_2 \propto \frac{q_1 q_2}{15^2}$$

$$\therefore \ \frac{F_2}{F_1} = \frac{(10)^2}{(15)^2} \Rightarrow F_2 = \frac{4}{9}F_1$$

96. *(d)* Let P luminous intensity of source, then amount of light diverging from source in 1 s is $4\pi P$.

Light escapes in air through a cone of semi-vertical angle i_c,

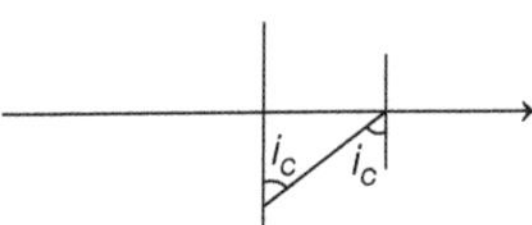

So, fraction of light that escapes in the air is

$$f = \frac{2\pi P(1 - \cos i_c)}{4\pi P} = \frac{1}{2}(1 - \cos i_c)$$

$$= \frac{1}{2}(1 - \sqrt{1 - \sin^2 i_c})$$

This fraction is independent of depth h, of sources below the surface.

97. *(b)* In first process,

$$\Delta Q = \Delta U + 0 \Rightarrow C_V \Delta T = \frac{R}{\gamma - 1} \cdot \Delta T$$

Also, $pV = RT$ and $3pV = R(T + \Delta T)$

$\Rightarrow \ 2pV = R\Delta T$

$\therefore \ 2RT = R\Delta T \Rightarrow \Delta T = 2T$

Hence, final temperature is $3T$.

In second process,

$$(3p)V = R(3T)$$

$$\therefore \quad T_{\text{Final}} = \frac{6RT}{R} = 6T$$

and $\quad \Delta Q = C_V(6T - 3T) + 3\phi(2V - V)$

$$\Rightarrow \quad \Delta Q = C_V 3T + 3pV$$

$$= \frac{R(3T)}{\gamma - 1} + 3RT$$

So, total heat given is

$$Q_{\text{Total}} = \frac{R(2T)}{\gamma - 1} + \frac{R(3T)}{\gamma - 1} + 3RT$$

$$= \frac{5RT}{\gamma - 1} + 3RT$$

As, $\ Q_{\text{Total}} = C(6T - T)$

We have,

$$5C = \frac{5R}{\gamma - 1} + 3R$$

$$= \frac{R}{7/5 - 1} + \frac{3}{5}R = \frac{31}{30}R$$

98. *(a)* $v = 3t^2 - 24t + 36$

$$= 3(t^2 - 8t + 12)$$

$$= 3(t^2 - 6t - 2t + 12)$$

$$= 3(t - 6)(t - 2)$$

$$v = 0 \text{ at } t = 6\,\text{s and } t = 2\,\text{s}$$

At $t = 2$ s, $x = 32$ and at $t = 6$ s, $x = 0$

At $t = 8$ s, $x = 32$.

99. *(c)*

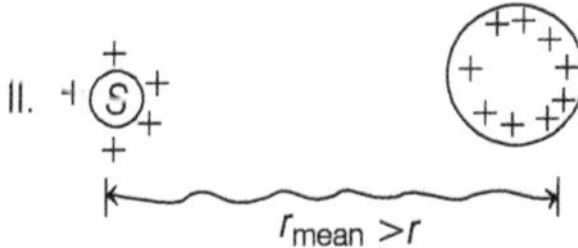

In II case, mean distance between charges is more due to mutual repulsion.

$$\therefore \quad E_2 < E_1$$

100. *(b)* $v_{\text{outflow}} = \frac{\text{Volume flow rate}}{\text{Area}}$

$$= \frac{5 \times 10^{-4}}{1 \times 10^{-4}} = 5\ \text{ms}^{-1}$$

For vertical motion of spray,

$$u_y = u \cdot \sin\theta = 5 \times \frac{1}{\sqrt{2}}$$

[For range maximum, $\theta = 45°$]

$$h = -5\,\text{m}$$

Using $h = ut + \frac{1}{2}at^2$, we have

$$\Rightarrow \quad -5 = \frac{5}{\sqrt{2}} \cdot t - \frac{1}{2} \times 10 \times t^2$$

$$\Rightarrow \quad 5t^2 - \frac{5}{\sqrt{2}}t - 5 = 0$$

$$t^2 - \frac{t}{\sqrt{2}} - 1 = 0$$

Applying Sridharacharya rule, we get

$$\Rightarrow \quad t = \frac{\frac{1}{\sqrt{2}} \pm \sqrt{\frac{1}{2} - 4(1)(-1)}}{2 \times 1} = \frac{\frac{1}{\sqrt{2}} \pm \frac{3}{\sqrt{2}}}{2}$$

$$= \frac{4}{2\sqrt{2}} = \sqrt{2}\ \text{s}$$

During this time water keeps on moving in forward horizontal direction with speed $u\cos\theta$.

So, water spray can cover a distance is $u\cos\theta \times t$.

Radius of circular region of spray

$$= 5 \times \frac{1}{\sqrt{2}} \times \sqrt{2} = 5\text{ m}$$

Hence, Area irrigated $= \pi R^2 = 25\pi\text{ m}^2$

101. *(b, d)* (a) The unit cell of CsCl has bcc arrangement of ions in which each ion has eight oppositely charged ions around it, in the nearest neighbours as shown below :

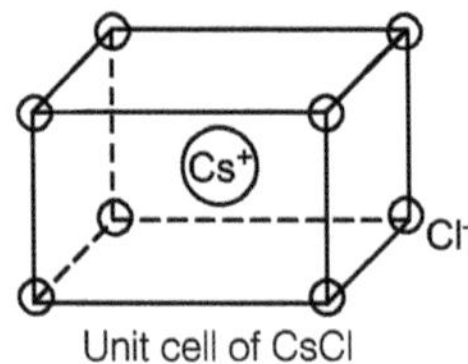

Unit cell of CsCl

(b) In bcc, coordination number of metal is 8.

(c) In an unit cell, a corner is shared in eight unit cells and a face centre is shared between two adjacent unit cells.

(d) In NaCl unit cell; $2(r_{Na^+} + r_{Cl^-}) = a$

$\Rightarrow \quad a = 2(95 + 181) = 552$ pm

Thus, incorrect statements in given options (b) and (d).

102. *(a)* Optical isomerism arises in those compounds where the mirror images are non-superimposable.

Complexes like $[M(AA)_2X_2]$ (only *cis* form) and $[M(AA)_3]$ exhibit optical isomerism.

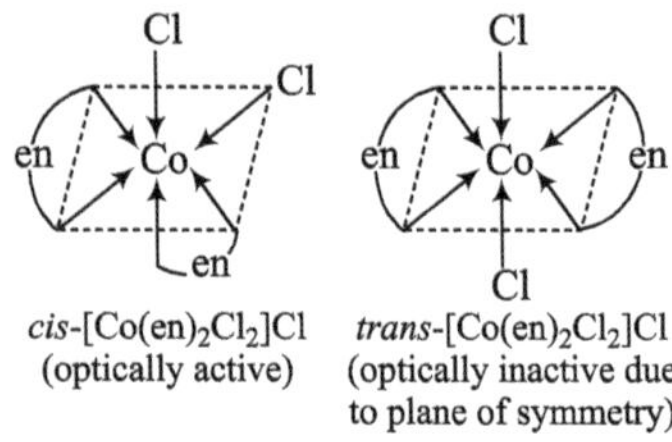

cis-$[Co(en)_2Cl_2]Cl$ (optically active) *trans*-$[Co(en)_2Cl_2]Cl$ (optically inactive due to plane of symmetry)

$[Co(NH_3)_4Cl_2]Cl$ can exist in both *cis* and *trans* forms that are given below:

trans-$[Co(NH_3)_4Cl_2]Cl$ (optically inactive) *cis*-$[Co(NH_3)_4Cl_2]Cl$ (optically inactive)

$[Co(NH_3)_4Cl_3]$ exists in *fac* and *mer*-isomeric forms and both are optically inactive.

fac-isomer (optically inactive) *mer*-isomer (optically inactive)

Thus, the correct option is (a).

103. *(c)* NH_2 (aniline) $\xrightarrow[\text{Acetylation}]{1.CH_3COCl}$ $NHCOCH_6$ Acetanilide

$NHCOCH_3$ $\xrightarrow{Br_3}$ p-Br-acetanilide (Major) + o-Br-acetanilide (Minor)

$\downarrow KOH/C_2H_5OH$

NH_2 / Br — *p*-bromo aniline

Note NH_2 group is highly activating group, hence reaction can take place at all the *ortho*/*para* positions. However, if the ring is deactivated by acetylation then mono haloaniline derivative (*para*) is obtained as major product.

104. *(d)* Molarity *(M)*

$$= \frac{\text{Number of moles of solute}}{\text{Volume of solution (in L)}}$$

$\therefore$ Number of moles of complex

$$= \frac{\text{Molarity} \times \text{volume (in mL)}}{1000}$$

$$= \frac{0.1 \times 100}{1000} = 0.01\text{ mol}$$

Number of moles of ions precipitate

$$= \frac{1.2 \times 10^{22}}{6.02 \times 10^{23}}$$

$$= 0.02\text{ mol}$$

$\therefore$ Number of Cl^- present in ionisation sphere

$$= \frac{\text{Number of moles of ions precipitated}}{\text{Number of moles of complex}}$$

$$= \frac{0.02}{0.01} = 2$$

$\therefore 2\ Cl^-$ are present outside the square brackets, i.e. in ionisation sphere. Thus, the formula of complex is

$$[Co(H_2O)_5Cl]Cl_2 \cdot H_2O$$

105. *(b)* $CH_3—CH_2—CH_2Br \xrightarrow[80°C]{\text{Alc· NaOH}(X)} CH_3—CH{=}CH_2$

$\xrightarrow{HBr(Y)} CH_3—\underset{Br}{CH}—CH_3$

Step 1 Haloalkanes on reaction with alcoholic NaOH undergoes elimination reaction to give alkenes.

Step 2 The formed alkene then reacts with HBr to give monobrominated product. This product is in accordance markonwnikoff addition.

106. *(c)* The van der Waals' equation of state is

$$\left(p + \frac{n^2a}{V^2}\right)(V - nb) = nRT$$

For one mole and when $b = 0$, the above equation condenses to

$$\left(p + \frac{a}{V^2}\right)V = RT$$

$$\Rightarrow \quad pV = RT - \frac{a}{V} \quad \ldots\text{(i)}$$

Eq. (i) is a straight line equation between pV and $\frac{1}{V}$ whose slope is '$-a$'. Equating with slope of the straight line given in the graph.

$$-a = \frac{20.1 - 21.6}{3 - 2} = -1.5$$

$$\Rightarrow \quad a = 1.5$$

107. *(a)* $\Lambda^c_m = 40.1\text{ S cm}^2\text{ mol}^{-1}$

$$\Lambda^\circ_m = \Lambda^\circ_{m(HCOO^-)} + \Lambda^\circ_{m(H^+)}$$

$$= (54.6 + 349.6)\text{ S cm}^2\text{ mol}^{-1}$$

$$= 404.2\text{ S cm}^2\text{ mol}^{-1}$$

$$\alpha = \frac{\Lambda^c_m}{\Lambda^\circ_m} = \frac{40.1}{404.2} = 0.0992$$

Now, $HCOOH \overset{H_2O}{\rightleftharpoons} HCOO^-(aq) + H^+(aq)$

Initial conc.	C	0	0
Equil. conc.	$C(1-\alpha)$	$C\alpha$	$C\alpha$

$$K_a = \frac{C\alpha^2}{1-\alpha}$$

$$= \frac{0.25 \times (0.0992)^2}{(1 - 0.0992)}$$

$$= 2.73 \times 10^{-3}\text{ mol L}^{-1}$$

108. *(a)* Val-Gly-Phe-Val-Ala-Val

Random hydrolysis gives Ala-Val dipeptide fragment, which indicates the connected presence of Ala and Val. Thus, the correct sequence is given in option (a).

109. *(d)* Molecular formula $C_3H_8O(C_nH_{2n+2}O)$ suggests that the organic compound is either alcohol or ether. Since, the compound on reaction with HI gives two different compounds, it must be an unsymmetrical ether and its formula must be $CH_3OC_2H_5$ (methoxyethane).

The reactions are as follows:

$$\underset{\text{Methoxyethane (A)}}{CH_3OC_2H_5} + 2HI \longrightarrow \underset{X}{CH_3I} + \underset{Y}{C_2H_5I} + H_2O$$

$$\underset{Y}{C_2H_5I} + \underset{\text{Aqueous}}{2NaOH} \longrightarrow \underset{Z}{C_2H_5OH} + NaI$$

$$\underset{(Z)}{C_2H_5OH} + \underset{\text{Aqueous}}{6NaOH} + 3I_2 \longrightarrow \underset{\text{Iodoform}}{CHI_3} + HCO\overset{-}{O}\overset{+}{Na} + H_2O + 5NaI$$

Thus, the compound '*A*' is methoxyethane.

110. *(c)*

$$C_6H_{11}-C(CH_3){=}O \xrightarrow{(C_6H_5)_3P^+\ddot{C}HCH_2CH_3} C_6H_{11}-C(CH_3){=}CHCH_2CH_3 \xrightarrow[H_2O_2,\ OH^-]{B_2H_6} C_6H_{11}-\underset{CH_3}{CH}-\underset{CH}{CHCH_2CH_3}$$

Carbonyl compounds reacts with $(C_6H_5)P^+\ddot{C}HCH_2CH_3$ to form phosphoniumylide, which then gives alkenes. The formed alkenes then reacts with B_2H_6 to give addition product, alkane which is in accordance to antimarkownikoff's rule.

111. *(a)*

Column I	Column II
Sap vacuole	Stores and concentrates mineral salts and nutrients
Contractile vacuole	Osmoregulation
Food vacuole	Contains digestive enzyme
Air vacuole	Stores metabolic gases
Spherosome	Stores lipids

112. *(d)* Normal allele

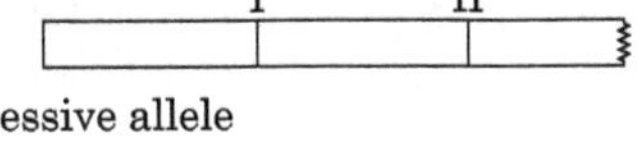

Recessive allele

II

If an individual is a carrier he will have 1 copy of the normal allele and 1 copy of the recessive allele. When digested with restriction enzyme pst I, there will be 5 restriction fragments formed

1 copy
1 copy
1 copy
2 copies

Gel electrophoresis is used for separating the mixture of fragments obtained through a gel in an electrical field. The negatively charged DNA molecules migrate toward the positive electrode at a rate inversely proportional to their length. Thus, the largest fragment will be closest to the negative electrode, while the 3 smaller fragments which are similar in length are closer to the positive electrode.

113. *(a)* Since, all the species of finches differ in their feeding habits. This may be due to the fact that finches developed different kinds of beak in order to feed on different kinds of food.

114. *(a)* The diagram and graph show that newborn deer calves that fall within the 6-8 kg range have the lowest mortality. This results in stabilising selection, where a selection pressure towards a particular range exists.

115. *(d)* Oxidative phosphorylation contributes 34 ATP out of the 38 ATP produced per glucose molecule, which is 89.98%. Glycolysis and the Krebs' cycle contribute 4 ATP.

116. *(a)* Woman *X* will never inherit the BRCA 1 allele since her father has inherited only normal genes from his parents.

117. *(a)* The enzyme *V* is responsible for making amino acid 2 from 1, whilst *Z* is responsible for the conversion of 4 to 5. These two enzymes *V* and *Z* were thus absent in mutant strain.

118. *(c)* In the DNA molecule, the width between the 2 backbones is equal to the width of 1 base pair, i.e. 1 purine and 1 pyrimidine.

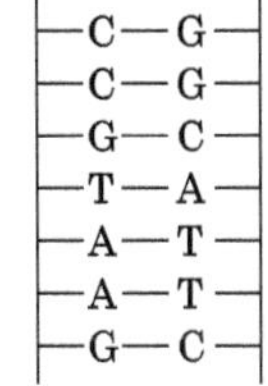

In DNA, the bases pair with complementary bases of the opposite chain *via* hydrogen bonds, adenine to thymine and cytosine to guanine.

Thus, the number of G = C, number of A = T.

Hence, when the percentage concentration of cytosine is similar to guanine, this suggests that source is DNA.

119. *(c)* The rate of reaction can be determined from the gradient of the graph, the faster the rate of reaction, the steeper the gradient while the slower the rate of reaction the more gentle the gradient.

Decreasing the temperature decreases the rate of enzyme controlled reaction, resulting in a graph with a more gradual gradient with 10 arbitrary units of product formed.

120. *(d)* The correct match are

Crustacea – Class
Hominidae – Family
Dermaptera – Order
Ctenophora – Phylum
Archaea – Domain

PRACTICE SET 4

Stream : SB/SX

MM : 160

Instructions

1. There are 120 questions in this paper.
2. The question paper contains two parts; Part I (1 Mark Questions) and Part II (2 Marks Questions).
3. There are four sections in each part; Mathematics, Physics, Chemistry and Biology.
4. Out of the four options given with each question, only one is correct.

PART-I (1 Mark Questions)

MATHEMATICS

1. If in ΔABC, $A = (1, 10)$, circumcentre $= \left(-\frac{1}{3}, \frac{2}{3}\right)$ and orthocentre $\left(\frac{11}{3}, \frac{4}{3}\right)$. The coordinates of the mid-point of the sides opposite to A is

(a) (1, 5) (b) $\left(1, -\frac{11}{3}\right)$ (c) (1, – 3) (d) (1, 6)

2. If $f(x) \cdot f(y) = f(x) + f(y) + f(xy) - 2 \; \forall \; x, y \in R$ and if $f(x)$ is not a constant function, then the value of $f(1)$ is equal to

(a) 1 (b) 2 (c) 0 (d) – 1

3. $\lim_{n \to \infty} (-1)^{n-1} \sin\left(\pi \sqrt{n^2 + \frac{n}{2} + 1}\right)$ is equal to

(a) 0 (b) 1 (c) $\frac{1}{\sqrt{2}}$ (d) $-\frac{1}{\sqrt{2}}$

4. Number of values of x satisfying the equations $5\{x\} = x + [x]$ and $[x] - \{x\} = \frac{1}{2}$ is

(a) 1 (b) 2 (c) 3 (d) 4

5. Given a, b, c are positive integers forming an increasing GP, $b - a$ is a perfect square of a natural number and $\log_6 a + \log_6 b + \log_6 c = 6$, then $a + b + c$ is equal to

(a) 108 (b) 109 (c) 110 (d) 111

6. Let set 'A' has 7 elements and set B has 5 elements. If one function is selected from all possible defined function from A to B, then the probability that it is onto is

(a) $\frac{7! \times 2}{3 \times 5^6}$ (b) $\frac{7!}{10 \times 5^6}$ (c) $\frac{7!}{5^6}$ (d) $\frac{7!}{5^7}$

7. If $I_n = \int_1^e (\ln x)^n \, dx, n \in N$, then $I_{10} + 10 I_9$ is equal to

(a) e^{10} (b) $\frac{e^{10}}{10}$ (c) e (d) $e - 1$

8. If $\frac{a_0}{n+1} + \frac{a_1}{n} + \frac{a_2}{n-1} + \ldots + \frac{a_n - 1}{2} + a_n = 0$, then the equation $a_0 x^n + a_1 x^{n-1} + a_2 x^{n-2} + \ldots + a_{n-1} x + a_n = 0$ has

(a) exactly one root in (0, 1)
(b) no root in (0, 1)
(c) at least one root in (0, 1)
(d) at the most one root in (0, 1)

9. In a right angle ΔABC, the hypotenuse $AB = p$, then $\mathbf{AB} \cdot \mathbf{AC} + \mathbf{BC} \cdot \mathbf{BA} + \mathbf{CA} \cdot \mathbf{CB}$ equals to

(a) $2p^2$ (b) p^2 (c) $\frac{p^2}{2}$ (d) $2p$

10. Two mutually perpendicular tangents of the parabola $y^2 = 4ax$ meet its axis in P_1 and P_2. If S is the focus of the parabola, then $\frac{1}{SP_1} + \frac{1}{SP_2}$ is equal to

(a) $\frac{4}{a}$ (b) $\frac{2}{a}$ (c) $\frac{1}{a}$ (d) $\frac{1}{4a}$

11. Number of ordered pairs (a, x) satisfying the equation $\sec^2(a+2)x + a^2 - 1 = 0; -\pi < x < \pi$ is

(a) 1 (b) 2 (c) 3 (d) 4

12. In a ΔABC, if $\angle A = 30°$ and $BC = 2 + \sqrt{5}$, then the distance of the vertex A from the orthocentre of the ΔABC is

(a) $2(2+\sqrt{5})$ (b) $\sqrt{3}(2+\sqrt{5})$ (c) $\frac{\sqrt{5}+2}{2\sqrt{2}}$ (d) $\frac{2+\sqrt{5}}{\sqrt{3}}$

13. $\int_0^2 \sqrt{x + \sqrt{x + \sqrt{x + \dots \infty}}}\, dx$ is equal to

(a) $\frac{19}{6}$ (b) $\frac{17}{6}$

(c) $\frac{13}{6}$ (d) Cannot be determined

14. The number of solution of the equation $\log(-2x) = 2\log(x+1)$ is

(a) 0 (b) 1 (c) 2 (d) 3

15. The number of ways of arranging the letters *AAAAA*, *BBB*, *CCC*, *D*, *EE* and *F* in a row of the letters *C* are separated from one another is

(a) ${}^{13}C_3 \frac{12!}{5!3!2!}$ (b) $\frac{13!}{5!3!3!2!}$

(c) $\frac{14!}{3!3!2!}$ (d) None of these

16. Let $S = \left\{x \in (-\pi, \pi) : x \neq 0, \pm \frac{\pi}{2}\right\}$. The sum of all distinct solution of the equation $\sqrt{3}\sec x + \operatorname{cosec} x + 2(\tan x - \cot x) = 0$ in the set S is equal to

(a) $-\frac{7\pi}{9}$ (b) $-\frac{2\pi}{9}$ (c) 0 (d) $\frac{5\pi}{9}$

17. Let C be the circle with centre at (1, 1) and radius = 1. If T is the circle centred at $(0, y)$ passing through origin and touching the circle C externally, then the radius of T is equal to

(a) $\frac{\sqrt{3}}{\sqrt{2}}$ (b) $\frac{\sqrt{3}}{2}$ (c) $\frac{1}{2}$ (d) $\frac{1}{4}$

18. The locus of the foot of perpendicular drawn from the centre of the ellipse $x^2 + 3y^2 = 6$ on any tangent to it is

(a) $(x^2 - y^2)^2 = 6x^2 + 2y^2$ (b) $(x^2 - y^2)^2 = 6x^2 - 2y^2$

(c) $(x^2 + y^2)^2 = 6x^2 + 2y^2$ (d) $(x^2 + y^2)^2 = 6x^2 - 2y^2$

19. If $10^9 + 2(11)^1(10)^8 + 3(11)^2(10)^7 + \dots + 10(11)^9 = k(10)^9$, then k is equal to

(a) $\frac{121}{10}$ (b) $\frac{441}{100}$ (c) 100 (d) 110

20. If $\log_{3x-1}(x-2) = \log_{(9x^2-6x+1)}(2x^2 - 10x - 2)$, then x equals to

(a) $9 - \sqrt{15}$ (b) $3 + \sqrt{15}$ (c) $2 + \sqrt{5}$ (d) $6 - \sqrt{5}$

PHYSICS

21. Inner surface of a cone is coated with a reflecting layer forms a conical mirror. A small point source S is placed over the axis of this cone. Minimum angle α of the cone for which rays emitted by source S will be reflected from conical surface only once is

(a) $\alpha = 45°$ (b) $\alpha = 90°$ (c) $\alpha = 120°$ (d) $\alpha = 135°$

22. Consider the following reaction.

$$^{235}_{92}\text{U} + ^{1}_{0}n \longrightarrow ^{83}_{32}\text{Ge} + ^{153}_{60}\text{Nd}$$

Now, the ratio of kinetic energies of Ge and Nd is

(a) 83 : 153 (b) 1 : 1 (c) 11 : 6 (d) 60 : 32

23. A system consists of three coins that can come up either head or tail. Coins are tossed and that results in two tails and one head. Entropy of three coins system will be

(a) $\frac{k_B}{3}$ (b) $3k_B$ (c) $\frac{k_B}{\ln 3}$ (d) $k_B \cdot \ln 3$

24. Position vector of a particle is $\mathbf{r} = 3\cos\omega t\,\hat{\mathbf{i}} + 4\sin\omega t\,\hat{\mathbf{j}}$. Now, choose the correct option.

(a) Particle moves along circular path
(b) Particle moves along parabolic path
(c) Particle moves along elliptical path
(d) Particle moves along straight line

25. Two sound waves A and B have intensities of $10\frac{\mu W}{cm^2}$ and $500\frac{\mu W}{cm^2}$. Difference in their intensity levels is

(a) 490 dB (b) 50 dB (c) 19 dB (d) 17 dB

26. A ball of mass m, moving with a speed of $2v_0$ collides inelastically with an identical ball at rest.

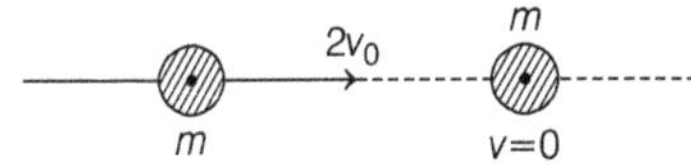

Now, choose the correct option.

(a) For a general head on collision first ball retraces its path backwards
(b) For a general head on collision first ball stops after collision
(c) For a general head on collision angle between velocities of scattered balls is 90°
(d) For a general head on collision angle between velocities of scattered balls is less than 90°

27. An object is placed 40 cm in front of a lens as shown below.

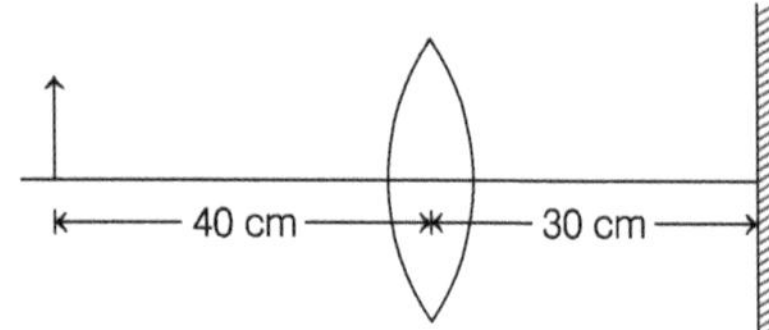

A plane mirror is placed 30 cm behind the lens. Focal length of lens is $f = +8.0$ cm.

I. There are in all three images.
II. Two of images are erect.
III. Two of images are inverted.
IV. Two of images are of same size.

Which of the above statements are correct?

(a) Statements I and IV are correct
(b) Statements II and IV are correct
(c) Statements I, II and IV are correct
(d) Statements I, III and IV are correct

28. Angular position θ of a body changes with time as shown below.

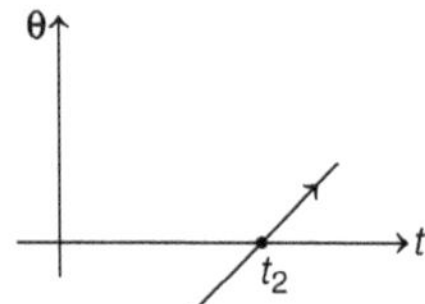

Now, choose the correct option.

(a) Body is initially rotating anti-clockwise, then it starts rotating clockwise
(b) Body is rotating anti-clockwise
(c) Body is rotating clockwise
(d) Cannot be said

29. A particle of mass m is confined to a narrow tube of length L.
If $L = 0.5$ nm, then possible energies of electrons are

(a) 1.5 eV, 6 eV, 13.5 eV......, etc
(b) 2 eV, 4 eV, 6 eV ......, etc
(c) 1.5 eV, 3 eV, 6 eV......, etc
(d) 3 eV, 5 eV, 7 eV......, etc

30. Angular momentum of earth with time varies as

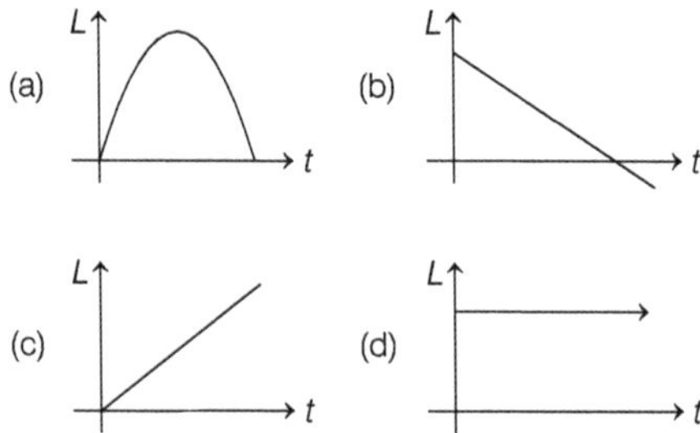

31. Electric field in a region is $\mathbf{E} = \dfrac{a(x\hat{\mathbf{i}} + y\hat{\mathbf{j}} + z\hat{\mathbf{k}})}{x^2 + y^2 + z^2}$

where, a is a positive constant. Charge enclosed in a sphere of radius R and centre at origin is

(a) $4\pi\varepsilon_0 aR$ (b) $4\pi\varepsilon_0 a \log R$
(c) $4\pi\varepsilon_0 aR^2$ (d) $\dfrac{4\pi\varepsilon_0 a}{R}$

32. A triangular lamina is given,

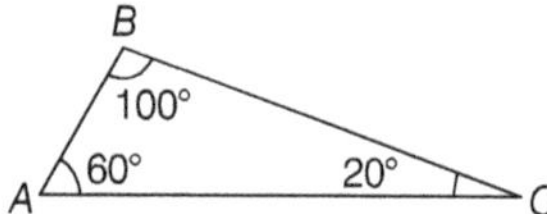

Let, I_{AB}, I_{BC} and I_{AC} are the moments of inertia of lamina about AB, BC and AC.
Now, choose the correct option.

(a) $I_{AB} + I_{BC} = I_{AC}$ (b) $I_{AC} > I_{AB} < I_{BC}$
(c) $(I_{AB})^2 + (I_{BC})^2 = (I_{AC})^2$ (d) $I_{AB} > I_{BC} > I_{AC}$

33. Water is filled to a height H in a dam-with a stopping wall of width ω.

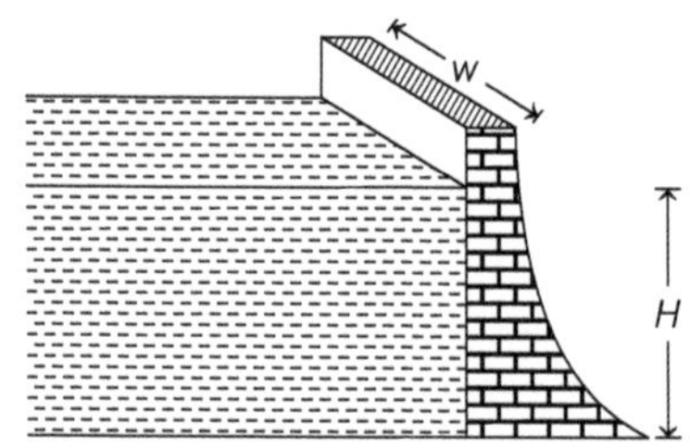

Force on face of dam is (take, g = 10 ms^{-2})

(a) $\omega\rho g H^2$ (b) $\dfrac{2}{3}\omega\rho g H^2$ (c) $\dfrac{1}{2}\rho g \omega H^2$ (d) $\dfrac{1}{3}\omega\rho g H^2$

34. Consider two rectangular wave pulses each of which is moving towards the other with a speed of 1 mms^{-1}.

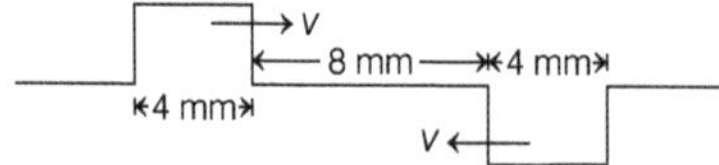

Now, choose the correct option.

(a) After 8s energy is 100% kinetic
(b) After 16s energy is 100% kinetic
(c) After 6s energy is 100% kinetic
(d) After 12s energy is 100% kinetic

35. A parallel plate capacitor has a plate area $A = 200\ \text{cm}^2$ and a plate separation of 2 mm.

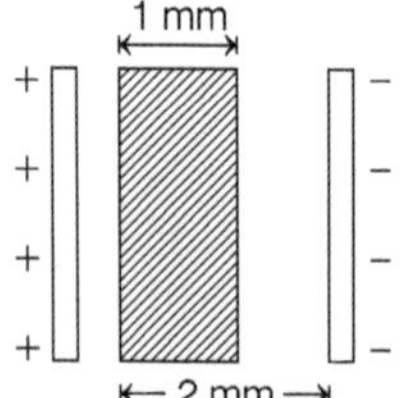

A dielectric slab of dielectric constant $k = 7$ and width 1 mm fills the capacitor. Percentage of energy stored in dielectric is

(a) 12.5% (b) 25% (c) 50% (d) 100%

36. A positively charged ion is accelerated by using a voltage V and then it is allowed to enter a region of perpendicular magnetic field. Frequency of rotation of ion is x. If accelerating voltage is increased four times, then frequency becomes

(a) $2x$ (b) $\frac{x}{2}$ (c) $4x$ (d) x

37. Stationary wave over a stretched string is $y(x, t) = 0.01 \sin 62.8x \cdot \cos 628t$. If at mid point of string there is an anti-node, then

(a) fundamental frequency = 10 Hz
(b) fundamental frequency = 15 Hz
(c) fundamental frequency = 20 Hz
(d) fundamental frequency = 25 Hz

38. A mass m is connected to four ideal springs of same force constant k as shown below.

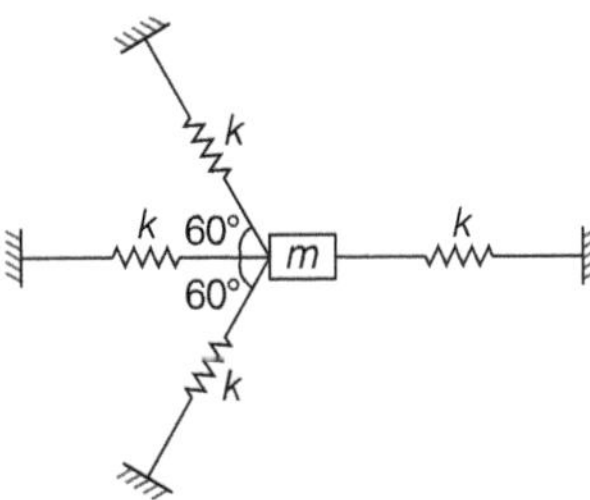

This system is placed on a horizontal frictionless surface. Time period of oscillation when mass is slightly displaced is

(a) $2\pi\sqrt{\frac{m}{k}}$ (b) $2\pi\sqrt{\frac{m}{3k}}$
(c) $2\pi\sqrt{\frac{3m}{4k}}$ (d) $2\pi\sqrt{\frac{2m}{5k}}$

39. Two particles are thrown from top of a tower from same point in opposite directions as shown below.

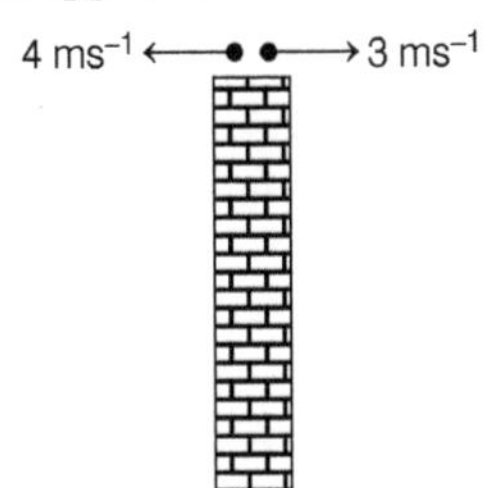

Their velocities are perpendicular after a time interval of

(a) $\sqrt{3}$ s (b) $\frac{\sqrt{3}}{2}$ s (c) $\frac{\sqrt{3}}{5}$ s (d) $\frac{\sqrt{3}}{7}$ s

40. A bead is free to slide down a smooth wire tightly stretched between two points over a vertical wire loop (as shown below).

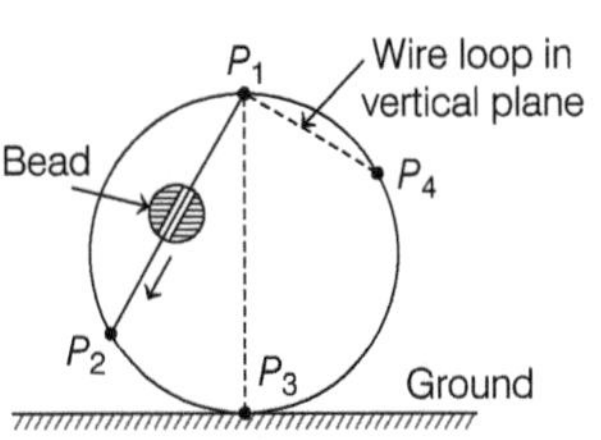

Now, choose the correct option.

(a) Time to slide down along P_1P_4 is maximum
(b) Time to slide down along P_1P_3 is minimum
(c) Time to slide down along P_1P_2 is more than time to slide along P_1P_3
(d) All are incorrect

CHEMISTRY

41. Which of the following molecules has the maximum dipole moment ?

(a) CO_2 (b) CH_4 (c) NH_3 (d) NF_3

42. Which one of the following does not undergo iodoform reaction?

(a) $C_2H_5—CH(OH)—CH_3$ (b) $CH_3—CH(OH)—CH_3$
(c) $C_2H_5—C(=O)—C_2H_5$ (d) $CH_3—CH_2—OH_3$

43. Correct order of bond angles for the following species is

(a) $NH_3 > PCl_3 > BCl_3$ (b) $BCl_3 > NH_3 > PCl_3$
(c) $BCl_3 > PCl_3 > NH_3$ (d) $PCl_3 > BCl_3 > NH_3$

44. The complexes $[Co(NH_3)_6][Cr(CN)_6]$ and $[Cr(NH_3)_6][Co(CN)_6]$ are the examples of which type of isomerism?

(a) Ionisation isomerism (b) Coordination isomerism
(c) Geometrical isomerism (d) Linkage isomerism

45. Which of the following compound is aromatic?

(a) H, + (b) (c) (d) H, +

46. Which of the following represents physical adsorption?

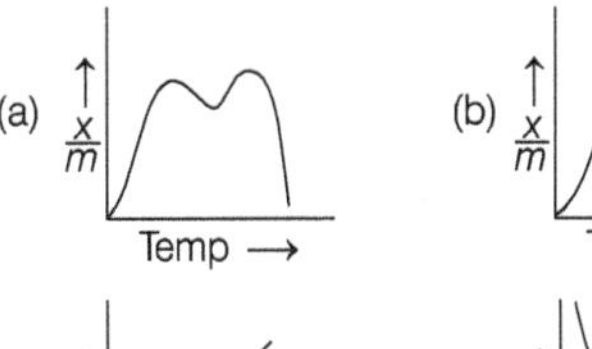

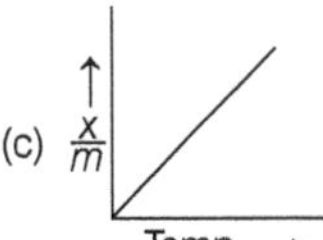

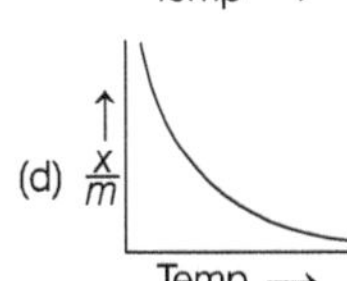

47. In the reversible reaction,

$2NO_2 \underset{k_2}{\overset{k_1}{\rightleftharpoons}} N_2O_4$ the rate of disappearance of NO_2 is equal to

(a) $\frac{2k_1}{k_2}[NO_2]^2$ (b) $2k_1[NO_2]^2 - 2k_2[N_2O_4]$

(c) $2k_1[NO_2]^2 - k_2[N_2O_4]$ (d) $(2k_1 - k_2)[NO_2]$

48. XeF_4 reacts with water at – 80°C to give

(a) $XeOF_2$ (b) $XeOF_4$ (c) XeO_3 (d) XeO_2F_2

49. Calculate the emf of the cell in which the following reaction takes place.

$Ni(s) + 2Ag^+ (0.002\text{ M}) \longrightarrow Ni^{2+} (0.160\text{ M}) + 2Ag(s)$

Given that $E^\circ_{cell} = 1.05\text{ V}$

(a) 0.61 V (b) 0.91 V (c) 0.82 V (d) 0.023 V

50. The structure of diborane (B_2H_6) contains

(a) four 2C – $2e^-$ bonds and four 3C – $2e^-$ bonds
(b) two 2C – $2e^-$ bonds and two 3C – $3e^-$ bonds
(c) two 2C – $2e^-$ bonds and four 3C – $2e^-$ bonds
(d) four 2C – $2e^-$ bonds and two 3C – $2e^-$ bonds

51. Acetophenone when reacted with a base, C_2H_5ONa, yields a stable compound, which has the structure.

(a) $C_6H_5-C(CH_3)=CH-C(=O)-C_6H_5$

(b) $C_6H_5-C(CH_3)-CH_2-C(=O)-C_6H_5$

(c) $C_6H_5-C(CH_3)(OH)-C(CH_3)(OH)-C_6H_5$

(d) $C_6H_5-CH(OH)-CH(OH)-C_6H_5$

52. If one strand of DNA has the sequence ATGCTTGA, the sequence in the complementary strand would be

(a) TCCGAACT (b) TACGTAGT
(c) TACGAACT (d) TACGTAGT

53. Two Faraday of electricity is passed through a solution of $CuSO_4$. The mass of copper deposited at the cathode is (at. mass of Cu = 63.5 u)

(a) 0 g (b) 63.5 g
(c) 2 g (d) 127 g

54. Which of the following is a redox reaction?

(a) $NaCl + KNO_3 \longrightarrow NaNO_2 + KCl$
(b) $CaC_2O_4 + 2HCl \longrightarrow CaCl_2 + H_2C_2O_4$
(c) $Ca(OH)_2 + 2NH_4Cl \longrightarrow CaCl_2 + 2NH_3 + 2H_20$
(d) $2K[Ag(CN)_2] + Zn \longrightarrow 2Ag + K_2[Zn(CN)_4]$

55. Which of the following complexes has the least wavelength of light in visible region?

(a) $[Co(CN)_6]^{3-}$ (b) $[Co(H_2O)_6]^{3+}$
(c) $[Co(NH_3)_6]^{3-}$ (d) $[CoF_6]^{3-}$

56. CH_3CH_2OH can be converted into CH_3CHO by

(a) catalytic hydrogenation
(b) treatment with $LiAlH_4$
(c) treatment with pyridinium chlorochromate
(d) treatment with $KMnO_4$

57. Which of the following represents the correct order of decreasing number of S = O bonds?

(a) $H_2SO_4 > H_2SO_3 > H_2S_2O_8$ (b) $H_2S_2O_8 > H_2SO_3 > H_2SO_4$
(c) $H_2S_2O_8 > H_2SO_4 > H_2SO_3$ (d) $H_2SO_3 > H_2S_2O_8 > H_2SO_4$

58. A unit cell with edge length $a \neq b \neq c$ and axial angles $\alpha = \beta = \gamma = 90°$ is called

(a) cubic (b) tetragonal
(c) orthorhombic (d) hexagonal

59. For the reaction,

$$CaO + 2HCl \longrightarrow CaCl_2 + H_2O$$

1.23 g of CaO is reacted with excess of hydrochloric acid and 1.85 g of $CaCl_2$ is formed. What is the per cent yield?

(a) 76.1 (b) 86.3
(c) 95.1 (d) None of these

60. The major product in the given reaction is

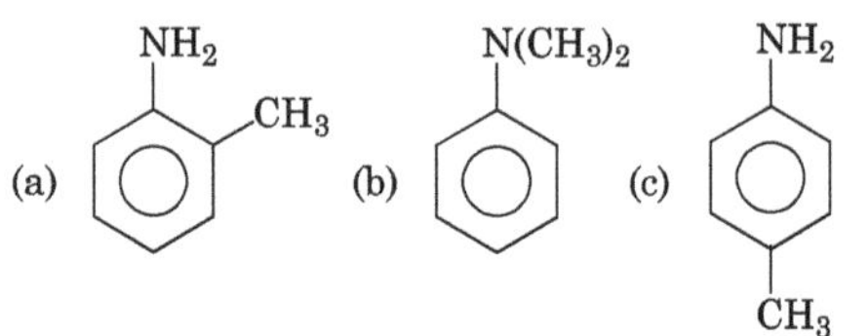

(a) $C_6H_4(NH_2)(CH_3)$ (2-methyl; NH_2, CH_3 ortho) (b) $C_6H_5N(CH_3)_2$ (c) $p\text{-}CH_3C_6H_4NH_2$ (d) $p\text{-}CH_3C_6H_4\overset{+}{N}H_3$

BIOLOGY

61. Single step large mutation leading to speciation is also called

(a) founder's effect (b) saltation
(c) branching descent (d) natural selection

62. Which of the following is correct for immunomodulators?

(a) They always suppress immune system
(b) They never suppress immune system
(c) They always stimulate immune system
(d) Specific immunomodulators stimulate the immunoresponse of immune system, whereas some other immunomodulators inhibit it

63. A transplant between individuals of the same species but with different MHC/HLA alleles is
(a) autograft (b) isograft
(c) xenograft (d) allograft

64. Which of the following is not used as a biopesticide?
(a) *Bacillus thuringiensis*
(b) *Trichoderma harzianum*
(c) Nuclear Polyhedrosis Virus (NPV)
(d) *Xanthomonas compestris*

65. Which is correct regarding genetically engineered insulin using *E.coli*?
(a) Difficult to purify
(b) Obtained in large unlimited quantities
(c) Possibility of transmission of animal disease
(d) Insulin obtained varies in chemical structure

66. If birth rate is 100, death rate is 10 and number of individuals in population is 1000, then what will be the percentage of natural growth rate?
(a) 0.09% (b) 9.0% (c) 0.9% (d) 90%

67. Which of the following is expected to have the highest value ($gm/m^2/yr$) in a grassland ecosystem?
(a) Secondary Production (SP)
(b) Tertiary Production (TP)
(c) Gross Production (GP)
(d) Net Production (NP)

68. Recent reports of acid rains in big industrial cities are due to the effect of atmospheric pollution by
(a) more release of NO_2 and SO_2 by burning of fossil fuels
(b) more release of CO_2 by burning of coal/wood cutting of forests and increasing populations
(c) excessive release of NH_3 by coal gas/ industries
(d) excessive release of CO by incomplete combustion of carbonaceous fuels

69. Identify the bacterium that appears violet after Gram staining.
(a) *Salmonella enterica*
(b) *Escherichia coli*
(c) *Mycobacterium tuberculosis*
(d) *Rhizobium meliloti*

70. The nitrogenous metabolic waste in *Hydra* is mostly
(a) ammonia and is removed from whole surface of body
(b) urea and is removed mainly by tentacles
(c) urea and is removed from whole surface of body
(d) uric acid and is removed from whole surface of body

71. Halophitic archaebacterium, e.g. *Halobacterium salinarum* found in great salt lake and dead sea cannot live in
(a) less than 3 M NaCl concentration
(b) less than 5 M NaCl concentration
(c) more than 4 M NaCl concentration
(d) more than 3 M NaCl concentration

72. In the sieve elements which one of the following is the most likely function of P-proteins?
(a) Deposition of callose on sieve plates
(b) Providing energy for active translocation
(c) Autolytic enzymes
(d) Sealing mechanism on wounding

73. Autonomic genome system is present in
(a) ribosomes and Golgi bodies
(b) Golgi bodies and mitochondria
(c) mitochondria and chloroplast
(d) chloroplasts and ribosomes

74. Which statement regarding coenzyme is incorrect?
(a) Every coenzyme is a cofactor and every cofactor is a coenzyme
(b) Every coenzyme is a cofactor but every cofactor is not a coenzyme
(c) Most of the coenzymes are nucleotides and are composed of vitamins
(d) Coenzymes are the active constituents of enyzmes

75. Most cytogenic activities occur during
(a) interphase (b) telophase
(c) prophase (d) anaphase

76. Which of the following statements regarding cyclic flow of electrons during light reactions is false?
(a) This process takes place in the stromal lamellae
(b) ATP synthesis takes place
(c) NADPH + H^+ is synthesised
(d) Takes place only when light of wavelength beyond 680 nm is available for excitation

77. Enzyme enolase catalyses the conversion of 2 PGA to phosphoenol pyruvic acid in the presence of which of the following cofactors?
(a) Mn^{2+} (b) Fe^{2+} (c) Mg^{2+} (d) Zn^{2+}

78. In human beings, the three pairs of salivary glands and numerous buccal glands produce about
(a) 1.0 dm^3 of saliva per day
(b) 1.5 dm^3 of saliva per day
(c) 2.0 dm^3 of saliva per day
(d) 2.5 dm^3 of saliva per day

79. Coronary heart disease is due to
(a) Streptococci bacteria
(b) inflammation of pericardium
(c) weakening of the heart valves
(d) insufficient blood supply to the heart muscles

80. At their resting stage, the body cells exhibit a potential of –5 to –100 mV known as
(a) polarisation
(b) resting potential
(c) repolarisation
(d) depolarisation

PART-II (2 Marks Questions)

MATHEMATICS

81. The tangent to the curve $y = e^x$ drawn at the point (c, e^c) intersects the line joining the points $(c-1, e^{c-1})$ and $(c+1, e^{c+1})$ is

(a) on the left of $x = c$ (b) on the right of c
(c) at not point (d) at all points

82. Let $g(x) = \dfrac{(x-1)^n}{\log \cos^m (x-1)}$; $0 < x < 2$, m and n are integers, $m \neq 0$, $n > 0$, and let P be the left hand derivatives of $|x-1|$ at $x = 1$. If $\lim_{x \to 1^+} g(x) = P$, then

(a) $n = 1, m = 1$ (b) $n = 1, m = -1$
(c) $n = 2, m = 2$ (d) $n > 2, m = n$

83. The locus of orthocentre of the triangle formed by lines $(1+p)x - py + p(1+p) = 0$, $(1+q)x - qy + q(1+q) = 0$ and $y = 0$ where $p \neq q$ is

(a)a hyperbola (b) a parabola
(c) an ellipse (d) a straight line

84. The number of 3×3 matrices A whose entries are either 0 or 1 and for which the system $A\begin{bmatrix} x \\ y \\ z \end{bmatrix} = \begin{bmatrix} 1 \\ 0 \\ 0 \end{bmatrix}$ has exactly two distinct solutions, is

(a) 0 (b) $2^9 - 1$ (c) 168 (d) 2

85. Let w be a complex cube root of unity with $w \neq 1$. A fair die is thrown three times. If r_1, r_2 and r_3 are the numbers obtained on the die, then the probability $w^{r_1} + w^{r_2} + w^{r_3} = 0$ is

(a) $\frac{1}{18}$ (b) $\frac{1}{9}$ (c) $\frac{2}{9}$ (d) $\frac{1}{36}$

86. Let P, Q, R and S be the points on the plane with position vectors $-2\mathbf{i} - \mathbf{j}, 4\mathbf{i}, 3\mathbf{i} + 3\mathbf{j}$ and $-3\mathbf{i} + 2\mathbf{j}$ respectively. The quadrilateral $PQRS$ must be

(a) parallelogram, which is neither a rhombus nor a rectangle
(b) square
(c) rectangle but not a square
(d) rhombus but not a square

87. Let (x_0, y_0) be the solution of the following equations $(2x)^{\ln 2} = (3y)^{\ln 3}, 3^{\ln x} = 2^{\ln y}$, then x_0 is

(a) $\frac{1}{6}$ (b) $\frac{1}{3}$ (c) $\frac{1}{2}$ (d) 6

88. Let $f : (0, 1) \to R$ be defined by $f(x) = \dfrac{b - x}{1 - bx}$, where b is constant such that $0 < b < 1$. Then,

(a) f is not invertible in (0, 1)
(b) $f \neq f^{-1}$ on (0, 1) and $f'(b) = \dfrac{1}{f(0)}$
(c) $f = f^{-1}$ on (0, 1) and $f'(b) = \dfrac{1}{f'(0)}$
(d) f^{-1} is differentiable in (0, 1)

89. For a real number x let $[x]$ denote the largest integer less or equal to x and $\{x\} = x - [x]$. The smallest integer value of x for which $\int_1^n [x]\{x\}\, dx$ exceeds 2020 is

(a) 63 (b) 64 (c) 90 (d) 91

90. If g is the inverse of a function f and $f'(x) = \dfrac{1}{1 + x^5}$, then $g'(x)$ is equal to

(a) $1 + x^5$ (b) $5x^4$
(c) $\dfrac{1}{1 + (g(x))^5}$ (d) $1 + (g(x))^5$

PHYSICS

91. There is a hole in centre of a thin circular biconvex lens of focal length 4 cm. Diameter of hole is half of diameter of the lens. A point source of light is placed 9 cm from the wall and lens is placed in between, such that a single circular illuminated spot with sharp bright edges is formed on the wall. Distance of lens and source must be

(a) 1 cm (b) 3 cm
(c) 5 cm (d) 4 cm

92. A tumor on a person's leg has a mass of 3g. What is the minimum activity a radiation source can have, if it is to deliver a dose of 10 Gy (Gray) to the tumor in 14 min. Assume each disintegration provides an energy of 0.7 MeV to the tumor.

(a) 3.7×10^5 Bq (b) 3.7×10^8 Bq
(c) 3.7 Bq (d) 3.7×10^{10} Bq

93. An electron typically spends about 10^{-8} s in an excited state before it drops to a lower state by emitting a photon.

Number of revolutions made by an electron in $n = 2$, Bohr's orbit in 1.00×10^{-8} s is

(a) about 10^6 (b) about 10^{14}
(c) about 10^{15} (d) about 10^2

94. A system undergoes a cycle consiting of three processes.

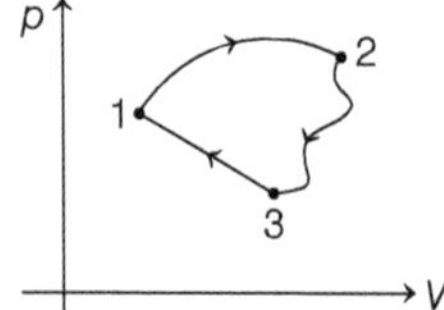

Process involved are listed as,

Process	ΔQ (kJ)	ΔW (kJ)	ΔU (kJ)
$1 \to 2$	a	100	100
$2 \to 3$	b	−50	c
$3 \to 4$	100	d	−200

Values of a, b, c and d, respectively are

(a) 200 kJ, 50 kJ, 100 kJ, 300 kJ
(b) 200 kJ, 50 kJ, 300 kJ, 100 kJ
(c) 50 kJ, 100 kJ, 200 kJ, 300 kJ
(d) 300 kJ, 200 kJ, 200 kJ, 50 kJ

95. Position of a particle is $s = t^3 - 6t^2 + 9t$ m. Distance travelled by the particle in first 5 s is

(a) 20 m (b) 24 m
(c) 28 m (d) 23 m

96. Consider a partially filled capacitor.

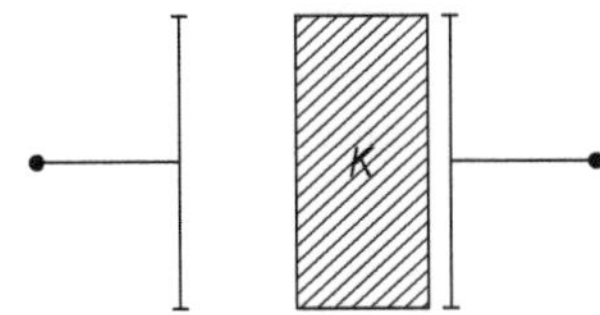

Half of the volume is filled with a dielectric of succeptibility $X_E = 3$. Percentage of total energy stored in the dielectric is

(a) 100% (b) 60% (c) 20% (d) 50%

97. Let R = radius of a spherical refracting surface and l = least distance between conjugate focii. Then, refractive index of medium of refracting surface is

(a) $\dfrac{R+l}{R-l}$ (b) $\left(\dfrac{R+l}{R-l}\right)^2$

(c) $\dfrac{2(R-l)}{R+l}$ (d) $2\left(\dfrac{R-l}{R+l}\right)^2$

98. A heavy block is immersed in water, then

(a) it is easier to lift block at 0° C
(b) it is easier to lift block at room temperature
(c) it is easier to lift block at 4° C
(d) lifting block is equally difficult at every temperature

99. For an ideal gas, let $C_p = a + cT$ and $C_V = b + cT$. where, a, b and c are constants.

T-V relation that holds in adiabatic expansion is

(a) $V^{a-b} \cdot T^b \cdot e^{cT}$ = constant
(b) $T^{a-b} \cdot V^b \cdot e^{cT}$ = constant
(c) $T^{a-b} \cdot V^a \cdot e^{cT}$ = constant
(d) $V^{a-b} \cdot T^a \cdot e^{cT}$ = constant

100. A driver is caught crossing a red light. The driver claims to the judge the colour she actually saw was green ($f_G = 5.6 \times 10^{14}$ Hz) and not red ($f_R = 4.80 \times 10^{14}$ Hz) because of the doppler effect. The judge accepts this explaination and instead fines her for speeding at the rate of 1 ₹ for each kilometer per hour she exceeded the speed limit of 100 km/h. Find charged will be

(a) ₹ 164, 999,999 (b) ₹ 165, 000, 000
(c) ₹ 174, 999, 900 (d) ₹ 164, 999, 900

CHEMISTRY

101. H_3C–(cyclopentane)–$C(=O)CH_3$ $\xrightarrow[\text{(ii) NaOH}]{\text{(i) } CF_3COOH}$

Find the organic acid produced from the above reaction.

(a) H_3C–(cyclopentane)–OH

(b) H_3C–(cyclopentane)–COO^-Na^+

(c) H_3C–(cyclopentane)–CH_2–COO^-Na^+

(d) None of the above

102. In a decay series, $^{206}_{82}Pb$ is obtained at the end from $^{238}_{92}U$. How many alpha and beta particles must have been emitted?

(a) 7α, 10β (b) 8α, 6β (c) 5α, 6β (d) 6α, 8β

103. In the given reaction identify X and Y.

Chlorobenzene (Cl) $\xrightarrow{\text{(i) NaOH/H}^+} X \xrightarrow{CHCl_3 + aq.\ NaOH} Y$

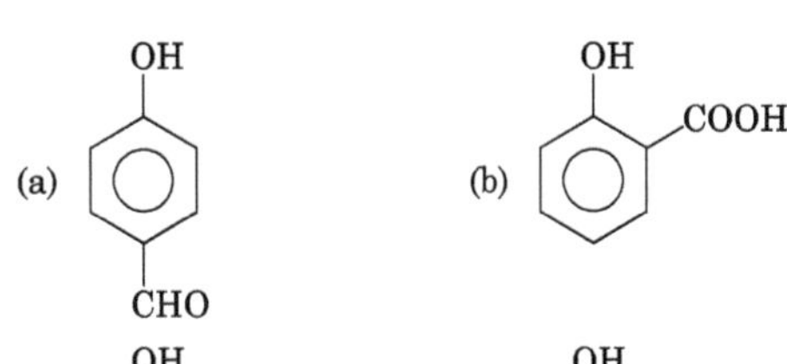

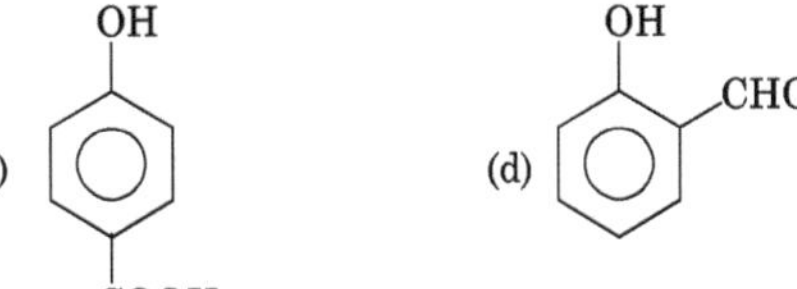

104. The variation of concentration of the prodcut P with time in the reaction, $A \to P$ is shown in following graph.

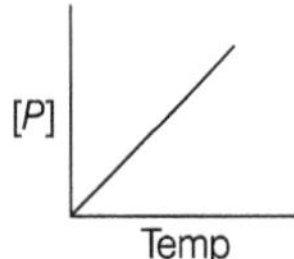

The graph between $\frac{-d[A]}{dt}$ and time will be of the type

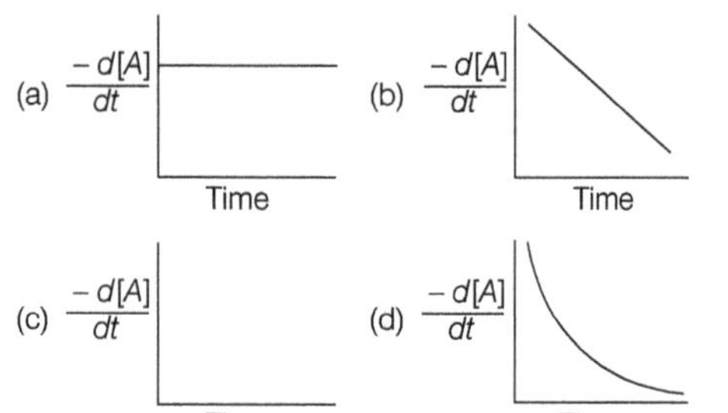

105. 75.2 g of C_6H_5OH (phenol) is dissolved in a solvent of $K_f = 14$. If the depression in freezing point is 7 K, then find the percentage of phenol that dimerises.

(a) 62% (b) 98%
(c) 55% (d) 75%

106. The standard enthalpies of formation of $CO_2(g)$, $H_2O(l)$ and glucose(*s*) at 25°C are – 400 kJ/mol, – 300 kJ/mol and – 1300 kJ/mol, respectively. The standard enthalpy of combustion per gram of glucose at 25°C is

(a) + 2900 kJ (b) – 2900 kJ
(c) – 16.11 kJ (d) + 16.11 kJ

107. Chromium metal crystallises with a body centred cubic lattice. The length of the unit edge is found to be 287 pm. Calculate the atomic radius . What would be the density of chromium in g/cm^3?

(a) 7.3 (b) 8.1
(c) 5.1 (d) 11.3

108. The hybridisation and magnetic moment of complex $[Cr(H_2O)_6]Cl_2$ is

(a) sp^3d^2, 4.89 BM (b) d^2sp^3, 4.89 BM
(c) dsp^2, 5.89 BM (d) sp^3d^2, 3.39 BM

109. Calculate the standard free energy change for the formation of methane at 298 K. The value of $\Delta_r H°$ for $CH_4(g)$ is –74.81 kJ mol^{-1} and *S* values for *C*(graphite) $H_2(g)$ and $CH_4(g)$ are 5.70, 130.7 and 186.3 JK^{-1} mol^{-1} respectively.

(a) – 80.8 (b) – 98.91
(c) – 50.74 (d) – 40.4

110. Compound(s) that on hydrogenation produce(s) optically inactive compound(s) is/are

(a)
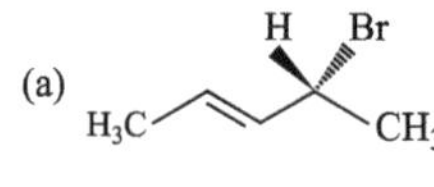

(b) H Br, H_2C, CH_3

(c) H Br, H_2C, CH_3, CH_3

(d)
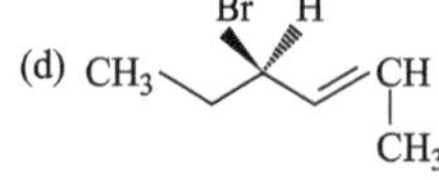

BIOLOGY

111. Given below is a matrix of possible interactions [beneficial (+), harmful (–), neutral (0)] between species 1 and 2. The names of interactions *P*, *Q*, *R* and *S* respectively, are

		Species 1		
		+	–	0
Species 2	+	*R*	*Q*	
	–		*P*	*S*

(a) Predation, Competition, Mutualism, Commensalism
(b) Mutualism, Competition, Amensalism, Commensalism
(c) Competition, Predation, Mutualism, Amensalism
(d) Competition, Mutualism, Commensalism, Predation

112. Which of the following gives the correct human disease-causal microbe match for each?

Human disease	Causal microbe
A. Chronic gastritis	1. *Borrelia burgdorferi*
B. Lyme disease	2. *Helicobacter pylori*
C. Scarlet fever	3. *Rickettsia prowazekii*
D. Typhus	4. *Streptococcus pyogenes*

Codes

	A	B	C	D		A	B	C	D
(a)	2	1	4	3	(b)	2	3	1	4
(c)	4	1	2	3	(d)	4	3	1	2

113. A gene that was 5055 base pairs long resulted in the expression of functional proteins that were 350, 450 and 1500 amino acids long. What could be most likely reason for this?

(a) *m*RNA degradation
(b) Gene amplification
(c) Alternative splicing
(d) Mutations resulting in early termination signals

114. Which of the following statements about the life cycle of the HIV (Human Immunodeficiency Virus) are false?

I. The HIV particles recognise the host cell through the sialic acid containing proteins or lipids on the membrane of the host cell.

II. Upon entry of the HIV into the host cell, a drop of pH will result in a conformational change in the HIV structure, consequently releasing the viral genome into the host cell.

III. The viral DNA which enters the host cell's nucleus will be integrated into the genetic material of the host cell using the host cell's enzyme, integrase.

IV. The viruses are released from the host cell by exocytosis.

(a) I and III (b) II and IV
(c) I, III and IV (d) All of these

115. Genes P, Q, R and S occur on the same chromosome. Investigation of a large population produced the following cross over values between pairs of genes.

P and R	34%	P and Q	59%
R and S	12%	S and Q	37%

Which one of the following sequences of letters represents the sequence of genes on the chromosome?
(a) PRSQ (b) PSRQ (c) QSPR (d) RQSP

116. In cats, the genes controlling coat-colour are codominant (incompletely dominant) and are carried on the X-chromosomes. When a black female was mated with a ginger male, the resulting littre consisted of black male and tortoise shell female kittens. What phenotypic ratio would be expected in the F_2-generation?

(a) 1 balck male : 1 ginger male : 2 tortoise shell females
(b) 1 black male : 1 ginger male : 1 tortoise shell female : 1 black female
(c) 2 black males : 1 tortoise shell female : 1 ginger female
(d) 1 black male : 1 tortoise shell male : 1 ginger female : 1 black female

117. The diagram shows the movement of substances into and out of a chloroplast.

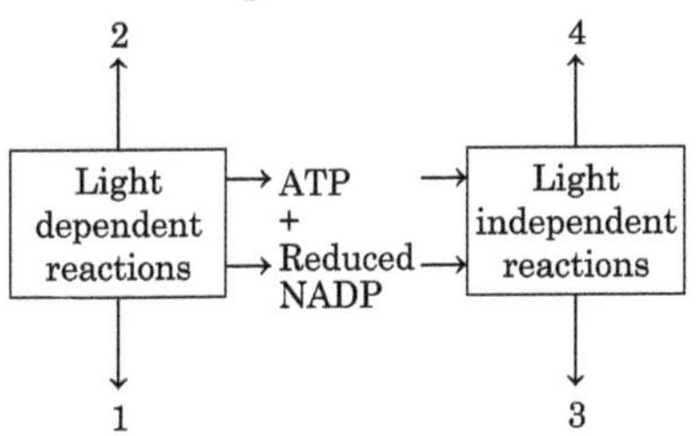

What do labels 1 to 4 represent?

	1	2	3	4
(a)	CO_2	ATP	H_2O	Starch
(b)	CO_2	H_2O	Sugars	O_2
(c)	H_2O	O_2	CO_2	Sugars
(d)	Sugars	H_2O	ATP	O_2

118. In an investigation into the properties of membranes, washed beet root cells were shaken for five minutes in different solvents. The intensity of the red colour of the solvent was then measured. Which of the following is possible?

Tube conditions	Intensity of colour (arbitrary units)
(a) Tap water at 20°C	10
(b) Ethanol at 20°C	35
(c) Tap water at 35° C	35
(d) Distilled water at 45°C	50

119. The chemiosmotic coupling hypothesis of oxidative phosphorylation proposes that Adenosine Triphosphate (ATP) is formed because

(a) high energy bonds are formed in mitochondrial proteins
(b) ADP is pumped out of the matrix into the intermembrane space
(c) a proton gradient forms across the inner membrane
(d) there is a change in the permeability of the inner mitochondrial membrane towards Adenosine Diphosphate (ADP)

120. The graph below shows the effect of radiation on two small inbreeding populations of *Drosophila,* derived from the same parental stock.

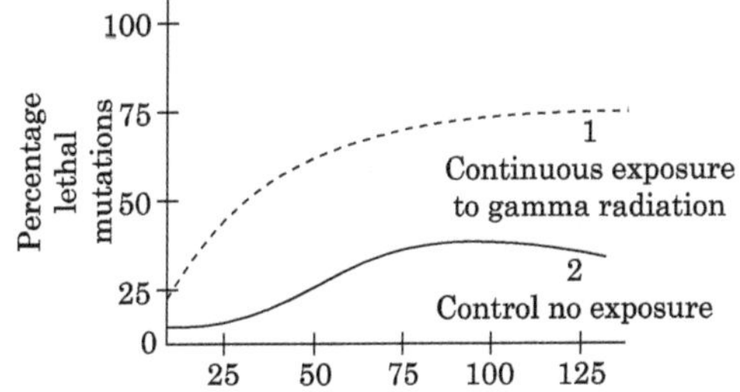

What causes the rise in graph 2 ?

(a) Spontaneous mutation in a small gene pool
(b) Natural selection
(c) A decrease in genetic drift
(d) An increase in the background radiation

Answers

PART-I

1 (b)	2 (b)	3 (d)	4 (a)	5 (d)	6 (a)	7 (c)	8 (c)	9 (b)	10 (c)
11 (c)	12 (b)	13 (a)	14 (b)	15 (a)	16 (c)	17 (d)	18 (c)	19 (c)	20 (b)
21 (c)	22 (c)	23 (d)	24 (c)	25 (d)	26 (d)	27 (d)	28 (b)	29 (a)	30 (d)
31 (a)	32 (d)	33 (c)	34 (c)	35 (a)	36 (d)	37 (c)	38 (b)	39 (c)	40 (d)
41 (c)	42 (c)	43 (b)	44 (b)	45 (d)	46 (d)	47 (b)	48 (a)	49 (b)	50 (d)
51 (a)	52 (c)	53 (b)	54 (d)	55 (a)	56 (c)	57 (c)	58 (c)	59 (a)	60 (b)
61 (b)	62 (d)	63 (d)	64 (d)	65 (b)	66 (b)	67 (c)	68 (a)	69 (c)	70 (a)
71 (a)	72 (d)	73 (c)	74 (a)	75 (a)	76 (c)	77 (c)	78 (b)	79 (d)	80 (b)

PART-II

81 (a)	82 (c)	83 (d)	84 (a)	85 (c)	86 (a)	87 (c)	88 (a)	89 (d)	90 (d)
91 (b)	92 (b)	93 (a)	94 (a)	95 (c)	96 (c)	97 (b)	98 (c)	99 (a)	100 (d)
101 (a)	102 (b)	103 (d)	104 (a)	105 (d)	106 (c)	107 (a)	108 (a)	109 (c)	110 (b)
111 (c)	112 (a)	113 (c)	114 (d)	115 (b)	116 (b)	117 (c)	118 (b)	119 (c)	120 (a)

Solutions

1. *(b)* We know that centroid divides the orthocentre and circumcentre in the ratio 2 : 1.

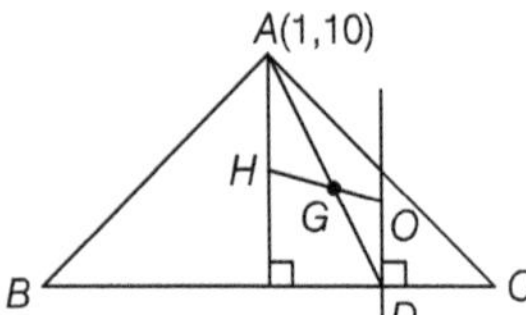

Centroid 2 1

$H\left(\frac{11}{3},\frac{4}{3}\right)$ G $O\left(-\frac{1}{3},\frac{2}{3}\right)$

$$G=\left(\frac{-\frac{2}{3}+\frac{11}{3}}{3},\frac{\frac{4}{3}+\frac{4}{3}}{3}\right)$$

$$G=\left(1,\frac{8}{9}\right)$$

Also, $AG:GD=2:1$

2 1

$A(1,10)$ $G\left(1,\frac{8}{9}\right)$ $D(h,k)$

$$1=\frac{2h+1}{3},\frac{8}{9}=\frac{2k+10}{3}$$

$$h=1,k=-\frac{11}{3}$$

$\therefore$ Coordinate of mid-point of opposite side of A is $\left(1,-\frac{11}{3}\right)$.

2. *(b)* We have,

$f(x)\cdot f(y)=f(x)+f(y)+f(xy)-2$

Put $x=y=1$, we get

$$(f(1))^2=3f(1)-2$$

$\Rightarrow$ $f(1)=1$ or 2

Let $f(1)=1$, then put $y=1$

$\therefore$ $f(x)\cdot f(1)=f(x)+f(1)+f(x)-2$

$\Rightarrow$ $f(x)=2f(x)+1-2$

$\Rightarrow$ $f(x)=1$ constant function,

then $f(1)=2, f(1)\neq 1$

3. *(d)* We have,

$$\lim_{n\to\infty}(-1)^{n-1}\sin\left(\pi\sqrt{n^2+\frac{n}{2}+1}\right)$$

$$=\lim_{n\to\infty}\sin\left(n\pi-\pi\sqrt{n^2+\frac{n}{2}+1}\right)$$

$[\because \sin x=(-1)^{n-1}\sin(n\pi-x)]$

$$=\lim_{n\to\infty}\sin\left(\pi\left(n-\sqrt{n^2+\frac{n}{2}+1}\right)\right)$$

$$=\lim_{n\to\infty}\sin\left[\frac{\pi\left(n^2-n^2-\frac{n}{2}-1\right)}{n+\sqrt{n^2+\frac{n}{2}+1}}\right]$$

$$=\lim_{n\to\infty}\sin\left[\frac{\pi\left(-\frac{1}{2}-\frac{1}{n}\right)}{1+\sqrt{1+\frac{1}{2n}+1}}\right]$$

$$\sin\left(-\frac{\pi}{4}\right)=-\frac{1}{\sqrt{2}}$$

4. *(a)* We have,

$$5\{x\}=x+[x] \quad \text{...(i)}$$

$$[x]-\{x\}=\frac{1}{2} \quad \text{...(ii)}$$

$\therefore$ $0\le\{0\}<1$

$\Rightarrow$ $[x]-\frac{1}{2}=\{x\}$

$$0\le[x]-\frac{1}{2}<1\Rightarrow[x]=1,\{x\}=\frac{1}{2}$$

$\therefore$ $\frac{5}{2}=x+1\Rightarrow x=\frac{3}{2}$ only one solution

5. *(d)* We have, a, b, c are in GP.

$\therefore$ $b^2=ac$

and $\log_6 a+\log_6 b+\log_6 c=6$

$\Rightarrow$ $\log_6(abc)=6\Rightarrow abc=6^6$

$\Rightarrow$ $b^3=6^6\Rightarrow b=36$

$\Rightarrow$ $ac=36\times36=2^4\times3^4$

$\Rightarrow$ $b-a=N^2$

$\Rightarrow$ $36-a=N^2$

a is a factor of $2^4\times3^4$

$a=27$ is possible value

$\therefore a=27, b=36, c=48$

$a+b+c=27+36+48=111$

6. *(a)* Given, $n(A)=7, n(B)=5$

Total number of function from A to $B=5^7$

Total number of onto function from A to B is

$$n(E)=\frac{7!}{3!4!}5!+\frac{7!}{3!2!}\times\frac{1}{2!2!}\times5!$$

$$=\frac{7!\times20}{6}$$

$\therefore$ Required probability $=\dfrac{\frac{7!\times20}{6}}{5^7}=\dfrac{7!\times2}{3\times5^6}$

7. *(c)* We have,

$$I_n=\int_1^e(\ln x)^n\,dx$$

$$\Rightarrow \quad I_n=[(\ln x)^n x]_1^e-\int_1^e\frac{n(\ln x)^{n-1}}{x}.x\,dx$$

$\Rightarrow$ $I_n=e-nI_{n-1}$

$\Rightarrow$ $I_n+nI_{n-1}=e$

$\therefore$ $I_{10}+10I_9=e$

8. *(c)* Consider the function

$$f(x)=\frac{a_0x^{n+1}}{n+1}+\frac{a_1x^n}{n}+\frac{a_2x^{n-1}}{n-1}+\ldots+\frac{a_{n-1}x^2}{2}+a_nx$$

$\Rightarrow$ $f(0)=f(1)=0$

$\Rightarrow$ $f'(x)=a_0x^n+a_1x^{n-1}+a_2x^{n-2}+\ldots+a_{n-1}x+a_n$

By Rolle's theorem

$f'(x)=0\, x\in(0,1)$

Hence, at least one solution in (0, 1).

9. *(b)* ABC is right angle triangle

AB is hypotenuse and $AB=p$

$\therefore$ $\angle C=90°$

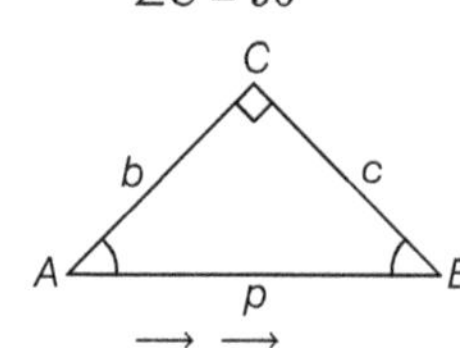

$\Rightarrow$ $\overrightarrow{CA}\cdot\overrightarrow{CB}=0$

$\Rightarrow$ $\mathbf{AB}\cdot\mathbf{AC}=|\mathbf{AB}||\mathbf{AC}|\cos A$

$\Rightarrow$ $\mathbf{BC}\cdot\mathbf{BA}=|\mathbf{BC}||\mathbf{BA}|\cos B$

$\Rightarrow \mathbf{AB}\cdot\mathbf{AC}+\mathbf{BC}\cdot\mathbf{BA}+\mathbf{CA}\cdot\mathbf{CB}$

$\Rightarrow |\mathbf{AB}||\mathbf{AC}|\cos A+|\mathbf{BC}||\mathbf{BA}|\cos B+0$

$\Rightarrow |\mathbf{AB}|(|\mathbf{AC}|\cos A+|\mathbf{BC}|\cos B)$

$\Rightarrow$ $|\mathbf{AB}||\mathbf{AB}|=|\mathbf{AB}|^2=p^2$

$[\because p=b\cos B+c\cos A]$

10. *(c)* Tangents at P_1 and P_2 are mutually perpendiculars

$\therefore$ $t_1t_2=-1$

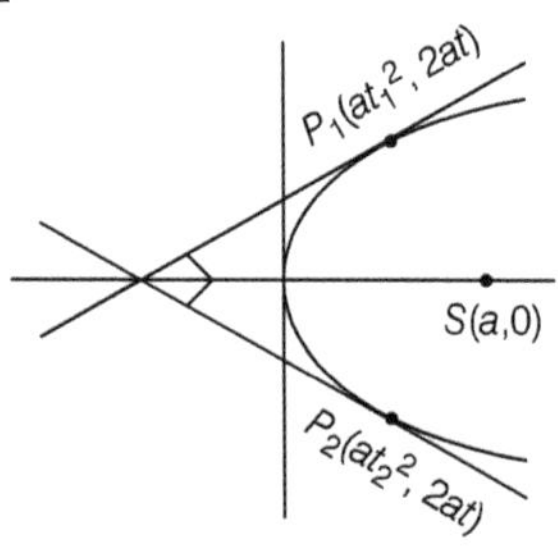

$$SP_1=\sqrt{(a-at_1^2)^2+(2at_1)^2}$$

$$SP_1=a+at_1^2$$

Similarly, $SP_2=a+at_2^2$

$\Rightarrow \quad SP_2 = a + a\left(\frac{1}{t_1}\right)^2 \quad \left[\because t_2 = -\frac{1}{t_1}\right]$

$\Rightarrow \quad \frac{1}{SP_1} + \frac{1}{SP_2} = \frac{1}{a + at_1^2} + \frac{t_1^2}{at_1 7a}$

$= \frac{1}{a}\left(\frac{1 + t_1^2}{1 + t_1^2}\right) = \frac{1}{a}$

11. *(c)* We have,

$\sec^2 (a + 2)x + a^2 - 1 = 0$

$\Rightarrow \quad \tan^2(a + 2)x + a^2 = 0$

$\therefore \quad \tan(a + 2)x = 0$ and $a = 0$

$\Rightarrow \quad \tan(2x) = 0$ and $a = 0$

$\Rightarrow \quad 2x = n\pi$ and $a = 0$

$\Rightarrow \quad x = \frac{n\pi}{2}$ and $a = 0$

$x \in (-\pi, \pi)$

$\therefore \quad x = 0, \frac{\pi}{2}, -\frac{\pi}{2}$

Number of ordered pair of (a, x) is 3

i.e.$(0,0)\left(0, \frac{\pi}{2}\right)$ and $\left(0, -\frac{\pi}{2}\right)$.

12. *(b)* In ΔABC, $\angle A = 30°$

$BC = 2 + \sqrt{5}$

$\frac{a}{\sin A} = 2R \Rightarrow \frac{2 + \sqrt{5}}{\sin 30°} = 2R$

$R = \frac{2 + \sqrt{5}}{2 \times \frac{1}{2}} = 2 + \sqrt{5}$

Now, $AH = 2R\cos A$

$\because \quad AH = 2(2 + \sqrt{5}) \cos 30°$

$\Rightarrow \quad AH = 2(2 + \sqrt{5}) \frac{\sqrt{3}}{2} = (2 + \sqrt{5})(\sqrt{3})$

$\therefore \quad AH = \sqrt{3}(2 + \sqrt{5})$

13. *(a)* We have,

$I = \int_0^2 \sqrt{x + \sqrt{x + \sqrt{x + \ldots\infty}}}\, dx$

Let $\quad y = \sqrt{x + \sqrt{x + \sqrt{x + \ldots}}}$

$y = \sqrt{x + y} \Rightarrow y^2 - y - x = 0$

$\Rightarrow \quad y = \frac{1 \pm \sqrt{4x + 1}}{2}$

$\Rightarrow \quad y = \frac{1 + \sqrt{4x + 1}}{2} \quad y \geq 1$

$\because \quad I = \int_0^2 \frac{1 + \sqrt{4x + 1}}{2} dx$

$\Rightarrow \quad I = \frac{1}{2}\left[x + \frac{(1 + 4x)^{3/2}}{6}\right]_0^2$

$\Rightarrow \quad I = \frac{1}{2}\left[\left(2 + \frac{27}{6}\right) - \left(0 + \frac{1}{6}\right)\right]$

$\Rightarrow \quad I = \frac{1}{2}\left(\frac{12 + 27 - 1}{6}\right) = \frac{19}{6}$

14. *(b)* We have,

$\log(-2x) = 2\log(x + 1)$

$-2x > 0 \quad x < 0$ and $x + 1 > 0 \quad x > -1$

$\therefore \quad x \in (-1, 0)$

$-2x = (x + 1)^2$

$\Rightarrow \quad x^2 + 4x + 1 = 0$

$\Rightarrow \quad x = \frac{-4 \pm 2\sqrt{3}}{2} = -2 \pm \sqrt{3}$

So, $x = -2 + \sqrt{3}$ only one solution lies in $(-1, 0)$.

15. *(a)* All *AAAAABBBDEEF* can be arranged in $\frac{12!}{5!3!2!}$ ways.

Between the gaps *C* can be arranged $^{13}C_3$ ways

Total ways $= {^{13}C_3} \times \frac{12!}{5!3!2!}$

16. *(c)* We have,

$\sqrt{3}\sec x + \operatorname{cosec} x + 2(\tan x - \cot x) = 0$

$\because \quad \frac{\sqrt{3}}{\cos x} + \frac{1}{\sin x} + 2\left(\frac{\sin x}{\cos x} - \frac{\cos x}{\sin x}\right) = 0$

$\Rightarrow \sqrt{3}\sin x + \cos x + 2(\sin^2 x - \cos^2 x) = 0$

$\Rightarrow \quad \sqrt{3}\sin x + \cos x = 2\cos 2x$

$\Rightarrow \quad \cos\left(x - \frac{\pi}{3}\right) = \cos 2x$

$\Rightarrow \quad \cos 2x = \cos\left(x - \frac{\pi}{3}\right)$

$\Rightarrow \quad 2x = 2n\pi \pm \left(x - \frac{\pi}{3}\right)$

$\Rightarrow \quad x = (6n - 1)\frac{\pi}{3}$ or $(6n + 1)\frac{\pi}{9}$

$\Rightarrow \quad x = -\frac{\pi}{3}, \frac{\pi}{9}, \frac{7\pi}{9}$ and $-\frac{5\pi}{9}$ in $(-\pi\ \pi)$

$\therefore$ Sum $= \frac{-\pi}{3} + \frac{\pi}{9} + \frac{7\pi}{9} - \frac{5\pi}{9} = 0$

17. *(d)* In ΔABC

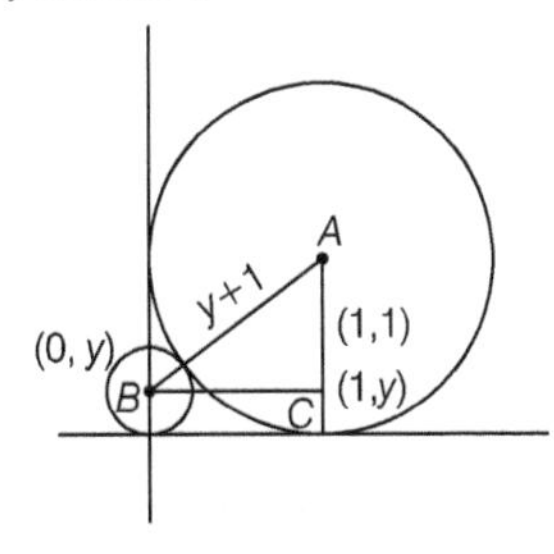

$AB = y + 1$

$AC = 1 - y$

$BC = 1$

$AB^2 = AC^2 + BC^2$

$(y + 1)^2 = (1 - y)^2 + (1)^2$

$(1 + y)^2 - (1 - y)^2 = 1$

$4y = 1 \Rightarrow y = \frac{1}{4}$

Hence, radius of circle $T = \frac{1}{4}$

18. *(c)* Equation of ellipse,

$x^2 + 3y^2 = 6$

Equation of tangent of ellipse with slope *m* is

$y = mx \pm \sqrt{6m^2 + 2}$

$m = -\frac{h}{k}$

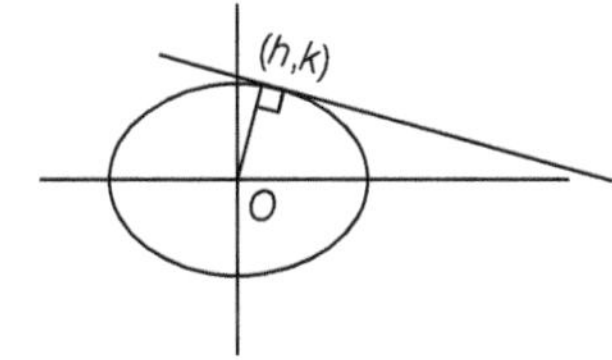

and tangent passes through (h, k)

$\therefore \quad k = h\left(-\frac{h}{k}\right) \pm \sqrt{\frac{6h^2}{k^2} + 2}$

$\Rightarrow \quad k^2 + h^2 = \pm\sqrt{6h^2 + 2k^2}$

$\Rightarrow (h^2 + k^2)^2 = 6h^2 + 2k^2$

So, locus is $(x^2 + y^2)^2 = 6x^2 + 2y^2$.

19. *(c)* We have,

$(10)^9 k = 10^9 + 2(11)(10)^8 + 3(11)^2(10)^7 + \ldots + 10(11)^9$

$\Rightarrow k = 1 + 2\left(\frac{11}{10}\right) + 3\left(\frac{11}{10}\right)^2 + \ldots + 10\left(\frac{11}{10}\right)^9$

$\Rightarrow \frac{11}{10}k = \frac{11}{10} + 2\left(\frac{11}{10}\right)^2 + \ldots 9\left(\frac{11}{10}\right)^9 + \left(\frac{11}{10}\right)^{10} 10 - \frac{k}{10}$

$= 1 + \frac{11}{10} + \left(\frac{11}{10}\right)^2 + \ldots \left(\frac{11}{10}\right)^9 - \left(\frac{11}{10}\right)^{10} 10$

$-\frac{k}{10} = \frac{\left(\frac{11}{10}\right)^{10} - 1}{\frac{11}{10} - 1} - \left(\frac{11}{10}\right)^{10} 10$

$\therefore \quad k = 100$

20. *(b)* We have,

$\log_{(3x-1)}(x - 2)$

$= \log_{(9x^2 - 6x + 1)}(2x^2 - 10x - 2)$

$\Rightarrow \log_{(3x-1)}(x-2)$
$= \log_{(3x-1)^2}(2x^2 - 10x - 2)$
$\Rightarrow 2\log_{(3x-1)}(x-2)$
$= \log_{(3x-1)}(2x^2 - 10x - 2)$
$\Rightarrow \log_{(3x-1)}(x-2)^2$
$= \log_{(3x-1)}(2x^2 - 10x - 2)$
$\Rightarrow (x-2)^2 = 2x^2 - 10x - 2$
$\Rightarrow x^2 - 4x + 4 = 2x^2 - 10x - 2$
$\Rightarrow x^2 - 6x - 6 = 0$
$\Rightarrow x = 3 \pm \sqrt{15}$
$\therefore x = 3 + \sqrt{15}, x > 2$

21. *(c)* Cone with angle α and source S is as shown below.

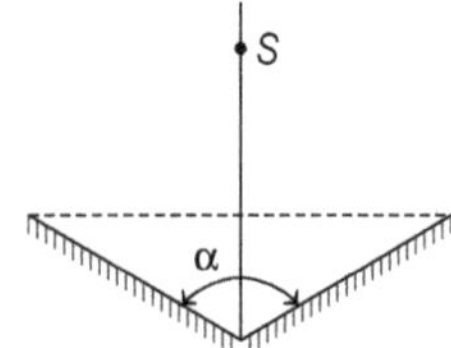

Now, we draw reflected rays as shown below.

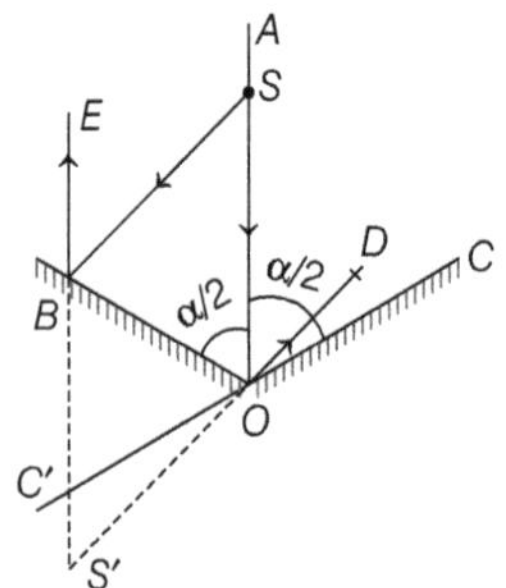

From above construction, image of S is S'. Now, for rays to reflect only once point S' must be below line OB and OCC'.

So, $\angle S'OB + \angle SOB + \angle AOC = \frac{3\alpha}{2} \geq 180°$

or $\alpha \geq \frac{2 \times 180}{3} \Rightarrow \alpha \geq 120°$

22. *(c)* Both kinetic energy and linear momentum are conserved.

So, $KE_{Ge} + KE_{Nd} = 0$

and $m_{Ge}v_{Ge} + m_{Nd}v_{Nd} = 0$

Combining both, we get

$$K_{Ge} = \left(\frac{m_{Nd}}{m_{Nd} + m_{Ge}}\right) \times Q$$

$$K_{Nd} = \left(\frac{m_{Ge}}{m_{Nd} + m_{Ge}}\right) \times Q$$

So, the ratio is $\frac{K_{Ge}}{K_{Nd}} = \frac{m_{Nd}}{m_{Ge}}$

$= \frac{153}{83} \approx \frac{11}{6}$

23. *(d)* We use Boltzmann relation $S = k_B \cdot \ln \Omega$, where Ω is the number of ways the state can occur.

Here two tails and one head can occur in three different ways.

So, entropy, $S = k_B \cdot \ln 3$

24. *(c)* Position of particle is

$$\mathbf{r} = 3\cos\omega t\,\hat{\mathbf{i}} + 4\sin\omega t\,\hat{\mathbf{j}}$$

So, $x = 3\cos\omega t$, $y = 4\sin\omega t$

$\Rightarrow \frac{x}{3} = \cos\omega t$ and $\frac{y}{4} = \sin\omega t$

So, we have $\frac{x^2}{9} + \frac{y^2}{16} = \cos^2\omega t + \sin^2\omega t$

or $\frac{x^2}{9} + \frac{y^2}{16} = 1$

This is an ellipse.

25. *(d)* Sound levels of A and B are

$$\beta_A = 10\log\left(\frac{I_A}{I_O}\right)$$

$$\beta_B = 10\log\left(\frac{I_B}{I_O}\right)$$

$$\therefore \beta_A - \beta_B = 10\left(\log\frac{I_A}{I_o} - \log\frac{I_B}{I_O}\right)$$

$$= 10\log\left(\frac{I_B}{I_A}\right) = 10\log\left(\frac{500}{10}\right)$$

$$\Rightarrow \beta_A - \beta_B = 10\log 50$$
$$= 10 \times 1.70 = 17 \text{ dB}$$

26. *(d)* By conservation of momentum,

$$p_i = p_1 + p_2$$

For a general collision, some energy is lost

$$\Rightarrow \frac{p_i^2}{2m} > \frac{p_1^2}{2m} + \frac{p_2^2}{2m} \Rightarrow p^2 > p_1^2 + p_2^2$$

$\therefore \theta < 90°$

27. *(d)* Ray diagram for given situation is

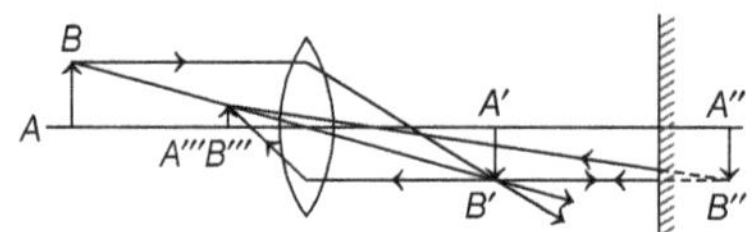

First image $A'B'$ is formed by lens. Image of $A''B''$ is formed in plane mirror as $A''B''$. Image of $A''B''$ is formed by lens as $A'''B'''$.

Note *That this is real and erect.*

28. *(b)* As $\omega = \frac{d\theta}{dt}$ = slope of $\theta - t$ line which is positive in given case so, body is rotating anti-clockwise.

29. *(a)* de-Broglie waves will resonate with a node at each end of tube.

$\therefore L = \frac{\lambda_1}{2}, \lambda_2, \frac{3}{2}\lambda_3 \ldots \frac{n\lambda_n}{2}$, etc

So, $\lambda_n = \frac{2L}{n}$, $n = 1, 2, 3..$

As, $\lambda_n = \frac{h}{p_n}$ and $(KE)_n = \frac{p^2}{2m}$

We have, $(KE)_n = \frac{n^2h^2}{8L^2m}$, $n = 1, 2, 3, \ldots$, etc

With $m = 9.1 \times 10^{-31}$ and $L = 5 \times 10^{-10}$ m

$(KE)_n = 2.4 \times 10^{-19}\, n^2 \text{J} = 1.5\, n^2 \text{eV}$

$\therefore$ KE = 1.5 eV, 6 eV, 13.5 eV, ……, etc

30. *(d)* Angular momentum of earth remains constant with time. So, correct option is (d).

$$\tau_{ext} = 0 = \frac{dL}{dt} \Rightarrow L = \text{constant}$$

31. *(a)* $\mathbf{E} = \frac{a(x\hat{\mathbf{i}} + y\hat{\mathbf{j}} + z\hat{\mathbf{k}})}{x^2 + y^2 + z^2} = \frac{a\mathbf{r}}{r^2} = \frac{a}{r} \cdot \hat{\mathbf{r}}$

$$d\phi = \mathbf{E}\, d\mathbf{s} = \left(\frac{a}{R} \cdot \hat{\mathbf{r}}\right)(ds \cdot \hat{\mathbf{r}})$$

$$\Rightarrow \phi = \int d\phi = 4\pi aR$$

$\therefore Q_{enclosed} = \varepsilon_0\phi = 4\pi\varepsilon_0 aR$

32. *(d)* Moment of inertia about shortest side is maximum because effective distance of mass distribution about this side is maximum.

33. *(c)* Force $= P_{avg} \times$ Area

$$= \left(\frac{P_{top} + P_{botton}}{2}\right) \times \text{Area} = \frac{1}{2}\rho g\omega H^2$$

34. *(c)* After 6 s, both pulses overlapps completely.

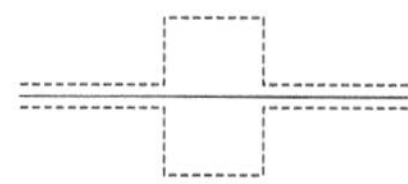

So, there is no potential energy as $y_{net} = 0$. Hence, energy is 100% kinetic after 6 s.

35. *(a)* $\frac{U_D}{U_{Total}} = \frac{U_D}{U_{gap} + U_D}$

$$= \frac{u_D \times (\text{Vol})_D}{\{u_{gap} \times (\text{Vol})_{gap} + (u_D)(\text{Vol})_D\}}$$

$$= \frac{\frac{1}{2}\varepsilon_0 \frac{E_{gap}^2}{K} A \cdot l}{\frac{1}{2}\varepsilon_0 E_{gap}^2 A(d-l) + \frac{1}{2}\varepsilon_0 \frac{E_{gap}^2}{K} A \cdot l}$$

$$= \frac{l/K}{(d-l) + l/K} = \frac{l}{(d-l)K + l}$$

$$= \frac{1}{1 \times 7 + 1} = \frac{1}{8}$$

$\therefore$ Per centage of energy stored

$$= \frac{1}{8} \times 100 = 12.5\%$$

36. *(d)* Frequency, $f = \frac{2\pi m}{qB}$ and it is independent of accelerating voltage or velocity.

37. *(c)* From given equation,

$$\omega = 2\pi f = 628$$

$$\Rightarrow \quad f = \frac{628}{2\pi} = 100\,\text{Hz}$$

This may be third, fifth or seventh harmonic as there is an anti-node at mid-point.

So, fundamental frequency may be

$$\frac{100}{3}, \frac{100}{5}, \frac{100}{7}, \ldots, \text{etc.}$$

38. *(b)* For displacement x,

Force on mass m is

$$F_{net} = ma = -\left(kx + kx + 2k\left(\frac{x}{2}\right)\right)$$

$$\Rightarrow \quad a = -\frac{3k}{m} \cdot x$$

So, $\quad T = 2\pi\sqrt{\frac{m}{3k}}$

39. *(c)* Velocities of the particle after time t seconds is

$$\mathbf{v}_1' = \mathbf{v}_1 + \mathbf{g}t_1$$

$$\mathbf{v}_2' = \mathbf{v}_2 + \mathbf{g}\,t_2$$

$$\mathbf{v}_1' \perp \mathbf{v}_2' = \mathbf{v}_1' \cdot \mathbf{v}_2' = 0$$

As, v_1 and v_2 are in opposite directions

$\therefore \quad \mathbf{v}_1 \cdot \mathbf{v}_2 = -\mathbf{v}_1\mathbf{v}_2$

$$(\mathbf{v}_1' + \mathbf{g}t_1') \cdot (\mathbf{v}_2' + \mathbf{g}\,t_2) = 0$$

$$\Rightarrow \quad -v_1 v_2 + g^2 t^2 = 0$$

$$\Rightarrow \quad t = \frac{\sqrt{v_1 v_2}}{g} = \frac{\sqrt{4 \times 3}}{10} = \frac{\sqrt{3}}{5}$$

40. *(d)* Acceleration of the bead down the wire is $g\cos\theta$.

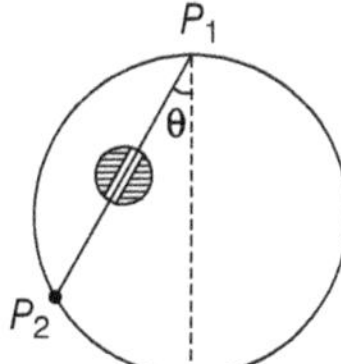

Also, length P_1P_2 is $2R\cos\theta$.

So, $v^2 = u^2 + 2as$

Gives, $v^2 = 2g\cos\theta \times 2R\cos\theta$

$$= 2\sqrt{gR}\,\cos\theta$$

Time, $\quad t = \frac{v}{a} = \frac{2\sqrt{gR}\,\cos\theta}{g\cos\theta} = 2\sqrt{\frac{R}{g}}$

Which is same regardless of where P_2 is located.

Time of free fall,

$$2R = \frac{1}{2}gt^2 \Rightarrow t^2 = \frac{4R}{g} \Rightarrow t = 2\sqrt{\frac{R}{g}}$$

41. *(c)* CO_2 and CH_4 have zero dipole moment as these are symmetrical in nature. Between NH_3 and NF_3, NH_3 has greater dipole moment, though in NH_3 and NF_3 both, N possesses one lone pair of electrons. This is because in case of NH_3, the net N—H bond dipole is in the same direction as the direction of dipole of lone pair, but in case of NF_3, the direction of net bond dipole moment of three —N—F bonds is opposite to that of the dipole moment of the lone pair.

(a) O=C=O, $\mu_{net} = 0$ (b) CH_4 (μ_{res}, μ_1, μ_2, μ_3, μ_4), $\mu_{net} = 0$

(c) NH_3 (μ_1, μ_2, μ_3, μ_4), $\mu_{net} = 1.5$ D

(d) NF_3 (μ_1, μ_2, μ_3, μ_4), $\mu_{net} = 0.2$ D

42. *(c)* Iodoform reaction is given by only those compounds that have either

CH_3—C(=O)— or CH_3CH—OH group.

(a) C_2H_5—CH(OH)—CH_3
(s-butyl alcohol gives iodoform reaction)

(b) CH_3—CH(OH)—CH_3
(Isopropyl alchohol gives iodoform reaction)

(c) C_2H_5—C(=O)—C_2H_5
Diethyl ketone
(does not give iodoform reaction)

(d) CH_3CH(H)OH
Ethyl alcohol
(gives iodoform reaction)

43. *(b)* BCl_3 is trigonal in shape where bond angle is 120°. In NH_3 and PCl_3 (both are having pyramidal geometry), though the central atom has equal number of lone pairs and bond pairs, the valence shell of P is relatively bigger, so electrons suffers lesser repulsion and thus bond angle is lower.

Thus, the correct order of bond angle is $BCl_3 > NH_3 > PCl_3$.

44. *(b)* The complexes $[Co(NH_3)_6][Cr(CN)_6]$ and $[Cr(NH_3)_6][Co(CN)_6]$ are the examples of coordination isomerisms. This isomerism occurs only in those complexes in which both cation and anion are complex. It occurs due to exchange of ligands between cation and anion.

45. *(d)* The compounds which have all sp^2-hybridised carbon atoms and follows Huckel's rule [$(4n + 2)\pi$ electrons] are aromatic compounds.

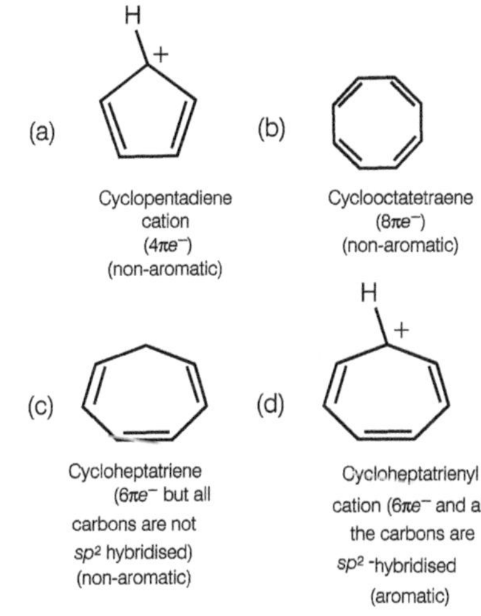

(a) Cyclopentadiene cation ($4\pi e^-$) (non-aromatic)
(b) Cyclooctatetraene ($8\pi e^-$) (non-aromatic)
(c) Cycloheptatriene ($6\pi e^-$ but all carbons are not sp^2 hybridised) (non-aromatic)
(d) Cycloheptatrienyl cation ($6\pi e^-$ and all the carbons are sp^2-hybridised) (aromatic)

46. *(d)* The physical adsorption isobar at constant pressure shows a decrease in x/m with rise in temperature.

Thus, the correct plot is given in option (d).

47. *(b)* $2NO_2 \underset{k_2}{\overset{k_1}{\rightleftharpoons}} N_2O_4$

Rate of reaction $= -\frac{1}{2}\frac{d[NO_2]}{dt}$

$$= k_1[NO_2]^2 - k_2[N_2O_4]$$

$\therefore$ Rate of disappearance of NO_2.

i.e. $\frac{-d[NO_2]}{dt} = 2k_1[NO_2]^2 - 2k_2[N_2O_4]$

48. *(a)* Xenon tetrafluoride reacts with water at – 80°C to give xenon oxyfluoride and hydrogen fluoride.

$$XeF_4 + H_2O \xrightarrow{-80°C} \underset{\text{Xenon oxyfluoride}}{XeOF_2} + 2HF$$

49. *(b)* From the given cell reaction and Nernst equation,

$$E_{cell} = E_{cell}^{\circ} - \frac{0.0591}{n}\log\frac{[Ni^{2+}]}{[Ag^+]^2}$$

$$= 1.05V - \frac{0.0591}{2}\log\frac{[0.160]}{[0.002]^2}$$

$$= 1.05 - \frac{0.0591}{2}\log(4\times10^4)$$

$$= 1.05 - \frac{0.0591}{2}(4.6021)$$

$$= 1.05 - 0.14 = 0.91\text{V}$$

$$E_{cell} = 0.91\text{V}$$

50. *(d)* The structure of diborane (B_2H_6) has four $2C-2e^-$ bonds and two $3C-2e^-$ bonds.

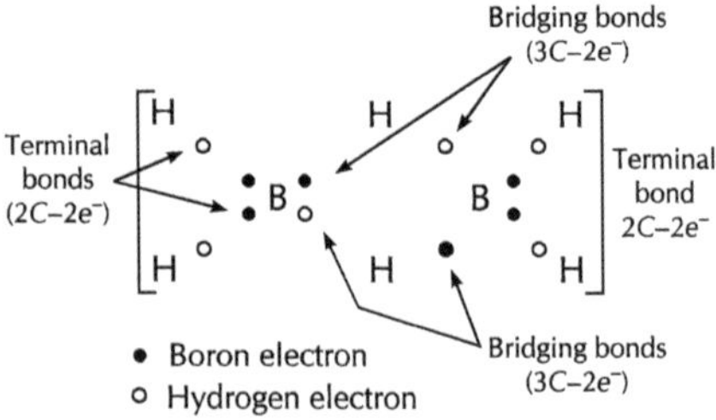

51. *(a)* Aldehydes or ketones with α-H-atoms, in presence of dilute base, undergoes aldol condensation to give β-hydroxy aldehyde or β-hydroxy ketone. On heating, aldols eliminate water molecule to form α, β-unsaturated compound.

$$C_6H_5-\overset{O}{\overset{\|}{C}}-CH_3 + HCH_2-\overset{O}{\overset{\|}{C}}-C_6H_5$$

Acetophenone

$$\xrightarrow[\text{Base}]{C_2H_5ONa} C_6H_5-\underset{CH_3}{\overset{OH}{C}}-CH_2-\overset{}{\underset{O}{\underset{\|}{C}}}-C_6H_5 \quad \text{Aldol}$$

$$\xrightarrow{-H_2O} C_6H_5-\underset{CH_3}{C}=CH-\underset{O}{\underset{\|}{C}}-C_6H_5$$

This is an example of self aldol condensation.

52. *(c)* In DNA, adenine (*A*) and thymine (*T*) are joined by two hydrogen bonds while guanine (*G*) and cytosine (*C*) are joined by three hydrogen bonds. No other combination of four bases is possible.

Hence, the sequence of complementary strand is TACGAACT.

53. *(b)* According to Faraday's second law

Given, $Q = 2F$

Atomic mass of Cu = 63.5u

Valency of the metal $Z = 2$

We have, $CuSO_4 \longrightarrow Cu^{2+} + SO_4^{2-}$

$$\underset{1\,mol}{Cu^{2+}} + \underset{\substack{2\,mol\\2F}}{2e^-} \longrightarrow \underset{1\,mol\,=\,63.5g}{Cu}$$

Alternatively.

$$W = ZQ = \frac{E}{F}\cdot 2F = 2E = \frac{2\times63.5}{2} = 63.5\text{ g}$$

54. *(d)* (a) $\overset{+1}{Na}\overset{-1}{Cl} + \overset{+1}{K}\overset{-1}{NO_3} \longrightarrow \overset{+1}{Na}\overset{-1}{NO_3} + \overset{+1}{K}\overset{-1}{Cl}$

(b) $\overset{+2}{Ca}\overset{-2}{C_2O_4} + 2\overset{+1}{H}\overset{-1}{Cl} \longrightarrow \overset{+2}{Ca}\overset{-1}{Cl_4} + \overset{+1}{H_2}\overset{-2}{C_2O_4}$

(c) $\overset{+2}{Ca}(\overset{-1}{OH})_2 + 2\overset{+3}{N}\overset{+1}{H_4}\overset{-1}{Cl} \longrightarrow \overset{+2}{Ca}\overset{-1}{Cl_2} + 2\overset{-3}{N}\overset{+1}{H_3} + 2\overset{+1}{H_2}\overset{-2}{O}$

In all these cases during reaction, there is no change in oxidation state of ion or molecule or constituent atom, thus these are simply ionic reactions.

(d) $2K[Ag(CN)_2] + Zn \longrightarrow 2Ag + K_2[Zn(CN)_4]$

$Ag^+ \longrightarrow Ag$ gain of e^-, reduction

$Zn \longrightarrow Zn^{2+}$ loss of e^-, oxidation.

As both oxidation and reduction takes place simultaneously.

Thus, it is a redox reaction.

55. *(a)* As we know that, strong field ligand split the five degenerate energy levels with more energy separation than weak field ligand, i.e. as strength of ligand increases crystal field splitting energy increases. Also,

$$\Delta E = \frac{hc}{\lambda}$$

$$\Rightarrow \quad \Delta E \propto \frac{1}{\lambda}$$

$$\Rightarrow \quad \lambda \propto \frac{1}{\Delta E}$$

As energy separation increases, the wavelength decreases.

Thus, the correct order of increasing CFSE and decreasing wavelength is

$$[CoF_6]^{3-} > [Co(H_2O)_6]^{3+} > [Co(NH_3)_6]^{3+} > [Co(CN)_6]^{3-}$$

Here, strength of ligand increases, ΔE increases, CFSE increases and λ absorbed decreases.

Hence, correct option is (a).

56. *(c)* Ethanal (CH_3CHO) is an oxidised product of ethanol.Pyridinium chlorochromate ($C_5H_5\overset{+}{N}HCl\overset{-}{C}rO_3$), PCC oxidises primary alcohols to aldehydes. Strong oxidising agents such as $KMnO_4$ are used for getting carboxylic acid from alcohols.

The oxidation process can be stopped at the aldehyde stage, if Cr(VI) reagents such as Collin's reagent ($CrO_3\cdot 2C_5H_5N$), Corey's reagent or pyridinium chlorochromate are used PCC and pyridinium dichromate $[(C_5H_5NH)_2^{2+}Cr_2O_7^{2-}]$ in anhydrous medium are used as the oxidising agent. So, the correct option is (c).

$$\underset{\text{Ethanol}}{CH_3CH_2OH} \xrightarrow{\text{PCC}} \underset{\text{Ethanal}}{CH_3CHO}$$

57. *(c)* The structures of are as follows:

$H_2S_2O_8$, H_2SO_4, H_2SO_2

$$HO-\underset{O}{\overset{O}{\overset{\|}{\underset{\|}{S}}}}-O-O-\underset{O}{\overset{O}{\overset{\|}{\underset{\|}{S}}}}-OH$$

4S $=$O bonds
$H_2S_2O_8$

$$HO-\underset{O}{\overset{O}{\overset{\|}{\underset{\|}{S}}}}-OH \qquad HO-\overset{O}{\overset{\|}{\ddot{S}}}(OH)-OH$$

2S $=$O bonds H_2SO_4 1S $=$O bonds H_2SO_2

Thus, the correct order of decreasing number of S$=$O bonds is

$$H_2S_2O_8 > H_2SO_4 > H_2SO_3$$

58. *(c)* A unit cell with edge length $a \neq b \neq c$ and axial angles $\alpha = \beta = \gamma = 90°$ is called orthorhombic, e.g. rhombic sulphur, KNO_3 and $BaSO_4$.

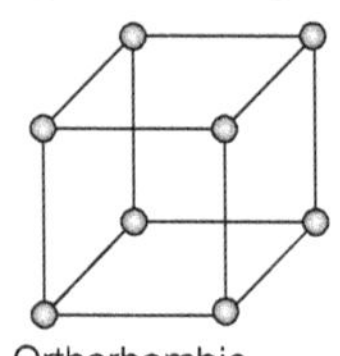

Orthorhombic

59. *(a)* The balanced equation is

$$\underset{\substack{1\text{ mol}\\56\text{ g}}}{CaO} + 2HCl \longrightarrow \underset{\substack{1\text{ mol}\\111\text{ g}}}{CaCl_2} + H_2O$$

56 g of CaO produces $CaCl_2$ = 111 g

1.23 g of CaO will produce

$$CaCl_2 = \frac{111}{56}\times 1.23$$

$$= 2.43\text{ g}$$

Thus, theoretical yield = 2.43 g

Actual yield = 1.85 g

$$\text{Per cent yield} = \frac{1.85}{2.43}\times 100 = 76.1$$

60. *(b)* When nitrobenzene reacts with Sn/HCl, it undergoes reduction to form aniline ,which on reaction with methyl iodide gives N, N-dimethyl aniline.

NO_2 — Benzene $\xrightarrow{\text{(i) Sn/HCl}}$ NH_2 — Aniline $\xrightarrow{(CH_3I)\ (2\ mol)}$ $N(CH_3)_2$ — N, N-dimethyl aniline

61. *(b)* Single step large mutation leading to speciation is called **saltation**. It is the occurrence of a major mutation in single generation, bringing about significant change.

62. *(d)* Immunomodulators are natural or synthetic substances that help to regulate or normalise the immune system. There are two specific types of immunomodulators *viz* immuno suppressants, which cause immuno suppression or inhibit the immune system and immunostimulants (immunostimulators) which stimulate immune system by increasing activity of its components.

63. *(d)* Allograft is transplant between the individuals of same species, but with different genetical background. It is generally used as a temporary measure until the damaged tissue is able to repair itself.

64. *(d)* The bacterium *Xanthomonas compestris* is not used as biopesticide. It is the causative agent of a plant disease called black rot of cabbage. Rest all are biopesticides.

65. *(b)* The only correct option about genetically engineered insulin using *E.coli* is that it can be obtained in huge amounts and unlimited quantities. The other options given are not appropriate for recombinant insulin as it is not difficult to purify. It does not allow transmission of animal diseases and this insulin obtained does not vary in chemical structure.

66. *(b)* Birth rate = 100

Death rate = 10

Number of individuals in a population = 1000

Natural growth rate = 100 – 10 = 90

So, percentage of growth rate

$$= \frac{90}{1000} \times 100$$

= 9.0%

67. *(c)* The rate of total capture of energy or the rate of total production of organic material is Gross Primary Productivity (GPP), while the balance or biomass remaining after meeting the cost of respiration of producers is Net Primary Productivity (NPP). Hence, gross productivity has highest value in grassland ecosystem.

68. *(a)* More release of NO_2 and SO_2 cause acid rain. The acid rain is a mixture of H_2SO_4 and HNO_3. The SO_2 and NO_2 produced during the combination of coal and petroleum reacts with water vapour and formed H_2SO_4 and HNO_3, respectively. These acids combine with air droplets/water vapour causing acid rain.

69. *(c)* *Mycobacterium tuberculosis* is a Gram positive bacterium that appears violet after Gram staining. This is because it retains the colour crystal violet dye due to the presence of a thicker peptidoglycan layer.

70. *(a)* In *Hydra*, the exchange of oxygen and carbon dioxide and the excretion of waste nitrogenous matter (chiefly ammonia) occur directly by diffusion through cell membrane to outside.

71. *(a)* Archaebacterium cannot live in less than 3 M NaCl concentration as their survival requires very high salt concentrations.

72. *(d)* Sieve elements are a component of phloem tissue and are responsible for conduction of food in plants. A sieve tube element has peripheral layer of cytoplasm without any nucleus. The central part is occupied by a network of canals which contain fibrils of P-protein. The protein is vibratite and is believed to actively participate in the transport of nutrients. One general property of it is its ability to form a gel and it functions as a puncture repair substance forming a plug at any site of damage in sieve element, thus preventing loss of food materials being translocated by the phloem. Thus, it is believed to have sealing function on wounding.

73. *(c)* Autonomic genome system is present in mitochondria and chloroplast. These are the autonomous bodies. In these, small circular DNA particles are present which can duplicate and express themselves.

74. *(a)* Option (a) is incorrect as there are three types of cofactors, i.e. prosthetic group, coenzymes and metalions. Thus, every coenzyme is a cofactor but every cofactor is not a coenzyme.

75. *(a)* Interphase is the period between the end of one cell division to the begining of the next cell division. It is the phase where most of the cytogenetic activities take place. The cell becomes metabolically very active and prepares itself for both cell growth and DNA replication in an orderly manner.

76. *(c)* Cyclic flow of light reaction is the process of photosynthesis in which light of wavelength more than 680 nm is required. It takes place in the stromal lamella of the chloroplast. During cyclic photophosphorylation only the synthesis of ATP takes place (i.e. not NADPH + H^+) because the stroma lamellae do not possess enzyme NADP reductase due to which the excited electrons in the cyclic flow do not pass on to NADP instead get back to the PS-I complex only.

77. *(c)* Conversion of 2 PGA to PEP is the 8th step of glycolysis. This conversion is catalysed by the enzyme enolase and takes place in the presence of cofactor Mg^{2+} and R^+ with the loss of a water molecule.

78. *(b)* The saliva is a colourless, opalascent and sticky liquid produced by the salivary glands. It contains salivary amylase for digestion of carbohydrate. In quantity it is about 1.5 dm^3 (cubic decimeter), i.e 1.5 L/day.

79. *(d)* Coronary heart disease occurs due to the insufficient blood supply of blood to the heart muscle. It occurs due to the hardening of arteries and arteriole because of thickening of fibre tissue and the consequent loss of elasticity. It is also referred to as atherosclerosis.

80. *(b)* Resting potential is the difference in electrical potential that exists across the membrane of nerve cells. The resting potential is maintained with the help of sodium-potassium pump.

81. *(a)* Slope of line joining the points $(c - 1, e^{c-1})$ and $(c + 1, e^{c+1})$ is equal to

$$\frac{e^{c+1} - e^{c-1}}{2} > e^c$$

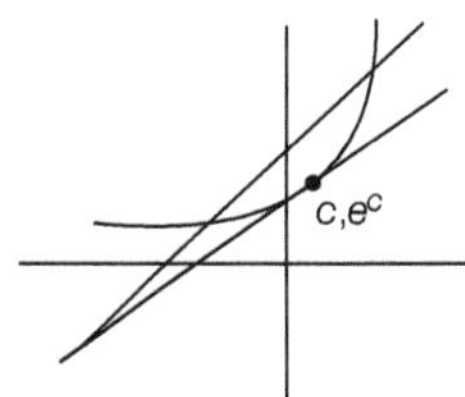

$\Rightarrow$ tangent to the curve $y = e^x$ will intersect the given line to the left of the line $x = c$.

82. *(c)* Given, $g(x) = \dfrac{(x-1)^n}{\log \cos^m (x-1)}$

P is left hand derivative of $|x-1|$

$\therefore \quad P = -1$

$$\lim_{x \to 1^+} g(x) = P$$

$$\Rightarrow \lim_{h \to 0} g(1+h) = -1$$

$$\Rightarrow \lim_{h \to 0} \frac{h^n}{\log \cos^m h} = -1$$

$$\Rightarrow \lim_{h \to 0} \frac{nh^{n-1}}{m(-\tan h)} = -1$$

$$\Rightarrow \frac{-n}{m} \lim_{h \to 0} \frac{h^{n-1}}{\tan h} = -1$$

Which holds of $n - 1 = 1 \Rightarrow n = 2$

and $-\dfrac{n}{m} = -1 \Rightarrow n = m = 2$

83. *(d)* We have,

$$(1+p)x - py + p(1+p) = 0 \quad \text{...(i)}$$
$$(1+q)x - qy + q(1+q) = 0 \quad \text{...(ii)}$$
$$y = 0 \quad \text{...(iii)}$$

From Eqs. (i) and (ii), we get

$$x = pq,\ y = (p+1)(q+1)$$

From Eqs. (i) and (iii), we get

$$x = -p,\ y = 0$$

From Eqs. (ii) and (iii), we get

$$x = -q,\ y = 0$$

C(pq,(1+p)(1+q))
(h,k)
A(−q,0) y=0 B(−p,0)

Equation of altitude from C to AB is

$$x = pq$$

Altitude from B to AC is

$$y = \frac{-q}{1+q}(x+p)$$

Put $x = pq$, we get $y = -pq$

$\because \quad h = pq,\ k = -pq$

$\therefore \quad h + k = 0$

Locus of orthocentre $x + y = 0$ which represent the equation of straight line.

84. *(a)* System of equation represent the equation of three planes.

But three plane cannot intersect at exactly two points.

$\therefore$ No solution exists.

85. *(c)* $r_1, r_2, r_3 \in \{1, 2, 3, 4, 5, 6\}$

r_1, r_2, r_3 are forms of $(3m, 3m+1, 3m+2)$

$\therefore$ Required probability

$$= \frac{3! \times {}^2C_1 \times {}^2C_1 \times {}^2C_1}{6^3} = \frac{2}{9}$$

86. *(a)* $\mathbf{PQ} = 6\mathbf{i} + \mathbf{j}$

$\mathbf{QR} = -\mathbf{i} + 3\mathbf{j}$

$\mathbf{SR} = 6\mathbf{i} + \mathbf{j}$

$\mathbf{PS} = -\mathbf{i} + 3\mathbf{j}$

$\mathbf{PQ} \parallel \mathbf{BR}$ and $\mathbf{QR} \parallel \mathbf{PS}$

S(−3i+2j) R(3i+3j)
P(−2i − j) Q(4i)

$\because PQRS$ is a parallelogram

$$\mathbf{PQ} \cdot \mathbf{QR} = (6\mathbf{i} + \mathbf{j})(-\mathbf{i} + 3\mathbf{j})$$
$$= -6 + 3$$
$$= -3 \neq 0$$

$\therefore PQRS$ is not a rectangle.

$$|\mathbf{PQ}| = \sqrt{36+1} = \sqrt{37}$$
$$|\mathbf{QR}| = \sqrt{1+9} = \sqrt{10}$$
$$|\mathbf{PQ}| = |\mathbf{QR}|$$

$\therefore$ It is not a rhombus.

87. *(c)* We have,

$$(2x)^{\ln 2} = (3y)^{\ln 3}$$

$$\Rightarrow \ln 2(\ln 2 + \ln x) = \ln 3(\ln 3 + \ln y)$$

$$\Rightarrow (\ln 2)^2 + \ln 2 \cdot \ln x = (\ln 3)^2 + \ln 3 \cdot \ln y \quad \text{...(i)}$$

$$\Rightarrow 3^{\ln x} = 2^{\ln y}$$

$$\Rightarrow \ln x \ln 3 = \ln y \ln 2 \quad \text{...(ii)}$$

From Eqs. (i) and (ii), we get

$$(\ln 2)^2 + \ln 2 \cdot \ln x = (\ln 3)^2 + \ln 3\left(\frac{\ln x \cdot \ln 3}{\ln 2}\right)$$

$$\Rightarrow (\ln 2)^3 + (\ln 2)^2 \ln x = (\ln 3)^2 \ln 2 + (\ln 3)^2 \ln x$$

$$\Rightarrow (\ln x)[(\ln 3)^2 - (\ln 2)^2] = \ln 2[(\ln 2)^2 - (\ln 3)^2]$$

$$\Rightarrow \ln x = -\ln 2$$

$$\therefore \quad x = \frac{1}{2}$$

88. *(a)* We have,

$$f(x) = \frac{b-x}{1-bx},\ x \in (0, 1) \text{ and } b \in (0,1)$$

Let $f(x) = y$

$$\therefore \quad y = \frac{b-x}{1-bx}$$

$$\Rightarrow y - ybx = b - x$$

$$\Rightarrow \quad x = \frac{b-y}{1-by}$$

Range $f(x) \neq R$

$\therefore f(x)$ is not invertible in $(0, 1)$.

89. *(d)* We have,

$$\int_1^n [x]\{x\}\, dx > 2020$$

$$\Rightarrow \int_1^2 \{x\}\, dx + 2\int_2^3 \{x\} dx + \dots + (n-1)\int_{n-1}^n \{x\} dx > 2020$$

$$\Rightarrow (1 + 2 + 3 \dots + n - 1)\int_0^1 x\, dx > 2020$$

$$\Rightarrow \frac{(n-1)(n)}{2}\left[\frac{n^2}{2}\right]_0^1 > 2020$$

$$\Rightarrow \frac{n(n-1)}{2 \times 2} > 2020$$

$$\Rightarrow n(n-1) > 8080$$

Minimum of x is 91.

90. *(d)* We have,

$$f'(x) = \frac{1}{1+x^5}$$

$g(x)$ is inverse of $f(x)$.

$\therefore \quad f(g(x)) = x$

On differentiating, we get

$$f'(g(x) \cdot g'(x)) = 1$$

$$\Rightarrow g'(x) = \frac{1}{f'(gx)}$$

$$\Rightarrow g'(x) = \frac{1}{\dfrac{1}{1+(g(x)^5)}}$$

$$\Rightarrow g'(x) = 1 + (g(x))^5$$

91. *(b)* To form a sharp image, the light cone that passes the circular hole and light cone that is formed by the refracted light have same base area on screen. This is shown below.

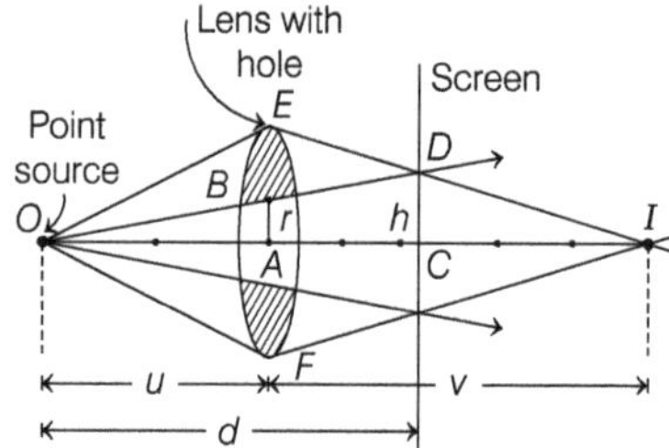

Now, from lens equation, we have

$$\frac{1}{v} - \frac{1}{u} = \frac{1}{f}$$

$$\frac{1}{v} = \frac{1}{f} + \frac{1}{-x} = \frac{x-f}{fx}$$

Image distance $= v = \dfrac{fx}{x-f} = \dfrac{fu}{u-f}$

Let, h = radius of image on screen,

r = radius of hole in lens

and R = radius of lens.

Now, from similar triangles ΔOAB and ΔOCD, we have

$$\frac{h}{r} = \frac{d}{u} \quad \ldots(i)$$

***Note** That no sign convention is required here.*

And from similar triangles ΔDCI and ΔAEI, we have

$$\frac{h}{R} = \frac{u+v-d}{v} \quad \left(\because R = 2r \text{ and } v = \frac{fu}{u-f}\right)$$

$$\Rightarrow \frac{h}{2r} = \frac{u + \left(\frac{fu}{u-f}\right) - d}{\left(\frac{fu}{u-f}\right)} \quad \ldots(ii)$$

From Eqs. (i) and (ii), we have

$$\Rightarrow \frac{d}{2u} = \frac{u(u-f) + fu - d(u-f)}{fu}$$

$$\Rightarrow 2u^2 - 2du + df = 0$$

$$\Rightarrow u = \frac{2d \pm \sqrt{4d^2 - 8df}}{4}$$

$$\Rightarrow u = \frac{d}{2} \pm \left(\frac{1}{2}\sqrt{d(d-2f)}\right)$$

Here, $f = 4$ cm, $d = 9$ cm.

So, $u = \frac{9}{2} \pm \frac{1}{2}\sqrt{9(9-8)} = \frac{9}{2} \pm \frac{3}{2}$

$\therefore$ $u = 6$ cm or $u = 3$ cm

So, both are possible positions.

92. *(b)* A 10 Gy dose of radiation corresponds to 10 J of radiation energy. As tumor mass is 0.003 kg energy for 10 Gy dose is

$$E_1 = 0.003 \times 10 = 0.03 \text{ J}$$

Each disintegration provides 0.7 MeV.

$$E_2 = 0.7 \text{ MeV} = 0.7 \times 10^6 \times 1.6 \times 10^{-19} \text{ J}$$
$$= 1.12 \times 10^{-13} \text{ J}$$

To provide 0.03 J number of disintegrations required is

$$N = \frac{0.03}{1.12 \times 10^{-13}} = 2.68 \times 10^{11} \text{ decays}$$

These must occurs in 14 min,

$\therefore$ Number of decays per second $= n = \frac{N}{t}$

$$= \frac{2.68 \times 10^{11}}{14 \times 60} = 3.7 \times 10^8 \text{ decays s}^{-1}$$

$\therefore$ Source activity required is

$$R = 3.7 \times 10^8 \text{ Bq}$$

93. *(a)* Frequency of revolution in a Bohr's orbit is

$$f_n = \frac{-E_1}{h}\left(\frac{2}{n^3}\right)$$

where, E_1 = energy in $n = 1$ state.

So, $f_2 = \frac{-E_1}{h} \times \frac{2}{2^3} = 0.823 \times 10^{15}$ rps

$= 8.23 \times 10^{14}$ rps

Number of revolutions made by electron in 1×10^{-8} s is

$$N = f_2 \times \Delta t = 8.23 \times 10^{14} \times 1 \times 10^{-8}$$
$$= 8.23 \times 10^6 \text{ revolutions}$$

94. *(a)* From first law of thermodynamics,

we have, $\Delta Q - \Delta W = \Delta E$

Applied to process $1 \to 2$, we get

$$a - 100 = 100 \Rightarrow a = 200 \text{ kJ}$$

For process $3 \to 1$,

$$100 - d = -200$$
$$d = 300 \text{ kJ}$$

For entire cycle, $\Sigma Q = \Sigma W$

$\Rightarrow 200 + b + 100 = 350 \Rightarrow b = 50$ kJ

For a cyclic process, $\Sigma \Delta U = 0$

$\Rightarrow \quad 100 + c - 200 = 0$

$\Rightarrow \quad c = 100$ kJ

95. *(c)* Position of a particle is

$$s = 3t^2 - 12t + 9 = 3(t-1)(t-3)$$

So, path of particle is

$t = 0$, $x = 0$ $\quad$ $t = 1$, $x = 4$ $\quad$ $t = 5$, $s = 20$

$\therefore$ Total distance covered

$$s = 4 + 4 + 20 = 28 \text{ m}$$

96. *(c)* $\frac{U_D}{U_{\text{Total}}} = \frac{U_D}{U_{\text{gap}} + U_D}$

$$= \frac{U_D \text{ Vol}_D}{U_{\text{gap}}\text{Vol}_{\text{gap}} + U_D \text{ Vol}_D}$$

$$= \frac{\frac{1}{2}\epsilon_0 \frac{E_0^2}{K}\frac{V}{2}}{\left(\frac{V}{2} \times \frac{1}{2}\epsilon_0 E^2 + \frac{V}{2} \times \frac{1}{2}\frac{\epsilon_0 E^2}{K}\right)}$$

$$= \frac{\frac{1}{K}}{1 + \frac{1}{K}} = \frac{1}{1+K} = \frac{1}{5}$$

$\Rightarrow \quad U_D = 20\%U \quad [\because K = 1 + X = 1 + 3 = 4]$

97. *(b)*

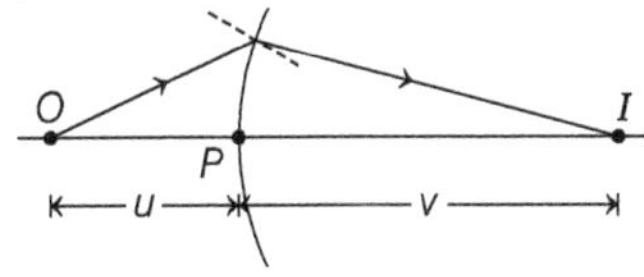

For a spherical refracting surface,

$$\frac{\mu}{v} - \frac{1}{(-u)} = \frac{\mu - 1}{R}$$

$$\Rightarrow \quad \frac{\mu}{v} + \frac{1}{u} = \frac{\mu - 1}{R}$$

Let $u + v = x$

$$\Rightarrow \quad \frac{\mu}{x - u} + \frac{1}{u} = \frac{\mu - 1}{R}$$

$$\Rightarrow \quad x = u + \frac{\mu R u}{(\mu - 1)u - R}$$

$$\therefore \quad \frac{du}{dx} = \frac{\mu R^2}{((\mu - 1)u - R)^2}$$

l is minimum, when $\frac{du}{dx} = 0$.

$$\therefore \quad u = \frac{R}{\sqrt{(\mu)} - 1}$$

From Eq (i), we have

$$v = \frac{\sqrt{\mu}\, R}{\sqrt{\mu} - 1}$$

As $l = u + v$, $l = \frac{R}{\sqrt{\mu} - 1} + \frac{\sqrt{\mu} \cdot R}{\sqrt{\mu} - 1}$

$$\mu = \left(\frac{R + l}{R - l}\right)^2$$

98. *(c)* Density of water is maximum at 4° C. So, upthrust is maximum at 4° C.

99. *(a)* For adiabatic process,

$$\Delta Q = \Delta U + \Delta W = 0$$

$$\therefore \quad C_V dT + p dV = 0$$

$$\Rightarrow \quad C_V dT + \frac{RT}{V} \cdot dV = 0$$

$$\Rightarrow \quad C_V dT + (C_p - C_V)T \cdot \frac{dV}{V} = 0$$

$$(b + cT)dT + (a - b)T \cdot \frac{dV}{V} = 0$$

$$\Rightarrow \quad b\frac{dT}{T} + c\,dT + (a - b)\frac{dV}{V} = 0$$

Integrating above equation, we have

$$b \ln T + cT + (a - b)\ln V = \text{a constant}$$

So, $\ln T^b + cT + \ln V^{a-b}$ = a constant

$\Rightarrow V^{a-b} \cdot T^b \cdot e^{cT}$ = constant

100. *(d)* Longitudinal Doppler effect in light gives,

$$f = f_0\left(\frac{1 + v/c}{1 - v/c}\right)^{1/2}$$

where, f = frequency observed,

f_0 = frequency emitted by source,

v = speed of observer moving towards source

and $\quad c$ = speed of light.

$\therefore$ We have, $v = c\left(\frac{f^2 - f_0^2}{f^2 + f_0^2}\right)$

$$v = 3 \times 10^8\left(\frac{(5.6)^2 - (4.8)^2}{(5.6)^2 + (4.8)^2}\right)$$
$$= 4.59 \times 10^7 \text{ m/s}$$
$$= 1.65 \times 10^8 \text{ km/h}$$

So, $\quad$ fine $= (1.65 \times 10^8 - 100)$

$= ₹\, 164, 999, 900$

101. *(a)* The complete reaction can be represented as:

H_3C–(cyclopentyl)–C(=O)–CH_3 $\xrightarrow[\text{(Epoxide formation)}]{\text{(i) } CH_3COOH}$ H_3C–(cyclopentyl)–O–C(=O)–CH_3 $\xrightarrow[\text{(ii) NaOH}]{\text{Hydrolysis}}$ H_3C–(cyclopentyl)–OH + CH_3COONa

Reaction (i) is known as Baeyer's Villiger oxidation in which ketone group gets oxidised to ester group in presence of peroxide.

102. *(b)* ${}^{238}_{92}U \longrightarrow {}^{206}_{82}Pb$

Number of α-particles

$$= \frac{\text{change in mass number}}{4}$$

$$= \frac{238-206}{4}$$

$$= \frac{32}{4} = 8\alpha$$

Number of β-particles $= 2 \times$ number of α-particles – change in atomic number

$$= 2 \times 8 - (92 - 82)$$

$$= 16 - 10 = 6\beta$$

103. *(d)* Chlorobenzene (Cl) $\xrightarrow[\text{Dow's process}]{\text{(i) NaOH, (ii) } H^+}$ Phenol (OH) X $\xrightarrow[\text{(ii) } H^+ \text{ Reimer-Tiemann reaction}]{\text{(i) } CHCl_3 + aq. \text{ NaOH}}$ Salicylaldehyde (OH, CHO) Y

Here, reaction (i) is known as Dow's process, where chlorobenzene gets converted into phenol in presence of NaOH. The formed phenol then reacts with chloroform in aqueous NaOH to give salicylaldehyde. This reaction is known as Reimer-Tiemann reaction.

104. *(a)* As from the given graph, it is clear that the concentration of the product increases linearly with time t, thus order of the reaction

$$= 0 \text{ and } -\frac{d[A]}{dt} = \text{constant}.$$

Thus correct plot is given option (a).

105. *(d)* Molar mass of solute

$$(M_B) = \frac{1000 \times K_f \times W_B}{W_A \times \Delta T_f}$$

$$\Rightarrow \quad M_B = \frac{1000 \times 14 \times 75.2}{1000 \times 7}$$

$$M_B = 150.4 \text{ g per mol}$$

Actual molar mass of phenol = 94 g/mol

Now, van't Hoff factor,

$$i = \frac{\text{Calculated molar mass}}{\text{Observed molar mass}}$$

$$\therefore \quad i = \frac{94}{150.4} = 0.625$$

Dimerisation of phenol can be shown as :

	$2C_6H_5OH \rightleftharpoons$	$(C_6H_5OH)_2$
Initial	1	0
At equilibrium	$1-\alpha$	$\frac{\alpha}{2}$

Total number of moles at equilibrium,

$$i = 1 - \alpha + \frac{\alpha}{2}$$

$$i = 1 - \frac{\alpha}{2}$$

But $i = 0.625$, thus, $0.625 = 1 - \frac{\alpha}{2}$

$$\frac{\alpha}{2} = 1 - 0.625$$

$$\alpha = 0.75$$

Thus, the percentage of phenol that dimerises is 75%.

106. *(c)* Given

$$H^\circ_{CO_2} = \Delta_f H^\circ(CO_2) = -400 \text{ kJ mol}^{-1}$$

$$H^\circ_{H_2O} = \Delta_f H^\circ(H_2O) = -300 \text{ kJ mol}^{-1}$$

$$H^\circ_{\text{glucose}} = \Delta_f H^\circ \text{(glucose)}$$

$$= -1300 \text{ kJ mol}^{-1}$$

$$H^\circ_{O_2} = \Delta_f H^\circ(O_2) = 0.00$$

$$C_6H_{12}O_6(s) + 6\,O_2(g) \longrightarrow 6\,CO_2(g) + 6H_2O(l)$$

$\Delta_c H^\circ$ (glucose)

$$= 6[\Delta_f H^\circ(CO_2) + \Delta_f H^\circ(H_2O)] - [\Delta_f H^\circ(C_6H_{12}O_6) + 6\Delta_f H^\circ(O_2)]$$

$$= 6[-400 - 300] - [-1300 + 6 \times 0]$$

$$= -2900 \text{ kJ mol}^{-1}$$

Molar mass of $C_6H_{12}O_6 = 180 \text{ g mol}^{-1}$

Thus, standard heat of combustion of glucose per gram

$$= \frac{-2900}{180} = -16.11 \text{ kJ g}^{-1}$$

107. *(a)* In bcc unit cell, $4r = \sqrt{3}a$

$$\Rightarrow r(\text{Cr}) = \frac{\sqrt{3}a}{4} = \frac{\sqrt{3}}{4} \times 287 \text{ pm}$$

$$= 124.3 \text{ pm}$$

$$\text{Density of solid} = \frac{NM}{N_A \cdot a^3}$$

N = number of atoms per unit cell,

M = molar mass

a^3 = volume of cubic unit cell,

N_A = Avogadro's number

For bcc $Z = 2$

$$= \frac{2 \times 52 \text{ g}}{6.023 \times 10^{23}} \times \left(\frac{1}{2.87 \times 10^{-8} \text{ cm}}\right)^3$$

$$= 7.3 \text{ g/cm}^3$$

108. *(a)* Cr is in +2 oxidation state in $[Cr(H_2O)_6]Cl_2$. The electronic configuration of Cr^{2+} is $3d^4 4s^0$.

	3d					4s	4p		
Cr	1	1	1	1					

As H_2O is a weak field ligand, so pairing will not occur.

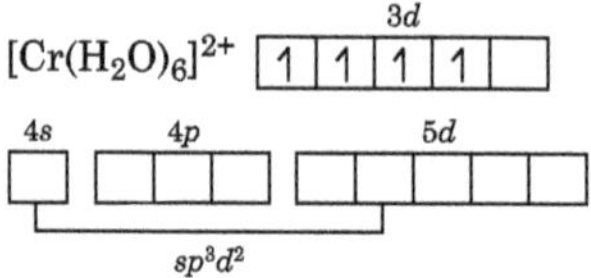

Thus, the hybridisation of $[Cr(H_2O)_6]^{2+}$ is sp^3d^2.

Magnetic moment $= \sqrt{n(n+2)}$

For this complex, $n = 4$

$$\mu = \sqrt{4(4+2)}$$

$$= \sqrt{24}$$

$$= 4.89 \text{ BM}$$

109. *(c)* For the reaction;

$$C + 2H_2 \longrightarrow CH_4$$

Given, $\Delta H^\circ = -74.81 \text{ kJ mol}^{-1}$

ΔS°_m can be calculated as:

$$\Delta S^\circ_m = S^\circ_{m\,(\text{product})} - S^\circ_{m\,(\text{reactant})}$$

$$= S^\circ_m CH_4(g) - [S^\circ_m C_{(\text{graphite})} + 2S^\circ_m H_2(g)]$$

$$= [186.3 - (5.70 + 2 \times 130.7)] \text{ JK}^{-1} \text{ mol}^{-1}$$

$$= -80.8 \text{ JK}^{-1} \text{ mol}^{-1}$$

$$= -80.8 \times 10^{-3} \text{ JK}^{-1} \text{ mol}^{-1}$$

Since, $\Delta G^\circ = \Delta H^\circ - T\Delta S^\circ$

$= -74.81 - [(298) \times (-80.8 \times 10^{-3})]$

$= -74.81 + 24.07$ kJ mol^{-1}

$= -50.74$ kJ mol^{-1}

110. *(b)* In option (b) hydrogenation of olefinic bond will render compound achiral as two identical ethyl group will come at the α-carbon, which was earlier chiral carbon. Thus, it gives optically inactive compound on hydrogenation. However, in (a), (c) and (d) chirality will be retained even after hydrogenation.

(a) H_3C–CH=CH–CH_2–C(H)(Br)–H_2C $\xrightarrow{H_2}$ H_3C–… C(H)(Br)–CH_3 Chiral

(b) H_2C=… C(H)(Br)–CH_3 $\xrightarrow{H_2}$ H, H_2C, C(H)(Br)–CH_3, H Achiral

(c) H_2C=C(CH_3)–C(H)(Br)–CH_3 $\xrightarrow{H_2}$ H_3C–CH(CH_3)–C(H)(Br)–CH_3 Chiral

(d) H_3C–CH_2–C(Br)(H)–CH=CH–CH_3 $\xrightarrow{H_2}$ CH_3–C(Br)(H)–CH_2–CH_3 Chiral

111. *(c)* *P* is competition as in this both organisms are harmed.

Q is predation as in this prey is harmed and predotor is benefitted.

R is mutualism as in this both organisms are benefitted.

S is amensalism as in this one organism is harmed and other is neutral.

112. *(a)* Chronic gastritis– *Helicobacter pylori*

Lyme disease– *Borrelia burgdorferi*

Scarlet fever– *Streptococcus pyogenes*

Typhus– *Rickettsia prowazekii*

113. *(c)* Alternative RNA splicing is a form of regulation of gene expression where different *m*RNA molecules are produced from the same primary transcript, depending on this some RNA segments are treated as exons and some as introns. This enables the single gene of 5055 base pairs to code for more than one kind of polypeptide, depending on which segments are spliced together during RNA processing.

114. *(d)* All of the statements are incorrect. These can be corrected as

Statement I and II The glycoprotein 120 on the HIV envelope must attach to both a CD_4 molecule and a chemokine receptor. The interaction between gp 120 and chemokine receptor brings about a conformation change in gp 41 on the HIV virion, allowing the fusion of HIV envelope and host cell membrane. The fusion allows the nucleocapsid to enter the host cell. The capsid is removed by cellular enzymes and the viral RNA genome and various enzymes are released into the host cell.

Statement III The viral DNA which enters the host cell's nucleus will be integrated into the genetic material of the host cell using the HIV virus enzyme, integrase and not the host cell's enzyme.

Statement IV The viruses bud out from the host cell's plasma membrane and are not released by exocytosis.

115. *(b)* In a linakge group, the relative frequency of crossing over is a measure of the distance apart of genes. The higher the percentage, the further is the distance between the two genes. Therefore, the sequence of the genes is PSRQ.

P S R Q

P ← 34 → R

P ← 59 → Q

S ← 12 → R

S ← 37 → Q

116. *(b)* B : allele for black coat

G : allele for ginger coat

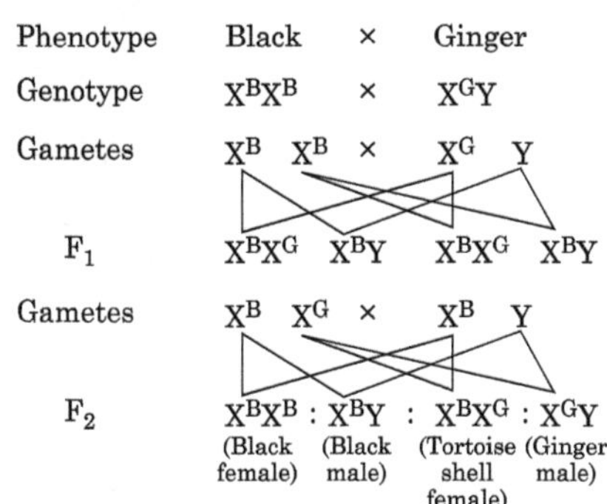

117. *(c)* In the light dependent reactions, photolysis of water occurs to release electrons, hydrogen ions and oxygen. Oxygen is released as oxygen gas. In the light independent reactions, carbon dioxide is fixed, producing carbohydrates.

118. *(b)* Option (b) is correct. Ethanol is a non-polar solvent anel the phospholipid bilayer of the beet root cell membrane is also non-polar. Non-polar solutes dissolve in non-polar solvents and the lipids dissolve in alcohol (ethanol is an organic solvent that can dissolve non-polar and even some polar substance), thus the cell membrane will be disrupted as it is dissolved by the ethanol solution. This means that the structure will be altered and there will be more gaps in the phospholipid bilayer, thus the membrane will be more permeable and more of the beet root's red pigment will leak out into the alcohol solvent, causing the solution to become a more intense red colour.

119. *(c)* According to chemiosmotic hypothesis ATP synthetase becomes active in ATP formation only where there is a protein gradient having higher concentration of H^+ or protons on the inner mitochondrial as compared to outer membrane.

120. *(a)* Spontaneous mutations are those that occur without a known cause. They may be truly spontaneous, resulting from an inherent low level of metabolic errors (i.e. mistakes) during DNA replication or because of mutagenic agents in the environment which is probably not the cause in this case.

PRACTICE SET 5

Stream : SB/SX

MM : 160

Instructions

1. There are 120 questions in this paper.
2. The question paper contains two parts; Part I (1 Mark Questions) and Part II (2 Marks Questions).
3. There are four sections in each part; Mathematics, Physics, Chemistry and Biology.
4. Out of the four options given with each question, only one is correct.

PART-I (1 Mark Questions)

MATHEMATICS

1. Suppose $\log_a b + \log_b a = c$. The smallest positive integer value of c for all, $a, b > 1$

(a) 4 (b) 3 (c) 2 (d) 1

2. If a, b, c are odd integers, then equation $ax^2 + bx + c = 0$ cannot have

(a) imaginary roots (b) real roots
(c) irrational roots (d) rational roots

3. When $1! + 2! + 3! + 4! + \dots + 2020!$ is divided by 35, then remainder will be

(a) 31 (b) 32 (c) 33 (d) 34

4. The probability that the length of a randomly chosen chord of a circle lies between $\frac{2}{3}$ and $\frac{5}{6}$ of its diameter is

(a) $\frac{1}{4}$ (b) $\frac{5}{12}$ (c) $\frac{1}{16}$ (d) $\frac{5}{16}$

5. Range of $y = \cos\left(\pi \sin\left(\frac{\pi}{2}\cos(\pi \sin x)\right)\right)$, where $x \in R$ is

(a) $[-\pi, \pi]$ (b) $[-1, 1]$
(c) $[0, 1]$ (d) $[-1, 0]$

6. If the product of the intercepts made by the line $x\tan\alpha + y\sec\alpha = 1$ on the coordinate axes is $\sin\alpha$, then α is equal to

(a) $\frac{\pi}{4}$ (b) $\frac{\pi}{2}$ (c) π (d) 0

7. Let $g(x) = \int_0^{|x|^{3/4}} t^{2/3} \sin\frac{1}{t}\, dt$, for all real x, then $\lim_{x \to 0} \frac{g(x)}{x}$ is equal to

(a) ∞ (b) $-\infty$ (c) 0 (d) $\frac{3}{4}$

8. A circle is inscribed into a rhombus $ABCD$ with one angle 60°. The distance from the centre of the circle to the nearest vertex is equal to 1. If P is any point of the circle, then $|PA|^2 + |PB|^2 + |PC|^2 + |PD|^2$ is equal to

(a) 9 (b) 10 (c) 11 (d) 12

9. A parabola $y = ax^2 + bx + c$ crosses the X-axis at $(\alpha, 0)$, $(\beta, 0)$ both to the right of the origin. A circle also passes through those two points. The length of a tangent from the origin to the circle is

(a) $\sqrt{\frac{bc}{a}}$ (b) ac^2 (c) $\frac{b}{a}$ (d) $\sqrt{\frac{c}{a}}$

10. Let P be an $m \times m$ matrix such that $P^2 = P$. Then $(I + P)^n$ equals (where I is identity matrix of order m)

(a) $I + P$ (b) $I + nP$
(c) $I + 2^n P$ (d) $I + (2^n - 1)P$

11. Let $[n]$ denotes the greatest integer less than or equal to λ. How many real numbers x satisfy the equation $x^2 + 10000[x] = 10000x$

(a) 198 (b) 199 (c) 200 (d) 201

12. The value of $\int_0^\infty \frac{1}{(1 + x^{2020})(1 + x^2)} dx$ is

(a) π (b) $\frac{\pi}{2}$ (c) $\frac{\pi}{4}$ (d) 0

13. In a ΔXYZ, let a, b and c be the lengths of the sides opposite to the angles X, Y and Z, respectively. If $1 + \cos 2X - 2\cos 2Y = 2\sin X \sin Y$, then $\frac{a}{b}$ is equal to

(a) 1 (b) 2 (c) 3 (d) 4

14. Let the harmonic mean of two positive real numbers a and b be 4. If q is positive real number such that $a, 5, q, b$ is an arithmetic progression, then the values of q is equal to

(a) 3 (b) 4 (c) 5 (d) 6

15. Let **a**, **b** and **c** be three non-coplaner unit vectors such that the angle between every pair of them is $\frac{\pi}{3}$. If $\mathbf{a} \times \mathbf{b} + \mathbf{b} \times \mathbf{c} = p\mathbf{a} + q\mathbf{b} + r\mathbf{c}$, where p, q and r are scalars, then the value of $\frac{p^2 + 2q^2 + r^2}{q^2}$ is

(a) 6 (b) 10 (c) 8 (d) 4

16. How many 3×3 matrices M with entries from $\{0, 1, 2\}$ are there, for which the sum of diagonal entries of $M^T M$ is 5?

(a) 198 (b) 162 (c) 126 (d) 135

17. Three randomly chosen non-negative integers x, y and z are found to satisfy the equation $x + y + z = 10$. Then, the probability that z is even

(a) $\frac{1}{2}$ (b) $\frac{36}{55}$ (c) $\frac{6}{11}$ (d) $\frac{5}{11}$

18. The function $f : [0, 3] \to [1, 29]$, defined by $f(x) = 2x^3 - 15x^2 + 36x + 1$, is

(a) one-one and onto (b) onto but not one-one
(c) one-one but not onto (d) neither one-one nor onto

19. The ellipse $E_1 = \frac{x^2}{9} + \frac{y^2}{4} = 1$ is inscribed in a rectangle R whose sides are parallel to the coordinate axes. Another ellipse E_2 passing through the point (0, 4) circumscribes the rectangle R.

The eccentricity of ellipse E_2 is

(a) $\frac{\sqrt{2}}{2}$ (b) $\frac{\sqrt{3}}{2}$ (c) $\frac{1}{2}$ (d) $\frac{3}{4}$

20. Let f be a non-negative function defined on the interval [0, 1]. If $\int_0^x \sqrt{1 - (f'(t))^2}\, dt = \int_0^x f(t)\, dt$ $0 \le x \le 1$ and $f(0) = 0$, then

(a) $f\left(\frac{1}{2}\right) < \frac{1}{2}$ and $f\left(\frac{1}{3}\right) > \frac{1}{3}$ (b) $f\left(\frac{1}{2}\right) > \frac{1}{2}$ and $f\left(\frac{1}{3}\right) > \frac{1}{3}$

(c) $f\left(\frac{1}{2}\right) < \frac{1}{2}$ and $f\left(\frac{1}{3}\right) < \frac{1}{3}$ (d) $f\left(\frac{1}{2}\right) > \frac{1}{2}$ and $f\left(\frac{1}{3}\right) < \frac{1}{3}$

PHYSICS

21. A conductor is charged in a room, its potential varies with time (considering all practical factors) will be

(a)
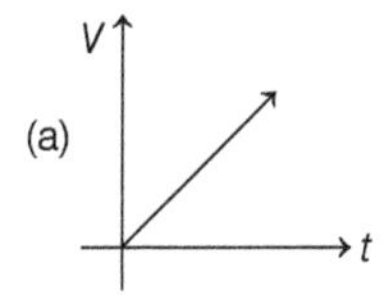

(b)
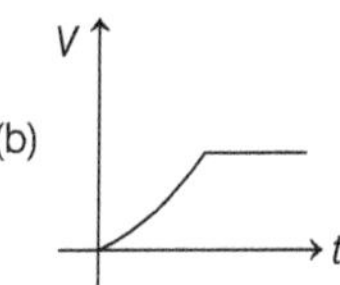

(c)
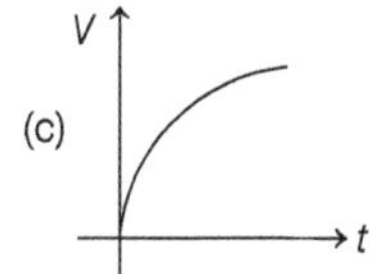

(d)
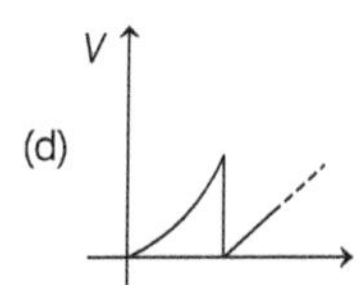

22. Which of these is a path for a long range projectile such as an intercontinental ballistic missile?

(a) Parabolic (b) Semi-circular
(c) Elliptical (d) Straight line

23. Surface tension of a liquid at room temperature is 6.99×10^{-2} Nm^{-1}. If vapour pressure of the liquid is 2.33×10^3 Pa, then radius of smallest drop which can be formed without evaporating at room temperature is

(a) 0.06 mm (b) 0.4 mm
(c) 0.6 mm (d) 0.32 mm

24. A ball is dropped over an inclined plane of inclination 45°.

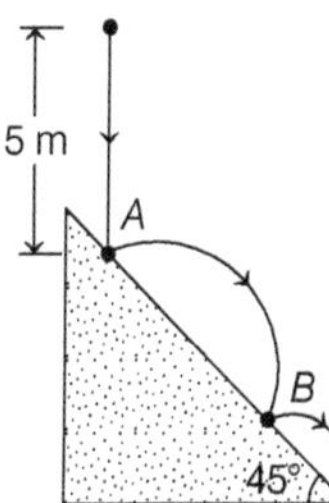

If the ball is dropped from a height of 5 m above the inclined plane and collision is elastic, then the length AB over the plane is (take, $g = 10$ ms^{-2})

(a) $20\sqrt{2}$ m (b) $10\sqrt{2}$ m
(c) $5\sqrt{2}$ m (d) $15\sqrt{2}$ m

25. Consider a uniform semi-circular wire of radius r placed, such that its centre coincides with its centre. Then, coordinates of its centre of mass are best located in

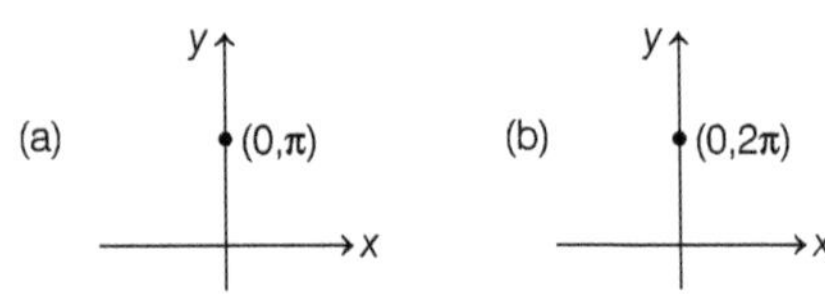

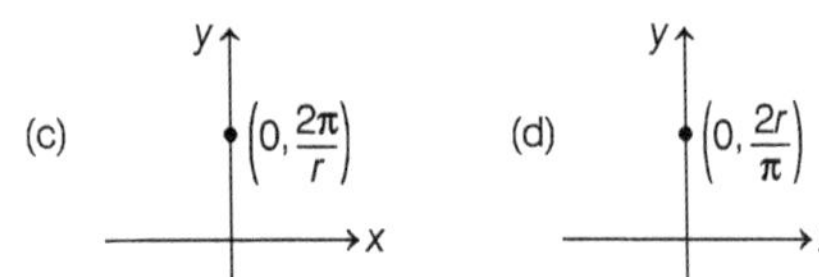

26. Two radio antennas separated by 300 m are simultaneously broadcasting signals of same wavelengths.
A car is moving over a parallel path at a distance of 1000 m from line joining antennas as shown below.

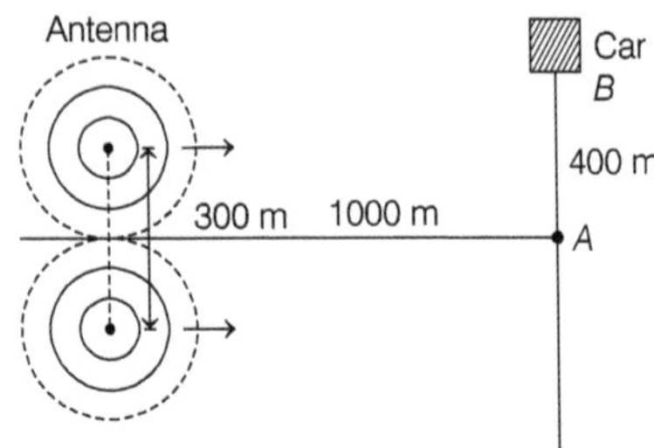

At 400 m from the centre line of antenna, a radio fitted in car registers a second minima. Wavelength of transmission is around

(a) 110 m (b) 45 m (c) 65 m (d) 55 m

27. A Michelson interferometer invented to form an interference pattern by splitting a light beam into two parts and recombining them after both parts covers unequal distances. This interferometer is used in LIGO appratus. This appratus detects

(a) matter waves
(b) electromagnetic waves of high frequencies
(c) radio waves from distant galaxies
(d) gravitational waves

28. A capacitor is connected to a cell and after a long time, a dielectric slab is allowed to fall through the gap of capacitor plates with out touching them.

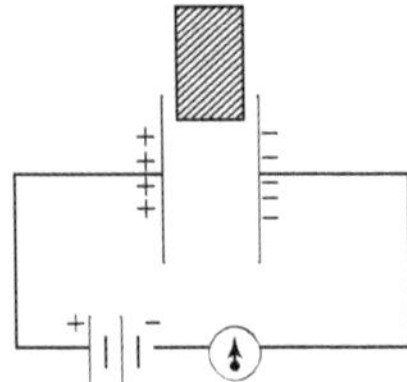

As slab passes through the gap needle of a sensitive galvanometer shows

(a) no deflection at all
(b) deflection towards left hand side
(c) deflection towards right hand side
(d) initial left then towards right hand side deflection

29. A ball is rolled over a horizontal ground making an angle of 45° with a mirror wall in an art gallery.

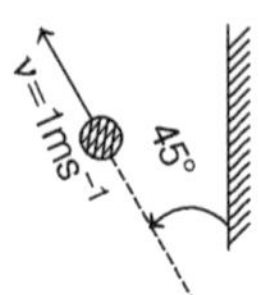

If velocity of ball is 1 ms^{-1}, then velocity of image of ball with respect to ball will be

(a) $\sqrt{2}$ ms^{-1} (b) $\sqrt{3}$ ms^{-1} (c) $\frac{1}{\sqrt{2}}$ ms^{-1} (d) $\frac{1}{\sqrt{3}}$ ms^{-1}

30. A thin rod of length l, coefficient of linear expansion α is heated, so that its temperature changes by ΔT. Fractional increase in moment of inertia of rod about an axis perpendicular to its length and passing through one of its end is

(a) $\alpha\Delta T$ (b) $\frac{3}{2}\alpha\Delta T$ (c) $\frac{2}{3}\alpha\Delta T$ (d) $\frac{4}{3}\alpha\Delta T$

31. A transverse propagating wave is described by

$$y = A\sin 2\pi\left(Bt - \frac{x}{C}\right)$$

If maximum particle speed is exactly equal to wave speed, then value of $\frac{C}{A}$ is

(a) π (b) 2π (c) $\frac{\pi}{2}$ (d) $\frac{B}{2\pi}$

32. A sound source emitting sound with frequency f_0 is moving towards a fixed wall. Now, consider three observers O_1, O_2 and O_3 as shown below.

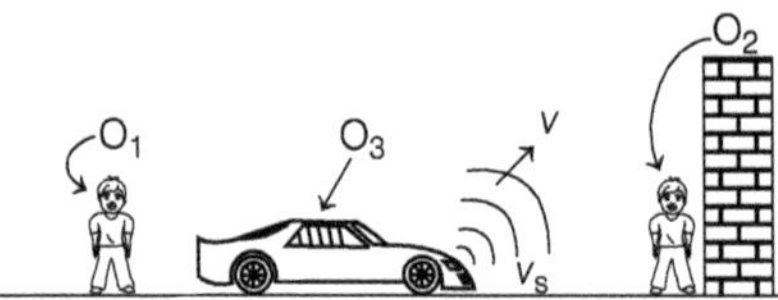

Ratio of frequencies observed by observers $f_1 : f_2 : f_3$ is (take, v_s = speed of source, v = speed of sound in air and assume that O_3 receives only reflected sound and $v_s = \frac{v}{2}$)

(a) 1 : 2 : 3 (b) 6 : 3 : 2 (c) 3 : 2 : 1 (d) 2 : 6 : 9

33. Choose the correct option.

(a) For opaque solids, Eg > 1.8 eV
(b) For coloured solids, Eg > 3.1 eV
(c) For transparent solids, Eg > 3.1 eV
(d) For semi-conductors, Eg > 5 eV

34. A block of wood is floating in water in a vessel. This vessel is kept in a moving lift.

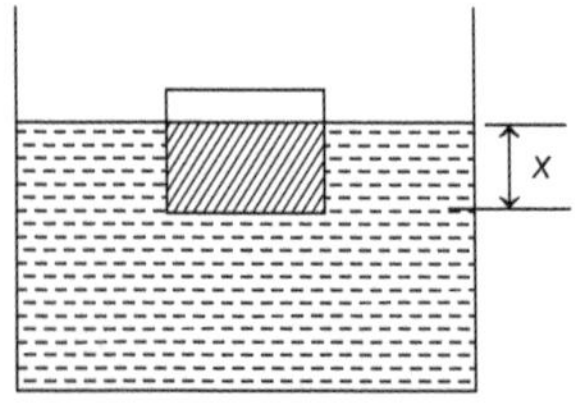

Portion of block inside water is let x.

(a) When lift is moving with a constant velocity then portion of block inside water is x

(b) When lift is accelerating up then portion of block inside water is x

(c) When lift is accelerating down then portion of block inside water is x

(d) All of the above

35. For a particle, potential energy $U(x)$ varies with position x as shown below.

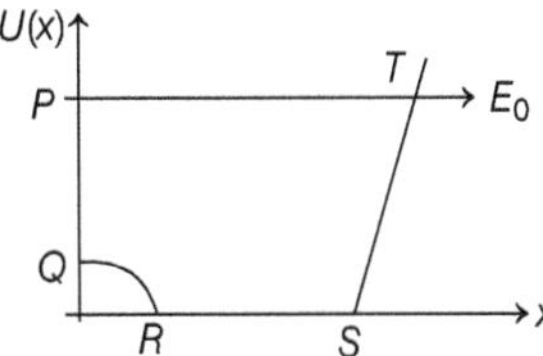

In above figure, E_0 is total energy of the particle. Velocity-position of the particle is

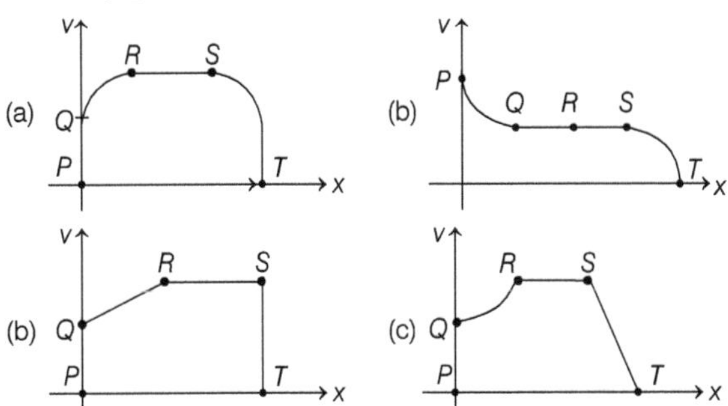

36. A particle of mass m collides elastically with a rigid wall as shown below.

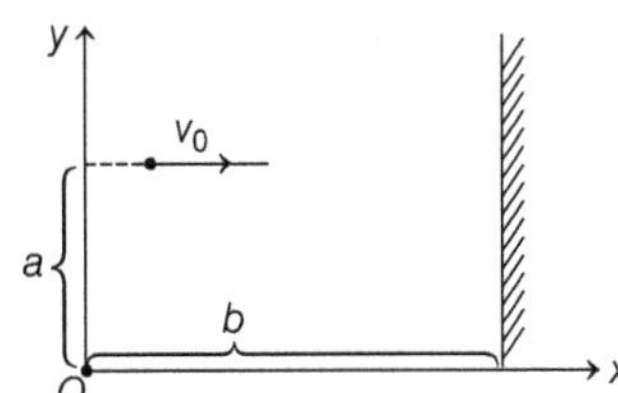

If duration of collision is Δt, then torque provided by the wall about origin is

(a) 0 (b) $\frac{mva}{\Delta t}$ (c) $\frac{2mva}{\Delta t}$ (d) $\frac{2mvb}{\Delta t}$

37. A billiard ball is stuck with a cue stick horizontally as shown below.

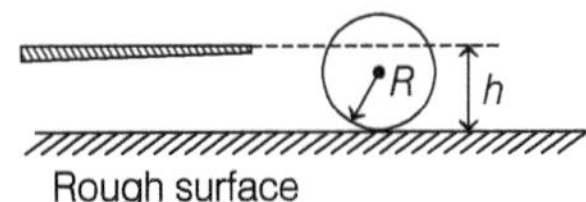

Now, choose the incorrect option.

(a) When $h = \frac{R}{2}$, sphere spins anti-clockwise and looses energy by friction

(b) When $h = R$, sphere translates without rotation and looses energy by friction

(c) When $h = \frac{3R}{2}$, sphere rolls without and slip and no loss of energy

(d) When $h = \frac{7R}{2}$, sphere rolls perfectly over the surface

38. Velocity-time graph for a moving object is as shown below.

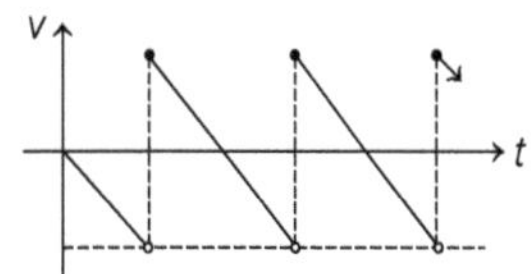

Its acceleration time is

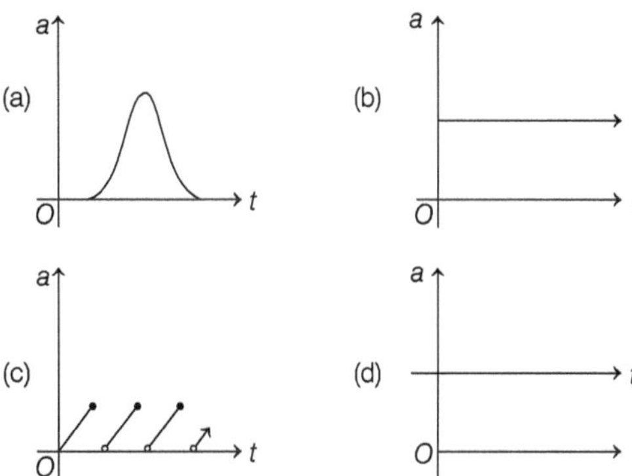

39. Your mobile phone is operating in which of these frequency bands?

(a) 380 to 1900 MHz (b) 30 to 40 MHz
(c) Less than 30 MHz (d) 30 to 300 GHz

40. Let a radioactive sample undergoes α, β and γ decays (not necessarily in any particular order), then correct order of half-lives of these decays is

(a) $\alpha > \beta > \gamma$ (b) $\gamma > \beta > \alpha$
(c) $\beta > \alpha > \gamma$ (d) not following any order

CHEMISTRY

41. Conductivity of a saturated solution of a sparingly soluble salt AB at 298 K is 1.85×10^{-6} Sm^{-1}. Solubility product of the salt AB at 298 K is

[given, $\Lambda^{\circ}_{m}(AB) = 140 \times 10^{-4}$ Sm^{-2} mol^{-1}]

(a) 5.7×10^{-12} (b) 1.32×10^{-12}
(c) 7.5×10^{-12} (d) 1.74

42. The number of $p\pi$-$d\pi$ 'pi' bonds present in XeO_3 and XeO_4 molecules, respectively

(a) 3, 4 (b) 4, 2 (c) 2, 3 (d) 3, 2

43. The Langmuir adsorption isotherm is deduced using the assumption

(a) the adsorption takes place in multilayers

(b) the adsorption sites are equivalent in their ability to adsorb the particles

(c) the heat of adsorption varies with coverage

(d) the adsorbed molecules interact with each other

44. By which of the following processes, pure nitrogen gas is prepared?

(a) $(NH_4)_2Cr_2O_7 \xrightarrow{\Delta}$ (b) $NH_4Cl + NaNO_2 \xrightarrow{\Delta}$

(c) $NH_3 + NaNO_2 \xrightarrow{\Delta}$ (d) $N_2O + Cu \xrightarrow{\Delta}$

45. $CH_3CH{=}CH_2 \xrightarrow{X} \xrightarrow{Y} CH_3CH_2\overset{O}{\overset{\|}{C}}OH$

X and Y are respectively

(a) H_3O^+, MnO_4^-/H^+

(b) H_3O^+, CrO_3/pyridine

(c) $BH_3 \cdot THF/H_2O_2$, CrO_3/pyridine

(d) $BH_3 \cdot THF/H_2O_2$, OH^-, CrO_3/H_3O^+

46. In acidic H_2O_2 reacts with dichromate to give coloured complex of Cr. Oxidation state of Cr in this complex is

(a) + 5 (b) +3

(c) +6 (d) −10

47. Ge (II) compounds are powerful reducing agents, whereas Pb (IV) compounds are strong oxidants. It can be because

(a) Pb is more electropositive than Ge

(b) ionisation potential of lead is less than that of Ge

(c) ionic radii of Pb^{2+} and Pb^{4+} are larger than those of Ge^{2+} and Ge^{4+}

(d) more pronounced inert pair effect in lead than Ge

48. Vapour pressure of pure benzene is 119 torr and that of toluene is 37.0 torr at the same temperature. Mole fraction of toluene in vapour phase which is in equilibrium with a solution of benzene and toluene having a mole fraction of toluene 0.50, will be

(a) 0.137 (b) 0.237

(c) 0.435 (d) 0.205

49. In the conversion of

$$C_6H_5{-}CH{=}CH{-}CH_2{-}\overset{O}{\overset{\|}{C}}{-}CH_3 \xrightarrow{X} C_6H_5{-}CH{=}CH{-}CH_2{-}\underset{CH_3}{\underset{|}{\overset{OH}{\overset{|}{CH}}}}$$

X is

(a) H_2 / Pt (b) Zn—Hg / HCl

(c) Li / NH_3 (d) $NaBH_4$

50. During change of O_2 to O_2^- ion, the electron adds on which one of the following orbitals?

(a) $\overset{*}{\pi}$-orbital (b) π-orbital (c) $\overset{*}{\delta}$-orbital (d) σ-orbital

51. Total number of voids in 0.5 mole of a compound forming hexagonal closed packed structure are

(a) 6.022×10^{23} (b) 3.011×10^{23}

(c) 9.033×10^{23} (d) 4.516×10^{23}

52. Out of the following electronic arrangements for outer electronic configuration.

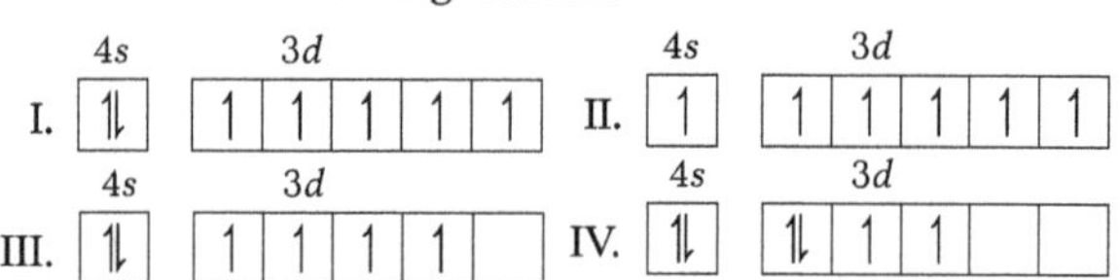

The most stable arrangement is

(a) Only I (b) Only II (c) Only III (d) Only IV

53. Which of the following is the most reactive towards electrophilic reagent?

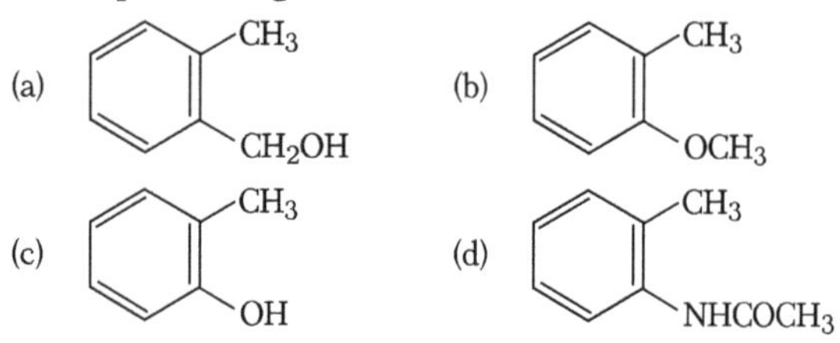

54. A 0.150 mole sample of an ideal gas is allowed expand at 294 K from 10.00 atm to 1.00 atm.

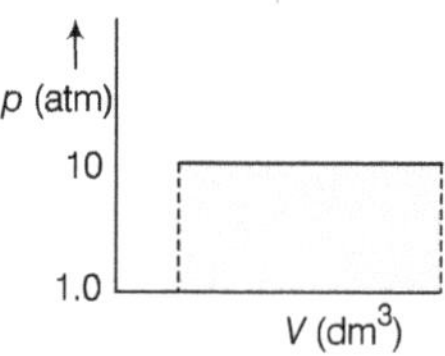

If external pressure is kept constant at 1.00 atm work done is

(a) −330 J (b) −3.26 dm^3 atm

(c) Both (a) and (b) (d) None of these

55. Consider the following gaseous equilibria with equilibrium constant K_1 and K_2 respectively.

$$SO_2(g) + \frac{1}{2}O_2(g) \rightleftharpoons SO_3(g)$$

$$2SO_3(g) \rightleftharpoons 2SO_2(g) + O_2(g)$$

The equilibrium constant is related as:

(a) $2K_1 = K_2^2$ (b) $K_1^2 = \frac{1}{K_2}$ (c) $K_2^2 = \frac{1}{K_1}$ (d) $K_2 = \frac{2}{K_1^2}$

56. A magnetic moment of 1.73 BM will be shown by one among the following.

(a) $[Cu(NH_3)_4]^{2+}$ (b) $[Ni(CN)_4]^{2-}$ (c) $TiCl_4$ (d) $[CoCl_6]^{4-}$

57. Monomer of $\left[-\underset{CH_3}{\underset{|}{\overset{CH_3}{\overset{|}{C}}}}-CH_2- \right]_n$ is

(a) 2-methylpropene (b) styrene

(c) propylene (d) ethene

58. The reaction,

$$CH_3{-}\overset{CH_3}{\overset{|}{CH}}{-}CH_2{-}O{-}CH_2{-}CH_3 + HI \xrightarrow{\text{Heated}} \ldots.$$

Which of the following compounds will be formed?

(a) $CH_3—\underset{\displaystyle |}{\overset{\displaystyle CH_3}{CH}}—CH_2—I + CH_3CH_2OH$

(b) $CH_3—\underset{\displaystyle CH_3}{\underset{|}{CH}}—CH_3 + CH_3CH_2OH$

(c) $CH_3 — \underset{\displaystyle CH_3}{\underset{|}{CH}} — CH_2OH + CH_3CH_3$

(d) $CH_3 — \overset{\displaystyle CH_3}{\overset{|}{CH}} — CH_2OH + CH_3 — CH_2 — I$

59. The following sequence of reactions of A gives

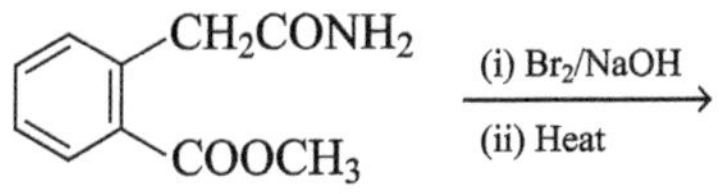

(benzene ring with ortho CH_2CONH_2 and $COOCH_3$) $\xrightarrow[\text{(ii) Heat}]{\text{(i) } Br_2/NaOH}$

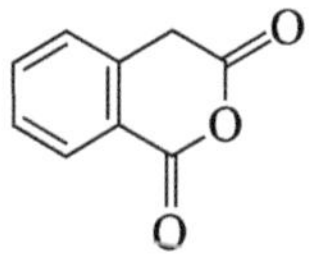

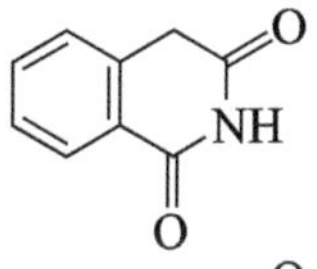

(a) (benzene fused six-membered ring: CH_2–C(=O)–O–C(=O))

(b) (benzene fused six-membered ring: CH_2–C(=O)–NH–C(=O))

(c) (benzene fused five-membered ring: CH_2–NH–C(=O))

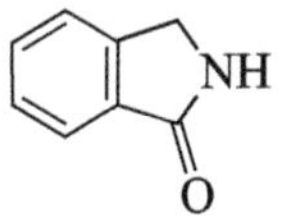

(d) (benzene fused six-membered ring: CH_2–C(=O)–CH_2–C(=O))

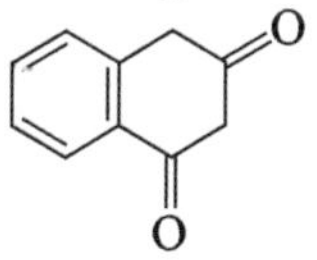

60. $C_3H_6Cl_2$ on reaction with NaOH forms C_3H_6O which gives yellow precipitate on heating with NaOH and I_2. Thus, $C_3H_6Cl_2$ is

(a) 1,1-dichloropropane (b) 1,2-dichloropropane
(c) 2,2-dichloropropane (d) 1,3-dichloropropane

BIOLOGY

61. If sexual reproduction takes place between the filaments of *Rhizopus* of different strains, one with 80 nuclei and another with 24 nuclei, what would be the total number of spores of different strains put together?

(a) 24 (b) 48 (c) 96 (d) 114

62. All mammals without any exception are characterised by

(a) viviparity and biconcave red blood cell
(b) extra abdominal testis and four-chambered heart
(c) heterodont teeth and 12 pairs of cranial nerves
(d) a muscular diaphragm and milk producing glands

63. Flocculation or coagulation of protoplasm is the

(a) interchangeability between sol and gel states
(b) ability to scatter that beam of light
(c) erratic *zig-zag* movement of protoplasmic particles
(d) ability of protoplasm to increase in size when they lose charges

64. Which one of the following precedes reformation of the nuclear envelope during M-phase of the cell cycle?

(a) Decondensation from chromosome and reassembly of the nuclear lamina
(b) Transcription from chromosomes and reassembly of the nuclear lamina
(c) Formation of the contractile ring and formation of the phragmoplast
(d) Formation of the contractile ring and transcription from chromosomes

65. Lipids, which can be found in oil based salad dressings and ice-cream, during digestion are splitted into

(a) fatty acids and glycerol
(b) glycerol and amino acids
(c) glucose and fatty acids
(d) glucose and amino acids

66. A twig kept in water having some salt remains fresh for longer period due to

(a) decrease in bacterial degradation
(b) exosmosis
(c) decrease in transpiration rate
(d) absorption of more water

67. Farmers in a particular region were concerned that premature yellowing of leaves of a pulse crop might cause decrease in the yield. Which treatment could be most beneficial to obtain maximum seed yield?

(a) Frequent irrigation of the crop
(b) Treatment of the plants with cytokinins along with a small dose of nitrogenous fertiliser
(c) Removal of all yellow leaves and spraying the remaining green leaves with 2, 4, 5-trichlorophenoxy acetic acid
(d) Application of iron and magnesium to promote synthesis of chlorophyll

68. Which of the following is maximum in chloroplast?

(a) RuBP carboxylase (b) Hexokinase
(c) Phosphatase (d) Nuclease

69. When breastfeeding is replaced by less nutritive food low in proteins and calories, the infants below the age of one year are likely to suffer from

(a) marasmus (b) rickets
(c) kwashiorkor (d) pellagra

70. A large proportion of oxygen is left unused in the human blood even after its uptake by the body tissues. This O_2

(a) raises the pCO_2 of blood to 75 mm of Hg
(b) is enough to keep oxyhaemoglobin
(c) helps in releasing more O_2 to the epithelial tissues
(d) acts as a reserve during muscular exercises

71. In humans, blood passes from the postcaval to the diastolic right atrium of heart due to

(a) pushing open of the venous valves
(b) suction pull
(c) stimulation of the sinoauricular node
(d) pressure difference between the caval and atrium

72. Which is a bridge between nervous system and endocrine system?

(a) Thalamus (b) Hypothalamus
(c) Limbic system (d) Parietal lobe

73. Steroid hormones work as

(a) they enter into target cells and bind with specific receptor and activate specific genes to form protein
(b) they bind to cell membrane
(c) they catalyse formation of cAMP
(d) None of the above

74. In Mendelian dihybrid cross when heterozygous round yellow are self-crossed, round green offsprings are represented by the genotype

(a) RrYy, RrYy and RRYy (b) Rryy, RRyy and rryy
(c) rrYy and rryY (d) Rryy and RRyy

75. If there are 120 adenine molecules in a B-DNA double helical structure showing 20 coils, what is the number of pyrimidine nucleotides forming three hydrogen bonds in it?

(a) 80 (b) 100 (c) 120 (d) 140

76. Which antibiotic inhibits interaction between *t*RNA and *m*RNA during bacterial protein synthesis?

(a) Erythromycin (b) Neomycin
(c) Streptomycin (d) Tetracycline

77. An insect bite may result in inflammation of that spot. This is triggered by the alarm chemicals such as

(a) histamine and dopamine
(b) histamine and kinins
(c) interferons and opsonin
(d) interferons and histones

78. Palaeontologists unearthed a human skull during excavation. A small fragment of the scalp tissue was still attached to it. Only little DNA could be extracted from it. If the genes of the ancient man need to be analysed, the best way of getting sufficient amount of DNA from this extract is by

(a) hybridising the DNA with a DNA probe
(b) subjecting the DNA to polymerase chain reaction
(c) subjecting the DNA to gel-electrophoresis
(d) treating the DNA with restriction endonucleases

79. If the mean and the median pertaining to a certain character of a population are of the same value, the following is most likely to occur

(a) a normal distribution
(b) a bimodal distribution
(c) a T-shaped curve
(d) a skewed curve

80. Which one of the following statements is not correct during protein synthesis?

(a) UAA codon codes for lysine
(b) UGG codon codes for tryptophan
(c) Cysteine is coded by UGU and UAC codons
(d) Tyrosine is coded by UAU and UAC codons

PART-II (2 Marks Questions)

MATHEMATICS

81. If $\log_{2x} 1944 = \log_x 486\sqrt{2}$, and $x^6 = 2^a \cdot 3^b$, then $a + b$ is equal to

(a) 24 (b) 26 (c) 28 (d) 30

82. The number of solution of the equation $\sqrt{3x^2 + 6x + 12} + \sqrt{5x^2 + 10x + 9} = 4 - 2x - x^2$ is equal to

(a) 1 (b) 2 (c) 3 (d) 4

83. The value of $\lim\limits_{n \to \infty} \sum_{r=0}^{n} \left(\frac{1}{4r+1} - \frac{1}{4r+3} \right)$ is equal to

(a) $\frac{\pi}{2\sqrt{2}}$ (b) $\frac{\pi}{2}$ (c) $\frac{\pi}{4}$ (d) $\frac{\pi}{8}$

84. ΔABC is such that a circle touches AB at B and passes through centroid of ΔABC and C, if $AB = 6$, $BC = 4$, then AC is equal to

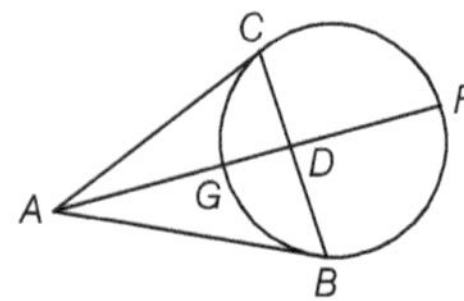

(a) $2\sqrt{2}$ (b) $3\sqrt{2}$ (c) $2\sqrt{14}$ (d) $2\sqrt{13}$

85. Area enclosed by $y = g(x)$, $x = 1$ and $x = 15$, where $g(x)$ is inverse of $f(x) = x^3 + 3x + 1$ is

(a) $\frac{9}{4}$ (b) 18 (c) $\frac{3}{4}$ (d) $\frac{9}{2}$

86. The remainder when $\left(\sum_{k=1}^{5} {}^{20}C_{2k-1} \right)^6$ is divided by 11, is

(a) 1 (b) 3 (c) 5 (d) 7

87. Let A, B, C be the angles of ΔABC with vertex $A(4, -1)$ and $x - 1 = 0$ and $x - y = 1$ are internal angle bisectors through B and C respectively. Let D, E, F be points of contact of sides BC. CA and AB with incircle of ΔABC. If D', E', F' are images of D, E and F in internal angle bisector of A, B, C, then equation of circumcircle of $\Delta D'E'F'$ is

(a) $(x-1)^2 + y^2 = 5$ (b) $x^2 + (y-1)^2 = 25$
(c) $(x-1)^2 + (y-1)^2 = 5$ (d) $x^2 + y^2 = 25$

88. Initially there is 50 gm of salt in tank with 100 L of water present A liquid at rate of 5 L/min with 2 gm/L of salt is coming into tank. After proper mixing in tank it is running out with 4 L/min. The amount of salt present in tank after time $t = 100$ min is

(a) 300 gm (b) $\frac{1525}{4}$ gm (c) $\frac{1555}{4}$ gm (d) $\frac{3125}{8}$ gm

89. Let $\mathbf{a} = \hat{\mathbf{i}} + \hat{\mathbf{j}} + \hat{\mathbf{k}}$, $\mathbf{b} = \hat{\mathbf{i}} - \hat{\mathbf{j}} + \hat{\mathbf{k}}$, $\mathbf{a} \times \mathbf{b} = \mathbf{b} + \lambda\, \mathbf{a}$ and $\mathbf{a} \cdot \mathbf{c} = 1$, then which of the following is true

(a) $[\mathbf{a}\ \mathbf{b}\ \mathbf{c}] = -\frac{8}{3}$ and $\lambda = -\frac{1}{3}$

(b) $[\mathbf{a}\ \mathbf{b}\ \mathbf{c}] = \frac{8}{3}$ and $\lambda = -\frac{1}{3}$

(c) $[\mathbf{a}\ \mathbf{b}\ \mathbf{c}] = -\frac{8}{3}$ and $\lambda = -\frac{2}{3}$

(d) $[\mathbf{a}\ \mathbf{b}\ \mathbf{c}] = -\frac{8}{3}$ and $\lambda = \frac{2}{3}$

90. Let $f(x)$ be a twice differentiable function all real values of x and satisfies $f(1) = 1$, $f(2) = 4$, $f(3) = 9$, then which of the following is definitely true?

(a) $f''(x) = 2$, $x \in (1, 3)$

(b) $f''(x) = f'(x) = 5$, for some $x \in (2, 3)$

(c) $f''(x) = 3$, $x \in (2, 3)$

(d) $f''(x) = 2$ for some $x \in (1, 3)$

PHYSICS

91. A metal bar (length L, mass m) can slides over two connected rails fitted over an inclined plane without friction. Rails are of very less resistance and metal bar is of resistance R.

When a magnetic field **B** is switch ON perpendicular to ground, then terminal speed of the metal bar obtained is

(a) $\frac{Rmg}{B^2L^2}$ (b) $\frac{Rmg}{B^2L^2} \cdot \sec\theta \cdot \tan\theta$

(c) $\frac{Rmg}{B^2L^2} \cdot \tan\theta$ (d) $\frac{Rmg}{B^2L^2} \cdot \cot\theta$

92. Two identical cylindrical rods of radii 10 cm each and refractive index $\sqrt{3}$ rests over a flat horizontal plane mirror.

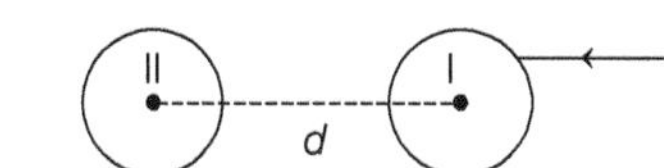

A horizontal light ray is made incident over right side glass rod I, such that it leaves the rod at a height of 10 cm from the plane mirror. Final emergent ray is found leaving glass rod parallel to the mirror.

Distance d between glass rods is nearly

(a) 31.5 cm (b) 32.5 cm
(c) 37.5 cm (d) 38.5 cm

93. v-x graph of an object is

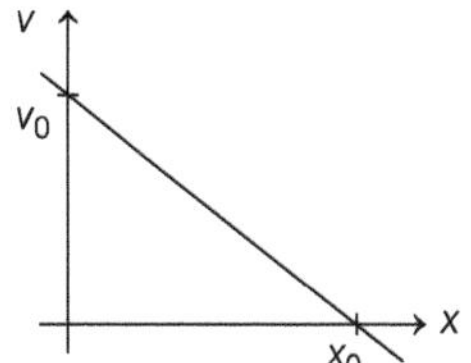

a-x graph of body is

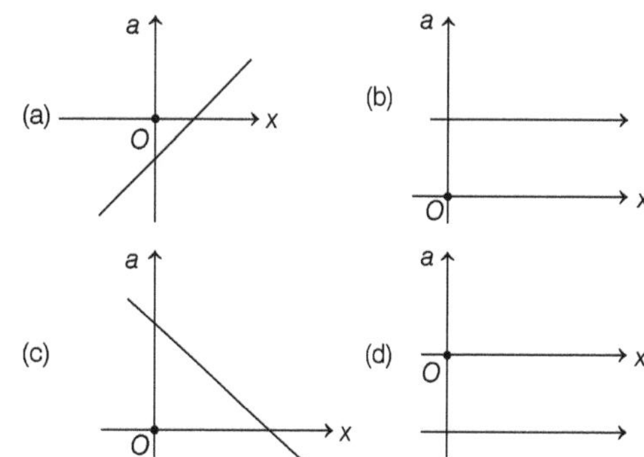

94. For a charged circular ring, maximum value of electrical potential is at

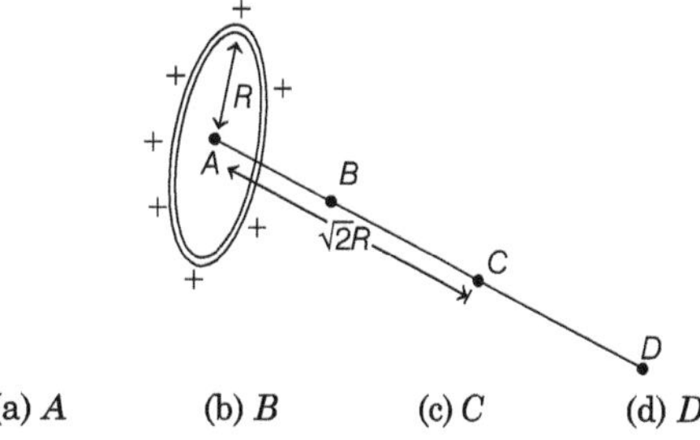

(a) A (b) B (c) C (d) D

95. Five identical rods are connected between two large reservoirs of temperatures 100° C and 0° C. Each rod has thermal conductivity k, length l, and area A. There is no heat transfer through sides of rods.

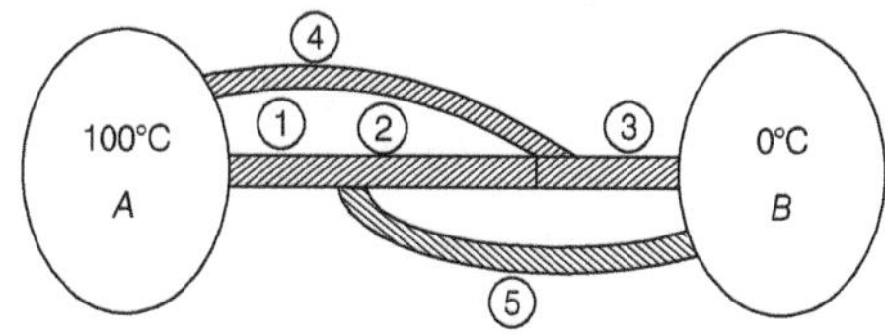

Heat flow rate from reservoir A to B is

(a) $\frac{100\,kA}{l}$ (b) $\frac{100\,kA}{7\,l}$ (c) $\frac{100\,k\,A}{3\,l}$ (d) $\frac{100\,kA}{13\,l}$

96. A resistance R is measured using a voltmeter and an ammeter as shown below.

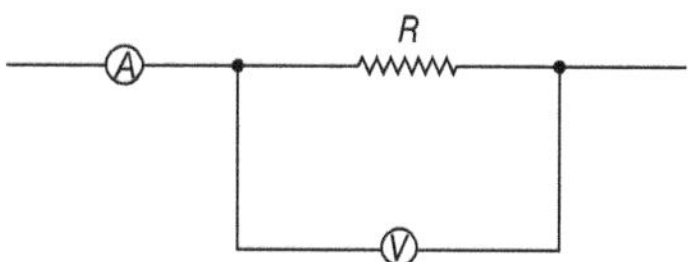

Ammeter shows 2A and voltmeter shows 120V. Internal resistance of voltmeter is 3000 Ω. Error in measurement of R compared to the reading taken with an ideal voltmeter will be

(a) 3.2 Ω (b) 4.2 Ω (c) 1.2 Ω (d) 0.2 Ω

97. A free neutron decays into a proton and an electron as, ${}_0^1n \longrightarrow {}_1^1p + {}_{-1}^{0}e + \nu$. If neutron-hydrogen atom mass difference is 840 μu, then the maximum possible kinetic energy of electron will be (1 u = 932 MeV)

(a) 0.783 MeV (b) 0.840 MeV
(c) 0.589 MeV (d) 0.687 MeV

98. A chord ACB, 5m long is attached at points A and B to the vertical walls 3 m apart.

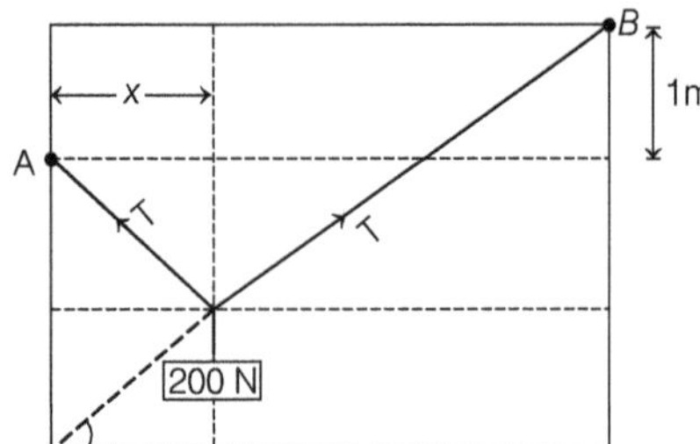

A pulley of negligible mass and negligible radius carries 200 N load is free to roll over chord without friction. Dimension x in figure, when pulley is in equilibrium is

(a) $\frac{9}{8}$ m (b) $\frac{7}{5}$ m (c) $\frac{4}{3}$ m (d) $\frac{3}{4}$ m

99. A uniform solid sphere rolls up an inclined plane.

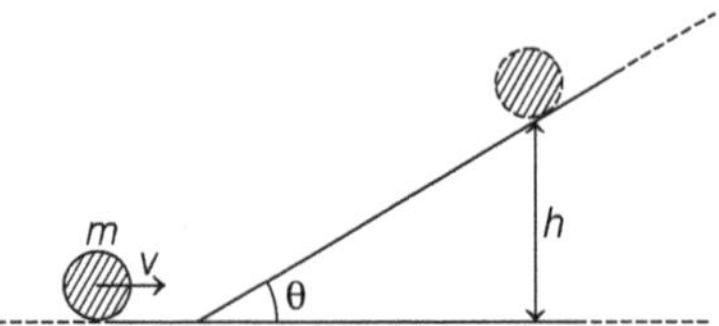

Now, choose the correct option.

(a) $h \propto \frac{1}{m}$ (b) $h \propto \frac{1}{\tan\theta}$

(c) $h \propto v^2$ (d) $h \propto \frac{1}{r^2}$

100. Water is filled in a tank upto height h. Bulk modulus of water is B and its density at surface is ρ_0. Density at depth h is

(a) $\frac{\rho_0}{(1-\rho_0 gh/B)}$ (b) $\frac{\rho_0}{(1-B/\rho_0 gh)}$

(c) $\rho_0\left(1-\frac{\rho_0 gh}{B}\right)$ (d) $\rho_0\left(1-\frac{B}{\rho_0 gh}\right)$

CHEMISTRY

101. The heat of atomisation of $PH_3(g)$ and $P_2H_4(g)$ are 953 and 1485 kJ mol^{-1} respectively. The P—P bond energy in kJ mol^{-1} is

(a) 214 (b) 426
(c) 318 (d) 1272

102. An organic compound undergoes first order decomposition. The time taken for its decomposition to 1/8 and 1/10 of its initial concentration are $t_{1/8}$ and $t_{1/10}$ respectively.

What is the value of $\frac{[t_{1/8}]}{[t_{1/10}]} \times 10$? ($\log_{10}2 = 0.3$)

(a) 4 (b) 8
(c) 9 (d) 11

103. The given graph represents the variation of compressibility factor $(Z) = \frac{pV}{nRT}$, for three real gases A, B and C. Identify the only incorrect statement.

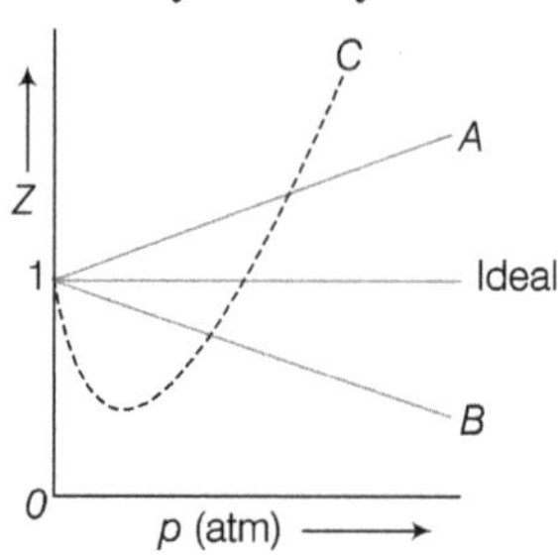

(a) For the gas A, $a = 0$ and its dependence on p is linear at all pressure

(b) For the gas B, $b = 0$ and its dependence on p is linear at all pressure

(c) For the gas C, which is typical real gas for which neither a nor $b = 0$. By knowing the minima and point of the intersection, with $Z = 1$, a and b can be calculated

(d) At high pressure, the slope is positive for all real gases

104. $E^\circ_{Fe^{3+}/Fe} = -0.336$ V, $E^\circ_{Fe^{2+}/Fe} = -0.439$ V. The value of standard electrode potential for the charge, $Fe^{3+}(aq) + e^- \longrightarrow Fe^{2+}(aq)$ will be

(a) – 0.072 V (b) 0.385 V (c) 0.770 V (d) – 0.270 V

105. Among Ga ($Z = 64$), Lu ($Z = 71$), La ($Z = 103$), Ta ($Z = 73$) the elements having half-filled f-orbital is

(a) Ga (b) Lu (c) La (d) Ta

106. Identify the major products A and B in the following reaction.

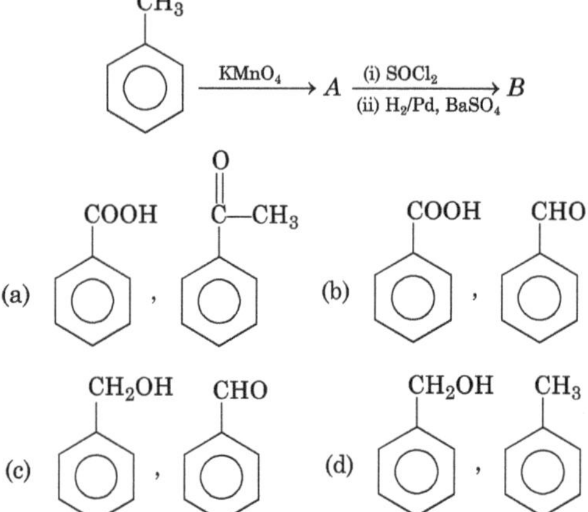

107. The correct statement with respect to the complexes $[Ni(CO)_4]$ and $[Ni(CN)_4]^{2-}$ is

(a) nickel is in the same oxidation state in both

(b) both are paramagnetic in nature

(c) have square planar and tetrahedral geometry respectively

(d) have tatrahedral and square planar geometry respectively

108. Compound *A*, $C_5H_{10}O_5$ gives a tetraacetate with Ac_2O and oxidation of '*A*' with Br_2 / H_2O gives an acid, $C_5H_{10}O_6$. Reduction of '*A*' with HI gives *iso*-pentane. What is the possible structure of *A*?

(a) COOH | HO—C—CH_2OH | CHOH | CH_2OH

(b) CHO | HO—C—CH_2OH | CHOH | CH_2OH

(c) CHO | HO—C—H | $(CHOH)_2$ | CH_2OH

(d) HO—C=O | HO—C—CH_2OH | CHOH | CH_2OH

109. Identify the correct statement about the following pairs of compounds.

OH, OH (A, B); Ph, CO_2Me, O (C); Ph, CO_2Me, O (D)

(a) *A* and *B* diastereomers ; *C* and *D* diastereomers
(b) *A* and *B* enantiomers ; *C* and *D* diastereomers
(c) *A* and *B* diastereomers ; *C* and *D* enantiomers
(d) *A* and *B* enantiomers ; *C* and *D* enantiomers

110. Given

(I) H_3C, CH_3, CH_3 (II) H_3C, CH_2, CH_3 (III) H_2C, CH_2, CH_2

The enthalpy of hydrogenation of these compounds will be the order as:

(a) I > II > III (b) III > II > I
(c) II > III > I (d) II > I > III

BIOLOGY

111. The diagram shows the changes in temperature of the reaction mixture during the Polymerase Chain Reaction (PCR).

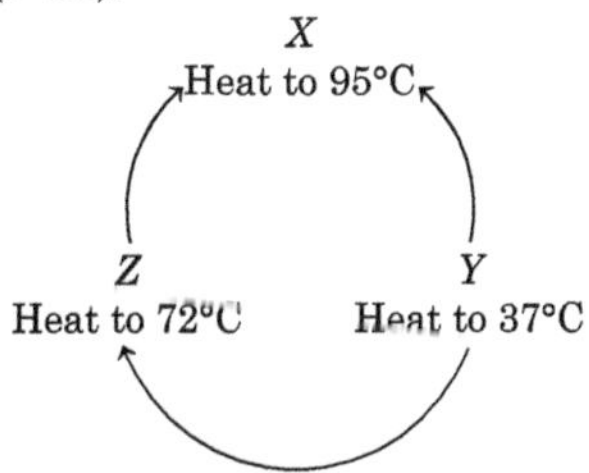

The main events during one cycle of the reaction are listed below.

I. Binding of RNA primers
II. DNA synthesis
III. Separation

Which combination correctly matches each event with the temperature in the reaction mixture?

	X	*Y*	*Z*		*X*	*Y*	*Z*
(a)	I	II	III	(b)	II	III	I
(c)	III	I	II	(d)	III	II	I

112. Gene therapy is used for the treatment of

I. Cystic fibrosis
II. Severe Combined Immunodeficiency (SCID)

The difference between them is

(a) I is a transient treatment while II can possibly be a permanent treatment
(b) I usually uses the *ex vivo* approach while II usually uses the *in vivo* approach
(c) I usually uses adenoviruses as vectors while (II) usually uses liposomes as vectors
(d) I treats a recessive disorder while (II) treats a dominant disorder

113. Strips of plant tissue were immersed in a range of sucrose solutions of different concentrations. Their lengths were measured before immersion and after 30 minutes in the different solutions. The graph shows the ratio of initial length to final length.

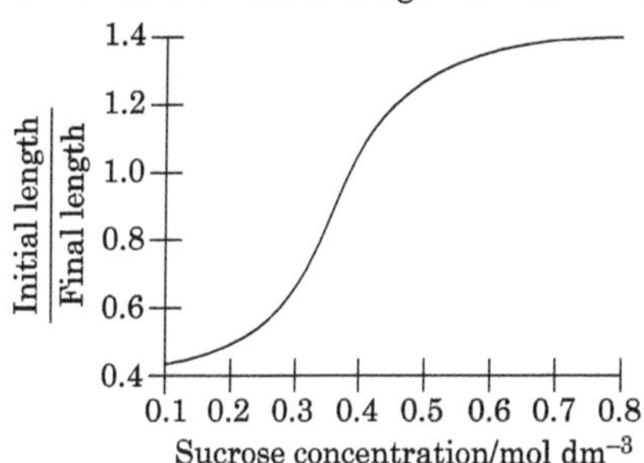

Which concentration of sucrose solution has the same water potential as the cell sap?

(a) 0.1 mol dm^{-3} (b) 0.25 mol dm^{-3}
(c) 0.45 mol dm^{-3} (d) 0.6 mol dm^{-3}

114. What is the approximate ratio of ATP synthesised during anaerobic respiration compared with aerobic respiration?

(a) 1 : 2 (b) 1 : 10 (c) 1 : 20 (d) 1 : 30

115. A plant is heterozygous for a pair of alleles that are codominant. This plant is self-pollinated and the resulting seeds are germinated and allowed to grow. Which ratios are expected in the offspring?

	Ratio of phenotypes	Ratio of genotypes
(a)	1 : 2 : 1	1 : 2 : 1
(b)	1 : 2 : 1	3 : 1
(c)	3 : 1	1 : 2 : 1
(d)	3 : 1	3 : 1

116. In fruitflies, a sex-linked gene controls the development of eye colour. The eyes are either red or white. The male is the heterogametic sex.

What will be the expected percentage of eye colours in the progeny when a heterozygous red-eyed female is crossed with a white-eyed male?

	Red eyes		White eyes	
	Male	**Female**	**Male**	**Female**
(a)	25.0	25.0	25.0	25.0
(b)	37.5	37.5	12.5	12.5
(c)	12.5	12.5	37.5	37.5
(d)	0	50.0	50.0	0

117. The *m*RNA triplet UGA acts as a stop codon to terminate the synthesis of a polypeptide. The diagram shows a strand of DNA coding for 4 amino acids. Where would a mutation, involving the insertion of a thymine nucleotide, result in the termination of translation?

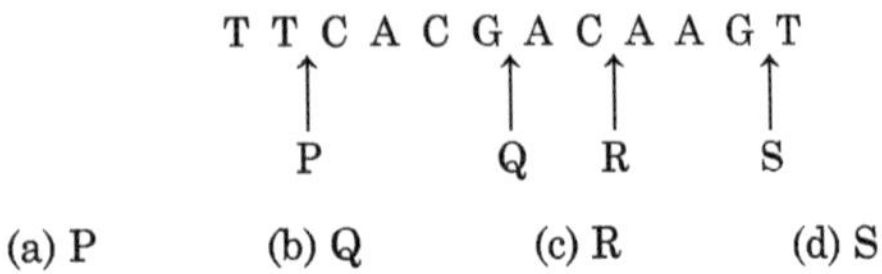

(a) P (b) Q (c) R (d) S

118. Match major events in the history of life with earth's geological period.

Event	Geological Period
A. First reptiles	1. Quaternary
B. First mammals	2. Triassic
C. First humans	3. Carboniferous
D. First amphibians	4. Devonian

Codes

	A	B	C	D		A	B	C	D
(a)	1	2	3	4	(b)	2	3	4	1
(c)	4	1	2	3	(d)	3	2	1	4

119. The sequence of bases on a messenger RNA molecule is shown.

AUCGAAGUUCGU

It is transcribed from one strand of DNA.

What is the sequence of bases on the complementary, non-transcribed strand?

(a) ACGAACTTCGAT (b) ATCGAAGTTCGT
(c) TAGCTTCAAGCA (d) UGCUUGAAGCUA

120. Albinism in humans is controlled by a recessive allele. How many copies of this allele will be found at one of the poles of a cell at telophase-I of meiosis in an albino person?

(a) 23 (b) 4 (c) 2 (d) 1

Answers

PART-I

1 (c)	2 (d)	3 (c)	4 (a)	5 (b)	6 (a)	7 (c)	8 (c)	9 (d)	10 (d)
11 (b)	12 (c)	13 (a)	14 (b)	15 (d)	16 (a)	17 (c)	18 (b)	19 (c)	20 (c)
21 (d)	22 (c)	23 (a)	24 (a)	25 (d)	26 (d)	27 (d)	28 (d)	29 (a)	30 (c)
31 (b)	32 (d)	33 (c)	34 (d)	35 (a)	36 (c)	37 (c)	38 (d)	39 (a)	40 (a)
41 (d)	42 (a)	43 (a)	44 (b)	45 (c)	46 (c)	47 (d)	48 (b)	49 (d)	50 (a)
51 (c)	52 (a)	53 (c)	54 (c)	55 (b)	56 (a)	57 (a)	58 (d)	59 (c)	60 (c)
61 (c)	62 (c)	63 (a)	64 (a)	65 (a)	66 (c)	67 (d)	68 (a)	69 (a)	70 (d)
71 (d)	72 (b)	73 (a)	74 (d)	75 (a)	76 (d)	77 (b)	78 (b)	79 (a)	80 (a)

PART-II

81 (b)	82 (a)	83 (c)	84 (c)	85 (b)	86 (b)	87 (a)	88 (d)	89 (b)	90 (d)
91 (b)	92 (a)	93 (a)	94 (b)	95 (a)	96 (c)	97 (a)	98 (a)	99 (c)	100 (a)
101 (a)	102 (c)	103 (b)	104 (c)	105 (a)	106 (b)	107 (d)	108 (b)	109 (d)	110 (b)
111 (c)	112 (a)	113 (c)	114 (c)	115 (a)	116 (a)	117 (c)	118 (d)	119 (b)	120 (c)

Solutions

1. *(c)* We have,

$$\log_a b + \log_b a = c$$

$$\Rightarrow \quad \log_a b + \frac{1}{\log_a b} = c$$

By $\quad AM \geq GM$

$$\therefore \quad \log_a b + \frac{1}{\log_a b} \geq 2$$

$$\therefore \quad c \geq 2$$

2. *(d)* We have,

Let $\frac{p}{q}$ is roots of $ax^2 + bx + c = 0$

$$\therefore \quad a\left(\frac{p}{q}\right)^2 + b\left(\frac{p}{q}\right) + c = 0$$

$ap^2 + bpq + cq^2 = 0$ is never possible

a, b, c are odd.

3. *(c)* $1! + 2! + 3! + 4! + 5! + 6! + 7! + 8! \ldots + 2020!$

$$1 + 2 + 6 + 24 + 120 + 720 + 35k$$

[$\because$ 7! onward are divisible by 35]

$$\Rightarrow \quad 873 + 35k$$

$$\Rightarrow \quad 840 + 33 + 35k$$

$$\Rightarrow \quad 35 \times 24 + 35k + 33$$

$$\Rightarrow \quad 35(24 + k) + 33$$

$\therefore$ Remainder = 33

4. *(a)* $AB = \frac{5}{6}r \Rightarrow CD = \frac{2}{3}r$

$$\Rightarrow \quad OB = \sqrt{r^2 - \frac{25}{36}r^2}$$

$$\Rightarrow \quad OB = \frac{\sqrt{11}}{6}r$$

Similarly, $OD = \sqrt{r^2 - \frac{4}{9}r^2} = \frac{\sqrt{5}}{3}r$

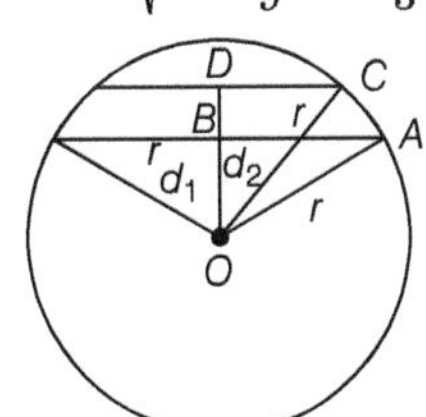

Required probability

$$= \frac{\pi\left(\frac{5}{9}r^2 - \frac{11}{36}x^2\right)}{\pi r^2} = \frac{20 - 11}{36} = \frac{1}{4}$$

5. *(b)* We have,

$$y = \cos\left(\pi \sin\left(\frac{\pi}{2}\cos(\pi \sin x)\right)\right)$$

$$-1 \leq \sin x \leq 1$$

$$\therefore \quad -\pi \leq \pi \sin x \leq \pi$$

$$\Rightarrow \quad -1 \leq \cos(\pi \sin x) \leq 1$$

$$\Rightarrow \quad -1 \leq \sin\left(\frac{\pi}{2}\cos(\pi \sin x)\right) \leq 1$$

$$\Rightarrow \quad -\pi \leq \pi \sin\left(\frac{\pi}{2}\cos(\pi \sin x)\right) \leq \pi$$

$$\Rightarrow \quad -1 \leq \cos\left(\pi \sin\left(\frac{\pi}{2}\cos(\pi \sin x)\right)\right) \leq 1$$

$$\Rightarrow \quad -1 \leq y \leq 1$$

$\therefore$ Range of $y \in [-1, 1]$

6. *(a)* We have,

$$x \tan\alpha + y \sec\alpha = 1$$

$$\frac{x}{\cot\alpha} + \frac{y}{\cos\alpha} = 1$$

intercept are $\cot\alpha$ and $\cos\alpha$.

Given, $\quad \cot\alpha \cdot \cos\alpha = \sin\alpha$

$$\tan^2\alpha = 1$$

$$\Rightarrow \quad \tan\alpha = \pm 1$$

$$\alpha = \pm\frac{\pi}{4} \Rightarrow \alpha = \frac{\pi}{4}$$

7. *(c)* We have,

$$g(x) = \int_0^{|x|^{3/4}} (t)^{2/3} \sin\left(\frac{1}{t}\right) dt$$

$$g'(x) = |x|^{\frac{1}{2}} \sin\frac{1}{|x|^{3/4}} \quad \lim_{x \to 0} \frac{g(x)}{x}$$

$$\lim_{x \to 0} \frac{g'(x)}{1} = \lim_{x \to a} |x|^{1/2} \sin\frac{1}{|x|^{3/4}} = 0$$

8. *(c)* $r = \sqrt{3}\sin 30° = \frac{\sqrt{3}}{2}$

$$|PA|^2 + |PB|^2 + |PC|^2 + |PD|^2$$

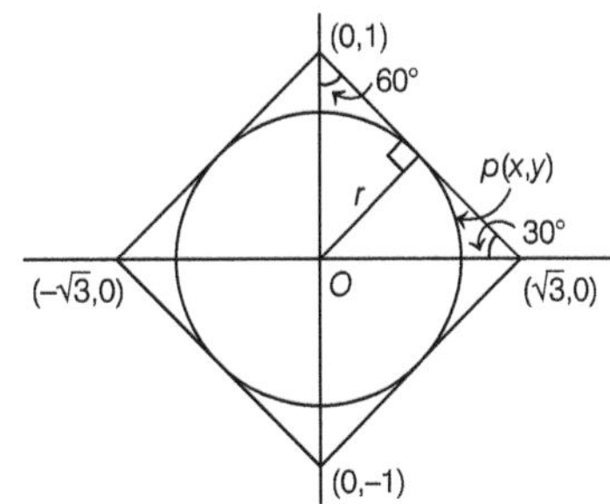

$$\Rightarrow \quad (x - \sqrt{3})^2 + y^2 + x^2 + (y - 1)^2 + (x + \sqrt{3})^2 + y^2 + x^2 + (y + 1)^2$$

$$= 4(x^2 + y^2 + 2)$$

$$= 4(r^2 + 2) \qquad [\because x^2 + y^2 = r^2]$$

$$= 4\left(\frac{3}{4} + 2\right) = 11$$

9. *(d)* α, β are the roots of

$$ax^2 + bx + c = 0$$

$$\therefore \quad \alpha + \beta = -b/a$$

$$\text{and} \quad \alpha\beta = c/a$$

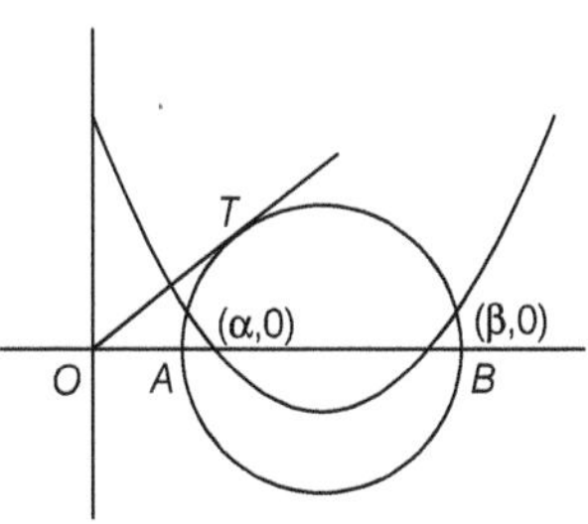

Circle passes through $(\alpha, 0)$ and $(\beta, 0)$

$$OA = \alpha, OB = \beta$$

$\because$ OT is tangent of circle

$$\therefore \quad OT^2 = OA \cdot OB$$

$$\Rightarrow \quad OT^2 = \alpha\beta$$

$$\Rightarrow \quad OT = \sqrt{\alpha\beta}$$

$$\Rightarrow \quad OT = \sqrt{c/a}$$

10. *(d)* We have,

$$P^2 = P$$

$$\Rightarrow \quad P^{-1}P^2 = P^{-1}P$$

$$\Rightarrow \quad P = I$$

$$\Rightarrow \quad (I + P)^n = (P + P)^n$$

$$= (2P)^n = 2^n P^n = 2^n P$$

$$= P + 2^n P - P$$

$$= I + (2^n - 1) P$$

11. *(b)* We have,

$$x^2 + 10000[x] = 10000x$$

$$\Rightarrow \quad x^2 + 10000(x - \{x\}) = 10000x$$

$$\Rightarrow x^2 + 10000x - 10000\{x\} = 10000x$$

$$\Rightarrow \quad \{x\} = \frac{x^2}{10000} \Rightarrow 0 \leq \{x\} < 1$$

$$\therefore \quad 0 \leq \frac{x^2}{10000} < 1$$

$$-100 < x < 100$$

Total number of x is 199.

12. *(c)* Let

$$I = \int_0^\infty \frac{1}{(1 + x^{2020})(1 + x^2)} dx$$

Put $\quad x = \frac{1}{t}\ dx = \frac{-1}{t^2} dt$

$$\therefore \quad I = \int_0^\infty \frac{1}{\left(1 + \frac{1}{t^{2020}}\right)\left(1 + \frac{1}{t^2}\right)} - \frac{dt}{t^2}$$

$$\Rightarrow \quad I = \int_0^\infty \frac{t^{2020} dt}{(t^{2020} + 1)(t^2 + 1)}$$

$$\Rightarrow \quad I = \int_0^\infty \frac{x^{2020} dx}{(1 + x^{2020})(1 + x^2)}$$

$$\Rightarrow \quad 2I = \int_0^\infty \frac{x^{2020}+1}{(1+x^{2020})(1+x^2)} dx$$

$$\Rightarrow \quad 2I = \int_0^\infty \frac{1}{(1+x^2)} dx \Rightarrow 2I = [\tan^{-1} x]_0^\infty$$

$$\Rightarrow \quad 2I = (\tan^{-1}\infty - \tan^{-1} 0)$$

$$\Rightarrow \quad 2I = \frac{\pi}{2} \Rightarrow I = \frac{\pi}{4}$$

13. *(a)* We have,

$$1 + \cos 2X - 2\cos 2Y = 2\sin X \sin Y$$

$$\Rightarrow 1 + 1 - 2\sin^2 X - 2(1 - 2\sin^2 Y) = 2\sin X \sin Y$$

$$\Rightarrow 4\sin^2 Y - 2\sin^2 X - 2\sin X \sin Y = 0$$

$$\Rightarrow \quad 4b^2 - 2a^2 - 2ab = 0$$

$$\left[\because \frac{a}{\sin X} = \frac{b}{\sin Y} = \frac{c}{\sin Z} = k\right]$$

$$\Rightarrow \quad \frac{a^2}{b^2} + \frac{a}{b} - 2 = 0$$

$$\Rightarrow \quad \left(\frac{a}{b}\right)^2 + \left(\frac{a}{b}\right) - 2 = 0$$

$$\therefore \quad \frac{a}{b} = 1$$

14. *(b)* Given, harmonic mean of a and b is 4

$$\therefore \quad 4 = \frac{2ab}{a+b} \quad \ldots(\text{i})$$

and $a, 5, q, b$ are in AP

$$\therefore \quad q = 10 - a, q = \frac{b+5}{2}$$

$$\Rightarrow \quad a = 10 - q, b = 2q - 5$$

On putting, the values of a and b in Eq. (i), we get

$$4 = \frac{2(10-q)(2q-5)}{10 - q + 2q - 5}$$

$$\Rightarrow \quad 2q^2 - 23q + 60 = 0$$

$$\Rightarrow \quad 2q^2 - 15q - 8q + 60 = 0$$

$$\Rightarrow \quad (2q - 15)(q - 4) = 0$$

$$\Rightarrow \quad q = 4, \frac{15}{2}$$

15. *(d)* We know,

$$[\mathbf{a}\ \mathbf{b}\ \mathbf{c}]^2 = \begin{vmatrix} \mathbf{a}\cdot\mathbf{a} & \mathbf{a}\cdot\mathbf{b} & \mathbf{a}\cdot\mathbf{c} \\ \mathbf{b}\cdot\mathbf{a} & \mathbf{b}\cdot\mathbf{b} & \mathbf{b}\cdot\mathbf{c} \\ \mathbf{c}\cdot\mathbf{a} & \mathbf{c}\cdot\mathbf{b} & \mathbf{c}\cdot\mathbf{c} \end{vmatrix}$$

$$[\mathbf{a}\ \mathbf{b}\ \mathbf{c}]^2 = \begin{vmatrix} 1 & \frac{1}{2} & \frac{1}{2} \\ \frac{1}{2} & 1 & \frac{1}{2} \\ \frac{1}{2} & \frac{1}{2} & 1 \end{vmatrix}$$

$$[\mathbf{a}\ \mathbf{b}\ \mathbf{c}]^2 = \frac{5}{4} - \frac{3}{4} = \frac{1}{2}$$

$$[\mathbf{a}\ \mathbf{b}\ \mathbf{c}] = \frac{1}{\sqrt{2}}$$

as given $\mathbf{a} \times \mathbf{b} + \mathbf{b} \times \mathbf{c} = p\mathbf{a} + q\mathbf{b} + r\mathbf{c}$

Take dot product with $\mathbf{a}$ is

$$\mathbf{a}\cdot(\mathbf{a}\times\mathbf{b}) + \mathbf{a}\cdot(\mathbf{b}\times\mathbf{c}) = p(\mathbf{a})^2 + q\mathbf{a}\cdot\mathbf{b} + \mathbf{r}(\mathbf{a}\cdot\mathbf{c})$$

$$0 + \frac{1}{\sqrt{2}} = p + \frac{q}{2} + \frac{r}{2} \quad \ldots(\text{i})$$

Similarly take dot product with $\mathbf{b}$ and $\mathbf{c}$

$$0 = \frac{p}{2} + q + \frac{r}{2} \quad \ldots(\text{ii})$$

and

$$\frac{1}{\sqrt{2}} = \frac{p}{2} + \frac{q}{2} + r \quad \ldots(\text{iii})$$

From Eqs. (i) and (iii), we get

$p = r$ and $p + q = 0$

$$\therefore \quad \frac{p^2 + 2q^2 + r^2}{q^2} = \frac{p^2 + 2p^2 + p^2}{p^2} = 4$$

16. *(a)* Let the matrix

$$M = \begin{bmatrix} a_{11} & a_{12} & a_{13} \\ a_{21} & a_{22} & a_{23} \\ a_{31} & a_{32} & a_{33} \end{bmatrix}$$

$$\therefore \quad M^T = \begin{bmatrix} a_{11} & a_{21} & a_{31} \\ a_{12} & a_{22} & a_{32} \\ a_{13} & a_{23} & a_{33} \end{bmatrix}$$

Sum of diagonal entries of

$$M^T M = \Sigma a_{ij}^2 = 5$$

Possible cases are

1, 2, 0, 0, 0, 0, 0, 0, 0 which gives $\frac{9!}{7!}$ matrices = 72

and 1, 1, 1, 1, 1, 0, 0, 0, 0 which gives $\frac{9!}{5!4!}$ matrices = 126

$\therefore$ Total matrices = 72 + 126 = 198

17. *(c)* We have, $x + y + z = 10$

$\therefore$ Total number of sample space

$$= {}^{12}C_2 = 66$$

Number of possibilities of z is even

$$z = 0 \Rightarrow {}^{11}C_1$$

$$\Rightarrow \quad z = 2 \Rightarrow {}^{9}C_1$$

$$\Rightarrow \quad z = 4 \Rightarrow {}^{9}C_1$$

$$\Rightarrow \quad z = 6 \Rightarrow {}^{5}C_1$$

$$\Rightarrow \quad z = 8 \Rightarrow {}^{3}C_1$$

$$\Rightarrow \quad z = 10 \Rightarrow {}^{1}C_1$$

Total = 11 + 9 + 7 + 5 + 3 + 1 = 36

Required probability $= \frac{36}{66} = \frac{6}{11}$

18. *(b)* We have,

$$f(x) = 2x^3 - 15x^2 + 36x + 1$$

$$\Rightarrow \quad f'(x) = 6x^2 - 30x + 36$$

$$\Rightarrow \quad f'(x) = 6(x-2)(x-3$$

$f(x)$ is increasing in [0, 2] and decreasing in [2, 3].

$\therefore f(x)$ is many one

$$f(0) = 1, f(2) = 29, f(3) = 28$$

Range = [1, 29]

$\therefore f(x)$ is onto but not one-one.

19. *(c)* Let the ellipse $E_2 = \frac{x^2}{a^2} + \frac{y^2}{b^2} = 1$ as it is passing through (0, 4) $b^2 = 16$

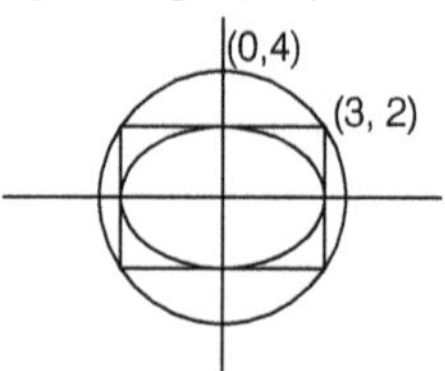

Also it is passes through (3, 2).

$$\therefore \quad a^2 = 12$$

$$\therefore \quad e = \sqrt{1 - \frac{a^2}{b^2}} = \sqrt{1 - \frac{12}{16}}$$

$$e = \frac{1}{2}$$

20. *(c)* We have,

$$\int_0^x \sqrt{1 - (f'(t))^2}\, dt = \int_0^x f(t)\, dt$$

On differentiating, we get

$$\sqrt{1 - (f'(x))^2} = f(x)$$

$$\Rightarrow \quad f'(x) = \pm\sqrt{1 - (f(x))^2}$$

$$\therefore \quad f(x) = \sin x$$

$$f(x) \neq -\sin x$$

$$x > \sin x$$

$$\therefore \quad \frac{1}{2} > \sin\frac{1}{2}$$

$$\Rightarrow \quad \frac{1}{2} > f\left(\frac{1}{2}\right) \text{ and } \frac{1}{3} > f\left(\frac{1}{3}\right)$$

21. *(d)* Conductor is charged till breakdown of air, then its potential is reduced to zero.

22. *(c)* For a particle thrown over very large distance (range), all acceleration g vectors points towards earth's centre and their magnitudes are different. Hence, path is elliptical in shape.

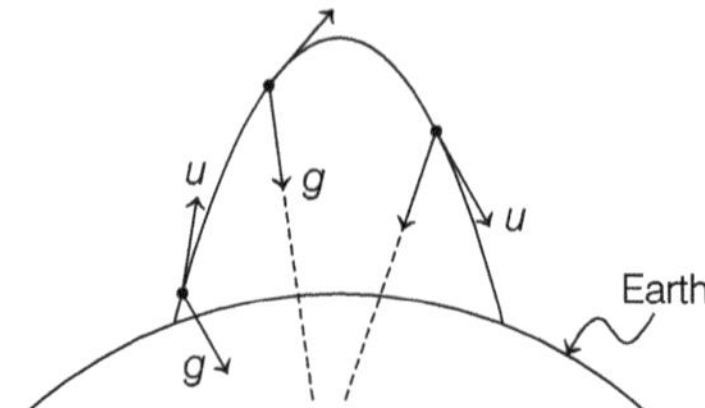

23. *(a)* Drop evaporates if pressure inside drop is greater than its vapour pressure. So, to avoid evaporation,

$$p_{\text{excess}} \leq p_{\text{vapour}} \Rightarrow \frac{2S}{R} = p_{\text{vapour}}$$

$$R = \frac{2S}{p}$$

$$= \frac{2 \times 6.99 \times 10^{-2}}{2.33 \times 10^3}$$

$$= 6 \times 10^{-5} \text{ m} = 0.06 \text{ mm}$$

24. *(a)*

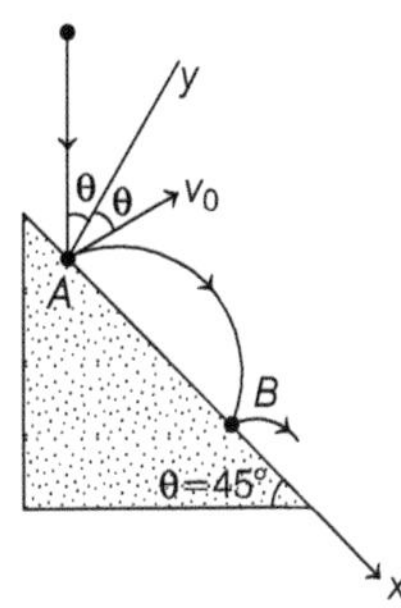

Speed of ball with which it strikes the plane is

$$v_0 = \sqrt{2gh} = \sqrt{2 \times 10 \times 5} = 10 \text{ ms}^{-1}$$

The ball bounces with same speed at an angle θ = 45° with *Y*-axis (shown in above figure), taken perpendicular to plane.

Now, from *A* to *B*

$$u_y = v_0 \cos\theta = 10 \times \cos 45° = \frac{10}{\sqrt{2}} \text{ ms}^{-1}$$

$$a_y = -g\cos\theta = -\frac{10}{\sqrt{2}} \text{ ms}^{-2}$$

Now, $y = u_y t + \frac{1}{2} a_y t^2$

As $y = 0$, (for entire motion of projectile from *A* to *B*)

$$\therefore \quad 0 = u_y t + \frac{1}{2} a_y t^2 \Rightarrow t = \frac{2v_0 \cos\theta}{g\cos\theta}$$

or time of flight, $t = \frac{2v_0}{g}$

$$\Rightarrow \quad t = \frac{2 \times 10}{10} = 2 \text{ s}$$

Now, for motion along *X*-axis,

$$x = L = AB$$

$$u_x = v_0 \sin\theta = \frac{10}{\sqrt{2}} \text{ ms}^{-1}$$

$$a_x = g\sin\theta = \frac{10}{\sqrt{2}} \text{ ms}^{-1}$$

$$t = 2 \text{ s}$$

So, $x = AB = u_x t + \frac{1}{2} a_x t^2$

$$= \frac{10}{\sqrt{2}} \times 2 + \frac{1}{2} \times \frac{10}{\sqrt{2}} \times (2)^2$$

$$= \frac{10}{\sqrt{2}} \times 4 = 20\sqrt{2} \text{ m}$$

25. *(d)*

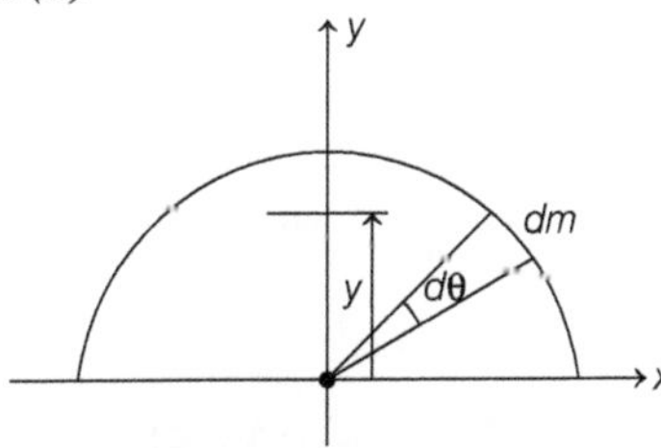

By symmetry, $x_{CM} = 0$

$$dm = \frac{m}{\pi} d\theta$$

$$y_{CM} = \frac{\int y\,dm}{\int dm} = \int_0^{\pi} \frac{(r\sin\theta)\left(\frac{m}{\pi} d\theta\right)}{\int_0^{\pi} dm}$$

$$= \frac{\frac{rm}{\pi} \cdot \int_0^{\pi} \sin\theta\, d\theta}{m} = \frac{2r}{\pi}$$

26. *(d)* As distance between antennas is large, we are not using approximation

$$\theta \approx \sin\theta \approx \tan\theta$$

Now, from figure

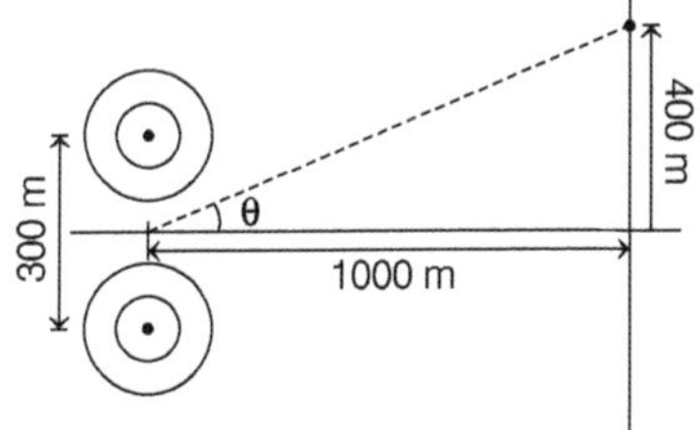

$$\tan\theta = \frac{400}{1000} = 0.4$$

So, $\quad \sin\theta = \frac{4}{\sqrt{116}}$

Now, wavelength, $\lambda = \frac{d\sin\theta}{m}$

$$= \left(\frac{300 \times \frac{4}{\sqrt{116}}}{2}\right) = 55.65 \text{ m}$$

So, $\lambda \approx 55$ m

27. *(d)* LIGO stands for Laser Interferometer gravitational wave observatory. These waves propagate from the site of any gravitational disturbance, such as explosion of a supernova or rotation of a double star. It recently (in March 2019) detects gravitational waves.

28. *(d)* As dielectric slab is entring the capacitor, the capacity increases and so galvanometer shows a deflection towards left. When dielectric slab is moving away from the capacitor, charge returns to cell and this causes a right hand side deflection.

29. *(a)* Velocity components of object and image are as shown below.

Clearly, $v_{\text{object}} = v\left(-\frac{1}{\sqrt{2}}\hat{\mathbf{i}} + \frac{1}{\sqrt{2}}\hat{\mathbf{j}}\right)$

and $\quad v_{\text{image}} = v\left(\frac{1}{\sqrt{2}}\hat{\mathbf{i}} + \frac{1}{\sqrt{2}}\hat{\mathbf{j}}\right)$

Relative velocity of image with respect to object is $v_{\text{rel}} = v_{\text{image}} - v_{\text{object}}$

$$= \sqrt{2}\, v\hat{\mathbf{i}} = \sqrt{2}\,\hat{\mathbf{i}} \text{ ms}^{-1}$$

30. *(c)* Initially, $I_1 = \frac{ML^2}{3}$

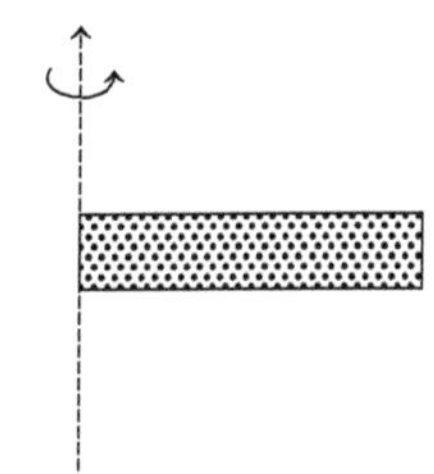

When temperature is increased, length of rod becomes

$$l' = l(1 + \alpha\Delta T)$$

So, moment of inertia will be

$$I_2 = \frac{M}{3}(l(1 + \alpha\Delta T))^2$$

$$= \frac{M}{3}(l^2 + l^2(\alpha\Delta T)^2 + 2l^2\alpha\Delta T)$$

Neglecting $l^2\alpha^2\Delta T^2$ as it is very small, we have

$$I_2 = \frac{Ml^2}{3} + \frac{2}{3}\alpha\Delta T$$

So, $\quad I_2 - I_1 = \frac{2}{3} Ml^2 \alpha\Delta T$

and $\quad \frac{I_2 - I_1}{I_1} = \frac{2}{3}\alpha\Delta T$

31. *(b)* Comparing given equation with

$$y = A\sin 2\pi\left(\frac{t}{T} - \frac{x}{\lambda}\right)$$

We have, $T = \frac{1}{B}$ and $\lambda = C$

So, f = frequency $= \frac{1}{T} = B$

Hence, wave speed, $v_{\text{wave}} = f\lambda$

$$\Rightarrow \quad v_{\text{wave}} = BC$$

Also, particle's maximum speed,

$$v_{\max} = A\omega = A\left(\frac{2\pi}{T}\right)$$

$$\Rightarrow \quad v_{\max} = A(2\pi B)$$

Now given, $v_{\text{wave}} = v_{\max}$

$$\Rightarrow \quad BC = 2\pi AB$$

$$\Rightarrow \quad \frac{C}{A} = 2\pi$$

32. *(d)* For O_1, observed frequency

$$f_1 = f_0\left(\frac{v}{v + v_s}\right)$$

Also, $\quad f_2 = f_0\left(\frac{v}{v - v_s}\right)$

and $\quad f_3 = \left(\frac{v + v_s}{v - v_s}\right) f_0$

So, $f_1 : f_2 : f_3 :: 2 : 6 : 9$

33. *(c)* Visible photon energies are from 1.8 eV to 2.8 eV.

$\therefore$ If Eg < 1.8 eV, all visible light is absorbed. Also, 1.8 < Eg < 3.1 eV, solid will be coloured.

If Eg > 3.1 eV, solid will be transparent.

For Eg > 3 eV, material is an insulator.

34. *(d)* Fraction of block submerged depends on weight of block and buoyant force. So, fraction of block submerged does not depends on acceleration of frame of reference. So, depth inside water remains same in all cases.

35. *(a)* v is zero at P, at R and S it is maximum, at T it is again zero. So, correct option is (a).

36. *(c)* Applied torque, τ

= rate of change of angular momentum

$$= \frac{\Delta L}{\Delta t} = \frac{2mva}{\Delta t}$$

37. *(c)*

Conservation of angular momentum gives, $mv(h - R) = I\omega = \frac{2}{5} mR^2 \cdot \frac{v}{R}$

$$\Rightarrow \quad mv(h - R) = \frac{2}{5} mvR$$

$$\Rightarrow \quad h - R = \frac{2}{5} R \Rightarrow h = \frac{7}{5} R$$

So, option (d) is correct.

Torque due to applied force about centre of mass = $\tau = F(h - R)$

For $\tau = 0$, $h = R$ and no rotation exists. So, option (b) is correct.

For $h > R$, sphere spins clockwise and looses energy. So, option (c) is incorrect.

For $h < R$, $\tau < 0$ and sphere spin anti-clockwise. So, option (a) is correct.

38. *(d)* Acceleration = Slope of $v - t$ graph which is negative and constant. Such v-t graph exists for a bouncing ball.

39. *(a)* Mobile phones are using 380 to 1900 MHz frequency bands.

40. *(a)* If half-life of α-decay is in years then half-life of β-decay is in hours and half-life of γ-decay is in nano-seconds.

41. *(d)* $\Lambda_m^\circ = \frac{\kappa \times 1000}{S}$

$$S = \frac{\kappa \times 1000}{\Lambda_m^\circ}$$

$$= \frac{1.85 \times 10^{-6} \times 1000}{140 \times 10^{-4}} = 1.321 \text{M}$$

$$AB \rightleftharpoons A^+ + B^-$$

$$K_{sp} = [A^+][B^-] = S^2 = (1.321)^2$$
$$= 1.74 \text{ mol}^2 \text{ L}^{-2}$$

42. *(a)* The structure of XeO_3 and XeO_4 are follows:

(XeO_3)

Hence, $3p\pi$-$d\pi$ bonds are present.

(XeO_4)

Hence, 4 $p\pi$-$d\pi$ bonds are present.

43. *(a)* The main postulates of Langmuir's theory of adsorption are as:

(i) Adsorption takes place on the surface of the solid only till the whole of the surface is completely covered with a unimolecular layer of the adsorbed gas. This suggests all the adsorption sites present on solids surface are equivalent in their ability to adsorb the particles.

(ii) Adsorption consist of two opposing processes (a) condensation and (b) evaporation.

(iii) The rate of condensation depend upon the uncovered surface of the adsorbent available for condensation.

44. *(b)* In laboratory, pure N_2 gas is prepared by heating NH_4Cl with sodium nitrite.

$$NH_4Cl + NaNO_2 \xrightarrow{\Delta} NH_4NO_2 + NaCl$$
$$NH_4NO_2 \xrightarrow{\Delta} N_2 + 2H_2O$$

$$(NH_4)_2Cr_2O_7 \xrightarrow{\Delta} N_2 + Cr_2O_3 + 4H_2O$$

This reaction will not give pure N_2 as it contain traces of Cr_2O_3.

45. *(c)* $CH_3CH{=}CH_2 \xrightarrow[H_2O_2]{BH_3 \cdot THF(X)} CH_3-CH_2-CH_2OH \xrightarrow{CrO_3 / \text{pyridine } (Y)} CH_3CH_2-\overset{\overset{O}{\|}}{C}-H$

Propene undergoes hydroboration in presence of peroxide, to give propanol. This formed alcohol undergoes oxidation with oxidising agent, CrO_3 in pyridine to give propanal.

46. *(c)* When H_2O_2 in added to an acidified solution of a dichromate, $Cr_2O_7^{2-}$, a deep blue coloured complex, chromic peroxide, CaO_3 [or $CrO(O_2)_2$] is formed.

$$Cr_2O_7^{2-} + 2H^+ + 4H_2O_2 \longrightarrow \underset{\text{Chromic peroxide}}{2CrO(O_2)_2} + 5H_2O$$

This deep blue coloured complex has the following structure.

Oxidation state of Cr is + 6 due to the presence of two peroxide linkages, which can be calculated as ln $CrO(O_2)_2$, let the oxidation state of Cr be x.

$$x + (-1)4 + (-2) = 0$$
$$x - 6 = 0$$
$$x = +6$$

47. *(d)* Inert pair effect is more pronounced in heavier members, like Pb. Hence, Pb (IV) compounds act as strong oxidising agents and are reduced to more stable Pb (II) compounds.

48. *(b)* From Raoult's law, for ideal solution,

$$p = p_B^0 \chi_B + p_T^0 \chi_T = 119 \times 0.5 + 37 \times 0.5 \quad (\because \chi_B = 1 - \chi_T)$$
$$= 59.5 + 18.5$$
$$= 78 \text{ torr}$$

Mole fraction of toluene in vapour phase

$$(\chi_T)_v = \frac{p_T^o \chi_T}{p} = \frac{18.5}{78}$$
$$= 0.237$$

49. *(d)*

$$C_6H_5-CH{=}CH-CH_2-\overset{\overset{O}{\|}}{C}-CH_3 \xrightarrow{NaBH_4(X)} C_6H_5-CH{=}CH-CH_2-\underset{\underset{CH_3}{|}}{\overset{\overset{OH}{|}}{CH}}$$

$NaBH_4$ is a selective reducing agent, it reduces $C{=}O$ group to alcohol. It does not reduce olefin (alkene) double bond.

50. *(a)* During change of O_2 to O_2^-, the electronic configuration of O_2 (total $e^- = 16$) is

$$\sigma(1s^2), \overset{*}{\sigma}(1s^2), \sigma(2s^2), \overset{*}{\sigma}(2s^2), \sigma(2p_z^2)$$
$$\pi(2p_x^2) \approx \pi(2p_y^2), \overset{*}{\pi}(2p_x^1) \approx \overset{*}{\pi}(2p_y^1)$$

and configuration of O_2^-, (total e^- =17)

$\sigma(1s^2), \overset{*}{\sigma}(1s^2), \sigma(2s^2), \overset{*}{\sigma}(2s^2)\ \overset{*}{\sigma}(2p_z^2), \pi(2p_x^2)$
$\approx \pi(2p_y^2), \overset{*}{\pi}(2p_x^2) \approx \overset{*}{\pi}(2p_y^1)$

Thus, electron goes in the $\overset{*}{\pi}$-orbital.

51. *(c)* In hexagonal close packed structure, there are 6 atoms per unit cell.

∴ Number of octahedral voids = 6

∴ Number of tetrahedral voids

$= 2 \times 6 = 12$

Total number of voids $= 6 + 12 = 18$

∴ Total number of voids per atom $= \frac{18}{6} = 3$

∵ In 1 mole of a compound, the total number of voids $= 3 \times 6.023 \times 10^{23}$

∴ In 0.5 mole compound, the total number of voids

$= 3 \times 0.5 \times 6.023 \times 10^{23}$
$= 9.034 \times 10^{23}$

52. *(a)* Most stable arrangement is

4s		3d				
⥮		1	1	1	1	1

Because, (I) according to Aufbau principle, electrons enter in the subshell of an atom in the increasing order of energy. (II) according to Hund's rule of maximum multiplicity, pairing of electrons in the orbitals of a subshell does not take place until all orbitals of a subshell are singly occupied.

53. *(c)* + *R*-effect of OH is more pronounced than OCH_3 group.

Hence, [benzene ring with CH_3 and OH] 2-methyl phenol is more reactive towards electrophilic reagent.

54. *(c)* $V_{(initial)} = \frac{nRT}{p}$

$= \frac{0.150 \times 0.0821 \times 294}{10} = 0.362\ dm^3$

$V_{(final)} = \frac{0.150 \times 0.0821 \times 294}{1} = 3.62\ dm^3$

∴ $W = -p\Delta V = -1 \times (3.62 - 0.362)$
$= -3.26\ dm^3$ atm

Thus, (b) is correct.

dm^3 atm can be converted into joules as follows

$= \frac{-3.26 \times 8.314}{0.0821}\ J = -330\ J$

Thus (c) is correct.

55. *(b)* For the reaction,

$SO_2(g) + \frac{1}{2} O_2(g) \rightleftharpoons SO_3(g)$

Equilibrium constant,

$$K_1 = \frac{[SO_3]}{[SO_2][O_2]^{1/2}} \quad ...(i)$$

For the reaction,

$2SO_3(g) \rightleftharpoons 2SO_2(g) + O_2(g)$

Equilibrium constant,

$$K_2 = \frac{[SO_2]^2[O_2]}{[SO_3]^2} \quad ...(ii)$$

On squaring both sides in Eq.(i), we get

$$K_1^2 = \frac{[SO_3]^2}{[SO_2]^2[O_2]} \quad ...(iii)$$

Eqs. (ii) × (iii), we get

$K_1^2 \times K_2 = 1$

$K_2 = \frac{1}{K_1^2} \Rightarrow K_1^2 = \frac{1}{K_2}$

56. *(a)* Magnetic moment, μ is related with number of unpaired electrons as:

$\mu = \sqrt{n(n+2)}$ BM

$(1.73)^2 = n(n+2)$

On solving, $n = 1$

Thus, the complex/compound having one unpaired electron exhibit a magnetic moment of 1.73 BM.

(a) In $[Cu(NH_3)_4]^{2+}$ the electronic configuration of Cu^{2+} = [Ar] $3d^9$

⥮	⥮	⥮	⥮	1

(Although in the presence of strong field ligand NH_3, the unpaired electrons gets excited to higher energy level, but it still remains unpaired.)

(b) In $[Ni(CN)_4]^{2-}$ the electronic configuration of Ni^{2+} = [Ar] $3d^4$

⥮	⥮	⥮	1	1

But CN^- being strong field ligand pair up the unpaired electrons and hence, in this complex, number of unpaired electrons = 0

(c) In $[TiCl_4]$ the electronic configuration of Ti^{2+} = [Ar], no unpaired electron.

(d) In $[CoCl_6]^{4-}$ Co^{2+} = [Ar]$3d^7$

⥮	⥮	1	1	1

It contains three unpaired electrons. [Thus, $[Cu(NH_3)_4]^{2+}$ is the complex that exhibits a magnetic moment of 1.73 BM.]

57. *(a)* Monomer of

$\left[-\underset{CH_3}{\overset{CH_3}{C}}-CH_2- \right]_n$ polymer is 2-methyl propene or *iso*-butene, i.e. $H_3C-\underset{CH_3}{C}=CH_2$.

2-methylpropene

58. *(d)* When conc. HI or HBr reacts with ether, the corresponding alcohol and iodide are formed. When there is a case of mixed ethers, the halogen atom attaches to the smaller alkyl group, due to steric effect.

$CH_3-\underset{}{\overset{CH_3}{CH}}-CH_2-O-CH_2-CH_3 + HI \xrightarrow{\Delta}$

$CH_3-\overset{CH_3}{CH}-CH_2OH + CH_3CH_2I$

59. *(c)*

[benzene ring with CH_2CONH_2 and $COOCH_3$] $\xrightarrow{(i)\ Br_2/NaOH}$ [benzene ring with CH_2NH_2 and $COOCH_3$] $\xrightarrow{(ii)\ \Delta}$ [cyclic lactam with NH and C=O]

Reaction (i) is known as Hofmann bromamide reaction. In reaction (ii) the product formed in reaction (i) undergoes elimination reaction.

60. *(c)* C_3H_6O (carbonyl compound) gives iodoform test (yellow ppt. with I_2/NaOH). This suggest that, it contains $-\overset{O}{\overset{||}{C}}-CH_3$

$-\overset{O}{\overset{||}{C}}-$ group is due to hydrolysis of $-\underset{Cl}{\overset{Cl}{C}}-$.

Then $C_3H_6Cl_2$ will be

$CH_3-\underset{Cl}{\overset{Cl}{C}}-CH_3 \xrightarrow[-H_2O]{2NaOH} CH_3-\overset{O}{\overset{||}{C}}-CH_3$

2,2-dichloropropane

61. *(c)* Sexual reproduction in *Rhizopus* takes place by gametangial copulation. The fusing gametangia are multinucleate and genetically different. The resulting structure is a diploid zygospore.

Here 24(n) nuclei from different filaments combine to form 24($2n$) nuclei which undergo meiosis to form 96 nuclei. (or spores) through sexual reproduction.

62. *(c)* In mammals, the teeth are heterodont, i.e. dentition consists of incisors, canines, premolar and molar; and thecodont (in sockets of jaw bones). The brain has 12 pairs of cranial nerves in all mammals.

63. *(a)* Flocculation or coagulation of protoplasm is interchangeability between sol and gel states. Protoplasm is generally found in two states, i.e. peripheral gel-like ectoplasm and central sol-like endoplasm.

64. *(a)* At telophase stage, nuclear membrane vesicles associate with the surface of individual chromosome and fuse to reform the nuclear membranes, which partially enclose a cluster of chromosomes before coalescing to reform the complete nuclear envelope. During this process, the nuclear pores reassemble and reassociate to form the nuclear lamina. One of the lamina protein (lamina-B) remains with the nuclear membrane fragments throughout mitosis and may help nuclear reassembly. After the nucleus reforms, the pores pump in nuclear proteins, the chromosomes decondense and RNA synthesis resumes, causing the nucleolus to reappear. Thus, option (a) is correct.

65. *(a)* All lipids digestion takes place in the small intestine. The fats found in foods like salad dressing, ice-creams, etc., during digestion get split into fatty acids and glycerol which act as the digestive products of lipids.

66. *(c)* On addition of salt into water, the gradient of water becomes more negative. So, transpiration rate decreases, hence the cut flowers or twigs remain fresh for longer period.

67. *(d)* If a pulse crop is undergoing premature yellowing of leaves and decrease in yield then application of iron and magnesium to promote synthesis of chlorophyll may become most beneficial to overcome the problem and to obtain maximum seed yield.

68. *(a)* RuBP Carboxylase (RuBisCO) is the most abundant enzyme or protein in chloroplasts and probably on earth. It catalyses the carboxylation of ribulose 1, 5-bisphosphate to form two molecules of 3-phosphoglyceric acid in C_3 cycle of photosynthesis.

69. *(a)* When breastfeeding is replaced by less nutritive food, low proteins and calories, the infants below one year of age will suffer from marasmus. Whereas kwashiorkor is caused by the deficiency of proteins, rickets by the deficiency of vitamin-D and pellagra by the deficiency of vitamin-B_3.

70. *(d)* The tissues in our body are able to utilise only 25% of O_2 carried by arterial blood. The venous blood is still 75% saturated with O_2. This O_2 acts as a reserve during muscular exercise.

71. *(d)* Due to the different pressure between the caval and atrium, the blood passes from the postcaval to the diastolic right atrium of human heart.

72. *(b)* Hypothalamus acts as a bridge between nervous system and endocrine system. It is a centre for hunger, thirst, sweating, sleep, fatigue, temperature, anger, pleasure, love, hate and satisfaction. It releases regulatory factors for endocrine glands, controls autonomic nervous system and regulates parasympathetic activity.

73. *(a)* Steroid hormones are lipid soluble. So, they can pass freely across the lipid bilayer of plasma membrane. After getting into the cytoplasm, molecules of steroid hormones bind to receptor molecules, located within the cytoplasm of target cell and thus a hormone receptor complex is formed. Now, this complex moves into the nucleus of the cell and activates specific genes that ultimately produce specific proteins.

74. *(d)* Heterozygous round yellow has genotype RrYy. On selfing, round green offsprings produced and are represented by Rryy and RRyy [genotypes on (R = Round, yy = Green)].

75. *(a)* There are 80 nucleotides which contain 3 hydrogen bonds. B-DNA shows 10 nucleotides per turn (coil) of helix, if there are 20 coils then total number of nucleotides is 200 out of which 120 are adenine (equal amount of thymine). So, the number of guanine (equal amount of cytosine) nucleotides is 80.

76. *(d)* Tetracycline interferes with the attachment of *t*RNA carrying the amino acid to the *m*RNA ribosome complex preventing the addition of amino acids to the growing polypeptide chain. Streptomycin interferes with the initial steps of protein synthesis by changing the shape of 30S portion of the 70S prokaryotic ribosome.

77. *(b)* An insect bite may result in inflammation of that spot, that is triggered by the release of histamine and kinins. These act as allergy medication.

78. *(b)* Polymerase Chain Reaction (PCR) is a technique by which small samples of DNA can be amplified. Starting with a very small piece of DNA, this technique is used to make literally billions of copies in only a few hours.

79. *(a)* For a normal distribution, the mean, median and mode are actually equivalent.

80. *(a)* UAA (ochre) is a termination or non-sense codon and does not code for lysine (lys). It is the codon of a functional gene, helps in termination of polypeptide chain. It does not code for any amino acid.

81. *(b)* We have,

$$\log_{2x} 1944 = \log_x 486\sqrt{2}$$

$$\Rightarrow \frac{\log_2 1944}{\log_2 2 + \log_2 x} = \frac{\log_2 486\sqrt{2}}{\log_2 x}$$

$$\Rightarrow \log_2 x \cdot \log_2 1944 = \log_2 486\sqrt{2}(1 + \log_2 x)$$

$$\Rightarrow \log_2 x(\log_2 1944 - \log_2 486\sqrt{2}) = \log_2 486\sqrt{2}$$

$$\Rightarrow \log_2 x\left(\log_2 \frac{1944}{486\sqrt{2}}\right) = \log_2 486\sqrt{2}$$

$$\Rightarrow \log_2 x \log_2 2\sqrt{2} = \log_2 486\sqrt{2}$$

$$\Rightarrow \frac{3}{2}\log_2 x = \log_2 486\sqrt{2}$$

$$\Rightarrow \log_2 x^{3/2} = \log_2 486\sqrt{2}$$

$$\Rightarrow x^{3/2} = 486\sqrt{2}$$

$$\Rightarrow x^6 = (486\sqrt{2})^4$$

$$\Rightarrow x^6 = 2^6 \times 3^{20}$$

$$\therefore a = 6, b = 20$$

$$a + b = 26$$

82. *(a)* We have,

$$\sqrt{3x^2 + 6x + 12} + \sqrt{5x^2 + 10x + 9} = 4 - 2x - x^2$$

$$3x^2 + 6x + 12 = 3(x + 1)^2 + 9$$

$$\therefore \sqrt{3x^2 + 6x + 12} \geq 3$$

$$5x^2 + 10x + 9 = 5(x + 1)^2 + 4$$

$$\sqrt{5x^2 + 10x + 9} \geq 2$$

$$\therefore \text{LHS} \geq 3 + 2 = 5$$

RHS $4 - 2x - x^2 = 5 - (x + 1)^2$

RHS ≤ 5

$\therefore$ Only equality is possible.

$\therefore$ Equality is possible if $x = -1$

$\therefore x = -1$ is the only solutions.

83. *(c)* We have,

$$\lim_{n \to \infty} \sum_{r=0}^{n} \left(\frac{1}{4r + 1} - \frac{1}{4r + 3}\right)$$

$$= \lim_{n \to \infty} \sum_{r=0}^{n} \left(\int_0^1 x^{4r} - x^{4r+2}\right) dx$$

$$= \int_0^1 (1 + x^4 + x^8 + \ldots) - (x^2 + x^6 + x^{10} \ldots) dx$$

$$= \int_0^1 \left(\frac{1}{1 - x^4} - \frac{x^2}{1 - x^4}\right) dx$$

$$= \int_0^1 \frac{1 - x^2}{1 - x^4} dx = \int_0^1 \frac{dx}{1 + x^2}$$

$$= [\tan^{-1} x]_0^1 = \frac{\pi}{4}$$

84. *(c)* Let $AD = 3x$

$\therefore \quad AG = 2x$

$GD = x$

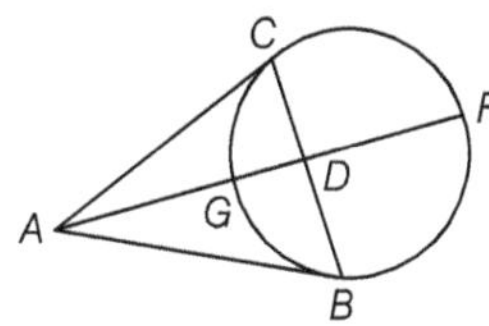

$DF = y$

Given, $AB = 6$

$BC = 4$

$BD = DC = \frac{1}{2} BC = 2$

$GD \times DF = BD \times CD$

$xy = 4$ and $AB^2 = AG \times AF$

$36 = 2x(3x + y)$

$36 = 6x^2 + 2xy$

$\Rightarrow \quad 6x^2 = 36 - 8$

$\Rightarrow \quad x^2 = \frac{28}{6}$

In ΔABC, by Apollonius law,

$$AC^2 + AB^2 = 2\left(AD^2 + \left(\frac{BC}{2}\right)^2\right)$$

$\Rightarrow \quad AC^2 + 36 = 2(9x^2 + 4)$

$\Rightarrow \quad AC^2 + 36 = 2\left(9 \times \frac{28}{6} + 4\right)$

$\Rightarrow \quad AC^2 = 2(42 + 4) - 36$

$\Rightarrow \quad AC^2 = 56$

$\Rightarrow \quad AE = \sqrt{56} = 2\sqrt{14}$

85. *(b)* We have,

$y = g(x)$

$\Rightarrow \quad f(x) = x^3 + 3x + 1$

$\therefore \quad fog(x) = x$

$[\because g(x)$ is inverse of $f(x)]$

Area enclosed between $y = g(x)$, $x = 1$ and $x = 15$ is

$$\text{Area} = \int_1^{15} g(x)\, dx = \int_1^{15} f^{-1}(x)\, dx$$

$\therefore$ Let $f^{-1}(x) = t$

$x = f(t)$

$dx = f'(t)\, dt$

When $x = 1, t = 0\, x = 15, t = 2$

$\therefore \quad \text{Area} = \int_0^2 tf'(t)\, dt = \int_0^2 t(3t^2 + 3)\, dt$

$[\because f(t) = t^3 + 3t + 1$

$\therefore f'(t) = 3t^2 + 3]$

$$= \left[\frac{3t^4}{4} + \frac{3t^2}{2}\right]_0^2 = 12 + 6 = 18$$

86. *(b)* We have,

$$\left(\sum_{k=1}^{5} {}^{20}C_{2k-1}\right)^6$$

$= ({}^{20}C_1 + {}^{20}C_3 + {}^{20}C_5 + {}^{20}C_7 + {}^{20}C_9)^6$

$= (2^{18})^6$

$= 8 \times (32)^{21} = 8(33 - 1)^{21}$

$= 8(11\lambda - 1) = 8 \times 11\lambda - 11 + 3$

$= 11(8\lambda - 1) + 3$

$\therefore$ Remainder is 3 when divided by 11.

87. *(a)* Mirror image of $A(4, -1)$ with respect to line $x - 1 = 0$ and $x - y - 1 = 0$ are respectively $(-2, -1)$ and $(0, 3)$ which lie on BC

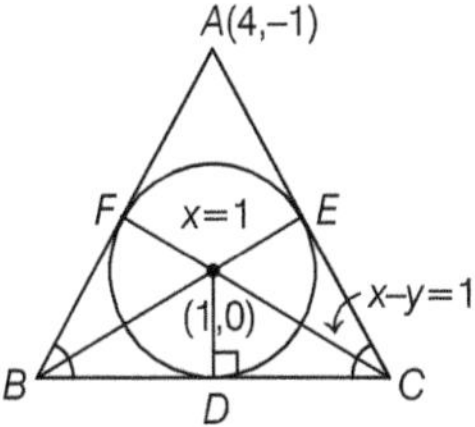

$\therefore$ Equation of $BC = 2x - y + 3 = 0$

$$r = \left|\frac{2 - 0 + 3}{\sqrt{5}}\right| = \frac{5}{\sqrt{5}} = \sqrt{5}$$

Now image of D, E, F are also lies on circumcircle with respect to the diameter.

$\therefore$ D', E', F' are lie on in circle of ΔABC

Equation of incircle of

$\Delta ABC = (x - 1)^2 + y^2 = 5$

88. *(d)* Let the amount of salt present in tank is 'm' gm

$$\frac{dm}{dt} = 10 - \left(\frac{m}{100 + t}\right)^4$$

$\Rightarrow \quad \frac{dm}{dt} + \frac{4m}{100 + t} = 10$

$\Rightarrow \quad (100 + t)^4 m = \frac{10(100 + t)^5}{5} + c$

$\Rightarrow \quad m(100 + t)^4 = 2(100 + t)^5 + c$

At $t = 0$, $m_0 = 50$ gm

$50(100)^4 = 2(100)^5 + c$

$c = -150 \times 100^4$

$\therefore \quad m_t = \frac{2(100 + t)^5}{(100 + t)^4} - \frac{150 \times 100^4}{(100 + t)^4}$

$m_{100} = 2(100 + 100) - \frac{150 \times 100^4}{(200)^4}$

$m_{100} = 400 - \frac{150}{16}$

$m_{100} = 400 - \frac{75}{8} = \frac{3200 - 75}{8} = \frac{3125}{8}$

89. *(b)* We have,

$\mathbf{a} = \hat{\mathbf{i}} + \hat{\mathbf{j}} + \hat{\mathbf{k}}, \mathbf{b} = \hat{\mathbf{i}} - \hat{\mathbf{j}} + \hat{\mathbf{k}}$

$\Rightarrow \quad \mathbf{a} \times \mathbf{b} = \mathbf{b} + \lambda \mathbf{a}$

$\Rightarrow \quad \mathbf{a} \cdot (\mathbf{a} \times \mathbf{c}) = \mathbf{a} \cdot \mathbf{b} + \lambda |\mathbf{a}|^2$

$0 = 1 + 3\lambda$

$\Rightarrow \quad \lambda = -\frac{1}{3}$

$\Rightarrow \quad \mathbf{a} \times (\mathbf{b} \times \mathbf{c}) = \mathbf{a} \times \mathbf{b} + \lambda(\mathbf{a} \times \mathbf{a})$

$\Rightarrow \quad |\mathbf{a}|^2 \mathbf{c} - (\mathbf{a} \cdot \mathbf{c})\, \mathbf{a} = \mathbf{b} \times \mathbf{a}$

$\Rightarrow \quad \mathbf{c} = \frac{\mathbf{b} \times \mathbf{a} + (\mathbf{a} \cdot \mathbf{c})\mathbf{a}}{|a|^2}$

$\Rightarrow \quad (\mathbf{a} \times \mathbf{b}) \cdot \mathbf{c} = \frac{|\mathbf{a} \times \mathbf{b}|^2 + \mathbf{a} \cdot (\mathbf{a} \times \mathbf{b})}{|a|^2}$

$\Rightarrow \quad [\mathbf{a}\ \mathbf{b}\ \mathbf{c}] = \frac{|a \times b|^2}{|a|^2} = \frac{8}{3}$

90. *(d)* We have,

$f(1) = 1, f(2) = 4, f(3) = 9$

$\therefore \quad f(x) = x^2$

Let $\quad g(x) = f(x) - x^2$

$\therefore \quad g(1) = g(2) = g(3) = 0$

From Rolle's theorem on $g(x)$

$g'(x) = 0\, x \in (1, 2)$

Similarly, $g'(x) = 0$ for at least $x \in (2, 3)$

$g''(x) = f''(x) - 2$

$g''(x) = 0$ if $f''(x) = 2$ for some $x \in (1, 3)$

91. *(b)* Due to motion of bar in magnetic field,

emf induced, $E = Blv = BLv\cos\theta$

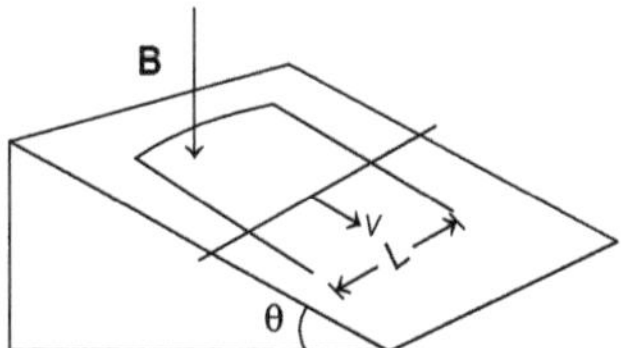

Current in metal bar, $I = \frac{E}{R} = \frac{BLv}{R}\cos\theta$

Force on bar up the inclined plane is

$$F = BIL\cos\theta = \frac{B^2L^2v}{R}\cos^2\theta$$

When this force is equal to downwards gravitational pull, bar reaches terminal speed.

$$\frac{B^2L^2v\cos^2\theta}{R} = mg\sin\theta$$

$$v = \frac{Rmg}{B^2L^2} \cdot \sec\theta \cdot \tan\theta$$

92. *(a)* As, $\mu = \frac{\sin i}{\sin r}$

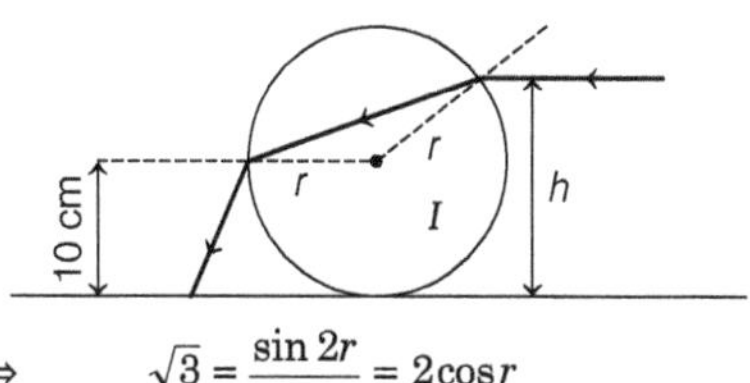

$\Rightarrow \quad \sqrt{3} = \frac{\sin 2r}{\sin r} = 2\cos r$

$\therefore \quad r = 30^\circ \Rightarrow i = 60^\circ$

So, $h = 10 + 10\sin 60° = 18.66\,\text{cm}$

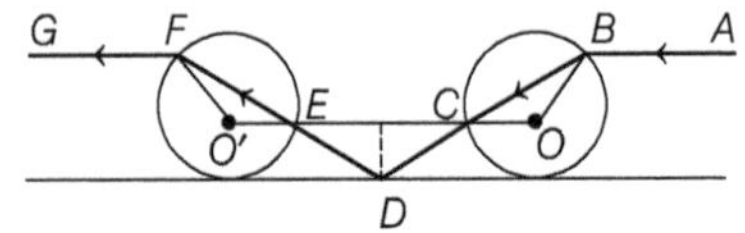

From above ray diagram,

$$OO' = O'E + EC + CO$$
$$= 10 + 2 \times 10\cot 60° + 0$$
$$= 10 + 11.55 + 10$$
$$= 31.55\,\text{cm}$$

93. *(a)* From v-x graph, equation is

$$v = \frac{-v_0}{x_0} \cdot x + v_0$$

Differentiating w.r.t. t, we get

$$a = \frac{-v_0}{x_0} \cdot v = \frac{-v_0}{x_0}\left(-\frac{v_0}{x_0} \cdot x + v_0\right)$$

$$\Rightarrow \quad a = \frac{v_0^2}{x_0^2} \cdot x - \frac{v_0^2}{x_0}$$

[Positive slope and negative intercept]

94. *(b)* $E_{\text{axis}} = \dfrac{kxQ}{(x^2 + R^2)^{3/2}}$

$$E_{\max} \Rightarrow \text{when } \frac{dE}{dx} = 0 \Rightarrow x = \pm \frac{R}{\sqrt{2}}$$

$$E_{\max} = \frac{Q}{6\sqrt{3}\,\pi \epsilon_0}$$

From $\dfrac{d\mathbf{V}}{dr} = -\mathbf{E}$, we have

$$V_{\max} = \frac{Q}{2\sqrt{6}\,\pi \epsilon_0}$$

Which occurs at B.

95. *(a)* Rods are connected as,

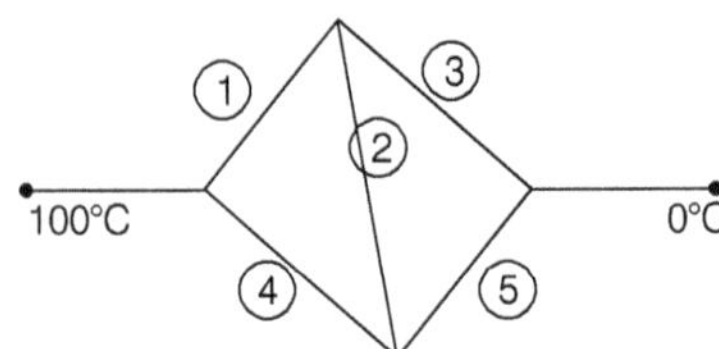

Following electrical analogy no heat current flows through rod 2.

So, equivalent thermal resistance of network is

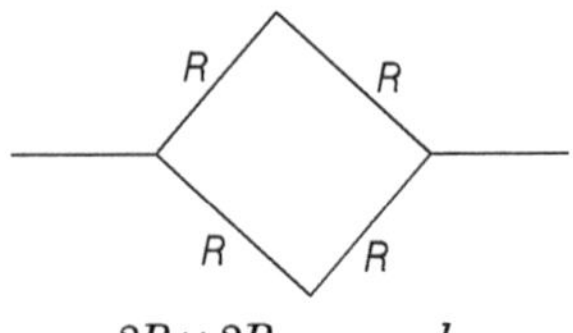

$$R_{eq} = \frac{2R \times 2R}{2R + 2R} = R = \frac{l}{kA}$$

So, heat flow rate through network is

$$H = \frac{\Delta T}{R_{eq}} = \frac{100\,kA}{l}$$

96. *(c)* Let I = ammeter reading and V = voltmeter reading.

$$V = \frac{R_V \cdot R}{R_V + R} \cdot I$$

where, R_V is the voltmeter resistance.

or $$R = \frac{VR_V}{IR_V - V}$$

$$\Rightarrow \quad R = \frac{120 \times 3000}{2 \times 3000 - 120} = 61.2\,\Omega$$

As, $$R = \frac{VR_V}{IR_V - V} = \frac{V}{I - (V/R_V)}$$

when $R_V = \infty$, i.e. voltmeter is ideal,

$$R = \frac{V}{I} = \frac{120}{2} = 60\,\Omega$$

So, error in $R = 61.2 - 60 = 1.2\,\Omega$

97. *(a)* Electron's energy is maximum if no neutrino is emitted. Then,

$$K_{\max} = (m_n - m_p - m_e) \cdot c^2$$
$$= \{m_n - (m_p + m_e)\} \cdot c^2$$

As, $m_p + m_e$ = mass of H-atom.

We have,

$$K_{\max} = (m_n - m_H) \cdot c^2 = (840\,\mu\text{u}) \cdot c^2$$
$$= 840 \times 10^{-6}\,\text{u} \times 932\,\frac{\text{MeV}}{\text{u}} = 0.783\,\text{MeV}$$

98. *(a)* Geometry of given arrangement is as shown below.

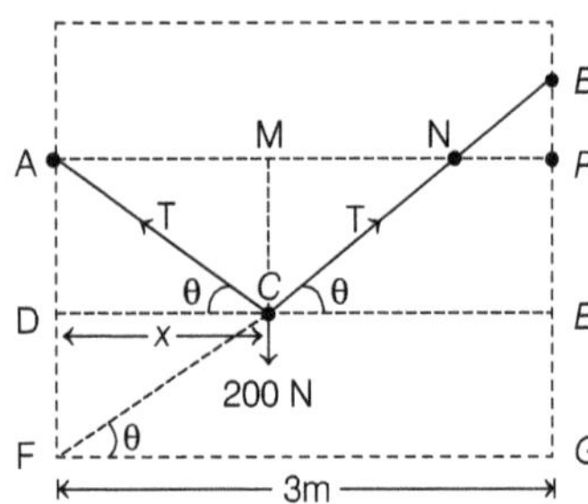

Triangles ΔADC, ΔBEC and ΔBGF are similar triangles. Also, ΔACD is congruent to ΔFCD. So, $AC = FC$.

Hence, $FB = 5$ m.

In ΔBGF,

$$\cos\theta = \frac{\text{base}}{\text{hypotenuse}} = \frac{3}{5}$$

So, $\sin\theta = \dfrac{4}{5}$ and $\tan\theta = \dfrac{4}{3}$

Now, $\Delta AMC \cong \Delta NMC$ (by AAS rule)

$$AM = MN$$

Also, $AM + MN + ND = 3$ m.

$\Rightarrow \quad NP = 3 - 2x \quad (\because AM = x \text{ and } MN = x)$

Now in ΔBPN,

$$\tan\theta = \frac{BP}{NP} \Rightarrow \frac{4}{3} = \frac{1}{3 - 2x} \Rightarrow x = \frac{9}{8}\,\text{m}$$

99. *(c)* Equating rotational KE at bottom with potential energy at top, we have

$$\frac{1}{2}mv^2 + \frac{1}{2}I\omega^2 = mgh$$

$$\frac{1}{2}mv^2 + \frac{1}{2} \times \frac{2}{5}mr^2 \times \frac{v^2}{r^2} = mgh$$

$$v^2 \times \left(\frac{1}{2} + \frac{1}{5}\right) = gh$$

So, h is independent of mass of ball or angle of inclination.

100. *(a)* Let mass m has volume V_0 at surface and volume $V_0 - \Delta V$ at depth h.

So, density at depth h is

$$\rho = \frac{m}{V_0 - \Delta V}$$

$$\Rightarrow \quad \frac{\rho}{\rho_0} = \frac{V_0}{V_0 - \Delta V} = \frac{1}{1 - \Delta V/V_0}$$

As, bulk modulus B is

$$B = P\frac{\Delta V}{V_0} \Rightarrow \frac{\Delta V}{V_0} = \frac{P}{B}$$

So, $\dfrac{\rho}{\rho_0} = \dfrac{1}{1 - \dfrac{P}{B}}$ or $\rho = \dfrac{\rho_0}{\left(1 - \dfrac{\rho_0 gh}{B}\right)}$

101. *(a)* $PH_3(g) \longrightarrow P(g) + 3H(g)$

$$\Delta H = 953\text{ kJ mol}^{-1}$$

$\therefore$ BE of (P—H) bond $= \dfrac{953}{3}$

$$P_2H_4(g) \longrightarrow 2\,P(g) + 4H(g)$$

This involve breaking of one (P—P) bond and four (P—H) bonds.

$$\therefore \quad \Delta H = \frac{953}{3} \times 4 + \text{BE (P—P)}$$

$$1485 = \frac{953}{3} \times 4 + \text{BE (P—P)}$$

$$\therefore \quad \text{BE (P—P)} = 214.3\,\text{kJ}$$

102. *(c)* For a first order process,

$$kt = \ln\frac{[A]_0}{[A]}$$

where, $[A]_0$ = initial concentration.

$[A]$ = concentration of reactant remaining at time t.

$$kt_{1/8} = \ln\frac{[A]_0}{[A]_0/8} = \ln 8 \quad \ldots\text{(i)}$$

$$kt_{1/10} = \ln\frac{[A]_0}{[A]_0/10} = \ln 10 \ldots\text{(ii)}$$

Therefore,

$$\frac{t_{1/8}}{t_{1/10}} = \frac{\ln 8}{\ln 10} = \log 8 = \log 2^3 \; 3\log 2$$

$$\frac{t_{1/8}}{t_{1/10}} = 3 \times 0.3 = 0.9$$

$$\frac{t_{1/8}}{t_{1/10}} \times 10 = 0.9 \times 10 = 9.0$$

103. *(b)* From the graph, it is clear that the value of Z decreases with increase of pressure. We can explain this on the basis of van der Waals' equation.

At high pressure, when 'p' is large, V will be small and one cannot ignore 'b' in comparison to V. However, the term $\frac{a}{V^2}$ may be considered negligible in comparison to 'p' in van der Waals' equation.

$$\left(p + \frac{a}{V^2}\right)(V - b) = nRT$$

$$p(V - b) = nRT$$

$$\Rightarrow pV - pb = nRT$$

or $$\frac{pV}{nRT} = 1 + \frac{pb}{nRT}$$

or $$Z = 1 + \frac{pb}{nRT}$$

Thus, Z is greater than 1. As pressure is increased (at constant T), the factor $\frac{pb}{nRT}$ increases. This explains why after minima in the curves, Z increases continuously with pressure. Hence, the only incorrect statement is (b).

104. *(c)* Given,

$$Fe^{3+} + 3e^- \longrightarrow Fe;$$

$$E_1^\circ = -0.336\text{ V} \quad \ldots\text{(i)}$$

$$Fe^{2+} + 2e^- \longrightarrow Fe; E_2^\circ = -0.439\text{ V} \quad \ldots\text{(ii)}$$

We need to calculate

$$Fe^{3+} + e^- \longrightarrow Fe^{2+}, E_3^\circ = ? \quad \ldots\text{(iii)}$$

We can obtain the Eq. (iii) by subtracting Eq. (ii) from Eq. (i), but E_3°, cannot be obtained by this method because electrode potential is intensive property. Thus we can it by determine

$$\Delta G_3 = \Delta G_1 - \Delta G_2$$

(ΔG is an extensise property)

$$\Delta G_3 = 3 \times 0.336\text{ V} - 2 \times 0.439\text{ V}$$

$$\Delta G_3 = 0.108\text{ V} - 0.878\text{ V}$$

$$-1 \times \text{V} \times E_3^\circ = -0.770\text{ V}$$

$$\Rightarrow E_3^\circ = 0.770\text{ V}$$

105. *(a)* Gadolinium ($Z = 64$)

$[Xe]\,4f^7, 5d^1, 6s^2$

Lutetium ($Z = 71$) $[Xe]\,4f^{14}, 5d^1, 6s^2$

Lawrencium ($Z = 103$) $[Rn]\,5f^{14}, 6d^1, 7s^2$

Tantalum ($Z = 73$) $[Xe]\,4f^{14}, 5d^3, 6s^2$

Hence, gadolinium has got incompletely half-filled f-orbital.

106. *(b)* The reaction takes place as follows:

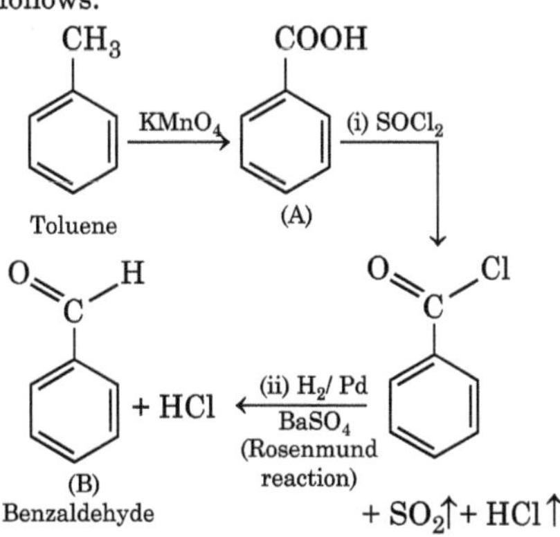

107. *(d)* In $[Ni(CO)_4]$, Ni is in zero oxidation state. Its electronic configuration is $[Ar]\,3d^4 4s^2$. Pairing of electrons occurs due to strong field ligand CO.

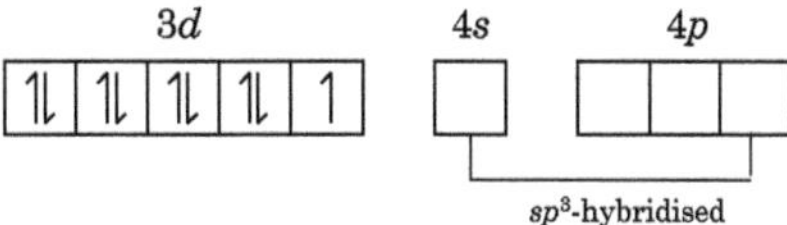

It has tetrahedral geometry and is diamagnetic (as all the electrons are paired).

The oxidation state of Ni in $[Ni(CN)_4]^{2-}$ is + 2. It's electronic configuration is $3d^8 4s^0$. Pairing of electrons will also occur in this case as CN^- is a strong field ligand.

$[Ni(CN)_4]^{2-}$

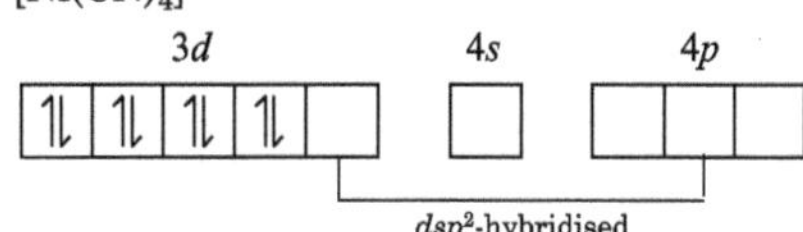

It has square planar geometry and is diamagnetic in nature.

108. *(b)*

$$HO-\overset{\displaystyle COOH}{\underset{\displaystyle \underset{\displaystyle CH_2OH}{CHOH}}{C}}-CH_2OH \xleftarrow[H_2O]{Br_2} HO-\overset{\displaystyle CHO}{\underset{\displaystyle \underset{\displaystyle CH_2OH}{CHOH}}{C}}-CH_2OH \;(A)$$

$$A \xrightarrow{HI\,/\,\text{Red P}} H_3C-\overset{\displaystyle CH_3}{CH}-CH_2CH_3 \;(\textit{iso}\text{-pentane})$$

$$A \xrightarrow[NaOAc]{Ac_2O} AcO-\overset{\displaystyle CHO}{\underset{\displaystyle \underset{\displaystyle CH_2OAc}{CHOAc}}{C}}-CH_2OAc$$

109. *(d)* Diastereomers are the optically active isomers in which the mirror images are non-superimpossable on each other whereas enantiomers are those optical isomers in which the mirror images are superimposable on each other.

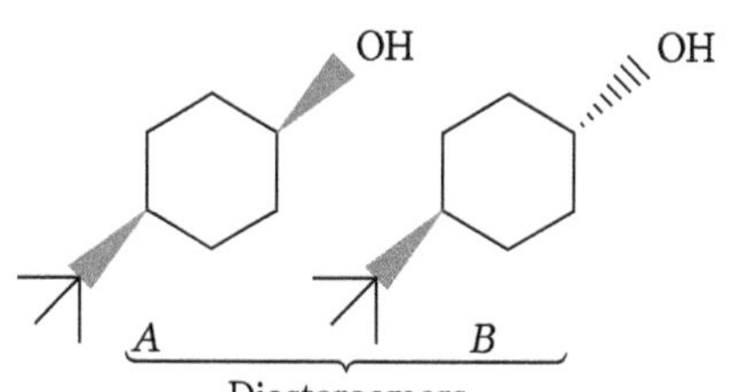

Because A and B are non-superimposable mirror images of each other.

Hence, these are diastereomers.

Ph CO₂Me (C) — Ph CO₂Me (D) epoxides: Enantiomers

Both C and D are the superimposable mirror images of each other hence, are enantiomers.

110. *(b)* The enthalpy of hydrogenation of given compounds is inversely proportional to stability of alkene.

III (less stable resonance form) ; II → H_3C, CH_3, CH_3 substituted benzene (Stable)

I: H_3C, CH_3, CH_3 = Already aromatic compound because of $6\pi e^-$s highly stable.

Hence, correct order of enthalpy of hydrogenation is

$$III > II > I.$$

111. *(c)* There are three major steps in a PCR.

Step 1 : Denaturation The thermal cycler heats the mixture of primer and DNA fragment to about 95°C. Hydrogen bonds holding the double-stranded DNA fragment break and the double-stranded DNA separates and dissociates into single strands.

Step 2 : Annealing of primers The solution is allowed to cool to 37°C. As it cools, primers base pair with complementary sequences at the 3' end of the single-stranded DNA.

Step 3 : Primer extension The thermal cycler then raises the temperature to 72°C. Using the primer, the polymerase synthesises the rest of the fragment resulting in the new strand of DNA.

112. *(a)* Gene therapy for both the Cystic Fibrosis (CF) and SCID uses the *in vivo* approach. Gene therapy for CF uses liposomes while gene therapy for SCID uses retrovirus as vectors.

Both CF and SCID are recessive disorders.

113. *(c)* When the concentration of sucrose solution has the same water potential as the cell sap, there would be no net movement of water across the cell membranes of the plant tissue and there would be no change in the length of the strip. For the **0.45 mol dm^{-3}** sucrose solution, the ratio of initial length: final length was 1.0, which means that the length of the strip of plant tissue was unchanged.

114. *(c)* 36 molecules of ATP are produced during aerobic respiration, whilst only 2 ATP molecules are released in aerobic respiration. Thus, approximate ratio is 1 : 20.

115. *(a)* For a plant with codominant alleles, both characteristics will show, unless the plant has identical alleles. In this case, both the phenotypic and genotypic ratio will be the same. Therefore, the answer could be either (a) or (d). In case of the option (d), there is only 2 types of outcome, but with 2 alleles, the outcome should be 3 different types. Hence, the answer should be (a).

116. *(a)*

Phenotype	**Red-eyed female (F)**	×	**White-eyed male (M)**
Genotype	X^RX^r	×	X^rY
Gametes	(X^R) (X^r)		(X^r) (Y)

F_1 genotypes : X^RX^r X^rX^r X^RY X^rY

F_1 phenotype : red F, white F, red M, white M

117. *(c)* If the T is inserted into the position R, the DNA strand will be read as

TTC ACG ACT AAG T

When transcripted, the *m*RNA coded for will be

AAG UGC UGA UUC → UGA is a stop codon and will result in termination of translation.

118. *(d)* First reptiles – Carboniferous

First mammals – Triassic

First humans – Quaternary

First amphibians – Devonian

119. *(b)* The DNA sense strand would have the bases TAGCTTCAAGCA. Therefore, the non-transcribed strand would be ATCGAAGTTCGT, where thymine is incorporated instead of uracil.

120. *(c)* Since the allele is recessive, both homologous chromosomes in a somatic cell of an albino person would have the allele. After meiosis-I, each end would have a homologous chromosome with the allele. Since the chromosome is existing as a pair of sister chromatids at this stage, each chromosome and hence each end would have 2 copies of the allele.

www.ingramcontent.com/pod-product-compliance
Ingram Content Group UK Ltd.
Pitfield, Milton Keynes, MK11 3LW, UK
UKHW061547190726
13845UKWH00030B/1068

9 789324 195388